SANDSTONE DIAGENESIS:
Recent and Ancient

AF695085

REPRINT SERIES VOLUME 4 OF THE INTERNATIONAL ASSOCIATION OF SEDIMENTOLOGISTS

SANDSTONE DIAGENESIS: Recent and Ancient

EDITED BY

Stuart D. Burley and Richard H. Worden

Blackwell Publishing

© 2003 International Association of Sedimentologists
and published for them by
Blackwell Publishing Ltd

350 Main Street, Malden, MA 02148-5018, USA
108 Cowley Road, Oxford OX4 1JF, UK
550 Swanston Street, Carlton South, Melbourne, Victoria 3053, Australia
Kurfürstendamm 57, 10707 Berlin, Germany

The rights of Stuart Burley and Richard Worden to be identified as the Authors of the Editorial Material in this Work have been asserted in accordance with the UK Copyright, Designs, and Patents Act 1988.

All rights reserved. No part of this publication may be reproduced, stored in a retrieval system, or transmitted, in any form or by any means, electronic, mechanical, photocopying, recording or otherwise, except as permitted by the UK Copyright, Designs, and Patents Act 1988, without the prior permission of the publisher.

First published 2003

Library of Congress Cataloging-in-Publication Data

Sandstone diagenesis: recent and ancient/edited by Stuart D. Burley and Richard H. Worden.
p. cm. — (Reprint series volume 4 of the International Association of Sedimentologists)
Includes bibliographical references and index.
ISBN 1-40510-897-5 (pbk.: alk. paper)
1. Sandstone. 2. Diagenesis. I. Burley, Stuart D. II. Worden, Richard H. III. Reprint series . . . of the International Association of Sedimentologists; v. 4.

QE471 .15.S25 S245 2003
552′.5—dc21 2002038289

A catalogue record for this title is available from the British Library.

Set in 9.5/12pt Melior
by Graphicraft Limited, Hong Kong

For further information on
Blackwell Publishing, visit our website:
http://www.blackwellpublishing.com

MIX
Paper from responsible sources
FSC® C013604

Contents

Introduction

Sandstone diagenesis: the evolution of sand to stone

R.H. WORDEN* *and* S.D. BURLEY†,‡

**Department of Earth Sciences, University of Liverpool, Brownlow Street, Liverpool L69 3GP, UK*

†BG Group, Thames Valley Park Drive, Reading, Berkshire RG6 1PT, UK

‡Basin Dynamics Research Group, School of Geology and Geography, University of Keele, Keele ST5 5BG, UK

INTRODUCTION TO DIAGENESIS

Diagenesis comprises a broad spectrum of physical, chemical and biological post-depositional processes by which original sedimentary assemblages and their interstitial pore waters react in an attempt to reach textural and geochemical equilibrium with their environment (Curtis, 1977; Burley *et al.*, 1985). These processes are continually active as the ambient environment evolves in terms of temperature, pressure and chemistry during the deposition, burial and uplift cycle of basin history. As such, diagenesis encompasses a broad spectrum of post-depositional modifications to sediments. It ranges from weathering in subaerial environments and oxidation in the water column, includes compaction and lithification of sediments during burial, and eventually grades through a continuum into low-temperature metamorphism. In both humid and arid climates where geochemical reactions approach completion, gibbsite, kaolin group minerals and smectites form from aluminosilicate precursors. In cooler, temperate climates a greater variety of clay minerals occurs in weathering profiles reflecting metastable, intermediate breakdown products of aluminosilicates. Diagenesis is differentiated from metamorphism by a variety of mineral and thermal-history indices (Fig. 1; Frey, 1987; Slater *et al.*, 1994), but broadly a temperature transition of 180–250°C is thought to separate the two regimes. The classic transition from diagenesis to metamorphism was described eloquently in the Salton Sea geothermal field, south-east California, where sandstones of broadly similar composition are present over a temperature interval of 100°C to 350°C (McDowell & Elders, 1980). Here, clearly diagenetic, non-equilibrium mineral assemblages

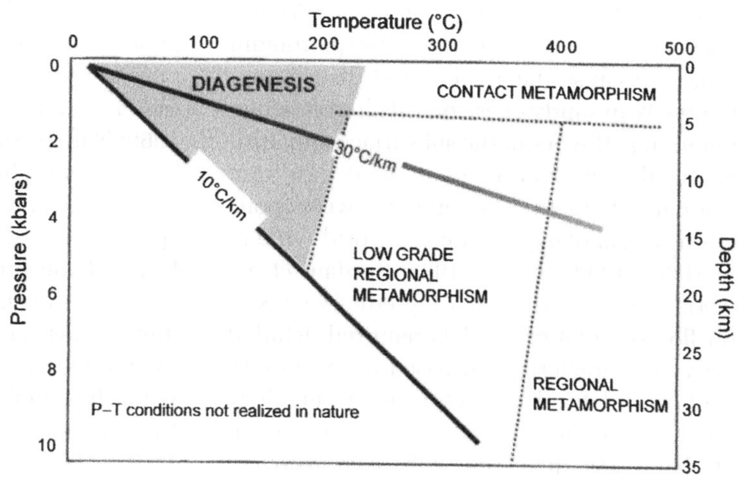

Fig. 1 Pressure–temperature diagram relating diagenesis to metamorphic regimes and typical P–T gradients in Earth's crust. The crustal geotherm of 10°C km^{-1} is representative of stable cratons, whereas a value of 30°C km^{-1} is typical of rifted sedimentary basins. Low-pressure (shallow)–high-temperature conditions are realized only in geothermal systems or in the vicinity of igneous contacts (so-called contact diagenesis: McKinley *et al.*, 2001).

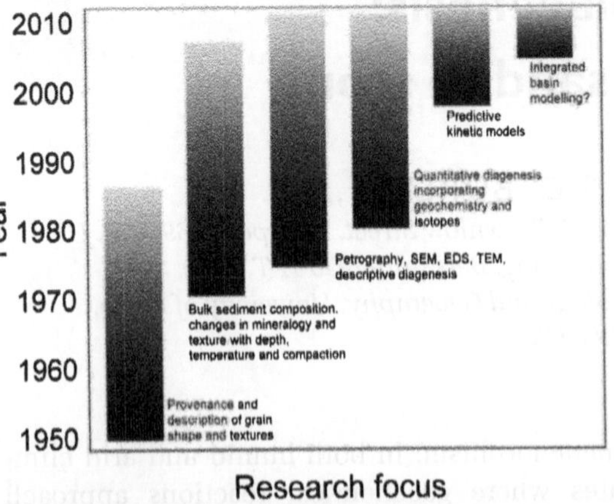

Fig. 2 The history of the study of diagenesis and probable future directions of the subject.

are replaced by an equilibrium metamorphic albite, chlorite and quartz assemblage. In the broadest sense, therefore, diagenesis can be considered as everything that contributes to making a sediment into a sedimentary rock from its weathering through to metamorphism during deep burial.

The study of sandstone diagenesis is relatively new, having grown from the description of grain shapes and textures coupled with the analysis of the evolution of bulk sediment composition with increasing depth and temperature of burial (Fig. 2). The importance of sandstone diagenesis as a subject is evidenced by the explosive growth of both pure and applied research in this subject through the 1980s and 1990s. This was largely driven by the petroleum industry because the amount and distribution of porosity in sandstones controls hydrocarbon migration pathways in the subsurface and, ultimately, the production of oil and gas from reservoirs. Prediction of porosity 'sweet spots' became a goal of explorationists world-wide in the 1980s–1990s (Curtis, 1983; Surdam *et al.*, 1984). The construction of predictive reservoir-flow-simulation models required detailed reservoir characterization based on an understanding of detrital and authigenic mineralogy (Hurst, 1987). Furthermore, as the demand for petroleum has increased, enhanced recovery techniques have developed that require the injection of reactive chemicals into the pore space of sandstones and potentially result in damage to reservoir formations (Pittman & King, 1986; Kantorowicz *et al.*, 1992). There is thus a need to understand chemical reactions with the host sandstone through the introduction of steam, surfactants, polymers or acids into reservoirs.

This introductory account provides a framework for the papers included in this compilation, sets the geochemical theme and provides definitions of the terms generally used in diagenesis. Wherever possible we have referred to the papers included in this volume to illustrate the points made. For more comprehensive reviews of characteristics of sandstone diagenesis the reader is referred to Morad (1998), Morad *et al.* (2000) and Worden & Morad (2000).

WHY DO DIAGENETIC REACTIONS TAKE PLACE?

The fundamental driving mechanisms for diagenetic reactions are changes in one or more of ambient pore-fluid chemistry, temperature and pressure. Empirical evidence suggests that this sequence is the order of importance of controlling parameters for diagenetic reactions. Indeed, it has long been recognized in studies of pedogenesis that it is the amount of average rainfall, i.e. the extent of flushing with water, that controls the rate of weathering. In dry temperate climates average saprolite formation rates of 5 mm per 1000 yr are an order of magnitude less than in wet, tropical climates (Tardy, 1969; Nahon, 1991). It is the presence of unstable or metastable mineral components bathed in aqueous pore waters charged with dissolved species that defines the diagenetic system. Without the presence of aqueous pore fluids, diagenesis effectively ceases. The sedimentary mineral assemblage reacts through water–rock interaction via pore fluids towards equilibrium with the ambient geochemical environment. Diagenesis is thus a dynamic suite of processes; as the burial history of a sedimentary basin develops, and pore fluids evolve through time, the diagenetic fabric and mineralogy must change in response.

Time is a critical component in diagenetic reactions. Early diagenetic reactions can be extremely rapid, with marine cementation in carbonates and sandstones taking place over mere decades (Taylor & Illing, 1969; Pye *et al.*, 1990; Al-Agha *et al.*, 1995), although few mineralogical changes take place in most silicates during sediment transport and deposition. More typical is reddening of desert sands, which demonstrably takes place between 5000 and 20 000 yr (Gardener, 1983; Pye, 1983). Mature lateritic soil profiles are known to develop over time-scales of 10 000 to 1 000 000 yr (Valeton, 1983), and calcretes form over comparable time intervals, although possibly much shorter (Wright & Tucker, 1991). Considerably greater time intervals are available during burial, where associated higher temperatures and pressures increase reaction rates and so favour true chemical equilibrium. Even here, some evidence suggests that deep burial diagenetic reactions may be quite rapid, at least in a geological sense, occurring on a scale of tens of thousands of years (Walderhaug, 1994; Duddy *et al.*, 1998; Worden *et al.*, 2000b) to a few million years (Boles, 1987).

Diagenetic assemblages are a function of chemical thermodynamics and kinetics. In this context, phase equilibrium diagrams are useful for predicting the direction of diagenetic mineral reactions (Fig. 3; Garrels & Mackenzie, 1971), but are much less effective at predicting aqueous pore solution compositions in geological systems containing diverse complex minerals. Free energy calculations can be used to identify unstable or metastable systems and predict relative mineral stability (Curtis, 1978). Indeed, it is the bond strength of minerals and free energy of mineral surfaces that define mineral reactivity (Helgeson *et al.*, 1978; Hurst, 1981). However, differential rate phenomena in many systems dominate over thermodynamics, especially in systems at low temperatures in early diagenesis. During diagenesis it is typically the difference between chemical reaction rates and transport rates that is the critical rate-limiting factor in the resultant mineral reaction.

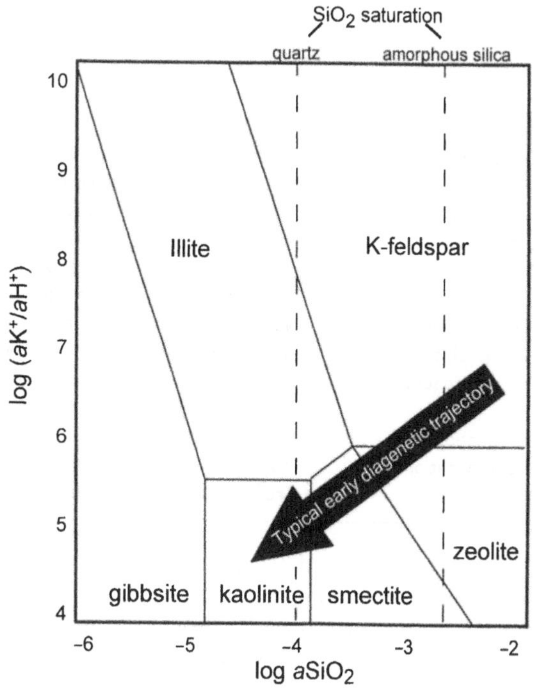

Fig. 3 Phase diagram for the system $\log(aK^+/aH^+)$–$\log aSiO_2$ (where 'a' indicates chemical activity of species) illustrating how changing isothermal water geochemistry can lead to diagenetic reactions. For example, if water in contact with feldspathic sandstones is flushed with fresh water (e.g. meteoric water) then the water will progressively evolve to lower aqueous potassium and silica concentrations. This explains why advanced eogenesis (and telogenesis) leads to kaolinite growth at the expense of feldspar minerals.

DEFINING DIAGENESIS

A variety of terms are used to describe diagenetic processes that have acquired a specific meaning. Important terms used in this text, and commonly applied to diagenesis, are defined in Table 1 and a brief commentary on their meanings is given below.

Authigenesis literally means 'generation *in situ*' and is usually applied to describe all diagenetic mineral formation in sediments.

Table 1 Common terms used in diagenesis.

Term	Definition
Authigenesis	*In situ* mineral growth
Cementation	Growth or precipitation of minerals in pore spaces
Compaction	Suite of processes resulting in the collapse of pore space in a sandstone
Decarboxylation	Loss of CO_2 from organic matter in response to increased temperature
Dehydration	Loss of H_2O from minerals and organic matter as a result of increased temperature
Dissolution	Process whereby a mineral is destroyed by interaction with a fluid leaving behind a cavity
Lithification	The process of indurating a loose or friable sediment through compactional and cementation processes
Neoformation	New growth of minerals during diagenesis
Neomorphism	Transformation of a mineral typically involving changes in crystal chemistry
Paragenetic sequence	The order in which diagenetic processes occur in sediments as recorded or inferred by petrographic, geochemical or isotopic methods
Precipitation	Crystallization of a mineral from solution
Recrystallization	Dissolution followed by precipitation involving changes in crystal size or habit of a given specific mineral
Replacement	Growth of a chemically different authigenic mineral within the body of a pre-existing mineral

Authigenic minerals are thus distinct from detrital (transported) minerals and formed *in situ* within the host sediment in which they now occur.

Cementation is the diagenetic process by which authigenic minerals are precipitated in the pore space of sediments which thereby become lithified. Compaction typically includes simple grain rearrangement during shallow burial as well as the ductile deformation of soft sand grains and intergranular matrix. This is quite different from the process known as chemical compaction, which involves the chemically induced dissolution of grains at intergranular contacts and reprecipitation of the dissolved material on grain surfaces facing open pores. Dissolution is the diagenetic process by which a solid component in the host sediment is dissolved by an aqueous pore solution leaving behind a space or cavity within the host sediment (see Schmidt & MacDonald, 1979; Burley & Kantorowicz, 1986).

Neomorphism describes the processes of replacement and recrystallization of one mineral by a related mineral but involving changes in the details of the mineral chemistry, excluding simple pore filling processes (Folk, 1965). The term has been applied widely to limestones and dolostones in which it is commonly used to describe the coarsening of aragonitic micrite into calcite microspar and is equally applicable to sandstones. Examples of neomorphism in sandstones are the conversion of aragonite to low Mg calcite cement or the evolution of sand-grain-coating green clays (e.g. Odin, 1990) into chlorite (e.g. Ehrenberg, 1993). Neomorphism commonly preserves textural evidence (ghost fabrics) of the previous phase.

Recrystallization, in contrast, is the change in crystal size or shape resulting from thermodynamic instability (such as the reprecipitation of finely grained calcite by coarse grained calcite cement; Hendry *et al.*, 1996; or kaolinite by dickite; Ehrenberg *et al.*, 1993) when kinetic barriers are exceeded to allow the reaction to proceed. In the diagenetic realm both recrystallization and neomorphism always require and involve the presence of an aqueous medium. These processes are distinct from replacement whereby an authigenic mineral occupies the place of another former mineral (either detrital or authigenic) via a dissolution–precipitation process (e.g. carbonate cements replacing detrital quartz; Hesse, 1987; or the albitization of detrital K-feldspar grains; Ramseyer *et al.*, 1992). It has been suggested replacement must take place via a 'thin film' mechanism (Pettijohn *et al.*, 1972).

A paragenetic sequence is the interpreted order in which diagenetic processes occurred.

The sequence is constructed on the basis of observations about the order of mineral growth and, typically, some degree of interpretation. Thus, for example, if quartz cement encloses illite crystals, the quartz is interpreted to have grown after the illite. A paragenetic sequence includes all the processes listed above as well as compaction, dissolution, recrystallization and precipitation. So very high 'minus cement volumes' (the sum of pore space and the space occupied by cements) in carbonate cemented sandstones, for example, is commonly taken to indicate that the carbonate cementation took place before compaction, and thus before significant burial. A paragenetic sequence is a simple way of relating a potentially complex series of events in a time series. However, it always should be remembered that a paragenetic sequence constructed for any sandstone is largely an interpretation of textural data. In some instances, a paragenetic sequence can be constrained by the use of absolute age determinations from radiometric dating of minerals (such as the K–Ar technique for illite) or via a burial curve using an independent temperature (or pressure) determination (such as apatite fission-track dating or fluid-inclusion microthermometry).

REGIMES OF DIAGENESIS

The concept of diagenetic regimes is a broad framework that relates diagenetic processes to the evolution of sedimentary basins (Fig. 4). Three conceptual regimes are commonly recognized: early diagenesis (eogenesis), burial diagenesis (mesogenesis) and uplift-related diagenesis (telogenesis). This terminology was adopted from a scheme developed initially by Choquette & Pray (1970) to describe limestone diagenetic processes, but is now more generally applied: correctly so, as the same fundamental processes and controls operate in clastic diagenesis and in carbonate diagenesis. Alternative schemes (e.g. the Russian system including

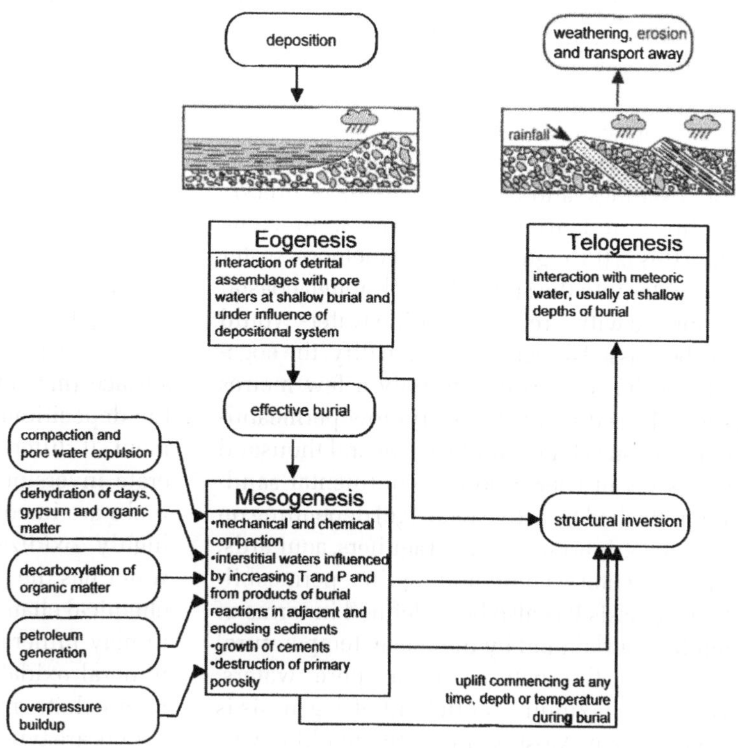

Fig. 4 Flow chart showing the links between the regimes of diagenesis. The change from mesogenesis (burial diagenesis) to telogenesis can occur at any stage during burial. Telogenesis (uplift-related processes) only happens when surface waters penetrate into the inverted basin and cause mineral reactions.

such terms as catagenesis and epigenesis; see Prozorovich, 1970, for example) have been used but are less commonly applied now. This is because systems and classifications defined by the maximum temperature of burial run into the difficulty of the effect of varying time spent at a given temperature—a direct consequence of the kinetic control on the rate of diagenetic reactions. Moreover, classifications defined by the equivalent stage of petroleum generation in source rocks are not useful for sandstones because these contain only sparse organic matter and their different thermal and poro-mechanical properties compared with fine grained, organic-matter-rich sediments results in quite different responses to ambient geochemical environments. Recognition must be made of the arbitrary nature of any diagenetic classification, but the three-fold scheme adopted here is simple and inclusive.

Eogenesis equates broadly to early diagenesis but is defined as including all processes that occur at or near the surface of the sediments where the chemistry of the interstitial waters is controlled mainly by the depositional environment (Fig. 4; Berner, 1980; Chapelle, 1993). Strictly speaking, this is the regime where the influence of the original depositional pore water dominates, and so includes weathering and soil development in continental depositional settings, and bacterially mediated redox reactions in marine environments. In some cases this is the regime in which meteoric water penetrates into the subsurface, although coastal sediments with a reflux of marine waters would also be classed as eogenetic. In reality, the eogenetic realm may extend to only a few metres below the sediment surface (in low permeability mudstones for example) or several thousand metres (as in coarse porous continental sandstones flushed by active recharge) depending on the geometric arrangement of aquifers, aquitards, synsedimentary faults and aquifer permeability. Eogenesis has also been defined in terms of depth of burial (and by inference, temperature) instead of the nature of the pore waters, where the maximum depth limit of eogenesis is about 1–2 km. Most basins of the world have a geothermal gradient between 20 and 30°C km^{-1} so that the limit of eogenesis (for a mean surface temperature of 10°C) lies between 30 and 70°C according to this scheme (Morad *et al.*, 2000).

Mesogenesis occurs during burial and is defined as those diagenetic processes occurring once the sediment has passed from the influence of the depositional environment through to the earliest stages of low-grade metamorphism. Boundaries may not be sharp, and may be difficult to define by textural, mineralogical or isotopic means. In many cases, this regime therefore includes sediments of between about 100–1000 m burial and those at depths with equivalent temperatures of up to 200–250°C. Mesogenesis is often termed burial diagenesis because it happens during burial. However, mesogenesis can continue following burial in inverted sedimentary basins that have experienced a degree of uplift and cooling. The main factors that influence mesogenesis include the time–temperature history, the primary mineralogy and fabric, loss and gain of material to neighbouring lithologies (so-called 'mass-transfer'), the geochemistry of the pore water and the presence of petroleum-related fluids (including oil, hydrocarbon gas, organic-derived CO_2 and H_2S). The boundary between eogenesis and mesogenesis can be defined in depth and temperature terms instead of the connectivity of pore water with surface waters (Morad *et al.*, 2000; and see above). In this case mesogenesis begins at 1–2 km burial (temperatures of between 30 and 70°C).

Telogenesis occurs in uplifted and exhumed rocks that have been exposed to the influx of surface (meteoric) water that is not related to the depositional environment of the host sediment. It differs from mesogenesis during moderate inversion and uplift simply because the rocks are in contact with flowing, low salinity, highly oxidized, CO_2-charged waters. Such water has the capacity to cause significant geochemical changes, including feldspar alteration to clay minerals (usually kaolinite) and ferric mineral oxidation (including alteration of ferroan calcite and dolomite), even though these waters are typically of low ionic strength. Many early studies of diagenesis were on outcrop samples that had inevitably undergone variable degrees of telogenetic alteration.

DIAGENETIC MINERAL CEMENTS IN SANDSTONES

A host of minerals occur as cements in sandstones. The most common mineral cements are quartz (and related chalcedonic silica varieties), the carbonate minerals, and a variety of aluminosilicate clay minerals. Less common, although locally important, cements include anhydrite (and gypsum), pyrite, feldspars, zeolites, haematite, apatite and Ti-rich minerals (such as sphene).

Quartz cement (Figs 5 & 6) is mineralogically the most simple of cements but occurs in a variety of forms. Quartz overgrowths are approximately equal thickness rinds that form on detrital quartz grains (Waugh, 1971). These are usually optically continuous with the substrate minerals (forming a syntaxial fabric) revealing that the two types of quartz are in perfect crystal continuity. In the case of detrital polycrystalline quartz grains, or strained quartz, the overgrowths typically adopt the crystallographic orientation of the detrital quartz immediately adjacent to the cement. Microcrystalline

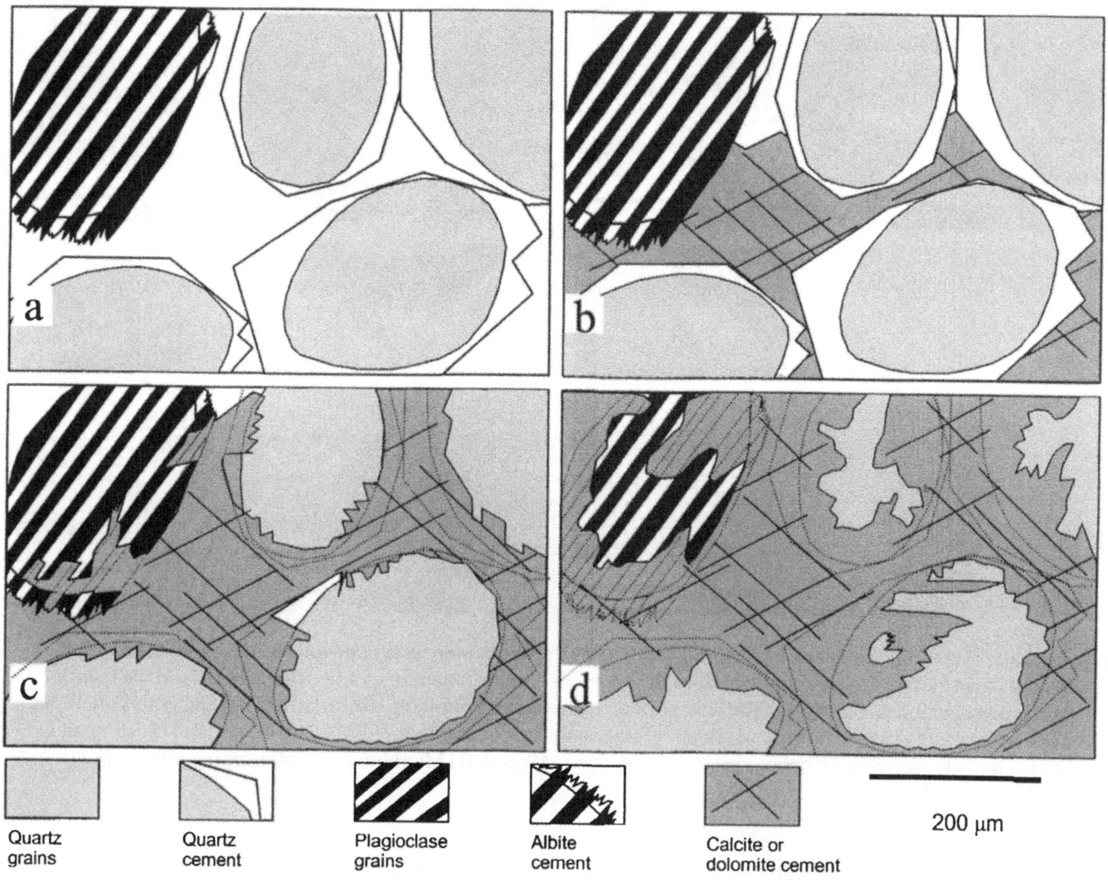

Fig. 5 Schematic cartoon illustrating types of pore-filling cementation in a paragenetic sequence. (a) syntaxial overgrowth cements, (b) passive pore-filling cements post-dating overgrowths, (c) peripherally grain-replacive cements post-overgrowths and (d) extensive grain replacement cements.

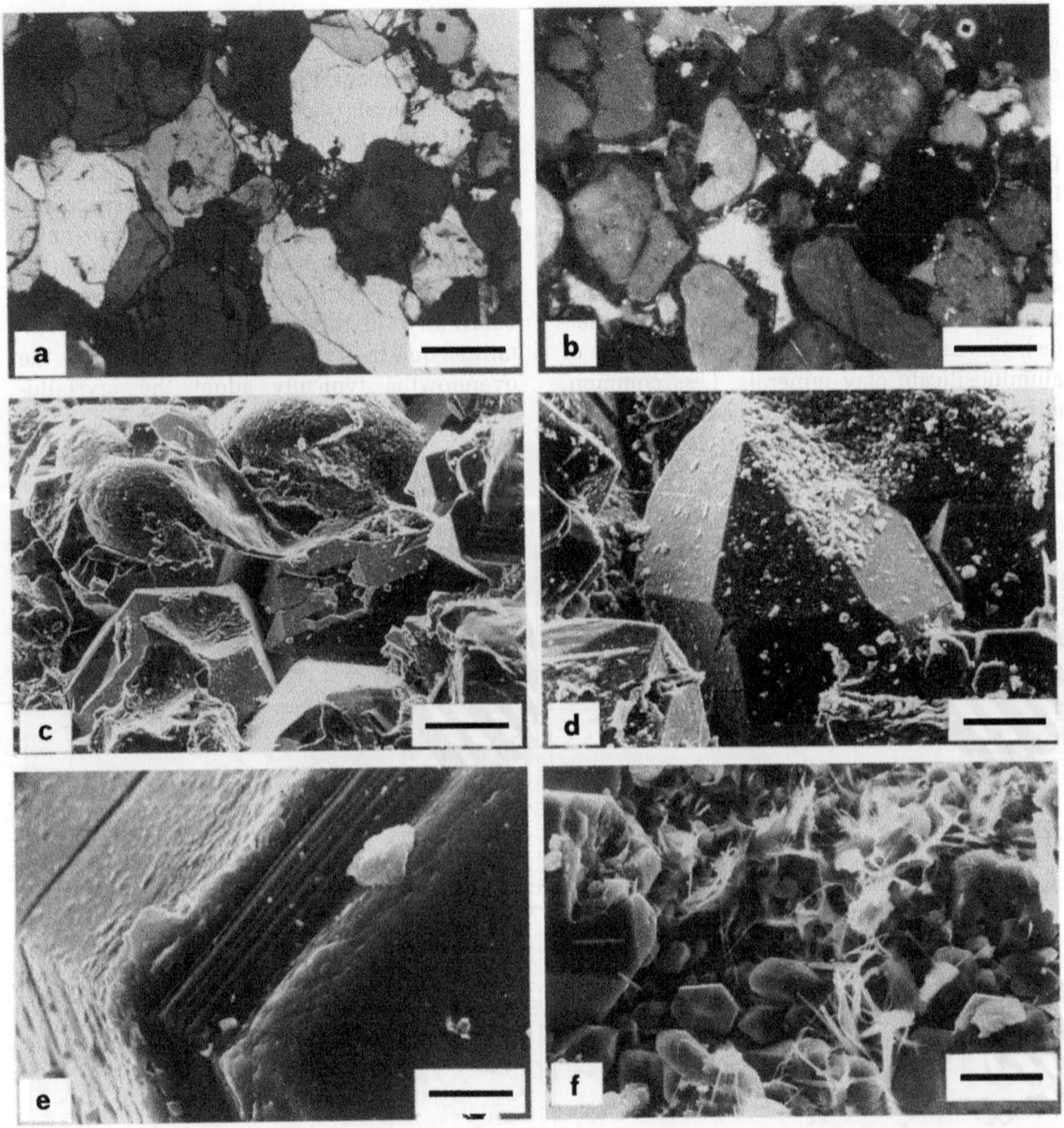

Fig. 6 Quartz cement in sandstones. (a) Optical micrograph of sandstone with large quartz overgrowths (crossed polars). (b) Same field of view under cathodoluminescence showing extent of quartz overgrowths (scale 300 μm). (c) A scanning election microscopy (SEM) image of the same sample showing well-rounded detrital grains enclosed in quartz overgrowths (scale 180 μm). (d) Detail of large euhedral overgrowth, SEM (scale 60 μm). (e) Close-up of edge of overgrowth showing successive growth zones (scale 10 μm). (f) Microquartz crystals lining intergranular pores, SEM (scale 20 μm).

quartz cement grows as a multitude of crystals of less than 10 μm length. These are often in optical continuity with detrital quartz but can also grow on detrital or diagenetic clay minerals. Quartz not in optical continuity with its host is termed epitaxial. Microcrystalline quartz is often observed to have replaced siliceous fossils (such as *Rhaxella*; Vagle *et al.*, 1994),

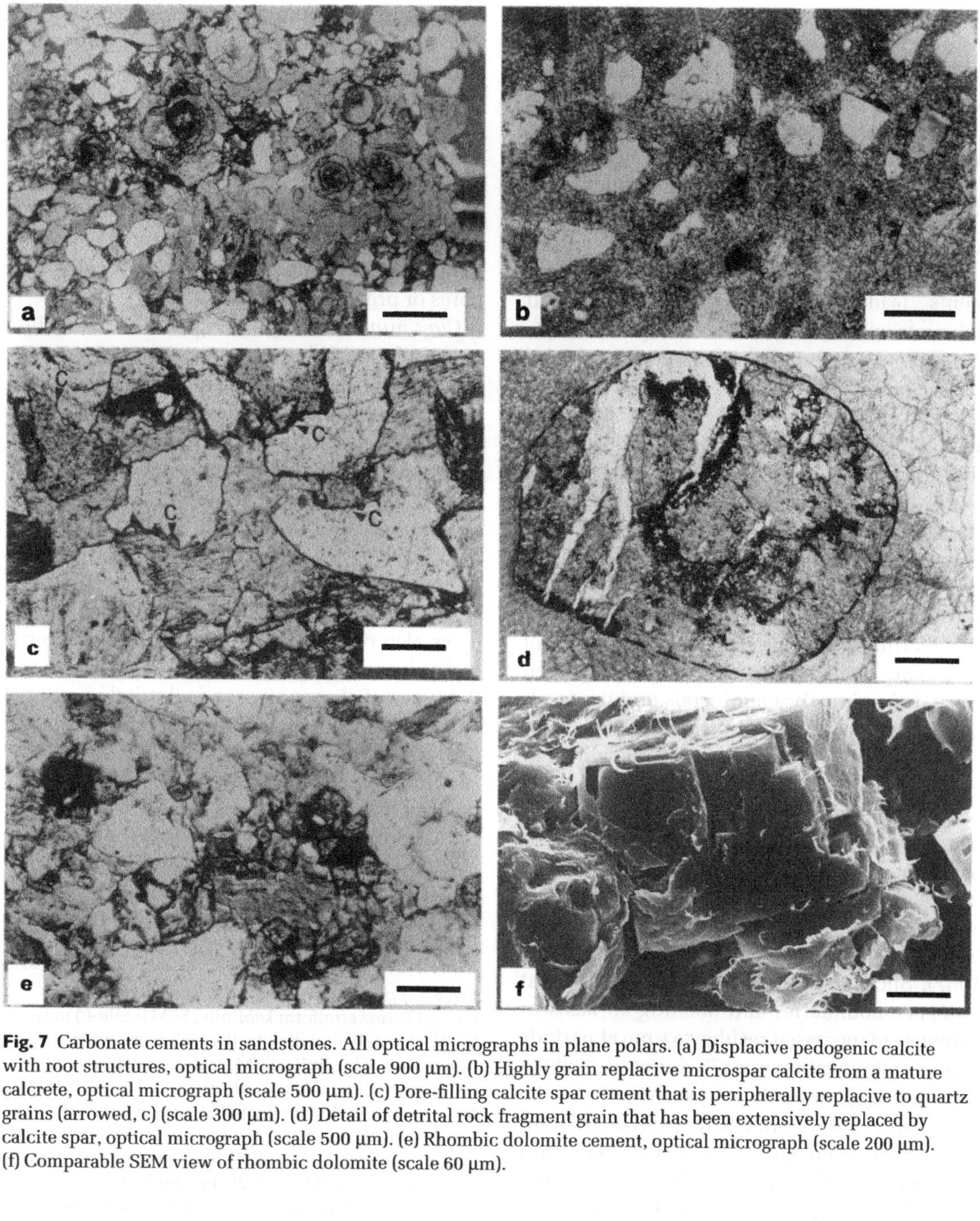

Fig. 7 Carbonate cements in sandstones. All optical micrographs in plane polars. (a) Displacive pedogenic calcite with root structures, optical micrograph (scale 900 μm). (b) Highly grain replacive microspar calcite from a mature calcrete, optical micrograph (scale 500 μm). (c) Pore-filling calcite spar cement that is peripherally replacive to quartz grains (arrowed, c) (scale 300 μm). (d) Detail of detrital rock fragment grain that has been extensively replaced by calcite spar, optical micrograph (scale 500 μm). (e) Rhombic dolomite cement, optical micrograph (scale 200 μm). (f) Comparable SEM view of rhombic dolomite (scale 60 μm).

which it is generally inferred to have replaced former biogenic Opal A (Williams *et al.*, 1980). Quartz cement tends to occur during burial diagenesis at temperatures above 70°C (Bjørlykke and Egeberg, 1993) although silcrete formation can also result in quartz overgrowths (Thiry *et al.*, 1988).

Carbonate cements (Figs 5 & 7) include calcite ($CaCO_3$), dolomite ($MgCa(CO_3)_2$) and siderite ($FeCO_3$). Calcite and dolomite cements occur as both ferroan and non-ferroan compositions. This is clearly revealed by the sensitivity of the mineral to Fe-detecting stains (Dickson, 1965). All three main carbonate mineral cements found in clastic rocks can develop during eogenesis as well as mesogenesis. Eogenetic cementation results in pore-filling cement fabrics composed of fine microspar crystals in quantities ≤ 40%.

Pedogenesis in arid continental environments leads to non-ferroan calcrete and dolocrete with an abundance of biogenic calcite and dolomite cement fabrics (Purvis & Wright, 1991; Spötl & Wright, 1992). The presence of calcrete intraclasts results in carbonate recementation at the bases of fluvial channels in red bed and other arid or semi-arid successions (Burley, 1984). Shallow marine sandstones are often cemented with nodules or discrete layers of eogenetic calcite (Wilkinson, 1991) owing to the dissolution and reprecipitation of shell detritus (Hendry *et al.*, 1996) as non-ferroan calcite. Marine sandstones develop a wide range of carbonate cements through reaction between detrital aluminosilicate minerals and the products of the breakdown of organic matter (Hein *et al.*, 1979). Eogenetic siderite develops in iron-rich, partially reduced systems with a minimal marine influence (Baker *et al.*, 1996; compare with Pye *et al.*, 1990). Complete reduction in natural systems with a significant marine influence leads to sulphide (initially greigite, ageing to pyrite) growth owing to the reduction of sulphate as well as ferric iron (Love, 1967). Such early cements commonly form at sequence boundaries and flooding surfaces in response to increased residence time at periods of low sedimentation rate (Taylor *et al.*, 2000).

Burial diagenesis is characterized by the recrystallization of pre-existing carbonate minerals (calcite and dolomite) in a ferroan form and results in cement typically at the 2–10% level. Burial diagenesis also results in a coarser grained fabric of locally pore-filling rhombohedral crystals of ferroan dolomite (and less often ferroan calcite). Many crystals exhibit zoning, with an increasing Fe content evident during diagenesis. Late diagenetic carbonate crystals (usually ferroan dolomite) can also develop in the apparent absence of pre-existing carbonate minerals. These probably result from an influx of source-rock-derived CO_2 and alkaline earth elements (and iron) from reactive-sand grain decomposition processes and from neighbouring mudstones or evaporites. Eogenetic and burial diagenetic carbonate cements frequently have an exaggerated effect upon the flow properties of sandstone because they locally fill pores or preferentially block pore throats.

Clay mineral cements (Figs 8 & 9) are volumetrically small but important components of sandstones because of the enormous effect they have on permeability. Clay minerals are thus a vital consideration in studies involving the movement of fluids through pore spaces. The most common clay minerals are kaolinite ($Al_2Si_2O_5(OH)_4$), illite ($KAl_3Si_3O_{10}(OH)_2$) and chlorite ($[Fe–Mg]_5Al_2Si_3O_{10}(OH)_8$). The smectite (typical dioctohedral clay; montmorillonite, typical trioctahedral clay; saponite) family of clays also occur as cements but these are generally less well documented than the others. All types of clay can occur as detrital components although hydrodynamic sorting tends to prevent co-deposition of sand grains and clays. Bioturbation, mass flow and soft-sediment deformation are the most likely mechanisms for introducing detrital clays into the fabric of marine sandstones, whereas mechanical infiltration is the dominant mechanism in continental sandstones (Walker *et al.*, 1979). Detrital

Fig. 8 (*opposite*) Clay cements in sandstones. (a) Eogenetic vermiform kaolinite, SEM (scale 40 μm). (b) Blocky, euhedral dickite, the high-temperature polymorph of kaolinite, SEM (scale 5 μm). (c) Grain-coating smectite preferentially nucleated on a detrital ferromagnesian (FM) grain, but absent on quartz (Q) and feldspar (F) grains, optical micrograph (scale 80 μm). (d) Comparable SEM view of grain-coating radial smectite, SEM (scale 20 μm). (e) Post-quartz overgrowth pore bridging 'hairy' illite, SEM (scale 80 μm). (f) Detail of authigenic illite showing the lath-shaped nature of the illite crystals, SEM (scale 4 μm). (g) Pre-quartz overgrowth grain-coating chlorite cement, SEM (scale 60 μm). (h) Detail of authigenic chlorite enclosed by but partially inhibiting quartz overgrowths, SEM (scale 4 μm).

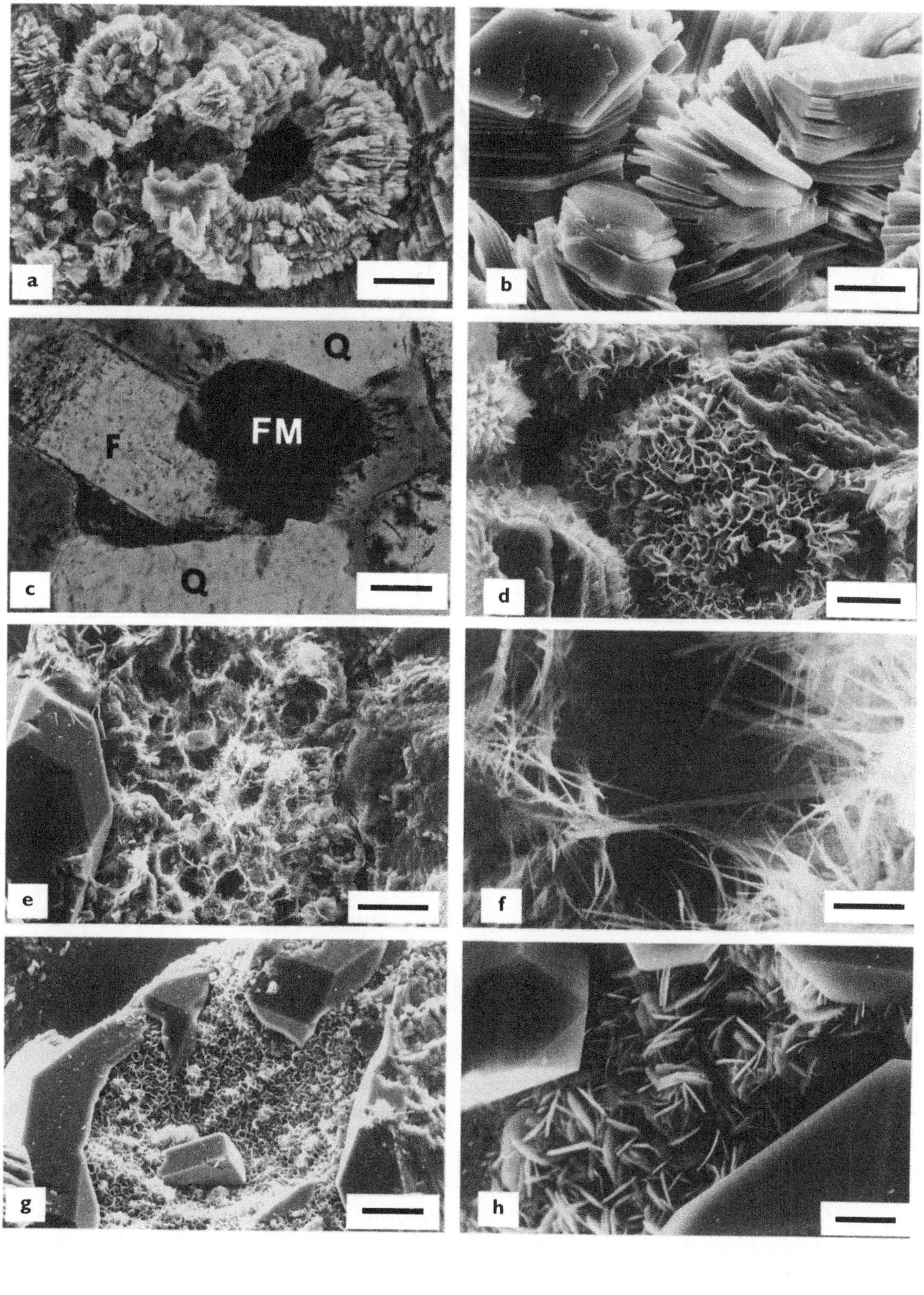
a
b
Q
FM
F
Q
c
d
e
f
g
h

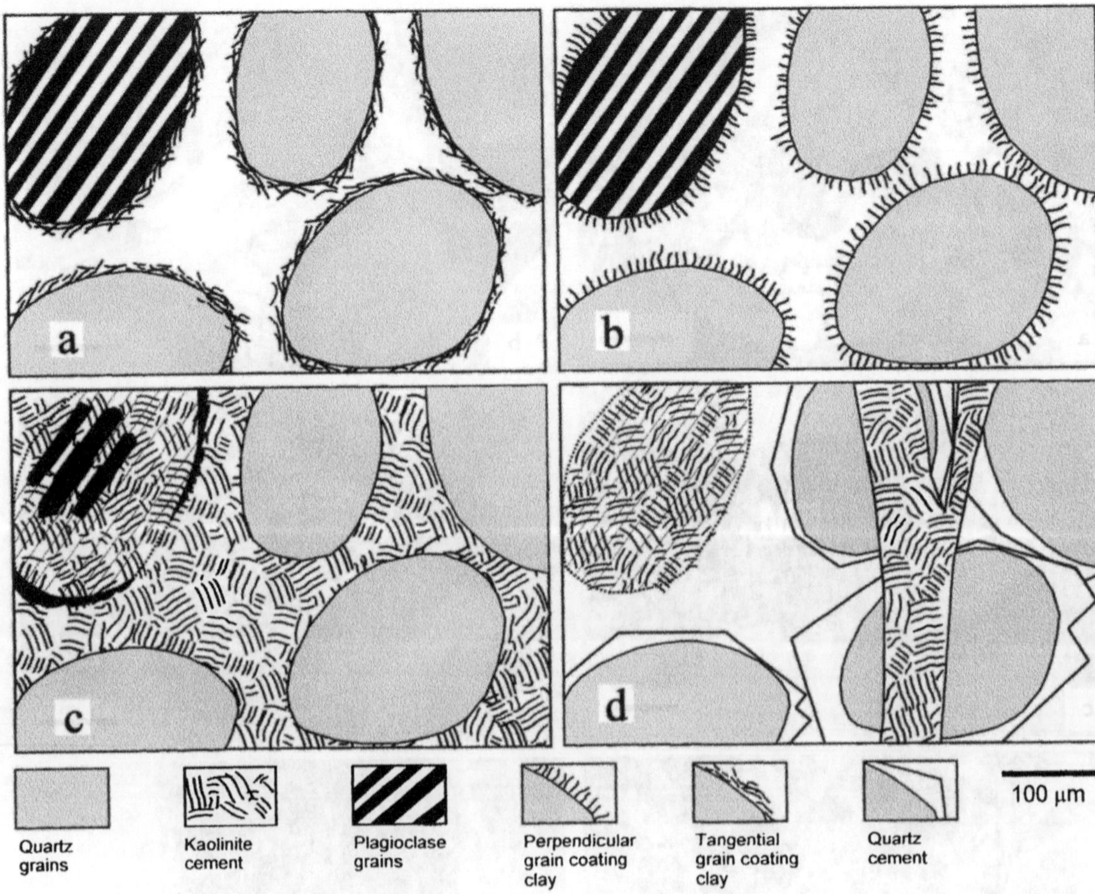

Fig. 9 Schematic cartoon illustrating types of clay morphologies in sandstones. (a) Grain coating, tangential clays typified by detrital and mechanically infiltrated clays, (b) grain coating, perpendicular clays typified by smectite, illite and chlorite, (c) pore-filling and feldspar-replacive clays typified by kaolinite, (d) grain replacive and fracture filling cements (typically kaolinite).

clay, of whatever mineral chemistry, occurs as tiny, ragged, abraded crystals and naturally accumulates in pore spaces forming tangential grain-coating and pore-bridging fabrics. Formerly, it was widely held that clay mineral neoformation took place when detrital clays, derived from continental weathering, entered the marine environment (see Millot, 1970). However, it is now documented that detrital clays do not undergo significant chemical or structural alteration during transport or deposition. Flocculation leads to a sedimentary fractionation of clays in passing from the fluvial into an estuarine or open marine environment (Jeans, 1989). Only cation exchange reactions with the surface of clays take place during transport and deposition. In the late 1960s one of the great debates of the time centred on whether the clay matrix characteristic of argillaceous Palaeozoic sandstones (so-called greywackes) was detrital or diagenetic in origin. This debate was resolved as diagenetic by Whetten & Hawkins (1970) in one of the first detailed petrographic and mineralogical studies of diagenesis.

Smectites, kaolinite, and chlorite all occur as eogenetic cements dependent on the composition of the detrital sediment supply, rate of *in situ* weathering of detrital minerals versus subsequent sediment supply, climate and the composition of eogenetic pore waters. Smectites and related minerals (such as palygorskite) form in Na–K-rich, alkaline brines under oxidizing conditions (Jeans, 1978). By contrast, kaolinite and the kaolin family of aluminosilicates require low pH, low ionic strength waters and so are typical of weathering profiles and early diagenesis in fluvial and deltaic environments. Chlorite often occurs in reduced, marine waters whereas glauconite cement develops in Fe-rich, oxidized shallow marine environments (Odin & Matter, 1981). The habit of eogenetic clays varies with their mineralogy and chemistry (Wilson & Pittman, 1979) but has not been investigated systematically. Smectites tend to occur as very small crystals that line intergranular pores and replace grains that contain Mg and Fe required for their development, resulting in the so-called 'grain localization' effect of Burley (1984). Kaolinite usually occurs as pore-filling aggregates (termed 'books') of euhedral hexagonal crystals.

Kaolinite, illite and chlorite all occur as burial diagenetic cements. Kaolinite forms vermiform masses that occupy, and locally fill, pores. It can also occur as a pseudomorphic replacement of altered detrital feldspar minerals. Much of what is described as kaolinite in mesogenetic clay assemblages is most probably the polymorph dickite, which forms as a replacement of kaolinite as temperature increases (Ehrenberg *et al.*, 1993; McAulay *et al.*, 1994). Illite is an exclusively burial authigenic clay that forms at temperatures exceeding approximately 70°C in potassium-bearing formation waters and has been shown to be of phengitic composition (Warren & Curtis, 1989). It typically grows as fine, ribbon-like filaments on the surface of detrital grains and on the surface of earlier burial cements and replaces kaolinite at depth (Hancock & Taylor, 1978). Chlorite cement typically grows as radial detrital grain coatings composed of euhedral, interlocking plates that typically shield detrital grains from subsequent diagenetic processes.

Feldspar mineral cements are common although typically much less abundant than quartz, carbonates and clays. Early diagenetic K-feldspar (Morad *et al.*, 1989; Fig. 10) and burial diagenetic albite (Waugh, 1978; Ramseyer *et al.*, 1992) routinely are found in sandstone. Both minerals can form as overgrowths on detrital grains or as pore-filling cements. The diagenetic minerals typically have subtly different mineral chemistry and crystallography from the substrate grains often leading to an absence of optical continuity (i.e. non-uniform optical extinction positions). These differences also lead to differential diagenetic reactivity. Diagenetic feldspar minerals are more thermodynamically stable in the diagenetic environment than detrital feldspars. A consequence of this is that in some cases detrital feldspars are dissolved away leaving a rind of the earlier diagenetic feldspar overgrowth. Such reactions are temperature dependant and so depth-defined decrease in detrital feldspar abundance is common in many sedimentary basins (e.g. Harris, 1989). Albitization occurs when detrital K-feldspar or calcic plagioclase are replaced by albite (Ramseyer *et al.*, 1992). The first of these can occur during increasing salinity of formation water (usually dominated by Na in terms of the aqueous cation population). The second can occur at any stage during diagenesis because the anorthite feldspar end-member is unstable at all conditions found in sedimentary basins. The second stage releases Al and thus leads to clay mineral growth as well as feldspar replacement.

Pyrite cement (FeS_2) occurs as both eogenetic and burial diagenetic cements and is common although often at the $\leq 1\%$ bulk sandstone volume level. The eogenetic form is typically microcrystalline framboids (like submicrometre sized raspberries) and results from the microbial reduction of detrital ferric iron and typically the presence of seawater sulphate during earliest burial (Love, 1967). The burial diagenetic form of pyrite is somewhat coarser than the framboids and is usually at least subhedral. It is often found to be one of the last cements to form and is especially associated with the reduction of haematite in the presence of hydrocarbons (Elmore *et al.* 1987).

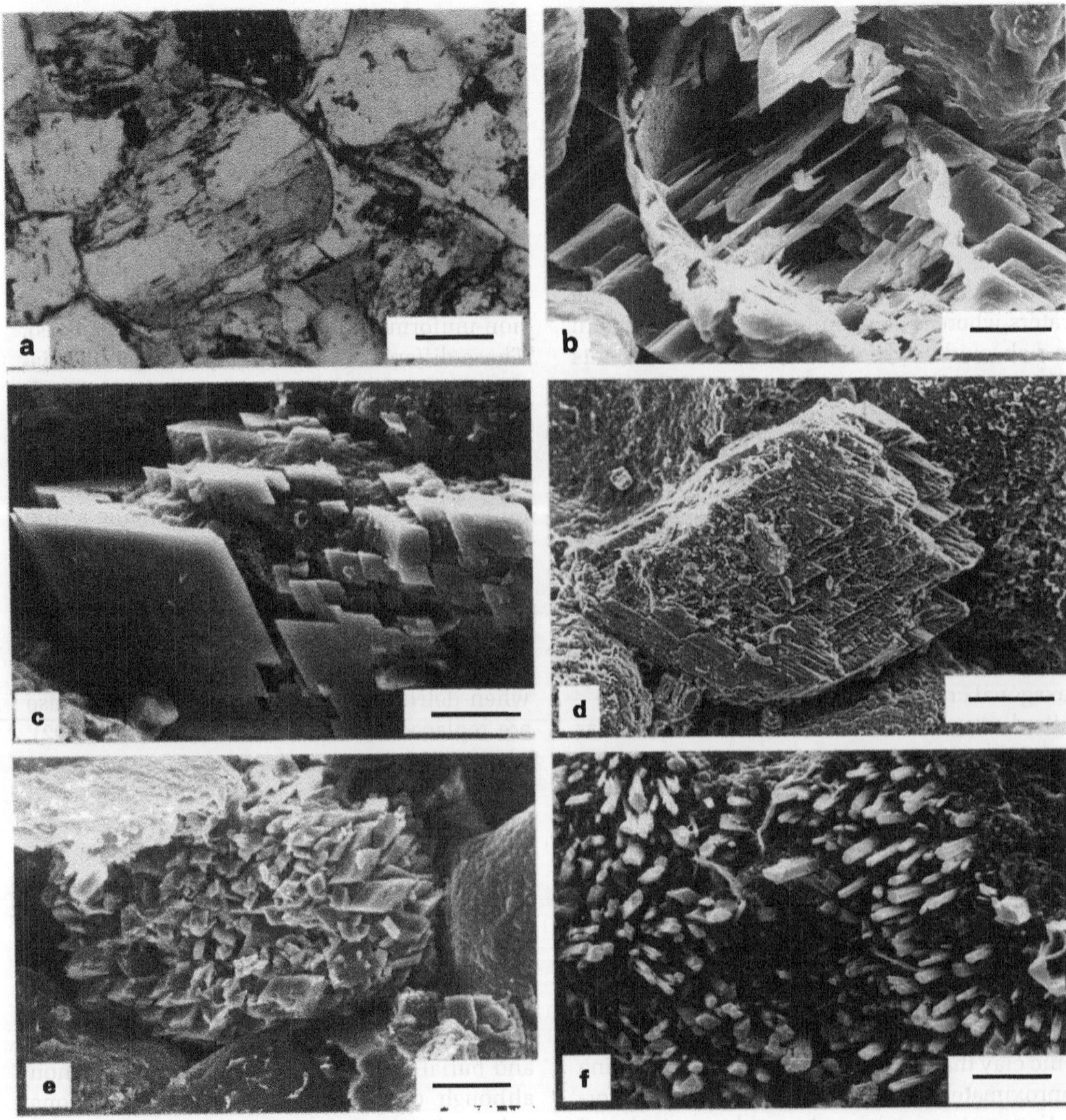

Fig. 10 Feldspar cement in sandstones. (a) Partially dissolved detrital feldspar grain with overgrowth, optical micrograph, plane polars (scale 60 μm). (b) Comparable view with the SEM showing dissolved grain interior and euhedral overgrowths (scale 40 μm). (c) Early stage of K-feldspar overgrowth development, SEM (scale 60 μm). (d) Complete, euhedral K-feldspar overgrowth, SEM (scale 120 μm). (e) Cluster of euhdral pore filling authigenic K-feldspar crystals, SEM (scale 60 μm). (f) Epitaxial K-feldspar overgrowth on detrital grain, SEM (scale 40 μm).

Anhydrite cement ($CaSO_4$) and other sulphates (e.g. barite $BaSO_4$) also occur as burial cements in typically minor but locally important quantities (Sullivan *et al.*, 1990; Baines *et al.*, 1991). They result from circulation of evaporite-related pore waters and occur as subhedral laths and poikilotopic crystals in similar amounts as the burial carbonate cements.

EOGENESIS (DEPOSITIONAL-ENVIRONMENT RELATED DIAGENESIS)

Diagenetic processes that fall into the eogenetic realm (dominated by interstitial waters that are connected to the surface of the sediment pile) are controlled by the overall physical, biological and geochemical characteristics of the depositional system. The gross depositional environment of the sand controls most eogenetic processes and imparts a distinctive diagenetic 'fingerprint', or assemblage, on the sandstone. Complex diagenetic assemblages and paragenetic sequences can result if the depositional environment changes as a result of relative sea-level fluctuations prior to burial and isolation of the sand unit (Curtis & Coleman, 1986).

On deposition, the primary sand comprises a mixture of minerals formed under a wide range of conditions of temperature, pressure, oxidation state, water composition and pH. Much depositional sand contains unstable grains that have not had the opportunity to react during initial weathering, erosion and transport before deposition. Therefore, a detrital mineral assemblage is typically inherently unstable and during eogenesis will tend to react with the ambient water and atmosphere. In many cases (especially in non-marine environments), these processes are simply a continuation of processes initiated during weathering.

The inherent instability of detrital sand during eogenesis results in the subsequent sandstones being 'fingerprinted' by the nature of the interstitial waters. These waters, directly influenced by the environment of deposition, leave their mark on the sandstone if any of the eogenetic minerals are preserved into the mesogenetic or telogenetic regimes. Although there are few index minerals that uniquely define a particular environment of deposition, this is often evidenced by the assemblage and specific chemistry of mineral suites. There are three broad subdivisions that frequently result in distinct eogenetic assemblages.

1 Subaerial, arid, hot (tropical to subtropical) environments where sediment hinterlands often undergo little chemical weathering because of the reduced rainfall. The resulting interstitial waters are alkaline and concentrated and dominated by Ca^{2+}, Mg^{2+} and HCO_3^- species.

2 Subaerial, humid, warm (subtropical to temperate) verdant environments where sediment hinterlands are subject to intense weathering in the presence of very dilute (less than a few 100 ppm) interstitial waters. These pore waters are dominated by aqueous Na^+ and HCO_3^- and are slightly to moderately acidic owing to the presence of decaying organic material.

3 Marine environments characterized by slightly alkaline waters (pH of seawater 8.3) and dominated by aqueous Na^+ and Cl^- (with subordinate SO_4^{2-}, HCO_3^-, Ca^{2+} and Mg^{2+}) with a salinity of 35‰.

Hot, dry, continental environments (Fig. 11) are highly oxidizing and have a paucity of reducing organic matter and low water tables. All iron remains in the ferric state and coats minerals in a hydroxide or sesquioxide form leading to a characteristic red coloration of continental sediments (and hence the term: red beds). These oxides age with time and increasing temperature to the mineral haematite (Walker *et al.*, 1979). Carbonate minerals that precipitate in these environments are thus iron-free. Evaporation often exceeds meteoric influx, leading to an upward flux of groundwater, evaporation and the consequent development of caliche (mostly calcrete, dolocrete or gypcrete). Diagenetic alunite ($KAl_3(SO_4)_2(OH)_6$) also develops under these conditions (Khalaf, 1990). The minimal weathering in the hinterland and minimal flux of water in the eogenetic zone leads to the common accumulation or growth of various smectite clays (e.g. montmorillonite, saponite, palygorskite), because these typically represent the results of the earliest stages of aluminosilicate mineral weathering and reaction in pore waters with high Ca and Mg activity. Eogenetic silica cement is not uncommon in arid environments and can occur in silcrete nodules on bedding surfaces on the distal parts of braid plains (Leckie & Cheel, 1990).

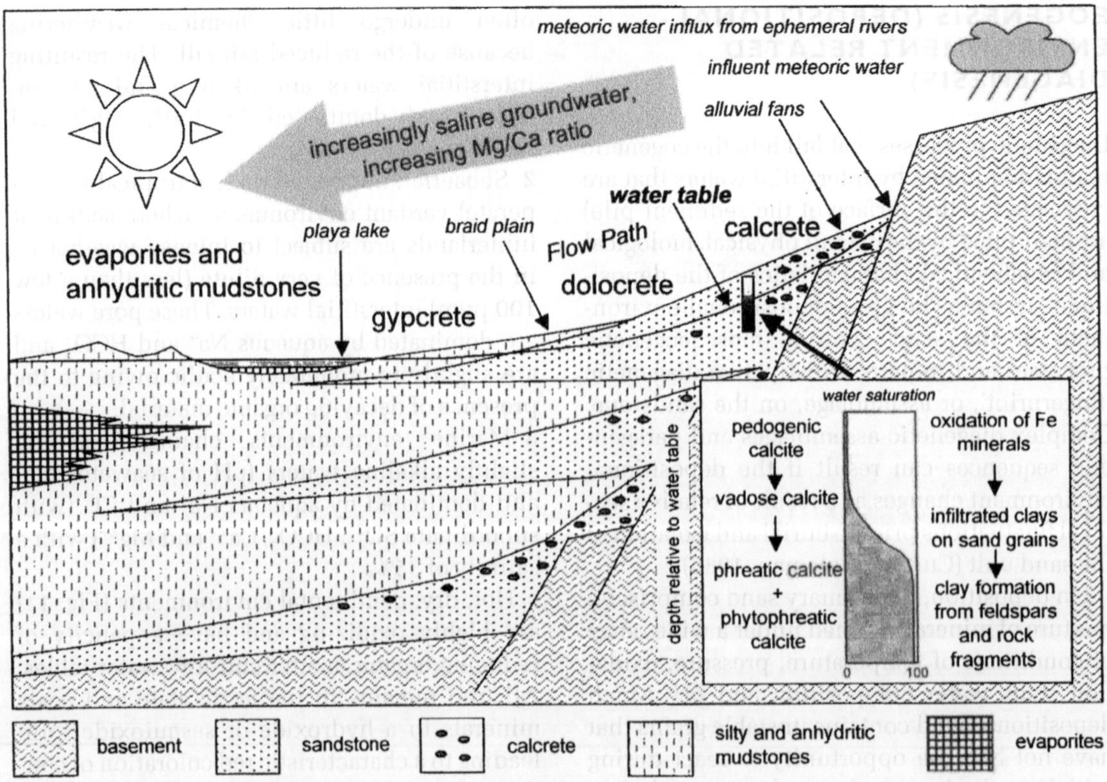

Fig. 11 Schematic diagram summarizing reactions in the fluvial–alluvial, hot and dry eogenetic regime. Calcretes develop closest to the rift shoulder, dolocretes form some distance down the flow path and gypcrete only forms in the vicinity of playa lakes. Waters have increasing Mg/Ca ratios down the flow path as calcite growth leaves Mg in the water. The groundwaters become increasingly saline owing to evaporation and weathering/alteration of detrital grains. Pedogenic calcretes form near to the surface although groundwater calcretes (phreatic) form deeper in the section. Plant activity can raise local HCO_3^- levels and cause localized phytophreatic calcrete formation. Dolocretes can also form in soils and in groundwaters.

Warm, wet, typically verdant, subaerial eogenetic environments (Fig. 12) have an abundance of reducing organic matter that undergoes bacterially mediated decay (Curtis *et al.*, 1986). Any ferric iron in the solid sediment is soon reduced to the Fe^{2+} form by the redox processes and is often available for incorporation in siderite. Siderite-dominated eogenetic carbonates occur owing to local enrichment of iron in the system, resulting from the alteration of Fe-rich detrital clays (Baker *et al.*, 1996; Browne & Kingston, 1993). However, the relative absence of SO_4^{2-} and Mg^{2+} leads to a different suite of minerals than found in the marine eogenetic realm. Sulphide minerals are relatively scarce; carbonate minerals include the nearly universal eogenetic mineral calcite but also ferroan dolomites and siderite, including the iron-rich dolomite pistomesite. In the relative absence of sulphur (in the form of sulphide), any iron is free to associate with carbonate or aluminosilicate clay minerals. The clays in this environment are typified by kaolinite as this mineral represents the last (i.e. advanced) stage of silicate weathering prior to the development of oxides and hydroxides (e.g. bauxite and goethite, as in laterites). Andrews & Turner (1991) documented a case of dolomite cement forming during non-marine eogenesis, assigning the abundance of Mg to sporadic temporary inundations by seawater,

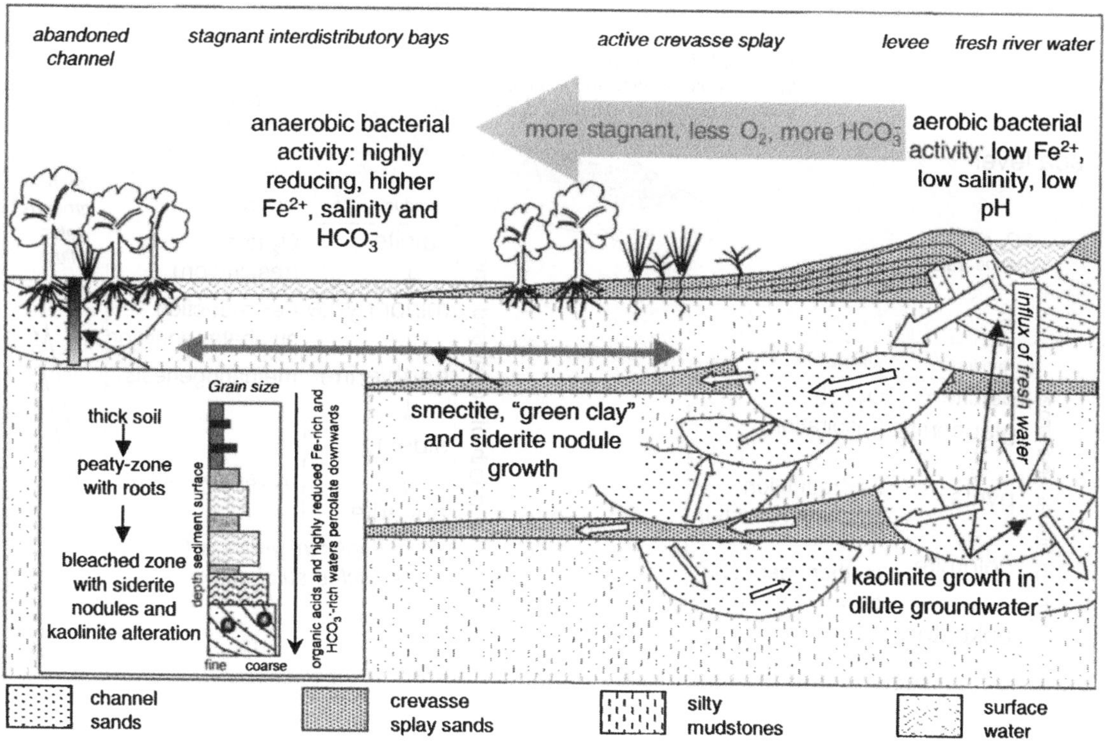

Fig. 12 Schematic diagram summarizing reactions in the fluvial, humid eogenetic regime. Influent, fresh, oxidized water from the river will lead to low salinity, low pH and low Fe-waters in the underlying sands and thus kaolinite growth with minimal carbonate mineral formation. Flow from rivers (magnitude represented by size of arrow) decreases away from the main river axis, with flow rates depending on connectivity of sandbodies. Distal to the main channel axis, groundwaters become stagnant, and in the presence of decaying organic matter, promote the activity of anaerobic bacteria, leading to elevated aqueous Fe, HCO_3^- and higher pH values. Stagnant waters result in higher aqueous potassium and silica concentrations and thus result in smectite growth (instead of kaolinite growth as under the main channel) or the growth of various 'green clays' (Fe-rich smectites). Vegetated abandoned channels are filled with organic-rich silt and detrital clay that promote anaerobic conditions. The abundance of humic acids from resulting peat development lead to bleaching of sands (so-called 'seat-earths', effectively advanced 'weathering') and kaolinite growth and the development of siderite nodules deeper in the section.

weathering of a local source of Mg in the hinterland, or even contemporary volcanism.

Bacterially catalysed processes characterize marine eogenesis (Fig. 13). The interaction of reducing organic matter and oxidizing inorganic solutes (e.g. SO_4^{2-}) and minerals (e.g. Fe^{3+} minerals) causes rapid eogenetic alteration of shallow buried sediments (Irwin *et al.*, 1977; Hesse, 1986). The bacteria and a potent mix of oxidizing and reducing material leads to an authigenic suite of minerals, including pyrite (and other sulphides), various carbonate cements and a range of typically green sheet-silicates ('glaucony') including glauconite, illite and smectite. The unusual glauconite–phosphate assemblage is typical of suboxic zones and may be related to sulphide oxidation (Coleman, 1985). Colonization of shallow-marine sandstones by shelly invertebrates (especially during breaks in sedimentation) can lead to an abundant supply of the components needed for carbonate authigenesis. The development of calcite nodules and sheets is often characteristic of (especially shallow) marine eogenesis. Eogenetic dolomite cements can develop (see the carbonate literature, e.g. Tucker & Bathurst,

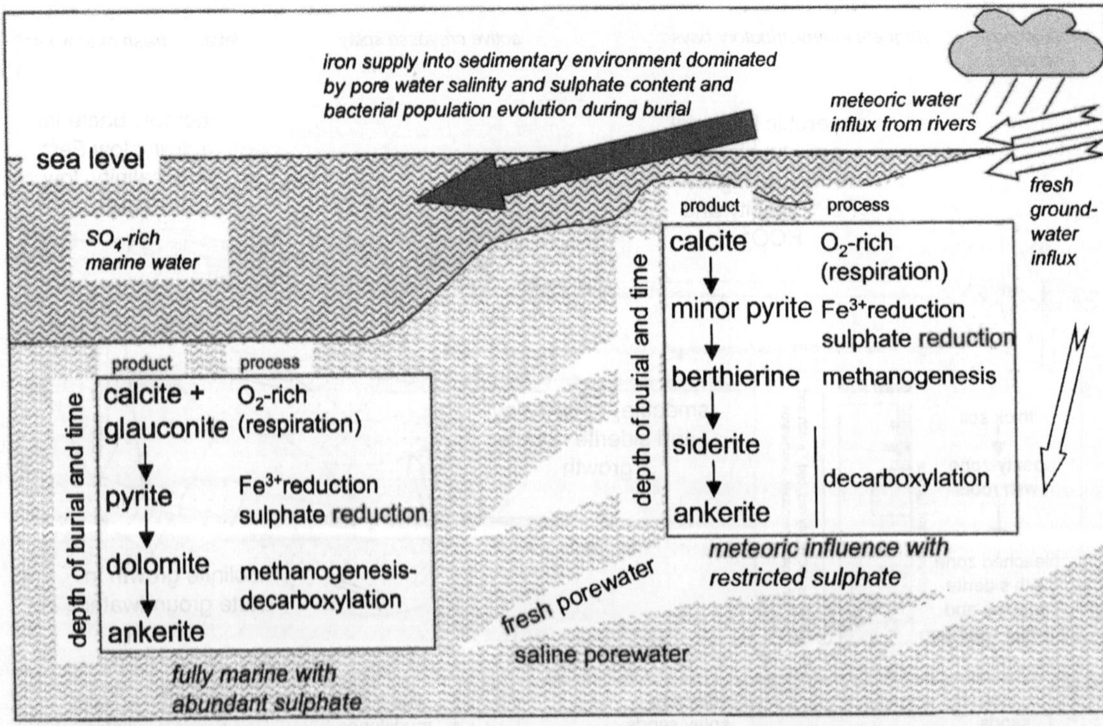

Fig. 13 Schematic diagram summarizing reactions in the marine eogenetic regime. Seawater-saturated sediments undergo a series of bacterially mediated processes: respiration, Fe and then sulphate reduction, methanogenesis and decarboxylation. These lead to a series of depth-related eogenetic facies: calcite +/– glauconite, pyrite and dolomite (+/–ankerite). Deltaic, fresh or brackish water eogenetic facies are different mainly because there is much less sulphate available (leading to less sulphide and thus less pyrite). The Fe that remains, owing to the lack of pyrite growth, instead forms 'green clays' such as berthierine because methanogenesis destroys CO_2 and prevents Fe-carbonate mineral growth. Decarboxylation reactions take over from methanogenesis with further burial, and result in siderite growth.

1990) although they are less abundant than calcite.

COMPACTION

This is the process of volume reduction and consequential pore-water expulsion within sediments. Normally this takes place in response to vertical shear–compressional stresses owing to increasing weight of overburden, but the same processes operate under tectonic compressional forces. Compaction can be expressed as a percentage of the original porosity of the sediment or by specific compressibility values based on strength or rigidity. All of these parameters are influenced by the lithology of the sandstone (Pittman & Larese, 1991) and relative porosity values commonly used by sedimentologists are a useful way of expressing compaction during burial (Fig. 14).

Moderately sorted, coarse-grained clastic sediments have depositional porosity values of about 40% (Beard & Weyl, 1973), which corresponds to a random packing fabric of uniform spheres. In fact initial porosity values are strongly dependent on grain packing arrangements and so are much more affected by sorting than grain size. Compaction takes place as the sediment package is gradually buried by younger sediment and the overburden load increases. The overburden load operates as an effective stress, defined as the difference between the lithostatic pressure and fluid pres-

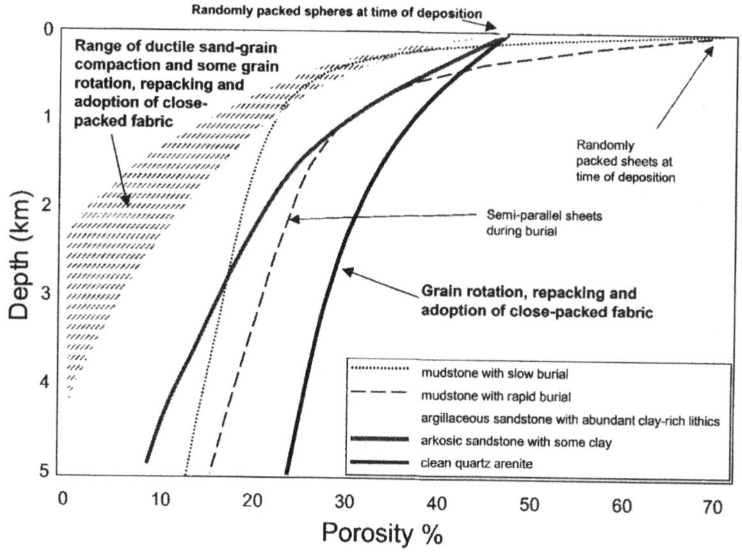

Fig. 14 Representative family of compaction curves for sandstones of different primary lithologies and, for comparison, mudstone compaction curves for rapid and slow sedimentation rates. Note that these curves represent only the effects of compaction, and not the effects of cementation. Depositional sands have porosities of about 45%, the value typical for randomly packed, well-sorted spheres. Depositional muds have higher porosities of up to 70% owing to electrostatically bound water and the platy nature of mudflakes. Compacting sandstones free of ductile (clay-rich) grains reach minimum porosity values of about 26% (the value typical for close-packed spheres) simply as a result of grain rearrangement. Increasing quantities of ductile grains result in lower porosities for a given depth of burial because the clay-rich grains get squashed and ultimately form pseudomatrix.

sure (Gretener, 1976). In normally pressured systems, the fluid pressure is the hydrostatic pressure and depends only on the fluid density and depth of burial. The resulting degree of compaction is largely dependent on the ratio of brittle to ductile sand grains (Worden *et al.*, 2000a). Compaction in sandstone occurs in response to four processes: grain rearrangement, plastic deformation of ductile components, dissolution and brittle fracturing (Fig. 15). It is important to understand the relative effects of compaction and cementation on overall sandstone porosity-loss. Based on the assumption that sandstones start with a porosity of 40%, this can be achieved by determining the volume of cement and the intergranular volume (IGV; sum of remaining primary pore space and volume of pore filling cement; see Fig. 16).

In sandstones composed predominantly of competent grains (e.g. quartz, feldspars) the potential for grain reorganization is much less than for mudstones. Physical compaction and the adoption of a close packed geometric arrangement should lead to a theoretical porosity value of about 26% (Beard & Weyl, 1973). This mechanical style of compaction is dominant in the first 1000 m or so of burial and typically results on porosities in the range 25–30%, conforming to the theoretical value (Worden *et al.*, 1997) when continued to depths of 2000–4000 m burial.

In sandstones with an abundance of less competent (ductile) sand grains such as clay-rich weathered volcanic fragments, mud intraclasts or metapelite rock fragments, there is the opportunity for intense plastic deformation. The ductile grains become extruded between the more rigid grains. Much porosity loss results from this style of mechanical compaction and negligible porosities may be found in ductile-rich sandstones even at 2000 m burial (Pittman & Larese, 1991).

After the initial physical rearrangement of grains, chemical compaction can continue

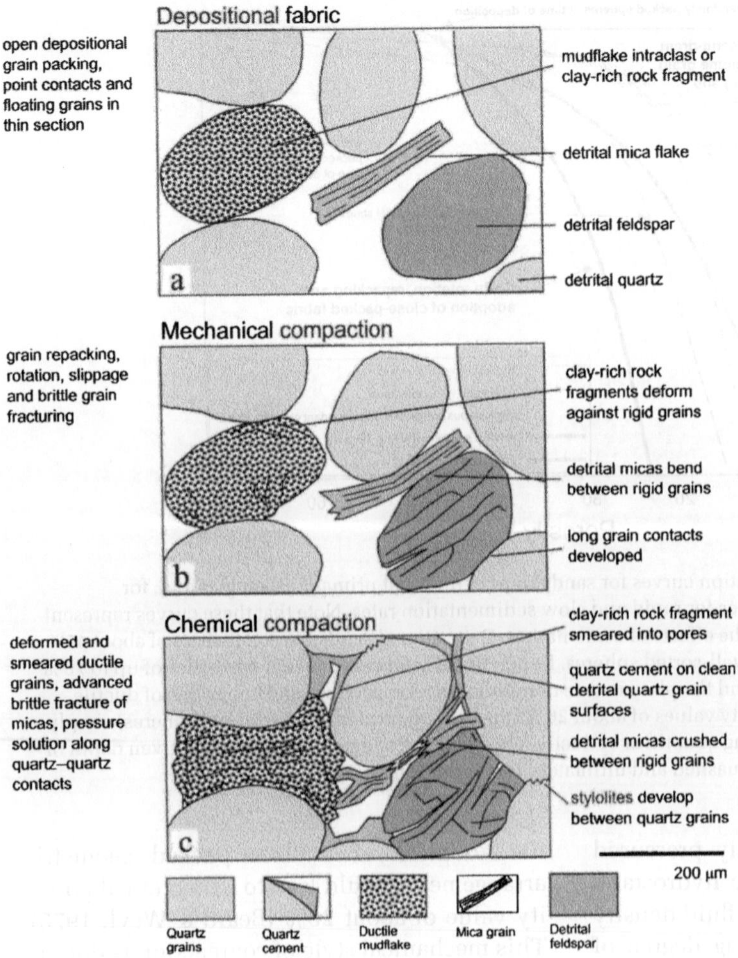

Fig. 15 Schematic cartoon illustrating compaction processes in sandstones. (a) Typical high-porosity fabric at deposition, (b) fabric after mechanical compaction, (c) fabric after mechanical and chemical compaction.

during burial, resulting in further porosity loss (Fig. 15; Thompson, 1959). The solubility of silicate minerals tends to increase with increasing pressure and temperature (Robin, 1978). Pressure dissolution is thus a compactional response of sandstones during burial to increase the surface area of grain–grain contacts, thus spreading the load (effective stress owing to overburden) over a greater area (Robin, 1978; cf. Bjørkum, 1996). Thus point-contacts between grains evolve through straight-elongate to concavo-convex and ultimately interpenetrative or even sutured contacts (Taylor, 1950; Plate 1, facing p. 24). Such depth-related compactional pressure dissolution processes are well documented in many sedimentary basins (e.g. Houseknecht, 1984; Ramm & Bjørlykke, 1994). Mullis (1992) argued the contrast in grain size at lithological boundaries is responsible for the heterogeneity of pressure dissolution fabrics recorded from many sequences. Pressure dissolution is strongly influenced by mineralogy, mineral chemistry and structure, as evidenced by differential amounts of dissolution between pressure dissolved minerals (Matter & Ramseyer, 1985), even between detrital quartz grains of differential solubility (Plate 1).

If the fluid pressure is greater than hydrostatic (i.e. if the fluid is overpressured), then the effective stress is reduced. Consequently, the effects of both mechanical and chemical compaction are reduced relative to normally

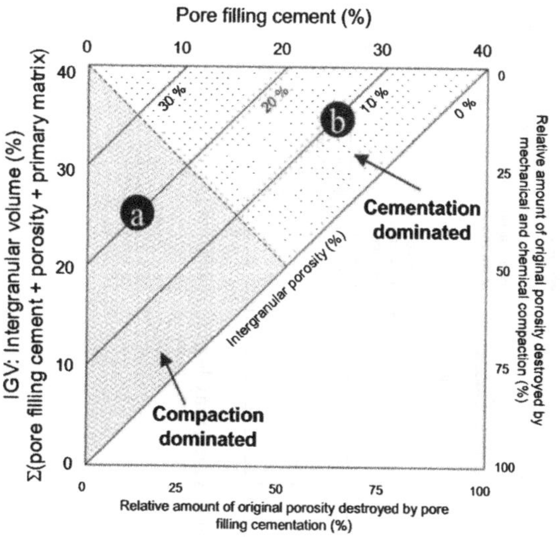

Fig. 16 Diagram illustrating contrasting styles of porosity loss in sandstones (eponymous after Houseknecht, 1984). IGV is the intergranular volume equal to the sum total of remaining intergranular porosity and the volume of pore-filling cements and matrix. The IGV does not include cements that have replaced detrital grains or secondary porosity. The plot assumes an initial (deposional) porosity of about 40%. The region marked 'a' represents sandstones dominated by compactional porosity loss. The region marked 'b' represents sandstones dominated by cementational porosity loss. Sandstone 'a' has 5% cement but 20% intergranular porosity, suggesting that 15% porosity loss was the result of compaction. Conversely, sandstone 'b' has 25% cement but 10% intergranular porosity, suggesting that only 5% porosity loss resulted from compaction.

pressured sandstones (Osborne & Swarbrick, 1999). Compaction is thus a combined function of depth of burial, lithology (rigid ductile grain ratio, mineralogy) and fluid pressure. Although less well documented, compaction is also probably a function of strain rate, as a consequence of changes in burial history or evolution of overpressure development.

MESOGENESIS (BURIAL DIAGENESIS)

This regime was originally defined by Schmidt & Macdonald (1979). It includes the physical, chemical and biological processes that act upon a sediment during burial as it is gradually removed from the influence of the depositional environment and persists until the onset of metamorphism, or structural uplift and exposure to meteoric water. During burial, a multitude of physical and chemical reactions are initiated in response to changes in the ambient geochemical environment. Subsequent to eogenenesis, the sediment comprises a mixture of detrital components that were initially stable, or metastable, in the eogenetic system, and new authigenic minerals that reached stability during eogenesis. Further changes in pressure, temperature and pore-water chemistry effect changes in the stable mineralogy and cause diagenetic reactions to occur. In response to changes in pressure, temperature and pore-water composition, many mineral assemblages, previously stable under eogenetic conditions, become unstable and thus react to produce new stable authigenic minerals. Reaction rate increases during burial following kinetic principles. However, the equilibrium conditions also shift during progressive burial. Diagenesis is thus a dynamic process. Geochemical equilibrium, rarely achieved by the residual mineral matrix, constantly shifts as pressure, temperature and pore-water composition change.

The physical processes of compaction combine to reduce porosity with increasing burial. Initially, porosity is reduced by mechanical compaction, principally through grain reorientation and, less commonly, brittle fracturing. At greater depths, chemical compaction becomes important (Thompson, 1959) as pressure dissolution operates. As a direct result of compaction, the volume of intergranular pore space is greatly reduced. The amount of physical compaction is most intense during the first 500 m of burial, but continues until the closest grain packing arrangement for the particular sandstone grain framework is reached. For well-sorted sandstones this equates to an intergranular volume of around 26%, with decreasing percentages for progressively less well-sorted sandstones.

In parallel with the physical processes of compaction, both inorganic and organic constituents of clastic rocks (including sandstones

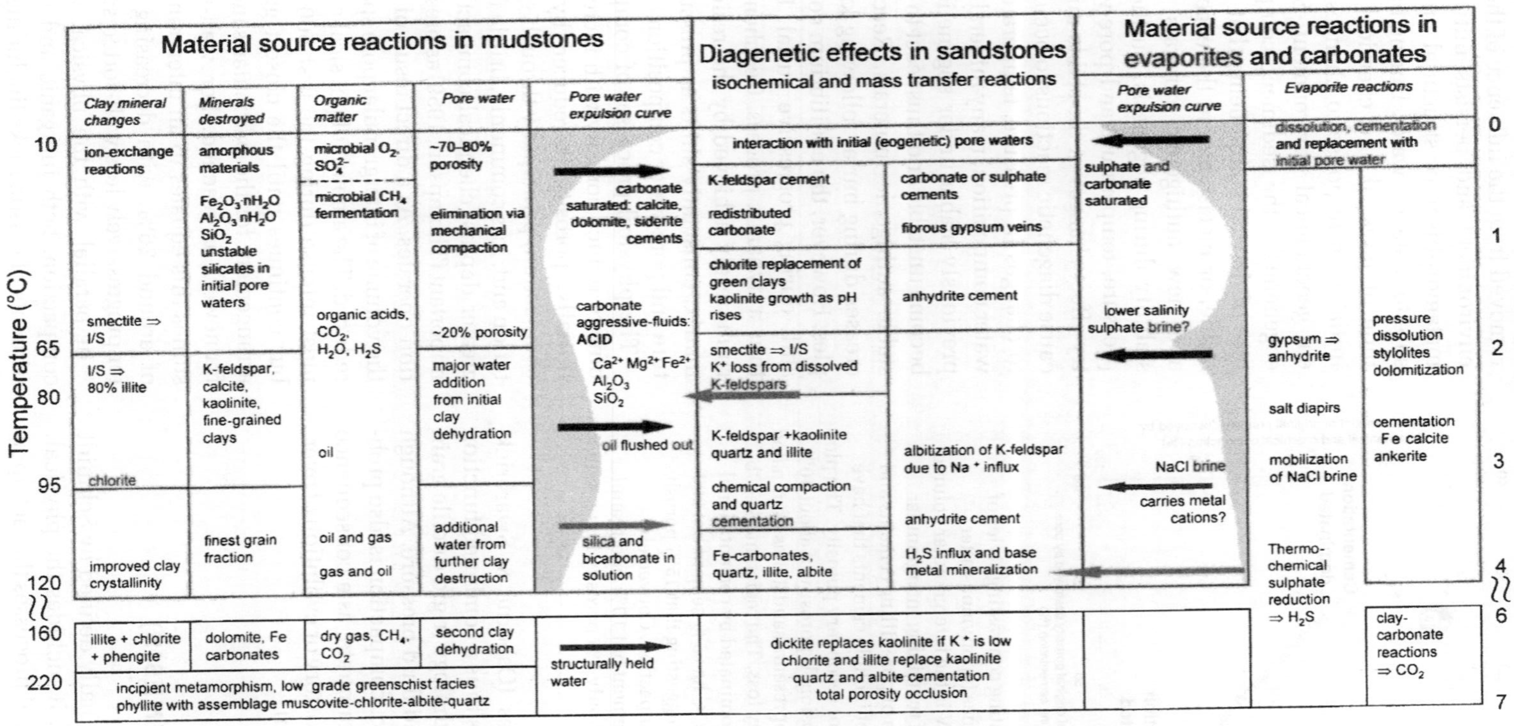

Fig. 17 Summary of the key depth-related diagenetic processes in mudstones, evaporites and carbonates and which may lead to reactions in sandstones (modified from Burley *et al.*, 1985; Burley, 1993).

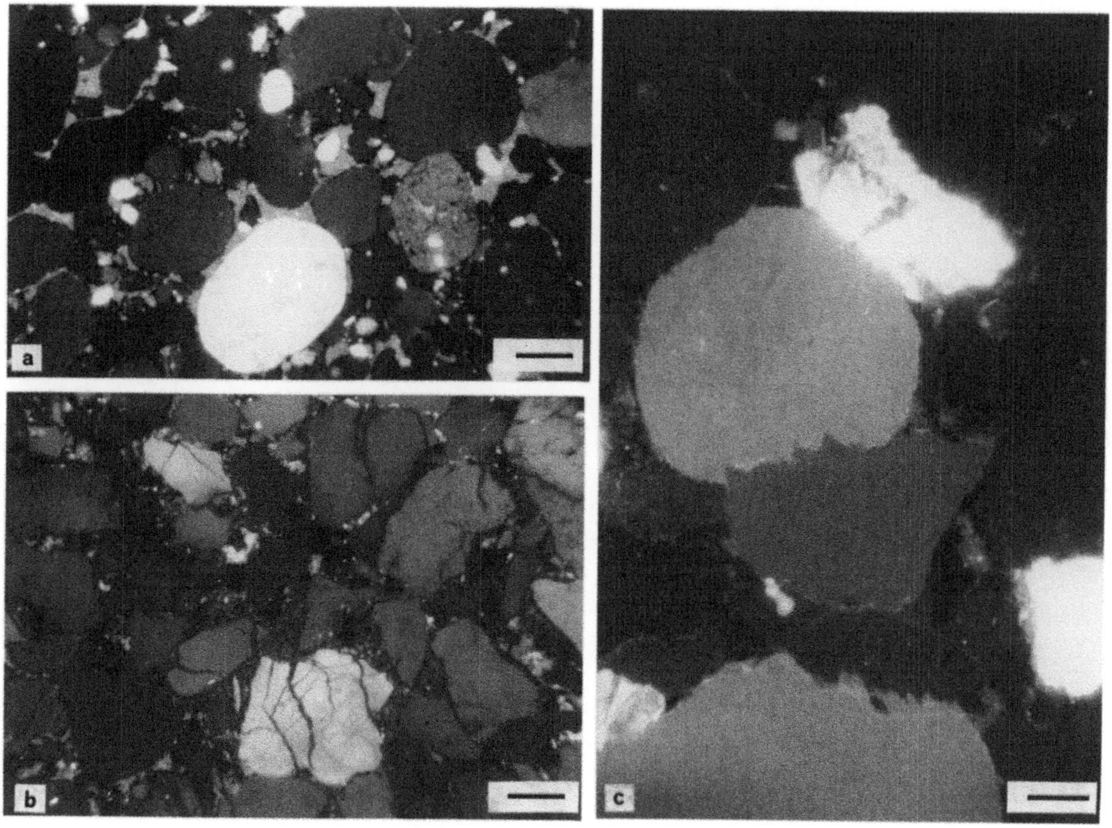

Plate 1 Compaction fabrics in sandstones revealed by cathodoluminesence. Blue grains are plutonic quartz, pink grains are volcanic quartz, white grains are K-feldspar in all images. (a) Initial stages of chemical compaction (after mechanical compaction) showing close packing grain fabric and initial development of concavo-convex grain contacts (scale 500 μm). (b) Extreme mechanical compaction accomplished by grain reorientation and brittle fracturing, accompanied by quartz cementation (scale 400 μm). (c) Later stages of chemical compaction showing development of interpenetrative grain contacts between quartz grains. Note the differential solubility of quartz types at grain contacts (scale 1600 μm).

and mudstones) undergo a variety of chemical depth-related changes in response to increases in pressure and temperature. Most diagenetic reactions are strongly temperature dependent. Increasing temperature during burial promotes many reactions for several reasons. First, increasing ambient temperature adds energy to the reacting system. Reaction rates thus increase and reaction barriers are more easily overcome: diagenetic processes occur more quickly with increasing temperature. Second, the stable mineral assemblage changes with increasing temperature leading to a series of prograde diagenetic reactions (analogous to prograde metamorphism and metamorphic 'index' minerals: Walther & Wood, 1984). Thus, whereas kaolinite and smectite minerals may be stable at low temperature, illite and quartz (and other minerals) become stable at higher diagenetic temperatures.

Many burial diagenetic reactions are dehydration reactions. Releasing the volatile components of minerals (–OH, $-CO_3$) increases the entropy of a mineral system and will be favoured with increasing temperature (Yardley, 1989). With increasing depth of burial, less hydrous minerals tend to be most stable (anhydrite over gypsum, illite over smectite, feldspars over zeolites). There is also a tendency for burial environments to have increasingly lower redox potential, so that reactions incorporating reduced valence metals are favoured, especially in the presence of hydrocarbons. The increase in effective stress during burial may favour the precipitation of more dense minerals in pressure sensitive reactions (such as calcite to dolomite).

It is the reactivity of the sandstone, at times influenced by large-scale sediment–water interaction, that characterizes burial diagenesis. All the various lithological components of a gradually subsiding sedimentary basin will respond to the evolving physico-chemical regime. Minerals may dissolve, precipitate, or be replaced. Pore waters will be modified or replaced dependent upon the relative rates of compaction of neighbouring rock units. The constituents of a subsiding volume of sandstone therefore should not be considered in isolation. An integration of the processes in the whole sediment column is required to fully understand burial diagenetic processes in a given sand body (Curtis, 1983; Burley *et al.*, 1985; see Fig. 17).

Analysis of large data sets from sandstones that have undergone widely variable degrees of burial and heating have allowed the synthesis of numerous 'generic rules' regarding the temperatures at which different diagenetic reactions take place. Figure 17 summarizes these reactions for sources of reactive components (mudstones, carbonates and evaporites) alongside sandstone 'sinks'. The key reactions are detailed in the following synthesis. Although this is not a rigorous analysis of all the kinetic data and reactions, it provides an overview of some of the main burial diagenetic reactions compiled from many studies.

1 Dioctahedral smectite undergoes progressive replacement by illite at temperatures greater than about 70–90°C (Boles & Franks, 1979; McKinley *et al.*, 2003). The replacement of smectite by illite in sandstones is broadly concomitant with earlier stages of oil generation in mudstones adjacent to sandstones. In sandstones, it is possible that illite with minor smectite typically results from smectite precursors whereas pure illite (at least those that have not been heated beyond about about 100–120°C) results from direct precipitation from pore waters, with K derived from the breakdown of other aluminosilicate minerals.

2 K-feldspar dissolution in sandstones occurs over depth ranges of 1.5 to 4.5 km (temperature ranges from 50 to 150°C; Wilkinson *et al.*, 2001) but is commonly extensive by 2.5 km burial producing so called 'diagenetic quartz arenites' (Harris, 1989).

3 The reaction of kaolinite with K-feldspar to produce illite and quartz is important in both sandstones and mudstones (Bjørkum & Gjelsvik, 1988). The reaction is prevalent at temperatures greater than about 70°C but becomes pervasive at temperatures greater than 130°C. According to equilibrium thermodynamic models, K-feldspar and kaolinite are universally unstable together and their coexistence at lower diagenetic temperatures results from either the presence of a kinetic reaction barrier or their slow rate of reaction at low temperatures.

4 Quartz cementation is a significant process only at temperatures greater than about 70–80°C (e.g. Giles *et al.*, 2000). Most quartz cementation seems to occur in sandstones over the temperature range 80–100°C.

5 Grain-coating chlorite forms primarily by the transformation of Fe-rich clays (primarily berthierine) at burial depths greater than about 3 km and temperatures greater than 90–100°C (Ehrenberg, 1993; Aagaard *et al.*, 2000). The precursor clay mineral was described by Aagaard *et al.* (2000) as an amorphous Fe-bearing, berthierine-dominated mixed-layer clay coating the surface of sand grains. Odinite has been suggested as a precursor clay mineral for chlorite formation by Ryan & Reynolds (1996).

6 Carbonate cementation during mesogenesis is dominated by ferroan dolomite, which grows at similar or somewhat higher temperatures than illite and quartz cements (about 100°C). However, a wide range of carbonate cements (ferroan calcite, non-ferroan dolomite, siderite) can grow over a wide range of diagenetic conditions in sandstones (see references in Morad, 1998).

7 Increasing temperature promotes recrystallization of carbonate cements, particularly calcite, to cements of increasing crystal size.

8 At burial depths between about 3.0 and 4.5 km (T = 90–130°C), eogenetic kaolinite is subjected to severe dissolution and reprecipitation as blocky crystals of ordered and disordered dickite (Ehrenberg *et al.*, 1993; McAulay *et al.*, 1994; Beaufort *et al.*, 1998).

9 Kaolinite can be replaced by chlorite, especially in systems that have little available potassium (thus preventing the occurrence of illite growth at the expense of kaolinite). Boles & Franks (1979) recorded that the chloritization of kaolinite occurs at burial depths between about 3.5 and 4.5 km (T = 165–200°C). In contrast, chloritization of kaolinite in Triassic reservoir sandstones of the Lunde Formation, offshore Norway occurs close to the oil–water contact at depths of about 2.5 km (T = 100°C; Ehrenberg, 1993).

10 Gypsum dehydrates to anhydrite at depths of between 1.5 and 4 km (T = 50–120°C) depending upon pore-fluid salinity, pressure and thermal gradient (Hardie, 1967; Jowett *et al.*, 1993).

On top of this general framework of sandstone mineral composition moving towards thermodynamic equilibrium with enclosing pore waters, is superimposed the migration of pore waters in the subsurface under a compactional pressure drive. Pore waters tend to migrate from low permeability to high permeability rocks (from high compressibility, low permeability, low porosity rocks—such as mudstones—to more porous, less compressible rocks—such as sandstones). The transfer of dissolved species in such pore fluids—especially Na^+, K^+, Ca^{2+}, Mg^{2+} Fe^{2+}, $HCO^-{}_3$, SO^{2-}_4 and Cl^-—is known as 'mass solute transfer' and has provoked widespread controversy in the diagenetic community. Mass solute transfer is potentially an extremely important process during burial. Fluid and thus solute migration tend to take place from evaporites, carbonates and organic mudstones, all rich in solutes and base metals, to more porous units (sandstones and some carbonates) where cementation and mineralization can result (Jackson & Beales, 1967; Boles & Franks, 1979; Baines *et al.*, 1991). However, the importance of mass transfer in sandstone diagenesis is still the subject of much debate. Arch proponents of isochemical diagenesis claim that mass transfer is an irrelevance (Bjørkum & Gjelsvik, 1988), whereas the allochemical diagenetic proponents advocate that regional patterns of diagenesis and porosity/mineral anomalies can be explained by large-scale fluid circulation models (Sullivan *et al.*, 1990; de Caritat & Barker, 1992; Gluyas & Coleman, 1992), especially where faults form part of the migration system (Burley *et al.*, 1989).

Basin-scale fluid flow and sandstone diagenesis

Basin-scale fluid flow is the mass convective movement of fluid either within one (typically permeable) formation or across formation boundaries (known as cross-formational flow). The fluids may be aqueous (either water or brine), non-hydrocarbon gases (CO_2, H_2S, N_2 or the noble gases) or petroleum fluids (oil, gas or condensate). Basin-wide fluid flow can occur at any stage in the history of a basin, from its inception through to its terminal stages following inversion and uplift (Crossey *et al.*, 1996; Moñtanez *et al.*, 1997).

Background aqueous fluid pressure ordinarily increases as a function of depth (height of fluid column) and fluid density (Gretener, 1976). This fluid pressure is known as hydrostatic pressure because the fluid is at equilibrium and does not flow in response to this pressure. Aqueous fluid movement occurs in response to increased pressure gradients, with the fluid moving away from zones of elevated fluid pressure (overpressure). Increased fluid pressure can develop as a result of: (i) inhibited compaction during rapid burial (too much fluid remains); or (ii) localized generation of fluid. Processes that cause fluid generation include source-rock maturation producing petroleum (and water, CO_2, H_2S, or N_2), gypsum dehydration and clay mineral dehydration reactions (e.g. smectite to illite reaction) (see Fig. 17). In contrast to aqueous fluids, hydrocarbon migration is largely caused by the inherent buoyancy of hydrocarbons in water (Schowalter, 1979). Consequently, hydrocarbon migration is decoupled from aqueous fluid migration.

Basin-scale water movement is thought to be responsible for a variety of diagenetic effects in sandstones (see Fig. 17). An influx of low salinity water may result in dissolution of feldspars and carbonates in sandstones (Bjørkum *et al.*, 1990; cf. McLaughlin *et al.*, 1994). Conversely, widespread precipitation of authigenic cements may occur if the incoming water is highly saline (e.g. Sullivan *et al.*, 1994) and mixing of different formation waters takes place (e.g. Burley *et al.*, 1989). The characteristic feature of large-scale fluid migration is the potential for mixing zones to occur that are preferentially the loci of intense diagenetic reaction (Giles, 1987). Widespread flow of aqueous fluids from mudstones has been thought to be responsible for the growth of (locally zoned) ankerite in sandstones (Gawthorpe, 1987; Mozley & Joernle, 1990).

Basin-scale water flow can take place at any stage during the evolution of a sedimentary basin (Fig. 18). In the early stages of basin formation large-scale water flow is usually driven by topographic head (Bethke & Marshack, 1990). It is well documented from continental basins bounded by mountain ranges where groundwater influx can penetrate for kilometres into sedimentary sequences. This flow results in the infiltration of fine grained sedimentary materials (typically clays bearing hydrous oxides), and causes replacive alteration of detrital aluminosilicate minerals, growth of calcretes or dolocretes and promotes the oxidation of detrital Fe-minerals (Walker *et al.*, 1979). However, equally remarkable is the displacive power of different density waters. Present-day fresh water, for example, discharges tens of kilometres off the south-eastern coast of the USA from a hinterland with effectively no topographic head (Manheim & Paull, 1981).

During burial and mesogenesis large-scale water movement occurs, initially in response to compactional processes in sandstones and mudstones (Gretener, 1976). As pores collapse, fluids either remain in the reduced pore volume leading to elevated fluid pressure, or flow out of the compacting sediments towards rock volumes of close to hydrostatic pressure. During subsequent burial, thermally induced events such as source-rock maturation, gypsum dehydration, the smectite to illite reaction or pressure dissolution become the dominant compactional processes (see Fig. 17). Most fluid flow is upwards and towards basin margins because the fluid pressure is always hydrostatic approaching the surface. However, localized downward flow can occur if fluid pressures decrease (towards hydrostatic values) deeper in the section. Basin-scale fluid flow is typically limited by the existence of permeable escape routes and flow may be prevented (fluids become trapped) where low permeability rocks (cap rocks) overlie porous and permeable sandstones. Faults commonly exhibit a duality of

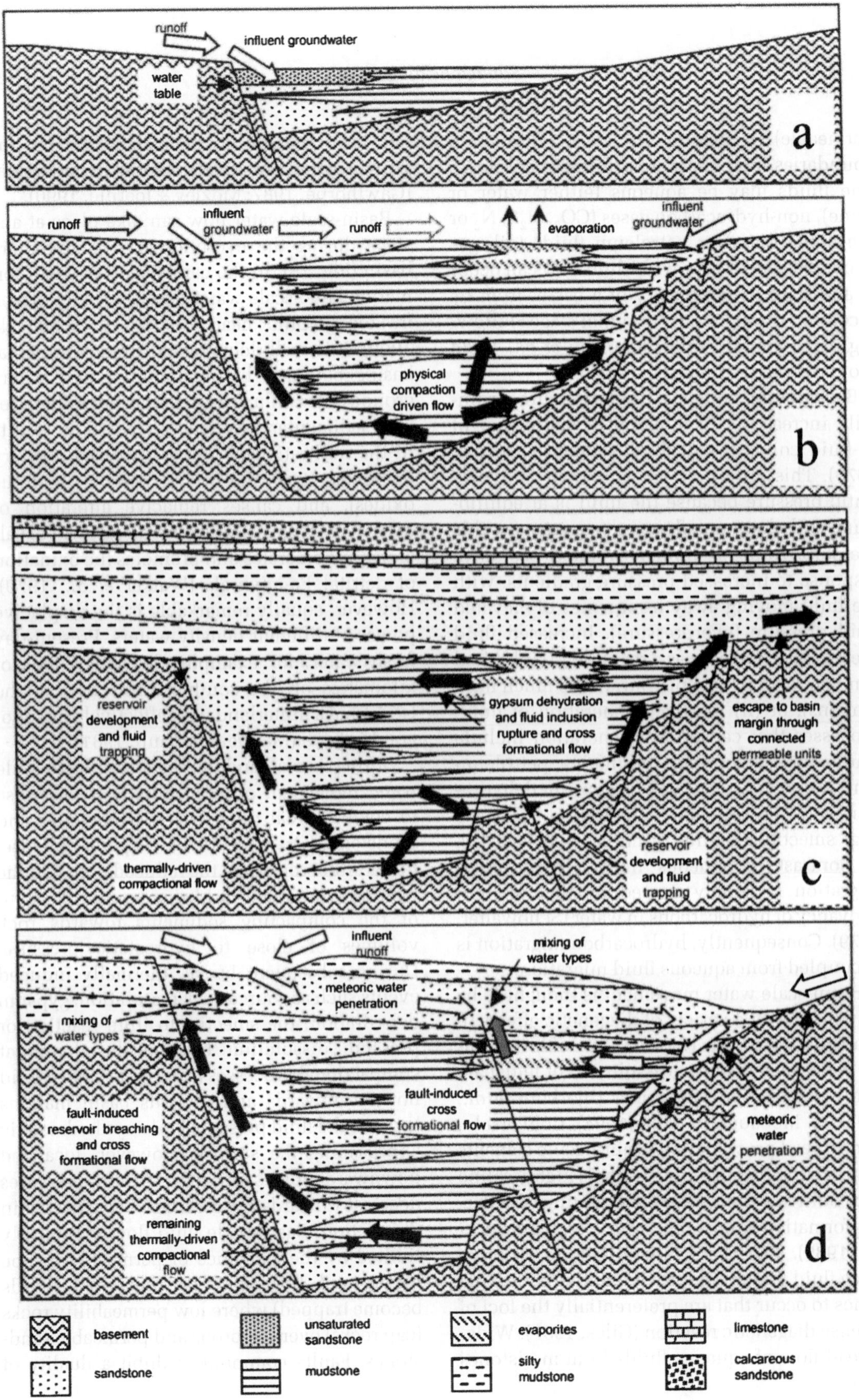
runoff
influent groundwater
water table
a
runoff
influent groundwater
runoff
evaporation
influent groundwater
physical compaction driven flow
b
reservoir development and fluid trapping
gypsum dehydration and fluid inclusion rupture and cross formational flow
escape to basin margin through connected permeable units
thermally-driven compactional flow
reservoir development and fluid trapping
c
influent runoff
mixing of water types
meteoric water penetration
mixing of water types
fault-induced reservoir breaching and cross formational flow
fault-induced cross formational flow
meteoric water penetration
remaining thermally-driven compactional flow
d
basement
sandstone
unsaturated sandstone
mudstone
evaporites
silty mudstone
limestone
calcareous sandstone

behaviour, acting as seals for large parts of the basin history, but as preferential conduits at times of overpressure development or tectonic stress (Knipe, 1993). The movement of fluids from evaporites seems to be locally important to sandstones in many basins, resulting in high-salinity (especially Na^+, Ca^{2+}, Cl^-, SO_4^{2-} and, in some cases, K^+) water that is capable of causing anhydrite cementation, albitization of K-feldspar, illitization of kaolinite, (de Caritat & Barker, 1992; Sullivan *et al.*, 1994; Worden *et al.*, 1999) and, in the presence of hydrocarbons, sulphide mineralization (Burley *et al.*, 1989; Baines *et al.*, 1991; Rowe & Burley, 1997).

Large-scale movement of formation waters certainly occurs in the subsurface. However, the importance of fluid-mixing processes, the volumes of fluids available for diagenetic reaction, and their geochemical character, remain controversial, compared with isochemical diagenetic processes, as a means of causing basin-wide patterns of diagenesis. Large-scale fluid movement is hard to predict in terms of the amounts, direction, water chemistry and controlling factors. This has led to some diagenetic researchers tacitly adopting an assumption that sandstone diagenesis neither involves nor requires basin-scale fluid movement. This stance certainly makes prediction of diagenesis and porosity-loss simpler but it is likely to be incomplete at least in detail, and possibly in general. Remarkably, despite many years of diagenetic and hydrological research, the importance of basin-scale fluid flow as a cause of diagenetic reaction remains the root of much vigorous and lively debate.

Fig. 18 (*opposite*) Schematic diagram of basin-scale fluid migration in a rifted basin that undergoes burial and inversion. (a) Synrift sands have a well-developed water table with extensive meteoric water influx and groundwater flow. (b) Later synrift sandstones have meteoric water influx at the basin margins but mechanical compaction results in water expulsion from buried sediments and flow along high permeability routes. (c) During burial, the higher temperature causes petroleum generation and migration. Post-rift sedimentation seals fluid escape from the synrift sediment packages allowing petroleum accumulation. Chemical compactional processes result in continued water expulsion from buried sediments and flow along high permeability routes. Evaporites lose saline water through gypsum dehydration and rupture of aqueous fluid inclusions in primary halite. (d) Minor inversion ceases sedimentation, with faulting allowing breaching of traps and upwards fluid loss. Erosional unconformities allow meteoric water influx (and local telogenesis). Mixing of the meteoric water and the basinal water leads to diagenetic processes (mostly mineral precipitation).

TELOGENESIS (UPLIFT AND EXPOSURE RELATED DIAGENESIS)

The telogenetic regime is where the waters associated with early diagenetic processes or with burial diagenetic processes are displaced by a subsequent influx of meteoric water. The inward flux of meteoric water must be driven by a pressure head associated with rainfall on upland regions (Fig. 18). The magnitude of the pressure head is given by the height of input zone above the output zone, directly analogous to a siphon. Influx is usually permitted and driven by tectonic uplift, sea-level fall (and thus sediment surface exposure) or through seaward progradation of a coastline. The main controls on the occurrence and magnitude of meteoric water influx include the density difference between fresh and sea water, the amount of opposing upward flux of compaction-related water, the bulk permeability of the formation and the geometry of the input and output points. Telogenetic processes are seldom found in deep basin centres typical of petroleum exploration, but are common in outcropping sedimentary rocks. Telogenetic processes are thus commonly reported in outcrop studies of diagenesis but are more rarely reported in core-based studies of diagenesis, except where ancient unconformities are encountered (Al-Gailani, 1981; Shanmugam, 1988). However, the scale over which meteoric water ingress occurs can be enormous; freshwater lenses are reported 320 km from shoreline in the Bahamas (Bjørlykke, 1989), and deeply buried sandstones in the Canadian Alberta Basin are argued to have experienced extensive flushing with meteoric water (Bethke & Marshack, 1990; Longstaffe & Ayalon, 1987).

Meteoric waters are commonly very dilute, oxidizing, saturated with CO_2 and are thus potentially acidic in nature. This contrasts with most burial-diagenesis-associated waters, which are typically (although variably) saline, reducing and with rock-limited CO_2 concentrations and a near neutral pH. An influx of meteoric water is thus often responsible for characteristic changes in the host coarse siliciclastic rock. These include oxidation of reduced iron-bearing cements including ferroan carbonates (most commonly ankerite but also siderite and calcite), dissolution of feldspars and chert grains (Shanmugam & Higgins, 1988), alteration of feldspar minerals to clay minerals (e.g. the development of kaolinite in Brent Group and Kimmeridge Clay Formation sandstones below the Cimmerian unconformity; Bjørkum *et al.*, 1990; Emery *et al.*, 1990).

Most telogenetic processes occur within the first few metres or tens of metres of the sediment surface. Although large volumes of meteoric water can flow deep into aquifers over protracted periods of geological time, the consequence is often muted in siliciclastic sequences because meteoric waters are commonly of low ionic strength and will tend to reach saturation with the major aqueous species very rapidly during influx, limiting the geochemical consequences. The rock dominates the water–rock system and it is geochemically inappropriate to ascribe deep (hundreds of metres or more) processes to meteoric water influx. Pervasive modification of sandstone mineralogy by meteoric water influx is restricted to the soil or weathering zone where it is responsible for lateritic profiles in subtropical to tropical regions, ultimately leading to bauxite development (Esteoule-Choux, 1983). It is highly unlikely that telogenesis can lead to major movement of alumina and silica as these are so sparingly soluble and are so abundant in the rock. There is, however, greater potential for carbonate and sulphate cementation associated with influx of marine waters from above regional unconformities (as in the Zechstein for example; Sullivan *et al.*, 1990; Stanislavsly & Gvirtzmann, 1999).

Basin inversion and associated fault movement, uplift, erosion and the formation of mountains exposed to meteoric water influx lead to significant changes in the types of basin-scale fluid flow. Fault movement may breach overpressured fluid traps, allowing fluid release to shallower formations. Newly formed mountain ranges can lead to artesian groundwater systems and allow oxidized freshwater influx deep into the basin (Fig. 18). Mixing of artesian meteoric water with geochemically incompatible fault-released deep (highly reduced and possibly saline) formation water has been cited as the cause of a range of diagenetic events in inverted basins (Lee & Bethke, 1994; Chan *et al.*, 2000).

DIAGENESIS IN A TIME FRAMEWORK

Textural relationships and their interpretation define a paragenetic sequence, but provide no information on the specific age or duration of diagenetic events and processes. Sometimes, textural relationships can be used to relate diagenetic events to eogenetic, mesogenetic or telogenetic regimes, especially in relation to the amount of compaction that can be determined from thin-section.

Two approaches have been adopted to date diagenetic events. So-called 'absolute dating' uses radiometric techniques. The K–Ar and Rb–Sr decay schemes have been applied extensively to authigenic clay minerals, principally illites (Hamilton *et al.*, 1989), and occasionally chlorites, as well as authigenic feldspars (Girard *et al.*, 1988), whereas Ar–Ar step heating and laser ablation techniques have been applied less commonly.

The second approach is less direct, involving an independent determination of diagenetic mineral formation temperature (from fluid inclusions or oxygen isotopes), which is then used to date the reaction via a modelled thermal history and burial curve usually obtained from basin modelling. This can be a powerful approach as modelling techniques enable the sensitivity of the 'age' of mineral formation to be investigated in relation to uncertainty in the geology, such as sea-level fluctuations or stratigraphy (Burley & Cornford, 1998; Worden *et al.*, 2000b).

Despite considerable uncertainty regarding the Ar retention of illites (Hamilton *et al.*, 1989), and the reliability of fluid inclusion homogenization temperatures for recording precipitation temperatures (Osborne & Haszeldine, 1993), these techniques have been used successfully to date authigenic mineral precipitation, deduce reaction mechanisms and infer fluid migration events. For example, the timing of diagenetic events in the Rotliegend of the Southern North Sea (Robinson *et al.*, 1993), North Sea Jurassic Brent Sandstone (Liewig *et al.*, 1987) and in the Texas Gulf Coast (Aronson & Hower, 1976; Morton, 1985) have all been constrained by radiometric dating of authigenic minerals. It is this type of quantitative approach that makes a significant contribution to understanding diagenetic processes and reactions.

STYLES OF MESOGENETIC REACTIONS IN SANDSTONES

Much has been written on the possible controls on diagenetic reactions during burial. The following is a brief synthesis of some of the main controls on diagenesis in the mesogenetic realm represented by papers compiled in this volume. The list is by no means exhaustive and some of the items discussed are highly controversial. We highlight some of the more topical issues from our own perspective.

The influence of mineralogical maturity

Sandstones are traditionally described in terms of their relative proportions of detrital quartz, feldspars and lithic grains (Folk, 1965). In terms of chemical diagenesis, the three main feldspar minerals (albite, K-feldspar and anorthite/plagioclase) behave quite differently. Similarly, the mineralogical content of rock fragments is a significant, although commonly ignored, parameter. Cathodoluminescence petrography indicates that the type of detrital quartz—igneous volcanic, plutonic, metamorphic or vein—also influences grain framework stability.

In general, sandstone authigenesis may be grouped in three ways with reference to detrital mineralogy (Primmer *et al.*, 1997). Quartzose sandstones and lithic sandstones dominated by polycrystalline quartz tend to be characterized by quartz cement with subordinate clay (kaolinite and illite) cements. K-feldspar-dominated arkoses and lithic sandstones dominated by granitic rock fragments, tend to have abundant clay cement (kaolinite and illite) as well as quartz cement. Plagioclase-dominated arkoses and lithic sandstones dominated by basaltic, mudstone and metamorphic rock fragments, tend to have abundant clay cement (smectite, chlorite and illite) with zeolite, carbonate and local quartz cement. The degree of clay and other aluminosilicate mineral diagenesis is controlled strongly by the original lithotype of the sandstone as well as the geothermal gradient and the time of sediment deposition relative to heat flow history (Lee & De Vries Klein, 1986).

Allogenic cements, including eogenetic carbonate cements, are naturally independent of the primary mineralogy of the host sandstone. However, the chemical 'inertness' of quartzose sandstones renders them relatively immune to changes in formation water chemistry (redox state, CO_2 content). Conversely, the chemical reactivity of plagioclase-dominated arkoses and many lithic sandstones will make them highly sensitive to fluxes from external sources.

Rate of burial versus rate of heating

One of the key attributes of the diagenetic realm is the dominance of kinetics over equilibrium. On a hand-specimen scale, many sandstones are not at thermodynamic equilibrium: they are effectively metastable. Thermodynamically they are still able to react with their ambient environment but the rate of reaction is so slow that equilibrium is not attained even over geological time. However, geochemical processes speed up with increasing temperature (with an approximate rule of thumb of doubling the rate for every 10°C rise in temperature). The rate of burial and the consequent heating is thus a potentially important issue in sandstones. When rocks are buried and slowly heated they may more nearly reach equilibrium than rocks that are heated up quickly. This means that sandstones in active rift basins or in basins with large amounts of accommodation space (both with very high sedimentation rates) are more likely to be far from equilibrium than those in stable, slowly subsiding intracratonic basins. Thermal disequilibrium is likely to be normal in basins with high sedimentation rates. However, active rift basins tend to have a much higher heat flow than intracratonic basins, so that the former are much hotter at a given depth than the latter. The balance between rate of burial and heating and the heat flux tends to be dominated by the latter. The higher temperatures associated with rifting will often cause extreme diagenesis despite the rapid burial rate.

Pressure regime

The presence of inter-quartz grain dissolution seams is often called pressure dissolution although the role of pressure in the process of stylolite formation has been questioned recently (Bjørkum, 1996). Temperature and the rate of quartz precipitation are now thought by many to be the main control on the rate of stylolite formation. However, pressure (or, more accurately, the effective stress) is apparently important in controlling diagenesis. Elevated porosity is commonly reported for overpressured sandstones. This is routinely thought to be the result of reduced compaction as the effective stress is equivalent to that for a less deeply buried sandstone. However, overpressured sandstones are also reported to contain less quartz cement than their normally pressured equivalents (Osborne & Swarbrick, 1999). This suggests that overpressure not only limits compaction but also restricts quartz cement growth. The interplay between pressure (and pressure dissolution) and cementation is not yet fully understood.

Formation water chemistry and presence of petroleum

Formation waters in sandstones are often saline but the salinity varies from a few per cent dissolved solids to 30% (Hanor, 1987). Significant temporal variations in formation water salinity have been reported for sandstones deep in sedimentary basins over protracted geological time-scales. These are typically associated with halite deposits and convective water flux (e.g. Sullivan *et al.*, 1994; Worden *et al.*, 1999). Chloride is almost invariably the dominant anion in formation water. Natural waters are necessarily electrically neutral so that the total sum of the anions and cations (weighted by the number of charges per atom) must be the same. If the aqueous chloride concentration increases, then so must the amount of aqueous metals (Hanor, 1994). Practically no minerals in clastic rocks contain chloride so that an increase in aqueous chloride does not directly affect sandstone diagenesis. However, the common aqueous metals: sodium, potassium, calcium and magnesium are reactive aqueous species. Their relative proportions routinely affect clay, carbonate and feldspar mineral stability, leading to, for example, albitization of K-feldspar and illitization of smectite. Mineral systems often limit (or buffer) the freedom of a formation water to vary in terms of the ratio of sodium to other metals. Thus sodium cannot simply increase to match an increase in aqueous chloride even though halite dissolution typically controls overall salinity. Charge balance must be maintained, leading to increasing concentrations of potassium and other metals. Diagenetic processes may thus be caused, in some circumstances, simply by an increase (or decrease) in formation water salinity. The rock buffering of formation-water metal ratios is manifest in clay–feldspar–carbonate reactions. An example of this could be an increase in salinity, owing to convective halite-saturated waters, leading first to K-feldspar albitization. This releases potassium to the formation water, resulting in illitization of kaolinite. The overall reaction would be:

$$Na^+ + KAlSi_3O_8 \Rightarrow K^+ + NaAlSi_3O_8$$
K-feldspar albite

$$3Al_2Si_2O_5(OH)_4 + 2K^+ \Rightarrow 2KAl_3Si_3O_{10}(OH)_2 + 3H_2O + 2H^+$$
kaolinite illite

with increasing salinity causing growth of albite and illite at the expense of K-feldspar and kaolinite under uniform conditions of pressure and temperature.

Oil-bearing sandstones are commonly observed to have undergone less diagenesis than the underlying aquifer (e.g. Yurkova, 1970). This is used as evidence that sandstone diagenesis is inhibited in the presence of oil. The effect of hydrocarbons on diagenesis depends on the wettability of the oil–water–rock system and the specific rate controlling step (Worden *et al.*, 1998). Adding oil to water-wet sandstones that have a surface control on rate of diagenesis has a minimal effect. In contrast, diagenesis in oil-wet sandstones is strongly affected by the presence of oil (Worden & Morad, 2000). Oil can have other effects, including the inducement of feldspar alteration and carbonate cementation if the oil is undergoing biodegradation and oxidation (Ehrenberg & Jakobsen, 2001)

POROSITY

Porosity in sandstones is the aggregate total of all the openings or interstices in a rock framework and within grains. The interstices are individually referred to as pores. Porosity is normally expressed as a percentage (or fraction) of the bulk rock volume.

Pores are the discrete spaces within the rock fabric and are connected together by pore throats. Collectively, pores and pore throats constitute the pore network. There are different ways of classifying pores dependent upon the specific objective of the classification (Fig. 19). Porosity that is only visible with a powerful (e.g. electron) microscope is known as microporosity. Porosity that is visible with the naked eye is termed macroporosity.

Porosity that occurs between primary sand grains and that was present at deposition of the sediment is known as primary porosity. Porosity that develops during diagenetic processes such as mineral dissolution is termed secondary porosity. Porosity that develops at a site formerly occupied by a mineral that dissolved and reprecipitated locally elsewhere in the rock is known as redistributional secondary porosity (as no new porosity is created).

Porosity between grains is defined as intergranular porosity (Choquette & Pray, 1970). Porosity within grains (e.g. micropores within diagenetically reacting sands grains) is intragranular porosity. Such porosity is typically microporosity simply by virtue of the fact that the pores must be much smaller than the sand grains. Porosity can also develop in rocks undergoing brittle deformation and is termed fracture porosity. The previous types of porosity are thus matrix porosity relative to fracture porosity.

Porosity in sandstones is a complex function of the depositional porosity, depth of burial and compaction, and the growth of mineral cements in the pore network. Depositional porosity depends solely upon the fabric of the sandstone, with the most pronounced influence being the sorting. The presence of a range of grain sizes in poorly sorted sandstones diminishes the depositional porosity simply because the interstices between the larger grains become filled with the smaller grain size material (Beard & Weyl, 1973). For perfectly sorted sandstone, grain size has no impact on porosity owing to the scale-independence of grain packing.

SECONDARY POROSITY

The development of secondary pores during burial diagenesis has always been, and remains, a controversial topic. Secondary porosity is defined as new porosity created by the dissolution of detrital grains, matrix, or even earlier formed cement coupled with the *removal* of the dissolution products from the sandstone, thus adding to the sum total porosity (Schmidt & MacDonald, 1979). The generation of secondary

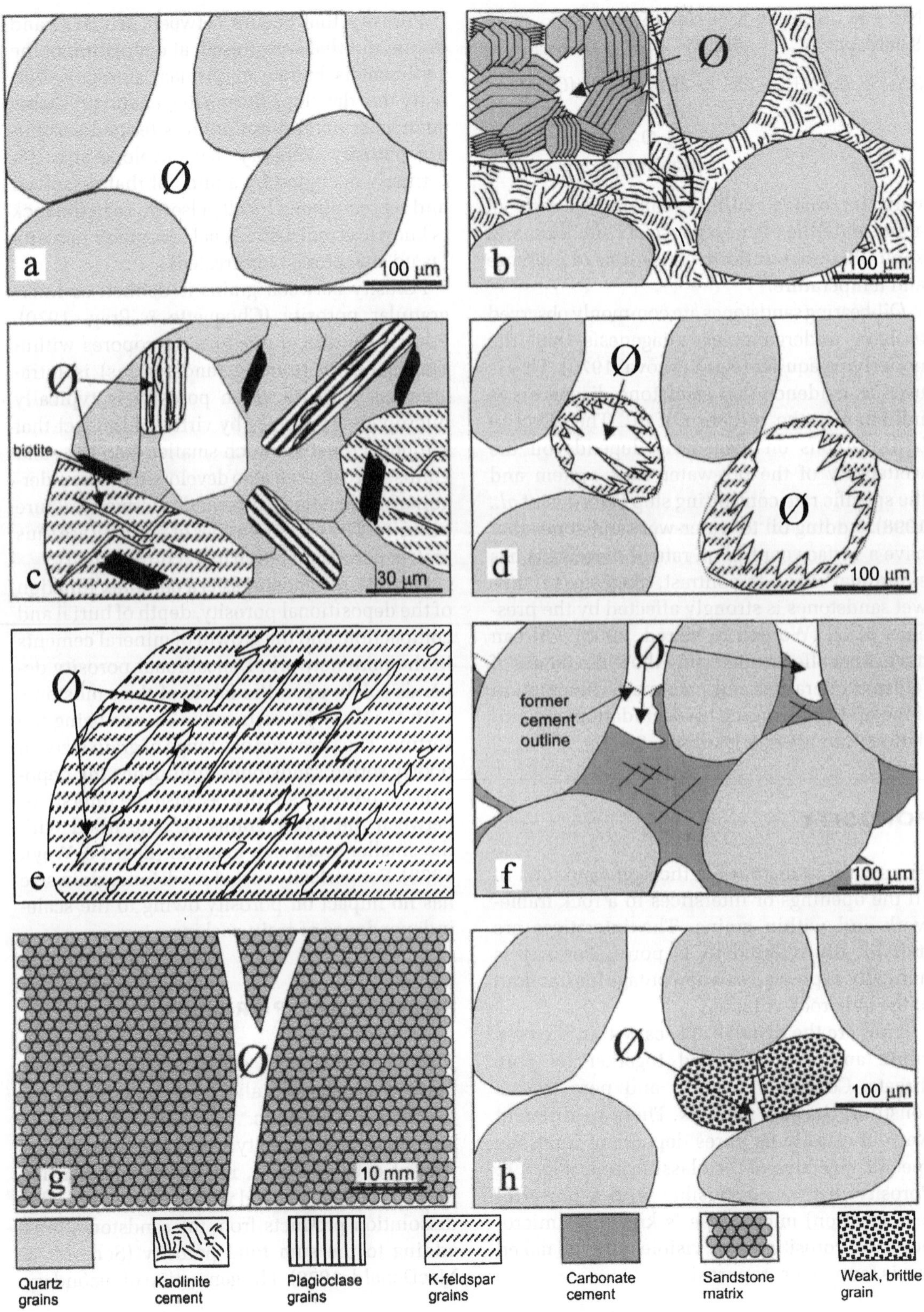
Ø
a
100 µm
b
100 µm
biotite
c
30 µm
d
100 µm
e
former
cement
outline
f
100 µm
g
10 mm
h
100 µm
Quartz grains
Kaolinite cement
Plagioclase grains
K-feldspar grains
Carbonate cement
Sandstone matrix
Weak, brittle grain

porosity can result in an apparently undercompacted fabric, oversized pores, hollowed-out grains, and systematic changes of the detrital rock mineralogy and composition. There are, however, serious problems in dissolving and flushing out rock-forming materials (aluminosilicates, quartz or carbonates) in sufficient quantities to add significantly to the overall porosity of sandstones, given the known solubilities of most sedimentary rock-forming minerals, the composition of pore fluids and the volumes of formation waters available (Giles & Marshall, 1986). Despite this, secondary porosity formed during burial has been suggested to be the cause of anomalously high porosity values in some North Sea oil fields (e.g. Burley, 1986; Harris, 1989; Wilkinson *et al.*, 1997) and oil fields onshore USA (Surdam *et al.*, 1984). The secondary pores in these cases are ascribed to the dissolution of carbonate cements and K-feldspar respectively, together with the mass removal of the dissolution products out of the sandstone.

More commonly developed is redistributional secondary porosity whereby the products of dissolution reactions are locally precipitated (Giles & de Boer, 1990). This involves the dissolution of reactive detrital grains (e.g. Morton *et al.*, 1989) or matrix (or even earlier formed cements) and the local reprecipitation of reaction products. This can lead to new porosity being being created within the dissolving mineral but the occlusion of intergranular porosity by the newly formed minerals. An example of this is the creation of porosity within dissolving detrital K-feldspar grains and the local precipitation of illite in primary pores and quartz on the free surfaces of detrital quartz grains. Redistributional secondary porosity is probably common in sandstones during burial, although the importance of additional secondary porosity remains unclear.

Fig. 19 (*opposite*) Schematic diagram of porosity types in sandstones. (a) Primary intergranular macroporosity, (b) intergranular microporosity between authigenic kaolin crystals (detail enlarged inset), (c) primary intragranular microporosity in a weathered rock fragment, (d) secondary intragranular macroporosity (e.g. within weathered feldspar grains), (e) secondary intragranular microporosity (e.g. weathered out plagioclase zones in detrital perthitic feldspar), (f) secondary intergranular macroporosity (e.g. dissolved eogenetic calcite cement), (g) macrofracture porosity, (h) microfracture porosity.

FUTURE CHALLENGES FOR DIAGENETIC RESEARCH

Perhaps the main aim of diagenetic research, and certainly one of the most practical applications, is to predict the occurrence and amounts of authigenic mineral assemblages in sandstones of a given burial (i.e. temperature and pressure) history. Such a capability would allow geologists to predict reservoir quality ahead of exploration drilling in aquifers and oil reservoirs, to predict the distribution of deposits such as 'China clays', or, for example, the concentration of base metals at palaeo-unconformities associated with lateritization. One possible approach would be to use so-called 'basin modelling' applications that provide a spatial and temporal three-dimensional framework for modelling basin-scale processes and reactions. These models incorporate temperature and pressure evolution to predict compactional processes and the generation, migration and accumulation of petroleum (Schnider *et al.*, 1996; Burley *et al.*, 2000). Most silicate reactions can be modelled in the same framework using kinetic schemes to define reaction thresholds and rates; indeed some basin models already incorporate quartz cementation, and the step to clay mineral diagenesis is not large. The next major advance may be to incorporate equilibrium reaction-path models available for predicting the speciation of state of dissolved components and saturation states (Ben Baccar & Fritz, 1993). Such an integrated modelling scheme for diagenetic reactions would enable prediction of the distribution of reservoir quality and its evolution through time. This long-term goal still has many obstacles to overcome. Significant gaps remain in our fundamental understanding of diagenetic processes and mineral properties that require research focused on low-temperature geochemistry for their resolution. Some ideas for areas of potentially fruitful research are indicated in this concluding section.

As our understanding of eogenesis improves, the importance of time as a key control on diagenesis has become apparent (Coleman *et al.*, 1979), opening the potential to place eogenetic processes in a sequence stratigraphical framework. As eogenesis is influenced strongly by residence time, the controls that define the sequence stratigraphical character of sediments (absolute sea-level, climate, rate of sedimentation, rate of subsidence) must also govern early diagenetic assemblages and the extent to which they are developed. It may become possible to routinely predict eogenetic facies based simply on an understanding of sequence stratigraphy (McKay *et al.*, 1995; Taylor *et al.*, 2000). This may also make the recognition of 'diagenetic facies' possible from seismic data. As seismic stratigraphy is evolving towards seismic sedimentology, so a further advance in understanding reservoir quality may derive from 'seismic diagenetic' studies. Regional mapping of lithology, porosity and diagenetic facies will become possible as a result of the dramatic improvement in seismic acquisition, imaging and inversion techniques, coupled with progress in rock physics that links sediment elastic properties with bulk properties (Dvorkin *et al.*, 2002).

The question of the scale of a diagenetic system and its components is belatedly being acknowledged. However, still too few diagenetic studies consider the extent of diagenetic processes and flux of components, particularly when mass solute transfer is proposed (but see Giles, 1987; Milliken *et al.*, 1989; McLaughlin *et al.*, 1994). It is meaningless talking about open versus closed systems without defining the size of the system in question. This is especially pertinent as the scale of a given diagenetic system may be radically different for different species. A diagenetic system may, for example, operate on the scale of tens of kilometres for CO_2 (advective influx into sandstones), but only tens of centimetres for SiO_2 (diffusive influx into sandstones; see, for example, Worden & Barclay, 2000). Determination of the relative extent of movement of individual species within a diagenetic system may be the key step that resolves the isochemical versus allochemical debate.

Image analysis offers the possibility of removing much of the subjectivity from the quantification of diagenetic fabrics (Cooper *et al.*, 2000). Point counting, routinely used to determine the abundance of sandstone components and porosity, but fraught with problems of resolution, recognition and experience (Pettijohn *et al.*, 1972), may be superceded by integration of different types of digital images, including light optical, backscattered electron, cathodoluminescence (CL), X-ray map and UV fluorescence. Integration of backscattered electron and CL images has been used to quantify the amounts of quartz cement in some studies (e.g. Hogg *et al.*, 1992), and image analysis techniques have been automated to capture modal components from thin-sections (Erlich *et al.*, 1991). Already this approach is being extended to determine permeability from thin-section determination of porosity (Lander & Walderhaug, 1999); further integration with petrophysical techniques, particularly micro-imaging tools, is inevitable if modal analyses on the thin-section scale are to be effectively related to the reservoir scale.

As oil and, especially, gas exploration extends to deeper and hotter sandstones, so the need to understand the later stages of diagenesis will become paramount. 'Tight gas sand' is the term routinely used to describe low permeability, clay-rich sandstones, but the study should include clay-poor, deep sandstones where quartz and carbonate cements occlude most of the porosity and authigenic clays drastically reduce permeability. Understanding the terminal stages of diagenesis and focusing upon the processes that close the last pores in a sandstone during advanced heating will help place a genuine economic basement in the deepest and hottest sedimentary basins.

The role of bacteria during eogenesis is now well established (e.g. Irwin *et al.*, 1977; Claypool & Kvenvolden, 1983). However, there are bacterial communities being found in ever deeper sandstone formations, suggesting that bacterially mediated reactions during mesogenesis are a real possibility (Spark *et al.*, 2000; Ehrenberg & Jacobsen, 2001). Bacteria not only influence what reactions take place but also catalyse the rate at which they occur. This is especially important for sulphur-bearing compounds (sulphides and sulphates) and carbonate minerals, although it is also possible that bacteria influence the progress of silica and aluminosilicate reactions, as suggested by the presence of ammonium in illite–smectite clays (Lindgreen, 1994) and authigenic feldspar (Ramseyer *et al.*, 1993). Advancing our understanding of bacterial activity in the mesogenetic realm may yet revolutionize prediction of rates of diagenetic processes.

Finally, still with the theme of understanding the mechanisms for diagenetic processes, prediction of the occurrence and amounts of cement will rest ultimately with a better knowledge of the rates of diagenetic processes. It is relatively easy to predict what the equilibrium mineral assemblage should be (see, for example, Garrels & Mackenzie, 1971), but to predict the rate at which equilibrium is approached is much more difficult. Rate data are not routinely available for most diagenetic reactions. They can be acquired in various ways.

1 Experimental approaches have been used successfully in metamorphic petrology but the relatively low temperatures encountered in diagenesis mean that reactions cannot easily be synthesized under realistic conditions in a laboratory. The usual route is to increase the temperature (beyond the diagenetic realm) and then extrapolate rate data back to lower temperatures (e.g. Huang *et al.*, 1993). This requires the dangerous assumption that the mechanism is identical at high and low temperatures. Over and above the mechanism issue, extrapolation is always a risky business with ever greater uncertainty away from the control region.

2 Theoretical rate data can be acquired based on an *ab initio* approach. This is also dangerous since the fundamental understanding of rates of bond breakage, diffusion in grain boundary regions are poorly known.

3 Empirical rate data can be generated by examination of large data sets (of, for example, cement types and volumes) placed in a time and temperature context (e.g. Walderhaug, 1994). This, however, has the risk of aggregating a range of processes, inputs and variables, but has the advantage that the output is based on real data of interest and relevance to those seeking to predict sandstone reservoir quality at the basin scale.

Obtaining meaningful predictive rate data may come from a combination of these three approaches. Using one approach in isolation is likely to lead to abberrant rate data. As ever in science, it pays to use multiple working hypotheses rather than sticking blindly to a single path. The next generation of step changes in understanding diagenetic processes and product are most likely to derive from an integration of methodological approaches, techniques and scientific disciplines.

ACKNOWLEDGEMENTS

Laura Crossey and Ian Jarvis are thanked for constructive reviews of the manuscript.

REFERENCES

Aagaard, P., Jahren, J., Harstad, A.O., Nilsen, O. & Ramm, M. (2000) Formation of grain-coating chlorite in sandstones; laboratory synthesized vs. natural occurrences. *Clay Minerals*, **35**, 261–269.

Al-Agha, M.R., Burley, S.D., Curtis, C.D. & Esson, J. (1995) Complex cementation textures and authigenic mineral assemblages in Recent concretions from the Lincolnshire Wash (east coast, UK) driven by Fe(0) to Fe(II) oxidation. *Journal of the Geological Society*, **152**, 157–171.

Al-Gailani, M.B. (1981) Authigenic mineralisations at unconformities: implications for reservoir characteristics. *Sedimentary Geology*, **29**, 89–115.

Andrews, J.E. & Turner, M.S. (1991) The anatomy of an early Dinantian terraced floodplain: palaeoenvironment and early diagenesis. *Sedimentology*, **38**, 271–287.

Aronson, J.L. & Hower, J. (1976) Mechanism of burial metamorphism of argillaceous sediment. 2. Radiogenic argon evidence. *Geological Society of America Bulletin*, **87**, 738–744.

Baker, J.C., Kassan, J. & Hamilton, P.J. (1996) Early diagenetic siderite as an indicator of depositional environment in the Triassic Rewan Group, Southern Bowen Basin, eastern Australia. *Sedimentology*, **43**, 77–88.

Baines, S.J., Burley, S.D. & Gize, A.P. (1991) Sulphide mineralisation and hydrocarbon migration in North Sea oilfieds. In: *Source, Transport and Deposition of Metals* (Eds, M. Pagel & J. Leyroy), 507–511. Balkema, Rotterdam.

Beard, D.C. & Weyl, P.K. (1973) Influence of texture on porosity and permeability of unconsolidated sand. *American Association Petroleum Geologists Bulletin*, **57**, 348–369.

Bethke, C.M. & Marshack, S. (1990) Brine migration across North America—the plate tectonics of groundwater. *Annual Review of Earth Sciences*, **18**, 287–315.

Beaufort, D., Cassagnabere, A., Petit, S., Lanson, B., Berger, G., Lacharpagne, J.C. & Johansen, H. (1998) Kaolinite-to-dickite reaction in sandstone reservoirs. *Clay Minerals*, **33**, 297–316.

Ben Baccar, M. & Fritz, B. (1993) Geochemical modelling of sandstone diagenesis and its consequences on the evolution of porosity. *Applied Geochemistry*, **8**, 285–295.

Berner, R.A. (1980) *Early Diagenesis: A Theoretical Approach*. Princeton Series in Geochemistry, Princeton University Press, Princeton, NJ.

Bjørkum, P.A. (1996) How important is pressure in causing dissolution of quartz in sandstones? *Journal of Sedimentary Research*, **66**, 147–154.

Bjørkum, P.A. & Gjelsvik, N. (1988) An isochemical model for the formation of authigenic kaolinite, K-feldspar and illite in sandstones. *Journal of Sedimentary Petrology*, **58**, 506–511.

Bjørkum, P.A., Mjos, R., Walderhaug, O. & Hurst, A. (1990) The role of the late Cimmerian unconformity for the distribution of kaolinite in the Gullfaks Field, northern North Sea. *Sedimentology*, **37**, 395–406.

Bjørlykke, K.O. (1989) Sandstone diagenesis and porosity during basin evolution. *Geologische Rundschau*, **78**, 169–205.

Bjørlykke, K. & Egeberg, P.K. (1993) Quartz cementation in sedimentary basins. *American Association of Petroleum Geologists Bulletin*, **77**, 1536–1548.

Boles, J.R. (1987) Six million year diagenetic history, North Coles Levee, San Joaquin Basin, California. In: *Diagenesis of Sedimentary Sequences* (Ed. J.D. Marshall). pp. 191–200. Spec. Publ. Geol. Soc. London, No. 36, Blackwell Scientific Publications, Oxford.

Boles, J.R. & Franks, S.G. (1979) Clay diagenesis in Wilcox sandstones of southwest Texas: implications of smectite diagenesis on sandstone cementation. *Journal of Sedimentary Petrology*, **49**, 55–70.

Browne, G.H. & Kingston, D.M. (1993) Early diagenetic spherulitic siderites from Pennsylvanian palaeosols in the Boss Point Formation, Maritime Canada. *Sedimentology*, **40**, 467–474.

Burley, S.D. (1984) Patterns of diagenesis in the Sherwood Sandstone Group (Triassic), United Kingdom. *Clay Minerals*, **19**, 403–440.

Burley, S.D. (1986) The development and destruction of porosity in Upper Jurassic reservoir sandstones of the Piper and Tartan oilfields, Outer Moray Firth, North Sea. *Clay Minerals*, **21**, 706–720.

Burley, S.D. (1993) Burial diagenesis. In: *The Encyclopedia of the Solid Earth Sciences* (Ed. P. Keary), pp. 72–76. Blackwell Scientific Publications Oxford.

Burley, S.D. & Cornford, C. (1998) Carbonate cements constrain the burial history of the Portledge-Peppercombe Permian outlier, North Devon. *Geoscience in South West England*, **9**, 188–202.

Burley, S.D. & Kantorowicz, J.D. (1986) Thin section and SEM textural criteria for the recognition of cement dissolution porosity in sandstones. *Sedimentology*, **33**, 587–604.

Burley, S.D., Kantorowicz, J.D. & Waugh, B. (1985) Clastic diagenesis. In: *Sedimentology: Recent and Applied Aspects* (Eds P. Brenchley & B.P.B. Williams). Spec. Publ. Geol. Soc. London, No. 18, 189–226. Blackwell Scientific Publications, Oxford.

Burley, S.D., Mullis, J. & Matter, A. (1989) Timing diagenesis in the Tartan Reservoir (UK North Sea): constraints from combined cathodoluminesence microscopy and fluid inclusion studies. *Marine and Petroleum Geology*, **6**, 98–120.

Burley, S.D., Clarke, S., Dodds, A., *et al.* (2000) New insights on petroleum migration from the applicaton of 4D basin modelling in oil and gas exploration. *Journal of Geochemical Exploration*, **69**, 465–470.

Chan, M.A., Parry, W.T. & Bowman, J.R. (2000) Diagenetic hematite and manganese oxides and fault-related fluid flow in Jurassic sandstones, South-eastern Utah. *American Association Petroleum Geologists Bulletin*, **84**, 1281–1310.

Chapelle, F.H. (1993) *Ground-water Microbiology and Geochemistry*. John Wiley & Sons, New York, 448 pp.

Choquette, P.W. & Pray, L. (1970) Geologic nomenclature and classification of porosity in sedimentary carbonates. *American Association Petroleum Geologists Bulletin*, **54**, 207–250.

Claypool, G.E. & Kvenvolden, K.A. (1983) Methane and other hydrocarbon gases in marine sediment. *Annual Review of Earth Sciences*, **11**, 299–327.

Coleman, M.L. (1985) Geochemistry of non-silicate minerals: kinetic considerations. *Philosophical Transactions of the Royal Society, London*, **315**, 39–56.

Coleman, M.L., Curtis, C.D. & Irwin, H. (1979) Burial rate: a key to source and reservoir potential. *World Oil*, **188**, 83–92.

Cooper, M.A., Evans, I.J.E., Flint, S.S., Hogg, A.J.C. & Hunter, R.H. (2000) Quantification of detrital, authigenic and porosity components in sandstones: a quantitative image analysis approach. In: *Quartz Cementation in Sandstones* (Eds R.H. Worden & S. Morad). Spec. Publs Int. Assoc. Sediment., No. 29, pp. 89–101. Blackwell Science, Oxford.

Crossey, L.J., Loucks, R. & Totten, M.W. (1996) *Siliciclastic Diagenesis and Fluid Flow: Concepts and Applications*. Spec. Publ. Soc. Econ. Paleont. Miner., Tulsa, OK, **55**, 220 pp.

Curtis, C.D. (1977) Sedimentary geochemistry: environments and processes dominated by involvement of an aqueous phase. *Philosophical Transactions of the Royal Society, London*, **286**, 353–372.

Curtis, C.D. (1978) Possible links between sandstone diagenesis and depth-related geochemical reactions occurring in enclosing mudstones. *Journal of the Geological Society, London*, **135**, 107–117.

Curtis, C.D. (1983) Geochemistry of porosity enhancement and reduction on clastic sediments. In: *Petroleum Geochemistry and Exploration of Europe* (Ed. J. Brooks). Spec. Publ. Geol. Soc. London, No. 12, 113–125. Blackwell Scientific Publications, Oxford.

Curtis, C.D. & Coleman, M.L. (1986) Controls on the precipitation of early diagenetic calcite, dolomite and siderite concretions in complex depositional sequences. In: *Roles of Organic Matter in Sediment Diagenesis* (Ed. D.C. Gautier). Spec. Publ. Soc. Econ. Paleont. Miner. Tulsa, OK, **38**, 23–33.

Curtis, C.D., Coleman, M.L. & Love, L.G. (1986) Pore water evolution during sediment burial from isotopic and mineral chemistry of calcite, dolomite and siderite concretions. *Geochimica et Cosmochimica Acta*, **50**, 2321–2334.

de Caritat, P. & Baker, J.C. (1992) Oxygen isotope evidence for upward cross formational flow in a sedimentary basin near maximum burial. *Sedimentary Geology*, **78**, 155–164.

Dickson, J.A.D. (1965) A modified staining technique for carbonates in thin section. *Nature*, **205**, 587.

Duddy, I.R., Green, P.F., Hegarty, K.A., Bray, R.J. & O'Brien, G.W. (1998) Dating and duration of hot fluid flow events determined using AFTA and vitrinite reflectance-based thermal history reconstruction. In: *Dating and Duration of Fluid Flow and Fluid Rock Interaction* (Ed. J. Parnell). Spec. Publ. Geol. Soc. Longon, No. 144, pp. 41–51. Geological Society Publishing House, Bath.

Dvorkin, J., Gutierrez, M.A. & Nur, A. (2002) On the universality of diagenetic trends. *The Leading Edge*, **January**, 40–43.

Ehrenberg, S.N. (1993) Preservation of anomalously high porosity in deeply buried sandstones by grain coating chlorite: examples from the Norwegian continental shelf. *American Association of Petroleum Geologists Bulletin*, **77**, 1260–1286.

Ehrenberg, S.N. & Jacobsen, K.G. (2001) Plagioclase dissolution related to biodegradation of oil in Brent Group sandstones (Middle Jurassic) of Gullfaks Field, northern North Sea. *Sedimentology*, **48**, 703–722.

Ehrenberg, S.N., Aagaard, P., Wilson, M.J., Fraser, A.R. & Duthie, D.M.L. (1993) Depth-dependant transformation of kaolinite to dickite in sandstones of the Norwegian continental shelf. *Clay Minerals*, **28**, 325–352.

Elmore, R.D., Engel, M.H., Crawford, L., Nick, K., Imbus, S. & Sofer, Z. (1987) Evidence for a relationship between hydrocarbon migration and authigenic magnetite. *Nature*, **325**, 428–430.

Emery, D. Myers, K.J. & Young, R. (1990) Ancient subaerial exposure and freshwater leaching in sandstones. *Geology*, **18**, 1178–1181.

Erlich, R., Crabtree, S.J., Horrowitz, K.O. & Horrowitz, J.P. (1991) Petrography and reservoir physics: objective classification of reservoir porosity. *American Association of Petroleum Geologists Bulletin*, **75**, 1547–1562.

Esteoule-Choux, J. (1983) Kaolinitic weathering profiles in Brittany: genesis and economic importance. In: *Residual Deposits: Surface Related Weathering Processes and Materials* (Ed. R.C.L. Wilson). Spec. Publ. Geol. Soc. London, No. 11, pp. 33–38. Blackwell Scientific Publications, Oxford.

Folk, R.L. (1965) Some aspects of recrystallisation in ancient limestones. In: *Dolomitization and Limestone Diagenesis* (Eds L.C. Pray & R.G. Murray). Spec. Publ. Soc. Econ. Paleont. Miner., Tulsa, OK, **13**, 14–48.

Frey, M. (1987) Very low grade metamorphism of clastic sedimentary rocks. In: *Low Temperature Metamorphism* (Ed. M. Frey), pp. 9–58. Blackie, London.

Gardener, R. (1983) Reddening of tropical coastal dune sands. In *Residual Deposits: Surface Related Weathering Processes and Materials* (Ed. R.C.L. Wilson). Spec. Publ. Geol. Soc. London, No. 11, pp. 103–115. Blackwell Scientific Publications, Oxford.

Garrels, R.M. & Mackenzie, F.T. (1971) *Evolution of Sedimentary Rocks*. Norton, New York, 397 pp.

Gawthorpe, R.L. (1987) Burial dolomitization and porosity development in a mixed carbonate–clastic sequence: an example from the Bowland Basin, northern England. *Sedimentology*, **34**, 533–558.

Giles, M.R. (1987) Mass transfer and problems of secondary porosity creation in deeply buried hydrocarbon reservoirs. *Marine and Petroleum Geology*, **4**, 188–204.

Giles, M.R. & de Boer, R.B. (1990) Origin and significance of redistributional secondary porosity. *Marine and Petroleum Geology*, **7**, 378–397.

Giles, M.R. & Marshall, J.D. (1986) Constraints on the development of secondary porosity in the subsurface —re-evaluation of processes. *Marine Petroleum Geology*, **3**(3), 243–255.

Giles, M.R., Indrelid, S.L., Beynon, G.V. & Amthor, J. (2000) The origin of large-scale quartz cementation: evidence from large datasets and coupled heat-fluid mass transport modelling. In: *Quartz Cementation in Sandstones* (Eds R.H. Worden & S. Morad). Spec. Publs Int. Assoc. Sediment. No. 29, pp. 21–38. Blackwell Science, Oxford.

Girard, J.P., Aaronson, J.L. & Savin, S.M. (1988) Separation, K–Ar dating and $O^{18}/^{16}O$ ratio measurement of diagenetic K-felspar overgrowths: an example from the Lower Cretaceous arkoses of the Angola margin. *Geochimica Cosmochimica Acta*, **52**, 2207–2214.

Gluyas, J.G. & Coleman, M.L. (1992) Material flux and porosity changes during sediment diagenesis. *Nature*, **356**, 52–54.

Gretener, P.E. (1976) Pore pressure: fundamentals, general ramification and implications for structural geology. *American Association of Petroleum Geology, Continuing Education Course Notes Series*, **4**.

Hamilton, P.J., Kelly, S. & Fallick, A.E. (1989) K–Ar dating of illite. I: hydrocarbon reservoirs. *Clay Minerals*, **24**, 215–232.

Hancock, N.J. & Taylor, A.M. (1978) Clay mineral diagenesis and oil migration in the Middle Jurassic Brent Group. *Journal of the Geological Society, London*, **135**, 69–72.

Hanor, J.S. (1987) *Origin and Migration of Subsurface Sedimentary Brines*. SEPM Short Course Notes, No. 21. Tulsa, OK. Society of Economic Paleontologists and Mineralogists.

Hanor, J.S. (1994) Physical and chemical controls on the composition of waters in sedimentary basins. *Marine and Petroleum Geology*, **11**, 31–46.

Hardie, L.A. (1967) The gypsum–anhydrite equilibrium at one atmopshere pressure. *American Mineralogist*, **52**, 171–200.

Harris, N.B. (1989) Diagenetic quartz arenite and destruction of secondary porosity: an example from the Middle Jurassic Brent Sandstone of northwest Europe. *Geology*, **17**, 361–364.

Hein J.R., O'Neill, J.R., Jones, M.G. (1979) Origin of authigenic carbonates in sediment from the deep Bering Sea. *Sedimentology*, **26**, 681–705.

Helgeson, H.C., Delany, J.M., Nesbitt, H.W. & Bird, D.K. (1978) Summary and critique of the thermodynamic properties of rock forming minerals. *American Journal of Science*, **278A**, 1–229.

Hendry, J.P., Trewin, N.H. & Fallick, A.E. (1996) Low-Mg-calcite marine cement in Cretaceous turbidites: origin, spatial distribution and relationship to seawater chemistry. *Sedimentology*, **43**, 877–900.

Hesse, R. (1986) Early diagenetic pore-water/sediment interaction: modern offshore basins. In: *Diageneis. Geoscience Canada*, **13**, 165–196.

Hesse, R. (1987) Selective and reversible carbonate–silica replacements in Lower Cretaceous carbonate-bearing turbidites of the Eastern Alps. *Sedimentology*, **34**, 1055–1077.

Hogg, A.J.C., Sellier, E. & Jourdain, A.J. (1992) Cathodoluminescence of quartz cements in Brent Group sandstones, Alwyn South, UK North Sea. In: *Geology of the Brent Group* (Eds A.C. Morton, R.S. Haszeldine, M.R. Giles & S. Brown). Spec. Publ. Geol. Soc. London, No. 61, pp. 421–440. Geological Society Publishing House, Bath.

Houseknecht, D.W. (1984) Influence of grain size and temperature on intergranular pressure dissolution quartz cementation and porosity in a quartzose sandstone. *Journal of Sedimentary Petrology*, **54**, 348–361.

Huang, W.L., Longo, J.M. & Pevear, D.R. (1993) An experimentally derived kinetic model for smectite to illite conversion. *Geological Society of America Bulletin*, **87**, 162–177.

Hurst, A.R. (1981) A scale of dissolution for quartz and its implications for diagenetic processes in sandstones. *Sedimentology*, **28**, 451–459.

Hurst, A.R. (1987) Problems of reservoir characterisation in some North Sea sandstone reservoirs solved by the application of microscale geological data. In: *North Sea Oil and Gas Reservoirs* (Eds J. Kleppe, E.W. Berg, A.T. Buller, O. Hjelmeland & O. Torsaeter), pp. 153–167. Norwegian Petroleum Directorate, Graham and Trotman, London.

Irwin, H., Curtis, C.D. & Coleman, M.L. (1977) Isotopic evidence for source of diagenetic carbonate formed during the burial of organic rich sediments. *Nature*, **269**, 209–213.

Jackson, S.A. & Beales, F.W. (1967) An aspect of sedimentary basin evolution: the concentration of Mississippi-Valley type ores during the late stages of diagenesis. *Canadian Petroleum Geology Bulletin*, **15**, 383–433.

Jeans, C.V. (1978) The origin of the Triassic clay mineral assemblages of Europe with special reference to the Keuper Marl and Rhaetic of parts of England. *Philosophical Transactions of the Royal Society, London*, **289**, 549–639.

Jeans, C.V. (1989) Clay diagenesis in sandstones and shales: an introduction. *Clay Minerals*, **24**, 127–136.

Jowett, E.C., Cathles, L.M. & Davis, B.W. (1993) Predicting the depths of gypsum dehydration in evaporitic sedimentary basins. *American Association of Petroleum Geologists Bulletin*, **77**, 402–413.

Kantorowicz, J.D., Eigner, M.R.P., Livera, S., Van Schijndel-Goester, F.S. & Hamilton, P.J. (1992) Integration of petroleum engineering studies of producing

Brent Group fields to predict reservoir properties in the Pelican Field, UK North Sea. In: *Geology of the Brent Group* (Eds A.C. Morton, R.S. Haszeldine, M.R. Giles & S. Brown). Spec. Publ. Geol. Soc. London, No. 61, pp. 453–469.

Khalaf, F.I. (1990) Diagenetic alunite in clastic sequences, Kuwait, Arabian Gulf. *Sedimentology*, **37**, 155–164.

Knipe, R. (1993) The influence of fault zone processes and diagenesis on fluid flow. In: *Diagenesis and Basin Development* (Eds A.D. Horbury & A.G. Robinson). American Association Petroleum Geologists, Tulsa, Studies in Geology, **36**, 135–154.

Lander, R.H. & Walderhaug, O. (1999) Predicting porosity through simulating sandstone compaction and quartz cementation. *American Association of Petroleum Geologists Bulletin*, **83**, 433–456.

Leckie, D.A. & Cheel, R.J. (1990) Nodular silcrete in the Cypress Hills Formation (Upper Eocene to middle Miocene) of southern Saskatchewan, Canada. *Sedimentology*, **37**, 445–454.

Lee, M.K. & Bethke, C. (1994) Groundwater flow, late cementation and petroleum accumulation in the Permain Lyons sandstone and its role in oil accumulation. *American Association of Petroleum Geologists Bulletin*, **78**, 217–237.

Lee, Y.I. & DeVries Klein, G. (1986) Diagenesis of sandstones in the back-arc basins of the Western Pacific Ocean. *Sedimentology*, **33**, 651–669.

Liewig, N., Clauer, N. & Sommer, F. (1987) Rb–Sr and K–Ar dating of clay diagenesis in Jurassic sandstone oil reservoir, North Sea. *American Association of Petroleum Geologists Bulletin*, **71**, 1467–1474.

Lindgreen, H. (1994) Ammonium fixation during illite–smectite diagenesis in Upper Jurassic shale, North Sea. *Clay Minerals*, **29**, 527–537.

Longstaffe, F.J. & Aylalon, A. (1987) Oxygen isotope studies of clastic diagenesis in the Lower Cretaceous Viking Formation, Alberta: implications for the role of meteoric water. In: *Diagenesis of Sedimentary Sequences* (Ed. J. Marshal). Spec. Publ. Geol. Soc. London, No. 36, pp. 277–296. Blackwell Scientific Publications, Oxford.

Love, L.G. (1967) Early diagenetic iron sulphide in Recent sediments of the Wash, England. *Sedimentology*, **9**, 327–352.

Manheim, F.T. & Paull, C.K. (1981) Patterns of groundwater salinity changes in a deep continental transect off the southeast Atlantic coast of USA. *Journal of Hydrology*, **54**, 95–105.

Matter, A. & Ramseyer, K. (1985) Cathodoluminescence microscopy as a tool for provenance studies of sandstones. In: *Provenance of Arenites* (Ed. G.G. Zuffa) Reidell, Dordrecht, pp. 191–211.

McAulay, G.E., Burley, S.D., Fallick, A.E. & Kusznir, N.J. (1994) Palaeohydrodynamic fluid flow regimes during diagenesis of the Brent Group in the Hutton–NW Hutton reservoirs: constraints from oxygen isotopic studies of authigenic kaolin and reverse flexural modelling. *Clay Minerals*, **29**, 609–626.

McDowell, S.D. & Elders, W.A. (1980) Authigenic layer silicate minerals in Borehole Elmore 1, Salton Sea geothermal Field, California, USA. *Contributions to Mineralogy and Petrology*, **74**, 293–310.

McKay, J.L., Longstaffe, F.J. & Plint, A.G. (1995) Early diagenesis and its relationship to depositional environment and relative sealevel fluctuations (Upper Cretaceous Marshybank Formation, Alberta and British Columbia) *Sedimentology*, **42**, 161–190.

McKinley, J.M., Worden, R.H. & Ruffell, A.H. (2001) Contact diagenesis: the effect of an intrusion on reservoir quality in the Triassic Sherwood Sandstones, N. Ireland. *Journal of Sedimentary Research*, **71**, 484–495.

McKinley, J.M., Worden, R.H. & Ruffell, A.H. (2003) Smectite in sandstones: a review of the controls on occurrence and behaviour during diagenesis. In: (Worden, R.H. Morad, S., eds.) *Clay Mineral Cements in Sandstones* (Eds R.H. Worden & S. Morad). Spec. Publs Int. *Assoc. Sediment.*, No. 34, pp. 109–128. Blackwell Publishers, Oxford.

McLaughlin, O.M., Haszeldine, R.S., Fallick, A.E. & Rogers, G. (1994) The case of the missing clay, aluminium loss and secondary porosity. *Clay Minerals*, **29**, 651–664.

Milliken, K.L., McBride, E.F. & Land, L.S. (1989) Numerical assessment of dissolution versus replacement in the subsurface destruction of detrital feldspars, Oligocene Frio Formation, South Texas. *Journal of Sedimentary Petrology*, **59**, 740–757.

Millot, G. (1970) *Geology of Clays: Weathering, Sedimentology, Geochemistry*. Springer-Verlag, New York.

Moñtanez, I.P., Gregg, J.M. & Shelton, K.L. (1997) *Basin-wide Diagenetic Patterns: Integrated Petrologic, Geochemical, and Hydrologic Considerations*. Soc. Econ. Paleont. Miner. Spec. Publ., Tulsa, OK, **57**, 302 pp.

Morad, S. (1998) *Carbonate Cementation in Sandstones*. Spec. Publs Int. Assoc. Sediment., No. 26, 511 pp. Blackwell Science, Oxford.

Morad, S., Marfil, R. & de la Pena, J.A. (1989) Diagenetic K-feldspar pseudomorphs in the Triassic Buntsandstein sandstones of the Iberian Range, Spain. *Sedimentology*, **36**, 635–650.

Morad, S., Ketzer, J.M. and De Ros, L.F. (2000) Spatial and temporal distribution of diagenetic alterations in siliciclastic rocks: implications for mass transfer in sedimentary basins. *Sedimentology*, **47** (Millenium Reviews), 95–120.

Morton, A.C., Borg, G., Hansley, P.L., Haughton, P.D.W., Krinsley, D.H. & Trusty, P. (1989) The origin of faceted garnets in sandstones: dissolution or overgrowth. *Sedimentology*, **36**, 927–942.

Morton, J.P. (1985) Rb–Sr evidence for punctuated illite/smectite diagenesis in the Oligocene Frio Formation, Texas Gulf Coast. *American Association Petroleum Geologists Bulletin*, **96**, 114–122.

Mozley, P.S. & Joernle, K. (1990) Geochemistry of carbonate cements in the Sag River and Shublik Formations (Triassic/Jurassic), North Slope, Alaska: implications for the geochemical evolution of formation water. *Sedimentology*, **17**, 817–836.

Mullis, A.M. (1992) A numerical model for porosity modification at a sandstone–mudstone boundary by quartz pressure solution and diffusive mass transfer. *Sedimentology*, **39**, 99–107.

Nahon, D.B. (1991) *Introduction to the Petrology of Soils and Chemical Weathering*. Wiley Interscience, New York, 313 pp.

Odin, G.S. (1990) Clay mineral formation at the continental–ocean boundary: the verdine facies. *Clay Minerals*, **25**, 477–483.

Odin, G.S. & Matter, A. (1981) De glauconiarium originae. *Sedimentology*, **28**, 614–641.

Osborne, M. & Haszeldine, S. (1993) Evidence for resetting fluid inclusion temperatures from quartz cements in oilfields. *Marine and Petroleum Geology*, **10**, 271–278.

Osborne, M. & Swarbrick, R.E. (1999) Diagenesis in North Sea HPHT reservoirs—consequences for porosity and overpressure prediction. *Marine and Petroleum Geology*, **16**, 337–353.

Pettijohn, F.J., Potter, P.E. & Siever, R. (1972) *Sand and Sandstone*. Springer-Verlag, New York.

Pittman, E.D. & King, G.E. (1986) Petrology and formation damage control, Upper Cretaceous sandstone, offshore Gabon. *Clay Minerals*, **21**, 781–790.

Pittman, E.D. & Larese, R.E. (1991) Compaction of lithic sands: experimental results and application. *American Association of Petroleum Geologists Bulletin*, **75**, 1279–1299.

Primmer, T.J., Cade, C.A., Evans, I.J., *et al.* (1997) Global patterns in sandstones diagenesis: application to reservoir quality prediction for petroleum exploration. *American Association of Petroleum Geologists Memoir*, **69**, 61–78.

Prozorovich, G.E. (1970) Determination of the time of oil and gas accumulation by epigenesis studies. *Sedimentology*, **15**, 41–52.

Purvis, K. & Wright, V.P. (1991) Calcretes related to phreatophytic vegetation from the Middle Triassic Otter sandstone of south west England. *Sedimentology*, **38**, 539–551.

Pye, K. (1983) Post-depositional reddening of late Quaternary coastal dune sands, north-eastern Australia. In: *Residual Deposits: Surface Related Weathering Processes and Materials* (Ed. R.C.L. Wilson). Spec. Publ. Geol. Soc. London, No. 11, pp. 117–130. Blackwell Scientific Publications, Oxford.

Pye, K., Dickson, J.A.D., Schiavon, N., Coleman, M.L. & Cox, M. (1990) Formation of siderite–Mg-calcite–iron concretions in intertidal marsh and sandflat sediments, north Norfolk, England. *Sedimentology*, **37**, 325–343.

Ramm, M. & Bjørlykke, K. (1994) Porosity/depth trends in reservoir sandstones: assessing the quantitative effects of varying pore pressure, temperature history and mineralogy, Norwegian Shelf data. *Clay Minerals*, **29**, 475–490.

Ramseyer, K., Boles, J.R. & Lichtner, P.C. (1992) Authigenic K–NH_4–feldspar in sandstones: a fingerprint of the diagenesis of organic matter. *Journal of Sedimentary Petrology*, **62**, 349–356.

Ramseyer, K., Diamond, L.W. & Boles, J.R. (1993) Mechanism of plagioclase albitisation. *Journal of Sedimentary Petrology*, **63**, 1092–1099.

Robin, P.-Y.F. (1978) pressure solution at grain-to-grain contacts. *Geochimica Cosmochemica Acta*, **42**, 1383–1389.

Robinson, A.G., Coleman, M.L. & Gluyas, J.G. (1993) The age of illite cement growth, Village Fields area, Southern North Sea: evidence from K–Ar ages and $^{18}O/^{16}O$ isotope ratios. *American Association of Petroleum Geologists Bulletin*, **77**, 68–80.

Rowe, J.E. & Burley, S.D. (1997) Fault-related diagenetic cementation in Triassic Sandstones at Alderley Edge, north-eastern Cheshire. In: *Petroleum Geology of the Irish Sea and its Margins* (Eds N. Meadows & G. Cowan). Spec. Publ. Geol. Soc. London, No. 124, pp. 325–352. Geological Society Publishing House, Bath.

Ryan, P.C. & Reynolds, R.C. Jr. (1996) The chemical composition of serpentine/chlorite in the Tuscaloosa Formation, U.S. Gulf Coast: EDX vs. XRD determinations, implications for mineralogic reactions, and the origin of anatase. *Clays and Clay Minerals*, **45**, 339–352.

Schmidt, V. & Macdonald, D.A. (1979) The role of secondary porosity in the course of sandstone diagenesis. In: *Aspects of Diagenesis* (Eds P.A. Scholle & P.R. Schuldger). Soc. Econ. Paleont. Miner. Spec. Publ., Tulsa, OK, **29**, 175–207.

Schnider, F., Potdevin, J.L., Wolf, S. & Faille, I. (1996) Mechanical and chemical compaction model for sedimentary basin simulators. *Tectonophysics*, **263**, 307–317.

Schowalter, T.T. (1979) Mechanics of secondary hydrocarbon migration and entrapment. *American Association of Petroleum Geologists Bulletin*, **63**, 723–760.

Shanmugam, G. (1998) Origin, recognition and importance of erosional unconformities in sedimentary basins. *New Perspectives in Basin Analysis* (Eds K.L.

Kleinspehn & C. Paola), pp. 83–108. Springer-Verlag, New York.

Shanmugam, G. & Higgins, J.B. (1988) Porosity enhancement from Chert dissolution beneath Neocomian unconformity, Ivishak Formation, North Slope, Alaska. *American Association of Petroleum Geologists Bulletin*, **72**, 523–535.

Slater, D.J., Yardley, B.W.D., Spiro, B. & Knipe, R.J. (1994) Incipient metamorphism and deformation in the Variscides of SW Dyfed, Wales: first steps towards isotopic equilibrium. *Journal of Metamorphic Geology*, **12**, 237–248.

Spark, I., Patey, I., Duncan, B., Hamilton, A., Devine, C. & McGovern-Traa, C. (2000) The effects of indigenous and introduced microbes on deeply buried hydrocarbon reservoirs, North Sea. *Clay Minerals*, **35**, 5–12.

Spötl, C. & Wright, V.P. (1992) Groundwater dolocretes from Late Triassic of the Paris Basin, France: a case study of an arid continental diagenetic facies. *Sedimentology*, **39**, 1119–1136.

Stanislavsky, E. & Gvirtzmann, H. (1999) Basin scale migration of continental rift brines: paleo-hydrologic modelling of the Dead Sea basin. *Geology*, **27**, 792–794.

Sullivan, M.D., Haszeldine, R.S. & Fallick, A.E. (1990) Linear coupling of carbon and strontium isotopes in Rotliegende sandstone, North Sea: evidence for cross formational flow. *Geology*, **18**, 1215–1218.

Sullivan, M.D., Haszeldine, R.S., Boyce, A.J., Rogers, G. & Fallick, A.E. (1994) Late anhydrite cements mark basin inversion: isotopic and formation water evidence, Rotliegend sandstone, North Sea. *Marine and Petroleum Geology*, **11**, 46–54.

Surdam, R.C., Boese, S.W. & Crossey, L.J. (1984) The chemistry of secondary porosity. *Clastic Diagenesis* (Ed. R.C. Surdam). *American Association Petroleum Geologists Memoir*, **37**, 127–150.

Tardy, Y. (1969) Geochemie des alterations. *Mem. Serv. Carte. Geol. Alsace Lorraine, Strasbourg*, **31**, 199 pp.

Taylor, J.M. (1950) Pore space reduction in sandstones. *American Association of Petroleum Geologists Bulletin*, **34**, 701–716.

Taylor, J.C.M. & Illing, L.V. (1969) Holocene intertidal calcium carbonate cementation, Qatar, Persian Gulf. *Sedimentology*, **12**, 69–108.

Taylor, K.G., Gawthorpe, R.L., Curtis, C.D., Marshall, J.D. & Awwiller, D.N. (2000) Carbonate cementation in a sequence stratigraphic framework: Upper Cretaceous Sandstones, Book Cliffs, Utah Colorado. *Journal of Sedimentary Research*, **70**, 360–372.

Thiry, M., Ayrault, M.B. & Grisoni, J.-C. (1988) Groundwater silicification and leaching in sands: example from the Fontainebleau Sand (Oligocene) in the Paris Basin. *Bulletin of the Geological Society of America*, **100**, 1283–1290.

Thompson, A. (1959) Pressure solution and porosity. In: (Ireland, H.A., ed.) *Silica in Sediments* (Ed. H.A. Ireland). Spec. Publ. Soc. Econ. Paleont. Miner., Tulsa, OK, **7**, 92–110.

Tucker, M. & Bathurst, R.L. (1990) *Carbonate Diagenesis*. International Association of Sedimentologists Reprint Series No. 1, 320 pp. Blackwell Scientific Publications, Oxford.

Vagle, G.B., Hurst, A. & Dypvik, H. (1994) Origin of quartz cements in some sandstones from the Jurassic of the Inner Moray Firth. *Sedimentology*, **41**, 363–377.

Valeton, I. (1983) Palaeoenvironment of lateritic bauxites with vertical and lateral differentiation. In: *Residual Deposits: Surface Related Weathering Processes and Materials* (Ed. R.C.L. Wilson). Spec. Publ. Geol. Soc. London, No. 11, pp. 77–90. Blackwell Scientific Publications, Oxford.

Walderhaug, O. (1994) Precipitation rates for quartz cement in sandstones determined by fluid inclusion microthermometry and temperature history modelling. *Journal of Sedimentary Research*, **A64**, 324–333.

Walker, T.R., Waugh, B. & Crone, A.J. (1979) Diagenesis in first cycle desert alluvium of Cenozoic age, southwestern United States and northwestern mexico. Bulletin, Geological Society of America, **89**, 19–32.

Walther, J.V. & Wood, B.J. (1984) Rate and mechanism in prograde metamorphism. *Contributions to Mineralogy and Petrology*, **88**, 246–259.

Warren, E.A. & Curtis, C.D. (1989) The chemical composition of authigenic illite within two sandstone reservoirs as analysed by ATEM. *Clay Minerals*, **24**, 137–156.

Waugh, B. (1971) Formation of quartz overgrowths in the Penrith Sandstone (Lower Permian) of northwest England as revealed by scanning electron microscopy. *Sedimentology*, **17**, 309–320.

Waugh, B. (1978) Authigenic feldspar in British Permo-Triassic sandstones. *Journal of the Geological Society*, **135**, 51–56.

Whetten, J.T. & Hawkins, J.W. (1970) Diagenetic origin of graywacke matrix minerals. Sedimentology, **15**, 347–361.

Wilson, M.D. & Pittman, E.D. (1979) Authigenic clays in sandstones: recognition and influence on reservoir properties and palaeoenvironmental analysis. *Journal of Sedimentary Petrology*, **47**, 3–31.

Williams, L.A., Park, G.A. & Crerar, D.A. (1980) Silica diagenesis, I. Solubility controls. *Journal of Sedimentary Petrology*, **55**, 301–311.

Wilkinson, M. (1991) The concretions of the Bearreraig Sandstone Formation: geometry and geochemistry. *Sedimentology*, **38**, 899–912.

Wilkinson, M., Darby, D., Haszeldine, R.S. & Couples, G.D. (1997) Secondary porosity generation during deep burial associated overpressure leak-off: Fulmar Formation, United Kingdom Central Graben. *American Association of Petroleum Geologists Bulletin*, **81**, 803–813.

Wilkinson, M., Milliken, K.L. & Haszeldine, R.S. (2001) Systematic destruction of K-feldspar in deeply buried rift and passove margin sandstones. *Journal of the Geological Society, London*, **158**, 675–683.

Worden, R.H. & Barclay, S.A. (2000) Internally-sourced quartz cement due to externally-derived CO_2 in sub-arkosic sandstones, Northern North Sea, UKCS. *Journal of Geochemical Exploration*, **69–70**, 645–649.

Worden, R.H. & Morad, S. (2000) *Quartz Cementation in Sandstones*. Spec. Publs. Int. Assoc. Sediment., No. 29, 342 pp. Blackwell Science, Oxford.

Worden, R.H., Mayall, M.J. & Evans, I.J. (1997) Predicting reservoir quality during exploration: lithic grains, porosity and permeability in Tertiary clastics of the south China Sea basin. In: *Petroleum Geology of South East Asia* (Eds A.J. Fraser, S.J. Matthews & R.W. Murphey). Spec. Publ. Geol. Soc. London, No. 124, pp. 107–115. Geological Society Publishing House, Bath.

Worden, R.H., Oxtoby, N.H. & Smalley, P.C. (1998) Can oil emplacement prevent quartz cementation in sandstones? *Petroleum Geoscience*, **4**, 129–139.

Worden, R.H., Coleman, M.L. & Matray, J.-M. (1999) Basin scale evolution of formation waters: a diagenetic and formation water study of the Triassic Chaunoy Formation, Paris Basin. *Geochemica et Cosmochimica Acta*, **63**, 2512–2528.

Worden, R.H., Mayall, M.J. & Evans, I.J. (2000a) The effect of ductile–lithic sand grains and quartz cement on porosity and permeability in Oligocene and Lower Miocene clastics, South China Sea: prediction of reservoir quality. *American Association of Petroleum Geologists Bulletin*, **84**, 345–359.

Worden, R.H., Ruffell, A.H. & Cornford, C. (2000b) Surface temperature, sequence stratigraphy and deep burial diagenesis. *Clay Minerals*, **35**, 11–21.

Wright, V.P. & Tucker, M.E. (1991) *Calcretes*. International Association of Sedimentologists Reprint Series No. 2, 360 pp. Blackwell Science, Oxford.

Yardley, B.W.D. (1989) *An Introduction to Metamorphic Petrology*. Longman, New York, 248 pp.

Yurkova, R.M. (1970) Comparison of post-sedimentary alterations of oil-, gas- and water-bearing rocks. *Sedimentology*, **15**, 53–68.

Eogenesis (early diagenesis)

Marine eogenesis

Marine waters are a complex solution of dissolved atmospheric gases, organic products, inorganic anions and cations together with suspended organic matter, shell debris and particulate detritus. Diagenetic alteration in sediments bathed in marine waters can follow a number of paths, resulting in the growth of sulphide minerals, carbonates, clays and complex silicates.

The unstable mixture of geochemically reduced detrital organic matter and oxidized inorganic species (predominantly sulphate and ferric iron) suggest that pyrite growth is inevitable, albeit through a complex series of intermediate species of ferric hydroxides and monosulphides (Love, 1967). However, bacteria mediate practically all reactions in marine sediments and bacterial reaction rates are governed by a complex series of factors. Pye *et al.* (1990) demonstrate that siderite can grow instead of pyrite if the rate of bacterial sulphate reduction is slower than the rate of bacterial iron reduction. Even porous marine sediments are capable of sustaining geochemically distinct microenvironments.

Marine carbonate cements are commonplace although they result from a spectrum of processes. Hein *et al.* (1979) report a wide variety of carbonate minerals (calcite, dolomite, siderite, rhodochrosite) resulting from the reaction of aluminosilicate minerals with the products of organic matter oxidation. Conversely, carbonate cements can also form as a result of recrystallization of biogenic shell detritus (Hendry *et al.*, 1996). Between these two extremes, Wilkinson *et al.* (1991) illustrate a carbonate cementation scheme involving first sulphate reduction of organic matter and then dissolution of detrital shells producing a second generation of calcite cementation.

Clay minerals are not particularly common as a depositional component of high-energy sands from delta channel, beach and parallic environments. However, bioturbation is capable of intimately mixing mud into initially clean sands. Odin & Matter (1981) unravel some of the mysteries surrounding the origin of glaucony and clearly differentiated the relatively reduced green clay, berthierine, and the more oxidized glauconite. The latter clay, whether as a veneer on detrital grains or as a grain replacement, forms in moderately deep seawater (60–500 m) in areas of slow sedimentation, whereas berthierine forms in shallower water with a faster rate of sedimentation. Glauconite precipitates initially as an expandable mineral (smectite–glauconite) and transforms into a non-expandable form (mica–glauconite) with time.

Reprinted from Sedimentology (1967) **9**, *327–352*

EARLY DIAGENETIC IRON SULPHIDE IN RECENT SEDIMENTS OF THE WASH (ENGLAND)

L. G. LOVE

Department of Geology, University of Sheffield, Sheffield (Great Britain)

(Received June 6, 1967)

SUMMARY

Microscopical and chemical examination of intertidal flats of the Wash indicate that phases of iron corresponding to ferric hydroxide and iron monosulphide are involved in the early diagenetic formation of pyrite, the first being an important source of the iron. Sedimentological data are tested against analyses of total iron extractable by hydrochloric acid and concentrated nitric acid, and of the pyritic fraction of this iron; ignition data for residual organic matter are considered. The occurrence of the pyrite as isolated grains and spherules of framboidal texture is recorded, their size ranges given, and iron monosulphide recognised as the precursor of both forms. Ordering in the pyrite framboidal spherules is recorded and internal coalescence confirmed as a feature of primary growth which obscures the ordering.

INTRODUCTION

The intertidal flats of the Wash in eastern England represent a good example of an environment in which pyrite is forming in sediments at the present day. The area is possibly in more complex a situation than those in which extensive geological deposits of shale and mudstone formed in the past because, twice daily, tides alternately cover the flats and re-expose them to the atmosphere while between high and low tide levels there are several sub-environments of sedimentation; but the counterweighing factor of accessibility makes it appropriate for a primary study. The work here was stimulated by Van Straaten's now classic report on the composition and structure of Recent marine sediments of the Netherlands Wadden Sea (VAN STRAATEN, 1954) and by more recent geochemical work by BERNER (1963, 1964a, b) and KAPLAN et al. (1963) concerning the formation of iron monosulphide and pyrite in sediments and the origin of the sulphide itself through bacterial anaerobic decomposition of sulphate. The observations also illuminate those already made by the author on early diagenetic pyrite in shales (e.g., LOVE and AMSTUTZ, 1966).

The objects of the present work were the forms exhibited by iron hydroxide, monosulphide and pyrite in such a Recent sediment, and what might be learnt of the physical stages involved in the production of pyrite, because up to the present microscopical observation has lagged behind the geochemical. Indeed in rocks, microscopic evidence is much relied upon for the determination of the diagenetic history of sulphides and yet the origin of some of the textures is still obscure. Chemical and sedimentological data are also briefly presented to enable comparison to be made with other areas and at the same time to illustrate some possible causative relationships.

The intertidal area of the Wash studied is that for which the sediments and environments of deposition have been described by EVANS (1958, 1965), who did not, however, discuss these aspects of the iron compounds in detail.

The intertidal flats—origin of samples

Collection was made from the intertidal area along a line seaward from Freiston Shore (N.G. TF 397425) 7 km (4–5 miles) east-southeast of Boston on the eastern coast of England. EVANS (1965), generalising on the predominant character of the sediment, distinguished six sedimentological sub-environments respectively between high and low watermarks; his interpretation of their stratification is discussed later: salt marsh, higher mud flats, inner sand flats, *Arenicola* sand flats, lower mud flats, and lower sand flats.

The superficial sediments are pale brown to brownish red, but at depths from a few cm to 1 m and with an intervening mixed zone, an intensely black colouration starts. At a depth of 2 m in the middle area of the flats this fades to grey. The colour zonation, due to the state of the iron in the sediment, was featured in Van Straaten's account as: hydroxide zone—brownish, monosulphuric zone—intensely black, bisulphuric or pyrite zone—more or less grey. This arrangement is tied to the present surface of the sediments.

The samples used analytically in the present study were taken from the salt marsh, higher mud flats and *Arenicola* sand flats, as tabulated in Fig.1 to indicate to their possible stratigraphical inter-relationship following EVANS (1965; see p.331). In the salt marsh the upper sample lay in brown mud just below the root zone but the lower ones were from beneath the sloping bank of the creek side. The higher mud flats, intersected by creeks, provided the deepest section and here sampling over a depth of 3 m was possible on the outer side of a sharp bend in the creek where a clean and steep section could be dug out, while in the sides, behind 2–5 cm of brown material, the brown–black–grey zonation parallel with the top of the flats was well preserved. About 40 cm into this face, where the samples were taken, there was no apparent abnormal oxidation as a result of the presence of the creek. The section was continued to a depth of about 25 cm below the creek level at its lowest. In the *Arenicola* sand flats the depth of collection was limited by difficulty of excavation below the very high water table.

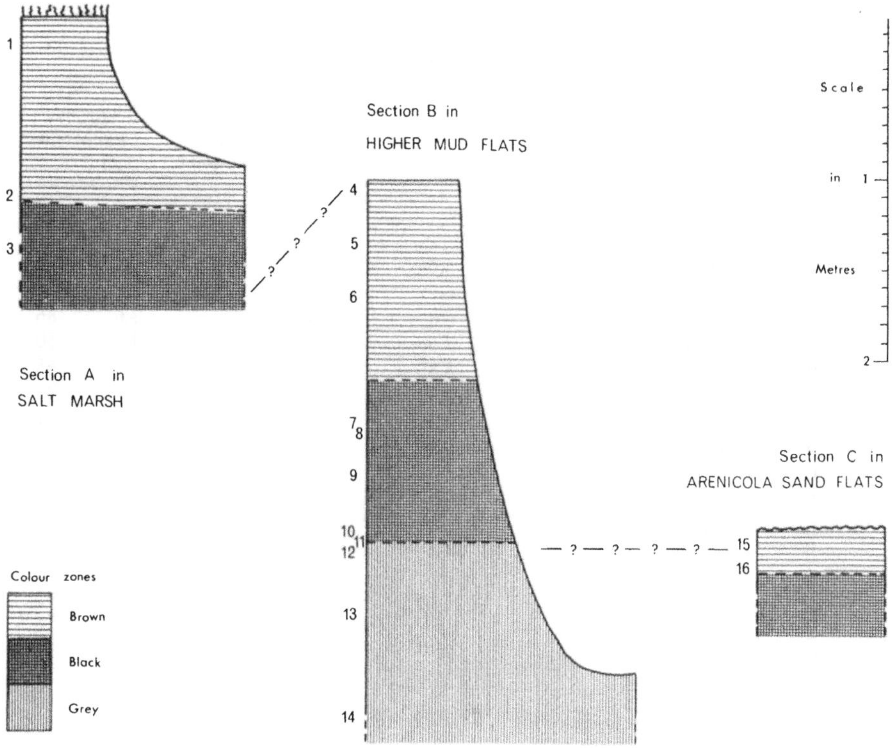

Fig.1. Sections of salt marsh, higher mud flats and *Arenicola* sand flats in the intertidal sediments of the Wash near Freiston. Sample numbers and possible correlations are indicated.

As will be seen from Evans' illustration of cores the sediments are not homogeneous so samples about 5 cm in thickness were collected, a narrow section of each being placed immediately in 5*N* hydrochloric acid for estimation of the non-pyrite acid-soluble iron, but the jar not sealed until effervescence was complete, while an adjacent narrow section was sealed in a small tube for drying and total acid-soluble iron determination.

PYRITE

Pyrite is the dominant stable sulphide present in the sediment; quantitative data concerning its abundance are given in Table I. After acid treatment of heavy residue, X-ray analysis revealed no other sulphide; in particular marcasite was sought (if greigite occurs, as suspected from scanty microscopic evidence, it must be in a very low proportion to the magnetite present). Traces of a second sulphide mineral have been seen, tentatively regarded as a variety of pyrite. The pyrite occurs as both framboidal spherules and single crystals. Both may be found within such enclosed spaces

TABLE I

PHYSICAL AND CHEMICAL DATA OF INTERTIDAL SEDIMENTS FROM THE WASH (ENGLAND)[1]

Number	Section	Colour	Total acid-soluble Fe as Fe_2O_3	Acid-soluble Fe in pyrite (%)	Combustible organic material (% dry weight)	% of whole sediment finer than 5 μ	% of whole sediment finer than 76 μ
1	A	brown	3.22	5.0	0.39	38	93
2	A	black	3.02	6.2	0.29	21	86
3	A	black	3.65	6.8	0.32	33	86
4	B	brown	2.24	3.58	1.62	14	50
5	B	brown	2.74	8.86	2.12	20	68
6	B	black	1.90	22.10	1.08	6	48
7	B	black	1.59	9.7	0.55	3	31
8	B	black	2.25	9.01	0.85	5.5	27
9	B	black	1.34	6.06	0.54	3	30
10	B	black	1.36	17.50	0.36	2	24
11	B	black	2.38	23.0	0.66	9	47
12	B	grey	2.27	24.68	0.76	7	86
13	B	grey	2.02	19.0	0.67	14	65
14	B	grey	2.00	15.05	0.91	6	32
15	C	brown	1.39	1.8	0.48	3.5	8.5
16	C	brown/black	1.64	4.7	0.76	8.5	37

[1] For location of samples and sections see Fig.1. Relationships of data are plotted in Fig.3–6.

as foraminiferid tests, diatom frustules and plant cells, as noted by several authors concerning various Recent sediments. The forms of pyrite are described in detail in the light of observations made by LOVE and AMSTUTZ (1966) on spherules and isolated crystals in the Chattanooga Shale where a full account is given of the history of investigation of the early diagenetic microscopic pyrite of sediments. In particular in the present section the phenomenon of ordering in framboids is described from Recent unconsolidated sediment for the first time, apart from a record by PAPUNEN (1966, fig.4).

External appearance of pyrite spherules

Shape and outline

The spherules are framboidal, composed, that is, of numerous small grains of pyrite: the overall shape varies from spheroid to ovoid with profiles from smooth to ragged; the latter is due to the irregularity in the outer two or three grains and could have been increased during separation. A deviation from the standard shape given by a constriction around the middle is noticeable but not common among larger spherules; although a pollen species of the same shape but usually larger size is present in the sediments the spherule is more likely to have originated from intergrowth of two adjacent blobs of sulphide.

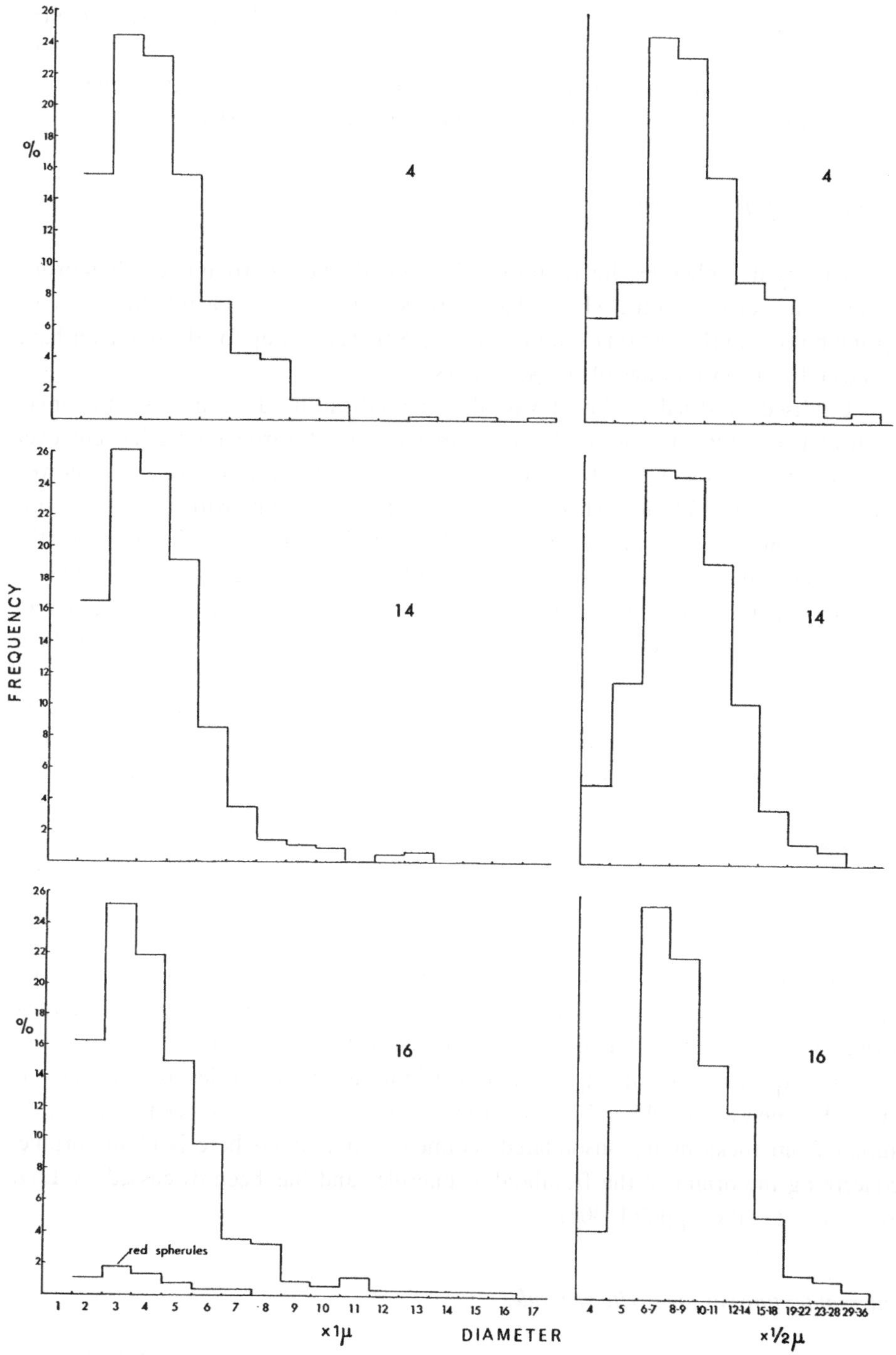

Fig.2. Size distributions of pyrite spherules from three samples (as numbered) on linear and logarithmic size-class scales. For No.16 is also indicated data for "red spherules" from the same count.

Usually, however, when in groups the outlines of framboids are straightened or even recessed against each other without any merging having occurred. In sections of both internally unordered and ordered framboids (see below) a rare tendency to polygonal outline is shown, as also observed in the Chattanooga Shale.

Size range of pyrite spherules

The pyrite spherules lie in positively skewed and approximately lognormal unimodal size distributions, whose characteristics show no significant shift between top and bottom of the section examined (Fig.2). Spherules of up to 70 μ diameter have been seen from concentrates of coarser grades.

Oxidised (air dried) sediment was taken in pin-head sized quantities and spread evenly and very thinly to about a 2 cm square in a drop of warmed glycerine jelly over a warmed microscopic slide without a cover glass after which the glycerine was allowed to cool and set for 30 min. This medium caused no problems with the microscope immersion oil and used thinly still allowed distinctive reflection of light from the pyrite. Much attention was paid to the best method of preparing material for microscopic examination and separated concentrates of pyrite were rejected in favour of use of specks of the untreated sediment, because significant regular loss from the finer part of the pyrite itself during separation was occurring, partly perhaps due to the difficulty of sufficiently dehydrating the fresh clay to make it wholly "wettable" with bromoform for centrifuging. Repeated washing with acetone had been used in this process, to avoid the necessity for baking and crushing. The problem has not arisen in the case of indurated rocks, where, anyway, checks against polished sections are possible.

The logarithmic classes used in Fig.2 accommodate the original data recorded with a $\frac{1}{2}$ μ scale, with the minimum of distortion. They correlate approximately with $^1/_4$ φ classes except at the fine end of the scale where the classes of actual measurement are too coarse and lead to difficulties. With the mode so far to the left, and only two smaller measurement divisions in use, particular attention was paid to this critical area to confirm that the decrease indicated here is actually present.

The question of an exponential distribution for spherules rather than an unimodal one was raised by VALLENTINE (1963, p.22); such has not been found by the author from rocks or unconsolidated sediment. That found here is of importance concerning the origin of the framboidal spherules and has been discussed by LOVE and AMSTUTZ (1966, pp.281, 303).

Internal textures of the pyrite spherules

The internal textures of pyrite spherules are only to be seen in polished section and in many ways are similar to those described by LOVE and AMSTUTZ (1966) from the Chattanooga Shale; the same terminology will be used.

Crystal size within framboids

The crystals, many of them euhedral, range in size up to 3 μ but examples as coarse as the latter are rare, and on the whole are smaller than those of the Chattanooga Shale framboids. With up to 25–50 % more crystal grains lying along a diameter, the texture of the framboids appears finer.

Internal homogenisation

More striking in this Recent sediment than in the Chattanooga Shale is the effect whereby in polished sections of some framboids the internal detail is blurred by continuity of the pyrite between adjacent grains so that the latter do not appear clearly defined or even become indistinguishable. Because of the irregular indicence in sections, smearing of the mineral during polishing is rejected in favour of an original condition in the framboids and, furthermore, the phenomenon is known from older rocks where it cannot be suspected that the sulphide is in some way immature and not fully hardened. Referred to as "infilling" by LOVE and AMSTUTZ (1966) the entry of new pyrite was favoured on balance but, seen now to occur so early in the history of the sediment, continued original growth of the grains towards each other until unified is the most likely cause. In this way, less fully grown areas, more often than not on the perimeter if such differentiation is seen, are likely to be unaffected, but at the other extreme a smooth outline can be produced (Plate I, contrast D with G). Secondary sulphide is also known but appears to be distinguishable, and is described below.

Ordering

The pattern of crystal grains seen in sections of framboids varies from random and unordered to distinctly ordered. Due to the dominant small grain size and more extensive homogenisation, detail of the ordering is not so clear as in instances described by LOVE and AMSTUTZ (1966, p.284) and therefore further detailed descriptions are not given here. In fact where the homogenisation is intense ordering can only be detected by the regular arrangement of any minute intergranular spaces unfilled, while unordered forms likewise show a trend towards a convolute or cerebral appearance when grains in close proximity are fully joined. At least 15% ordering or partial ordering has been detected. Square patterns are found, some of which are due to cube crystals in a simple cubic packing: oblique cuts give rhomboidal grain sections in oblique lines. Pavements of hexagonal grains are also seen and complex layered patterns involving six sided sections reminiscent of patterns thought by Love and Amstutz to be due to pyritohedra in open and close packings. Evidence is seen also that pyrite tetrahedra might be involved in some textures, as might have been expected but which has not so far been confirmed from the framboids of the Chattanooga Shale. Curious is the occurrence of bilaterally symmetrical ordering patterns as if twinning were also involved in the crystallographic basis of the ordering (Plate IG). Certain ordering patterns are essentially polygonal— one excellent example is comparable with that illustrated from Rammelsberg (Devonian) by RAMDOHR (1953,

PLATE I

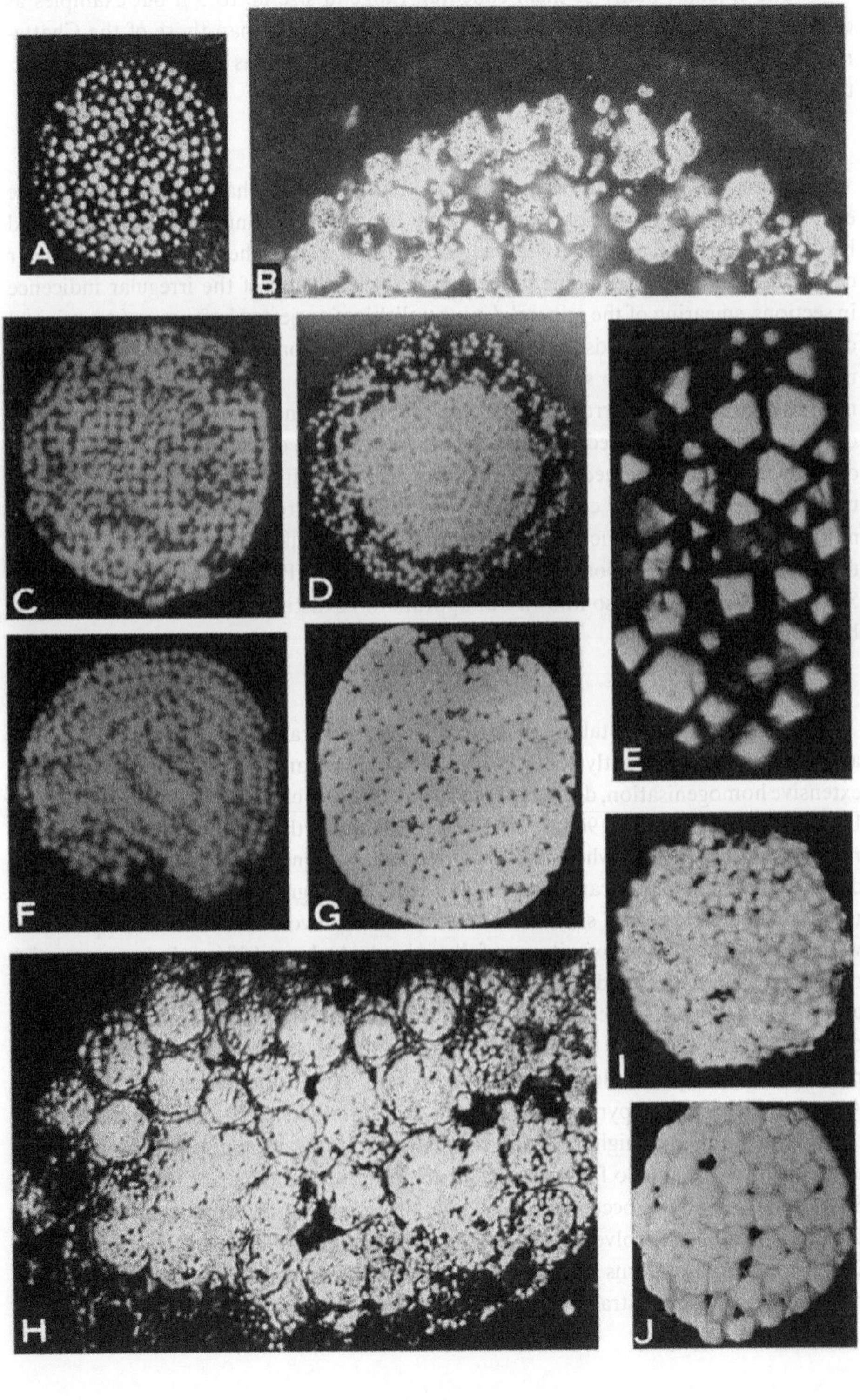

p.428, Abb.69)—some are to be explained as being in sections oblique to all axes of isometric and other regular packings, whilst others on closer examination appear to be due to the dominance of different oblique lineations of regular networks of grains in different areas of the framboid.

PAPUNEN (fig.4, p.119) illustrated pyrite framboids showing "rectilinear and circular ordering", from a layer in a Quaternary peat bog of Finland; the framboids are surrounded by a porous, less reflective and easily tarnished iron sulphide. Ordering in framboidal pyrite has further been observed by the present author from bottom sediments of the English Channel and from estuarine and marine sediments from The Netherlands, all of post-Pleistocene age.

Secondary sulphide

What is interpreted as a later generation of sulphide surrounding framboids and even filling in between the grains within framboids has been observed in rare instances (Plate IH, I, J). With a slightly weaker reflectance than the normal pyrite the mineral survives treatment in hydrochloric acid and appears to be the type referred to by RAMDOHR (1953) as Melnikovite-pyrite. Although found affecting framboids within microfossil tests in a post-Pleistocene sediment of the English Channel, no such proof of formation in situ has yet been obtained for the much younger Wash specimens, and as similar secondary sulphide has been recorded from Coal Measures shale (LOVE 1965, p.195 and Pl.14, fig.18), although not so far seen from coal particles from the Wash sediments, such a source cannot yet be ruled out.

Non-framboidal pyrite

Simple, non-framboidal pyrite grains are, very rarely, as large as 11 μ, and do clearly exceed the modal diameter of the spherules, an empirical limit suggested earlier by the author (LOVE, 1964, p.12), although even at this level they are very much less common. For progressively smaller sizes their abundance increases greatly.

PLATE I

All material sectioned and polished: photographed under incident light.

A, C, D, F, G. Framboidal pyrite spherules showing styles of internal ordering and degrees of infilling. Approx. × 2,000.

B. Sector of 70 μ diatom containing numerous pyrite framboids, of which the largest are 4–5 μ. Note pores in the siliceous test. Approx. × 1,500.

E. Loose cluster of mainly euhedral pyrite grains, some showing granular internal texture. Approx. × 2,000.

H, I, J. Framboidal pyrite with secondary "melnikovite-pyrite". H: × 1,500; I, J. × 2,000. All approx.

Reliable data are not available from untreated material, but in concentrates, where proportions of coarser simpler grains are probably increased as for framboids, those in the sub 2 μ category are none the less disctinctly the most abundant. For the Chattanooga Shale the author's counts from polished sections of rock showed single crystals to lie in a distribution approaching exponential (LOVE and AMSTUTZ, 1966, p.293; Fig.3) and the indications so far from the Wash are that such would best fit the general observations made.

Typically euhedral in transmitted light the grains commonly show 4-, 5-, 6- and 8-sided profiles while under incident light, and especially the Ultropak system on untreated mounts, simple cubes and octahedra have been recognised as well as more complex habits some of which must be pyritohedra. Both in polished section and from

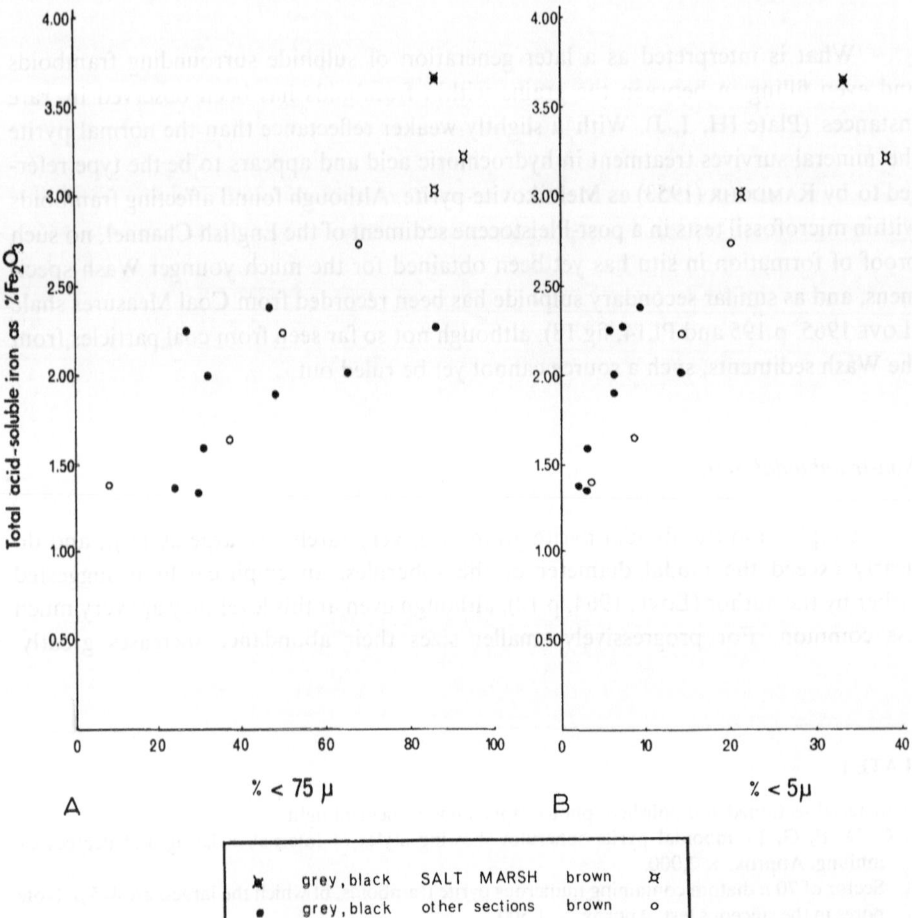

Fig.3. Total iron (as % Fe_2O_3 of dry sediment weight) extracted by HCl and HNO_3 from samples.
A. Against proportion of sediment finer than 75 μ.
B. Against proportion of sediment finer than 5 μ.

external appearance in the solid under incident light a small proportion of single crystals give some indication of being composite, nearly framboidal, despite clearly outlined square or triangular faces or sections (Plate I, E). There appears to be transition, therefore, between single crystals and some framboidal spherules which show a tendency to polygonal outline.

Sulphide in detrital coal

Coal fragments, as already noted by EVANS (1965, p.217), form a minor component of the sediment but one which here necessitates special attention on account of its contribution to the total analysed contents of organic material and iron in the sediments as well as to the pyrite.

The coal fragments are present in all samples taken down the principal section and are not more abundant in the surface layers so they must be present due to natural erosion of coal-bearing strata and not solely as industrial waste. Simple sedimentation tests do, however, reveal that the coal fragments appear to be progressively concentrated in the coarser fractions of the sediment and are not at all common below the 200 mesh level. Possibly this is a consequence of the mechanics of coastal or tidal drift of sediment and of the lower than average specific gravity of the partly weathered coal as it is found in the sediment.

The coal fragments must contribute to the results of the combustion analyses of organic material given later but as the sediment now stands, the coal is part of the residue of resistant organic detritus which in due course of time would appear as the bituminous or carbonaceous fraction of the indurated rock unless in the meantime it becomes oxidised.

In polished thin section, some fragments, although considerably weathered at the margins, still contained pyrite, as framboidal spherules, as small single crystals and as minute veinlets in the cleat. Including partly oxidised pyrite framboids and assuming for this purpose that the whole area of the framboid is pyrite, point counting gave a maximum of only 0.1% by volume and the effect of this on the total analysed pyrite content of the whole sediment is not great. By transmitted light, red-brown blobs of limonite or iron hydroxide may be seen, presumably from the oxidation during weathering of pyrite, for the same can be seen in and around pyrite itself. A small contribution to the soluble iron content of the sediment will be available from this source, too.

Although the pyrite grains and spherules in the coal are not individually distinguishable from those in the rest of the sediment, the size distribution as a whole showed a much broader spread than those given in Fig.2 (for which only loose framboids were measured, so few if any were derived from this coal). A clear modal value did not emerge but might be predicted to lie between 5 and 6 μ and the rounded red-brown clots referred to lay in the same size range, evidence of their origin as oxidised framboids.

Coal of Carboniferous age in the Midlands and east and northeast of England (on spore evidence, the source of the fragments here) varies greatly in its pyrite content, from virtually nil up to 5% by weight. The proportion observed, allowing for addition of individual grain and vein pyrite, is well below what might be expected: the balance is certainly to be accounted for by oxidation in outcrop and in transit. Evans has suggested offshore glacial deposits as an intermediate repository.

BLACK IRON MONOSULPHIDE, "FeS"

Iron monosulphide gives rise to the black colour of unconsolidated sediments when they are in the reduced state, and in the intertidal sediments of the Wash it does so starting at depths from 5 cm to 1 m. Some of its characteristics are easily recognised—it readily dissolves in dilute hydrochloric acid, evolving hydrogen sulphide; on exposure to the atmosphere it turns into reddish brown iron hydroxide; and under the microscope it is black, opaque and weekly reflective. Beyond this, its instability in air makes it elusive and some of the observations made here are only circumstantial.

Exposed surfaces of sediment containing the monosulphide do not immediately turn totally brown-red by oxidation but the process takes between a few minutes and a few hours, occurring soonest superficially and appearing to be most rapid on drying of the material. In uncovered glycerine jelly microscope slides of fresh sediment the pyrite was to be distinguished by its lack of change while red spherules developed from other black spherules, and red specks and small particles came also from black forms. These were of comparable size with pyrite grains of the same size so it can be concluded that in the original reduced sediment spherules and single particles of iron monosulphide existed side by side with spherules and single crystal grains of pyrite. The monosulphide spherules showed some degree of composition from individual particles, so resembling framboidal pyrite; red spherules, after the change, showed granularity in the red material and a few showed a suggestion of internal ordering but as these factors could have emerged during the change they are not at present valid evidence of the physical state of the monosulphide. Other red particles, formed from the black, tended to be irregular in form and texture. The presence of the monosulphide inside such enclosed spaces as foraminerid tests and diatoms has been inferred from black contents turning red: foraminiferid tests with red spherules have also been isolated.

On the exact composition and mineralogy of the black iron monosulphide (FeS) accounts have been confusing (see BERNER, 1962, 1964a), but some clarity has been introduced by BERNER (1967b). In the laboratory the action of H_2S on iron solutions has yielded products which have been identified by X-ray diffraction and analysis: (*1*) mackinawite, tetragonal FeS; (*2*) greigite, cubic Fe_3S_4, magnetic; and (*3*) non crystalline FeS. Mackinawite often appears from descriptions—under the term hydrotroilite—to be hydrated and has been referred to as $FeS.nH_2O$. BERNER (1967a) suggests this to be an adsorption effect due to the very fine grained nature of the material. Magnetic forms have been termed melnikovite, with some of the properties of greigite indicated. Mixtures of mackinawite, greigite and non-crystalline FeS with the first two in various degrees of crystallisation would appear to be important constituents of the early product of the action of H_2S on iron in natural sediments.

In the present work X-ray studies of the unstable sulphide have not been made. With reference to the trend in analyses made by various authors, and the probability that the tetragonal FeS is dominant, it is referred to as "black iron monosulphide (FeS)".

RED FERRIC HYDROXIDE

The material which gives the oxidised sediment its characteristic red-brown colour, here termed "ferric hydroxide", is seen microscopically in undried sediment to be translucent, red and birefringent in transmitted light, and weakly reflective in red to orange shades. It has not been examined here by X-ray methods so distinction cannot be made between goethite, lepidocrocite, limonite or other poorly defined oxides, but most likely hydrated forms are dominant. Berner (1962) indicated that wet tetragonal iron sulphide rapidly oxidises in air to give lepidocrocite (and orthorhombic sulphur) and in most specimens examined, therefore, this must be present.

Ranging from red spherules to irregular blobs and specks, not usually greater than the accompanying pyrite, the hydroxide is most abundant in the uppermost red-brown parts of the sections. Larger particles are free but the smaller ones are as likely on the surface of any other mineral or scattered in and tinting the poorly defined clots of clay mineral characteristic of the sediment.

It was observed microscopically that in uncovered fresh black sediment the amount of red hydroxide increases on standing by red specks and spherules appearing in place of black ones. Some red material is also seen at first, either resulting from transformation during opening and handling of the sediment sample or originally present in the black sediment. In air-oxidised sediment, the size distribution of red spherules roughly corresponds with that of pyrite spherules as in Fig.2 (no.16) but this is based only on a small sample and on the highly subjective selection of well rounded red blobs and the exclusion of more ragged ones. It is thought that the majority of red spherules is secondary to monosulphide (and, may be to pyrite) spherules, but a proportion of more irregular and mainly hydroxide particles could represent a primary origin as gel produced by hydrolisation of iron in river water as it enters saline water, and subsequent aggregation of the colloidal material into charged masses attached to clays and other minerals.

Partially pyritic spherules

Red spherules have been observed in transmitted light to include black grains reflective like pyrite if near the upper surface. Also, pyrite framboids with reddish fringes have been seen. These are either pyrite spherules in the course of oxidation, when progressively fewer grains of pyrite remain unaltered, or the oxidised result of mixed monosulpide–pyrite spherules, presumably part way through the conversion to pyrite. On treatment with acid which dissolves the hydroxide (or in fresh material, the monosulphide) the pyrite remnant would presumably fall apart and add to the loose small grains present.

SEDIMENTOLOGICAL MEASUREMENTS

Having described the visual evidence concerning the occurrence of pyrite, iron monosulphide and iron hydroxide in the sediment, and before going on to consider factors relating to their origin, it is appropriate to refer to physical aspects of the sediments as a whole. The sedimentological data shown in Table I and in relation to

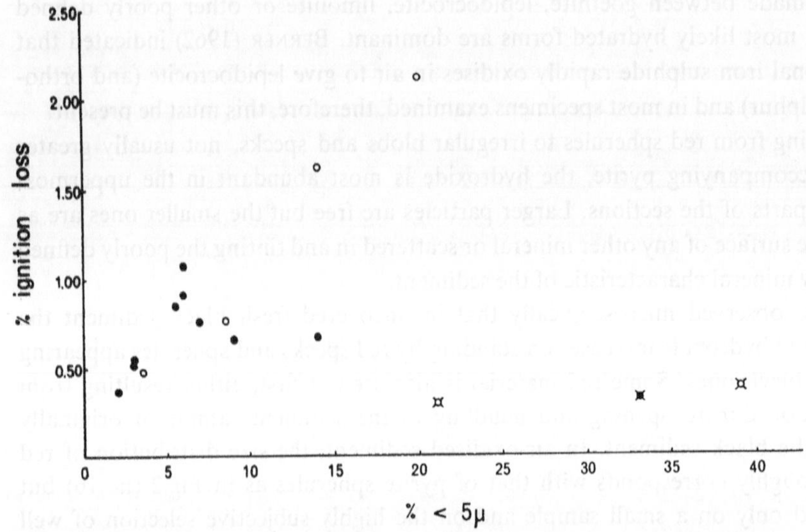

Fig.4. Proportion of dry sediment lost on heating at 375°C against proportion of sediment finer than 5 μ. Legend see Fig.3.

chemical factors in Fig.3–6 were obtained following Ackroyd (1957, p.73) in hydrometric analysis of the portion of weighed dry material washed after dispersion through a 200 mesh B.S. sieve (75 μ). In all the samples the values for the fractions finer than 5 μ and 10 μ appear roughly proportional to those of the total fraction passing the 75 μ mesh sieve. Due to the increased uncertainties of determination of the particles within the clay range, both from experimental error and the theoretical aspects of the sedimentation method data at 2 μ are not quoted; they do, however, appear to show the same trends between the specimens as do the clay-with-silt grades.

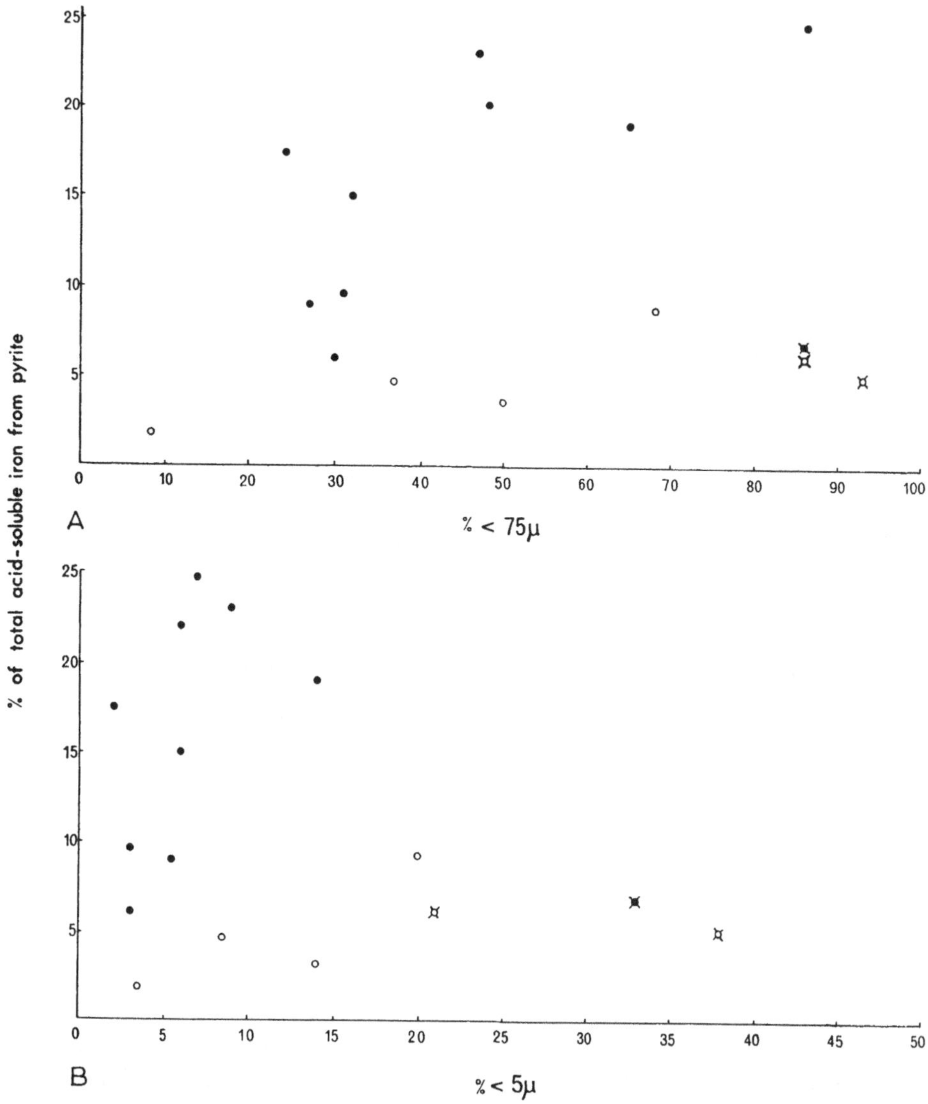

Fig.5. Proportion of total acid-extractable iron due to pyrite.
A. Against proportion of sediment finer than 75 μ.
B. Against proportion of sediment finer than 5 μ.
Legend see Fig.3.

Sedimentological history of the region

In the proportions of clay, silt and sand present, the sediments of the three sections studied show some support for EVANS' (1965, fig.4, p.214) differentiation of the superficial sediment types on the intertidal flats. Sediment from the salt marsh is altogether the finest, with that from the higher mud flat deposits next. Moreover, in

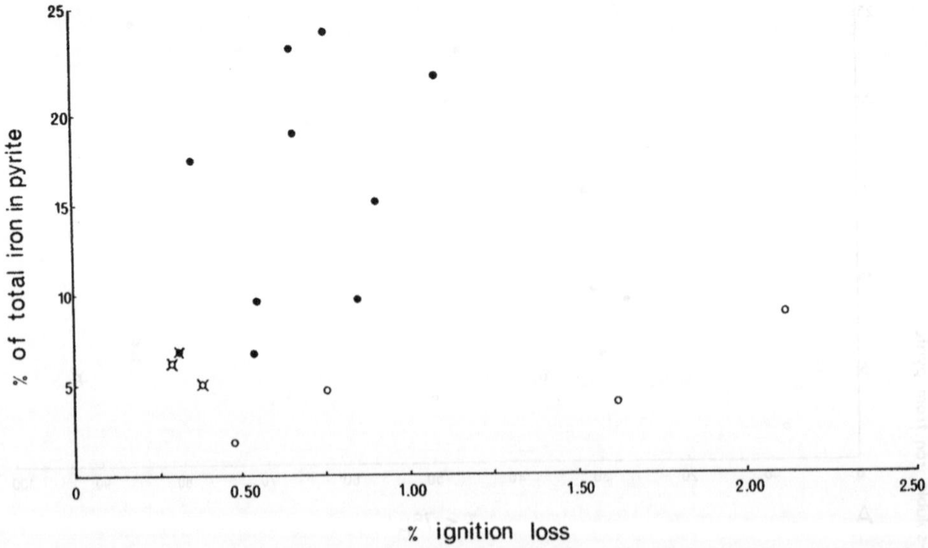

Fig.6. Proportion of acid-extractable iron due to pyrite against loss on heating at 375°C. Legend see Fig.3.

the longer section in the higher mud flats the clay–silt–sand proportions support the stratification of deposits suggested by EVANS (1965, p.283), whose contention was that the facies now represented by the principal zones on the intertidal surface pass laterally inland each beneath the one next landward, with a very slight dip towards the land. To yield such a succession of diachronous, sedimentologically distinct horizons, steady seaward and upward migration of the different sub-environments would be required. Joysey (in EVANS, 1965, p.241), suggested, on the contrary, that the sediments of the upper part of the sequence, occupying for each sub-environmental zone only a relatively small vertical range on any part of the shore, were more likely to be interlayered within the "higher mud" and "inner sand" horizons following fluctuations up and down the intertidal area of these types of deposition. Some alteration is not inconsistent with the evidence available. Nonetheless it does appear established in general that accretion and building out of the intertidal sediment has occurred over a long period. Although currently in the area under study the *Arenicola* sand flats appear to show no overall accretion but, rather, reworking in the top 1.4 cm, the salt marsh may receive an addition of between 1 and 6 cm of material within a year, and the higher mud flats between 1 and 1.5 cm (EVANS, 1965, p.233). Accretion of fine sediment on the upper parts of the intertidal flats clearly follows the now well-known principle of a progressive increase in resistance to erosion by water movement as the grain size of a sediment falls below 180 μ. For the Dutch Wadden Sea POSTMA (1961) showed how very fine sediment is deposited during the period when the incoming tide has lost its velocity but may only be resuspended by much higher currents during the most vigorous part of the ebb; if achieved at all this will be so late that effectively the material is now in suspension in water that is relatively further inshore than that which previously deposited it. In this way there is growth of the tidal flats to high tide level; removal of material only tends to occur in creek erosion.

CHEMICAL ANALYSES

Iron analyses of the sediments

Method

Table I gives some indication of the extent to which iron is available in the sediments for the formation of iron sulphide, the extent to which it has, in fact, been converted to pyrite and the role in these processes of grain size and residual organic matter. Of direct concern here are pyrite, iron monosulphide and hydroxide; siderite was not found present by EVANS (1965, p.227) using X-ray methods. The assumption is made that neither 5*N* hydrochloric acid nor nitric acid significantly dissolves lattice iron from clay minerals or from other silicates, or from heavy oxides if present, and that 5*N* hydrochloric acid does fully dissolve iron hydroxide and monosulphide but leaves the pyrite which may subsequently be dissolved by nitric acid. To the extent that this is incorrect so perhaps is this extra iron available for precipitation with sulphides.

Iron soluble in these acids was determined from dried sediment but the proportions of "pyrite–iron" to "non-pyrite–iron" was from a fresh sample acidified immediately on collection lest, during delay, the monosulphide might be converted to pyrite or vice-versa. The evident change, on oxidation, from black monosulphide to hydroxide does not, however, affect the result in analysis of dried, weighed samples for total acid-soluble iron.

The sediments analysed here were stood in excess 5*N* hydrochloric acid for 12 h, with boiling for 10 min at some period after the completion of solution of monosulphide and calcium carbonate. Microscopic examination of the residues from test specimens indicated that in fresh undried samples solution was completed by the boiling but in oven-dried material removal of all the red-brown specks was slower. It is much easier, however, than in iron-stained indurated rock. Hydrochloric acid solutions, after filtration, were neutralised with ammonia and the iron hydroxide precipitate redissolved and then converted to perchlorate for spectrophotometric analysis: in the case of nitric acid solutions from the first residue, when also organic matter had probably been attacked, that acid was fumed off with perchloric acid before analysis.

Testing for the rate of change of the pyrite proportion, a number of samples collected from two horizons, each about a 5 cm cube, were left undisturbed in thin polythene bags for varying periods before analysis of a specimen from the centre. No significant increase or decrease in the pyrite proportion was detected. Although within a short time black samples can go brown, and some surface samples can go black in the centre after a few days if kept from the air and not dried, these changes appear only to be affecting the unstable forms of iron but none the less, first treatment for analysis within a few hours of collection, or, best, at the time of collection, seems advisable.

Total soluble iron

The content of total soluble iron lies, as would be expected, below the maximum of the ranges given by VAN STRAATEN (1954) and EVANS (1965) who presumably estimated total iron, but the content obtained of HCl-soluble iron is comparable with data provided by Van Straaten from the Wadden Sea flats.

Fig.3 shows linear relationship between the total acid soluble iron (HCl, HNO_3) and the overall fineness of the sediment given as the proportion of material finer than 75 μ and 5 μ. The same trend is in fact seen when a limit at 2 μ is taken, and this implies that it is due to some factor in the finest fraction of the deposit. Such a proportionality with clay has been observed by CARROLL (1958) and others. The relationship holds as well for the salt marsh section as for the others and also irrespectively of the state of oxidation of the sediment, confirming that the same fraction of iron is involved in the conversions from hydroxide to monosulphide and on to pyrite, or back. It is also to be deduced, however, that iron contents of 0.5% and 1% respectively characterise the silt and sand grades. Evans observed a similar value (personal communication) and Van Straaten demonstrated about 0.5% of HCl-soluble "Fe_2O_3" when the proportion of sub-16 μ material is likewise extrapolated to zero. This iron must come from hydroxide carried into the sediment as particles on minerals other than clays or as independent clots (even the coarsest sediments here do show a brown iron colouration when fully oxidised) or from other minerals. Van Straaten's data did not include iron from pyrite which certainly does exist in the coarser sediment of the sections studied here, but correspondingly he did not use nitric acid to release iron.

Proportion of total acid-soluble iron as pyrite

No consistent trend with depth is shown by the pyrite fraction of the iron but a number of the samples indicate a simple proportionality (Fig.5) with the average fineness of the sediment. Those not conforming are, firstly, the oxidised ones, brown or partly brown, where pyrite, although present, is much less than would otherwise be expected, and, secondly, the very fine clays of the salt marsh. Although Fig.5 might only indicate that specimens richer in acid-soluble iron contain more pyrite, the former parameter being itself in relation to grain size it will, however, be seen that among the oxidised sediments the same trend to a greater proportion of pyrite with finer grain size is separately shown.

The black salt marsh specimen appeared in the field to be from below the level of black–brown colour transition but its repeated association graphically with the two oxidised members of the same section suggests some property of this sediment as a whole consistently leading to lower pyritisation. The much greater content of finest material in the salt marsh deposits reducing permeability below an optimum value, or less residual organic material, might be significant. In the five samples "from the surface of the tidal flats at Dollard" quoted by VAN STRAATEN (1954, p.38) the ratio of pyrite iron to the total HCl-soluble iron and pyrite iron, when plotted against the proportion of sediment less than 16 μ, lies in the same part of the field as do those of the salt marsh specimens plotted for this grain size although the pyrite content of the Dollard samples is about double. Presumably they are from the hydroxide zone. Their sub-16 μ grain size content is higher than all but the salt marsh samples from the Wash. Looking towards the subsequent progress of this part of diagenesis, perhaps, Van Straaten shows how in two Holocene "Old Sea Clays" of similar grain size content at 16 μ, the pyrite proportion is about 30% and 45%, suggesting what might be found in deeper, older, sediments in the Wash succession.

Organic material in the sediment: its relation to pyrite formation

Siderite being absent and preliminary heating to a constant weight at 110° C serving to decompose FeS to oxide, during which stage grey and black samples turned brown, the reproducible loss of weight on ignition between 110° C and 375° C was taken to represent the proportion present of organic material. This is often discussed in relation to pyrite and the reduced condition.

Microscopy makes it clear that much of the organic matter is partly degraded, but resistant, cellulosic plant material and pollen exines together with thin tissue such as that associated with foraminiferid tests: some of the spores and pollen are of Carboniferous type and are possibly derived from weathered coal particles which are more easily seen, and are concentrated, in the coarse fraction (see p.327). All this material is significant mainly in the extent to which it was first accompanied in the sediment by rapidly hydrolised substances such as polysaccharides and proteins whose easy decomposition can lead to rapid using up of free oxygen in the sediment soon after deposition.

In Fig.4, with the exclusion of the salt marsh samples, a concentration appears of combustible organic material in the finer sediments, irrespective of the present state of oxidation of the iron, and the superficial, oxidised sediments of the higher mud flats and the *Arenicola* sand flats conform to this. Nonetheless it was experimentally verified that on tight packing in thick polythene of fresh, moist undisturbed samples of any of the oxidised sediments, after two weeks the inner parts were turning black; the fraction of the organic material necessary for this reaction, is therefore present in adequate quantity. Although much organic material does appear to be of size above 75 μ, the relationship inferred here probably represents the degree of preservation during the history of the sediment so far, the averagely finer sediment being the better preserving medium perhaps due to decreased diffusion of oxygen. Possibly, however, the generally very fine sediment such as in the salt marsh originally had less of the bulky organic debris to start with[1].

Excluding oxidised specimens, some very broad relationship may be seen in Fig.6 between the proportions of available iron present as pyrite and of combustible organic material. Unlike the relationship of pyritisation to grain size of the sediment the salt marsh specimens are included with the others.

Water associated with the sediments

In the higher mud flats water was sampled at low tide from the stream in the creek bottom, seepage from excavations in the black sediments in the creek side (about 1.5 m from the top), and a shallow pool in the surface of the mud (replenished at high tide and not at the date of sampling affected by recent rain). In none was the iron content detectable by spectrophotometric methods in unconcentrated samples. Values of pH determined by use of short range universal indicator paper at the time of collection were: surface pond 7.5, seepage at 1.5 m 7.9, creek 8.2. While emphasis must not be placed on the absolute values from this method, that for the surface water is what might be expected for seawater, while the relative differences for the other two samples were noticeable.

DISCUSSION

Formation of pyrite

In recent years it has been made much clearer how sulphide originates in unconsolidated sediments. EMERY and RITTENBERG (1952) demonstrated that within the top few centimetres or metres of submarine sediment, changes in pH can accompany those equally startling changes in Eh which signal the using up of free oxygen in the decomposition of the least resistant organic material buried with the sediment, while under the new physico-chemical regime so developed anaerobic bacteria become intensely active, in place of the previously dominant aerobes. Amongst them sulphate reducers are important. According to many authors the normal proportion of sulphur contained within the organic matter of sediments is quite inadequate as a source of sulphide for the precipitation of metals but more positively KAPLAN et al. (1963) have shown that in the various processes of the sulphur cycle in sediments reduction of sulphates is "the single most important process", and that the overlying sea water provides an inexhaustible reservoir of sulphate for downward diffusion. Control by

[1] On a flat surface on the salt marsh a 25 cm cube of surface material left lying upon an adjacent part of the undisturbed surface was found nearly 2 months later (October–December 1965) intact and when lifted revealed that the marsh top beneath had turned from a dull red-brown colour to black. Any water covering the flat top of the marsh had clearly done so very gently.

sedimentological factors must therefore create a sulphate gradient (BERNER, 1964b) an expression of which is the restriction of the sulphate reduction in ordinary circumstances to the uppermost part of the sediment mass. The extent of transfer of sulphate can be calculated; assuming a water content in the freshly deposited sediment of 75–25% by volume, and taking the mean content of sulphate in sea water as about 2.6 parts per thousand (RANKAMA and SAHAMA, 1950, p.287), then the sulphur content due to pyrite alone in the richest sample of Wash sediment (No.12 in Table I—about a quarter of the iron is pyritised) would be such as would exhaust of sulphate two to twenty volumes respectively of water as originally in association.

Throughout the sediment the pyrite does not appear to have been formed as a coating upon other iron yielding minerals. In particular within plant cells, to which entrance is possible only at the molecular level, or in foraminiferid and diatom shells where the pores may be seen as distinctly smaller than much of the pyrite (Plate IB), then the iron must have been available in solution for the reaction (and the pyrite is conclusively of early diagenetic origin).

With much of the iron originally brought into the area as forms of ferric hydroxide adsorbed on clay minerals or present as clots independently or on other minerals, it may be supposed that its mobility follows the changes in Eh resultant upon the reduction of the sediment, and before precipitation of the monosulphide the iron enters solution in the ferrous state in which it is more readily soluble than as ferric iron. This process has recently been emphasized by CURTIS (1967) who also showed how, under the combined influence of reducing conditions and precipitation by sulphide, iron will be drawn from other iron species than limonite, for instance chlorite, in a ratio dependent upon the activation energies of solution of the various iron phases. Clearly then, in early diagenetic pyrite formation, the hydroxide cannot be regarded exclusively as the source of iron, and other minerals play their part. Even with iron at so low a level in the seepage water analysed it may be noted that characteristic of the stream bottoms in the creeks are the pavements of lamellibranch shells washed from the sediments in which they burrowed during life; these shells are very strongly iron stained indicating some deposition or oxidation in the water. It may be noted that even within the sediment mass the pH is not lowered sufficiently to dissolve calcite extensively (which begins at pH 7.8), for both large and microscopic shells are to be found at all the levels examined but within the sediment, from which most of these shells were washed, in the darker layers some were found partly blackened and etched; this is still under mineralogical investigation.

The relationship of the monosulphide to the pyrite has been examined by BERNER (1963, pp.112–113) who showed on theoretical grounds that strictly a change by oxidation is required to give the pyrite; and from analytical evidence that "the transformation involves the addition of sulphur and not the subtraction of iron".

The reaction might develop in two stages:

$$FeS + H^+ \rightleftharpoons HS^- aq. + Fe^{2+}$$
$$Fe^{2+} + S^\circ + HS^- \rightleftharpoons FeS_2 + H^+$$

Berner pointed out that elemental sulphur is capable of serving in both the capacities of a source of sulphur and as "one of the few oxidising agents found in reduced sediments that contain H_2S", while the amount of elemental sulphur needed to form a saturated S^{2-} or HS^- solution at the pH of most sediments is very small. This is discussed further in BERNER (1967a, b).

Altogether half the total amount of sulphur found in the pyrite must enter by this reaction and the process may exercise a degree of control over the pyritisation. The only adequate source of elemental sulphur must be the hydrogen sulphide released bacterially from sulphate.

In submarine sediments from the Gulf of California BERNER (1964b, p.138) records that the black monosulphide is rare. On balance this is held to be due to some combination of two factors: sufficiently slow deposition to allow completion of pyrite formation at the surface, and at the sediment–water interface an abundant supply of reactive sulphur (from his own preceding discussion the additional oxidising agent postulated would appear unnecessary). In contrast over a depth of up to 3 m in the intertidal sediments of the Wash, the monosulphide is readily found and no more than a quarter of the total available iron is fully converted to pyrite. In seeking an explanation, environmental factors must be examined. In a situation where re-oxidation is a strong trend, and free hydrogen sulphide absent, it would seem likely that the availability of elemental sulphur for the pyritisation of the monosulphide would be low; also while we do not know to what extent equilibrium is approached, and so cannot make predictions from free energy data, it might seem unlikely that the phases Fe_2O_3 and FeS can co-exist with pyrite in greater concentration than found.

In the intertidal sediments of the Wash, the depth of the oxidised zone and the absence in its sediment of free hydrogen sulphide are roughly predictable in terms of distance from drainage creeks and the permeability of the sediment, and are clearly the resultants in these respects of the movement of the tidally based water table. Attainment of the generally reduced condition appears to be tied to more or less permanent saturation with water, where the regular flushing does not occur. In the middle of an extensive area of the *Arenicola* sand flats this is within 10–15 cm of the surface: here *Arenicola* burrows penetrate the black sediment although the worms ingest material from the surface. Outward diffusion of oxygen from their burrows is usually just enough to give a brown colouration to the sand for one to a few mm all round. In the higher mud flats the water table may fall by 1–2 m in the vicinity of creeks and here reddish brown crusts of iron hydroxide or oxide around burrows and cracks are evident, in some cases with the formation of small nodules (such were avoided in sampling for analysis). In the deeper waters of the Black Sea (VOLKOV, 1960), there is no oxidised zone, but an excess of free hydrogen sulphide which in the more extremely anaerobic areas may extend above the sediment–water interface; a balance between sediment type and bottom currents is found so that in shallower parts a thin upper zone of oxidised sediment is present without free H_2S. In the intertidal sediments of the Wash, most of all in the deeply drained higher mud flats, it is therefore probable that the reactions postulated are occurring over a much greater depth range than is recorded by various authors for purely submarine sediment. None the less, despite these environmental differences, the same essential processes are at work and the ultimate pyrite distributions probably indistinguishable.

While the colour zonation of the sediments, brown–black–grey, is helpful in understanding the inter-relationships of the different forms of iron present, and indeed does correspond to obvious stages in the formation of pyritous sediment, it must be remembered that it does not represent the exclusive occurrences of the hydroxide, monosulphide and pyrite, nor from the presence in the uppermost brown sediment of both iron monosulphide and pyrite is it necessary to conclude that this is therefore originally sediment of a black monosulphide zone, or of the grey pyrite even deeper, which is becoming oxidised during erosion of the surface and downward progression of the zone of oxidation; indeed Evans' record that there is distinct evidence of accretion of sediment in the upper parts of the intertidal zone has already been stated. Some restricted spaces in which pyrite formed are evident localities of the reduced state attained well in advance of that in the surrounding material. It is not difficult further to envisage small aggregates of clay or organic debris also defining microscopic regions where similar changes can first occur. Clearly one factor in this is the rate of diffusion, particularly of oxygen but also of other ions to and from any point of reaction. The relationships to be found by general observation and analysis represent gross averaging of vast numbers of physico-chemical "cells" in what is microscopically a most inhomogeneous medium.

The framboidal texture in pyrite

The pyrite, resulting from conversion of monosulphide in situ, has essentially the same distribution as that of the monosulphide. In size, taking framboidal spherules and isolated grains together, it is approximately exponential if correctly inferred as similar to that found from polished sections of the Chattanooga Shale by LOVE and AMSTUTZ (1966). They argued that such a distribution might be expected from precipitation "over a period at any one level within the uncompressed and unconsolidated sediment at multitudinous points" with "irregular rates of formation and of coagulation between adjacent particles in a medium allowing limited diffusion at molecular sizes". It was further pointed out that many authors have taken the rounded form of the spherules to imply that at the earliest stage the material was in the form of blobs of gel with surface tension effects of particular importance. At some stage, however, the forms of the monosulphide in sediments give X-ray diffraction patterns indicating some crystallinity.

The outward form of the spherules, natural enough for blobs of colloidal material but equally an inheritance from the monosulphide stage is unnatural for the crystallised pyrite in the framboid. Its powerful effect on the surface can sometimes indeed be seen to dominate the euhedral crystallisation of the outermost grains of the pyrite, when otherwise cubic grains are rounded off at the outer edge of the framboid (LOVE and AMSTUTZ, 1966, Pl.I, fig.8). This is also evidence that framboids are not formed by the gathering up into balls under surface tension effects of the clouds of more dispersed pyrite grains, in the sediment, a trick readily performed on a larger scale in the laboratory with a suspension of pyrite powder in water. Crystallisation of pyrite from the blob of sulphide appears to have proceeded to one of two products: the single more or less euhedral crystal or the framboidal spherule. In the former no remnant form of the blob is left while in the latter it is dominant. The framboid obviously develops from a number of centres of crystallisation whose crystals remain distinct as the separate grains of the framboid, unless finally they merge to a homogeneous whole.

The statistical distribution of the framboids against the single euhedral crystals suggested to Love and Amstutz that with increasing size an increasing proportion of sulphide blobs ultimately yielded the framboidal rather than the single crystal form of pyrite. It is obviously an oversimplification to regard this as evidence of a limiting factor in the possible size of the blob before fragmentation occurs, especially because when it does the ordering of the pyrite grains in the framboid appears as an extension of the crystallography of the pyrite through the whole framboid. This phenomenon also appears briefly to have been observed by KALLIOKOSKI (1965) who suggested that framboids represent "macrocrystals" of pyrite. To the extent that single crystals can be regarded as having several centres of formation, mutually in crystallographic harmony, or otherwise, then this is a useful concept. If the starting point of the crystallisation is in blobs already isolated and not accreting with further material, but to the contrary, under strong surface tension effects in relation to the surrounding phase, and colloidal moreover, being highly metastable with respect to the crystalline state, then perhaps some of the rules for crystallisation from solutions are not immediately relevant. In a study of microscopic pyrite in veins in two andesites LOVE and AMSTUTZ (in press) observe some superficially comparable phenomena developed from what would be regarded as a colloidal stage of the sulphide.

CONCLUSIONS

(*1*) Acid soluble forms of iron in the Recent sediments of the Wash exist in the phases referrable to ferric hydroxide, iron monosulphide and pyrite. It is confirmed that this is the progression in the formation of pyrite during the early stages of diagenesis.

(*2*) The downward sequence of brown, black and grey colour zones in the sediment indicates the dominance of each phase in turn but pyrite and iron monosulphide are found higher in localised micro-environments more advanced in the chemical sequence.

(*3*) The thickness of the hydroxide zone is closely related to the depth of fluctuation of the tidally-based water table and associated oxygenation; in the latter the intertidal flats probably differ from the source of most pyritic shales and silty

mudstones but texturally, if not quantitatively, the final effects of early diagenetic pyrite formation will be indistinguishable.

(*4*) Part of the source of the iron hydroxide is an association with the clay, part independent of it as specks colouring other sediment grains.

(*5*) The pyrite is present as single isolated grains and framboidal spherules: the former inferred to have increasing abundance with smaller size, and the spherules in unimodal, logarithmic distributions whose modal value, 3–4 μ, does not significantly vary down the 3 m section.

(*6*) The framboidal pyrite spherules show surface and internal features already found characteristic of this texture in rocks. Similar grades of internal ordering are revealed but are much more obscured here by a coalescence of the grains within the framboid which is therefore a primary and not a secondary process.

(*7*) Transformation of the monosulphide to pyrite is in situ: occurrence of both in enclosed spaces and independently of other iron-bearing minerals indicates that iron has passed into solution before precipitation as monosulphide. It is inferred that inheritances from the monosulphide stage include the size distribution of the pyrite and the spherule form—the latter surviving framboidal crystallisation of the pyrite, being from an original gel condition; spherules and isolated small grains of monosulphide are detected microscopically.

(*8*) Conversion of acid-soluble iron to pyrite is not complete in the higher mud flats at 3 m depth and shows no consistent trend with depth. A proportionality here to the fineness of the sediment is not applicable to the much finer sediment of the salt marsh where perhaps reduced permeability is a contrary factor. Residual organic matter, represented by ignition loss, follows the same pattern but the link between pyritisation and ignition loss is not conclusively demonstrable on data available: it would probably be reflecting some more fundamental relationship rather than a causal one in itself.

(*9*) Sedimentological evidence supports the generalised structure of the tidal flats being that of the passage of each environmental sediment zone beneath its shoreward neighbour with perhaps some evidence of alternation represented in the intermediate strata examined.

ACKNOWLEDGEMENTS

The work described was mainly carried out in the Department of Geology of the University of Sheffield. Acknowledgement is particularly made to Mr. V. Samogyi for performing spectrophotometric analyses and immediate preparations for these, and for discussion on other stages; and to Dr. P. E. Brown for laboratory facilities. Polished sections were kindly made by Mr. Lämmler of the Institute of Petrology and Mineralogy, University of Heidelberg, by arrangement with Prof. Dr. G. C. Amstutz, whose encouragement is also acknowledged. Financial support for field-work and the collection of samples was received in Research Fund Grant No.518

from the University of Sheffield. Colleagues of the author kindly discussed aspects of the work in progress and read the manuscript but must not be held responsible for the paper and the balance of ideas in it.

REFERENCES

ACKROYD, T. N. W., 1957. *Laboratory Testing in Soil Engineering*. Soil Mechanics, Ltd., London, 263 pp.

BERNER, R. A., 1962. Tetragonal iron sulphide. *Science*, 137 : 669.

BERNER, R. A., 1963. Experimental studies in the formation of sedimentary iron sulphides. In: M. L. JENSEN (Editor), *Biogeochemistry of Sulfur Isotopes—Natl. Sci. Found., Symp. 1962, Yale Univ., New Haven, Conn.*, pp.107–120.

BERNER, R. A., 1964a. Iron sulphides formed from aqueous solution at low temperatures and atmospheric pressure. *J. Geol.*, 72 : 293–306.

BERNER, R. A., 1964b. Distribution and diagenesis of sulfur in some sediments from the Gulf of California. *Marine Geol.*, 1 : 117–140.

BERNER, R. A., 1967a. Diagenesis of iron sulfide in Recent marine sediments. In: *Estuaries—Am. Assoc. Advan. Sci., Publ.*, 83 : 268–272.

BERNER, R. A., 1967b. Thermodynamic stability of sedimentary iron sulfides. *Am. J. Sci.*, 265 : 773–785.

CARROLL, D., 1958. Role of clay minerals in the transportation of iron. *Geochim. Cosmochim. Acta*, 14 : 1–27.

CURTIS, C. D., 1967. Diagenetic iron minerals in some British Carboniferous sediments. *Geochim. Cosmochim. Acta*, 31 : 2109–2133.

DEGENS, E. T., 1965. *Geochemistry of Sediments*. Prentice Hall, Englewood Cliffs, N.J., 342 pp.

EMERY, K. O. and RITTENBERG, S. C., 1952. Early diagenesis of California Basin sediments in relation to the origin of oil. *Bull. Am. Assoc. Petrol. Geologists*, 36 : 735–806.

EVANS, G., 1958. Some aspects of Recent sedimentation in the Wash. *Eclogae Geol. Helv.*, 51 : 508–515.

EVANS, G., 1965. Intertidal flat sediments and their environments of deposition in the Wash. *Quart. J. Geol. Soc. London*, 121 : 209–245.

KALLIOKOSKI, J., 1965. Framboids—macrocrystals of colloidal pyrite. *Econ. Geol.*, 60 : 1562 (abstract).

KAPLAN, I. R., EMERY, K. O. and RITTENBERG, S. C., 1963. The distribution and isotopic abundance of sulphur in Recent marine sediments off southern California. *Geochim. Cosmochim. Acta*, 27 : 297–331.

LOVE, L. G., 1964. Early diagenetic pyrite in fine grained sediments and the genesis of sulphide ores. In: G. C. AMSTUTZ (Editor), *Sedimentology and Ore Genesis*. Elsevier, Amsterdam, pp.11–17.

LOVE, L. G., 1965. Micro-organic material with diagenetic pyrite from the Lower Proterozoic Mount Isa Shale and a Coal Measures shale. *Proc. Yorkshire Geol. Soc.*, 35 : 187–202.

LOVE, L. G. and AMSTUTZ, G. C., 1966. Review of microscopic pyrite from the Devonian Chattanooga Shale and Rammelsberg Banderz. *Fortschr. Mineral.*, 43 : 273–309.

LOVE, L. G. and AMSTUTZ, G. C., in press. Framboidal pyrite in two andesites.

PAPUNEN, H., 1966. Framboidal texture of the pyritic layer found in a peat bog in S. E. Finland. *Compt. Rend. Soc. Geol.Finlande*, 38 : 117–125.

POSTMA, H., 1961. Transport and accumulation of suspended matter in the Dutch Wadden Sea. *Neth. J. Sea Res.*, 1 : 148–190.

RAMDOHR, P., 1953. Mineralbestand, Strukturen und Genesis der Rammelsberg-Lagerstätte. *Geol. Jahrb.*, 67 : 367–494.

RANKAMA, K. and SAHAMA, T. G., 1950. *Geochemistry*. Univ. Chicago Press, Chicago, Ill., 912 pp.

VALLENTINE, J. R., 1963. Isolation of pyrite spherules from Recent sediments. *Oceanog. Limnol.*, 8 : 16–30.

VAN STRAATEN, L. M. J. U., 1954. Composition and structure of Recent marine sediments in the Netherlands. *Leidse Geol. Mededel.*, 19 : 1–110.

VOLKOV, I. I., 1960. The distribution of free hydrogen sulphide in the sediments of the Black Sea. *Dokl. Akad. Nauk SSSR*, 134 : 676–679 (in Russian).

Reprinted from Sedimentology (1990) **37**, *325–343*

Formation of siderite–Mg-calcite–iron sulphide concretions in intertidal marsh and sandflat sediments, north Norfolk, England

K. PYE*, J. A. D. DICKSON†, N. SCHIAVON†, M. L. COLEMAN*‡ *and* M. COX‡

**Postgraduate Research Institute for Sedimentology, University of Reading, Whiteknights, Reading RG6 2AB; †Department of Earth Sciences, Downing Street, Cambridge CB2 3EQ; ‡Geochemistry Branch, BP Research Centre, Chertsey Road, Sunbury-on-Thames, Middlesex TW16 7LN, England*

ABSTRACT

Concretions cemented mainly by siderite, Mg-calcite and iron monosulphide are common in late Holocene marsh and sandflat sediments on parts of the north Norfolk coast. Field experiments have shown that the concretions are actively forming in reduced sediments in which sulphate-reducing bacteria are active. $\delta^{13}C$ values ranging from -3 to $-11{\cdot}8$‰ (mean $-5{\cdot}9$‰) suggest that the carbonate in the concretions is derived partly from marine sources and partly from microbial degradation of organic matter. $\delta^{18}O$ values ranged from $-6{\cdot}4$‰ to $+0{\cdot}8$‰ (mean $-1{\cdot}0$‰) suggesting that carbonate precipitated in porewaters ranging from pure seawater to seawater diluted with meteoric water. Chemical analysis of porewaters showed no evidence of significant sulphate depletion at the depth of concretion formation. Some concretions have formed around fragments of wood or metal, but others contain no apparent nucleus. In field experiments siderite, FeS and Mg-calcite were precipitated around several different nuclei within a period of six months. We suggest that siderite may form wherever the rate of iron reduction exceeds the rate of sulphate reduction, such that insufficient dissolved sulphide is available to precipitate all the available dissolved ferrous iron.

INTRODUCTION

Concretions cemented mainly by siderite, Mg-calcite and iron monosulphides are actively forming in intertidal marsh and sandflat sediments near Warham on the north Norfolk coast. These concretions are of considerable geological interest for several reasons. First, there are few reports of siderite precipitation under marine conditions, although its formation in fresh and brackish water environments is well-documented (Ho & Coleman, 1969; Postma, 1977, 1982; Stoops, 1983). Pyrite, Mg-calcite and aragonite are the early diagenetic minerals most frequently encountered in temperate shallow marine sediments (Van Straaten, 1957; Coleman & Gagliano, 1965; Wiedemann, 1972; Jorgensen, 1976; Al-Hashimi, 1977; Adams & Schofield, 1983). Ferrous iron in the pore waters would normally be expected to react quickly with H_2S produced by bacterial sulphate reduction, thereby precluding the formation of ferroan carbonates, and siderite is widely regarded as a diagnostic indicator of methanic or post-oxic environments which contain little or no H_2S (e.g. Berner, 1981; Maynard, 1982; Gautier, 1982, 1986). The occurrence of siderite-bearing concretions in marine sediments at Warham therefore appears to challenge a number of well-established ideas in sedimentary geochemistry. Second, the Warham concretions are remarkable for the rapidity of their formation. Model calculations (e.g. Berner, 1968) have suggested that carbonate concretions take several hundreds or thousands of years to form, but along the north Norfolk coast some concretions up to 0·4 m in diameter have clearly formed in sediments deposited since the Second World War. Third, shell material and plant remains show varying degrees of replacement and preservation by authigenic minerals in the Norfolk concretions and may aid interpretation of important fossil-bearing siderite concretions in the geological record, such as those of the Francis Creek Shale in the Mazon Creek area (Schopf, 1979; Woodland & Stenstrom, 1979; Baird *et al.*, 1986; Marples, 1986).

The work described in this paper was undertaken to provide a better understanding of the processes responsible for concretion formation at Warham, following preliminary investigations by Pye (1981, 1984). The programme of experimental and analytical work was designed particularly to establish whether the formation of siderite reflects (1) depletion of porewater sulphate at shallow depth due to a high sedimentation rate, as postulated by Pye (1981), (2) dilution by freshwaters draining from the land, (3) locally high Fe/Ca ratios in the porewaters generated by corrosion of metal artifacts, and (4) local or short-term excesses of iron reduction relative to sulphate reduction and H_2S generation.

STUDY AREA

The north Norfolk coast (Fig. 1) is fringed by an extensive series of salt marshes and sand flats which are partly protected by barrier islands and spits. In places the zone of Holocene coastal sediments is more than 3 km wide. These sediments onlap an eroded boulder clay and chalk surface which dips to the north (Funnell & Pearson, 1984). Freshwater peats which accumulated on this surface during the early Holocene were subsequently inundated by rising sea level, and a sequence of intertidal muds and sands grading up into salt marsh was deposited. The major geomorphological features such as Blakeney Point and Scolt Head Island were probably established some 4000 yr ago but have been modified during historical times. On the basis of historical, archaeological and air photo evidence, Pethick (1980) concluded that the North Norfolk marshes fall into three broad age groups: (a) pre-Romano-British (i.e. 2000+BP), (b) mediaeval (400+BP) and (c) recent (post 1850). Marsh development since 1850 has occurred mainly in areas protected by prograding shingle laterals at the western end of Blakeney Point and Scolt Head Island. The only development of open coast marsh occurred at Warham after 1950.

At Warham a pre-Romano-British Upper Marsh (average elevation 2·7 m O.D.) is separated from the post-1950 Lower Marsh (average elevation 2·0 m) by a low beach ridge capped by a thin veneer of aeolian sand (Fig. 2). The Upper Marsh is covered only by the highest spring tides (6% of all tides), whereas the Lower Marsh is covered by about 50% of tides. Mean neap tidal range is approximately 2·5 m and mean spring tidal range 5·5 m. Air photographs of the Warham Lower Marsh area taken in 1948 show it was largely unvegetated with a poorly defined creek system, but by 1960 vegetation was well established and the creek system more developed. The seaward edge of the Lower Marsh has recently suffered erosion, forming a number of muddy residuals on the landward side of the intertidal sand flats.

The upper 2 m of Upper Marsh sediment consists of greyish brown, silty sands and clayey silts which fine upwards towards the surface. Lamination is weakly preserved and root channels penetrate throughout. Interbedded gravel and sand layers are present, especially close to creeks. A permanent water table is present about 1·8 m below the surface, corresponding approximately with the level of standing water in Stonemeal Creek at low tide. Adjacent to the major creeks the sediments below the water table, and in places up to 0·4 m above it, are coloured black by iron monosulphides. Away from the creeks the sediments are usually light brown or grey in colour, even below the water table.

On the Lower Marsh 0·6–0·8 m of brownish, mottled silty sands overlie at least 1 m of sands and gravelly sands. A permanent groundwater table is present approximately 1·0–1·2 m below the surface, corresponding to the standing water level in the creeks at low tide (Fig. 3). As on the Upper Marsh, black sediments occur below and just above the water table close to the creeks, but elsewhere the predominant sediment colour is brown or grey. In some places a discontinuous zone of black sediment (0·1–0·3 m thick) occurs above the water table, often at depths of 0·4–0·5 m below the marsh surface.

Pools of standing water, known as 'pans', are common on both the Upper and Lower Marshes. Some dry out during summer due to seepage and evaporation, but others, developed above a perched water table, persist throughout the year. The sediments on the bottom of the permanent pans are organic-rich and coloured black or dark greyish brown by organic matter and iron monosulphides.

On the intertidal sand flats the water table is located 0·2–0·4 m below the surface at low tide. The saturated sand immediately below the water table, and in some areas up to 0·2 m of unsaturated sand above it, is black. The thickness of the surface oxidized zone varies according to local patterns of erosion and accretion. In places black sands with oxidized *Arenicola* burrows are exposed at the surface. The thickness of the black monosulphide zone varies from 0·4 m near the Lower Marsh front to only 0·15 m near low

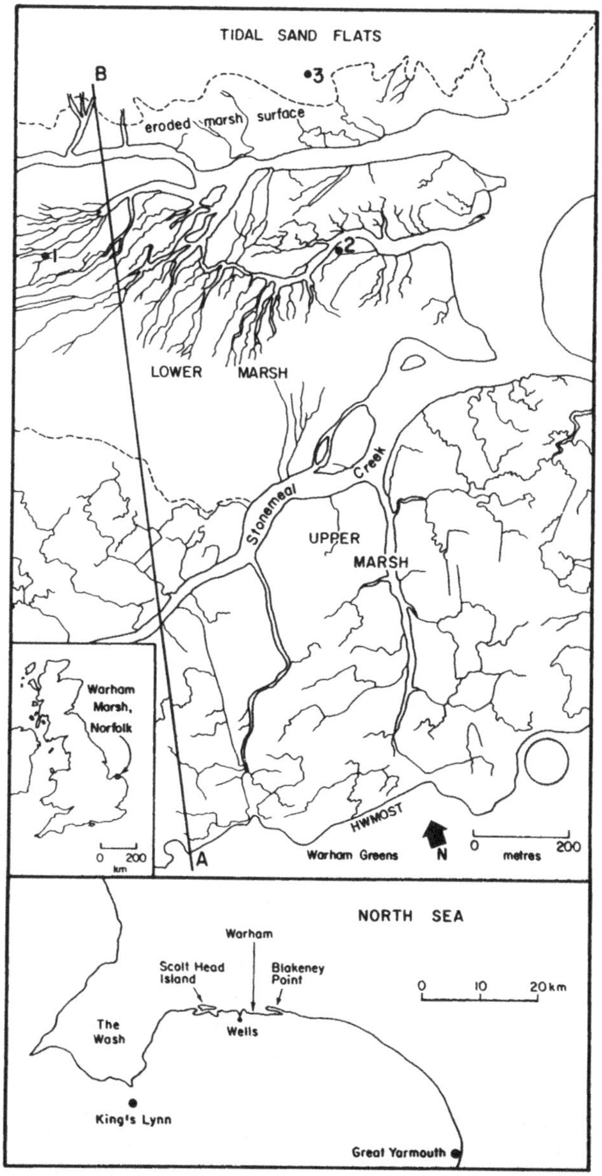

Fig. 1. The study area, showing location of the three experimental sites. The cross-section A–B is shown in Fig. 2.

water mark. With increasing depth there is a sharp transition to greyish or brownish sands.

Black concretions are most commonly seen in the sediments below the water table on the Lower Marsh and the more landward sand flats (Fig. 4). They also occur in thin reduced layers of local reduction spots within the brown, partially oxidized Lower Marsh sediments, and in black, organic-rich pan-bottom and creek-bottom sediments (Fig. 5). Shell and gravel lag deposits in some creek beds are also cemented. The concretions are best exposed where creek migration and bed scouring has caused bank collapse, along the eroded Lower Marsh front, and along the margins of major creeks on the Upper Marsh. Laminations can be traced through some concretions into the surrounding sediment, testifying to their *in situ* development. Reworked, partially oxidized concretions are found in the creek beds and on the surface of the sand flats. A

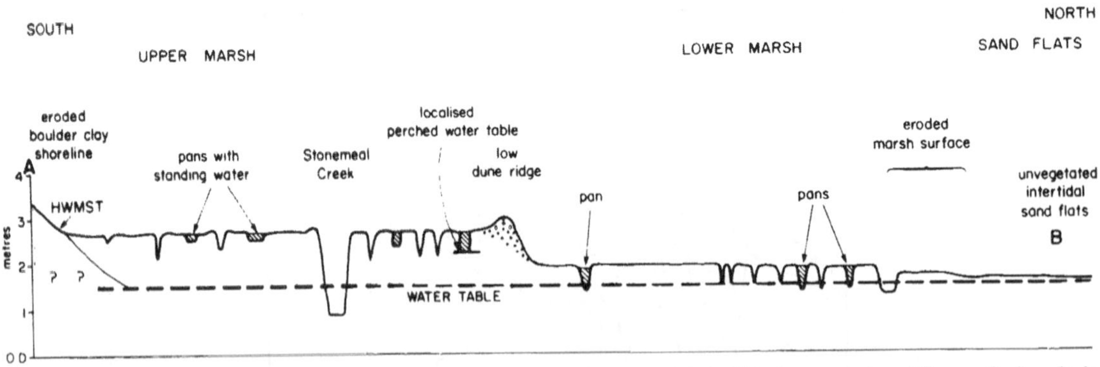

Fig. 2. Schematic cross-section showing the main morphological features of the Warham marshes. The vertical scale is exaggerated.

Fig. 3. Pit section in the Lower Marsh, showing weakly laminated muddy sands overlying cleaner sands of the former intertidal flat facies. A concretion can be seen in black reduced sediments just above the water table.

few concretions exceed 0·3 m in diameter, but 0·1–0·2 m is more typical. The shape is variable, reflecting the form of the nucleus (where one is present) and, in some instances, coalescence of neighbouring concretions. Cemented root tubules, many still containing woody material, frequently project from the concretions. Many concretions contain shells and pieces of cemented gravel. The concretions are black when fresh, but brown or orange oxidation rims develop rapidly on exposure to air.

INVESTIGATIVE METHODS

More than forty concretions were collected from creek bank exposures and specially dug pits. Immediately after collection the concretions were placed in sealed containers filled with black mud to minimize oxidation. Selected examples were sawn into parallel slices for petrographic and geochemical analysis. One slice was vacuum impregnated with Spurr low-viscosity epoxy resin to facilitate preparation of polished thin sections. These were then examined using optical microscopy, cathodoluminescence microscopy (CL), and backscattered scanning electron microscopy (BSEM). The chemical composition of the cements was determined quantitatively by electron microprobe analysis (EMPA). A second slice from each concretion was cut into small chips for stable isotope analysis and mineralogical determination by X-ray diffraction (XRD).

All sub-samples chosen for stable isotope analysis were treated by low-temperature plasma ashing to remove organic matter before analysis. Approximately 30 mg quantities of each sample were reacted with 100% phosphoric acid at 60°C for several days

Fig. 4. Two freshly collected concretions.

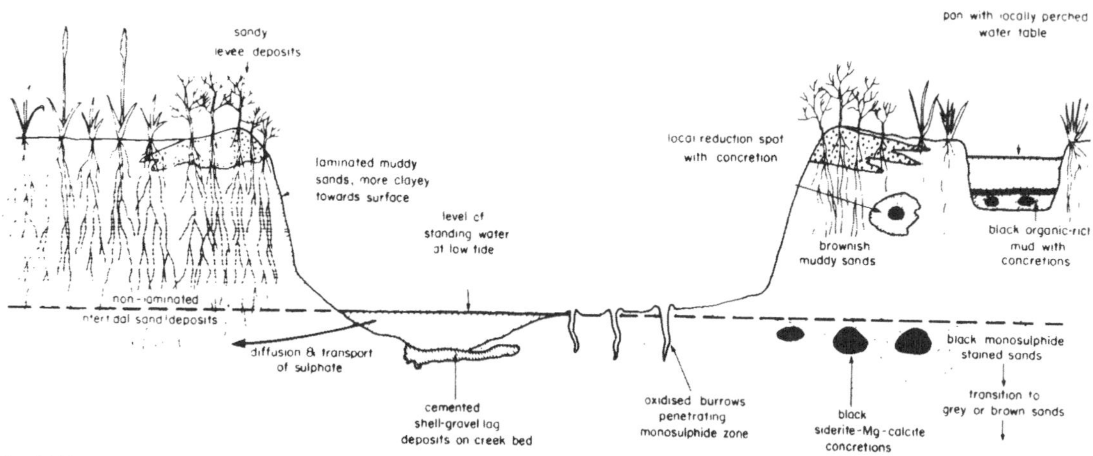

Fig. 5. Diagram showing the main situations where concretions are found in the marsh sediments.

until there was no sign of further reaction. The carbon dioxide was then transferred to vessels containing lead acetate to remove any H_2S which had been released by the acid. The samples were then analysed by mass spectrometry using a VG 903 or a VG SIRA 12 instrument. The corrections for fractionation of the oxygen on reaction with the phosphoric acid assumed all siderite reacted at 60°C (using a value of $\alpha = 1{\cdot}01046$, extrapolated from the data of Carothers *et al.*, 1988). Sub-samples from one concretion were analysed before and after removal of the Mg-calcite component using 10% acetic acid in order to check whether the siderite and Mg-calcite cements have similar isotopic characteristics.

Chemical analysis by inductively coupled plasma spectrometry (ICP) was performed on a dilution of the phosphoric acid used to extract the gas for isotope analysis (Coleman *et al.*, 1989). Since iron oxides and sulphides were partly dissolved by the acid, the chemical data may not give an accurate indication of the ratios of different carbonate minerals present. However, the aluminium concentrations indicate whether significant dissolution of silicate minerals, which affect the measured chemical compositions, has taken place.

Sediment and water samples were also collected from several sites for particle size, mineral and chemical analysis. The grain size distributions of the sediments were determined by wet-sieving, the mineralogy by X-ray powder diffraction, and the major element composition by X-ray fluorescence spectrometry (XRF). Anions in the water samples were determined using a Dionex ion chromatograph; cations were determined after HCl acidification by atomic absorption spectrophotometry (AAS) or ICP.

The numbers and activities of sulphate-reducing bacteria in two sediment cores were determined using the radio-assay techniques described by Maxwell & Hamilton (1986).

RESULTS

Composition of the host sediments

The upper 0·4 m of sediments on the Lower Marsh contain 30–50% silt and clay and have a median size of 70–135 μm (Table 1). They can generally be classified as muddy sands. The underlying sand flat sediments contain 2–8% fines and have a median diameter of 180–212 μm. There is a corresponding difference in the bulk chemical composition, with the quartz-rich sandy sediments containing more Si and less Al, Fe, Ca, Mg, Na and K than the more clayey and silty marsh sediments (Table 2).

Quartz comprises over 90% of the intertidal sands, with subsidiary K-feldspar, plagioclase, mica, kaolinite, chlorite, calcite and dolomite (Table 3). Aragonite was also detected in some samples. Mica, chlorite, kaolinite, calcite and dolomite are more abundant in the finer-grained marsh sediments. With the exception of aragonite, all of these minerals occur in the eroded boulder clay cliffs to the east, from which much of the sediment is thought to be derived. Aragonite is present in small amounts both as shell debris and authigenic crystals in the black intertidal flat and marsh sediments. Pyrite framboids are rare, although they occur occasionally, as does euhedral pyrite, in organic-rich microenvironments.

Composition of the concretions

Approximately a third of the concretions collected were found to contain a metal nucleus, a quarter contained fragments of wood and in the remainder no nucleus could be identified. Many of the metal nuclei represent fragments of military shells from wartime training exercises.

Concretions from all parts of the marsh–sandflat system contain siderite, Mg-calcite and iron monosulphide as the main cements, although their relative proportions vary considerably between, and even within, individual concretions. Authigenic aragonite is also present in some Mg-calcite-rich concretions, often as overgrowths on shells. The presence of griegite was suggested in some XRD traces, but mackinawite and pyrite were not detected. Most of the iron sulphide is X-ray amorphous, acid-volatile and easily oxidized in air. Reworked concretions exposed on creek beds and the intertidal flats contain varying amounts of goethite, akaganeite, amorphous iron oxyhydroxides and gypsum (Pye, 1988).

The texture of the concretions differs depending on whether they have formed in sandy or muddy host sediments. Two extremes are shown in Figs 6 and 7.

Table 1. Grain size of sediments on the Lower Marsh and Intertidal Sand Flats, determined by wet-sieving.

	Sample (% weight in each size class)										
	LM1A	LM1B	LM1C	LM1D	LM2A	LM2B	LM2C	LM2D	LM3	LM4	LM5
Depth (cm)	0–5	20–25	60–65	75–80	0–5	20–25	55–60	90–95	50–60	50–60	50–60
> 1000 μm	0·28	0·30	0·44	0·11	0·26	0·24	0·54	0·15	0·06	0·08	0·83
500–1000 μm	0·42	0·30	0·59	0·23	0·21	0·64	0·07	0·26	0·08	0·17	0·21
250–500 μm	4·10	1·98	27·47	29·94	1·80	1·84	16·48	30·96	19·26	10·32	25·73
125–500 μm	42·63	41·89	68·37	65·99	54·80	50·53	70·69	62·20	78·23	64·75	66·04
63–125 μm	5·93	7·46	1·10	0·52	8·93	6·30	3·64	1·84	1·09	3·58	1·87
< 63 μm	46·61	48·04	2·01	3·17	33·95	40·41	8·56	4·55	1·25	21·07	5·29
Median (μm)	90	70	212	210	135	132	186	183	204	160	205

NB: LM3—experimental site 3, Intertidal Flats; LM4—experimental site 1, Lower Marsh; LM5—experimental site 2, Lower Marsh.

Table 2. Major element composition of sediments from the Warham Lower Marsh and Intertidal Sand Flats, determined by X-ray fluorescence spectrometry.

	Sample (weight % oxide)										
	LM1A	LM1B	LM1C	LM1D	LM2A	LM2B	LM2C	LM2D	LM3	LM4	LM5
Depth (cm)	0–5	20–25	60–65	75–80	0–5	20–25	55–60	90–95	50–60	50–60	50–60
SiO_2	78·89	71·06	89·68	91·60	73·88	66·77	94·56	94·79	93·22	87·64	92·94
Al_2O_3	5·69	7·98	3·16	2·03	6·88	9·08	1·41	1·40	1·32	3·36	1·85
TiO_2	0·37	0·48	0·22	0·14	0·51	0·60	0·10	0·09	0·13	0·21	0·10
Fe_2O_3	2·78	3·78	1·53	1·00	3·59	4·44	0·62	0·62	0·80	1·80	0·83
MgO	1·04	1·40	0·47	0·41	1·24	1·58	0·16	0·20	0·26	0·47	0·27
CaO	2·37	3·43	0·70	1·07	2·84	3·63	0·38	0·53	0·85	1·13	0·74
Na_2O	0·94	1·09	0·57	0·55	0·95	1·16	0·54	0·40	0·50	0·57	0·55
K_2O	1·37	1·64	1·08	0·87	1·49	1·83	0·69	0·69	0·64	1·02	0·81
MnO	0·04	0·06	0·01	0·02	0·06	0·06	0·01	0·01	0·01	0·01	0·01
P_2O_5	0·12	0·16	0·07	0·03	0·15	0·20	0·03	0·03	0·04	0·07	0·05
LOI	6·68	9·20	2·23	2·07	8·13	11·03	1·32	1·53	1·71	3·29	1·62
Total	100·29	100·28	99·72	99·79	99·73	100·37	99·82	100·27	99·47	99·57	99·77

The finer grained marsh sediments contain numerous fragments of organic matter whose cellular structures are partially preserved by authigenic cement crystals (Figs 8 and 9).

In the sandy concretions, where cement crystals have grown unhindered into the pore spaces, siderite displays a number of morphologies, the two most common being globular (Fig. 10A & B) and acicular (Fig. 10C & D). Botryoids of radial-fibrous siderite occur around the edges of larger voids. It is possible that these may be replacive after aragonite. Mg-calcite occurs mainly as equant rhombs and elongated prismatic crystals (Fig. 11A & B). It is brightly luminescent in CL micrographs, whereas the siderite is non-luminescent.

BSEM examination revealed complex intergrowths of siderite, Mg-calcite and FeS in many concretions (Fig. 11C & D). Several had an outer zone dominated by Mg-calcite cement and a core cemented mainly by siderite, but others showed a more patchy distribution of the two minerals. This is clearly seen when sawn surfaces oxidize on exposure to air. Areas cemented mainly by calcite and FeS remain black, while the predominantly siderite-cemented areas turn brown. For reasons which cannot yet be explained, the FeS in the siderite areas is more easily oxidized than that in the Mg-calcite areas.

The concretions formed in muddy sediments (i.e. $>50\%$ of the material <63 μm) contain many shrinkage cracks, root pipes filled with cemented sediment, faecal pellets and fragments of preserved organic matter. Framboidal pyrite and vivianite crystals are present in small numbers (Fig. 11E & F).

Microprobe analysis showed that the siderite contains less than 5 mol% Ca and Mg (Fig. 12), while the Mg-calcite contains 7–14 mol% Mg and 0–7 mol% Fe. Probe analyses of iron sulphide coatings and crack fillings indicated the presence of 8–20% S and 17–35% Fe. The presence of impurities and the fine-grained, porous nature of the material makes accurate analysis and interpretation of this material difficult. During SEM examination, FeS was distinguished from pyrite by differences in grey level in backscattered electron images and on the basis of Fe/S peak height ratios determined from the EDS spectra.

Table 3 and Fig. 13 summarize the isotopic and chemical compositional data obtained from three typical concretions collected from the intertidal sand flat, mid-Lower Marsh and mid-Upper Marsh. The chemical data give only a general indication of the relative proportions of siderite and Mg-calcite present due to contamination from iron monosulphides and oxyhydroxides. Due to the fine intergrowth of these minerals, complete physical separation was not possible. However, the isotopic data display a consistency despite the mineralogical variation between individual sub-samples. As a further check, duplicate samples from one concretion (W7A from the intertidal flats) were run before and after removal of the Mg-calcite component with 10% acetic acid. Only a small variation in the results was obtained ($\delta^{13}C$ values of $-3{\cdot}0$ and $-3{\cdot}3‰$, respectively), suggesting that differences in the carbon and oxygen isotopic signatures of the siderite and Mg-calcite are small.

The majority of $\delta^{18}O$ values lie in the range 0 to $-3‰$. Based on the fractionation data of Carothers

Table 3. Isotopic and elemental composition of sub-samples from a number of Warham concretions (see also Fig. 13). Elemental data in ppm. $\delta^{13}C$ and $\delta^{18}O$ in ‰PDB. (−) means not determined.

Concretion W5A (Intertidal Sand Flats)

Sub-sample	$\delta^{13}C$	$\delta^{18}O$	Mg	Ca	Fe	Mn	Sr	Al
5·1.a	−5·4	−6·4	0·06	0·15	0·79	<0·01	<0·01	0·01
5·1.b	−1·7	−4·2	—	—	—	—	—	—
5·3	−3·1	−1·8	0·08	0·46	0·45	<0·01	<0·01	0·49
5·5	−3·7	−1·3	0·04	0·13	0·83	<0·01	0·03	0·06
5·7	−2·5	−0·5	—	—	—	—	—	—
5·9	−1·3	+0·2	—	—	—	—	—	—
5·11	−3·0	−0·4	—	—	—	—	—	—
5·13.a	−3·8	−4·1	0·01	0·12	0·98	0·01	0·01	0·03
5·13.b	−0·3	+0·4	—	—	—	—	—	—
5·14.a	−1·9	−0·6	0·06	0·33	0·61	0·01	0·01	0·01
5·14.b	−1·6	−0·4	—	—	—	—	—	—
5·18	−2·7	−1·6	0·04	0·45	0·51	0·01	0·01	0·43

Concretion W4A (Lower Marsh)

Sub-sample	$\delta^{13}C$	$\delta^{18}O$	Mg	Ca	Fe	Mn	Sr	Al
4·3	−7·1	−0·3	—	—	—	—	—	—
4·4.a	−8·9	−0·8	0·06	0·09	0·85	0·01	<0·01	2·72
4·4.b	−8·4	+0·8	—	—	—	—	—	—
4·5	−8·2	+0·6	—	—	—	—	—	—
4·6.a	−5·4	−0·5	0·09	0·48	0·42	0·01	<0·01	2·72
4·6.b	−5·4	−0·5	—	—	—	—	—	—
4·11	−6·8	+0·1	0·08	0·31	0·60	0·06	0·01	0·17
4·12.a	−11·8	−3·8	0·04	0·04	0·92	<0·01	<0·01	1·14
4·12.b	−9·8	+0·1	—	—	—	—	—	—
4·13	−8·9	−0·3	—	—	—	—	—	—
4·14	−6·5	+0·1	0·08	0·34	0·57	0·01	<0·01	0·63
4·20	−10·5	−0·3	—	—	—	—	—	—
4·21	−9·0	−0·2	0·04	0·06	0·89	0·01	<0·01	2·42
4·22	−6·4	+0·1	0·04	0·09	0·87	0·01	<0·01	0·77
4·25	−4·9	−0·4	0·10	0·27	0·62	0·01	<0·01	5·22

Concretion W6A (Upper Marsh, close to Stonemeal Creek)

Sub-sample	$\delta^{13}C$	$\delta^{18}O$	Mg	Ca	Fe	Mn	Sr	Al
6·5.a	−6·5	−2·6	0·11	0·57	0·31	0·01	0·01	5·92
6·5.b	−6·2	−2·2	—	—	—	—	—	—
6·5.c	−5·8	−1·4	—	—	—	—	—	—
6·7.a	−5·7	0·0	0·09	0·11	0·79	0·01	0·01	5·95
6·7.b	−5·2	−0·3	—	—	—	—	—	—
6·7.c	−5·7	−1·6	—	—	—	—	—	—
6·9	−5·9	−0·8	0·01	0·46	0·45	0·01	0·01	3·27
6·15	−5·0	−1·2	—	—	—	—	—	—
6·17	−6·4	−0·3	0·04	0·04	0·91	0·01	0·01	5·34
6·19	−5·5	−1·1	—	—	—	—	—	—
6·21	−5·5	+0·3	—	—	—	—	—	—
6·26.a	−6·9	−2·4	0·11	0·49	0·39	0·01	0·01	6·02
6·26.b	−6·7	−0·6	—	—	—	—	—	—
6·27	−7·4	−0·4	—	—	—	—	—	—
6·28.a	−7·2	−2·0	0·08	0·12	0·79	0·01	0·01	7·51
6·28.b	−7·2	−2·0	—	—	—	—	—	—

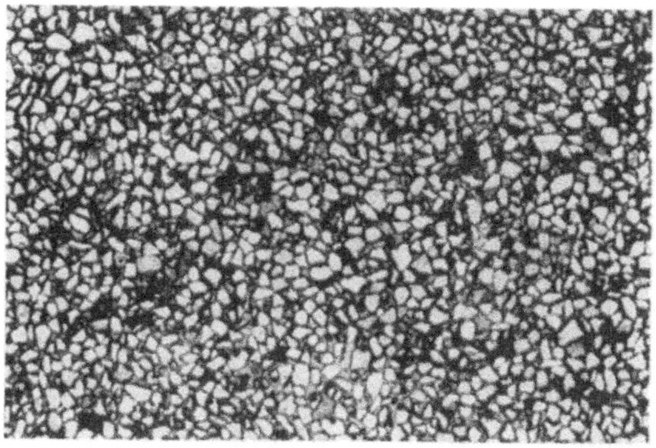

Fig. 6. Transmitted light micrograph showing typical fabric of a concretion developed in the intertidal sand facies (picture width = 1·3 mm).

et al. (1988), and assuming a maximum formation temperature of 25°C, the most negative $\delta^{18}O$ value expected for precipitation of siderite is −0·5‰ (the lowest measured porewater value is +0·4‰). Given the same assumptions, the most negative recorded $\delta^{18}O$ value for a siderite could have been precipitated from porewater representing the most negative $\delta^{18}O$ value of meteoric water which might have fallen in the area (−7·5‰, G. Darling, B. G. S., pers. comm.).

Water analyses

Table 4 summarizes the composition of water samples collected from creeks, pans and pits dug following relatively rain-free periods in September 1985 and May 1986. The surface waters were found to be similar in composition to seawater. Groundwater samples taken from the intertidal flats and two sites near the seaward edge of the Lower Marsh also had a near-seawater composition, although one sample, which was collected after a period of rain and neap tides, showed evidence of freshwater dilution.

Samples taken from a pit in the mid-Lower Marsh in September showed concentrations of chloride, sulphate, Ca, Mg, Na and K higher than those present in seawater, probably due to evapotranspiration by the marsh vegetation. Only one sample (W1/3), collected from a depth of 0·6 m near the seaward edge of the Lower Marsh, showed any evidence of sulphate depletion. The groundwater samples showed higher alkalinity and Ca levels compared with seawater, particularly in the Lower Marsh sediments, where microbial activity and shell dissolution are most evident. Levels of dissolved Fe were found to be low, only exceeding 1 ppm in three of the sixteen samples analysed.

Sulphur-reducing bacteria counts and activity measurements

Sediment cores were taken in June 1987 from the landward edge of the intertidal sand flats and from a well-vegetated area on the Lower Marsh for sulphate-reducing bacteria (SRB) counts and measurement of their activity by radioassay. At the first site, black

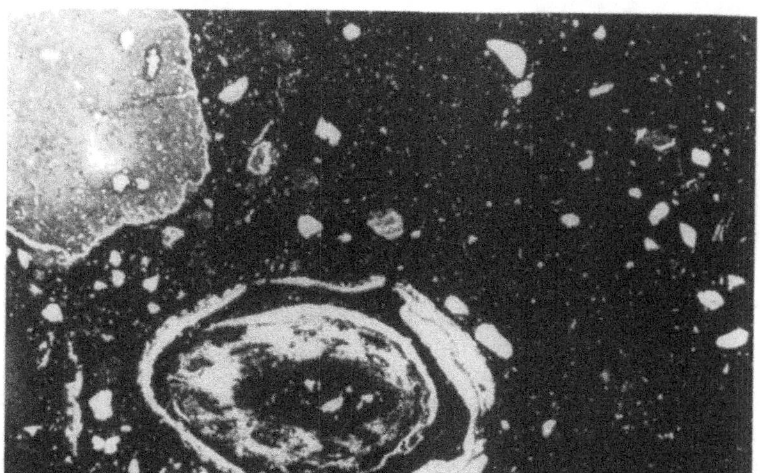

Fig. 7. Transmitted light micrograph showing typical fabric of a concretion formed in fine-grained salt pan sediments (picture width = 0·65 mm).

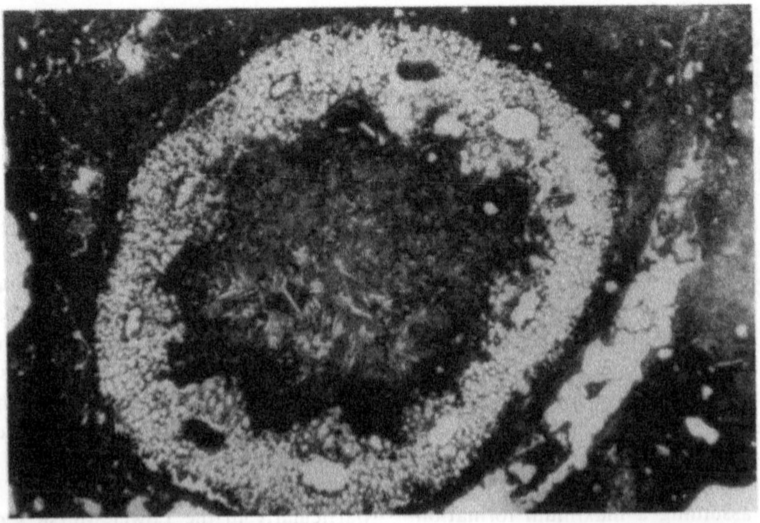

Fig. 8. Transmitted light micrograph showing preservation of woody material in a fine-grained concretion. The centre of the root system is infilled by acicular siderite (picture width = 0·65 mm).

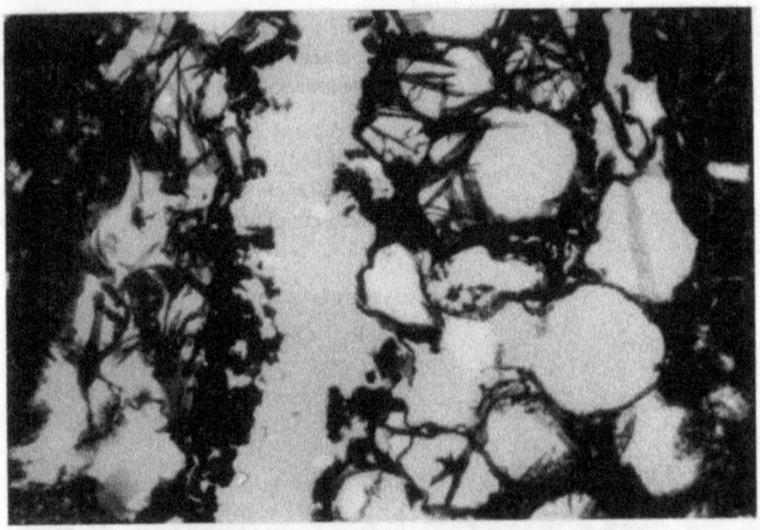

Fig. 9. Higher magnification view of part of Fig. 7, showing cell walls partly replaced by FeS and lined with acicular crystals of siderite (picture width = 0·16 mm).

sands were present beneath a thin (1 mm) surface oxidized layer. At the second site the upper 0·1 m of sediment, including the vegetation mat, was removed to expose greyish brown sandy mud before the core was taken. A concretion from pan bottom sediments on the Lower Marsh was also collected for bacterial activity determination. The results (Table 5) show a close agreement between SRB numbers and activities. Both numbers and activities of SRB decrease with depth in the cores, as observed elsewhere (e.g. Nedwell & Abram, 1978). The SRB counts and activities were two orders of magnitude higher at the vegetated Lower Marsh site than at the unvegetated sandflat site or in the pan concretion.

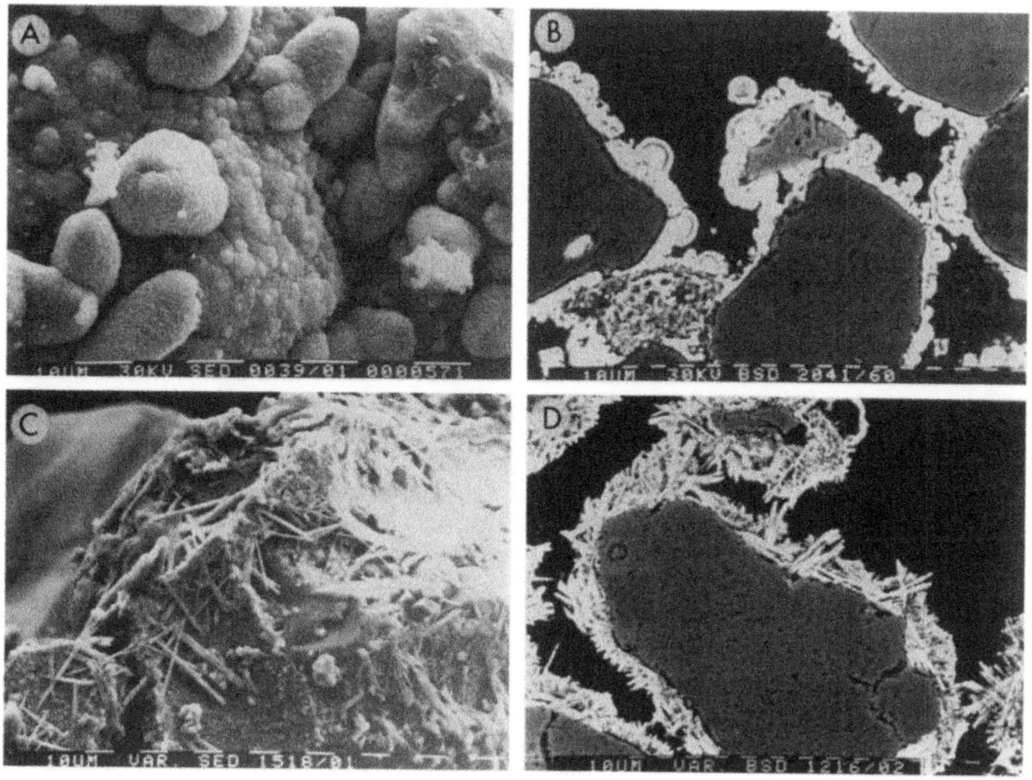

Fig. 10. Scanning electron micrographs showing the nature of globular (A and B) and acicular (C and D) siderite cement. B and D are backscattered images of polished sections. Scale bars = 10 μm.

Field experiments

In an attempt to establish whether siderite, Mg-calcite and FeS are precipitated at the same time in different microenvironments, or whether there is a seasonal pattern of precipitation possibly related to cycles of microbial activity, a number of experimental nuclei were placed at three different sites in April 1986 and September 1987. Two sites were on the mid-Lower Marsh and the third on the upper intertidal sand flats (see Fig. 1 for locations). The latter site was eroded by a storm tide in October 1987, but devices from the two remaining sites were recovered in late January 1988 and examined for evidence of mineral precipitation. The devices consisted of 250 g of pure quartz sand contained in a fine plastic mesh bag. One of five different nuclei [sawn hardwood, aluminium strip, galvanized (Zn-coated) steel strip, phosphor bronze mesh and polycarbonate sheet] was placed in each to establish what effect nucleus composition may have on the rate of authigenic mineral precipitation and on the composition of the precipitates formed. The importance of substrate control has been highlighted by Raiswell (1988), who suggested that concretion formation is often controlled by surface-reaction rates rather than by rates of ionic diffusion and fluid transport.

The devices were placed approximately 0·2 m apart just below the permanent groundwater table in the monosulphide zone and the pits backfilled with reduced sediment.

After recovery the mesh bags were dried, impregnated with blue epoxy resin, and sectioned for BSEM examination and X-ray microanalysis.

Significant precipitation of authigenic carbonate and sulphide, forming a zone 2–10 mm wide, occurred around all of the phosphor bronze and galvanized steel nuclei. Limited precipitation (forming a zone 0·1–2 mm wide) occurred around the aluminium and wood nuclei but none was observed around the polycarbonate nuclei. This suggests that nucleus composition has an effect on the potential for concretion development, but the precise mechanism

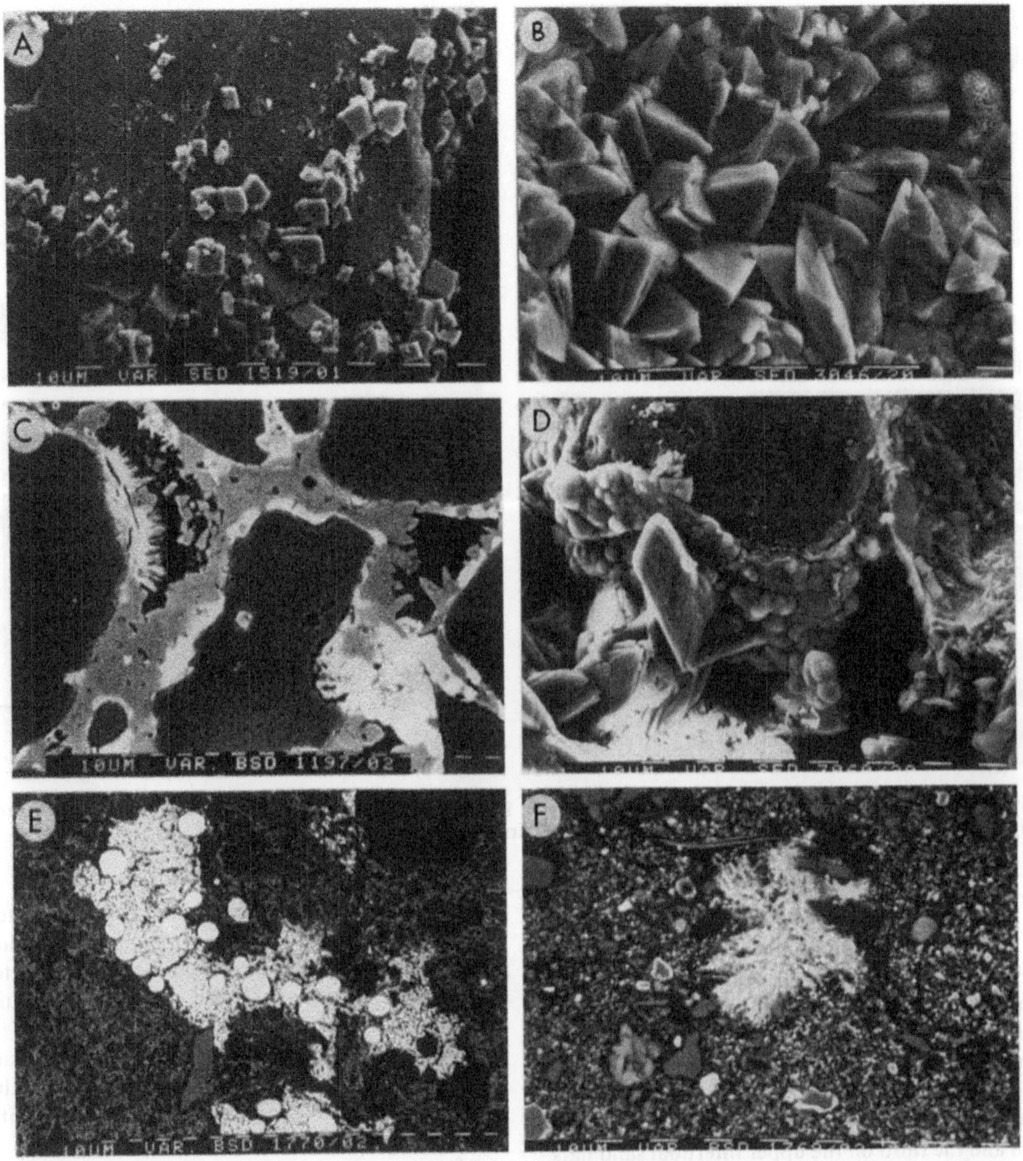

Fig. 11. Scanning electron micrographs showing the nature of the Mg-calcite cement: (A) equant rhombs; (B) elongate prismatic crystals; (C) backscattered image of polished section showing elongated prismatic crystals; (D) Mg-calcite overgrowing globular siderite coating on quartz grain; (E) framboidal pyrite cluster in a fine-grained concretion cemented mainly by Mg-calcite; (F) a rare example of iron phosphate crystals (vivianite?). Scale bars = 10 μm.

of the control remains uncertain. There appear to be two main possibilities which require further evaluation: (1) inorganic corrosion of metallic substrates may directly enhance mineral precipitation, and (2) mineral precipitation may be governed by the varying suitability of different substrates for bacterial colonization.

A thin organic-rich zone containing P, S, Cl and Sn formed adjacent to the corroded edge of the phosphor bronze nuclei. Outside this were thin zones cemented by Cu sulphide, Cu–Fe sulphide and Mg-calcite/Fe sulphide with minor patches of siderite (Fig. 14A & B). The precipitates around the galvanized steel nuclei included an inner zone of Zn carbonate surrounded

Table 4. Chemical composition of surface and sub-surface water samples from Warham. Elemental data in ppm. $\delta^{18}O$ and δD in ‰SMOW.

Sample	Depth (cm)	pH	Alkalinity (mg/l $CaCO_3$)	δD	$\delta^{18}O$	SO_4^{2-}	Cl^-	S^{2-}	PO_4	Ca	Mg	Fe	Mn	Sr	Al	Na	K	Cl^-/SO_4^{2-}
W1/1	0	7·5	—	−2·5	−0·4	2630	18400	—	1	393	1180	<1	0·1	7	0·1	10800	485	7·0
W1/3	60	7·5	—	−1·8	−0·3	2380	18500	—	1	397	1180	2	0·1	7	0·1	10800	510	7·9
W1/6	29	7·0	—	3·2	0·9	2680	19100	—	20	474	1190	4	0·2	8	20·0	10700	540	7·1
W3/8	0	6·9	—	−5·3	−0·5	2660	20200	—	1	592	1290	<1	0·1	8	0·1	11500	500	7·6
W3/9	13	7·8	—	−2·4	−0·2	3480	24400	—	1	747	1670	3	1·4	10	0·7	14300	560	7·0
W3/10	23	8·1	—	−4·2	−0·5	3320	23300	—	1	764	1490	<1	0·1	10	3·6	13300	570	7·0
W3/11	32	7·7	—	−5·2	−0·2	3450	23000	—	1	787	1420	<1	0·1	10	0·2	13000	760	6·5
W3/12	40	8·1	—	−3·2	0·6	3740	23000	—	1	781	1440	<1	0·1	10	1·6	13300	825	6·2
W1J	30	8·0	280	—	—	2520	17600	<1	—	395	1100	<1	—	—	—	—	—	7·0
W2J	0	7·9	135	—	—	2680	19600	<1	—	330	1000	<1	—	—	—	—	—	7·3
W3J	50	7·4	335	—	—	2720	18200	<1	—	500	1100	<1	—	—	—	—	—	6·7
LM8	0	7·7	140	—	—	2686	18800	—	—	408	1410	<1	—	—	—	10900	590	7·0
LM9	120	7·4	445	—	—	2597	17000	—	—	499	1220	<1	—	—	—	10100	560	6·6
LM10	0	7·6	127	—	—	2566	19100	—	—	403	1390	<1	—	—	—	11000	580	7·4
ITF1	30	7·5	265	—	—	2179	15500	—	—	378	1110	<1	—	—	—	9000	510	7·1
ITF2	30	7·3	190	—	—	2655	18700	—	—	432	1450	<1	—	—	—	11400	610	7·0

Location of samples:
W1/1: surface water from creek at edge of Lower Marsh; W1/3 & W1/6: from a pit near seaward edge of Lower Marsh; W3/8: surface water from pan, mid- Lower Marsh; W3/9–W3/12 from a pit in mid-Lower Marsh; W1J & ITF1: from pits near the landward edge of the intertidal sand flats; ITF2: from a pit near the seaward edge of the intertidal sand flats; W3J and LM9: from pits in the mid- Lower Marsh; LM8: surface water from a creek near the landward edge of the Lower Marsh; LM10: surface water from a pan, landward edge of Lower Marsh; W2J: surface water from a creek in mid- Lower Marsh. Samples W1/1–W1J were collected in late summer (September), samples LM8–ITF2 in spring (May). (—) means not determined.

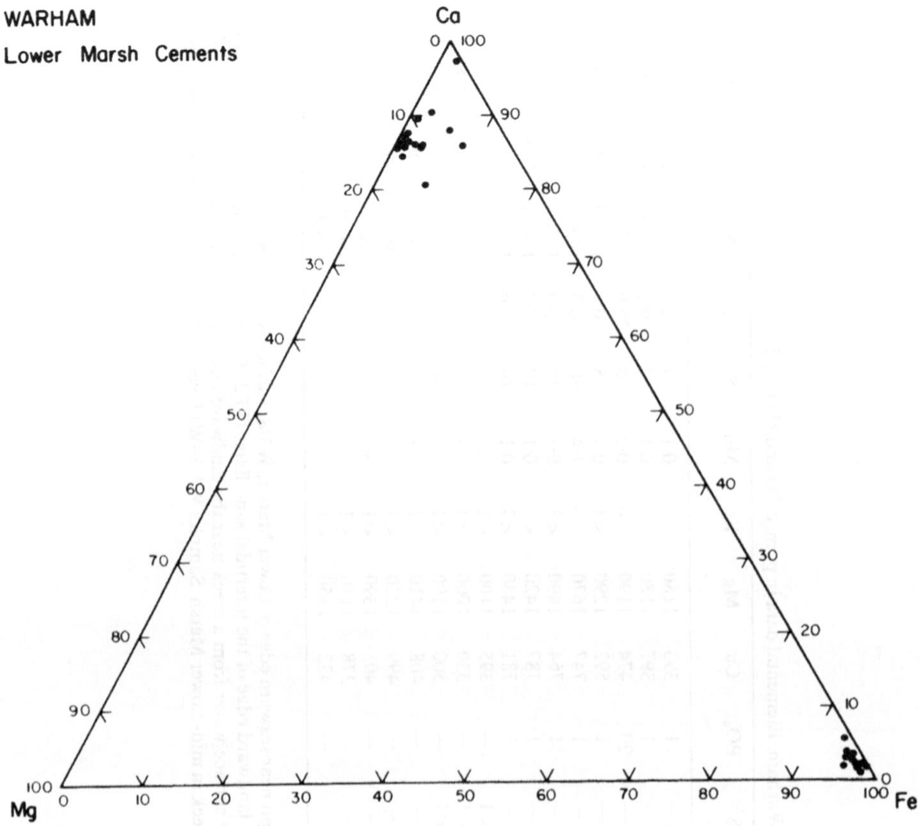

Fig. 12. Ionic ratios of siderite and Mg-calcite cements determined by electron microprobe analysis.

by zones of Zn sulphide and mixed Zn–Fe sulphide. These in turn were surrounded by a diffuse zone cemented by Mg-calcite, Fe sulphide and siderite (Fig. 14C). The zinc coatings on the nuclei were clearly corroded and clearly provided the source of Zn; some Fe must also have been derived from the underlying steel. Only limited cementation occurred around the aluminium and wood nuclei, but the precipitates included siderite as well as Mg-calcite and FeS (Fig. 14D), indicating that an intrinsic source of Fe is not a prerequisite for siderite precipitation.

Surprisingly, four concretions up to 40 mm in diameter formed in the sediments adjacent to the mesh bags at experimental site 2 (care had been taken to ensure no concretions were present when the experimental pits were backfilled with sediment). On

Table 5. Numbers and activities of sulphate-reducing bacteria in Lower Marsh and Intertidal Sand Flat sediment cores.

Location	Mean depth from surface (cm)	SRB count (MPN)	SRB activity $^{35}SO_4^{2-}$ ($nm\ SO_4^{2-}\ ml^{-1}\ d^{-1}$)
Intertidal Sand Flats (eroded Lower Marsh)	2	9×10^2	4·15
	8	$4{\cdot}5 \times 10^2$	5·78
	14	$2{\cdot}5 \times 10^2$	4·38
Mid-Lower Marsh	12	$9{\cdot}5 \times 10^5$	786·09
	18	$4{\cdot}5 \times 10^4$	619·87
Concretion from pan on Lower Marsh	20	$4{\cdot}5 \times 10^2$	30·49

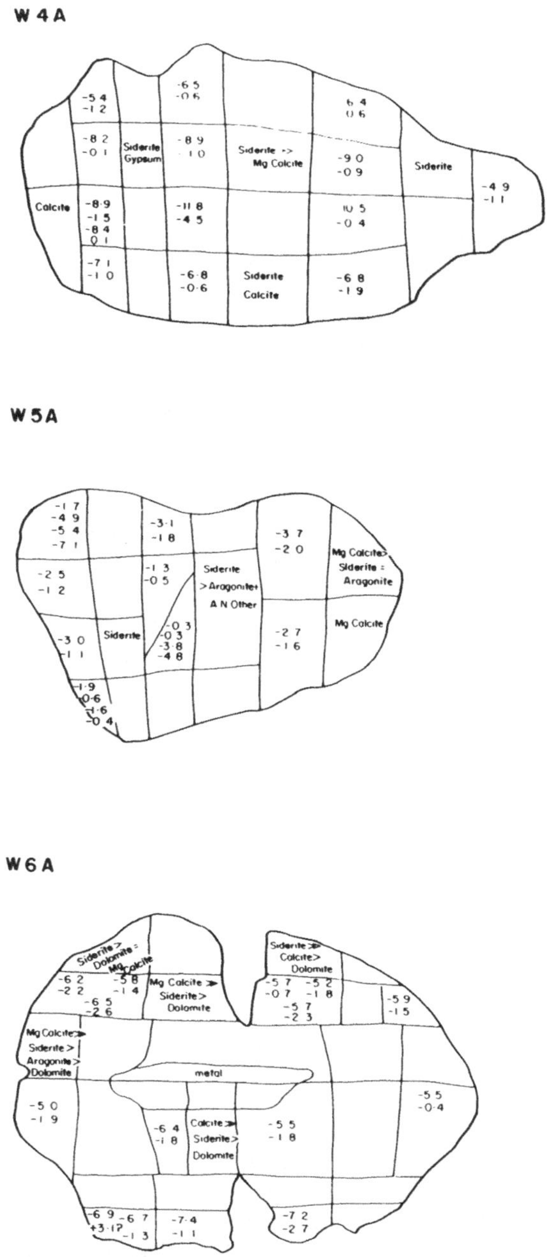

Fig. 13. Mineralogical and isotopic composition of subsamples taken from three Warham concretions (W4A—Lower Marsh; W5A—Intertidal Flats; W6A—Stonemeal Creek, Upper Marsh). $\delta^{13}C$ values above $\delta^{18}O$ values.

sectioning these concretions were found to contain no obvious nucleus, although a number of small shells had been incorporated within their margins. BSEM, EDS and XRD analysis showed that the cements consisted of complex intergrowths of Mg-calcite, siderite and FeS. These polymineralic concretions must have formed within a period of six months, pointing either to very local (possibly substrate) controls on mineral nucleation, or to very short-term fluctuations in pore water chemistry.

Aragonite shells contained within these concretions were found to be partly or wholly replaced by siderite (Fig. 15A, B, C), with preservation of fine structural detail. Included foraminifera and *Hydrobia* shells displayed complete replacement by siderite (Fig. 15D), even in areas where the surrounding cement consisted of Mg-calcite. In these instances it is not possible to tell whether the siderite replacement occurred before, during or after the episode of Mg-calcite cementation.

Laboratory experiments

Black sand collected from below the water table on the intertidal sand flats was placed in a five gallon plastic container, covered with 0·2 m of seawater, and taken complete with in-fauna to the laboratory for further experimentation. The level of water in the containers was maintained by periodically adding more seawater to compensate for slight losses due to evaporation. Steel washers, pieces of aluminium strip and pieces of sawn wood were suspended by nylon threads and immersed in the sediment for periods of 7, 14 and 28 days before being withdrawn for SEM and XRD analysis. After 14 and 28 days crystals of authigenic Mg-calcite were observed around the aluminium nuclei, while both Mg-calcite and siderite were observed adjacent to the steel nuclei. No significant precipitation was observed around the wood nuclei or in the sediment a few centimetres away from the nuclei. These results suggest that carbonate precipitation is especially favoured around metallic nuclei. Although steel is rapidly attacked by H_2S forming FeS and releasing H^+, this acidity may be offset by the alkalinity generated by bacterial biofilms which commonly colonize corroding metal surfaces (Chandler, 1985; Hamilton, 1985; Pankhania *et al.*, 1986). In this experiment corrosion of steel appeared to favour siderite formation, possibly by locally raising the dissolved Fe/Ca ratio of the porewaters.

DISCUSSION AND CONCLUSIONS

A number of authors have pointed out previously that siderite can only form where insufficient H_2S is

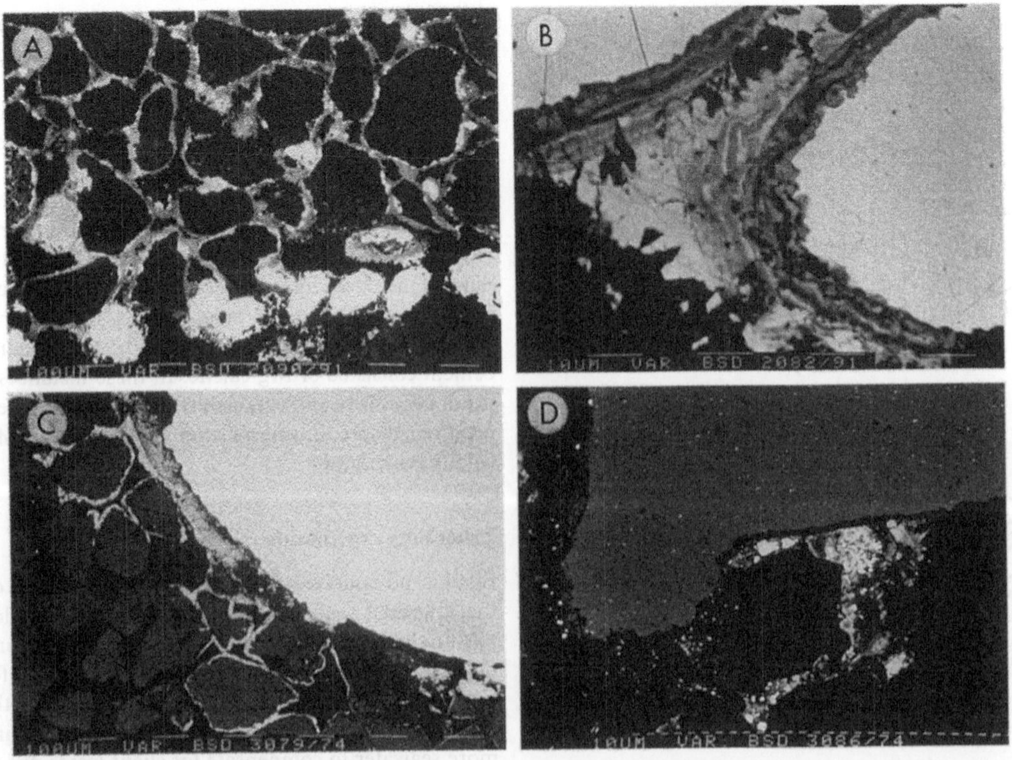

Fig. 14. Backscattered scanning electron micrographs showing precipitates formed around experimental nuclei. (A) Mg-calcite, siderite and FeS precipitated around phosphor-bronze mesh; (B) zoned copper sulphide precipitated immediately adjacent to the bronze mesh; (C) zinc sulphide and zinc carbonate precipitated around a bronze nucleus; (D) siderite precipitated adjacent to an aluminium nucleus. Scale bars in A = 100 μm; scale bars in B–D = 10 μm.

available to precipitate all the available reduced iron as FeS or pyrite (e.g. Curtis & Spears, 1968). In freshwater sediments, production of H_2S is normally limited by the low levels of dissolved sulphate available for reduction by sulphate-reducing bacteria, leading to the creation of 'methanic' conditions and siderite precipitation at shallow depth (Berner, 1981; Maynard, 1982; Gautier, 1982). In marine sediments with similar permeability and organic carbon content, sulphate depletion may be expected to occur at a greater depth due to a much higher initial sulphate content of the porewaters, although cases of sulphate depletion at shallow depth have been reported (e.g. Murray *et al.*, 1978; Devol, 1983). Production of H_2S in near-surface marine sediments is more commonly limited by the availability of reactive organic matter rather than sulphate (Berner, 1981; Westrich & Berner, 1984). Formation of iron sulphides in marine sediments which contain a large terrigenous component is not usually limited by iron availability, since ferric hydroxide coatings on detrital grains are reduced relatively easily (Berner, 1985). As a result, iron sulphides (either FeS or pyrite) should be the dominant authigenic iron minerals formed in the zone of sulphate reduction. Accordingly, the view is widely held that siderite and other ferroan carbonates form only at greater depth in diagenetic zones below the zone of sulphate reduction. This view is expressed by Gautier (1986, p. 56), who states that 'in marine sediments which contain abundant organic matter, siderite is a dependable indicator of the methanic diagenetic zone'. However, our results suggest that this is not necessarily the case. The $\delta^{13}C$ values strongly suggest that the carbonate in the Warham concentrations is partly of marine origin and partly derived from biogenic degradation of organic matter, principally in the zone of sulphate reduction. The intertidal sand flat concretions contain a larger marine component than the concretions in the more organic-rich marsh sediments. Little CO_2 is likely to have been contributed by

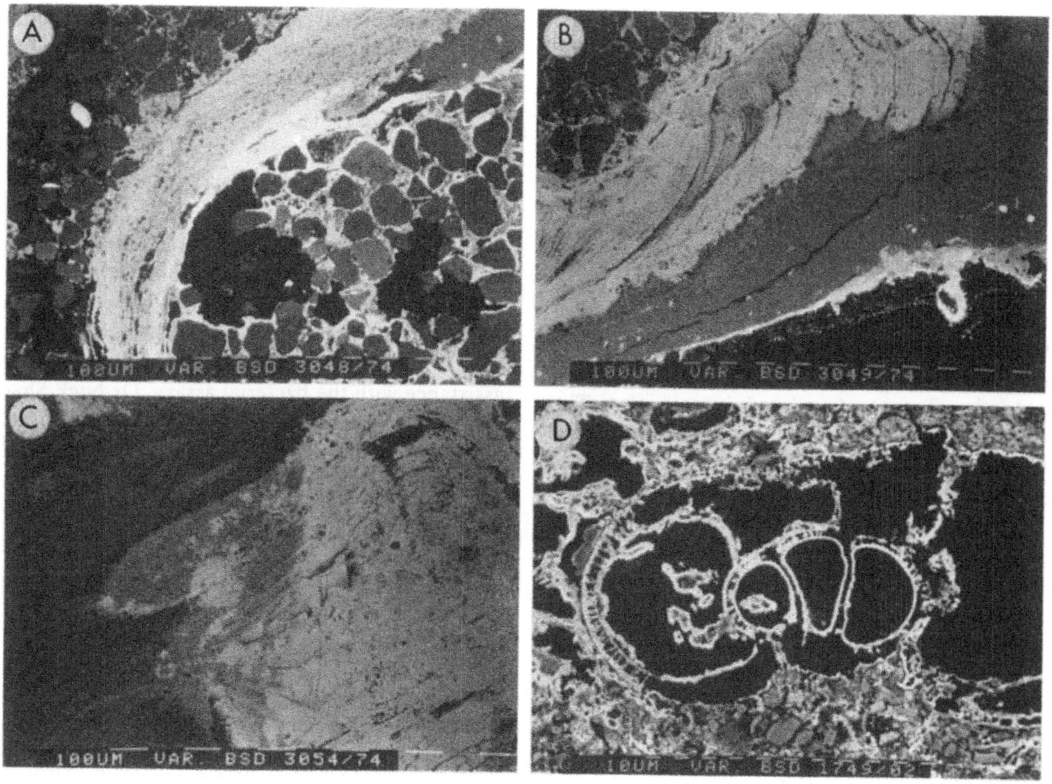

Fig. 15. Backscattered SEM micrographs showing replacement of an aragonite shell (A–C) and a calcite shell (D) by siderite. Scale bars in A–C = 100 μm; in D = 10 μm.

methane oxidation (Reeburgh, 1980) since carbonates formed from such a source typically show much more strongly negative $\delta^{13}C$ values (typically −20 to −60‰ PDB) (Roberts & Whelan, 1975; Nelson & Lawrence, 1984; Hovland *et al.*, 1987). Although interpretation of the oxygen isotope data suggests dilution of some samples by rainwater, sulphate determinations showed no evidence of sulphate depletion at the depth of concretion formation. It is therefore unlikely that the activity of sulphate-reducing bacteria is limited by sulphate availability.

The formation of authigenic siderite in the zone of sulphate reduction at Warham may result from a deficiency of bacterially produced H_2S relative to the rate of iron reduction, at least locally, within the sediments. Coleman (1985) and Curtis (1987) have pointed out that, where iron (III) reduction exceeds sulphate reduction, iron-rich authigenic carbonate minerals may be expected to co-precipitate with iron sulphide. Both sulphate reduction and iron reduction introduce HCO_3^- into the porewater system, thereby raising the likelihood of authigenic carbonate precipitation, particularly if H_2S escapes upwards to the surface (Coleman, 1985; Curtis, 1987).

The relative rates of iron reduction and sulphate reduction are controlled by several factors including the availability of organic matter, availability of reactive iron, and the activity of various bacteria. Nutrient availability plays a significant role in determining the distribution and rate of bacterial activity (Postgate, 1984). Our measurements showed that sulphate reducing bacteria are most active in the upper 0·2 m of the vegetated marsh sediments where biological productivity is high and organic nutrients are abundant. These sediments are typically mottled, with black patches around decaying root stems and larger patches of organic matter in which sulphate-reducing bacteria are likely to be concentrated. A great deal of the H_2S produced in this zone is probably oxidized or escapes to the surface, creating an opportunity for ferroan carbonate precipitation. The observed decline in numbers and activities of sulphate-reducing bacteria with depth is probably attributable to a reduction in the availability of suitable organic nutrients, since there is no shortage of sulphate. As a consequence of the reduced SRB activity with increasing depth, the rate of H_2S generation must also decline with depth, increasing the likelihood that there will be an excess of iron reduction.

Further evidence that nutrient availability is an important control on the rate of bacterial sulphate reduction is provided by the observation that, away from creeks, which act as pathways for the transfer of dissolved and suspended nutrients in and out of the marsh system, the sediments at depth are brownish due to the absence of reduced iron species. The porewaters in these sediments contain slightly higher concentrations of sulphate and chloride than seawater, but are slightly undersaturated with respect to calcite.

The near-surface sand flat sediments contain fewer and less active sulphate-reducing bacteria than the near-surface marsh sediments, probably because the sediments contain less organic matter. Close to the marsh front, where the sandflats are flushed by nutrient-rich waters from the vegetated marsh, the zone of black sediment containing FeS and authigenic carbonate is 0·3–0·4 m thick. With increasing distance from the marsh front the thickness of this zone diminishes in parallel with the expected availability of organic nutrients.

It has been shown in previous studies that the activity of sulphate-reducing bacteria is strongly influenced by temperature and shows marked seasonal variation (Nedwell & Abram, 1978; Postgate, 1984). The generation of H_2S is therefore also likely to vary seasonally, with possible implications for the dissolved Ca/Fe ratio and porewater alkalinity. However, our experimental data indicate no simple seasonal pattern of mineral precipitation, since siderite, Mg-calcite and FeS were found to co-precipitate within a single season.

According to Berner (1971) a dissolved Ca/Fe ratio of <20 is required for siderite formation, but porewaters in the Warham marsh system generally have a much higher Ca/Fe ratio. Conditions favourable for siderite formation may therefore exist only in certain microenvironments. The field experiments demonstrated that an iron-bearing nucleus is not essential for primary siderite precipitation, although corrosion of steel did appear to favour siderite relative to Mg-calcite, possibly by lowering the Ca/Fe ratio of the porewaters close to the steel nucleus.

This study has demonstrated the complexity of mineral authigenesis in a modern intertidal system. Much further work is required to clarify the relationships between patterns of microbial activity, porewater chemistry and authigenic mineral composition.

ACKNOWLEDGMENTS

This work was supported by NERC Grant GR3/6016. KP also acknowledges financial support provided by the Royal Society. Permission to work at Warham was granted by the National Trust and the Nature Conservancy Council. We thank Dr Ali Moosavi for assistance with the bacterial assays, Dr Nick Shackleton for some of the isotope analyses and Gillian Foreman for sample preparation. The paper also benefited from critical reviews by Rob Raiswell and John Hudson. University of Reading PRIS Contribution No. 030.

REFERENCES

ADAMS, A.E. & SCHOFIELD, K. (1983) Recent submarine aragonite, magnesian calcite, and hematite cements in a gravel from Islay, Scotland. *J. sedim. Petrol.*, **53**, 417–421.

AL-HASHIMI, W.S. (1977) Recent carbonate cementation from seawater in some weathered dolostones, Northumberland, England. *J. sedim. Petrol.*, **47**, 1375–1391.

BAIRD, G.C., SROKA, S.D., SHABICA, C.W. & KUECHER, G.J. (1986) Taphonomy of Middle Pennsylvanian Mazon Creek area fossil localities, northwast Illinois: significance of exceptional fossil preservation in syngenetic concretions. *Palaios*, **1**, 271–285.

BERNER, R.A. (1968) Rate of concretion growth. *Geochim. Cosmochim. Acta*, **32**, 477–483.

BERNER, R.A. (1971) *Principles of Chemical Sedimentology*. McGraw Hill, New York.

BERNER, R.A. (1981) A new geochemical classification of sedimentary environments. *J. sedim. Petrol.*, **51**, 359–365.

BERNER, R.A. (1985) Sulphate reduction, organic matter decomposition and pyrite formation. *Phil. Trans. R. Soc. A*, **315**, 25–38.

CAROTHERS, W.W., ADAMI, L.H. & ROSENBAUER, R.J. (1988) Experimental oxyten isotope fractionation between siderite-water and phosphoric acid liberated CO_2 siderite. *Geochim. Cosmochim. Acta*, **52**, 2445–2450.

CHANDLER, K.A. (1985) *Marine and Offshore Corrosion*. Butterworths, London, 413 pp.

COLEMAN, J.M. & GAGLIANO, S.M. (1965) Sedimentary structures: Mississippi River deltaic plain. In: *Primary Sedimentary Structures and Their Hydrodynamic Interpretation* (Ed. by G. V. Middleton). *Spec. Publs Soc. econ. Paleont. Miner, Tulsa*, **12**, 133–148.

COLEMAN, M.L. (1985) Geochemistry of diagenetic non-silicate minerals: kinetic considerations. *Phil. Trans. R Soc. A*, **286**, 353–372.

COLEMAN, M.L., WALSH, J.N. & BENMORE, R.A. (1989)

Determination of both chemical and stable isotope composition in milligramme size carbonate samples. *Sediment. Geol.*, in press.

CURTIS, C.D. (1987) Mineralogical consequences of organic matter degradation in sediments: inorganic/organic diagenesis. In: *Marine Clastic Sedimentology* (Ed. by J. K. Leggett & G. G. Zuffa), pp. 108–123. Graham and Trotman, London.

CURTIS, C. D. & SPEARS, D.A. (1968) The formation of sedimentary iron minerals. *Econ. Geol.*, **63**, 257–270.

DEVOL, A.H. (1983) Methane oxidation rates in the anaerobic sediments of Saanich Inlet. *Limnol. Oceanogr.*, **28**, 738–742.

FUNNELL, B.M. & PEARSON, I. (1984) A guide to the Holocene geology of North Norfolk. *Bull. geol. Soc. Norfolk*, **34**, 123–148.

GAUTIER, D.L. (1982) Siderite concretions: indicators of early diagenesis in the Gammon Shale (Cretaceous). *J. sedim. Petrol.*, **52**, 859–871.

GAUTIER, D.L. (1986) Interpretation of early diagenesis in ancient marine sediments. *Soc. econ. Paleont. Miner, Short Course Notes*, **17**, 6–78.

HAMILTON, W.A. (1985) Sulphate-reducing bacteria and anaerobic corrosion. *Ann. Rev. Microbiol.*, **39**, 195–217.

HO, C. & COLEMAN, J.M. (1969) Consolidation and cementation of Recent sediments in the Atchafalaya Basin. *Bull. geol. Soc. Am.*, **80**, 183–192.

HOVLAND, M., TALBOT, M.R., QVALE, H., OLAUSSEN, S. & AASBERG, L. (1987) Methane-related carbonate cements in pock-marks of the North Sea. *J. sedim. Petrol.*, **57**, 881–892.

JORGENSEN, B.B. (1983) The microbial sulphur cycle. In: *Microbial Geochemistry* (Ed. by W. E. Krumbein), pp. 91–124. Blackwell Scientific Publications, Oxford.

JORGENSEN, N.O. (1976) Recent high magnesian calcite/aragonite cementation of beach and submarine sediments from Denmark. *J. sedim. Petrol.*, **46**, 940–951.

MAPLES, C.G. (1986) Enhanced paleoecological and paleoenvironmental interpretations result from analysis of early diagenetic concretions in Pennsylvanian shales. *Palaios*, **1**, 512–516.

MAXWELL, S. & HAMILTON, W.A. (1986) Modified radiorespirometric assay for determining the sulphate reduction activity of biofilms on metal surfaces. *J. Microbiol. Methods*, **5**, 83–91.

MAYNARD, J.B. (1982) Extension of Berner's 'New Geochemical Classification of Sedimentary Environments' to ancient sediments. *J. sedim. Petrol.*, **52**, 1325–1331.

MURRAY, J.W., GRUNDMANIS, V. & SMETHIC, M.S. JR. (1978) Interstitial water chemistry in the sediments of Saanich Inlet. *Geochim. Cosmochim. Acta*, **42**, 1011–1026.

NEDWELL, D.B. & ABRAM, J.W. (1978) Bacterial sulphate reduction in relation to sulphur geochemistry in two contrasting areas of saltmarsh sediment. *Estuar. Coast. Mar. Sci.*, **6**, 341–351.

NELSON, C.S. & LAWRENCE, M.F. (1984) Methane-derived high-Mg calcite submarine cement in Holocene nodules from the Fraser Delta, British Columbia, Canada. *Sedimentology*, **31**, 645–654.

PANKHANIA, I.P., MOOSAVI, A.N. & HAMILTON, W.A. (1986) Utilization of cathodic hydrogen by *Desulfovibrio vulgaris* (Hildenborough). *J. Gen. Microbiol.*, **132**, 3357–3365.

PETHICK, J.S. (1980) Salt-marsh initiation during the Holocene transgression: the example of the North Norfolk marshes, England. *J. Biogeogr.*, **7**, 1–9.

PETHICK, J.S. (1981) Long-term accretion rates on tidal marshes. *J. sedim. Petrol.*, **51**, 571–577.

POSTGATE, J.R. (1984) *The Sulphate-reducing Bacteria*, 2nd edn. Cambridge University Press, Cambridge.

POSTMA, D. (1977) The occurrence and chemical composition of Recent Fe-rich mixed carbonates in a river bog. *J. sedim. Petrol.*, **47**, 1089–1098.

POSTMA, D. (1982) Pyrite and siderite formation in brackish and freshwater swamp sediments. *Am. J. Sci.*, **282**, 1151–1183.

PYE, K. (1981) Marshrock formed by iron sulphide and siderite cementation in saltmarsh sediments. *Nature*, **294**, 650–652.

PYE, K. (1984) SEM analysis of siderite cements in intertidal marsh sediments, Norfolk, England. *Mar. Geol.*, **56**, 1–12.

PYE, K. (1988) An occurrence of akaganéite (β-FeOOH·Cl) in Recent oxidized carbonate concretions, Norfolk, England. *Min. Mag.*, **52**, 125–126.

RAISWELL, R. (1987) Non-steady state microbiological diagenesis and the origin of concretions and nodular limestones. In: *Diagenesis of Sedimentary Sequences* (Ed. by J. D. Marshall), pp. 41–54. Blackwell, Oxford.

RAISWELL, R. (1988) Evidence for surface reaction-controlled growth of carbonate concretions in shales. *Sedimentology*, **35**, 571–575.

REEBURGH, S. (1980) Anaerobic methane oxidation rate depth distribution in Skan Bay sediments. *Earth Planet. Sci. Lett.*, **47**, 345–352.

ROBERTS, H.H. & WHELAN, T. III. (1975) Methane-derived carbonate cements in barrier and beach sands of a subtropical delta complex. *Geochim. Cosmochim. Acta*, **39**, 1085–1089.

SCHOPF, J.M. (1979) Evidence of soft-sediment cementation enclosing Mazon plant fossils. In: *Mazon Creek Fossils* (Ed. by M. H. Nitecki), pp. 105–128. Academic Press, New York.

STOOPS, G. (1983) SEM and light microscopic observations of minerals in bog-ores of the Belgian Campine. *Geoderma*, **30**, 179–186.

VAN STRAATEN, L.M.J.U. (1957) Recent sandstones on the coasts of the Netherlands and of the Rhone Delta. *Geol. Mijnb.*, **19**, 196–213.

WESTRICH, J.T. & BERNER, R.A. (1984) The role of sedimentary organic matter in bacterial sulphate reduction: the G model tested. *Limnol. Oceanogr.*, **29**, 236–249.

WIEDEMANN, H.U. (1972) Shell deposits and shell preservation in Quaternary and Tertiary estuarine sediments in Georgia, USA. *Sediment. Geol.*, **7**, 103–125.

WOODLAND, B.G. & STENSTROM, R.C. (1979) The occurrence of siderite concretions in the Francis Creek Shale (Pennsylvanian) of northeastern Illinois. In: *Mazon Creek Fossils* (Ed. by M. H. Nitecki), pp. 69–104. Academic Press, New York.

Reprinted from Sedimentology (1979) **26**, *681–705*

Origin of authigenic carbonates in sediment from the deep Bering Sea

JAMES R. HEIN, JAMES R. O'NEIL *and* MARJORIE G. JONES

U.S. Geological Survey, 345 *Middlefield Road, Menlo Park, California* 94025, *U.S.A.*

ABSTRACT

Forty beds of authigenic carbonate were identified from the deep Bering Sea in cores taken on Leg 19 of the Deep Sea Drilling Project. Carbonate minerals were mainly high-magnesium calcite and protodolomite, less commonly siderite, rhodochrosite, low-magnesium calcite, and manganosiderite. Authigenic carbonates cement and replace diatom ooze, ash and bentonite beds, and, less commonly, clastic beds. Replacement zones are as much as 60 cm thick. Eighty-five per cent of carbonate beds occurred below 400 m sub-bottom depth and 70% in sediment older than 4 m.y. $\delta^{13}C$ values averaged $-17{\cdot}20‰$ PDB and $\delta^{18}O$ ranged from 18·59 to 34·11‰ SMOW. The carbon was derived from oxidation of organic matter under anaerobic conditions during bacterial reduction of sulphate, or from CO_2 produced in concert with CH_4 during degradation of organic matter. The cations (Ca, Mg, Fe, Mn) were derived from alteration of ash beds. In Bering Sea deposits, ash beds altered to smectite within about 3–5 m.y. Carbonate precipitated simultaneously at different stratigraphic levels within the 627–1057 m sections at temperatures of 7–85°C. No apparent calcite precursor of biogenic origin was found for these authigenic carbonates.

INTRODUCTION

Most limestone and carbonate sediment in the deep sea are biogenic deposits or diagenetic products originating from calcareous biogenic debris. Lithification of carbonate deposits in the deep sea may occur in several ways. Upon burial, biogenic calcite or aragonite of foraminiferal, nannofossil, and pteropod oozes may be diagenetically altered to chalk or limestone by local dissolution and reprecipitation (Packham & van der Lingen, 1973; Matter, 1974; Schlanger & Douglas, 1974). Chalk and limestone, as well as nonskeletal shallow water carbonates (e.g. ooids, grapestones, and carbonate mud) have $\delta^{13}C$ compositions that range from about 0 to $+5‰$ (Milliman & Müller, 1977). Lithification can also result from changes in the carbonate equilibria of the interstitial water; for example, changes in the pH, HCO_3^-, or Ca^{2+} of the pore water resulting from diagenesis of volcanic debris may cause

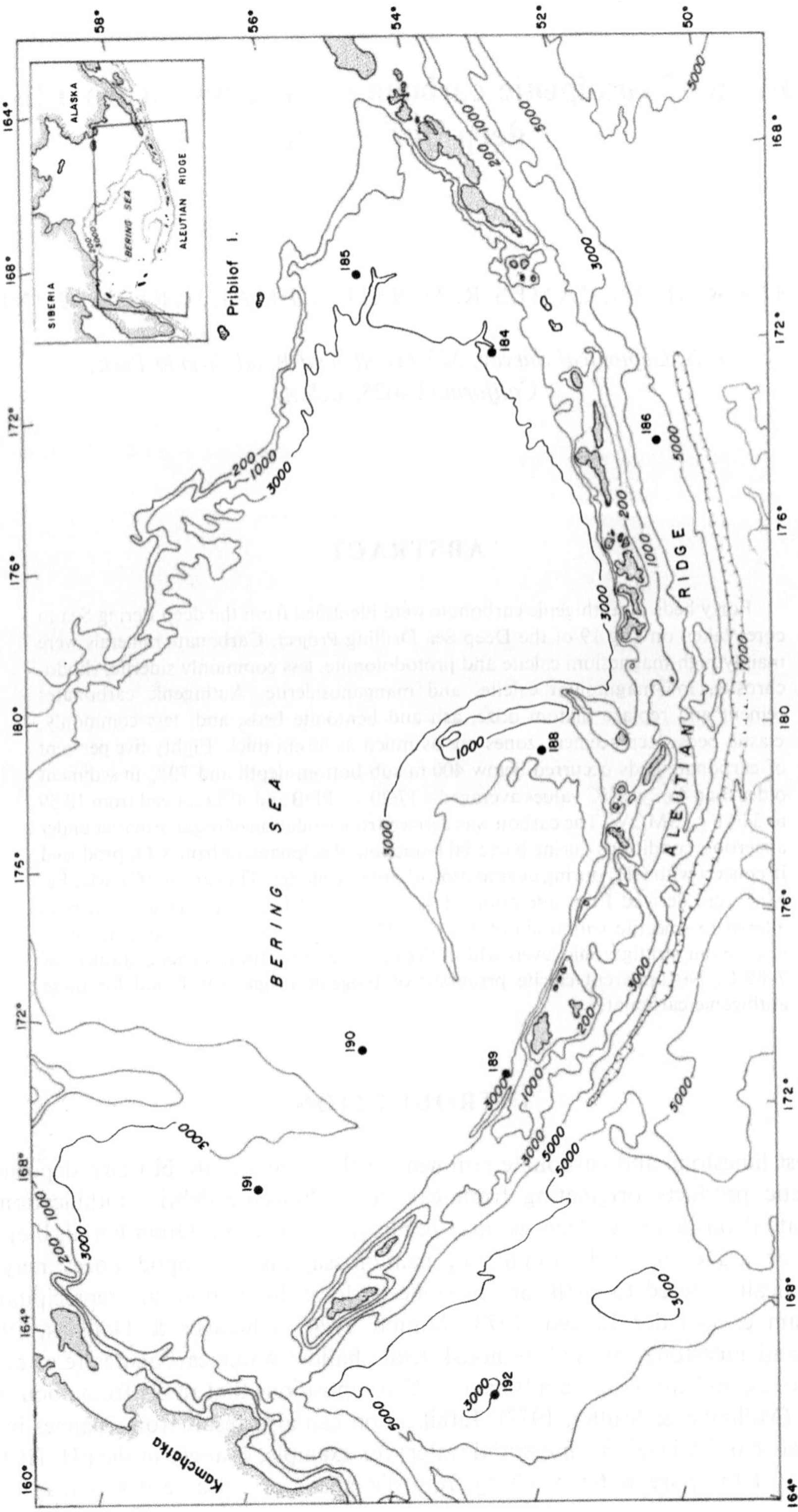
BERING SEA
ALEUTIAN RIDGE
Pribilof I.
Kamchatka
SIBERIA
ALASKA
184
185
186
188
189
190
191
192
200
1000
3000
5000
58°
56°
54°
52°
50°
160°
164°
168°
172°
176°
180°

dissolution and reprecipitation of biogenic carbonate. Metamorphism accompanying intrusion of lava can lithify carbonate sediments (Engel & Engel, 1970; Thompson, 1972; Garrison, Hein & Anderson, 1973).

In addition to these deep-sea biogenic carbonates, rare carbonate veins, crusts, nodules, and isolated crystals, all of apparent nonbiogenic origin, have been described in the literature (Lancelot & Ewing, 1972). In contrast to the low-magnesium calcite or aragonite typical of the deep sea biogenic deposits, inorganic carbonates are commonly siderite, dolomite, rhodochrosite, or high-magnesium calcite. Nonbiogenic carbonates are characterized by highly negative $\delta^{13}C$ values (-15 to -60), suggesting that the carbon was derived from the bacterial degradation of organic matter or from oxidation of CH_4 produced by bacterial processes (Russel *et al.*, 1967; Hathaway & Degens, 1969; Lancelot & Ewing, 1972). In contrast, authigenic carbonates formed in a deep-sea hypersaline environment have $\delta^{13}C$ compositions that ran gemainly from 0 to $+5‰$ (Milliman & Müller, 1977). Most investigators report that either the calcium or the carbon of deep sea carbonates was derived from biogenically formed calcite or aragonite.

We describe forty authigenic carbonate beds from the southern Bering Sea and far northwest Pacific in cores drilled during Leg 19 of the Deep Sea Drilling Project (DSDP). Carbonate deposits form beds, replacement zones and cements, and were probably formed entirely inorganically of high-magnesium calcite, siderite, rhodochrosite, protodolomite, and manganosiderite. The carbon is derived from bacterial degradation of organic matter. The cations, Ca, Mg, Fe and Mn, are considered to be derived from alteration of ash beds and possibly from weathering of oceanic basaltic basement. There was apparently no biogenic calcite precursor.

Authigenic carbonate deposits are from DSDP Sites 184, 185, 188, 189, 190, and 191 in the southern Bering Sea, Site 186 in Atka Basin, midway between the Aleutian Islands and the Trench, and Site 192 atop Meiji Guyot in the far northwest Pacific (Fig. 1). The Bering Sea is a marginal oceanic basin bounded on the south by the volcanically active Aleutian arc. In the deep Bering Sea, the sedimentary section is typically more than 3 km thick and at some places as much as 10 km thick (Cooper *et al.*, 1977a). All stratigraphic sections drilled on Leg 19 consist mainly of two lithologic units, an upper Miocene to Holocene diatom ooze or diatomaceous mudstone (370–700 m thick) and an underlying mudstone. Sediments drilled at Site 192 include 100 m of Eocene and Upper Cretaceous chalk that overlies an alkali-basaltic basement. The change from terrigenous to dominantly biogenic sedimentation occurred during the late Miocene (Fullam *et al.*, 1973; Scholl & Creager, 1973). Minor to moderate amounts of sand, volcanic ash, bentonite, limestone, and ice-rafted debris occur in these sections (Fig. 2). Authigenic carbonate deposits make up less than 0·5% of the 627–1057 m sections.

METHODS

Forty-nine samples were analysed by X-ray diffraction (CuKα) for general and carbonate mineralogy. Relative percentages of carbonate minerals within each sample were determined by comparing peak heights of (100) reflections; values listed in Fig. 2 and Table 1 are, therefore, only approximate. Assuming no interference from non-carbonate reflections, the mole percent $MgCO_3$ (includes $FeCO_3$ and $MnCO_3$) was

Fig. 2. Lithology and distribution of carbonates and pyroclastic deposits in Leg 19 cores. The code for carbonate X-ray mineralogy means, for example, 100 D (47) is 100% protodolomite containing 47 mol % $MgCO_3$.

Table 1. Characteristics of Bering Sea carbonates, Leg 19, Deep Sea Drilling Project

Sample (Site Core Sec)	Age (m.y.)	Depth below sea-floor (m)	Bed thick-ness (cm)	Lithology	Petrography	X-ray mineralogy*	$MgCO_3$ (mole per cent)	$\delta^{18}O$†	$\delta^{13}C$	T°C‡	EDAX Chemistry
184-14-4	3·8	388	n.d.	Calcareous ash	Calcite cemented and replaced ash	100% calcite	3·0	n.d.	n.d.	n.d.	n.d.
184-16-1	4·2	427	23	Limestone	Calcite cemented and replaced diatom ooze	100% Mg-calcite	5·0	n.d.	n.d.	n.d.	n.d.
184-20-5	5·7	545	n.d.	Siliceous limestone	Silica cemented and calcite cemented and replaced diatom ooze	92% calcite 8% rhodochrosite	1·5	25·16	−19·91	39·0	Si, Ca, Al, Fe, K, Mg
184-20-5	5·7	546	10	Limestone	Calcite cemented and replaced diatom ooze	100% Mg-calcite	5·0	26·75	−21·09	30·7	Ca, Si, Al, Na
184-22-6	6·5	599	28	Limestone (top of bed)	Calcite cemented and replaced tuffaceous diatom ooze	100% calcite	3·0	n.d.	n.d.	n.d.	n.d.
184-22-6	6·5	599	28	Tuffaceous limestone (bottom of bed)	Calcite cemented and replaced vitric-crystal ash	100% calcite	4·0	n.d.	n.d.	n.d.	n.d.
184B-1-2	7·5	671	10	Tuffaceous sandstone	Calcite cemented sandstone	50% Mg-calcite 25% protodolomite 25% siderite	23·0 42·0	n.d.	n.d.	n.d.	n.d.
184B-1-2	7·5	672	10	Calcareous black shale	n.d.	100% calcite	0·0	n.d.	n.d.	n.d.	n.d.
184B-9-3	10·0	897	7	Sandstone and calcareous shale	Calcite cemented and replaced carbonaceous shale and sparry calcite cemented feldspathic sandstone	87% calcite 13% rhodochrosite	3·0	n.d.	n.d.	n.d.	n.d.
184B-10-3	10·2	907	60	Limestone	Micrite and microspar replaced and recrystallized diatom ooze?	100% calcite	3·0	n.d.	n.d.	n.d.	n.d.

Table 1 *continued on page* 686

Table 1—*continued*

Sample (Site Core Sec)	Age (m.y.)	Depth below sea-floor (m)	Bed thick-ness (cm)	Lithology	Petrography	X-ray mineralogy*	$MgCO_3$ (mole per cent)	$\delta^{18}O$†	$\delta^{13}C$	T°C‡	EDAX Chemistry
184B-11-2	10·3	918	n.d.	Calcite veins in ash	Sparry calcite veins, crystals up to ~ 0·5 mm	100% calcite	3·5	n.d.	n.d.	n.d.	n.d.
185-8-4	2·4	174	5	Dolostone	Micritic dolomite cemented and replaced diatom ooze, rare foraminifers	100% protodolomite	47·0	n.d.	n.d.	n.d.	n.d.
185-16-2	6·3	519	50	Dolostone	Micritic dolomite cemented and replaced diatom ooze	100% protodolomite	47·0	34·11	−7·16	−0·7	Ca, Si, Mg, Al, K, Fe, S
185-17-3	7·1	582	40	Limestone	Microspar cemented and replaced diatom ooze	100% Mg-calcite	9·0	28·98	+2·04	20·1	n.d.
185-18-2	7·8	650	7	Limestone	Calcite cemented and replaced diatomaceous silty claystone	86% Mg-calcite 14% siderite	7·0	n.d.	n.d.	n.d.	n.d.
186-12-1	1·2	250	n.d.	Calcareous diatom ooze	Micritic calcite cemented pyritized silty diatom ooze	42% Mg-calcite 38% Mg-calcite 20% rhodochrosite	5·0 24·0	n.d.	n.d.	n.d.	n.d.
186-18-3	2·4	445	5	Tuffaceous sandstone	Calcite cemented tuffaceous sandstone	100% Mg-calcite	10·0	n.d.	n.d.	n.d.	n.d.
186-21-3	2·6	510	5	Limestone	n.d.	60% Mg-calcite 40% siderite, rhodochrosite, and protodolomite (?)	9·0	n.d.	n.d.	n.d.	n.d.
188-9-1	1·8	283	5	Dolostone	Micritic dolomite cemented and replaced burrowed diatom ooze	100% protodolomite	46·0	n.d.	n.d.	n.d.	n.d.
188-12-1	2·5	425	20	Dolostone	Micritic dolomite cemented and replaced burrowed diatom ooze	100% protodolomite	46·0	31·70	−16·44	7·0	Ca, Si, Mg, Al, Fe, S. K, Mn
188-17-1	4·3	602	20	Dolostone	Microspar dolomite replaced porcelanite	100% protodolomite	47·0	27·15	−13·77	28·7	Ca, Mg, Si, Al, Fe, Mn

Table 1 *continued on page* 687

Table 1—*continued*

Sample (Site Core Sec)	Age (m.y.)	Depth below sea-floor (m)	Bed thick-ness (cm)	Lithology	Petrography	X-ray mineralogy*	$MgCO_3$ (mole per cent)	$\delta^{18}O$†	$\delta^{13}C$	T°C‡	EDAX Chemistry
189-11-2	4·2	644	n.d.	Calcite veins in bentonite	Large (to 2·0 mm) calcite crystals oriented perpendicular to vein margins	100% calcite	1·5	21·13	−13·60	63·7	n.d.
189-12-2	4·9	711	6	Limestone	Calcite cemented and replaced diatomaceous, tuffaceous, pelagic clay	95% Mg-calcite 5% manganosiderite	7·0	19·70	−18·73	74·0	Ca, Si, Al, Mn, Fe, Mg, K, S
189-14-1	5·4	743	21	Breccia	Calcite cemented (30%) breccia; mostly volcanic rock fragments, chert, and plagio-clase clasts that are cement supported	100% Mg-calcite	5·0	18·59	−20·54	82·7	n.d.
189-15-4	5·7	776	10	Calcareous mudstone (top of bed)	Calcite replaced silty pelagic or hemipelagic carbonaceous mudstone	100% Mg-calcite	5·0	n.d.	n.d.	n.d.	n.d.
189-15-4	5·7	776	10	Limestone (bottom of bed)	95% micrite, diatom ghosts	100% calcite	4·0	26·42	−19·95	32·4	Ca, Si, Al, Fe, Mg, Mn, Ba, S
189-19-1	6·6	848	5	Marble (?)—limestone	Large patches (0·5 mm) calcite with indistinct crystal boundaries	100% calcite	1·5	18·67	−20·67	82·0	n.d.
190-15-cc	6·8	618	n.d.	Limestone	Calcite cemented and replaced diatom ooze and diatoma-ceous siltstone	100% Mg-calcite	7·5	26·64	−20·39	31·3	Si, Ca, Al, Mg, Fe, Mn, K, S
191-7-2	0·8	233	12	Calcareous mudstone	n.d.	32% Mg-calcite 27% rhodochrosite 41% protodolomite	9·0 43·0	33·57	−14·38	1·3	n.d.
191-8-1	1·1	281	12	Limestone	Micritic calcite replaced tuffaceous diatom ooze	100% Mg-calcite	9·0	31·87	+ 0·53	7·8	Ca, Si, Fe, K, Al, S, Mg, Mn

Table 1 *continued on page* 688

Table 1—*continued*

Sample (Site Core Sec)	Age (m.y.)	Depth below sea-floor (m)	Bed thickness (cm)	Lithology	Petrography	X-ray mineralogy*	$MgCO_3$ (mole per cent)	$\delta^{18}O$†	$\delta^{13}C$	T°C‡	EDAX Chemistry
191-12-cc	2·4	527	n.d.	Sandstone	Poorly sorted volcanic sandstone cemented with calcite that forms in large indistinct patches and replaces all clast types, cement supported	100% Mg-calcite	6·0	n.d.	n.d.	n.d.	n.d.
192-22-2	6·6	525	3	Limestone	Calcite cemented and replaced diatom ooze; rare foraminifers	100% Mg-calcite	11·5	31·38	−16·92	9·8	Ca, Si, Al, Mg, K, S, Mn, Fe
192-27-3	13·9	750	17	Limestone	Indistinct patches of calcite, with undulatory extinction, that replaces carbonaceous porcelanite	100% Mg-calcite	6·0	n.d.	n.d.	n.d.	n.d.
192-29-1	16·2	794	n.d.	Calcareous claystone	n.d.	100% calcite	1·0	n.d.	n.d.	n.d.	n.d.
192-29-1	16·2	794	n.d.	Chalk	n.d.	100% calcite	1·0	n.d.	n.d.	n.d.	n.d.
192-33-3	~21·9	920	760	Chalk	90% micritic calcite, nannofossil plates, barite veins	100% calcite	1·0	n.d.	n.d.	n.d.	n.d.
192-33-3	~21·9	920	760	Chalk	90% micritic calcite, nannofossil plates, burrowed, foraminifers	100% calcite	1·0	n.d.	n.d.	n.d.	n.d.
192-34-3	~22·4	925	125	Chalk	Carbonaceous, siliceous, micrite-nannofossils	100% calcite	1·0	n.d.	n.d.	n.d.	n.d.
192-34-3	~22·4	926	25	Mudstone	Nannofossil, siliceous mudstone	100% calcite	1·0	n.d.	n.d.	n.d.	n.d.
192A-1-1	~38·0	943	700	Chalk (top of bed)	Micritic-nannofossil calcite, burrowed, carbonaceous foraminifers	100% calcite	0·0	n.d.	n.d.	n.d.	n.d.
192A-1-1	~38·0	944	700	Chalk (middle upper part of bed)	Micritic-nannofossil calcite, burrowed, carbonaceous, foraminifers	100% calcite	2·0	n.d.	n.d.	n.d.	n.d.

Table 1 *continued on page* 689

Table 1—*continued*

Sample (Site Core Sec)	Age (m.y.)	Depth below sea-floor (m)	Bed thick-ness (cm)	Lithology	Petrography	X-ray mineralogy*	$MgCO_3$ (mole per cent)	$\delta^{18}O$†	$\delta^{13}C$	T°C‡	EDAX Chemistry
192A-1-3	~38·0	947	700	Chalk (middle of bed)	Carbonaceous, siliceous, micritic-nannofossil calcite, burrowed	100% calcite	3·0	n.d.	n.d.	n.d.	n.d.
192A-1-6	~38·0	950	42	Chalk	Silty, siliceous, carbonaceous, micritic-nannofossil calcite; siliceous microfossils are compressed-deformed	100% calcite	1·0	n.d.	n.d.	n.d.	n.d.
192A-2-3	~42·0	954	40	Chalk (top of bed)	Carbonaceous, silty, siliceous, micritic-nannofossil calcite, foraminifers	97% calcite 3% rhodochrosite	2·0	27·94	+2·64	24·9	Ca, Si, Al, Fe, Mn, S, Ba
192A-2-3	~42·0	954	40	Chalk (bottom of bed)	Carbonaceous, silty, siliceous, micritic-nannofossil calcite; foraminifers	96% calcite 4% rhodochrosite	1·0	n.d.	n.d.	n.d.	n.d.
192A-2-6	~42·0	957	105	Chalk (top of bed)	Tuffaceous, micritic-nannofossil calcite; foraminifers	97% calcite 3% rhodochrosite	0·0	n.d.	n.d.	n.d.	n.d.
192A-2-6	~42·0	959	105	Chalk (bottom of bed)	Micritic-nannofossil calcite, foraminifers, rhodochrosite rhombs	98% calcite 2% rhodochrosite	0·0	n.d.	n.d.	n.d.	n.d.
192A-4-4	~45·0	1023	50	Chalk	Micrite recrystallized to microspar (95%) calcite; foraminiferal walls replaced by quartz which in turn is partly replaced by calcite, sparry calcite fills chambers	98% calcite 2% rhodochrosite	1·0	n.d.	n.d.	n.d.	n.d.
192A-5-1		1044	n.d.	Calcite vein in basalt	n.d.	100% calcite	0·0	29·85	−0·07	16·2	Ca, Si, Al, K, Fe, Mg

* All carbonates summed to 100%.
† $\delta^{18}O$ values corrected for mol percentage of $MgCO_3$ (Tarutani, Clayton & Mayeda, 1969).
‡ Calculated from $\Delta^{18}O$ ($CaCO_3$—H_2O) (O'Neil, Clayton & Mayeda, 1969). We assumed sea-water values for $\delta^{18}O$ of pore waters.
n.d. is not determined.

determined by measuring the 2θ shift in the calcite (or dolomite) peaks and comparing the values with calibration curves of Goldsmith, Graf & Heard (1961) and Goldsmith & Graf (1958). This technique has an accuracy of about 1 mol %. A silicon standard was used to measure *d* spacings precisely. Forty-three thin sections were examined with the petrographic microscope to study textures. Texture, mineralogy, and chemical composition of some samples were also studied using techniques of scanning electron microscopy (SEM) and energy-dispersive analysis by X-rays (EDAX). Details of the X-ray, SEM, and EDAX techniques have been given by Hein, Scholl & Gutmacher (1976) and Hein & Scholl (1978). Standard techniques were used to determine carbon and oxygen isotope ratios in the carbonates. All $^{13}C/^{12}C$ and $^{18}O/^{16}O$ ratios are reported in the δ-notation relative to PDB and SMOW standards, respectively. $\delta^{18}O$ values are adjusted for the mole percentage of $MgCO_3$ in the calcite and protodolomite by subtracting 0·06‰ times the mole percentage of $MgCO_3$ from the measured $\delta^{18}O$ (Tarutani, Clayton & Mayeda, 1969).

RESULTS

Authigenic carbonate layers range in thickness from 3 to 60 cm (Fig. 2, Table 1). The biogenic chalk beds are much thicker. Sites 184 and 189 have the most authigenic zones, 11 and 7 respectively, Site 190 the least, with 2. The stratigraphically highest deposit is 174 m below the sea floor (Site 185), the lowest 918 m (Site 184). Eighty-five per cent of the authigenic beds occur below 400 m. The stratigraphically highest bed occurs in deposits 1·2 m.y. old, the lowest beds in deposits 16 m.y. old; 70% occur in deposits older than 4 m.y. An Upper Cretaceous and Eocene nannofossil chalk section occurs at the base of Site 192; nannofossils and calcareous foraminifers, however, are rare or absent at all Bering Sea sites. Biogenic calcite is absent below about 150–250 m and occurs only in trace amounts in the upper deposits. Foraminifers occur below 250 m but are non-calcareous, mostly agglutinated benthonic forms. Authigenic limestone occurs as commonly in the upper diatomaceous sections, as in the lower mudstone units (Fig. 2). Whereas carbonate primarily replaces diatomaceous ooze in the upper sections, it cements sand or replaces ash or bentonite beds in the lower mudstone deposits.

Authigenic limestone beds were not deposited by sedimentation at the sea floor but rather are replacement zones (see section on Petrography). Most commonly, carbonate replaces and cements diatom ooze or ash beds; more rarely it is a cement for sandstone and breccia (Table 1). These deposits might, therefore, be more properly called replacement zones rather than beds. We will, however, generally refer to them as limestone, dolostone, or carbonate beds. The lateral extent of the carbonate beds is not known, as they cannot be correlated between sites; some samples may be fragments of concretions.

Carbonate mineralogy and chemistry

The dominant carbonate mineral in Bering Sea deposits is calcite with a $MgCO_3$ content that ranges from 0·0 to 24·0 mol %. Of secondary importance is protodolomite, with 42–47 mol % $MgCO_3$; rhodochrosite, siderite, and manganosiderite occur in

lesser amounts (Table 1). Authigenic carbonates are composed of high-magnesium calcite (more than 4 mol % $MgCO_3$), protodolomite, iron and manganese carbonates; they are rarely low-magnesium calcite. In contrast, chalk (biogenic calcite) is composed of low-magnesium calcite and minor secondary rhodochrosite. Most authigenic limestone beds are monomineralic, but those that contain much clay and tuffaceous debris also contain two or three carbonate minerals. Calcite dominates the multi-mineralic samples except in sample 191-7-2 where protodolomite is most abundant. X-ray reflections of ordered dolomite are present but of low intensity and broadened. Protodolomite has 3–8 mol % excess $CaCO_3$. Siderite is a minor carbonate mineral mostly at Sites 184, 185, and 186, where it occurs with Mg-calcite and protodolomite and makes up 25% or less of the limestone. Rhodochrosite is present in ten samples, representing as much as 27% of the total carbonate; more commonly, it makes up only 3%, five of the ten samples are from the basal chalk sequence at Site 192. Manganosiderite occurs in trace amounts.

The chemical composition of the limestones, as determined by EDAX (Table 1), suggests that there is at least minor substitution of Fe and Mn for Mg in the Mg-calcite and protodolomite. The substitution of Fe for Mg is difficult to detect by X-ray diffraction and is, therefore, not usually reported. Fe and Mn may be quite common in authigenic carbonates (Nash & Pittman, 1975). Also, manganosiderite, siderite, and rhodochrosite may occur in trace amounts in most samples, which would account for the Fe and Mn detected by EDAX. Barium occurs in two samples. Petrographic study shows the presence of barite veins in these samples and in several other samples for which chemical data are not available. All other elements are attributed to biogenic silica, detrital minerals, pyroclastic debris, and authigenic smectite and silica in the limestone beds (Table 1).

Petrography

Three distinct carbonate lithologies are apparent: (1) nannofossil chalk, (2) carbonate-cemented and replaced diatomaceous, tuffaceous, and clastic beds, and (3) micritic limestone.

Nannofossil chalk

Site 192, drilled on the crest of Meiji Guyot, is the only site studied that is south of the Aleutian Trench (Fig. 1) and is the only site where beds rich in nannofossils were found (Fig. 2). These chalks consist of nannofossils and disaggregated nannofossil plates that together make up more than 80% of the deposit (Fig. 3a and b). Recrystallized foraminiferal tests and sparse diatoms, commonly replaced by sparry calcite, occur in the chalks. Pyrite is common. Clay minerals in the chalk are variable, but consist mostly of smectite (Hein *et al.*, 1976). Much of the chalk is burrowed; burrows are commonly marked by coarser grained calcite (microspar) and trains or pockets of detrital grains. Rhodochrosite was not identified in thin section but is shown on X-ray diffractograms (Table 1). Diagenesis of the chalk section is typical of numerous other sections described in the Initial Reports of the DSDP (for details, see Schlanger *et al.*, 1973; Matter, 1974). Nannofossils occur in many states of preservation and were commonly overgrown by calcite or had calcite rhombs formed at their centres (Fig. 3a).

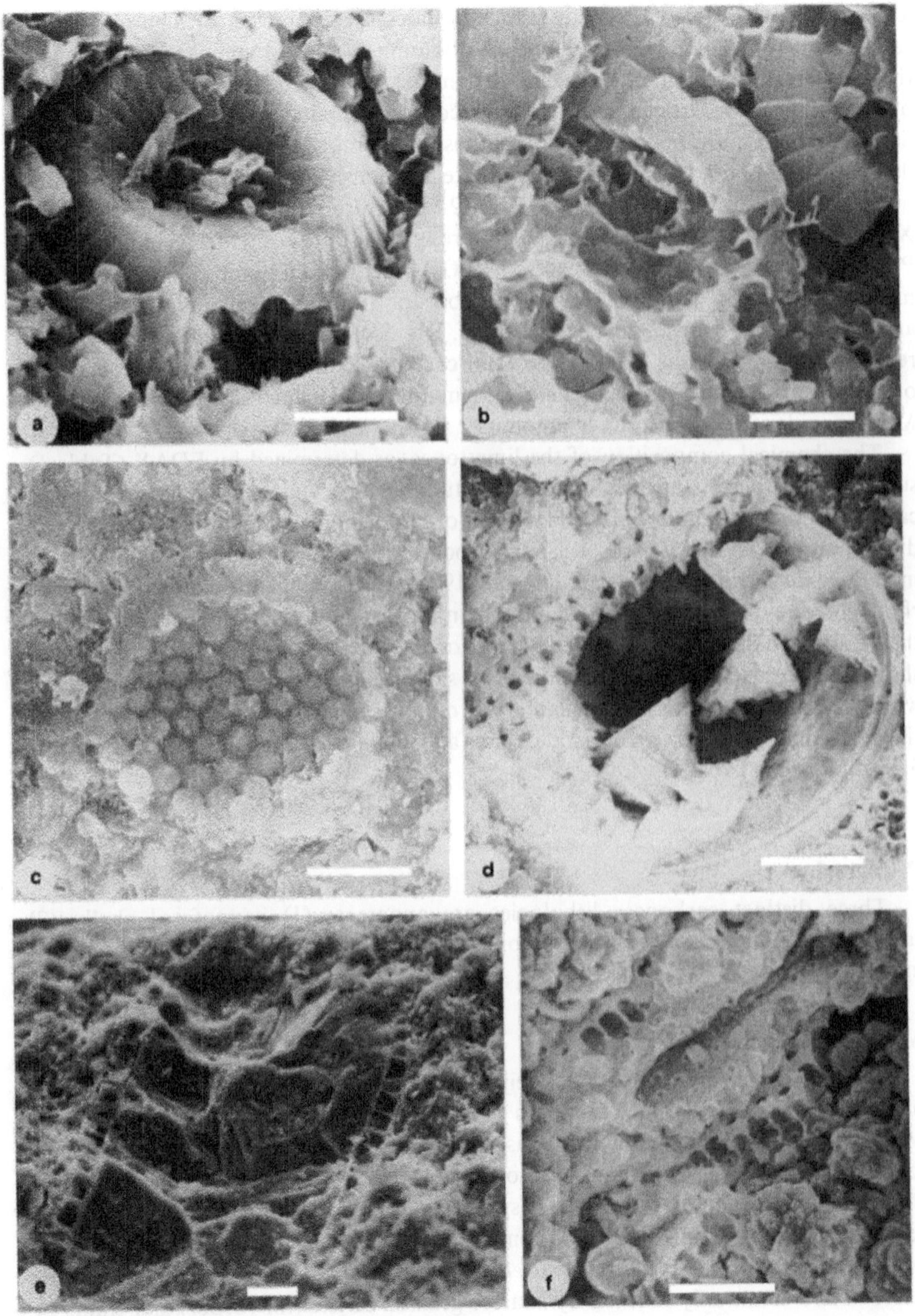

Fig. 3. SEM (scanning electron microscopy) photographs of authigenic carbonates. (a) Broken and partly dissolved coccolith; note how the definition of the nannofossil plate boundaries is obscured by diagenesis. The matrix is clay minerals and disaggregated nannofossil plates. (b) Partial disaggregation of nannofossil plates a and b are from sample 192A-2-3. (c) Diatom replaced by calcite, sample 185-18-2. (d) Void fill calcite precipitated in a diatom frustule, sample 191-8-1. (e) Sparry protodolomite in diatom frustule and micritic protodolomite matrix, sample 185-16-5. (f) Granular calcite replaced diatom ooze, sample 184-20-5. Scales represent 3, 2, 13, 10 and 29 μm, respectively.

Carbonate-cemented or replaced deposits

Calcite- and protodolomite-cemented and replaced diatom ooze is the most common authigenic carbonate lithology. Siliceous microfossils are partly to completely replaced by carbonate (Fig. 4a and b) but, in most samples, are fresh or show only slight replacement. Some frustule structure may be retained after replacement by

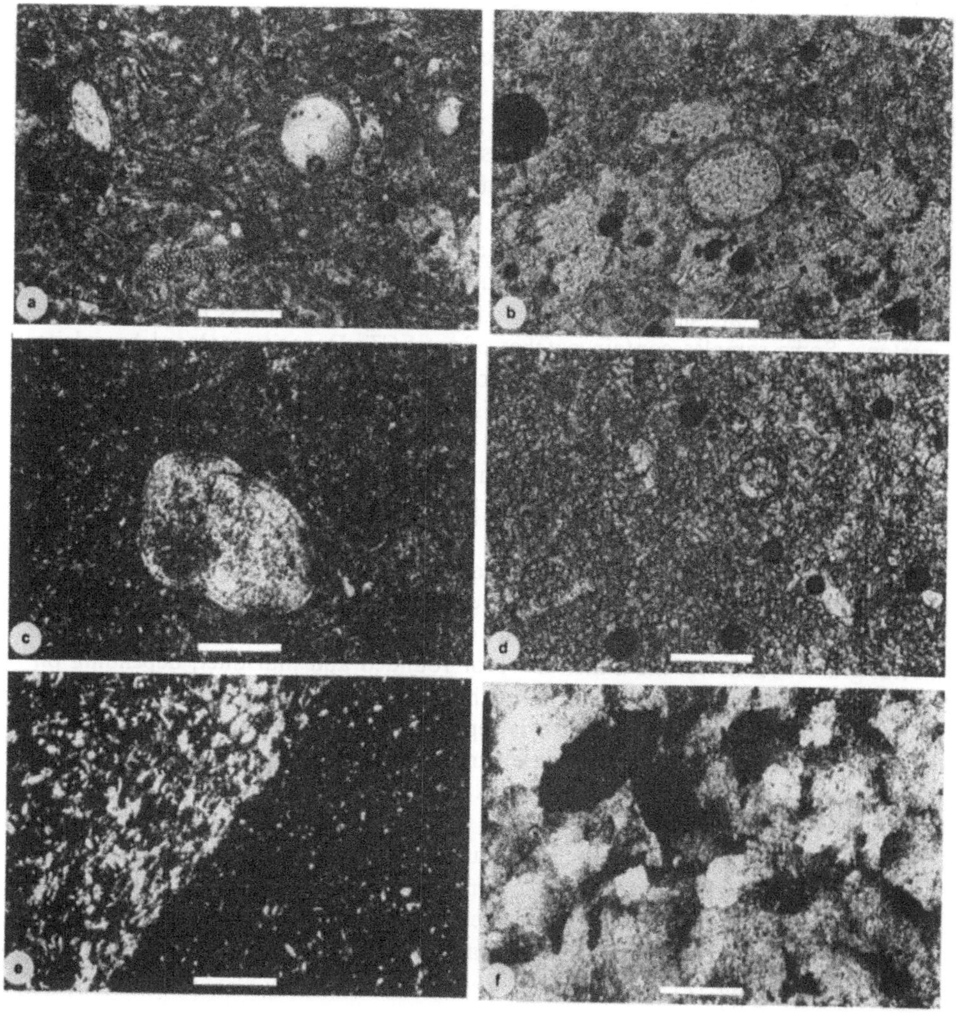

Fig. 4. Photographs of thin sections of authigenic carbonates. (a) Calcite- and rhodochrosite cemented diatom ooze; sample 188-12-1, plane light, scale represents 0·5 mm. (b) Protodolomite cemented diatom ooze, 188-12-1, plane light, scale represents 0·5 mm. (c) Recrystallized foraminifers in calcite- and siderite-cemented and replaced diatomaceous silty claystone; sample 185-18-2, crossed nicols, scale represents 0·5 mm. (d) Protodolomite microspar replaced porcelanite; note diatom in upper right centre. Dark circles are bubbles in the thin section; sample 188-17-1, plane light, scale represents 0·5 mm. (e) Calcite replaced silty pelagic or hemipelagic carbonaceous mudstone, replacement contact diagonally crosses the view; sample 189-15-4, crossed nicols, scale represents 2·0 mm. (f) Recrystallized or hydrothermally altered limestone. Dark areas may be pyritized diatoms; sample 189-19-1, crossed nicols, scale represents 2·0 mm.

carbonates. The more deeply buried the carbonate zone, the greater the degree to which carbonate replaced opal-A (biogenic silica) or opal-CT (Fig. 3c); because opal-A transforms to opal-CT after 400–500 m burial, opal-CT is the most commonly replaced silica polymorph. Siliceous microfossils originally made up 60–90% of deposits that are now 60–95% carbonate. Foraminifers are rare, but are as abundant in the limestones as in surrounding deposits. All foraminiferal tests in the limestones were recrystallized (Fig. 4c).

Particle sizes of the carbonates range from 1 to 4 μm micrite. Diatoms were infilled with microspar or sparry calcite with crystal sizes as great as 200 μm (Fig. 3c and d). Rarely, patches of neoformed carbonate occur that have grain sizes of 8–10 μm. In several samples, frustules were filled with both quartz and calcite, quartz having formed first. All samples contain pyrite as frustule infillings, coatings on microfossils, lenses, patches, or framboids. Limestones have various amounts of admixed terrigenous, pyroclastic, and organic debris. SEM photographs show that no nannofossils occur in any of the limestones or adjacent deposits (e.g., Fig. 3f). Many beds were burrowed. Burrows contain more relatively coarser grained detrital minerals and less, but coarser grained calcite, than the enclosing limestone; diatoms are more concentrated in burrows that also contain opal-CT and less calcite than the limestone. Some samples contain patches of opal-CT. Deposits replaced by calcite show no petrographic differences from those replaced by protodolomite. In some samples, diatom ooze was transformed by dissolution and reprecipitation to porcelanite before the opal-CT was replaced by carbonate (Fig. 4d). And in some samples, the replaced diatom ooze recrystallized to microspar leaving diatom ghosts and more coarsely crystalline patches of calcite as the only signs of a previous history (see next subsection).

Some ash and bentonite beds were cemented and replaced by carbonates. These beds are about 50% calcite and 50% volcanic glass, feldspar, volcanic rock fragments, and opaque minerals. Calcite commonly replaced glass and feldspar. Glass and mineral grain boundaries are either sharp or rounded and indistinct, depending on the amount of replacement calcite. Two or three carbonate minerals occur where ash, tuffaceous diatom ooze, or mudstone is replaced. Calcite veins occur in some ash and bentonite beds. Calcite crystals grew perpendicularly to the vein walls or formed interlocking mosaics with grain sizes up to 2 mm.

Sparry calcite cement makes up 25–60% of some volcanic sandstone and breccia. Calcite grains are as large as 2 mm; their boundaries can be either sharp or indistinct. Volcanic rock fragments and plagioclase grains were partly replaced. Relict grains are seen in the cement. Framework grains are cement supported; clasts float in cement.

Rarely, mudstone containing organic matter, pyrite, and barite veins was partly replaced by carbonate (Fig. 4e).

Micritic limestone

Micritic limestone occurs in a few places; it shows little or no evidence of a precursor as it consists of uniform mosaics of micrite (or microspar in sample 189-17-1) made up of pure carbonates, 80–98% calcite or protodolomite (Fig. 4d). Some diatom ghosts or patches of microspar and sparry calcite (to 100 μm) occur in the micrite, which is 2–8 μm in grain size. In sample 189-17-1, protodolomite grains are 3–18 μm and average about 10–12 μm. Detrital minerals, volcanic glass, pyrite, and organic matter occur. Coarser-grained patches indicate secondary recrystallization.

Micrite may represent recrystallized varieties of the replaced diatomaceous and tuffaceous beds, most commonly occurring at depths of burial more than 750 m.

Sample 189-19-1 is texturally unusual; it looks much like an impure marble composed of large optically coherent patches with diffuse boundaries interlocking to form a mosaic (Fig. 4f). Large pyrite patches may represent replaced diatoms, but no other biogenic remains occur (Fig. 4e).

Carbon and oxygen isotopes

Oxygen and carbon-isotopic compositions were determined for seventeen samples (Table 1). $\delta^{13}C$ is highly variable, ranging from +2·64 to −21·09‰. Values for four of these samples are close to zero, which is typical for biogenic carbonates. The remaining thirteen samples have highly negative $\delta^{13}C$ values that average −17·20‰. $\delta^{18}O$ ranges from 18·59 to 34·11‰ (Table 1). The four samples with $\delta^{13}C$ values close to 0·0 have the relatively narrow $\delta^{18}C$ range of 27·94–31·87‰ (Fig. 5). Considering the isotopic data for all sites collectively, $\delta^{13}C$ values generally decrease with age of enclosing sediment, with depth of burial, and with decreasing mole percentage of $MgCO_3$ in the calcites. These correlations are not well developed but are best defined if the four biogenic-carbonate values are omitted. The best developed relation is a decrease in $\delta^{18}O$ with increasing depth of burial and increasing age. This correlation is true at each DSDP site as well as for all sites considered together. There is an overall poorly defined decrease in $\delta^{18}O$ with decreasing mole percentage of $MgCO_3$ in the calcites.

DISCUSSION

Authigenic carbonates in deep-sea sediment may include isolated dolomite

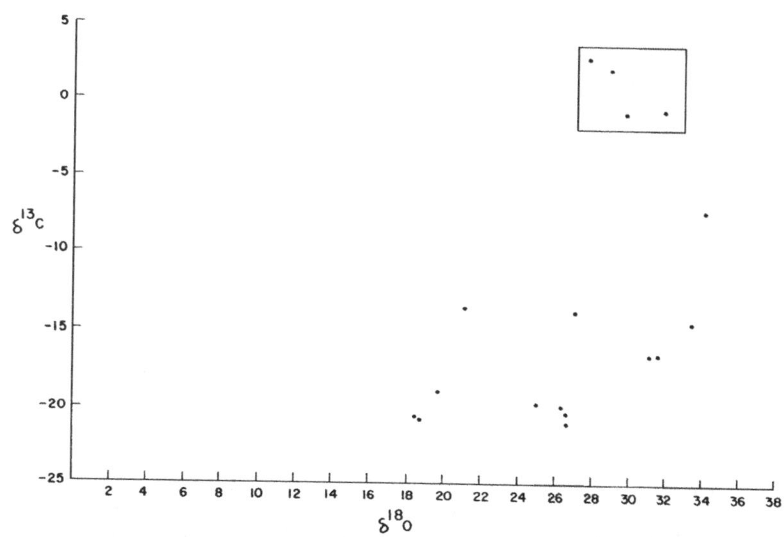

Fig. 5. Plot of $\delta^{13}C$ relative to $\delta^{18}O$ for Leg 19 carbonates. The enclosed area represents range of marine biogenic calcite and recent biogenic carbonate sediment.

rhombs, carbonate nodules, cements, crusts, and veins in mudstone, volcanic, and pyroclastic rocks, and, less commonly, layers (Peterson *et al.*, 1970; Lancelot & Ewing, 1972; Berger & von Rad, 1972; Irwin, Curtis & Coleman, 1977). Sedimentary sections in the Bering Sea contain at least forty authigenic carbonate beds or replacement zones that are as much as 60 cm thick. Bering Sea type deposits may not be unusual, however, as the sediment components from most DSDP legs have not been reported in detail. We speculate that some beds from the deep sea, described simply as limestone, may, upon close examination, be found to be similar to the authigenic carbonates discussed here.

The occurrence of abundant authigenic carbonate implies a major redistribution and reorganization of elements within the sedimentary sections and coexisting pore waters. To understand the magnitude and significance of this chemical mobility, several questions need to be answered: (1) what is the origin of the carbon and oxygen? (2) what is the origin of the cations, Ca, Mg, Fe and Mn? (3) what are the chemical and physical conditions in which carbonates precipitate? and (4) what is the time of formation? In general, conditions most conducive to carbonate precipitation are high pH and abundant alkaline-earth cations and carbonate species in the pore water.

Origin of the carbon

The possible sources of carbon and their corresponding $\delta^{13}C$ values are: (1) sea-water bicarbonate (including that in pore waters), $\delta^{13}C$ close to zero (Table 2, Fig. 5), (2) sea water influenced by volcanogenic CO_2, $\delta^{13}C - 7{\cdot}0$°/$_{oo}$ (Pineau, Javoy & Bottinga 1976); calcites closely associated with volcanic rocks have $\delta^{13}C$ values that range from

Table 2. $\delta^{13}C$ (PDB) of various carbonates, gases and organic compounds

	$\delta^{13}C$* (°/$_{oo}$) Average	Range
Authigenic carbonates (this study)	−17·20	−7·16 to −21·09
Biogenic carbonates (this study)	+1·29	−0·70 to +2·64
Recent biogenic sediments and marine biogenic calcite	n.d.	−2·0 to −3·0
Organic matter		
Terrestrial	n.d.	(~)−23 to −28
Marine plankton	n.d.	(~)−18 to −23
Crude oil	n.d.	(~)−26 to −30
DSDP gases		
CO_2 Biogenic	−18·8	−7·5 to −32·0
CH_4	−72·3	−47·1 to −90·0
Gases experimenta		
CO_2 Biogenic	−14·0	−9·2 to −18·5
CH_4	−83·9	−70·6 to −94·3

* Data from: Cheney & Jensen (1965); Degens (1969); Letolle & Martin (1970); Claypool, Presley & Kaplan (1973).

n.d. = not determined.

about +1 to +5‰ (Milliman & Müller, 1977), (3) biogenic calcite and aragonite, $\delta^{13}C$ close to zero, (4) CO_2 derived from metabolic oxidation of organic matter, $\delta^{13}C$ −18 to −23‰, or about the same as the $\delta^{13}C$ of the organic matter used as a food by the bacteria (Presley & Kaplan, 1968; Degens, 1969; Table 2), (5) CO_2 derived from oxidation of CH_4, $\delta^{13}C$ about −18 to −23‰; bacteriogenic CH_4 *per se* has values of −50 to −90‰ (Claypool, Presley & Kaplan, 1973).

Because methane production results in the selective removal of ^{12}C, the $\delta^{13}C$ of the remaining CO_2 can be shifted to values as heavy as +18‰ in sediments where methane is being generated (Nissenbaum, Presley & Kaplan, 1972); thus, if limestone formed early during this process they would contain light carbon (−20‰; Claypool, 1974), whereas, limestones precipitated late, when primarily heavy residual CO_2 was present, would have correspondingly heavy carbon isotopic composition (+15‰; Irwin *et al.*, 1977). Through mixing of carbon produced by these various processes, marine authigenic carbonates can form with $\delta^{13}C$ values between about +20 and −60‰ (Hathaway & Degens, 1969; Murata, Friedman & Madsen, 1969; Deuser, 1970).

Biogenic calcite can be eliminated as a possible precursor for the authigenic carbonates because nannofossils and foraminifers occur only in trace amounts in surface deposits (see Lisitsyn, 1966); no nannofossils were observed in thin sections or with the SEM in the limestones or surrounding deposits. The chemical and physical environments in Bering Sea deposits have not been sufficiently adverse to the preservation of biogenic calcite to cause the complete removal of nannofossils (Lisitsyn, 1966; Scholl & Creager, 1973). The relatively heavy ^{13}C composition of biogenic calcite and sea water bicarbonate suggests that neither is the carbon source for the limestones.

Methane, present in many areas where carbonates occur, was probably not oxidized to CO_2; the ubiquitous occurrence of pyrite and ferrous carbonate argues against the occurrence of an aerobic environment believed necessary for the conversion of CH_4 to CO_2. Carbon dioxide produced by metabolic oxidation of organic matter during bacterial reduction of sulphate, or CO_2 used early in the carbonate reduction stage appears to be the major carbon source for carbonates recovered during Leg 19. $\delta^{13}C$ values in the range −20 to −15‰ support this conclusion (Table 2).

Similar limestones and carbonate nodules were described by Irwin *et al.* (1977) from the Jurassic Kimmeridge Clay, but in contrast to Bering Sea carbonates, they also found ^{13}C-rich limestones ($\delta^{13}C$ to +9‰). They suggested that the heavy carbonates formed during a diagenetic stage characterized by bacterial fermentation, a stage apparently not represented (or of limited extent) in Bering Sea deposits. Instead, Leg 19 carbonates apparently formed in the stratigraphically higher diagenetic zone of Irwin *et al.*, a zone characterized by bacterial sulphate reduction and $\delta^{13}C$ values near −15 to −25 (see also Curtis, 1978).

Origin of cations

We suggest that alteration of andesitic and more silicic ash beds provided the bulk of the cations required to form the authigenic carbonate deposits. Hein & Scholl (1978) showed that ash beds recovered during Leg 19 alter primarily to smectite clay minerals and minor clinoptilolite. It takes about 3–5 m.y. for the ash to be transformed to smectite at sub-bottom depths greater than 300 m. As stated, 85% of the limestones occur at sub-bottom depths greater than 400 m, and 70% in sediment older than 4 m.y. Moreover, 72% of the limestones are closely associated with ash or bentonite layers (Fig. 2). Clay in the insoluble residue of the carbonate is commonly, but not invariably, smectite-rich.

Alteration of volcanic debris to smectite in the marine environment involves the removal of Ca, K, Fe, Mn, Ni, and Cu and the uptake of Mg (Lawrence, Gieskes & Broecker, 1975; Gieskes, Kastner & Wanter, 1975; Bischoff & Dickson, 1975). Much of the Ca released probably comes from alteration of plagioclase, to a lesser degree from ferromagnesium silicates. Hein & Scholl (1978) showed that plagioclase is a minor part of bentonite beds as compared with ash layers. Stewart (1977) found that sand layers in the lower 400 m at Site 191 contain abundant authigenic smectite. He further found that after 700 m of burial, pyroxene, plagioclase, and other reactive minerals were replaced by smectite. Fe and Mn were derived from the ash beds and alteration of ferromagnesium silicates. The Mn that forms rhodochrosite at Site 192 is probably derived from weathering of the immediately underlying basaltic basement (Natland, 1973).

The source of Mg in the protodolomite and magnesium calcite is less easily understood, because the Mg released to the pore waters during alteration of the ash was removed by formation of smectite. Possibly the altered ash is sufficiently abundant to provide Mg for both the smectite and carbonate. This possibility is suggested by the fact that at some sites ash beds outnumber carbonate beds by as much as twenty to one. Since the variety of smectite in the bentonite beds is montmorillonite (dioctahedral) rather than saponite (trioctahedral; Hein & Scholl, 1978), the smectite may contain a relatively small amount of structural Mg. And some of the 'Mg' content determined by X-ray diffraction in these carbonates may actually be Fe (see Nash & Pittman, 1975; Table 1).

Alternatively, the protodolomite may have formed from a calcite precursor; as calcite replaced silicates containing Mg, the freed Mg would have been available to substitute for Ca in the carbonates. There appear, however, to be only minor recrystallization textures for all but the stratigraphically lowest. On the basis of textural evidence, we would favour primary precipitation of protodolomite. The evidence is not entirely clear, and dolomitization of limestone or Mg incorporated during recrystallization may have played a role in Mg enrichment. Recrystallization would have to have been on a sub-micron to micron scale. At least in biogenic deposits, recrystallization of low-magnesium calcite to high-magnesium calcite in sea water solutions has not been demonstrated (Sayles & Manheim, 1975) although it has been proposed to occur under conditions of high salinity (Milliman, Ross & Ku, 1969; Müller & Fabricius, 1974). If the protodolomites were not precipitated directly from solution as suggested by their textures, then dolomitization is a more likely process contributing to Mg enrichment. If dolomitization takes place by solution and redeposition in pore water rather than by solid-state diffusion (which is unlikely at low temperatures), the ^{18}O content of the dolostone would increase; lower temperatures and equilibrium considerations predict this. Thus, because the carbonates of this study most enriched in ^{18}O are protodolomites, direct precipitation, or dolomitization by solution-redeposition mechanisms are favoured.

Environment of deposition

Sediment cored at Leg 19 sites is primarily diatomaceous with various amounts of admixed terrigenous and volcanic debris. In general, sedimentation rates were high,

Table 3. Oxidation–reduction reactions of metabolic processes and reactions for simultaneous methane production and carbonate precipitation*

(A) Sulphate reduction

$$2CH_2O + 2H_2O \rightarrow 2CO_2 + 4H_2$$
$$4H_2 + SO_4^{2-} \rightarrow S^{2-} + 4H_2O$$

$$2CH_2O + SO_4^{2-} \rightarrow S^{2-} + 2CO_2 + 2H_2O$$

(B) Carbonate reduction

$$2CH_2O + 2H_2O \rightarrow 2CO_2 + 4H_2$$
$$4H_2 + HCO_3^- + H^+ \rightarrow CH_4 + 3H_2O$$
$$CO_2 + H_2O \rightarrow HCO_3^- + H^+$$

$$2CH_2O \rightarrow CH_4 + CO_2$$

(C) Methane production—Carbonate precipitation

$$HCO_3^- + 8H \rightarrow CH_4 + 2H_2O + OH^-$$
$$HCO_3^- + OH^- \rightarrow CO_3^{2-} + H_2O$$
$$Me^{2+} + CO_3^{2-} \rightarrow MeCO_3$$

$$Me^{2+} + 2HCO_3^- + 8H \rightarrow CH_4 + MeCO_3 + 3H_2O$$
where Me = Ca, Mg, Fe, etc.

* From Claypool (1974).

mostly greater than 100 m/m.y. which prevented complete oxidation of organic matter. During early diagenesis, therefore, bacterial reduction of sulphate in association with the anaerobic oxidation of organic matter resulted in the production of CO_2 and H_2S (Table 3, eqn A). Subsequently, precipitation of iron sulphide occurred with consequent depletion of sulphate, and probably Fe, in the pore waters in the upper deposits (Claypool *et al.*, 1973; Sayles, Waterman & Manheim, 1973; Claypool, 1974). Claypool (1974, p. 144) found that prior to complete removal, up to eight times as much sulphate can flow into the sediment column by diffusion as was originally buried with the sediment: '. . . organic diagenesis is not limited to the original amount of sulphate buried with sediment and more bicarbonate is generated which can ultimately be used for CH_4 production or precipitation of diagenetic carbonates' (Table 3, eqn C). High alkalinities are found in the upper deposits at Leg 19 sites (Sayles *et al.*, 1973). During relatively later diagenesis, after completion of sulphate reduction, CH_4 was produced by bacterial fermentation and reduction of CO_2 in the sulphate-free pore waters (Table 3, Eqn B; Claypool *et al.*, 1973). Thermocatalytic generation of CH_4 and CO_2 may have occurred at depth at some sites. Methane probably occurred over larger stratigraphic intervals than those shown in Fig. 2. Its previous existence is indicated by vertical clay-filled fractures that were probably gas migration or escape channels.

Dissolution of diatomaceous debris was coincident with, or later than, CH_4 generation, although these two diagenetic processes are not genetically related. Silica dissolution began after 400–500 m of burial and became widespread at about 550–650 m (Hein *et al.*, 1978). At the same time that major reconstitution of biogenic silica was taking place (opal-A → opal-CT + minor clinoptilolite), ash deposits altered to bentonite (ash → montmorillonite + minor clinoptilolite; Hein & Scholl, 1978). Finally, at great depths of burial, terrigenous detrital minerals such as feldspar and pyroxene altered to smectite or were replaced by calcite.

Within this continuing series of diagenetic reactions, it appears likely that most carbonate beds did not form during very early diagenesis, that is, during sulphate reduction; authigenic carbonates are not found in near-surface sediment where sulphate reduction occurs, as carbonate cannot compete with sulphide for Fe. This does not mean that the CO_2 for the carbonate deposits was not derived from the metabolic oxidation of organic matter; it probably was, but, as was mentioned earlier, it can not be determined from our data whether the CO_2 was produced early, during the sulphate reduction stage, or later, during the beginning of carbonate reduction and production of CH_4.

If our suggestion for the origin of the cations is correct, then the limestone beds formed during or after alteration of ash beds. Both alteration of ash and production of CH_4 create pore waters with relatively high pH, a condition favourable for carbonate precipitation. Milliman *et al.* (1969) suggested that precipitation of high-magnesium calcite resulted from a sudden rise in pH, which in turn resulted from bacterial reduction processes. Formation of carbonate during late diagenesis is also supported by evidence that authigenic carbonates commonly form below the depth of the pore water alkalinity maximum (Anderson & Lawrence, 1976) which is deeper than 200 m for sites drilled during Leg 19 (Sayles *et al.*, 1973). Relatively late diagenetic formation is further indicated by the several carbonate beds that replaced diatomaceous deposits that had reached a relatively advanced stage of diagenesis. Considering all these facts, abundant carbonate formation in Leg 19 deposits probably did not begin until after at least 400–500 m of burial. Carbonates occur abundantly below 400–500 m and uncommonly in sediment above this level. Oxygen isotopic compositions (see next section) suggest that carbonates can form at several stratigraphic levels at the same time. Once the solid carbonate forms, it does not undergo isotopic exchange with pore water or mix with previously formed carbonates unless dissolution–reprecipitation occurs (Claypool, 1974).

Carbonate is not evenly dispersed through the sediment but rather concentrates into beds as cement and replacement zones. Most commonly, nearly pure diatom ooze is cemented and replaced. Abundant pore spaces and high permeability characterize the diatom ooze, partly because compaction of the ooze is insignificant at Leg 19 sites (Lee, 1973). When pure diatom ooze is absent, ash or diatomaceous ooze with some terrigenous debris is the locus of precipitation. The main factors that determine where carbonate will precipitate are: (1) porosity and permeability of the host sediment, (2) proximity to ash beds that release cations and create high pH environments during alteration, and (3) perhaps proximity to organic-rich layers where CH_4 generation and high pH conditions prevail.

Time and temperature of formation

If we assume precipitation from pore fluids of known oxygen isotopic composition under conditions of equilibrium, then we can determine temperatures of formation from the oxygen-isotopic compositions of the carbonates. Using oxygen-isotope temperatures calculated from the expression of O'Neil *et al.* (1969) and reasonable estimates for thermal conductivity and geothermal gradients, the depth at which the carbonates formed can be calculated (Table 4). Down-hole temperature measurements were made at Sites 184 and 185 during drilling operations. Therefore, calculated depths of formation for carbonates from these two sites are likely to be

Table 4. Determination of depth of formation of authigenic carbonates

Sample No.	Sub-bottom depth of bed (m)	$\delta^{18}O$ T°C of formation	Geothermal gradient T°C/m*	Geothermal gradient measured T°C/m†	Geothermal gradient from heat flow T°C/m‡	T°C at dpeth of bed (present)	Probable depth of formation (m)§	Increased burial since formation (m)
184-20-5	545	39	0·072	0·082	0·088	47	476	69
184-20-5	546	31	0·057	0·082	0·088	47	378	168
185-17-3	582	20	0·034	⩾0·050	0·088	29	400	182
188-12-1	425	7	0·016	n.d.	0·083	35	84	341
188-17-1	602	29	0·048	n.d.	0·083	50	349	253
189-11-2	644	64	0·099	n.d.	0·132	85	485	159
189-12-2	711	74	0·104	n.d.	0·132	94	561	150
189-14-1	743	83	0·112	n.d.	0·132	98	629	114
189-15-4	776	32	0·041	n.d.	0·132	102	242	534
189-19-1	848	82	0·097	n.d.	0·132	112	621	227
190-15-CC	618	31	0·050	n.d.	0·064	40	484	134
192-22-2	525	10	0·053	n.d.	0·055	29	182	343

* Geothermal gradient that would exist if the carbonate formed at the depth where it now occurs in equilibrium with pore waters.

† Geothermal gradient determined from down-hole temperature measurements during coring operations at Leg 19 Sites 184 and 185 (Erickson, 1973).

‡ Geothermal gradient calculated from nearby heat-flow measurements assuming a thermal conductivity of 0·022 ± 0·003. (Data from Watanabe, Langseth & Anderson, 1977; Cooper, Marlow & Scholl, 1977b; Cooper, unpublished data.)

§ Possible error estimated to be ± 50 m, probably much more for sample 188-12-1.

n.d. = not determined.

most reliable. On the basis of comparison of down-hole temperature measurements with those determined from heat flow data, calculated depths of formation listed in Table 4 can be considered reasonable to within ± 50 m. In general, formation temperatures calculated for protodolomite are not so realistic as those calculated for calcite. This implies that something is wrong with the assumptions used and may tell us something more about the formation conditions of protodolomite. The data in Table 4 show that the carbonates are in isotopic equilibrium under thermal conditions that now exist 69–534 m higher in the respective cores. This can be interpreted to mean that sample 184-20-5, for example, formed less than a million years ago, when the depth of burial was 69 m less than it is now, assuming a sedimentation rate of 80 m/m.y. These data also show that carbonates do not form over a narrow depth range characterized by a particular temperature or pressure but rather at several levels simultaneously with correspondingly different temperatures at each level. The high temperatures recorded at Site 189 are reasonable values considering its location at the base of the volcanically active Aleutian Ridge. Such temperatures explain the unusual texture of sample 198-19-1 (Table 1), clearly from a deposit either being modified by circulating hydrothermal fluids or undergoing major recrystallization. Further, thermocatalytic production of CH_4 (important at $>50°C$) is likely at Site 189. Oxygen isotopic compositions suggest that sample 189-15-4 at Site 189 formed relatively soon after burial compared with the surrounding carbonates. In other words, samples 189-11-21, 189-12-2, 189-14-1 and 189-19-1 probably formed at about the same time and after much deeper burial than sample 189-15-4 (Tables 1 and 4).

CONCLUSIONS AND SUMMARY

From our analysis of carbonate material taken from DSDP cores of Leg 19, we conclude the following: (1) Authigenic carbonates, high-magnesium calcite, protodolomite, siderite, rhodochrosite, and manganosiderite formed at depth in the sediment of the Bering Sea. (2) Carbon for the carbonates was produced by oxidation of organic matter under anaerobic conditions, an origin supported by low $\delta^{13}C$ values that average $-17{\cdot}2‰$. (3) Ca, Mg, Fe, and Mn were derived from the alteration of andesitic ash, and by the weathering of sea-floor basalt at least at Site 192. Alteration of ash and possibly generation of CH_4 provided the high pH and abundant carbonate species in pore waters necessary for carbonate precipitation. (4) The main phases of alteration of ash, diagenesis of biogenic siliceous debris, CH_4 generation, and carbonate precipitation all took place at about the same time, after about 3–4 m.y. and 400–550 m of burial, but over a wide range of temperatures, 7–85°C. (5) Owing to its high porosity, carbonate most commonly precipitated in pure diatom ooze, and at deep levels in the section, some diatom frustules were actually replaced by carbonate. (6) Limestone recrystallized after about 750 m of burial. (7) In organic-rich clay or tuffaceous deposits, two or three different carbonate minerals occur together. (8) There was no apparent calcite precursor of biogenic origin for the authigenic carbonates.

ACKNOWLEDGMENTS

We appreciate helpful discussion with Miriam Kastner, Scripps Institution of

Oceanography, James L. Bischoff, George Claypool, Alan K. Cooper, and David W. Scholl, U.S. Geological Survey. Robert E. Garrison, University of California at Santa Cruz, Phillip W. Choquette, Marathon Oil Company, Colorado, J. D. Milliman, Woods Hole Oceanographic Institution, H. C. Jenkyns, Oxford University, and J. D. Hudson, University of Leicester, critically reviewed the report. Alan K. Cooper kindly furnished unpublished heat-flow data. Technical assistance was given by Christine Gutmacher, Elaine Alexander, Teresa Long, C. Robin Ross, Lanford Adami, and Jeanne Henning. R. Oscarson helped with the EDAX chemistry; R. E. Garrison arranged for us to use the Jelco SEM at the University of California at Santa Cruz. This work would not have been possible without the efforts of the Deep Sea Drilling Project and the scientists and crew of the GLOMAR CHALLENGER sponsored in part by the National Science Foundation.

REFERENCES

ANDERSON, T.F. & LAWRENCE, J.R. (1976) Stable isotope investigation of sediments, basalts, and authigenic phases from Leg 35 cores. In: *Initial Reports of the Deep Sea Drilling Project*, **XXXV** (Ed. by C. D. Hollister, C. Craddock *et al.*) pp. 497–505. U.S. Government Printing Office, Washington.

BERGER, W.H. & VON RAD, A. (1972) Cretaceous and Cenozoic sediments from the Atlantic Ocean. In: *Initial Reports of the Deep Sea Drilling Project*, **XIV** (Ed. by D. E. Hays, A. C. Pimm *et. al.*), pp. 787–954. U.S. Government Printing Office, Washington.

BISCHOFF, J.L. & DICKSON, F.W. (1975) Sea water–basalt interaction at 200°C and 500 bars; implications as to origin of sea floor heavy metal deposits and regulation of sea water chemistry. *Earth Planet. Sci. Letts*, **25**, 385–397.

CHENEY, E.C. & JENSEN, M.L. (1965) Stable carbon isotopic composition of biogenic carbonates. *Geochim cosmochim. Acta*, **29**, 1331–1346.

CLAYPOOL, G.E. (1974) *Anoxic diagenesis and bacterial methane production in deep sea sediments.* Unpublished Ph.D. thesis, University of California, Los Angeles.

CLAYPOOL, G.E., PRESLEY, B.J. & KAPLAN, I.R. (1973) Gas analyses in sediment samples from Legs 10, 11, 13, 14, 15, 18, and 19. In: *Initial Reports of the Deep Sea Drilling Project*, **IXX** (Ed. by J. S. Creager, D. W. Scholl *et al.*), pp. 879–884. U.S. Government Printing Office, Washington.

COOPER, A.K., CHILDS, J.R., MARLOW, M.S., RABINOWITZ, P.D., SCHOLL, D.W. & LUDWIG, W.J. (1977a) Preliminary isopach and structure maps of the Bering Sea Basin. *U.S. geol. Surv. Map MF*-907.

COOPER, A.K., MARLOW, M.S. & SCHOLL, D.W. (1977b) The Bering Sea—a multifarious marginal basin. In: *Island Arcs, Deep Sea Trenches, and Back-Arc Basins* (Ed. by M. Talwani and W. C. Pitman III), pp. 437–450. *Am. geophys. Union, Maurice Ewing Series I*, Washington, D.C.

CURTIS, C.D. (1978) Possible links between sandstone diagenesis and depth-related geochemical reactions occurring in enclosing mudstones. *Q.J. geol. Soc. London*, **135**, 1–11.

DEGENS, E.T. (1969) Biogeochemistry of stable carbon isotopes. In: *Organic Geochemistry* (Ed. by G. Eglinton and M. T. J. Murphy), pp. 304–329, Springer-Verlag, Berlin.

DEUSER, W.G. (1970) Extreme $^{13}C/^{12}C$ variation in Quaternary dolomites from the continental shelf. *Earth Planet. Sci. Letts*, **8**, 118–124.

ENGEL, A.E.J. & ENGEL, G.C. (1970) Igneous rock samples recovered on Leg 2. In: *Initial Reports of the Deep Sea Drilling Project*, **II** (Ed. by M. N. A. Peterson, N. T. Edgar *et al.*), p. 387. U.S. Government Printing Office, Washington.

ERICKSON, A. (1973) Initial Report on downhole temperature and shipboard thermal conductivity measurements, Leg 19, Deep Sea Drilling Project. In: *Initial Reports of the Deep Sea Drilling Project*, **IXX** (Ed. by J. S. Creager, D. W. Scholl *et al.*), pp. 643–656. U.S. Government Printing Office, Washington.

FULLAM, T.J., SUPKO, P.R., BOYCE, R.E. & STEWART, R.J. (1973) Some aspects of Late Cenozoic sedimentation in the Bering Sea and North Pacific Ocean. In: *Initial Reports of the Deep Sea Drilling Project*, **IXX** (Ed by J. S. Creager, D. W. Scholl *et al.*), pp. 887–896. U.S. Government Printing Office, Washington.

GARRISON, R.E., HEIN, J.R. & ANDERSON, T.F. (1973) Lithified carbonate sediment and zeolitic tuff in basalts, Mid-Atlantic Ridge. *Sedimentology*, **20**, 399–410.

GIESKES, J.M., KASTNER, M. & WANTER, T.B. (1975) Evidence for extensive diagenesis, Madagascar Basin, Deep Sea Drilling Site 245. *Geochim. cosmochim. Acta*, **39**, 1385–1394.

GOLDSMITH, J.R. & GRAF, D.L. (1958) Relation between lattice constants and composition of the Ca–Mg carbonates. *Am. Miner.* **43**, 84–101.

GOLDSMITH, J.R., GRAF, D.L. & HEARD, H.C. (1961) Lattice constants of calcium-magnesium carbonates. *Am. Miner.* **46**, 453–457.

HATHAWAY, J.C. & DEGENS, E.T. (1969) Methane-derived marine carbonates of Pleistocene age. *Science*, **165**, 690–692.

HEIN, J.R. & SCHOLL, D.W. (1978) Diagenesis and distribution of Late Cenozoic volcanic sediment in the southern Bering Sea. *Bull. geol. Soc. Am.* **89**, 197–210.

HEIN, J.R., SCHOLL, D.W., BARRON, J.A., JONES, M.G. & MILLER, J. (1978) Diagenesis of Late Cenozoic diatomaceous deposits and formation of the bottom simulating reflector in the southern Bering Sea. *Sedimentology*, **25**, 155–181.

HEIN, J.R., SCHOLL, D.W. & GUTMACHER, C.E. (1976) Neogene clay minerals of the far N.W. Pacific and southern Bering Sea. In: *AIPEA Proceedings*, 1975, *International Clay Conf., Mexico City* (Ed. by S. W. Bailey), pp. 71–80. Applied Publishing Ltd, Wilmette, Illinois.

IRWIN, H., CURTIS, C.D. & COLEMAN, M. (1977) Isotopic evidence for source of diagenetic carbonates formed during burial of organic-rich sediments. *Nature*, **269**, 209–213.

LANCELOT, Y. & EWING, J.I. (1972) Correlation of natural gas zonation and carbonate diagenesis in Tertiary sediments from the north-west Atlantic. In: *Initial Reports of the Deep Sea Drilling Project*, **XI** (Ed. by C. D. Hollister, J. I. Ewing *et al.*), pp. 791–799. U.S. Government Printing Office, Washington.

LAWRENCE, J.R., GIESKES, J.M. & BROECKER, W.S. (1975) Oxygen isotope and cation composition of DSDP pore waters and the alteration of layer II basalts. *Earth Planet. Sci. Letts*, **27**, 1–10.

LEE, H.J. (1973) Measurements and estimates of engineering and other physical properties, Leg 19. In: *Initial Reports of the Deep Sea Drilling Project*, **IXX** (Ed. by J. S. Creager, D. W. Scholl *et al.*), pp. 701–719. U.S. Government Printing Office, Washington.

LETOLLE, R. & MARTIN, J.M. (1970) Carbon isotope composition of suspended organic matter in two European estuaries. *Mod. Geol.* **1**, 275–278.

LISITSYN, A.P. (1966) Recent sedimentation in the Bering Sea. *Acad. Sci. USSR Inst. Oceanology* (Israel program for Sci.—Trans.; *U.S. Dept. Commerce, Nat. Sci. Found.* 1969).

MATTER, A. (1974) Burial diagenesis of pelitic and carbonate deep-sea sediments from the Arabian Sea. In: *Initial Reports of the Deep Sea Drilling Project*, **XXIII** (Ed. by R. W. Whitmarsh, O. E. Weser, *et al.*), pp. 421–469. U.S. Government Printing Office, Washington.

MILLIMAN, J.D. & MÜLLER, J. (1977) Characteristics and genesis of shallow-water and deep-sea limestones. In: *The Fate of Fossil Fuel CO_2 in the Oceans* (Ed. by N. R. Anderson and A. Malahoff), pp. 655–672, Plenum Publishing Corporation, N.Y.

MILLIMAN, J.D., ROSS, D.A. & KU, T. (1969) Precipitation and lithification of deep-sea carbonates in the Red Sea. *J. sedim. Petrol.* **39**, 724–736.

MÜLLER, J. & FABRICIUS, F. (1974) Magnesian–calcite nodules in the Ionian deep sea: an actualistic model for the formation of some nodular limestones. In: *Pelagic Sediments: on Land and Under the Sea* (Ed. by K. J. Hsu and H. C. Jenkyns). *Spec. Publ. Int. Ass. Sedim.* **1**, 235–247. Blackwell Scientific Publications, Oxford.

MURATA, K.J., FRIEDMAN, I. & MADSEN, B.M. (1969) Isotopic composition of diagenetic carbonates in marine Miocene formations of California and Oregon. *Prof. Pap. U.S. geol. Surv.* **614-B.**

NASH, A.J. & PITTMAN, E.D. (1975) Ferro-magnesian calcite cement in sandstones. *J. sedim. Petrol.* **45**, 258–265.

NATLAND, J.H. (1973) Basal ferromanganoan sediments at DSDP Site 183, Aleutian Abyssal Plain, and Site 192, Meiji Guyot, Northwest Pacific, Leg 19. In: *Initial Reports of the Deep Sea Drilling Project,* **IXX** (Ed. by J. S. Creager, D. W. Scholl *et al.*), pp. 629–636. U.S. Government Printing Office, Washington.

NISSENBAUM, A., PRESLEY, B.J. & KAPLAN, I.R. (1972) Early diagenesis in a reducing fjord, Saanich Inlet, British Columbia. I. Chemical and isotopic changes in major components of interstitial water. *Geochim. cosmochim. Acta.* **36,** 1007-1027.

O'NEIL, J.R., CLAYTON, R.N. & MAYEDA, T.K. (1969) Oxygen isotope fractionation in divalent metal carbonates. *J. chem. Phys.* **51,** 5547–5558.

PACKHAM, G.H. & VAN DER LINGEN, G.J. (1973) Progressive carbonate diagenesis at Deep Sea Drilling Sites 206, 207, 208, and 210, in the southwest Pacific and its relationship to sediment physical properties and seismic reflectors. In: *Initial Reports of the Deep Sea Drilling Project,* **XXI** (Ed. by R. E. Burns, J. E. Andrews *et al.*), pp. 495–521. U.S. Government Printing Office, Washington.

PETERSON, M.N.A., EDGAR, N.T., VAN DER BORCH, C.C. & REX, R.W. (1970) Cruise Leg Summary and Discussion. *Initial Reports of the Deep Sea Drilling Project,* **II** (Ed. by M. N. A. Peterson, N. T. Edgar, *et al.*), pp. 413–427.

PINEAU, F., JAVOY, M. & BOTTINGA, Y. (1976) $^{13}C/^{12}C$ ratios of rocks and inclusions in popping rocks of the mid-Atlantic Ridge and their bearing on the problem of isotopic compositions of deep-seated carbon. *Earth Planet. Sci. Letts,* **29,** 413–421.

PRESLEY, B.J. & KAPLAN, I.R. (1968) Changes in dissolved sulfate, calcium and carbonate from interstitial water of nearshore sediments. *Geochim. cosmochim. Acta,* **32,** 1037–1048.

RUSSEL, K.L., DEFFEYES, K.S., FOWLER, G.A. & LLOYD, R.M. (1967) Marine dolomite of unusual isotopic composition. *Science,* **155,** 189–191.

SAYLES, F.L. & MANHEIM, F.T. (1975) Interstitial solutions and diagenesis in deeply buried marine sediments: Results from the Deep Sea Drilling Project. *Geochim. cosmochim. Acta,* **39,** 103–127.

SAYLES, F.L., WATERMAN, L.S. & MANHEIM, F.T. (1973) Interstitial water studies and small core samples, Leg 19. In: *Initial Reports of the Deep Sea Drilling Project,* **IXX** (Ed. by J. S. Creager, D. W. Scholl *et al.*), pp. 871–874. U.S. Government Printing Office, Washington.

SCHLANGER, S.O. & DOUGLAS, R.G. (1974) The pelagic-chalk-limestone transition and its implications for marine stratigraphy. In: *Pelagic Sediments: on Land and Under the Sea* (Ed. by K. J. Hsü and H. C. Jenkyns). *Spec. Publ. Int. Ass. Sedim.* **1,** 117–148. Blackwell Scientific Publications, Oxford.

SCHLANGER, S.O., DOUGLAS, R.G., LANCELOT, Y., MOORE, T.C. & ROTH, P.H. (1973) Fossil preservation and diagenesis of pelagic carbonates from the Magellan Rise, central North Pacific Ocean. In: *Initial Reports of the Deep Sea Drilling Project,* **XVII,** pp. 407–427. U.S. Government Printing Office, Washington.

SCHOLL, D.W. & CREAGER, J.S. (1973) Geological Synthesis of Leg 19 (DSDP) Results: Far north Pacific and Aleutian Ridge and Bering Sea. In: *Initial Reports of the Deep Sea Drilling Project,* **IXX** (Ed. by J. S. Creager, D. W. Scholl *et al.*), pp. 891–913. U.S. Government Printing Office, Washington.

STEWART, R.J. (1977) Neogene trubidite and sedimentation in Komandorskiy Basin, Western Bering Sea. *Bull. Am. Ass. Petrol. Geol.* **61,** 192–206.

TARUTANI, T., CLAYTON, R.N. & MAYEDA, T.K. (1969) The effect of polymorphism and magnesium substitution on oxygen isotope fractionation between calcium carbonates and water. *Geochim. cosmochim. Acta.* **33,** 987–996.

THOMPSON, G. (1972) A geochemical study of some lithified carbonate sediments from the deep-sea. *Geochim cosmochim. Acta.* **36,** 1237–1253.

WATANABE, T., LANGSETH, M.G. & ANDERSON, R.N. (1977) Heat flow in back-arc basins of the western Pacific. In: *Island-arcs, Deep-sea Trenches, and Back-arc Basins* (Ed. by M. Talwani and W. C. Pitman III), pp. 137–161. *Am. geophys. Union, Maurice Ewing Series I,* Washington, D.C.

(Manuscript received 25 *August* 1978 *revision received* 5 *February* 1979*)*

Reprinted from Sedimentology (1981) ***28****, 611–641*

De glauconiarum origine

GILLES S. ODIN* *and* ALBERT MATTER†

**Département de géologie dynamique, Université P.-et-M. Curie,* 75230 *Paris Cédex* 05, *France and* †*Geologisches Institut, Universität Bern, Sahlistrasse* 6, 3012 *Bern, Switzerland*

ABSTRACT

The glauconitic facies is widespread on present-day continental shelves from 50° S to 65° N and at water depths between 50 and 500 m, and is in particularly great abundance on the upper slope and outer shelf between 200 and 300 m. It is also common in many ancient rocks of post-late Precambrian age.

It occurs as sand- to pebble-sized, essentially green particles (granular facies) but also as a surface coating on particles and hardgrounds and as a diffuse impregnation (film and diffuse facies). We suggest the replacement of the term 'glauconite', which has been interchangeably used to designate a morphological form and a specific mineral, by glaucony (facies) and glauconitic smectite and glauconitic mica as end members of the glauconitic mineral family.

The widely accepted model of Burst and Hower for glauconitization requires a degraded, micaceous (2:1 layer lattice structure) parent clay mineral. However, detailed analysis of numerous samples of Recent glaucony reveals that such a parent substrate is exceptional. The model therefore requires modification. Generally the parent material is carbonate particles, argillaceous (kaolinitic) faecal pellets, infillings of foraminiferal tests, various mineral grains and rock fragments, that pass gradually into the commonly occurring green grains.

We show that the process of glauconitization is achieved by *de novo* authigenic growth of automorphous crystallites in the pores of the substrate, accompanied by progressive alteration and replacement of the substrate. It is this two-fold evolution that causes the 'verdissement' of granular substrates, macrofossils and hardgrounds. The authigenic mineral is an iron-rich and potassium-poor glauconitic smectite. While new smectites are growing into the remaining pore space the earlier smectites are modified by incorporation of potassium, producing decreasingly expandable minerals with a non-expandable glauconitic mica as the end member. This mineralogical diversity of the glauconitic mineral family explains the highly variable physical and chemical properties of glaucony. Four categories, nascent, little-evolved, evolved and highly-evolved glaucony are distinguished.

Glauconitization appears to be controlled by a delicate balance between degree of physical confinement of a particle and the amount of ionic exchange between the micro-environment and ambient open marine sea water. The optimum conditions for glauconitization are those of semi-confinement. As a result the interior of a grain is more glauconitized than its less confined periphery. Similarly, for identical substrate types, large grains (500 μm) provide more favourable substrates for glauconitization than lesser confined small grains.

On a larger scale the formation of glaucony is governed by the availability of iron and potassium and the balance between detrital influx and winnowing. Low accumulation rates expose grains to the open marine environment for sufficiently long times (10^5–10^6 years for highly-evolved glaucony).

RÉSUMÉ

Depuis une douzaine d'années des glauconies sont récoltées sur les marges continentales. Nous avons étudié par divers procédés analytiques des séries d'échantillons ou de fractions d'échantillon afin de saisir la *dynamique* de leur évolution. Celle ci est double et concerne d'une part un support en déséquilibre avec son milieu de dépôt, il s'altère; d'autre part des minéraux authigènes qui se développent à l'abri de ce support. Nous insistons sur la différence fondamentale qui existe entre **la glauconie** mot qui désigne un faciès comportant deux types de composants évoluant séparément et **les minéraux glauconitiques** qui désignent l'un de ces composants: l'authigène, lui-même varié minéralogiquement.

Les minéraux glauconitiques sont des phyllites vertes dioctaédriques, riches en fer dès leur apparition sous la forme d'une smectite glauconitique de composition chimique spécifique. Au cours de l'évolution du grain vert, ces smectites recristallisent en minéraux plus riches en potassium et, par ce procédé répété, se mettent en équilibre avec un milieu ouvert sur le milieu marin général. Une composition chimique quasiment constante subsiste excepté pour le potassium qui devient abondant et atteint 9 % (K_2O) dans le mica-glauconitique (la glauconite, terme réservé au minéralogiste). L'homogénéité chimique des minéraux authigènes, cristallographiquement variés, est le reflet de l'homogénéité d'ensemble du milieu marin.

Les supports granulaires de cette évolution sont de 4 types principaux: les remplissages de micro-tests, les coprolithes, les débris carbonatés biogéniques, les débris minéraux. Tous montrent des propriétés physiques communes: ils sont ou deviennent poreux par altération, ces pores déterminent une grande surface de réaction et un milieu *semi-confiné par rapport à l'eau de mer* susjacente caractère fondamental permettant les synthèses minérales et l'altération. Le support est un carrefour où s'échangent lentement les ions de l'eau de mer, de l'eau interstitielle, du support lui-même. La nature chimique du support, si elle influe sur son altérabilité n'a aucune influence primordiale sur la nature minéralogique de l'authigenèse. Ce fait capital est en contradiction avec l'hypothèse de la transformation d'une phyllite T.O.T. dégradée héritée comme support nécessaire de la genèse des minéraux glauconitique comme l'exige l'hypothèse la plus souvent admise. Lorsque le milieu, au lieu d'être marin franc, est sub-deltaique (pH plus bas) les mèmes 4 types de support donnent naissance sous des apparences quasi identiques au *faciès berthiérine* dont les minéraux authigènes sont des phyllites ferrifères de structure voisine de la kaolinite.

La glauconitisation est donc d'abord le résultat d'une croissance cristalline *de novo* de smectite à l'abri d'un support poreux de nature et de forme variées; c'est la glauconie 'naissante' dont la teneur en potassium peut être fixée entre 2 et 4 % (K_2O). La glauconie 'peu-évoluée' est le stade suivant pendant lequel le support disparaît en grande partie (K_2O = 4–6 %). La glauconie 'évoluée' (K_2O = 6–8 %) se caractérise par des recristallisations dans des grains d'où la texture héritée du support disparait peu à peu alors que des fissures apparaissent. Dans la glauconie 'très-évoluée' l'origine n'est plus reconnaissable et le minéral authigène pur à une structure proche d'un mica et une composition en potassium élevée caractéristique du mica glauconitique pôle d'évolution de la famille.

INTRODUCTION

The glauconitic facies, consisting mainly of green grains but also present as thin films or in a diffuse form, occurs in many ancient rocks. Glauconitization has taken place since the earliest formation of soils and clay minerals, i.e. since the last third of the Precambrian. The glauconitic facies today covers wide areas of the continental shelves.

Numerous attempts have been made to explain the process of glauconitization. These can be grouped into three schools of thought: (1) precipitation of gelatinous silica followed by 'hydration of silica and subsequent absorption of bases' (Takahashi, 1939, p. 501); (2) derivation from a biotite or iron mica parent (Galliher, 1935). This theory was later modified into (3) the 'layer lattice theory' (Burst, 1958a, b; Hower, 1961) involving the absorption of potassium and iron by a degraded layer lattice mineral.

However, studies of glauconitic facies of present-day continental shelves reveal that glauconitization cannot be explained by the existing theories. We have studied green grains from both the present-day continental shelves and from ancient sediments. The new model we propose for glauconitization fits the existing and new data better than the 'layer lattice theory'.

VARIABILITY OF GLAUCONY

The term 'glauconite' has been commonly used to designate green grains, but also as a name of a mineral species. However, most of the green grains are not composed of the mineral species glauconite but of various other components. This double meaning has created confusion, as recognized by previous authors (e.g. McRae, 1972). This confusion masks variations that are important in two fields of research: (1) the understanding of the process of glauconitization; (2) the isotopic dating of glauconitic sediments. Numerous solutions have been proposed, introducing the terms 'glauconite', 'glauconite *sensu stricto*' and 'protoglauconite'. To avoid further confusion we recommend that the term 'glauconite' be discontinued and instead, for the facies the term **glaucony** (pl. glauconies) and, for the minerals *glauconitic smectite* and *glauconitic mica* be used as the end members of the glauconitic mineral family.

The most characteristic feature of glaucony is the variability of its morphology and of its physical and chemical properties. In order to reconstruct the modes of origin of glaucony, all of the different appearances and the range and limits of their variabilities have to be taken into account.

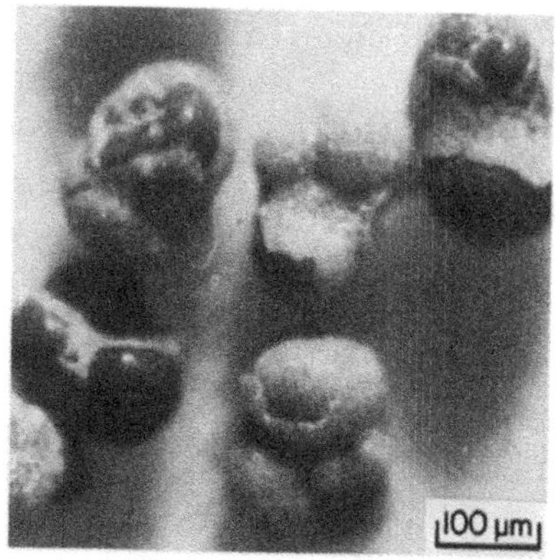

Fig. 1. Different stages of glauconitization of globigerinid Foraminifera illustrated by the amount of dissolution of the tests and the darkening (verdissement) of infillings.

Fig. 1. Différentes étapes d'évolution de moules de Foraminifères: Plateau continental N.O. espagnol, échantillon M. Lamboy. La couleur plus ou moins foncée et la dissolution du test indiquent un stade d'évolution plus ou moins avancé.

Morphological variability

The morphological diversity was recognized by early workers and was used by Cayeux (1916) to establish the first classification of glaucony.

However, the many different morphologies of the materials which cause the greenish aspect of what has been called greensand or greenearth can schematically be grouped into only two facies, the granular and the film facies. The granular facies consists of sand-sized, more or less green grains whose longest diameter ranges from several tens of microns to several millimetres.

The film facies is characterized by a pellicular penetration of a substrate by finely divided glauconitic minerals. The depth of penetration, which may vary from several hundred microns to a few centimetres, appears to be a function of the alteration of the substrate and the resulting porosity.

The type of substrate includes particles of diverse origin and size such as detrital minerals, calcareous macrofossil debris, pebbles and even blocks of sedimentary rocks up to several metres in diameter. Hardgrounds, especially those on carbonate rocks, are often marked by a glaucony film which may extend over several square kilometres.

A few investigators e.g. Collet (1908) and Millot (1964) envisaged yet a third facies: a diffuse impregnation of the entire rock mass without any recognizable polarity towards the sediment–water interface. The physicochemical nature of this pigment is very difficult to define due to difficulty in its separation. A thorough description and analysis of this facies has yet to be made.

Although it is convenient for descriptive purposes to regard granular and film facies as separate, in the granular facies a continuous morphological sequence from large and thoroughly glauconitized grains to small pebbles with only a surface green coating (film) occurs. Hence there is no fundamental break between the two facies. Often the particle which has been replaced can still be identified, whereas in other cases no particular structure is recognizable. Detailed investigations of Recent 'green grains' found in their pre-burial stage have revealed a continuous sequence ranging from whitish to dark green grains whose origin would be unidentifiable had the less altered members not been observed (see Odin, 1975a for numerous references). Several authors have observed and analysed different examples demonstrating that carbonate debris, infillings of foraminiferal tests (Fig. 1), feldspar and quartz grains and kaolinitic and other argillaceous coprolites pass gradually into the commonly occurring, fissured and cracked green grains. The changes in internal texture which accompany the external morphological change were studied by scanning electron microscopy and are discussed below.

The observations made on granular glaucony from ancient formations and from present-day continental shelves suggest that four kinds of substrate are common (Fig. 2):

(1) *Internal moulds* (or casts of predominantly carbonate microfossil tests (e.g. Foraminifera). Internal moulds sometimes dominate in a sample or

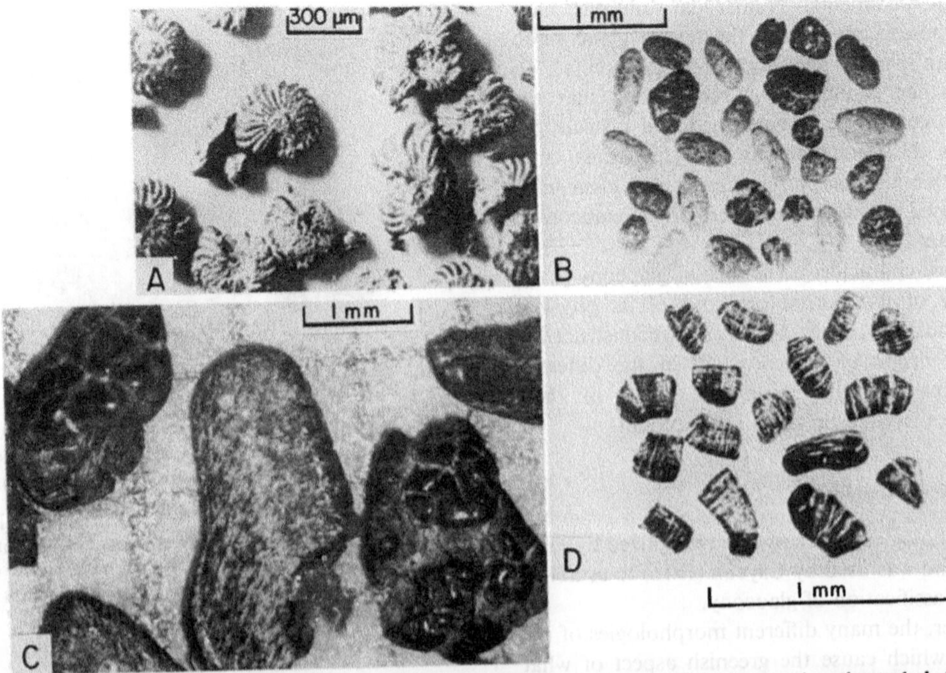

Fig. 2. Main types of glauconitized substrates. (A) Dark green casts of foraminiferal tests (continental shelf of N.E. Spain). (B) Light to dark green coprolites reflecting different stages of glauconitization (subsurface Quaternary off Senegal). (C) Green shell debris with original shell texture (zebra texture) preserved on lower surface of particle (centre) and cracked and bulbous upper surfaces (Lutetian, Paris Basin). (D) Accordion habit developed by glauconitization of mica grains (shelf off N.W. Spain).

Fig. 2. Les principaux types de supports glauconitisés: (A) Remplissages de microtests vert foncé (plateau continental N.O. espagnol). (B) Coprolithes à différents stades d'évolution, plus ou moins verts (Quaternaire enfoui au large du Sénégal). (C) Débris coquilliers verdis (Lutétien du Bassin de Paris) noter les stries caractéristiques de ces grains zébrés qui conservent localement la texture originale. (D) Debris minéraux verdis (micas du plateau continental N.O. espagnol).

are at least present in small amounts (Murray & Renard, 1891, pp. 378–391; Collet, 1908; Caspari, 1910; Wermund, 1961; Ehlmann, Hulings & Glover, 1963; Bjerkli & Östmo-Saeter, 1973).

(2) *Faecal pellets*, which are more or less argillaceous or limy and contain variable amounts of organic matter. They frequently occur in great abundance in ancient and Recent glauconitic sediments (Takahashi & Yagi, 1929; Moore, 1939; Bell & Goodell, 1967; Porrenga, 1967a; Tooms, Summerhayes & McMaster, 1970; Giresse & Odin, 1973). Most pellets are more or less ellipsoidal with a longest diameter of 50–500 µm. Pryor (1975) shows that most of the pellets are true faecal pellets produced in large quantities by filter feeding organisms in present-day shallow seas. They also occur at greater depths (Moore, 1939).

(3) *Biogenic carbonate debris*, formed either by disarticulation after disintegration of the organic tissue or by biological and mechanical break-up, is frequently found replaced by glaucony (Dangeard, 1928; Houbolt, 1957; Lamboy, 1974).

(4) *Mineral grains and rock fragments*, whether they contain iron and/or silica and aluminium as major ions, or whether or not they are phyllitic, may become glauconitized (Cayeux, 1916; Wermund, 1961; Ojakangas & Keller, 1964; Odin, 1972a; Hein, Allwardt & Griggs, 1974). We have observed glauconitized quartz, feldspar, biotite and muscovite, calcite, dolomite, phosphates, volcanic glass shards, volcanic and plutonic rock fragments as well as chert grains.

A rough appraisal of the observed substrates and chemistries indicates that none is dominant nor, *a fortiori*, required as a starting material for glauconitization. Nevertheless, carbonate appears to represent an especially favourable substrate, as noted

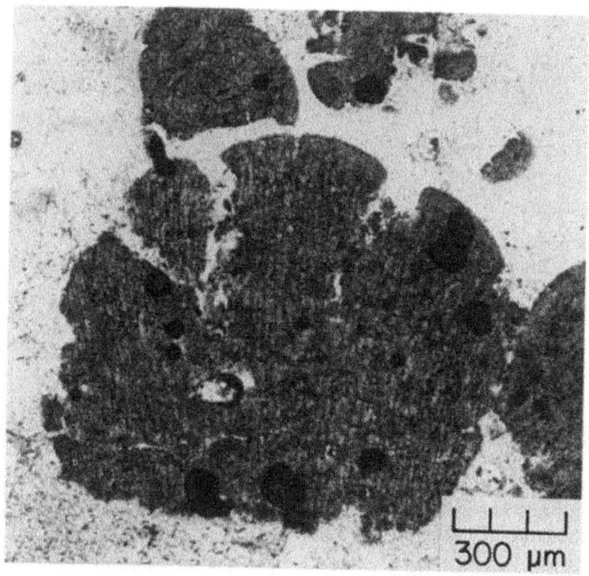

Fig. 3. Aspect of glaucony as seen under the optical microscope: green grain (as in Fig. 2C) with original shell architecture preserved.

Fig. 3. Aspects de la glauconie au microscope optique: grains verts du Lutétien (voir Fig. 2C) montrant l'origine biogène du support de verdissement.

by previous workers (Cayeux, 1916; Millot, 1964; Lamboy, 1976).

The glauconitic grains within a bed may be derived from different parent materials, each of which may itself be represented by several instances of the glauconitization process; it is important to bear in mind that the bulk analysis of a sample, even purified, is likely to be a mixture of initial substrates and authigenic glauconitic materials. In addition, the numerous fissures, cracks, hollows and the surface coatings are sites of trapped impurities.

Variability of physical properties

Viewed in thin-section between crossed polars, most glaucony grains are cryptocrystalline and show aggregate polarization. The crystallites are smaller than the thickness of the thin-section (0·02 mm) which renders optical work difficult.

Sometimes the crystallites are oriented. They may be disposed fibroradially in the peripheral part of a grain or they may show a zebra structure resulting from the replacement of the crossed-lamellar structure of bivalves as shown on Fig. 3. In plane-polarized light, different shades of green are observed ranging from pale green, sometimes yellowish, to dark green. Most grains are weakly pleochroic.

Refractive indices, which range from 1·59 to 1·63, increase particularly with higher amounts of iron and with higher ordering (less expandable layers) of the glauconitic mineral (Bentor & Kastner, 1965; Cimbalnikova, 1970; Velde & Odin, 1975). The birefringence is difficult to estimate as the natural colours generally mask the interference colours. Hence visual identification of glauconitic materials other than by their general aspect remains tentative (Fig. 3).

On the other hand, investigations of glauconitic grains with the scanning electron microscope at high magnifications reveal that the tiny glaucony crystals may take on different habits. Furthermore the textural evolution caused by progressive glauconitization can be seen and documented.

Borst & Keller (1969) were the first to illustrate the surface texture of a glaucony grain using SEM, unfortunately without reference to its mineralogical nature.

Authigenic minerals of Recent granular glaucony in its initial stage of formation often exhibit tiny, ill-defined globules less than 0·5 μm in diameter. They eventually become attached to each other forming 'caterpillar' structures 2–3 μm long, as shown on Fig. 4(A). Small blades about 1–3 μm wide have been observed, especially in glauconitized mica

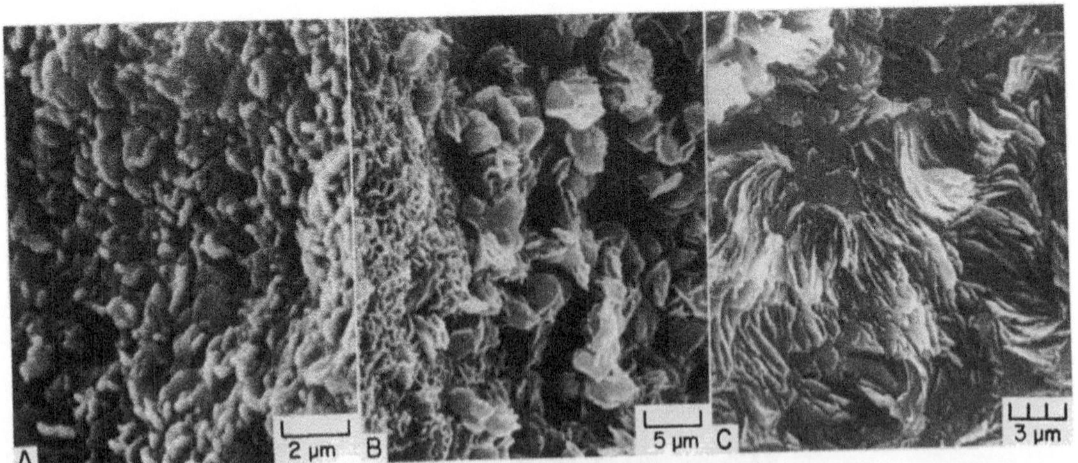

Fig. 4. SEM photomicrographs showing the main types of nanostructure of glaucony. (A) Ill-defined globules and caterpillar structures typical of nascent glaucony (4 % K_2O), (B) Boxwork and rosette structures of evolved glaucony (6·5 % K_2O), (C) Lamellar structure of highly-evolved glaucony (8 % K_2O).

Fig. 4. Aspects de la glauconie au microscope à balayage: (A) Les globules et chenilles caractérisent des minéraux glauconitiques peu évolués. (B) Les lamelles parfois arrangées en réseau ou en petites roses des sables se recontrent partout notamment dans des microgéodes pour les dernières qui dénotent une automorphie certaine. (C) Les lames caractérisent les grains les plus évolués où les minéraux sont riches en potassium, elles peuvent atteindre 5–10 µm de long.

flakes. The blades cover lattice lamellae or cracks in the micas in a boxwork-like fashion or make up minute lepispheres 3–4 µm in diameter as shown by Odin (1972a). In more evolved grains, which no longer show any trace of the initial substrate, the blades form up to 4–5 µm aggregates resembling rosettes (Fig. 4B). Well-developed lamellae up to 5–10 µm long are found in ancient glaucony as well as in dark green, mature grains from present-day continental shelves. These lamellae are always slightly sinuous and show sub-parallel alignment (Fig. 4C). This habit is characteristic of the most evolved glauconitic grains (step 4 of Fig. 14). However, identification of glauconitic minerals from scanning electron micrographs should be undertaken with caution because other phyllitic and non-phyllitic minerals may show the same habits.

Obviously the nanostructure of authigenic glauconitic minerals is subject to considerable variation, and it may be either xenomorphic or automorphic (Fig. 5). The glauconitized bivalve shell shown in Fig. 3 still shows the fine details of the former aragonitic architecture even at high magnifications, which indicates a piecemeal replacement of the individual aragonite crystals (zebra structure). As a rule the growth habits are more organized in more evolved grains. They are also similar to authigenic growth forms of other minerals, including other micaceous minerals. On the other hand, the inherited substrates show evidence of alteration, which is particularly well illustrated by carbonate substrates. Several degrees of recrystallization of the glauconitic minerals are observed during the evolutionary process. It is important to realize that it is the *sequence of nanostructures* which enables us to understand the evolutionary nature of glauconitization.

The specific gravity of glauconitic grains varies from 2·2 to 2·9 according to different authors (Cimbalnikova, 1970, Shutov *et al.*, 1970, and others). Lloyd & Fuller (1965) reported values as high as 2·9–3·1. Statistically, the greener the glaucony the higher the specific gravity. Moreover the spread of the values is wide for low-density glauconitic grains as opposed to a narrow clustering of those with a specific gravity greater than that of quartz.

The paramagnetic susceptibility of glaucony varies. It permits an effective separation of glaucony from other grains and the different kinds of glaucony from each other.

With progressive replacement of the initial substrates and subsequent evolution, the grains become increasingly green, and have a higher specific gravity and magnetic susceptibility. Correspondingly, the originally wide scatter of the values (heterogeneous

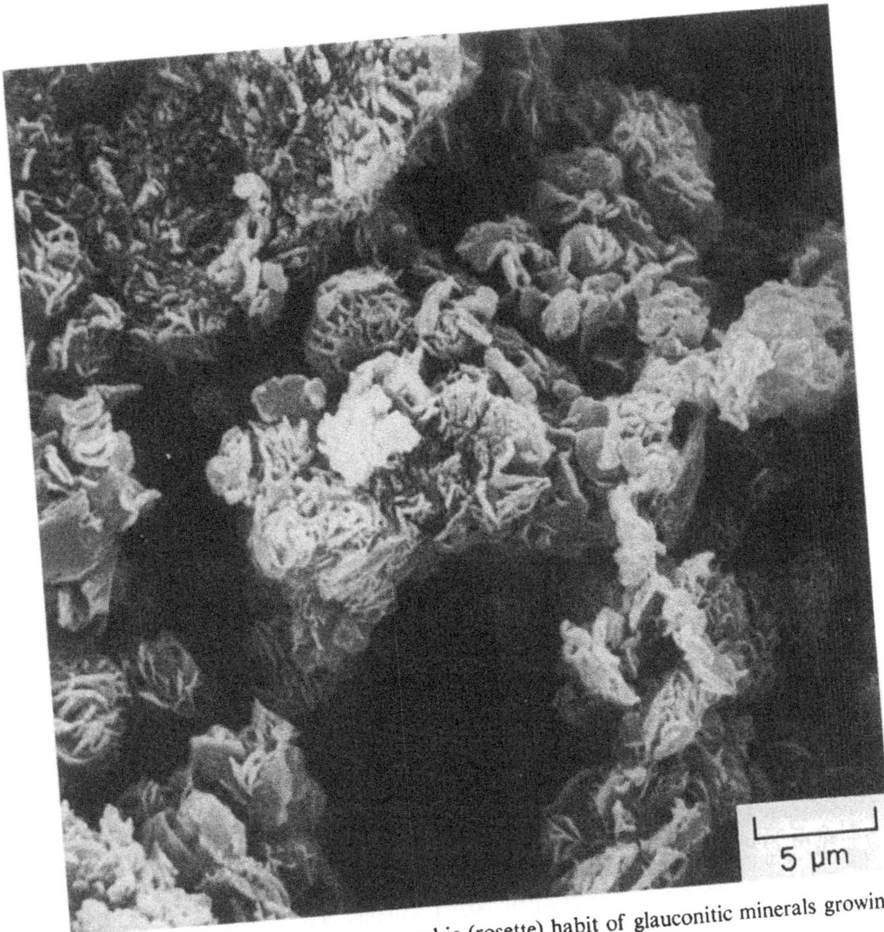

Fig. 5. SEM photomicrograph showing the automorphic (rosette) habit of glauconitic minerals growing in an echinodermal stereom.

Fig. 5. Automorphie des minéraux glauconitiques dans des supports de glauconitisation biogéniques (stéréome d'Echinodermes).

sample) narrows, indicating a more homogeneous sample.

Mineralogical variability

From the discussion above it follows that it is essential to recognize and distinguish the inherited minerals of the substrate from the authigenic glauconitic material which has grown on or within this substrate and to unravel the mineralogical evolution of both separately. The granular facies is used to show the mineralogical variability.

Results of X-ray diffraction analysis

For the mineralogist the term 'glauconite' refers to a particular mineral, a hydrated ferric mica with less than 5 % expandable layers (Burst, 1958a, b; Hower, 1961; Bentor & Kastner, 1965). This perfect glauconitic mica ('mineral glauconite' of Burst; Wermund, 1961; 'glauconite *sensu stricto*' of McRae, 1972) however has been observed so far in only a small number of samples (Fig. 6). The geologist uses the term in a broader morphological sense and not everything that he calls 'glauconite' has the diffraction characteristics of the mineral defined above.

Another green mineral, celadonite, which is found in volcanic rocks, is closely related to glauconitic mica but is slightly richer in SiO_2 (52–56 %) and contains 5–9 % MgO instead of 2–3 % (Hendricks & Ross, 1941; Foster, 1969). Ferric illite is also mineralogically and chemically quite similar to glauconitic mica, differing only by its much lower

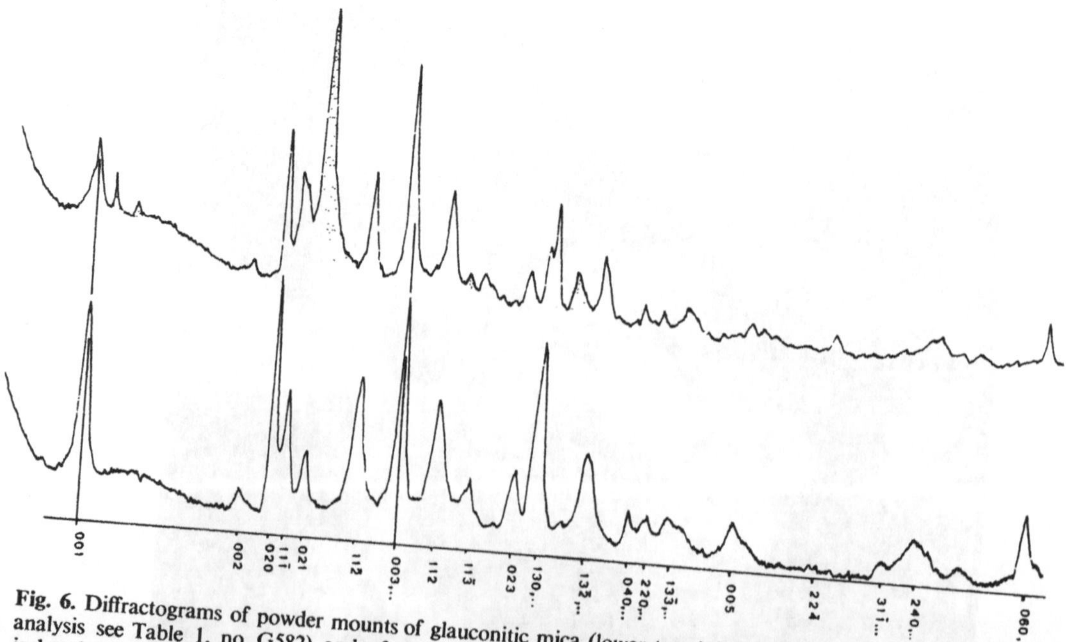

Fig. 6. Diffractograms of powder mounts of glauconitic mica (lower trace) from the Ordovician of Estonia (chemical analysis see Table 1, no. G582) and of celadonite from volcanic rocks, Cyprus. The peaks of glauconitic mica are indexed according to Warshaw (1957). Stippled peaks in celadonite pattern (upper trace) result from the presence of another mineral.

Fig. 6. Diffractogrammes de poudre de mica glauconitique (en bas) et de céladonite (en haut): L'analyse chimique du mica glauconitique (pur) est donné Tableau 1–1; ce pôle de la famille des minéraux glauconitiques est rarement atteint et seulement après une longue évolution. La céladonite a été récoltée sous forme de bloc bleu vert pâle à Chypre et n'est pas pure. L'indexation des raies est celle proposée par Warshaw (1957) pour le mica glauconitique; on notera que le pic le plus haut est toujours dû à la diffraction (020) sur nos diffractogrammes de poudre mais il est plus fin que les deux raies (001) et (130) auxquelles Warshaw donne l'intensité 100. Le caractère ferrifère de ces minéraux est visible dans le rapport très élevé des hauteurs de pics (001/002). Céladonite et mica glauconitique se distinguent d'après ces exemples comme d'après les fiches ASTM par des intensités relatives différentes des pics (001 plus faible dans la céladonite) et le dédoublement net de certains pics tels (130) dans la céladonite.

iron content. According to Kossovskaya & Drits (1970) this mineral is indicative of restrictive hypersaline conditions.

On X-ray diffraction patterns of green grains, especially of those in their early stage of glauconitization, it is sometimes difficult to know which peak to assign to the substrate and which to the authigenic material. An analysis of a total glaucony sample is of little use, whereas *an ordered sequence of X-ray patterns of different fractions reveals genetic information* by separating the different stages of evolution. Our use of powder mounts rather than the usual oriented specimens illustrates that X-ray diffraction patterns of the evolution of glauconitic smectite to glauconitic mica are not identical to those of smectite to illite.

Variability of the first-order basal reflection. On randomly oriented diffraction patterns of powder mounts the first-order basal reflection (001) may be located anywhere between 14 and 10 Å. Upon glycol solvation it hardly shifts position except after preliminary cationic exchange of K. The peak generally shows a fairly broad base with asymmetrical sides (Fig. 7). Only when the peak is situated at 10 Å does it possess a sharp appearance.

Variability of other diffraction peaks. Although the chronological evolution of glauconitic minerals goes from less-ordered to well-ordered varieties, it is easier to start with the latter in description of the diffraction patterns from random powder mounts.

Starting from a mica-type glauconitic material, first the $(11\bar{3})$ peak disappears, then $(11\bar{1})$ and (021) and finally also (023) (Fig. 8). In the absence of hydrated oxides of iron in the grains, the $(11\bar{1})$ and (021) reflections are the most characteristic; the (002) peak is always subdued which indicates a high amount of iron in octahedral positions. The size of the (112) and $(11\bar{2})$ reflections is a very reliable measure of the ordering of the layer silicate lattice

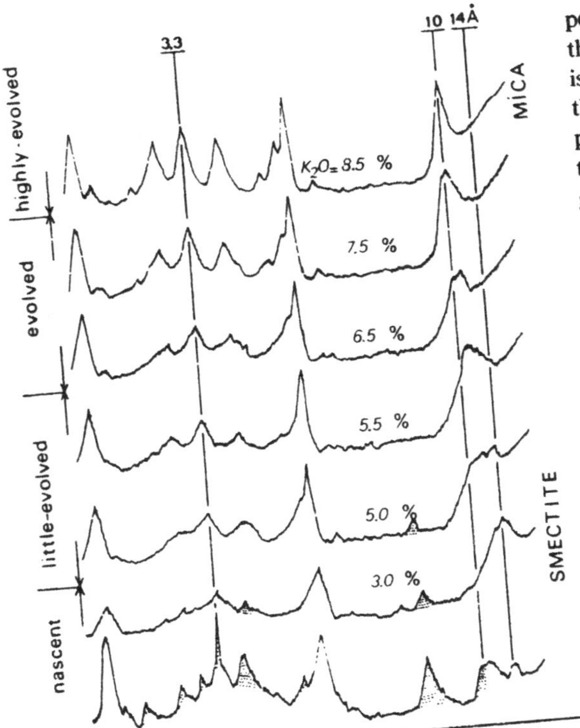

Fig. 7. Powder diffractograms representing stages of glauconitization. The substrate (stippled peaks) is still recognizable in diffraction patterns of nascent and little-evolved glaucony. Note the change in shape and position of the (001) peak (between 14 and 10 Å) and the increase in height of the (hkl) peaks of the authigenic component during evolution of glaucony.

Fig. 7. Diffractogrammes de poudre de glauconies à différents stades d'évolution: tous les grains analysés ont été séparés, sont paramagnétiques et de couleur verte. Les minéraux du support donnent encore des diffractions dans les glauconies naissantes et peu-évoluées. Les minéraux authigènes montrent une série continue où l'on peut distinguer 2 types de variation: la position de la raie de diffraction (001), la présence et la hauteur des raies (h, k, l). La raie (020) est un repère constant. Le pic de diffraction (001) des minéraux authigènes des glauconies naissantes et peu évoluées 'gonfle' difficilement sous l'action de l'éthylène glycol (cations interfoliaires potassiques), mais passe aisément à 10 Å après chauffage. *Noter que l'aspect d'ensemble d'un diffractogramme est largement lié à la teneur en potassium.* L'analyse des différentes fractions glauconitiques d'un sédiment, notamment peu-évolué, permet de saisir et de reconstituer une partie de la séquence d'évolution minéralogique.

(Bentor & Kastner, 1965). In a diffraction pattern of a 10 Å mica-type lattice these two peaks are approximately of the same size as the (003) peak. They are subdued domes when the basal reflection is near 12 Å. When $(11\bar{2})$ and (112) finally become indistinct a concomitant decrease in size of the (003) peak is observed. The position of the summit of these peaks remains stationary. The (020) reflection is the most constant in both position and size and therefore serves as a reference. The variation of the peaks mentioned above permits a determination of the stage of evolution of green grains in the process of glauconitization (Bentor & Kastner, 1965, pp. 19–22).

Briefly, the X-ray diffraction data discussed above define a crystallographic family ranging from a green smectite carrying no particular mineral name, but which we here call *glauconitic smectite*, to a micaceous mineral called *glauconitic mica*. Apart from a few details (size of (002) peak, behaviour upon glycolation of (001)], this family is crystallographically identical to the illite family which evolve during diagenesis from smectite to illite.

Chemical aspects of the glauconitic minerals

The examination of our own data (Table 1), and of the analyses reported in the literature reveal that a certain number of common, and even specific, chemical characteristics can be recognized.

The *silica* content is fairly constant and generally comprises 47·5–50 % SiO_2 according to our data (Table 1) as well as those compiled by Hendricks & Ross (1941) and Smulikowski (1954).

The *aluminium* varies in general between 5·0 and 8·0 % Al_2O_3. All values lie between 3·5 and 11 % in our data. Similar results have been obtained by Hendricks & Ross (1941), Smulikowski (1954), Foster (1969) and Cimbalnikova (1971a). We believe that the extremely high values of > 20 % Al_2O_3 measured by Shutov *et al.* (1970) from Palaeozoic glaucony are due to post-depositional evolution.

The content of *iron* is fairly uniform and varies from 19 to 27 % Fe_2O_3, again with only a few samples higher in iron (Table 1). These abnormally high values are almost certainly caused by 'impurities' such as oxidized grain surfaces. The values reported in the literature are consistent with ours. Foster (1969) noted that the few samples with 'extremely high' aluminium content (12 % Al_2O_3) have a rather low iron content of < 19 % Fe_2O_3 total. These results confirm that the glauconitic mineral family consists of a number of iron-rich varieties whose iron content varies within a relatively narrow range.

Although the amount of ferrous iron can generally only be determined approximately, our values are rather constant in the range 1·0–3·2 % FeO. Hendricks & Ross (1941) report systematically higher values between 2·0 and 4·0 % FeO.

14962395-1-Text.pdf TPS: 246x189 Plates: 0 Spine

Table 1. Chemical analyses of cleaned glauconies of: (1) Ordovician, (2) Jurassic, (3) Upper Cretaceous, (4) Eocene, (5) Oligocene, (6) Ypresian of Monts des Flandres, (7) relict glaucony from the shelf of north-west Spain, (8) Pleistocene-Holocene glaucony from borehole representing a mean of several analyses. Letters along sample numbers indicate. (A)160–500 μm fraction, (B) 100–160 μm, (FF) < 20 μm and (L) denser than 2·89 (bromoform) fraction. The sample numbers in 6 indicate depths in borehole. GL-O reference sample for isotopic and chemical analyses; values given are means of data from 15 laboratories. (+) oxidized sample. (m) mean of several independent analyses

Tableau 1. Analyses chimiques par voie humide de grains verts purifiés (G.) de séries sédimentaires marines. 1, Ordovicien; 2, Jurassique; 3, Crétacé supérieur; 4, Eocène; 5, Oligocène; 6, Yprésien des Monts des Flandres; 7, Mio-Pliocène relique de surface du plateau N.O. Espagnol; 8, Pléistocène-Holocène de surface du Golfe de Guinée; Pleistocène enfoui. La fréquence relative des teneurs en potassium élevées n'est pas répresentative, et due à une sélection préalable des échantillons évolués en vue de datations isotopiques. Quelques échantillons ont été analysés plusieurs fois, on donne la moyenne (m). 'A' désigne la fraction granulométrique 160–500 μm; 'B' désigne la fraction granulométrique 100–160 μm; 'F.F.' désigne la fraction fine inférieure à 20 μm; 'L' désigne une fraction plus dense que le bromoforme. Pour les échantillons de l'Yprésien, les glauconies sont désignées par leur profondeur de prélèvement en sondage (Odin, 1975a). GL-O est un matériel de référence d'analyse chimique et isotopique; *les données de cette colonne sont une moyenne des données d'une quinzaine de laboratoires*. La donnée la plus élevée en Fe_2O_3(+) a été obtenue sur une glauconie oxydée probablement peu avant son enfouissement. On notera l'homogénéité des teneurs en fer (fortes), en magnésium et aluminium (faibles), la faible teneur en calcium et sodium comparée à celle en potassium. Cette homogénéité d'ensemble est liée selon nous à l'homogénéité du milieu et des conditions de formation en relation avec le milieu marin.

La variation large des teneurs en potassium est compensée par celle des teneurs en eau (total K_2O+H_2O = constant); cette variation définit une suite spécifique de composés minéraux authigènes dans ce milieu homogène

	1	2	3													
	G.582A	G.482A	G.9A	G.16A m	G.17A	G.18A	G.60A m	G.61A	G.63A	G.71A m	G.71B	G.73A	G.151	G.279A	GL-O m	G563A
SiO_2	49·2	45·9	52·1	52·6	48·0	51·5	46·6	50·6	48·4	50·0	49·2	47·8	50·9	46·3	50·9	50·2
Al_2O_3	7·2	5·5	6·8	5·2	7·4	7·6	4·5	5·5	5·8	6·6	16·1	10·7	5·9	4·4	7·55	7·3
Fe_2O_3	19·1 }	27·1	{ 21·0	19·8 }	21·0	{ 17·6	26·0+	21·0	24·7	20·0	14·4	21·4	18·8 }	26·8	{ 17·2	17·8
	4·4 }		{ 2·15	2·2 }		{ 2·2	3·3	3·0	0·4	2·8	1·3	1·5	3·0 }		{ 2·2	2·9
TiO_2	0·05	0·05	—	—	0·05	—	—	—	—	—	—	—	—	00·5	0·05	0·07
CaO	0·4	1·5	0·7	0·5	0·7	0·4	—	0·6	0·8	0·8	—	—	0·8	0·4	0·95	0·6
MgO	3·6	3·0	1·5	4·1	4·2	4·9	1·7	3·6	3·8	3·7	3·1	3·3	4·5	4·5	4·45	4·3
Na_2O	0·05	—	0·2	0·15	—	0·1	0·15	0·15	0·2	0·1	0·4	0·4	0·2	—	0·05	0·05
K_2O	8·3	8·15	6·8	8·7	8·9	8·5	9·0	7·6	7·6	8·8	8·1	7·2	8·5	7·8	7·95	8·8
P_2O_5	—	—	0·05	0·10	—	0·10	—	0·10	0·10	0·10	0·05	—	0·2	—	0·35	—
H_2O^-	1·3	1·3 }		{ 1·6	2·2	1·3 }		{ 2·2	2·3	1·8	3·1	3·1	1·8 }		{ 2·5	1·7
H_2O^+	5·5	7·2 }	8·05	{ 5·7	6·9	5·5 }	8·0	{ 5·5	5·9	5·3	4·5	5·2	3·2 }	8·2	{ 5·6	5·8
	99·1	99·7	99·35	100·65	99·35	99·7	99·25	99·85	100·0	100·0	100·2	100·6	97·8	98·45	99·75	99·5

	4															5
	G.42A m	G.49A m	G.58A	G.66	G.78A m	G.79A	G.98	G.101A	G.110A	G.128A	G.132A	G.134A	G.440A	G.153A	G.154A	G.139
SiO_2	45·9	48·8	49·3	50·0	48·6	48·0	49·0	46·9	49·1	49·6	49·3	49·2	47·5	49·6	49·3	49·2
Al_2O_3	9·6	10·1	8·3	7·2	7·4	5·4	5·2	4·5	5·6	5·5	3·9	3·7	4·8	6·1	6·4	6·5

Fe_2O_3	21·2	18·3	19·2	24·5	22·1	22·8	22·3	26·8	21·3	21·6	23·7	23·1	25·9	21·0	20·9	22·7
FeO	1·9	1·3	2·8	1·1	2·2	1·7	3·6	2·3	2·7	1·5	1·8	2·4	1·9	1·1	1·1	0·6
TiO_2	—	—	—	—	—	—	—	—	—	—	—	—	0·15	—	—	—
CaO	—	—	—	—	0·5	1·0	0·7	0·8	0·9	0·8	0·8	0·8	0·8	0·8	0·8	0·8
MgO	5·0	3·9	2·2	—	2·9	4·3	3·5	2·6	3·5	4·6	4·5	4·6	3·5	4·5	4·5	3·9
Na_2O	0·4	0·35	0·2	—	0·2	0·2	0·2	0·15	0·15	0·15	—	—	0·1	0·1	0·15	0·15
K_2O	7·5	7·6	7·5	6·4	8·4	8·3	8·1	6·9	7·6	7·6	7·8	8·0	6·8	7·6	7·5	7·0
P_2O_5	0·05	0·05	—	—	0·15	—	0·2	0·2	0·2	0·1	0·05	0·1	—	0·05	0·05	0·1
H_2O^-	3·2	5·0 }	10·3	—	{ 2·0	2·1	2·3	3·8	3·2	3·5	2·9	2·3	2·8	3·7	3·9	2·9
H_2O^+	5·5	4·6 }		—	{ 5·8	6·1	5·2	4·4	5·3	5·1	4·9	5·8	5·4	5·4	5·6	6·5
	100·25	100·0	99·5		100·25	99·9	100·3	99·35	99·55	100·05	99·65	100·0	99·65	99·95	100·2	100·35

	6															
	–41.9A	–41·9A	–41·9B	–49·0A	–54·0A	–59·8A	–67·3A	–73·0A	–76·0A	–82·6A	–82·6B	–83·7A	–92·8B	–95·2	–96·1	–97·2B
SiO_2	48·6	48·5	47·8	—	49·8	—	49·1	51·1	51·7	48·5	48·0	49·8	48·6	47·9	48·4	48·4
Al_2O_3	5·7	5·2	6·3	—	6·2	—	5·2	5·9	7·0	7·1	7·4	6·8	6·4	6·3	7·6	6·2
Fe_2O_3	22·0	21·1	20·9	24·0	21·9	23·75	21·4	19·2	18·4	20·3	20·3	20·3	21·8	21·0	20·0	20·4
Fe O	1·9	1·7	1·7	—	1·3	—	0·75	1·3	1·5	1·9	1·7	1·7	1·8	1·8	1·5	1·2
TiO	—	0·1	0·1	—	—	—	0·4	0·5	0·4	0·05	0·1	0·5	0·5	0·5	0·5	0·6
CaO	0·7	—	—	—	1·4	—	1·5	2·1	1·5	—	—	1·5	1·5	2·7	1·8	1·5
MgO	4·1	5·1	4·9	—	3·5	—	3·5	2·6	2·9	3·2	3·0	3·0	2·8	2·1	2·7	3·0
Na_2O	0·2	0·1	0·1	0·25	0·1	0·3	0·6	0·6	0·3	0·3	0·4	0·3	0·2	0·6	0·5	1·8
K_2O	8·7	8·0	7·9	7·9	7·1	7·2	6·4	5·8	5·7	6·7	6·3	5·9	5·4	5·0	4·6	4·6
P_2O_5	—	0·05	0·1	—	—	—	0·15	0·15	0·15	0·2	0·15	0·15	0·2	0·2	0·1	0·2
H_2O^-	2·0	3·3	3·2	—	4·0	—	4·8	4·6	4·7	4·3	5·1	4·8	4·9	5·4	6·1	6·7
H_2O^+	5·8	6·8	6·9	—	5·5	—	6·3	6·1	6·0	6·8	6·9	5·6	5·7	6·3	6·0	5·5
	99·0	99·95	99·9	—	100·8	—	98·6	99·95	98·75	99·35	99·15	98·85	98·3	99·8	99·8	98·6

	7					8								9	
	G.242A	G.243A	G.245A	G.250A	G.252A	G.312A	G.313A	G.318A	G.319A	G.362A	G.490cA	G.490hA	G.448FF	G.486A	G.600A
Si O_2	45·7	46·9	47·6	45·4	48·7	46·7	45·7	47·6	47·5	45·9	46·6	46·9	40·8	46·1	47·3
Al_2O_3	6·2	6·5	4·3	5·8	7·8	9·7	9·8	6·6	6·4	6·4	6·1	6·9	21·8	4·0	9·3
Fe_2O_3	25·1	23·1	24·6	26·4	19·9 }	20·9	{ 21·6	22·9 }	23·8	23·8	{ 23·8 }	22·0	11·7	{ 24·9	18·7
FeO	2·1	1·6	1·7	1·6	2·6 }		{ 1·1	1·1 }			{ 1·2 }			{ 2·3	1·4
TiO_2	—	—	—	—	—	0·7	0·8	0·8	0·4	0·4	0·3	0·5	1·1	0·1	0·4
CaO	0·7	0·8	0·7	0·7	0·7	1·4	1·4	1·4	1·4	1·4	1·4	1·4	2·6	1·5	3·1
MgO	4·1	3·8	3·9	3·5	3·8	3·0	2·9	2·8	2·8	3·5	4·2	3·6	1·7	4·4	3·6
Na_2O	0·2	0·2	0·3	0·3	0·2	0·2	0·5	0·4	0·15	0·1	0·2	0·3	0·15	0·2	0·2
K_2O	8·8	8·9	8·9	8·2	8·1	2·3	3·0	5·1	6·6	6·6	4·0	3·4	1·5	6·4	3·7
P_2O_5	—	—	—	—	—	0·15	0·25	0·2	0·15	0·05	—	—	—	—	—
H_2O^-	1·4	1·4	1·4	1·6	1·8	5·0	4·8	4·6	3·4	2·3	5·8	6·6	3·6	3·7	4·8
H_2O^+	6·3	6·1	6·1	6·2	6·2	10·0	8·5	6·6	6·9	9·1	6·5	8·3	14·1	5·9	6·6
	99·9	99·3	99·8	99·0	99·1	98·65	98·95	98·7	98·1	98·15	100·1	99·9	99·05	99·5	99·1

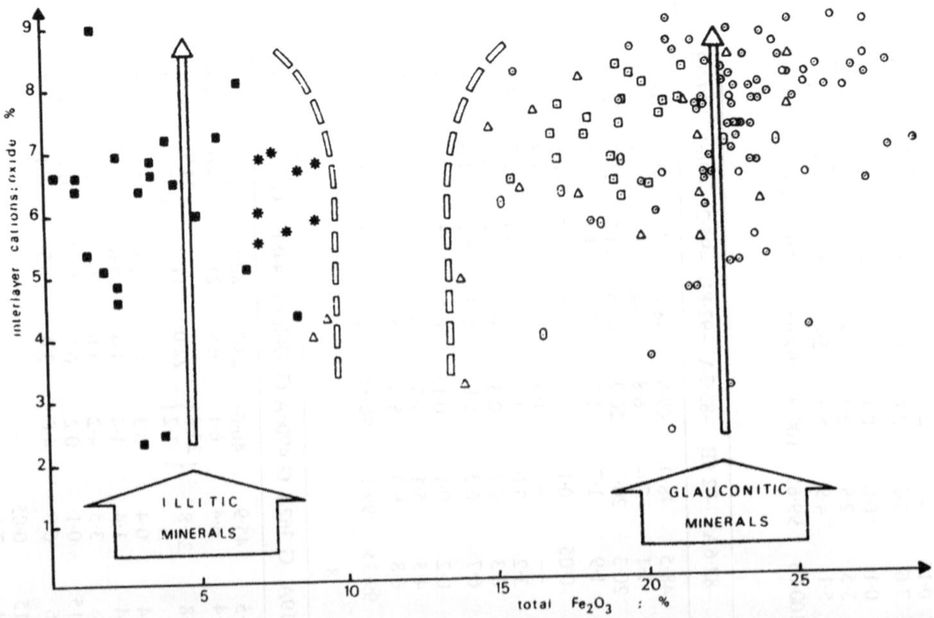

Fig. 8. Diagram showing the position of the illitic and the glauconitic mineral families with respect to their iron and interlayer cation content. Data from ○ Table 1, □ Cimbalnikova (1971a), ʘ Parry & Reeves (1966), △ Hower (1961) —these samples appear to be mixtures of clay minerals of authigenic and of substrate origin—★ Kossovskaya & Drits (1970), ■ Hower & Mowatt (1966). Lack of values between about 10–15% Fe_2O_3 suggests absence of a mineralogical transition between illitic and glauconitic minerals. Arrows show evolution from a smectitic to a mica-type clay mineral within both families. Glaucony does not increase in iron content with increasing amounts of interlayer cations.

Fig. 8. Diagramme d'abondance des cations en fonction de la teneur en fer (Fe_2O_3 total: deux séries distinctes sont mises en évidence de part et d'autre d'une coupure entre 10 et 15 % de fer. Cette coupure permet de distinguer parmi les minéraux T.O.T. dioctaédriques une famille glauconitique dont une des originalités est la teneur élevée en fer. Les 2 flèches symbolisent des variations chimiques entre des minéraux de type smectite et de type mica, généralement atteints après une longue évolution géochimique. Les minéraux de la famille glauconitique ne montrent pas, sur ce diagramme, de tendance à un enrichissement en fer lors de l'enrichissement en cations interfoliaires. La reconstitution de l'évolution minérale probable est effectuée ici grâce à l'analyse d'échantillons de niveaux variés qui permet de se prémunir contre des cas particuliers.

It appears that the percentage of *magnesium* is very constant because the range obtained by us (2·6–4·6 % MgO) compares well with the ranges given by Hendricks & Ross (1941) and Foster (1969).

Of all the cations found in glauconitic minerals *potassium* is the most variable. Although obviously the lower limit of the potassium content in glauconitic minerals remains somewhat vague we suggest, on the basis of X-ray diffraction patterns, that pure glauconitic smectite contains at least 3 % K_2O. Glauconitic clay mineral lattices with less potassium may eventually exist, but they would display uncharacteristic X-ray diffraction patterns with very subdued reflections. We have measured up to 8·5 % potassium (K_2O) in grains of Cenomanian and Pliocene age (Table 1) and several other workers (e.g. Foster, 1969; Lamboy, 1974) have found exceptional values of close to 9 %. Small amounts of other cations are also found incorporated in glauconitic minerals.

Starting from the same chemical data different structural formulae of glauconitic minerals can be constructed depending on the assumption of cation sites as well as the calculation itself (Kelley, 1945). For example A1 and Fe may occupy octahedral and tetrahedral sites whereas Mg may be in octahedral or interlayer positions.

A simplified representative structural formula for glauconitic minerals is

$$(Si_{(4-x)}Al_x)_{\sim 4}(Fe^{+++},\ Al,\ Fe^{++},\ Mg)_{\sim 2}\,O_{10}(OH)_2K_{(x+y)}$$

whereby x varies from 0·2 to 0·6 and y ranges from

0·4 to 0·6. Heavy letters indicate the major cations in the tetrahedral layers.

The high proportion of potassium is related to the marine origin of glaucony. Iron contributes more than half of the octahedral charge. These are the salient chemical features of the glauconitic mineral family which distinguishes it from the other three layer lattice silicates as noticed by Kossovskaya & Drits (1970) and shown on Fig. 8. Takahashi & Yagi (1929) insisted on the presence of a high and constant proportion of iron in the first stage of glauconitization which was later confirmed by Ehlmann *et al.* (1963), Pratt (1963) and Seed (1968, p. 230).

There is no correlation between interlayer cations (mainly potassium) and iron in octahedral positions because incorporation of the two cations takes place by different processes and at different times during the glauconitization process. The fixation of iron precedes the incorporation of potassium (Foster, 1969; Velde & Odin, 1975; Birch, Willis & Rickard, 1976). The general aspect of non-oriented X-ray diffraction patterns is closely dependant on the potassium content (Fig. 7).

The amount of expandable layers (Velde & Odin, 1975), the density of the grains (Shutov *et al.*, 1970) the index of refraction (Cimbalnikova, 1970) and the ion exchange capacity (Cimbalnikova, 1971b) all vary with the amount of potassium. These properties characterize more or less precisely the stage of evolution of the glauconitic minerals and their position within the glauconitic family.

Summing up, the glauconitic minerals represent a *specific family* whose unique chemical and crystallographic characteristics include a minor substitution of tetrahedral silicon, a dominance of iron in the octahedral sites and a high proportion of interlayer potassium. A potassium-poor glauconitic smectite and a potassium-rich glauconitic mica are the two end members of the family. We interpret the lack of data points between *c.* 10–15 % total Fe_2O_3 shown on Fig. 8 as evidence that there is no chemically continuous series between these minerals and the illitic clay minerals. This characteristic of glauconitic minerals reflects a certain homogeneity of the environment of their formation.

The minerals inherited from the substrate

As mentioned above (p. 613) almost *any kind of mineral particle* can provide a substrate for authigenesis of glauconitic clay minerals. According to our observations it appears that kaolinite (1:1 layer lattice) is by far the most important initial material of Recent glauconitized faecal pellets. It is noteworthy that among the large number of grains from different ancient and Recent settings we came across one single example with a 2:1 (smectitic) clay mineral forming most of the substrate (Eocene glaucony of the Monts des Flandres, Odin, 1972b). This at least raises doubts as to the validity of the glauconitization model proposed by several authors since Burst (1958a). This involves as parent material a degraded 2:1 clay mineral lattice that is gradually transformed to glauconitic mica by substitution of iron for aluminium and fixation of potassium. The data presented in Fig. 8 do not support such an evolution; on the contrary they show a natural gap between the illitic and the glauconitic minerals. The basis of this smectitic-host model is either merely theoretical or rests on a rather few coincidental occurrences of glaucony with smectitic substrate.

The X-ray diffractograms of Recent grains in their initial stages of evolution show a gradual disappearance of the peaks of the substrate. This is the most obvious mineralogical feature related to the developing green colouration. The green authigenic minerals are poorly crystallized and do not give any characteristic peaks. Therefore the nature of the authigenic minerals often remains uncertain.

Although few smectitic initial substrates have been observed whose ions and layer silicate lattices may participate in the formation of the authigenic glauconitic mineral, their role has been grossly overemphasized by the existing theory which takes into account neither the mineralogical variability of the substrates nor the general lack of inherited clay minerals in authigenic glauconies.

Before discussing the distribution and mode of formation of glaucony it is useful briefly to outline below those ferriferous mineral parageneses which may occur together and which are closely related to the glauconitic paragenesis.

Some aspects of other green iron-silicate facies

Ochrous to green granules of quite similar aspect, and therefore visually often indistinguishable from glaucony, have been described especially by Von Gaertner & Schellmann (1965), Porrenga (1967b), Rohrlich, Price & Calvert (1969), Giresse (1969) and Hardjosoesastro (1971) from marine environments. X-ray diffraction analysis has revealed that the colour of these grains is caused by a 7 Å layer lattice silicate identified as berthierine. According to

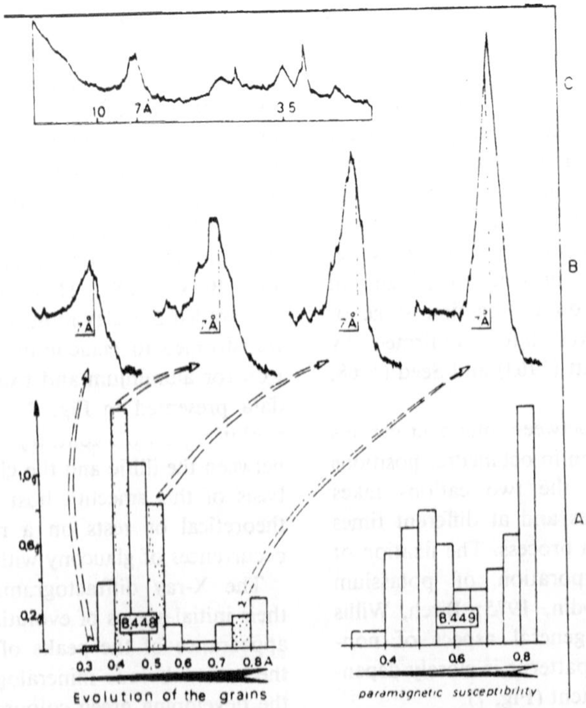

Fig. 9. Characteristic X-ray diffraction peaks with respect to the paramagnetic behaviour of berthierine. (A) Histogram classes represent abundances of grains separated at different currents; (B) (001) diffraction peak of four different fractions displaying an evolution from nascent berthierine showing only a peak of a kaolinite substrate to evolved berthierine showing a small peak of a poorly crystallized authigenic mineral; (C) complete diffractogram of an evolved berthierine. Note presence of inherited quartz.

Fig. 9. Caractéristiques diffractométriques et paramagnétiques de grains verts de type berthiérine les deux histogrammes de distribution du paramagnétisme sont établis en fonction de l'intensité (A) du courant d'alimentation du séparateur Frantz. Les grains sont d'autant plus verts et homogènes qu'ils sont plus attirables (intensité faible). L'analyse diffractométrique des 4 fractions repérées a donné les 4 raies de diffraction (001) reproduites. Malgré l'évolution évidente de la coloration, de la forme, de la densité et du paramagnétisme aucun nouveau pic ne semble créé sur le diffractogramme mais le pic à 7 Å évolue nettement: diminution d'intensité, léger décalage vers une épaisseur moyenne des feuillets de l'ordre de 7·3 Å. Au pic aigu de la kaolinite héritée dans les coprolithes se substitue un pic caractéristique de la berthiérine. Seule *l'étude dynamique de diverses fractions reliées à une évolution morphologique* a permis de mettre en évidence la nature de l'authigenèse étant donné les caractéristiques diffractométriques voisines de l'hérité et de l'authigène. Nous ne concluons pas pour cela à une filiation nécessaire: kaolinite-berthiérine car l'apparition du pic caractéristique de la berthiérine sédimentaire a été suivie de la même manière dans des grains micacés de Guyane et des remplissages de microfaune au large du Sénégal.

Brindley *et al.* (1968) the term berthierine for the 7 Å iron layer lattice silicate has historical priority over that of chamosite.

This iron-rich mineral has a kaolinite-type (1:1) structure (Fig. 9) with the iron occupying octahedral sites. The chemical data of some Recent berthierines (Table 2) indicate a low ferrous and high ferric iron content.

The mode of occurrence of berthierine (infillings of various microfossils, faecal pellets, pellicular alteration of carbonate bioclasts) as well as its mineralogical evolution shown in Fig. 9 suggest a mode of formation, however difficult to observe, similar to that of glaucony. Berthierine minerals are often poorly crystallized, have uncharacteristic X-ray patterns and a nearly constant basal spacing close to 7·5 Å. In Recent sediments they are often associated with detrital kaolinite.

In general, Recent berthierine occurs in tropical seas (Fig. 10B), which explains its association with kaolinite. All berthierine occurrences found so far are located near river mouths. The influence of temperature can be seen in the general lack of berthierine on shelves swept by cold currents or

Table 2. Chemical analyses of evolved berthierine from: (1) Gulf of Guinea (*grains partly oxidized), (2) shelf of Guyana. Amount of Fe^{3+} exceeds amount of $Fe^{2+}+Mg^{2+}$ indicating a mainly dioctahedral mineral. Note anomalously high content of K for a 1:1 clay mineral

Tableau 2. Analyses chimiques de fractions purifées de grains verts évolués de type berthiérine: 1—berthiérines du Golfe de Guinée partiellement oxydées: Fe_2O_3 excessif (*), 2—berthiérines du plateau continental guyanais. Noter l'abondance du fer ferrique un peu supérieure à celle du total 'fer ferreux plus magnésium', lui-même supérieur à la teneur en aluminium. Remarquer la teneur en potassium non négligeable de ces grains verts

	1		2			
	$448A_1$	1150bA	508102B	508169B	508172B	508175A
SiO_2	32·7	31·6	36·9	39·1	37·5	38·2
Al_2O_3	11·7	9·5	11·0	11·8	10·8	10·8
Fe_2O_3	30·2*	30·7*	17·9	18·3	19·5	18·8
FeO	4·1	4·7	6·5	5·7	6·1	6·0
TiO_2	0·4	0·4	0·4	0·5	0·5	0·5
CaO	1·2	1·1	0·6	0·6	0·5	0·7
MgO	4·1	7·0	11·0	8·3	8·9	10·6
Na_2O	0·2	0·2	0·2	0·2	0·2	0·3
K_2O	1·1	1·0	1·4	1·3	1·3	1·2
H_2O^-	3·8	3·2	3·0	3·5	3·8	3·1
H_2O^+	10·5	10·2	10·2	10·1	10·3	9·2
	100·0	99·6	99·1	99·4	99·4	99·4

areas of upwelling (Fig. 10B). It appears that low temperatures counterbalance the effects of the deltaic environment.

Celadonite described from altered volcanic rocks is yet another one of the greenish minerals that is strikingly similar to mature glauconitic minerals for which it was occasionally mistaken. However, the chemical composition is slightly different and celadonite is always associated with volcanic materials. We therefore strongly recommend the retention of the terms glaucony and glauconitic minerals for the marine green sedimentary facies and celadonite for the authigenic mineral formed in volcanic rocks. Consequently, the term glaucony keeps its genetic significance.

Other materials which render a green stain to a sediment, such as chlorite formed through alteration of biotite, have been mistaken for glaucony (Odin & Létolle, 1978).

OCCURRENCE OF GLAUCONY

Glaucony in ancient rocks

Glaucony occurs in sedimentary rocks of all ages. Nevertheless, it appears that during certain periods such as the Albian to Cenomanian, the early Tertiary and the Quaternary to Recent, glauconitization has been particularly common on a basin-wide or even world-wide scale. Glaucony is known from Precambrian sediments older than 10^9 years (Polevaya, Murina & Kazakov, 1961; Webb, McDougall & Cooper, 1963), but we are not aware of any glaucony older than 2×10^9 years.

The occurrences of 'continental glaucony' have been reviewed by Kossovskaya & Drits (1970) who concluded that these minerals are chemically not strictly identical to glauconitic minerals as defined above but are ferric illites with a lower iron content (Fig. 8). Hence it seems that granular glaucony, bearing the typical mineralogical and chemical attributes of glaucony, is restricted to the marine environment.

Environment of Recent glaucony

Green grains are common on modern continental shelves (for detailed discussion and references see Odin, 1975a). As shown in Fig. 10, Recent glaucony occurs on both sides of the Atlantic. Berthierine (which is also present in a Scottish loch, Rohrlich *et al.*, 1969) dominates on the shelf, whereas glaucony is forming in deeper water on the outer shelf and upper slope. Berthierine is abundant off river deltas in Senegal, the Gulf of Guinea (Giresse & Odin, 1973) and from the Orinoco delta to the Amazon mouth,

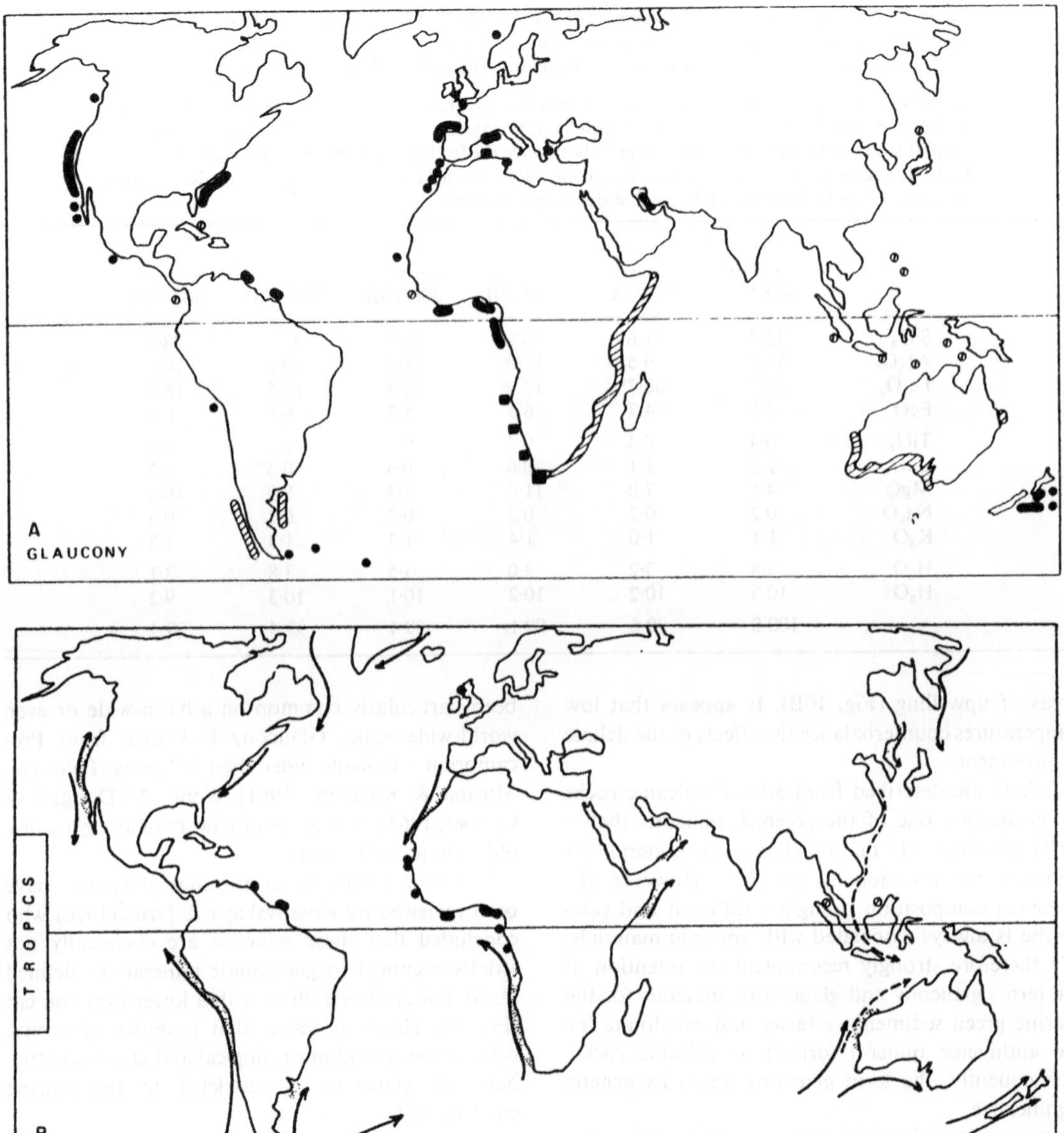

Fig. 10. Distribution of glaucony (A) and berthierine (B) on the present seafloor. In (A) solid round symbol: glaucony identified by X-ray diffraction; hatched areas are occurrences of unidentified green grains, squares are glaucony remanié. In (B) solid symbol: berthierine identified by X-ray diffraction. Note absence of berthierine in areas of upwelling (stippled) and cold currents (arrows). Hatched circles: berthierine identified by visual inspection only.

Fig. 10. Répartition comparée des berthiérines et glauconies récoltées sur les fonds actuels: carte A: On distingue les glauconies reconnues par analyse minéralogiques (taches) des grains verts non encore analysés (hachures) qui pourraient être de la berthiérine. Carte B (berthiérines): On a noté en noir la répartition des berthiérines pour lesquelles nous disposons de diffractogrammes; les courants longitudinaux froids principaux auxquels sont parfois liée les up-wellings (zones pointillées) sont indiqués pour montrer l'absence de berthiérine dans ces régions. Au contraire les berthiérines correspondent à des zones où des courants océaniques chauds sont connus. Les glauconies sont réparties à des latitudes diverses n'excluant que les plus hautes tandis que les berthiérines paraissent essentiellement intertropicales.

and glaucony is forming farther offshore in these areas.

Glaucony is also known from the Mediterranean Sea, the Persian Gulf and the Indian Ocean. Green grains which proved to be berthierine were found between New Guinea and the Indonesian Archipelago (Fig. 10B).

In the West Pacific, glaucony occurs near Japan and at water depths between 250 and 2000 m off Chatham Island, New Zealand. On the east side of the Pacific, numerous studies have dealt with the nature and origin of glaucony and so far no berthierine has been found. Our own geochronological results on samples off California (obtained through courtesy of G. H. Curtis) confirm the Recent age of the glauconitization process of various initial substrates. Farther to the south, off Mexico, we have discovered glaucony in Quaternary to Recent sediments (DSDP Leg 66) which are in water depths between 1200 and 1900 m.

In this paper, we distinguish, following Porrenga (1967a), between green grains of the glaucony and of the berthierine type; in fact with increased use of X-ray diffraction techniques an increasing number of 'glauconite' grains are identified as berthierine.

It is commonly assumed that abundant primary berthierine is restricted to tropical seas, shallow water depths and areas shorewards of glaucony occurrences. Figure 10(B) shows that berthierine is absent on the shelves crossed by cold currents. This confirms Porrenga's (1967b) observation that temperatures greater than about 20 °C are essential for berthierine formation.

Holocene glauconies locally contain the continuous spectrum of authigenic green minerals the basal spacing (001) of which ranges from 14 to 11 Å. Well-ordered glauconitic mica is only found among relict grains (pre-Pleistocene) which have never been buried.

The chemical variability of the different glauconitized substrate components of Recent sediments clearly shows that it is not a determinant factor in the geochemical reactions which cause glauconitization, and therefore *no particular starting material is required.*

As to the accurate time of genesis of glaucony, the occurrences described above present various possibilities. In the Mediterranean Sea, glauconitization occurred at different times during the Würm (Robert & Odin, 1975). Along the west African shelf, the pre-Holocene transgression, dated by C-14 at 18,000 years, caused an oxidation of the pre-existing glaucony grains which now occur at 100–120 m water depth in a 'red'-belt. We have recognized this red-belt from south Spain to the mouth of the Congo river. Glaucony found below this water depth is older than 18,000 years and glaucony above it is younger (Giresse, 1975; Odin & Giresse, 1976).

On the present-day seafloors, glaucony grains occur mainly in water depths between 100 and 500 m. They are *abundant*, comprising 10–90% of the sediment on the outermost shelf and upper slope between 200 and 300 m water depth. However, abundant glaucony has also been recorded at much greater depths (400–800 m) on some submarine highs (Bell & Goodell, 1967; Cullen, 1967).

At some localities, for example off the Congo river mouth, no evolution of grains has been observed in piston cores, but Martin (1973) mentions that only the uppermost few decimetres of the sediment of the Ivory Coast contain glauconitized grains.

Glaucony occurs from 50° south, on the shelf of the Southern Antilles, to 65° north, off the Norwegian coast. To us this indicates that no particular climatic conditions are required for glaucony formation. Under present conditions very little detrital material reaches the areas of active glaucony formation. However, glaucony has also been observed in areas usually known for their high sedimentation rates such as the last glacial (Würm) muds in the Golfe du Lion and Aegean Sea (Mediterranean Sea).

Because of eustatic sea-level changes which may sometimes enhance glauconitization, the grains may experience a complex life history. For example, after glauconitization at about 100 m water depth, a regression may occur and grains may become exposed to highly oxidizing and agitated waters (e.g. west African margin). This set of changing environmental conditions leaves a recognizable imprint on the grains. The view of numerous authors that a strongly oxidizing and agitated environment is essential for glauconitization is incorrect; it is *after* the formation of the glaucony grains that these conditions prevail.

THE ORIGIN OF GLAUCONY

Various theories have been proposed to explain glauconitization and the mineralogical as well as the morphological aspects of glaucony. They range from the early theory involving co-precipitation of Mg-, Fe-, Al-, and Si-gels which subsequently absorbed K (Murray & Renard, 1891) to the widely

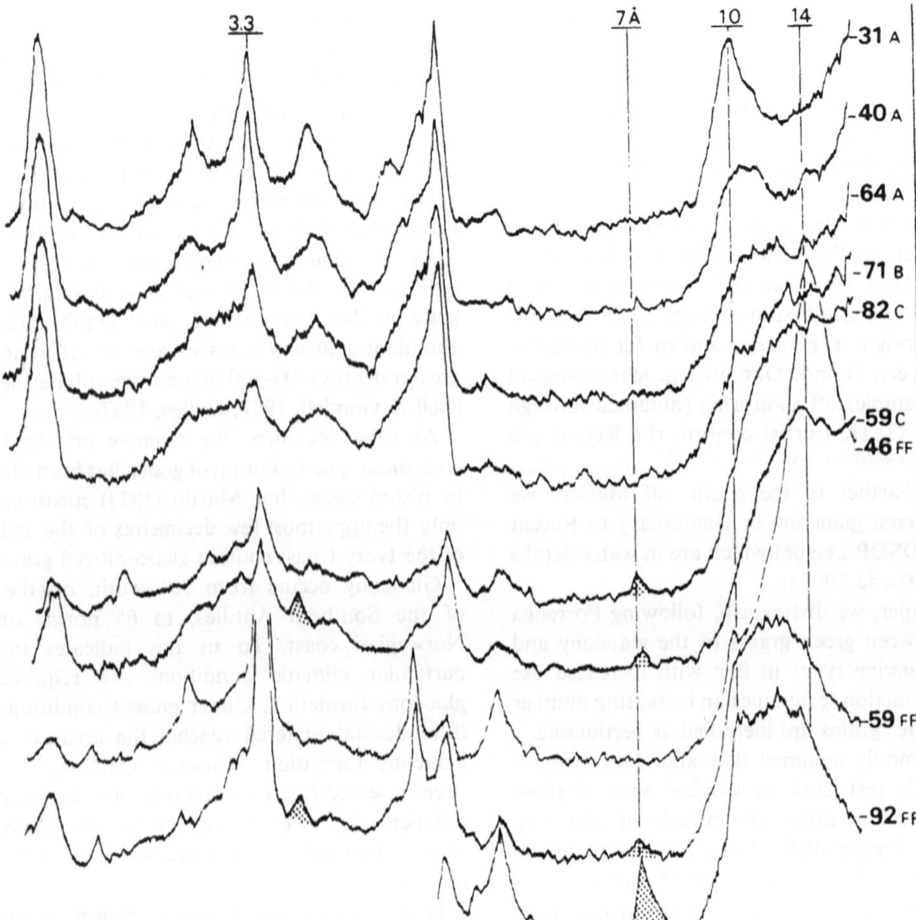

Fig. 11. X-ray diffractograms of various size fractions of borehole samples from the Monts des Flandres (Belgium). Numbers on right refer to depth in borehole in metres. FF grey, non-glauconitized matrix < 10 μm (oriented slides). Glaucony grains (powder mounts): (C) 50–100 μm, (B) 100–160 μm, (A) 160–500 μm. The least evolved (smallest) grains show peaks similar to the matrix, notably traces of kaolinite (stippled). Evolved grains contain only authigenic glauconitic minerals.

Fig. 11. Diffractogrammes de diverses fractions de sédiments de l'Yprésien des Monts des Flandres: les échantillons sont prélevés en sondage, la profondeur est donnée du côté droit. FF signifie fraction fine non glauconitisée, grise; préparation orientée. (C) Grains petits (50–100 μm) vert pâle préparation non orientée; (B) Grains moyens glauconitisés (100–160 μm). (A) Grains gros (160–500 μm) glauconitisés. Les grains les moins évolués, les plus petits, montrent des pics proches de ceux de la matrice notamment des traces de kaolinite. Les grains évolués ne sont plus constitués *que* de minéraux glauconitiques authigènes nettement plus fermés que la fraction fine du même échantillon. Il y a apparemment passage cristallographique continu depuis la matrice jusqu'aux grains très évolués. En réalité les minéraux authigènes sont chimiquement très différents des minéraux de la matrice dès leur apparition et il n'y a pas filiation mais coincidence de structure.

accepted, almost classic 'transformation of degraded layer lattice silicate theory' (Burst, 1958a,b). The 'layer lattice theory' will be discussed first and checked against the new results already presented.

The 'layer lattice theory'

Since Burst (1958a,b) explained the formation of glaucony by transformation of generally degraded layer lattice silicates and Hower (1961) proposed a precise mechanism for this process, most students of glaucony have accepted this theory and tried to find further supporting evidence. Essentially, Burst (1958a, b) had ascribed to glauconitization the transformation process which is generally accepted by clay mineralogists as the mechanism responsible for the evolution of inherited clays during diagenesis. Lucas (1962) has identified a similar process for clay minerals in saline deposits. In the latter case, according to Grim & Bradley (1955, cited in Millot, 1964, pp. 369–374), a *memory of the past structure* allows rebuilding of the old crystal lattice in the new clay minerals.

In Burst's view, glauconitization involves the more or less simultaneous incorporation of iron and potassium into the lattice of a clay mineral derived from the continent. The implications of this theory are therefore: (1) that the substrate of glauconitization is a high alumina smectite, an illite or a degraded mica, and (2) that the main process causing glauconitization is ionic exchange in the permanent lattice structure of detrital clay minerals. This theory therefore takes into account the X-ray diffraction patterns of green grains and explains the clearly recognizable evolution during glauconitization from open, expandable, more or less impure minerals to closed, non-expandable minerals.

Some authors note that this theory cannot possibly explain glauconitization of other than the 2:1 clay mineral substrates (micaceous materials), such as, for example, carbonate, quartz grains etc. (pp. 613–614) and therefore neomorphic replacement has been suggested as an alternative, accessory mechanism. The necessity of two hypotheses involving two completely different mechanisms which operate in an identical homogeneous environment (open sea-water) and convert various substrates to chemically homogeneous specific minerals, is at least logically not satisfying.

The Ypresian sandy clay series of the Belgium Basin provides an excellent example of the above-mentioned apparent transformation (Odin, 1972b). As shown in Fig. 11, there is apparently a continuum of the X-ray characteristics from the clay fraction to the most evolved grains. Very small (50 μm) grey-green and small (100 μm) green pellets contain smectites with a trace of kaolinite. Coarse pellets (160 μm) contain no kaolinite at all. Even coarser pellets (about 500 μm) consist of a mica-like mineral. In the coarser pellets a similar sequence can be observed from bottom to top in the borehole. These results show that the initial substrate was pelletal mud and that small pellets are always less different from the clay matrix than the coarser ones which are therefore more evolved.

In addition, chemical analyses do not support the postulated cationic exchanges of the 'layer lattice theory'. We have analysed two different samples of the matrix (< 20 μm) which was originally the initial substrate of the pellets. These samples contain 8·3 and 12·9 % Fe_2O_3 whereas of 20 samples of glaucony with various K contents (Table 1–6) the lowest value is > 20 % Fe_2O_3. Thus the green minerals, as soon as they are recognizable as such, are chemically different to the substrate. We are convinced that chemical composition intermediate between inherited and authigenic clay could only be obtained by mixing the two chemically distinct components.

The 'layer lattice theory' encounters other difficulties, enumerated below in attempting to explain several important, new observations.

(1) SEM studies by Odin (1972a) show that glauconitization of detrital micas (biotite and muscovite) takes place by authigenic crystal growth in the open pore space between mica sheets which act as substrates. As mica sheets remain unaltered, neither their crystal architecture nor their ions are used or are necessary.

(2) We observe that it is the non-micaceous, particularly carbonate, materials that provide most examples of a gradual evolutionary process; the old theory does not account for this.

(3) The fact that, given the same parent material, in some areas berthierine and in others glaucony is forming suggests that the general environment of formation is more important than the mineralogy of the parent material.

(4) In virtually all of the studied examples of glauconitic hardgrounds, glauconitization proceeded by neomorphic growth of green glauconitic minerals in the pores of limestones (Aubry & Odin, 1973).

(5) Most of the clay matrix of shales with glaucony shows no trace of glauconitization. Assuming that

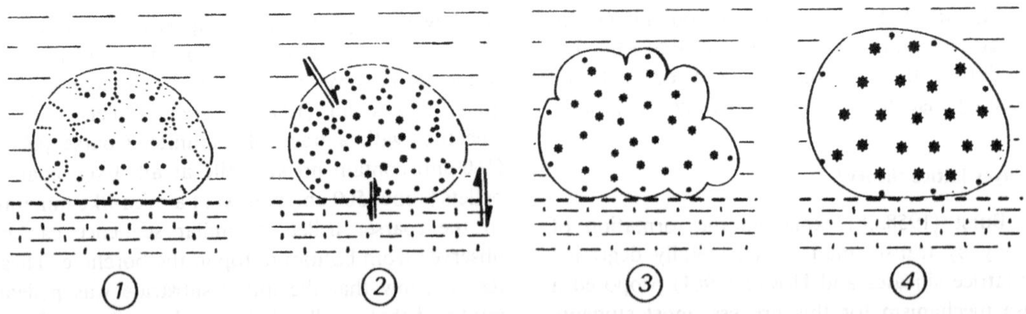

Fig. 12. Schematic representation of the evolution of a grain during glauconitization. Four stages from nascent (1) to highly-evolved (4) are selected from the continuum of changes observed in grains from the present shelf floor. Asterisks represent growth of glauconitic minerals. For discussion, see text.

Fig. 12. Schéma de l'évolution d'un grain au cours de sa glauconitisation: reconstitution synthétique d'après des observations dans divers gisements des fonds marins. 4 moments ont été choisis dans une suite continue de changements. Cette évolution peut être arrêtée et le stade atteint fossilisé, à n'importe quel moment par l'enfouissement du grain. L'évolution complète se déroule dans un milieu original schématiquement intermédiaire entre l'eau de mer renouvelée, au-dessus, et le sédiment confiné sous-jacent: cette 'interface' est en réalité épaisse de quelques centimètres, décimètres ou plus en fonction des possibilités d'échange réelles. (1) Le premier moment symbolise la phase altération du support, assez rapide, qui entraîne la porosité et la création du semiconfinement capital pour des premières synthèses. (2) La porosité étant réalisée les échanges ioniques ont lieu entre l'eau de mer l'eau interstitielle du sédiment et le grain qui prend une coloration verte et acquiert un paramagnétisme typique des minéraux glauconitiques engendrés (petites étoiles). Le grain peu évolué perd sa structure mais a encore la forme générale du support. (3) Les restes du support ont entièrement disparu, au centre du grain les croissances cristallines et recristallisations se développent entraînant un craquellement de la zone externe, de même que se craquelle la surface du pain qui lève en cuisant. Le grain a grossi. (4) Le grain est entièrement 'rempli' de minéraux verts évolués. Ceux-ci peuvent encore recristalliser sous l'influence du potassium de l'eau de mer disponible. Les craquelures et fissures de surface deviennent à leur tour un lieu de formation de minéraux glauconitiques moins évolués. Peut être à l'occasion d'un confinement d'ensemble de l'environnement du grain (enfouissement) une pellicule de minéraux glauconitiques tardifs se développe parfois arrondissant le grain.

matrix and pellets consisted of the same clay mineral assemblage, it becomes obvious that the micro-environment within the pellets is a much more important factor than the specific nature of the substrate. In fact most of the glauconitized clay coprolites we analysed had been formed from mainly kaolinitic clays instead of 2:1 clay minerals. Hower's (1961) priorities for glauconitization are: (*a*) a 2:1 structure, (*b*) the environment, and (*c*) the presence of Fe; we show in this paper that (*a*) is neither necessary nor especially favourable while (*b*) is the most important factor at the very local scale as well as at a general scale. Concerning (*c*), Fe is frequently not present in the substrate itself.

(6) If glauconitization did, in general, start on aluminium-rich clay minerals as stated by the 'layer lattice theory', minerals of composition intermediate between the illite family and the glaucony mineral family should not be missing (Fig. 8). We believe that the example of the Ypresian clays of Belgium described above is conclusive. In addition, we analysed the clay fraction and the green grains from Recent sediments of the Gulf of Guinea. The Fe_2O_3 concentrations of four samples of the clay matrix are 6·7, 8·3, 8·7 and 11·7 % whereas the lowest value measured on the glaucony pellets was over 20 % total Fe_2O_3. These data clearly demonstrate that there is a gap instead of a gradual passage between the inherited and degraded aluminium-rich clay minerals and the ferriferous glauconitic smectite which, to us, is yet another indication of its authigenic origin.

(7) We find it difficult to admit, as do Hower (1961) and Seed (1968), that the same layer lattice silicate structure allows *simultaneously and in the same environment* the departure of the aluminium ion from the octahedral layer while the trivalent iron ions enter it. These ions are geochemically alike and occur in comparable concentrations in sea-water and interstitial water.

In summary the classic 'layer lattice theory' and the various attempts to give it a sound base by suggesting precise mechanisms for the process encounter numerous difficulties in explaining the observations. If the diagenetic change from a degraded clay mineral to glaucony should occur at all, it would, according to our opinion, represent a specific case of local importance compared with the large number of examples demonstrating authigenic growth of glauconitic minerals.

The mechanism of formation of glaucony

The 'verdissement' of the grains

The formation of glaucony, as we envisage it to date, is best illustrated by the granular facies. First of all, if glaucony so commonly occurs in its granular facies it is because the initial substrate *was itself granular*. As reported in the previous sections, initial granular substrates are always highly porous either because of primary intraparticle porosity (boring or solution of biogenic particles) or the presence of fractures and fissures (Fig. 12, stage 1). It is in these pores of 5–10 μm size, which may extend across the entire grain, that growth of the first green glauconitic crystals is inferred. They are iron and potassium rich glauconitic smectites. The first blades grow on edge, perpendicular to the surface of the substrate, and develop into a boxwork-like fabric by coalescence (Fig. 4). Finally the entire pore space is filled (Fig. 12, stage 2), at which stage the grains already appear green.

The initial substrate disappears, often by dissolution. While the last smectites are growing into the remaining pore space of the substrate grain, the earlier smectites are being recrystallized. Chemical analyses of this stage already reveal a high iron content whereas the potassium content just begins to increase.

The substrate will persist for different durations depending on its ease of alteration in the marine environment (Lamboy, 1976). Initial carbonate substrates alter more quickly and are therefore more favourable to glauconitization than are, for example, silicate substrates which are resistant to alteration. Therefore glauconitization of silicate substrates will be slow and incomplete.

The disappearance of the substrate may be observed by SEM and X-ray diffraction. Nevertheless pieces of the substrate may still be present even if they cannot be detected by SEM and X-ray diffraction. We have measured the argon content of faecal pellets from the Gulf of Guinea at different stages of evolution (Odin *et al.*, 1979). There is no visible trace of the substrate remaining when pellets contain > 4–5 % K_2O. However, in spite of this, there is still nearly 10 % of the inherited argon trapped in the pellet with K_2O as high as 6·6 % (Fig. 13). To

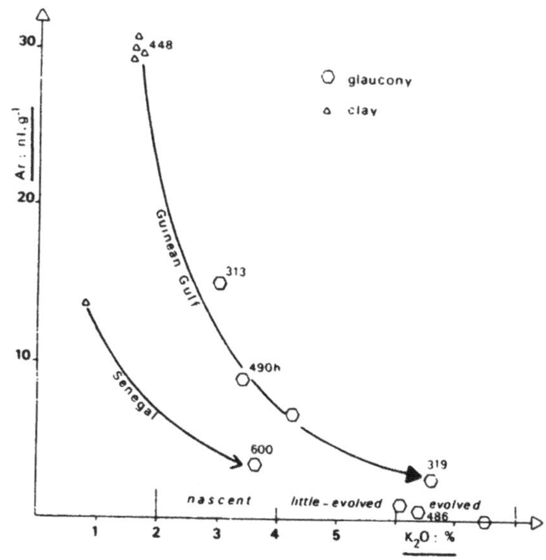

Fig. 13. Relationship between inherited argon and the stage of evolution of glaucony. For chemical analyses of these samples see Table 1.

Fig. 13. Teneur en argon radiogénique hérité dans des glauconies en voie de formation: les analyses isotopiques montrent qu'il reste de l'argon hérité du support (coprolithes kaoliniques) bien longtemps après que toute trace de ce support soit disparue sur les diffractogrammes.

us this shows that, even in quite well glauconitized substrates, some pieces of the initial minerals may have retained their argon.

When, according to the X-ray diffraction pattern, the parent material has become a minor admixture in the grain, important textural changes are observed in the grain under the SEM: (1) progressive loss of the primary texture; (2) growth of larger and better shaped crystallites which form lamellae or lepispheres resembling rosettes; and finally (3) deformation of the grain which takes on a bulbous, cracked habit. This corresponds to stage 3 of Fig. 12.

The deformation of the initial substrate grain results from the displacive growth of glauconitic crystals. In its last stage of evolution (stage 4, Fig. 12), the grain is generally embedded beneath an outer crust of little-evolved glauconitic minerals. At the end of this process a smooth crust is formed over the initially irregular surface, increasing the roundness of the grains (Lamboy & Odin, 1975). This sequence of events was observed on grains collected from the modern shelf floor; these grains have been *in permanent contact with the marine environment*.

The duration of each of the phases was plotted

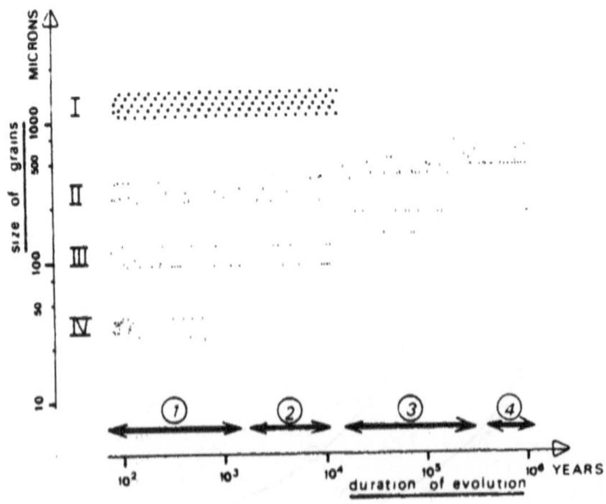

Fig. 14. Statistical evolution of different size classes of a population of substrate grains during glauconitization. (I) Large substrate particles whose peripheral parts only are glauconitized; (II) substrate grains whose adequate size permits complete glauconitization (note the two growth episodes and possible grain size decrease); (III) small substrate grains which remain in the little-evolved stage; (IV) substrate grains that are too small to provide shelter for growth of authigenic glauconitic minerals. The sizes indicated were measured on coprolites from many locations on the shelves of the Atlantic and would vary for different substrates (see text). Note that the four stages from 1 (nascent) to 4 (highly-evolved) are not of same duration. The time-scale was obtained from palaeontological information.

Fig. 14. Evolution statistique de la taille d'un population de supports au cours de la glauconitisation: (I) Supports gros dont le centre trop confiné ne permet pas le verdissement. (II) Supports de taille adéquate subissant une évolution très longue. (III) Supports petits insuffisamment confinés, l'évolution s'arrête à un stade peu-évolué. (IV) Supports délimitant un milieu trop petit, trop ouvert, s'altérant sans abriter de minéralogenèse. Les tailles indiquées sont celles observées dans les coprolithes récoltés en divers fonds de l'Océan Atlantique. Elles varient d'un type de support à l'autre; plus grandes pour les débris carbonatés: milieu plus poreux (plus ouvert); plus petite pour les loges de tests de microfaune: milieu plus confiné. Les 4 phases: (1) naissante; (2) peu-évoluée; (3) évoluée; (4) très-évoluée, sont notées en abscisse afin d'en souligner la durée très différente estimée d'après l'observation de diverses glauconies récentes complétées par quelques glauconies enfouies au Plio-Quaternaire.

on Fig. 14 using data compiled by Giresse, Lamboy & Odin 1980). Two growth episodes can be recognized in faecal pellets of the size range II, which appears to be the most favourable. The first and more important is the formation by displacive crystal growth of larger, bulbous and fissured grains. Sometimes the fissures penetrate deep enough to cause complete fracturing of the grains leading to grain diminution. The second growth episode corresponds to the formation of the outer crust which measures some tens of microns in thickness (Fig. 14).

If a small coprolite is being glauconitized, the evolution of glauconitic minerals is arrested in an early stage (Fig. 14, case III), whereas those formed in larger coprolites of the same sediment reach a more advanced stage. Because argillaceous faecal pellets of at least 300 μm lead to the most evolved glaucony, we concluded that this size is more favourable for glauconitization than smaller pellets. The optimal grain size is slightly larger for carbonate particles, smaller for mica and smallest for quartz grains.

Figure 15 shows the evolutionary stages of an echinoid spine and of a test of *Orbulina* sp. It is quite clear that different substrate materials all end up more or less alike as fissured glaucony grains. It is generally difficult to determine the precursor of these grains, but Giresse *et al.* (1980) give some criteria that at least permit the determination of from which group of substrates a glaucony grain was derived. Briefly, glaucony infilling chambered microfossil tests will end up as small grains > 100 μm in size; micas will form small elongated grains. Although small faecal pellets may look similar, the data from Recent sediments show that they give rise to a variable group of grains generally ranging in size from 2 to 500 μm. Carbonate particles form the largest glaucony grains. For example, flat-shaped shell debris up to several millimetres in size retains its flat shape for a long time during glauconitization.

According to SEM observations, the evolution from nascent (smectitic) minerals to highly-evolved (micaceous) minerals takes place by a recrystallization process. A thermodynamic study of Odin, Velde & Bonhomme (1977) also shows that recrystallization occurred under experimental conditions.

The influence of confinement

We have suggested, in the previous section, a new model for the formation of glaucony. Some observations are now presented which allow a deduction of the major geochemical factors that control the mechanisms of glauconitization with regard to the proposed model in the granular and film facies. The term *confinement* refers to the partial chemical isolation from sea-water of pore fluids within various substrates.

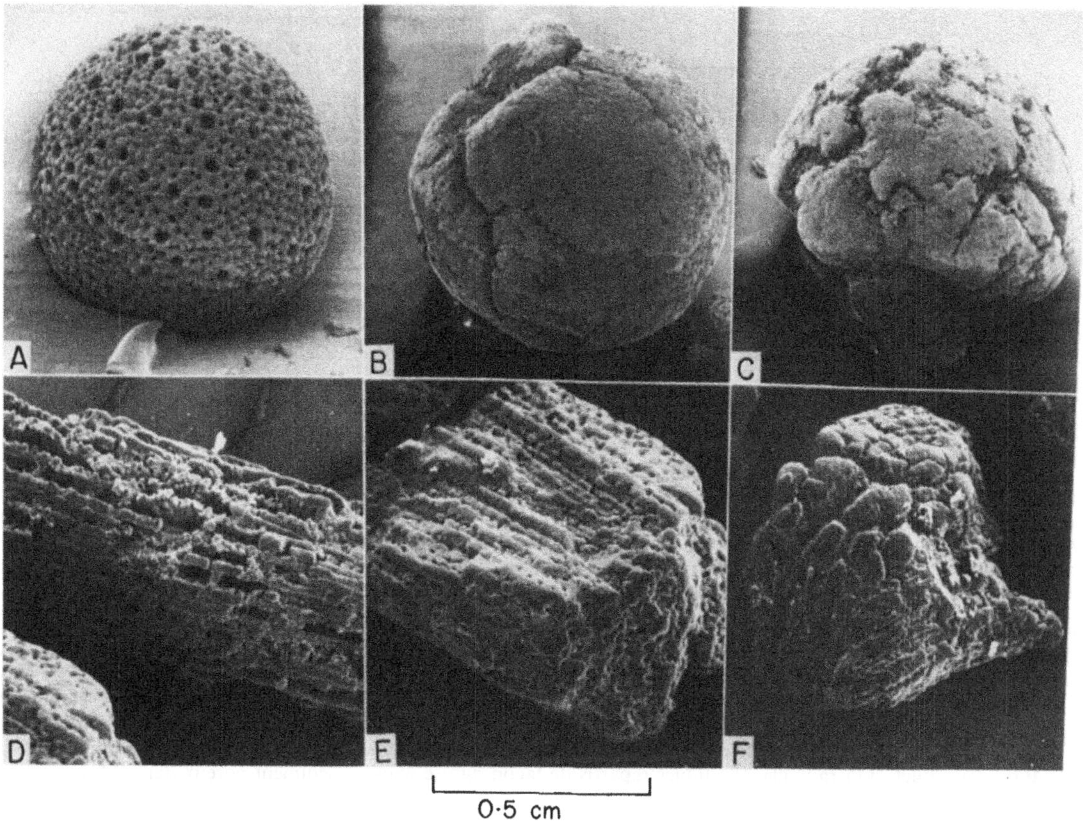

Fig. 15. SEM photomicrographs of Orbulina tests (A, B, C) and echinoderm fragments (D, E, F) showing progressive deformation during glauconitization. (A) Orbulina with pale green interior; (B) green mould with development of cracks; (C) dark green mould with large cracks leading eventually to splitting. Sequence (D) to (F) shows loss of original texture of echinoderm stereom during replacement by authigenic glauconitic crystals and development of a typical cracked grain.

Fig. 15. Exemple de déformation du support au cours de la glauconitisation. A: Orbuline, intérieur vert pâle, B: remplissage vert clair, C: remplissage vert foncé. En même temps que le grain apparaît plus foncé à la loupe binoculaire, les fissures se développent entraînant parfois la cassure du grain à droite. D–F: débris d'Echinoderme: spicule à 3 stades d'évolution, de gauche à droite le grain est de plus en plus foncé; en même temps la texture disparaît peu à peu pendant que croissent les minéraux glauconitiques entraînant des craquelures en surface, photo de droite. L'aspect original bien reconnaissable au début disparaît peu à peu.

The association of authigenic glauconitic smectite with microfossil tests, pores and fissures in particles and burrows and borings in hardgrounds was observed by other authors and taken as an indication that glauconitization requires a certain *confinement*. Confinement produces a micro-environment which is different from the surrounding marine environment. This explains the observations made in several cases studies, e.g. the Ypresian of Belgium, the Miocene of the Aquitaine Basin in France (Odin, Hunziker & Lorenz, 1975) that small grains > 100 μm are less evolved than larger grains 200–400 μm in size. The interiors of small grains are relatively unconfined leading to excessive exchange with ambient sea-water and no glauconitization takes place (case I, Fig. 16). Grains which are larger provide more confined microenvironments suitable for glauconitization in their centres (case II, Fig. 16). It is remarkable that, statistically, glaucony grains are essentially found in the 100–1000 μm size range.

Furthermore, the more pronounced formation of glauconitic minerals in the centre of the grain rather than on its periphery, which leads to large and cracked grains described in the previous section, is another indication of the importance of the confined environment. Within the confinement of the grain centre a micro-environment exists whose chemical conditions are more favourable to crystal growth than

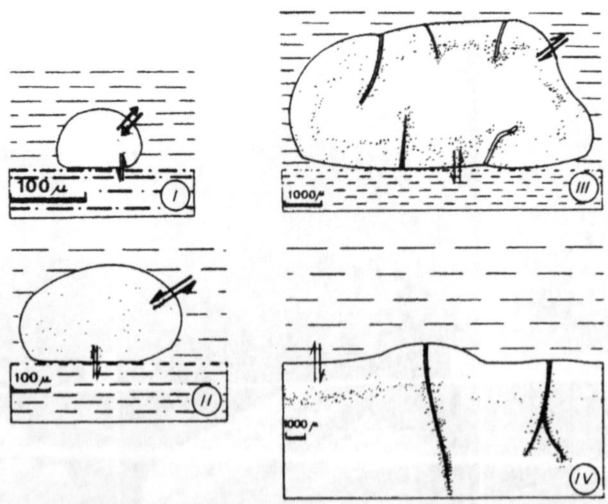

Fig. 16. Schematic representation of the role of confinement in glauconitization of various substrates. There are three types of confinement related to the size of the substrate which control the degree of glauconitization. Light stipple—unconfined environment leading to incomplete glauconitization; dense stipple—semi-confined environment promoting complete glauconitization; blank—lack of glauconitization due to insufficient exchange with ambient water.

Fig. 16. L'équivalence des différents types de glauconitisation en terme de semi-confinement. (I, II) correspondent à la glauconie en grains. (III) à la glauconie pelliculaire sur galet, (IV) à la glauconie pelliculaire sur surface durcie Les tailles sont indicatrices et seront plus faibies pour des supports peu altérables: quartz et plus fortes pour les supports aisément altérés: carbonate. Le facteur commun à ces supports est un milieu intermédiaire entre sédiment confiné et mer ouverte; il est poreux et permet des échanges lents entre le grain et l'extérieur tout en protégeant les croissances cristallines de l'effet diluant de l'eau de mer. Le cas IV peut être épaissi jusqu'à quelques décimètres pour interpréter la formation d'éventuels minéraux glauconitiques répartis de facon diffuse dans un sédiment poreux dans son ensemble.

those of the environment in the periphery of the grain. This differential growth in different parts of the grains causes the cracks. However, if the degree of confinement is such as to inhibit a certain minimum exchange of ions, e.g. due to low porosity and permeability it may become an obstacle to glauconitization (Fig. 16, case III, granular substrate or case IV, hardground).

Various observations that are unequivocal proof of the reality of ionic exchange are briefly recalled: (1) Since the nature of the substrate grain has no influence on the kind of layer lattice silicate formed, yet this product is positively correlated with the general environmental conditions (berthierine in deltaic and glauconitic minerals in open marine environments), an ionic exchange between the general environment and the interstitial waters of the substrate grain must take place. In the granular facies, ions are exchanged between the interstitial water of the particles and sea-water and probably also with the interstitial water of the sediment. (2) The fact that argillaceous pellets in shales are glauconitized but not the clay matrix is thought to reflect insufficient ionic exchanges between the underlying sediment and sea-water, i.e. too confined conditions compared with those of the pellets on the clayey seafloor. On the other hand larger grains (2–3 mm) and pebbles are glauconitized on the surface only. The interiors of the grains remain more or less unaltered because ionic exchange does not reach deep enough. Ionic exchange is often facilitated in carbonate grains by Thallophyte borings (Fig. 17). Moreover they dissolve more easily which is another reason for suggesting that the largest glaucony grains are derived from carbonate particles.

From the discussion above, it follows that *a particular semi-confined environment exists in granular substrates* because of their porosity. The large surface area determined by the pores is a catalyst accelerating the reactions and the growth of glauconitic minerals. This peculiar micro-environment, which is different from that of sea-water above and from the interstitial environment of the underlying sediments, is a common characteristic of all the known examples of Recent glauconitization.

Within mica flakes the porosity and the semi-confined micro-environment are found between individual sheets which split along the plane of

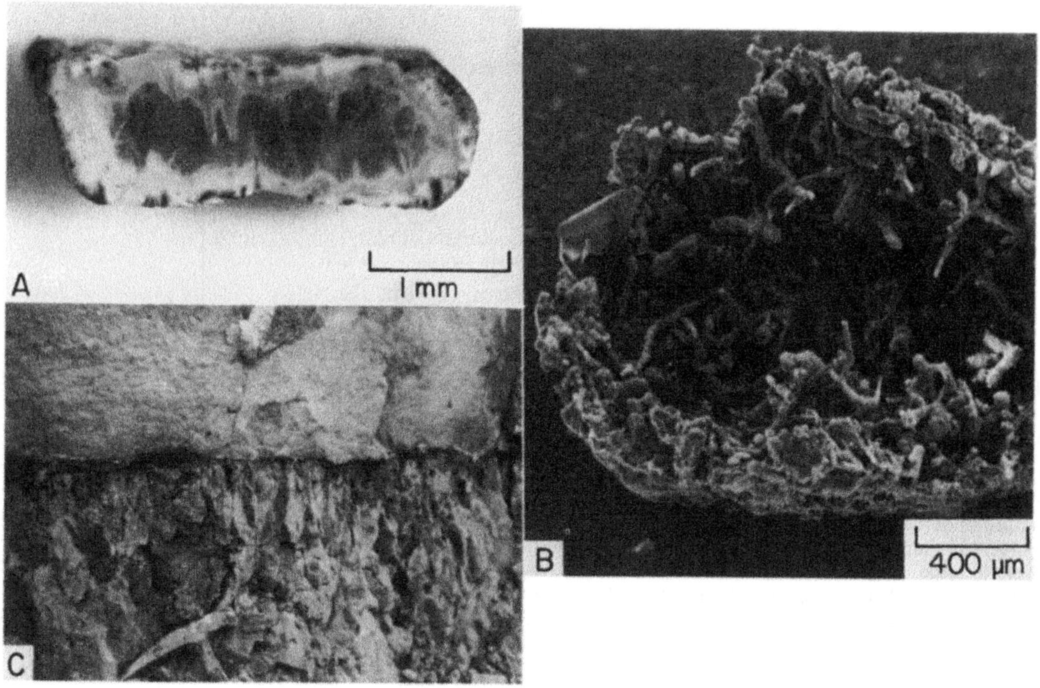

Fig. 17. Glauconitization of large particles and hardgrounds. (A) Broken carbonate particle with unaltered centre, altered periphery penetrated by glauconitized thallophyte borings shown in (B). (B) SEM photomicrograph of acid-treated carbonate grain with glaucony envelope typical of the film facies (sample by M. Lamboy). (C) Film facies on hardground of Campanian chalk, Limburg. See hammer head for scale.

Fig. 17. Exemples de glauconitisation de supports très grands: (A) Une particule carbonatée de plusieurs millimètres de long montre un centre non altéré gris, une périphérie en voie d'altération, blanche, pénétrée par des perforations de Thallophyte, noires, qui ont été remplies de glauconie. (B) Une particule équivalente a été traitée à l'acide chlorohydrique par M. Lamboy (microscopie électronique à la balayage) noter l'abondance de filaments, verts, glauconieux, à la périphérie; leur absence au centre du grain qui était peu altéré; ces deux échantillons du plateau N.O. espagnol. (C) Glauconie sous son faciès pelliculaire le plus développé: le verdissement d'une surface durcie au sommet du Campanien du Limbourg, le marteau donne l'échelle.

primary cleavage, parallel to (001). Here the glauconitic minerals start growing and further displace the sheets until finally an accordion-like pellet is fermed.

The semi-confined environment within microfossil tests is created by the wall of the test which forms a semi-permeable barrier for migrating ions. It seems that, given the same sediment (e.g. Moroccan and Senegalese shelves), this barrier is more effective in tests than in faecal pellets whose optimal size is generally larger. In the case of large clasts of very low primary porosity, the secondary porosity formed by dissolution and various kinds of boring organisms permits glauconitization of the periphery of the particle (Lamboy, 1968). Chert pebbles are hardly altered on the seafloor and develop only a glauconitic film several hundreds of microns thick, whereas pebbles and blocks of chalk often have a glauconitized outer zone thicker than 1 cm. Hardgrounds also develop an upper more porous zone with a semi-confined environment in which glauconitic minerals form wherever there is a sufficiently long period of non-sedimentation. Thus glauconitic horizons mark the different periods of non-deposition within a sequence as described for example by Juignet (1974) from the chalk of the Paris Basin.

Thus the semi-confined environment of the pores acts as a passageway for the ions derived from the sea-water, from the initial substrate of the grain and from the underlying sediments. These three

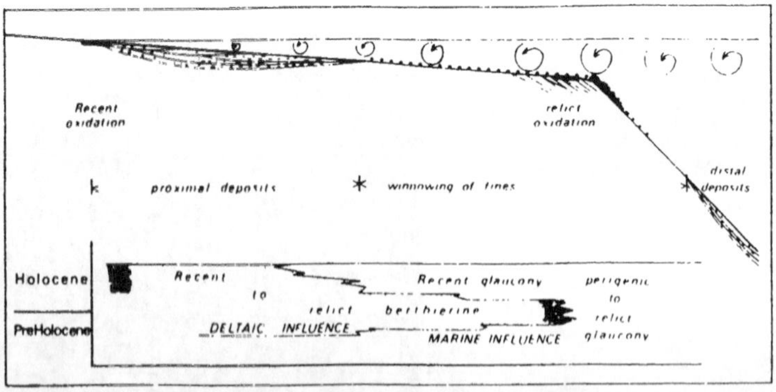

Fig. 18. Schematic diagram showing the environments of formation of glaucony and berthierine as a function of detrital influx and energy levels on a passive continental margin. Cross-section below shows facies relationships of glaucony and berthierine as a result of the Holocene transgression.

Fig. 18. Schéma synthétique de la situation du phénomène de glauconitisation sur une marge passive: la zone de glauconitisation se situe au-delà des dépôts fins proximaux dans une zone d'absence de depôt fins entretenue en partie par l'action profonde de la houle. Celle-ci, et d'autres courants, permettent de renouveler les ions à l'interface eau-sédiment où se situent les supports favorables. La zone d'accumulation distale des grains verts est due soit à un lent déplacement vers le large des particules qui s'accumulent lorsqu'ils arrivent hors de portée des courants, soit à une zone spécialement favorable à une époque de régression (soit aux deux raisons). On a noté un niveau d'oxydation relique qui représente ce niveau de régression. La glauconie située au-dessus de ce niveau d'altération est plus récente que la régression.

sources combine to supply the ions required for the geochemical reactions of glauconitization.

Environment of glauconitization

Depth of formation

At the metre or kilometre scale the specific environment which is favourable for the formation of glauconitic minerals can be ascertained from occurrences on the present continental shelves, provided that the relict grains originating from previous environments are distinguished. Glaucony is not a reliable depth indicator. It is present today on shelves between 60 and 500 m. At greater depths, up to 800 m, abundant glauconitic grains have been recovered from submarine highs which may have been tectonically subsiding and from active continental margins. At shallow depths the turbulent, and therefore well-oxygenated, waters are unfavourable to glauconitization. It appears that from 50 to 500 m water depth, a layer of granular or film glaucony is able to develop.

This limited zone of formation of glaucony on passive margins can be tentatively explained by the balance between detrital influx and winnowing by marine currents as shown on Fig. 18. In the zone close to the shore the detrital influx exceeds erosion. Despite the presence of suitable substrates (e.g. faecal pellets) glaucony cannot form because of relatively high net accumulation. Beyond about 50 m water depth, detrital influx is less and, in addition, winnowing causes a continual redistribution of sediment beyond the shelf edge. The progradational features recognizable on seismic profiles across the outer shelf confirm the operation of this process. It is in this zone near the shelf edge that the substrates are exposed for sufficiently long times to be glauconitized. At greater depths the energy is less, the winnowed sediment accumulates and suitable substrates are not glauconitized.

The local geochemical conditions are those of the sediment/sea-water interface beyond the zone of fluvial influence. The glauconitic facies is therefore characteristic of the *open-marine environment*. This environment, which is moderately alkaline (pH 7–8), favours the formation of 2:1 (T.O.T.) clay minerals in general. The Eh is not particularly oxidising, as pointed out in various publications, thereby precluding the degradation of glauconitic grains that occurred on numerous occasions in the past after falls in sea-level (regressions).

As shown above, the rounded aspect of the grains is no evidence for a highly turbulent environment. However, this does not exclude the possibility of bottom currents which cause moderate transport of the grains and bring new ions into contact with the green particles, facilitating ionic exchanges between them and sea-water. The existence of such bottom

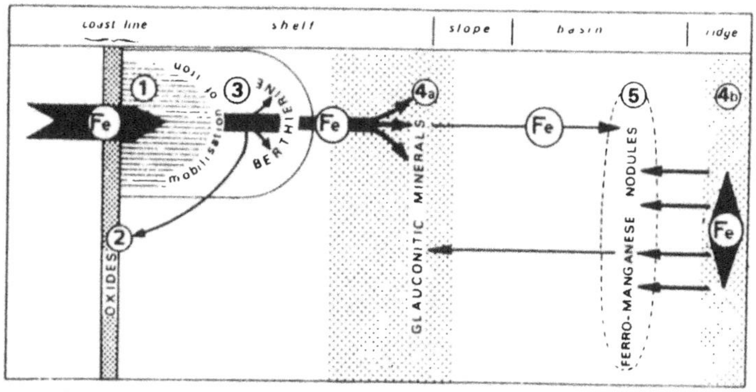

Fig. 19. Position of green sheet silicates in the geochemical path of iron in the sea. Zone 1, accumulation of detrital iron; zone 2, chemical precipitation of iron and oxidation of detrital iron minerals; zone 3, authigenic growth of berthierine; zones 4a, b, authigenic growth of glauconitic minerals; zone 5, growth of ferromanganese nodules. Modified from Odin (1975b).

Fig. 19. Place des phyllites vertes dans l'itinéraire géochimique du fer: le fer, élément cardinal des authigenèses silicatées marines, provient en dernier ressort du continent mais aussi des couches internes. Cet élément est géochimiquement actif dans des environnements variés: très oxydant et côtier (2), subdeltaique non oxydant et acide (3), marin ouvert non oxydant et basique (4 a et b); marin probablement confiné (5). Ce schéma est très souvent déformé par un courant parallèle à la côte qui étale ce spectre d'un côté de l'embouchure.

currents on the shelves and most other sea floors is well-documented by modern oceanographic work.

A marine transgression provides favourable conditions for glauconitization on a wide scale. Granular detritus, formed in abundance near-shore, is submerged to greater depths and thus into less turbulent water. At the same time the zone of accumulation of clastic detritus is shifted landwards, leaving the earlier formed grains in a zone of no deposition, exposed for sufficiently long periods for glauconitization to take place. The fact that initial substrates for glaucony grains are initially formed near shore, such as in the Lutetian series of the Paris Basin (Odin, 1969), does not imply that glauconitization took place in this environment. It was achieved *later* when the grains were in at least 50 m water depth. A transgression is only one of several possible situations favouring glauconitization. There are sources of abundant substrates other than submergence of a shoreline and other causes of non-deposition.

The recognition of the kind of initial substrate may be used to estimate the depth difference between its site of deposition and its site of glauconitization. For example, a population of glaucony grains derived from shell debris implies a large change of water depth whereas a glaucony infilling of planktonic tests implies none.

Glauconitization in the geochemical path of iron in the sea

Iron is the crucial element in the glauconitization process. Figure 19 represents an attempt to locate the formation of glauconitic minerals with regard to this element according to our observations of the ferriferous minerals in present-day seas. Iron is supplied to the sea by streams as detrital iron and by exhalations from volcanoes and rifts as true juvenile iron. Five zones of iron activity are distinguished on Fig. 19.

Zone 1 is the area of passive iron, characterized by rapid deposition of particulate and colloidal iron by gravity settling flocculation, precipitation or filter-feeding organisms (Glangeaud, 1941; Pryor, 1975). An important fraction of the continent-derived iron is thus largely immobilized near river outflows and because of high accumulation rates has no chance to react with sea-water. However, in the presence of organic matter or hydrogen sulphide in the sediment, iron is reduced and becomes soluble as Fe^{2+}. This explains the much greater concentrations of iron in the interstitial waters of marine sediments than in the supernatant sea-water (Goldhaber & Kaplan, 1974). The iron is either precipitated in the sediment, e.g. as pyrite or migrates into the sea-water.

Zone 2 appears to be restricted to warm climates. In fact, on the shores and the nearshore seafloor of

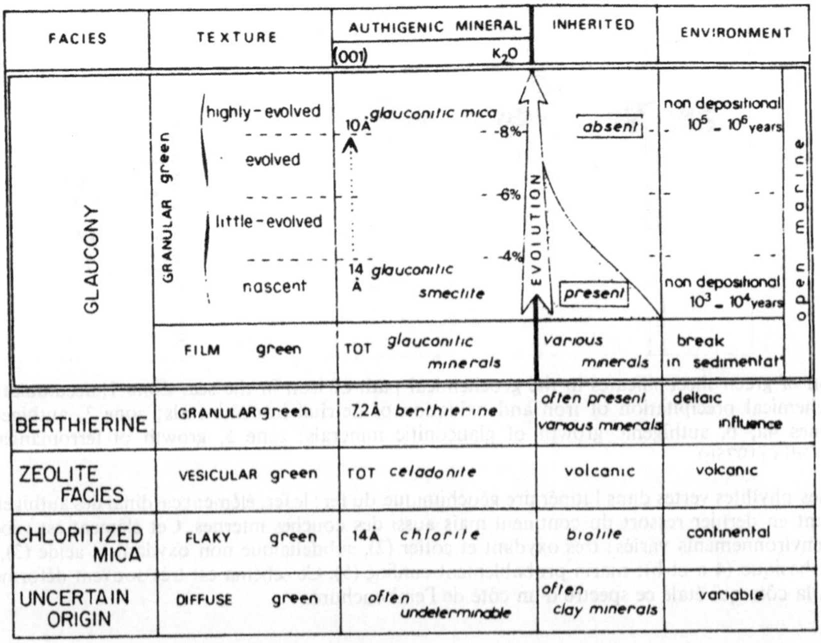

Fig. 20. Genetic classification of glaucony and related facies.

Fig. 20. Classificatson génétique des glauconies et faciès verts parfois confondus. Dans les glauconies vraies: faciès marin à authigenèses T.O.T. le plus souvent en grains, on doit distinguer fondamentalement 2 évolutions opposées celle du support qui s'altère, celle des minéraux authigènes que l'on caractérise soit par leur teneur en potassium, soit par l'épaisseur moyenne de leurs feuillets. Il n'y a pas de lien chimique ni minéralogique entre support et minéraux authigènes à aucun moment de ces évolutions.

the tropics, traces of chemical or biochemical iron precipitates have been observed in less than 10 m water depth. In addition, quartz, oolites and pellet coated by iron oxides, hydroxides or carbonates are observed (Giresse, 1969; Bouysse, Kudrass & Le Lann, 1977). On a global scale these precipitates and oxidized particles are today of minor importance, but this might not always have been the case during the geological past (Borchert, 1965).

Zone 3 is characterized by the presence of authigenic, poorly crystallized, green berthierine grains The formation of berthierine occurs within the influence of rivers and seems to be restricted today to the tropical seas. During burial diagenesis the 7 Å berthierine evolves rather quickly into a 14 Å chlorite.

Zone 4 consists of two sub-zones and covers a much wider area than zones 1–3. Its primary geochemical feature is the formation of glauconitic minerals. One sub-zone encompasses the outer continental shelf and upper slope which are still reached by continental iron. The other sub-zone of glaucony formation comprises some oceanic highs and mid-ocean ridges and rises which receive essentially juvenile iron. The glauconitization in active margin areas remains to be documented.

Finally *zone 5* covers the vast expanses of the deep-sea floors where iron is found incorporated into ferromanganese nodules and encrustations and also as iron smectites. Most of this iron is derived from juvenile sources.

Obviously iron is present in different forms in each of the five zones. When dealing with ancient formations this information provides a rough indication as to the palaeogeographic framework within which an iron-bearing marine sediment was deposited.

CONCLUSIONS

Glaucony is a green marine facies which develops today on continental margins and on oceanic highs. Its highly variable aspect is the result of replacement of mineralogically different initial substrates by authigenic minerals. Within the shelter of the initial substrate a semi-confined micro-environment (different from that of the supernatant sea-water and of the underlying sediment) is established. During the process of glauconitization the initial substrate component tends to disappear with the authigenic component taking its place. The specific authigenic minerals form the series of the *glauconitic mineral family* which is crystallographically comparable to the smectite-illite family from which it differs mainly by the very high amount of iron over aluminium in the octahedral layers. Glaucony initially forms by *de novo* crystal growth of *glauconitic smectite* within the pores of chemically different substrates. As glauconitization proceeds this glauconitic smectite evolves into a *glauconitic mica* end product by a recrystallization process.

The widespread glauconitic facies represents the more or less advanced stage of evolution of sea bottoms covered with suitable substrates during periods of no active deposition. The break of sedimentation and the time required for the formation of the initial glauconitic smectite in altered materials is estimated to be 10^3–10^4 years, and about 10^5–10^6 years for highly-evolved glauconitic mica.

The glauconitic minerals occur in film and granular facies. The latter can be further subdivided (into four different stages) according to the stage of evolution of the authigenic glauconitic minerals as shown in Fig. 20. Other green mineral facies, which are sometimes confused with the glauconitic facies, have been added to this scheme. The evolution from nascent to highly evolved glaucony seems to be achieved by a recrystallization process and is related to a progressive equilibrium with bottom sea-water. A clear distinction must be made between the alteration of the substrate and of the authigenic mineral phase. This duality of significance of the components of glaucony was not recognized in the earlier classifications proposed by Burst (1958a, b), Hower (1961) and Bentor & Kastner (1965). Consequently, the substrate was mistaken by previous authors as impurities and the authigenic nascent smectite as an inherited degraded component. According to Burst (1958b) and Hower (1961) glauconitization proceeds by the slow substitution of iron displacing aluminium with a synchronous increase in potassium. We have shown that this is erroneous.

We suggest the following criteria for glauconitization: (1) the necessity of a semi-confined micro-environment within a physical substrate, (2) the chemical suitability of the open sea-water and absence of deposition which permit iron and potassium but also Si, Al, Mg to be incorporated into the geochemical reaction.

While we believe we have clarified the problem of the origin of glaucony, it is obviously far from being fully understood and should stimulate further research in the thermodynamical and physico-chemical aspects of this marine authigenesis.

ACKNOWLEDGMENTS

This paper has greatly benefited from the detailed comments of Miriam Kastner. S. Calvert also provided helpful criticism. We have no space fully to acknowledge the results and help of numerous geologists, in particular G. Birch, J. C. Faugères, M. Gennesseaux, P. Giresse, M. Lamboy, L. Leclaire, J. P. Masse, L. Martin, R. Mathieu, G. Moguedet and J. Pinson. Thanks are also due to Philip Allen for helpful suggestions and to S. Damatha (Paris) and E. Schmid (Berne) for typing various versions of the manuscript since February 1979.

REFERENCES

Aubry, M.P. & Odin, G.S. (1973). Sur la nature minéralogique du verdissement des craies: formation d'une phyllite apparentée aux glauconies en milieu semi-confiné poreux. *Bull. Soc. géol. Normandie*, **LXI**, 11–22.

Bell, D.L. & Goodell, H.G. (1967). A comparative study of glauconite and the associated clay fraction in modern marine sediments. *Sedimentology*, **9**, 169–202.

Bentor, Y. K. & Kastner, M. (1965). Notes on the mineralogy and origin of glauconite. *J. sedim. Petrol.* **35**, 155–166.

Birch, G.F., Willis, J.P. & Rickard, R.S. (1976). An electron microprobe study of glauconites from the continental margin off the west coast of S. Africa. *Mar. Geol.* **22**, 271–384.

Bjerkli, K. & Östmo-Saeter, J.S. (1973). Formation of glauconie in foraminiferal shells on the continental shelf off Norway. *Mar. Geol.* **14**, 169–178.

Borchert, H. (1965). Formation of marine sedimentary iron ores. In: *Chemical Oceanography*, *2* (Ed. by J.P. Riley and G. Skirrow), pp. 159–201. Academic Press, London.

Borst, R.L. & Keller, W.D. (1969). Scanning electron micrographs of API reference clay minerals and other selected samples. *Proc. Int. Clay Conf.* Tokyo, **1**, 871–891.

Bouysse, P., Kudrass, H.B. & Le Lann, F. (1977). Reconnaissance sédimentologique du plateau continental de la Guyane française. *Bull. B.R.G.M.* **2**, 141–179.

Boyer, P.S., Guinness, E.A., Lynch-Blosse, M.A. & Stolzman, R.A. (1977). Greensand fecal pellets from New Jersey. *J. sedim. Petrol.* **47**, 267–280.

Brindley, G.W., Bailey, S.W., Faust, G.T., Forman, S.A. & Rich, C.I. (1968). Report of the Nomenclature Committee (66–67) of the Clay Minerals Society. *Clays Clay Miner.* **16**, 322–324.

Burst, J.F. (1958a) 'Glauconite' pellets: their mineral nature and applications to stratigraphic interpretations. *Bull. Am. Ass. Petrol. Geol.* **42**, 310–327.

Burst, J.F. (1958b) Mineral heterogeneity in glauconite pellets. *Am. Miner.* **43**, 481–497.

Caspari, W.A. (1910). Contributions to the chemistry of submarine glauconite. *Proc. R. Soc. Edinb.* **30**, 364–373.

Cayeux, L. (1916). *Introduction à l'étude pétrographiques des roches sédimentaires*, pp. 241–252. Imprimerie Nationale, Paris. 524 pp.

Cimbalnikova, A. (1970). Index of refraction and density of glauconites. *Cas. Miner. Geol. Českosl.* **15**, 335–345.

Cimbalnikova, A. (1971a) Chemical variability and structural heterogeneity of glauconites. *Am. Miner.* **56**, 1385–1392.

Cimbalnikova, A. (1971b) Cation exchange capacity of glauconites. *Čas. Miner. Geol. Českosl.* **16**, 15–21.

Collet, L.W. (1908) *Les dépôts marins*, pp. 132–194. Doin, Paris, 325 pp.

Cullen, D.J. (1967) The age of glauconite from the Chatham Rise, East of New-Zealand. *N.Z. J. mar. Freshwat. Res.* **1**, 399–406.

Dangeard, L. (1928) Observations de géologie sous-marine et d'océanographie relatives à la Manche. *Ann. Inst. océanogr.* **VI**, 1, chap. XI, 199–211.

Ehlmann, A. J., Hulings, N.C. & Glover, E.D. (1963) Stages of glauconite formation in modern foraminiferal sediments. *J. sedim. Petrol.* **33**, 87–96.

Foster, M.D. (1969) Studies of celadonite and glaucnite. *Prof. Pap. U.S. geol. Surv.* **614-F**, 17 pp.

Galliher, E.W. (1935) Geology of glauconite. *Bull. Am. Ass. Petrol. Geol.* **19**, 1569–1601.

Giresse, P. (1969) Etude des différents grains ferrugineux authigènes des sédiments sous-marins au large du delta de l'Ogooué (Gabon). *Sciences Terre*, **XIV**, 27–62.

Giresse, P. (1975). Essai de chronométrie de la glauconitisation dans le Golfe de Guinée. *C.r. somm. Séanc. Soc. géol. Fr.* **5**, 163–164.

Giresse, P., Lamboy, M. & Odin, G.S. (1980) Evolution géométrique des supports de glauconitisation, reconstitution de leur paléo-environnement. *Ocean. Acta*, **3**, 251–260.

Giresse, P. & Odin, G.S. (1973) Nature minéralogique et origine des glauconies du plateau continental du Gabon et du Congo. *Sedimentology*, **20**, 457–488.

Glangeaud, L. (1941) Sur la formation et la répartition des faciès vaseux dans les estuaires. *C. r. hebd. Séanc. Acad. Sci. Paris*, 1022–1024.

Goldhaber, M.B. & Kaplan, I.R. (1974). The sulfur cycle. In: *The Sea* (Ed. by E. G. Goldberg), **5**, 569–655. Wiley, New York.

Hardjosoesastro, R. (1971) Note on chamosite in sediments of the Surinam shelf. *Geologie Mijnb.* **50**, 29–33.

Hein, J.R., Allwardt, A.O. & Griggs, G.B. (1974) The occurrence of glauconite in Monterey Bay, California. Diversity, origins and sedimentary environmental significance. *J. sedim. Petrol.* **44**, 562–571.

Hendricks, S. B. & Ross, C.S. (1941) Chemical composition and genesis of glauconite and celadonite. *Am. Miner.* **26**, 683–708.

Houbolt, J.J.H.C. (1957) *Surface sediments of the Persian Gulf near the Qatar Peninsula.* Ph. D. thesis. University of Utrecht. Mouton, Den Haag, 113 pp.

Hower, J. (1961) Some factors concerning the nature and origin of glauconite. *Am. Miner.* **46**, 313–334.

Hower, R.J. & Mowatt, T.C. (1966) The mineralogy of illite and mixed-layer illite montmorillonite. *Am. Miner.* **51**, 825–854.

Juignet, P. (1974) *La transgression crétacée sur la bordure orientale du Massif armoricain* (Aptien-Albien-Cénomanien de Normandie et du Maine: le stratotype du Cénomanien). Thèse Sciences. University of Caen, 806 pp.

Kelley, W.P. (1945) Calculating formulas for fine grained minerals. *Am. Miner.* **30**, 1–26.

Kossovskaya, A.G. & Drits, V.R. (1970) Micaceous minerals in sedimentary rocks. *Sedimentology*, **15**, 83–101.

Lamboy, M. (1968) Sur un processus de formation de la glauconie en grains à partir des débris coquilliers. Rôle de organismes perforants. *C.r. hebd. Séanc. Acad. Sci., Paris*, **266**, 1937–1940.

Lamboy, M. (1974) La glauconie du plateau continental au Nord-Ouest de l'Espagne dérive d'anciens débris coquilliers. *C.r. hebd. Séanc. Acad. Sci., Paris*, **280**, 157–160.

Lamboy, M. (1976) *Géologie marine du plateau continental au N.O. de l'Espagne.* Thèse Doctorat d'Etat, University of Rouen, 283 pp.

Lamboy, M. & Odin, G.S. (1975) Nouveaux aspects concernant les glauconies du plateau continental Nord-Ouest espagnol. *Rev. Géogr. phys. Géol. dyn.* **XVII**, 99–120.

Lloyd, A.F. & Fuller, A.O. (1965) Glauconite from shallow marine sediments of the South African coast. *S. Afr. J. Sci.* **61**, 442–448.

Lucas, J. (1962) *La transformation des minéraux argileux dans la sédimentation.* Etudes sur les argiles du Trias. *Mém. Serv. Carte géol. Als.-Lorr.* **23**, 202 pp.

McRae, S.G. (1972) Glauconite. *Earth Sci. Rev.* **8**, 397–440.

Martin, L. (1973) *Morphologie, sédimentologie et paléogéographie au Quaternaire récent du plateau continental ivoirien.* Thèse Doctorat d'Etat, Paris, 339 pp.

Millot, G. (1964) *Géologie des Argiles.* Masson, Paris, 499 pp.

Moore, M.B. (1939) Faecal pellets in relation to marine deposits. In: *Recent Marine Sediments. A symposium* (Ed. by P.D. Trask), pp. 516–524. Am. Ass. Petrol. Geol., Tulsa.

Murray, J. & Renard, A.F. (1891) *Report on Deep-Sea Deposits Based on the Specimens Collected During the Voyage of H.M.S. 'Challenger' in the years 1872–1876*, H.M.S.O., London, 525 pp.

Odin, G.S. (1969) Méthode de séparation des grains de glauconie, intérêt de leur étude morphologique et structurale. *Revue Géogr. phys. Géol. dyn.* **XI**, 171–174.

Odin, G.S. (1972a) Observations nouvelles sur la structure

de la glauconie en accordéon; description du processus de genèse par néoformation. *Sedimentology*, **19**, 285–294.

ODIN, G.S. (1972b) Modalités du passage continu du sédiment argileux au minéral glauconite dans les formations éocènes du Rodeberg (Flandres occidentales). *C.r. hebd. Séanc. Acad. Sci., Paris*, **274**, 660–663.

ODIN, G.S. (1975a) *De glauconiarum constitutione, origine, aetateque*. Unpublished Thése Doctorat d'Etat, Paris, 280 pp.

ODIN, G.S. (1975b) Migrations du fer des eaux continentales jusqu'aux eaux océaniques profondes. *C.r. hebd. Séanc. Acad. Sci. Paris*, **281**, 1665–1668.

ODIN, G.S., DODSON, M., HUNZIKER, J.C. & KREUZER, H. (1979) Radiogenic argon in glauconies during their genesis. *Bull. Liaison Inform. Projet PICG, 133*, **6**, 7–8.

ODIN, G.S. & GIRESSE, P. (1976) Essai de chronométrie de la glauconitisation dans le golfe de Guinée, compléments et remarques. *C.r. somm. Séanc. Soc. géol. Fr.* **3**, 108–111.

ODIN, G.S., HUNZIKER, J.C. & LORENZ, C.R. (1975) Age radiométrique du Miocène inférieur en Europe occidentale et centrale. *Geol. Rdsch.* **64**, 570–592.

ODIN, G.S. & LÉTOLLE, R. (1978) Les glauconies et aspects voisins ou confondus: signification sédimentologique. *Bull. Soc. géol. Fr.* **XX**, 553–558.

ODIN, G.S., VELDE, B. & BONHOMME, M. (1977) Radiogenic argon in glauconies as a function of mineral recrystallization. *Earth planet. Sci. Lett.* **37**, 154–158.

OJAKANGAS, R.W. & KELLER, W.D. (1964) Glauconitization of rhyolite sand grains. *J. sedim. Petrol.* **34**, 84–90.

PARRY, W.T. & REEVES, C.C. (1966) Lacustrine glauconititic mica from pluvial Lake Mound, Lym and Terry counties, Texas. *Am. Miner.* **51**, 229–235.

POLEVAYA, N.I., MURINA, G.A. & KAZAKOV, G.A. (1961) Glauconite in absolute dating. *Ann. N.Y. Acad. Sci.* **91**, 298–310.

PORRENGA, D.H. (1967a) Clay mineralogy and geochemistry of recent marine sediments in tropical areas. *Publ. Fysisch-Geographisch Lab. Univ. Amsterdam*, **9**, 145 pp. Dort-Stolk, Amsterdam.

PORRENGA, D.H. (1967b) Glauconite and chamosite as depth indicators in the marine environment. *Mar. Geol.* **5**, 495–501.

PRATT, W.L. (1963) Glauconite from the sea floor off Southern California. In: *Essays in Marine Geology in Honor of K.O. Emery* (Ed. by R.L. Miller), pp. 97–119. University of California Press, Los Angeles.

PRYOR, W.A. (1975) Biogenic sedimentation and alteration of argillaceous sediments in shallow-marine environments. *Bull. geol. Soc. Am.* **86**, 1244–1254.

ROBERT, C. & ODIN, G.S. (1975) Niveaux glauconieux dans les sédiments récents du seuil Nord-Egéen. *Bull. Grpe fr. Argiles*, **27**, 1–11.

ROHRLICH, V., PRICE, N.B. & CALVERT, S.E. (1969) Chamosite in the recent sediments of Loch Etive (Scotland). *J. sedim. Petrol.* **39**, 624–631.

SEED, D.P. (1968) The analysis of the clay content of some glauconitic oceanic sediments. *J. sedim. Petrol.* **38**, 229–231.

SHUTOV, V.D., KATZ, M.Y., DRITS, V.A., SOKOLOVA, A.L. & KASAKOV, G.A. (1970) Crystallochemical heterogeneity of glauconite as depending on the conditions of its formation and post-sedimentary change. *Int. Clay Conf., Madrid*, **1**, pp. 327–339.

SMULIKOWSKI, K. (1954) The problem of glauconite. *Polska Akad. Nauk. Komitet. Geol., Arc. Min. Warsaw* **18**, 21–120.

TAKAHASHI, J.I. (1939) Synopsis of glauconitization. In: *Recent Marine Sediments. A symposium* (Ed. by P.D. Trask), pp. 502–512. Am. Ass. Petrol. Geol., Tulsa.

TAKAHASHI, J.I. & YAGI, T. (1929) The peculiar mud grains in the recent littoral and estuarine deposits,with special reference of the origin of glauconite. *Econ. Geol.* **24**, 838–852.

TOOMS, J.S., SUMMERHAYES, C.P. & MCMASTER, R.L. (1970) Marine geological studies on the north west African margin: Rabat-Dakar. In: *The Geology of the East Atlantic Continental Margin* (Ed. by F.M. Delany). *Inst. Geol. Sci. Rep. No.* **70/16**, 9–25.

VELDE, B. & ODIN, G.S. (1975) Further information related to the origin of glauconite. *Clays Clay Miner.* **23**, 376–381.

VON GAERTNER, H.R. & SCHELLMANN, W. (1965) Rezente Sedimente im Küstenbereich der Halbinsel Kaloum, Guinea. *Tschermaks miner. petrogr. Mitt.* **10**, 349–367.

WARSHAW, C.M. (1957) *The mineralogy of glauconite.* Unpublished Ph.D. Thesis. Pennsylvania State University, University Park, 155 pp.

WEBB, A.W., MCDOUGALL, I. & COOPER, J. A. (1963) Retention of radiogenic argon in glauconites from Proterozoic sediments, Northern Territory Australia. *Nature* **199**, 270–271.

WERMUND, E.G. (1961) Glauconite in early Tertiary sediments of Gulf Coastal Province. *Bull. Am. Ass. Petrol. Geol.* **45**, 1667–1696.

(*Manuscript received* 2 *June* 1980; *revision received* 5 *November* 1980)

Reprinted from Sedimentology (1996) **43**, *877–900*

Low-Mg calcite marine cement in Cretaceous turbidites: origin, spatial distribution and relationship to seawater chemistry

JAMES P. HENDRY,* NIGEL H. TREWIN† and ANTHONY E. FALLICK‡

**Department of Geology, School of Geosciences, The Queen's University, Belfast BT7 1NN, UK (j.hendry@qub.ac.uk)*

†Department of Geology & Petroleum Geology, University of Aberdeen, King's College, Aberdeen AB9 2UE, UK

‡Isotope Geosciences Unit, Scottish Universities Research and Reactor Centre, East Kilbride, Glasgow G75 0QF, UK

ABSTRACT

Lower Cretaceous (Hauterivian) bioclastic sandstone turbidites in the Scapa Member (North Sea Basin) were extensively cemented by low-Mg calcite spars, initially as rim cements and subsequently as concretions. Five petrographically distinct cement stages form a consistent paragenetic sequence across the Scapa Field. The dominant and pervasive second cement stage accounts for the majority of concretions, and is the focus of this study. Stable-isotope characterization of the cement is hampered by the presence of calcitic bioclasts and of later cements in sponge spicule moulds throughout the concretions. Nevertheless, trends from whole-rock data, augmented by cement separates from synlithification fractures, indicate an early calcite $\delta^{18}O$ value of+0·5 to −1·5‰ PDB. As such, the calcite probably precipitated from marine pore fluids shortly after turbidite deposition. Carbon isotopes ($\delta^{13}C$=0 to −2‰ PDB) and petrographic data indicate that calcite formed as a consequence of bioclastic aragonite dissolution. Textural integrity of calcitic nannoplankton in the sandstones demonstrates that pore fluids remained at or above calcite saturation, as expected for a mineral-controlled transformation.

Electron probe microanalyses demonstrate that early calcite cement contains <2 mol% $MgCO_3$, despite its marine parentage. Production of this cement is ascribed to a combination of an elevated aragonite saturation depth and a lowered marine Mg^{2+}/Ca^{2+} ratio in early Cretaceous 'calcite seas', relative to modern oceans. Scapa cement compositions concur with published models in suggesting that Hauterivian ocean water had a Mg^{2+}/Ca^{2+} ratio of ≤ 1. This is also supported by consideration of the spatial distribution of early calcite cement in terms of concretion growth kinetics. In contrast to the dominant early cement, late-stage ferroan, ^{18}O-depleted calcites were sourced outwith the Scapa Member and precipitated after 1–2 km of burial.

Our results emphasize that bioclast dissolution and low-Mg calcite cementation in sandstone reservoirs should not automatically be regarded as evidence for uplift and meteoric diagenesis.

INTRODUCTION

Reservoir properties of sandstones are profoundly influenced by the spatial distribution of carbonate cements, which may be dispersed through individual beds, aggregated into discrete nodules, or form extensive and impermeable stratabound layers (Hurst, 1987). Efficient exploitation of clastic hydrocarbon reservoirs requires development and production strategies that take account

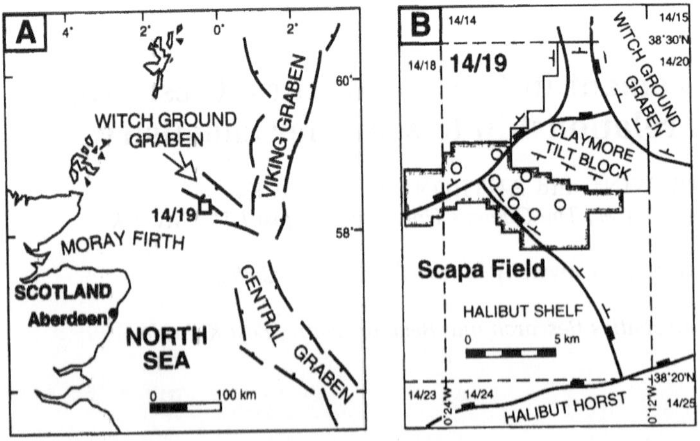

Fig. 1. (A) Location map of North Sea Block 14/19; (B) enlargement showing Scapa Field and positions of sampled wells (open circles).

of geological controls on carbonate cement distributions, particularly with respect to sedimentary environments and tectonic settings. Recent research on shallow-marine sandstones has shown that depositional facies, detrital composition, burial history and palaeohydrological conditions can all influence carbonate cementation patterns (e.g. Bjørkum & Walderhaug, 1990a,b; Wilkinson & Dampier, 1990; Pirrie & Marshall, 1991; Walderhaug & Bjørkum, 1992). Cements are frequently low-Mg calcite, formed by redistribution of biogenic carbonate during ingress of calcium-carbonate-undersaturated meteoric groundwaters, sourced from adjacent shorelines or uplifted basin margins (e.g. Haszeldine *et al.*, 1992; Wilkinson, 1993a; Prosser *et al.*, 1994).

Carbonate authigenesis in deep-water marine sandstones is less comprehensively understood, and few previous studies have documented massive early cementation. Such sandstones have variable biogenic carbonate contents, and commonly remain below the range of coastal groundwater aquifers and of eustatic sea-level fluctuations. Carbonate authigenesis is frequently mesogenetic, and solutes tend either to be allochthonous or related to silicate and organic matter degradation reactions. Cements are consequently dominated by ferromagnesian carbonate minerals that possess strong 'organic' carbon isotopic signatures (e.g. Boles & Ramseyer, 1987; Macaulay *et al.*, 1993; Haszeldine, 1994; Watson *et al.*, 1995). Cement volumes are typically less than in shallow-marine sandstones, and spatial distributions may reflect proximity to extraformational solute sources and/or subsurface flow pathways (e.g. Burley, 1993).

This paper presents detailed petrographic and compositional data from abundant sparry calcite cements in Lower Cretaceous, calcarenaceous sandstone turbidites of the Scapa Member (Valhall Formation, North Sea Basin). The cements are Mg-poor and form concretionary layers and nodules analogous to those described from shallow-marine sandstones, yet meteoric fluids were apparently not involved in their formation. By demonstrating that cementation resulted from redistribution of biogenic carbonate in depositional marine pore fluids, it is possible to re-assess and extend the diagenetic models developed for shallow-marine sandstones. The implications of long-term oceanographic changes for sandstone cementation are also examined, by evaluating the calcite composition and fabrics in the context of published estimates for early Cretaceous ocean chemistry.

SAMPLING AND METHODS

More than 180 polished, blue-epoxy-impregnated thin sections of porous and calcite-cemented sandstones were prepared from 3·8-cm-diameter core plugs, taken from eight wells that penetrated the main reservoir unit ('SD' sand) of the Scapa Member across the Scapa Field (Fig. 1). Stained acetate peels were also prepared for all calcite-cemented samples. Thin sections were examined in transmitted light and cathodoluminescence (CL). Selected examples were studied using blue-light epifluorescence (with incident light of λ=560 nm), and in backscatter mode in a scanning electron microscope (BSEM) equipped with an energy-dispersive X-ray analyser (EDXA).

Sampling strategies for calcite stable isotope analysis depended upon the degree of fine-scale petrographic complexity within calcite-cemented layers. Whole-rock samples of about 50–100 mm^3 were milled from core plugs using a masonry drill, whilst a hand-held microdrill was used to separate individual cement stages from fractures. Aliquots containing 5·0–10·0 mg of calcite were plasma-ashed for at least 6 h to remove volatile organic matter, before being reacted overnight with 100% H_3PO_4 *in vacuo* at 25°C. Isotope ratios of purified CO_2 were measured on a V.G. Isogas SIRA-10 triple collector mass spectrometer, against a calibrated internal standard. Results are expressed in δ notation as ‰ deviation from the PDB international standard. Analytical precision was better than 0·1‰ (1σ) for both $\delta^{13}C$ and $\delta^{18}O$.

Wavelength-dispersive electron-probe microanalyses (EPMA) of calcite cement were made using a CAMECA Camebax probe. Operating conditions were: 10-kV accelerating potential; 10·6-nA specimen current; 40° take-off angle. Count times were 60 s for Mg and Sr, and 30 s for Ca, Fe and Mn. Backgrounds on each analysis point were measured for 15 s either side of the peak positions. In order to minimize sample volatilization, a focused beam was rastered over an area of about 12 μm diameter. Sample location involved distinction of major calcite types based on direct observation of CL through the optics of the microprobe, and qualitative compositional mapping using an on-line BSEM/EDXA system. EPMA detection limits were as follows (mol%): $FeCO_3$, 0·052; $MgCO_3$, 0·034; $MnCO_3$, 0·049; $SrCO_3$, 0·027 (~235 p.p.m.). Analytical precision was monitored by 50 analyses of a relatively homogeneous low-Mg calcite of independently determined composition ('DIS'='Derbyshire Iceland Spar'). Precisions at 95% confidence are as follows: $FeCO_3$, ±6%; $MgCO_3$, ±7%; $MnCO_3$, ±22%. The Sr content of DIS was close to the detection limit, giving an unrealistically low precision; no homogeneous standard of similar Sr content to the Scapa calcites was available.

SCAPA SANDSTONES

Sedimentology

The Scapa Field is located in the Outer Moray Firth (Block 14/19, North Sea Basin). It occupies a half-graben situated north and east of the Halibut Horst, a long-lived topographic high (Fig. 1). Scapa SD sandstones were sourced from an intervening, high-energy, shallow-marine shelf (Harker & Chermak, 1992; Riley *et al.*, 1992) and are Hauterivian in age. Concentrated, cohesionless sediment gravity flows reworked sands rich in shallow-water macrofossil debris but poor in detrital clay, and deposited them as an extensive base-of-slope apron in the adjacent half-graben (Boote & Gustav, 1987). SD sandstones are mainly fine to very fine grained, and poorly stratified to planar laminated. Subordinate pebbly sandstones and sand-matrix fossiliferous conglomerates are generally massive, but individual beds occasionally display normal, coarse-tail or reverse (basal) grading. Nonreservoir facies include burrowed heterolithic sandstones or thin hemipelagic siltstones and marls between discrete turbidite deposits, and sporadic debris-flow conglomerates with basement-derived clasts (McGann *et al.*, 1991). Beds are frequently amalgamated, and vertical and lateral facies organization is poorly structured, typical of shelf-fed turbidite systems (e.g. Heller & Dickinson, 1985; Surlyk, 1987).

Petrology

The majority of SD sandstones are fossiliferous quartz arenites, with subordinate fossiliferous feldspathic or lithic subarenites. They contain a complex and diverse authigenic mineralogy, including several stages of calcite and quartz cementation (Fig. 2). Calcite is present as tightly cemented layers and nodules, fracture fills, and as rim and syntaxial cements in porous sandstones. Quartz cements formed from precursor opal-CT, and ultimately from sponge spicules which were abundant throughout the sandstones at deposition (Hendry & Trewin, 1995). Although the spicules have dissolved, CL reveals numerous pseudomorphs within calcite-cemented sandstone (see below). Early diagenetic pyrite forms sparsely disseminated framboids in porous and calcite-cemented sands, particularly within microfossil chambers and vascular plant fragments. Mesogenetic pyrite, ankerite, kaolinite, illite and additional calcite and quartz formed during deep burial of the Scapa Member.

Aragonitic shell material has been pervasively dissolved from the sandstones, although micritized outlines of coral, bivalve and green algal fragments can be locally identified. In contrast, calcitic bioclasts such as red algae, brachiopod, echinoderm, bryozoan and crustacean fragments are well preserved in both calcite-cemented and porous sandstones. CL reveals some texturally retentive stabilization of echinoderm grains and foraminifera. However, there is no evidence

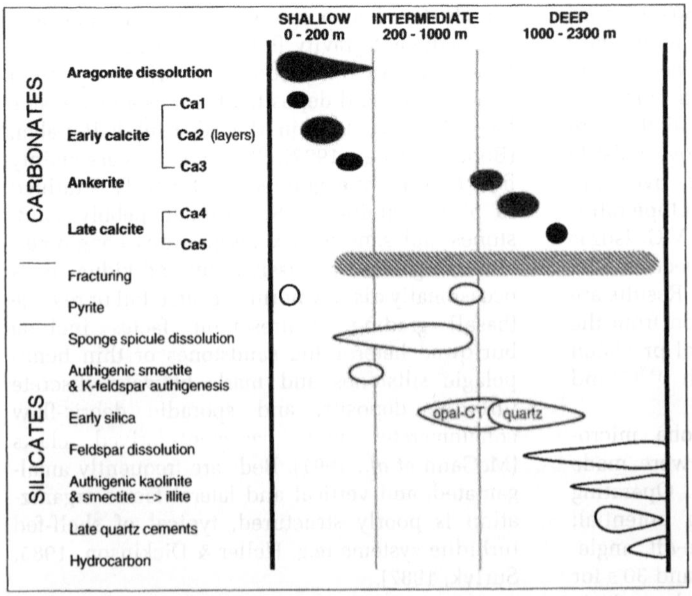

Fig. 2. Simplified paragenetic sequence for Scapa SD reservoir sandstones; shading highlights carbonate phases.

for any dissolution of low-Mg calcites, even on delicate microfossils such as coccolith plates. Preserved faunal diversity is greatest in the coarsest sand-rich lithofacies, whereas fine- to very fine-grained sandstones are dominated by echinoderm plates, foraminifera and comminuted bivalve fragments. This suggests a degree of hydrodynamic sorting in the shallow source area prior to redeposition in sediment gravity flows.

CONCRETIONS

Calcite-cemented concretions of tabular to ovoid/spherical form are abundant within the SD sandstones of all cored Scapa Field wells (except in heterolithic sandstones, which are tightly cemented by microquartz and kaolinite). Concretions account for about 25% of the sandstones in individual wells and thereby constitute a significant reduction in the reservoir net-to-gross sand ratio. The concretions are tightly cemented (porosity <5%, permeability <0·2 mD), but their morphology and spacing are highly variable (e.g. Fig. 3). In fine- to very fine-grained sandstones, concretions range from about 2 cm to 1 m in thickness, and are present at decimetre- to metre-scale intervals. Concretion margins are usually sharp; they may be planar to curved, and bedding-parallel to discordant. The outer few millimetres to centimetres of decimetre-scale concretions (and entire centimetre-thick nodules) are oil stained (e.g. Fig. 3). Pebbly sandstones and sand-matrix conglomerates tend to exhibit a gradation between patchy and full cementation rather than well-defined concretions.

Small nodules terminating within the cores possess an elliptical to irregular outline. In laminated sandstones the depositional fabric is deflected around these nodules, implying a precompactional origin. Some decimetre-scale concretions in poorly stratified or massive sandstones possess slightly curved, convex-outward margins, compatible with a spherical or elliptical shape. Analogous concretions in laminated sandstones may possess bedding-parallel margins, suggesting a more tabular morphology. Highly irregular concretions are locally present in poorly structured sandstones, sometimes in association with soft sediment deformation.

Where distinct sandstone beds can be discriminated, they typically contain concretions at one, or a maximum of two, levels within the beds. These concretions are too thin to resolve seismically, and their lateral extent is indeterminate. Formation pressure data argue against field-wide development of individual cemented layers (Harker & Chermak, 1992), but are unable to discriminate randomly distributed concretions from stratabound zones of partial to total cementation (Bjørkum & Walderhaug, 1990a).

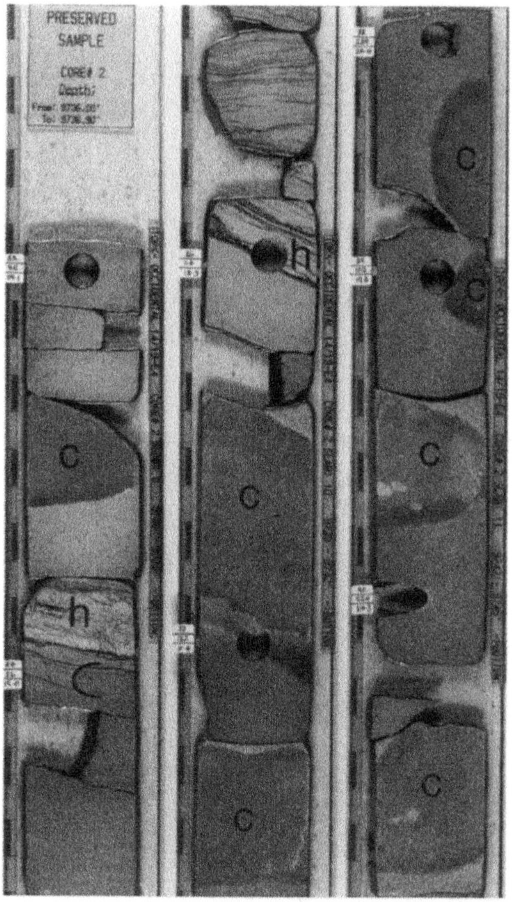

Fig. 3. Abundant concretions (c) in poorly stratified Scapa Member sandstone. Most of the sands are clean-washed; minor, bioturbated heterolithic sands (h) are present at the top of some beds. Scale divisions are ~3 cm.

CALCITE CEMENTS

Diagenetic fabrics within concretions appear deceptively simple in thin section, but staining and CL reveal considerable paragenetic complexity. CL patterns may vary between individual concretions, but five calcite cement stages are recognized to form a regionally consistent paragenetic sequence (Table 1). Eogenetic calcites (Ca1–3) enclose smooth quartz and feldspar grains with no authigenic overgrowths, whilst mesogenetic calcites (Ca4–5) post-date precipitation of early microquartz and K-feldspar overgrowths (Hendry & Trewin, 1995), mesogenetic euhedral pyrite, ankerite and blocky kaolinite, and incipient dissolution of alkali and plagioclase feldspars.

Early calcite cements

The first calcite cements (Ca1) are infrequent, volumetrically minor and nonferroan. They are brightly fluorescent in blue light and are the only cement stage to possess prominent concentric zonation in CL (Fig. 4A,B). The cements are limited to epitaxial, columnar to equant isopachous fringes on polycrystalline allochems (e.g. bivalve and brachiopod grains), and inclusion-rich syntaxial overgrowth crystals on monocrystalline allochems (principally echinoderm plates and disaggregated inoceramid prisms). CL zonation sequences are locally consistent but vary between samples. A typical pattern is nonluminescent (± thin bright subzones), passing to bright then dull or very dull outer zones. Zonation is also visible with BSEM, and EDXA analyses suggest that is it largely controlled by variable Mn substitution. Zonal morphologies tend to be anhedral, suggesting that crystal faces were disrupted by poisoning or rapid growth (Braithwaite & Heath, 1989). Organic contamination in the cements (and the host grains) could account for epifluorescence of the calcite. Inclusion trails in syntaxial overgrowths are orientated normal to the allochem surface and appear from CL and BSEM to represent replacement of a metastable spired overgrowth by later (post-Ca1) cement (Fig. 4A). This metastable phase may have been Mg-calcite, given that similar CL patterns are seen in the otherwise well-preserved echinoderm grains.

Fine- to very fine-grained Ca2 spars are the principal early calcite cements. They account for the majority of concretions (Fig. 4C,D), as well as forming discrete overgrowths on bioclasts (and Ca1 cements) in the porous sandstones (Fig. 4E,F). Pore-filling Ca2 cement forms a tightly interlocking and drusy mosaic of limpid, equant spar. Smallest crystals are located on microcrystalline calcite allochems, which presumably served as nucleation sites. The cement is weakly ferroan, poorly fluorescent, and very dull brown in CL. Subtle sectoral zonation may be present in the coarsest crystals, but luminescence is generally uniform (Fig. 4D). Ca2 contains disseminated single-phase fluid and calcareous solid inclusions.

Ca2 spar frequently fills intergranular pore space within concretions, but is sometimes syntaxially overgrown by more coarsely crystalline Ca3; for example in biomoulds and relatively large intergranular voids of pebbly sandstones and sand-matrix conglomerates. Transition from Ca2 to Ca3 cement is marked by an overall

Table 1. Petrographic characteristics of calcite cement phases in the Scapa SD sandstones.

Cement stage	Grain size	Distribution	Stain	CL	Epifluorescence	Inclusions
Ca1	Usually 250–1000 μm	Localized epitaxial fringes and syntaxial overgrowths on bioclasts	Nonferroan	Complex bright/dark banded	Bright	Calcite, rare pyrite
Ca2	Usually 100–500 μm, locally ≤100 μm in fractures	Concretions, pervasive overgrowths on bioclasts and on Ca1 cements	Weakly ferroan	Dark brown, rarely zoned	Dull to moderate	Rare pyrite, single phase fluid
Ca3	Mostly >500 μm, may be several mm if poikiloptic	Syntaxial overgrowth on Ca2 in biomoulds, locally forms concretion rims and fills fractures	Variable, tends to be nonferroan	Variable, usually moderate or dull orange–brown, typically unzoned	Very dull	Casts of microborings in biomoulds, rare pyrite
Ca4	Mostly >500 μm, rarely 250–500 μm	Fractures, biomoulds (especially sponge spicules) in concretions, also cements some concretion margins, thin overgrowths on Ca2 in porous sands	Ferroan	Moderate to dull orange–brown with prominent sectoral zonation	Mostly very dull, locally can be bright	Microquartz, calcitized opal-CT, rare single phase fluid inclusions, pyrite
Ca5	Mainly 100–250 μm	Veinlets and altered bioclasts, pseudomorphs after opal-CT	Nonferroan	Bright yellow	Variable	Microquartz

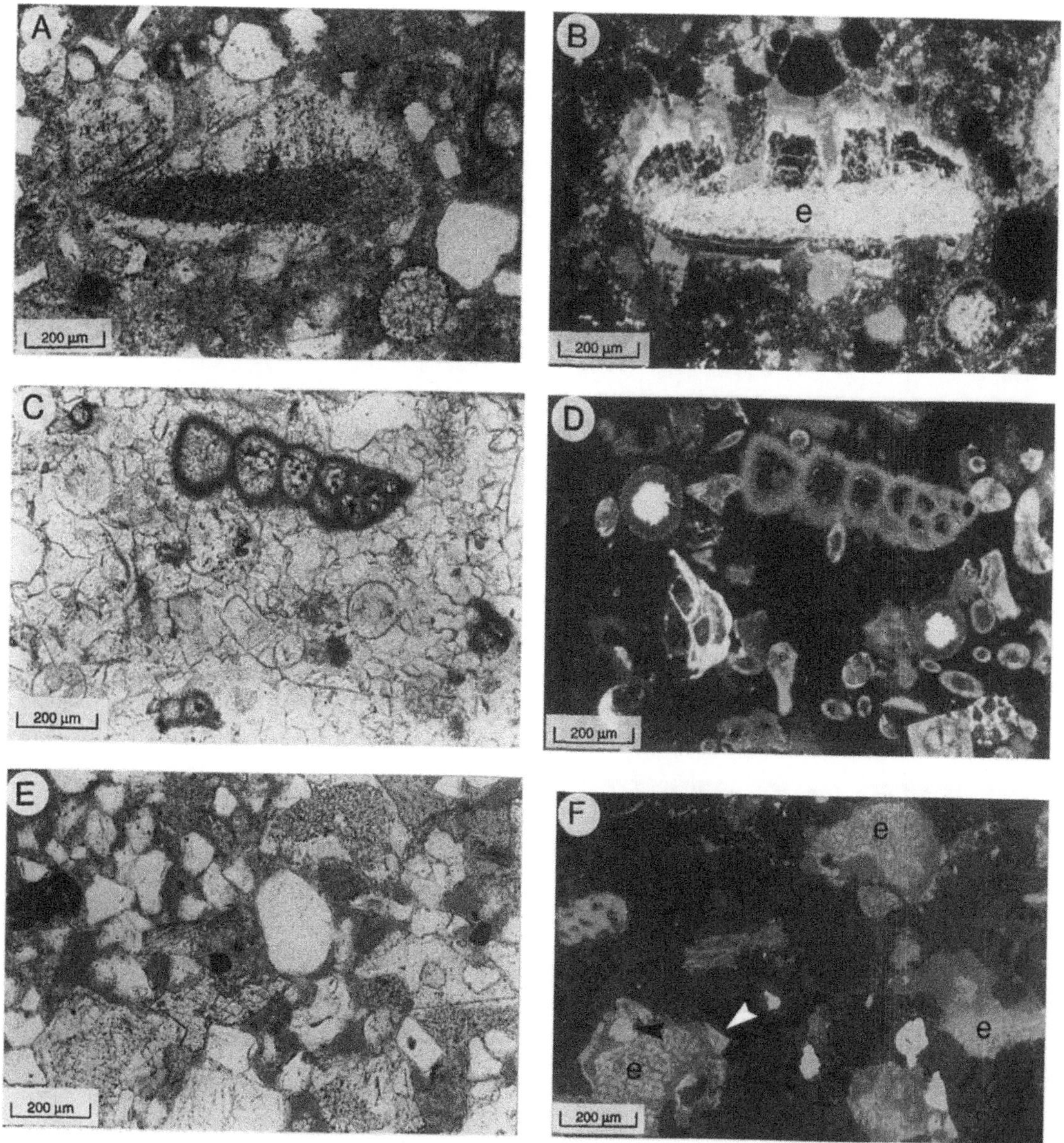

Fig. 4. Paired transmitted light and CL photomicrographs. (A, B) Zoned Ca1 overgrowth on an echinoderm plate (e). Inclusion trails in the cement are interpreted as relics of a former bladed or acicular fringing cement. (C,D) Very dark luminescent Ca2-cemented concretion; note the well-preserved foraminifer and common circular to elliptical sponge spicule moulds filled by later and brighter (Ca4–5) cements. The latter are not clearly visible in transmitted light. (E, F) Ca2 overgrowth on echinoderm plates (e) in a porous sandstone. Note Ca2 growth around sponge spicules (e.g. black arrow) which were subsequently dissolved and replaced by Ca4 cement. Ca4 locally restored a euhedral crystal face in these cases (e.g. white arrow). Dark background in (E) is blue-resin-impregnated porosity. The image in (F) is slightly over-exposed to distinguish very dull luminescent detrital quartz from nonluminescent microquartz.

reduction in Fe content and increase in luminescence. However, Ca3 is very variable in CL, ranging from dull orange–brown to bright yellow, and the boundary with Ca2 may be sharp, subzoned or gradational. Biomouldic Ca3 (and Ca4) locally contains silicified casts of endolith borings and/or collapsed micrite envelopes from previously dissolved aragonite bioclasts (Fig. 5A). Poikilotopic

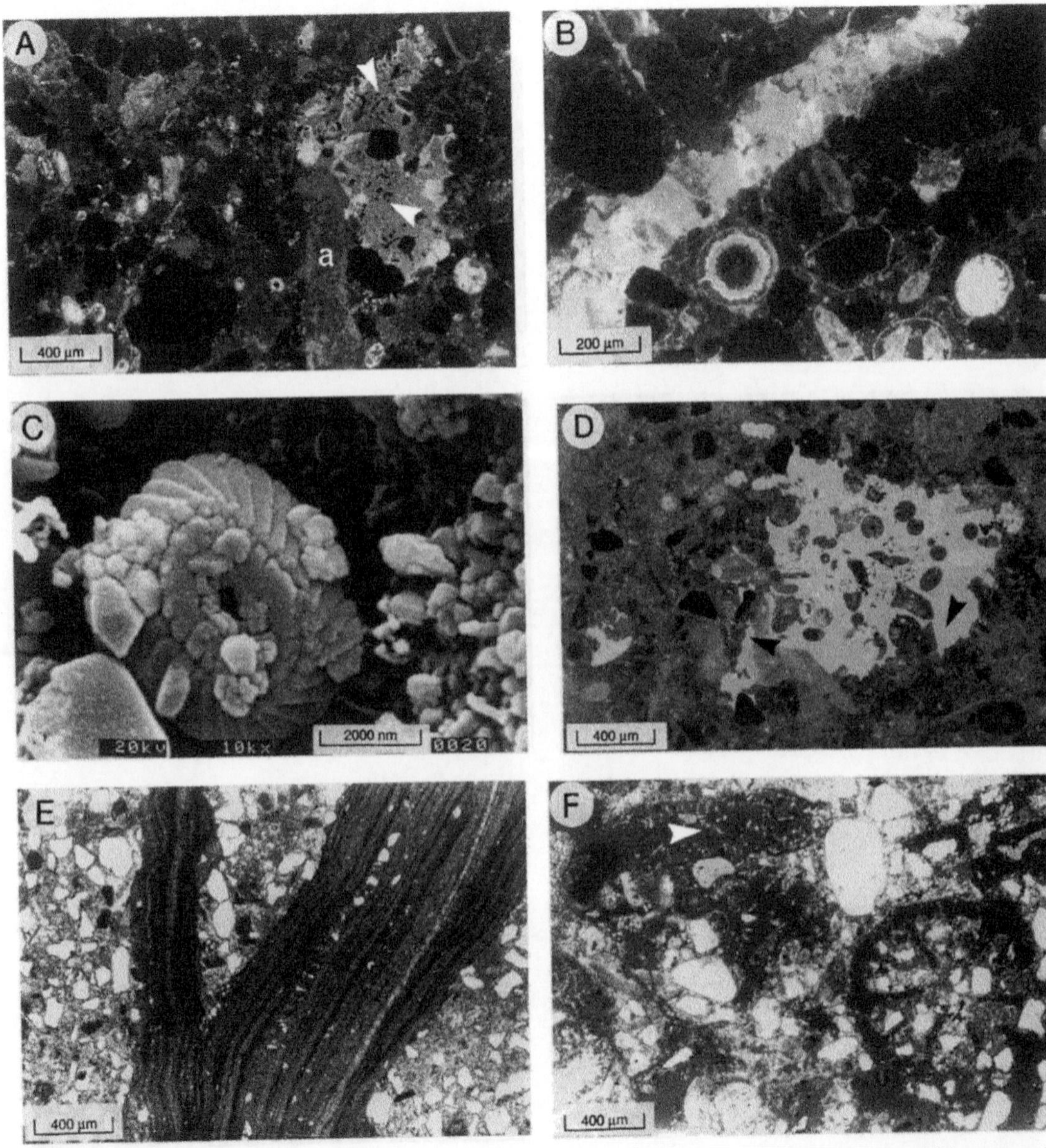

Fig. 5. (A) Poorly luminescent Ca2 cement passes to moderate luminescent Ca3 within a bivalve mould (quartz grains are nonluminescent). Tiny black inclusions in Ca3 (white arrows) are silicified casts of endolith borings from the original shell fragment. Note micritized algal fragment (a) and Ca4 in spicule moulds. (B) CL photograph of fracture-filling, sector-zoned Ca4 spar cutting a Ca2-cemented concretion. (C) SEM photomicrograph of a typically well-preserved coccolith in a porous sandstone. Calcareous nannoplankton were probably entrained into sand-prone sediment gravity flows from underlying hemipelagites. (D) Bright luminescent Ca3-cemented region within a dull luminescent Ca2 concretion; note the absence of corrosion textures on surrounding Ca2 crystal terminations (arrowed). (E) Early, synlithification fracture in a concretion, containing multiple generations of microsparitic calcite cement. CL and staining indicate that this microspar is paragenetically equivalent to coarser Ca2–3 spars. (F) Biomoulds at the edge of a concretion; surrounding cement is Ca2–3 and dark background is blue-resin-filled porosity. Note quartz-filled endolith borings (arrowed); compare with (A).

Ca3 is also the principal intergranular cement at the top and/or bottom margins of some concretions with Ca2-dominated cores.

Late calcite cements

The prevailing late calcite cement (Ca4) has a strongly ferroan stain, moderate to dull orange–brown luminescence with prominent sectoral zonation, and moderate to dull epifluorescence. Ca4 spar most commonly fills secondary (fracture and biomouldic) porosity within Ca2 concretions (Figs 4D and 5B), and sometimes forms outer rims on earlier syntaxial overgrowths in porous sandstones. Ca4 replaces many sponge spicules, restoring euhedral crystal faces where Ca2 overgrowths surrounded the spicules in porous sandstones (Fig. 4F). In addition, Ca4 is the dominant cement in outer margins of some concretions, and (rarely) accounts for entire centimetre-scale nodules. Pore-filling Ca4 is medium- to coarse-grained (sometimes poikilotopic), anhedral, equant spar. It contains sparse single-phase fluid inclusions and sporadic solid inclusions, mostly aggregates of microquartz and illite/smectite that precipitated between Ca2 and Ca4 authigenesis (Hendry & Trewin, 1995).

The final calcite stage (Ca5) is volumetrically minor, and present either in hairline fractures through earlier concretions, or as a replacement of opal-CT lepispheres (cf. Hendry and Trewin, 1995; Fig. 8B). It is nonferroan and bright yellow luminescent.

Textural relationships

Ca2-cemented concretions appear deceptively under-packed in transmitted light, with 'floating' siliciclastic grains and as much as 60% intergranular spar. Whilst this would appear to fulfil criteria for a neomorphic replacement of a micritic matrix (Bathurst, 1975), such an origin is inconsistent with the sedimentary facies, and with the textural integrity of microcrystalline peloids, foraminifera and coccoliths (Fig. 5C) throughout the sandstones. Instead, staining and CL demonstrate that the loosely packed detrital fabric reflects an abundance of calcitized sponge spicules, which are otherwise poorly resolved from the surrounding cement (e.g. Figs 4C–F and 5B). Up to 40% of the SD unit detrital assemblage consisted of highly inequant, monaxon and triaxon megascleres, now mainly replaced by Ca4–5 calcites. Some perforate microscleres were also present.

An unusual feature in several Ca2-cemented sandstones is the presence of Ca3–4 cements as isolated, 500-μm to several-millimetre-sized patches of intergranular spar (Fig. 5D). Polyhedral terminations of surrounding Ca2 crystals argue against corrosion or dissolution, and there is no evidence that Ca3–4 replaces a previous cement phase. Furthermore, these 'islands' of Ca3–4 spar contain K-feldspar overgrowths, plus microquartz and/or calcitized opal-CT lepispheres (Hendry & Trewin, 1995), suggesting that these regions of sandstone remained porous subsequent to encasement by Ca2. Intriguingly, these features are commonly centred on pyritized plant fragments.

Ca4-cemented samples contain deformed micas and display a similar packing to the porous sandstones, reflecting the dissolution of sponge spicules and mechanical compaction/grain reorientation prior to late calcite precipitation.

Fracture-filling calcites

Concretions in the SD sandstones are commonly dissected by dilational fractures that host a variety of authigenic mineral fills. Fractures rarely extend into adjacent porous sandstones and may have originated by stress concentration and/or hydraulic fracturing processes (cf. Lloyd *et al.*, 1982; Marin *et al.*, 1993). There are two distinct types which permit physical separation of Ca2–3 and Ca4 cements for isotopic characterization:

1 irregular fractures with splayed terminations and laminated calcite microspar fills, suggesting repeated episodes of rupture and cementation (Fig. 5E). These are interpreted to have formed during lithification, under relatively low confining pressures. From CL and staining, the microspar cement can be assigned to Ca2–Ca3;

2 subsequent, relatively straight, brittle cracks, with a passive Ca3 or Ca4 spar infill (e.g. Fig. 5B). These may also contain pyrite, blocky kaolinite and late-stage quartz.

CALCITE COMPOSITION

Stable isotopes

Concretions in the SD sandstones typically contain a complex mixture of at least three calcite phases on a sub-millimetre scale. For example, early Ca2–Ca3-cemented sandstones contain calcitic bioclasts, Ca4 cement within sponge spicule moulds and local veinlets of Ca5 (Fig. 4D). Ca4-cemented samples also contain bioclasts,

Table 2. 'Whole-rock' and fracture-fill calcite stable isotope results. See Fig. 6 for identification of concretion sample traverses.

Whole-rock calcites				Fracture-fill cements		
Dominant cement	Traverse (Fig. 6)	$\delta^{13}C$ (‰PDB)	$\delta^{18}O$ (‰PDB)	Cement stage	$\delta^{13}C$ (‰PDB)	$\delta^{18}O$ (‰PDB)
Ca2	A	−0·8	−3·1	Ca2	−0·3	−1·3
Ca2	A	−0·8	−3·2	Ca2	−0·5	+0·5
Ca2	A	−0·7	−2·9	Ca2	−1·3	+0·5
Ca2	A	−0·9	−2·8	Ca2	−0·9	−0·0
Ca2	B	−1·3	−1·9	Ca3	−4·5	−1·4
Ca2	B	−1·2	−1·9	Ca3	−5·4	−0·6
Ca2	B	−1·0	−2·7	Ca3	−1·5	−2·7
Ca2	B	−0·9	−2·8	Ca3	−1·5	−3·1
Ca2	B	−0·5	−3·5	Ca4	−0·3	−10·8
Ca2	B	−1·3	−5·7	Ca4	−0·3	−11·4
Ca2	B	−0·7	−1·9	Ca4	−1·2	−9·8
Ca2	B	−1·3	−2·2	Ca4	−1·1	−11·4
Ca2	B	−1·2	−3·2	Ca4	−1·5	−9·1
Ca2	C	−0·9	−0·7	Ca4	−0·3	−8·9
Ca2	C	−1·6	−1·3	Ca4	−2·6	−11·3
Ca2	C	−2·5	−2·6	Ca4	−2·7	−10·7
Ca2	C	−1·1	−1·5	Ca4	−2·6	−11·2
Ca2	—	−1·2	−2·7			
Ca2	—	−1·3	−4·2			
Ca2	—	−1·2	−5·0			
Ca2	—	−0·4	−1·2			
Ca3	A	−2·2	−3·0			
Ca3	A	−1·8	−4·0			
Ca3	C	−3·1	−2·8			
Ca3	C	−2·2	−1·8			
Ca3	—	−6·3	−2·2			
Ca3	—	−2·5	−1·6			
Ca3	—	−2·4	−2·1			
Ca3	—	−4·2	−2·5			
Ca3	—	−1·8	−1·6			
Ca4	—	−2·7	−5·5			
Ca4	—	−7·0	−7·7			
Ca4	—	−5·7	−8·2			
Ca4	—	−1·1	−9·0			
porous sst	A	−0·0	−5·6			
porous sst	B	−0·8	−4·5			
porous sst	C	−2·6	−5·7			
porous sst	—	−0·2	−5·4			
porous sst	—	−1·7	−7·3			
porous sst	—	−1·3	−7·4			

sometimes with earlier Ca2 and/or Ca3 rim cements. Most of the analyses carried out were of a 'whole-rock' nature, sample powders being taken from petrographically screened samples and being of sufficient size to ensure representative proportions of each calcite phase (for simplicity, samples with Ca5 veinlets were avoided). Stable isotope traverses were made across three concretions, and several additional samples were extracted from cemented sandstones with a predominant Ca2, 3 or 4 intergranular cement (Table 2). Data trends were interpreted in terms of multicomponent mixing, based on visual approximation of the relative proportions of the different calcite phases. These trends were confirmed by microsampling individual cement phases from fracture fills (Table 2).

Results of the concretion traverses are plotted on Fig. 6, and the data are combined with the remaining whole-rock and fracture cement analyses on Fig. 7. The results suggest that:

1 Ca2–Ca3 cements possess similar $\delta^{18}O$ values, ranging between +0·5 and −3·1‰ PDB; in contrast Ca4 cement has much more negative $\delta^{18}O$ values,

between −8·9‰ and −11·4‰ PDB. Because of common Ca4-replaced sponge spicules, most of the Ca2-cemented sandstone 'whole-rock' samples fall between −1 and −4‰ PDB;

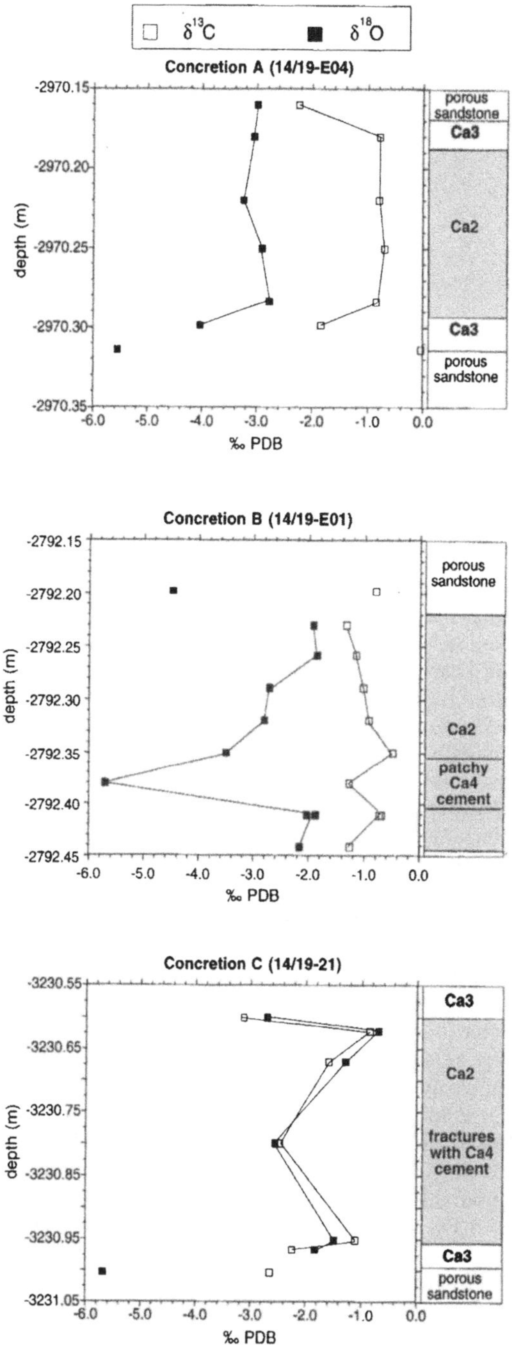

2 Ca2 cements have $\delta^{13}C$ values of about 0 to −2‰ PDB, but Ca3 values are several per mil more negative (to ≤ −6·3‰ PDB, based on a Ca3-cemented 'whole-rock' sample). $\delta^{13}C$ values of Ca4 cement are between −0·3 and −2·7‰ PDB;

3 the gross trends between Ca2-, Ca3- and Ca4-cemented sandstone samples are generally analogous to those of the fracture cement separates (Fig. 7), showing that it is reasonable to assume consistent proportions of calcitic bioclasts and Ca4-replaced sponge spicules within the 'whole-rock' sample suite. An exception is the two Ca4-cemented sandstone samples, which have much more negative $\delta^{13}C$ values (−5·7 and −7·1‰ PDB) than the pure Ca4 fracture cements. Although CL examination suggests that these samples also contain some Ca3 cement pre-dating the Ca4 spar, the data indicate that there must be some local variation in Ca4 $\delta^{13}C$ composition. One of these samples is from the same well as concretion 'C'. Covariation of oxygen and carbon isotope compositions in the 'C' traverse (Fig. 6) suggests that the Ca4 cement within it is depleted in both ^{18}O and ^{13}C with respect to surrounding Ca2 spars;

4 bulk carbonate in typical concretions contains roughly 60–70% intergranular Ca2 cement with subequal calcitic bioclasts and biomouldic Ca4, whereas porous sandstones contain more bioclastic than authigenic calcite. However, 'whole-rock' porous sandstones sampled adjacent to concretions 'A–C' are up to 3‰ more negative in $\delta^{18}O$ than the concretion samples. This suggests that the bioclast grains may, on average, be more depleted in ^{18}O than the Ca2 cement.

Trace elements

Four thin sections from different concretions were probed for Ca2 and Ca4, and a single thin section was probed for Ca1 (Table 3). All of the sampled cements are low-Mg calcites (<2 mol% $MgCO_3$), although Ca1 and Ca2 are relatively enriched in Mg compared with Ca4. They also have relatively more Sr, but less Fe and Mn than

Fig. 6. 'Whole-rock' stable isotope sample traverses through three SD sandstone concretions. CL and stained peel petrography shows that the intergranular cement of concretion 'A' is mainly Ca2, with Ca3 on the upper and lower rims. Concretion 'B' is also dominated by Ca2 but contains some patches of intergranular Ca4. Concretion 'C' contains patches of Ca3 cement and some fracture-hosted Ca4, surrounded by dominant Ca2 spars. All concretions contain calcitic bioclasts and Ca4-filled sponge spicule moulds.

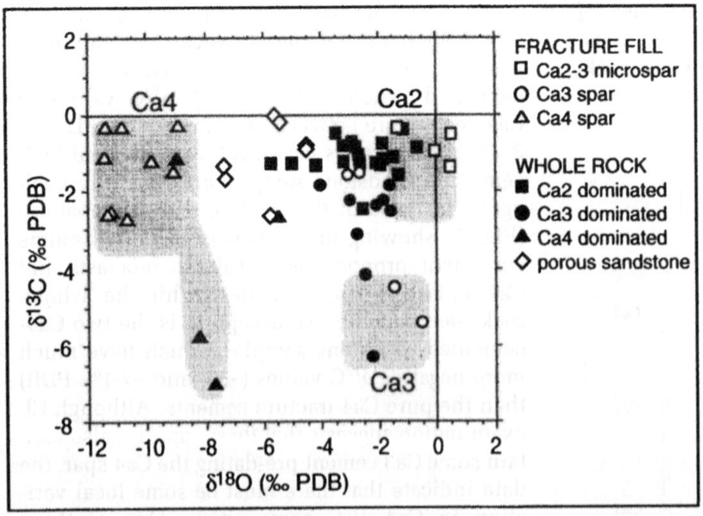

Fig. 7. $\delta^{13}C$–$\delta^{18}O$ scatter plot of calcite-cemented SD sandstone samples and fracture-hosted cement separates. Shaded fields represent extrapolated approximate end-member compositions of Ca2, Ca3 and Ca4 (based on whole-rock and cement data).

the later cement (Fig. 8). There is clearly some intersample variability in Ca2 chemistry, and in the Fe and Mn compositions of Ca4; the relatively large range in the latter is compatible with pronounced sectoral zonation visible in CL (cf. Hendry & Marshall, 1991). Scatter in the Ca2 analyses may also reflect sectoral zonation, particularly of Mg (which would not be recorded by CL). Very dull luminescence of Ca2 is surprising given its relatively low Fe content and Mn/Fe ratio >1; brighter luminescence in Ca4 presumably reflects higher absolute Mn concentrations, although the influence of other trace cations may also be important (Savard *et al.*, 1995).

DISCUSSION

Cementation environments

Early cements

Ca2 cement accounts for the majority of concretions within the Scapa SD unit sandstones. The abundance of calcitized sponge spicules and other highly inequant bioclasts prohibits estimation of cementation depth from minus-cement porosities. However, Ca2 cementation took place prior to opaline sponge spicule dissolution (Figs 2 and 4D), indicating relatively shallow burial (cf. Hendry & Trewin, 1995). The Scapa member was deposited in an open-marine basin (Boote & Gustav, 1987), and provided that reasonable assumptions are made about depositional (seafloor) palaeotemperatures and isotopic equilibrium attainment, oxygen isotopes provide a means of estimating cement precipitation depths. Increased temperature during shallow to moderate (<3 km) sediment burial typically results in increasingly negative cement $\delta^{18}O$ values. Pore fluid ^{18}O-enrichment is highly unlikely in high water-to-rock ratio conditions pertaining to clay-poor sandstones (cf. Marshall, 1982). Consequently, pure Ca2 cements are unlikely to have a much more negative $\delta^{18}O$ than the Ca2–3 microspars in early fractures which cross-cut them. On this basis, a reasonable estimate for the $\delta^{18}O$ of pure Ca2 cement is +0·5 to −1·5‰ PDB (Fig. 7).

Assuming equilibrium fractionation, the $\delta^{18}O$ of a calcite precipitate (δ_c) from parent water (δ_w) varies as a function of temperature according to the Craig (1965) relationship, modified for current international standards by Anderson & Arthun (1983):

$$T\,(^\circ C) = 16{\cdot}0 - 4{\cdot}14(\delta_c - \delta_w) + 0{\cdot}13(\delta_c - \delta_w)^2.$$

Applying this relationship to Ca2 suggests precipitation from Cretaceous nonglacial oceanid water (*c.* −1‰ SMOW; Hudson & Anderson, 1989) over a temperature range of about 10–18 °C. Bearing in mind a Hauterivian palaeolatitude of the Outer Moray Firth of about 45°N (Smith *et al.*, 1981), and given estimated water depths of about 150 m (O'Driscoll *et al.*, 1990), this compares favourably with Hauterivian palaeotemperatures of ~12–15 °C from benthic foraminifera in the north-western Pacific (Hudson & Anderson, 1989), and 13–15 °C from belemnites in Germany (Spaeth *et al.*, 1971). Furthermore, the data clearly preclude the involvement of meteoric water at

Table 3. Statistical summary of EPMA results from Ca1, Ca2 and Ca4 cements.

Cement stage	Sample		$MgCO_3$ (mol%)	$MnCO_3$ (mol%)	$FeCO_3$ (mol%)	$SrCO_3$ (mol%)	*m*Mg/*m*Ca
Ca1	14/19-E02	max.	1·848	0·347	1·717	0·112	0·0189
	2769·9m	min.	0·421	0·000	0·014	0·000	0·0043
		avg.	1·300	0·079	0·469	0·050	0·0133
		SD	0·394	0·082	0·547	0·027	0·0040
Ca2	14/19-E01	max.	1·087	1·062	0·082	0·088	0·0111
	2792·4m	min.	0·200	0·498	0·000	0·036	0·0020
		avg.	0·664	0·720	0·042	0·060	0·0068
		SD	0·253	0·177	0·019	0·015	0·0026
	14/19-E02	max.	1·911	0·479	0·146	0·153	0·0196
	2792·5m	min.	0·613	0·218	0·002	0·048	0·0062
		avg.	1·319	0·334	0·078	0·099	0·0135
		SD	0·318	0·063	0·039	0·024	0·0033
	14/19-E09	max.	1·437	0·764	0·097	0·126	0·0146
	3374·6m	min.	0·278	0·241	0·000	0·028	0·0028
		avg.	0·773	0·502	0·055	0·059	0·0079
		SD	0·337	0·155	0·026	0·020	0·0035
	14/19–25	max.	1·654	0·963	0·068	0·175	0·0170
	2742·9m	min.	0·866	0·557	0·000	0·045	0·0088
		avg.	1·209	0·746	0·036	0·107	0·0124
		SD	0·207	0·113	0·022	0·038	0·0022
	All Ca2	max.	1·910	1·060	0·150	0·180	0·0200
		min.	0·200	0·220	0·000	0·030	0·0020
		avg.	0·986	0·576	0·053	0·081	0·0100
		SD	0·396	0·215	0·032	0·034	0·0041
Ca4	14/19-E01	max.	0·411	2·182	1·743	0·062	0·0043
	2792·4m	min.	0·050	0·639	0·610	0·015	0·0005
		avg.	0·179	1·424	1·284	0·034	0·0019
		SD	0·107	0·524	0·333	0·013	0·0011
	14/19-E02	max.	0·436	2·286	2·532	0·052	0·0045
	2792·5m	min.	0·088	0·355	0·290	0·011	0·0009
		avg.	0·247	1·172	1·614	0·031	0·0026
		SD	0·130	0·646	0·510	0·012	0·0014
	14/19-E09	max.	0·380	1·918	2·998	0·086	0·0039
	3374·6m	min.	0·029	0·235	0·322	0·016	0·0003
		avg.	0·191	1·179	1·166	0·037	0·0020
		SD	0·107	0·586	0·697	0·018	0·0011
	14/19–25	max.	0·850	2·030	0·836	0·064	0·0086
	2742·9m	min.	0·101	0·047	0·202	0·018	0·0010
		avg.	0·279	1·241	0·352	0·037	0·0029
		SD	0·234	0·507	0·153	0·016	0·0024
	All Ca4	max.	0·850	2·290	3·000	0·090	0·0090
		min.	0·030	0·050	0·200	0·010	0·0000
		avg.	0·224	1·254	1·124	0·034	0·0023
		SD	0·154	0·565	0·652	0·014	0·0016
DIS std		max.	0·379	0·259	0·230	0·055	0·0038
		min.	0·253	0·105	0·105	0·000	0·0025
		avg.	0·333	0·178	0·180	0·025	0·0034
		SD	0·025	0·034	0·028	0·011	0·0002

realistic temperatures, particularly as regional early Cretaceous continental runoff was probably at least as negative as –8‰ SMOW (Hendry, 1993a). Likewise, precipitation at elevated temperatures during burial can be discounted as requiring unrealistically ^{18}O-enriched pore fluids (e.g. ≥ +5·5‰ SMOW for 50 °C).

Although the isotopic composition of Ca1 cements could not be directly measured, they must have formed prior to significant burial to comply with the Ca2 data. Transition from Ca1 to Ca2 involves a drop in Fe concentration (Fig. 8) and it is tempting to equate Ca2 with the zone of near-surface bacterial sulphate reduction (cf.

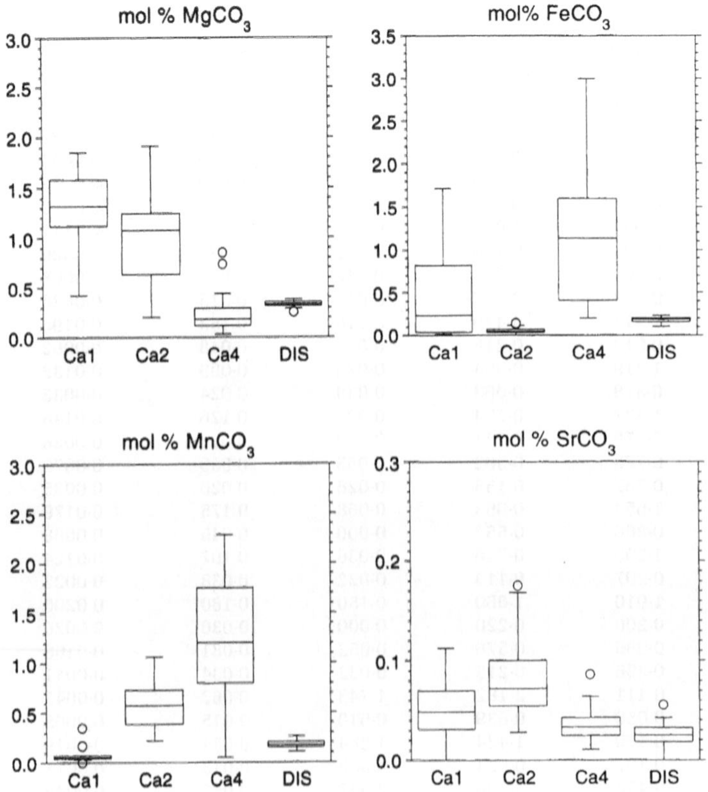

Fig. 8. Box-and-whisker plots of EPMA data from Ca1, Ca2 and Ca4 cements, plus Derbyshire Iceland spar (DIS) standard. The plots are based on a cumulative frequency distribution for each data population. Boxes include 50% of the data, from the 25th to 75th percentiles, with the median values indicated by horizontal bars. The whiskers extend to the upper and lower ranges of the data, with statistical outliers denoted by open circles. Determination of absolute EPMA accuracy is impossible, but the following error estimates are based on comparison of DIS data with pooled results of several independent, wet-chemical analyses of the same standard: $FeCO_3$, $\pm 0{\cdot}06$ mol%; $MgCO_3$, $\pm 0{\cdot}05$ mol%; $MnCO_3$, $\pm 0{\cdot}07$ mol%; $SrCO_3$, $\pm 0{\cdot}02$ mol%.

Hendry, 1993b). Banded non- to bright or dull luminescence in Ca1 would then be compatible with low and fluctuating Eh conditions in the

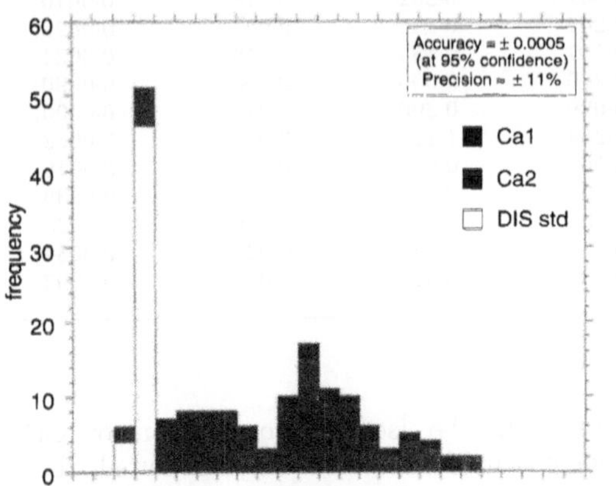

Fig. 9. Histogram of Mg/Ca ratios in Ca1 and Ca2 cements, plus Mg/Ca ratios from repeated analysis of Derbyshire Iceland spar standard.

oxic to post-oxic environments (e.g. Barnaby & Rimstidt, 1989; Coniglio, 1989). Post-depositional Eh decline in marine sediment pore fluids is governed by the progressive microbial oxidation of organic matter (Froelich *et al.*, 1979; Curtis, 1987). However, the episodic depositional nature of turbidite systems disrupts O_2 and SO_4^{2-} concentration gradients in the uppermost sediment, and allows for a more complex redox history during early burial (e.g. Wilson *et al.*, 1985; Buckley & Cranston, 1988; Burdige, 1993). This would agree with the local consistency but interbed variability of Ca1 zonation sequences.

There is no evidence for a major break in precipitation between Ca2 and Ca3, and the most isotopically light Ca3 spar ($\delta^{18}O = -3{\cdot}1$‰ PDB) would have formed in equilibrium with marine pore fluid at 25 °C. Ca2–3 data therefore imply that early cementation forming the majority of concretions took place very shortly after deposition of the Scapa sandstones, with minor cementation continuing during several hundred metres burial (assuming an average 30 °C km^{-1} geothermal gradient based on present-day bottom hole temperatures). The presence of Ca2–3 in irregular fractures which must have formed at shallow burial depths supports this interpretation.

Late cement

The $\delta^{18}O$ shift of 5–7‰ between Ca3 and Ca4 suggests a much deeper burial origin for the late cement, as does the textural and paragenetic data. Depletion in ^{18}O could also be interpreted in terms of an influx of meteoric waters, but this is untenable on the basis of reconstructed subsidence curves for the Scapa Field (Hendry, unpublished data). Furthermore, the Scapa Member is encased within an argillaceous marine succession and is unlikely to have hosted any regional groundwater aquifers, such as those inferred from the diagenesis of Tertiary turbidites adjacent to the East Shetland Platform (Watson *et al.*, 1995). An isotopically invariant pore fluid $\delta^{18}O$ of −1‰ SMOW yields Ca4 precipitation temperatures of 50–70 °C, equivalent to between 1 and 2 km burial.

Formation of early calcite cements

A marine eogenetic origin for early calcite cement in the Scapa Member places tight constraints upon possible sources of calcium and bicarbonate. Low solubility of $CaCO_3$ means that between 10^3 and 10^4 pore volume exchanges of supersaturated pore fluid are required to cement a sandstone (Bjørkum & Walderhaug, 1990a). The hydrodynamics required for external sourcing of the calcite cement would only be accomplished with a long-lived meteoric aquifer, the existence of which has already been ruled out. Fluid flux is less important if the cementation represented a mineral-controlled transformation, in which dissolution of aragonitic bioclasts and precipitation of calcite cement is thermodynamically driven (James & Choquette, 1990). Furthermore, aragonitic bioclasts provide the only sufficient source of Ca^{2+} for cementation, as early calcite cements pre-dated any feldspar dissolution.

Slightly negative $\delta^{13}C$ values of Ca2 cement (Fig. 7) argue for a predominantly marine bicarbonate source, although the absence of any $\delta^{13}C$ values greater than zero may point to a subsidiary component of organogenic carbon in the pore fluids. The coincidence of early calcite cementation with the shallow zones of microbial organic matter oxidation has already been suggested on compositional and petrographic grounds. The Scapa Member sandstones are unlikely to have been highly enriched in organic material at deposition, but only a minor contribution of bicarbonate from sulphate reduction (*c.* −25‰ PDB; Curtis, 1987) would be required to produce the Ca2 cements with $\delta^{13}C$ values of 0 to −2‰ PDB.

Ca3 cements have more negative $\delta^{13}C$ values than Ca2, suggesting either an increased organic carbon component due to more efficient oxidation of organic matter and/or a relative diminution of the marine carbon (aragonite) supply. The latter is supported by the fact that Ca3 cements are much less abundant than Ca2, and generally much coarser grained (Table 1), both of which suggest relatively lower supersaturation of the pore fluids (cf. Wilkinson, 1993a). Ca2–3 spars were unable to occlude relatively large intergranular and biomouldic voids, particularly in the coarsest pebbly sandstones and sand-matrix conglomerates (Fig. 5F). Precipitation presumably ceased when the supply of aragonite-derived $CaCO_3$ was exhausted. Likewise, it is reasonable to assume that volumetrically minor Ca1 cements precipitated at the earliest stage of aragonite dissolution, with carbonate supplied from the most unstable bioclasts and possibly from microbial oxidation of the most reactive organic matter.

A wide variety of stenohaline calcitic taxa are present in the SD sandstones, but formerly aragonitic fossils are not preserved. There is direct evidence for their former presence in pebbly sandstones and sand-matrix conglomerates (e.g. Fig. 5A,F), but little unequivocal evidence in finer grained sandstones. Nevertheless, Cretaceous shallow-marine faunas are likely to have been mollusc-dominated and relatively aragonite-rich, and ample detrital aragonite can be assumed to have been present throughout the sandstones at deposition. Dissolution of finely comminuted aragonitic debris would leave no textural record in an unlithified sandstone matrix, because the packing of detrital silicate grains could readjust during burial and compaction.

Origin of late calcite cements

The source of Ca4 cement is uncertain, but can be assumed to have been external to the sandstones given that no aragonite bioclasts survived intact beyond Ca2–3 authigenesis and none of the calcitic bioclasts or early cements display evidence of dissolution (e.g. Fig. 5C). Resurgence of calcite Fe^{2+} and Mn^{2+} contents (Fig. 8) also argues for an extraformational fluid source. The variable, but generally small, negative $\delta^{13}C$ values of Ca4 suggest that late-stage bicarbonate had an ultimately marine parentage, with minor organogenic input. The Scapa Member is surrounded by hemipelagic marls of the Valhall Formation and underlain by organic-rich Kimmeridge Clay mudrocks, both of which may have exported Ca^{2+} and bicarbonate into the sands during burial dewatering.

SPATIAL DISTRIBUTION OF EARLY CEMENT

In addition to forming tightly cemented concretions, Ca2 cements are present as epitaxial overgrowths on calcitic bioclasts throughout the intervening porous sandstones. Understanding the origin of this cementation fabric is important for the prediction of overall reservoir quality and potential subsurface flow barriers. In simple terms the relationship implies that hydrological or geochemical heterogeneities localized the precipitation of Ca2 some time after it had nucleated on all the available substrates.

Spatial confinement of calcite cement in sandstones is influenced by three main factors (Irwin & Hurst, 1983): (i) detrital mineral/bioclast distribution; (ii) primary poroperm variations; (iii) location of organic-rich mudrocks relative to the sandstones. Only the first of these is important for eogenetic and autochthonous cementation exemplified by Ca2. This is because mass transport during shallow burial would have been dominated by pore-scale to bed-scale ionic diffusion, the rates of which are independent of permeability over the range typically encountered in sandstones (Wilkinson & Dampier, 1990). Consequently, bioclast distribution is likely to have controlled the spatial distribution of cement. Patterns of calcite cementation under hydrodynamically quiescent conditions are dependent upon the distribution of suitable calcitic nuclei and/or on localized and transient fluctuations in pore fluid supersaturation (Bjørkum & Walderhaug, 1990a; Wilkinson, 1993a). In either case, cementation should be volumetrically concentrated in the most bioclastic regions of a sandstone. For this reason, calcitic concretions in shallow-marine sandstones are frequently present at sedimentary heterogeneities such as condensed horizons, storm deposits and channel lags (Bryant *et al.*, 1988; Bjørkum & Walderhaug, 1990b; Gibbons *et al.*, 1993).

Most of the SD sandstones are poorly stratified and, where they are laminated, the sedimentary layering is one or two orders of magnitude finer in scale than the Ca2 concretions. Calcitic shell fragments are evenly distributed through the sandstones (within limits of visual estimation), and there is no reason to doubt that aragonitic bioclasts were dispersed in a similar fashion. Carbonate supply and suitable nucleation sites would consequently have been available throughout the sandstone beds, and the development of spatially segregated concretions was most probably influenced by kinetic factors during cementation.

Eogenetic concretion growth in sediments can be modelled as taking place under transport and/or surface reaction control (Berner, 1980; Wilkinson & Dampier, 1990). In the first case, the rate-limiting step is diffusion of ionic species to the growth interface, whilst in the second it is the various physiochemical processes involved in dehydration of the ions and incorporation into the calcite lattice. Once concretions have started to grow under transport-control, development of a $CaCO_3$ concentration gradient may reduce supersaturation sufficiently to curtail precipitation on available substrates within a 'range of influence' around the fastest growing concretions (Bjørkum & Walderhaug, 1990a). The radius of this non-precipitation zone increases rapidly at first, and then much more slowly, governed by the rate of growth of the central concretion. Whilst the initial cement nucleation sites are probably dictated by substrate availability (Wilkinson, 1993b), cement growth (and formation of concretions) in this situation would be expected to show a non-random distribution (Bjørkum & Walderhaug, 1990a).

Geochemical modelling by Wilkinson & Dampier (1990) suggests that elevated Mg^{2+}/Ca^{2+} ratios characteristic of marine pore fluids favour concretion growth under surface reaction rather than transport controls. In this situation, concentration gradients are not developed and cementation would continue on all the initial substrates. Similar patchy cementation would be expected under hydrodynamically active conditions, in which pore fluid flow outpaced ionic diffusion rates (Walderhaug & Bjørkum, 1992). However, shallow sedimentary basins typically have compactional fluid flow rates of *c.* 10^{-2}–10^{-3} cm yr^{-1}, which exert negligible influence on diffusion-controlled mass transport over typical bed to formation scales (Bjørkum & Walderhaug, 1990a; Bjørlykke, 1993).

In the absence of active pore fluid flow, the pattern of Ca2 distribution apparently records a switch from initial surface-reaction to subsequent transport-controlled kinetics during precipitation. This is not unreasonable, given that aragonite dissolution usually outpaces calcite precipitation during early transformation stages (Mucci, 1983; Walter, 1986). High supersaturations would therefore have been present throughout the sandstone, allowing Ca2 (± preceding Ca1) cementation to occur on all suitable nuclei. As net supersaturation fell with progressive calcite precipitation, transport-controlled kinetics became dominant and concentration gradients developed around the fastest-growing nuclei. These would then have developed into concretions. Subtle variations in bioclast distribution within individual beds would have been sufficient to determine which concretion nuclei were 'successful'. Such variations probably resulted from subtle depositional heterogeneities within the generally well-mixed turbidite sandstones.

Because of the abundance of calcitic bioclasts with Ca2 overgrowths, it is likely that concretions developed by local impingement of cement clusters, rather than in a radially directed fashion. This mode of concretion growth has been recognized in mudrocks (Feistner, 1989; Mozley, 1989), and may account for some patches of Ca2-free sandstone within concretions, which were subsequently filled by Ca3 or Ca4 (Fig. 5D). Alternatively, the presence of partially pyritized plant fragments within many of these areas could indicate that highly localized pH changes during microbial reactions temporarily impeded Ca2 precipitation around the plant fragments.

The vertical distribution of concretions within the SD cores suggests a maximum 'range of influence' on the centimetre–decimetre scale during cementation. Whilst concretion thicknesses are similar to those in many shallow-marine sandstones, the frequency per unit vertical thickness (e.g. Fig. 3) is somewhat greater (cf. Walderhaug *et al.*, 1989). This probably reflects a greater relative abundance of labile carbonate in the Scapa Member than in typical shelf sandstone deposits. A relatively uniform bioclast distribution in the turbidites would be predicted to produce dispersed concretions, none of which would persist laterally over a distance significantly greater than its thickness. A corresponding absence of areally extensive, stratiform, cemented horizons is in full agreement with reservoir performance data (Harker & Chermak, 1992).

Variability in concretion thickness most probably reflects changes in bed-to-bed aragonite content. This would imply mass transfer on a very local scale, with calcite precipitation taking place only millimetres or centimetres away from the sites of aragonite dissolution. Supporting evidence includes a paucity of early calcite cement in bioclast-poor heterolithic sands, its relative abundance in bioclast-rich conglomeratic sands and the spatial variability of Ca3 $\delta^{13}C$ values which suggest a geochemically closed system on a bed-to-bed scale.

Concretion shape is difficult to assess in core. The sedimentary facies and concretion growth model discussed above predicts approximately spherical morphologies. Many examples in the SD sands possess slightly curved upper and lower margins which are compatible with a quasispherical shape. More irregularly shaped or tabular concretions are also present, and suggest that a subtle stratification of bioclast content promoted relatively higher lateral than vertical growth rates (in a diffusional mass-transport system the effect of fine-scale permeability heterogeneities would be unimportant). Likewise, irregular concretions probably represent lateral merging of individual nodules in beds with a heterogeneous bioclast distribution. As net supersaturation fell, concentration gradients around individual concretions were diminished, allowing eventual impingement of adjacent growth centres. In contrast, irregular and patchy cementation in pebbly and conglomeratic sandstones is probably related to significantly higher concentrations of bioclastic aragonite, and correspondingly prolonged episodes of surface-reaction controlled cementation in these beds.

SIGNIFICANCE OF MARINE LOW-Mg CALCITE CEMENT

Abiotic, marine calcites are generally regarded as having elevated Mg contents. For example, marine calcite cement formed in tropical shoal and reefal environments contains between 12 and 19% $MgCO_3$ (Tucker & Wright, 1990). Deep-sea calcite cements related to methane seeps also contain high Mg concentrations (Hovland *et al.*, 1987). In contrast, calcites with less than 2 mol% $MgCO_3$ are more typical of cementation from low-Mg^{2+}/Ca^{2+} meteoric waters or deep burial fluids. Petrographic and isotopic evidence indicates that low Mg concentrations in Ca2 are probably primary, and therefore result from one or more factors specific to the timing and environment of cementation. Several possibilities can be considered (Morse & Mackenzie, 1990; Burton & Walter, 1991; Meyers, 1991; Burton, 1993): (i) recrystallization of high-Mg calcites; (ii) elevated PCO_2 levels; (iii) low temperatures; (iv) anomalous seawater chemistry.

Biogenic Mg-calcites commonly undergo a fabric-retentive loss of Mg during early diagenesis; however, if any Mg-calcite cements were present in the samples examined they probably formed the spired overgrowths on echinoderm grains which were encased by Ca1 as described above. It is also noteworthy that porous sandstones which contained a relatively far greater proportion of Mg-calcite echinoderm grains to Ca2 than concretions yet have more negative 'bulk' $\delta^{18}O$ values (Fig. 7). This suggests that echinoderm grains have been stabilized during diagenesis, whilst Ca2 is a primary low-Mg calcite precipitate.

High PCO_2 levels can be generated by organic matter oxidation, and may partially account for 'marine' low-Mg calcite precipitation in mudrock-hosted concretions. Negative $\delta^{13}C$ values (of Ca3 in particular) suggest comparable processes in the Scapa sandstones, but the relatively high permeability of the sediments mitigates against build up of PCO_2 to the same extent as in organic-rich muds. Inverse temperature dependence of Mg partitioning in calcite is well documented, and abiotic low-Mg calcite cements are present in Tertiary to Recent deep-water periplatform environments (Schlager & James, 1978; Malone *et al.*, 1990). However, precipitation temperatures were considerably less than 10–18 °C estimated for Ca2. Synsedimentary calcite cements in high-latitude Oligocene shelf sediments contain at least 8 mol% $MgCO_3$ despite precipitation at 10 °C (James & Bone, 1992), and marine calcite cements recovered from 500 m bathymetry (about 10 °C) in the Bahamas contain 10–18 mol% $MgCO_3$ (Major & Wilber, 1991). Therefore, formation of marine low-Mg calcite cement in the Scapa Member most probably reflects a lower seawater Mg^{2+}/Ca^{2+} ratio than at present.

Long-term variation in ocean chemistry has been recognized from mineralogical changes in abiotic shallow-water precipitates, and is believed to be ultimately controlled by plate-tectonic processes (Sandberg, 1983; Wilkinson *et al.*, 1985). The late Mesozoic was characterized by relatively high global rates of plate movement coupled with eustatic maxima. As a consequence there was widespread carbonate deposition, and diagenetic–metamorphic reactions at subduction zones increased oceanic PCO_2 levels, promoting dolomitization and Mg^{2+} scavenging from seawater (Wilkinson & Algeo, 1989). Increased hydrothermal alteration at mid-ocean ridges further reduced the net oceanic Mg^{2+}/Ca^{2+} ratio. Various empirical and theoretical studies suggest early Cretaceous atmospheric PCO_2 levels were 4–12 times current values (Berner, 1992, and references therein; Berner, 1994). This alone would be insufficient to account for precipitation of marine calcite with <2 mol% $MgCO_3$ (Burton & Walter, 1991), but a combination of diminished seawater Mg^{2+}/Ca^{2+} and elevated PCO_2 in the early Cretaceous 'calcite seas' was probably adequate to account for precipitation of low-Mg Ca2 cement at <20 °C. Build up of pore-fluid Mg^{2+}/Ca^{2+} during cementation may have been curbed by diffusion of Mg^{2+} into the subadjacent water column.

The Mg^{2+}/Ca^{2+} ratio of the Scapa pore fluids can be estimated in a simple calculation. The mMg/mCa value of Ca2 cement ranges between 0·002 and 0·020 (Fig. 10), and the mean value is 0·010. Assuming equilibrium partitioning of Mg^{2+} in an open system, the calcite and pore fluid compositions are related by the homogeneous distribution coefficient, D_{Mg}:

$$[mMg^{2+}/mCa^{2+}]_{calcite}/D_{Mg}=[mMg^{2+}/mCa^{2+}]_{fluid}.$$

Upper and lower limits on D_{Mg} can be assumed to be 0·014, determined for low-Mg calcite at 15 °C by Oomori *et al.* (1987), and 0·048, determined for the aragonite to low-Mg calcite transformation by Katz (1973) and extrapolated to 15 °C. This range includes the value of 0·030 for direct precipitation of low-Mg calcite at 25 °C determined by Howson *et al.* (1987). These values yield aqueous Mg^{2+}/Ca^{2+} ratios of 0·2–0·6 based on the mean Ca2 composition. An upper value of 1·4 is obtained using the most Mg-rich analysis and the lower of the two D_{Mg} values. Despite uncertainties inherent in the calculation, the Ca2 compositions strongly suggest that early Cretaceous seawater had a Mg^{2+}/Ca^{2+} ratio of ≤ 1. This value is very close to the theoretical prediction made by Wilkinson & Algeo (1989) for the early Cretaceous.

MARINE ARAGONITE DISSOLUTION

An eogenetic origin for early calcite cements in the Scapa Member implies that aragonite dissolution began shortly after deposition and in marine pore fluids. Aragonite solubility in seawater increases with decreasing temperature

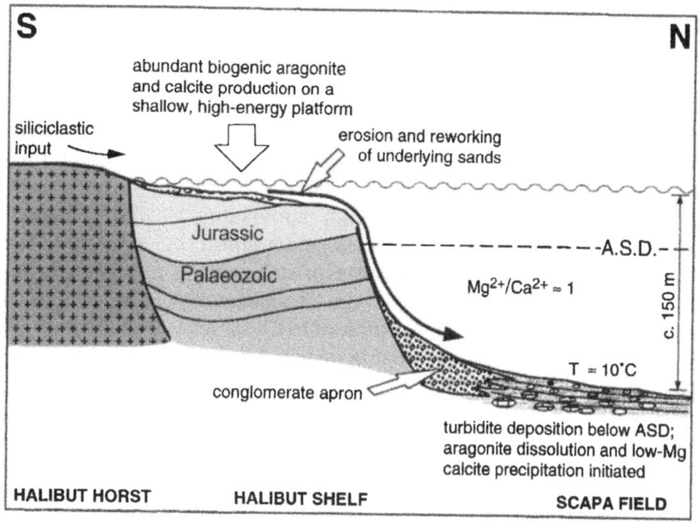

Fig. 10. Simplified model for deposition and early carbonate diagenesis of Scapa Member sandstones. A.S.D.= aragonite saturation depth. See text for discussion.

and increasing pressure and PCO_2. Consequently, ocean waters become progressively less saturated with respect to aragonite with increasing depth until the aragonite compensation depth (ACD), below which there is no net aragonite accumulation on the sea floor. The depth at which ocean water becomes undersaturated with respect to aragonite (aragonite saturation depth or ASD) depends ultimately upon oceanic circulation and chemistry, and currently varies from an average of about 500 m in the Pacific Ocean to almost 3000 m in the Atlantic Ocean (Morse & Mackenzie, 1990). Rapid aragonite to calcite transformation occurs during early diagenesis of modern periplatform sediments deposited in the Indian Ocean at about 540 m below sea level, going to completion by about 160 m of burial and in less than 6 Myr (Malone *et al.*, 1990). A similar or shorter time-scale is likely for aragonite dissolution in the SD sandstones, particularly if Ca2 precipitation took place in the sulphate reduction zone.

A relative shallowing of calcite compensation depths in late Mesozoic ocean basins (Berger & Winterer, 1974; Van Andel, 1975) attests to the control of the plate tectonic processes discussed above upon carbonate saturation levels in global seawater. A major consequence of the early Cretaceous 'calcite-sea' chemistry (particularly increased PCO_2) would have been a lower net saturation with respect to aragonite, and hence elevation of the aragonite saturation depths with respect to modern oceans. This did not prevent metabolic biomineralization of aragonite, but almost certainly increased its susceptibility to *post-mortem* dissolution on the sea floor or during very shallow burial. Low-Mg^{2+}/Ca^{2+} seawater would have been particularly favourable to the aragonite–calcite transformation, which is catalysed in the presence of aqueous NaCl but inhibited by Mg^{2+} (Morse & Mackenzie, 1990). 'Calcite seas' extended through the Cretaceous (Sandberg, 1983), and there is recent evidence for marine aragonite dissolution in late Cretaceous reefs (Opdyke & Wilson, 1993; Opdyke *et al.*, in press). Seawater in shallow marine environments can be chemically modified by water–rock interaction, evaporation, photosynthetic activity and other biological processes. However, early Cretaceous deep-seawater would probably have retained a composition close to the global average at the time. The Scapa depositional system reworked biogenic material from the shallow Halibut shelf into deeper water, probably close to or below the ambient ASD. Consequently, aragonite dissolution would have begun immediately, and the temperature, PCO_2 and Mg^{2+}/Ca^{2+} of the ambient pore water were such that low-Mg calcite was the thermodynamically stable precipitate (Fig. 10).

Aragonite to calcite transformations are rapid on geological time-scales (e.g. Malone *et al.*, 1990). In this respect, the CL homogeneity of Ca2 is compatible with rapid precipitation relative to shallow-burial microbial organic–inorganic interactions (cf. Hendry, 1993b). Once calcite begins to precipitate, the process is driven to completion by its solubility contrast relative to dissolving aragonite, and the pore fluid is buffered between aragonite and calcite saturation levels (Berner, 1980). Consequently, biogenic low-Mg calcites in the SD sands, such as coccoliths, display no evidence of dissolution (Fig. 5C). Because diverse aragonitic bioclasts encompass a range of solubilities, they fit the requirement of a *finite* $CaCO_3$ source which could maintain high supersaturations throughout the Ca1–Ca2–Ca3 cementation episode. Thus, early calcite cement was sufficient to occlude intergranular porosity in fine-grained sandstone concretions, but insufficient to fill large intergranular pores and biomoulds of pebbly and conglomeratic sandstones (Fig. 5F).

IMPLICATIONS FOR CALCITE CEMENTATION MODELS

Dissolution of bioclasts and precipitation of sparry low-Mg calcite in shallow-marine sandstone reservoirs is commonly cited as evidence for uplift and meteoric diagenesis. This study shows that these processes can occur in marine pore fluids, particularly with sediments deposited during 'calcite sea' times (e.g. many of the mid- to late Jurassic and Cretaceous reservoirs of the North Sea). There is no reason why marine aragonite–low-Mg calcite transformations should be restricted to deep-water sediments. Low-Mg calcite concretions in Jurassic shelf sandstones of southern England were interpreted by Bryant *et al.* (1988) as having formed from shallow burial aragonite dissolution in marine fluids, and there is palaeoecological evidence that the aragonite saturation depth onlapped shallow-marine platforms in the Middle Jurassic (Palmer *et al.*, 1988). Aragonite dissolution in these conditions may be further catalysed by biogeochemical interactions (Hendry, 1993a; Hendry *et al.*, 1995).

Low Sr concentrations in the calcite cements are also compatible with a marine redistributional origin. Although the accuracy of Sr microprobe data is less well constrained than for Mg, Ca2 cement clearly contains <0·2 mol% $SrCO_3$. The Sr budget during precipitation will have been dependent on the relative timing of aragonite dissolution, and the effective water–rock ratio. Calcite precipitating in a closed system should theoretically adopt an Sr^{2+}/Ca^{2+} ratio approaching that of the dissolving aragonite. However, Sr partitioning in calcite is particularly sensitive to kinetic modification. The lowest experimentally determined Sr partition coefficients (<0·06) are obtained by aragonite to calcite transformation experiments, so low-Sr calcites can potentially form by aragonite recrystallization in marine waters (Katz *et al.*, 1972). Additionally, it has recently been demonstrated that Sr incorporation into calcite is directly correlated with Mg content (Carpenter *et al.*, 1991).

The results of this study underscore the potential influence of seawater chemistry on spatial patterns of calcite cementation in sandstones. Wilkinson & Dampier (1990) demonstrated that the switch from surface-reaction to transport-controlled growth kinetics is strongly influenced by pore-fluid Mg^{2+}/Ca^{2+}. Surface reaction is the rate-limiting step throughout growth of concretions from modern seawater ($Mg^{2+}/Ca^{2+}=5$), producing patchy or dispersed cements (depending on the distribution of suitable nuclei). For Mg-free (fresh) water, transport-controlled kinetics take over for micrometre- to centimetre- sized concretion nuclei, leading to spatially segregated concretions which may subsequently impinge to form stratabound cemented layers (Bjørkum & Walderhaug, 1990b; Wilkinson, 1993a). Therefore, the pattern of Ca2 cementation recorded in this study confirms that early Cretaceous seawater had a very low Mg^{2+}/Ca^{2+} ratio.

CONCLUSIONS

1 Early diagenetic calcite formed grain-rimming overgrowths and subsequent tightly cemented concretions in Hauterivian age Scapa Member sandstones. Petrographic and $\delta^{13}C$ data indicate that the calcite was sourced by remobilization of bioclastic aragonite on a local (bed) scale.

2 Interpretation of 'whole-rock' calcite $\delta^{18}O$ values is hampered by mixing with later diagenetic phases in sponge spicule moulds. However, a combination of whole-rock data with fracture-hosted cements shows that early calcites precipitated from marine pore fluids at shallow burial depths. Calcite cementation began in the post-oxic environment, but mostly took place in anoxic conditions, possibly contemporaneous with microbial sulphate reduction.

3 Eogenetic calcite cementation in the Scapa sandstones resulted from the instability of metabolically produced aragonite in Late Mesozoic seawater. Similar processes occur today at greater depths, and have been recognized from Jurassic shallow-platform carbonates.

4 The chemistry of early Cretaceous 'calcite seas' accounted for low-Mg marine calcite precipitation. A rough calculation suggests that ambient seawater had an Mg^{2+}/Ca^{2+} ratio of ≤ 1, in close agreement with theoretical estimates of palaeo-ocean composition.

5 Aragonite dissolution initially outpaced calcite precipitation, generating high supersaturation with respect to calcite. Earliest cements consequently formed overgrowths on all suitable (biogenic calcite) substrates. As pore fluid supersaturation fell to lower levels, precipitation kinetics switched from surface-reaction to transport control, leading to the formation of discrete concretions at roughly decimetre-scale spacings.

6 Because the $CaCO_3$ was locally redistributed, spacing and shape of concretions reflects the original bioclast distribution. In general, the most bioclastic sand packages received the most cement. The SD sands were deposited from high-concentration turbulent flows which appear compositionally well mixed. Consequently, early calcite concretions are of limited lateral extent and frequently quasispherical. More elongate/tabular concretions are also present, suggesting that some of the 'massive' sandstones are in fact compositionally stratified. However, there is no indication that any concretions were extensive enough to affect reservoir performance.

7 Unless there is incontrovertible isotopic evidence for meteoric diagenesis, extreme caution should be exercised in using calcite cement compositions or fabrics as palaeohydrological and stratigraphic tools. Patchy low-Mg calcite cements may result from precipitation in an active meteoric palaeoaquifer (Walderhaug & Bjørkum, 1992), but also from stagnant marine pore fluids during times of low oceanic Mg^{2+}/Ca^{2+} ratio (this study). Consequently, palaeoceanography should be routinely considered when evaluating carbonate cementation patterns in sandstone hydrocarbon reservoirs.

ACKNOWLEDGMENTS

This paper forms part of a wider study of the Scapa reservoir by J.P.H., which was generously funded by Elf Enterprise Caledonia Ltd. We are grateful to E.E.C. and their operating partners for making cores, logs and other data available, and for permission to publish our results. We particularly thank Nigel Langridge for valuable discussions, and appreciate the helpful review comments from Kevin Taylor and Ian Jarvis. EPMA analyses were carried out at Edinburgh University Geology Department and isotopic analyses were performed at the Scottish Universities Research and Reactor Centre; we thank the support staff at both institutions for their advice and assistance. S.U.R.R.C. is funded by the Natural Environment Research Council and the Universities of the Scottish Consortium. Tony Dickson kindly supplied us with the DIS standard.

REFERENCES

Anderson, T.F. and Arthur, M.A. (1983) Stable isotopes of oxygen and carbon and their application to sedimentologic and palaeoenvironmental problems. In: *Stable Isotopes in Sedimentary Geology, Soc. econ. Palaeont. Miner. Short Course* 10 (Ed. by M. A. Arthur, M. A., Anderson, T. F., Kaplan, I. R., Veizer, J. and Land, L. S), pI.1–I.151.

Barnaby, R.J. and Rimstidt, J.D. (1989) Redox conditions of calcite cementation interpreted from Mn and Fe contents of authigenic calcites. *Bull. geol. Soc. Am.*, **101**, 795–804.

Bathurst, R.G.C. (1975) *Carbonate Sediments and Their Diagenesis. Developments in Sedimentology Vol. 12.* Elsevier, Amsterdam.

Berger, W.H. and Winterer, E.L. (1974) Plate stratigraphy and the fluctuating carbonate line. In: *Pelagic Sediments On Land and Under The Sea* (Ed. by K. J. Hsu and H. C. Jenkyns, H. C), *Spec. Publ. int. Ass. Sediment.*, **1**, 11–48.

Berner, R.A. (1980) *Early Diagenesis: A Theoretical Approach.* Princeton University Press, Princeton.

Berner, R.A. (1992) Palaeo-$CO_2$and climate. *Nature*, **359**, 114.

Berner, R.A. (1994) GEOCARB II: a revised model of atmospheric CO_2 over Phanerozoic time. *Am. J. Sci.*, **294**, 56–91.

Bjørkum, P.A. and Walderhaug, O. (1990a) Geometrical arrangement of calcite cementation within shallow marine sandstones. *Earth-Sci. Rev.*, **29**, 145–161.

Bjørkum, P.A. and Walderhaug, O. (1990b) Lateral extent of calcite-cemented zones in shallow marine sandstones. In: *North Sea Oil and Gas Reservoirs-II* (Ed. by A. T. Buller, E. W. Berg, O. Hjelmeland, J. Kleppe, O. Torsaeter and J. O. Aasen), pp. 331–336. Graham and Trotman, London.

Bjørlykke, K. (1993) Fluid flow in sedimentary basins. *Sediment. Geol.*, **86**, 137–158.

Boles, J.R. and Ramseyer, K. (1987) Diagenetic carbonate in Miocene sandstone reservoir, San Joaquin basin, California. *Bull. Am. Ass. petrol. Geol.*, **71**, 1475–1487.

Boote, D.R.D. and Gustav, S.H. (1987) Evolving depositional systems within an active rift, Witch Ground Graben, North Sea. In: *Petroleum Geology of North West Europe* (Ed. by J. Brooks and K. Glennie), pp. 819–833. Graham and Trotman, London.

Braithwaite, C.J.R. and Heath, R.A. (1989) Inhibition, distortion and dissolution of overgrowth cements on pelmatozoan fragments. *J. sedim. Petrol.*, **59**, 267–271.

Bryant, I.D., Kantorowicz, J.D. and Love, C.F. (1988) The origin and recognition of laterally continuous carbonate-cemented horizons in the Upper Lias Sands of southern England. *Mar. petrol. Geol.*, **5**, 108–133.

Buckley, D.E. and Cranston, R.E. (1988) Early diagenesis in deep sea turbidites: the imprint of paleo-oxidation zones. *Geochim. Cosmochim. Acta*, **52**, 2925–2939.

Burdige, D.J. (1993) The biogeochemistry of manganese and iron reduction in marine sediments. *Earth-Sci. Rev.*, **35**, 249–284.

Burley, S.D. (1993) Models of burial diagenesis for deep exploration plays in Jurassic fault traps of the Central and Northern North Sea. In: *Petroleum Geology of Northwest Europe: Proceedings of the 4th Conference* (Ed. by J. R. Parker), pp. 1353–1375. Geol. Soc., London.

Burton, E.A. (1993) Controls on marine carbonate mineralogy: review and reassessment. *Chem. Geol.*, **105**, 163–179.

Burton, E.A. and Walter, L.M. (1991) The effects of PCO_2 and temperature on magnesium incorporation in calcite in seawater and $MgCl_2$–$CaCl_2$ solutions. *Geochim. Cosmochim. Acta*, **55**, 777–785.

Carpenter, S.J., Lohmann, K.C., Holden, P., Walter, L.M., Huston, T.J. and Halliday, A.N. (1991) $\delta^{18}O$ values, $^{87}Sr/^{86}Sr$ and Sr/Mg ratios of Late Devonian abiotic marine calcite: implications for the composition of ancient seawater. *Geochim. Cosmochim. Acta*, **55**, 1991–2010.

Coniglio, M. (1989) Neomorphism and cementation in ancient deep-water limestones, Cow Head Group (Cambro-Ordovician), western Newfoundland, Canada. *Sediment. Geol.*, **65**, 15–33.

Craig, H. (1965) The measurement of oxygen isotope palaeotemperatures. In: *Stable Isotopes in Oceanographic Studies and Palaeotemperatures* (Ed. by E. Tongiorgi), pp. 161–182. Consiglio Nazionale delle Richerche, Labotratorio di Geologia Nucleare, Pisa.

Curtis, C.D. (1987) Mineralogical consequences of organic matter degradation in sediments: Inorganic/organic diagenesis. In: *Marine Clastic Sedimentology* (Ed. by J. K. Leggett and G. G. Zuffa), pp. 108–123. Graham and Trotman, London.

Feistner, K.W.A. (1989) Petrographic examination and re-interpretation of concretionary carbonate horizons from the Kimmeridge Clay, Dorset. *J. geol. Soc. Lond.*, **146**, 345–350.

Froelich, P.N., Klinkhammer, G.P., Bender, M.L., Luedtke, N.A., Heath, G.R., Cullen, D., Dauphin, P., Hammond, D., Hartman, B. and Maynard, V. (1979) Early oxidation of organic matter in pelagic sediments of the eastern equatorial Atlantic: suboxic diagenesis. *Geochim. Cosmochim. Acta*, **43**, 1075–1090.

Gibbons, K., Hellem, T., Kjemperud, A., Nio, S.D. and Vebenstad, K. (1993) Sequence architecture, facies development and carbonate-cemented horizons in the Troll Field reservoir, offshore Norway. In: *Advances in Reservoir Geology* (Ed. by M. Ashton), *Spec. Publ. geol. Soc. Lond.*, **69**, 1–31.

Harker, S.D. and Chermak, A. (1992) Detection and prediction of Lower Cretaceous sandstone distribution in the Scapa Field, North Sea. In: *Exploration Britain: Geological Insights For The Next Decade* (Ed. by R. F. P. Hardman), *Spec. Publ. geol. Soc. Lond.*, **67**, 221–246.

Haszeldine, R.S. (1994) Comparing diagenesis in the Magnus, South Brae, Andrew Fans: North Sea (abstract). *Reservoir Characterisation of Deep Marine Clastic Systems* (Ed. by Hartley, A. J. and Prosser, D. J.), p. 14. University of Aberdeen.

Haszeldine, R.S., Brint, J.F., Fallick, A.E., Hamilton, P.J. and Brown, S. (1992) Open and restricted hydrologies in Brent Group diagenesis: North Sea. In: *Geology of the Brent Group* (Ed. by A. C. Morton, R. S. Haszeldine, M. R. Giles and S. Brown), *Spec. Publ. geol. Soc. Lond.*, **61**, 401–419.

Heller, P.L. and Dickinson, W.R. (1985) Submarine ramp facies model for delta-fed, sand-rich turbidite systems. *Bull. Am. Ass. petrol. Geol.*, **69**, 960–976.

Hendry, J.P. (1993a) Geological controls on regional subsurface carbonate cementation: An isotopic-palaeohydrologic investigation of Middle Jurassic limestones in central England. In: *Diagenesis and Basin Development* (Ed. by A. G. Robinson and A. Horbury), *A.A.P.G. Studies in Geology*, **36**, 231–260.

Hendry, J.P. (1993b) Calcite cementation during bacterial manganese, iron and sulfate reduction in Jurassic shallow marine carbonates. *Sedimentology*, **40**, 87–106.

Hendry, J.P., Ditchfield, P.W. and Marshall, J.D. (1995) Two stage neomorphism of Jurassic aragonitic bivalves; implications for early diagenesis. *J. sedim. Res.*, **A65**, 214–224.

Hendry, J.P. and Marshall, J.D. (1991) Disequilibrium trace element partitioning in Jurassic sparry calcite cements: implications for crystal growth mechanisms during diagenesis. *J. geol. Soc. Lond.*, **148**, 835–848.

Hendry, J.P. and Trewin, N.H. (1995) Authigenic quartz microfabrics in Cretaceous turbidites; evidence for silica transformation processes in sandstones. *J. sedim. Res.*, **A65**, 380–392.

Hovland, M., Talbot, M.R., Qvale, H., Olaussen, S. and Asaberg, L. (1987) Methane-related carbonate cements in pockmarks of the North Sea. *J. sedim. Petrol.*, **57**, 881–892.

Howson, M.R., Pethybridge, A.D. and House, W.A. (1987) Synthesis and distribution coefficient of low-magnesium calcites. *Chem. Geol.*, **64**, 79–87.

Hudson, J.D. and Anderson, T.F. (1989) Ocean temperatures and isotopic compositions through time. *Trans. Roy. Soc. Edinburgh: Earth Sci.*, **80**, 183–192.

Hurst, A. (1987) Problems of reservoir characterization in some North Sea sandstone reservoirs solved by the application of microscale geological data. In: *North Sea Oil and Gas Reservoirs* (Ed. by J. Kleppe, E. W. Berg, A. T. Buller, O. Hjelmeland and O. Torsaeter), pp. 153–167. Graham and Trotman, London.

Irwin, H. and Hurst, A. (1983) Applications of geochemistry to sandstone reservoir studies. In:

Petroleum Geochemistry and the Exploration of Europe (Ed. by J. Brooks), *Spec. Publ. geol. Soc. Lond.*, **12**, 127–145.

James, N.P. and Bone, Y. (1992) Synsedimentary cemented calcarenite layers in Oligo-Miocene cool-water shelf limestones, Eucla Platform, Southern Australia. *J. sedim. Petrol.*, **62**, 860–872.

James, N.P. and Choquette, P.W. (1990) Limestones–The meteoric diagenetic environment. In: *Diagenesis; Geoscience Canada Reprint Series 4* (Ed. by I. A. McIlreath and D. W. Morrow), pp. 35–74. Geol. Assoc. Can.

Katz, A. (1973) The interaction of magnesium with calcite during crystal growth at 25–90 °C and one atmosphere. *Geochim. Cosmochim. Acta*, **37**, 1563–1586.

Katz, A., Sass, E., Starinsky, A. and Holland, H.D. (1972) Strontium behaviour in the aragonite-calcite transformation: an experimental study at 40–98 °C. *Geochim. Cosmochim. Acta*, **36**, 481–496.

Lloyd, G.E., Ferguson, C.C. and Reading, K. (1982) A stress-transfer model for the development of extensional fracture boudinage. *J. Struct. Geol.*, **4**, 355–372.

Macaulay, C.I., Haszeldine, R.S. and Fallick, A.E. (1993) Distribution, chemistry, isotopic composition and origin of diagenetic carbonates: Magnus Sandstone, North Sea. *J. sedim. Petrol.*, **63**, 33–43.

Major, R.P. and Wilber, R.J. (1991) Crystal habit, geochemistry, and cathodoluminescence of magnesian calcite marine cements from the lower slope of Little Bahama Bank. *Bull. geol. Soc. Am.*, **103**, 461–471.

Malone, J.M., Baker, P.A., Burns, S.J. and Swart, P.K. (1990) Geochemistry of periplatform carbonate sediments, leg 115, site 716 (Maldives Archipelago, Indian Ocean). In: *Proceedings of the Ocean Drilling Program, Scientific Results*, **115**, 647–659.

Marin, B.A., Clift, S.J., Hamlin, H.S. and Laubach, S.E. (1993) Natural fractures in Sonora Canyon sandstones, Sonora and Sawyer Fields, Sutton County, Texas. *Soc. petrol. Eng. Pap.*, **25895**.

Marshall, J.D. (1982) Isotopic composition of displacive fibrous calcite veins: reversals in pore-water composition trends during burial diagenesis. *J. sedim. Petrol.*, **52**, 615–630.

McGann, G.J., Green, S.C.H., Harker, S.D. and Romani, R.S. (1991) The Scapa Field, Block 14/19, UK North Sea. In: *United Kingdom Oil and Gas Fields, 25 Years Commemorative Volume* (Ed. by I. L. Abbotts), *Mem. geol. Soc. Lond.*, **14**, 369–376.

Meyers, W.J. (1991) Calcite cement stratigraphy: An overview. In: *Luminescence Microscopy: Quantitative and Qualitative Aspects* (Ed. by C. E. Barker and O. C. Kopp), *Soc. econ. Paleont. Miner. Short Course*, **25**, 133–148.

Morse, J.W. and Mackenzie, F.T. (1990) *Geochemistry of Sedimentary Carbonates*. Elsevier, Amsterdam.

Mozley, P.S. (1989) Complex compositional zonation in concretionary siderite: implications for geochemical studies. *J. sedim. Petrol.*, **59**, 815–818.

Mucci, A. (1983) The solubility of calcite and aragonite in seawater at various salinities, temperatures, and one atmosphere total pressure. *Am. J. Sci.*, **283**, 780–799.

O'Driscoll, D., Hindle, A.D. and Long, D.C. (1990) The structural controls on Upper Jurassic and Lower Cretaceous reservoir sandstones in the Witch Ground Grbane, UK North Sea. In: *Tectonic Events Responsible For Britain's Oil and Gas Reserves* (Ed. by R. F. P. Hardman and J. Brooks), *Spec. Publ. geol. Soc. Lond.*, **55**, 299–323.

Oomori, T., Kaneshima, H. and Maezato, Y. (1987) Distribution coefficient of Mg^{2+} ions between calcite and solution at 10–50 °C. *Mar. Chem.*, **20**, 327–336.

Opdyke, B.N. and Wilson, P.A. (1993) Geochemical constraints on Late Cretaceous tropical surface water temperature and chemistry based on relatively pristine carbonate from Leg 144, site 877 A, Wodejebato Guyot. Proceedings of the American Geophysical Union Fall Meeting, San Fransisco. *EOS*, **74** (suppl.), 353.

Opdyke, B.N., Wilson, P.A. and Enos, P. (in press) Geochemistry, diagenesis, and petrology of an Upper Cretaceous rudist reef from Site 877 A, Wodejebato Guyot. *Proceedings of the Ocean Drilling Program Leg 144, Part B- Scientific Results.* College Station, TX.

Palmer, T.J., Hudson, J.D. and Wilson, M.A. (1988) Palaeoecological evidence for early aragonite dissolution in ancient calcite seas. *Nature*, **335**, 809–810.

Pirrie, D. and Marshall, J.D. (1991) Field relationships and stable isotope composition of concretions from James Ross Island, Antarctica. *Sediment. Geol.*, **71**, 137–150.

Prosser, D.J., Daws, J.A., Fallick, A.E. and Williams, B.P.J. (1994) Geochemistry and diagenesis of stratabound calcite cement layers within the Rannoch Formation of the Brent Group, Murchison Field, North Viking Graben (northern North Sea). *Sediment. Geol.*, **87**, 139–164.

Riley, L.A., Harker, S.D. and Green, S.C.H. (1992) Lower Cretaceous palynology and sandstone distribution in the Scapa Field, U.K. North Sea. *J. petrol. Geol.*, **15**, 97–110.

Sandberg, P.A. (1983) An oscillating trend in Phanerozoic non-skeletal carbonate minerology. *Nature*, **305**, 19–22.

Savard, M.M., Veizer, J. and Hinton, R. (1995) Cathodoluminescence at low Fe and Mn concentrations: a SIMS study of zones in natural calcites. *J. sedim. Res.*, **A65**, 208–213.

Schlager, W. and James, N.P. (1978) Low-magnesian calcite limestones forming at the deep-sea floor, Tongue of the Ocean, Bahamas. *Sedimentology*, **25**, 675–702.

Smith, A.G., Hurley, A.M. and Briden, J.C. (1981) *Phanerozoic Palaeocontinetal World Maps.* Cambridge University Press, Cambridge.

Spaeth, C., Hoefs, J. and Vetter, U. (1971) Some aspects of isotopic composition of belemnites and related paleotemperatures. *Bull. geol. Soc. Am.*, **82**, 3139–3150.

Surlyk, F. (1987) Slope and deep gully sandstones, Upper Jurassic, East Greenland. *Bull. Am. Ass. petrol. Geol.*, **71**, 464–475.

Tucker, M.E. and Wright, V.P. (1990) *Carbonate Sedimentology*. Blackwell Scientific Publications, Oxford.

Van Andel, T.H. (1975) Mesozoic/Cenozoic calcite compensation depth and the global distribution of calcareous sediments. *Earth planet. Sci. Lett.*, **26**, 187–194.

Walderhaug, O. and Bjørkum, P.A. (1992) Effect of meteoric water flow on calcite cementation in the Middle Jurassic Oseberg Formation, well 30/3–2, Veslefrikk Field, Norwegian North Sea. *Mar. petrol. Geol.*, **9**, 308–318.

Walderhaug, O., Bjørkum, P.A. and Bolås, H.M.N. (1989) Correlation of calcite-cemented layers in shallow-marine sandstones of the Fensfjord Formation in the Brage Field. In: *Correlation in Hydrocarbon Exploration* (Ed. by J. D. Collinson), pp. 367–375. Graham and Trotman, London.

Walter, L.M. (1986) Relative efficiency of carbonate dissolution and precipitation during diagenesis: a progress report on the role of solution chemistry, In: *Roles of Organic Matter in Sediment Diagenesis* (Ed. by D. L. Gautier), *Spec. Publ. Soc. econ. Paleont. Miner.*, **38**, 1–11.

Watson, R.S., Trewin, N.H. and Fallick, A.E. (1995) The formation of carbonate cements in the Forth and Balmoral Fields, Northern North Sea: a case for biodegradation, carbonate cementation and oil leakage during early burial. In: *Reservoir Characterisation of Deep Marine Clastic Systems* (Ed. by A. J. Hartley and D. J. Prosser), *Spec. Publ. geol. Soc. Lond.*, **94**, 177–200.

Wilkinson, B.H. and Algeo, T.J. (1989) Sedimentary carbonate record of calcium–magnesium cycling. *Am. J. Sci.*, **289**, 1158–1194.

Wilkinson, B.H., Owen, R.M. and Carroll, A.R. (1985) Submarine hydrothermal weathering, global eustacy, and carbonate polymorphism in Phanerozoic marine oolites. *J. sedim. Petrol.*, **55**, 171–183.

Wilkinson, M. (1993a) Concretions of the Valtos Sandstone Formation of Skye: geochemical indicators of palaeo-hydrology. *J. geol. Soc. Lond.*, **150**, 57–66.

Wilkinson, M. (1993b) Geometrical arrangement of calcite cementation within shallow marine sandstones–Comment. *Earth-Sci. Rev*, **34**, 47–51.

Wilkinson, M. and Dampier, M.D. (1990) The rate of growth of sandstone-hosted calcite concretions. *Geochim. Cosmochim. Acta*, **54**, 3391–3399.

Wilson, T.R.S., Thomson, J., Colley, S., Hydes, D.J., Higgs, N.C. and Sorensen, J. (1985) Early organic diagenesis: the significance of progressive subsurface oxidation fronts in pelagic sediments. *Geochim. Cosmochim. Acta*, **49**, 811–822.

Manuscript received 29 August 1995; revision accepted 16 February 1996.

Reprinted from Sedimentology (1991) **38**, *899–912*

The concretions of the Bearreraig Sandstone Formation: geometry and geochemistry

MARK WILKINSON*

Department of Geology, University of Leicester, University Road, Leicester LE1 7RH, UK

ABSTRACT

The sandbodies of the Bearreraig Sandstone Formation (Inner Hebrides, UK) are cemented by two generations of calcite. The first generation, an inhomogeneous ferroan calcite (0·05–3·28 mol% $FeCO_3$) formed during sulphate reduction ($\delta^{13}C = -24$ to -32‰ PDB) in marine porewaters ($\delta^{18}O$ of cement from -1 to -4‰ PDB) at very shallow burial depths (a few centimetres). These cements are rare but form millimetre-scale clusters of crystals which acted as nuclei to the later, concretionary cements.

The second generation of cements are more homogeneous ferroan calcites (mean 1·58% mol% $FeCO_3$) which evolve to progressively higher Fe/Mg ratios. They are sourced by shell dissolution ($\delta^{13}C$ of cement from $+1$ to -3‰ PDB) into meteoric ($\delta^{18}O$ of cement from -6 to -10‰ PDB) or mixed marine–meteoric waters ($\delta^{18}O$ of cement from -4 to -6‰ SMOW). These were introduced into the formation either during Bathonian times as a freshwater lens, or, subsequent to partial inversion, by confined aquifer flow. Corroded feldspars within the concretions suggest that an interval of at least 8 Ma separated the deposition of the sediments from the onset of concretion growth.

Abundant concretions are preferentially developed at certain horizons within the sandbodies, where the early generation of ferroan calcite cements provided nuclei. The latter formed close to the sediment–water interface, the concentration of cement within the sediment being related to sedimentation rate. The relatively high concentrations of the first generation of cement, upon which the concretionary horizons are nucleated, formed during periods of minimal sedimentation.

INTRODUCTION

The distribution of calcite-cemented intervals within sandstones has been recognized as a potentially important factor in the development of hydrocarbon reservoirs (e.g. Kantorowicz *et al.*, 1987; Bryant *et al.*, 1988; Saigal & Bjørlykke, 1988). The degree to which calcite concretions affect fluid flow is primarily a function of how effectively they coalesce laterally to form impermeable sheets. Individual concretions will not significantly influence porefluid flow, although more laterally persistent concretionary beds may divide a reservoir into discrete units (Kantorowicz *et al.*, 1987).

The sandstone-hosted calcite concretions in the Jurassic of the Inner Hebrides of Scotland are believed to be analogous to the calcite-cemented intervals encountered in rocks of similar age in the subsurface of the North Sea. The geometry of the carbonate cements in North Sea reservoir sandstone intervals remains obscure, because, although they can be detected where they intersect drill holes, either in core or on wire-line logs, their degree of lateral connectivity is difficult to ascertain. Because the North Sea sandstones are important petroleum reservoirs, the prediction of the geometry of calcite cement-bodies is a problem of commercial importance (Kantorowicz *et al.*, 1987).

Although shale-hosted concretions have been the subject of intensive study by a large number of workers (e.g. Irwin *et al.*, 1977; Raiswell, 1987 and references therein), sandstone-hosted concretions have received less attention. Recently, a number of papers have been published concerning sandstone-hosted concre-

*Present address: Department of Earth Sciences, Liverpool University, Brownlow Street, PO Box 147, Liverpool L69 3BX, UK.

tions (e.g. Pirrie, 1987; Dickinson, 1988; McBride, 1988; Johnson, 1989), reflecting an increase in interest amongst the academic community. This paper presents the results of a study of the concretions of the Bearreraig Sandstone Formation of Trotternish, Skye, Inner Hebrides (Fig. 1).

GEOLOGICAL SETTING

The Aalenian–Bajocian Bearreraig Sandstone Formation of Trotternish, Skye, was deposited in a marine environment within the Hebrides Basin; it comprises six members (Fig. 2) of which four are concretion-bearing (Morton, 1976, and references therein). The bio- and lithostratigraphy is well established (Morton, 1976) although no detailed sedimentological study of the sequence has been published. The sequence is dominated by the deposits of migrating sandsheets on a storm- and tide-swept shelf (Morton, 1983).

Three members are considered here: the Ollach Sandstone Member, a siltstone which coarsens into a sandstone; the Udairn Shale Member, which is actually a poorly-sorted sandy shale and silt forming a gradationally coarsening-upwards unit; and the Holm Sandstone Member, a silty sandstone which coarsens up into a medium-grained sandstone (Morton, 1987).

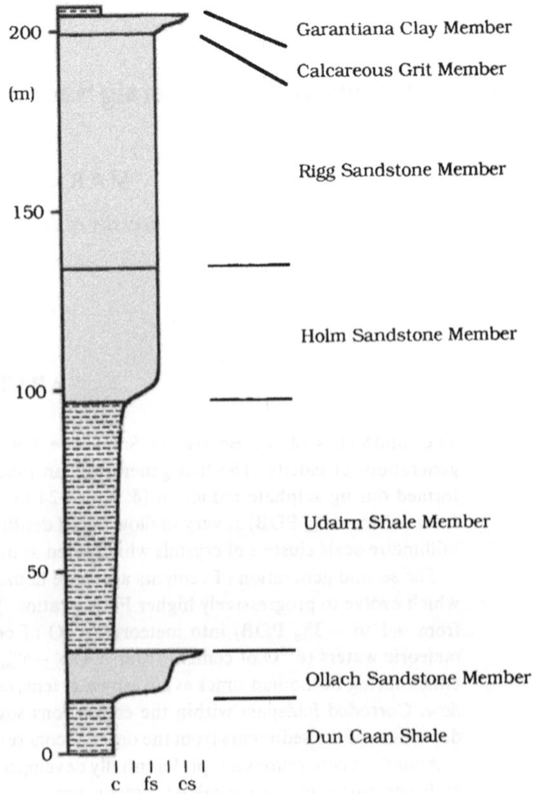

Fig. 2. Stratigraphy of the Bearreraig Sandstone Formation. From data in Morton (1987).

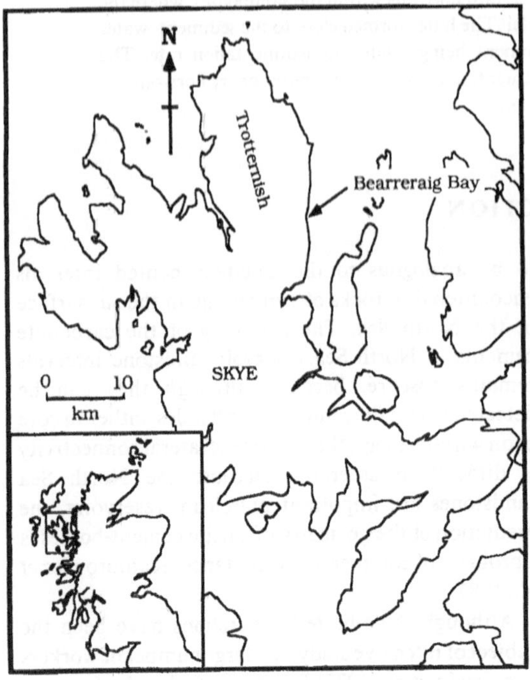

Fig. 1. Map showing location of Bearreraig Bay. The inset shows the position of Skye on the north-west coast of Scotland, UK.

METHODOLOGY

Cement mineralogy

Carbonate cement mineralogy was determined by staining thin-sections using the method of Dickson (1966) and by X-ray diffraction studies of powdered samples. All thin-sections were examined by cathodoluminescence microscopy. Non-carbonate cements were examined on a Hitachi S-520 scanning electron microscope, with qualitative energy-dispersive analysis to aid mineral identification. Samples were pretreated using a 1% (w/w) solution of EDTA to remove the carbonate cement.

Minor element geochemistry of carbonates

Minor element concentrations within the carbonate cements were determined by wavelength-dispersive analysis using a Jeol JXA-8600S microprobe. The electron beam was de-focused to a diameter of 15 μm to minimize specimen damage. Counting times were 2 × 10 s for all elements (Ca, Mg, Mn, Fe). Two sigma is approximately ±5% of any result. Complete data are tabulated in Wilkinson (1989b). Whole cement analyses were produced by ICP analysis after extraction from whole-rock samples using 0·1 M HCl. Reaction time was minimized to prevent leaching of cations from aluminosilicate phases. Two sigma is approximately ±10% of any result. Data are tabulated in Table 1, as mole percent carbonate.

Stable isotopes

Samples containing between 2·5 and 3·5 mg of calcite were digested in anhydrous phosphoric acid under vacuum and the carbon dioxide was collected and cleaned. The product was analysed on a V.G. Sira-12 triple-collector, dual-inlet mass spectrometer against a Lincolnshire Limestone internal laboratory stand-

Table 1. Stable isotope (‰, PDB) and minor element (mol%) data.

Sample	$\delta^{13}C$	$\delta^{18}O$	$FeCO_3$	$MgCO_3$	$MnCO_3$	$SrCO_3$
Ollach Sandstone Member						
M87201	−1·36	−4·19	1·115	1·780	0·102	0·052
M87202	−0·37	−4·52	1·673	1·536	0·106	0·053
M87203	−0·36	−5·36	1·929	1·424	0·156	0·052
M87204	−0·47	−5·42	2·149	1·277	0·333	0·055
M87205	−3·73	−3·22	0·211	2·189	0·164	0·049
M87206	−8·21	−2·60	0·254	2·018	0·152	0·053
M87207	−8·65	−5·55	0·426	1·742	0·151	0·062
M87208	−7·31	−4·80	0·801	1·456	0·175	0·069
M87209	−0·85	−4·16	1·311	2·277	0·058	0·058
M87210	−0·11	−4·76	1·676	2·160	1·929	0·059
M87211	0·30	−4·47	1·810	1·938	0·079	0·058
M87212	−1·30	−3·65	0·974	2·433	0·056	0·057
Udairn Shale Member						
M8737	−8·75	−5·79				
M87101	−0·03	−8·17				
M87104	−0·21	−7·38				
M87105	−1·15	−8·05				
M87106	−0·25	−7·70				
M87107	−0·37	−8·20				
M87108	−31·49	−1·03				
M87109	−0·37	−8·32·				
M87110	−0·25	−6·00				
M87111	−26·23	−3·53				
M87112	−27·24	−3·01				
M87113	−0·23	−8·28				
M87114	−0·11	−8·30				
M87115	0·02	−6·02				
M87116	0·11	−8·39				
M87117	−0·04	−6·58				
M87119	−6·98	−6·01				
·Holm Sandstone Member						
M87254	−24·04	−2·70	0·909	2·225	0·136	0·084
M87254(rpt)	−24·07	−2·72	—	—	—	—
M87255	0·69	−8·44	1·476	1·438	0·062	0·091
M87256	0·67	−8·61	1·354	1·320	0·065	0·094
M87257	0·45	−8·62	1·290	1·077	0·051	0·089

ard. The results were corrected using standard techniques (Craig, 1957) and an alpha value of 1·01025 (for calcite-CO_2 fractionation for reaction with phosphoric acid). They are presented in standard delta notation in per mil difference from the PDB standard.

Precipitation temperatures for the measured calcites are derived from (Craig, 1957, modified by Anderson & Arthur 1983):

$$T(°C) = 16{\cdot}0 - 4{\cdot}14(\delta_c - \delta_w) + 0{\cdot}13(\delta_c - \delta_w)^2, \quad (1)$$

where δ_w and δ_c are the $\delta^{18}O$ values for the water and the calcite respectively. The former is expressed relative to SMOW, the latter to the international PDB isotopic standard. The results are given in Table 1.

Sediment size distribution

Sediments decalcified using dilute hydrochloric acid were sized by wet sieving, residues being collected at 0·5 ϕ unit intervals from 0·0 to 4·0 ϕ. Selected sieve residues were examined with the aid of a Hitachi S-520 scanning electron microscope to ensure that adequate sediment disaggregation had been achieved. The finer portion of the sediment was sized, after separation from the bulk sample by dry sieving, using a Sedigraph 5000ET particle size analyser. Data are tabulated in Table 2.

RESULTS

The Ollach Sandstone Member

The Ollach Sandstone Member hosts metre-scale calcite concretions with irregular shapes. They appear elongate horizontally when observed in vertical cliff exposures, and are preferentially developed at certain horizons within the sandbody (Fig. 3). The concretions commonly coalesce laterally to form extensive cement sheets covering up to 75% of some bedding planes. The host sandstone has little calcite cement, but has a well-developed quartz cement in the form of overgrowths on framework grains. Because the quartz cement is either not present, or is only very poorly developed within the concretions, it must generally post-date the concretionary calcite (Gluyas, 1984). Corroded feldspars have been observed overgrown by concretionary calcite cement.

Two calcite cement generations have been identified within the concretions of the Ollach Sandstone Member. The first generation (OC1) is a chemically inhomogeneous, although generally ferroan, calcite (0·05–3·28 mol% $FeCO_3$, Fig. 4) characterized by preserving a high minus-cement porosity within the host sediments where developed (80–90%, visual estimate). This cement generation has irregular crystal boundaries and a high concentration of finely divided opaque material (Fig. 5). In thin-section it appears as subspherical to irregular shaped patches up to 5 mm in diameter, frequently with a high content of shell material. Some of the cement patches exhibit rounded margins suggesting reworking by sedimentary processes, while others include concentrically arranged shell debris and resemble burrow fills. The cement is only a minor volumetric component of the concretions (<1%), and is invariably surrounded by the second cement generation described below.

The second cement generation (OC2) within the concretions is an equant ferroan calcite spar (mean 1·58 mol% $FeCO_3$, Fig. 6) with no discernable zonation. It is volumetrically the most important carbonate cement, occluding all available porosity. Crystal boundaries are more regular than those of the first generation and the minus-cement porosity is lower. There are few included opaque minerals. Figure 6 shows the element contents determined for four samples from a single concretion. Each sample shows only a limited range of Fe/Mg ratios, with a centre-to-edge trend for the concretion of increasing Fe/Mg ratio.

It has not proved possible to separate the two cement generations for isotopic analysis. Whole-cement values (Fig. 7) form a poorly defined cluster around $\delta^{13}C = -1‰$ and $\delta^{18}O = 5‰$, although analyses are present with lighter carbon and heavier oxygen (down to $\delta^{13}C = 8{\cdot}2‰$ and $\delta^{18}O = -2{\cdot}6‰$).

Minor element trends (determined by ICP, Table 1) across two Ollach Sandstone Member concretions (OB1 and OB2) are presented in Fig. 8. The results suggest that porewater iron and manganese levels increased during precipitation, while magnesium levels decreased (assuming centre-to-edge growth).

Udairn Shale Member

The Udairn Shale Member, actually a poorly sorted sandy shale and silt, hosts discrete septarian nodules 0·5 m in diameter. The cement in these concretions has $\delta^{13}C$ values of −26·2 to −31·5‰ and $\delta^{18}O$ values of −1·0 to −3·5‰ (Fig. 7); this is refered to here as cement UC1. A second cement (UC2) forms horizontal sheets 0·5 m thick which are laterally continuous for tens of metres, and which surround the septarian concretions. The cemented bands are frequently less

Table 2. Sediment grain size.

Sample	Type*	Unit†	Percentage of sample finer than indicated size (in ϕ units)													
			0·0	0·5	1·0	1·5	2·0	2·5	3·0	4·0	5·0	6·0	7·0	8·0	9·0	10·0
M8807	C	H	100	100	99·9	99·7	99·0	96·7	58·1	14·0	10·0	6·3	4·6	3·5	2·8	2·4
M8808	U	H	100	100	99·9	98·7	94·8	87·7	49·2	16·9	11·8	7·8	6·1	5·1	4·1	3·4
M8809	C	H	100	100	100	96·8	93·1	88·3	55·3	20·9	15·0	9·4	7·1	5·2	4·2	3·1
M8810	U	H	100	100	100	98·6	95·2	91·9	66·7	19·3	14·1	8·7	6·4	4·6	3·5	2·7
M8811	U	H	100	100	99·9	99·8	99·5	95·9	59·6	18·0	13·9	9·9	6·9	4·5	3·2	2·3
M8801	U	R	100	100	100	99·8	99·2	97·3	65·9	22·9	19·5	11·5	8·8	6·9	5·7	4·4
M8802	C	R	100	100	100	99·7	98·8	73·7	29·3	6·1	4·6	2·9	2·3	1·8	1·3	1·0
M8803	U	R	100	99·1	98·1	—	90·0	58·5	31·1	7·8	5·4	2·9	1·5	0·85	0·58	0·39
M8804	C	R	100	100	100	98·5	95·8	57·7	30·7	12·0	8·4	5·2	3·6	0·58	1·9	1·6
M8882	C	S	100	100	99·9	99·7	98·4	96·3	85·9	45·1	36·5	24·8	17·6	11·7	8·6	6·3
M8883	U	S	100	99·9	99·8	99·7	99·1	97·4	89·8	42·8	32·1	21·0	14·5	9·8	6·8	4·7
M8884	C	S	100	99·9	99·6	99·3	98·5	97·1	78·2	41·3	33·0	21·9	14·9	10·3	7·2	5·8
M8885	U	S	100	99·9	99·7	99·4	99·1	93·4	79·8	39·5	31·6	20·9	13·4	8·7	5·1	4·0

*Sample collected from concretionary horizon (C) or non-concretionary horizon (U).
†Sample from Holm Sandstone Member (H), Rigg Sandstone Member (Bearreraig Sandstone Formation; R) or Scalpa Sandstone Formation (S).

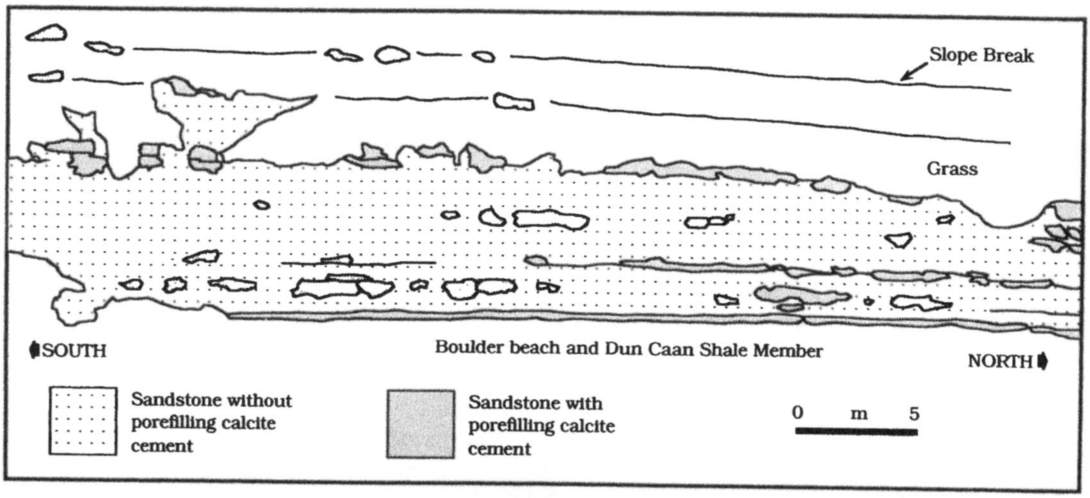

Fig. 3. Sketch of cliff-face at Bearreraig Bay, Skye, showing the Ollach Sandstone Member of the Bearreraig Sandstone Formation. Drawn from a photographic mosaic.

than fully cemented. Isotopically, they are distinct from the cements of the septarian concretions, with $\delta^{18}O$ values of $-5{\cdot}7$ to $-9{\cdot}3$‰ and $\delta^{13}C$ values of 0·1 to $-8{\cdot}7$‰ (Fig. 7).

Holm Sandstone Member

The most conspicuous feature of the Holm Sandstone Member at outcrop is a single layer of concretions approximately 2 m from the base (Fig. 9). These are of metre scale, and distinguished from those of the Ollach Sandstone Member by a spherical or subspherical geometry. Many of the concretions are compound, i.e. are composed of a cluster of mutually interfering bodies, sufficiently spaced so that the individual growths are readily distinguishable.

Two cement types have been identified, and are petrographically indistinguishable from the OC1 and OC2 cements of the Ollach Sandstone Member. They are here denoted HC1 and HC2. Cement HC1

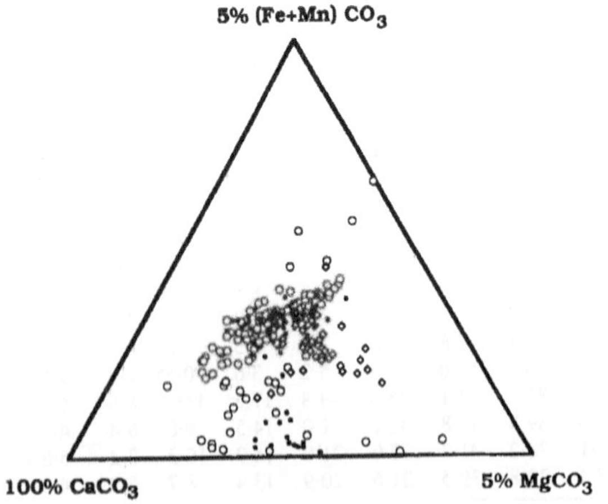

Fig. 4. Ca–(Fe+Mn)–Mg plot of microprobe data from cement OC1 of the Ollach Sandstone Member (as mole percent). The symbols differentiate analyses from different thin-sections. All are from a single concretion.

($\delta^{18}O = -2{\cdot}6‰$ and $\delta^{13}C = -24{\cdot}0‰$, Fig. 7) has been seen in only a single thin-section from the centre of a concretion, the remainder of the concretion being composed of cement HC2 ($\delta^{18}O = -8{\cdot}4$ to $-8{\cdot}6‰$ and $\delta^{13}C = 0{\cdot}5$ to $0{\cdot}7‰$, Fig. 7). The isotope data are similar to those collected by Marshall (1981) from the infillings of ammonite chambers from the same formation, which are included on Fig. 7 for comparison.

Figure 8 shows centre-to-edge minor element and isotope trends for a single Holm Sandstone Member concretion (OH1). The centre of the concretion is composed primarily of cement HC1, and is geochemically distinct from the outer regions. The host sediments, which are friable at outcrop, lack a pore-filling calcite cement, but are partially cemented by quartz overgrowths which largely post-date the concretionary calcite cements (Gluyas, 1984). Corroded feldspars (Fig. 10) are found overgrown by cement HC2.

DISCUSSION

Isotopic data

The three Members of the Bearreraig Sandstone Formation described all contain two distinct cement generations. The early cement generations (OC1, UC1 and HC1) have petrographic and geochemical features in common which suggest a common mode of origin.

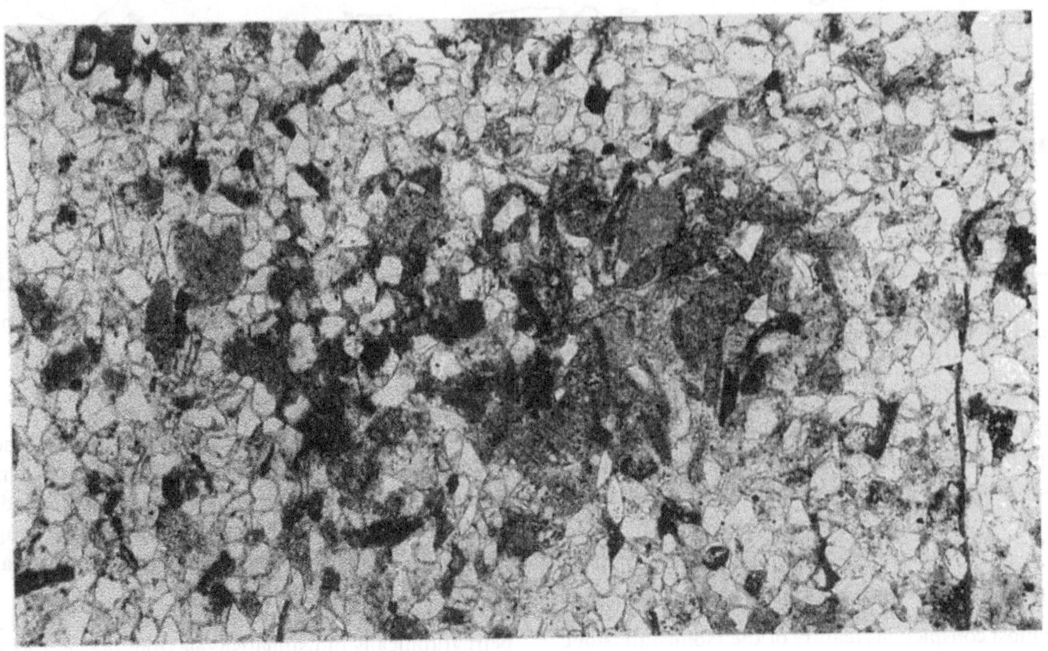

Fig. 5. Photomicrograph of a ring of early marine cement (dark) enclosed within later (meteoric) cement from an Ollach Sandstone Member concretion. The field of view is 4 mm across.

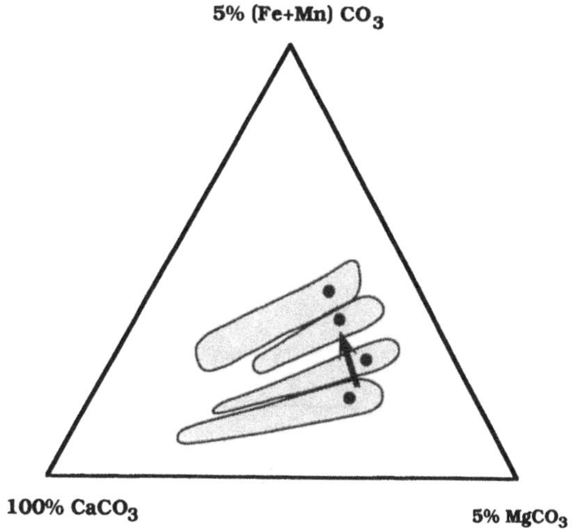

Fig. 6. Ca–(Fe+Mn)–Mg plot of microprobe data from cement OC2 of the Ollach Sandstone Member (as mole percent). The data are from a single concretion. The fields contain between 123 and 174 analyses each. The plotted points are the means for the data. The arrow indicates the centre-to-edge trend of increasing Fe/Mg ratio within the sampled concretion.

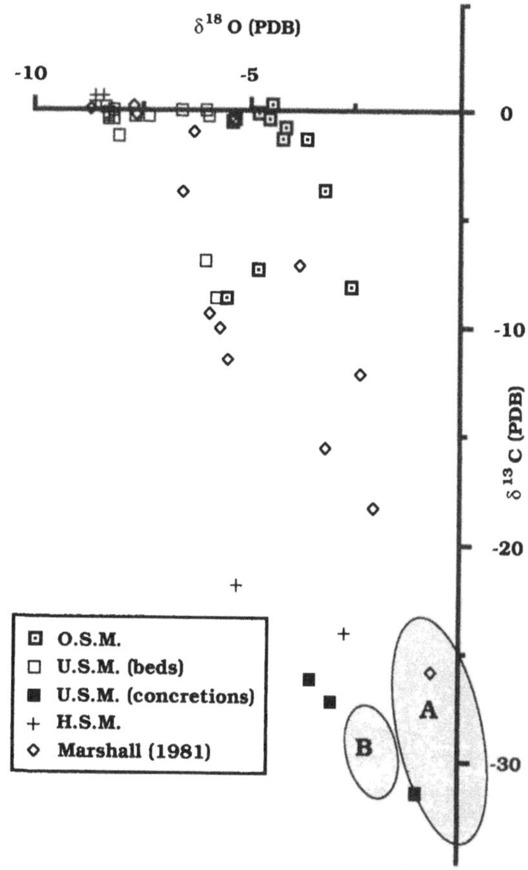

Fig. 7. Stable isotope cross-plot (‰) for 'whole-rock' cement samples from the Bearreraig Sandstone Formation. Data are shown from the Ollach Sandstone Member (O.S.M.), the cemented beds of the Udairn Shale Member (U.S.M. (beds)), the concretions of the Udairn Shale Member (U.S.M. (concretions)) and the Holm Sandstone Member (H.S.M.). Also shown are data of Marshall (1981) from the Holm Sandstone Member, and the 'early cements' of Kantorowicz *et al.* (1987) (area A) and Walderhaug *et al.* (1989) (area B).

Oxygen isotope data (from UC1 and HC1 only; δ^{18}O $-1{\cdot}0$ to $-3{\cdot}5$‰) suggest precipitation from marine waters (δ^{18}O $= -1$‰; Shackleton & Kennet, 1975) at shallow burial depths and temperatures of 16–27°C. This is consistent with surface temperature estimates (for the overlying Great Estuarine Group; Hudson & Andrews, 1987), the high minus-cement porosity of the host sediment containing the cement, and the possible reworking of the cement by sedimentary processes.

Carbon isotope data (δ^{13}C $= -24{\cdot}0$ to $-31{\cdot}5$‰) are interpreted to indicate carbonate formation as a by-product of sulphate reduction, acting upon either *in situ* organic material or methane derived from greater burial depths (e.g. Raiswell, 1987). The inclusion of finely divided pyrite within the cements supports this. Calcite cements associated with sulphate reduction and pyrite production are frequently non-ferroan (e.g. Irwin *et al.*, 1977), as pyrite is highly insoluble and its rapid formation can buffer either the porewater sulphide or iron concentrations to very low values. Ferroan calcite may be a precipitation product provided that iron is present in excess.

Marshall (1981) presented isotopic data for cements within ammonite chambers from the Holm Sandstone Member, which show the same trend as the concretionary cements. However, the early cements within the ammonite chambers were originally aragonite, and have undergone recrystallization. Marshall (1981) ascribed the isotopic trend to either progressive porewater evolution, or to a variable degree of alteration, but was unable to identify the controlling factor. The isotopic data from the concretions require only two end-member porewater compositions to be involved, suggesting that the trend of Marshall's (1981) data is due to variable degrees of isotopic resetting during recrystallization.

Shallow-marine sandstones from the Fensfjord Formation in the Brage Field of the North Sea (Norwegian Sector; Walderhaug *et al.*, 1989) contain calcites with almost identical isotopic compositions to

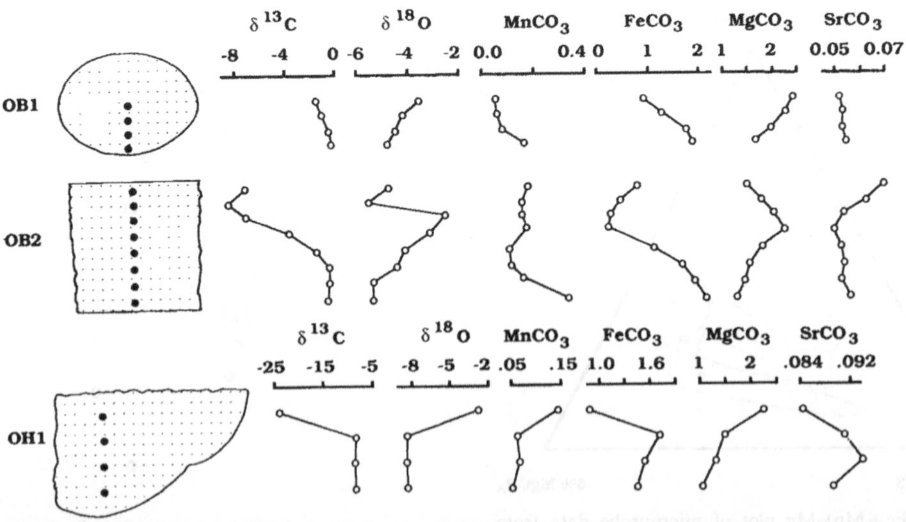

Fig. 8. 'Whole-cement' isotopic (‰) and minor element (mol%) trends across concretions from the Ollach Sandstone and Holm Sandstone members.

Fig. 9. A line of concretions from within the Holm Sandstone Member (Bearreraig Sandstone Formation) at Bearreraig Bay, 2 m above the transitional contact with the Udairn Shale Member. The line of concretions lies 1·5 m above the grass covered ledge.

those of the UC1 and HC1 cements, within decimetre-scale concretions. Walderhaug *et al.* (1989) interpret this to indicate that the cements contain a significant proportion of biogenic carbonate. The Viking Group sandstones of the Troll Field (North Sea, UK Sector) host isotopically similar calcites within fully cemented non-bioclastic beds (Fig. 7; Kantorowicz *et al.*, 1987). These are interpreted as incorporating bicarbonate generated by sulphate-reducing bacterial processes at or close to the sediment–water interface.

Using the two examples above as analogues, cements OC1, HC1 and UC1 are interpreted to be the

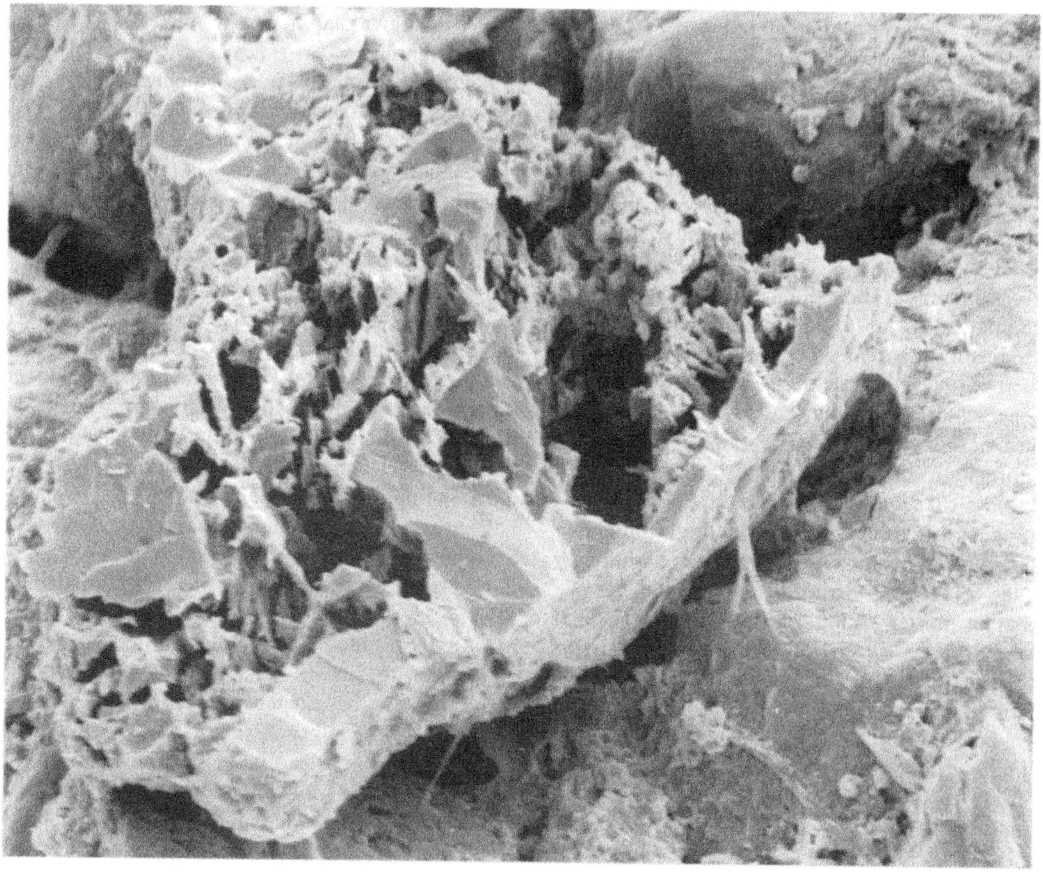

Fig. 10. Skeletal feldspar from within concretion OH1 of the Holm Sandstone Member. The feldspar was originally overgrown by cement OC2, which was removed as described in the text.

by-product of bacterial sulphate reduction at minimal depths of burial (a few centimetres). As a proportion of the cement OC1 includes concentrically arranged shell fragments and resembles burrow cross-sections, some of the early cements are interpreted to have formed within organic-matter-rich burrows, although the majority of the burrows were only partly cemented. The large range of observed cement compositions (Fig. 4) suggests limited connection between the various sites of precipitation. The distribution of these early cements is probably a function of sedimentation rate, a prolonged pause in sedimentation allowing time for cement precipitation (e.g. Raiswell, 1987).

Cements UC2 and HC2 are similar isotopically, $\delta^{13}C=0{\cdot}1$ to $-8{\cdot}7$‰ for UC2 and 0·5 to 0·7‰ for HC2, $\delta^{18}O=-5{\cdot}7$ to $-9{\cdot}3$‰ for UC2 and $-8{\cdot}4$ to $-8{\cdot}6$‰ for HC2 (Fig. 7). They are interpreted to indicate derivation from a marine carbon source (shell dissolution, $\delta^{13}C$ of approximately 0‰), and precipitation from meteoric waters ($\delta^{18}O=-5$ to -6‰; Hudson & Andrews, 1987) at burial depths of less than 600 m. Precipitation from fully marine pore-waters can be discounted. A minority of the analyses (e.g. $\delta^{18}O=-6{\cdot}0$‰, $\delta^{13}C=-0{\cdot}25$‰) imply precipitation from isotopically heavier water, interpreted as mixed marine–meteoric water. All the analyses of cement OC2 are interpreted to involve precipitation from mixed marine–meteoric water. The mixing line between cement OC1 and OC2 is only poorly defined due to the dispersed nature of cement OC1. As previously stated, it has not proved possible to separate a sufficient amount of cement OC1 for isotopic analysis. The radial decrease in $\delta^{18}O$ within concretion OB1 of the Ollach Sandstone Member (Fig. 8) suggests increasing temperatures during growth, presumably due to progressive burial.

Timing and depth of concretion formation

The timing and depth of formation of the concretions is important in constraining solute fluxes during growth. If the cements form close to the sediment–water interface, then diffusion of seawater-sourced components may contribute significantly to concretion formation (Wilkinson, 1989a). Although it is clear that, as above, cements OC1, HC1 and UC1 formed within centimetres of the sediment surface, the depth of precipitation of the second generation of cements is not so readily apparent. The relative dating criteria of Raiswell (1971) for shale-hosted concretions are of little use in sandstones (Gluyas, 1984). Several lines of evidence, however, suggest that the Bearreraig Sandstone Formation concretions formed at some depth.

The sediments within the concretions show the same range of structures as the surrounding sediments, and there are no (known) reworked concretions. Hence the concretions formed below the influence of bioturbation and reworking. The late cements (OC2, HC2 and UC2) are invariably ferroan, indicating reducing conditions. Because the bottom waters were oxic, this indicates formation at depths below the diffusive range of oxygen. Hudson & Andrews (1987) used similar criteria to demonstrate the burial diagenetic origin of similar concretions from the Valtos Sandstone Formation (Great Estuarine Group) of Skye.

The skeletal remains of feldspars have been observed within the concretions of both the Ollach Sandstone and Holm Sandstone members (Fig. 10). Assuming this dissolution took place after deposition of the sediment, then an estimate of the time required to dissolve the feldspar provides a minimum estimate of the time interval between sedimentation and the onset of concretion growth. Busenburg & Clemency (1976) describe feldspar dissolution as a four-stage process, of which only the last stage could be geologically significant. A linear relationship between dissolution rate and surface area was proposed, which will hold provided the minimum particle dimension significantly exceeds the lattice defect spacing (Holdren & Speyer, 1985). Surface reaction can be assumed to be the rate-determining step, as even the relatively insoluble aluminium will diffuse away in less than 1 year (Stoessel, 1987). For an idealized, spherical, feldspar, simple integration yields a proportional relationship between total reaction time (t) and feldspar radius (r):

$$t = r/(v_m \cdot k), \qquad (2)$$

where v_m is the molar volume of the feldspar, and k a rate constant.

The rate constant (k) has been determined experimentally, and is the subject of some controversy (e.g. Paces, 1983, and references therein). As the conditions of reaction of the majority of the experiments described in the literature are not directly applicable to the subsurface, the results of Paces (1983) are used; these are derived from mass-balance studies of hydrological catchment areas. The most probable value of k is 1×10^{-14} mol m s^{-1}, such that a feldspar 0·5 mm in diameter will undergo complete dissolution in approximately 8 Ma. Uncertainty in the value of k leads to a range of possible reaction times between 120 000 years and 16 Ma. Experimental data (e.g. Busenburg & Clemency, 1976) support the more rapid reaction times.

The extent to which the feldspars of the Bearreraig Sandstone Formation were corroded prior to deposition is unknown. Bjørlykke *et al.* (1986) related the variable leaching of feldspars within the Brent Group to the extent of incipient leaching within a weathering zone prior to burial. Passaratti & Eslinger (1987) demonstrated the existence of corroded feldspars within a Holocene beach sand. It is concluded that the corroded felspars are at least consistent with the existence of a significant time interval between deposition of the host sands and the onset of concretion formation. The best estimate of the minimum duration of this interval is 8 Ma; however, this figure is subject to a high degree of uncertainty.

Oxygen stable isotope data (above and Fig. 7) are interpreted to indicate that cements UC2 and HC2 precipitated from meteoric waters, and cement OC2 from mixed marine–meteoric water. As the faunas of the Bearreraig Sandstone Formation indicate deposition in marine waters (e.g. Morton, 1983), the original porewaters must have been at least partly replaced by meteoric waters before the onset of concretion growth. The timing of meteoric water introduction into the sands is problematic; Emery *et al.* (1987) suggest four mechanisms by which this may be achieved. The Great Estuarine Group, which overlies the Bearreraig Sandstone Formation, was deposited in waters of variable but reduced salinity (Hudson & Andrews, 1987). Hence, during Bathonian times, a meteoric lens may have penetrated down to the Bearreraig Sandstone Formation. Meteoric water could also have been introduced when the area suffered partial inversion between the end of the early Kimmeridgian and the later Cretaceous (Hudson & Andrews, 1987), possibly by confined aquifer flow.

The presence of corroded feldspars within the concretions suggest that a time period of significant duration separated the onset of concretion growth from the deposition of the host sands. The best estimate for the length of this is 8 Ma (from feldspar corrosion rates), consistent with concretion growth in a freshwater lens of Bathonian age, developed from the reduced-salinity lagoons within which the overlying Great Estuarine Group was deposited. It is highly unlikely that the concretionary cements (OC2, HC2 and UC2) were precipitated with diffusive contact to the sediment–water interface.

Geometry

Sandstone-hosted concretions have commonly been observed to lie along certain bedding surfaces (e.g. Pirrie, 1987; Wilkinson, 1989b; Bjørkum & Walderhaug, 1990a). The following explanations are considered here:

(1) concretion distribution mirrors shell distribution, as calcite nucleation occurs most rapidly within shell-rich strata (Bjørkum & Walderhaug, 1990b);
(2) concretions grow where porefluid flow is most rapid, i.e. in high-permeability strata. Concretion distribution hence mirrors permeability distribution within a sandbody;
(3) concretionary cements nucleate upon early sulphate-reduction cements, which form strata-bound accumulations during sedimentary stillstands (Kantorowicz *et al.*, 1987).

Bjørkum & Walderhaug (1990b) present a model for the formation of strata-bound concretions in which concretion distribution is controlled by the initial distribution patterns of shell material within the host sandbody. They suggest that, at very low degrees of supersaturation, the concretions cannot grow upon detrital or fossil calcite because of surface poisoning, for example by magnesium. They hence propose that the nucleation of calcite is driven by the differing solubilities of aragonite, high-Mg calcite and low-Mg calcite. Although homogeneous nucleation cannot be driven by the low degrees of supersaturation under consideration (De Boer, 1977), heterogeneous nucleation may well take place upon detrital shell material. As Bjørkum & Walderhaug (1990b) suggest that porewaters will become supersaturated with respect to calcite more rapidly within shell-rich horizons, it follows that these will be the sites of concretion growth.

However, it is unlikely that the timing of development of supersaturation is significant in the formation of concretionary nuclei. Calculations (not shown) indicate that the supersaturation state of the porewaters within a typical sandbody will be homogenized by diffusion within approximately 100 years. To grow a concretion of 1 m diagenetic redistribution takes around 9 Ma in stationary porewaters (Wilkinson & Dampier, 1990), so that the timing of the development of supersaturation will not significantly influence concretion growth.

Concretions form more rapidly in flowing fluids than in stationary ones (Nielson, 1961; Berner, 1968, 1980, p. 119; Wilkinson & Dampier, 1990), hence, in a sandbody with an abundance of nuclei, concretionary horizons could develop within beds with high flow rates. If this were the case, concretions should lie long the most permeable horizons of a sandbody. Unconsolidated sand samples collected from both within and between concretionary horizons were subjected to size analysis, on the assumption that, in poorly cemented sands, permeability increases with grain size (e.g. Beard & Weyl, 1973). The results obtained clearly do not support the hypothesis that there is a permeability control on the formation of concretionary horizons (Fig. 11).

Kantorowicz *et al.* (1987) described carbonate-cemented intervals within reservoir sandstones from the Troll Field of the North Sea (UK Sector). Fringing cements formed locally during periods of non-deposition or emergence, and formed the nuclei to more extensive mosaic calcites. By analogy, it is proposed that the early cements of the Ollach Sandstone and Holm Sandstone members (OC1 and HC1) acted as nuclei to the later, concretionary cements (OC2 and HC2). As there is no evidence that the sands of the Bearreraig Sandstone Formation became emergent at any time during deposition, cements OC1 and HC1 must have developed in unusual concentrations during periods of non-deposition (cf. Raiswell, 1987). The distribution of concretionary cement is hence controlled by the distribution of the early marine cements which developed in response to changes in sedimentation rates.

CONCLUSIONS

The sandbodies upon which this study is based underwent two generations of calcite precipitation. The first, which was of only minor volumetric significance, was a ferroan calcite interpreted to be

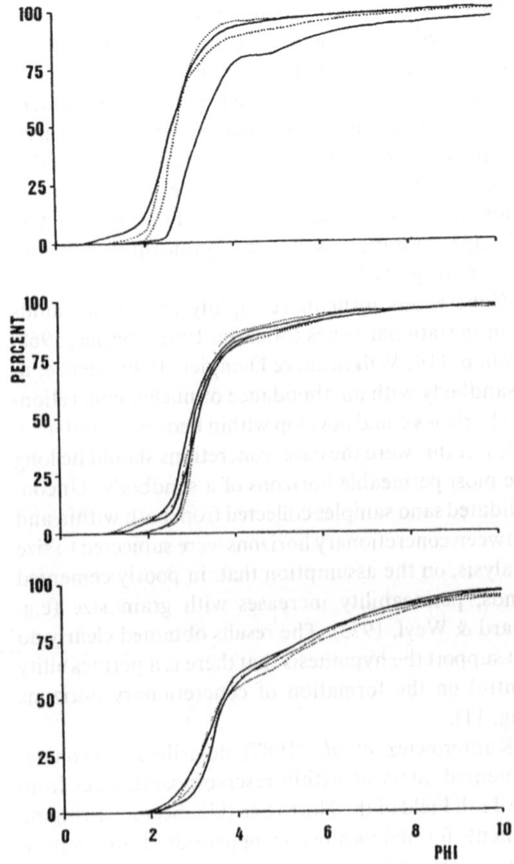

Fig. 11. Grain-size distributions from concretionary horizons (dotted line) and interbedded non-concretionary horizons (solid lines). Because of the limited size of the data set for the study rocks (Holm Sandstone Member, upper plot), data are also included for the Rigg Sandstone Member (Bearreraig Sandstone Formation, centre plot), and the underlying Scalpa Sandstone Formation (lower plot). There is no significant difference between the concretionary and non-concretionary sediments, ruling out a permeability control on the siting of the concretionary horizons.

the product of sulphate reduction. Although the porewaters within the burrows were of marine origin, cement precipitation caused modification leading to a wide range of geochemical compositions. The quantity of this cement produced was a function of sedimentation rate, relatively high cement concentrations being associated with sedimentary stillstands.

The second calcite cement generation was nucleated upon the first and formed metre-scale concretions. These laterally coalesced to occupy up to 75% of some bedding planes. The cement is a ferroan calcite which displays an increasing Fe/Mg ratio with time. It was sourced by the dissolution of shell material into porewaters of meteoric, or mixed marine–meteoric origin. Strongly corroded feldspars within the concretions suggest that a significant time interval elapsed (best estimate of 8 Ma) between the precipitation of the two cement generations.

The timing of freshwater penetration into the sequence is poorly constrained, although the reduced-salinity faunas of the Great Estuarine Group, which overlies the Bearreraig Sandstone Formation, suggest that a freshwater lens may have been established during Bathonian times. Confined aquifer flow could only have been important after partial basin inversion; this occurred between the early Kimmeridgian and the late Cretaceous. The influence of mixed water is restricted to the lowermost sandstone member of the formation (the Ollach Sandstone Member), suggesting that this was close to the lower limit of freshwater penetration within the sequence.

Hence, although most of a concretionary band may be comprised of meteoric cement, its degree of lateral persistence is a function of the distribution of early marine cements. If such cements (or similar nuclei) are rare, or absent, any concretions formed will be distributed randomly throughout the sandbody, and they are unlikely to coalesce laterally.

ACKNOWLEDGMENTS

I would like to acknowledge the receipt of a research grant from Shell UK Exploration and Production, and the contribution to this paper by Dr J. D. Hudson of Leicester University who supervised the work. Thanks also to Dr J. D. Marshall of Liverpool University for use of the stable isotope laboratory, and to T. Astin and R. Raiswell for constructive and helpful reviews. Thanks to M. Allen, T. Wilkinson, M. Williams and M. Wakefield for help in the field.

REFERENCES

ANDERSON, T.F. & ARTHUR, M.A. (1983) Stable isotopes of oxygen and carbon and their application to sedimentologic and paleoenvironmental problems. In: *Stable Isotopes in*

Sedimentary Geology (Ed. by M.A. Authur), *Soc. econ. Paleont. Miner., Short Course, Tulsa*, **10**, 1.1–1.151.

BEARD, D.C. & WEYL, P.K. (1973) Influence of texture on porosity and permeability of unconsolidated sand. *Bull. Am. Ass. petrol. Geol.*, **57**, 349–369.

BERNER, R.A. (1968) Rate of concretion growth. *Geochim. Cosmochim. Acta*, **32**, 477–483.

BERNER, R.A. (1980) *Early Diagenesis, A Theoretical Approach*. Princetown University Press, Princetown, 241 pp.

BYØRKUM, P.A. & WALDERHAUG, O. (1990a) Lateral extent of calcite-cemented zones in shallow marine sandstones. In: *North Sea Oil and Gas Reservoirs II* (Ed. by A.T. Buller *et al.*), pp. 331–336. The Norwegian Institute of Technology, Graham and Trotman, London.

BYØRKUM, P.A. & WALDERHAUG, O. (1990b) Geometrical arrangement of calcite cementation within shallow marine sandstones. *Earth Sci. Rev.*, **29**, 145–161.

BJØRLYKKE, K., AAGAARD, P., DYPVIK, H., HASTINGS, D.S. & HARPER, A.S. (1986) Diagenesis and reservoir properties of Jurassic sandstones from the Haltenbanken area, offshore mid Norway. In: *Habitat of Hydrocarbons on the Norwegian Continental Shelf* (Ed. by A. M. Spencer *et al.*), pp. 275–286. Graham and Trotman, London.

BRYANT, I.D., KANTOROWICZ, J.D. & LOVE, C.F. (1988) The origin and recognition of laterally continuous carbonate-cemented horizons in the Upper Lias Sands of southern England. *Mar. petrol. Geol.*, **5**, 108–133.

BUSENBERG, G. & CLEMENCY, C.V. (1976) The dissolution of feldspars at 25° C and 1 atmosphere CO_2 partial pressure. *Geochim. Cosmochim. Acta*, **40**, 41–50.

CRAIG, H. (1957) Isotopic standards for carbon and oxygen factors for mass-spectrometric analysis of carbon dioxide. *Geochim. Cosmochim. Acta*, **12**, 133–149.

DE BOER, R.B. (1977) Influence of seed crystals on the precipitation of calcite and aragonite. *Am. J. Sci.*, **277**, 38–60.

DICKINSON, W.W. (1988) Isotopic and petrographic evidence for carbonate diagenesis in non-marine sandstones, Green River Formation, Wyoming. *J. sedim. Petrol.*, **58**, 327–338.

DICKSON, J.A.D. (1966) Carbonate identification and genesis as revealed by staining. *J. sedim. Petrol.*, **36**, 491–505.

EMERY, D., HUDSON, J.D., MARSHALL, J.D. & DICKSON, J.A.D. (1987) The origin of late spar cements in the Lincolnshire Limestone, Jurassic of central England. *J. geol. Soc. London*, **145**, 621–633.

GLUYAS, J.G. (1984) Early carbonate diagenesis within Phanerozoic shales and sandstones of the NW European Shelf. *Clay Miner.*, **19**, 309–321.

HOLDREN, G.R. & SPEYER, P.M. (1985) Reaction rate-surface area relationships during early stages of weathering. I. Initial observations. *Geochim. Cosmochim. Acta*, **49**, 675–681.

HUDSON, J.D. & ANDREWS, J.E. (1987) The diagenesis of the Great Estuarine Group, Middle Jurassic, Inner Hebrides, Scotland. In: *Diagenesis of Sedimentary Sequences* (Ed. by J. D. Marshall), pp. 259–276. Blackwell Scientific Publications, Oxford, 360 pp.

IRWIN, M., CURTIS, C.D. & COLEMAN, M.L. (1977) Isotopic evidence for source of diagenetic carbonates formed during burial of organic-rich sediments. *Nature*, **269**, 209–213.

JOHNSON, M.R. (1989) Paleogeographic significance of oriented calcareous concretions in the Triassic Katberg Formation, South Africa. *J. sedim. Petrol.*, **59**, 1008–1010.

KANTOROWICZ, J.D., BRYANT, I.D. & DAWANS, J.M. (1987) Controls on the geometry and distribution of carbonate cements in Jurassic sandstones: Bridport Sands, southern England and Viking Group, Troll Field, Norway. In: *Diagenesis of Sedimentary Sequences* (Ed. by J. D. Marshall), pp. 103–118. Blackwell Scientific Publications, Oxford, 360 pp.

MARSHALL, J.D. (1981) Zoned calcites in Jurassic ammonite chambers: trace elements, isotopes and neomorphic origin. *Sedimentology*, **28**, 867–887.

MCBRIDE, E.F. (1988) Contrasting diagenetic histories of concretions and host-rock, Lion Mountain Sandstone (Cambrian), Texas. *Bull. geol. Soc. Am.*, **100**, 1803–1810.

MORTON, N. (1976) Bajocian (Jurassic) stratigraphy in Skye, Western Scotland. *Scott. J. Geol.*, **12**, 23–33.

MORTON, N. (1983) Paleocurrents and paleo-environments of part of the Bearreraig Sandstone (Middle Jurassic) of Skye and Raasay, Inner Hebrides. *Scott. J. Geol.*, **19**, 87–95.

MORTON, N. (1987) Jurassic subsidence history in the Hebrides, North-west Scotland. *Mar. petrol. Geol.*, **4**, 226–242.

NIELSON, A.E. (1961) Diffusion controlled growth of a moving sphere. The kinetics of crystal growth in potassium perchlorate precipitation. *J. Phys. Chem.*, **65**, 46–49.

PACES, T. (1983) Rate constants of dissolution from the measurement of mass-balance in hydrological catchments. *Geochim. Cosmochim. Acta*, **47**, 1855–1863.

PASSARATTI, M. L. & ESLINGER, E. V. (1987) Dissolution and relic textures in framework grains of Holocene sediments from the Brazos River and Gulf Coast of Texas. *J. sedim. Petrol.*, **57**, 94–97.

PIRRIE, D. (1987) Oriented calcareous concretions from James Ross Island, Antarctica. *Bull. Br. Antarctic Surv.*, **75**, 41–50.

RAISWELL, R. (1971) The growth of Cambrian and Liassic concretions. *Sedimentology*, **17**, 147–171.

RAISWELL, R. (1987) Non-steady state microbiological diagenesis and the origin of concretions and nodular limestones. In: *Diagenesis of Sedimentary Sequences* (Ed. by J. D. Marshall), pp. 41–54. Blackwell Scientific Publications, Oxford, 360 pp.

SAIGAL, G. C. & BJØRLYKKE, K. (1987) Carbonate cements in clastic reservoir rocks from offshore Norway—relationships between isotopic composition, textural development and burial depth. In: *The Diagenesis of Sedimentary Sequences* (Ed. by J. D. Marshall), pp. 313–324. Blackwell Scientific Publications, Oxford, 360 pp.

SHACKLETON, N. J. & KENNET, J. P. (1975) Paleotemperature history of the Cenozoic and the initiation of Antarctic glaciation: oxygen and carbon isotope analyses in the DSDP sites 277, 279 and 281. In: *Initial Reports of the Deep Sea Drilling Project* (Ed. by J. P. Kennett and R. E. Hontz), Vol. 29, pp. 743–55. US Government Printing Office, Washington.

STOESSEL, R. K. (1987) Mass transport in sandstones around dissolving plagioclase grains. *Geology*, **15**, 295–298.

WALDERHAUG, O., BJØRKUM, P.A. & NORGÅRD BØLAS, H.M. (1989) Correlation of calcite-cemented layers in shallow-marine sandstones of the Fensfjord Formation in the

Brage Field. In: *Correlation in Hydrocarbon Exploration* (Ed. by J. D. Collisson *et al.*), pp. 367–375. Norwegian Petroleum Society.

WILKINSON, M. (1989a) Discussion: evidence for surface reaction-controlled growth of carbonate concretions in shales. *Sedimentology*, **36**, 951–953.

WILKINSON, M. (1989b) *Sandstone-hosted concretionary cements of the Hebrides, Scotland*. PhD thesis, University of Leicester, UK.

WILKINSON, M. & DAMPIER, M.D. (1990) The rate of growth of sandstone-hosted calcite concretions. *Geochim. Cosmochim. Acta*, **54**, 3391–3399.

(Manuscript received 3 December 1990; revision received 15 March 1991)

Non-marine eogenesis I: warm and wet environments

Diagenesis in fluvial and floodplain environments commonly involves the growth of carbonate minerals. The types of carbonates that can grow include calcite, dolomite and siderite and each can contain appreciable quantities of manganese. Siderite is cited as being diagnostic of non-marine environments, especially when it contains little Ca or Mg and has low $\delta^{18}O$ values, as the relative absence of aqueous sulphate (and thus sulphide) allows iron to enter carbonates instead of sulphide minerals. However, Andrews *et al.* (1991) document a case of dolomite cement forming during non-marine eogenesis, assigning the abundance of Mg to (i) sporadic temporary inundations by sea water, (ii) weathering of a local source of Mg in the hinterland, or (iii) contemporary volcanism. McKay *et al.* (1995) also describe a dolomite (ankerite) cemented non-marine sandstone but assigned the source of Mg to detrital dolomite-rich lithic fragments and the relative paucity of siderite to the growth of pyrite from brackish, sulphate-rich, groundwaters. Siderite-dominated eogenetic carbonates are described by Baker *et al.* (1995) and Browne & Kingston (1993), who consign the enrichment of iron in the system to the deposition and subsequent alteration of chlorite and smectite.

The carbon in these eogenetic carbonates has several potential sources but excludes bacterial sulphate reduction because of the lack of sulphate in this groundwater type. Oxidation of organic matter in general, and methane in particular, is cited as a source of low $\delta^{13}C$ values ($\leq -20‰$), whereas methanogenesis is cited as a source of high $\delta^{13}C$ values ($\geq 0‰$) (Andrews *et al.*, 1991; Baker *et al.*, 1995). $\delta^{13}C$ values can thus be used to trace the origin of the carbonate but it is clear that the carbonate enrichment is fundamentally related to the abundance of organic matter in sediment formed in warm and wet environments.

Kaolinite is reported to be the predominant eogenetic aluminosilicate product by McKay *et al.* (1995) although the clay eogenetic story is assumed to be subordinate to the carbonate and sulphide stories.

Reprinted from Sedimentology (1991) ***38****, 271–287*

The anatomy of an early Dinantian terraced floodplain: palaeo-environment and early diagenesis

JULIAN E. ANDREWS, MICHAEL S. TURNER, GHULAM NABI *and* BARUCH SPIRO*

School of Environmental Sciences, University of East Anglia, Norwich NR4 7TJ, UK

**Natural Environment Research Council Isotope Geoscience Laboratories, British Geological Survey, Keyworth, Nottingham NG12 5GG, UK*

ABSTRACT

The lowermost Carboniferous rocks in the Cockburnspath area of east Berwickshire (southern Scotland) are interpreted as coastal floodplain sediments. A lower mudstone-dominated unit is composed of silty mudstones and shales with subordinate sandstones and argillaceous ferroan dolomites (cementstones). These are interpreted as distal floodplain sediments with periodic crevasse-splay deposition. The dark grey colour of the mudrocks suggests deposition in reducing conditions, probably in floodplain lakes. Most of the cementstones are concretionary, some with septarian cracks, suggesting an early diagenetic origin. An immature palaeosol suggests periodic pedogenesis under improved drainage. A synsedimentary erosion surface indicates incision of a valley into the floodplain, presumably in response to base-level lowering.

An upper sandstone-dominated unit starts with fine-grained rippled sandstones, cut by small channel sandstones. These are interpreted as floodplain lake deposits fed by crevasse channels. A distinctive conglomerate with cementstone clasts, wood fragments and fish remains is interpreted as a major overbank deposit, dumped into a pre-existing floodplain lake. A bivalve fauna was established in the overlying mudstones, followed by a thin limestone with a restricted marine fossil assemblage, showing that seawater flooding of the lakes occurred at times.

Mudrocks throughout the sequence contain no pyrite, except for the marine band which has an organic-carbon/sulphur ratio and degree of pyritization value typical of marine sediments. The concretionary cementstones have $\delta^{13}C$ values around -4 to -6‰ PDB which are interpreted as indicative of anaerobic oxidation of organic matter. The combined geochemical data suggest a significant involvement of iron reduction in cementstone formation, although the $\delta^{13}C$ values are ambiguous in assessing the relative involvement of methanogenesis and methane oxidation. Limited seawater inundation of the floodplains might have supplied magnesium and calcium ions for dolomite formation assuming that any H_2S derived as a result of sulphate reduction was oxidized by iron reduction. Alternatively a weathering source for solutes might have been involved.

INTRODUCTION

Ancient alluvial sediments are generally studied with emphasis on the clastic sedimentology, and comparison with modern river systems has led to a good understanding of their depositional processes and products (e.g. see Bridge, 1984; Walker & Cant, 1984; Collinson, 1986, for reviews). Carbonates in alluvial sediments are usually volumetrically small, restricted to calcretes in floodplain sediments (e.g. Allen, 1974), although where floodplain lakes occur, impure bedded carbonates may form (e.g. Friend & Moody-Stuart, 1970). Despite their restricted occurrence, carbonates and palaeosols in alluvial sediments are powerful indicators of palaeoclimate, floodplain drainage and pauses in sedimentation which might be linked to channel migration or incision and terrace formation (Leeder, 1975; Kraus, 1987). In contrast, the study of concretionary carbonates in sediments has mainly been the concern of sedimentary geochemists who have concentrated on marine sedimentary systems (e.g. Raiswell, 1988), or authigenic carbonates, typically siderites, in ancient deltaic and floodplain deposits (e.g. Curtis & Coleman, 1986).

In this study we have tried to understand the depositional system and history of a Carboniferous coastal floodplain and have then used this framework, combined with geochemistry, to explain the formation of concretionary authigenic carbonates, locally called cementstones. The cementstones (impure, argillaceous dolostones) are common throughout the Lower Carboniferous of Scotland and northern England (Francis, 1965). The general facies models interpret these sediments as coastal alluvial plain deposits (Leeder, 1974; Anderton, 1985). Within this province the cementstones are often associated with evaporite minerals (e.g. Scott, 1986) and some are laminated and clearly primary (Belt, Freshney & Read, 1967), probably formed in playa-like lakes. Others, however, are concretionary and of early diagenetic origin. Limited sulphur isotope data from anhydrite and gypsum in cementstone-bearing facies (Ballagan Formation) of the Midland Valley of Scotland (Pattrick, Coleman & Russell, 1983; B. Spiro & M. Turner, unpublished data) suggest that the evaporites are seawater derived, which has led to the general supposition that the dolomite-promoting fluid was, at least in part, seawater (Anderton, 1985; Scott, 1986).

In the sequence that we have studied the geochemical data were, in part, intended to test whether seawater was involved in cementstone formation, and, in particular, we tried to constrain the early diagenetic processes involved in cementstone formation by studying their carbon isotope composition.

Stratigraphy and field relations

The studied sequence is part of the Calciferous Sandstone Measures (old terminology) of the southern Scottish Lower Carboniferous (Fig. 1). The sediments accumulated in the Oldhamstocks Basin (*sensu* Lagios, 1983) which is situated on the NE margin of the Southern Uplands massif. Mudstones in the studied interval yield miospores of the CM Zone (Clayton, 1971; Neves *et al.*, 1973) which indicate the Ivorian Stage. These rocks belong to the Tyninghame For-

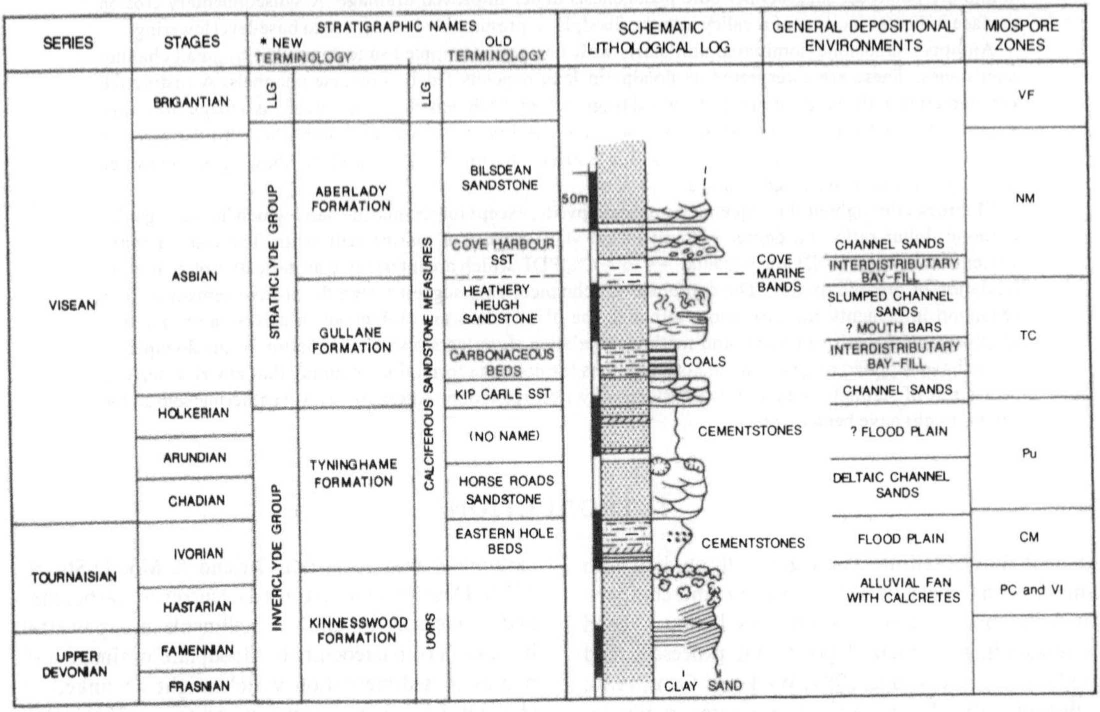

Fig. 1. Stratigraphy of the Upper Devonian–Lower Carboniferous strata of the Cockburnspath area (Berwickshire). * New terminology after Patterson & Hall (1986) and Chisholm *et al.* (1990). UORS = Upper Old Red Sandstone, LLG = Lower Limestone Group. Black and white scale bars = 50 m. See key to Fig. 3 for lithological symbols.

mation of the Inverclyde Group (Patterson & Hall, 1986; Chisholm, McAdam & Brand, 1990), and are notable for the cementstones (argillaceous ferroan dolostones) and a distinctive conglomerate with cementstone clasts (the Eastern Hole Conglomerate, e.g. see Clarkson, 1986). We refer to these rocks as the Eastern Hole beds (Fig. 1) and our discussion concentrates on the lower part of them.

The basal Eastern Hole beds overlie a distinctive red-weathering bed with an irregular bedding surface approximately 500 m NNW of Pease Sands (Fig. 2; grid reference NT 7910 7155), which is taken as the top of the Old Red Sandstone (Kinnesswood Formation of Patterson & Hall, 1986). In the field this bed appears to be a coarse sandstone (Greig, 1988, p. 42), but is in fact a silty mudstone with intergrowth of coarse dolomite crystals. Petrographic examination of chert nodules from this bed reveals relict textures of soil-zone fabrics and we interpret this bed as a coalesced nodular calcrete (now dolomite). It is texturally similar to the nodular calcretes (locally called cornstones) which characterize the Kinnesswood Formation. The upper part of the Eastern Hole beds (not further discussed here) are dominantly fluvial-derived sandstones, probably proximal crevasse-splay deposits, interbedded with mudrocks and a few thin cementstones. A distinct erosion surface marks the base of the thickly bedded Horse Roads Sandstone at Horse Roads Rock (Fig. 2; grid reference NT 7910 7165).

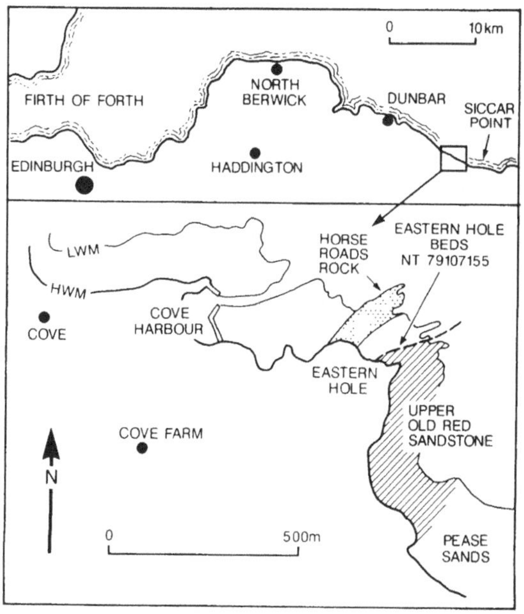

Fig. 2. Location of the studied sequence. The dashed line marks the boundary between the Upper Old Red Sandstone and the Eastern Hole beds.

METHODS

Stable isotope analyses were performed on CO_2 derived from *c.* 10-mg samples which were reacted with anhydrous H_3PO_4 at 55°C for about 16 h. Isotope ratios were measured using a VG Sira 10 mass spectrometer at the NERC Isotope Geoscience Laboratories and a VG Sira Series II mass spectrometer at the University of East Anglia. Results are expressed relative to the PDB standard after correction for ^{17}O following the method of Craig (1957). A fractionation factor of 1·0100 was applied for dolomite–phosphoric-acid fractionation (based on the data of Rosenbaum & Sheppard, 1986). Replicate analyses of the laboratory standard ($n = 15$) gave a precision of $\pm 0{\cdot}06‰$ for carbon and $\pm 0{\cdot}11‰$ for oxygen.

Elemental analyses were made on either a Pye Unicam SP9 or Varian Spectr AA-10 atomic absorption spectrophotometer (AAS), after dissolution of 1 g of sample powder in 15% HCl at 120°C. In addition three samples of shale were leached and analysed under the same conditions to monitor contributions from the insoluble residues. Replicate analyses gave a precision of 4–6%. An elemental traverse across one cementstone sample was performed on a Jeol JXA-86100 microprobe where reproducibility was around 1% for Ca and Mg, and 5–10% for Fe and Mn. Insoluble residues (%IR) were calculated from the AAS leaches, dried on filter paper (blank corrected) and weighed.

Carbon-to-sulphur (C/S) ratios were determined on samples which met the criteria of Berner & Raiswell (1984). The ground mudrock samples were leached in cold 10% HCl for 4 h to remove trace $CaCO_3$, and then *c.* 15 mg of sample was combusted at 1020°C in a Carlo-Erba 1108 CHNS elemental analyser. Percent carbon and percent sulphur values were calculated by comparison with a sulphanilamide standard, and repeats indicated a reproducibility of $\pm 0{\cdot}03\%$ for both carbon and sulphur. The %C and %S values are assumed to represent %organic carbon and %pyrite sulphur only. Degree of pyritization (DOP) was calculated after the method of Raiswell *et al.* (1988).

Clay mineral identification was based on standard techniques (e.g. Andrews, 1987) and the relative proportions of minerals were calculated using the Biscaye (1965) method.

FACIES ANALYSIS—DESCRIPTION

The sequence is broadly divisible at the base of bed 21 (Fig. 3) into a lower two-thirds (unit) dominated by mudrocks and thin sandstones, and an upper third (unit) dominated by sandstones.

Lower mudrock-dominated unit

This unit (beds 1–20) is mainly composed of grey shales, mudstones and muddy siltstones (75% of compacted lithologies), interbedded with silty, fine- to medium-grained sandstones (11% of compacted lithologies). Except for wood and macroscopic plant remains the mudrocks are unfossiliferous, and the content of finely divided organic carbon is low (0·2–0·6% C). No trace of pyrite was found during optical petrography or C/S analysis. (C/S ratios of the Cove Marine Bands (Fig. 1) are consistent with the data of Berner & Raiswell (1984) (G. Nabi, unpublished data) so we are confident that weathering effects have not severely affected the C and S contents.) The clay mineralogy of the <2-µm fraction (Fig. 3) indicates an average composition for beds 6–20 dominated by illite (85%) with attendant kaolinite (14%, but increasing lower in the section) and trace chlorite (1%), in general agreement with Wilson *et al.* (1972). The mudrocks are either intercalated with silt–fine-sandstone lenticles (e.g. bed 4), or cut by tabular sandstone beds which pinch or thin laterally over a few metres. Ripple cross-lamination and low-angle planar cross-bedding are the only sedimentary structures observed, with well-preserved, weakly asymmetric, sharp-crested ripple marks seen on the top of bed 11. Cementstones (impure ferroan dolostones) comprise the rest of the unit. They generally have a coalesced nodular appearance with compaction of the underlying shale laminae around them, and two beds (9 and 17) show septarian cracks infilled by sparry dolomite and calcite cements.

Upper sandstone-dominated unit

A scoured contact at the base of bed 21 marks the base of the unit. The scour is infilled by coarse- to medium-grained sandstones with low-angle trough cross-bedding, low-angle erosion surfaces and decimetre-scale planar cross-bedding. Laterally over 10 m this unit pinches out, showing scoured contacts cutting into micaceous silty fine-grained sandstones. These are overlain by 2 m of similar purple silty sandstones (bed 22). Ripple cross-lamination with bidirectional foresets is abundant throughout the bed, and poorly preserved, sinous-crested ripple marks with 'tuning-fork' bifurcations are seen in places. One surface contains burrows 0·5 m in diameter orientated oblique to bedding.

A less obvious scour surface truncates bed 22 and is followed by a poorly sorted conglomerate with flat-pebble clasts (bed 23). The matrix is a red, coarse-grained sandstone, whilst the clasts, up to 20 cm in diameter, are generally yellow-weathering cementstone with primary lamination preserved. The clasts lie in horizontal to subhorizontal position, but are not imbricated. Fish scales and spines, plant and wood fragments, mainly *Lepidodendron calamopsides* Long (Scott & Rex, 1987), are also common. Above the conglomerate are micaceous silty sandstones (bed 24), packed with plant fragments.

The uppermost beds of the sequence are dark mudstones interrupted by a well-cemented sandstone (bed 27) with abundant coalified plant fragments of *Lepidodendron calamopsides* at the base, succeeded by a limestone bearing bivalves, *Sanguinolites* sp. (Clough, Barrow & Crampton, 1910), rare ostracodes and fish scales in the upper 6 cm. Petrography of the limestone reveals a rich fauna, not visible in the field, including crinoid, brachiopod, foraminifer, fenestellid bryozoan, calcisphere and (?) red calcereous algal fragments. The clay mineralogy of these upper mudrocks is more kaolinite-rich (33%) and illite-poor (60%) than the average values for the mudrocks between beds 6 and 20 (Fig. 3). Finely divided organic matter attains 1%, and up to 0·57% pyrite sulphur is present, with a DOP value of 0·26.

FACIES ANALYSIS—INTERPRETATION

The general interpretation of the mudrock-dominated unit (beds 1–20) is deposition in an alluvial plain

Fig. 3. Graphic log and mineralogical data for the base of the Eastern Hole beds. Note that the log can be divided into a lower mudrock-dominated unit and an upper sandstone-dominated unit. Palaeocurrent data are corrected for tectonic dip; north is towards the top of the page. ORS = Old Red Sandstone. Black and white scale bars = 1 m. The dominance of illite is interpreted as an inherited clay mineral, i.e. a weathering product of the surrounding catchment. Increases in the amount of kaolinite in the siltier beds, and those intercalated with sandstones, suggest that kaolinite is largely a diagenetic product of feldspar breakdown.

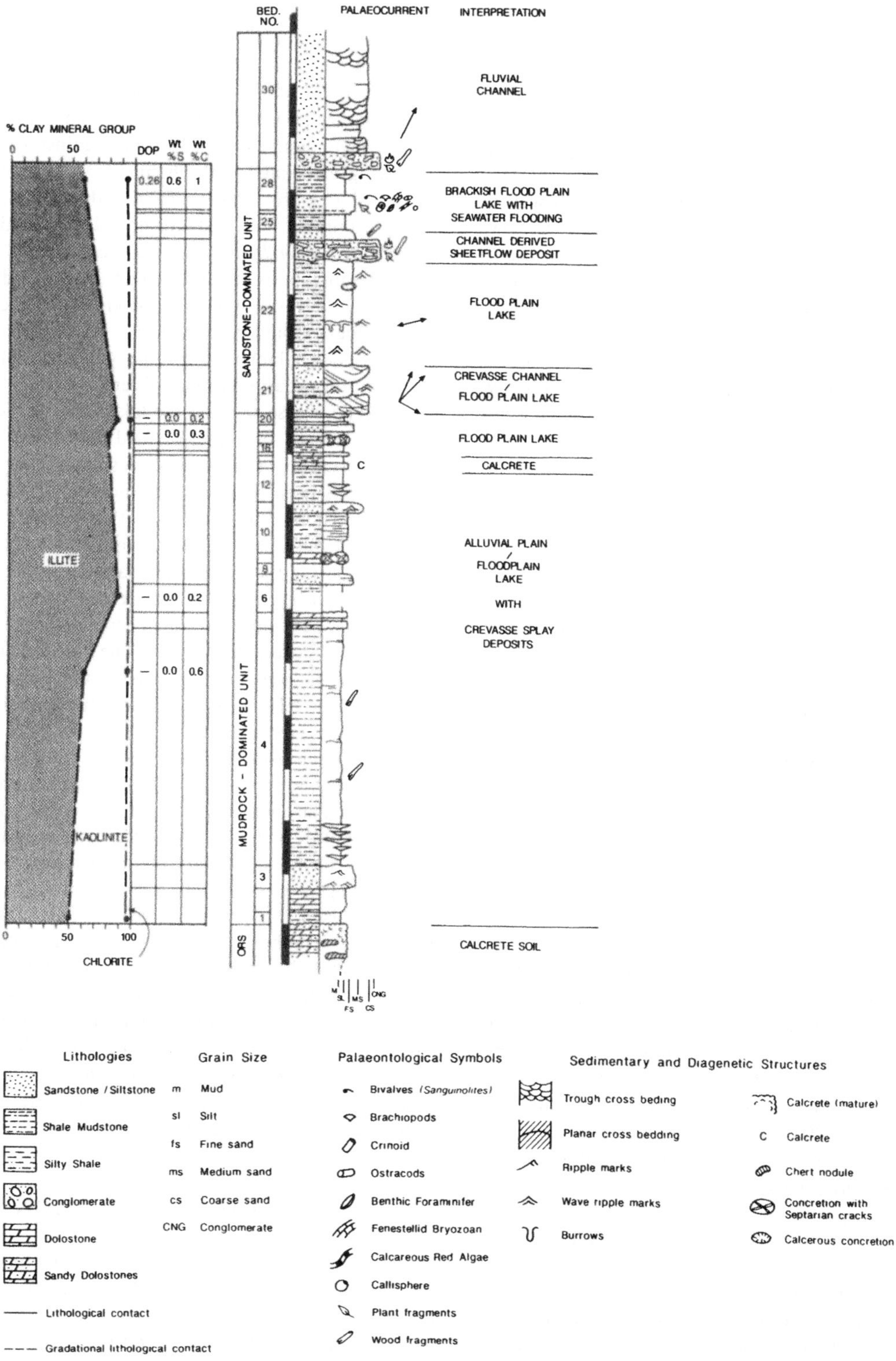
% CLAY MINERAL GROUP
0
50
DOP
Wt %S
Wt %C
0.26
0.6
1
ILLITE
KAOLINITE
CHLORITE
0.0
0.2
0.0
0.3
0.0
0.2
0.0
0.6
100
BED. NO.
SANDSTONE-DOMINATED UNIT
MUDROCK - DOMINATED UNIT
ORS
30
28
25
22
21
20
16
12
10
8
6
4
3
1
PALAEOCURRENT
INTERPRETATION
FLUVIAL CHANNEL
BRACKISH FLOOD PLAIN LAKE WITH SEAWATER FLOODING
CHANNEL DERIVED SHEETFLOW DEPOSIT
FLOOD PLAIN LAKE
CREVASSE CHANNEL / FLOOD PLAIN LAKE
FLOOD PLAIN LAKE
CALCRETE
C
ALLUVIAL PLAIN / FLOODPLAIN LAKE
WITH
CREVASSE SPLAY DEPOSITS
CALCRETE SOIL
M
SL
FS
MS
CS
CNG
Lithologies
Sandstone / Siltstone
Shale Mudstone
Silty Shale
Conglomerate
Dolostone
Sandy Dolostones
Lithological contact
Gradational lithological contact
Grain Size
m Mud
sl Silt
fs Fine sand
ms Medium sand
cs Coarse sand
CNG Conglomerate
Palaeontological Symbols
Bivalves (Sanguinolites)
Brachiopods
Crinoid
Ostracods
Benthic Foraminifer
Fenestellid Bryozoan
Calcareous Red Algae
Callisphere
Plant fragments
Wood fragments
Sedimentary and Diagenetic Structures
Trough cross beding
Planar cross bedding
Ripple marks
Wave ripple marks
Burrows
Calcrete (mature)
C Calcrete
Chert nodule
Concretion with Septarian cracks
Calcerous concretion

environment, the mudrocks representing fine-grained overbank deposits, relatively remote from active channels. The lack of desiccation cracks and rootlets seems to imply that the sediment rarely dried out, and was unvegetated. Excepting the basal mudstones (beds 1 and 4), red coloration is uncommon, which argues against strong aerobic oxidation of the sediment. This suggests that the floodbasin was waterlogged, probably indicating a floodplain of very low slope with rivers running close to base level (Collinson, 1986). The lack of pyrite is usually taken as evidence against a significant contribution of seawater-derived sulphate (Berner, 1984).

The siltstone lenticles and laterally thinning, rippled sandstones are interpreted as distal crevasse-splay deposits. Preservation of sharp-crested ripples suggests a lack of reworking of the ripple tops, and might imply formation in standing water, for example a floodplain lake.

The nodular appearance, lack of primary sedimentary structures (except one bed), and compaction of shale around the dolomite nodules implies that they formed as concretions in the sediment during early diagenesis (see Raiswell, 1976), at times coalescing to form more persistent dolostone beds. Coalescence of nodules during early diagenesis requires a 'freezing' of the geochemical conditions which promote concretion growth, and as these are depth related (Curtis, 1977), implies significant pauses in sedimentation (Raiswell, 1987).

In general, we interpret the dolostones to have formed in the reducing muds of coastal floodplain lakes, which rarely, if ever, dried out. A similar depositional environment has been invoked for these rocks by Scott & Rex (1987) and Greig (1988), and for their lateral equivalents in Berwickshire (Anderton, 1985) and the Midland Valley of Scotland (Belt *et al.*, 1967). It is, however, important to stress that no trace of evaporite minerals has been found in the Eastern Hole beds, suggesting that highly saline waters were not associated with these cementstones. This is in contrast to the Tweed Embayment cementstones which formed 20 km to the SE (Scott, 1986).

The basal scour-surface-confined sandstones of the sandstone-dominated unit (beds 21–28) are clearly channel deposits and their small size suggests a crevasse origin. The surrounding and overlying silty sandstones with bidirectional ripple foresets appear to have been formed by wave action; however, the lack of marine indicators militates against tidal marine conditions. We interpret these as proximal crevasse-splay sands, fed and cut by the channels, and accumulating in floodplain lakes. The rippling and bioturbation occurred during pauses in crevasse deposition.

The integrity of the cementstone clasts, lack of imbrication, presence of plant stems and sandy matrix in the conglomerate (bed 23) argue against strong winnowing associated with channel-lags. The presence of fish remains implies a fairly stable floodplain lake environment with well-developed food chains, whereas *Lepidodendron* fragments were clearly derived, presumably from vegetated areas surrounding the lake. We think the conglomerate represents catastrophic overbank flooding from a river, probably as a sheet-flood, which fed into an existing floodplain lake, dumping robust fragments of cementstone, plant material and coarse sand on the lake floor. As flood velocity decreased, only finer-grained sands and plant fragments were deposited (bed 24). A similar sequence of events has been proposed by Anderton (1985) to explain sandstone intercalations in the Foulden Fish Bed, part of the Lower Carboniferous Cementstone Group bordering the Northumberland Trough. The yellow-weathering cementstone clasts in the conglomerate are distinct from the bedded cementstones seen lower in the Eastern Hole beds. The conglomerate clasts imply that some primary cementstones existed, perhaps around lake margins as dolomitized mudflats (cf. Eugster & Hardie, 1975).

The overlying beds represent quieter-water conditions in the lake with a bivalve fauna becoming established. The abundant marine fossils in the upper few centimetres of bed 27 record a brief seawater-flooding event in the history of the floodplain and prove that open-sea conditions were never far away to the east (Wilson, 1989). The presence of pyrite sulphur in bed 28, above the marine-fossil-bearing limestone, suggests limited input of seawater sulphate, i.e. brackish conditions, because the C/S ratio is not indicative of fully marine conditions (Berner & Raiswell, 1984). This might imply a gradual freshening of the lake following seawater flooding. The DOP values indicate normal, aerobic bottom water and bottom sediment (Raiswell *et al.*, 1988).

LATERAL FACIES VARIATION

Lateral lithological changes over tens of metres are common throughout the Eastern Hole beds, especially in the fluvial sandstones, but here we only describe the lower mudrock-dominated unit (beds 1–20). Lateral facies changes in ancient fine-grained alluvial plain deposits are not well documented, and in this respect, these rocks are unusual. At low tide, the outcrops can be traced along strike for about 80 m, and at about 40 m (horizontally) above MLWM there is a marked lineament which cuts across beds 3–10. This lineament truncates the beds on the SW side, whilst those on the NE side lap out against it. The beds on the SW side of this feature are laterally continuous, except for local pinching-out of the crevasse-splay sandstones (Fig. 4, columns 1 & 2). Immediately NE of the lineament, the stratigraphy changes (Fig. 4, column 3), although all the lithological elements are retained (mudrocks, sandstones and cementstones). The thick cementstone (bed 2, Figs 3 & 4) is not cut by this lineament, and the mudrock sequences either side of the feature have a different bedding arrangement (they are also thinner on the NE side), which argues against the lineament being a synsedimentary fault line. At MLWM (Fig. 4, column 4) the lithologies of beds 3 and 4 change suddenly to medium- and coarse-grained sandstones about 2 m thick. The outcrop is encrusted with barnacles and seaweed, which obscure sedimentary structures, but there can be little doubt that these are channel sandstones.

We interpret this feature as a synsedimentary erosion surface which cuts into the floodplain sediments. However, the strata immediately NE of the erosion surface are not channel sandstones, but rather, floodplain muds and crevasse-splay sandstones with concretionary cementstones. If early diagenetic cementstone formation requires pauses in sedimentation (see below), perhaps in the order of hundreds of thousands of years (Berner, 1968), then column 3 (Fig. 4) should also represent alluvial plain sediments. Therefore, the erosion surface is not a channel- but a valley-margin (terrace), cut into the pre-existing floodplain sediments. The channel sandstones occur 30 m NE of the erosion surface (Fig. 4, column 4).

Sequences NE of the erosion surface (Fig. 4, columns 3 & 4) are slightly thinner than those on the SW side. This suggests that renewed floodplain aggradation at the lower level never quite filled the valley topography. The upper beds (11–20, Fig. 4)

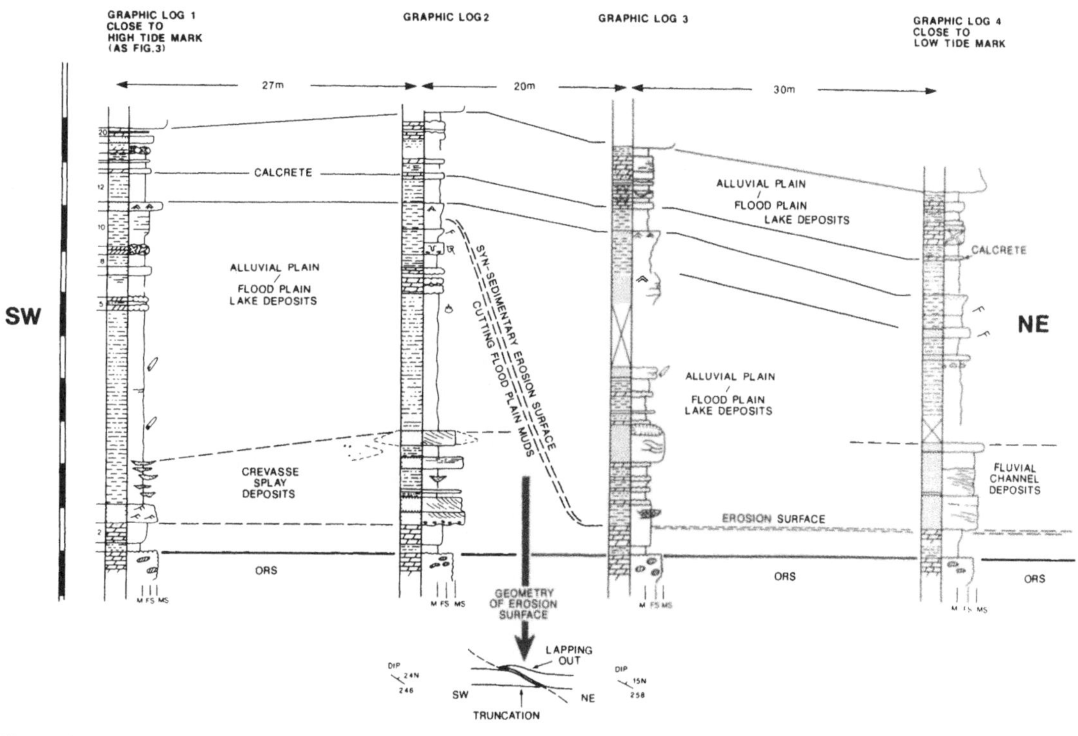

Fig. 4. Graphic logs taken along strike in the mudrock-dominated unit of the Eastern Hole beds (see Fig.3). Log 1 is a reproduction of the lower part of Fig. 3. Note the location of the erosion surface, its geometry and the small change of dip and strike. Palaeoenvironmental interpretations are given between logs. Black and white scale bars = 1 m. See key to Fig. 3 for lithological symbols.

mark a return to floodplain sedimentation characterized by both cementstones and a palaeosol (see below).

There are few comparable ancient floodplain sequences which show valley incision, the Lower Devonian example of Allen & Williams (1982) being a notable exception. Their sequence 5 shows a similar scale of downcutting (about 10 m) with thin channel sandstone fill around 3–4 m thick, similar to our example. The causes of incision could be intrinsic (channel diversion) or extrinsic (climate change, sea-level change or tectonism) (Schumm, 1977; Allen & Williams, 1982) and on the basis of our limited data it is not possible to decide between them.

PALAEOSOL

Description

One cementstone, bed 13 (Figs 3 & 4) is distinct due to its lateral continuity and red–yellow mottled appearance. Slabbed surfaces show that the mottling is caused by cementstone clasts, apparently breccia fragments, set in a red matrix with coarse dolomite crystals (Fig. 5). Petrography shows that the clasts are composed of a dense brown dolomicrite with abundant silt-sized detrital grains and patches rich in clay and/or organic matter. The micrite has a clotted fabric with peloids, lumps and wisps common. Anastomosing fractures dissect the larger lumps, and these are partially infilled with peloids and fragments of micrite (Fig. 6a). Some of the micrite lumps and some of the crack fills are replaced or cemented by a lighter-brown inclusion-rich microsparry dolomite cement which in places has a spherulitic texture (Fig. 6b). The dolomite crystals in the matrix were probably originally euhedral, but are now partially dissolved (Fig. 6c), and petrography shows that the red matrix is in fact a later-stage cement (see below). Under cathodoluminescence (CL) the micrites luminesce dull-orange, whilst the inclusion-rich cements, filling cracks, have either identical dull-orange luminescence, or are non-luminescent.

Fig. 5. A polished slab of cementstone bed 13 showing the brecciated macro-fabric. Light clasts are composed of dolomicrite and formed the cementstone body prior to brecciation. Dark areas are reddened matrix composed of silt-sized dolomite crystals. Note the pseudo-lamination in the top 2 cm of the slab which is caused by re-cementation of the breccia fragments. The overall features of this bed are consistent with formation by soil-zone processes (see text). Scale bar = 2 cm.

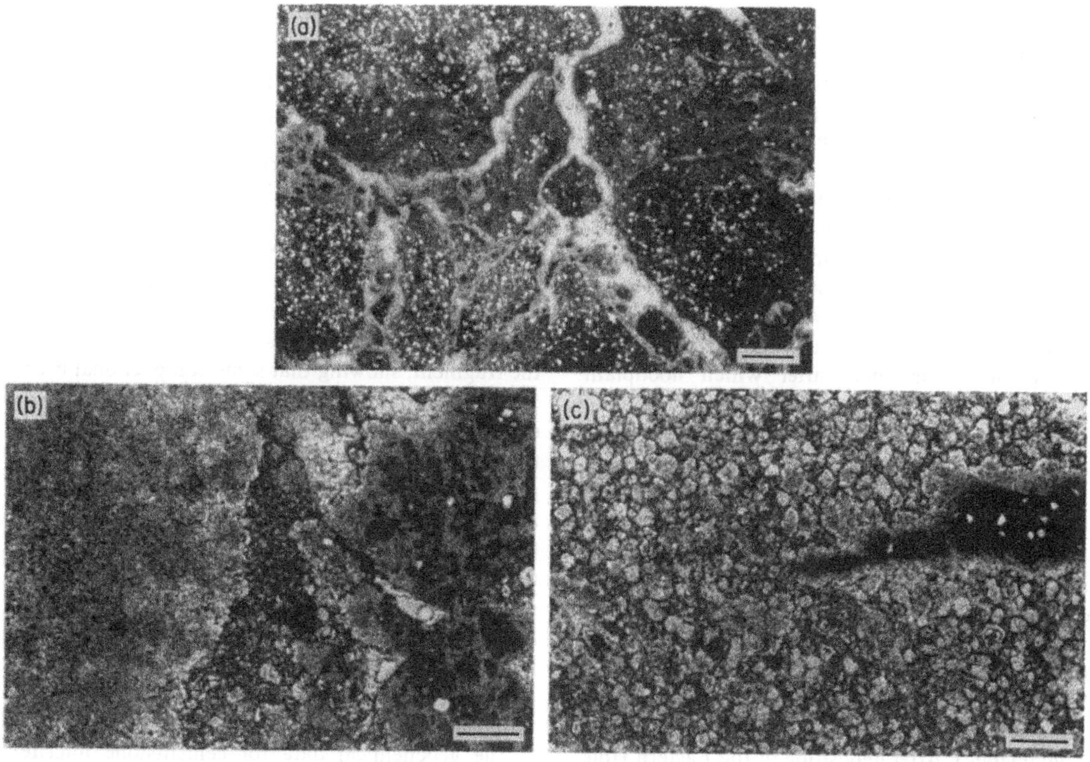

Fig. 6. Photomicrographs of cementstone bed 13. (a) Anastomosing fractures (light) in the cementstone body. The fragments are largely re-fittable suggesting that brecciation occurred *in situ* and did not involve transport. The lumps of cementstone (peds) are silty, whilst the fractures are healed by dolomite and calcite cements. Scale bar = 1 mm. (b) The brecciated cementstone texture (right half) and clotted texture (left half). The clotted texture is composed of interlocking spherulites of inclusion-rich dolomite cement which replaces original cementstone fabrics. Scale bar = 500 μm. (c) Dolomite crystals of the red silty matrix (see Fig. 5). The dolomite crystals were originally euhedral but are now partly dissolved. The dark clast is composed of dolomicrite and is fringed by an early diagenetic, inclusion-rich dolomite cement (same as the spherulitic cements seen in b). Scale bar = 500 μm.

Interpretation

We interpret this bed as a calcrete horizon, which developed on the alluvial plain when the floodplain lakes dried out. The calcrete is continuous across both the terrace and valley-fill sediments (Fig. 4), implying that calcrete formation was not related to improved drainage on the terrace following valley incision.

The overall structure of the lumps (peds) is similar to the angular blocky structure described by Fitzpatrick (1980, Plate 3B). The fractures in the peds, and general brecciation were probably caused by shrinkage (possibly of expandable clays; Fitzpatrick, 1980) during desiccation, although we cannot rule out brecciation by plant roots. The similarity of CL colour between the micrite and inclusion-rich cement suggests that the cement formed early in diagenesis, probably only centimetres below the sediment surface. Clotted, peloidal and spherulitic fabrics are probably indicative of bacterial activity in early diagenetic soil-zone environments and occur in other ancient calcretes (Andrews, 1985) and cyanobacterial mats (Andrews, 1986). The matrix cements are interpreted below (Origin of the Cementstones).

SUMMARY OF EVENTS AT EASTERN HOLE

Deposition of the Eastern Hole beds began after a significant period of non-deposition during which mature calcretes had formed in the Upper Old Red Sandstone sediments (see also Leeder, 1976). Mud-dominated coastal floodplain lake sediments accumulated, punctuated by crevasse-splay sand deposition. Pauses in floodplain aggradation allowed authigenic cementstone nucleation and growth to occur in the top 1 m of floodplain sediment. After deposition of bed 10 (Fig. 3), the floodplain sediments were incised by a river channel which cut down several tens of metres to regain base level. The channel then re-accumulated floodplain sediments, and the presence of cementstones again suggests significant periods of non-deposition. The valley was gradually filled by younger floodplain sediment and its topography was more or less eliminated, after which floodplain aggradation continued across both the terrace and the former valley suggesting a significant period of basin subsidence.

The floodplain experienced pauses in this overall subsidence-dominated period, during which an immature calcrete formed. Renewed aggradation buried the soil but further breaks in sedimentation allowed cementstone formation. These conditions were disrupted by an episode of crevasse-channel and crevasse-splay deposition probably feeding a floodplain lake. Catastrophic overbank events occasionally led to sediment-dumping of conglomerate (Eastern Hole Conglomerate). After deposition of the Eastern Hole Conglomerate more quiescent conditions ensued in the coastal floodplain lakes allowing a brackish bivalve fauna to establish. Soon after this, seawater inundated the floodplain establishing a short-lived saline lagoon in which a restricted marine fauna lived. Gradually, the lagoon was isolated from the sea and proximal channellized floodplain sedimentation was re-established.

ORIGIN OF THE CEMENTSTONES

Petrography and geochemistry

The dolostones are usually homogeneous with anhedral crystals around 50 μm in diameter. Quartz and feldspar silt, clay patches and plant fragments are the only common detrital materials, and total acid-insoluble residues vary between 8 and 21%. The matrix of the septarian concretions shows a gradation in crystal size from submicrometre in the clay-rich core to 200 μm diameter toward the outer edge. Sedimentary lamination and brecciation occur within the concretion body of bed 9 (Fig. 3), which is inherited either from the pre-existing mudstone, or a carbonate mudstone precursor. The filling of septarian cracks is by three stages of carbonate cement. The first stage is a fibrous, inclusion-rich, brown, ferroan dolomite, which luminesces dull-orange, identical to the dolomicrite of the concretion body. (This is the same as the spherulitic cement which heals fractures in the palaeosol.) This inclusion-rich cement is overgrown by a clear non-ferroan dolospar, which is non-luminescent, but partly replaced by later diagenetic cements. (This cement is the same as the partially dissolved dolomite crystals in the palaeosol matrix.) Shards of brecciated concretion body are found in the cracks and may retain these cements on *one* side of the fragment, showing that some compactional brecciation post-dated first-stage cementation. The second-stage dolospar cement begins with a 2–5-μm-thick, bright-orange luminescing zone (not always present), followed by a dull red–orange luminescing zone of variable thickness, usually greater than 100 μm (this cements the partly dissolved dolomite crystals in the palaeosol). In most cases this cement is partially replaced by blue–purple luminescing kaolinite, and it is followed by stage-three calcite spar, usually the pore-filler, which is bright-orange to red luminescing and vein fed. Stage-two and stage-three cements are also seen in the palaeosol matrix.

The geochemical data for cementstone micrite/microspar, cements and a six-point traverse across one septarian concretion are shown in Table 1.

Petrography and geochemistry—interpretation

Cementstone dolomicrite/microspar

The insoluble residue contents of the cementstones could be interpreted as indicative of pre-carbonate sediment porosities of between 80 and 90%, if it were assumed that carbonate passively displaced pore fluids (Lippman, 1955; Raiswell, 1971). However, the crystal sizes, especially toward the outer edge of the concretions (up to 200 μm), are unlikely to represent pores of a similar size (Feistner, 1989), indicating some replacive or displacive growth. We suggest that concretion growth began where porosities were high, probably within 1 m of the sediment surface (Baldwin & Butler, 1985). Oxygen isotope values also suggest carbonate precipitation before significant burial (see below). This reasoning combined with the high Fe and Mn values for cementstone dolomicrospars (Table 1) fixes dolomite nucleation within the first few metres of sediment burial under reducing conditions.

The $\delta^{13}C$ values for these dolomicrospars are low (−4·5 to −6·9‰, Table 1) which could be caused in a

Table 1. Geochemical data.

Sample	Ca (wt%)	Mg (wt%)	Fe (wt%)	Mn (ppm)	Sr (ppm)	%IR	$\delta^{13}C$ (‰)	$\delta^{18}C$ (‰)	Ca (mol% X_{CO_3})	Mg (mol% X_{CO_3})	Fe (mol% X_{CO_3})
Cementstones											
Bed 2[1]	19·06	11·11	1·94	2672	102	8·4	−5·42	−4·91	47·6	45·7	3·5
Bed 5[1]	20·43	10·04	3·13	2836	136	15·7	—	—	51·0	41·3	5·6
Bed 9[2]	20·39	10·85	2·54	2816	86	14·6	—	—	50·9	44·6	4·6
Bed 13[1]	19·86	10·28	2·95	3679	119	13·8	—	—	49·6	42·3	5·3
Bed 15[1]	19·72	10·90	3·49	4398	86	18·8	−6·69	−2·97	49·2	44·8	6·3
Bed 17[1]	19·57	10·87	2·62	2916	119	20·7	−5·52	−3·50	48·8	44·7	4·7
Bed 19[1]	19·87	10·59	3·19	5022	98	12·1	−6·89†	−4·04†	49·6	43·6	5·8
Bed 20[1]	19·83	10·28	3·46	3536	96	13·4	—	—	49·5	42·3	6·3
Bed 20[1]	21·00	10·07	2·88	3510	111	10·4	−6·02	−4·66	52·3	41·4	5·2
Dolomite cements from fracture fills											
Bed 20	19·74	10·98	2·95	2510	62	1·6	−13·26	−1·16	49·2	45·2	5·3
Bed 9	20·57	11·42	1·24	3429	80	0·0	—	—	51·3	47·0	2·2
Palaeosol (bed 13)											
Micritic clast	—	—	—	—	—	—	−6·93†	−5·79†	—	—	—
Peloidal micrite	—	—	—	—	—	—	−7·30†	−6·40†	—	—	—
Spherulitic dolomite	—	—	—	—	—	—	−7·73†	−7·03†	—	—	—
Red matrix	—	—	—	—	—	—	−8·31†	−3·32†	—	—	—
Traverse across septarian nodule (equivalent to bed 4[3])											
0·5‡	20·45	10·80	2·06	9858	—	—	−5·56	−4·87	51·0	44·4	3·7
1·5	21·43	10·51	2·39	6209	—	—	−6·22	−4·59	53·4	43·2	4·3
2·5	21·86	9·87	4·09	4502	—	—	−6·78	−4·89	54·5	40·6	7·4
3·5	21·42	9·40	4·08	7645	—	—	−6·77	−4·94	53·4	38·7	7·4
5·0	21·73	11·23	1·96	7490	—	—	−5·81	−4·46	54·2	46·2	3·5
6·0	21·76	11·99	1·94	7529	—	—	−5·50	−5·39	54·3	49·3	3·5
Septarian cements in septarian nodule											
n.a	21·23	11·34	1·30	9786	—	—	−6·23	−6·36	53·0	46·6	2·3
n.a	21·22	11·23	1·03	8215	—	—	−6·67	−6·27	52·9	46·1	1·8

Samples: [1] Internally homogeneous silty cementstone. [2] Septarian concretion with internal lamination. [3] Septarian concretion with fine-grained centre. NB cement analyses are average of cement stages 2 & 3 described in the text.

† Isotope analyses determined at the University of East Anglia.

‡ Distance from top rim (cm).

number of ways. Modern meteoric-derived waters show an enormous variation in their dissolved inorganic carbon (DIC) $\delta^{13}C$ values (−25 to 0‰ or heavier), depending on the relative involvement of: (i) CO_2 contribution by organic matter degradation, (ii) contribution of HCO_3^- from the weathering of pre-existing carbonate and acid hydrolysis of silicates, (iii) the extent of equilibration of standing water with atmospheric CO_2 and (iv) the extent of photosynthetic withdrawal of ^{12}C into plants (Anderson & Arthur, 1983). Theoretically, the observed $\delta^{13}C$ values could simply result from equilibrium precipitation of dolomite from bicarbonate-dominated waters, where at 20°C, the precipitated dolomite would be approximately 4·5‰ heavier than the $\delta^{13}C$ of the DIC (see Emrich, Ehhalt & Vogel, 1970; Sheppard & Schwarcz, 1970). This is unlikely, however, given that most authigenic carbonates in sediment need a mechanism to increase alkalinity such that carbonate precipitation is forced (Raiswell, 1987). This is usually caused by oxidation of organic matter buried in the sediment. In this case the organic matter would be of terrestrial origin, which has an average $\delta^{13}C$ around −26‰ (e.g. see Deines, 1980).

Given that dolomite nucleation occurred under reducing conditions (see above), oxidation of organic matter could theoretically have occurred by a number of microbially mediated anaerobic processes including: manganese-, nitrate-, iron- and sulphate reduction (see Froelich *et al.*, 1979; Berner, 1981). Complete oxidation of organic matter by any of these processes would result in a DIC $\delta^{13}C$ of the pore waters as light as $-26‰$. If methanogenic reactions were involved the associated CO_2 is generally isotopically heavy (0 to $+10‰$; Claypool & Kaplan, 1974), unless oxidation of the methane occurred, in which case CO_2 with significantly lighter $\delta^{13}C$ values, down to $-65‰$ in freshwater sediments, would be supplied (Whiticar, Fabel & Schoell, 1986). The observed microspar values therefore seem to imply mixing of isotopically light DIC with an isotopically heavier source. Because the dolomite contains Fe^{3+}, we do not consider aerobic oxidation of organic matter to have been linked to dolomite formation and it is not discussed further.

The floodplain setting precludes sulphate reduction as a major organic matter oxidation mechanism, because meteorically-derived water contains little dissolved sulphate (Berner, 1984). This reasoning is confirmed by the general lack of pyrite sulphur in the mudrocks (see above). Manganese- and nitrate reduction probably occurred (the high Mn values in Table 1 point to available reduced Mn) but they are difficult to quantify. We doubt that they provided enough alkalinity to cause dolomite precipitation although Jones (1985) has shown that denitrification can be an important reaction in modern lake sediments, accounting for 20% of organic matter decomposition. This leaves iron reduction (the high Fe values in dolomite and the low DOP values show that iron availability was not limited) and methanogenesis as the most important organic matter oxidative processes.

We suspect that the mechanisms and by-products of organic matter oxidation, Fe^{2+} content of the cementstones and the alkalinity required to trigger cementstone nucleation, are linked. There is clearly a relationship between $FeCO_3$ and $\delta^{13}C$ in the cementstones, the lighter $\delta^{13}C$ values corresponding to higher $FeCO_3$ contents (Fig. 6). This suggests that the behaviour of reduced iron (Fe^{2+}) and isotopically light carbon was linked during cementstone growth. The role of iron reduction in early diagenesis is a relatively poorly studied process (but see Jones, Gardener & Simon, 1984; Jones, 1985; Aller, Mackin & Cox, 1986); however, reduced mineral phases of iron are common in modern sediments suggesting that the oxidizing capacity of iron is utilized (Reeburgh, 1983). The reducing agent consumed during iron reduction must be either reactive organic matter buried with the iron or methane derived from methanogenesis. Anaerobic oxidation of both these components will release isotopically light carbon and reduced iron into the pore water thus potentially explaining the coupling of $\delta^{13}C$ and $FeCO_3$ (Fig. 7). There are three possible mechanisms which could produce the alkalinity required for carbonate precipitation.

(1) Anaerobic microbial oxidation of organic matter using Fe^{3+} as the oxidant in a similar manner to that proposed for the explanation of isotopically light carbon signatures in banded iron formations (Walker, 1984), i.e.

$$3H_2O + 2Fe_2O_3 + CH_2O \rightarrow HCO_3^- + 4Fe^{2+} + 7OH^-.$$

(2) Anaerobic oxidation of isotopically light methane at the iron reducing/methanogenesis redox-boundary using Fe^{3+} as the oxidant, i.e. methanogenesis by acetate fermentation (accounts for about 70% of methane production in modern freshwater sediments; Whiticar *et al.*, 1986):

$$CH_3COOH \rightarrow CH_4 + CO_2,$$

followed by methane oxidation by iron reduction:

$$CH_4 + 2Fe_2O_3 \rightarrow HCO_3^- + 4Fe^{2+} + 3OH^-.$$

The latter reaction is analogous to that observed in marine sediments (Alperin & Reeburgh, 1985) where methane oxidation via sulphate reduction occurs at the base of the sulphate-reducing zone. While methane oxidation via sulphate reduction is generally accepted as an alkalinity promoting reaction (Reeburgh, 1983; Raiswell, 1987, 1988), anaerobic oxidation of methane via iron reduction

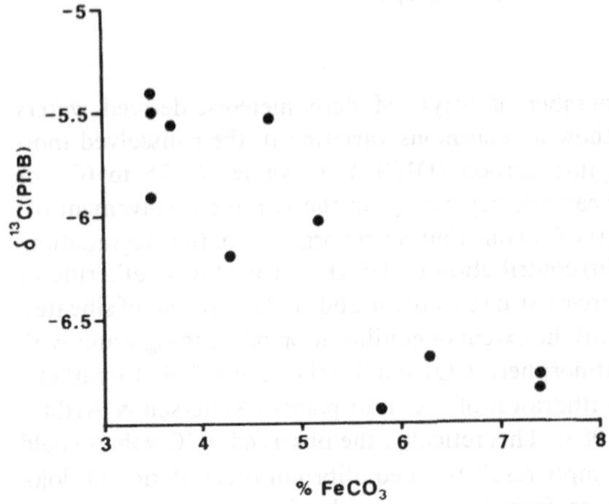

Fig. 7. Cross plot of $\delta^{13}C$ values vs. $\%FeCO_3$ for dolomitic cementstones. The relationship is weakly linear where high Fe values correspond with lighter $\delta^{13}C$ values.

is yet to be observed in modern biogeochemical systems. However, this hypothesis is reasonable for iron-rich, non-marine sediments where the methanogenic zone may begin only a few centimetres below the sediment–water interface in the absence of a major sulphate-reducing zone.

(3) Methanogenesis accompanied by iron reduction. Usually, methanogenesis produces CO_2, which lowers the pH (see Raiswell, 1987) unless significant iron reduction occurs simultaneously (Coleman, 1985), i.e.

$$13CH_2O + 2Fe_2O_3 + 3H_2O \rightarrow 6CH_4 + 7HCO_3^- + 4Fe^{2+} + OH^-.$$

Each mechanism promotes alkalinity although methane oxidation by iron reduction has the advantage of producing an alkalinity maximum close to the redox-boundary, i.e. a specific zone in the sediment is implicated (see Raiswell, 1987, 1988, for the case of methane oxidation by sulphate reduction). Moreover, it has the advantage of utilizing a mobile reduced carbon species, methane, which is not the case in iron reduction of organic matter (Coleman, 1985).

Choosing between these possible mechanisms should theoretically be helped by considering stable carbon isotopes. In case 1 isotopically light HCO_3^- will be produced, close to the value for terrestrial organic matter (−26‰, see above), whilst case 2 will produce extremely light HCO_3^- around −60‰ (see above). Clearly these are both lighter than the observed carbonate values around −6‰, which can only be produced by mixing these isotopically light species with isotopically heavier ones. We showed above that $\delta^{13}C$ values around 0‰ are about the heaviest likely values of meteoric water DIC. Hence, mixing HCO_3^- around 0‰ with HCO_3^- at −26‰ (iron reduction) implies incorporation of about 25% of the light HCO_3^- in the precipitated carbonate. Similarly, mixing HCO_3^- at 0‰ with HCO_3^- at −60‰ (methane oxidation) implies incorporation of about 10% of the light HCO_3^- in the precipitating carbonate.

Bicarbonate produced by methanogenesis accompanied by iron reduction (case 3) will be isotopically heavy (0 to +10‰, see above). Incorporation of this HCO_3^- in the precipitate would imply mixing with lighter sources. Clearly the carbon isotope data alone cannot discriminate between the possible mechanisms. Indeed the potential combinations are more complicated because isotopically light methane and isotopically heavy dissolved carbonate produced by methanogenesis may flux to the site of methane oxidation (the alkalinity maximum) at similar rates (Raiswell, 1987). In this case the resulting $\delta^{13}C$ value for HCO_3^- would then be close to that of the original matter, −26‰. Only disequilibrium in the fluxes will allow the formation of significantly lighter or heavier isotopic values.

Given these uncertainties we can only suggest a reasonable sequence of events to account for the $\delta^{13}C$ values of the cementstones.

(1) Pore waters in the Eastern Hole beds were derived from meteoric water which had already been modified by soil-zone reactions to produce alkalinity (DIC $\delta^{13}C$ is low negative). This water was, however, undersaturated with respect to dolomite.
(2) After a few centimetres burial, pore-water dissolved oxygen was exhausted by aerobic oxidation of organic matter and anaerobic manganese-, nitrate- and iron reduction of organic matter began. After a few more centimetres burial manganese and nitrate were exhausted, leaving iron reduction as the dominant mechanism. Bicarbonate produced as a by-product of this reaction began to build up in the pore waters causing near saturation with respect to dolomite. The bicarbonate derived from iron reduction had a $\delta^{13}C$ around −20‰, forcing DIC $\delta^{13}C$ values toward more negative values.
(3) At some 5 cm (or so) of sediment burial methanogenesis began, still accompanied by iron reduction. The bicarbonate produced by this reaction was isotopically heavy (≥0‰) and diffused upward adding its alkalinity to the isotopically light DIC, causing an increase in DIC $\delta^{13}C$ back toward more negative values. Saturation or supersaturation with respect to dolomite was now achieved.
(4) The methane produced by fermentation reactions also diffused upward and in the top 5 cm of sediment may have been oxidized by iron reduction to produce isotopically light bicarbonate (−60‰). The extra alkalinity produced may have been enough to force supersaturation with respect to dolomite, and ultimately nucleation. It is probable that only a small proportion of the available methane was oxidized, the rest fluxing out of the sediment as gas. This has the advantage of producing only a limited amount of bicarbonate, enough to cause dolomite nucleation whilst keeping the $\delta^{13}C$ values of the carbonate low (−6‰).

We are reasonably confident of steps 1 and 2. Steps 3

and 4 are considered likely, but we are less sure of their relative importance.

$FeCO_3$ and $\delta^{13}C$ profiling across an individual concretion body (Fig. 8) shows that the later precipitated dolomite is relatively depleted in $FeCO_3$ and heavier in $\delta^{13}C$. This suggests that as concretion growth progressed the supply of Fe^{2+} and alkalinity generated by iron reduction decreased to a point where concretion growth was not sustained. The small proportion of carbon left in these rocks (<0·6%) suggests that the reaction was limited by carbon availability or by an increase in sedimentation rate, which effectively diluted the alkalinity maximum (Raiswell, 1988).

The $\delta^{18}O$ values of the cementstones range from −5·39‰ to −2·97‰ (Table 1), a range which is consistent with precipitation from largely meteoric-derived water, as Lower Carboniferous marine dolomite has a predicted $\delta^{18}O$ between −1 and +2‰ (based on the calcite values of Popp, Anderson & Sandberg, 1986; Banner, Hanson & Meyers, 1988; Hudson & Anderson, 1989 and applying a 3‰ positive correction based on the Δ difference discussed in Land, 1980). The range in $\delta^{18}O$ values probably reflects varying amounts of evaporation of the surface- and pore-waters prior to dolomite precipitation, although the heavier values could indicate incorporation of a small amount of seawater (see below). The transect across a single concretion (Fig. 8) shows no significant variation in $\delta^{18}O$, which suggests that its total growth occurred at approximately the same temperature, and hence depth in the sediment.

Palaeosol and later diagenetic cements

The first stage of cement which fills septarian cracks has identical CL characteristics to the concretion body and is clearly of early diagenetic origin. This is overgrown by a clear, non-ferroan and non-luminescent dolospar, which suggests that both Fe and Mn (the latter a luminescence exciter in dolomite) were unavailable for incorporation into the precipitating cement.

The stage 2 and 3 carbonate cements are again luminescent, apparently indicating renewed Mn availability, and they are associated with kaolinite. The oxygen isotope values that we have measured for the cements are variable (−6·36 to −1·16‰, Table 1) and probably reflect the difficulty of separating different cement stages by sampling with a dental drill. However, we are reasonably confident that the heaviest value (−1·16‰) is from the stage 2 dolomite cement. This cement also has a distinctive $\delta^{13}C$ value of −13·20‰ implicating continued oxidation of organic matter during burial diagenesis. The red matrix of the palaeosol is composed of a mixture of non-luminescent dolomite crystals (end stage 1 cements) and luminescent stage 2 dolomite cements. Isotope values for this material are intermediate between early diagenetic and stage 2/3 cement values (Table 1) confirming the presence of mixed cements.

Why are the cementstones dolomitic?

Dolomite precipitation (or dolomitization) in any setting requires both a cation and carbonate supply and it is generally agreed that seawater is the most effective cation source in near-surface dolomite formation (e.g. see Given & Wilkinson, 1987). The dolomitic mineralogy of the cementstones thus poses a problem. We have shown that the general facies setting is that of an alluvial coastal floodplain. We see no evidence of marine influences until the obvious marine flooding event associated with bed 27 (see

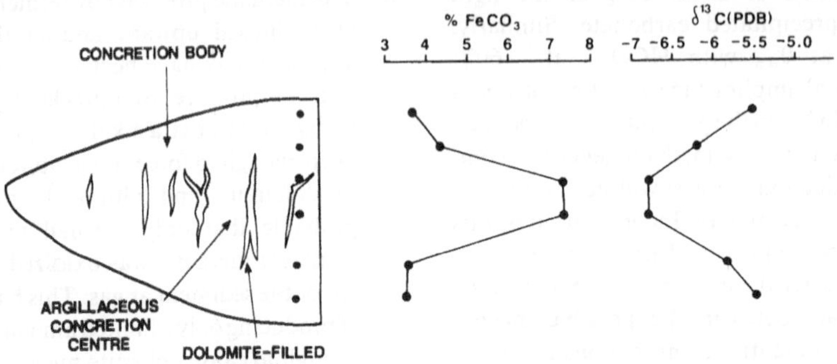

Fig. 8. Profile of $\delta^{13}C$ vs. $FeCO_3$ values across a single septarian concretion (first cementstone bed in graphic log 3, Fig. 4, which is the lateral equivalent of bed 4, Fig. 3). High Fe values correspond to lighter $\delta^{13}C$ values. Note the markedly high Fe values and light $\delta^{13}C$ values in the argillaceous concretion centre.

above; Fig. 3), which is not associated with the cementstone-bearing horizons. Moreover, we have not detected any pyrite sulphur either in the cementstones or their associated mudrocks, suggesting that pyrite formation was inhibited. This usually means that dissolved sulphate levels in the pore waters were very low. However, it *may not* rule out seawater as a cation source. Canfield (1989) has shown that in iron-rich sediments (offshore Mississippi delta) when Fe-oxides are present in high concentrations, dissolved sulphide is nearly absent from sediment pore water. In essence, any H_2S produced by sulphate-reducers is oxidized by iron reduction (see Aller *et al.*, 1986, for possible mechanisms). In this case, it is possible that periodic influxes of seawater (storms?) penetrated the coastal floodplain lakes. Small volumes of seawater would supply Mg^{2+} and Ca^{2+} ions required for dolomite formation, but limit sulphate reduction, the H_2S being buffered by the high iron content of the sediment. These brief flooding events would leave no marine faunal record.

The only other option is that solutes were supplied from a weathering source, as has been suggested for other ancient and Holocene non-marine dolomites (von der Borch, Lock & Schwebel, 1975; Eugster & Hardie, 1975). The Carboniferous rivers which fed these alluvial plains drained from the E and SE (based on palaeocurrent measurements in the overlying sandbodies), i.e. the Southern Uplands massif. During the Early Carboniferous much of the Southern Uplands area exposed a Lower Palaeozoic clastic sedimentary bedrock, punctuated by areas of mainly Lower Devonian volcanics (Greig, 1988). However, we have no evidence that these rocks supplied significant amounts of Mg or Ca.

It is also possible that contemporary volcanicity might have been a solute source, either by basalt weathering, or argillization of volcanic dust. Devonian–Carboniferous basalts (Kelso Traps) are known to outcrop 30 km to the SW of the basin (Greig, 1971). They are poorly dated, and there is no record of their former extent, although this could have been significantly greater than present outcrop (see Leeder, 1976). We cannot be sure, however, that they were exposed in the relevant catchment area. Similarly there is evidence of Early Carboniferous pyroclastic volcanicity in the Dunbar–Cockburnspath area (Davies, McCadam & Cameron, 1986; Francis, 1986), just a few kilometres to the north, but again the poor chronology means that we do not know whether these volcanoes were active during deposition of the Eastern Hole beds. Despite these uncertainities we do know that uppermost Old Red Sandstone sediments over much of southern Scotland were supplied with groundwater rich in both Ca and HCO_3^- because these rocks are characterized by pedogenic carbonate (Davies *et al.*, 1986). Although these cornstones probably formed during evaporative concentration of the soil-water a significant solute supply is indicated.

Whatever the mechanism of cation supply, the high levels of Ca and Mg in the pore waters, combined with low dissolved-sulphate levels, appear to have favoured dolomite formation (see Baker & Kastner, 1981) rather than calcite or siderite which are the most common authigenic carbonates in non-marine settings (Curtis & Coleman, 1986).

ACKNOWLEDGMENTS

J.E.A. thanks the School of Environmental Sciences, University of East Anglia (UEA), for field expenses, M.S.T. acknowledges the Natural Environment Research Council (NERC) and the British Geological Survey for a CASE studentship and G.N. was financed by a grant from the Ministry of Education, Government of Pakistan. Stable isotope analyses were made at the NERC Isotope Geoscience Laboratories, London, with the help of Peter Greenwood and Mike Fowler, and in the Stable Isotope Laboratory, UEA, with the help of Paul Dennis. We are grateful for the use of the microprobe at the Geology Department, University of Leicester, where Rob Wilson helped us with analyses. Steve Bennett helped with XRD and CL, Leo Reynolds, Richard Rix and Marcus Armes helped with AAS and Ian Marshall prepared the thin and polished sections. Philip Judge and Sheila Davies prepared the diagrams and photographs. Peter Friend, Rob Raiswell, and the reviewers, John Hudson and Mike Leeder, constructively criticised and improved our ideas.

REFERENCES

Allen, J.R.L. (1974) Studies in fluviatile sedimentation: implications of pedogenic carbonate units, Lower Old Red Sandstone, Anglo-Welsh outcrop. *Geol. J.*, **9**, 181–208.

Allen, J.R.L. & Williams, B.P.J. (1982) The architecture of an alluvial suite: rocks between the Townsend Tuff and Pickard bay tuff beds (early Devonian), southwest Wales. *Phil. Trans. R. Soc. Lond. B*, **297**, 51–89.

Aller, R.C., Mackin, J.E. & Cox, R.T. (1986) Diagenesis of Fe and S in Amazon inner shelf muds: apparent dominance of Fe reduction and implications for the genesis of ironstones. *Cont. Shelf Res.*, **6**, 263–289.

ALPERIN, M.J. & REEBURGH, W.S. (1985) Inhibition experiments on anaerobic methane oxidation. *Appl. Environ. Microbiol.*, **50**, 940–945.

ANDERSON, T.F. & ARTHUR, M.A. (1983) Stable isotopes of oxygen and carbon and their application to sedimentologic and palaeoenvironmental problems. In: *Stable Isotopes in Sedimentary Geology*. (Ed. by M. A. Arthur, T. F. Anderson, I. R. Kaplan, J. Veizer & L. S. Land), *Soc. econ. Paleont. Miner. Short Course Notes*, **10**, 1.1–1.151.

ANDERTON, R. (1985) Sedimentology of the Dinantian of Foulden, Berwickshire, Scotland. *Trans. R. Soc. Edinburgh, Earth Sci.*, **76**, 7–12.

ANDREWS, J.E. (1985) The sedimentary facies of a late Bathonian regressive episode: the Kilmaluag and Skudiburgh Formations of the Great Estuarine Group, Inner Hebrides, Scotland. *J. geol. Soc. London*, **142**, 1119–1137.

ANDREWS, J.E. (1986) Microfacies and geochemistry of Middle Jurassic algal limestones from Scotland. *Sedimentology*, **33**, 499–520.

ANDREWS, J.E. (1987) Jurassic clay mineral assemblages and their post-depositional alteration: upper Great Estuarine Group, Scotland. *Geol. Mag.*, **124**, 261–271.

BAKER, P.A. & KASTNER, M. (1981) Constraints on the formation of sedimentary dolomite. *Science*, **213**, 214–216.

BALDWIN, B. & BUTLER, C.O. (1985) Compaction curves. *Bull. Am. Ass. petrol. Geol.*, **69**, 622–626.

BANNER, J.L., HANSON, G.N. & MEYERS, W.J. (1988) Water rock interaction history of regionally extensive dolomites of the Burlington–Keokuk Formation (Mississipian): isotopic evidence. In: *Sedimentology and Geochemistry of Dolostones* (Ed. by V. Shukla & P. A. Baker), *Spec. publs Soc. Econ. Paleont. Miner.*, **43**, 97–115.

BELT, E.S., FRESHNEY, E.C. & READ, W.A. (1967) Sedimentology of Carboniferous cementstone facies, British Isles and eastern Canada. *J. Geol.*, **75**, 711–721.

BERNER, R.A. (1968) Rate of concretion growth. *Geochim. Cosmochim. Acta*, **32**, 477–483.

BERNER, R.A. (1981) A new geochemical classification of sedimentary environments. *J. sedim. Petrol.*, **51**, 359–365.

BERNER, R.A. (1984) Sedimentary pyrite formation: an update. *Geochim. Cosmochim. Acta*, **48**, 605–615.

BERNER, R.A. & RAISWELL, R. (1984) C/S method for distinguishing freshwater from marine sedimentary rocks. *Geology*, **12**, 365–368.

BISCAYE, P.E. (1965) Mineralogy and sedimentation of recent deep-sea clay in the Atlantic Ocean and adjacent seas and oceans. *Bull. geol. Soc. Am.*, **76**, 803–832.

VON DER BORCH, C.C., LOCK, D. & SCHWEBEL, D. (1976) Groundwater formation of dolomite in the Coorong region of South Australia. *Geology*, **3**, 283–285.

BRIDGE, J.S. (1984) Large-scale facies sequences in alluvial overbank environments. *J. sedim. Petrol.*, **54**, 583–588.

CANFIELD, D.E. (1989) Reactive iron in marine sediments. *Geochim. Cosmochim. Acta*, **53**, 619–632.

CHISHOLM, J.I., MCADAM, A.D. & BRAND, P.J. (1990) Lithostratigraphical classification of Upper Devonian and Lower Carboniferous rocks in the Lothians. *Br. geol. Surv., Technical Report WA/89/26, Onshore Geology Series.*

CLARKSON, E.N.K. (1986) Pease Bay to Cove. In: *Lothian Geology—an Excursion Guide* (Ed. by A. D. McAdam & E. N. K. Clarkson), pp. 140–145. Scottish Academic Press, Edinburgh, 221 pp.

CLAYPOOL, G.E. & KAPLAN, I.R. (1974) The origin and distribution of methane in marine sediments. In: *Natural Gases in Marine Sediments* (Ed. by I. R. Kaplan), pp. 97–139. Plenum Press, New York.

CLAYTON, G. (1971) A Lower Carboniferous miospore assemblage from the Calciferous Sandstone Measures of the Cockburnspath region of eastern Scotland. *Pollen & Spores*, **12**, 577–600.

CLOUGH, C.T., BARROW, G. & CRAMPTON, C. (1910) *Geology of East Lothian*, 2nd edn, *Mem. geol. Surv.*

COLEMAN, M.L. (1985) Geochemistry of diagenetic non-silicate minerals: kinetic considerations. *Phil. Trans. R. Soc. Lond. A*, **315**, 39–56.

COLLINSON, J.D. (1986) Alluvial sediments. In: *Sedimentary Environments and Facies*, 2nd edn (Ed. by H. G. Reading), pp. 20–62. Blackwell Scientific Publications, Oxford.

CRAIG, H. (1957) Isotopic standards for carbon and oxygen and correlation factors for mass-spectrometric analysis of carbon dioxide. *Geochim. Comochim. Acta.*, **12**, 133–149.

CURTIS, C.D. (1977) Sedimentary geochemistry; environments and processes dominated by an aqueous phase. *Phil. Trans. R. Soc. Lond. A*, **286**, 353–372.

CURTIS, C.D. & COLEMAN, M.L. (1986) Controls on the precipitation of early diagenetic calcite, dolomite and siderite concretions in complex depositional sequence. In: *Roles of Organic matter in Sediment Diagenesis* (Ed. by D. L. Gautier), *Spec. publs Soc. econ. Paleont. Miner.*, **38**, 23–33.

DAVIES, A., MCCADAM, A.D. & CAMERON, I.B. (1986). *Geology of the Dunbar District Mem. Br. geol. Surv.*, HMSO, London.

DEINES, P. (1980) The isotopic composition of reduced organic carbon. In: *Handbook of Environment Isotope Geochemistry, Vol. 1* (Ed. by P. Fritz & J. C. Fontes), pp. 329–406. Elsevier, Amsterdam.

EMRICH, K., EHHALT, D.H. & VOGEL, J.C. (1970) Carbon isotope fractionation during the precipitation of calcium carbonate. *Earth Planet. Sci. Lett.*, **8**, 363–371.

EUGSTER, H.P. & HARDIE, L.A. (1975) Sedimentation in an ancient playa-lake complex: the Williams Peak Member of the Green River Formation of Wyoming. *Bull. geol. Soc. Am.*, **86**, 319–334.

FEISTNER, K.W.A. (1989) Petrographic examination and re-interpretation of concretionary carbonate horizons from the Kimmeridge Clay, Dorset. *J. geol. Soc. Lond.*, **146**, 345–350.

FITZPATRICK, E.A. (1980) *Micromorphology of Soils. A Manual for the Preparation and Description of Thin Sections of Soils.* Department of Soil Science, Univeristy of Aberdeen.

FRANCIS, E.H. (1965) Carboniferous. In: *The Geology of Scotland* (Ed. by G. Y. Craig), pp. 309–357. Oliver & Boyd, Edinburgh.

FRANCIS, E.H. (1986) Dunbar. In: *Lothian Geology—an Excursion Guide* (Ed. by A. D. McAdam & E. N. K. Clarkson), pp. 119–132. Scottish Academic Press, Edinburgh, 221 pp.

FRIEND, P.F. & MOODY-STUART, M. (1970) Carbonate deposition on the river flood plains of the Wood Bay Formation (Devonian) of Spitzbergen. *Geol. Mag.*, **107**, 181–195.

FROELICH, P.N., KLINKHAMMER, G.P., BENDER, M.L., LUEDTHE, N.A., HEATH, G.R., CULLEN, D., DAUPHIN, P., HAMMOND, D., HARTMAN, B. & MAYNARD, V. (1979) Early oxidation of organic matter in pelagic sediments of the eastern equatorial Atlantic: suboxic diagenesis. *Geochim. Cosmochim. Acta*, **43**, 1075–1090.

GIVEN, R.K. & WILKINSON, B.H. (1987) Dolomite abundance and stratigraphic age: constraints on rates and mechanisms of Phanerozoic dolostone formation. *J. sedim. Petrol.*, **57**, 1068–1078.

GREIG, D.C. (1971) *British Regional Geology: The South of Scotland*, 3rd edn. HMSO, London.

GREIG, D.C. (1988) *Geology of the Eyemouth District. Mem. Br. geol. Surv.* HMSO, London.

HUDSON, J.D. & ANDERSON, T.F. (1989) Ocean temperatures and isotopic compositions through time. *Trans. R. Soc. Edinburgh Earth Sci.*, **80**, 183–192.

JONES, J.G. (1985) Microbes and microbial processes in sediments. *Phil. Trans. R. Soc. Lond. A*, **315**, 3–17.

JONES, J.G., GARDENER, S. & SIMON, B.M. (1984) Reduction of ferric iron by heterotrophic bacteria in lake sediments. *J. Gen. Microbiol.*, **130**, 45–51.

KRAUS, M.J. (1987) Integration of channel and floodplain suites. II. Vertical relations of alluvial paleosols. *J. sedim. Petrol.*, **57**, 602–612.

LAGIOS, E. (1983) A gravity study of the eastern Berwickshire Devonian basins, SE Scotland. *Scott. J. Geol.*, **19**, 189–203.

LAND, L.S. (1980) The isotopic and trace element geochemistry of dolomite: the state of the art. In: *Concepts and Models of Dolomitization* (Ed. by D. H. Zenger, J. B. Dunham & R. L. Ethington), *Spec. publs Soc. econ. Paleont. Miner.*, **28**, 87–110.

LEEDER, M.R. (1974) Lower Border Group (Tournaisian) fluvio-deltaic sedimentation and palaeogeography of the Northumberland Basin. *Proc. Yorks. geol. Soc.*, **40**, 129–180.

LEEDER, M.R. (1975) Pedogenic carbonate and flood sediment accretion rates: a quantitative model for alluvial, arid-zone lithofacies. *Geol. Mag.*, **112**, 257–270.

LEEDER, M.R. (1976) Palaeogeographic significance of pedogenic carbonates in the topmost Upper Old Red Sandstone of the Scottish border basin. *Geol. J.*, **11**, 21–28.

LIPPMAN, F. (1955) Ton, Geoden und Minerale des Barrene von Hoheneggelsen. *Geol. Rdsch.*, **43**, 475–503.

NEVES, R., GUEINN, K.J., CLAYTON, G., IONNIDES, N.S., NEVILLE, R.S.W. & KRUSZEWSKA, K. (1973) Palynological correlations within the Lower Carboniferous of Scotland and northern England. *Trans. R. Soc. Edinburgh*, **69**, 23–70.

PATERSON, I.B. & HALL, I.H.S. (1986) Lithostratigraphy of the late Devonian and early Carboniferous rocks in the Midland Valley of Scotland. *Rep. Br. geol. Surv.*, **18**, No. 3.

PATTRICK, R.A.D., COLEMAN, M.L. & RUSSELL, M.J. (1983) Sulphur isotopic investigation of vein lead–zinc mineralization at Tyndrum, Scotland. *Miner. Deposita*, **18**, 477–485.

POPP, B.N., ANDERSON, T.F. & SANDBERG, P.A. (1986) Brachiopods as indicators of original isotopic compositions in some Paleozoic limestones. *Bull. geol. Soc. Am.*, **97**, 1262–1269.

RAISWELL, R. (1971) The growth of Cambrian and Liassic Concretions. *Sedimentology*, **17**, 147–171.

RAISWELL, R. (1976) The microbiological formation of carbonate concretions in the Upper Lias of north east England. *Chem. Geol.*, **18**, 227–244.

RAISWELL, R. (1987) Non-steady state microbial diagenesis and the origin of concretions and nodular limestones. In: *Diagenesis of Sedimentary Sequences* (Ed. by J. D. Marshall), *Spec. publs geol. Soc.*, **36**, 41–54.

RAISWELL, R. (1988) Chemical model for the origin of minor limestone–shale cycles by anaerobic methane oxidation. *Geology*, **16**, 641–644.

RAISWELL, R., BUCKLEY, F., BERNER, R.A. & ANDERSON, T.F. (1988) Degree of pyritization of iron as a palaeoenvironmental indicator of botton water oxygenation. *J. sedim. Petrol.*, **58**, 812–819.

REEBURGH, W.S. (1983) Rates of biogeochemical processes in anoxic sediments. *Ann. Rev. Earth Planet. Sci.*, **11**, 269–298.

ROSENBAUM, J. & SHEPPARD, S.M.F. (1986) An isotopic study of siderites, dolomites and ankerites at high temperatures. *Geochim. Cosmochim. Acta*, **50**, 1147–1150.

SCHUMM, S.A. (1977) *The Fluvial System*. J. Wiley & Sons, New York, 338 pp.

SCOTT, W.B. (1986) Nodular carbonates in the Lower Carboniferous Cementstone Group of the Tweed Embayment, Berwickshire: evidence for a former sulphate evaporite facies. *Scott. J. Geol.*, **22**, 325–345.

SCOTT, A.C. & REX, G.M. (1987) The accumulation and preservation of Dinantian plants from Scotland and its borders. In: *European Dinantian Environments* (Ed. by J. Miller, A. E. Adams & V. P. Wright), pp. 329–344. John Wiley & Sons Ltd, Chichester, 402 pp.

SHEPPARD, S.M.F. & SCHWARCZ, H.P. (1970) Fractionation of carbon and oxygen isotopes and magnesium between coexisting calcite and dolomite. *Contrib. Miner. Petrol.*, **26**, 161–198.

WALKER, J.C.G. (1984) Suboxic diagenesis in banded iron formations. *Nature*, **309**, 340–342.

WALKER, R.G. & CANT, D.S. (1984) Sandy fluvial systems. In: *Facies Models*, 2nd edn (Ed. by R. G. Walker), *Geosci. Can.* Reprint Series, **1**, 71–103.

WHITICAR, M.J., FABER, E. & SCHOELL, M. (1986) Biogenic methane formation in marine and freshwater environments: CO_2 reduction vs. acetate fermentation-isotope evidence. *Geochim. Cosmochim. Acta.*, **50**, 693–709.

WILSON, M.J., BAIN, D.C., MCHARDY, W.J. & BERROW, M.L. (1972) Clay-mineral studies on some Carboniferous sediments in Scotland. *Sediment. Geol.*, **8**, 137–150.

WILSON, R.B. (1989) A study of the Dinantian marine macrofossils of central Scotland. *Trans. R. Soc. Edinburgh, Earth Sci.*, **80**, 91–126.

(Manuscript received 31 July 1990; revision received 5 October 1990)

Reprinted from Sedimentology (1995) **43**, *77–88*

Early diagenetic siderite as an indicator of depositional environment in the Triassic Rewan Group, southern Bowen Basin, eastern Australia

JULIAN C. BAKER,*†[1] JOCHEN KASSAN‡ and P. JOE HAMILTON*
**Australian Petroleum Cooperative Research Centre and Division of Petroleum Resources, CSIRO, PO Box 136, North Ryde, Sydney, New South Wales, 2113, Australia*
†Centre for Microscopy and Microanalysis, The University of Queensland, Brisbane, Queensland, 4072, Australia
‡34–36 Whistler Court, Greenbank, Brisbane, Queensland, 4124, Australia

ABSTRACT

Early concretionary and non-concretionary siderites are common in subsurface Triassic sandstones and mudrocks of the Rewan Group, southern Bowen Basin. A detailed petrological and stable isotopic study was carried out on these siderites in order to provide information on the depositional environment of the host rocks. The siderites are extremely pure, containing 85–97 mol% $FeCO_3$, and are commonly enriched in manganese. $\delta^{13}C$ (PDB) values are highly variable, ranging from $-18{\cdot}4$ to $+2{\cdot}9$‰, whereas $\delta^{18}O$ (PDB) values are very consistent, ranging from $-14{\cdot}0$ to $-10{\cdot}2$‰ (mean $= -11{\cdot}9 \pm 1{\cdot}0$‰). The elemental and oxygen isotopic composition of the siderites indicates that only meteoric porewaters were involved in siderite formation, implying that host rocks accumulated in totally non-marine environments. The carbon isotopic composition of the siderites is interpreted to reflect mixing of bicarbonate/carbon dioxide generated by methane oxidation and methanogenesis. Very low $\delta^{13}C$ values demonstrate that, contrary to current views, highly ^{13}C-depleted siderite can be produced at shallow burial depths in anoxic non-marine sediments.

INTRODUCTION

The geochemistry of early diagenetic carbonates can provide valuable information on depositional environment. In particular, the elemental and stable isotopic composition of early diagenetic siderite may be used to discriminate between marine and non-marine depositional environments (Mozley, 1989; Mozley & Wersin, 1992; Hart *et al.*, 1992) and thus has potentially useful application in situations where depositional environment cannot be resolved by more traditional means.

The Late Permian to Early Triassic Rewan Group constitutes part of the sedimentary fill of the Permo-Triassic Bowen Basin in east-central Queensland, Australia (Fig. 1). Although the Rewan Group is widely regarded as being non-marine (Dickins & Malone, 1973; Jensen, 1975; Cosgrove & Mogg, 1985; Harris, 1986; Fielding *et al.*, 1990a,b; Kassan, 1993, 1994), it has been suggested recently that, in the southern Bowen Basin, the Rewan Group may in part reflect a marine influence on sediment accumulation. Support for this notion could be given by the recently recorded occurrence of spinose acritarchs (Price, 1994) in the Rewan Group over the area. This evidence appears to contradict strong sedimentological and stratigraphic evidence for the Rewan Group being entirely non-marine.

Early diagenetic siderite is common in the Rewan Group over the southern Bowen Basin. In view of the relationship between early diagenetic siderite geochemistry and depositional environment demonstrated by Mozley (1989) and Mozley

[1]Present address: K. R. Martin Pty. Ltd, PO Box 22, Kenmore, Brisbane, Queensland, 4069, Australia.

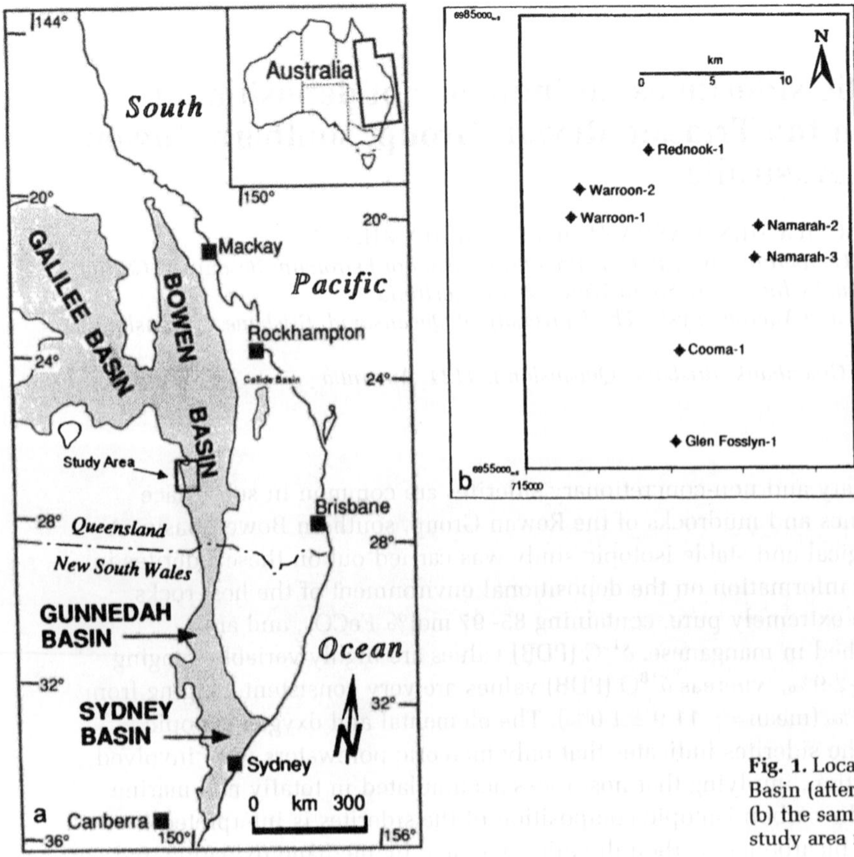

Fig. 1. Location of (a) the Bowen Basin (after Baker *et al.*, 1993) and (b) the sampled boreholes within study area marked in (a).

& Wersin (1992), an elemental and stable isotopic study was carried out on this siderite in an attempt to resolve the issue of depositional environment. The results highlight the potential of early diagenetic carbonates for providing information on depositional environment in situations where unequivocal sedimentological evidence is lacking. Moreover, the study provides additional data relevant to the use of siderite composition as an environmental indicator.

GEOLOGICAL SETTING

The Bowen Basin is a north-south elongate, back-arc extensional to foreland sedimentary basin that contains up to 10 km of variably deformed, marine and non-marine, Permo-Triassic siliciclastic sedimentary rocks. It constitutes the northern part of the much larger Bowen–Gunnedah–Sydney Basin system of eastern Australia (Fig. 1). The southern part of the Bowen Basin is concealed beneath relatively flat-lying Jurassic and Cretaceous sedimentary rocks of the Surat Basin. Details of the regional structure and stratigraphy of the Bowen Basin are given in Baker *et al.* (1993).

The Rewan Group occurs throughout the Bowen Basin, with the main NNW-elongate depocentre located near the eastern basin margin (Kassan, 1993). From here, the thickness of the Rewan Group progressively decreases from a maximum of over 3000 m to less than 70 m in the study area along the south-western flank of the Bowen Basin (Fig. 1). Constituent lithologies are mainly red, green, brown and grey volcanic-lithic sandstones, mudrocks and conglomerates that accumulated in continental (fluvial and alluvial fan) environments during rapid foreland basin subsidence. In the study area, the Rewan Group is entirely of Early Triassic age and includes quartzose sandstones and conglomerates of the lower Rewan Group and lithic sandstones and minor mudrocks of the upper Rewan Group (Kassan, 1994). The lower Rewan Group is of potential economic importance as a hydrocarbon

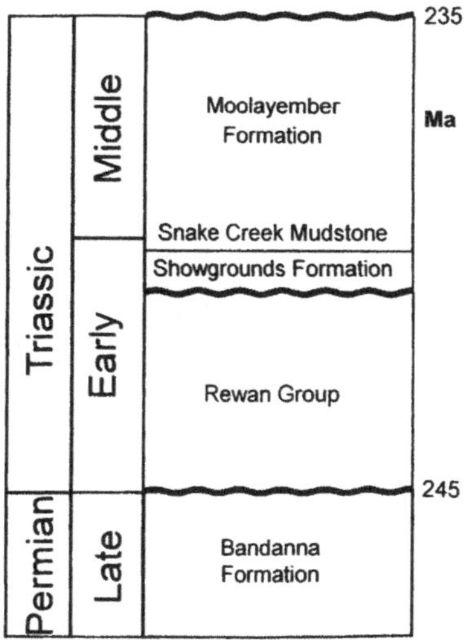

Fig. 2. Stratigraphic chart for latest Permian and Triassic of the south-western Bowen Basin (modified from Kassan, 1994).

reservoir. Over the margins of the southern Bowen Basin, the Rewan Group disconformably overlies coal measures of the Late Permian Bandanna Formation and is disconformably overlain by fluvial, quartz-rich sandstones of the Early Triassic Showgrounds Formation (Fig. 2).

SIDERITE OCCURRENCE

Early diagenetic siderite in the Rewan Group has only been recorded in the southern Bowen Basin, where it is common in sandstones and mudrocks constituting the upper Rewan Group. The siderite-bearing interval is up to 32 m thick and, based on facies relationships and sedimentary structures, is considered to have accumulated in a fluvial environment (Martin, 1992; Kassan, 1994). The siderite typically occurs as small spherulitic concretions, generally around 0·5–2·0 mm wide, that constitute up to 30% of the host rock. Locally, these sphaerosideritic concretions are so numerous that they coalesce to form irregular, siderite-cemented horizons. Single, irregular concretions with widths exceeding 1 cm are rare. Early, non-concretionary siderite locally forms sharply bounded, strongly cemented bands up to several centimetres wide. Commonly, siderite concretions are partly oxidized, imparting a spotty appearance to the host rocks.

Palaeosols are common in the Rewan Group in the northern Bowen Basin, but do not host siderite concretions and only occur within fine-grained successions (Kassan, 1993). The palaeosols typically show well-developed horizonation, are associated with rootlet penetration and have a variegated or mottled appearance. In the study area, the sideritic section lacks these features and the siderite concretions are commonly concentrated in sandy cross-bedded intervals. On this basis, it appears that, unlike some sphaerosideritic concretions elsewhere (e.g. Kantorowicz, 1990), the occurrence of siderite concretions in the Rewan Group in this area is not related to pedogenesis or the reworking of palaeosols.

SAMPLES AND METHODS

The siderite study involved 22 drill-core samples from depths between 2054·9 and 2284·3 m in seven boreholes located over an 85-km^2 area in the south-western part of the Bowen Basin (Fig. 1, Table 1). Nineteen samples are from the upper Rewan Group and three samples are from directly overlying intervals that had been tentatively logged as Showgrounds Formation. Samples consist of siderite-bearing sandstone and mudrock, and include both concretionary and non-concretionary siderite (Table 1).

Following routine thin-section examination of all samples, concretionary siderite in polished thin-sections of six samples was analysed using a JEOL JXA-8800L electron microprobe at an accelerating voltage of 15 kV. Fifty-nine quantitative elemental analyses were made on 2–4 concretions in each sample using a beam width of 1·0 μm. Qualitative elemental analyses were carried out on freshly exposed concretionary and non-concretionary siderite in another four samples using a Tracor Northern energy-dispersive spectrometer (EDS) attached to a JEOL 6400F SEM. Numerous EDS analyses were carried out on each of the four samples.

Concentrates of hand-picked siderite from crushed splits of all samples were reacted with 100% orthophosphoric acid for 3 days at 75°C to extract carbon dioxide for carbon and oxygen isotope analysis (McCrea, 1950). Isotope abundances in the extracted carbon dioxide were measured using a Micromass 602E mass spectrometer. Carbon and oxygen isotope data are presented in per mil (‰) with respect to Peedee belemnite (PDB). Table 1 also presents oxygen isotope data with respect to standard mean ocean water (SMOW). The reproducibility of standard $\delta^{13}C$ and $\delta^{18}O$ values was 0·2‰ or better.

Table 1. Stable isotope analyses for Rewan Group siderites.

Borehole	Depth (m)	Formation	Host lithology		$\delta^{13}C$ (‰ PDB)	$\delta^{18}O$ (‰ PDB)	$\delta^{18}O$ (‰ SMOW)
Cooma 1	2164·2	Rewan	Mudrock	A*	− 18·4	− 10·2	20·4
	2175·1	Rewan	Mudrock	A	− 13·8	− 11·5	19·0
	2182·2	Rewan	Mudrock	A	− 11·2	− 12·0	18·5
Glen Fosslyn 1	2090·6	Rewan	Sandstone	A	− 18·5	− 12·2	18·3
	2091·3	Rewan	Mudrock	A	− 13·0	− 11·4	19·1
	2091·5	Rewan	Mudrock	A	− 13·1	− 10·2	20·4
	2097·3	Rewan	Sandstone	B	− 5·1	− 14·0	16·4
Namarah 2	2252·5	Rewan	Mudrock	A	+ 2·9	− 11·4	19·1
	2258·9	Rewan	Sandstone	A	+ 2·8	− 12·7	17·8
	2266·5	Rewan	Sandstone	A	+ 0·8	− 12·8	17·7
	2266·6	Rewan	Sandstone	A	+ 0·3	− 12·6	17·9
	2266·9	Rewan	Sandstone	B	− 5·0	− 12·9	17·6
	2284·3	Rewan	Sandstone	A	+ 0·6	− 11·6	18·9
Namarah 3	2179·4	Rewan	Sandstone	A	− 4·6	− 12·4	18·1
	2184·4	Rewan	Sandstone	A	− 2·5	− 13·4	17·0
Rednook 1	2271·3	Rewan	Sandstone	A	− 4·2	− 12·4	18·1
	2272·5	Rewan	Sandstone	A	− 3·4	− 11·9	18·6
Warroon 1	2054·9	Showgrounds	Mudrock	A	− 16·7	− 11·2	19·3
	2057·5	Showgrounds	Siltstone	A	− 17·5	− 10·7	19·8
	2058·2	Rewan	Sandstone	B	− 6·4	− 11·9	18·6
	2058·4	Rewan	Mudrock	A	− 11·7	− 10·6	19·9
Warroon 2	2098·8	Showgrounds	Sandstone	A	− 0·9	− 12·2	18·3

*Mode of occurrence: A=concretionary; B =non-concretionary.

RESULTS

Host rock petrography

Sandstones

Rewan Group samples are mainly well-sorted, medium- to coarse-grained litharenites (Folk *et al.*, 1970) containing 10–50% quartz and up to 65% rock fragments. Showgrounds Formation samples are relatively quartzose, containing about 60–70% quartz. In all samples, feldspar (K-feldspar and plagioclase) contents are less than 2% and rock fragments are composed of felsic to intermediate volcanics, argillite and micaceous quartzite. Intraformational mudrock clasts are common. Accessory grains include biotite, tourmaline and zircon. Besides siderite, the main authigenic mineral is illite/smectite, which occurs as a late-stage alteration product of labile rock fragments. Late-stage calcite replaces framework grains outside siderite-cemented areas in Warroon 1 (2058·2 m). Kaolinite locally occurs as an alteration product of mica, and a few rock fragments have altered to green ferric illite in sandstones from above 2284·3 m in Namarah 2. Outside siderite-cemented areas, most micaceous and argillaceous grains have undergone deformation and dispersion, and grain-to-grain contact dissolution has occurred between juxtaposed quartz and quartzose grains to produce sutured grain contacts. Intergranular porosity has been eliminated by compaction effects and the formation of authigenic clay and siderite.

Mudrocks

Mudrock samples are well compacted and, except for the sample from Warroon 1 (2057·5 m), consist mainly of illitic clay matrix and scattered silt-sized quartz and mica grains. The sample from Warroon 1 (2057·5 m) is a micaceous, quartz-rich, coarse siltstone. Some mudrock samples contain trace amounts of anhedral to euhedral authigenic pyrite, most of which occurs inside siderite concretions.

Siderite petrography and composition

Spherulitic concretions in the Rewan Group typically consist of coarsely crystalline siderite that radiates outward from central masses of microcrystalline siderite (Fig. 3a,b). In the mudrocks, some spherulites have nucleated on (Fig. 3c) and enveloped (Fig. 3d) anhedral to euhedral authigenic pyrite crystals up to 0·8 mm wide. Crystal growth directions of this pyrite are commonly

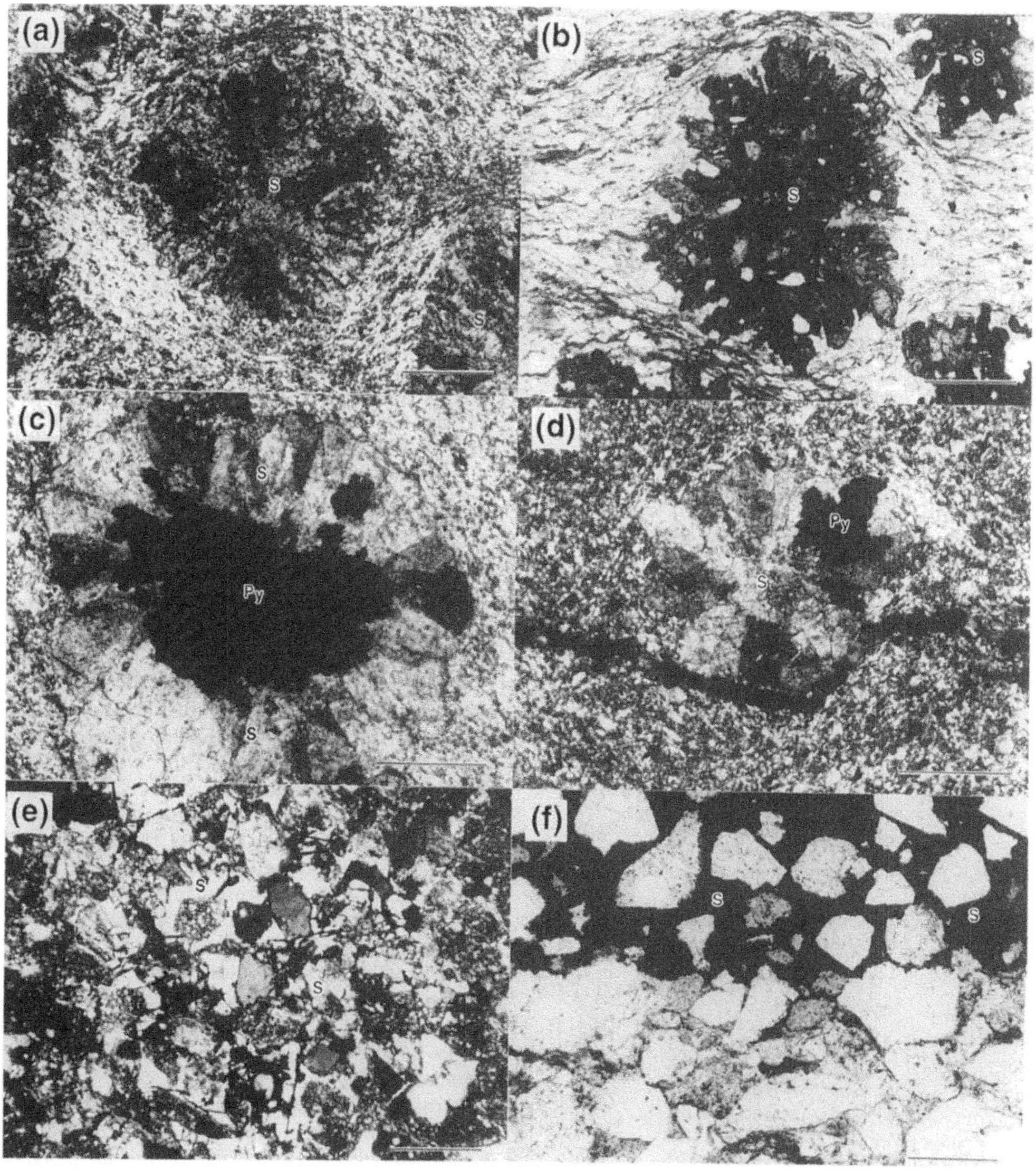

Fig. 3. (a) Typical siderite spherulites (S) bounded by compactionally deformed clay laminae; Warroon 1, 2058·4 m. Crossed polars. Scale bar=0·4 mm. (b) Siderite spherulites (S) in the Showgrounds Formation. Spherulites are texturally and compositionally similar to those in the underlying Rewan Group; Warroon 1, 2057·5 m. Plane polarized light. Scale bar=0·4 mm. (c) Siderite spherulite (S) with nucleus composed of euhedral pyrite (Py); Cooma 1, 2182·2 m. Crossed polars. Scale bar=0·2 mm. (d) Siderite spherulite (S) enclosing euhedral pyrite (Py) that apparently grew in the same direction as the spherulite following early spherulite formation; Cooma 1, 2182·2 m. Crossed polars. Scale bar=0·2 mm. (e) Concretionary siderite (S) poikilotopically enclosing uncompacted framework grains in sandstone. Framework grains outside cemented area are highly compacted: Namarah 2, 2266·5 m. Crossed polars. Scale bar=0·4 mm. (f) Sharp boundary between non-concretionary siderite band (S) and highly compacted, non-cemented sandstone (lower half of micrograph); Glen Fosslyn 1, 2097·3 m. Plane polarized light. Scale bar=0·4 mm.

Table 2. Microprobe analyses for Rewan Group siderites. Groups are for single concretions.

Sample	$FeCO_3$	$CaCO_3$	$MgCO_3$	$MnCO_3$
Cooma 1	94·5	0·4	0·4	4·7
2164·2 m	94·2	0·4	0·5	4·9
	94·6	0·4	0·4	4·6
	95·8	0·3	0·4	3·5
	93·7	0·4	0·6	5·3
	93·0	0·6	0·6	5·8
	94·1	0·4	0·4	5·1
	93·9	0·4	0·6	5·1
	92·6	0·7	0·9	5·8
	94·5	0·4	0·5	4·6
	93·9	0·4	0·4	5·3
	95·0	0·5	0·4	4·1
	90·3	1·0	0·5	8·2
Glen Fosslyn 1	98·8	0·3	0·1	0·8
2091·5 m	98·7	0·1	0·1	1·1
	99·2	0·1	0·1	0·6
	98·6	0·3	0·1	1·0
	98·6	0·3	0·1	1·0
	95·9	0·8	0·4	2·9
	98·9	0·1	0·1	0·9
	96·5	1·3	0·4	1·8
	96·2	0·4	0·7	2·7
	95·3	0·7	0·8	3·2
Namarah 2	87·0	0·3	0·2	12·5
2252·50 m	85·2	0·3	0·2	14·3
	85·4	0·3	0·2	14·2
	90·3	0·2	0·2	9·3
	89·7	0·3	0·2	9·8
	88·4	0·2	0·2	11·2
	86·3	0·2	0·2	13·3
	88·2	0·3	0·1	11·4
	87·6	0·3	0·2	11·9
	88·1	0·3	0·1	11·5
Namarah 3	96·4	0·6	0·3	2·5
2179·4 m	97·1	0·3	0·3	2·3
	96·4	0·3	0·4	2·9
	97·8	0·1	0·2	1·9
	97·3	0·2	0·2	2·3
	97·4	0·9	0·5	1·2
	97·4	0·2	0·2	2·2
	94·5	0·6	0·5	4·4
	94·2	0·7	0·7	4·4
	96·3	0·5	1·0	2·2
	96·2	0·9	1·1	1·8
	91·8	0·8	0·4	7·0
Rednook 1	96·7	0·5	0·2	2·6
2272·5 m	95·4	0·5	0·2	3·9
	96·8	0·5	0·2	2·5
	96·3	0·5	0·2	3·1
	96·5	0·6	0·2	2·7
Warroon 1	97·0	0·1	0·1	2·8
2058·4 m	96·9	0·2	0·1	2·8
	97·0	0·1	0·1	2·8
	97·5	0·3	0·3	1·9
	96·7	0·2	0·2	2·9
	97·0	0·1	0·1	2·8
	97·5	0·2	0·2	2·1
	97·4	0·2	0·2	2·2
	96·8	0·2	0·1	2·9

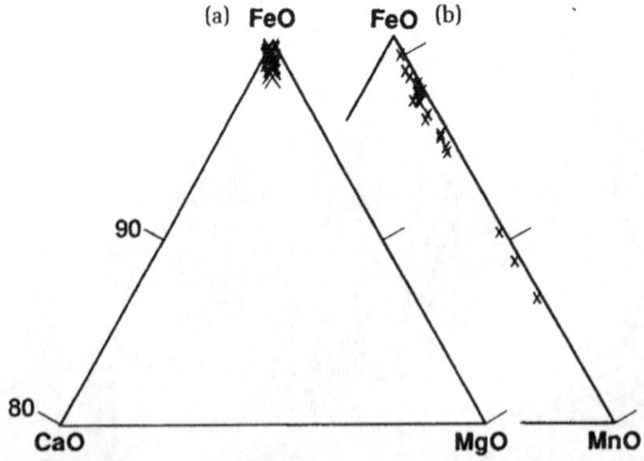

Fig. 4. Elemental composition of siderite in the Rewan Group in terms of (a) FeO, CaO and MgO, and (b) FeO, CaO and MnO. Each point is the average analysis for one concretion.

mimicked by those of the enclosing spherulite. Spherulites in the mudrocks rarely include detrital grains and are bounded by compactionally deformed clay laminae. Concretionary siderite in the sandstones forms irregular cement patches that poikilotopically surround uncompacted framework grains (Fig. 3e). Minus-cement-porosity in these siderite-cemented areas exceeds 35%. Relatively large (>1 cm) concretions in the sandstone intervals (sampled in Namarah 2, 2284·3 m) consist almost entirely of a combination of micritic siderite and a mosaic of fine and patchy coarse siderite crystals with widths between 20 and 100 μm. Non-concretionary siderite is generally uniformly finely crystalline (granular or rhombic) or coarsely crystalline and contains uncompacted framework grains (Fig. 3f). There are no visible indications that non-concretionary siderite bands are sheet concretions that formed by lateral coalescence of adjacent siderite concretions (e.g. Pirrie & Marshall, 1991; Coleman, 1993).

Microprobe analyses for concretionary siderites in six representative samples are given in Table 2 and are plotted in Fig. 4. The siderites are generally extremely pure, containing 85–97 mol% $FeCO_3$ and only 0·2–2·1 mol% $CaCO_3$+$MgCO_3$. Manganese contents range from 0·6 to 11·9 mol% $MnCO_3$ and are mainly greater than 2·0 mol% $MnCO_3$. Similar compositions are inferred from EDS spectra obtained for concretionary and non-concretionary siderite in Glen Fosslyn 1 (2091·3, 2097·3 m) and Namarah 2 (2266·5, 2284·3 m), although coarsely crystalline patches in the large concretion from Namarah 2 (2284·3 m) are slightly enriched in calcium and magnesium.

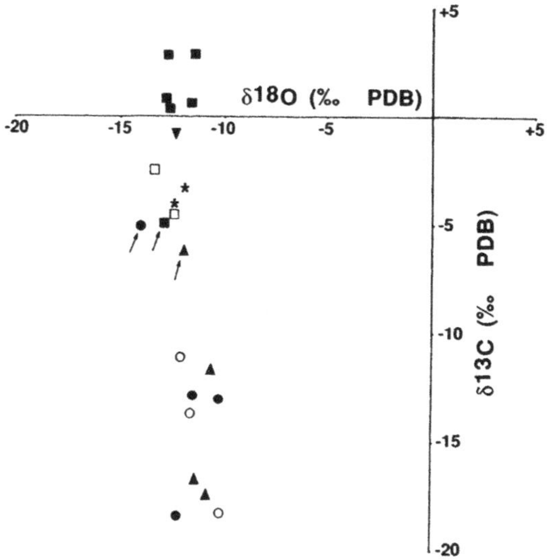

Fig. 5. Carbon and oxygen isotopic composition of concretionary and non-concretionary siderite in the Rewan Group and Showgrounds Formation. Key: open circle=Cooma 1; filled circle=Glen Fosslyn 1; filled square=Namarah 2; open square=Namarah 3; asterisk=Rednook 1; triangle=Warroon 1; inverted triangle=Warroon 2. Non-concretionary siderites indicated by arrows.

Otherwise, only very minor compositional variations were detected within single concretions and between concretions in the same sample (see Table 2).

Values of $\delta^{13}C$ and $\delta^{18}O$ for the siderites are given in Table 1 and are plotted in Fig. 5. Concretionary siderites have $\delta^{13}C$ values that range from $-18{\cdot}4$ to $+2{\cdot}9$‰ and are relatively consistent in each borehole. Values of $\delta^{13}C$ for non-concretionary siderites range from $-6{\cdot}4$ to $-5{\cdot}0$‰ and clearly differ from those of concretionary siderites in the same borehole. Values of $\delta^{18}O$ for concretionary and non-concretionary siderites are very consistent, ranging from $-14{\cdot}0$ to $-10{\cdot}2$‰, and averaging $-11{\cdot}9$‰ with a standard deviation of $\pm 1{\cdot}0$‰.

DISCUSSION

Mozley (1989) and Mozley & Wersin (1992) demonstrated that early diagenetic marine and non-marine siderites can be distinguished by their elemental and stable isotopic composition. Marine siderites are impure, with extensive substitution of magnesium (up to 41% $MgCO_3$) and lesser calcium (up to 15 mol% $CaCO_3$) for iron, and generally have lower $\delta^{13}C$ values than non-marine siderites. Non-marine siderites are relatively pure and manganese rich (commonly greater than 2 mol% $MnCO_3$), and can have much lower $\delta^{18}O$ values than marine siderites. Mozley & Wersin (1992) contended that geochemical characterization of early diagenetic siderite may help determine depositional environment in situations in which sedimentological evidence is ambiguous.

Compactionally deformed bounding clay laminae and the presence of uncompacted grains and high minus-cement-porosities within siderite-cemented areas provide classic textural evidence for formation of the concretionary and non-concretionary siderites prior to significant compaction. Furthermore, spherulites in the mudrocks are commonly composed entirely of siderite and show no evidence (e.g. grain ghosts or partly replaced grains/clay) of having replaced clay matrix. This suggests that these spherulites grew by displacive recrystallization within very soft sediment, and following the observations of Thyne & Boles (1989), Feistner (1989) and Huggett (1994), casts further doubt over the use of minus-cement-porosity as an accurate indicator of depth of concretionary cement development. In view of this textural evidence for a shallow burial origin for the siderites, it is assumed, for the purpose of this discussion, that the siderites formed from depositional porewaters at surface temperatures and that the conclusions of Mozley (1989) and Mozley & Wersin (1992) are applicable.

Siderite elemental composition

The very low calcium and magnesium content and relatively high manganese content of the siderites indicate siderite formation in a non-marine environment (cf. Mozley, 1989). Even siderite in the Namarah 2 section that contains spinose acritarchs is relatively pure and enriched in manganese, indicating that the presence of the acritarchs cannot be taken as evidence that the section accumulated in a marine-influenced environment. If the siderites had formed from both marine/brackish and fresh water, considerably more variation in calcium and magnesium content should be apparent as has been observed for coexisting marine and non-marine siderite in the Cretaceous Cardium Formation in the Western Canadian Sedimentary Basin (Hart *et al.*, 1992).

Coarsely crystalline, calcium- and magnesium-enriched siderite in the large concretion from Namarah 2 (2284·3 m) appears to be a microcavity-filling that formed at shallow depth after the main body of the concretion. This siderite probably reflects increased relative concentrations of porewater calcium and magnesium due to a decrease in reactive ferric iron availability following primary concretion formation in a closed system (Curtis *et al.*, 1975; Curtis & Coleman, 1986; Mozley & Carothers, 1992).

Siderite isotopic composition

Isotopic data for whole concretions can be correctly interpreted only if the concretions are isotopically homogeneous. In this case, except in the large concretion from Namarah 2 (2284·3 m), internal isotopic homogeneity is likely, based on the consistent elemental composition of the siderites and the lack of textural evidence for prolonged siderite development during progressive burial. In the large concretion from Namarah 2 (2284·3 m), the volume of relatively impure siderite, which may have an isotopic composition at slight variance with the bulk of the concretion, is less than 10%.

Oxygen isotopes

The consistent oxygen isotopic composition of the siderites accords with the textural and elemental evidence for siderite formation at very shallow burial depths from depositional porewaters that were consistently non-marine. This is because at very shallow burial depths and during accumulation of the sideritic section, burial temperature and the oxygen isotopic composition of meteoric water supplied to sediment accumulation sites by fluvial systems are likely to have remained relatively constant (siderite oxygen isotopic composition is mainly controlled by siderite formation temperature and the oxygen isotopic composition of the porewater from which the siderite is formed). If meteoric water was periodically mixed with marine water, variations in siderite oxygen isotopic composition would be expected as has been reported for early diagenetic siderites in Mississippi Delta sediments (Moore *et al.*, 1992) and the Cardium Formation (Hart *et al.*, 1992).

The measured siderite $\delta^{18}O$ values provide the strongest evidence that the siderites are of non-marine origin. Using the mean siderite $\delta^{18}O$ value of −11·9‰, a geologically realistic siderite precipitation temperature of 15°C (the southern Bowen Basin was at about 65°S during the Early Triassic; Embleton, 1984), and the siderite–water fractionation equation of Carothers *et al.* (1988), the $\delta^{18}O$ value of the porewater involved in siderite formation is calculated to be −15·6‰. Being strongly depleted in ^{18}O relative to SMOW, this porewater must have been of meteoric origin. Siderite formation from marine waters would be indicated by the siderite oxygen isotope data only if the siderites had formed at 95°C, an unlikely situation given the textural evidence for siderite formation at shallow depth.

Alteration of volcanic material to clay and precipitation of early ^{18}O-rich authigenic minerals can cause marine porewater $\delta^{18}O$ values to decrease with depth by at least 8‰ (see Mozley & Carothers, 1992). This mechanism has been invoked to explain anomalously low $\delta^{18}O$ values of early carbonate concretions and cements in marine sedimentary rocks (e.g. Pirrie & Marshall, 1991; Mozley & Carothers, 1992; Mozley & Burns, 1993; Morad & De Ros, 1994) and modern coastal sediments (Moore *et al.*, 1992). In the case of the Rewan Group siderites, even if marine water/sediment interaction did cause major ^{18}O depletion of the porewaters, it is improbable that the negative isotopic shift could account for the calculated porewater $\delta^{18}O$ value of −15·6%. Moreover, there is no evidence (e.g. presence of diagenetically altered grains within siderite cemented areas) that early (pre-siderite) mineral reactions took place in the host sandstones to cause ^{18}O depletion of porewaters during shallow burial. Finally, if marine water/sediment interaction was significant in causing porewaters to become depleted in ^{18}O, then siderites in the mudrocks should have higher $\delta^{18}O$ values than siderites in the sandstones because the mudrocks, consisting mainly of detrital clay, would have contained less reactive material (Pirrie & Marshall, 1991). The same argument also applies to siderite in the more quartzose samples. Lithology-dependent trends in $\delta^{18}O$ values do not occur, with siderite $\delta^{18}O$ values for all samples being very consistent.

If the host rocks did accumulate in marine environments, the low $\delta^{18}O$ values of the siderites could also be explained by early meteoric water incursion (e.g. Machemer & Hutcheon, 1988; Pirrie & Marshall, 1991; Mozley & Burns, 1993; Thyne & Gwinn, 1994). The consistently low siderite $\delta^{18}O$ values would imply extremely thorough early meteoric flushing of host mudrocks (which form beds up to 2·0 m thick), as well as host sandstones, a highly unlikely situation irrespective of depositional setting.

The siderite oxygen isotope values are unlikely to have been reset at higher temperatures due to siderite recrystallization. Siderite microfabrics are well preserved, suggesting that siderite recrystallization has not occurred. Moreover, siderite is stable at diagenetic temperatures (Mozley & Carothers, 1992) and the siderites have not experienced post-formational temperatures greater than 120–135°C as estimated from vitrinite reflectance data using the modified model of Barker & Pawlewicz (1986) ($\%R_{max}$=0·8–0·9 in the Rewan Group over the study area; Beeston & Smith, 1984).

Carbon isotopes

The concretionary siderites have $\delta^{13}C$ values that vary significantly between boreholes, but are relatively consistent in each borehole. Evidently, temporal changes in the carbon source and carbon dioxide/bicarbonate generation process, as well as porewater elemental and oxygen isotopic composition, were not significant at siderite formation depths.

The very ^{13}C-depleted siderites ($\delta^{13}C < -10‰$) clearly have an organic carbon source (e.g. Mozley & Carothers, 1992). Although some host mudrocks contain trace amounts of early diagenetic pyrite, the non-marine setting of the siderites, as interpreted from siderite elemental and oxygen isotopic compositions, implies that bacterially mediated sulphate and ferric iron reduction (Coleman, 1993) was not significant and thus not the process by which the very ^{13}C-depleted carbon dioxide and/or bicarbonate was generated from organic matter (Irwin *et al.*, 1977; Pye *et al.*, 1990; Coleman *et al.*, 1993). Moreover, siderite concretions in sulphidic coastal sediments of the Wash (UK), the occurrence of which has cast doubt over the widely held view that siderite cannot form in the sulphate-reducing zone (Pye *et al.*, 1990), have recently been shown to result from anthropogenic contamination (Al-Agha *et al.*, 1995).

Siderite spherulites and included pyrite have the same crystal growth directions (Fig. 3c,d), and, moreover, pyrite crystals are much larger inside the spherulites than in the host mudrocks. These relations are considered to indicate that siderite and pyrite growth were synchronous, which excludes the possibility that the siderites crystallized in a suboxic iron-reducing zone as has been interpreted for non-marine sphaerosideritic concretions elsewhere by Kantorowicz (1990) and Gibson *et al.* (1994). Although ^{13}C-depleted siderites can form in a suboxic iron-reducing zone, pyrite formation is impossible (e.g. Mozley & Carothers, 1992).

Anaerobic methane oxidation can produce carbonate with strongly negative $\delta^{13}C$ values (-60 to $-20‰$) (in Pye *et al.*, 1990) at very shallow depths (<1 m) and in situations of extremely low porewater sulphate concentrations (<2·0 mmol) (Raiswell, 1987). The reaction proceeds as a result of methane consumption by sulphate-reducing bacteria as follows:

$$CH_4 + SO_4^{2-} \rightarrow HS^- + HCO_3^- + H_2O.$$
(Raiswell, 1987)

Given that sulphide is also produced by this reaction, the presence of the associated euhedral pyrite is consistent with siderite formation in a methane oxidation zone (Raiswell, 1987). Moreover, the relatively ^{13}C-enriched concretionary siderites in Namarah 2, which must have formed largely within the zone of methanogenesis (Irwin *et al.*, 1977), show that methane was available for anaerobic oxidation. All concretionary siderite $\delta^{13}C$ values most likely indicate varying degrees of mixing between ^{13}C-depleted bicarbonate produced by methane oxidation and ^{13}C-enriched carbon dioxide produced by methanogenesis.

Values of $\delta^{13}C$ for non-concretionary siderites contrast with those of nearby concretionary siderites, confirming that non-concretionary siderite bands do not simply represent amalgamated siderite concretions. From the arguments presented above, the moderately negative $\delta^{13}C$ values of the non-concretionary siderites may reflect the presence of organic carbon made available by both methane oxidation and methanogenesis. Reasons for the contrasting $\delta^{13}C$ values between non-concretionary and concretionary siderites are unclear.

Comparison with siderite isotopic data of Mozley & Wersin (1992)

Based on a review of published isotopic compositions for marine and non-marine siderites, Mozley & Wersin (1992) suggested that siderites with $\delta^{13}C$ values less than $-8‰$ are most likely of marine origin, whereas those with $\delta^{18}O$ values less than $-13‰$ with positive $\delta^{13}C$ values are most likely of non-marine origin. Lacking independent evidence, concretionary siderites in

Namarah 2 would be correctly interpreted to have a non-marine origin on this basis. However, many of the siderites in the other boreholes have $\delta^{13}C$ values that are much lower than −8‰, placing them well within the siderite compositional field that has been defined exclusively by marine siderite carbon and oxygen isotopic values (see Mozley & Wersin, 1992; Fig. 4). Accordingly, the results of this study indicate that a non-marine origin for early diagenetic siderite cannot be discounted based on very low (< −8‰) siderite $\delta^{13}C$ values. Highly $\delta^{13}C$-depleted (−16·2‰), non-marine, sphaerosideritic concretions have also been reported by Kantorowicz (1990).

Palaeoclimatic implications of siderite isotope data

The analysed siderites crystallized from meteoric porewater with a $\delta^{18}O$ value that could not have been much higher than −15·6‰. Modern meteoric waters with this oxygen isotopic composition are restricted to high latitudes and altitudes (Sheppard, 1986) where average surface temperatures are around −5°C (Gregory *et al.*, 1989, Fig. 8). On this basis, the oxygen isotope data indicate that the water precipitated in cold climatic conditions, consistent with the reconstructed position of the southern Bowen Basin at about 65°S during the Early Triassic (Embleton, 1984). Given that fluvial systems drained a high mountain range to the east (Fielding *et al.*, 1990*a*; Elliott, 1993; Kassan, 1993), the strongly ^{18}O-depleted water composition may also reflect rain/snowfall at high altitude.

Mineral oxygen isotopic compositions have also been used to infer very low $\delta^{18}O$ values for Early Permian and latest Triassic to earliest Jurassic meteoric waters in the region. Kaolinites of weathering origin in the Gunnedah Basin indicate Early Permian meteoric water with a $\delta^{18}O$ value around −17‰, implying that most of this water precipitated as snow (Bird & Chivas, 1988). Kaolinites of diagenetic and weathering origin in the Bowen Basin indicate latest Triassic to earliest Jurassic meteoric water with a $\delta^{18}O$ value between −14 and −10‰ (Baker & Golding, 1992; Zhou & Dobos, 1994), and early diagenetic siderite in the Callide Coal Seam (Callide Basin) indicates latest Triassic meteoric waters with a $\delta^{18}O$ value around −15‰ (B. Faraj, unpubl. data). These data are consistent with the Bowen Basin region occupying a high-latitudinal position throughout the Permian and Triassic prior to northward drift of the Australian continent to middle to low latitudes since the Late Cretaceous (Embleton, 1984).

Sequence boundary implications

The contact between the Rewan Group and overlying Showgrounds Formation in the study area is disconformable (Kassan, 1993). In Warroon 1, the contact is marked by an abrupt change in sandstone composition from lithic-rich (upper Rewan Group) to quartzose (Showgrounds Formation). Siderite concretions occur below and above this contact and have consistent morphology (compare Fig. 3a and b), paragenesis and oxygen isotopic composition (Table 1). Moreover, the concretions have the same highly ^{13}C-depleted carbon isotopic signature, indicating that anaerobic methane oxidation was significant at shallow burial before and after development of the contact. The existence of identical shallow burial conditions before and after an episode (up to 3 Ma) of non-sediment accumulation and erosion, as implied by the siderites, is unlikely. It is considered therefore that the contact is conformable, implying that no hiatus exists between the Rewan Group, and Showgrounds Formation in Warroon 1, or that the quartzose sandstones above the contact are part of the upper Rewan Group.

CONCLUSIONS

Diagenetic siderites are abundant in the upper Rewan Group in the southern Bowen Basin, occurring mainly as small concretions that formed at shallow burial depths. Unlike marine siderites, these siderites are extremely pure and commonly enriched in manganese. Siderite $\delta^{18}O$ values are very consistent and indicate that the siderites formed from strongly ^{18}O-depleted porewater. On this basis, the siderites must have formed from porewaters that were entirely meteoric, confirming that host sedimentary rocks accumulated in freshwater environments that were not influenced by marine incursions.

Strongly negative to slightly positive siderite $\delta^{13}C$ values are interpreted to reflect mixing of bicarbonate/carbon dioxide generated by methane oxidation and methanogenesis. The strongly negative $\delta^{13}C$ values show that a freshwater origin for early diagenetic siderite should not be dismissed based on the siderite being highly ^{13}C depleted (cf. Mozley & Wersin, 1992).

The strongly ^{18}O-depleted composition of depositional porewaters, as implied by the siderite $\delta^{18}O$ values, indicates that the porewater precipitated in cold climatic conditions, consistent with the reconstructed position of the southern Bowen Basin at 65°S during the Early Triassic.

The occurrence of similar siderite concretions below and above the Rewan Group-Showgrounds Formation contact in Warroon 1 suggests that either this contact is at least locally conformable or that the contact represents a change in sandstone composition within the upper Rewan Group.

ACKNOWLEDGMENTS

We thank Oil Company of Australia N.L. for supporting the project and for permitting this paper to be published. The paper benefited from critical reviews by Chris Fielding and Journal referees Tim Young and Mark Hounslow. Skilled technical support was provided by Kim Baublys. The cooperation received from the Queensland Mines Department is greatly appreciated.

REFERENCES

Al-Agha, M.R., Burley, S.D., Curtis, C.D. and Esson, J. (1995) Complex cementation textures and authigenic mineral assemblages in Recent concretions from the Lincolnshire Wash (east coast, UK) driven by Fe(O) to Fe (II) oxidation. *J. geol. Soc. Lond.*, **152**, 157–171.

Baker, J.C., Fielding, C.R., de Caritat, P. and Wilkinson, M.M. (1993) Permian evolution of sandstone composition in a complex back-arc extensional to foreland basin: the Bowen Basin, eastern Australia. *J. sedim. Petrol.*, **63**, 881–893.

Baker, J.C. and Golding, S.D. (1992) Occurrence and palaeohydrological significance of authigenic kaolinite in the Aldebaran Sandstone, Denison Trough, Queensland, Australia. *Clays Clay Miner.*, **40**, 273–279.

Barker, C.E. and Pawlewicz, M.J. (1986) The correlation of vitrinite reflectance with maximum temperature in humic kerogen. In: *Palaeogeothermics* (Ed. by G. Buntebarth and L. Stegena), pp. 79–93. Springer-Verlag.

Beeston, J.W. and Smith R.J. (1984) Vitrinite reflectances from Bowen Basin strata below the Surat Basin, Queensland, *Geological Survey Record*, **1984/9**.

Bird, M.I. and Chivas, A.R. (1988) Stable-isotope evidence for low temperature kaolinitic weathering and post-formational hydrogen-isotope exchange in Permian kaolinites. *Chem. Geol.*, **72**, 249–265.

Carothers, W.W., Adami, L.H. and Rosenbauer, R.J. (1988) Experimental oxygen isotope fractionation between siderite–water and phosphoric acid liberated CO_2-siderite. *Geochim. Cosmochim. Acta*, **52**, 2445–2450.

Coleman, M.L. (1993) Microbial processes: Controls on the shape and composition of carbonate concretions. *Mar. Geol.*, **113**, 127–140.

Coleman, M.L., Hedrick, D.B., Lovley, D.R., White, D.C. and Pye, K. (1993) Reduction of Fe (III) in sediments by sulphate-reducing bacteria. *Nature*, **361**, 436–438.

Cosgrove, J.L. and Mogg, W.G. (1985) Recent exploration and hydrocarbon potential of the Roma Shelf, Queensland. *Australian Petrol. Expl. Ass. J.*, **25**, 217–234.

Curtis, C.D. and Coleman, M.L. (1986) Controls on the precipitation of early diagenetic calcite, dolomite and siderite concretions in complex depositional sequences. In: *Roles of Organic Matter in Sediment Diagenesis* (Ed. by D. L. Gautier), *Spec. Publ. Soc. econ. Palaeont. Miner.*, **38**, 23–33.

Curtis, C.D., Pearson, M.J. and Somogyi, V.A. (1975) Mineralogy, chemistry, and origin of a concretionary siderite sheet (clay-ironstone band) in the Westphalian of Yorkshire. *Miner. Mag.*, **40**, 385–393.

Dickins, J.M. and Malone, E.J. (1973) Geology of the Bowen Basin, Queensland. *Bull. Bureau Mineral Resources, Geology and Geophysics, Australia*, **130**.

Elliott, L.G. (1993) Post-Carboniferous tectonic evolution of eastern Australia. *Austr. Petrol. Expl. Ass. J.*, **33**, 215–236.

Embleton, B.J.J. (1984) Continental palaeomagnetism. In: *Phanerozoic Earth History of Australia. Oxford Monographs on Geology and Geophysics 2* (Ed. by J. J. Veevers), pp. 11–16. Clarendon Press, Oxford.

Feistner, K.W.A. (1989) Petrographic examination and re-interpretation of concretionary carbonate horizons from the Kimmeridge Clay, Dorset. *J. geol. Soc. Lond.*, **146**, 345–350.

Fielding, C.R., Falkner, A.J., Kassan, J. and Draper, J.J. (1990a) Permian and Triassic depositional systems in the Bowen Basin. In: *Bowen Basin Symposium Proceedings*, pp. 21–25. Geological Society of Australia.

Fielding, C.R., Gray, A.R.G., Harris, G.I. and Salomon, J.A. (1990b) The Bowen Basin and overlying Surat Basin. *Bull. Bureau Miner. Resources, Geology and Geophysics, Australia*, **232**, 107–118.

Folk, R.L., Andrews, P.B. and Lewis, D.W. (1970) Detrital sedimentary rock classification and nomenclature for use in New Zealand. *New Zealand J. Geol. Geophys.*, **13**, 947–968.

Gibson, P.J., Shaw, H.F. and Spiro, B. (1994) The nature and origin of ironstone bands in the Tertiary Lowmead and Duaringa Basins, Queensland. *Austr. J. Earth Sci.*, **41**, 255–263.

Gregory, R.T., Douthitt, C.B., Duddy, I.R., Rich, P.V. and Rich, T.H. (1989) Oxygen isotopic composition of carbonate concretions from the lower Cretaceous of Victoria, Australia: implications for the evolution of meteoric waters on the Australian continent in a paleopolar environment. *Earth planet. Sci. Lett.*, **92**, 27–42.

Harris, G.I. (1986) *Permo-Triassic stratigraphy of the Arcadia Valley, southeast central Queensland.* BSc(Hons) thesis, Queensland University.

Hart, B.S., Longstaffe, F.J. and Plint, A.G. (1992) Evidence for relative sea level change from isotopic and elemental composition of siderite in the Cardium Formation, Rocky Mountain Foothills. *Bull. Can. petrol. Geol.*, **40**, 52–59.

Huggett, J.M. (1994) Diagenesis of mudrocks and concretions from the London Clay Formation in the London Basin. *Clay Miner.*, **29**, 693–707.

Irwin, H., Curtis, C. and Coleman, M. (1977) Isotopic evidence for source of diagenetic carbonates formed during burial of organic-rich sediments. *Nature*, **269**, 209–213.

Jensen, A.R. (1975) Permo-Triassic stratigraphy and sedimentation in the Bowen Basin, Queensland. *Bull. Bur. Mineral Resources, Geology and Geophysics, Australia*, **154**.

Kantorowicz, J.D. (1990) Lateral and vertical variations in pedogenesis and other early diagenetic phenomena, Middle Jurassic Ravenscar Group, Yorkshire. *Proc. Yorks. geol. Soc.*, **48**, 61–74.

Kassan, J. (1993) *Basin analysis of the Triassic succession, Bowen Basin.* PhD thesis, Queensland University.

Kassan, J. (1994) *Facies analysis of drillcore from ATP 470P (Redcap Block) – Rewan Group, Showgrounds Formation, Snake Creek Mudstone Member.* Report to Oil Company of Australia.

Machemer, S.D. and Hutcheon, I. (1988) Geochemistry of early carbonate cements in the Cardium Formation, central Alberta. *J. sedim. Petrol.*, **58**, 136–147.

Martin, K.R. (1992) *Sedimentology, petrology and reservoir quality of the Bandanna Formation to Showgrounds Sandstone interval in Namarah no. 3, Bowen-Surat Basin.* Report to Oil Company of Australia.

McCrea, J.M. (1950) On the isotopic chemistry of carbonates and a palaeotemperature scale. *J. Chem. Phys.*, **18**, 849–857.

Moore, S.E., Ferrell, R.E. Jr and Aharon, P. (1992) Diagenetic siderite and other ferroan carbonates in a modern subsiding marsh sequence. *J. sedim. Petrol.*, **62**, 357–366.

Morad, S. and De Ros, L.F. (1994) Geochemistry and diagenesis of stratabound calcite cement layers within the Rannoch Formation of the Brent Group, Murchison Field, North Viking Graben (northern North Sea) – comment. *Sediment. Geol.*, **93**, 135–141.

Mozley, P.S. (1989) Relation between depositional environment and the elemental composition of early diagenetic siderite. *Geology*, **17**, 704–706.

Mozley, P.S. and Burns, S.T. (1993) Oxygen and carbon isotopic composition of marine carbonate concretions: an overview. *J. sedim. Petrol.*, **63**, 73–83.

Mozley, P.S. and Carothers, W.W. (1992) Elemental and isotopic composition of siderite in the Kuparuk Formation, Alaska: effect of microbial activity and water/sediment interaction on early pore-water chemistry. *J. sedim. Petrol.*, **62**, 681–692.

Mozley, P.S. and Wersin, P. (1992) Isotopic composition of siderite as an indicator of depositional environment. *Geology*, **20**, 817–820.

Pirrie, D. and Marshall, J.D. (1991) Field relationships and stable isotope geochemistry of concretions from James Ross Island, Antarctica. *Sediment. Geol.*, **71**, 137–150.

Price, P.L. (1994) *ATP 470P Redcap area, Mimosa Group, palynostratigraphic and palynofacies study.* Report to Oil Company of Australia.

Pye, K., Dickson, J.A.D., Schiavon, N., Coleman, M.L. and Cox, M. (1990) Formation of siderite – Mg-calcite – iron sulphide concretions in intertidal marsh and sandflat sediments, north Norfolk, England. *Sedimentology*, **37**, 325–343.

Raiswell, R. (1987) Non-steady state microbiological diagenesis and the origin of concretions and nodular limestones. In: *Diagenesis of Sedimentary Sequences* (Ed. by J. D. Marshall), *Spec. Publ. geol. Soc.*, **36**, 41–54.

Sheppard, S.M.F. (1986) Characterization and isotopic variations in natural waters. In: *Stable Isotopes in High Temperature Processes* (Ed. by J. W. Valley, H. P. Taylor Jr and J. R. O'Neil), *Miner. Soc. Am., Reviews in Mineralogy*, **16**, 165–183.

Thyne, G.D. and Boles, J.R. (1989) Isotopic evidence for origin of the Moeraki septarian concretions, New Zealand. *J. sedim. Petrol.*, **59**, 272–279.

Thyne, G.D. and Gwinn, C.J. (1994) Evidence for a paleoaquifer from early diagenetic siderite of the Cadium Formation, Alberta, Canada. *J. sedim. Res.*, **64**, 726–732.

Zhou, T. and Dobos, S.K. (1994) Stable isotope geochemistry of kaolinite from the 'White Section', Black Ridge, Clermont, central Queensland: implications for the age and origin of the 'White Section'. *Clays Clay Miner.*, **42**, 269–275.

Manuscript received 22 March 1995; revision accepted 14 June 1995

Reprinted from Sedimentology (1993) ***40****, 467–474*

Early diagenetic spherulitic siderites from Pennsylvanian palaeosols in the Boss Point Formation, Maritime Canada

G. H. BROWNE* *and* D. M. KINGSTON

Department of Geology, University of Western Ontario, London, Ontario, Canada N6A 5B7

ABSTRACT

The Boss Point Formation of Maritime Canada comprises alternating successions of braidplain sandstone, lacustrine mudstone, and lacustrine fluvio-deltaic sandstone and mudstone. These rocks were deposited within an active strike-slip basin during the early Pennsylvanian (Westphalian A), at a palaeolatitude of 8°S. Palaeosols comprise a small but distinct proportion of the formation, and typically occur at the tops of fining upward sandstone–siltstone cycles.

Spherulitic siderites from the Boss Point Formation occur within sandy and silty palaeosols. They occur as 'large' (>200 μm) and 'small' (<200 μm) diameter spherulitic nodules within calcrete, and show either a radially arranged ferroan calcite and siderite spar morphology, or contain a core of ferroan calcite, surrounded by a later formed sheath of siderite. Analytical data indicate that with respect to Ca–Mg–Mn, wide compositional variation occurs within the siderites. The Boss Point Formation siderites are similar to, but contain less Ca for a given Mg/Mn ratio, than previously published siderite compositional data. Published data, together with the results from the Boss Point Formation, indicate that a continuum exists in freshwater siderites, between the relatively Mn-depleted and the Mg-enriched analytical fields that were previously reported.

The relatively low Ca values determined in the Boss Point Formation samples reflect the general lack of calcium in the sedimentary basin during sedimentation. Both chlorite and smectite clays may have been important sources of iron. It is concluded that the spherulitic siderites are eogenetic, and formed within small anoxic ponds rich in organic matter, under reducing and low dissolved sulphate conditions. With time these pools dried up, ferroan calcite precipitating as calcrete nodules around the earlier formed siderite spherulites.

INTRODUCTION

The formation of siderite has been described from a range of geological environments. The general conditions necessary for its formation include a reducing environment, low dissolved sulphate concentration, near neutral pH and a CO_2 partial pressure $>0{\cdot}1$ Pa (Pearson, 1979; Postma, 1982; Bahrig, 1989). These physical and chemical conditions are common in anoxic environments in certain lakes (Postma, 1981, 1982; Bahrig, 1989) and in some paralic and marine settings (Pearson, 1979; Pye, 1984; Tasse & Hesse, 1984; Boles, 1987). In freshwater environments, production of H_2S is normally inhibited by the low levels of dissolved sulphate available for reduction by sulphate-reducing bacteria, thereby favouring siderite precipitation (Pye *et al.*, 1990). In marine settings, siderite can form either prior to sulphate reduction under suboxic conditions, or under strongly reducing conditions during methanogenesis (Coleman, 1985; Mozley & Carothers, 1992; Mozley & Wersin, 1992). In marine environments, however, the increase in dissolved sulphate levels will favour the precipitation of iron sulphides (pyrite) in preference to siderite.

Mozley (1989) has recently proposed that siderites formed in marine and non-marine environments can be differentiated geochemically using their elemental

* Present address: Institute of Geological & Nuclear Sciences, PO Box 30 368, Lower Hutt, New Zealand.

ratios, and has presented a series of ternary descrimination diagrams to characterize these domains (see also Matsumoto & Iijima, 1981). Freshwater siderites are typically Mn-rich and relatively pure (i.e. greater than 90 mol% $FeCO_3$), whereas siderites formed in marine environments are always impure, with extensive substitution of Mg, and to a lesser extent Ca, for Fe in the siderite lattice.

This study presents data from early diagenetic spherulitic siderites, which formed in palaeosols, of the Boss Point Formation in Maritime Canada. Both the palaeosols and the spherulitic siderites have not previously been recognized from the formation. The elemental composition of these siderites, which formed in an entirely continental environment, is compared with the analytical fields determined by Mozley (1989). Although spherulitic siderites have been noted from other settings, few studies have presented detailed descriptions or geochemical characterization of their morphology, or discussed their mode of formation.

Electron microprobe analyses were performed with a JEOL 8600 microanalyser equipped with Tracor-Northern TN-5500 automation. Carbonate analyses were carried out by wavelength dispersive spectrometry at an accelerating voltage of 15 kV, beam current of 10 nA and beam diameter of 5 μm. The following X-ray lines (and standards) were used: $Ca_{K\alpha 1}$, (calcite), $Mg_{K\alpha 1}$ (dolomite), $Mn_{K\alpha 1}$ (rhodonite) and $Fe_{K\alpha 1}$ (siderite). All analyses were done using 20 s counting times and data were corrected using (Tracor-Northern's ZAF program.

OCCURRENCE

The spherulitic siderites reported here are from the Boss Point Formation, a unit comprising several repeating successions (megacycles) of two main lithofacies associations: sandy braidplain, and lacustrine or fluvio-deltaic sandstone–siltstone. The formation is up to 800 m thick, and of Westphalian A (early Pennsylvanian) age. The outcrops examined occur in eastern New Brunswick and northern Nova Scotia; particularly good exposures occur around the shores of the northern Bay of Fundy (Fig. 1). The formation was deposited in the Cumberland Basin, which was bounded to the north-west and south by active strike-slip faults such as the Harvey–Hopewell Fault (Fig. 1).

The braidplain sandstones comprise erosively based, multi-storied channellized units up to 130 m thick, dominated by well sorted, fine to lower medium grained trough cross-bedded sandstones. They are intercalated with <60 m thick lacustrine mudstones and associated fluvio-deltaic sandstones. Mudstones consist of dark grey and reddish, massive and parallel laminated organic-rich siltstones and claystones, coals

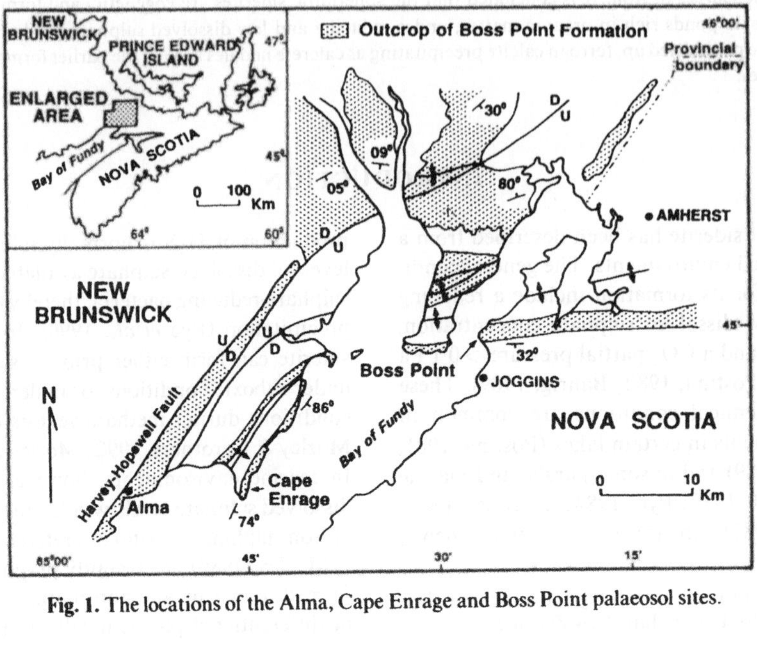

Fig. 1. The locations of the Alma, Cape Enrage and Boss Point palaeosol sites.

and ostracod-bearing, fine grained limestones (Browne, 1991). The ostracod-bearing limestones are <80 cm thick, and consist of richly carbonaceous micrite and wackestone, typically with crude to well developed horizontal stratification produced by the alignment of ostracod valve long axes. Bioclasts are almost exclusively of joined and disarticulated freshwater ostracod valves (*Carbonita*), with fewer bivalve, gastropods and fish scale fragments. Faunal details of the ostracods are presented by Copeland (1957). The deltaic rocks comprise centimetre to decimetre scale beds of coarsening upward mudstone to sandstone cycles that prograded into the lakes (Browne, 1991). Both sandstone and mudstone lithologies contain abundant carbonaceous debris ranging in size from comminuted material to large logs, the latter predominantly lycopods such as *Calamites*, *Lepidodendron* and *Sigillaria*. Ripple and plane laminated sandstone, pebbly sandstone and conglomerate are minor components in the formation.

Fining upward cycles 2–25 m in thickness are common throughout the formation and have been interpreted as having formed by the migration of fluvial channels (Browne, 1991). The palaeosols typically occur at the top of these fining upward cycles. There is no evidence for a marine influence during deposition of this succession. Marine sedimentation is in fact rare during the Carboniferous of Maritime Canada (Gibling *et al.*, 1992), known only from the Windsor Group (Mississippian) throughout Maritime Canada, and from the Sydney Mines Formation (Westphalian D) of Nova Scotia (Thibaudeau & Medioli, 1986).

The spherulitic siderites are components within indurated calcrete (caliche) nodules which occur within sandy and silty palaeosols (Fig. 2). Calcrete

Fig. 2. A silty palaeosol at Alma with a concentration of calcrete nodules in its middle portion. Scale is 1 m in length.

Fig. 3. Light-coloured calcrete nodules containing spherulitic siderites, from a silty palaeosol at Alma (cf. Fig. 2). Lens cap is 50 mm in diameter.

nodules are defined as secondary accumulations of carbonate in near surface settings, which result from cementation and/or replacement of host material by the precipitation of calcium carbonate from soil water or ground water (Netterberg, 1980; Goudie, 1983). In the Boss Point Formation the calcite within the calcrete is ferroan. Morphologically, the calcrete nodules in the formation equate with Stages II and III calcrete morphology of Machette (1985) i.e. they comprise isolated through to coalesced nodules 1–15 cm in diameter (Fig. 3).

At least 29 palaeosols have been recognized from the formation, and have a maximum thickness of 2·2 m. The palaeosols have been distinguished on the basis of colour mottling, and the presence of pedoturbation, organic debris and calcrete nodule development (e.g. Retallack, 1981; Francis, 1986).

Sandy palaeosols consist of green–grey (5G 5/2* and 10GY 5/2) and greyish brown (5YR 3/2), colour mottled, well sorted, very fine to lower medium grained sandstone. Sandstones include massive, laminated and ripple laminated textures, and in places may be bioturbated. Distinct horizons are only poorly developed; rather, profiles are typified by the presence of calcrete nodules. These nodules are generally scattered throughout individual pedogenic layers, but may also be concentrated into the upper parts of palaeosol profiles. Vertical and subvertical <3 cm wide calcareous tubes are common, and are interpreted either as rhizoliths or as calcified root casts (Browne, 1991).

Silty palaeosols consist of greyish green (5G 5/2 and 10GY 5/2) and reddish purple (5R 4/2, GY 5/2 and

* Colour designations follow the *Geological Society of America Rock Color Chart 1948*, reprinted 1979 (dry sample).

5RP 4/2), colour mottled, slightly sandy siltstone. The clay mineralogy of the palaeosols is dominated by kaolinite and illite, with less abundant albite, muscovite and rare smectite (Browne, 1991). The calcrete nodules are tan to dark coloured and disorthic (Brewer, 1964), with sharp outer margins. The nodules occur as both isolated and semi-continuous masses, comprising densely packed, erosion resistant intervals (Fig. 3). Bioturbation (or pedoturbation) is not evident in outcrop, but is typical in thin section. Many outcrops display arcuate, low to moderate angle shears, typically slickensided, and undulose shears (pseudo-anticlines), and are similar to structures described from ancient soils (Allen & Williams, 1979, Gray & Nickelsen, 1989; Driese & Foreman, 1992).

Although the palaeosols comprise only a small proportion of the total thickness of the formation (0·5%), they do consistently occur at approximately 60 m above the base of the formation. Palaeosols at this stratigraphic position can be traced through several sections (up to 40 km apart), covering an area of 900 km^2, and appear to form one pedogenic horizon.

The presence of colour mottling, little or no humus, pedogenic slickensides, pseudo-anticline structures and the calcrete nodules suggest that the palaeosols represent Vertisols and Aridisols (Ahmad, 1983; Nettleton & Peterson, 1983), and that they formed in a semi-arid palaeoclimatic setting (Browne, 1991). This interpretation is consistent with the early Pennsylvanian palaeolatitude of 8°S for this part of Maritime Canada (Roy & Morris, 1983).

In addition to the spherulitic siderite morphology, other petrographic varieties of calcrete morphology occur, including ped material, fragmentary nodules, siltstone aggregates and sparry calcite, and any of these types may occur within the same nodule (Browne, 1991). Concentration is given here to the spherulitic siderites.

Four spherulitic siderite samples from three localities were examined in this study (Fig. 1). Three samples (one from Alma and two from Cape Enrage) come from essentially the same stratigraphic interval between 60 and 70 m above the base of the formation. The fourth sample (Boss Point) comes from a stratigraphic position approximately 420 m above the base of the formation.

SPHERULITIC SIDERITE MORPHOLOGY AND COMPOSITION

Two distinct sizes of siderite nodule occur: 'large' spherulitic siderite nodules 200–500 μm in diameter, and 'small' spherulitic nodules 80–200 μm in diameter. Size variation is minor within a given sample. The siderite typically occurs as single spherulitic nodules, though they sometimes coalesce. 'Large' nodules typically consist of radially arranged ferroan calcite and sideritic spar (Fig. 4). 'Small' spherulitic siderites most typically consist of a core of mixed ferroan calcite (occupying as much as 70% of the grain) and siderite, surrounded by a sheath of later formed siderite (Fig. 5). 'Small' nodules lack a radial carbonate morphology. Nodules may contain several < 20 μm diameter quartz grains at their centres, but more commonly the quartz

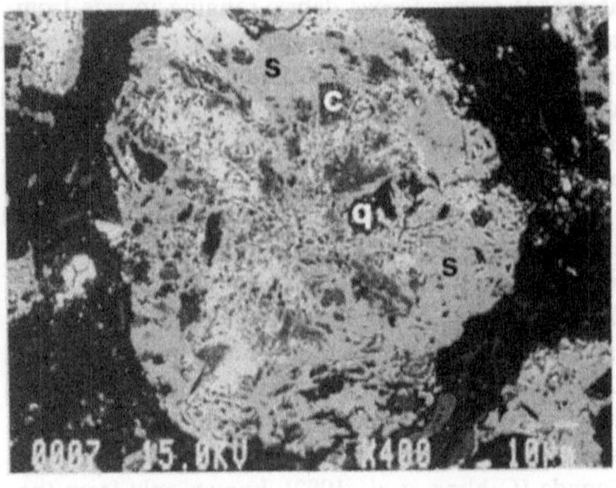

Fig. 4. A back-scattered electron microprobe image of a typical 'large' spherulitic siderite, showing dark coloured radially arranged ferroan calcite (c), and lighter coloured siderite (s). Black material is quartz (q). The grain is surrounded by a clay-rich matrix.

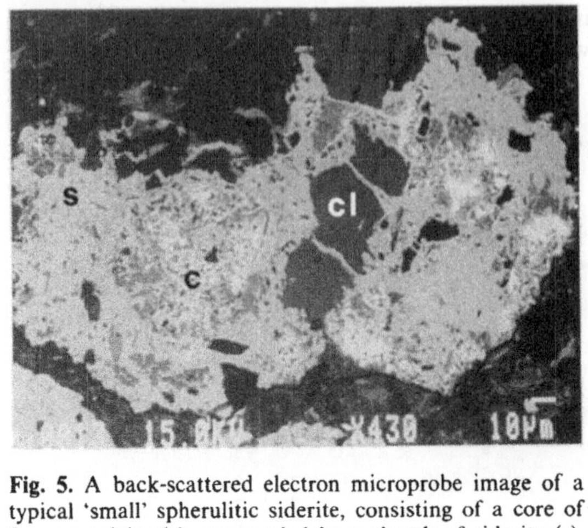

Fig. 5. A back-scattered electron microprobe image of a typical 'small' spherulitic siderite, consisting of a core of ferroan calcite (c), surrounded by a sheath of siderite (s). Dark areas to the right and top are K-rich clays (cl).

grains are scattered throughout the nodule (Figs 4 & 5). In addition, energy dispersive spectrometry (EDS) and X-ray diffraction (XRD) analyses indicate that several clay species are present within the nodules: Fe-rich clays (smectite) and K-rich clays (illite), as well as monazite ([Ce, La] PO_4) and albite. Some clays are surrounded by a thin (<2 µm thick) Mn-rich sideritic layer. A clay-rich matrix surrounds the spherulitic siderites, and consists of dark coloured Fe-rich (smectite) and K-rich (illitic) clays, as well as albite.

Electron microprobe determinations from 5 nodules (57 analyses) of both 'small' and 'large' spherulitic siderites have been plotted on discriminant $CaCO_3$–$MgCO_3$–$FeCO_3$ and $CaCO_3$–$MgCO_3$–$MnCO_3$ ternary plots (Fig. 6). These are compared with previously published freshwater siderite data from the Tyonek (Tertiary) and Ivishak (Permian–Triassic) formations (Mozley, 1989). The Ca–Mg–Fe ternary plot (Fig. 6a) shows that the Boss Point Formation samples are depleted in Fe with respect to the Tyonek Formation data, but fall within the compositional variation of

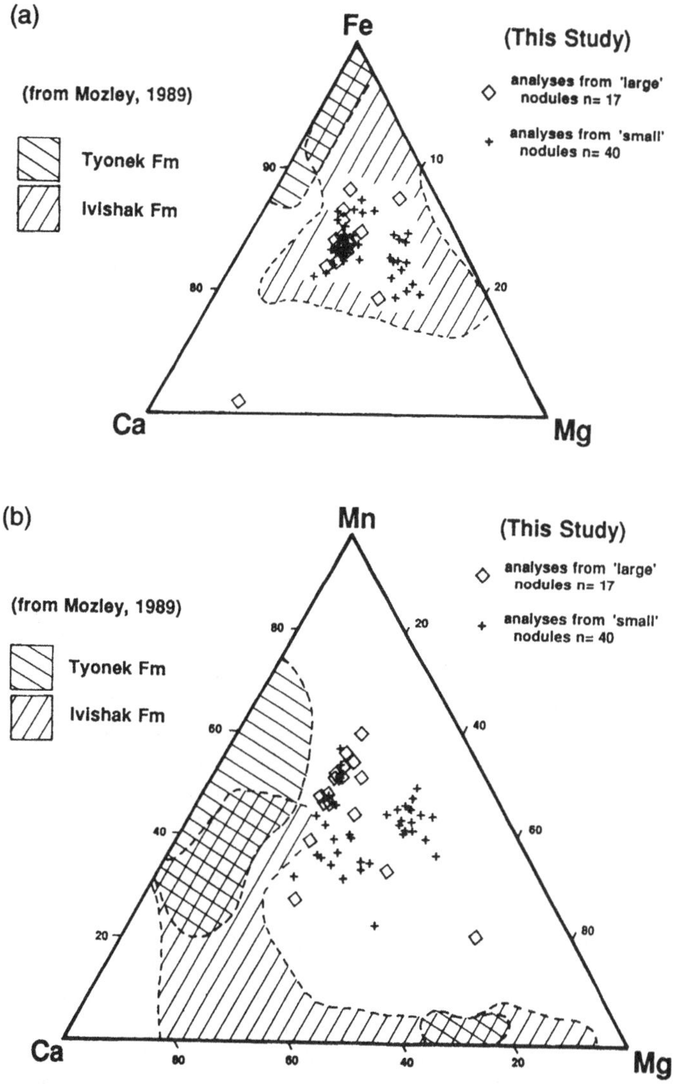

Fig. 6. Ternary $CaCO_3$–$MgCO_3$–$FeCO_3$ and $CaCO_3$–$MgCO_3$–$MnCO_3$ diagrams for Boss Point Formation spherulitic siderites. Data have been plotted on a mol% basis. Analyses from freshwater siderites in the Tyonek and Ivishak formations are taken from Mozley (1989, fig. 2).

the Ivishak Formation samples. Both the Ivishak and the Boss Point Formation samples are enriched in Mg. The high Mg siderite variants recognized by Mozley (1989, fig. 2) were not encountered in this study. 'Small' nodules appear to be more enriched in Mg than their 'large' counterparts, though there is considerable overlap (Fig. 6a). The Boss Point Formation samples plot between the extremes of Mg and Mn recognized by Mozley (1989, fig. 2), and in general are relatively depleted in Ca (Fig. 6b). Although there is considerable overlap, the smaller Boss Point Formation nodules tend to be more enriched in Mg than the larger nodules.

DISCUSSION

Previous work on other spherulitic siderites

Spherulitic siderites were first described by Teall (1899), and have since been described from a variety of freshwater and paralic settings. Their occurrence in pedogenic intervals is common (Fritz *et al.*, 1971; Leckie & Foscolos, 1986; Besly & Fielding, 1989; Hart, 1990; Kantorowicz, 1990; Lott & Humphreys, 1992). The Boss Point Formation occurrence is similar to the spherulitic siderite (equivalent to spherosiderite) of Spencer (1925), Fritz *et al.* (1971), Tucker (1981), Gibson (1989) and Gibson *et al.* (1990).

Besly & Fielding (1989) described spherulitic siderite from several localities in Britain, and argued that the association of siderite along vertical root structures was due to an upward movement of iron during lowering of the water table within swamp environments. Challis & Hornibrook (1989) indicated that spherulitic siderites occur as a replacement feature of foraminifera tests, and form principally by carbonation of chamositic clay which surrounds and partially infills the tests. Gibson *et al.* (1990) suggested that spherulitic siderites may form eogenetically, by growing in soft sediment under conditions of methanogenic fermentation in open lacustrine settings. Hart (1990) described $<0{\cdot}5$ mm diameter spherulitic siderites that comprised an Fe-rich core and a more Ca-rich margin. He interpreted that the siderites formed in a poorly drained palaeosol. Kantorowicz (1990) discussed their formation within waterlogged soils, and argued siderite formation from a bacterially derived bicarbonate source that supersaturated the pore waters during suboxic Fe^{3+} reduction.

Interpretation of the Boss Point Formation occurrence

The data suggest that with respect to Ca–Mg–Mn, the Boss Point Formation samples are relatively low in Ca for a given Mg/Mn ratio. The relatively low Ca values determined in the Boss Point Formation analyses reflect the general lack of calcium carbonate in the sedimentary basin, one that was instead dominated by siliciclastic sedimentation, eroded from the largely granitic and metamorphosed sedimentary rocks that surrounded the Cumberland Basin.

Mn values for the Boss Point Formation are in general high (cf. Mozley, 1989, fig. 2). The high Mn values (2·15–11·00 mol% $MnCO_3$) in Boss Point Formation samples is consistent with freshwater siderites, which typically have greater than 2 mol% $MnCO_3$ (Mozley, 1989).

The spherulitic siderite nodules are suggested to have formed via methanogenesis in small localized standing bodies of water (such as stagnant ponds), rich in organic matter, under reducing and low dissolved sulphate conditions (see Pearson, 1979; Postma, 1982; Gibson *et al.*, 1990). Methanogenesis is favoured in such low sulphate settings (Coleman, 1985; Mozley & Wersin, 1992). The siderite is suggested to have formed within the soft sediment of these ponds (as reported by Gibson *et al.*, 1990) although the possibility that they formed at the sediment–water interface cannot be ruled out. With time these anoxic ponds dried up, and became part of the pedogenic profile.

The Boss Point Formation spherulitic siderites occur within calcrete nodules, and are therefore considered to have formed early during diagenesis (eogenesis), at least over the period of pedogenesis. Once formed, the pore waters remained anoxic, but were overwhelmed by Ca-rich waters, promoting ferroan calcrete development. Calcrete growth after siderite reflects this change in pore water chemistry. The transition is suggested to reflect a change from small stagnant ponds in which the siderite developed, to a situation in which these ponds dried, and calcrete formed. The source of the calcium is uncertain. As part of this drier period, calcium may have been introduced from aeolian sources, the Mississippian Windsor Group being the most likely source. Alternatively, the calcium may have been derived from weathering of plagioclase or as a biological product of photosynthesis. CO_2 in soils is released by plant roots during respiration and during decay. The low partial pressure of CO_2 in soils promotes precipitation of $CaCO_3$, and is an important source of carbonate in modern calcretes (e.g. Salomons *et al.*, 1978). Whatever the source of calcium, the ferroan calcite nature of the calcrete nodules indicates that pore waters remained anoxic.

The question of the source of the iron should also be considered. Pearson (1979) concluded that siderite iron was derived from Fe-rich chloritic clays. Chlorite was detected by XRD in the mudstones and shales of the Boss Point Formation (Browne, 1991), but has not been found within the palaeosols. The only Fe-bearing clay species detected in the siderites is smectite. Smectite was also found in both the surrounding mudstone and shales of the Boss Point Formation, and in the form of Fe-rich smectites in the clay-rich matrix surrounding the spherulitic nodules. Both chlorite and smectite are thought to be important sources for iron. In addition, Carroll (1958) and Matsumoto & Iijima (1981) have shown that iron can be introduced from Fe-oxyhydroxide coatings on clay minerals, and this may also be a source of iron for the Boss Point Formation siderites. Alternatively, iron may have originated from Boss Point Formation river waters themselves, as has been suggested for siderite forming in some present day swamp deposits (Postma, 1982).

ACKNOWLEDGMENTS

The financial assistance of a New Zealand Department of Scientific and Industrial Research study award and G. F. Matthew Research Scholarship from the New Brunswick Museum to (G.H.B.), and NSERC operating grant A1917 (to A. G. Plint) are gratefully acknowledged. We wish to thank A. G. Plint and B. S. Hart for their encouragement and discussions. The manuscript was greatly improved by the critical comments of M. J. Kraus, J. D. Kantorowicz, M. G. Laird, J. McKay, K. A. Mertz and P. S. Mozley.

REFERENCES

AHMAD, N. (1983) Vertisols. In: *Pedogenesis and Soil Taxonomy II. The Soil Orders* (Ed. by L. P. Wilding, N. E. Smeck & G. F. Hall), pp. 91–123. Elsevier, Amsterdam.

ALLEN J.R.L. & WILLIAMS, B.P.J. (1979) Interfluvial drainage on Silurian–Devonian alluvial plains in Wales and Welsh Borders. *J. geol. Soc. Lond.*, **136**, 361–366.

BAHRIG, B. (1989) Stable isotope composition of siderite as an indicator of the palaeoenvironmental history of oil shale lakes. *Palaeogeogr. Palaeoclim. Palaeoecol.*, **70**, 139–151.

BESLY, B.M. & FIELDING, C.R. (1989) Palaeosols in Westphalian coal-bearing and redbed sequences, central and northern England. *Palaeogeogr. Palaeoclim. Palaeoecol.*, **70**, 303–330.

BOLES, J.R. (1987) Six million year diagenetic history, North Coles Levee, San Joaquin Basin, California. In: *Diagenesis of Sedimentary Sequences* (Ed. by J. D. Marshall). *Spec. Publ. geol. Soc. Lond.*, **36**, 191–200.

BREWER, R.C. (1964) *Fabric and Mineral Analysis of Soils*. Wiley, New York.

BROWNE, G.H. (1991) *The Sedimentology of the Boss Point Formation (Pennsylvanian) of eastern New Brunswick and northern Nova Scotia*. PhD thesis, University of Western Ontario, London, Ont.

CARROLL, D. (1958) Role of clay minerals in the transportation of iron. *Geochim. Cosmochim. Acta*, **14**, 1–28.

CHALLIS, G.A. & HORNIBROOK, N. DEB. (1989) Replacement of foraminifera by siderite in the Gore Lignite beds, Southland. *NZ Jl Geol. Geophys.*, **32**, 527–529.

COLEMAN, M.L. (1985) Geochemistry of diagenetic non-silicate minerals: kinetic considerations. In: *Geochemistry of Buried Sediments* (Ed. by G. Eglinton, C. D. Curtis, D. P. McKenzie & D. G. Murchison), pp. 39–54. The Royal Society, London.

COPELAND, M. J. (1957) The arthropod fauna of the Upper Carboniferous rocks of the Maritime Provinces. *Mem. geol. Surv. Canada*, **286**, 110 pp.

DRIESE, S. G. & FOREMAN, J. L. (1992) Paleopedology and paleoclimatic implications of late Ordovician vertic paleosols, Juniata Formation, southern Appalachians. *J. sedim. Petrol.*, **62**, 71–83.

FRANCIS, J.E. (1986) The calcareous paleosols of the basal Purbeck Formation (Upper Jurassic), Southern England. In: *Paleosols Their Recognition and Interpretation* (Ed. by V. P. Wright), pp. 112–138. Princeton University Press, Princeton, NJ.

FRITZ, P., BINDA, P.L., FOLINSBEE, F.E. & KROUSE, H.R. (1971) Isotopic composition of diagenetic siderites from Cretaceous sediments in Western Canada. *J sedim. Petrol.*, **41**, 282–288.

GIBLING, M.R., CALDER, J. H., RYAN, R., VAN DE POLL, H.W. & YEO, G.M. (1992) Late Carboniferous and Early Permian drainage patterns in Atlantic Canada. *Can J. Earth Sci.*, **29**, 338–352.

GIBSON, P.J. (1989) Petrology of two Tertiary oil shale deposits from Queensland, Australia. *J. geol. Soc. Lond*, **146**, 319–331.

GIBSON, P. J. SHAW, H. F. & SPIRO, B. (1990) The formation of eogenetic siderites in Tertiary lacustrine oil shales of Queensland, Australia. *Int. Ass. Sedimentol. Conf. Nottingham, England, August 1990*, abstract, pp. 494.

GOUDIE, A.S. (1983) Calcrete. In: *Chemical Sediments and Geomorphology* (Ed. by A. S. Goudie & K. Pye), pp. 93–131. Academic Press, London.

GRAY, M.B. & NICKELSEN, R.P. (1989) Pedogenic slickensides, indicators of strain and deformation processes in redbed sequences of the Appalachian foreland. *Geology*, **17**, 72–75.

HART, B.S. (1990) *Sedimentology and stratigraphy of the Upper Cretaceous Cardium Formation, in Northwestern Alberta and adjacent British Columbia*, PhD thesis, University of Western Ontario, London, Ont.

KANTOROWICZ, J.D. (1990) Lateral and vertical variations

in pedogenesis and other early diagenetic phenomena, Middle Jurassic Ravenscar Group, Yorkshire. *Proc. Yorks. geol. Soc.*, **48**, 61–74.

LECKIE, D.A. & FOSCOLOS, A.E. (1986) Paleosols and Late Albian sea level fluctuations: preliminary observations from the northwestern British Columbia foothills. In: *Current Research, Part B. Geol. Surv. Canada, Pap.*, **86–1B**, 429–441.

LOTT, G.K. & HUMPHREYS, B. (1992) The stratigraphy and petrology of Middle Jurassic (Ravenscar Group) sediments in cored boreholes from the north Yorkshire coast. *Proc. Yorks. geol. Soc.*, **49**, 23–40.

MACHETTE, M.N. (1985) Calcic soils of the southwestern United States. In: *Soils and Quaternary Geology of the Southwestern United States* (Ed. by D. L. Weide), *Spec. Pap. geol. Soc. Am.*, **203**, pp. 1–23.

MATSUMOTO, R. & IIJIMA, A. (1981) Origin and diagenetic evolution of Ca–Mg–Fe carbonates in some coalfields of Japan. *Sedimentology*, **28**, 239–259.

MOZLEY, P.S. (1989) Relation between depositional environment and elemental composition of early diagenetic siderite. *Geology*, **17**, 704–706.

MOZLEY, P.S. & CAROTHERS, W.W. (1992) Elemental and isotopic composition of siderite in the Kuparuk Formation, Alaska: effect of microbial activity and water/sediment interaction on early pore-water chemistry. *J. sedim. Petrol*, **62**, 681–692.

MOZLEY, P.S. & WERSIN, P. (1992) Isotopic composition of siderite as an indicator of depositional environment. *Geology*, **20**, 817–820.

NETTERBERG, F. (1980) Geology of south African calcretes: 1. Terminology. description, macrofeatures, and classification. *Trans. geol. Soc. S. Africa*, **83**, 255–283.

NETTLETON, W.D. & PETERSON, F.F. (1983) Aridisols. In: *Pedogenesis and Soil Taxonomy II. The Soil Orders* (Ed. by L. P. Wilding, N. E. Smeck & G. F. Hall), pp. 235–243. Elsevier, Amsterdam.

PEARSON, M.J. (1979) Geochemistry of the Hepworth Carboniferous sediment sequence and origin of the diagenetic iron minerals and concretions. *Geochim. Cosmochim. Acta*, **43**, 927–941.

POSTMA, D. (1981). Formation of siderite and vivianite and the pore-water composition of a recent bog sediment in Denmark. *Chem. Geol.*, **31**, 225–244.

POSTMA, D. (1982) Pyrite and siderite formation in brackish and freshwater swamp sediments. *Am. J. Sci.*, **282**, 1151–1183.

PYE, K. (1984) SEM analysis of siderite cements in intertidal marsh sediments, Norfolk, England. *Mar. Geol.*, **56**, 1–12.

PYE, K., DICKSON, J.A.D., SCHIAVON, N., COLEMAN, M.L. & COX, M. (1990) Formation of siderite–Mg—calcite–iron sulphide concretions in intertidal marsh and sandflat sediments, north Norfolk, England. *Sedimentology*, **37**, 325–343.

RETALLACK, G. J. (1981) Fossil soils: indicators of ancient terrestrial environments. In: *Palaeobotany, Palaeoecology and Evolution, Vol* 1 (Ed. by K. J. Niklas), pp. 55–102. Praeger Publishers, New York.

ROY, R.L. & MORRIS, W.A. (1983) A review of paleomagnetic results from the Carboniferous of North America: the concept of Carboniferous geomagnetic field horizon markers. *Earth planet. Sci. Lett.* **65**, 167–181.

SALOMONS, W., GOUDIE, A.S. & MOOK, W.G. (1978) Isotopic composition of calcrete deposits from Europe, Africa, and India. *Earth surf. Process.*, **3**, 43–57.

SPENCER, E. (1925) On some occurrences of spherulitic siderite and other carbonates in sediments. *J. geol. Soc. Lond.*, **81**, 667–705.

TASSE, N. & HESSE, R. (1984) Origin and significance of complex authigenic carbonates in Cretaceous black shales of the Western Alps. *J. sedim. Petrol.*, **54**, 1012–1027.

TEALL, J. (1899) *Summary of Progress of the Geological Survey for 1898*. Geological Survey (UK) Memoir.

THIBAUDEAU, S.A. & MEDIOLI, F.S. (1986) Carboniferous thecamoebians and marsh foraminifera: new stratigraphic tools for ancient paralic deposits. *Geol. Soc. Am., Abstracts with Programmes*, **18**, 771.

TUCKER, M.E. (1981) *Sedimentary Petrology. An Introduction.* Blackwell Scientific Publications, Oxford.

(Manuscript received 7 September 1992; revision accepted 29 January 1993)

Reprinted from Sedimentology (1995) **42**, *161–190*

Early diagenesis and its relationship to depositional environment and relative sea-level fluctuations (Upper Cretaceous Marshybank Formation, Alberta and British Columbia)

J. L. McKAY, F. J. LONGSTAFFE and A. G. PLINT
Department of Earth Sciences, University of Western Ontario, London, Ontario, Canada, N6A 5B7

ABSTRACT

Early diagenesis of the Upper Cretaceous (late Coniacian to early Santonian) Marshybank Formation was controlled by depositional environment (composition of depositional water, Fe and organic content of the sediment, sedimentation rate, proximity to the shoreline) and influx of meteoric water related to relative sea-level fall.

Five depositional environments, each characterized by a distinct early diagenetic mineral assemblage, have been recognized. Offshore shelf sediments that were deposited in a dysaerobic environment are characterized by abundant framboidal pyrite and rare septarian concretions, composed of 'early' calcite and siderite. Intense sulphate reduction, promoted by the dysaerobic depositional water, was the primary influence on early diagenesis. Offshore shelf sediments deposited under aerobic conditions are characterized by abundant concretions, composed of two generations of siderite (S1 and S2). In this environment, methanogenesis, rather than sulphate reduction, was more important. Early diagenesis of the inner shelf sands was generally limited. However, in sands deposited proximal to the shoreline, mixing of marine and meteoric waters promoted crystallization of Fe-rich chlorite and siderite. The shoreface was characterized by dissolution of detrital minerals in the upper portion, and precipitation of kaolinite or illite/smectite in the lower portion. In the coastal plain environment, brackish water and early reducing conditions resulted in formation of abundant euhedral pyrite. Ankerite, rather than siderite, is the typical early diagenetic carbonate.

The $\delta^{18}O$ values of the earliest cements (i.e. 'early' calcite, siderite S1, inner shelf siderite) indicate crystallization from a low-^{18}O, marine-derived porewater. Assuming crystallization at 25°C, a $\delta^{18}O$ value of about −7‰ (SMOW) can be estimated for the seaway during Marshybank Formation time. Similar calculations for the overlying Dowling Member (Puskwaskau Formation) suggest that the $\delta^{18}O$ value of the seaway increased to about −4% (SMOW), consistent with its transgressive nature. Very low $\delta^{18}O$ values are exhibited by siderite S2. These results indicate crystallization during intermediate diagenesis (≥60°C) from meteoric water (≥ −15‰ SMOW) that entered the Marshybank Formation during sea-level lowstand.

INTRODUCTION

Berner (1981) recognized that the composition of depositional water, in particular sulphur and oxygen content, is a major factor controlling early diagenesis of a sediment. Other factors are also important, such as degree of bioturbation (Maynard, 1982), sedimentation rate (Curtis, 1977, 1978; Maynard, 1982; Gautier & Claypool, 1984; Spears, 1989; Scotchman, 1991), organic content (Maynard, 1982), grain size and sorting, and detrital composition (Ashby & Pearson, 1979; Stonecipher & May, 1989).

In addition to these parameters, the effect of relative sea-level fluctuation on early porewater composition and early diagenesis must also be considered. The textural, mineralogical and chemical changes that occur in marine carbonate sediments as a result of early meteoric diagenesis have been described in numerous studies (e.g. Allan & Matthews, 1977, 1982; Magaritz *et al.*, 1979; Videtich & Matthews, 1980). Fewer studies have examined the effects of early meteoric diagenesis on marine clastic sediments. Segall *et al.* (1987) recognized subaerial exposure of marine rocks based on variations in clay mineralogy. Subaerial exposure surfaces have been identified on the basis of zones of potassium feldspar depletion and kaolinite enrichment (e.g. Al-Gailani, 1980, 1981; Huggett, 1984; Emery *et al.*, 1990). Precipitation of low-^{18}O, early diagenetic minerals in marine units has also been attributed to influx of meteoric water (e.g. Hutcheon *et al.*, 1985; Longstaffe & Ayalon, 1987; Machemer & Hutcheon, 1988; Bloch *et al.*, 1989; Bloch, 1990; Hart *et al.*, 1992; Longstaffe *et al.*, 1992).

The stratigraphic study of Plint & Walker (1987) prompted our interest in the diagenetic history of the Marshybank Formation. They suggested that the upper surface of the Marshybank Formation was an erosion surface formed during relative sea-level fall. Despite strong stratigraphic evidence for inferring subaerial emergence, physical evidence such as roots and palaeosols are lacking, presumably having been removed by transgressive erosion. Therefore, it was necessary to search for other evidence of subaerial exposure that might have been preserved beneath the erosion surface. Our attention was focused on two main questions: (1) how did the depositional environment influence early diagenesis? and (2) can early diagenesis resulting from the influx of meteoric water be recognized?

The study area extends for about 270 km along the Foothills thrust belt in north-western Alberta and north-eastern British Columbia, and about 80 km eastward into the Plains of north-western Alberta (Fig. 1). Nine outcrops and four cores were studied in detail.

GEOLOGICAL BACKGROUND

The Upper Cretaceous (late Coniacian–early Santonian) Marshybank Formation is predominantly a siltstone–sandstone unit that conformably overlies the siltstones and shales of the Muskiki Formation (Fig. 2). In the Foothills the Marshybank Formation is erosionally overlain by the Dowling Member shales of the Puskwaskau Formation, and in the Plains by siltstones and oolitic ironstone of the Bad Heart Formation (Fig. 2). Plint *et al.* (1990) have recently revised the stratigraphic nomenclature of the Muskiki, Marshybank and Bad Heart formations.

Stratigraphy and sedimentology of the Marshybank Formation

Stratigraphy

The stratigraphy of the Marshybank Formation was discussed by Plint (1990, 1991) and Plint & Norris (1991). The formation is subdivided into 12 allostratigraphic units (A–L; Plint, 1990; Plint & Norris, 1991; Fig. 2). The lowermost unit (unit A) of the Marshybank Formation and all of the underlying Muskiki Formation were deposited during relative sea-level rise and together form a transgressive systems tract (Plint, 1990). Units B–L of the Marshybank Formation downlap eastward on to the top of unit A and are interpreted as a highstand systems tract (Plint, 1990). This highstand systems tract is divided into seven, fourth-order progradational packages, or 'parasequences' (units B, C+D, E, F, H, I+J, K+L; Plint, 1991). In general, successive parasequences are increasingly progradational. The tops of most parasequences are marked by erosion surfaces overlain by thin conglomerates.

The top of the Marshybank Formation is a regional erosion surface that progressively truncates underlying strata from north-west to north-east. This surface has an erosional relief of about 50 m (Plint & Walker, 1987; Plint *et al.*, 1990), and can be traced from north-western Alberta into northern Montana (Plint, 1991). This erosion surface is believed to be the product of both eustatic and tectonic effects (Plint *et al.*, 1993).

Sedimentology

For the purposes of this study, five depositional environments are recognized: (1) dysaerobic offshore marine shelf, (2) aerobic offshore marine shelf, (3) inner marine shelf, (4) shoreface and (5) coastal plain environments. These environments correspond to facies 1 (environment 1); facies 3 and 4 (environment 2); facies 5, 6, 7 and 10 (environment 3); facies 8 and 9 (environment 4);

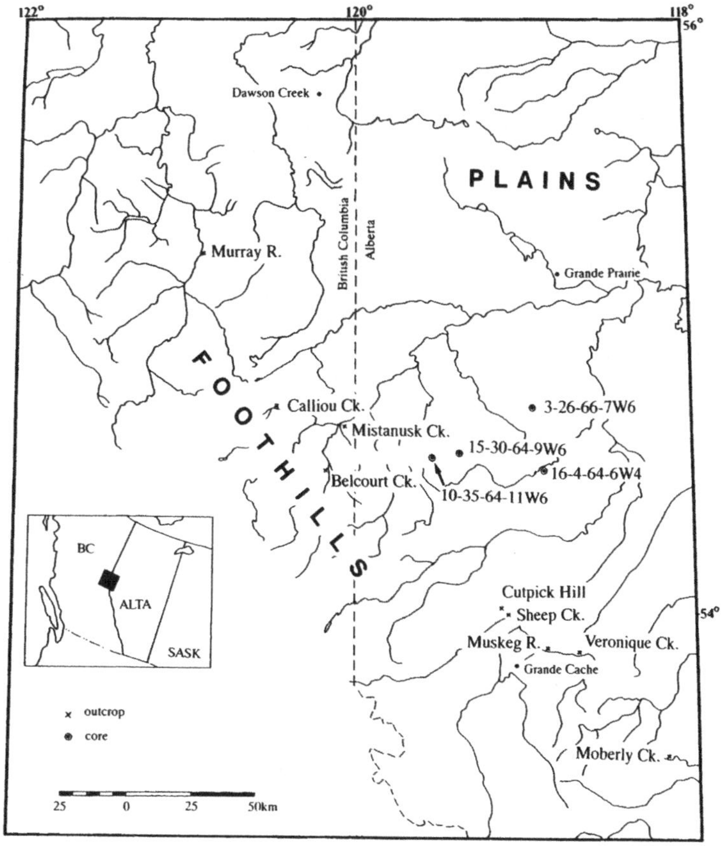

Fig. 1. Location of the study area and sample localities.

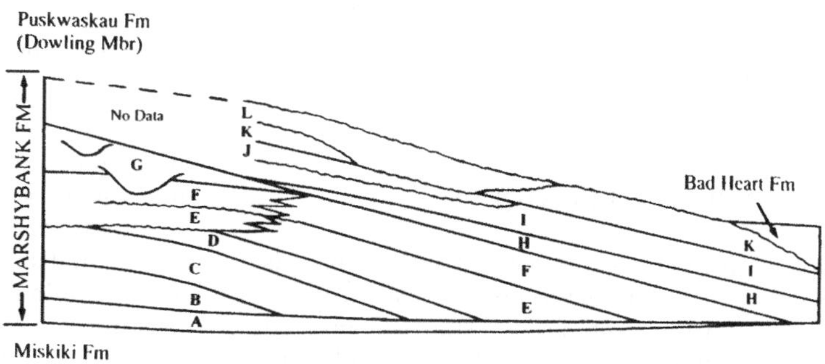

Fig. 2. Stratal geometry of the Marshybank Formation from west to east (left to right, respectively) after Plint (1990). In the Marshybank Formation, units A–D, the more basinward (i.e. eastern) portions of units E and F, and units H, I and K were deposited in the marine shelf environment. Shoreface sandstones characterize the western portion of units F and E, as well as units J and L. Coastal plain rocks are restricted to unit G.

and facies 13 and 14 (environment 5) of Plint & Norris (1991).

Within the study area, the dysaerobic offshore shelf is characterized by millimetre- to centimetre-scale bedded siltstones with negligible bioturbation. In contrast, pervasively bioturbated siltstones and sandy siltstones were deposited in the aerobic offshore shelf environment. According to Plint & Norris (1991), both dysaerobic and aerobic offshore facies were deposited more than 60 km from the shoreline. The presence (or lack) of bioturbation was controlled by the depth at which the pycnocline intersected the seafloor (Plint & Norris, 1991). Inner shelf facies are bioturbated silty sandstones and hummocky cross-stratified sandstones that gradationally overlie the offshore shelf facies. These rocks were deposited a few tens of kilometres seaward of the shoreface, in a storm-dominated environment (Plint & Norris, 1991). The shoreface environment (units L and J and the western portion of units E and F; Fig. 2) is characterized by swaley cross-stratified and parallel-laminated sandstones that were deposited in middle and upper shoreface environments. The upper shoreface is only preserved at the top of unit F in the western portion of the study area (Fig. 2). It is characterized by a thin (<20 cm), organic-rich 'black' sandstone and an underlying siliceous sandstone with abundant root traces. The coastal plain (unit G; Fig. 2) consists of interbedded sandstones, siltstones, shales and minor coals that represent lacustrine, lagoonal and floodplain deposits. Localized trough cross-bedded fluvial sandstones are also present.

Burial history

Maximum burial and subsequent uplift of the Marshybank Formation occurred in the early Eocene (Kalkreuth & McMechan, 1984). In the central portion of the study area (i.e. cores 11-35-64-11W6 and 15-30-64-9W6; Fig. 1) the estimated maximum depth of burial was about 4·3 km, which decreased to about 2·9 km in the west as a result of earlier uplift during the Laramide orogeny. Kalkreuth & McMechan (1984), based on measured vitrinite reflectance data, calculated that at the time of maximum burial the palaeogeothermal gradient in the study area was 27°C km^{-1}. Using this gradient and a palaeosurface temperature of 20°C (Kalkreuth & McMechan, 1984), the estimated maximum burial temperatures reached in the Marshybank Formation were approximately 100–140°C.

ANALYTICAL TECHNIQUES

Optical petrography, X-ray diffraction analysis, scanning electron microscopy and backscattered electron imaging were used to study the mineralogy of samples. The chemistry of whole rock samples was determined by X-ray fluorescence. Quantitative chemical analysis of minerals was performed using a JEOL JXA-8600 electron microprobe equipped with a backscattered electron detector and energy-dispersive system. Stable isotopic analysis of carbonate minerals and diagenetic quartz was also carried out. Preparation of carbonates for isotopic analysis involved hand-crushing of samples, sieving to obtain the <45-µm size fraction (Walters *et al.*, 1972), and overnight treatment in 3% sodium hypochlorite to remove organic matter. The samples were then reacted with anhydrous phosphoric acid under vacuum for the following reaction times and temperatures: (i) calcite for 1 h at 25°C and (ii) siderite for 2 h at 150°C. Isotopic compositions were calculated using the following phosphoric acid–CO_2 fractionation factors: (i) calcite (25°C)=1·01025 (Sharma & Clayton, 1965) and (ii) siderite (150°C)=1·00771 (Rosenbaum & Sheppard, 1986). Diagenetic quartz was separated following the method of Ayalon & Longstaffe (1992). Oxygen was extracted from this quartz using the BrF_5 method of Clayton & Mayeda (1963), and quantitatively converted to CO_2 over red-hot graphite. All isotopic data are reported in the standard δ-notation relative to the belemnite *Belemnitella americana* from the PeeDee Formation (PDB) for carbon (Craig, 1957) and standard mean ocean water (SMOW) for oxygen (Craig, 1961). An oxygen-isotope CO_2–H_2O fractionation factor of 1·0412 at 25°C was used to calibrate the mass spectrometer reference gas. The reproducibility of $\delta^{13}C$ and $\delta^{18}O$ values for a standard calcite analysed with each batch of samples was ±0·08‰ and ±0·11‰, respectively. Similar results for a standard siderite were ±0·20‰ and ±0·11‰, respectively. Reproducibility of duplicate analyses of unknowns was better than ±0·2‰ for both oxygen and carbon.

RESULTS

Detrital mineralogy

Marshybank marine sandstones and silty sandstones are classified as sublitharenites and litharenites after Folk (1974). Their detrital mineralogy is dominated by quartz (56–80%), with lesser chert (7–39%), dolomite (up to 13%, av. 2%) and shale fragments (2–13%). Feldspar is generally minor (av. 3%), although K-feldspar can constitute up to 8% in shoreface sandstones that outcrop in the northern portion of the study area. Marine sandstones are dominantly fine grained (125–250 µm), moderately well sorted and contain on average 7% matrix. Porosity, based on thin-section petrography, reaches a maximum of about 6% in shoreface sandstones.

Coastal plain sandstones and silty sandstones are classified as litharenites and lithic wackes (i.e. sublitharenites and litharenites with >15% matrix; Dott, 1964) with a detrital mineralogy similar to that of marine sandstones. However, quartz and chert are less abundant (24–57% and 5–12%, respectively) and both dolomite (up to

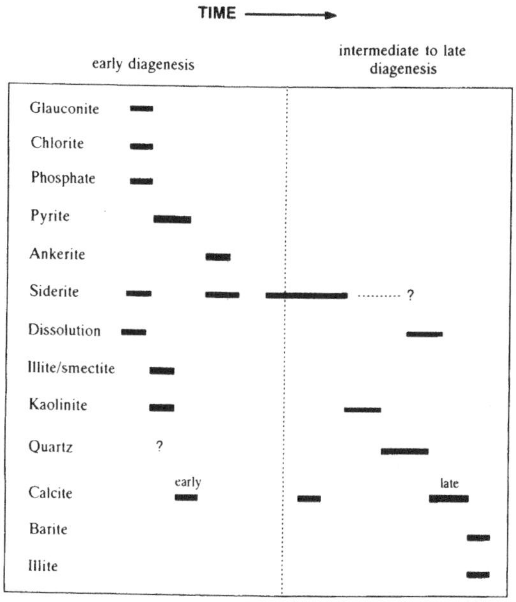

Fig. 3. Generalized paragenetic sequence for diagenesis of the Marshybank Formation. Many of the early diagenetic minerals are characteristic of a particular depositional environment (see Table 1). For example, glauconite and chlorite do not occur within the same sample, although both are early diagenetic and both pre-date pyrite. A question mark indicates that exact timing of precipitation with respect to other diagenetic minerals could not be determined. For example, the third stage of siderite crystallization, which began during early diagenesis, could have continued throughout intermediate diagenesis.

20%; av. 4%) and matrix (7–30%, av. 15%) are more abundant. The majority of the matrix is detrital illite and illite/smectite. These sandstones are fine- to medium-grained and more poorly sorted than the marine sandstones. Porosity is generally negligible.

The detrital mineralogy of siltstones is similar to the sandstones, except for the presence of abundant illitic matrix. Rare ooids are also present in offshore shelf sediments; these were originally phosphatic, although most have been replaced by calcite or siderite. Matrix- and clast-supported conglomerates are composed of well-rounded chert, shale and quartz clasts, as well as minor sideritic and phosphatic intraclasts.

Diagenetic mineralogy

A generalized paragenetic sequence for diagenesis of the Marshybank Formation is shown in Fig. 3. Early diagenesis was characterized by precipitation of glauconite, phosphate, pyrite, various clay minerals (Fe-rich chlorite, illite/smectite and kaolinite) and carbonates ('early' calcite and siderite). Unique assemblages of early diagenetic minerals, characteristic of the five depositional environments represented in the Marshybank Formation, are summarized in Table 1.

Precipitation of siderite appears to have continued during intermediate diagenesis, followed by precipitation of a second generation of kaolinite, as well as quartz, 'late' calcite, illite and barite during relatively late diagenesis. These diagenetic minerals (with the exception of siderite) are ubiquitous throughout the formation, reflecting the influence of basin-wide processes during relatively late burial and subsequent uplift. Relatively late diagenesis of the Marshybank Formation is similar to that of other Cretaceous sandstones located on the western margin of the Western Canada Sedimentary Basin. Readers are referred to Longstaffe & Ayalon (1987, 1991), Ayalon & Longstaffe (1988), Tilley & Longstaffe (1989) and Longstaffe *et al.* (1992) for further discussion. The discussion that follows focuses instead on early diagenesis of the Marshybank Formation.

Phosphate

Minor phosphate occurs within nodules, as laminae within ooids and as a pore-filling material, and is commonly enclosed by early diagenetic siderite. The occurrence of phosphate is closely associated with marine flooding surfaces (i.e. transgression and maximum flooding surfaces), although not every flooding surface is characterized by phosphate.

Glauconite

Pore-filling and peloidal glauconite is present in trace quantities in the Marshybank Formation. It occurs primarily in the offshore shelf siltstones and sandy siltstones.

Fe-rich chlorite

Fe-rich chlorite occurs as discontinuous grain-coatings that precipitated prior to siderite and pyrite (Fig. 4a). The occurrence of Fe-rich chlorite is restricted to inner shelf sandstones that were deposited proximal to the shoreline (i.e. Cutpick Hill and core 15-30-64-9W4; Fig. 1).

Pyrite

Pyrite is most abundant in the offshore shelf samples (up to 8%, av. 2·6%). In this

Table 1. Marshybank Formation depositional environments and associated detrital and early diagenetic phases.

Depositional environment	Sulphides	Carbonates	Clay minerals	Paragenetic sequence*
Dysaerobic offshore shelf	Abundant framboidal pyrite	Rare septarian concretions composed of a dominantly calcite interior (zone 1) and a siderite exterior (zone 2) Siderite is Fe-poor (av. 70 mol% Fe)	Detrital illitic clays	pyrite calcite zone 1 siderite zone 2 siderite
Aerobic offshore shelf	Framboidal and euhedral pyrite	Abundant siderite concretions composed of two generations of siderite, an early Fe-rich phase (av. 82 mol% Fe) and a later Fe-poor phase (av. 68 mol% Fe)	Dominantly detrital illitic clays; however, glauconite is present in some bioturbated siltstones	phosphate and/or glauconite framboidal pyrite euhedral pyrite siderite S1 siderite S2
Inner shelf	Trace framboidal pyrite	Minor disseminated siderite Only one phase of siderite characterized by a relatively low oxygen-isotope values (+18·3 to +24·6‰ SMOW) and relatively high Mn (av. 0·95 mol%)	Dominantly detrital illitic clays; however, diagenetic Fe-chlorite occurs in sands deposited proximal to the shoreline (i.e. core 15-30-64-9W4 and Cutpick Hill)	Basinward: pyrite siderite Basin margin: Fe-chlorite siderite pyrite
Shoreface (a) Lower shoreface	No pyrite	Nil to minor disseminated siderite Two siderite phases may be observed although generally only one is present Siderite chemistry is highly variable	Abundant early diagenetic kaolinite, except in core 10-35-64-11W6 where diagenetic illite/smectite (I/S) is present Kaolinite has a variable texture (block, platy and poorly crystallized) I/S occurs as crinkly grain coatings Minor detrital clay	feldspar dissolution kaolinite or illite/smectite siderite
(b) Upper shoreface	No pyrite	No detrital or early diagenetic carbonates are present	Minor early diagenetic kaolinite and illite/smectite No detrital clays	dissolution kaolinite illite/smectite
Coastal plain	Abundant euhedral pyrite and lesser framboidal pyrite Weathered pyrite is only observed in one sample	Pore-filling and replacive Ca-rich, ankerite (up to 9%) Minor disseminated, spherulitic and 'microconcretionary' siderite Siderite is characterized by >1 mol% Mn. Abundant detrital dolomite	Dominantly detrital illitic clays Abundant Fe-poor detrital chlorite Compact and ragged, pore-filling kaolinite	pyrite (and kaolinite?) ankerite and/or siderite

*Early diagenesis only; earliest phase or process listed first.

environment, disseminated framboidal pyrite is most common (Figs 4a–c and 5c), although pyramidal pyrite overgrowths on framboids, scattered and clustered pyrite octahedra, cubic pyrite (Fig. 4b), and pyrite pseudomorphs after organic matter (Fig. 4d) are also observed. In general, pyrite crystallization preceded siderite formation (Figs 4b and 5c), except in some inner shelf sandstones where pyrite followed both siderite and Fe-rich chlorite formation (Fig. 4a).

Pyrite is also abundant in the coastal plain rocks (up to 4%, av. 1·1%). A wide variety of pyrite textures are observed (Fig. 4e–h), including pyrite octahedra, dodecahedrons, pyrite aggregates and framboidal pyrite. However, in contrast to the marine rocks, euhedral pyrite is most common. In one coastal plain sample, pyrite has a pitted texture (Fig. 4h) typical of weathered pyrite (Mermut *et al.*, 1985). In the coastal plain environment, pyrite is the earliest diagenetic mineral, with the possible exception of rare, earlier formed, spherulitic siderite.

Calcite

Three types of diagenetic calcite are recognized in the Marshybank Formation. The earliest generation (i.e. 'early' calcite) is pore-filling and comprises 9–28% of the interior (zone 1) of septarian concretions (Figs 5a,b and 6). Precipitation of 'early' calcite preceded the crystallization of siderite in the outer zone (zone 2) of septarian concretions. A later generation of calcite, which precipitated after 'early' calcite and siderite, fills fractures in the septarian concretions (Figs 5a,b and 6). The exact timing of vein calcite formation, with respect to later diagenetic phases, could not be determined. The third type of calcite (i.e. 'late' calcite) is pervasive throughout the Marshybank Formation. This calcite is pore-filling and also replaces feldspar, bioclastic material and ooids. 'Late' calcite precipitated after most other diagenetic phases, with the exception of barite, microcrystalline quartz and possibly illite.

Elemental and stable isotopic data for diagenetic calcites are given in Table 2. Elementally, the three generations of calcite are similar, with Ca content ranging from 92 to 98 mol%, Mg content from <1 to 4 mol% and Fe from <1 to 5 mol%. However, the various generations of calcite are isotopically distinct (Fig. 7). The $\delta^{13}C$ values of 'early' calcite range from $-15{\cdot}4$ to $-9{\cdot}7$‰ (PDB), while the $\delta^{18}O$ values range from $+14{\cdot}8$ to $+20{\cdot}2$‰ (SMOW) and generally decrease toward the margin of zone 1. 'Late' calcite has a highly variable isotopic composition ($\delta^{13}C=-23{\cdot}6$ to $-1{\cdot}6$‰ PDB and $\delta^{18}O=+6{\cdot}3$ to $+18{\cdot}4$‰ SMOW). The isotopic composition of septarian fracture calcite ($\delta^{13}C=-12{\cdot}4$ to $-10{\cdot}5$‰ PDB and $\delta^{18}O=+8{\cdot}0$ to $+9{\cdot}6$‰ SMOW) is similar to the majority of 'late' calcite samples (Fig. 7). This similarity suggests that septarian vein calcite and 'late' calcite may be the same generation of cement and that septarian fractures were not filled by calcite until relatively late in diagenesis.

Ankerite

Ankerite (up to 9%) is present only in coastal plain samples, where it occurs as a pore-filling cement and as a replacement of detrital dolomite (Fig. 5d). Ankerite precipitated during early diagenesis apparently after pyrite formation (Fig. 4g).

The ankerite *d*(104) XRD peak occurs at 2·92 Å, rather than the 2·90 Å diffraction that is typical of many ankerites. This shift results from the high Ca content of the ankerite (57–59 mol% Ca). The average chemical formula of the ankerite is $(Ca_{1{\cdot}160}, Mg_{0{\cdot}472}, Mn_{0{\cdot}010}, Fe_{0{\cdot}358})(CO_3)_2$. The isotopic composition of ankerite could not be determined because of its low abundance and occurrence with detrital dolomite.

Siderite

Diagenetic siderite occurs in all environments of the Marshybank Formation, with the exception of the upper shoreface. However, its abundance, habit, texture and composition vary considerably, apparently as a function of depositional environment. Four distinct groups have been recognized: (i) rare septarian concretions composed of siderite and 'early' calcite are characteristic of the dysaerobic offshore shelf siltstones; (ii) abundant siderite concretions are characteristic of aerobic offshore shelf rocks; (iii) minor, disseminated siderite occurs in inner shelf and shoreface sandstones; and (iv) minor 'microconcretionary', disseminated and rare spherulitic siderite is present in the coastal plain rocks. Each group is discussed separately in the following section. Elemental and isotopic data for all types of siderite are provided in Tables 3 and 4, respectively.

Septarian concretion siderite. Septarian concretions only occur in the non-bioturbated siltstones of the Marshybank Formation. Detailed analyses were performed on two septarian concretions

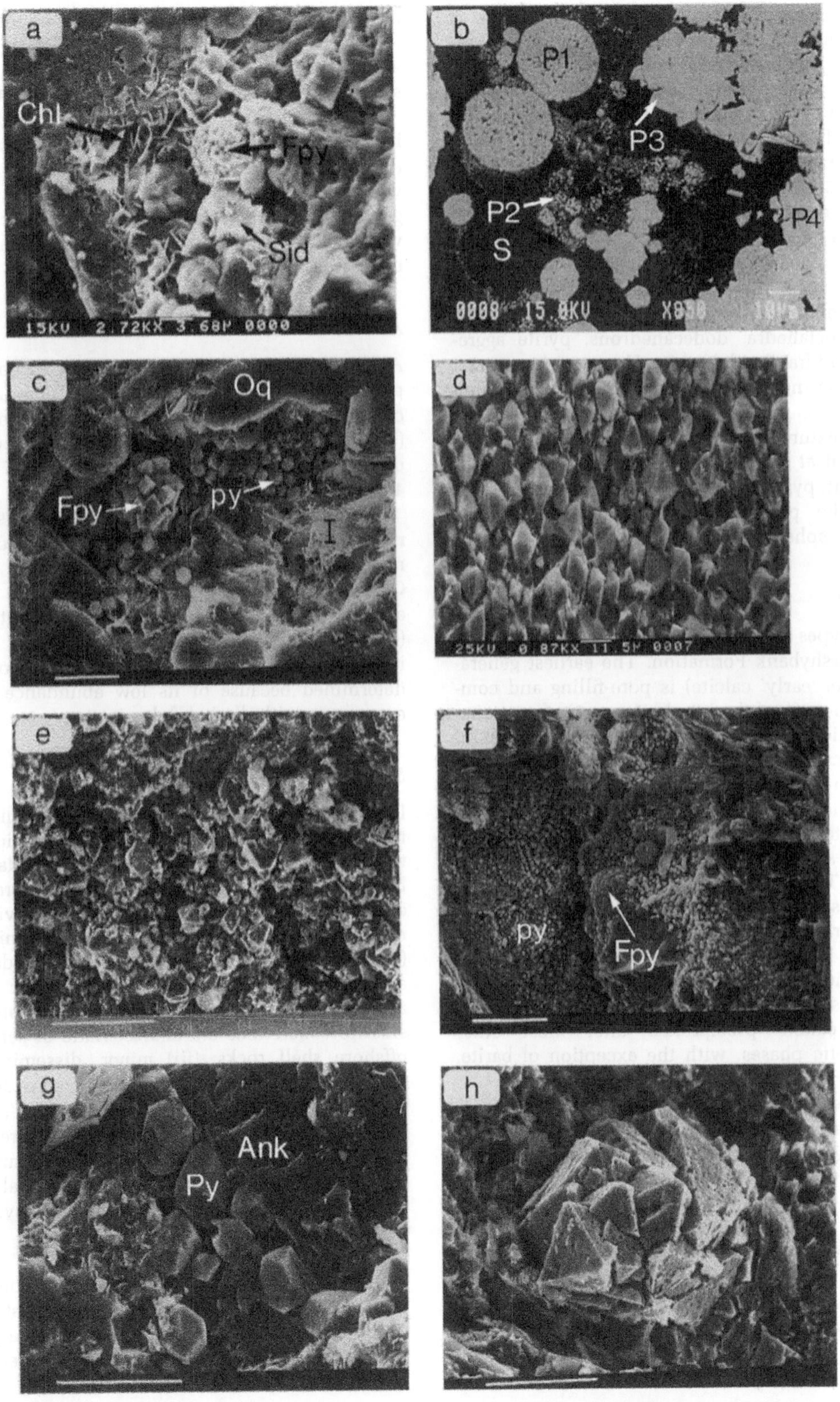
a
Chl
Fpy
Sid
b
P1
P3
P2
P4
S
c
Oq
Fpy
py
I
d
e
f
py
Fpy
g
Ank
Py
h

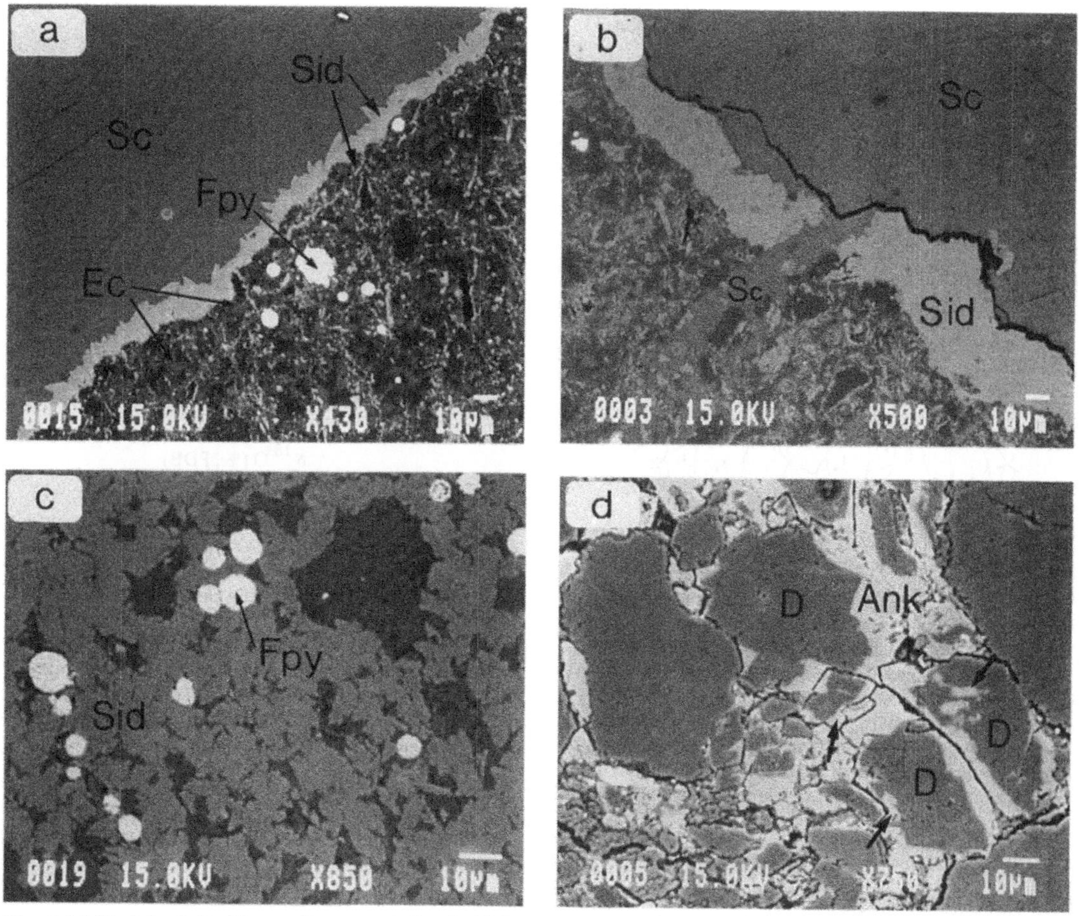

Fig. 5. (a) 'Early' calcite (Ec, medium grey), framboidal pyrite (Fpy, white) and siderite (Sid, light grey) within zone 1 of septarian concretion Mr-33. Zone 1 is cross-cut by a siderite-lined, calcite-filled septarian fracture (Sc). (b) Zone 1 in septarian concretion Mr-34 has two generations of calcite-filled septarian fractures (Sc). The earlier generation is siderite-lined (Sid), whereas the later generation is not. (c) Zone 2 of septarian concretions is characterized by abundant framboidal pyrite (Fpy) and later precipitated, although still early diagenetic, siderite (Sid, medium grey). (d) Ankerite (Ank)-cemented, sandy siltstone from the coastal plain environment. Note ankerite replacement (arrows) of detrital dolomite (D).

(Mr-33 and Mr-34) from the Muskeg River outcrop; the former was collected from the Dowling Member shale (Puskwaskau Formation), about 1 m above the top of the Marshybank Formation and the latter from about 36 m below the top of the Marshybank Formation. These concretions are

---

Fig. 4. Diagenetic pyrite from the marine (a–d) and coastal plain (e–h) environments. (a) The paragenetic sequence in this inner shelf sandstone is Fe-rich chlorite (Chl), siderite (Sid) and then framboidal pyrite (Fpy). (b) Four generations of pyrite occur in this sample. Early framboidal pyrite (P1), disseminated microcrystalline pyrite (P2), euhedral pyrite overgrowths (P3) and lastly blocky pyrite (P4). Pyrite crystallization was followed by the formation of siderite (S). (c) Framboidal pyrite (Fpy) composed of loosely intergrown pyrite cubes, and disseminated pyrite octahedra (py) (scale bar=10 µm). Later precipitated illitic clay (I) and quartz overgrowths (Oq) are also present. (d) Pyrite octahedra in a pyrite matrix. This texture was formed by pyrite replacement of organic material. (e) Large patches of euhedral pyrite are typical of the coastal plain environment (scale bar=40 µm). (f) A large mass of framboidal pyrite (Fpy) and non-framboidal, although somewhat ordered, pyrite octahedra (py) (scale bar=20 µm). (g) Pyrite dodecahedrons (Py) appear to be engulfed by ankerite (Ank) suggesting ankerite post-dates pyrite formation (scale bar=20 µm). (h) Weathered pyrite (scale bar=20 µm).

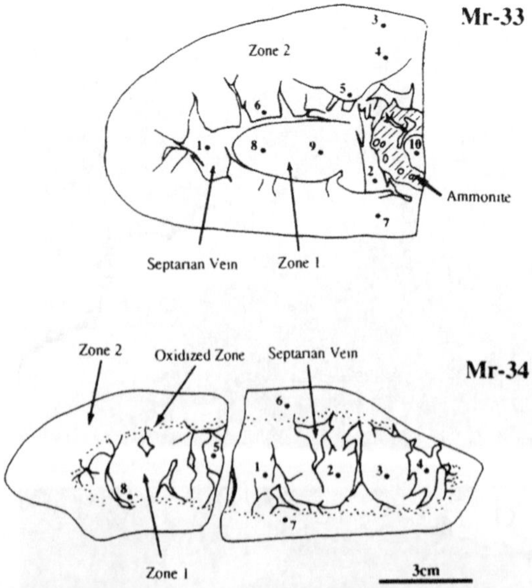

Fig. 6. Illustration of septarian concretions Mr-33 (Dowling Member, Puskwaskau Formation) and Mr-34 (Marshybank Formation). The concretions are divided into two zones: zone 1 is dominated by pyrite, 'early' calcite and minor siderite; zone 2 consists mainly of pyrite and siderite. In concretion Mr-34, the boundary between zones 1 and 2 is marked by an oxidized layer. Numbers indicate the positions of samples taken for stable isotopic analysis (data in Tables 2 and 4).

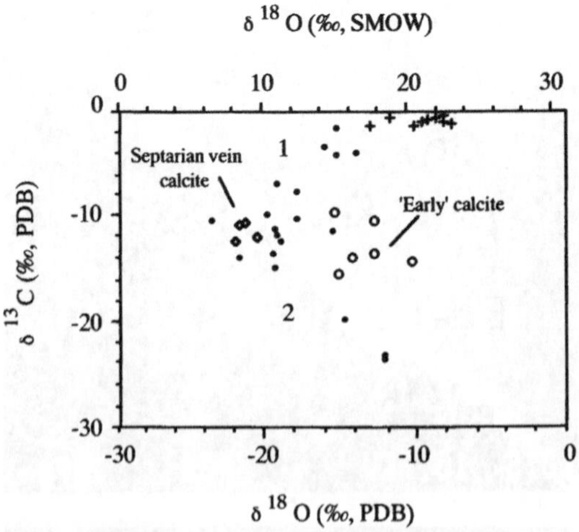

Fig. 7. Oxygen- versus carbon-isotope compositions of 'early' calcite (open circles), septarian vein calcite (open diamonds), 'late' calcite (small filled circles) and detrital dolomite (crosses). Two groups of 'late' calcite trend away from the main late calcite field. Group 1, which trends toward the isotopic field of detrital dolomite, were collected from the dolomite-rich coastal plain rocks. Dissolution of dolomite during late diagenesis influenced the composition of these samples. Group 2 samples were collected from the most basinward sample localities (i.e. cores 16-4-64-6W4 and 3-26-66-7W4; Fig. 1). It is believed that abiotic reactions, or possibly methane oxidation, was involved in the precipitation of this calcite, hence its low $\delta^{13}C$ values. 'Late' calcite data are discussed in more detail by McKay (1992).

ellipsoidal, relatively small (<20 cm long × <7 cm thick) and concentrically zoned (Fig. 6). The presence of septarian fractures suggests a very early diagenetic origin for these concretions (Raiswell, 1971; Gautier & Claypool, 1984; Astin, 1986; Scotchman, 1991).

The interior of the septarian concretions (zone 1; Fig. 6) contains abundant framboidal pyrite, 'early' calcite and up to 35% siderite (Fig. 5a,b). In concretion Mr-34 the outer margin of zone 1 is oxidized, but no oxidation rim was observed in concretion Mr-33. The exterior (zone 2; Fig. 6) contains pyrite and about 60–100% siderite (Fig. 5c).

Both zone 1 and zone 2 siderites are characterized by extensive substitution of Fe by Ca (11–19 mol%) and Mg (9–17 mol%); however, zone 2 siderite is slightly more Fe-rich (by 3–5 mol%) than zone 1 siderite in the same concretion. Zone 1 siderite has $\delta^{13}C$ values ranging from $-10{\cdot}5$ to $-8{\cdot}2$‰ (PDB), and $\delta^{18}O$ values ranging from $+15{\cdot}8$ to $+15{\cdot}9$‰ (SMOW). In concretion Mr-34, the $\delta^{13}C$ values of zone 1 siderite increase toward the outer edge of that zone. Zone 2 siderite has higher $\delta^{13}C$ ($-3{\cdot}8$ to $-0{\cdot}4$‰ PDB) and higher $\delta^{18}O$ ($+22{\cdot}4$ to $+27{\cdot}6$‰ SMOW) values than zone 1 siderite (Fig. 8). In concretion Mr-33, the isotopic values of zone 2 siderite decrease toward the concretion edge. One sample from the oxidized zone in concretion Mr-34 has an isotopic composition that is intermediate between zone 1 and zone 2 siderite (Fig. 8), perhaps suggesting that a mixture of the two siderite generations is present.

Non-septarian concretions. Non-septarian, concretionary siderite preferentially occurs in the aerobic offshore shelf rocks as single concretions and as laterally continuous sheets. Individual concretions are generally small (<10 cm thick) and round to ellipsoidal in shape, while concretionary sheets are up to 15 cm thick. Concretionary siderite is commonly located at or near the top of coarsening-up cycles, suggesting that coarser sediments provided a preferred pathway for porewaters from which the siderite precipitated (Hallam, 1969; Raiswell, 1971; Curtis *et al.*, 1975;

Table 2. Isotopic (‰) and elemental (mol%) data for diagenetic calcite.*

Sample	$\delta^{13}C$ (PDB)	$\delta^{18}O$ (SMOW)	$\delta^{18}O$ (PDB)	CaO	MgO	MnO	FeO	No. of analyses
'Early' calcite†								
Mr-34-1	−10·4	17·5	−13·0					
Mr-34-2	−15·4	15·3	−15·2					
Mr-34-3	−13·9	16·1	−14·3					
Mr-34-8	−9·7	14·8	−15·6					
Mr-33-9	−14·3	20·2	−10·1	95·23	1·71	0·23	2·84	9
Mr-33-8	−13·6	17·5	−13·0					
Septarian vein calcite								
Mr-34-4	−10·5	8·8	−21·5					
Mr-34-5	−10·9	8·4	−21·9	95·03	1·18	0·20	3·59	7
Mr-33-1	−12·0	9·6	−20·7					
Mr-33-2	−12·4	8·0	−22·2	94·30	1·69	0·28	3·73	4
'Late' calcite								
Vc-8	−15·0	10·8	−19·5					
Mr-41	−11·2	10·9	−19·4					
Mr-29	-	-	-	97·00	0·81	0·09	2·10	6
Mr-28	−10·2	12·5	−17·8					
Mr-27	−11·3	14·9	−15·5	97·06	0·70	0·10	2·13	8
Mr-16	−13·6	10·6	−19·7					
Bc-24	−4·0	16·4	−14·2	95·04	1·11	0·47	3·39	8
Bc-21	-	-	-	97·64	0·84	0·17	1·35	4
Bc-13	−4·1	15·1	−15·3					
Bc-8	−9·8	10·3	−20·0	95·00	1·23	0·07	3·70	14
Bc-11	−6·8	10·9	−19·5	97·17	1·12	0·09	1·62	9
Bc-7	-	-	-	95·81	1·02	0·04	3·13	2
Bc-3	-	-	-	94·73	1·11	0·14	4·01	2
Cal-19	−1·6	15·1	−15·4					
Cal-14	−3·3	14·3	−16·1					
Cal-13	-	-	-	97·92	1·24	0·10	0·74	7
Mur-8	−7·7	12·3	−18·0					
Mur-7	−10·4	6·5	−23·9	97·47	0·62	0·00	1·92	6
Mur-3	−13·9	8·4	−21·9					
11W6-19	−11·7	10·9	−19·4	96·55	1·00	0·21	2·24	7
11W6-12	-	-	-	96·43	0·93	0·37	2·26	4
9W6-11	-	-	-	97·37	0·66	0·16	1·80	5
9W6-8	−12·3	11·3	−19·0	96·67	0·82	0·10	2·42	10
9W6-5	-	-	-	97·39	0·69	0·14	1·78	2
7W6-4	−23·6	18·4	−12·2					
7W6-3	-	-	-	91·69	3·06	0·01	5·24	4
7W6-3	-	-	-	94·91	1·25	0·01	3·83	2
6W6-11	−19·9	15·6	−14·8					
6W6-10	−23·0	18·3	−12·2	96·34	0·79	0·20	2·67	2
6W6-10				94·54	4·10	0·17	1·19	3

*Mr-33 samples were collected from the Dowling Member (Puskwaskau Formation). All other samples are from the Marshybank Formation.

†Electron microprobe analyses of the 'early' calcite are less reliable because of the fine grain size (<10 μm) of this calcite.

Mr, Muskeg R.; Vc, Veronique Ck.; Bc, Belcourt Ck.; Cal, Calliou Ck.; Mur, Murray R.; 11W6, 10-35-64-11W6; 9W6, 15-30-64-9W6; 7W6, 3-26-66-7W6; 6W6, 16-4-64-6W4.

Raiswell & White, 1978; Pearson, 1979; Irwin, 1980). A similar origin is probable for the massive (up to 1 m thick) sideritized conglomerates present at the top of the Marshybank Formation.

Concretions are composed of 13–76% grain-coating (Fig. 9a) or pore-filling (Fig. 9b) siderite. The grain size, which ranges from micritic (1–10 μm) to medium grained (250–500 μm), is controlled by the grain size of the host sediment. Siderite cementation progressed through several stages, beginning with grain-coating and passive pore-filling, then displacive pore-filling and

Table 3. Elemental data (mol%) for siderite.*

Sample†	CaO	MgO	MnO	FeO	No. of analyses	Sample†	CaO	MgO	MnO	FeO	No. of analyses
					Septarian concretions (dysaerobic offshore shelf)						
	Zone 1 siderite						Zone 2 siderite				
Mr-33m	19·38	16·64	0·37	63·60	2	Mr-33t	14·52	15·49	0·46	69·53	6
						Mr-33b	17·97	15·40	0·45	66·19	8
Mr-34b	14·30	11·46	0·43	73·81	4	Mr-34t	11·38	9·26	0·29	79·07	1
					Siderite concretions (aerobic offshore shelf)						
	Siderite S1						Siderite S2				
7W6-3	11·37	8·84	0·48	79·32	3	7W6-3	13·86	20·30	0·20	65·64	6
6W6-3	9·59	3·63	2·12	84·66	7	6W6-3	11·64	15·35	0·20	72·82	4
9W6-8	10·00	0·44	0·69	88·87	8	9W6-8	5·89	21·20	0·63	72·28	5
Bc-11T	11·65	2·43	0·27	85·65	5	Bc-11T	15·84	24·66	0·17	59·33	7
Bc-11B						Bc-11B	14·73	16·63	0·52	68·12	11
Bc-9	10·33	2·84	0·28	86·56	3	Bc-9	16·57	22·78	0·02	60·63	4
Bc-8T	10·58	3·28	0·58	85·56	6	Bc-8T					
Bc-8Bt	11·11	3·16	0·64	85·10	6	Bc-8Bt	9·20	22·95	0·22	67·63	2
Bc-8Bm	9·21	5·07	0·58	85·15	2	Bc-8Bm	15·22	19·53	0·23	65·02	2
Bc-8Bb	10·48	3·12	0·66	85·75	3	Bc-8Bb	10·08	19·74	0·30	69·88	5
Bc-7T	9·65	3·94	0·68	85·73	3	Bc-7T	16·99	22·51	0·04	60·46	4
Bc-3T						Bc-3T	14·25	19·28	0·16	66·32	7
Bc-3B						Bc-3B	16·38	14·83	0·17	68·61	11
Cal-27Tt						Cal-27Tt	15·97	20·37	0·27	63·40	6
Cal-27Tm	11·95	11·46	0·28	76·32	6	Cal-27Tm	14·65	19·52	0·32	65·51	1
Cal-27Tb	10·89	10·87	0·25	77·99	8	Cal-27Tb	17·53	17·37	0·22	64·88	3
Cal-27Bm	10·44	10·01	0·25	79·30	3	Cal-27Bm					
Cal-27Bb	8·42	10·87	0·29	80·42	2	Cal-27Bb					
Cal-25Tt	10·25	6·82	0·43	82·50	2	Cal-25Tt	12·13	12·46	0·11	75·30	1
Cal-25Tb						Cal-25Tb	11·67	9·28	0·43	78·63	2
Cal-25Bt	10·35	4·68	0·50	84·47	1	Cal-25Bt	9·45	9·98	0·65	79·92	3
Cal-25Bb	11·31	4·61	0·50	83·58	2	Cal-25Bb	12·87	10·66	0·77	75·69	3
Mis-19	15·63	14·92	0·11	69·35	6	Mis-19	14·48	27·83	0·12	57·57	4
					Siderite (inner shelf)						
	Early siderite						Later siderite				
6W6-10	7·55	21·79	0·82	69·84	2	6W6-10					
9W6-13	7·66	14·59	1·25	76·50	6	9W6-13					
9W6-11	4·11	20·45	1·14	74·27	5	9W6-11	11·45	17·02	0·90	70·62	2
9W6-5	3·57	21·01	0·72	74·70	6	9W6-5					
11W6-19	4·76	24·59	1·24	69·41	5	11W6-19	12·70	25·72	0·55	61·03	11
					Siderite (shoreface)						
Mr-27	1·31	0·06	0·59	98·05	5	Mr-27	13·65	16·70	0·40	69·25	6
Mr-29	1·04	0·03	0·56	98·37	5	Mr-29	4·84	26·69	0·62	67·85	5
					Siderite (coastal plain)						
	Early siderite						Later siderite				
Bc-24	11·10	2·43	2·20	84·24	10	Bc-24					
Bc-21	5·42	2·61	2·84	89·13	3	Bc-21	7·22	23·75	1·32	67·71	6
Cal-13	4·70	0·77	3·70	90·83	4	Cal-13	4·11	20·75	1·38	73·75	2
11W6-12	4·62	24·20	1·58	69·60	8						
	8·81	24·89	1·58	64·72	3						

*Septarian concretion Mr-33 was collected from the Dowling Member (Puskwaskau Formation). All other samples are from the Marshybank Formation.
†T and B refer to samples from the top and bottom (respectively) of a concretion and t, m and b refer to the top, middle and base (respectively) of a particular thin section.

finally replacive growth. Because of the nature of siderite cementation, estimation of original sediment porosity by assuming that the amount of carbonate precipitated equals porosity at the time of precipitation (Raiswell, 1971; Oertel & Curtis, 1972; Gautier, 1982) is not valid. However, relatively early growth of these concretions is suggested by preservation of trace fossils and

Table 4. Isotopic data (‰) for siderite.*

Sample†	$\delta^{13}C$ (PDB)	$\delta^{18}O$ (SMOW)	$\delta^{18}O$ (PDB)	% Siderite	Remarks
Septarian concretions (dysaerobic offshore shelf)					
Mr-33-3	−3·8	22·4	−8·3	61	zone 2
Mr-33-4	−1·2	27·6	−3·2	63	zone 2
Mr-33-5	−1·6	27·5	−3·3	100	zone 2
Mr-33-7	−2·6	24·7	−6·0	100	zone 2
Mr-34-2	−10·5	15·8	−14·7	7	zone 1
Mr-34-3	−8·2	15·9	−14·6	16	zone 1
Mr-34-6	−0·4	25·4	−5·4	72	zone 2
Mr-34-7	−1·3	25·4	−5·3	61	zone 2
Mr-34-8	−5·8	20·8	−9·8	20	oxidized zone; mixture of zone 1 and 2 siderites
Siderite concretions (aerobic offshore shelf; two generations of siderite)					
7W6-4	−16·4	25·4	−5·4	54	sideritized horizon
7W6-3	−16·4	26·3	−4·5	43	sideritized burrows
6W6-4i	−11·6	26·8	−4·0	62	sideritized horizon
6W6-4iii	−9·8	27·0	−3·8	59	sideritized horizon
9W6-8	−6·3	23·1	−7·6	36	sideritized lag
9W6-1	−8·3	24·1	−6·7	13	sideritized burrows
11W6-1	−4·8	23·2	−7·5	23	sideritized burrows
Bc-11	−3·3	22·1	−8·5	25	sideritized lag
Bc-8ii	−11·8	23·0	−7·7	28	siderite horizon
Bc-8iii	−12·0	23·3	−7·3	28	siderite horizon
Bc-8iv	−10·6	22·5	−7·9	27	siderite horizon
Bc-3i	−8·9	14·1	−16·3	27	siderite horizon
Bc-3ii	−8·0	14·3	−16·1	35	siderite horizon
Bc-3iii	−5·5	16·9	−13·6	32	siderite horizon
Bc-3iv	−2·6	20·5	−10·1	59	siderite horizon
Bc-3v	−3·9	18·4	−12·1	49	siderite horizon
Mb-1i	−8·4	24·1	−6·6	51	siderite concretion
Mb-1ii	−5·9	27·7	−3·2	64	siderite concretion
Mb-1iv	−5·4	28·9	−2·0	76	siderite concretion
Mb-1v	−8·5	23·9	−6·8	45	siderite concretion
Mur-8	−8·7	20·9	−9·7	26	sideritized conglomerate
Mr-18	−10·7	25·0	−5·7	54	sideritized conglomerate
Cal-27iii	−6·4	26·6	−4·2	64	sideritized conglomerate
Cal-27iv	−8·0	26·6	−4·2	68	sideritized conglomerate
Cal-25i	−6·7	22·1	−8·5	59	sideritized conglomerate
Cal-25ii	−6·6	21·9	−8·7	46	sideritized conglomerate
Cal-25iii	−6·7	21·3	−9·3	55	sideritized conglomerate
Cal-25iv	−6·8	23·6	−7·1	50	sideritized conglomerate
Cal-25v	−7·9	24·8	−5·9	48	sideritized conglomerate
Siderite (inner shelf)					
6W6-10	−6·6	24·6	−6·1	11	small siderite nodule
9W6-14	−8·3	19·9	−10·7	40	thin (mm-scale) sideritized layer
9W6-13	−9·3	18·4	−12·2	2	disseminated siderite
Siderite (shoreface)					
11W6-5	−9·4	18·3	−12·2	6	disseminated siderite; one generation of siderite
Mr-27	−12·1	27·0	−3·8	22	disseminated siderite; two generations of siderite
Mr-28	−12·3	26·7	−4·1	19	disseminated siderite; two generations of siderite
Siderite (coastal plain)					
Bc-24 nod	−5·2	22·4	−8·3	54	small siderite nodule; two generations of siderite

*With the exception of septarian concretion Mr-33 all samples were collected from the Marshybank Formation. Mr-34 was collected from the Dowling Member (Puskwaskau Formation) ≈1 m above the Marshybank Formation.
†Roman numerals indicate subsamples taken vertically (top to bottom) through a concretion or concretionary layer.

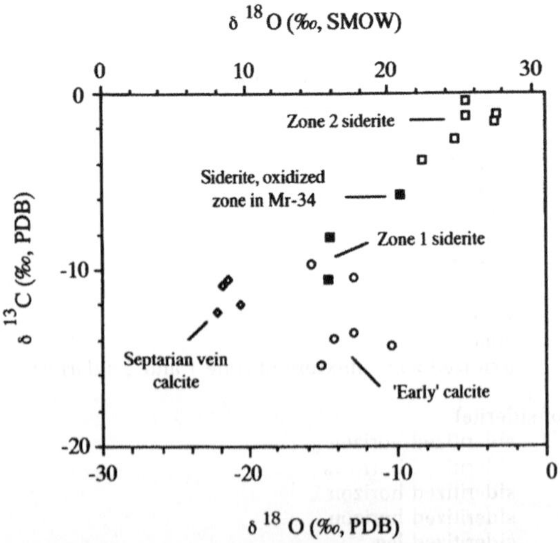

Fig. 8. Oxygen- versus carbon-isotope compositions of septarian concretion carbonates ('early' calcite, open circles; zone 1 siderite, filled squares; zone 2 siderite, open squares; septarian vein calcite, diamonds). One siderite sample from the oxidized zone in concretion Mr-34 appears to be a mixture of zone 1 and zone 2 siderites.

point contacts between detrital grains within the concretions.

The presence of two discrete varieties of siderite is suggested by a diffraction peak doublet at the siderite *d*(104) position (Fig. 9c). Electron microprobe analysis and backscattered electron imaging confirmed the presence of early Fe-rich siderite (S1) and paragenetically later Fe-poor siderite (S2) that commonly replaces siderite S1 (Fig. 9d). There is only one sample in which Fe-poor siderite apparently precipitated prior to Fe-rich siderite. In all samples pyrite formation preceded siderite formation.

Both generations of siderite exhibit extensive substitution of Fe by Ca (6–18 mol%) and Mg (up to 15 mol% in S1 and 28 mol% in S2); nevertheless, they are chemically distinct (Fig. 10). Siderite S1 is characterized by relatively high Fe (69–89 mol%, av. 82 mol%), and the Ca content (8–16 mol%) commonly exceeds Mg content (<1–15 mol%). Siderite S2 has significantly lower Fe (58–80 mol%, av. 68 mol%), higher Mg (9–28 mol%), and can have higher Ca (up to 18 mol%), although Mg content exceeds Ca content. The Mn concentration in both siderites is generally low (<1 mol% Mn).

The $\delta^{13}C$ and $\delta^{18}O$ values for the siderite concretions range from −16·4 to −3·3‰ (PDB) and +14·1 to +28·9‰ (SMOW), respectively. These values represent mixtures of siderites S1 and S2; the two generations could not be physically or chemically separated. The $\delta^{18}O$ values increase basinward from an average of 19·2‰ (SMOW) at Belcourt Creek to 26·9‰ (SMOW) in core 16-4-64-6W6 (Fig. 1). In many concretions there is a positive correlation among $\delta^{13}C$, $\delta^{18}O$ and the percentage of siderite S1 versus S2 (calculated using the *d*(104) XRD peak intensities). Using data from concretion Bc-3, the isotopic compositions of siderites S1 and S2 were calculated by plotting the isotopic composition of the siderite mixture versus the percentage of siderite S1 and S2. The resulting $\delta^{13}C$ and $\delta^{18}O$ values of siderite S1 are +3‰ (PDB) and +25‰ (SMOW), respectively (r^2=0·98 and 0·93, respectively). The $\delta^{13}C$ and $\delta^{18}O$ values of siderite S2 are −14‰ (PDB) and +9‰ (SMOW), respectively (r^2=0·98 and 0·93, respectively). In other concretions, correlation between the abundance of S1 and S2, and the isotopic composition is less well defined, probably because of varying degrees of replacement and incorporation of siderite S1 and detrital material by siderite S2 and variable porewater chemistry.

Inner shelf and shoreface siderite. Siderite in inner shelf and lower shoreface sandstones commonly occurs as grain-coatings on detrital dolomite (Fig. 11a) and as disseminated crystals (Fig. 11b). Two generations of siderite were identified in some samples (Fig. 11c). In the most basinward inner shelf sandstones, siderite precipitation post-dated pyrite formation, while in sandstones located proximal to the shoreline, siderite crystallized after Fe-rich chlorite but before pyrite (Fig. 4a). In shoreface sandstones timing of siderite precipitation with respect to illite/smectite formation could not be determined.

Inner shelf siderite has $\delta^{13}C$ and $\delta^{18}O$ values of −9·3 to −6·6‰ (PDB) and +18·4 to +24·6‰ (SMOW), respectively, and is characterized by high Mn (av. 1·0 mol%) relative to siderite from the offshore environment (av. 0·4 mol%). In the lower shoreface at Muskeg River, two generations of siderite were identified (Fig. 11c). The earliest phase is ragged in appearance and the most chemically pure siderite observed in the Marshybank Formation (>98 mol% Fe), while the second generation of siderite coats the first siderite and is relatively Fe-poor (68–69 mol% Fe). The $\delta^{13}C$ and $\delta^{18}O$ values of this siderite (mixture of the two generations) are −12‰ PDB and +27‰ SMOW, respectively. Only one generation of disseminated

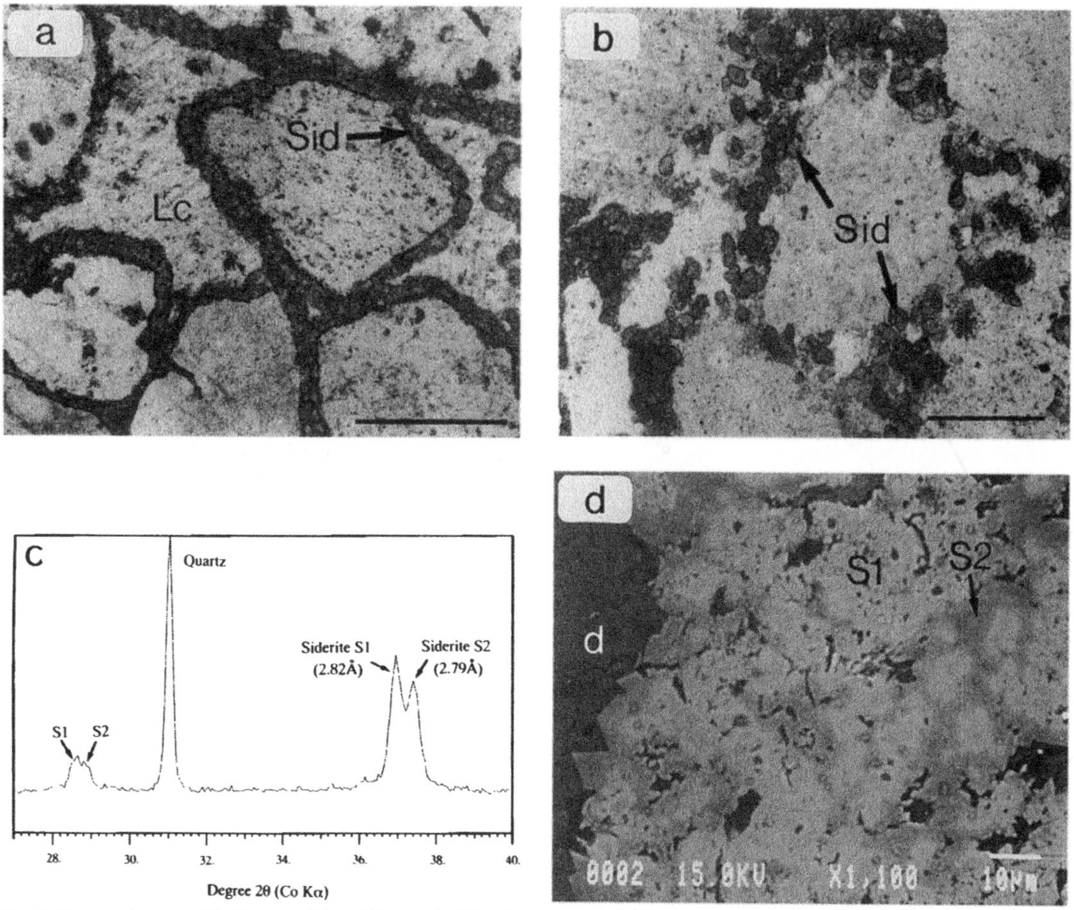

Fig. 9. Concretionary siderite, aerobic offshore shelf sediments. (a) Grain-coating siderite (Sid) and pore-filling 'late' calcite (Lc) in a sideritized horizon (scale bar=1 mm). Note the point contacts between detrital grains. (b) Pore-filling, granular to rhombic siderite (Sid) in a siderite concretion (scale bar=0·5 mm). (c) XRD showing the doubling of siderite diffraction peaks that suggests the presence of two generations of siderite (S1 and S2). (d) Backscatter photomicrograph showing two generations of siderite; a relatively Fe-rich variety (S1) and a later Mg-rich variety (S2). Siderite S2 has replaced both siderite S1 and detrital material.

siderite was present in the lower shoreface in core 10-35-64-11W6. Isotopically, this siderite is very similar to inner shelf siderite (δ^{13}C= − 9·4‰ PDB, δ^{18}O=+18·3‰ SMOW).

Coastal plain siderite. This siderite typically occurs as millimetre-scale 'microconcretions' (Fig. 11d), although minor disseminated and rare spherulitic siderite are also present. In general, siderite crystallization followed pyrite formation (Fig. 11d). The relative timing of siderite and ankerite precipitation is not known because these phases do not occur together.

The microconcretions comprise two generations of siderite, an earlier Fe-rich variety and later relatively Mg-rich variety. In contrast, only one generation of disseminated siderite is present. Both microconcretionary and disseminated siderites have a relatively high Mn content (1·6–3·7 mol% Mn). Isotopically, the siderite 'microconcretions' are indistinguishable from marine siderites (δ^{13}C= − 5·2‰ PDB and δ^{18}O=+22·4‰ SMOW). Insufficient disseminated siderite was available for isotopic analysis.

Kaolinite

Kaolinite is ubiquitous in the Marshybank Formation. It is least abundant in fine-grained marine lithologies and most abundant in shoreface sandstones, except in core 10-36-64-11W6

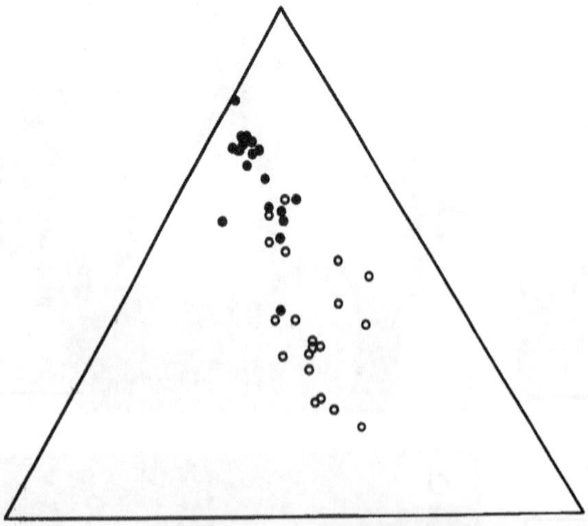

Fig. 10. Ternary plot comparing the elemental composition (mol%) of siderites S1 and S2 that comprise concretions in the aerobic offshore shelf facies. Within any one concretion, S1 (solid circles) is more Fe-rich than S2 (open circles).

where illite/smectite is dominant. The kaolinite is pore-filling; however, its texture varies depending on the environment of formation. Lower shoreface sandstones are characterized by uncompacted, pore-fillings of very fine-grained, blocky and platy kaolinite (Fig. 12a), while in upper shoreface and inner shelf sandstones, kaolinite is platy and may be somewhat ragged (Fig. 12b). In the coastal plain, kaolinite is typically ragged and tightly fills pore spaces (Fig. 12c), but minor platy kaolinite is also present (Fig. 12d).

Based on distribution and paragenetic relationships, two stages of kaolinite precipitation have been identified. The very fine-grained to blocky kaolinite, which is restricted to lower shoreface sandstones, is considered to be early diagenetic, having precipitated after feldspar dissolution and prior to any other diagenetic phase. Platy kaolinite, which is distributed ubiquitously throughout the formation, precipitated during late diagenesis following siderite, but prior to quartz formation.

Mixed-layer illite/smectite (I/S)

Two varieties of I/S were identified by X-ray diffraction. I/S with a distinct diffraction between 11·86 and 13·58 Å (i.e. 'peak' I/S; Fig. 13a) is characteristic of sandstones and occurs both as diagenetic, grain-coating crinkly plates (Fig. 13b), and as pore-filling to pore-bridging sheets. Precipitation of 'peak' I/S followed feldspar dissolution and siderite formation, but preceded quartz overgrowth formation.

Fine-grained samples or samples with abundant matrix are characterized by I/S with a broad *d*(001) diffraction (i.e. 'plateau' I/S; Fig. 13c) that lies between the *d*(001) diffractions of chlorite (14 Å) and illite (10 Å). 'Plateau' I/S appears as pasty-looking, ragged plates (Fig. 13d), suggesting a detrital origin.

Quartz

Diagenesis of the Marshybank Formation was dominated by quartz overgrowth formation (Fig. 4c), except where earlier siderite occluded most of the primary porosity. This quartz formed at or near maximum burial, after siderite, but before 'late' calcite and barite. Minor drusy quartz (Figs 11c and 13b) appears to have formed during relatively early diagenesis of inner shelf and shoreface sandstones.

The $\delta^{18}O$ values of eight samples of late diagenetic quartz range from +15·8 to +18·2‰ (SMOW).

DISCUSSION: DIAGENETIC HISTORY

Early diagenesis is considered here to include processes that occur shortly after deposition and at relatively low temperatures. As such, early diagenesis is strongly influenced by the chemistry of the depositional water and by microbially mediated processes.

Early diagenesis can be broadly described in terms of a depth zonation of chemical reactions

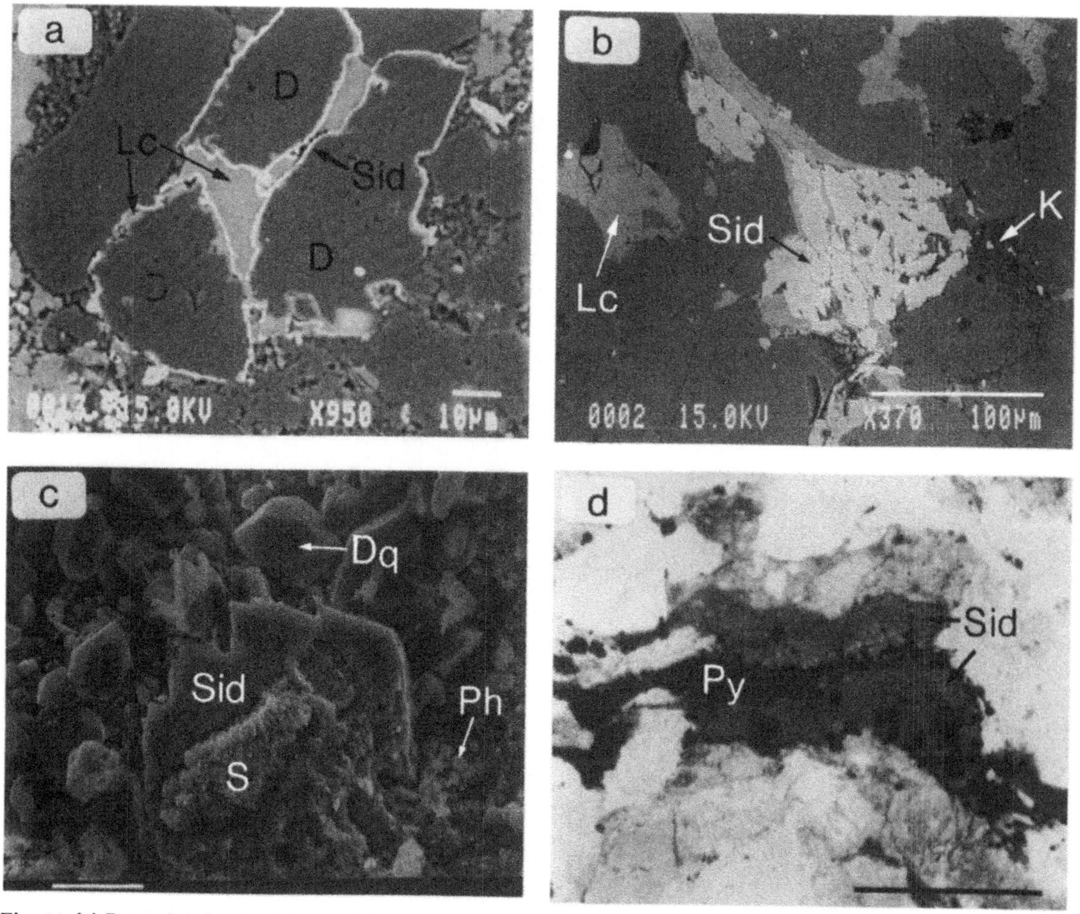

Fig. 11. (a) Detrital dolomite (D) coated by diagenetic siderite (Sid). This siderite habit is typical of inner shelf and lower shoreface sandstones. 'Late' calcite (Lc) is also present. (b) Siderite rhombs (Sid), pore-filling kaolinite (K) and 'late' calcite (Lc) in an inner shelf sandstone. (c) Two generations of siderite are present in the shoreface at Muskeg River. Ragged siderite (S) is overgrown by pristine siderite (Sid). Drusy quartz (Dq) and a phosphate mineral (Ph) are also present (scale bar=10 μm). (d) Early diagenetic pyrite (Py) and later formed (although still early diagenetic) 'micronodules' of siderite (Sid) are characteristic of the coastal plain environment (scale bar=1 mm).

(e.g. Curtis, 1977, 1978, 1980; Irwin *et al.*, 1977; Berner, 1981; Maynard, 1982; Coleman, 1985). The most important zones are: (i) the aerobic oxidation or oxic zone, (ii) the post-oxic or sub-oxic zone, (iii) the microbial sulphate reduction zone and (iv) the microbial methanogenic zone. In these zones, progressive degradation of organic matter, primarily by bacteria, is controlled by changing oxygen and sulphate content, decreasing reactivity and abundance of organic matter and Fe, and increasing temperature. At depth, the abiotic zone becomes important as temperature exceed 75°C (Irwin *et al.*, 1977).

During early diagenesis, the carbon-isotope composition of the porewater is commonly controlled by the method of organic degradation. For example, oxidation of organic matter yields bicarbonate with relatively low $\delta^{13}C$ values (about −25‰ PDB), whereas methanogenic processes (i.e. CO_2 reduction and acetate fermentation) yield bicarbonate with higher $\delta^{13}C$ values. Only during the very earliest stage of diagenesis will the $\delta^{13}C$ value of the depositional water be a significant factor, unless dissolution of early diagenetic carbonates and/or bioclastic material occurs during later diagenesis.

The oxygen-isotope composition of the earliest porewater is controlled primarily by the composition of the depositional water. Lowering of the porewater oxygen-isotope composition by up to

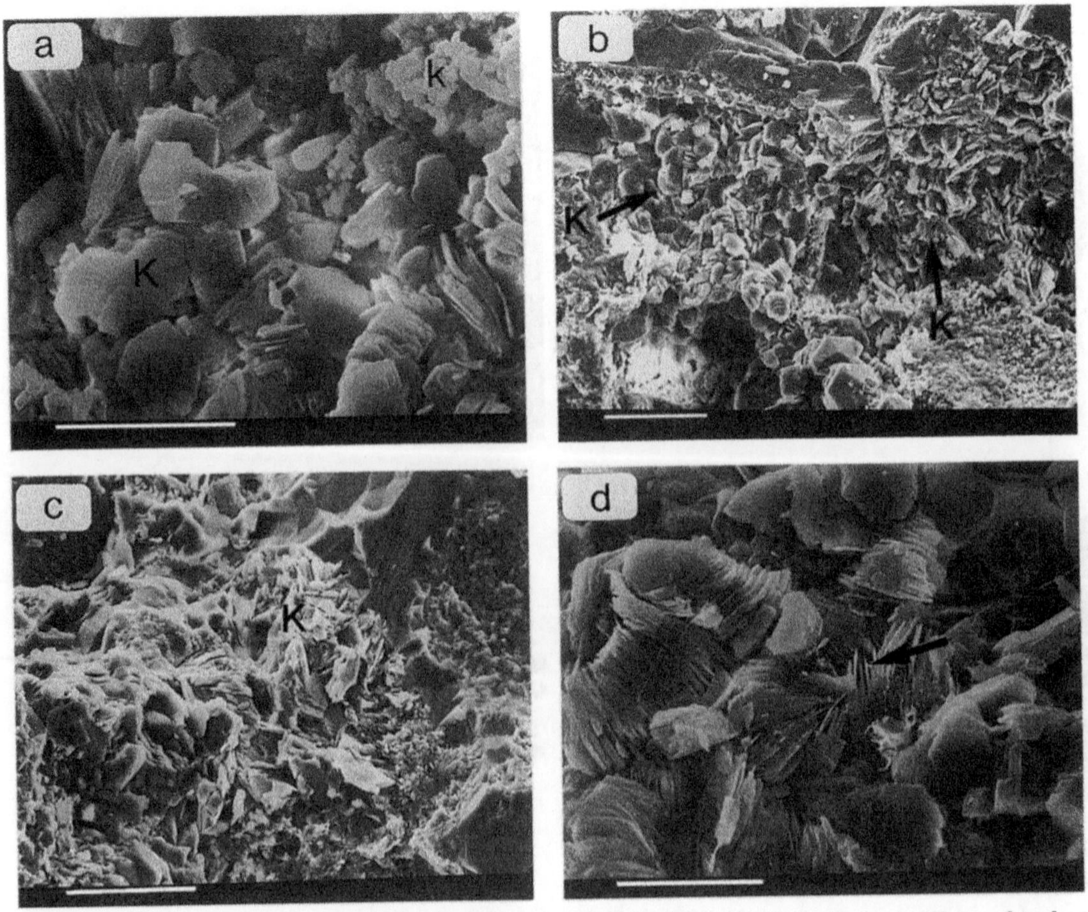

Fig. 12. (a) Mixtures of poorly crystallized kaolinite and blocky to platy kaolinite are characteristic of lower shoreface sandstones (scale bar=10 μm). (b) Platy, pore-filling kaolinite (K) in an inner shelf sandstone (scale bar=40 μm). (c) Compacted, pore-fillings of ragged kaolinite (K) are typical of coastal plain rocks. This texture suggests a detrital origin (scale bar=40 μm). (d) Rare, diagenetic platy kaolinite is also observed in the coastal plain environment. Note kaolinite dissolution (arrows) (scale bar=10 μm).

7‰ (SMOW) is possible within the first few 100 m of burial because of sediment–water interactions (Longstaffe, 1989). Such reactions can include the alteration of volcanic ash, with subsequent precipitation of ^{18}O-rich minerals (e.g. Lawrence *et al.*, 1975) and the release of ^{18}O-depleted oxygen from sulphate and organic matter during intense sulphate reduction (Sass *et al.*, 1991).

Because the Marshybank Formation was deposited in a range of environments (marine through coastal plain), the oxygen-isotope composition of the starting porewater may have ranged from 'normal' marine (0‰ SMOW) through meteoric (−15 to −10‰ SMOW; Rye & Sommer, 1980). To complicate the situation further, some evidence suggests that the Western Interior Seaway was '^{18}O-depleted', having oxygen-isotope compositions as low as −11 to −7‰ (SMOW) during parts of the Cretaceous (Kauffman, 1984; Wright, 1987; Eicher & Diner, 1989; Bloch, 1990; Kyser *et al.*, 1993).

Early diagenesis: dysaerobic offshore environment

The early diagenetic evolution of the dysaerobic, offshore siltstones is summarized in Fig. 14. Sedimentology indicates that these non-bioturbated siltstones were deposited under dysaerobic conditions. Accordingly, reactions in the oxic and post-oxic zones should be negligible, and highly

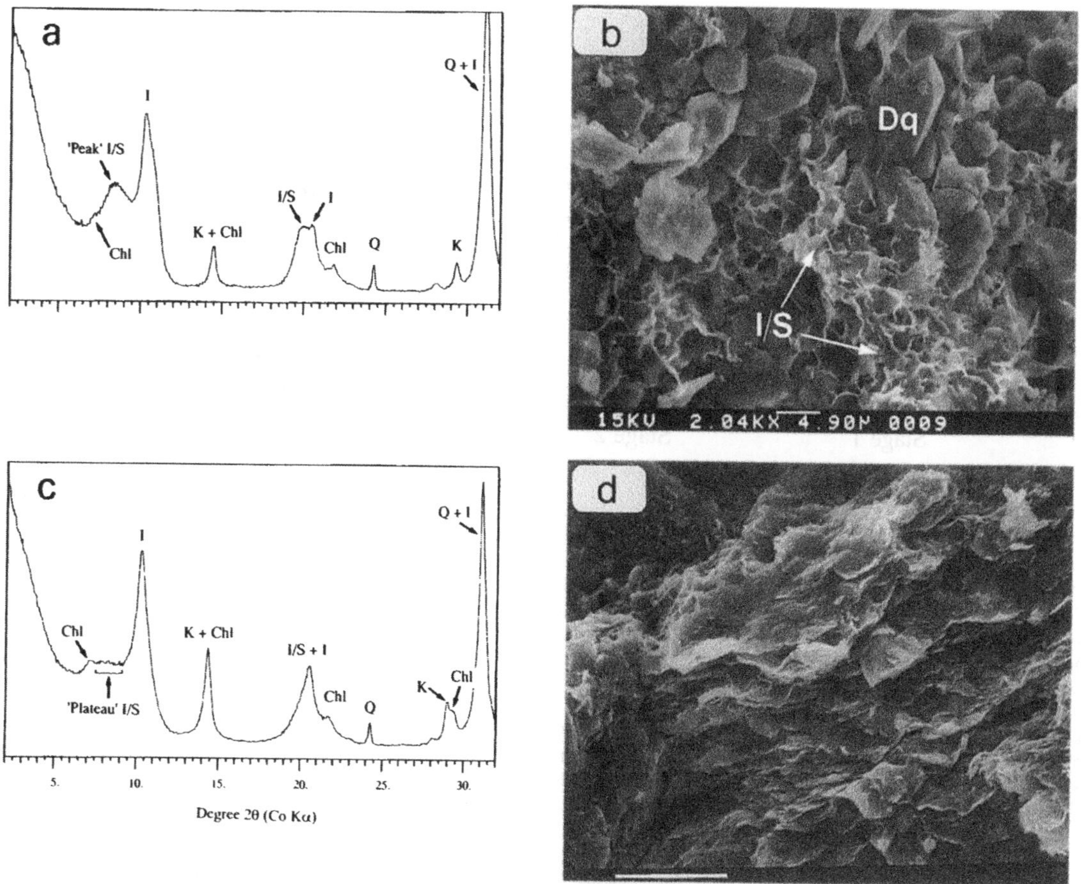

Fig. 13. (a) XRD pattern for the <2-μm size fraction of a shoreface sandstone. Distinct diffraction peaks are characteristic of 'peak' illite/smectite (I/S). Discrete illite (I), chlorite (Chl) and quartz (Q) are also present in this sample. (b) Pore-bridging illite/smectite (I/S) and drusy quartz (Dq) in a lower shoreface sandstone. (c) XRD pattern for the <2-μm size fraction of a marine siltstone. Detrital 'plateau' illite/smectite (I/S), as well as discrete illite (I), chlorite (Chl), kaolinite (K) and quartz (Q) are present. (d) Matted sheets of pasty to ragged detrital illite/smectite in a marine silty sandstone (scale bar=10 μm).

reactive Fe and organic matter would be available for intense sulphate reduction. The abundance of framboidal pyrite in the dysaerobic, offshore siltstones is consistent with this scenario (Stage 1a, Fig. 14).

As described earlier (see 'Septarian concretion siderite'), two septarian concretions were examined; Mr-34 from the Marshybank Formation, and Mr-33 from the base of the Dowling Member of the Puskwaskau Formation (Fig. 6). Intense sulphate reduction and pyrite formation resulted in bicarbonate supersaturation and precipitation of the 'early' calcite that forms zone 1 of these septarian concretions (Stage 1a, Fig. 14). The low $\delta^{13}C$ values of this 'early' calcite (−15·4 to −9·7‰ PDB) indicate that oxidation of organic matter during sulphate reduction was a major source of carbon, in addition to normal marine bicarbonate (0‰ PDB). The $\delta^{18}O$ values of this 'early' calcite (+14·8 to +20·2‰ SMOW) generally decrease from the interior to the margin of zone 1 in each concretion (Fig. 6). The values obtained for the interior samples should most closely approximate the depositional porewater, assuming precipitation temperatures equal to those at the sediment–water interface.

Crystallization from 'normal' marine water ($\delta^{18}O$=0‰ SMOW) would necessitate unrealistically high temperatures (i.e. 100 ± 25°C; calcite–water geothermometer of Friedman & O'Neil, 1977). Because the 'early' calcite is isotopically distinct from other diagenetic carbonates (Fig. 7),

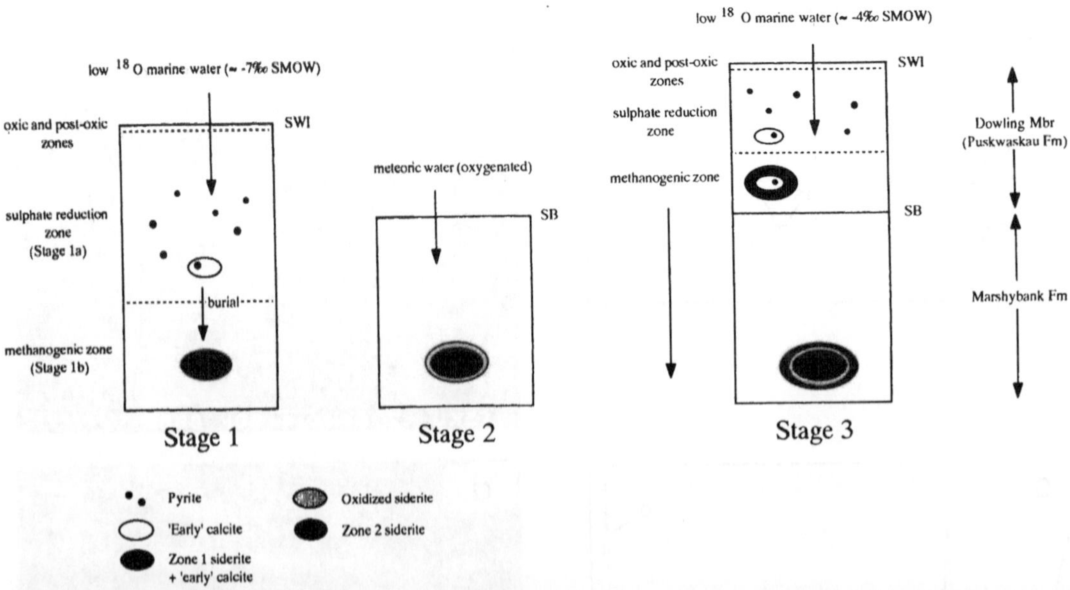

Fig. 14. Early diagenesis in the dysaerobic, offshore marine shelf environment. During Stage 1a, framboidal pyrite and 'early' calcite were precipitated in the sulphate reduction zone just below the sediment–water interface (SWI). As burial continued, the first generation of septarian fractures formed and sediments passed into the methanogenic zone where zone 1 siderite precipitated (Stage 1b). It is possible that during Stage 1b, influx of meteoric water began, related to relative sea-level fall. In Stage 2, sea-level lowstand occurred, a sequence boundary (SB) formed and zone 1 siderite was oxidized. During subsequent transgression and renewed burial (Stage 3), zone 2 siderite was precipitated, followed by formation of a second generation of septarian fractures. Diagenesis of the overlying Dowling Member shale (Puskwaskau Formation) was similar to Stage 1 diagenesis of the Marshybank dysaerobic, offshore shelf sediments.

it is also improbable that isotopic re-equilibration at higher temperatures occurred.

It appears likely that 'early' calcite formed from low-^{18}O porewater. An early diagenetic porewater with a δ^{18}O value of − 11‰ (SMOW) results in a more realistic formation temperature (25°C) for 'early' calcite. A brackish seaway at this time can explain the low δ^{18}O values. However, sediment–water interactions (e.g. precipitation of ^{18}O-rich minerals) and mixing with meteoric water can also yield low-^{18}O early diagenetic porewater (Mozley & Burns, 1993). Given the depositional environment (>60 km from the shoreline; Plint & Norris, 1991) and the relatively low permeability of siltstones, only sediment–water interactions are likely to have influenced the early porewater composition. Such reactions might explain the additional 3‰ decrease in the δ^{18}O values of 'early' calcite as precipitation progressed (i.e. towards the margin of zone 1). Assuming similar conditions of crystallization, the 3‰ difference in δ^{18}O values between 'early' calcite from the Marshybank Formation and the overlying Dowling Member may reflect a more ^{18}O-rich seaway during Dowling time. Higher δ^{18}O values are consistent with the transgressive nature of the Dowling sea.

Precipitation of 'early' calcite was followed by crystallization of siderite within the septarian concretions, once Fe was no longer being fixed by pyrite formation. Siderite precipitation requires porewater with very low dissolved sulphide (Pearson, 1979), high bicarbonate and high Fe^{2+} (Curtis & Spears, 1968). Degradation of organic matter is also essential since it is the principal source of bicarbonate (Curtis *et al.*, 1972; Hudson, 1978; Pearson, 1979; Coleman & Raiswell, 1981; Gautier, 1982; Raiswell, 1987). The extensive removal of Fe (and destruction of organic matter) during the early formation of abundant pyrite in the non-bioturbated siltstones may explain the low siderite content and the extensive substitution of Ca and Mg for Fe in this siderite.

The earliest siderite (zone 1 siderite, concretion Mr-34, Fig. 6) is characterized by low but increasing δ^{13}C values (− 10·5 to − 8·2‰ PDB). The relatively low values indicate that bicarbonate produced during sulphate reduction was still present in the porewater (Stage 1a, Fig. 14). However, the gradual enrichment in ^{13}C suggests introduction of bicarbonate produced by methanogenesis (Stage 1b, Fig. 14).

The $\delta^{18}O$ values of zone 1 siderite (+15·8 to +15·9‰ SMOW) are consistent with crystallization from −11‰ porewaters at 50°C (siderite–water geothermometer of Carothers *et al.*, 1988). However, if porewater $\delta^{18}O$ values had been further lowered by −5‰ (e.g. by introduction of meteoric water), crystallization temperatures similar to calcite result (about 25°C). Introduction of meteoric water would have occurred during relative sea-level fall at the end of Marshybank deposition (Stage 2, Fig. 14). Partial oxidation of zone 1 siderite in concretion Mr-34 (Fig. 6) probably occurred at sea-level lowstand after siderite precipitation had ceased (Stage 2, Fig. 14). By comparison, oxidation is not observed in concretion Mr-33 (Dowling Member, Fig. 6), which formed after sea-level lowstand.

Zone 2 siderite from both the Marshybank Formation and the Dowling Member septarian concretions has higher $\delta^{18}O$ values (+22·4 to +27·6‰ SMOW) than zone 1 siderite. Assuming initial siderite formation at about 25°C, the highest $\delta^{18}O$ values (inner portion on zone 2) are most simply explained by the influx of more ^{18}O-rich marine water following transgression of the Dowling sea (Stage 3, Fig. 14). This is consistent with the $\delta^{18}O$ values of 'early' calcite that suggest the Dowling sea was relatively ^{18}O-rich (about −4‰ SMOW) in comparison with the Marshybank sea. The decrease in zone 2 siderite $\delta^{18}O$ values toward the concretion margin probably reflects increasing temperature associated with progressively deeper burial. Other explanations, such as the influx of meteoric water and sediment–water interactions, do not correspond with the sedimentology, and normally $\delta^{18}O$ values increase during the later stages of burial diagenesis (Longstaffe *et al.*, 1992).

The $\delta^{13}C$ values of zone 2 siderite (−3·8 to −0·4‰ PDB) are higher than those of zone 1 siderite. This increase suggests that with burial, methanogenesis became the major carbon source, although Dowling seawater also may have supplied some bicarbonate (Stage 3, Fig. 14). The decrease in $\delta^{13}C$ values toward the margin of concretion Mr-33 may indicate that abiotic reactions began to supply low-^{13}C carbon as burial depth and temperature increased.

Both concretions Mr-34 and Mr-33 contain septarian fractures (Fig. 6). Formation of septarian fractures has been attributed to concretion dehydration and shrinkage (e.g. Raiswell, 1971) or tensile stress (Astin, 1986) related to transient overpressure development during shallow burial. Either way, these fractures indicate that concretion formation began prior to significant compaction and sediment dewatering. In septarian concretion Mr-34 (Marshybank Formation), two generations of calcite-filled fractures are observed. The first generation is lined by zone 1 siderite, and is cross-cut by the second generation of fractures which are filled only by calcite. By comparison, concretion Mr-33 (Dowling Member) contains only one generation of siderite-lined septarian fractures, similar to the first generation of fractures in the Marshybank concretion. Because the first generation of fractures in concretion Mr-34 pre-dates siderite formation (i.e. fractures are lined by siderite), it suggests that they formed during the initial burial of the Marshybank Formation (Stage 1, Fig. 14). The septarian fractures in concretion Mr-33 formed in a similar fashion shortly after deposition of the Dowling Member. We speculate that loading of the Marshybank Formation during Dowling Member deposition produced the second generation of fractures in concretion Mr-34. These fractures developed after siderite crystallization had ceased within the Marshybank Formation.

Based on their similar isotopic compositions (Fig. 7), the calcite which fills septarian fractures and 'late' calcite found throughout the Marshybank Formation are believed to be the same generation of cement. Therefore, it appears that septarian fractures were not filled by calcite until relatively late in diagenesis. Low $\delta^{18}O$ values of the 'late' calcite are typical of late diagenetic calcite in Mesozoic clastic rocks of western Alberta. This calcite formed during basin-wide influx of meteoric water that accompanied uplift of the Western Canada Sedimentary Basin during post-Laramide (Eocene ?) time (Longstaffe, 1993).

Early diagenesis: aerobic offshore environment

Sedimentological data suggest that the bioturbated, offshore marine sediments were deposited in an aerobic environment and, therefore, organic degradation in the oxic and post-oxic zones probably occurred. Glauconite and phosphate minerals formed at localities where relatively slow rates of sedimentation allowed sufficient reaction time within the post-oxic zone (Stage 1, Fig. 15).

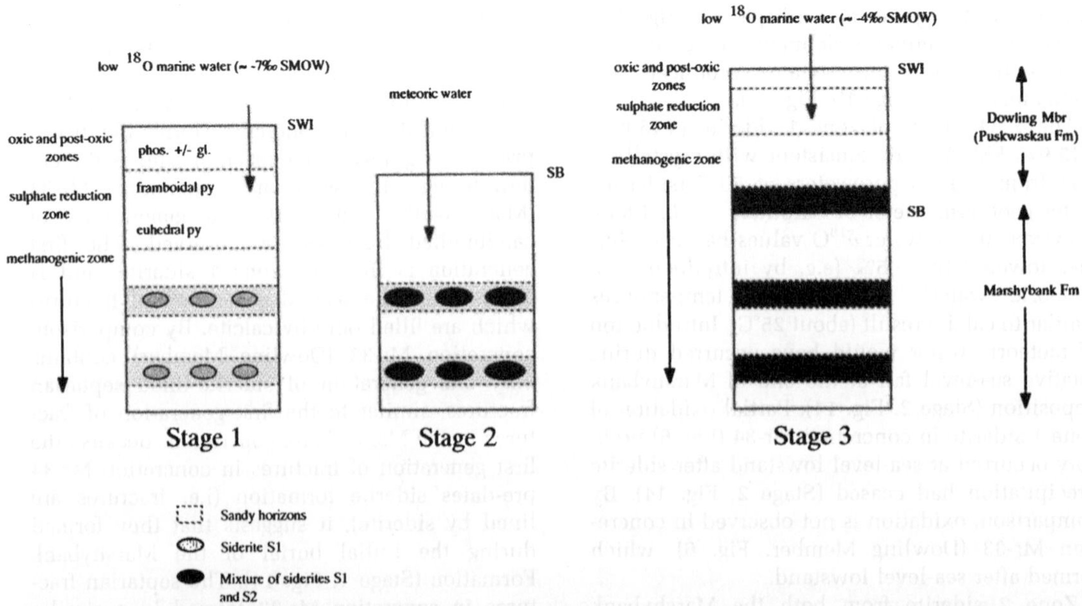

Fig. 15. Early diagenesis in the aerobic, offshore marine shelf environment. During Stage 1, phosphate and glauconite crystallized within the post-oxic zone, followed by pyrite formation within the sulphate reduction zone. Eventual depletion of porewater sulphide resulted in the switch from pyrite to siderite S1 formation in the methanogenic zone. Precipitation of siderite S1 occurred in relatively permeable horizons along which porewater movement was focused. Stage 2 was marked by influx of meteoric water related to relative sea-level fall. Influx of meteoric water was terminated in Stage 3 by transgression of the Dowling sea and renewed burial. The precipitation of siderite S2 may have been initiated during Stage 2; however, the bulk of siderite S2 precipitated during intermediate diagenesis (≥60°C), as meteoric water was expelled along permeable horizons (Stage 3).

Because of organic degradation in the oxic and post-oxic zones, less organic matter reached the sulphate reduction zone, resulting in lower sulphide supersaturation (Berner, 1978; Raiswell, 1982; Love *et al.*, 1983; Fisher, 1986; Spears, 1989). The change from framboidal to euhedral pyrite, and the succession of euhedral pyrite forms observed in the aerobic offshore environment (i.e. disseminated cubes to pyramidal overgrowth and lastly massive, cubic pyrite, Fig. 4b) are consistent with decreasing sulphide supersaturation with burial (Raiswell, 1982; Berner, 1984; Querol *et al.*, 1989; Dill & Kemper, 1990).

Relative to the dysaerobic sediments, the aerobic sediments are characterized by abundant siderite. This probably reflects that pyrite formation, which limited Fe availability for siderite formation in the dysaerobic environment, was less intense. In the aerobic environment siderite formed as concretions in permeable horizons along which porewater movement was focused. Concretions are characterized by two generations of siderite; early Fe-rich (S1) and later Mg-rich (S2) varieties. Concretions composed of two closely associated generations of siderite have been observed elsewhere (e.g. Pearson, 1974a,b, 1979; Curtis *et al.*, 1975, 1986; Postma, 1977; Jakobsen, 1988; Gibson, 1989; Mozley, 1989b; Larsen & Friis, 1991; Hart *et al.*, 1992; Mozley & Carothers, 1992). In an investigation of Carboniferous rocks from Yorkshire, Pearson (1979) suggested that earlier, Fe-rich siderite ($\delta^{13}C = +10$‰ PDB) precipitated during methanogenesis, whereas later Mg-rich siderite ($\delta^{13}C = -26$‰ PDB) formed during late diagenesis as the result of abiotic reactions. In contrast, Mozley & Carothers (1992) suggested that relatively Fe-rich siderite of the Kuparuk Formation (Alaska) precipitated in the post-oxic zone, whereas Mg-rich siderite crystallized during methanogenesis.

In the aerobic offshore facies of the Marshybank Formation, the calculated $\delta^{13}C$ value of siderite S1 (+3‰ PDB) is consistent with crystallization in the methanogenic zone (Stage 1, Fig. 15), although a contribution of marine bicarbonate cannot be ruled out. Interpretation of the oxygen-isotope compositions is more equivocal. If siderite S1 ($\delta^{18}O = +25$‰ SMOW) precipitated from a porewater derived from 'normal' marine water ($\delta^{18}O = 0$‰ SMOW), a crystallization temperature of about 60°C is required. However, much lower temperatures would have characterized the earliest stages of diagenesis. If siderite S1 crystallized at 25 ± 5°C, a $\delta^{18}O$ value of − 7 ± 1‰ (SMOW) is estimated for the early diagenetic porewater and hence the seaway. Because of the lower organic content of the aerobic offshore sediments, we speculate that ^{18}O depletion of the porewater from organic reactions during sulphate reduction was negligible. This may explain the apparent difference in early porewater compositions between dysaerobic (− 11‰ SMOW) and aerobic (− 7‰ SMOW) environments of the Marshybank Formation.

The $\delta^{13}C$ value calculated for siderite S2 (− 14‰ PDB) is not indicative of a particular carbon source. For example, microbial oxidation of organic matter during Fe- or S-reduction, or abiotic reactions during higher temperature diagenesis, all yield bicarbonate with such an isotopic composition. Soil-derived bicarbonate carried in meteoric water is another possible source. The $\delta^{18}O$ value calculated for siderite S2 is significantly lower than siderite S1 (+9 versus +25‰ SMOW) and suggests precipitation from a porewater that contained a sizable fraction of meteoric water, even if precipitation occurred at maximum burial (100–140°C). Meteoric water was introduced into the Marshybank Formation at least twice, once relatively early in diagenesis during relative sea-level fall, and then again during late diagenesis when regional uplift occurred.

Given that siderite S2 pre-dates crystallization of quartz during maximum burial, it is most probable that overprinting of the 'marine' isotopic signature (siderite S1) by a meteoric signature (siderite S2) resulted from relative sea-level fall at the end of Marshybank deposition (Stage 2, Fig. 15). However, the increase in bulk $\delta^{18}O$ values of concretionary siderite basinward (i.e. +19·2 to +26·9‰ SMOW) suggests that influx of meteoric water decreased basinward. A similar trend was observed for siderite from the Harmon Member of the Peace River Formation (Bloch *et al.*, 1989). Assuming a $\delta^{18}O$ value for meteoric water of − 15‰ (SMOW), the lowest possible crystallization temperature for siderite S2 is 60°C, indicating precipitation during burial (i.e. intermediate) diagenesis (Stage 3, Fig. 15). By maximum burial, when diagenetic quartz formed, porewater values had risen to − 2·5 ± 3·5‰ (SMOW) (using the quartz–water geothermometer of Clayton *et al.*, 1972). Similar porewater values have been obtained for other Upper Cretaceous sandstones in Alberta (Longstaffe, 1993). Subsequent to maximum burial, recharge by meteoric water during post-Laramide uplift once again lowered the porewater oxygen-isotope composition and low-^{18}O 'late' calcite formed.

Although its oxygen-isotope composition indicates crystallization from meteoric water, siderite S2 does not have the high Fe and Mn values typical of early diagenetic siderite of this origin (Carpenter *et al.*, 1988; Mozley, 1989a). Earlier removal of Fe and Mn by crystallization of pyrite and siderite S1 can explain this difference. That siderite S2 precipitated at all may have been directly related to the influx of Fe-rich meteoric water, given that Fe extraction during earlier diagenesis was extensive as the abundance of pyrite and siderite S1 suggest. Sediment–water interactions during burial diagenesis also probably supplied Fe to the porewater.

Early diagenesis: inner shelf environment

Early diagenesis in the most basinward shelf sandstones (i.e. core 16-4-64-6W6; Fig. 1) is similar to that of the aerobic offshore facies (i.e. pyrite followed by siderite; Table 1). However, pyrite and siderite are less abundant, possibly reflecting the lower organic content of sandstones. By comparison, sands deposited closer to the shoreline (i.e. Cutpick Hill and core 15-30-64-9W6; Fig. 1) underwent more intense early diagenesis, characterized by Fe-rich chlorite, followed by siderite and then pyrite (Table 1).

Fe-chlorite formation is favoured in brackish environments (Thomas, 1981; Huggett, 1984; Longstaffe, 1986; Segall *et al.*, 1987; Schwartz & Longstaffe, 1988; Tilley & Longstaffe, 1989). Proximal to the shoreline, chlorite formation probably resulted from mixing of depositional marine water and freshwater that penetrated these sands via their hydraulic connection with the shoreface. The absence of Fe-rich chlorite in the most basinward sandstones of this facies suggests that the brackish nature of the seaway alone was insufficient to promote chlorite precipitation.

Chlorite formation was followed by precipitation of siderite and then pyrite. This paragenetic sequence, and the relatively high Mn content of this siderite (>1 mol% Mn) relative to siderite from the dysaerobic and aerobic environments, suggest that crystallization occurred within the post-oxic zone. Therefore, the $\delta^{13}C$ values of this siderite (− 9·3 to − 6·6‰ PDB) probably reflect mixing of marine bicarbonate and bicarbonate formed by degradation of organic matter during microbial Fe-reduction.

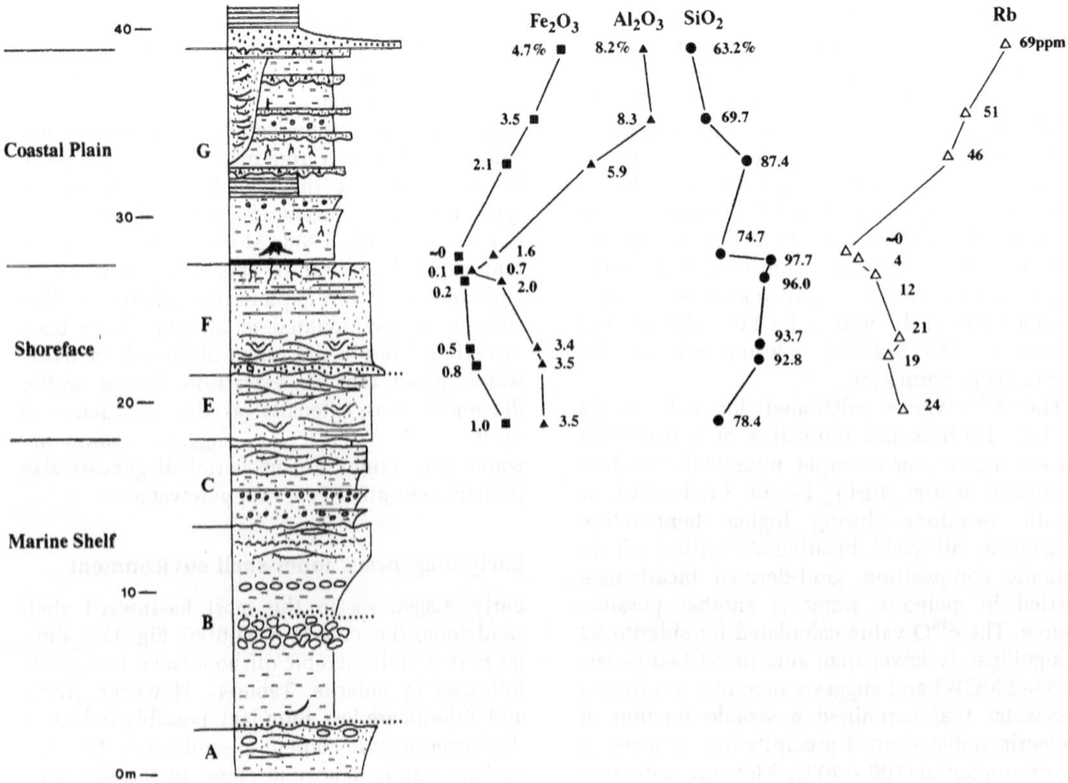

Fig. 16. Stratigraphic variations of various elements in the coastal plain and shoreface rocks at Belcourt Creek (Fig. 1). The uppermost portion of the upper shoreface in unit F is characterized by depletion of most elements (e.g. Fe, Al and Rb), and enrichment of Si and Zr (not shown) relative to the underlying shoreface. These elemental variations are the result of weathering and soil-forming processes.

The relatively low $\delta^{18}O$ values of siderite from the shoreline proximal sandstones (+18·9 to +19·9‰ SMOW) indicate that a significant fraction of meteoric water was present during its formation; crystallization at 25°C requires a porewater $\delta^{18}O$ value of $-12 \pm 0{\cdot}5$‰ (SMOW). By comparison, siderite from a more basinward sandstone has a higher $\delta^{18}O$ value (+24·6‰ SMOW) which, if precipitated at 25°C, requires a porewater of −7‰ (SMOW), the value that appears to be typical of the seaway at this time.

Early diagenesis: shoreface sandstones

The upper shoreface

The organic-rich 'black' sandstone that caps the upper shoreface of unit F is interpreted as the zone of organic accumulation in a podzolic soil (i.e. A_1 horizon). The underlying siliceous sandstone or ganister (i.e. A_2 horizon) was probably formed by oxidation, leaching and the physical translocation of clay-size material (Retallack, 1976; Percival, 1983a,b, 1986). Removal of material is indicated by leached feldspars and rock fragments, and depletion of many elements (e.g. Al, Fe, Ca, K, Mg, P, Zn, Cu, S, Rb, Ba) relative to the underlying portions of the shoreface (Fig. 16). Only Si and Zr are enriched in the leached zone, reflecting the accumulation of resistate minerals such as quartz and zircon, and the abundance of late diagenetic quartz.

A similar subaerial exposure surface probably developed at the top of the Marshybank Formation during sea-level lowstand. However, evidence of subaerial exposure was removed by subsequent transgressive erosion.

The lower shoreface

The lower shoreface at Belcourt Creek, Calliou Creek and Murray River (Fig. 1) is characterized by very fine-grained, blocky and poorly crystallized kaolinite. Kaolinite formation is favoured in acidic environments because of leaching of most cations and development of a high Al/Si ratio (Staub & Cohen, 1978; Hurst & Kundle, 1985). Such conditions are commonly associated with migration of freshwater through a sediment or rock (Hurst & Irwin, 1982; Stonecipher *et al.*, 1984; Longstaffe, 1984, 1986, 1987; Schwartz & Longstaffe, 1988). The very fine grain size of the kaolinite suggests that precipitation occurred rapidly from a highly supersaturated fluid (Curtis, 1978; Berner, 1981). By comparison, the coarser, platy kaolinite present throughout the Marshybank Formation probably formed more slowly at lower levels of supersaturation during late diagenesis.

In comparison, at Muskeg River and in core 10-35-64-11W6 (Fig. 1) illite/smectite rather than kaolinite is the dominant clay mineral. Both sample localities lie close to the maximum progradational edge of the shoreface. This location may favour crystallization of illitic clay from meteoric water enriched in K^+ and other ions as a result of leaching of the upper shoreface. The $\delta^{18}O$ value of early siderite (+18·3‰ SMOW) from the lower shoreface in core 10-35-64-11W6 suggests that the porewater was strongly meteoric (−13‰ SMOW at 25°C). The $\delta^{13}C$ values of the siderite (−9·4‰ PDB) is consistent with such an origin.

At Muskeg River, two generations of siderite cement occur in the lower shoreface sandstone located ≤1 m below the top of the Marshybank Formation. The earliest is chemically pure (>98 mol% Fe), a feature consistent with early precipitation from meteoric water (Mozley, 1989a). However, the later siderite has much higher Ca and Mg contents. Bulk isotopic results ($\delta^{18}O$=+26·7 to +27·0‰ SMOW) suggest that most of the siderite formed from porewaters that were too ^{18}O-rich to be solely meteoric (−6‰ SMOW at 25°C). The origin of these waters is equivocal; two possibilities are: (i) evaporation of Marshybank porewaters during subaerial exposure at sea-level lowstand and (ii) infiltration of Dowling seawater. In either case, the $\delta^{13}C$ values of the siderite mixture (−12·3 to −12·1‰ PDB) rule out normal marine bicarbonate and methanogenesis as major carbon sources.

Early diagenesis: coastal plain environment

Early diagenesis of the coastal plain sediments was characterized by formation of abundant pyrite, followed by ankerite and minor amounts of siderite (Table 1). In general, pyrite is more abundant in organic-rich, fine-grained marine rocks, rather than non-marine rocks, because of the high sulphate content of marine water (Berner, 1981; Gautier, 1982; Postma, 1982; Davison *et al.*, 1985). The occurrence of abundant pyrite in non-marine rocks is generally limited to coals where sulphur for pyrite formation is derived from organic matter.

The most probable explanation for abundant pyrite in the coastal plain rocks of the Marshybank Formation is that the depositional water was brackish, although addition of some organic sulphur cannot be ruled out. The predominantly euhedral texture of the coastal plain pyrite is typical of saltmarsh and estuarine sediments (Luther *et al.*, 1982), and results from direct pyrite formation at low sulphide supersaturation (Goldhaber & Kaplan, 1974; Howarth, 1979). Only one sample shows evidence of pyrite oxidation (Fig. 4h), which suggests that a high water table (and reducing conditions) existed throughout early diagenesis. In turn, this suggests that the Marshybank coastal plain sediments were deposited during relative sea-level rise.

Organic degradation during sulphate reduction, and subsequent pyrite formation, resulted in increased bicarbonate and reduced sulphide concentrations in the porewater, thus allowing crystallization of ankerite and siderite. Ankerite was favoured in this environment because of the presence of abundant detrital dolomite, which the ankerite partially replaces. As in the dysaerobic offshore shelf sediments, siderite precipitation was limited in the coastal plain environment because earlier pyrite formation extracted much of the Fe from the sediment.

SUMMARY

Early diagenesis in the Marshybank Formation was closely related to depositional environment. The main controls on diagenesis were: (i) composition of depositional water (especially oxygen and sulphate content), (ii) Fe and organic content of the sediment, (iii) rate of sedimentation and (iv) proximity to the shoreline. The influx of meteoric water associated with relative sea-level fall at the end of Marshybank deposition also had a major influence on diagenesis.

Five depositional–diagenetic environments have been recognized:

1 Non-bioturbated siltstones of the dysaerobic offshore shelf were characterized by intense sulphate reduction resulting in crystallization of abundant framboidal pyrite and 'early' calcite (zone 1 septarian concretions). Precipitation of 'early' calcite was followed by siderite crystallization within septarian concretions, once Fe was no longer being fixed by pyrite. The limited abundance of siderite in this environment and its high Ca and Mg content are the result of Fe and organic matter depletion during earlier pyrite formation.

'Early' calcite crystallized from a low $\delta^{18}O$ porewater (about −11‰ SMOW at 25°C). This porewater composition probably reflects the ^{18}O-depleted character of the seaway, plus a contribution of low-$\delta^{18}O$ water produced during water–sediment interactions. The $\delta^{18}O$ values of zone 1 siderite are consistent with crystallization from a similar porewater at higher temperatures (50°C). However, it is equally possible that porewater $\delta^{18}O$ values were lower because of meteoric water influx, and that zone 1 siderite formed at lower temperatures. Introduction of meteoric water is consistent with sea-level fall and lowstand that marked the end of Marshybank Formation deposition. Partial oxidation of zone 1 siderite also probably occurred at this time. Subsequent crystallization of zone 2 siderite from porewater with relatively high $\delta^{18}O$ values resulted from the influx of Dowling seawater ($\delta^{18}O = -4$‰ SMOW). The trend to lower $\delta^{13}C$ values observed for zone 2 siderite suggests that abiotic reactions became important as burial progressed.

2 In the aerobic offshore shelf environment, crystallization of minor phosphate and glauconite minerals occurred in the post-oxic zone at localities where sedimentation rates were sufficiently slow. This was followed by formation of pyrite in the sulphate reduction zone. Decreasing sulphide supersaturation (with burial) resulted in a change from framboidal to euhedral pyrite. Because sulphate reduction was less intense in the aerobic than dysaerobic environments, more Fe and organic matter were available to facilitate formation of abundant concretionary siderite (siderite S1 and S2).

The calculated $\delta^{18}O$ and $\delta^{13}C$ values of siderite S1 (+25‰ SMOW and +3% PDB, respectively) indicate precipitation from a low-$\delta^{18}O$ (marine-derived) porewater (about −7‰ SMOW) within the methanogenic zone. Siderite S2, which has significantly lower $\delta^{18}O$ and $\delta^{13}C$ values (+9‰ SMOW and −14‰ PDB, respectively), crystallized from evolved meteoric water at temperatures ≥60°C (intermediate diagenesis). The meteoric water entered the formation during sea-level fall and lowstand at the end of Marshybank Formation deposition.

3 Early diagenesis of the most basinward inner shelf sediments was generally minor, probably because of the limited availability of organic matter. However, in sands deposited proximal to the shoreline, mixing of marine and meteoric water promoted early diagenesis in the post-oxic zone and resulted in formation of Fe-rich chlorite and siderite. If this siderite crystallized at approximately 25°C, the $\delta^{18}O$ value of the earliest porewater ranged from about −13‰ (SMOW) nearest the shoreline to −7‰ (SMOW) in the most basinward sands. The latter composition is typical of seawater during Marshybank Formation deposition. The former, lower value suggests that a significant fraction of meteoric water had infiltrated the inner shelf sands close to the shoreline.

4 In the shoreface sandstones, early diagenesis was strongly influenced by the influx of meteoric water. Subaerial exposure and flushing of the upper shoreface sands by meteoric water resulted in feldspar and rock fragment dissolution, and the formation of a ganister. In the lower shoreface, kaolinite formed if porewaters were relatively dilute, whereas illite/smectite crystallized from evolved meteoric water that was enriched in K^+ and other ions.

5 In the coastal plain environment, abundant organic matter, brackish depositional water and anoxic conditions favoured the precipitation of abundant pyrite. Euhedral pyrite is more common than framboidal pyrite (the reverse is true in the marine sediments) because of low sulphide supersaturation in this environment. Once porewater was depleted in sulphide, ankerite and minor siderite crystallized. The high abundance of detrital dolomite in the coastal plain rocks and the extraction a significant amount of Fe during pyrite formation favoured ankerite rather than siderite crystallization.

ACKNOWLEDGMENTS

This paper summarizes the MSc thesis of J. McKay. Financial support was provided by NSERC grants A7387 (to F.J.L.) and A1917 (to A.G.P.). We also wish to acknowledge Canadian Hunter Exploration Ltd, Esso Resources Ltd and Unocal Canada Ltd for their financial and technical support. Special thanks go to Paul Middlestead for his help with isotopic analysis and to Avner Ayalon for his assistance with analytical techniques and discussion of the results. We would also like to thank Peter Mozley and Iain Scotchman for their helpful reviews of the manuscript.

REFERENCES

Al-Gailani, M.B. (1980) Geochemical identification of unconformities using semi-quantitative X-ray fluorescence analysis. *J. sedim. Petrol.*, **50**, 1261–1270.

Al-Gailani, M.B. (1981) Authigenic mineralizations at unconformities: implications for reservoir characteristics. *Sediment. Geol.*, **29**, 89–115.

Allan, J.R. and Matthews, R.K. (1977) Carbon and oxygen isotopes as diagenetic and stratigraphic tools: Surface and subsurface data, Barbados, West Indies. *Geology*, **5**, 16–20.

Allan, J.R. and Matthews, R.K. (1982) Isotope signatures associated with early meteoric diagenesis. *Sedimentology*, **29**, 797–817.

Ashby, D.A. and Pearson, M.J. (1979) Mineral distributions in sediments associated with the Alton Marine Band near Penistone, South Yorkshire. In: *International Clay Conference, Developments in Sedimentology 27* (Ed. by M. M. Mortland and V. C. Farmer), pp. 311–321.

Astin, T.R. (1986) Septarian crack formation in carbonate concretions from shales and mudstones. *Clay Miner.*, **21**, 617–631.

Ayalon, A. and Longstaffe, F.J. (1988) Oxygen-isotope studies of diagenesis and porewater evolution in the Western Canada Sedimentary Basin: evidence from the Upper Cretaceous basal Belly River sandstone, Alberta. *J. sedim. Petrol.*, **58**, 489–505.

Ayalon, A. and Longstaffe, F.J. (1992) Isolation of diagenetic silicate minerals in clastic sedimentary rocks for oxygen isotope analysis: a summary of methods. *Isr. J. Earth Sci.*, **39**, 139–148.

Berner, R.A. (1978) Sulfate reduction and the rate of deposition of marine sediments. *Earth planet. Sci. Lett.*, **37**, 492–498.

Berner, R.A. (1981) A new geochemical classification of sedimentary environments. *J. sedim. Petrol.*, **51**, 359–369.

Berner, R.A. (1984) Sedimentary pyrite formation: an update. *Geochim. Cosmochim. Acta*, **48**, 605–615.

Bloch, J. (1990) Stable isotopic composition of authigenic carbonates from the Albian Harmon Member (Peace River Formation): evidence of early diagenetic processes. *Bull. Can. petrol. Geol.*, **38**, 39–52.

Bloch, J., Hutcheon, I.E. and Krouse, H.R. (1989) Early diagenetic processes and fluid flow: Evidence from stable isotopic composition of authigenic carbonates, Harmon Member (Peace River Formation), Western Canada. In: *Proceedings of the 6th International Symposium of Water–Rock Interaction* (Ed. by D. L. Miles), pp. 91–94. Balkema, Rotterdam.

Carothers, W.W., Adami, L.H. and Rosenbauer, R.J. (1988) Experimental oxygen isotope fractionation between siderite–water and phosphoric acid liberated CO_2–siderite. *Geochim. Cosmochim. Acta*, **52**, 2445–2450.

Carpenter, S.J., Erickson, J.M., Lohmann, K.C. and Owen, R.R. (1988) Diagenesis of fossiliferous concretions from the Upper Cretaceous Fox Hill Formation, North Dakota. *J. sedim. Petrol.*, **58**, 706–723.

Clayton, R.N. and Mayeda, T.K. (1963) The use of bromine pentafluoride in the extraction of oxygen from oxides and silicates for isotopic analysis. *Geochim. Cosmochim. Acta*, **27**, 43–52.

Clayton, R.N., O'Neil, J.R. and Mayeda, T.K. (1972) Oxygen isotope exchange between quartz and water. *J. geophys. Res.*, **77**, 3057–3067.

Coleman, M.L. (1985) Geochemistry of diagenetic non-silicate minerals: Kinetic considerations. *Phil. Trans. R. Soc.*, **A315**, 39–56.

Coleman, M.L. and Raiswell, R. (1981) Carbon, oxygen and sulphur isotope variations in concretions from the Upper Lias of northeast England. *Geochim. Cosmochim. Acta*, **45**, 329–340.

Craig, H. (1957) Isotopic standards for carbon and oxygen and correction factors for mass-spectrometric analysis of carbon dioxide. *Geochim. Cosmochim. Acta*, **12**, 133–149.

Craig, H. (1961) Standards for reporting concentrations of deuterium and oxygen-18 in natural waters. *Science*, **133**, 1833–1834.

Curtis, C.D. (1977) Sedimentary geochemistry: environments and processes dominated by involvement of an aqueous phase. *Phil. Trans. R. Soc.*, **A268**, 353–372.

Curtis, C.D. (1978) Possible links between sandstone diagenesis and depth-related geochemical reactions occurring in enclosing mudstones. *J. geol. Soc. London*, **135**, 107–117.

Curtis, C.D. (1980) Diagenetic alteration in black shales. *J. geol. Soc. London*, **137**, 189–191.

Curtis, C.D., Coleman, M.L. and Love, L.G. (1986) Pore water evolution during sediment burial from isotopic and mineral chemistry of calcite, dolomite and siderite concretions. *Geochim. Cosmochim. Acta*, **50**, 2321–2334.

Curtis, C.D., Pearson, M.J. and Somogyi, V.A. (1975) Mineralogy, chemistry and origin of a concretionary siderite sheet (clay-ironstone band) in the Westphalian of Yorkshire. *Mineralog. Mag.*, **40**, 385–393.

Curtis, C.D., Petrowski, C. and Oertel, G. (1972) Stable carbon isotope ratios within carbonate concretions: a clue to place and time of formation. *Nature*, **235**, 98–100.

Curtis, C.D. and Spears, D.A. (1968) The formation of sedimentary Fe-minerals. *Econ. Geol.*, **63**, 257–270.

Davison, W., Lishman, J.P. and Hilton, J. (1985) Formation of pyrite in freshwater sediments: Implication for C/S ratios. *Geochim. Cosmochim. Acta*, **49**, 1615–1620.

Dill, H. and Kemper, E. (1990) Crystallographic and chemical variations during pyritization in the upper Barremian and lower Aptian dark claystones from the Lower Saxonian Basin (N.W. Germany). *Sedimentology*, **37**, 427–443.

Dott, R.H. (1964) Wacke, graywacke and matrix – what approach to immature sandstone classification? *J. sedim. Petrol.*, **34**, 625–632.

Eicher, D.L. and Diner, R. (1989) Origin of Cretaceous Bridge Creed Cycles in the Western Interior U.S. *Palaeogeogr. Palaeoclim. Palaeoecol.*, **74**, 127–146.

Emery, D., Myers, R.J. and Young, R. (1990) Ancient subaerial exposure and freshwater leaching in sandstones. *Geology*, **18**, 1178–1181.

Fisher, I.S. (1986) Pyrite formation in bioturbated clays from the Jurassic of Britain. *Geochim. Cosmochim. Acta*, **50**, 517–523.

Folk, R.L. (1974) *Petrology of Sedimentary Rocks*. Hemphill Publishing Co., Austin Texas.

Friedman, I. and O'Neil, J.R. (1977) Compilation of stable isotope fractionation factors of geochemical interest. In: *Data of Geochemistry*, 6th edn (Ed. by M. Fleischer), *United States Geological Survey Professional Paper*, **440-KK**.

Gautier, D.L. (1982) Siderite concretions: Indicators of early diagenesis in the Gammon Shale (Cretaceous). *J. sedim. Petrol.*, **52**, 859–871.

Gautier, D.L. and Claypool, G.E. (1984) Interpretation of methanic diagenesis in ancient sediments by analogy with processes in modern diagenetic environments. In: *Clastic Diagenesis* (Ed. by D. A. McDonald and R. C. Surdam), *Mem. Am. Ass. petrol. Geol.*, **37**, 111–123.

Gibson, P.J. (1989) Petrology of two Tertiary oil shale deposits from Queensland, Australia. *J. geol. Soc. London*, **146**, 319–331.

Goldhaber, M.B. and Kaplan, I.R. (1974) The sulfur cycle. In: *The Sea*, Vol. 5 (Ed. by E. D. Goldberg), pp. 569–655. Wiley (Interscience), London.

Hallam, A. (1969) Siderite- and calcite-bearing concretionary nodules in the Lias of Yorkshire. *Geol. Mag.*, **104**, 222–227.

Hart, B.S., Longstaffe, F.J. and Plint, A.G. (1992) Evidence for relative sea level change from isotopic and elemental composition of siderite in the Cardium Formation, Rocky Mountain Foothills. *Bull. Can. petrol. Geol.*, **40**, 52–59.

Hitchon, B. (1984) Geothermal gradients, hydrodynamics, and hydrocarbon occurrences, Alberta, Canada. *Bull. Am. Ass. petrol. Geol.*, **68**, 713–743.

Howarth, R.W. (1979) Pyrite: Its rapid formation in a salt marsh and its importance in ecosystem metabolism. *Science*, **203**, 49–51.

Hudson, J.D. (1978) Concretions, isotopes and the diagenetic history of the Oxford Clay (Jurassic) of central England. *Sedimentology*, **25**, 339–370.

Huggett, J.M. (1984) Controls on mineral authigenesis in Coal Measures sandstones of the East Midlands, U.K. *Clay Miner.*, **19**, 343–357.

Hurst, A.R. and Irwin, H. (1982) Geological modelling of clay diagenesis in sandstones. *Clay Miner.*, **17**, 5–22.

Hurst, V.J. and Kundle, A.C. (1985) Dehydroxylation, rehydration, and stability of kaolinite. *Clays Clay Miner.*, **33**, 1–14.

Hutcheon, I., Nahnybeda, C. and Krouse, H.R. (1985) The geochemistry of carbonate cements in the Avalon Sand, Grand Banks of Newfoundland. *Mineralog. Mag.*, **49**, 457–467.

Irwin, H. (1980) Early diagenetic carbonate precipitation and pore fluid migration in the Kimmeridge Clay of Dorset, England. *Sedimentology*, **27**, 577–591.

Irwin, H., Curtis, C.D. and Coleman, M. (1977) Isotopic evidence for source of diagenetic carbonates formed during burial of organic-rich sediments. *Nature*, **269**, 209–213.

Jakobsen, B.H. (1988) Accumulation of pyrite and Fe-rich carbonate and phosphate minerals in a lowland moor area. *J. Soil. Sci.*, **39**, 447–455.

Kalkreuth, W. and McMechan, M.E. (1984) Regional pattern of thermal maturation as determined from coal-rank studies, Rocky Mountain Foothills and Front Ranges North of Grande Cache, Alberta – Implications for petroleum exploration. *Bull. Can. petrol. Geol.*, **32**, 249–271.

Kauffman, E.G. (1984) Paleobiogeography and evolutionary response in the Cretaceous Western Interior Seaway of North America. In: *Jurassic – Cretaceous biochronology and paleogeography of North America* (Ed. by G. E. G. Westermann), *Spec. Pap. geol. Ass. Can.*, **27**, 273–306.

Kyser, T.K., Caldwell, W.G.E., Whittaker, S.G. and Cadrin, A.J. (1993) Paleoenvironment and geochemistry of the northern portion of the Western Interior Seaway during Late Cretaceous time. In: *Evolution of the Western Interior Basin* (Ed. by W. G. E. Caldwell and E. G. Kauffman), *Spec. Pap. geol. Ass. Can.*, **39**, 355–378.

Larsen, O.H. and Friis, H. (1991) Petrography, diagenesis and pore-water evolution of a shallow marine sandstone (Hasle Formation, Lower Jurassic, Bornholm, Denmark). *Sediment. Geol.*, **72**, 269–284.

Lawrence, J.R., Gieskes, J.M. and Broecker, W.S. (1975) Oxygen isotope and cation composition of DSDP pore waters and the alteration of layer II basalts. *Earth planet. Sci. Lett.*, **27**, 1–10.

Longstaffe, F.J. (1984) The role of meteoric water in diagenesis of shallow sandstones: Stable isotope studies of the Milk River Aquifer and Gas Pool, southeastern Alberta. In: *Clastic Diagenesis* (Ed. by D. A. McDonald and R. C. Surdam), *Mem. Am. Ass. petrol. Geol.*, **37**, 81–98.

Longstaffe, F.J. (1986) Oxygen isotope studies of diagenesis in the Basal Belly River sandstone, Pembina I-Pool, Alberta. *J. sedim. Petrol.*, **56**, 78–88.

Longstaffe, F.J. (1987) Stable isotope studies of diagenetic processes. In: *Short Course in Stable Isotope Geochemistry of Low Temperature Fluids* (Ed. by T. K. Kyser), *Mineralogical Association of Canada, Short Course*, **19**, 187–257.

Longstaffe, F.J. (1989) Stable isotopes as tracers in clastic diagenesis. In: *Short Course in Burial Diagenesis* (Ed. by I. E. Hutcheon), *Mineralogical Association of Canada, Short Course*, **15**, 201–277.

Longstaffe, F.J. (1993) Meteoric water and sandstone diagenesis in the western Canada sedimentary basin. In: *Diagenesis and Basin Development* (Ed. by A. D. Horbury and A. G. Robinson), *Am. Ass. petrol. Geol. Studies in Geology*, **36**, 49–68.

Longstaffe, F.J. and Ayalon, A. (1987) Oxygen-isotope studies of clastic diagenesis in the Lower Cretaceous Viking Formation, Alberta: implications for the role of meteoric water. In: *Diagenesis of Sedimentary Sequences* (Ed. by J. D. Marshall), *Spec. Publ. Geol. Soc. London,* **36,** 277–296.

Longstaffe, F.J. and Ayalon, A. (1991) Mineralogical and O-isotopic studies of diagenesis and porewater evolution in continental sandstones, Belly River Group, Alberta. *Appl. Geochem.,* **6,** 291–303.

Longstaffe, J.J., Tilley, B.J., Ayalon, A. and Connolly, C.A. (1992) Controls on porewater evolution during sandstone diagenesis, Western Canada Sedimentary Basin: An oxygen isotope perspective. *Spec. Publ. Soc. econ. Paleont. Miner.,* **47,** 13–34.

Love, L.G., Coleman, M.L. and Curtis, C.D. (1983) Diagenetic pyrite formation and sulfur isotope fractionation associated with a Wesphalian marine incursion, northern England. *Trans. R. Soc. Edinb. Earth Sciences,* **74,** 165–182.

Luther, G.W. 3rd, Giblin, A., Howarth, R.W. and Ryans, R.A. (1982) Pyrite and oxidized Fe mineral phases formed from pyrite oxidation in salt marsh and estuarine sediments. *Geochim. Cosmochim. Acta,* **46,** 2665–2669.

Machemer, S.D. and Hutcheon, I. (1988) Geochemistry of early carbonate cements in the Cardium Formation, central Alberta. *J. sedim. Petrol.,* **58,** 136–147.

Magaritz, M., Gavish, E., Bakler, N. and Kafri, U. (1979) Carbon and oxygen isotope composition – Indicators of cementation environment in Recent, Holocene and Pleistocene sediments along the coast of Israel. *J. sedim. Petrol.,* **49,** 401–412.

Maynard, J.B. (1982) Extension of Berner's "New geochemical classification of sedimentary environment's" to ancient sediments. *J. sedim. Petrol.,* **50,** 1325–1331.

McKay, J.L. (1992) *Diagenesis of the Upper Cretaceous Marshybank Formation, N.W. Alberta–N.E. British Columbia.* MSc thesis, University of Western Ontario, London, Ontario.

Mermut, A.R., Curtin, D. and Rostad, H.P. (1985) Micromorphological and sub-microscopial features related to pyrite oxidation in an inland marine shale from east central Saskatchewan. *J. Soil Sci. Soc. Am.,* **49,** 256–261.

Mozley, P.S. (1989a) Relationship between environment and the elemental composition of early diagenetic siderite. *Geology,* **17,** 704–706.

Mozley, P.S. (1989b) Complex compositional zonation in concretionary siderite: implications for geochemical studies. *J. sedim. Petrol.,* **59,** 815–818.

Mozley, P.S. and Burns, S.J. (1993) Oxygen and carbon istopic composition of marine carbonate concretions: an overview. *J. sedim. Petrol.,* **63,** 73–83.

Mozley, P.S. and Carothers, W.W. (1992) Elemental and isotopic composition of siderite in the Kuparuk Formation, Alaska: effect of microbial activity and water/sediment interaction on early pore-water chemistry. *J. sedim. Petrol.,* **62,** 681–692.

Oertel, G. and Curtis, C.D. (1972) Clay-ironstone concretion preserving fabrics due to progressive compaction. *Bull. geol. Soc. Am.,* **83,** 2597–2606.

O'Neil, J.R., Clayton, R.N. and Mayeda, T.K. (1969) Oxygen isotope fractionation in divalent metal carbonates. *J. chem. Phys.,* **51,** 5547–5558.

Pearson, M.J. (1974a) Sideritic concretions from the Westphalian of Yorkshire: a chemical investigation of the carbonate phase. *Mineralog. Mag.,* **39,** 696–699.

Pearson, M.J. (1974b) Magnesian siderite in carbonate concretions from argillaceous sediments in the Westphalian of Yorkshire. *Mineralog. Mag.,* **39,** 700–704.

Pearson, M.J. (1979) Geochemistry of the Hepworth Carboniferous Sediment Sequence and origin of the diagenetic iron minerals and concretions. *Geochim. Cosmochim. Acta,* **43,** 927–941.

Percival, C.J. (1983a) A definition of the term ganister. *Geol. Mag.,* **120,** 187–190.

Percival, C.J. (1983b) The Firestone Sill Ganister, Namurian, Northern England – A_2 horizon of a podzol or podzolic palaeosol. *Sediment. Geol.,* **36,** 41–49.

Percival, C.J. (1986) Paleosols containing an albic horizon: Examples from the Upper Carboniferous of northern England. In: *Paleosols, Their Recognition and Interpretation* (Ed. by V. P. Wright), pp. 87–111. Princeton University Press.

Plint, A.G. (1990) An allostratigraphic correlation of the Muskiki and Marshybank Formation (Coniacian-Santonian) in the Foothills and subsurface of the Alberta Basin. *Bull. Can. petrol. Geol.,* **38,** 280–306.

Plint, A.G. (1991) High-frequency relative sea level oscillations in Upper Cretaceous shelf clastics of the Alberta Foreland Basin: possible evidence for a glacio-eustatic control? In: *Sedimentation, Tectonics and Eustacy* (Ed. by D. I. M. MacDonald), *Spec. Publ. Int. Ass. Sedimentol.,* **12,** 409–428.

Plint, A.G., Hart, B.S. and Donaldson, W.S. (1993) Lithospheric flexure as a control on stratal geometry and facies distribution in Upper Cretaceous rocks of the Alberta foreland basin. *Basin Res.,* **5,** 69–77.

Plint, A.G. and Norris, B. (1991) Anatomy of a ramp margin sequence: facies successions, paleogeography and sediment dispersal patterns in the Muskiki and Marshybank Formations, Alberta Foreland Basin. *Bull. Can. petrol. Geol.,* **39,** 18–42.

Plint, A.G., Norris, B. and Donaldson, W.S. (1990) Revised definitions for the Upper Cretaceous Bad Heart Formation and associated units in the Foothills and Plains of Alberta and British Columbia. *Bull. Can. petrol. Geol.,* **38,** 78–88.

Plint, A.G. and Walker, R.G. (1987) Morphology and origin of an erosion surface cut into the Bad Heart Formation during major sea-level change, Santonian of west-central Alberta, Canada. *J. sedim. Petrol.,* **57,** 639–650.

Postma, D. (1977) The occurrence and chemical composition of recent Fe-rich mixed carbonates in a river bog. *J. sedim. Petrol.,* **47,** 1089–1098.

Postma, D. (1982) Pyrite and siderite formation in brackish and freshwater swamp sediments. *Am. J. Sci.,* **282,** 1151–1183.

Querol, X., Chinchon, S. and Lopeq-Soler, A. (1989) Iron sulfide precipitation sequence in Albian coals from the Maestrazgo Basin, southeastern Iberian Range, northeastern Spain. *Int. J. Coal Geol.*, **11**, 171–189.

Raiswell, R. (1971) The growth of Cambrian and Liassic concretions. *Sedimentology*, **17**, 147–171.

Raiswell, R. (1982) Pyrite texture, isotopic composition and the availability of iron. *Am. J. Sci.*, **282**, 1244–1263.

Raiswell, R. (1987) Non-steady state microbiological diagenesis and the origin of concretions and nodular limestone. In: *Diagenesis of Sedimentary Sequences* (Ed. by J.D. Marshall), *Spec. Publ. Geol. Soc. London*, **36**, 41–54.

Raiswell, R. and White, N.J.M. (1978) Spatial aspects of concretionary growth in the Upper Lias of northeast England. *Sediment. Geol.*, **20**, 291–300.

Retallack, G.J. (1976) Triassic palaeosols in the Upper Narrabeen Group of New South Wales. Part I: Features of the palaeosols. *J. geol. Soc. Aust.*, **23**, 383–399.

Rosenbaum, J. and Sheppard, S.M.F. (1986) An isotopic study of siderites, dolomites and ankerite at high temperature. *Geochim. Cosmochim. Acta*, **50**, 1147–1150.

Rye, D.M. and Sommer, M.A. (1980) Reconstructing paleotemperature and paleosalinity regimes with oxygen isotopes. In: *Skeletal Growth of Aquatic Organisms* (Ed. by D. C. Rhoads and R. A. Lutz), pp. 167–181. Plenum Press, New York.

Sass, E., Bein, A. and Almogi-Labin, A. (1991) Oxygen-isotope composition of diagenetic calcite in organic-rich rocks: Evidence for ^{18}O depletion in marine anaerobic pore water. *Geology*, **19**, 839–842.

Schwartz, F.W. and Longstaffe, F.J. (1988) Groundwater and clastic diagenesis. In: *The Geology of North America, Vol. O-Z, Hydrogeology* (Ed. by W. Back, J. S. Rosenhein and P. R. Seaber), pp. 413–434. Geological Society of America.

Scotchman, I.C. (1991) The geochemistry of concretions from the Kimmeridge Clay Formation of southern and eastern England. *Sedimentology*, **38**, 79–106.

Segall, M.P., Buckley, D.E. and Lewis, C.F.M. (1987) Clay mineral indicators of geological and geochemical subaerial modification of near surface Tertiary sediments on the northeastern Grand Banks of Newfoundland. *Can. J. Earth Sci.*, **24**, 2172–2187.

Sharma, T. and Clayton, R.N. (1965) Measurement of O^{18}/O^{16} ratios of total oxygen of carbonates. *Geochim. Cosmochim. Acta*, **29**, 1347–1354.

Spears, D.A. (1989) Aspects of iron incorporation into sediments with special reference to the Yorkshire ironstones. In: *Phanerozoic Ironstones* (Ed. by T. P. Young and W. E. G. Taylor), *Spec. Publ. geol. Soc. London*, **46**, 19–30.

Staub, J.R. and Cohen, A.D. (1978) Kaolinite-enrichment beneath coals: A modern analog, Snuggedy Swamp, South Carolina. *J. sedim. Petrol.*, **48**, 203–210.

Stonecipher, S.A. and May, J.A. (1989) Facies controls on early diagenesis: Wilcox Group, Texas Gulf Coast. In: *Prediction of Reservoir Quality Through Chemical Modeling* (Ed. by I. D. Meshri *et al.*), *Mem. Am. Ass. petrol. Geol.*, **49**, 25–44.

Stonecipher, S.A., Winn, R.D. Jr and Bishop, M.G. (1984) Diagenesis of the Frontier Formation, Moxa Arch: A function of sandstone geometry, texture and composition, and fluid flux. In: *Clastic Diagenesis* (Ed. by D. A. McDonald and R. C. Surdam), *Mem. Am. Ass. petrol. Geol.*, **37**, 289–316.

Thomas, J.B. (1981) Classification and diagenesis of clay minerals in tight gas sandstones: Case studies in which clay mineral properties are crucial to drilling fluid selection, formation evaluation, and completion techniques. In: *Clays and the Resource Geologist* (Ed. by F. J. Longstaffe), *Mineralogical Association of Canada, Short Course*, 104–118.

Tilley, B.J. and Longstaffe, F.J. (1989) Diagenesis and isotopic evolution of porewaters in the Alberta Deep Basin: the Falher Member and Cadomin Formation. *Geochim. Cosmochim. Acta*, **53**, 2529–2546.

Videtich, P.E. and Matthews, R.K. (1980) Origin of discontinuity surfaces in limestones: Isotopic and petrographic data, Pleistocene of Barbados, West Indies. *J. sedim. Petrol.*, **50**, 971–980.

Walters, L.J. Jr., Claypool, G.E. and Choquette, P.W. (1972) Reaction rates and $\delta^{18}O$ variation for the carbonate–phosphoric acid preparation method. *Geochim. Cosmochim. Acta*, **36**, 129–140.

Wright, E.K. (1987) Stratification of paleocirculation of the Late Cretaceous Western Interior Seaway of North America. *Bull. geol. Soc. Am.*, **99**, 480–490.

Manuscript received 9 February 1994; revision accepted 6 July 1994

Non-marine eogenesis 2: arid environments

Arid continental environments differ from humid environments in terms of the amount of organic debris in the sediment and by both the amount and geochemistry of the groundwater. Arid alluvial fans, braid plains and deserts preserve much less organic detritus than flood plains and their associated crevasse splays and channel fills. However, life still plays a role even in hostile environments. Rock varnish formed on sediment and rock surfaces in arid environments is shown by Nagy *et al.* (1991) and Krinsley (1998) to actively involve bacterial and fungal growth on the weathering surface. Organic filaments help to trap finely comminuted mineral particles into nanometre thick overgrowths on rock surfaces. The microstromatolite structure of rock varnish is the result of successive growth of bacteria and fungi and then trapping of minerals, followed by further growth of bacteria and fungi.

Calcretes and dolocretes are characteristic of arid environments and occur as soils or in the vadose and phreatic zones of unconfined aquifers. However, even some of the deeper forms of calcrete formation are likely to be influenced by organic activity. Purvis & Wright (1991) describe calcrete growth around deeply penetrating roots. The growth of this type of calcrete relates directly to the transpiration of the plant, as the uptake of water automatically concentrates pore solutions to the point of calcite saturation. The massive dolocretes described by Spötl & Wright (1992) are thought to have originated in the fully saturated region of an aquifer. These dolocretes also have silica cement associated with them, including microquartz and chalcedony. As dolocrete formation involves the pervasive dissolution of silicate grains, it is perhaps not surprising that some silica forms as an eogenetic cement. Indeed silcrete is not uncommon in arid environments and can occur as nodules formed on bedding surfaces (Leckie & Cheel, 1990) on the distal parts of braid plains. Silcrete is throught to be a characteristic manifestation of strongly seasonal and/or arid climates.

Zeolite is ordinarily not considered to be a common diagenetic mineral in clastic sediments, although it may be characteristic of the presence of highly saline continental brines, enriched in aqueous sodium and bicarbonate (Renaut, 1993). Highly alkaline lake and groundwaters lead to elevated silica solubility and encourage the stability of zeolite minerals at the expense of feldspars and clays. The zeolites that result are typically sodium-rich analcime and natrolite.

Alunites are unusual minerals that are rich in sulphate and either aluminium or ferric iron. They may be diagnostic of subsurface petroleum accumulations that leak to the surface (Khalaf, 1990). Along with petroleum can come any associated H_2S, a common non-hydrocarbon gas in some petroleum provinces. The seeping H_2S is oxidized, probably bacterially, upon reaching groundwater, resulting in highly concentrated sulphuric acid. This acid can then result in either gypsum or alunite growth.

Reprinted from Sedimentology (1990) **37**, *155–164*

Diagenetic alunite in clastic sequences, Kuwait, Arabian Gulf

FIKRY I. KHALAF

Environmental and Earth Sciences Division, Kuwait Institute for Scientific Research, PO Box 24885, 13109 Safat, Kuwait

ABSTRACT

Diagenetic alunite occurs with calcrete, gypcrete and silcrete deposits in a Mio-Pleistocene clastic sequence at several locations in southern Kuwait, Arabian Gulf. Based on their physical properties and textural characteristics the alunite deposits were divided into (1) chalky quartzitic alunite, (2) chalky quartz-free alunite, (3) hard pink alunitic sandstone. The chalky alunite is composed mainly of hydronium–alunite $(H_3O)Al_3(SO_4)_2(OH)_6$, while the hard pink alunite is composed solely of well developed potassium alunite $KAl_3(SO_4)_2(OH)_6$. These minerals resulted from the action of sulphuric acid on clays and K-feldspars in the muddy sandstone and mudstone host sediments. The sulphuric acid is most probably produced by the oxidation of hydrogen sulphide that might have seeped in from the oil fields of this area. Four diagenetic stages are suggested for the genesis of the studied alunites: gypsification, sulphuratization, silicification and alunitization. In oil field areas, the occurrences of alunite would serve as an indicator for the presence at depth of hydrocarbons.

INTRODUCTION

The term alunite is used for a group of iso-structural basic sulphate minerals that have a composition with the general formula $AR_3(SO_4)_2(OH)_6$, where A refers to a large cation in 12-fold coordination (Na, K, Pb, NH_4, Ag or H_3O) and R is either Fe^{3+} or Al^{3+} in octahedral coordinations. Alunite $KAl_3(SO_4)_2(OH)_6$, natroalunite $NaAl_3(SO_4)_2(OH)_6$, and basic aluminium sulphate or hydronium alunite $(H_3O)Al_3(SO_4)_2(OH)_6$, form an isomorphous series, but alunite is more abundant in nature. The mineral alunite is economically important since it is a potential source of aluminium.

Two genetic types of alunite are recognized, namely hydrothermal and diagenetic. Hydrothermal alunite is developed by the alteration of acidic volcanic rocks while diagenetic alunites are formed in potassium- and aluminium-rich sedimentary deposits. In both cases the genesis of alunite is attributed to reaction between potassium- and aluminium-rich rocks and sulphuric acid which is generated by the oxidation of gaseous or metallic sulphides (Lombardi & Sheppard, 1977; Hall, 1978). Diagenetic alunites have been reported from USA, Hungary, Australia, Greece and the Middle East (Stephenson, 1946; King, 1953; Keller, Gentile & Reesman, 1967; Ross *et al.*, 1968; El-Sharkawi & Khalil, 1977; Goldbery, 1978; Hall, 1978; Rouchy & Pierre, 1987).

The present paper reports, for the first time, the occurrence of diagenetic alunite from post-Eocene clastic deposits exposed in southern Kuwait and discusses the petrography and genesis of this mineral.

FIELD OCCURRENCE

The area of Kuwait constitutes a part of the interior homocline of the Arabian Peninsula where Eocene to Recent rocks are exposed. The surface of Kuwait is generally flat and is mostly carved into a partially calcretized sequence of clastic deposits of Mio-Pleistocene age, known locally as the Kuwait Group. Tectonically, Kuwait is located in the stable shelf region that covers eastern Saudi Arabia, UAE, Qatar, Bahrain and Southwestern Iraq, where sediments are comparatively thin and only mildly affected by the Alpine orogeny. The pretoliferous Cretaceous and

Tertiary deposits in Kuwait are exploited in several oil fields, namely; Burgan, Magwa, Ahmadi, Rawdhatin, Sabriyah, Bahra, Managish, Umm Ghudair, and Wafra and several offshore oil fields (Fig. 1). Most fields are located in the Kuwait Arch which is a large complex anticlinal structure running from Wafra through Burgan, Magwa, Ahmadi and Bahra to Sabriya in the north. Structures of Burgan and Wafra oil fields are reflected on the surface by the occurrence of small isolated hills which coincide with the anticlinal crests.

Alunite deposits occur with gypcretic and silcretic horizons in the upper clastic deposits of the Kuwait Group exposed at Warah, Burgan and Al-Wafra hills (Fig. 1). At Warah hill, alunite deposits occur as a massive lens about 2 m thick and about 30 m in lateral extent. It is associated with an iron-rich mudstone layer located at the contact between a calcrete profile, and silicified sandstones (silcrete) (Fig. 2). The alunite deposit at Warah hill is generally a white, chalky compact rock composed of sand grains floating in a chalky matrix. It breaks with conchoidal fractures and has a dull porcelain-like appearance. Patches and veins of various colours, mostly purplish, reddish and brownish, are also present. Fractured blocks of unaltered mudstone and silcrete are enclosed in the alunite deposits. Narrow fractures, less than 10 mm wide, are usually filled with displacive fibrous gypsum, while wider fractures are lined with gypsum and filled with alunite. The silcrete that caps the Warah hill is

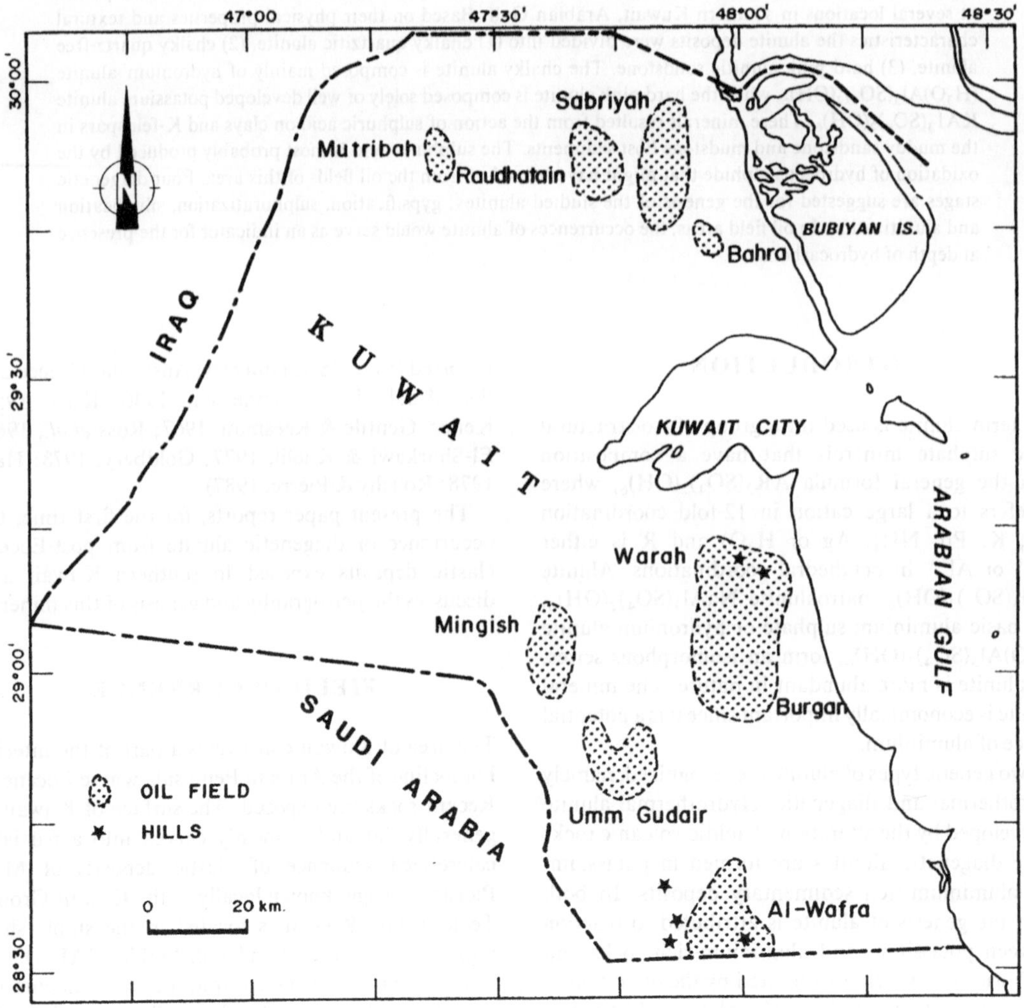

Fig. 1. Location map of the study area.

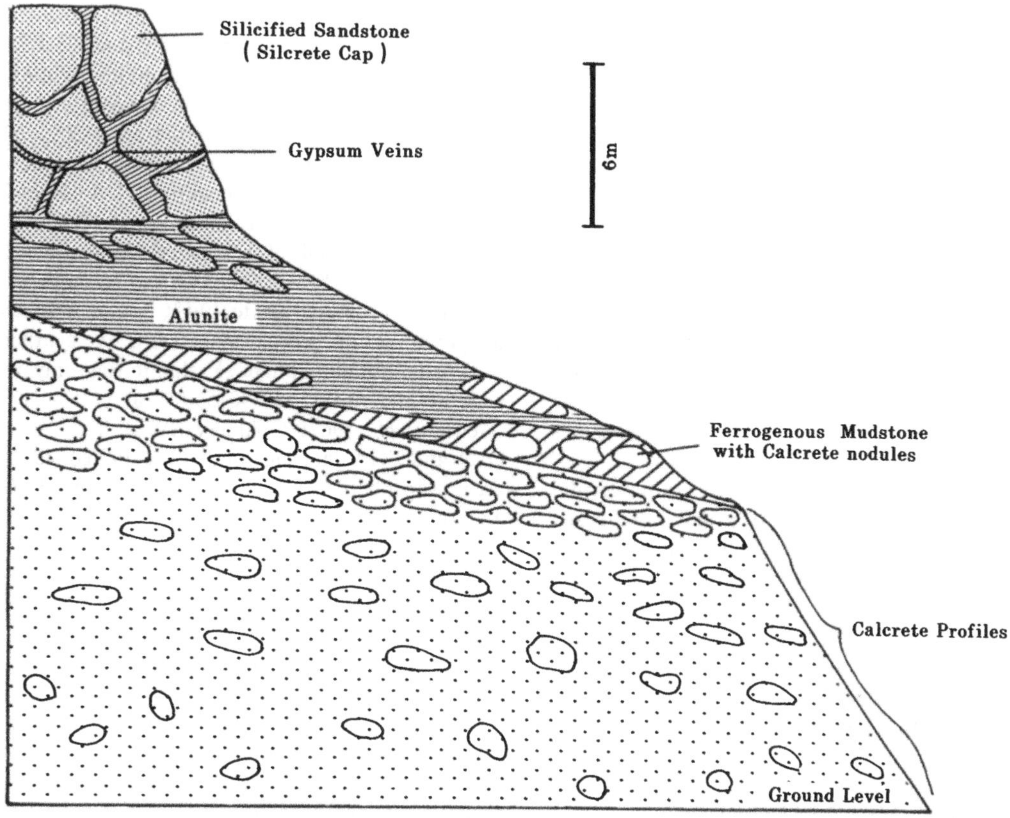

Fig. 2. Schematic diagram of the rock sequence exposed at Warah hill illustrating the mode of occurrence of alunite.

also fractured, the fractures being filled with gypsum and gypsiferous material.

Alunite deposits at Burgan hills are similar to that of Warah hill; however, they are harder and quartz grains are coarser and more frequent. Alunite deposits at Al-Wafra hills are very hard, vary in colour from creamy white to pink, and break with a conchoidal fracture. Alunite occurs as scattered cobbles on and around Al-Wafra hills and ridges. It was difficult to locate precisely the source of this alunite debris, which frequently occurs on hills capped with gypcrete and silcretic gypcrete.

PETROGRAPHY

The host sediment of alunite is calcretized fluviatile muddy sandstone capped with gypcrete and silcrete. Silcrete is composed of quartz grains cemented by syntaxial overgrowths and micro-crystalline quartz (Khalaf, 1988). Remnants of the original calcareous cement were recognized in the silcrete of Warah hill.

Gypcrete in Warah and Burgan hills is developed along fractures within the host sandstone. It is composed of detrital siliciclastic grains, mostly quartz and feldspars, cemented by gypsum. The contact between the host calcretic sandstone and the fractured gypcrete is gradational. Four zones were recognized along this contact, namely; (a) gypcrete, (b) gypcrete with carbonate remnants, (c) slightly gypcretized sandstone, and (d) the host calcretic sandstone. Gypcrete cap at Al-Wafra hills has a white saccharoidal appearance and is composed of detrital quartz grains cemented with coarsely crystalline limpid gypsum crystals. This gypcrete cap is silicified near its top, where a 0·2-m thick band of silcrete is developed.

Three types of alunitic material were recognized: (a) chalky quartzitic alunite, (b) chalky quartz-free alunite, and (c) hard pink alunitic sandstone. The chalky quartzitic alunite forms the main alunite deposits and is composed of well-sorted quartz grains of medium sand size floating in a matrix formed of aggregates of microcrystalline alunite. Alunite matrix has a very low birefrengence and displays a flow pattern around the quartz grains. It is very rich in spherical opaque particles of iron oxides that are usually less than 5 μm in diameter (Fig. 3A). The quartz and feldspar grains are severely corroded. Corrosion usually starts along fractures in quartz and cleavage planes in the feldspars (Fig. 3B and C).

Chalky quartz-free alunite occurs as lumps filling small pockets between gypsum veinlets which are usually less than 2 mm in width. These alunite lumps are formed of aggregates of microcrystalline alunite with abundant minute opaque particles of iron oxides (Fig. 3D).

The hard pink alunitic sandstone comprises detrital quartz grains floating in a massive matrix composed of a fine mosaic of microcrystalline aggregates of alunite; abundant opaque particles of iron oxides are responsible for the pink colour (Fig. 3E). Abundant quartz grains comprise more than 50% of the alunitic sandstone. The chalky quartzitic alunite is friable to semi-consolidated, and quartz grains constitute less than 50% of its bulk composition.

MINERALOGY

Identification of alunites was confirmed by XRD and Differential Thermal Analysis (DTA). X-ray (CuK_α) diffractograms of representative samples showed distinct peaks at 4·95Å, 3·50Å, 2·28Å and 1·90Å in addition to the main peak at 2·98Å, which are the characteristic diffraction peaks of alunites (Fig. 4). No significant difference occurs among the XRD data from the various types of alunites. Kaolinite is, however, present in quartzitic-alunite and quartz-free alunite. DTA curves of the alunites show three endothermic peaks, a very small peak at 300°C, a medium peak at 515°C and a small one at 750°C (Fig. 5). The small peak at 300°C is attributed to lattice interstitial water held as hydronium ion. The 515°C peak represents the dehydroxylation reaction which characterizes both K-alunite and (H_3O)-alunite, but it is closer to that of (H_3O)-alunite and is about 30°C less than that of the well-developed pure K-alunite. The 750°C peak represents an endothermic reaction that results from the breakdown of the strongly bonded sulphate groups and liberation of sulphur oxides.

Thermal Gravimetric Analysis curves (TGA) of the studied samples indicate that the dehydroxylation processes, which range between 475°C and 575°C, are responsible for 50% of the total weight loss, while the sulphate decomposition reaction ranging between 600°C and 900°C accounts for the other 50%. The total weight loss of the analysed samples is dependent on the purity of the alunitic material. TGA curves for chalky quartz free alunite indicate a weight loss of about 14% as a result of dehydroxylation and a total weight loss of 28·5% at 900°C. Davey, Lukaszewski & Scott (1963) reported up to 19·5% loss of weight at temperatures not greater than 600°C for synthetic basic aluminium sulphate. K-alunite has only two endothermic peaks at 540–560°C and 750–780°C (Keller *et al.*, 1967; Goldbery, 1978). The present DTA and TGA data suggest that the studied alunites are most probably a mixture of K-alunite and (H_3O)-alunite.

SEM investigations of the chalky quartzitic alunites show that most of the alunite matrix is formed of aggregates of minute needle-shaped crystals (2 μm long, 0·2 μm wide) with scattered individual rhombs that generally have a uniform size of about 1 μm (Fig. 6A, B). Energy dispersive X-ray analysis showed that the needle-shaped crystals and rhombs are rich in Al and S. The rhombs differ from the needle-shaped crystals by having considerable amounts of K. This suggests that the needle-shaped crystals are (H_3O)-alunite while the rhombs are K-alunite.

Many alunite rhombs in the quartzitic alunite are hollow (Fig. 6B). Cunningham *et al.* (1984) and Rouchy & Pierre (1987) interpreted similar hollow structures in alunite as moulds of gas or vapour bubbles. Incomplete alunitic rhombs were also observed (Fig. 6C). The incomplete nature of the K-alunite rhombs could be attributed to the exhaustion of available potassium during growth.

SEM investigations of the hard pink alunitic sandstones revealed that the matrix is composed largely of aggregates of well-developed K-alunite rhombs covered with iron-oxide particles (Fig. 6D, E). In the hard creamy alunite debris collected from Al-Wafra hills, alunite rhombs appeared more developed and twinned (Fig. 6F).

CHEMICAL ANALYSIS

Chemical analysis of the studied samples provided more support for the identification of alunite (Table 1). The average chemical composition of relatively pure alunite is about 11·0% K_2O, 38·0% Al_2O_3, 37·0% SO_3

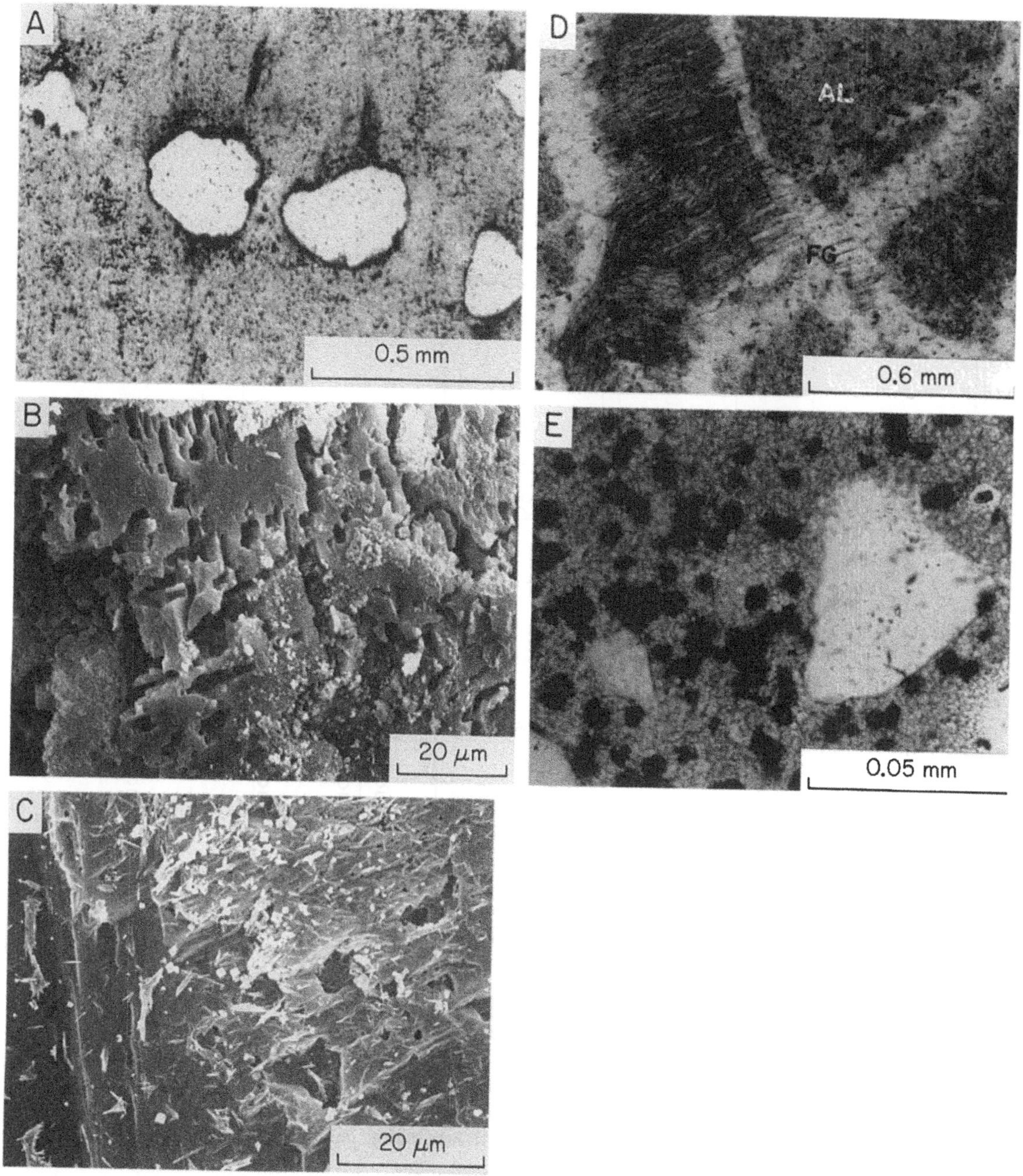

Fig. 3. Alunite. (A) Photomicrograph of chalky quartzitic alunite composed of detrital quartz grains floating in microcrystalline alunite matrix; note abundant minute spherical iron oxide particles. (B) SEM photograph of a skeleton of feldspar grain in chalky quartzitic alunite; note severe corrosion along the cleavage planes. (C) SEM photograph of the surface of a quartz grain embedded in chalky alunite showing triangular etching pits due to corrosion by sulphuric acid; note scattered minute K-alunite rhombs on the corroded surface. (D) Photomicrograph of chalky quartz-free alunite showing small pockets of microcrystalline alunite (AL) separated by micro-veins of fibrous gypsum (FG). (E) Photomicrograph of hard reddish alunitic sandstone composed of detrital quartz grains floating in alunite matrix, very rich in iron oxide particles.

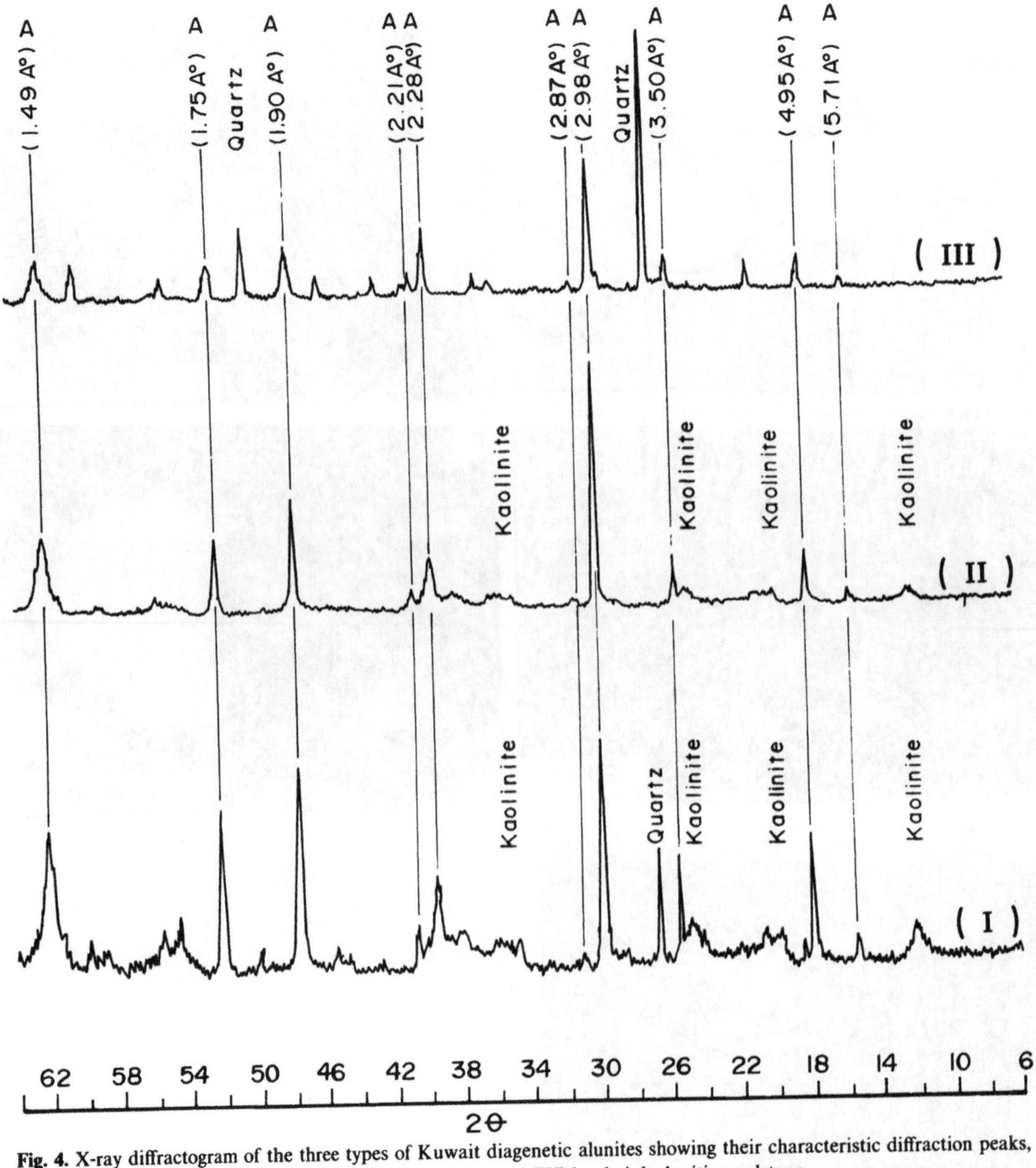

Fig. 4. X-ray diffractogram of the three types of Kuwait diagenetic alunites showing their characteristic diffraction peaks, (I) chalky quartzitic alunite, (II) chalky quartz-free alunite, and (III) hard pink alunitic sandstone.

and 14·0% H_2 (Palache, Berman & Frondel, 1951). The ratios of $Al_2O_3 : SO_3$ and $K_2O : Al_2O_3$ in the hard pink alunitic sandstone are similar to those in pure alunite. For the chalky alunites, the ratios are different because the amount of Al_2O_3 is approximately two times that of SO_3. This suggests that Al_2O_3 may be present in forms other than alunite. The presence of kaolinite in this type of alunite may be responsible for some of this excess of alumina. It is also possible that part of Al_2O_3 may be present as amorphous aluminium hydroxides.

DISCUSSION

Three types of diagenetic alunites occur in the Mio-Pleistocene fluviatile sequence of Kuwait, namely; (a)

Table 1. Chemical analysis of diagenetic alunites from Kuwait

	1	2	3	4	5	6	7
SiO_2	34·19	38·40	28·55	40·70	60·24	64·50	66·50
Al_2O_3	31·09	30·40	32·73	27·80	11·30	10·30	10·30
Fe_2O_3	1·29	2·39	1·55	4·40	4·00	3·40	0·51
MgO	0·04	0·06	0·12	0·11	0·22	0·08	0·09
CaO	0·24	0·40	0·52	0·32	1·30	0·67	0·30
K_2O	3·36	2·50	3·30	2·40	3·50	3·44	3·00
Na_2O	1·80	1·60	1·36	0·64	0·23	0·19	0·38
SrO	0·18	0·10	0·28	0·37	0·05	0·04	0·09
SO_3	15·50	14·75	15·50	14·20	12·80	10·30	10·78
H_2O	12·30	9·35	16·10	9·03	6·36	7·08	8·00

Note: 1, 2, 3 and 4 from chalky quartzitic alunite, at Warah; 5 and 6 from hard pink alunitic sandstone, at Al-Wafra; 7 from hard creamy alunitic sandstones, at Al-Wafra.

chalky quartzitic alunite, (b) chalky quartz-free alunite, and (c) hard pink alunitic sandstone. The first two types are present at Warah and Burgan hills while the third type occurs in the Al-Wafra hills. These alunite deposits are closely associated with gypcrete and silcrete. Detailed petrographic, mineralogical and geochemical analyses have shown that the Warah and Burgan chalky alunites are composed chiefly of basic aluminum sulphate or hydronium alunite with subordinate amounts of K-alunite, kaolinite and other amorphous aluminium compounds. The alunite in Al-Wafra alunitic sandstone is mostly present as well developed K-alunite.

Based on their lithological association and fabric, it is suggested that the alunites are the result of several consecutive diagenetic processes that took place in the host sediments. These diagenetic processes were responsible for the alteration of the fluviatile muddy sandstone to quartzitic alunites and alunitic sandstones. The mineral paragenesis probably encompassed gypsum, silica and alunite formation.

Gypsification was the first diagenetic process that affected the host sediment. Two stages of gypsification are recognized. Initially saccharoidal gypcrete developed by direct precipitation of gypsum from sulphate-rich groundwater in the porous sandstones. This was followed by a stage where sulphuric acid was produced by the oxidation of H_2S that could have permeated through the fractures of the calcareous sandstone in association with seeped petroleum gases. Kuwaiti crude oil is generally high in sulphur content (Aalund, 1983). During this stage, the calcite cement of the calcretic sandstone was replaced by gypsum. This may have proceeded with the dissolution of calcite by dilute sulphuric acid and the precipitation of gypsum.

Production of sulphuric acid probably occurred during periods of abundant hydrogen sulphide seepage which was oxidized and reacted with illitic clays and K-feldspars. This reaction could yield both silica and aluminium sulphates as follows:

$$\underset{\text{(illite)}}{KAl_2\{(OH)_2Al_2Si_3O_{16}\}} + H_2SO_4 \rightarrow \underset{\text{(silica)}}{3\,SiO_2} + \underset{\text{(alunite)}}{KAl_3(SO_4)_2(OH)_6}$$

Silicification started during or slightly after this

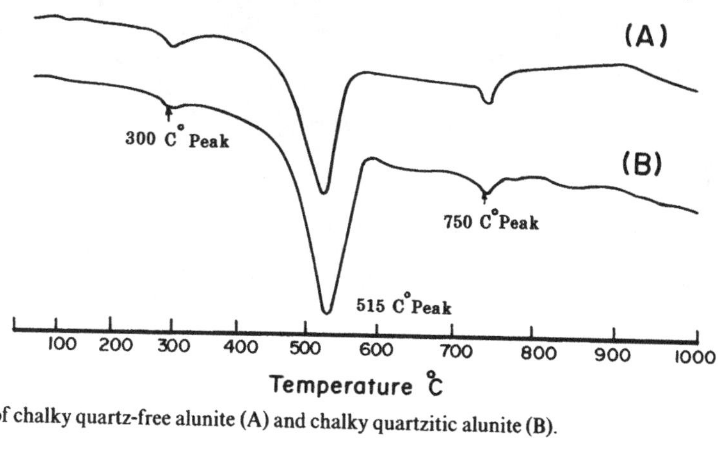

Fig. 5. DTA curves of chalky quartz-free alunite (A) and chalky quartzitic alunite (B).

Fig. 6. SEM photomicrographs of alunite. (A) General texture of chalky alunite; note needle-shaped crystals of hydronium alunite (HAL) and scattered minute rhombs of K-alunite (KAL). (B) Hollow in rhombs of K-alunite. (C) Incomplete rhombs of K-alunite. (D) Well-developed rhombohedral crystals of K-alunite in the hard pink alunite sandstone. (E) Well-developed K-alunite rhombs encrusted with iron-oxide particles. (F) Twinned rhombohedral crystals of K-alunite in the hard creamy alunitic sandstone.

second stage and lead to the precipitation of silica in the pore spaces of the host calcretic sandstones and gypcrete. This process was responsible for the genesis of silcrete associated with the alunite deposits. Two successive stages of silicification occurred: an early stage where syntaxial overgrowths developed due to slow precipitation of silica from a dilute solution and a later stage where microcrystalline quartz cement precipitated from a more concentrated solution or due to rapid influx of solutions (Heald & Renton, 1966). This suggests that sulphuric acid concentration was at first weak and then increased in strength.

The kaolinite associated with the chalky alunite material in the quartzitic alunite is authigenic and developed by direct precipitation from Al- and Si-rich solutions. The precipitation of considerable amounts of silica during the silicification process could have increased the Al:Si ratio in the solutions.

Alunite formation followed the stage of sulphuric acid production. The alunite replacing the illite matrix of the host sandstone was mostly hydronium alunite, followed by K-alunite precipitation. The absence of hydronium alunite and kaolinite in the hard pink alunitic sandstone and the occurrence of well-developed K-alunite suggest different diagenetic conditions for the formation of the chalky quartzitic alunite. The presence of K-alunite could be attributed to: (a) abundance of potassium in the host rock, (b) high temperature, as K-alunite is a higher temperature mineral than hydronium alunite, and (c) alteration of hydronium alunite in the presence of potassium ions (Adams & Hajeck, 1978, Sheridan & Royse, 1970, Zotov, 1971).

The fabric and mode of occurrence of the iron oxide particles in the hard pink alunitic sandstone indicate that they developed after K-alunite precipitation. This observation suggests that the sulphuric acid was not generated from pre-existing pyrites.

In conclusion, the alunites from Kuwait are late diagenetic and are a by-product of the reaction of Al- and K-rich clays and feldspars in the host sandstone and mudstone with sulphuric acid, which could have been produced by the oxidation of hydrogen sulphide associated with petroleum gases seeping along tectonic fractures within the area.

ACKNOWLEDGMENTS

The author thanks C. Schreiber, C. Pierre, C. Kahle and B. Jones for their constructive reviews of the original manuscript. Useful discussions were held with Professor M. El-Sharkawi, Kuwait University. XRD, SEM and chemical analysis were carried out by the Central Analytical Laboratory of the Kuwait Institute for Scientific Research.

REFERENCES

AALUND, L. (1983) Guide to export crudes, Middle East. *J. Oil and Gas*, **81**, 18, 204–215.

ADAMS, F. & HAJEK, B.F. (1978) Effects of solution sulphate, hydroxide and potassium concentrations on the crystallization of alunite, basalunite and gibbsite from dilute aluminum solutions. *Soil Sci.*, **126**, 169–173.

CUNNINGHAM, C.G., RYE, R.O., STEVEN, T.A. & MEHNERT, H.H. (1984) Origin and exploration significance of replacement and vein-type alunite deposits in the Marysvale volcanic Field, West Central Utah. *Econ. Geol.*, **79**, 50–71.

DAVEY, P.T., LUKASZEWSKI, G.M. & SCOTT, T.R. (1963) Thermal decomposition of the basic aluminum sulphate, $3Al_3O_3{\cdot}4SO_3{\cdot}9H_2O$. *Aust. J. appl. Sci.*, **14**, 137–154.

EL-SHARKAWI, M.A. & KHALIL, M.A. (1977) Glauconite, a possible source of iron for El-Gidida iron ore deposits, Bahariya Oases, Egypt. *Egypt. J. Geol.*, **21**, 1, 109–116.

GOLDBERY, R. (1978) Early diagenetic non-hydrothermal Na-alunite in Jurassic flint clays, Makhtesh Ramon, Israel. *Bull. geol. Soc. Am.*, **89**, 687–698.

HALL, R.B. (1978) World non-bauxite aluminium resources: alunite. *Prof. Pap. US Geol. Surv.*, **1076-A**.

HEALD, M.T. & RENTON, J.J. (1966) Experimental Study of Sandstone Concentration. *J. sedim Petrol.*, **36**, 977–991.

KELLER, W.D., GENTILE, R.J. & REESMAN, A.L. (1967) Allophane and Na-rich alunite from kaolinitic nodules in shale. *J sedim. Petrol.*, **37**, 215–220.

KHALAF, F.I. (1988) Petrography and diagenesis of silcrete from Kuwait, Arabian Gulf. *J. sedim. Petrol.*, **58**, 1014–1022.

KING, D. (1953) Origin of alunite deposits at Pidinga, South Australia. *Econ. Geol.*, **48**, 689–703.

LOMBARDI, G. & SHEPPARD, S.M.F. (1977) Petrographic and Isotopic Studies of the Altered Acid Volcanics of Tolfacerite Area, Italy; the Genesis of the Clays. *Class Clay Miner.*, **12(2)**, 147–162.

PALACHE, C., BERMAN, H. & FRONDEL, C. (1951) *The System of Mineralogy*, 7th edn, Vol. II. John Wiley, London. pp. 558.

ROSS, C.S., BERGQUIST, H.R, MONROE, W.H., FAHEY, J. & ROSS, M. (1968) Natroalunite in Upper Cretaceous sedimentary rocks, North-Central Texas. *J. sedim. Petrol.*, **38**, 4, 1155–1165.

ROUCHY, J.M. & PIERRE, C. (1987) Authigenic natroalunite in middle Miocene evaporites from the Gulf of Suez (Gemsa, Egypt). *Sedimentology*, **34**, 807–812.

STEPHENSON, L.W. (1946) Alunite at Woodbine–Engleford contact in northeastern Texas. *Bull. Am. Ass. Petrol. Geol.*, **3**, 1764–1770.

SHERIDAN, M.F. & ROYSE, C.F. (1970) Alunite: A new occurrence near Wickenburg, Arizona. *Am. Miner.*, **55**, 2016–2022.

ZOTOV, A.V. (1971) Dependence of the composition of alunite on the temperatures of its formations. *Geokhimiya*, **1**, 110–113.

(Manuscript received 16 December 1988; revision received 29 May 1989)

Reprinted from Sedimentology (1990) **37**, *445–454*

Nodular silcretes of the Cypress Hills Formation (upper Eocene to middle Miocene) of southern Saskatchewan, Canada

D. A. LECKIE* *and* R. J. CHEEL†

**Institute of Sedimentary and Petroleum Geology, Geological Survey of Canada, 3303 33rd St NW, Calgary, Alberta, Canada, T2L 2A7 and †Department of Geological Sciences, Brock University, St Catharines, Ontario, Canada, L2S 3A1*

ABSTRACT

Siliceous nodules in the upper Eocene to middle Miocene Cypress Hills Formation in southern Saskatchewan formed on the distal portions of an extensive braidplain. The nodules are similar to silcretes described elsewhere in the literature but their morphology and occurrence suggests that they are of a rare variety. The silcretes are discrete equant to disc-shaped nodules, 1–3 cm thick, and up to 15 cm long. The silicretes form horizontal, discontinous layers parallel to bedding within an unweathered profile. Void spaces and fractures within the nodules are lined or filled with drusy quartz. The silcretes contain less than 0·07% TiO_2. The $TiO_2/SiO_2/Al_2O_3$ and $TiO_2/SiO_2/Fe_2O_3$ ratios are similar to values obtained from nodules formed in an arid to semi-arid environment based on comparison with modern silcretes. This interpretation is confirmed by independent sedimentological and palaeontological climatic evidence for an arid to semi-arid climate throughout Oligocene time in the western Canadian plains.

INTRODUCTION

Horizons of siliceous nodules, interpreted here as silcretes, crop out in upper Eocene to middle Miocene, fine-grained sediments of the Cypress Hills Formation in southwestern Saskatchewan. The siliceous nodules are similar to silcretes described elsewhere but their unusual morphology and occurrence suggests that they are of a rare variety.

A definition of silcretes which we find appropriate is that suggested by Summerfield (1983b): 'Silcrete is an indurated product of surficial and penesurficial (near surface) silicification, formed by the cementation and/or replacement of bedrock, weathering deposits, unconsolidated sediments, soil or other materials and produced by low-temperature physio-chemical processes and not by metamorphic, volcanic, plutonic or moderate to deep burial diagenetic processes.' Silcrete commonly occurs as cement in sand and gravel (Langford-Smith, 1978; Wopner, 1978) but more rarely forms discrete nodules in finer-grained sediments (cf. Smale, 1973; Summerfield, 1982). Most silcretes are thought to indicate formation in 'arid to strongly seasonal' settings (Goudie, 1973; Stephens, 1978) although Summerfield (1983a, b) demonstrated that some recent to Cenozoic silcretes formed under humid coastal conditions. Summerfield (1983a, b) suggested that silcretes which formed under arid or humid conditions can be differentiated by their structure, texture, geochemistry and petrology, as well as the weathering characteristics of ambient sediments.

The purpose of this paper is to describe and interpret an uncommon variety of silcrete from the Cypress Hills Formation, constituting the first thorough description of this variety of nodular silcretes in the literature. Our interpretations provide further insight into the palaeoclimatic significance of these silcretes by providing independent palaeontological and sedimentological evidence for the palaeoclimatic conditions.

GEOLOGICAL SETTING

The Cypress Hills Formation outcrops in the dissected plateaus of the Cypress Hills in southwestern Saskatchewan and southeastern Alberta (Fig. 1). Vertebrate fossils indicate that the age of the formation

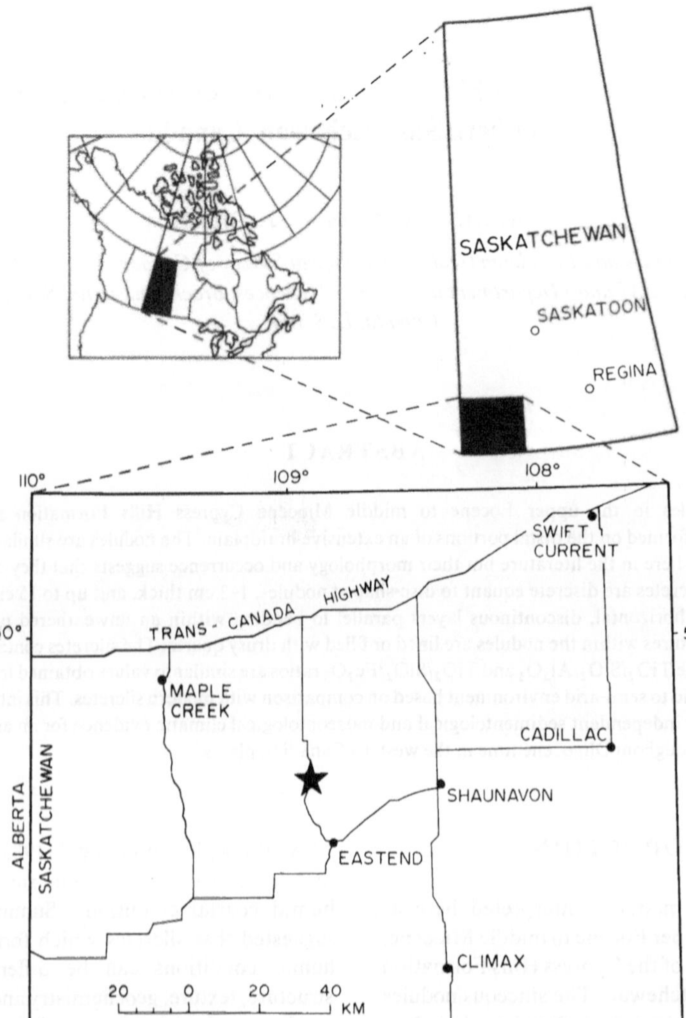

Fig. 1. Location of the roadcut outcrop containing the nodular silcretes (star). The Universal Transverse Mercator (UTM) grid location is 637050mE 5513750mN.

ranges from late Eocene to early Miocene (Russell, 1974; Storer, 1975, 1984). At the location described here the formation consists of predominantly boulder-size gravels deposited in braided channels which cut down into, and are interbedded with, finer-grained, interchannel deposits consisting of conglomeratic breccia, silts, sands, and lacustrine marls and clays. The detailed sedimentology of the Cypress Hills Formation is described by Leckie & Cheel (1989).

The nodular silcretes described herein occur in the eastern part of the Cypress Hills Formation, approximately 15 km northwest of the town of Eastend, Saskatchewan (Fig. 1). The outcrop is within 2 km of the Calf Creek Fossil Local, one of the most important and prolific Tertiary fossil sites in North America (Storer, 1978). The silcretes occur within fine-grained interchannel deposits approximately 17 m from the base of the exposure (Fig. 2).

DESCRIPTION

In outcrop, the silcrete nodules occur in two horizons, each 2–6 cm thick, within the upper 15 cm of a 1-m

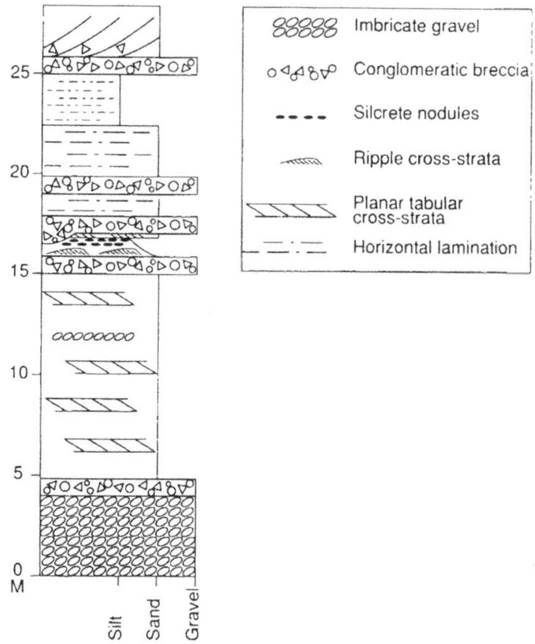

Fig. 2. Measured vertical section of outcrop containing the nodular silcrete.

thick, unconsolidated, upward-fining sand to silt unit (Figs 2 & 3). The sands in the unit are very fine-grained and current-ripple cross-laminated. The overlying silts are parallel-laminated with interbeds of rippled sand. The two horizons of silcrete nodules are superposed, 10 cm apart, within the silts. Both horizons can be traced laterally northwards for approximately 10 m, where they are erosionally truncated by a cross-bedded pebbly sand containing extraformational quartzite gravel and intraformational, brecciated marl and silcrete. Angular silcrete fragments comprise up to approximately 20% of the conglomeratic breccias (debris-flow or bank-collapse deposits; Leckie & Cheel, 1989) with fragments ranging from a few millimetres to several centimetres in size. Primary sedimentary structures in the sediments surrounding the silcrete nodules are well-preserved (Fig. 3) except where they are disrupted by the growth of the silcrete. The silcrete horizons are overlain by cross-stratified medium-grained sands.

The silcretes occur as discrete (i.e. not laterally connected), equant to disc-shaped nodules up to 15 cm long and 1–3 cm thick, lying parallel to horizontal bedding (Fig. 3). Individual nodules have a hardness of 7 on Moh's scale, are white to buff coloured and vary in overall morphology (Fig. 4). In cross-section, the nodules display an outer, white rind surrounding

Fig. 3. Nodular silcrete horizons (arrowed) within silts. Note the lack of weathering (pedogenesis) of surrounding sediments. The exposure is considered as a non-weathering profile in terms of Summerfield (1983a, b). Underlying sediments are ripple cross-laminated, very fine-grained sands. The scale rests on medium-grained, cross-bedded sand which cuts down and erodes the silcrete horizon laterally to the left.

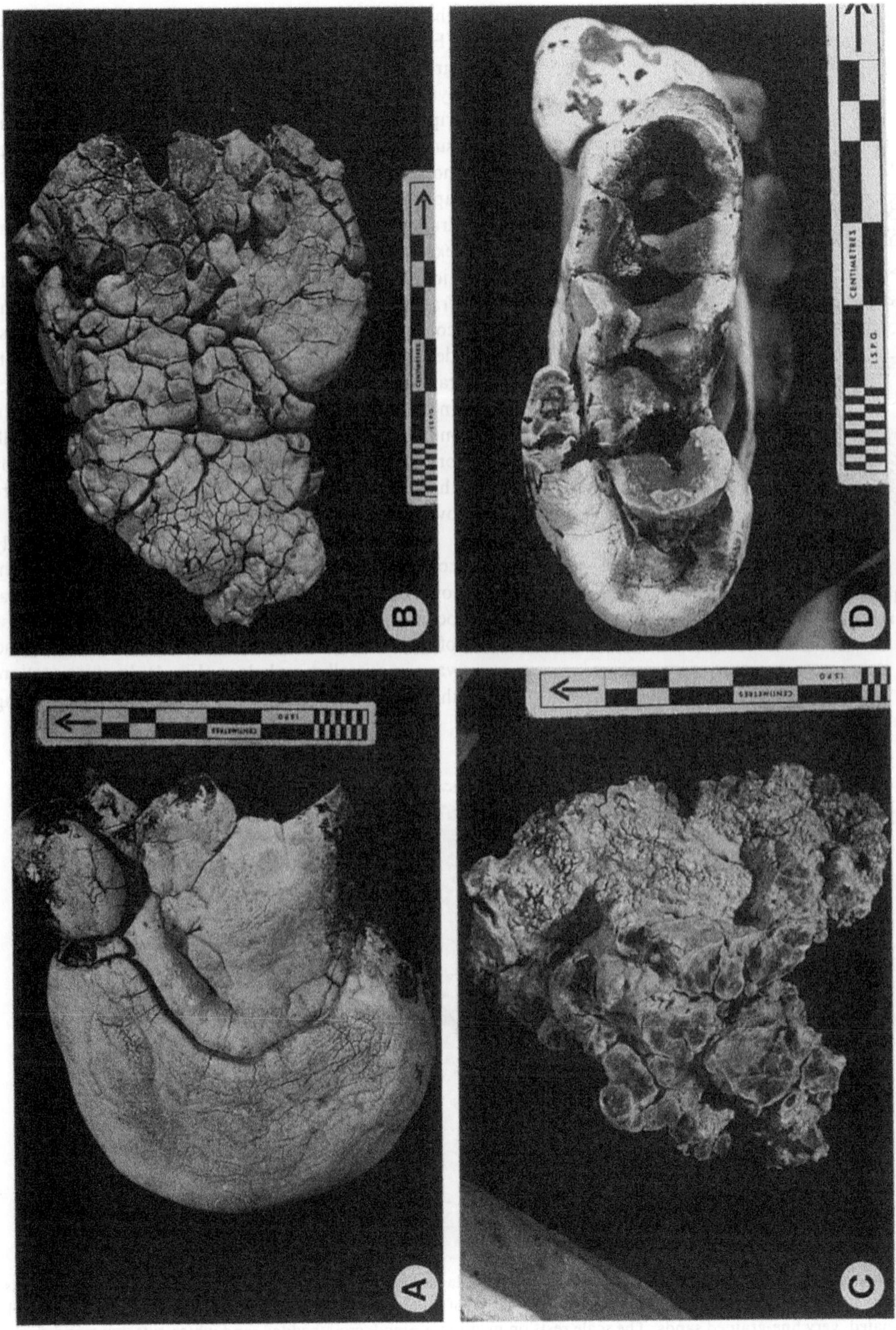

Fig. 4. (A, B, C) Examples of morphology of silcrete nodules. (D) Cross-section of silcrete nodule showing void space lined with drusy quartz.

darker material which may fill the nodule (Fig. 4D). Many of the nodules are geodes ('pedodes' of Brewer, 1976), having a hollow interior lined with drusy quartz (Fig. 4D), whereas others are completely filled. Nodule shape includes tuberose, reniform, lenticular and mamilliated forms (Fig. 4; Read, 1948). Outer surfaces are smooth to intensely fractured (Fig. 4). The surficial fractures commonly form polygonal patterns and are several millimetres deep, perpendicular to the nodule surface.

Petrographically, the silcretes consist largely of a massive microcrystalline quartz matrix which lacks organized fabric (Fig. 5, arrowed 'B'). The matrix contains minor amounts of clay and sesquioxides. In several samples, rounded to angular, larger grains of quartz (<5%; up to 100 μm in diameter) and rare calcite are widely dispersed within the matrix. Rounded to angular rock fragments, ranging in size from medium sand to granule, locally make up over 60% of the nodule material. However, areas rich in clastic fragments are of limited extent and occur adjacent to massive microcrystalline matrix. Circular voids and linear fractures are common within the geodes and are typically lined with drusy quartz (Fig. 5, arrowed 'A'), which may or may not fill the space. Crystal size increases towards the centre of the void or fracture. Where the voids and fractures are not filled, a thin layer of calcite may line the outermost quartz crystals. Several fractures are filled with equigranular quartz crystals, and lack the more common drusy form. The silcretes can generally be considered as massive, M- (matrix) fabric-types of Summerfield's (1983b) classification.

Scanning electron microscopy shows the form of the quartz in the silcrete, as well as the rare calcite (Fig. 6). Well-formed, rhombohedral quartz crystals, 50–200 μm wide, line the void spaces (Fig. 6A). Crystal size varies laterally and the quartz crystals also increase in size outwards from the precipitation surface which appears as interpenetrated-microcrystalline quartz crystals. Calcite locally coats the outer surface of quartz crystals (Fig. 6B).

Bulk chemical analyses of the Cypress Hills silcretes show that they contain 94·9–97·1% SiO_2. X-ray diffraction indicates that the silica is well-crystallized quartz with no indication of opaline silica. Titanium contents are low, ranging from 0·04 to 0·07%. Figure 7 shows plots of $TiO_2/SiO_2/Al_2O_3$ and $TiO_2/SiO_2/Fe_2O_3$ ratios (normalized to 100%) of nine silcrete samples. Oxide ratios of our silcretes fall into the geochemical fields for silcrete from non-weathering profiles (Fig. 7; Summerfield, 1983a, b). Oxides other than SiO_2 occur in minor amounts with Al_2O_3 (0·91–1·41%) and Fe_2O_3 (0·49–1·25%) predominating.

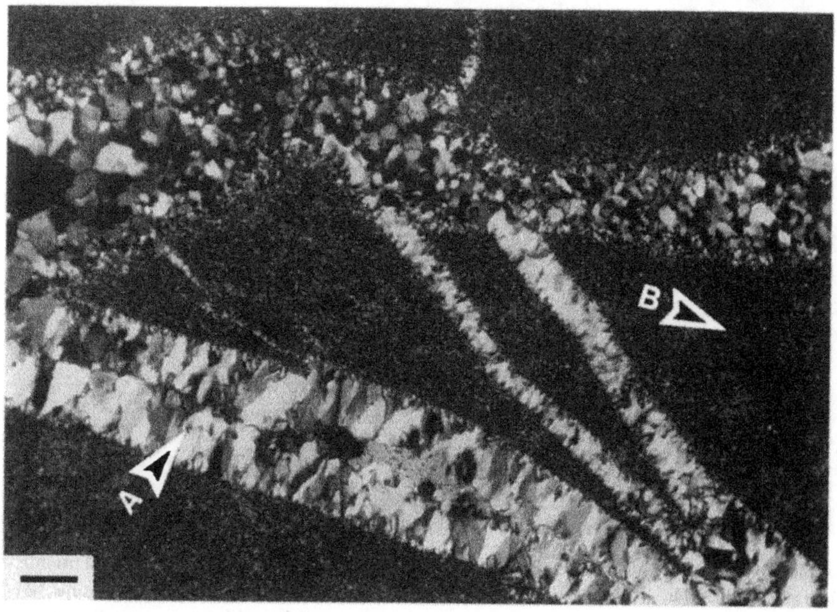

Fig. 5. Thin section of silcrete nodule. Massive microcrystalline quartz matrix having no organized fabric (arrowed 'B'). Drusy-quartz-filled fracture (arrowed 'A'). Scale bar = 0·2 mm.

Fig. 6. Scanning electron microscopy of silcrete nodules. (A) Void lined with quartz crystals, within a nodule. Note variation in grain size. The right of the photograph shows the microcrystalline quartz forming mass of the nodule. (B) Calcite coating (arrow) on quartz crystals within a nodule. The presence of calcite was confirmed by an energy-dispersive analyser on the SEM.

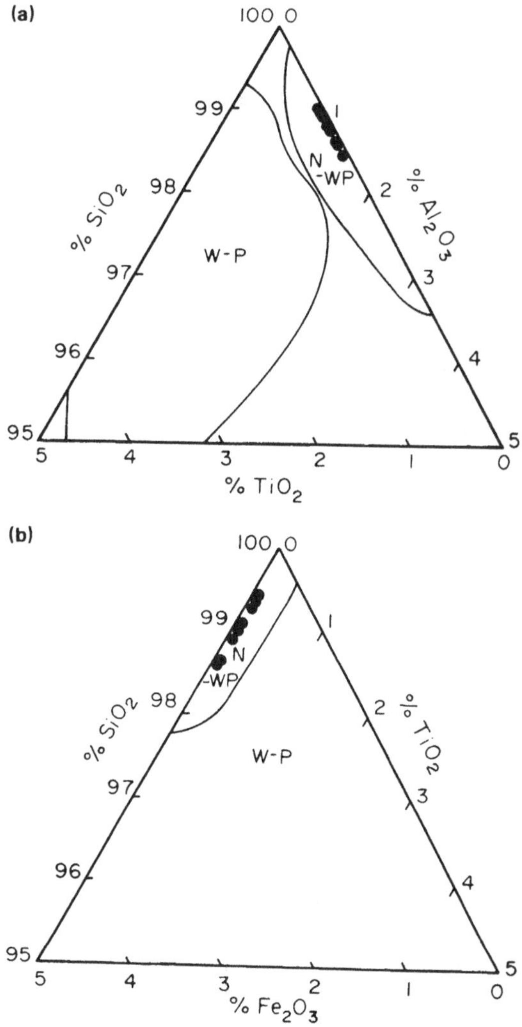

Fig. 7. Plots of (a) $TiO_2/SiO_2/Al_2O_3$ and (b) $TiO_2/SiO_2/Fe_2O_3$ from bulk chemical analyses (normalized to 100%) of the nine silcrete samples. N-WP and W-P refers to fields of non-weathering profile and weathering profile silcretes, respectively, described by Summerfield (1983a, b).

DISCUSSION

The silcrete nodules appear to have formed *in situ*, within the sediments of the interchannel areas of the distal braidplain which deposited the Cypress Hills Formation (Leckie & Cheel, 1989). The ripple cross-laminated sands below the silcretes may represent deposition within a shallow pond or lake. The size of the nodules, relative to associated sands, and their horizontal attitude (i.e. not imbricated) indicates that they have not undergone transport. However, transported silcrete fragments do occur in the overlying conglomeratic breccia. The material constituting the conglomeratic breccia was derived either from bank collapse into channels or was eroded by local floodwaters and sheetflow during torrential rainfalls passing on interchannel areas. In either case, the material which constitutes the conglomeratic breccias was derived by erosion of sediments on interchannel areas, including areas where silcrete had formed. The preservation of silcrete fragments within these deposits implies early silicification at shallow depths, possibly directly on the surface of interchannel areas.

Summerfield (1983a, b) showed that silcretes which formed under humid and arid conditions in southern Africa differ distinctively in terms of their geochemistry. Specifically, the difference lies in the relative concentration of TiO_2. Under humid conditions forming weathering-profile silcretes, TiO_2 becomes concentrated by removal of soluble minerals and localized remobilization of TiO_2 in very low-pH conditions. In arid environments SiO_2 undergoes absolute accumulation and the mechanism for TiO_2 concentration is lacking; consequently, the chemistry of the silcrete remains essentially the same as the ambient regolith. Low TiO_2 concentrations could have resulted from the mixing of silica-rich groundwater and surface water, concurrent with evaporation. The Cypress Hills Formation samples plot in fields (Fig. 7) delineated by silcretes from non-weathering profile settings under arid climatic conditions of the Kalahari desert (Summerfield, 1983a, b). The low TiO_2 concentration in the silcretes from the Cypress Hills Formation indicates that no titanium enrichment has taken place. However, the TiO_2 content alone is not an indicator of climate, because in some instances the host material may be devoid of TiO_2 (i.e. silt and sand typically have a TiO_2 content of $<0{\cdot}5\%$). Low TiO_2 concentrations within the silcrete may have been derived from deeper parent material through which groundwater percolated (cf. Thiry, Ayrault & Grisoni, 1988).

Summerfield (1983c) suggested that non-weathering profile silcretes developed under arid to semi-arid conditions from groundwater of possibly high (but fluctuating) pH. Silcrete formation appears to be favoured in areas where there is poor water drainage and fluctuating water-tables caused by marked variations in rainfall (Smale, 1973). Modern and late Cenozoic, non-weathering profile silcretes are generally associated with semi-arid climatic, calcrete and alkaline pan conditions (Summerfield, 1983c). The predominance of microcrystalline quartz (except where drusy quartz infilled void spaces) within the nodules may be the result of rapid silica precipitation from a supersaturated silica solution. Such silica precipitation could be induced by rainfall dilution or flooding which would cause a sharp shift of pH. During our studies, we did not observe any field or micromorphological evidence of caliche or calcrete in the Cypress Hills sediments. Calcite within the silcrete nodules comprises only a minor proportion and is largely limited to the outer surface of drusy quartz lining fractures, suggesting that calcite was precipitated after the silcrete nodules had formed. Thus, we do not feel that the silcrete nodules formed as a replacement of caliche.

Independent evidence for an arid to semi-arid climate is found in palaeontological evidence from the Cypress Hills Formation. The Calf Creek mammalian, amphibian and reptilian fauna, collected from within 2 km of the siliceous-nodule outcrops, have been inferred to be representative of a typical, balanced woodland savanna assemblage of the North America interior plains growing in a semi-arid climate which was only seasonally wet (Holman, 1969, 1972; J. Storer, pers. comm.). Skwara (1988) concluded that the Hemingfordian-aged Topham local fauna, also from near the occurrence of silcrete nodules, lived in a grassland plain, or dry savanna, with seasonal or intermittent streams. Furthermore, the absence of palynomorphs within the fine-grained sediments of the Cypress Hills Formation may also be the result of arid to semi-arid conditions by analogy with the poor recovery from arid climate, caliche-bearing sediments of Campanian and Maastrichtian rocks of southwestern Alberta (A. Sweet, pers. comm.). The absence of carbonaceous debris within the formation, in comparison with the underlying, coal-bearing Frenchman Formation, may be due to oxidation of organic matter in an arid climate with a low water-table. These climatic interpretations for the Cypress Hills Formation have been corroborated on a more regional basis for late Eocene to Oligocene sediments in the interior of western North America by Russell (1973), Thompson, Fields & Alt (1977, 1982) and Clark, Beerbower & Kietzke (1967).

Ambient sands are not considered the source of silica for the silcrete nodules. Hutton, Twidale & Milnes (1978) suggest that silcretes containing less than 1% TiO_2 form from silica which may have been derived from external sources. Potential sources of silica in solution may be volcanic ash interbedded with the finer-grained sediment of the Cypress Hills Formation or from groundwater enriched in silica. Silica within the groundwater may have been derived from the earlier, intense diagenetic alteration of potassium feldspar to kaolinite (Darragh *et al.*, 1966) in the kaolinite-rich sands of the underlying Whitemud Formation (Master, 1987). A third potential silica source could be biogenic. Twiss (1987) observed that significant amounts of biogenic silica were being derived from grasses in the dry climate of Kansas and concentrated near the surface by runoff and infiltration. In order for silica to have precipitated as quartz (or amorphous/opaline silica precipitation which was later altered to quartz), very slow rates of pore-water migration through the sediment would have been necessary.

A potential modern analogue for silcretes of the Cypress Hills Formation are those within pools along the Nata River, South Africa (Smale, 1973; Summerfield, 1982) although these were not described in detail. On the Nata River, the silcretes occur as horizons of white chalcedonic silica, up to 3 cm thick, within clayey pan (playa) deposits. Pools of water within the Nata River bed have high pH, high evaporation rates and, in some parts of the river bed, Smale (1973) observed flat nodules of silcrete which may be in the process of formation. However, Summerfield (pers. comm.) suggested that the silcretes in the Nata River may have previously formed as pan silcrete. Smale (1973) suggested that the Nata-River examples may have formed where upward-migrating groundwaters encountered downward-percolating less alkaline or more salty surface-waters.

Browne & Smale (1985) described silcretes of the Canterbury region of New Zealand which formed under warm, temperate to subtropical conditions. There, silcretes formed in the absence of weathering profiles but lacked the structures and textures which Summerfield (1983a, b) thought diagnostic of this setting. However, the New Zealand silcretes were relatively rich in TiO_2 (0·5–1·7%), characteristic of humid climatic settings. Similarly, the silcretes which we describe differ texturally and structurally from those of Summerfield (1983a, b) but are geochemically similar to those formed in an arid to semi-arid climate.

CONCLUSIONS

Siliceous nodules in the upper Eocene to middle Miocene Cypress Hills Formation in southern Saskatchewan formed on the distal portions of a widespread braidplain. The siliceous nodules differ from most previously described silcretes and provide an opportunity for a detailed description of this rare form. The silcretes are discrete, equant to disc-shaped nodules, 1–3 cm thick, and up to 15 cm long, forming horizontal layers parallel to bedding within an unweathered profile. Void and fracture spaces within the nodules are commonly lined or filled with drusy quartz. The silcretes contain less than 0·07% TiO_2 and $TiO_2/SiO_2/Al_2O_3$ and $TiO_2/SiO_2/Fe_2O_3$ ratios indicate that the nodules formed in an arid to semi-arid environment based on comparison with data of modern silcretes. This climatic interpretation for the Oligocene in the western Canadian plains is confirmed by independent sedimentological and palaeontological evidence in associated sediments.

ACKNOWLEDGMENTS

Financial support for this work by Energy, Mines and Resources Canada (Research Agreement No. 92) and the Natural Sciences and Engineering Research Council of Canada (Operating Grant No. A1618), to R.J.C., is gratefully acknowledged. We are also grateful to the following for their various contributions: A. Heinrich and J. Wong for conducting the geochemical analyses; J. Storer for interpretation and assessment of the palaeontological data from the Cypress Hills Formation; and U. Brand and G. Ross for their comments on earlier versions of this paper. The manuscript benefited from reviews by M. Summerfield and M. Thiry.

REFERENCES

BREWER, T. (1976) *Fabric and Mineral Analysis of Soils.* Kreiger, Huntington, NY, 482 pp.

BROWNE, G.H. & SMALE, D. (1985) An occurrence of silcrete from the Harper Hills, central Canterbury. *NZ geol. Rec.*, **8**, 128–134.

CLARK, J., BEERBOWER, J.R. & KIETZKE, K.K. (1967) Oligocene sedimentation, stratigraphy, paleoecology and paleoclimatology in the Big Badlands of South Dakota. *Fieldianda: Geol. Mem. Field Museum Nat. History*, **5**, 158 pp.

DARRAGH, P.J., GASKIN, A.J., TERRELL, B.C. & SANDERS, J.V. (1966) Origin of precious opal. *Nature*, **209**, 13–16.

GOUDIE, A. (1973) *Duricrusts in Tropical and Subtropical Landscapes.* Clarendon Press, Oxford, 174 pp.

HOLMAN, J.A. (1969) Lower Oligocene amphibians from Saskatchewan. *Q. J. Florida Acad. Sci.*, **31**, 273–289.

HUTTON, J.T., TWIDALE, C.R. & MILNES, A.R. (1978) Characteristics and origin of some Australian silcretes. In: *Silcrete in Australia* (Ed. by T. Langford-Smith), pp. 19–40. Department of Geography, University of New England.

LANGFORD-SMITH, T. (1978) A select review of silcrete research in Australia. In: *Silcrete in Australia* (Ed. by T. Langford-Smith), pp. 1–12. Department of Geography, University of New England.

LECKIE, D.A. & CHEEL, R.J. (1989) The Cypress Hills Formation (Upper Eocene to Miocene): a semi-arid braidplain deposit resulting from intrusive uplift. *Can. J. Earth Sci.*, **26**, 1918–1931.

MASTER, P.P. (1987) Evaluation and geology of kaolinitized sediments, Wood Mountain, Saskatchewan. In: *Economic Minerals of Saskatchewan* (ed. by C.F. Gilboy and L.W. Vigrass). *Spec. publs Saskatchewan Geol. Soc.*, **8**, 166–187.

READ, H.H. (1948) *Rutley's Elements of Mineralogy.* Thomas Murby and Co., London.

RUSSELL, L.S. (1973) Geological evidence on the extinction of some large terrestrial vertebrates. *Can. J. Earth Sci.*, **10**, 140–145.

RUSSELL, L.S. (1974) Revision of the fossil horses of the Cypress Hills Formation (Lower Oligocene) of Saskatchewan. *Can. J. Earth Sci.*, **12**, 636–648.

SKWARA, T. (1988) Mammals of the Topham Local Fauna: Early Miocene (Hemingfordian), Cypress Hills Formation, Saskatchewan. *Saskatchewan Museum of Natural History, Natural History Contribution* No. 9, 169 pp.

SMALE, D. (1973) Silcretes and associated silica diagenesis in southern Africa and Australia. *J. sedim. Petrol.*, **43**, 1077–1089.

STEPHENS, C.G. (1978) Silcrete. In: *The Encyclopedia of Sedimentology* (Ed. by R. W. Fairbridge and J. Bourgeois), pp. 741–742. Dowden, Hutchinson and Ross, Inc., Stroudsburg, PA.

STORER, J.E. (1975) Middle Miocene mammals from Cypress Hills, Canada. *Can. J. Earth Sci.*, **12**, 520–522.

STORER, J.E. (1978) Tertiary sands and gravels in Saskatchewan and Alberta: correlation of mammalian faunas. In: *Western and Arctic Canadian Biostratigraphy* (Ed. by C. R. Stelck and B. D. E. Chatterton), *Geol. Ass. Can. Spec. Paper*, **18**, 595–602.

STORER, J.E. (1984) Fossil land mammals of the Southfork Local Fauna (early Chadronian) of Saskatchewan. *Can. J. Earth Sci.*, **21**, 1400–1405.

SUMMERFIELD, M.A. (1982) Distribution, nature and probable genesis of silcrete in arid and semi-arid southern Africa. In: *Aridic Soils and Geomorphic Processes* (Ed. by D. H. Yaalon), pp. 37–65. Catena Supplement 1, Braunschweig.

SUMMERFIELD, M.A. (1983a) Silcrete as a paleoclimatic indicator: Evidence from southern Africa. *Palaeogeogr. Palaeoclim. Palaeoecol.*, **41**, 65–79.

SUMMERFIELD, M.A. (1983b) Silcrete. In: *Chemical Sediments and Geomorphology* (Ed. by A. S. Goudie and K. Pye), pp. 59–91. Academic Press, London.

SUMMERFIELD, M.A. (1983c) Petrography and diagenesis of silcrete from the Kalahari Basin and Cape coastal zone, southern Africa. *J. sedim. Petrol.*, **53**, 895–909.

THIRY, M., AYRAULT, M.B. & GRISONI, J.C. (1988) Groundwater silicification and leaching in sands: example of the Fountainbleau Sand (Oligocene) in the Paris Basin. *Geol. Soc. Am. Bull.*, **100**, 1283–1290.

THOMPSON, G.R., FIELDS, R.W. & ALT, D. (1977) Paleoclimatic interpretation of paleosols in the Northern Rockies. *Geol. Soc. Am.*, Rocky Mountain Section, Missoula, Montana, May 12–13, *Abstract Programs*, **9**, 768–769.

THOMPSON, G.R., FIELDS, R.W. & ALT, D. (1982) Land-based evidence of Tertiary climatic variations, northern Rockies. *Geology*, **10**, 413–417.

TWISS, P.C. (1987) Terrigenous biogenic silica as a possible source of late diagenetic chert in Permian Limestones in Kansas. *Geol. Soc. Am.*, 1987 Annual Meeting and Exposition, Phoenix, Arizona, October 26–29, *Program with Abstracts*, p. 873.

WOPNER, H. (1978) Silcretes of northern South Australia and adjacent regions. In: *Silcrete in Australia* (Ed. by T. Langford-Smith), pp. 93–141. Department of Geography, University of New England.

(Manuscript received 30 December 1988; revision received 30 October 1989)

Reprinted from Sedimentology (1991) ***38****, 1153–1171*

Rock varnish in the Sonoran Desert: microbiologically mediated accumulation of manganiferous sediments

BARTHOLOMEW NAGY, LOIS ANNE NAGY, MARK J. RIGALI, WILLIAM D. JONES

Laboratory of Organic Geochemistry, Department of Geosciences, The University of Arizona, Tucson, AZ 85721, USA

DAVID H. KRINSLEY

Department of Geology, Arizona State University, Tempe, AZ 85287, USA

NORVAL A. SINCLAIR

Department of Microbiology and Immunology, The University of Arizona, Tucson, AZ 85721, USA

ABSTRACT

Rock varnish occurs in virtually all environments, most commonly in arid and semi-arid climates, including Antarctica. Rock varnish consists of thin layers of intimately mixed aeolian and chemical sediments often showing botryoidal and more rarely stromatolite-like morphologies. Typical rock varnish samples collected at Twin Peak Mountain Park, near Phoenix, Arizona, consist of abundant quartz, with plagioclase, illite and a mixed layer, Fe-clay mineral, probably corrensite. EDS, SEM (BSE) and TEM analyses revealed that the typical Mn,Fe minerals occur as minute particles; some of these particles and other mineral grains are attached to filaments. XRD and electron diffraction showed that the Mn,Fe-bearing particles are poorly crystalline. The filaments, based on morphological criteria, are virtually indistinguishable from fungal filaments. Most filaments are fragments, probably broken by scraping during sample collection. Coccoid and rod-shaped forms, resembling cyanobacteria and other bacteria, respectively, are also present. Unlike definitive minerals, these filaments disintegrated in the concentrated energy of the SEM electron beam at the instrumental and experimental conditions used. In addition, no filamentous, rod-shaped or coccoid forms were observed in samples hydrolysed with 6 N HCl for 24 h at 100°C. Bacteria and fungi in powdered rock varnish were cultured on four media, incubated aerobically in the dark at 25°C. The culture media yielded dense growths of spore-forming bacteria and filamentous fungi. One fungus and two *Bacillus* isolates oxidized and concentrated manganese. Control experiments revealed that fungi and bacteria are present on and below the surfaces of rock varnish. Free and hydrolysed, peptide/protein-bound amino acids were identified in the rock varnish. Amino acids showed virtually no racemization with the exception of D/L asp = 0·1. Relatively high molecular weight humic matter was also separated from the rock varnish. High-resolution mass spectrometry revealed non-hydrocarbon moieties, similar to a Suwannee River (FL) humic acid standard. Micro-organisms and their original biochemical compounds do not seem to be preserved for long in the accreting varnish layer. The studies showed that the filaments helped to trap mineral particles of rock varnish, and that bacteria and fungi abetted Mn concentration. Some structures in the layers of rock varnish resemble stromatolites and present definitions would allow them to be termed as such.

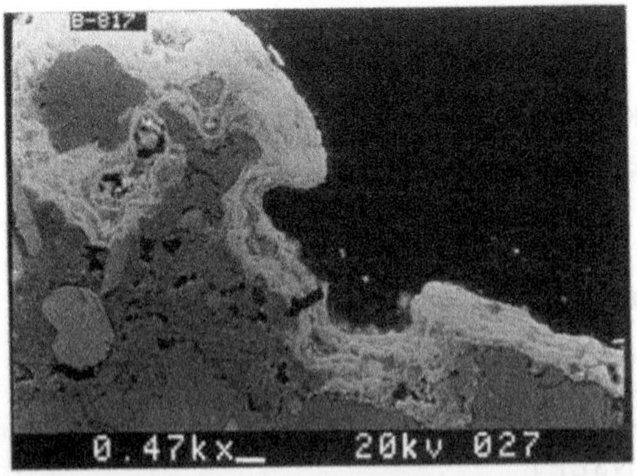

Fig. 1. Backscattered scanning electron micrograph of a polished cross-section of rock varnish (light coloured and laminated) on granite (grey, shown by arrow), Twin Peak Mountain Park, near Phoenix, Arizona. White scale bar at base next to 'kx' = 10 μm.

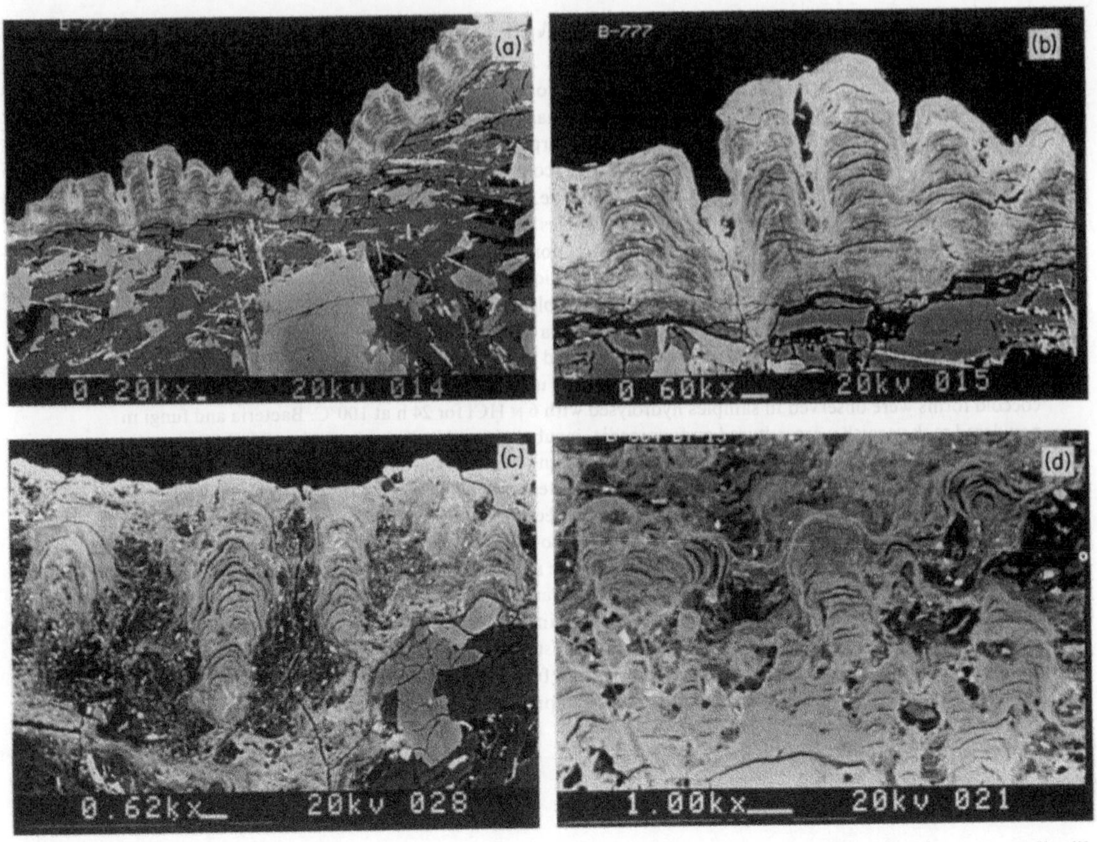

Fig. 2. Backscattered scanning electron micrographs of polished cross-sections of rock varnishes showing stromatolite-like morphologies: (a), (b) on an alluvial fan, Death Valley, California; (c) on a glacial moraine, White Mountains, Nevada; (d) on Bishop Tuff, Bishop, California. White scale bars at base next to 'kx' = 10 μm.

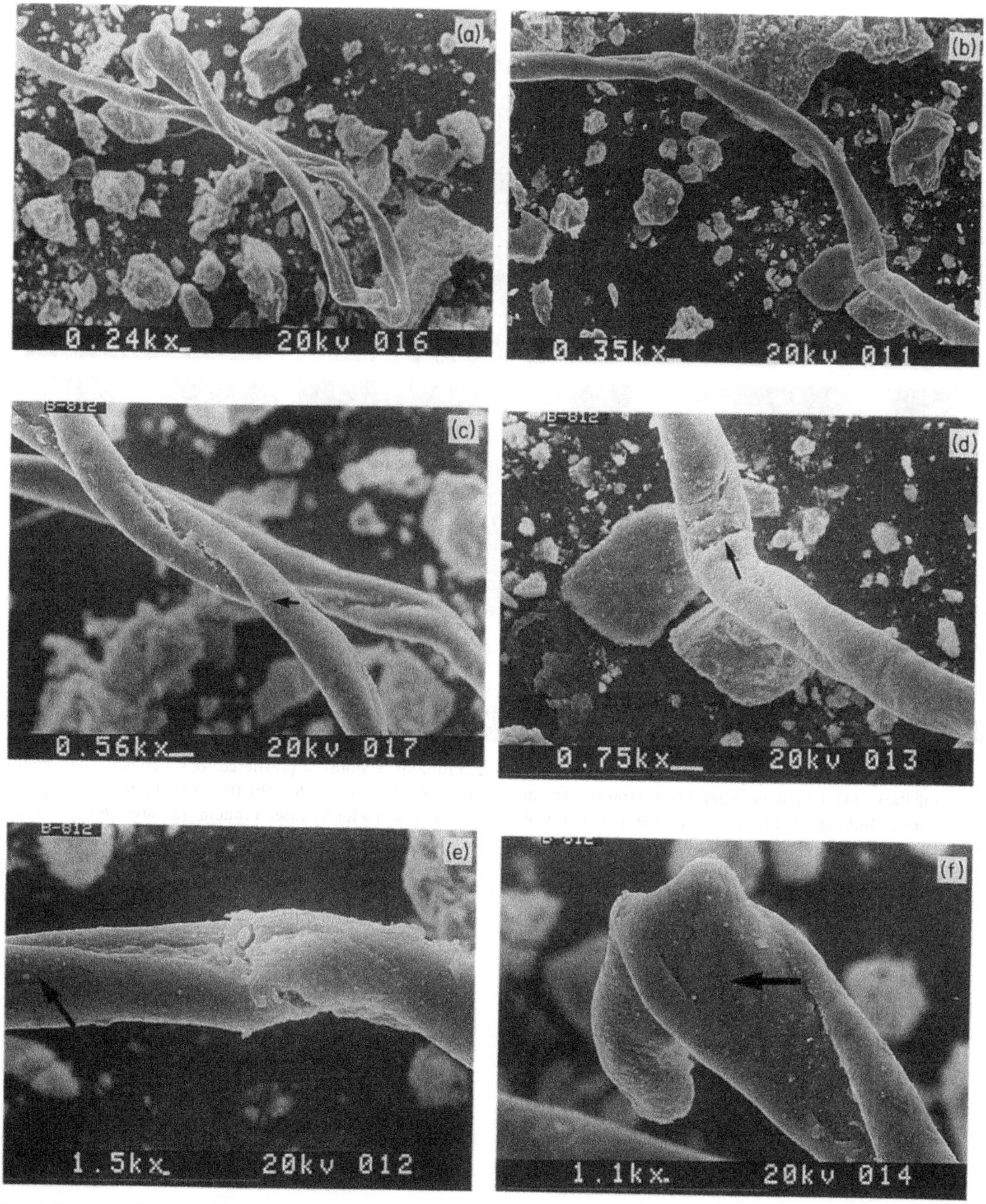

Fig. 4. Scanning electron micrographs of filaments (probably fungi) in powdered rock varnish, Twin Peak Mountain Park: (a)–(c) bent, twisted and twined filaments; (d), (e) fractures and poor preservation. Arrow on (d) points to what appears to be a brittle fracture. Arrows on (c), (e) and (f) point to the bound, cryptocrystalline Mn,Fe mineral, which occurs abundantly on the filament in (f). White scale bars at the base next to 'kx' = 10 μm in (a)–(d) and 1 μm in (e) and (f).

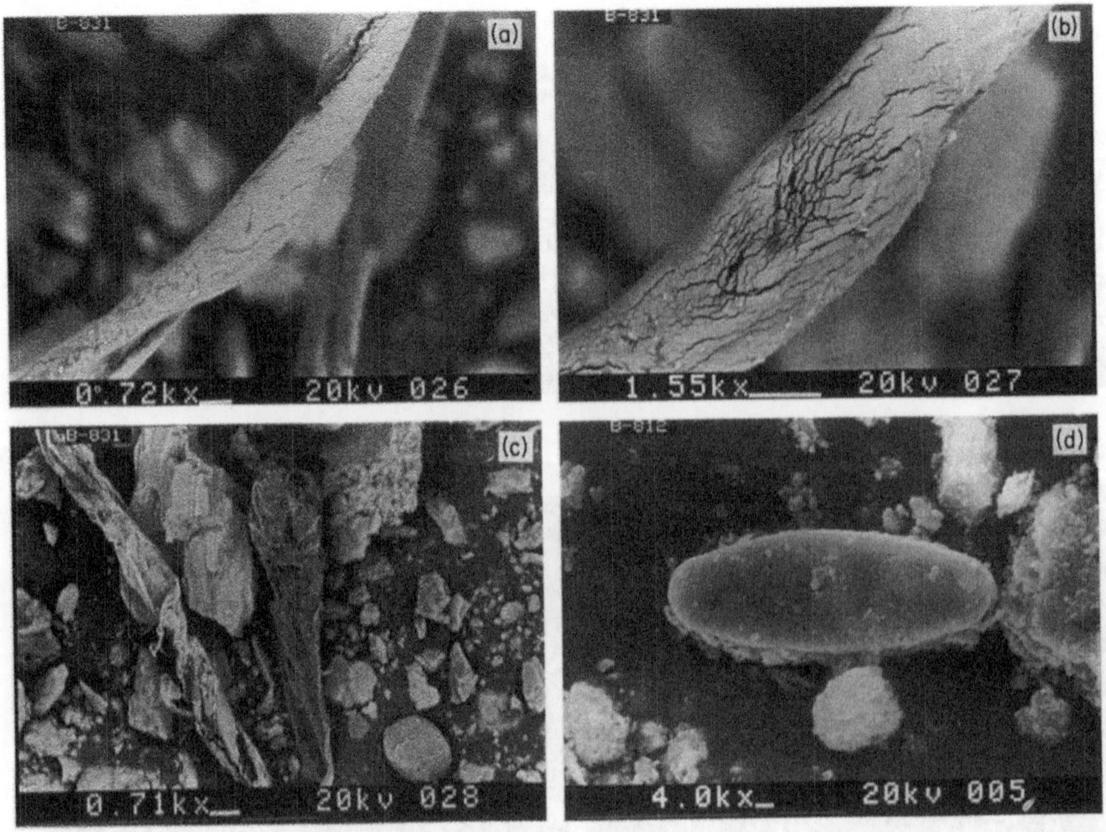

Fig. 5. Scanning electron micrographs of filaments and a particle (probably diatom) in powdered rock varnish, Twin Peak Mountain Park: (a) a filament after brief exposure to the electron beam of the SEM; (b) the same filament at higher magnification after exposure to more concentrated energy of the electron beam which caused random fractures on the wall of the filament—these fractures are absent in filaments shown in (a) and (c). The filament in (c) was substantially damaged, probably by scraping the rock varnish from the underlying granite during sample collection. Most filaments appear to be torn and broken in powdered preparations. The particle shown in (d) is apparently a diatom; note the pores around its edges. White scale bars at the base = 10 μm in (a)–(c) and 1 μm in (d).

Reprinted from Sedimentology (1998) ***45****, 711–725*

Models of rock varnish formation constrained by high resolution transmission electron microscopy

D. KRINSLEY
Department of Geological Sciences, University of Oregon, Eugene OR 97403–1272, USA
(E-mail: krinsley@oregon.uoregon.edu)

ABSTRACT

Nanometre-scale transmission electron microscope imagery of manganiferous rock varnishes from Death Valley, California, Peru, Antarctica, and Hawaii confirms prior infrared mineralogy studies. The building blocks of rock varnish are clay minerals that are cemented to the rock by oxyhydroxides of manganese and iron. Rock varnish is layered on the scale of nanometres, with the basic structure defined by the subparallel alignment of detrital clay minerals. Although only a few examples of bacteria were found, possible cell-wall encrustations are ubiquitous and aligned with the clay minerals. Mn–Fe oxides appear to be mobilized from bacterial casts and then reprecipitate on clay minerals that weather into monolayers. These observations have implications for varnish dating and palaeoenvironmental techniques: K–Ar and uranium-series dating of rock-varnish oxides can only yield minimum ages; however, this small spatial scale of mobilization would not alter the signal from micron-scale microchemical laminations.

Reprinted from Sedimentology (1991) ***38***, *539–551*

Calcretes related to phreatophytic vegetation from the Middle Triassic Otter Sandstone of South West England

K. PURVIS* *and* V. P. WRIGHT

Postgraduate Research Institute for Sedimentology, The University, PO Box 227, Whiteknights, Reading RG6 2AB, UK

ABSTRACT

Carbonate cementation in the Triassic Otter Sandstone of Budleigh Salterton, SW England, occurs in two distinct forms: large, vertical, concentrically zoned cyclinders, and thin subhorizontal sheets. The former represent rhizocretions of the tap roots of phreatophytic plants which colonized bars and abandoned channels on a large braidplain. The sheets represent cementation around ancient water-tables. The precipitation of the rhizocretions took place, at least in part, during the life of the plants and $\delta^{13}C$ and $\delta^{18}O$ values support the view that evapo-transpiration induced carbonate precipitation. Palaeosol profiles are rare in the Otter Sandstone, reflecting the geomorphological instability of the braidplain surfaces. In contrast, rhizocretionary and sheet calcretes, which formed several metres below the active depositional surface, are abundant because they had a high preservation potential.

INTRODUCTION

Palaeosols are important components of ancient alluvial sequences and have proved useful in a wide variety of ways for palaeo-environmental reconstruction. They have been well documented from ancient alluvial fan (e.g. Steel, 1974) and high-sinuosity fluvial systems (e.g. Allen, 1986) where confined channel belts, incision and terracing, and avulsion result in areas becoming isolated from net deposition or erosion, allowing prolonged pedogenesis. However, in low-sinuosity, braided channel systems, geomorphological surfaces are less stable and not surprisingly there are few descriptions of well-developed palaeosols from ancient deposits formed in such settings. In this paper we describe palaeocalcretes from a Triassic braided river system from South West England which represent carbonate precipitation around the tap roots of phreatophytic plants and rising water-tables. These carbonates formed well below the active land surface and had a higher preservation potential than soils *sensu stricto*.

* Present address: K.S.E.P.L., 2280 AB Rijswijk Z.H., The Netherlands.

TRIASSIC CALCRETES

Geological setting

The Middle Triassic Otter Sandstone crops out along the coast of southeast Devon between Budleigh Salterton (British Grid Reference SY056814) and Sidmouth (SY115869). The carbonate horizons analysed in this study are exposed at Budleigh Salterton (SY071819) and in the cliff to the east of the River Otter (SY080821) (Fig. 1a).

The Permo-Triassic sequence in the area consists of red beds (Fig. 1b). The Upper Permian Littleham Mudstone (a playa mudflat deposit; Mader, 1985) is overlain by the Lower Triassic Budleigh Salterton Pebble Beds, a thick braided stream conglomerate sequence (Mader, 1985, 1990). The Otter Sandstone is Anisian to Ladinian in age and is 118·8 m thick in a borehole at Great Well, near Ottery St Mary (SY105955) (Sellwood *et al.*, 1984); its thickness at Budleigh Salterton cannot be assessed because the outcrop is discontinuous. The junction between the Otter Sandstone and underlying Pebble Beds is marked by a layer of ventifacts overlain by cross-bedded aeolian sandstones which pass up into horizontally-stratified and cross-bedded fluvial sandstones.

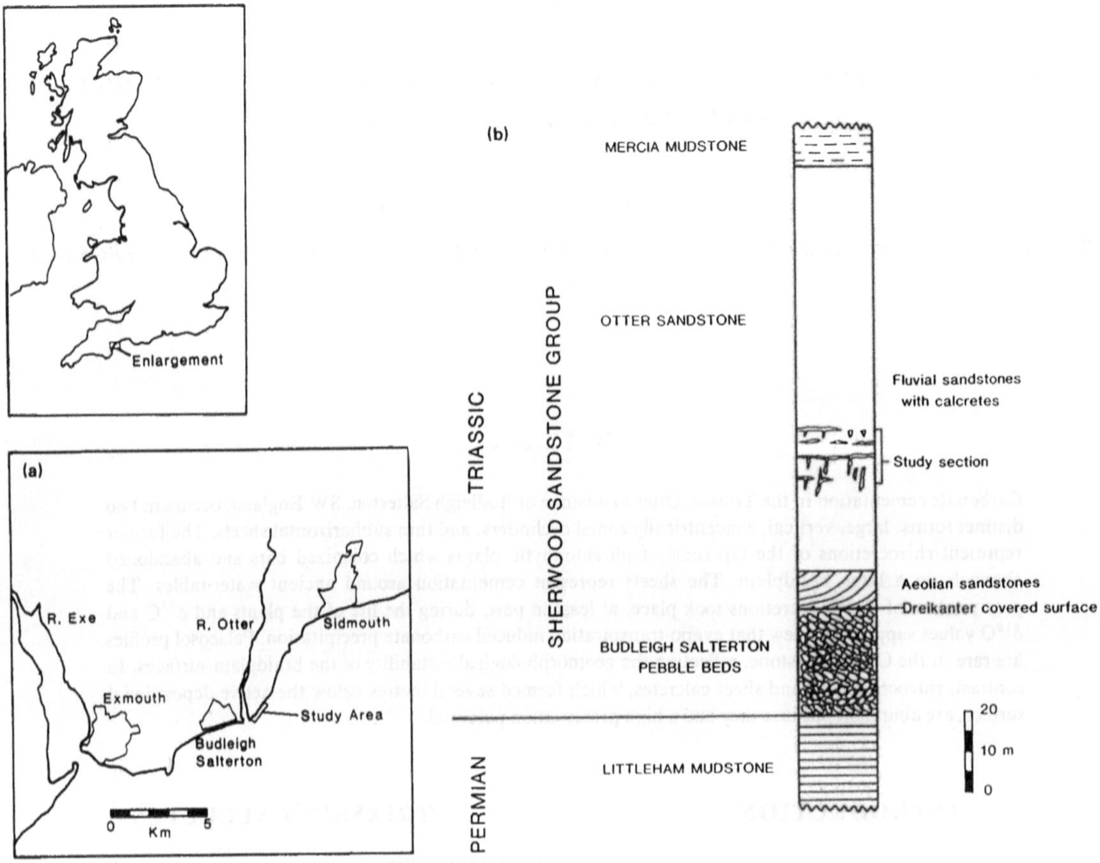

Fig. 1. (a) Study area on the south Devon coast, South West England. (b) Schematic sedimentary log of the Upper Permian and Triassic units exposed on the south Devon coast.

The Otter Sandstone consists of aeolian sandstones and braidplain deposits (Mader & Laming, 1985). The calcrete-bearing units consist of cycles, each with an erosional base, frequently with a channel-lag conglomerate, that grade up through cross-bedded fluvial sandstones into laminated siltstones and abandonment mudstones that are rarely preserved. Many of the intraformational conglomerates contain reworked calcrete clasts. Deposition was in low-sinuosity, braided streams with a palaeocurrent flow from the south (Henson, 1971). Carbonate cementation within the Otter Sandstone occurs as vertically elongate, cylindrical concretions and sheets (Fig. 2). The calcite sheets comprise laterally continuous calcite accumulations, orientated parallel to bedding surfaces, typically 0·1 m thick and up to 10 m in length. In comparison, the elongate, cylindrical concretions are orientated perpendicular to bedding surfaces, although they occasionally possess lateral offshoots parallel to bedding. The margins of the concretions are sharply defined, have an irregular outer surface, and commonly exceed 1 m in length with diameters of 0·1–0·15 m. They are commonly found concentrated in certain sandstone horizons which can be traced laterally for several tens of metres, and may be overlain by channel-lag conglomerates containing reworked concretionary material. Discrete palaeosol profiles have not been positively identified and it has not been possible to assess the relationships of the carbonates described above to any obvious subaerial surface.

Details of the sedimentological background to these deposits, and numerous field illustrations of the sandstones and carbonates, are to be found in Mader (1985, 1990, pp. 223–260).

Fig. 2. (a) Horizontal sheet calcretes and vertical, cylindrical concretions in sandstone at Budleigh Salterton. (b) Cylindrical concretions shown in (a).

MICROSTRUCTURE

Vertical structures

The vertical structures are composed of non-ferroan calcite and exhibit three distinct zones (Figs 3 & 4).

Zone 1 (Fig. 4). This consists mainly of calcite spar containing 'floating' micrite lumps (Fig. 5a) and silicate grains. The spar crystal sizes range from very fine to medium crystalline (in the sense of Folk, 1965) and are typically equant. The crystals (Fig. 5b) range from subhedral (with some euhedral rhombic crystals) to anhedral. Although the mosaics exhibit a high percentage of non-planar crystal boundaries, highly irregular, lobate boundaries are very rare. Areas of drusy calcite also occur.

Small extensions of this zone radiate outwards through zone 2 (Figs 3 & 4). Within these spar areas there are, centrally placed, laminated micrite-lined tubes (confirmed from transverse sections), filled by drusy calcite cement (Fig. 5c, d). The walls range in thickness from a few tens of micrometres to over 100 μm, and consist of from two to five or more micritic laminae (Fig. 5d, e). The longitudinal variations in the thickness and orientation of these structures suggest that the cement-filled tubes are moderately sinuous in form.

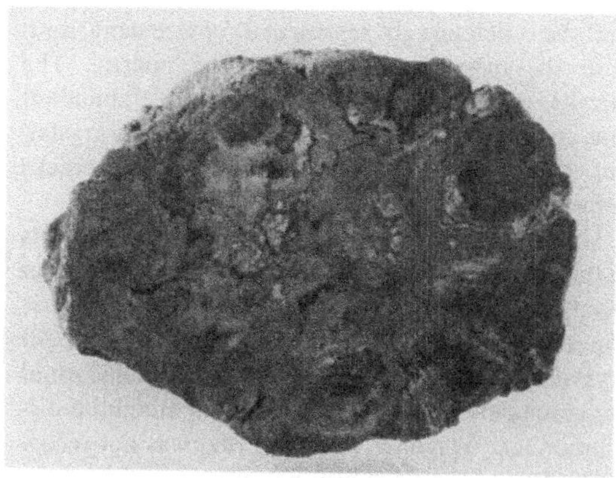

Fig. 3. Polished, transverse section through a cylindrical concretion. Note central carbonate zone with radiating extensions. Width of concretion = 6·8 cm.

The micrite lumps range from a few hundred micrometres to 10 mm in diameter (Fig. 5a). They

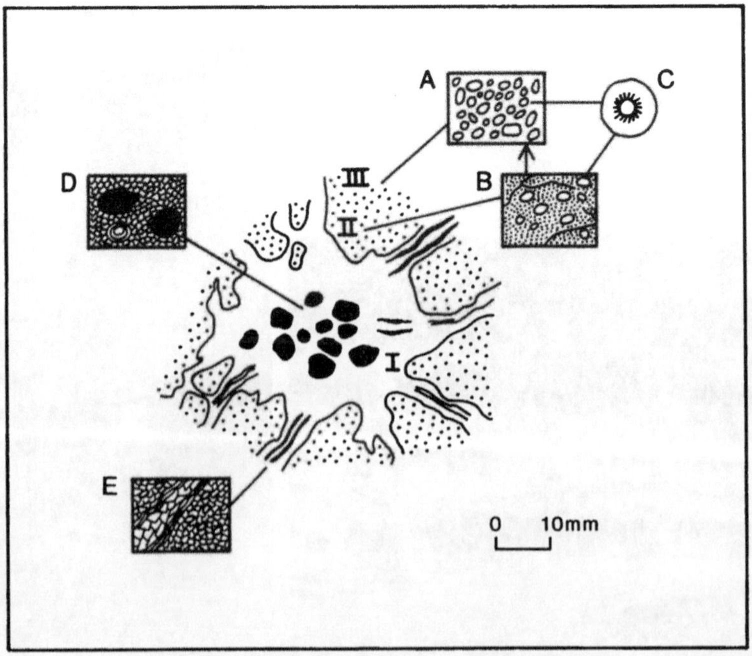

Fig. 4. Schematic showing the structure of a typical concretion, with zones 1–3 (see text). A: host sandstone cemented by sparry calcite (Fig. 9). B: cement-supported sandstone with dense micrite cement (Fig. 8). C: coronas on silicate grains (Fig. 8a). D: micro-pseudospar zones with micrite lumps (Fig. 5a). E: micrite-lined extension from zone 1 (Fig. 5c–e).

generally have sharp contacts with the surrounding spar but locally irregular contacts occur. This suggests that the contact might represent a replacement front, with the spar 'encroaching' on the micrite. The micrites contain floating silicate grains, abundant fine, commonly sinuous cracks, filled by finely crystalline calcite (Fig. 5a) and circum-granular cracks defining 'nodules'.

Under cathodoluminescence the spar exhibits highly varied zonation. Typically, the crystals have an inner weakly luminescent area in which the concentric zones are irregular, even wavy, in form and straight crystal faces are rare (Fig. 6). The individual zones are commonly non-isopachous although definite dissolution (cf. Wright & Peeters, 1989) was not recognized. Sector zoning is also not apparent. The outer parts of the crystals show a more regular zonation with clear, euhedral faces (Fig. 6b), of light and dark bands. The intercrystalline spaces are filled with brightly luminescing calcite which also fills cracks through the mosaics.

Straight compromise boundaries between crystals are very rare even where the later-stage, brightly luminescing cement is absent. This may be a function of the irregular zoning on the crystals (Fig. 7) and it appears that individual growth zones commonly thin where one crystal is in close proximity to another (Fig. 7), and preferred growth has occurred on faces furthest from approaching faces of other crystals.

A particular effort was made to find evidence of any displacive growth in these mosaics, such as impeded growth geometrics (Maliva, 1989), but none was found.

Zone 2 (Fig. 4) consists of a discontinuous zone, cut through by the radiating extensions of zone 1, composed of dense micrite (Fig. 8a). This micrite contains irregular, fine cracks filled with spar and floating silicate grains, which typically have coronas of fibrous calcite (Fig. 8). This zone passes outwards gradationally over several millimetres into zone 3.

Under cathodoluminescence this micrite, like that in zone 1, luminesces brightly with a very finely speckled pattern.

Zone 3 (Fig. 4) has a predominantly grain-supported fabric and a fine to medium crystalline spar cement (Fig. 9a). This zone passes out gradationally into the less well-cemented host rock. The spar cements are weakly luminescent.

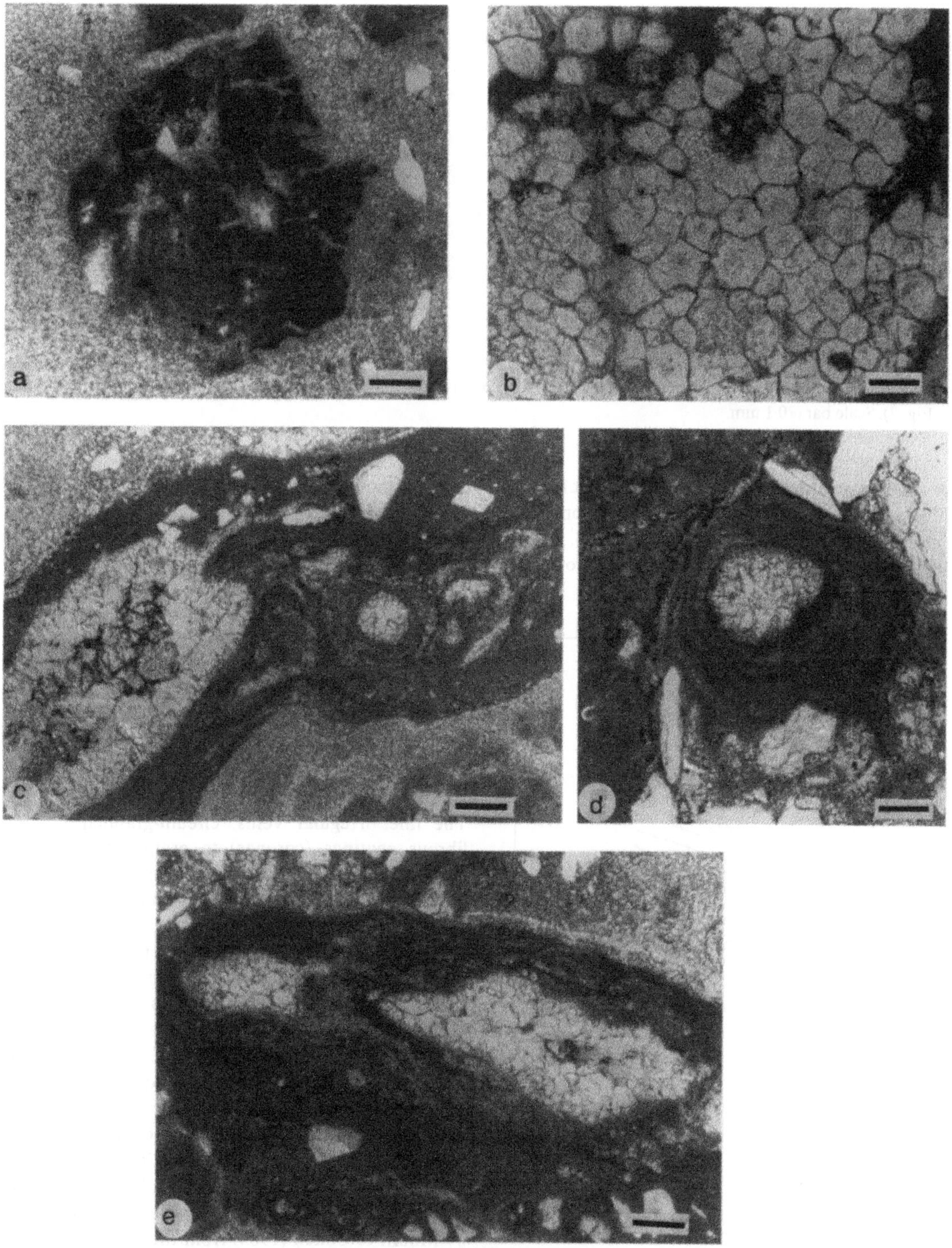

Fig. 5. Photomicrographs of zone 1. (a) Calcite spar with 'floating' lumps of micrite and silicate grains. Note the fine fracture pattern in the lump (scale bar = 0·2 mm). (b) Granular calcite mosaic (scale bar = 0·1 mm). (c) Micrite-lined walls to sinuous spar-filled extension of zone 1. Note traverse section showing tube-like micrite coatings, probably representing a rhizocretion (scale bar = 0·2 mm). (d) Detail of small rhizocretion with laminated walls (scale bar = 0·1 mm). (e) Oblique longitudinal section through one of the micrite-walled extensions, filled by drusy calcite cement (scale bar = 0·2 mm).

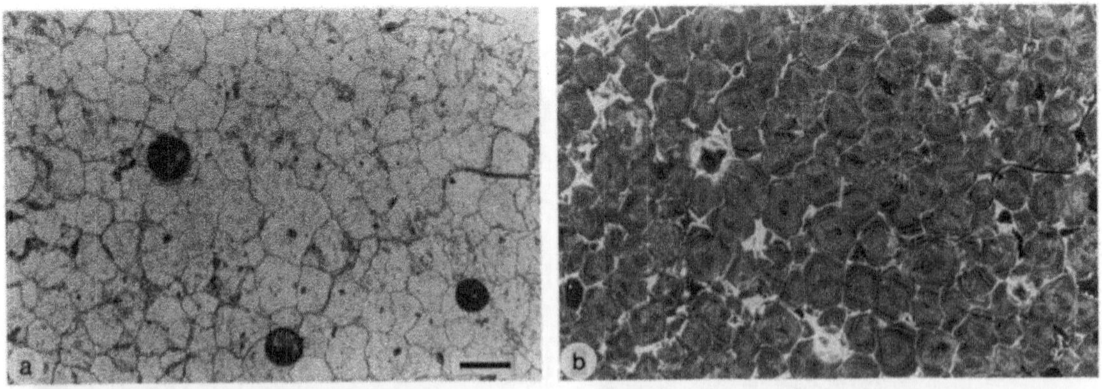

Fig. 6. Photomicrographic pair showing calcites from zone 1. (a) Granular mosaic under plane polarized light. (b) Growth zones illuminated under cathodoluminescence. Whereas solution discontinuities are rare, non-isopachous zonation is common (see Fig. 7). Scale bar = 0·1 mm.

A small percentage of silicate grains in each zone exhibit the effects of dissolution, but these appear to be more common in this outer zone (Fig. 9b). In the two outer zones most silicate grains have fibrous rim cements (coronas), both isopachous and non-isopachous.

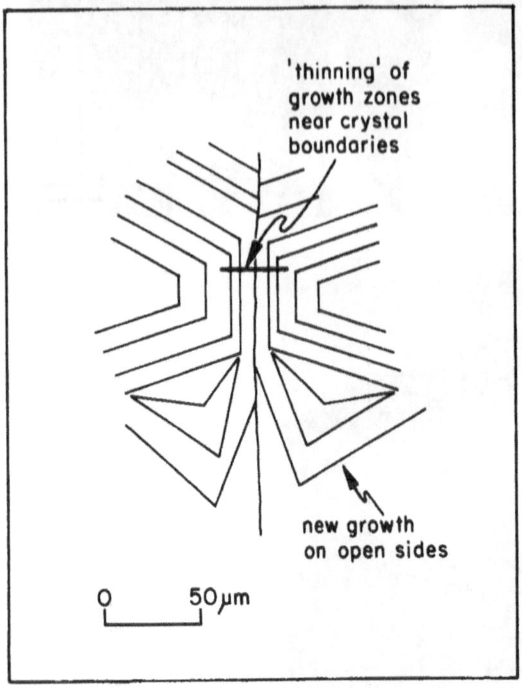

Fig. 7. Schematic representation of some features noted in zone 1 calcites under cathodoluminescence. As crystals have approached one another the areas closest to the adjacent crystal have thinner growth bands than other crystal faces. When crystals 'dock', new growth occurs on open sides.

Horizontal structures

The main fabric in these structures is a sparry calcite similar to that in zone 3. The spar generally exhibits less luminescent zonation than those in the vertical concretions. Micritic tubules are very rare as also are patches of micritic matrix like that in zone 2.

Interpretation

The fine, irregular veins, circum-granular cracks, fibrous coatings (coronas) to grains, floating and corroded silicate grains, micrite-walled tubules, and variable spar mosaics are typical of many modern and ancient calcretes (Wright, 1982; Esteban & Klappa, 1983; Goudie, 1983; Wright & Peeters, 1989).

The general shape of the vertical structures is clearly root-like and two other features support a root-related origin. First, the presence of a dense tubular micritic coating, passing gradationally into less well-cemented zones is a feature of many rhizocretions (Calvet *et al.*, 1975; Klappa, 1980; Mount & Cohen, 1984; Jones & Ng, 1988). Second, the presence of small tubules radiating from the central (root) cavity closely resemble rootlets arranged around a central tap root. These rhizocretions most closely resemble those described by Mount & Cohen (1984) from the Plio-Pleistocene Koobi Fora Formation of Kenya. In this example the rhizocretions were mainly composed of micrite, both as precipitated 'cylinders' around the roots and as later partial fills of the root moulds. The micrite in the

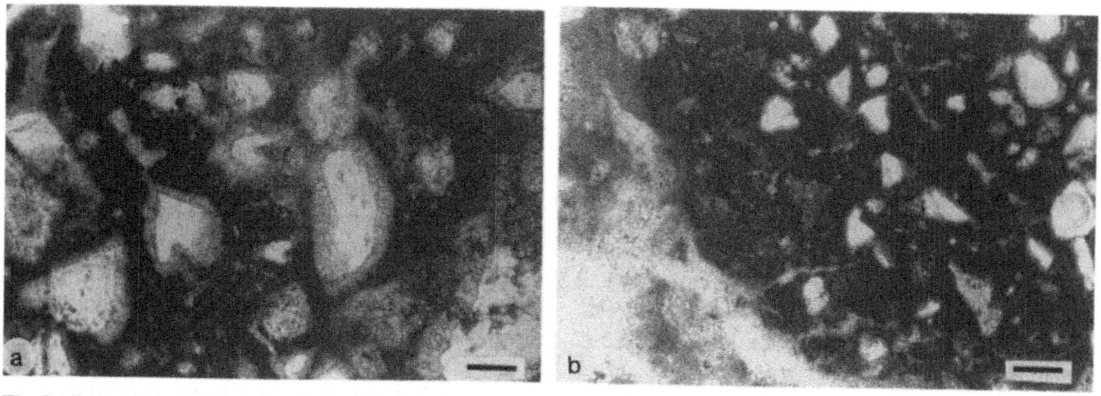

Fig. 8. Photomicrographs of zone 2. (a) Silicate grains with coronas of fibrous calcite (scale bar = 0·2 mm). (b) Edge of zone 2 shows coarser spar of zone 1 (bottom left corner) and dense micrite of zone 2 with floating silicate grains with coronas (scale bar = 0·2 mm).

Koobi Fora rhizocretions also occurs as laminated sheaths around smaller roots, as in the case of the radiating 'rootlets' in the Otter Sandstone (Fig. 5e), with spar cement as the final stage filling of the rhizocretions.

The earliest phase of cementation, post-dating the micrite and spar cements of zones 2 and 3, is represented by the fibrous grain-coating cements (coronas). This is also a common fabric in calcretes (Goudie, 1983) and occurs in the rhizocretions described by Mount & Cohen (1984). The non-isopachous nature of these Triassic coronas lends further support for their vadose origin. The dissolution of silicate grains is another feature also noted in the Koobi Fora rhizocretions (Cohen, 1982; Mount & Cohen, 1984).

The micrite 'lumps' in zone 1 possess displacive growth and desiccation-related features such as fine, irregular veins and circum-granular cracks. The occurrence of these features also supports a vadose site of formation (at least for some of the time). The vertical concretions exhibit no features suggestive of a burrow origin and no material resembling faecal pellets has been recognized in them.

The calcite sparry mosaics are a distinctive feature of these structures. They resemble neomorphic microspars and pseudospars in that they typically possess non-drusy fabrics, floating silicate grains and earlier-formed micritic calcrete lumps, and have non-planar intercrystalline boundaries. The apparently gradational contacts between the floating micrite and spar 'lumps' suggest that the spar mosaics may have locally

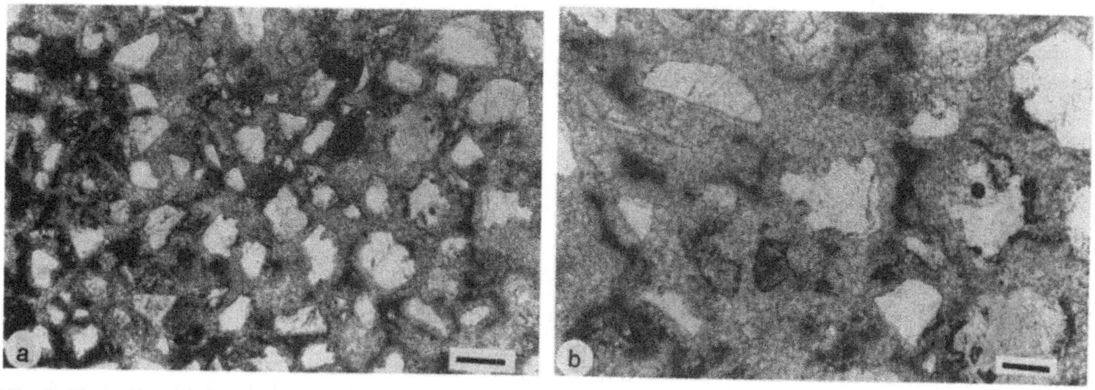

Fig. 9. Photomicrographs of zone 3. (a) Calcite-cemented silicate grains. Note irregular forms of many of the grains. The dissolution of many of the grains has probably resulted in the lack of grain-supporting framework (scale bar = 0·2 mm). (b) Corroded grain in centre of photograph (scale bar = 0·1 mm).

replaced the micrite. Many workers have noted similar neomorphic fabrics in Quaternary and pre-Quaternary calcretes (see review in Wright & Peeters, 1989). Recently, Tandon & Friend (1989) have described similar neomorphic spars from Upper Devonian–Lower Carboniferous calcretes of Arran, Scotland, which show distinctive cathodoluminescent patterns interpreted by those authors as reflecting replacement of a precursor micrite. Many of the rhombic, zoned calcites they observed contained relict luminescence patterns of ghost micrites, especially clear in the areas between masses of spar and micrite. However, most of the crystals they described apparently lack these relict luminescent features and it is difficult to envisage how the crystals were able to 'cleanse' themselves of the relict luminescences while still maintaining their fine zoning. An alternative interpretation for such relict patterns of 'micrite' luminescence is that the larger crystals had enclosed the smaller ones, not by some form of dissolution–precipitation, but by poikilotopic growth, encapsulating the earlier micritic-grade crystals. No such relict luminescence patterns were observed in the material described in this paper.

Wright & Peeters (1989) described similar neomorphic spars from Lower Carboniferous calcretes from South Wales, Belgium and Germany. They found little direct evidence of micrite replacement but favoured simple crystal growth. The zoned, rhombic to anhedral crystals they described exhibited prominent dissolution surfaces and this, together with the suggestion that growth had been in part displacive, was used to explain why the mosaics were so irregular, resembling 'classic' neomorphic spars.

Although the floating micrite 'lumps' may be regarded as supporting the replacement explanation, several other features appear to contradict it. First, the micrite of zone 2 has sharp contacts with the spar of zone 1 and shows no evidence of any neomorphism to spar. Second, the delicate laminar micrites of the small radiating tubules were also, seemingly, unaffected by the spar replacement (Fig. 5c, e).

Cannibilization of micrite by solution–precipitation may be an important mechanism whereby small solution-susceptible micritic crystals are replaced by coarser, secondary spars. Indeed, calcretes commonly exhibit evidence of phases of dissolution (Wright & Peeters, 1989). However, we see no direct evidence for it in the Otter Sandstone material. Non-isopachous growth is probably the cause of some of the irregular crystal forms and zonation, and for the paucity of straight compromise boundaries (Figs 5b & 6a). This may have been a function of 'competition' for available $CaCO_3$ between nearby crystals, or perhaps due to formation at higher supersaturations (above the critical saturation) or above the critical roughening temperature (apparently only 25°C for calcite (Tucker & Wright, 1990, p. 377). Such effects would promote random addition of atoms to the crystal faces. The latter two mechanisms have been invoked to explain some irregular mosaics in dolomites (Gregg & Sibley, 1984). A full evaluation of the causes of the unusual facrics in such crystalline calcretes is beyond the scope of this study, but convincing evidence of micrite neomorphism is lacking.

After the decay of the original root a void space must have been left, now filled by micritic lumps and sparry calcite. Both of these must have developed within some matrix which is no longer seen. The presence of floating silicate grains may reflect a precursor sand or finer grade sediment in which the carbonates formed.

The sequence of rhizocretion formation is strikingly similar to that noted by Mount & Cohen (1984) in the Koobi Fora Formation. This initial phase of cementation consisted of fibrous cements on the silicate grains. This was followed by extensive cementation by micrite around the root. On decay, the root cavity was, at least partly, filled by micrite (showing features such as fine, irregular cracks and circum-granular cracks) followed by the zoned spars and drusy cements.

The sheet structures exhibit few features indicative of a calcrete origin such as nodules, small rhizocretions or circum-granular cracks. Their different cathodoluminescence characteristics also supports a different origin to that of the rhizocretions. However, their occurrence with these root-related structures implies a similar site of formation, namely a near-surface environment. In particular, the lack of distinctive calcrete profiles suggests a non-pedogenic site of formation and, consequently, a groundwater origin is favoured. It is possible that the sheets developed at, or just above, the water-table in the capillary fringe zone, and the occurrence of numerous sheets reflects the episodic rises in the water-table as the sediment surface was aggraded. Groundwater cementation by carbonates is a common feature in many semi-arid to arid alluvial settings (Arakel & McConchie, 1982; Carlisle, 1983). Multiple sheet calcretes have been documented in zones of mobile water-tables by Seminiuk & Meagher (1981) from Quaternary dune systems in Western Australia. However, these sheets are the results of phreatophytic plant activity and are full of rhizocretions, which is not the case in the Otter Sandstone. Wright *et al.* (1988) have also documented multiple horizontal sheets of laminar, rhizocretionary calcrete from the Cretaceous of Spain related to root-mat calcification at water-tables, but again these contain clear evidence of a root origin. The lack of cathodoluminescent zonation may also reflect the stable Eh conditions in the groundwater zone as compared to the vadose accumulations in the rhizocretions.

STABLE ISOTOPE ANALYSES

Isotopic analyses were carried out on samples hand picked from rock slices. After powdering, the samples were reacted with 100% ortho-phosphoric acid in vacuuo and the CO_2 evolved was analysed on a VG Isogas SIRA 10 mass spectrometer. The carbon and oxygen isotopic results were corrected using standard procedures (Craig, 1957), and presented in standard PDB and SMOW notation. Standard NBS19 gave $\delta^{13}C=1{\cdot}92‰$ PDB, $\delta^{18}O=-2{\cdot}19‰$ and $\delta^{18}O=28{\cdot}65‰$ SMOW. Reproducibility was $\pm 0{\cdot}1‰$ for both carbon and oxygen.

The results are shown in Fig. 10. The majority of analyses are from zone 2 of the rhizocretions but the heaviest $\delta^{13}C$ came from the sheet calcrete.

Interpretation

Before discussing these Otter Sandstone analyses we wish to offer a word of warning as regards interpreting

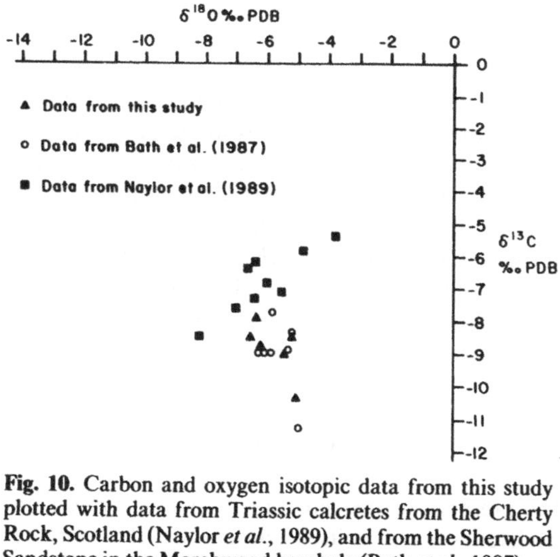

Fig. 10. Carbon and oxygen isotopic data from this study plotted with data from Triassic calcretes from the Cherty Rock, Scotland (Naylor *et al.*, 1989), and from the Sherwood Sandstone in the Marchwood borehole (Bath *et al.*, 1987).

soil carbonate stable isotope values, especially those in ancient calcretes. The factors influencing the isotopic chemistry of soil carbonate are extremely diverse. The isotopic composition of the oxygen or carbon is affected by latitude, elevation, temperature (including the degree of freezing), the vegetation type and its seasonal activity, isotopic composition of the rainfall, the degree of evaporation, the proximity to the sea, monsoonal effects, the degree of contamination by pre-existing carbonate in the soil, and many other factors (Cerling, 1984; Salomons & Mook, 1986; Dever *et al.*, 1987). Of special consideration is that most thick, mature calcretes require prolonged periods for their formation (Wright, 1990) and bulk analyses of such material must provide time-averaged values. This will be due in part to small-scale and larger-scale environmental change (especially as regards vegetation), but will also reflect changes in the depth of carbonate precipitation in the solum. There are significant vertical variations in the isotopic composition of carbonate within a soil, reflecting mixing and diffusion with the atmosphere, degassing and evaporation (Dever *et al.*, 1987; Quade *et al.*, 1989a). These depths will vary both due to changes in rainfall and evaporation budgets and especially as the soil aggrades or degrades. In addition, many calcretes are derived from groundwaters and little is known regarding their isotopic compositions. The distinction between pedogenic and groundwater types is not always an easy one to make (Wright, 1990).

However, the calcretes described here are of a relatively simple form, representing, apparently, rhizocretions. By analysing the early zone 2 micrites which presumably formed during root growth, the results should relate to carbonate precipitation in the active vadose zone. The $\delta^{13}C$ ($-7{\cdot}9$ to $-9{\cdot}03‰$ PDB) and $\delta^{18}O$ values ($-5{\cdot}1$ to $-6{\cdot}5‰$ PDB) are within the ranges of present-day soil carbonates given by Solomons *et al.* (1978) and Talma & Netterberg (1983), and are comparable to the Triassic calcretes documented by Bath *et al.* (1978) (Fig. 10). It must be stressed that we do not have any isotopic information on compositions of later carbonates in the Otter Sandstone. Burial cements in the unit are typically poikilotopic calcite spars (Leonard, 1982), quite unlike the calcite of zones 1–3. If isotopic re-equilibration of the zone 2 micrites took place during burial diagenesis, no petrographic evidence for the process has been recognized during this study. The similarities of the isotopic results to those from Quaternary and other Triassic calcretes further supports the view that these values are those of the original calcrete.

The carbon in soil carbonate is predominantly sourced from CO_2, an admixture of atmospheric and soil CO_2 (Quade *et al.*, 1989a). The carbon isotope values in soil carbonates can be used to assess the type (Quade *et al.*, 1989b) and degree of vegetation cover (Amundson *et al.*, 1988). The carbon values noted in the Otter Sandstone probably imply much input of light carbon (^{12}C) from biological sources. The carbon isotopic composition of soil carbonate in modern soils is typically much heavier than that of the coexisting vegetation (Cerling *et al.*, 1989). This is due to fractionation during the precipitation of the carbonate (soil carbonate is heavier by as much as +9·6‰ than soil CO_2; Friedman & O'Neil, 1977; Schlesinger, 1985), and because soil CO_2 is heavier than the associated vegetation (perhaps as much as 8–9‰; Solomons & Mook, 1986).

The light carbon values in the Otter Sandstone indicate a vegetation cover which was isotopically very light, perhaps with $\delta^{13}C$ values lower than −20‰. This is indicative of a C3 type vegetation (see Salomons & Mook, 1986; Quade *et al.*, 1989a), a group which includes virtually all modern trees and shrubs (Quade *et al.*, 1989b). Their values may be too light for CAM vegetation, such as desert succulents. In support of the idea that the vegetation cover consisted of trees or shrubs is the large size and depth of the tap roots.

However, the carbonate analysed here was formed adjacent to roots and very little is known of the isotopic composition of rhizocretionary carbonate. Roots apparently respire CO_2 enriched in the *heavier* isotope (Park & Epstein, 1960; Salomons & Mook, 1986), and so the microenvironment around the Triassic roots may also have been enriched in the heavier isotope. The lower values recorded also suggest a lack of mixing with atmospheric CO_2, which is probably due to the depths at which the rhizocretions formed, where most CO_2 would have been soil-sourced. The heaviest carbonate value ($\delta^{13}C$ of −7·93‰) came from a sheet calcrete sample which may indicate less light organic input.

The $\delta^{18}O$ values are also light and are similar to those noted by Bath *et al.* (1987) for Triassic pedogenic calcretes from the Sherwood Sandstone from Marchwood borehole. However, the values fall in a much narrower range than do those given by Naylor *et al.* (1989) for Permo-Triassic calcretes from the Moray Firth, Scotland (Fig. 10). These differences may reflect the different mechanisms involved in precipitating the calcretes. The co-variance of carbon and oxygen in the calcretes described by Naylor *et al.* (1989) (Fig. 10) is a typical feature of carbonate precipitated by evaporation (Salomons *et al.*, 1978; Schlesinger, 1985; Salomons & Mook, 1986). Evaporation is not the only drive for precipitation; degassing and evapo-transpiration needs also to be considered. Degassing causes a shift to heavier $\delta^{13}C$ values (Salomons & Mook, 1986; Dever *et al.*, 1987) and the light values recorded here might argue against this mechanism. This mechanism is unlikely to explain the preferential precipitation around the roots as roots respire CO_2. Evapo-transpiration is regarded as a major cause of carbonate precipitation in vegetated soils (Cerling, 1984; Salomons & Mook, 1986) and does not lead to fractionation of the isotopes.

The highly specific distribution of the early micrite cement supports such an origin; if evaporation or degassing were the cause of precipitation then carbonate would have been much more widely distributed. Microbial activity around roots is also an important factor in cementation but no petrographic evidence for such an origin has been noted (such as fungally mediated calcites; cf. Wright, 1986). The field, petrographic and stable isotope evidence favour an origin by evapo-transpiration but it is likely that more than one precipitation mechanism was operating at any one time.

The light $\delta^{18}O$ values probably reflect soil water of meteoric origin, although an input from evolved groundwater must also be considered. Little information is available on the isotopic compositions of groundwater-sourced calcrete but such groundwaters are typically evolved from meteoric waters depleted in the lighter isotopes by evaporation and biological fractionation (Manze & Brunnacker, 1977).

The Otter Sandstone rhizocretionary carbonate probably precipitated at some depth below the soil surface and hence any estimates of the palaeotemperatures based on the $\delta^{18}O$ values would be likely to be unrepresentative of the actual atmospheric temperatures. Indeed any palaeotemperature estimates using ancient calcretes must be based on a very careful study of the isotopic variations *within* a profile and temperature estimates based on single point analyses are unlikely to provide reliable values (Dever *et al.*, 1987; Quade *et al.*, 1989a; cf. Suchecki *et al.*, 1988).

DISCUSSION

The presence of rhizocretions in ancient braidplain sandstones is not in itself unusual, but the forms of these features in the Otter Sandstone are significant.

The sizes of the roots, as judged from the diameters in zone 1, where the original root was sited, range up to 0·10 m which is not large. However, their vertical extent over several metres is suggestive of deep tap roots. Tap roots are large vertical roots which descend into the substrate in order to reach deeper perched waters or the main water-table. Plants which use this water-gathering strategy are termed phreatophytes and rhizocretions attributed to such plants have been described by Cohen (1982) from the Plio-Pleistocene Koobi Fora Formation of Kenya. Such plants are typical of semi-arid to arid channel-bar or floodplain deposits (Cohen, 1982) and it is likely that the plants (probably trees) responsible for the rhizocretions in the Otter Sandstone were colonizing either abandoned river courses during periods of drought or perhaps more likely, grew on channel bars. The regional climate of the time was probably arid in nature, as reflected in the general depositional regime (Henson, 1971).

Regionally, the Otter Sandstone contains very few well-developed calcrete profiles (Leonard, 1982). However, well-developed palaeosols are known from slightly younger Triassic sequences in Britain from both alluvial-fan (Steel, 1974; Tucker, 1977) and playa settings (Wright *et al.*, 1988). Well-developed calcrete palaeosols on such alluvial-fan deposits reflect the occurrence of very stable geomorphological surfaces, caused by channel incision and/or stabilization following climatic changes (Steel, 1974; Wright & Alonso-Zarza, 1990). The palaeosols developed on playa settings were less well developed but had higher preservation potentials in the low-energy, actively aggrading playa centres (Wright *et al.*, 1988) (Fig. 11). The paucity of mature palaeosols in the Otter Sandstone probably reflects the lack of geomorphological stability and low preservation potential on the active braidplains. The rhizocretions and sheet carbonates formed several metres beneath the active land surface and, as such, were already 'fossilized', increasing their preservation potential.

CONCLUSIONS

Calcretes in the Triassic Otter Sandstone of south Devon represent calcification around the tap roots of phreatophytic plants or near the water-table. Stable isotopic analyses of the rhizocretions support such an origin, probably indicating precipitation by evapotranspiration. The geomorphologically unstable braidplain settings represented by the calcrete-bearing

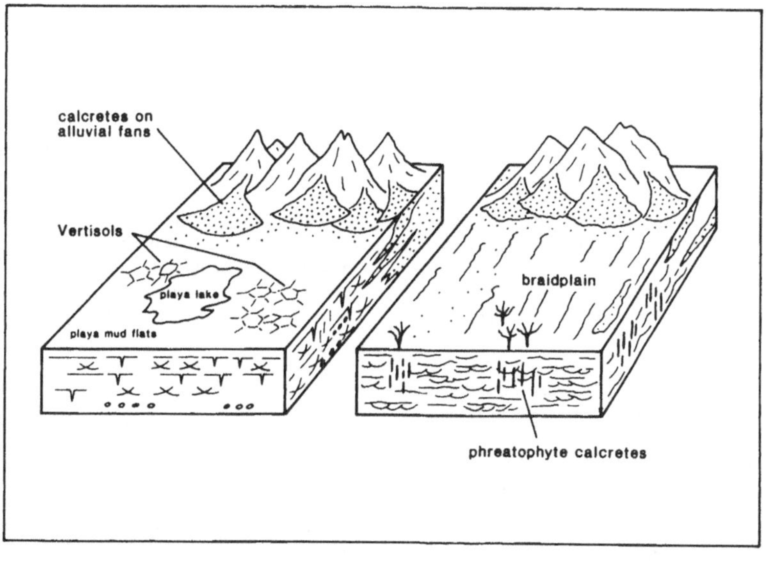

Fig. 11. Contrasting soil types in arid Triassic depositional systems. During the Late Triassic (left) calcretes developed on alluvial fan surfaces and vertisols on playas. During the deposition of part of the Otter Sandstone (right), the active braidplains were less stable and any soils which developed were eroded. The Otter Sandstone contains calcretes developed below the active land surface.

interval in the Otter Sandstone were not conducive to either the formation or preservation of soils. The calcretes described here formed beneath the active land surface and were pre-disposed to 'fossilization'.

ACKNOWLEDGMENTS

We especially thank Andrew Leonard (BP, Aberdeen) for access to unpublished thesis information on the Otter Sandstone. SURRC and Tony Fallick provided the isotopic analyses. The senior author thanks Shell and Esso for financial support for his studies of Triassic sandstones. The helpful comments of the reviewers, Hank Chafetz and John Hubert, were much appreciated. Alan Cross and James Watkinson provided assistance with the preparation of the manuscript and figures. University of Reading PRIS Contribution 116.

REFERENCES

ALLEN, R.L. (1986) Pedogenic calcretes in the Old Red Sandstone facies (late Silurian–early Carboniferous) of the Anglo-Welsh area, southern Britain. In: *Paleosols: their Recognition and Interpretation* (Ed. by V. P. Wright), pp. 58–86. Blackwell Scientific Publications, Oxford.

AMUNDSON, R.G., CHADWICK, O.A., SOWERS, J.M. & DONER, H.E. (1988) Relationship between climate and vegetation and the stable carbon isotope chemistry of soils in the eastern Mojave Desert, Nevada. *Quat. Res.*, **29**, 245–254.

ARAKEL, A.V. & MCCONCHIE, D. (1982) Classification and genesis of calcrete and gypsite lithofacies in paleodrainage systems of inland Australia and their relationship to carnotite mineralization. *J. sedim. Petrol.*, **52**, 1149–1170.

BATH, A.H., MILODOWSKI, A.E. & SPIRO, B. (1987) Diagenesis of carbonate cements in Permo-Triassic sandstones in the Wessex and East Yorkshire–Lincolnshire Basins, UK: a stable isotope study. In: *Diagenesis of Sedimentary Sequences* (Ed. by J. D. Marshall), *Spec. publs geol. Soc. London*, **36**, 173–190.

CALVET, F., POMAR, L. & ESTEBAN, M. (1975) Laz rizocreciones del Pleistoceno de Mallorca. *Inst. Invest. Geol. Univ. Barcelona*, **30**, 35–60.

CARLISLE, D. (1983) Concentration of uranium and vanadium in calcretes and gypcretes. In: *Residual Deposits* (Ed. by R. C. L. Wilson), *Spec. publs geol. Soc. London*, **11**, 185–195.

CERLING, T.E. (1984) The stable isotopic composition of modern soil carbonate and its relationship to climate. *Earth Planet Sci. Lett.*, **71**, 229–240.

CERLING, T.E., QUADE, J., WANG, Y. & BOWMAN, J.R. (1989) Carbon isotopes in soils and palaeosols as ecology and palaeoecology indications. *Nature*, **341**, 138–139.

COHEN, A.S. (1982) Paleoenvironments of root casts from the Koobi Fora Formation, Kenya. *J. sedim. Petrol.*, **52**, 401–414.

CRAIG, H. (1957) Isotopic standards for carbon and oxygen and correction factors for mass-spectrometric analysis of carbon dioxide. *Geochim. Cosmochim. Acta*, **12**, 133–149.

DEVER, L., FONTES, J. CH. & RICHÉ, J. (1987) Isotopic approach to calcite dissolution and precipitation in soils under semi-arid conditions. *Chem. Geol. (Isotope Geosci. Sec.)*, **66**, 307–314.

ESTABAN, M. & KLAPPA, C.F. (1983) Subaerial exposure environment. *Mem. Am. Ass. petrol. Geol.*, **33**, 1–54.

FOLK, R.L. (1965) Some aspects of recrystallization in ancient limestones. *Spec. publs Soc. econ. Paleont. Miner.*, **13**, 14–48.

FRIEDMAN, E. & O'NEIL, J. (1977) *Data of Geochemistry*, 6th edn. Compilation of stable isotope fractionation factors of geochemical interest. *Prof. Pap. US geol. Surv.*, **440-KK**.

GOUDIE, A.S. (1983) Calcrete. In: *Chemical Sediments and Geomorphology* (Ed. by A. S. Goudie & K. Pye), pp. 93–131. Academic Press, London.

GREGG, J.M. & SIBLEY, D.F. (1984) Epigenetic dolomitization and the origin of xenotopic dolomite textures. *J. sedim. Petrol.*, **54**, 908–931.

HENSON, M.R. (1971) *The Permo-Triassic rocks of south Devon*. PhD thesis, University of Exeter.

JONES, B. & NG, K-C. (1988) The structure and diagenesis of rhizoliths from Cayman Brac, British West Indies. *J. sedim. Petrol.*, **58**, 457–467.

KLAPPA, C.F. (1980) Rhizoliths in terrestrial carbonates: classification, recognition, genesis and significance. *Sedimentology*, **27**, 613–629.

LEONARD, A.J. (1982) *The diagenesis of the New Red Sandstone Deposits of South East Devon*. PhD thesis, University of Exeter.

MADER, D. (1985) Braidplain, floodplain and playa lake, alluvial fan, aeolian and palaeosol facies composing a diversified lithogenetical sequence in the Permian and Triassic of South Devon (England). In: *Aspects of Fluvial Sedimentation in the Lower Triassic Bundsandstein of Europe* (Ed. by D. Mader), pp. 15–64. Springer-Verlag, Berlin.

MADER, D. (1990) *Palaeoecology of the Flora in Bundsandstein and Keuper in the Triassic of Middle Europe*. Gustav Fischer Verlag, Stuttgart, 1650 pp.

MADER, D. & LAMING, D.J.C. (1985) Braidplain and alluvial fan environmental history and climatological evolution controlling origin and destruction of aeolian dune fields and governing overprinting of sand seas and river plains by calcrete pedogenesis in the Permian and Triassic of South Devon (England). In: *Aspects of Fluvial Sedimentation in the Lower Triassic Bundsandstein of Europe* (Ed. by D. Mader), pp. 519–528. Springer-Verlag, Berlin.

MALIVA, R.G. (1989) Displacive calcite syntaxial overgrowths in open marine limestones. *J. sedim. Petrol.*, **59**, 397–403.

MANZE, U. & BRUNNACKER, K. (1977) Uber das verhalten sauerstoff-und kohlenstoff-isotope in kalkkrusten und kalktuffen des mediterranean Raumes und der Sahara. *Z. Geomorph. N.F.*, **21**, 343–353.

MOUNT, J.F. & COHEN, A.S. (1984) Petrology and geochemistry of rhizoliths from Plio-Pleistocene fluvial and marginal lacustrine deposits, East Lake Turkana, Kenya. *J. sedim. Petrol.*, **54**, 263–275.

NAYLOR, H., TURNER, P., VAUGHAN, D.J. & FALLICK, A.E. (1989) The Cherty Rock, Elgin: a petrographic and isotopic study of a Permo-Triassic calcrete. *Geol. J.*, **24**, 205–221.

PARK, R. & EPSTEIN, S. (1960) Carbon isotope fractionation during photosynthesis. *Geochim. Cosmochim. Acta*, **21**, 110–126.

QUADE, J., CERLING, T.E. & BOWMAN, J.R. (1989a) Systematic variations in the carbon and oxygen isotopic composition of pedogenic carbonate along elevation transects in the southern Great Basin, United States. *Bull. geol. Soc. Am.*, **101**, 464–475.

QUADE, J., CERLING, T.E. & BOWMAN, T.R. (1989b) Development of Asian monsoon revealed by marked ecological shift during the latest Miocene in northern Pakistan. *Nature*, **342**, 163–166.

SALOMONS, W., GOUDIE, A.S. & MOOK, W.G. (1978) Isotopic composition of calcrete deposits of Europe, Africa and India. *Earth Surf. Proc.*, **3**, 43–57.

SALOMONS, W. & MOOK, W.G. (1986) Isotope geochemistry of carbonates in the weathering zone. In: *Handbook of Environmental Chemistry*, Vol. 2, *The Terrestrial Environment* (Ed. by P. Fritz & J. Ch. Fontes), pp. 239–270. Elsevier, Amsterdam.

SCHLESINGER, W.H. (1985) The formation of caliche in soils of the Mojave Desert, California. *Geochim. Cosmochim. Acta.*, **49**, 57–66.

SELWOOD, E.B., EDWARDS, R.A., SIMPSON, S., CHESHER, J.A., HAMBLIN, R.J.O., HENSON, M.R., RIDDOLLS, B.W. & WATERS, R.A. (1984) *Geology of the country around Newton Abbot.* Memoir for 1:50,000 geological sheet 339, new series. British Geological Survey, HMSO, London.

SEMINIUK, V. & MEAGHER, T.D. (1981) Calcrete in Quaternary coastal dunes in southwestern Australia: a capillary rise phenomenon associated with plants. *J. sedim. Petrol.*, **51**, 47–68.

STEEL, R.J. (1974) Cornstone (fossil caliche): its origin, stratigraphic and sedimentological importance in the New Red Sandstone, W. Scotland. *J. Geol.*, **82**, 351–369.

SUCHECKI, R.K., HUBERT, J.F. & BIRNEY DE WET, C.C. (1988) Isotopic imprint of climate and hydrogeochemistry on terrestrial strata of the Triassic–Jurassic Hartford and Fundy rift basin. *J. sedim. Petrol.*, **58**, 801–811.

TALMA, A.S. & NETTERBERG, F. (1983) Stable isotope abundances in calcretes. In: *Residual Deposits* (Ed. by R. C. L. Wilson), *Spec. publs geol. Soc. London*, **11**, 221–233.

TANDON, S.K. & FRIEND, P.F. (1989) Near-surface shrinkage and carbonate replacement processes, Aran Cornstone Formation, Scotland. *Sedimentology*, **36**, 1113–1126.

TUCKER, M.E. (1977) The marginal Triassic deposits of South Wales: continental facies and palaeogeography. *Geol. J.*, **12**, 169–188.

TUCKER, M.E. & WRIGHT, V.P. (1990) *Carbonate Sedimentology*. Blackwell Scientific Publications, Oxford, 482 pp.

WRIGHT, V.P. (1982) calcrete paleosols from the Lower Carboniferous Llanelly Formation, South Wales. *Sediment. Geol.*, **33**, 1–33.

WRIGHT, V.P. (1986) The role of fungal biomineralization in the formation of Early Carboniferous soil fabrics. *Sedimentology*, **33**, 831–838.

WRIGHT, V.P. (1990) Estimating rates of calcrete formation and sediment accretion in ancient alluvial deposits. *Geol. Mag.*, **127**, 273–276.

WRIGHT, V.P. & ALONSO-ZARZA, A. (1990) Pedostratigraphic models for interpreting alluvial fan deposits. *J. geol. Soc. London*, **147**, 8–10.

WRIGHT, V.P., NORTH, C.P., HANCOCK, P.L., CURTIS, M. & ROBINSON, D. (1988) Pedofacies variations across an arid alluvial basin: a case study from the Upper Triassic of S.W. Britain. *Int. Ass. Sedimentol. Eur. Mtg. Leuven 1988, Abstr.*, 227–228.

WRIGHT, V.P., PLATT, N.H. & WIMBLEDON, W.A. (1988) Biogenic laminar calcretes: evidence of root mats in paleosols. *Sedimentology*, **35**, 603–620.

WRIGHT, V.P. & PEETERS, C. (1989) Origins of some early Carboniferous calcrete fabrics revealed by cathodoluminescence: implications for interpreting the sites of calcrete formation. *Sediment. Geol.*, **65**, 345–353.

(Manuscript received 5 June 1990; revision received 31 December 1990)

Reprinted from Sedimentology (1993) ***40****, 271–301*

Zeolitic diagenesis of late Quaternary fluviolacustrine sediments and associated calcrete formation in the Lake Bogoria Basin, Kenya Rift Valley

R. W. RENAUT

Department of Geological Sciences, University of Saskatchewan, Saskatoon, Saskatchewan, Canada S7N OWO

ABSTRACT

Late Quaternary fluviolacustrine siltstones, mudstones and claystones (Loboi Silts) on the northern margins of the saline, alkaline Lake Bogoria in the Kenya Rift Valley contain up to *c*. 40% authigenic analcime and minor natrolite. The zeolitic sediments are reddish brown and up to 1 m thick. The amount of analcime increases upward in the profile, but decreases with distance from the lake. The altered sediments show many pedogenic features including zeolitic root mats, rootmarks, concretions and carbonate rhizoliths. Residual patches of calcrete locally cap the zeolitic rocks. The profile is interpreted as an exhumed palaeosol and land surface on the former margins of the lake.

The analcime occurs as submicroscopic (0·5–2·5 μm) subhedral and euhedral crystals, which have an average Si/Al ratio of 2·33 (as determined by X-ray microanalysis) or 2·18 (*d*-value of 639 analcime peak). The analcime formed in lake marginal sediments (soils) by reaction of silicate detritus with Na_2CO_3 rich pore waters concentrated close to the land surface by evaporative pumping and evapotranspiration. Poorly ordered clay minerals were probably the main reactants. Authigenic illite may have been a by-product of the reactions. Chemical analyses suggest that pore waters supplied some of Na^+, and possibly K^+ and SiO_2. The associated calcrete and rhizoliths were formed during or shortly after the main period of zeolitic alteration. The Ca^{2+} may have originated from infiltrating dilute runoff and groundwater. Authigenic smectite was precipitated in open porosity following analcime formation. The zeolitic alteration at Lake Bogoria provides a relatively recent analogue for lake marginal zeolites found in many ancient saline, alkaline lake sediments.

Reprinted from Sedimentology (1992) ***39****, 1119–1136*

Groundwater dolocretes from the Upper Triassic of the Paris Basin, France: a case study of an arid, continental diagenetic facies

CHRISTOPH SPÖTL* *and* V. P. WRIGHT†

**Geologisches Institut, Baltzerstrasse 1, CH-3012, Berne, Switzerland*
†Postgraduate Research Institute for Sedimentology, The University, PO Box 227, Whiteknights, Reading RG6 2AB, UK

ABSTRACT

Thick dolomite-cemented horizons (dolocretes) occur within a fluvial sandstone–mudstone sequence of Late Triassic age in the western part of the Paris Basin, France. Two types of dolomites can be distinguished: (a) nodular dolomitic beds less than a few metres thick, which formed within mottled overbank siltstones and mudstones; and (b) massive dolomite up to 16 m thick, which occurs in coarse grained channel sandstones and conglomerates.

The majority of the dolomite consists of a finely crystalline groundmass of dolomicrospar and, less commonly, dolomicrite. Glaebules, irregular spar-filled cracks, spheroidal dolomite, silicification and vuggy porosity are locally abundant in the massive dolomite. In contrast, biologically induced micromorphological features such as rhizocretions and alveolar–septal fabrics were observed in the thin, nodular dolomite beds.

The dolomite is near stoichiometric, well ordered and non-ferroan. $\delta^{13}C$ values range from $-7{\cdot}7$ to $-0{\cdot}4$‰ PDB and $\delta^{18}O$ values range from $-5{\cdot}1$ to $+1{\cdot}8$‰ PDB and no obvious difference in the stable isotopic composition between both types of dolomites was observed. Sr isotope ratios range from 0·7101 to 0·7126 and are invariably higher than the contemporary Triassic seawater.

A vadose–pedogenic origin for the thin dolocrete layers is indicated by the occurrence of rhizocretions and other biological structures. Several features, however, argue against a pedogenic origin for the massive carbonates, most notably the absence of biologically induced structures, the occurrence in coarse grained channel (and not overbank) deposits, and the great thickness. These units are thus interpreted as groundwater in origin. Phreatic calcretes of Quaternary age, widespread in inland Australia, are regarded as a modern analogue for the Triassic Paris Basin dolocretes.

Petrographic observations argue in favour of primary (proto)dolomite precipitation, although early diagenetic replacement of calcite by (proto)dolomite cannot be ruled out. Strontium and carbon isotope data of early diagenetic dolocrete cements and oxygen isotope data of early diagenetic silica indicate an entirely non-marine, continental origin for the groundwaters. The poorly ordered and non-stoichiometric protodolomite probably underwent stabilization upon further burial resulting in a near-stoichiometric, well ordered dolomite that clearly lacks evidence for pervasive recrystallization.

INTRODUCTION

Modern and ancient semi-arid to arid alluvial deposits contain a variety of authigenic carbonates. The most widely recognized are pedogenic forms, producing nodular to massive horizons within soil profiles. Such pedogenic calcretes or caliches have been extensively documented from the geological record. However, what appears to be less widely appreciated is that extensive carbonate precipitation occurs in some arid alluvial basins within the alluvial sediments as non-pedogenic calcretes (and dolocretes). These carbonate-cemented zones can cover areas of 100×10 km and

*Present address: US Geological Survey, Office of Energy and Marine Geology, National Center, M.S. 915, 12201 Sunrise Valley Drive, Reston, VA 22092, USA.

reach 10 m in thickness (Mann & Deutscher, 1978; Mann & Horwitz, 1979; Arakel & McConchie, 1982; Carlisle, 1983; Reeves, 1983; Arakel, 1986). Maizels (1987) records the presence of cemented Plio-Pleistocene alluvial fan gravels in Oman apparently over 200 m thick.

The carbonate in such occurrences is typically very finely crystalline, and locally highly porous. It displays a range of features indicating displacive and replacive growth, and is commonly associated with a variety of other authigenic minerals such as silica, clays (sepiolite and palygorskite) and gypsum.

The carbonates precipitate from mobile groundwaters that become progressively concentrated during down-dip flow such that a progressive increase in the Mg/Ca ratio occurs. The carbonate is precipitated mainly in the capillary fringe zone or just below the water table. Precipitation is caused by evaporation (and evapotranspiration), degassing and by the common ion effect where the groundwaters mix with playa groundwaters. A brief review of groundwater calcretes has been given by Wright & Tucker (1991).

Despite the occurrence of extensive groundwater dolocretes and calcretes in Quaternary drainage system sediments, there have been relatively few records in ancient sequences (but see Tandon & Narayan, 1981; Purvis & Wright, 1991). This paper describes thick (up to 16 m) finely crystalline and porous dolomites from Upper Triassic arid alluvial deposits of the Paris Basin, France, which are interpreted as groundwater dolocretes. Similar carbonates probably occur in other alluvial sequences and the criteria for their recognition are reviewed.

GEOLOGICAL SETTING

The study area is located south of Paris, in the western part of the Paris Basin, which is an intracratonic basin containing a combined Mesozoic–Tertiary sequence reaching nearly 4 km in maximum thickness (Fig. 1; Mégnien, 1980a; Cavelier & Lorenz, 1987). In the east of the basin, deposition commenced in the Early Triassic with Buntsandstein and Muschelkalk equivalents. In the west of the basin, sedimentation was initiated in the Late Triassic (Keuper). On a basin-wide scale the Keuper facies show a concentric pattern (Fig. 1; Mégnien, 1980b,c). Vast saline mud- and sandflats developed in the centre of the basin surrounded by a belt of alluvial fans at the margins. Marine evaporite deposition (halite) took place in the eastern part (Geisler-Cussey, 1986). The thick dolo-

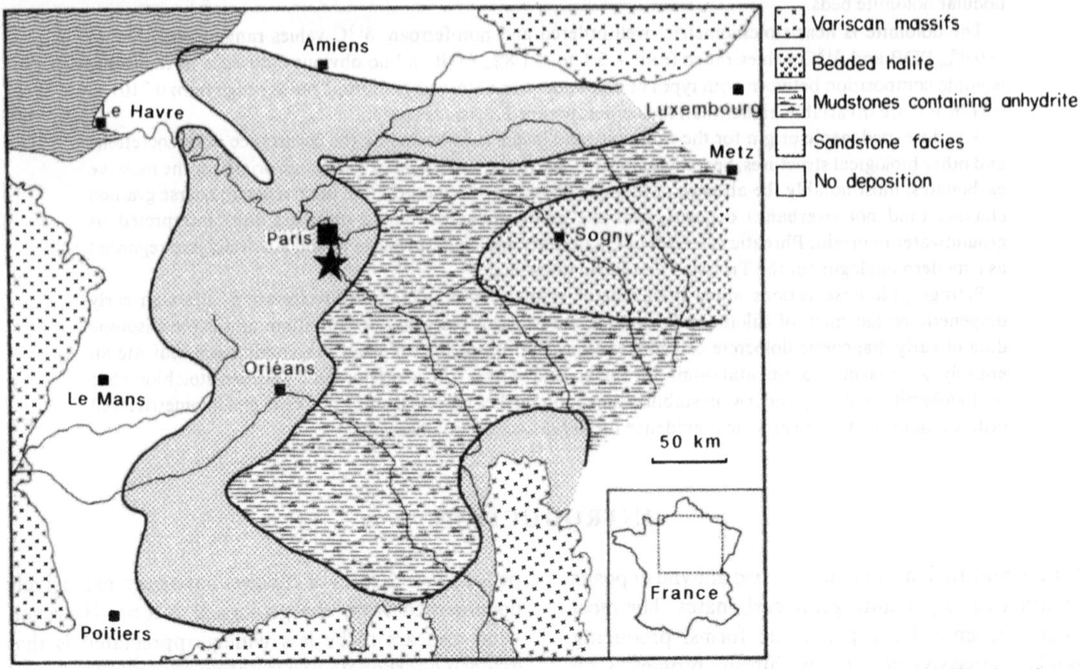

Fig. 1. General depositional pattern in the Keuper (Late Triassic) of the Paris Basin (after Mégnien, 1980b). The location of the wells studied is shown by a star. Bedded halite was deposited in the eastern part of the basin, surrounded by mudstones containing anhydrite.

crete horizons reported in this paper occur in the distal alluvial fan deposits interfingering with evaporitic mudflat deposits (Ricour, 1960; Mégnien, 1980b,c).

The Keuper in the western part of the basin, which contains major hydrocarbon plays (Perrodon & Zabek, 1991), is a red bed sequence consisting of several sandstone and mudstone horizons. Toward the top of the sequence is a sandstone unit (Grès de Chaunoy) up to 100 m thick. Capping this sandstone is a prominent dolomite-rich unit, the Dolomie sommitale ($\leqslant$20 m), which consists of several metre scale fluvial cycles and dolomite layers (Fig. 2). These are overlain by mudstones (Argile sommitale). The sandstones (Grès de Chaunoy) overlie a mudstone unit (Argile intermèdiaire) which represents a playa mudflat deposit.

Cores from 10 wells penetrating the upper part of the Keuper were examined, including the Vert le Grand field. The sandstones of the Grès de Chaunoy consist of stacked, fining upward cycles, ranging in thickness from 0·8 to 7 m, with a mean thickness of 2·3 m. A typical cycle consists of a channel lag horizon, grading into massive or crudely cross-stratified sandstones. The cycles are capped by laminated siltstones and mudstones, commonly truncated by the erosive base of the overlying cycle.

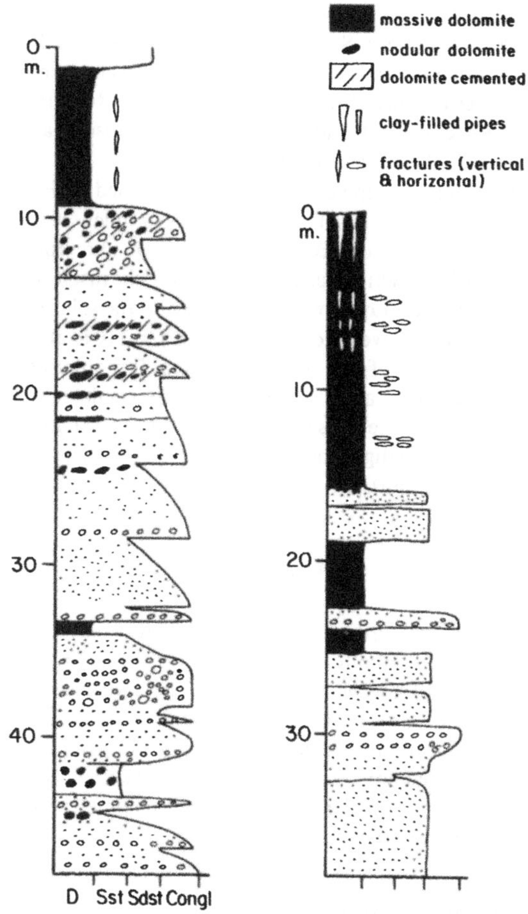

Fig. 2. Two lithological logs from cores through the top of the Grès de Chaunoy, Vert le Grand field. The thick dolomites were hosted in sandstone, little of which now remains due to replacive and displacive growth of the carbonate. D, dolomite; Sst, siltstone; Sdst, sandstone; Congl, conglomerate.

METHODOLOGY

Carbonate mineralogy and dolomite stoichiometry were determined using X-ray diffraction analysis (CuK_α radiation). Selected dolomite samples were also analysed in a Guinier camera (FeK_α radiation) and the unit cell dimensions were refined using silicon as an internal standard. All thin sections were examined by cathodoluminescence and UV/blue-light microscopy. Back-scattered electron imagery was used to recognize compositional zoning in the carbonate cements. Carbonate elemental composition was determined by electron microprobe analysis. Complete data and operating conditions are tabulated in Spötl (1991).

Stable isotope analyses of carbonate samples were performed on dolostone samples (whose compositional homogeneity was assessed by cathodoluminescence microscopy) and dolomite microsamples drilled from core slabs with a dental drill. Organic carbon was removed prior to acid treatment by oxidizing the sample powder in a plasma furnace. The powder was then reacted with phosphoric acid at 90°C and the resultant gas was measured with a VG Prism-II mass spectrometer. All analyses were corrected using the appropriate $\alpha_{CaCO_3\text{–phosphoric acid}}$ and are quoted relative to PDB. Reproducibility of duplicate samples is better than 0·1‰ for $\delta^{13}C$ and better than 0·2‰ for $\delta^{18}O$. No correction was applied to account for the differential fractionation factor of dolomite relative to calcite dissolving in phosphoric acid.

Six silicified samples were selected for oxygen isotope analysis of the authigenic silica. Special attention was paid not to include samples with minor siliciclastic detritus. All samples were crushed and white authigenic silica was hand-picked. After grinding in a tungsten carbide swing-mill the powder was treated with 3·3 M HCl in order to remove residual carbonate cement. Sample purity was checked by X-ray diffraction analysis. Oxygen isotope measurements were performed at the Scottish Universities Research and Reactor Centre at Glasgow and the results are reported relative to SMOW.

Strontium isotope analyses were performed by Elf-Aquitaine at Pau. The powdered samples were repeatedly washed in nanopure water and then dissolved in 1 M high purity HCl at 50°C. The insoluble residue was removed by centrifuging and the solution was then allowed to dry at 140°C. HCl at a concentration of 3 N was then used for dolomite dissolution (1 h at 80°C). All $^{87}Sr/^{86}Sr$ results were corrected for the presence of ^{87}Rb due to leaching during sample digestion in HCl using an initial age of 220 Ma (Norian).

DOLOMITES

Two principal types of carbonate horizons, consisting entirely of dolomite, were found in the Grès de Chaunoy: (a) less common, thinner nodular to massive dolomites which formed in fine grained overbank deposits, capping fining upward cycles; and (b) massively cemented bodies up to 16 m thick that occur predominantly in the coarser grained channel sandstones and cap the unit (Fig. 2).

Thin dolomite horizons

These dolomites occur in red siltstones and consist of a lower nodular horizon overlain by a more massively cemented one. Such profiles may reach thicknesses of a few metres. The mudstones beneath the carbonates are typically structureless and generally poorly preserved when cored. Mottling was observed locally. The nodules are 1–20 cm in size (Fig. 3a,c) and commonly form vertically aligned columns. The dolomite consists of dolomicrite and exhibits features such as 'floating grains', micritic grain coatings, rhizocretions and alveolar–septal fabrics. Rhizocretions are 60–120 μm wide, enveloped by organic-rich, dark micrite and can be traced over the scale of a thin section. The interior is filled by multiple generations of clear dolospar. Within the more massive horizons there are stringers and pockets of siltstone and mudstone between nodules, similar to a chicken-wire texture, suggesting that the horizon was formed by the coalescence of nodules.

The host siltstones and silty mudstones are indigo in colour, with minor grey and red mottling. Each of the dolomite profiles is truncated by the overlying sandstones, attesting to their pre-burial origins. No silicification was noted in these dolomites, in contrast to the massive dolomites, and similarly the distinctive porosity in the thicker dolomites was also absent.

Thick massive dolomites

Macroscopic appearance

The massive dolomites range in thickness from several metres to at least 16 m. Those that occur lower in the Grès de Chaunoy are rarely more than 3–4 m thick and cannot be correlated, but a thick dolomite, at least 16 m thick locally, occurs near the top of the sandstone and can be correlated within individual wells. It forms a crudely linear belt, orientated parallel to depositional dip.

These units commonly display a vertical evolution from carbonate-cemented conglomerates and pebbly sandstones at the base (Fig. 3b), grading upward into carbonate-cemented sandstones and finally into silty dolomites. The more massively cemented dolostones are generally light grey (Fig. 3e), medium grey or brownish grey in colour. Nodular textures, brecciation (Fig. 3d), including prominent circum-granular fractures and multiple parallel, horizontal veining, fracture and vuggy porosity (Fig. 3f), and patchy silicification are the only macroscopic features visible. The top of the dolomites is either sharp and erosive or gradational with overlying sandstones.

The porosity types range from millimetre scale vugs to larger horizontal forms up to 10 cm in length, arrayed in parallel sets (Fig. 2). Secondary dissolution features are locally abundant in the thick dolomite horizons (particularly near the top) and include (i) interconnected macroporosity (Fig. 3f) up to 15 cm in diameter, (ii) solution-enlarged pores partially filled by green clay-rich sediment and (iii) local *in situ* breccias. In one core a continuous downhole decrease in pore size and a decrease in the amount of illuviated soil-derived clay was observed within an individual massive dolomite layer. Capping the main dolomite is a zone of irregular, vertically orientated pipes filled with green, silty clay. Some of these pipes are clearly solutional in origin with silty clays probably having infiltrated into them. Elsewhere, however, the clays resemble remnant host sediment between columnar dolomite. The porosities are locally high and result in poor recovery of core.

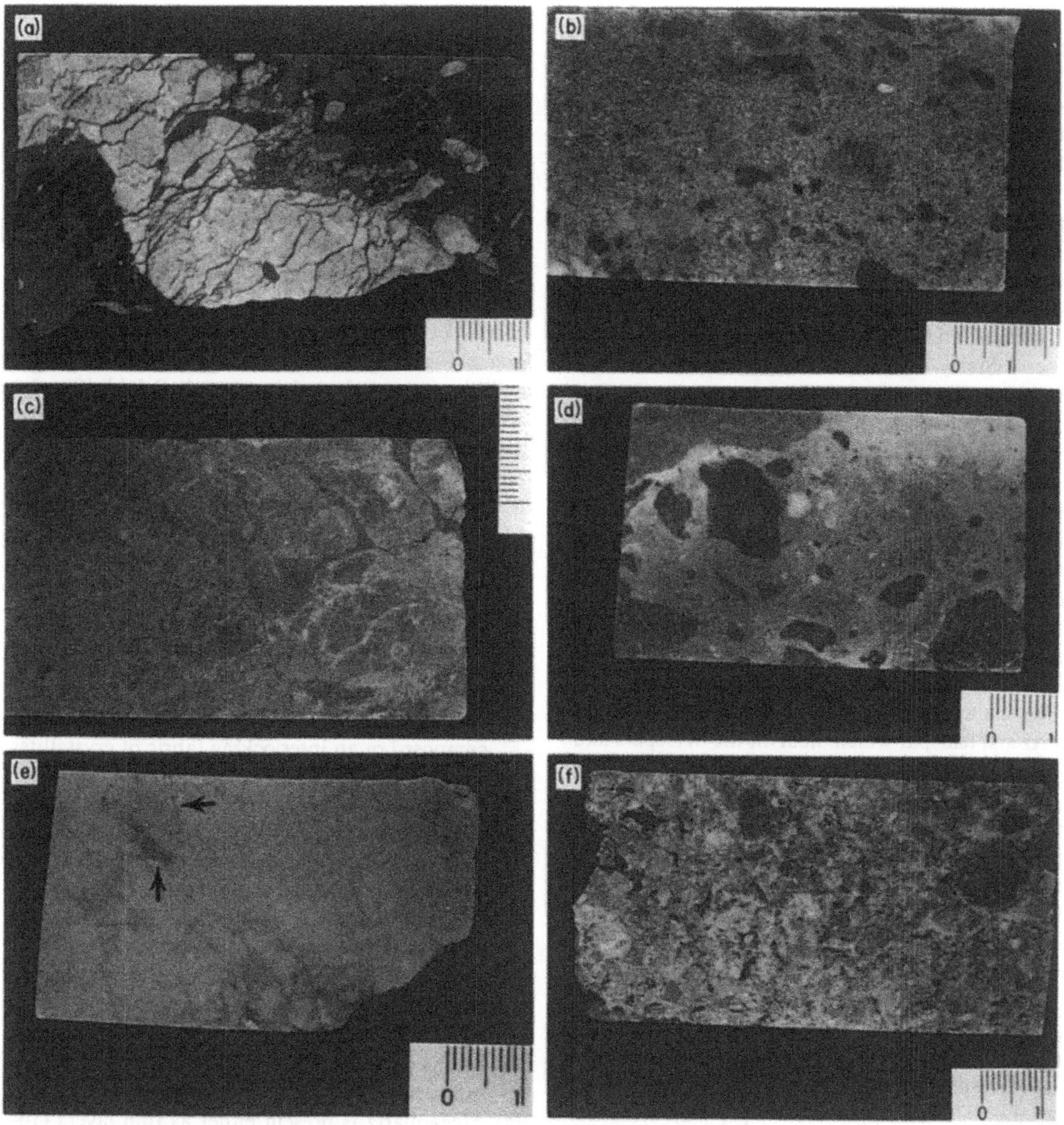

Fig. 3. Core slabs of macroscopic features in the Keuper dolomite. All small scale divisions are millimetres. (a) Dolomite nodules growing in a mottled argillaceous siltstone. This sample is taken from a thin dolomite horizon formed in an overbank deposit. The carbonate is usually finely crystalline and may contain biologically induced structures (e.g. alveolar–septal fabrics, rhizocretions). (b) A dolomite-cemented pebbly sandstone formed at the base of a carbonate-cemented fluvial cycle. (c) Nodular dolomite growing within a greenish sandstone and grading upward into more massive dolomite (toward the right). (d) A 'mud' supported carbonate breccia layer within a massively cemented dolomite, caused by brecciation and cementation during displacive carbonate growth. (e) Homogeneous, densely crystalline dolomite. Relict siliciclastic sediment can be seen (arrows) that has not been replaced by carbonate crystal growth. (f) Karstified dolocrete showing open dissolution porosity. The host rock resembles brecciated dolocrete as shown in (d).

Petrography

In thin section the thick dolomites consist of mixtures of dolomicrite (< 4 μm in diameter) and dolospar and detrital silicate material. Spötl (1991) was able to define a range of dolomites, including later burial ferroan dolomites. The descriptions here relate to the pre-burial cements.

The overwhelming majority of the samples examined in thin sections reveal a rather monotonous, dense crystalline groundmass (Fig. 4b) consisting of unimodal, planar-S and non-planar dolomite textures (in the sense of Sibley & Gregg, 1987). Planar-E dolomite textures are less common and appear to have formed as later diagenetic overgrowths on planar-S crystals. The dolospars exhibit moderately bright to bright orange to reddish brown luminescence under cathodoluminescence (Fig. 4a), whereas the dolomicrite luminesces moderately bright brown. Only a few dolomites showed visible fluorescence under a UV-fluorescence microscope (Fig. 4c). No relic or 'ghost' micromorphological structures were recognized in the massively cemented dolomite samples using cathodoluminescence and UV-fluorescence microscopy.

The main detrital grains in the groundmass are quartz and feldspar which 'float' in the dolospars and dolomicrites. These framework grains show evidence of strong replacement and etching by dolomite. Fractured quartz grains (Fig. 4a) and expanded micas, reflecting the displacive growth of the dolomite, are common in the more sandy levels. The amount of siliciclastic detritus tends to decrease as a function of dolocrete maturity. Mature dolocretes may be almost devoid of silicate grains (see Fig. 3e).

Apart from a dense crystalline dolomite groundmass, the dolomites exhibit a range of other features, including glaebules, complex cracks, crystallaria (Fig. 5a,b) and spheroidal dolomite (Fig. 4c,d). Dolospar and dolomicrite glaebules, up to several millimetres in diameter, are scattered in the groundmass. Many of these show complex, fine, dolospar-filled internal fractures, which rarely extend into the groundmass (Fig. 5a). The groundmass exhibits a wide range of dolospar-filled, planar to curved, millimetre scale fractures (Fig. 5b).

A distinctive component consists of spheroidal dolomite. Two types occur: type 1 consists of well developed spheroids with inclusion-rich concentric zones close to their margins (Fig. 4c). These spheroids are generally 30–90 μm in size but can reach 240 μm. They exhibit slightly undulose extinction. The spheroids commonly occur in coalesced masses. The larger spheroids are commonly fractured with fine radial cracks and reveal 'domain' extinction. Type 2 spheroids are up to 450 μm in diameter and consist of radiaxial dolomite; inclusion-rich concentric zones are poorly developed (Fig. 4d).

In a detailed study of one dolomite interval in Vert le Grand, a distinctive suite of fabrics was found only in the top few members of the unit. Such features were not recorded in the other thick dolomites. These include micrite-coated grains, locally with 'pendant' thickenings on the undersides of grains (Fig. 6a) and brace-like (meniscate) connections between adjacent grains (Fig. 6b). Irregular veins are also more common in the upper part of this dolomite unit. Peloids (Fig. 6c) are common in the top few metres. A particular distinctive feature of the upper part of this unit is the presence of strongly orientated clay domains (Fig. 7a), with a cross-cutting arrangement. Many of the coarser silicate grains have strongly orientated clay domains around them (Fig. 7b).

Local silicification is not uncommon and occurs mainly as passive, void-filling cement. Four types of textures can be distinguished.

(1) Randomly fibrous microquartz (in the sense of Milliken, 1979) is the most abundant type and is composed of an interlocking fabric of crystallites < 10 μm in size.
(2) Length-fast chalcedony occurs as well developed, radiaxial botryoidal layers, lining vugs (Fig. 4e,f). It nucleated as hemispherical aggregates 35–70 μm in diameter which resemble opal-CT lepispheres (cf. Maliva & Siever, 1988). However, they are far larger than those known from deep sea sediments and nodular cherts in ancient carbonates (5–20 μm).
(3) Length-slow chalcedony (quartzine) occurs only rarely, particularly as spherulites embedded in a groundmass of length-fast chalcedony.
(4) Megaquartz is present either as thin layers (10–20 μm) alternating with layers of length-fast chalcedony, or, most commonly, as an interlocking mosaic occluding the remaining pore space (Fig. 4e,f). Prismatic megaquartz was also noted on detrital quartz grains in the sandier dolocretes, growing into larger dissolution vugs. The megaquartz crystals show uniform extinction and are inclusion-poor, except for local crystallites of authigenic haematite.

Authigenic silica is often associated with opaque or brown masses that are tentatively identified as organic material, because of their generally bright fluorescence

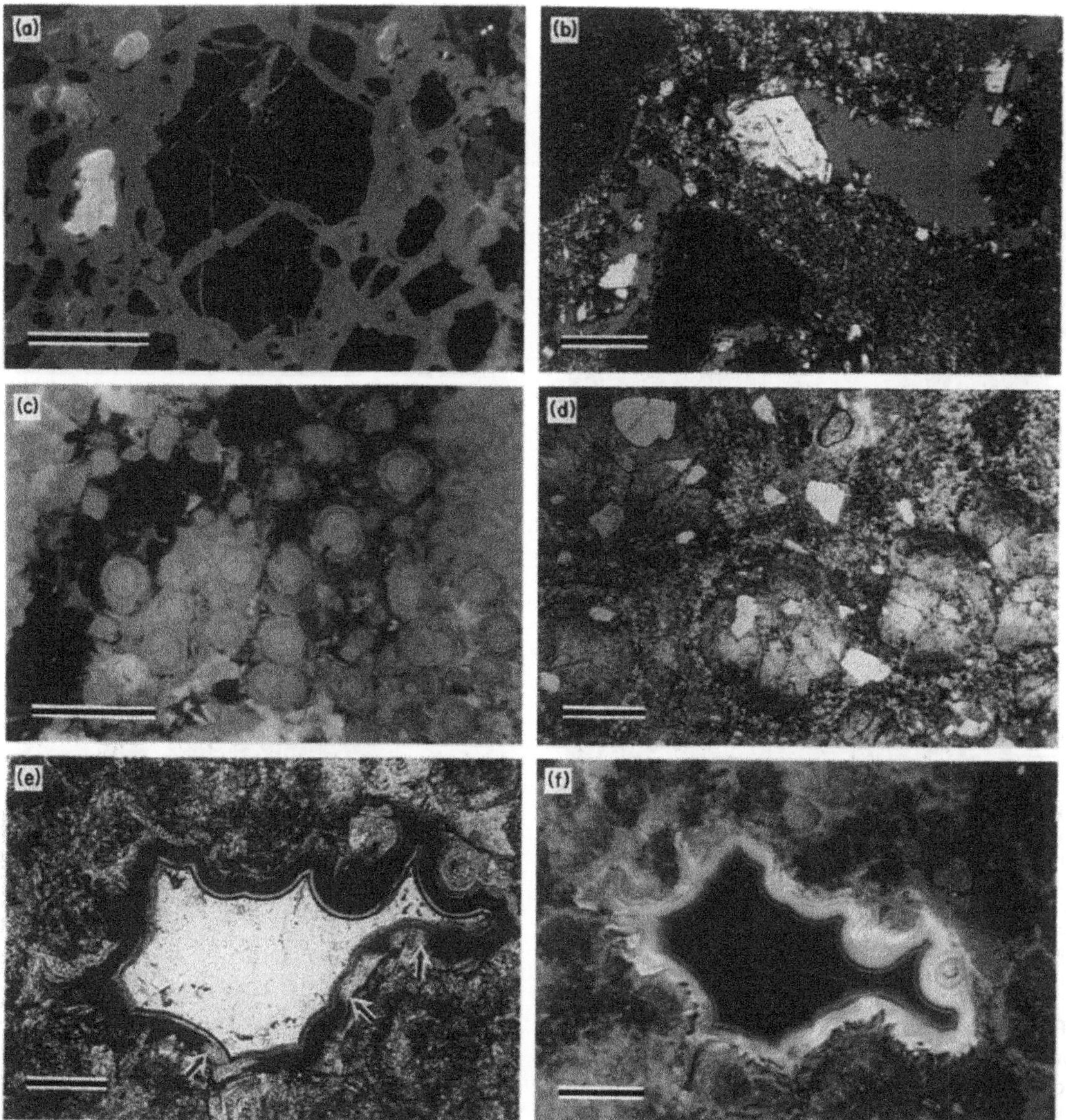

Fig. 4. Petrographic features of the Keuper dolomites. (a) A large broken polycrystalline quartz grain with clear fit of the pieces. Grain breakage due to displacive growth of dolomicrospar (orange CL colour) was a common process in these dolomites. Non-luminescent grains are detrital quartz and blue luminescent grains are alkali feldspar grains. CL photomicrograph, scale bar = 1 mm. (b) Karst macroporosity partially filled by late stage ferroan carbonate cements. The host dolomite is composed of dolomicrite and dolomicrospar. Plane-polarized light, scale bar = 250 μm. (c) Type 1 dolomite spheroids as seen under blue-light fluorescence. Note that ferroan dolomite filling the adjacent fracture is non-fluorescent. Scale bar = 500 μm. (d) Type 2 spheroidal dolomite growing in a dolomicrospar matrix. White spots are detrital quartz grains. Plane-polarized light, scale bar = 500 μm. (e), (f) Plane-polarized light and blue-light fluorescence pair illustrating silica cement stratigraphy in the Keuper dolomite. Botryoidal length-fast chalcedony is post-dated by translucent megaquartz. Note the discontinuous distribution of dark brown organic impurities in the chalcedony layer (arrows). Chalcedony exhibits bright greenish fluorescence, even in areas which appear transparent on (e). Host dolospar fluoresces moderately bright brown. Scale bars = 250 μm.

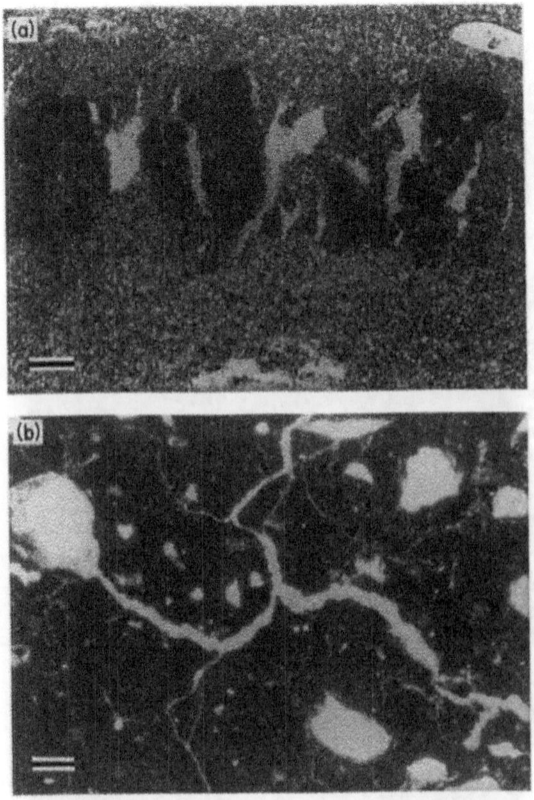

Fig. 5. (a) Parallel fractures in massive dolomites that formed within an elongate nodule of dolomicrite in a groundmass of dolomicrospar. Plane-polarized light, scale bar = 250 μm. (b) Irregular cracks and small vugs from massive dolomites, filled by dolospar. Plane-polarized light, scale bar = 100 μm. Both types of fractures are common in calcretes/dolocretes.

(Fig. 4f), and because EDS analyses showed no major peaks (e.g. Fe or Mn). Although brightly fluorescent zones coincide with impurity-rich layers (Fig. 4e), chalcedony layers that appear clear in transmitted light also exhibit moderately bright fluorescence (Fig. 4f).

The silica cement stratigraphy is variable but a general trend can be recognized from randomly fibrous microquartz, through radiaxial chalcedony to megaquartz (Fig. 4e). No relict evaporite minerals as inclusions in the silica were seen.

The silica cements post-date the early dolomites but pre-date all the ferroan, burial cements filling fractures. Silica cement clearly post-dates the formation of the irregular dissolution vugs in the dolomites, but there are localized occurrences of dissolution of early dolomite after silica precipitation.

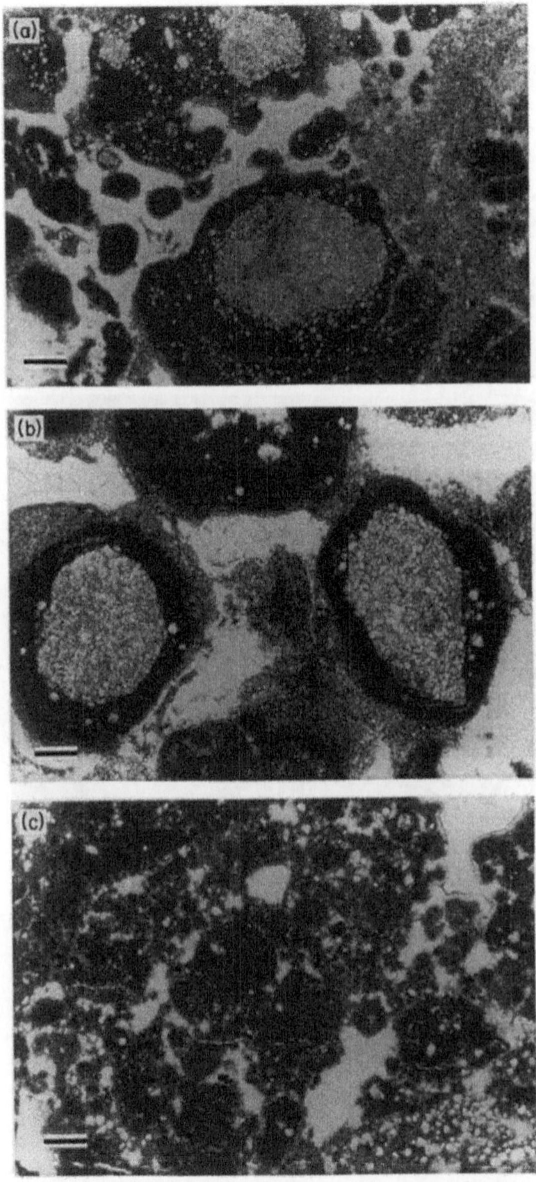

Fig. 6. Fabrics formed in the upper parts of massive dolomite profiles. (a) Nodules of dolomicrospar surrounded by dolomicrite coatings. Note irregular coatings, containing entrapped silicate and dolomite grains. The white areas are silica cements. Plane-polarized light, scale bar = 250 μm. (b) Coated grains with brace-like connections composed of dolomicrospar. The clear areas between are silica cements. Plane-polarized light, scale bar = 500 μm. (c) Peloidal fabric. Plane-polarized light, scale bar = 100 μm.

Carbonate mineralogy and geochemistry

Early diagenetic dolocrete cements are near stoichiometric and non-ferroan. The $CaCO_3$ content in the

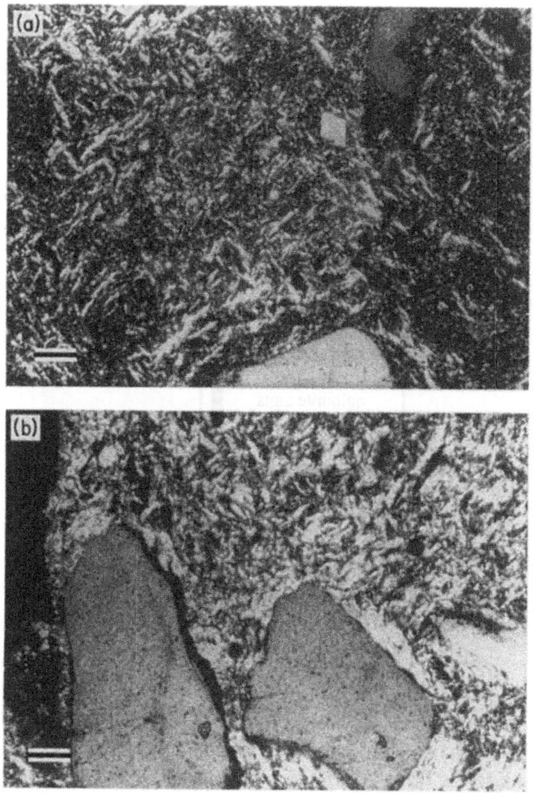

Fig. 7. Plane-polarized light photomicrographs of orientated clays from upper parts of massive dolocretes. (a) Cross-striated b-fabrics, reflecting orientated clay domains; (b) clay domains concentrated around sediment grains. Scale bars = 100 μm.

dolomite lattice ranges from 49·3 to 53·3 mol% with a mean of 51·4 ± 1·3 mol% (Table 1). No significant difference exists between samples from the two types of dolomite units. Fe and Mn are always < 0·1 wt%. Evaluation of X-ray diffraction traces showed the presence of sharp superstructure reflections at d_{101} = 22·1°, d_{015} = 35·3°, d_{021} = 43·8°, d_{030} = 67·4° and d_{0012} = 70·6° 2θ (using hexagonal indices), indicating well ordered dolomite. The degree of ordering (I_{015}/I_{110}) ranges from 0·52 to 1·30, with a frequency maximum at about 0·90. The results of the unit cell calculations are shown in Fig. 8 and Table 1. Dolocrete cements (dolomicrite and early diagenetic dolospar) have unit cells slightly expanded in the c_0 direction with respect to the 'ideal' dolomite. Dolomite from massive dolocrete units shows a slightly wider range in a_0 and c_0 values compared to dolomite from thin horizons.

Strontium concentrations in these samples vary between 27 and 813 ppm (mean of 40 samples = 254 ppm). No significant difference was found between samples from thin (238 ± 192 ppm) and massive dolocrete units (317 ± 229 ppm).

A total of 78 dolomite samples were analysed for their stable isotopic composition. The results are plotted in Fig. 9. $\delta^{13}C$ values range from −7·7 to −0·4‰ and $\delta^{18}O$ values range from −5·1 to +1·8‰. There is no significant difference between the stable isotopic composition of the thick dolomite units and the thinner ones (Fig. 9).

The strontium isotope data show a large spread in $^{87}Sr/^{86}Sr$ ratios and no significant differences between the dolomite types. Samples from the thin dolomite layers have ratios of 0·7103–0·7126 (mean 0·7110 ± 0·0006; n = 18) and samples from thick units have ratios of 0·7101–0·7119 (mean 0·7105 ± 0·0004; n = 16).

Six silica separates consisting of microquartz, megaquartz and chalcedony were analysed for their oxygen isotope composition. Their $\delta^{18}O$ values range from 28·3 to 31·6‰.

INTERPRETATION

Thin dolomite horizons

These dolomites formed on the alluvial surfaces represented by the silts and muddy silts. Their nodular to massive form and rhizocretions are typical features of pedogenic calcretes and dolocretes (Esteban & Klappa, 1983; Goudie, 1983; Wright & Tucker, 1991). The dolomite is interpreted as a primary precipitate as it lacks any relict unaltered precursors (see below).

Most modern and ancient pedogenic carbonates consist of low-Mg calcite and dolocretes are less common. The latter tend to develop on Mg-rich hosts such as basic or ultrabasic rocks or appear to be associated with evaporitic depositional environments (e.g. Spy-Anderson, 1981; Richter, 1985; Botha & Hughes, 1992).

Thick massive dolomites

Macroscopic and micromorphological features

Massive dolomite units in the Grès de Chaunoy contrast with most modern and ancient vadose (pedogenic) carbonates in that they are confined to zones of high primary porosities and permeabilities (channel sands and gravels). This might not be obvious in the case of thick, mature units, because of the almost complete replacement of silicate grains during carbonate cementation. However, the presence of

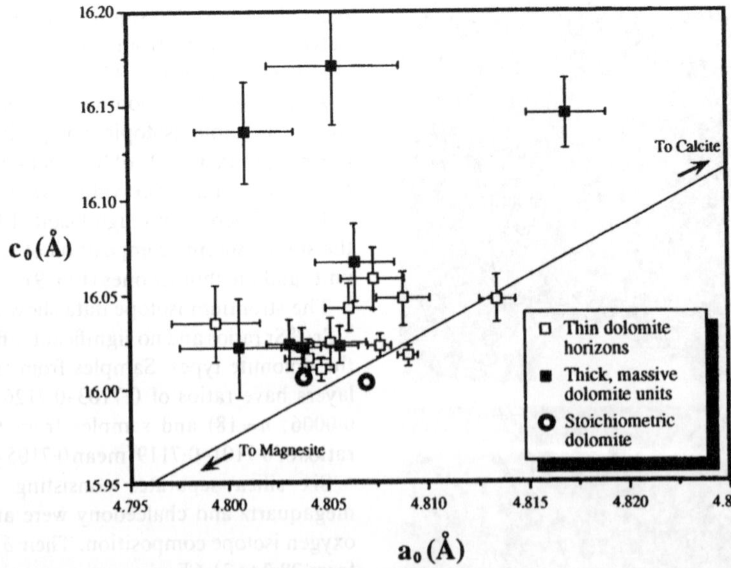

Fig. 8. Unit cell parameters for early diagenetic dolomicrite and dolospar cements. The line connects the unit cell dimensions of calcite and magnesite. Error bars show plus and minus one standard deviation. Natural stoichiometric dolomites are taken from Reeder (1983).

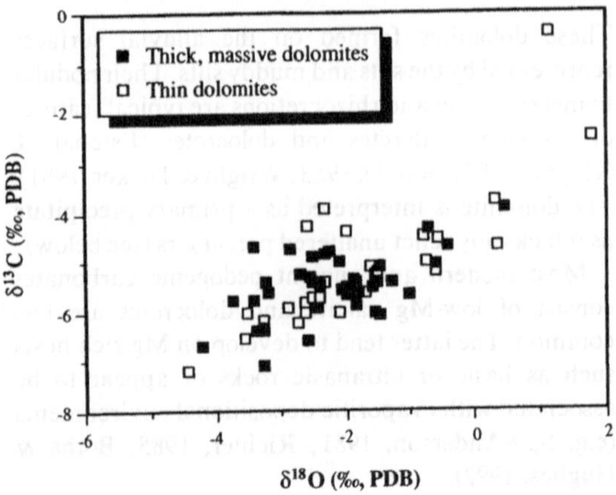

Fig. 9. Stable isotope composition of continental carbonates in the Keuper of the Paris Basin. Note covariance between oxygen and carbon isotope ratios in both types of dolomites.

scattered quartzitic pebbles, 'floating grain' textures and the general upward decrease in the amount of siliciclastic detritus (from carbonate-cemented gravels at the base through sandy dolomites up to almost pure dolostones near the top) indicate that carbonate formation occurred largely within a siliciclastic host sediment. In addition, most Quaternary and pre-Quaternary pedogenic calcretes and dolocretes are relatively thin, 1–3 m thick (Goudie, 1983; Wright & Tucker, 1991). The predominance of dolomite in the Triassic material would classify it as a dolocrete in the sense of Netterberg (1980), i.e. a duricrust having >90% of its carbonate as dolomite.

The massive dolomite possesses many features typical of calcretes and dolocretes such as nodular textures, brecciation, circum-granular spar-filled cracks, coated grains, orientated clays and 'floating' silicate grains. In contrast to the thin dolomite horizons, rhizocretions are absent, dense crystalline matrix dolomite is far more abundant and the dolomite crystals are uniformly larger.

A distinctive feature of the massive dolomite is the presence of spheroidal dolomite. This form has been recorded from groundwater dolocretes in Kuwait (Khalaf, 1990; El-Sayed *et al.*, 1991). Gunatilaka (1989) also described similar spheroids from a variety of lithologies from Kuwait. He correlated their occurrences with oil seepages, suggesting that gas bubbles, produced by bacteria feeding on the hydrocarbons, may have provided a nucleus for the spheroidal growth forms. However, El-Sayed *et al.* (1991) favoured split-growth as a process whereby the spherulites, and the closely related 'artichoke dolomites', had formed. There is no evidence for hydrocar-

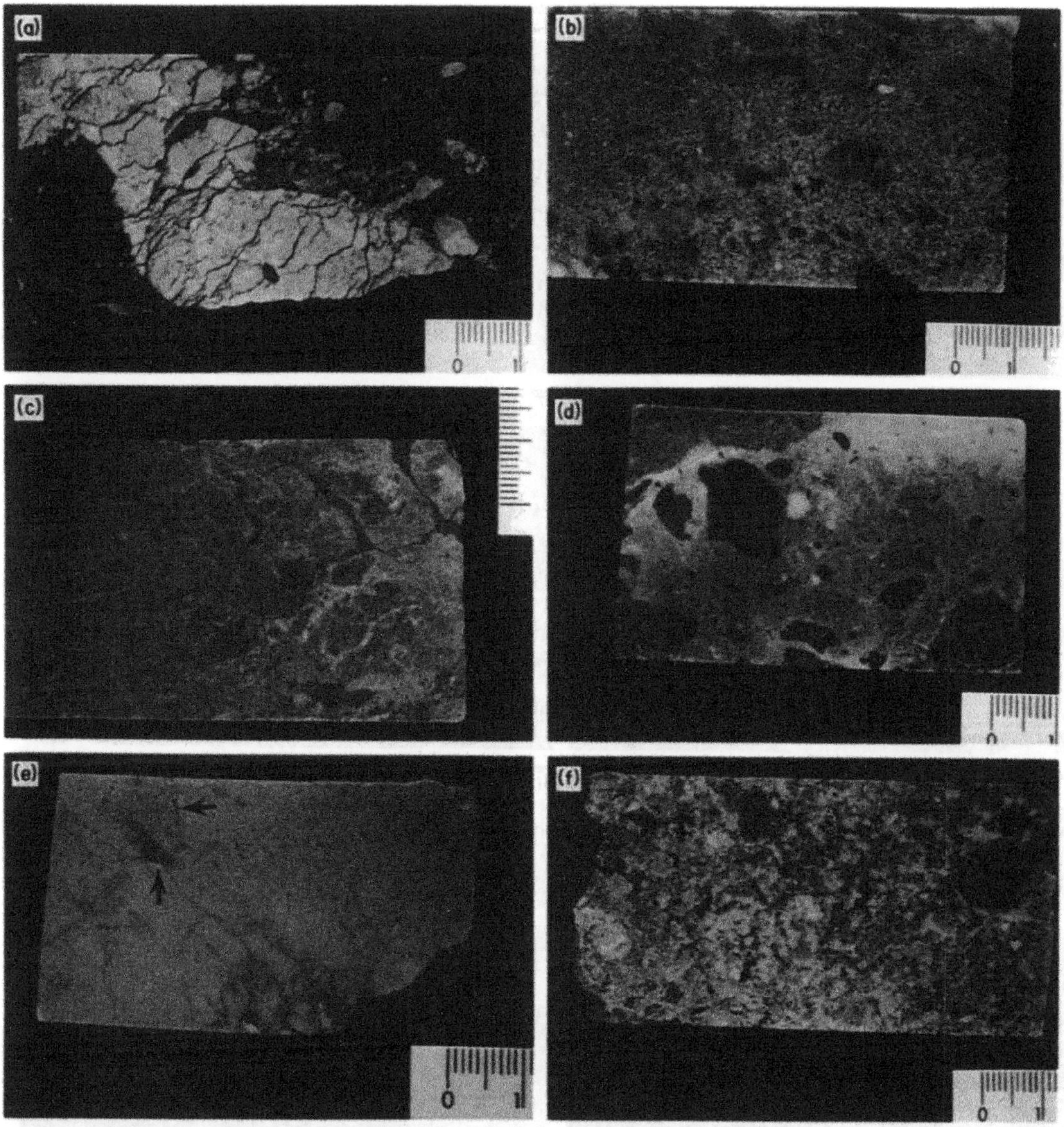

Fig. 3. Core slabs of macroscopic features in the Keuper dolomite. All small scale divisions are millimetres. (a) Dolomite nodules growing in a mottled argillaceous siltstone. This sample is taken from a thin dolomite horizon formed in an overbank deposit. The carbonate is usually finely crystalline and may contain biologically induced structures (e.g. alveolar–septal fabrics, rhizocretions). (b) A dolomite-cemented pebbly sandstone formed at the base of a carbonate-cemented fluvial cycle. (c) Nodular dolomite growing within a greenish sandstone and grading upward into more massive dolomite (toward the right). (d) A 'mud' supported carbonate breccia layer within a massively cemented dolomite, caused by brecciation and cementation during displacive carbonate growth. (e) Homogeneous, densely crystalline dolomite. Relict siliciclastic sediment can be seen (arrows) that has not been replaced by carbonate crystal growth. (f) Karstified dolocrete showing open dissolution porosity. The host rock resembles brecciated dolocrete as shown in (d).

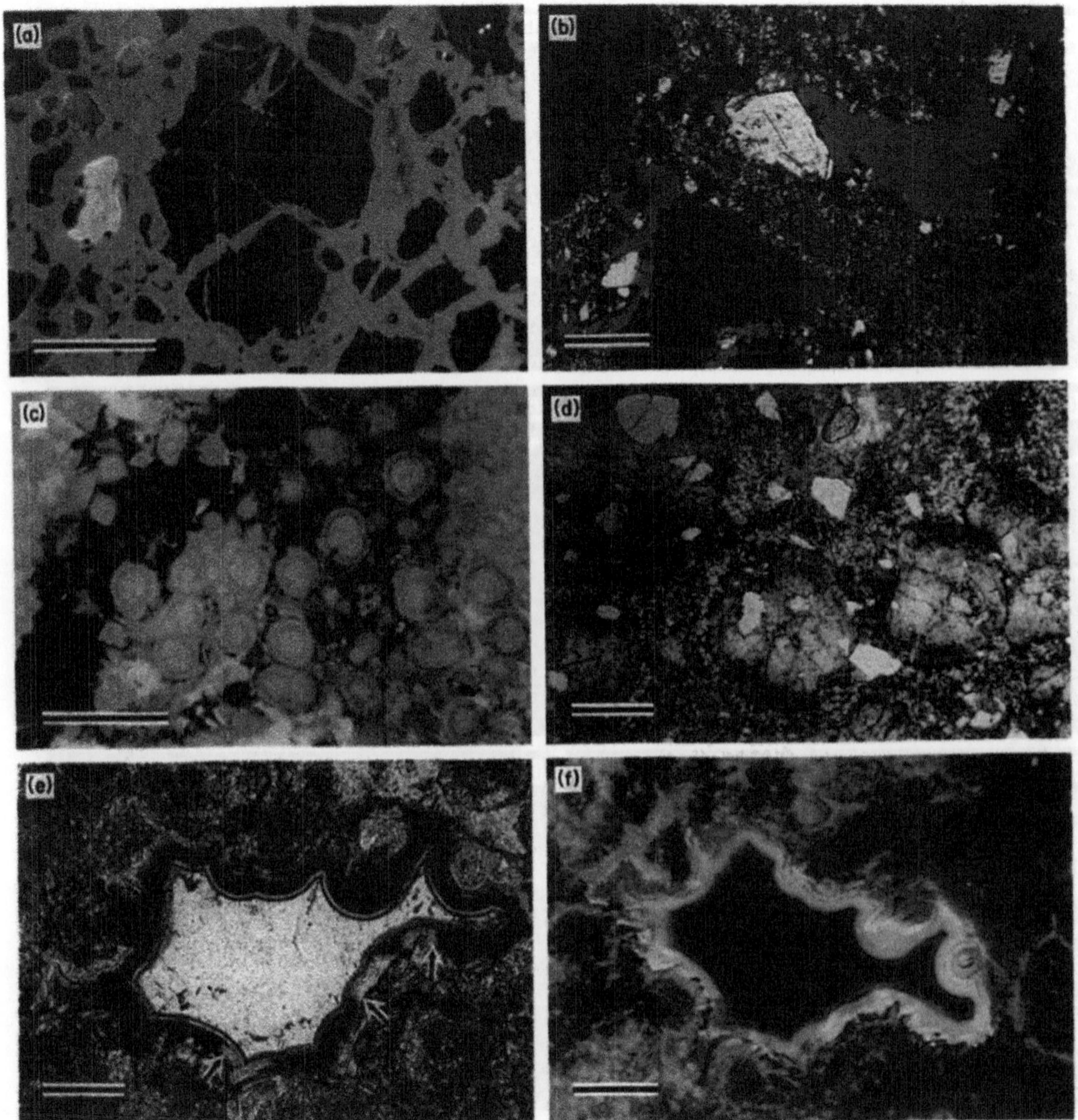

Fig. 4. Petrographic features of the Keuper dolomites. (a) A large broken polycrystalline quartz grain with clear fit of the pieces. Grain breakage due to displacive growth of dolomicrospar (orange CL colour) was a common process in these dolomites. Non-luminescent grains are detrital quartz and blue luminescent grains are alkali feldspar grains. CL photomicrograph, scale bar = 1 mm. (b) Karst macroporosity partially filled by late stage ferroan carbonate cements. The host dolomite is composed of dolomicrite and dolomicrospar. Plane-polarized light, scale bar = 250 μm. (c) Type 1 dolomite spheroids as seen under blue-light fluorescence. Note that ferroan dolomite filling the adjacent fracture is non-fluorescent. Scale bar = 500 μm. (d) Type 2 spheroidal dolomite growing in a dolomicrospar matrix. White spots are detrital quartz grains. Plane-polarized light, scale bar = 500 μm. (e), (f) Plane-polarized light and blue-light fluorescence pair illustrating silica cement stratigraphy in the Keuper dolomite. Botryoidal length-fast chalcedony is post-dated by translucent megaquartz. Note the discontinuous distribution of dark brown organic impurities in the chalcedony layer (arrows). Chalcedony exhibits bright greenish fluorescence, even in areas which appear transparent on (e). Host dolospar fluoresces moderately bright brown. Scale bars = 250 μm.

Table 1. Hexagonal unit cell dimensions (mean ± SD) and stoichiometry of dolomite cements.

Sample	a_0 (Å)	c_0 (Å)	V (Å)	n	$CaCO_3$ (mol%)
Thin dolomite units					
Well A 2322·90	4·8040 ± 0·0011	16·0168 ± 0·0097	320·1197	13	50·0
Well A 2335·20	4·8088 ± 0·0013	16·0477 ± 0·0138	321·3785	13	53·0
Well B 1982·10	4·8076 ± 0·0006	16·0236 ± 0·0061	320·7358	15	51·7
Well B 1982·50	4·8090 ± 0·0005	16·0179 ± 0·0056	320·8084	15	50·3
Well C 1916·60	4·8073 ± 0·0016	16·0584 ± 0·0164	321·3922	14	53·0
Well C 1926·90	4·7994 ± 0·0022	16·0351 ± 0·0201	319·8720	10	49·3
Well C 1935·40	4·8051 ± 0·0018	16·0255 ± 0·0120	320·4403	10	52·0
Well C 1935·70	4·8061 ± 0·0012	16·0434 ± 0·0124	320·9317	13	52·3
Well I 2282·40	4·8135 ± 0·0009	16·0474 ± 0·0124	322·0010	11	52·3
Well I 2285·90	4·8047 ± 0·0006	16·0108 ± 0·0064	320·0930	14	50·3
Thick dolomite units					
Well C 1893·25	4·8005 ± 0·0029	16·0221 ± 0·0264	319·7592	8	50·3
Well C 1895·45	4·8009 ± 0·0025	16·1356 ± 0·0265	322·0780	10	50·7
Well C 1899·60	4·8031 ± 0·0012	16·0239 ± 0·0134	320·1416	13	51·3
Well C 1934·50	4·8056 ± 0·0009	16·0238 ± 0·0094	320·4730	15	50·3
Well D 2130·55	4·8170 ± 0·0020	16·1454 ± 0·0186	324·4388	10	53·3
Well D 2131·85	4·8054 ± 0·0033	16·1709 ± 0·0312	323·3380	9	53·3
Well F 2233·50	4·8064 ± 0·0020	16·0667 ± 0·0207	321·4379	13	51·0
Well A 2312·95	4·8037 ± 0·0010	16·0221 ± 0·0103	320·1856	14	50·7

n is the number of lines used to calculate the unit cell parameters. The $CaCO_3$ content in the dolomite lattice was obtained from the measured d_{104} value (using a Guinier camera) and applying the equation of Lumsden & Chimahusky (1980).

bon seepages during dolomite growth in this Triassic example, and the hydrocarbon charge in the study area took place during the Late Mesozoic and Tertiary (Espitalié *et al.*, 1988).

The cross-cutting, strongly orientated clays found at the top of a single, thick dolocrete unit (see above) correspond to lattisepic or cross-striated b-fabrics found in soils (Brewer, 1976; Fitzpatrick, 1984; Bullock *et al.*, 1985). The orientated clay domains around the coarser silicate grains correspond to anisotropic aureoles or granostriated fabrics in the soil terminology (Fitzpatrick, 1984, p. 202; Bullock *et al.*, 1985). These fabrics reflect stress and movement in the soil matrix, typically resulting from wetting–drying and shrink–swell cycles. Within the massive dolomites these clays were only associated with solution piping and vadose features such as pendant cements.

Several macroscopic and micromorphological features argue against a pedogenic (vadose) origin for these massive carbonates, most notably the absence of biologically induced structures, their occurrence in coarse grained channel (and not overbank) deposits, and their great thickness. From the occurrence of calcrete/dolocrete-like features, and the presence of vadose-related features only at the tops of the thickest dolomites, we interpret them as mainly groundwater in origin. Such groundwater calcretes and dolocretes are best documented from central Australia where they form lenses and ribbons of well cemented alluvial deposits kilometres wide (up to 10 km), tens of kilometres long (over 100 km in some cases) and with an average thickness of 10 m (Mann & Deutscher, 1978; Mann & Horwitz, 1979; Arakel & McConchie, 1982; Carlisle, 1983; Arakel, 1986, 1991; Arakel & Wakelin-King, 1991). Similar dolocretes have been recorded from Kuwait by Khalaf (1990) and El Sayed *et al.* (1991).

These phreatic (groundwater or valley) calcretes and dolocretes are typically micritic and densely crystalline. They lack vadose features such as meniscus and pendant cements (Arakel & McConchie, 1982). The carbonates have been produced by cementation, displacement and replacement of the host sediments until almost pure carbonate bodies are produced. The carbonate may contain authigenic silica (Arakel *et al.*, 1989), various clays (typically Mg-rich forms such as sepiolite and palygorskite) and gypsum. Porosities in the range of 21–28% have been recorded in the massive zones of Quaternary groundwater calcretes (see Wright & Tucker, 1991).

On a local scale these carbonate zones may be lens shaped and show thickening to form domes which break through to the surface. The carbonates are sometimes overlain by uncemented alluvial deposits, with a gradational upper contact (cf. buried pedogenic calcretes or dolocretes) or may be capped by pedogenic carbonates. This may have been the case in this Triassic example where zones with orientated clays and other pedogenic features cap the main dolomite units. The possible karstic zones at the tops of some of the massive dolomites also have analogues in Quaternary duricrusts where topographic variations and exhumation of the duricrusts can result in extensive dissolution (Arakel, 1991).

The genesis of groundwater duricrusts has been reviewed by Wright & Tucker (1991) and the main mechanisms for carbonate precipitation are CO_2 degassing, evaporation/evapotranspiration and the common ion effect. Cementation preferentially occurs where groundwaters, flowing through the drainage system alluvium, are brought nearer the ground surface because of bedrock irregularities. Other sites for precipitation include areas where different groundwaters mix with more saline playa waters. The nature of the precipitation changes down the groundwater flow zone as the waters evolve, with the Mg/Ca ratio increasing down-dip as calcrete first forms, followed by dolomite (Arakel, 1986, 1988; Jacobsen *et al.*, 1988).

The silica in the massive dolocrete can be compared to silica in Quaternary duricrusts. Recently, Thiry & Milnes (1991) reviewed the occurrence of near surface continental silica formation (silcretes) and provided criteria to distinguish between pedogenic and groundwater silcretes. However, the complex history of changed environments recorded in Cainozoic silcretes shows that many silcretes are polyphase and may have been overprinted several times. No features indicative of pedogenic silcrete were noted in the massive dolocretes, suggesting a groundwater origin. The range of silica forms found in the thick dolocretes is similar to that from the Australian silcretes (Arakel *et al.*, 1989; Thiry & Milnes, 1991), which even include the brown and haematite-bearing forms. The complex interlayering of silica forms in the Triassic dolocrete probably reflects frequent fluctuations in the degree of supersaturation of the silica-bearing fluid; as with Quaternary silicified calcretes, this interlayering may mark the positions of original groundwater fluctuation zones (Arakel *et al.*, 1989). In general, the silicification post-dates dolocrete formation and probably relates to a major change in regional hydrology. In present day aquifers in Australia the silicification zone occurs above zones of more hypersaline waters (Arakel *et al.*, 1989). The sources of the silica are both dissolved silica derived from weathering horizons and, probably more importantly, silica released as a result of extensive replacement of detrital silicate grains during massive dolomite formation.

Mineralogy and chemical composition

The dolomites are well ordered and near stoichiometric. Unit cell dimensions of the Keuper dolocretes are slightly expanded compared to stoichiometric dolomite (Fig. 8). Reeder & Sheppard (1984) have demonstrated that unit cell dimensions of various non-ferroan sedimentary dolomites vary largely as a function of the amount of excess $CaCO_3$ in the crystal lattice. Keuper dolomites show only a moderate correlation between the unit cell volumes and $CaCO_3$ content ($R^2 = 0{\cdot}51$). This might reflect sample heterogeneity and the use of an indirect method to determine dolomite stoichiometry (X-ray diffractometry).

The strontium content varies considerably (27–813 ppm). Comparison with the Ca/Sr ratio of marine dolomite (Vahrenkamp & Swart, 1990) shows that the Triassic dolomites precipitated from fluids with highly variable Ca/Sr ratios and from fluids generally more enriched in strontium. Since meteoric water has very little Sr, the elevated values in the Triassic dolocretes suggest strong evaporative concentration of the groundwaters as they migrated toward the playa plain. El-Sayed *et al.* (1991) also noted a range of Sr values in dolocretes from Kuwait. They were generally somewhat lower than those recorded here (mostly 53–300 ppm).

Stable isotope geochemistry has been successfully applied in palaeoclimatic and palaeovegetational studies of ancient pedogenic calcretes, where the isotope fractionations are reasonably well understood (e.g. Cerling, 1984, 1991; Salomons & Mook, 1986; Mack *et al.*, 1991; Mora *et al.*, 1991). Very little, however, is known about stable isotopes in groundwater calcretes, including their possible use in differentiating pedogenic from non-pedogenic calcretes and dolocretes (Manze & Brunnacker, 1977; Talma & Netterberg, 1983).

Keuper dolocretes from the Paris Basin show similar carbon and oxygen isotope ratios for the possible groundwater and pedogenic forms (Fig. 9). Moreover, they compare well with many recent and ancient pedogenic calcretes as compiled by Talma & Netterberg (1983), taking into account a *c.* 3‰ difference in the $\delta^{18}O$ between coexisting calcite and dolomite (Land, 1983). They differ, however, from modern marginal marine dolocretes of the Persian Gulf (El-Sayed *et al.*, 1991), which have highly depleted carbon isotope signatures (-9 to $-10{\cdot}7$‰) and marine evaporitic oxygen isotope ratios (+0·6 to +3·3‰). These authors corrected their isotope data to account for the differential fractionation factor of dolomite dissolving in phosphoric acid. In order to compare their data with the results of the present study, a value of approximately 1·3‰ has to be added to their isotope ratios, making their oxygen isotope ratios even heavier.

Detailed interpretations of the carbon and oxygen stable isotope compositions of ancient calcretes are fraught with problems. The similarities between the $\delta^{13}C$ values of the two Triassic dolocretes are, perhaps, unexpected. It would be reasonable for the vadose dolocrete to exhibit lighter values, reflecting greater organic activity in the soil than lower in the substrate, unless the vegetation was one which produced soil CO_2 with relatively heavy values (such as C4 vegetation). Talma & Netterberg (1983) reported that the $\delta^{13}C$ values of phreatic calcretes from the Kalahari Desert tend to be lighter than the values of the more common pedogenic varieties of the same region. According to these authors, groundwater at greater depth may have been influenced by the deep roots of trees (C3 plants) compared to shallow soils where grass (C4 plants) will also have an effect. No such difference should be expected for the Triassic dolocretes from the Paris Basin, because of the exclusive C3 flora in the pre-Miocene (but see also Wright & Vanstone, 1991). These Keuper carbon values are heavier than those recorded by Purvis & Wright (1991) from Middle Triassic 'groundwater' rhizocretions from south-west England (ranging from −7 to −9‰ PDB), which would have received strong input of light, root-respired carbon.

The oxygen isotope composition of the groundwater varies mainly with climate, in particular with the mean annual temperature (Cerling, 1984). Evaporation will result in ^{18}O-enriched pore water. The oxygen isotope data of the most enriched Triassic groundwater dolocretes are consistent with the formation from ^{18}O-enriched meteoric groundwater. Assuming a groundwater temperature of 20°C (compatible with a low latitude setting of *c.* 30°N for the Paris Basin during the Triassic; Smith *et al.*, 1981), the $\delta^{18}O_{H_2O}$ in equilibrium with the Keuper dolomites is −8·6 to −1·7‰ SMOW (Land, 1983). The variation in the $\delta^{18}O$ of the Keuper groundwater dolocretes is probably due to evaporitive enrichment, probably in a downstream direction (cf. Sr results). Fluctuations in the temperature of carbonate precipitation could also have contributed to the spread in oxygen isotope ratios (cf. Schlesinger, 1985). Alternatively, variations in the degree of recrystallization (stabilization) of the Ca-rich precursor could also result in a spread of the oxygen isotope ratios.

It has been speculated that groundwater carbonates should be enriched in ^{18}O relative to their pedogenic counterparts, because they commonly form from evolved groundwater (Wright & Tucker, 1991). This is not substantiated by the present data. In addition, the few known occurrences of Quaternary groundwater calcretes from the Kalahari Desert tend to have $\delta^{18}O$ values similar or slightly lighter than their pedogenic counterparts (Talma & Netterberg, 1983). This phenomenon is probably due to the fact that the groundwaters are frequently only recharged by the heavier rainfalls which penetrate into the soil beyond the influence of evapotranspiration. Heavy rainfalls in this particular area tend to have lower $\delta^{18}O$ values than the mean annual rainfall (Vogel & Van Urk, 1975).

The dolocretes show a well developed linear trend of covariance of $\delta^{18}O$ and $\delta^{13}C$ (Fig. 9). This trend could have resulted from evaporation and degassing of CO_2 (Salomons *et al.*, 1978; Schlesinger, 1985; Salomons & Mook, 1986; Dever *et al.*, 1987), a trend also noted from Triassic calcretes by Naylor *et al.* (1989). The slightly heavier values recorded in the vadose (pedogenic) dolocretes probably reflect greater evaporation in the soil zone.

The strontium isotopic composition of the massive (groundwater) dolocretes is significantly more radiogenic than the contemporary Triassic seawater (0·7076; Koepnick *et al.*, 1990). Thus the pore water at the time of dolocrete formation was non-marine. In addition, the scatter of $^{87}Sr/^{86}Sr$ ratios is also consistent with a continental groundwater source. Present day continental surface and groundwaters typically show mixing phenomena, with widely varying $^{87}Sr/^{86}Sr$ ratios due to the complex lithological composition of the catchment areas (Faure, 1986). For instance, $^{87}Sr/^{86}Sr$ ratios in the dissolved load of modern Canadian rivers vary between 0·7025 and 0·7350 (Veizer, 1989). The mixing of two rivers draining catchment areas of different $^{87}Sr/^{86}Sr$ ratios, will ultimately produce a linear mixing line when plotted in an $^{87}Sr/^{86}Sr$ versus 1/Sr diagram (Faure, 1986). This was tested using the strontium isotope data of the early diagenetic dolocrete cements. The plot did not yield a linear relationship, and therefore the mixing responsible for the scatter of the values could not have been simply that of two different component solutions, each having a specific $^{87}Sr/^{86}Sr$ ratio and Sr concentration. Rather, it suggests a complex mixing system, as would be expected in a large alluvial system, draining a lithologically complex metamorphic hinterland.

The $\delta^{18}O$ ratios of the authigenic silica suggest a low temperature origin, assuming typical meteoric $\delta^{18}O$ values. Figure 10 shows the curves for minimum and maximum $\delta^{18}O$ values for authigenic silica in equilibrium with waters having various oxygen isotope compositions. The calculation is based on the data of Friedman & O'Neil (1977). Similar results were obtained using other fractionation factors (after Ligang *et al.*, 1989). If the samples equilibrated isotopically with water at a near surface temperature of 20°C, the corresponding range of water $\delta^{18}O$ values was -8.5 to -5.3‰ SMOW (Fig. 10). This is within the range of $\delta^{18}O_{H_2O}$ values inferred from the isotopic composition of the dolocretes.

The origin of these massive dolomites is difficult to constrain. Two models are largely consistent with the data set, primary dolomite precipitation and early diagenetic dolomitization of a calcitic precursor (calcrete).

A strong argument in favour of primary dolomite formation is the entire absence of calcite relics. X-ray diffraction, cathodoluminescence, UV/blue-light fluorescence and back-scattered electron microscopy failed to detect any calcite in these rocks. If these dolomites were primary, it might be assumed that the initial (proto)dolomite was non-stoichiometric and poorly ordered and has undergone recrystallization and stabilization during burial. The somewhat Ca-rich composition and slightly expanded unit cell dimensions of the Keuper dolomites are consistent with such a gradual stabilization process. The preservation of subtle calcrete-like textures and the similarity of the Triassic and Quaternary spheroidal dolomites, however, argues against a pervasive burial dolomitization/recrystallization event. Gunatilaka (1989) noted that the spheroidal dolomites from Kuwait ranged from near stoichiometric to $Ca_{56}Mg_{44}$, with the spheroids in some of the youngest units being the best ordered and less non-stoichiometric than forms in older host-rocks. El-Sayed *et al.* (1991) noted a range of 50·4–52·9 mol% $CaCO_3$ in other dolocretes from Kuwait that compare well with the Keuper example. In phreatic dolocretes formed in Tertiary clastic sediments of Kuwait, Khalaf (1990) reported the occurrence of minute dolomite crystals within the muddy matrix and the lack of any indications for a precursor calcite. These observations led him to suggest that the dolomite precipitated as a result of slow evaporation of Mg-rich porewater. Similarly, Spy-Anderson (1981) concluded that pedogenic dolocretes from the Keuper of south-east France formed without a calcium carbonate precursor.

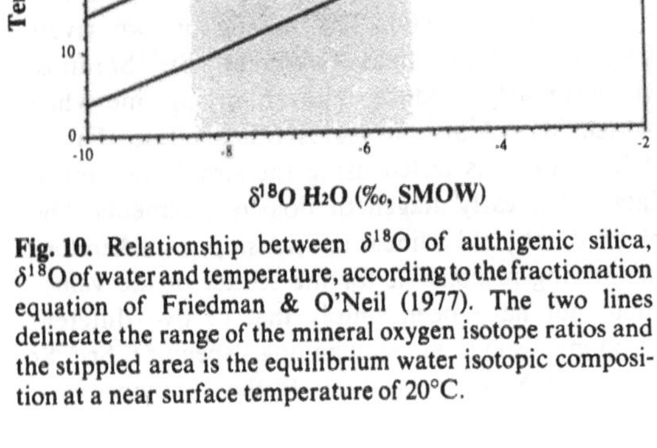

Fig. 10. Relationship between $\delta^{18}O$ of authigenic silica, $\delta^{18}O$ of water and temperature, according to the fractionation equation of Friedman & O'Neil (1977). The two lines delineate the range of the mineral oxygen isotope ratios and the stippled area is the equilibrium water isotopic composition at a near surface temperature of 20°C.

Alternatively, the Keuper dolocretes could have formed as initial calcretes and were dolomitized either penecontemporaneously with sedimentation or during burial (cf. Pelechaty & James, 1991). A deep burial dolomitization model can be ruled out, because (1) burial dolomite cements encountered in the Keuper dolocretes differ from early dolomicrite and dolospar in that they are ferroan, coarsely crystalline (partially saddle shaped) and show significantly depleted oxygen isotope compositions (Spötl, 1991), and (2) the common preservation of dolomicrite argues against a pervasive, late stage dolomitization event. Early diagenetic dolomitization of an initial calcium carbonate precursor, however, is a possible mechanism and seems to be the dominant process of dolomite formation in modern Australian calcretes (Arakel & McConchie, 1982). In this case, calcite would have been replaced by Ca-rich (proto)dolomite within probably less than a few tens of metres of burial. Although dolomitization cannot be entirely ruled out, primary (proto)dolomite precipitation is favoured because there is no petrographic evidence for extensive calcite-dominant replacement.

Regardless of which of the two dolomite formation processes actually operated in the Paris Basin, the fact that dolocrete and not calcrete occurs in this continental setting poses the question about the source for the high Mg^{2+} concentrations in the Keuper groundwaters. Seawater or seawater-derived brines are possible Mg^{2+} sources. Marine-derived evaporites are in fact common in the eastern part of the basin (Fig. 1; Geissler-Cussey, 1986) and saline groundwaters could have interacted with the sandstone and carbonate aquifers in the study area causing dolomite formation. Isotopic data, however, strongly argue against a marine component in the groundwaters of the study area. (1) Sr isotope ratios of the early dolocrete cements are incompatible with a marine (Triassic) seawater source. Considering the high salinities of the brines that precipitated sulphates and chlorides in the eastern Paris Basin, even small amounts of these brines mixing with the continental groundwaters in the study area would result in an overall $^{87}Sr/^{86}Sr$ ratio identical to that of marine waters. Such values were not measured in the Keuper dolocretes. (2) Dolocrete carbon isotope ratios are much lighter than values typical of marine evaporitic carbonates and are in perfect agreement with carbon isotope data reported from calcretes and dolocretes throughout the world (Talma & Netterberg, 1983). (3) Displacive anhydrite nodules and enterolitic anhydrite sampled in Keuper mudstones slightly to the east of the study area yielded low $\delta^{34}S$ and radiogenic $^{87}Sr/^{86}Sr$ ratios (Spötl, 1991). The $\delta^{34}S$ ratios range from $+11{\cdot}5$ to $+13{\cdot}9$‰ CDT and are significantly lower than that for Late Triassic seawater ($+18{\cdot}5$‰; Holser *et al.*, 1989). The strontium isotope ratios (0·7092–0·7104) are slightly lower than the values for the early diagenetic dolomite cements but still more radiogenic than the contemporary Triassic seawater value. Mixing calculations showed that these evaporites probably precipitated from evolved continental groundwaters (Spötl, 1991). A marine influence in the even more distal alluvial fan sediments (relative to the marine source in the east) is thus highly unlikely.

In summary, we feel confident that the Keuper dolocretes in the western Paris Basin are entirely continental in origin. Major exposures of (ultra)mafic rocks in the hinterland are considered unnecessary to yield high Mg^{2+} concentrations in the groundwaters draining these areas. The Australian examples clearly show that high Mg/Ca ratios in the groundwaters and extensive dolomite precipitation are common in continental arid settings and are caused by simply downstream progressive concentration of surface water and groundwater and precipitation of calcite in more proximal areas (Arakel & McConchie, 1982). Dolomite formation is favoured in the most distal areas adjacent to evaporitic playa flats. The overall geochemical similarity between the thin and massive dolocrete types suggests that both carbonates formed within the same hydrological system.

DISCUSSION

The thick, massive, porous dolomites in the Grès de Chaunoy are interpreted as Triassic groundwater dolocretes. The dolomites lack typical pedogenic features except at the tops of some profiles, have gradational tops, are anomalously thick compared to known pedogenic calcretes, and occur within channel rather than any floodplain lithofacies. Petrographic and mineralogical observations in conjunction with isotopic data suggest that the dolocretes represent early diagenetic dolomite precipitations from highly evolved, concentrated groundwater.

The dolomite strongly resembles Cainozoic groundwater dolocretes and calcretes which are associated with arid to semi-arid drainage basins. If the extent of similar duricrusts in areas such as central and Western Australia are any guide, zones of cementation like these may be widespread in ancient arid alluvial basins. Purvis & Wright (1991) have shown that non-pedogenic calcretes occur in the Middle Triassic Otter Sandstone of south-west England, and our observations suggest that similar cements are common in other Permo-Triassic alluvial sandstones on- and offshore in Western Europe; those of the Vert le Grand field are a very mature example.

It might be useful to offer criteria for distinguishing groundwater from phreatic dolocretes and calcretes; Thiry & Milnes (1991) and Wright *et al.* (1992) have provided criteria for differentiating pedogenic from groundwater silcretes and ferricretes respectively.

Groundwater carbonates do not typically form the multi-horizon profiles characteristic of pedogenic calcretes (and dolocretes; Watts, 1980; Arakel, 1982; Machette, 1985; Wright & Tucker, 1991). However, most groundwater carbonates form nodular, massive and brecciated horizons, but generally lack the capping laminar horizons of mature pedogenic calcretes. The presence or absence of such cappings is not a diagnostic criterion, because they only form in more mature pedogenic calcretes and some forms have been recorded on groundwater calcretes related to the activities of phreatophytic plants (Semeniuk & Meagher, 1981). Groundwater calcrete and dolocrete profiles are likely to have gradational tops unless they have been exhumed and eroded.

It is likely that because most groundwater calcretes and dolocretes form below the main zone of biological activity, they will possess little evidence of extensive rooting in the form of rhizocretions. Where such features occur they tend to represent vertical tap roots rather than smaller, shallow rooting forms (Purvis & Wright, 1991). The reduced level of biological activity favours the formation of simple crystalline calcretes (alpha calcretes of Wright, 1990) and not the varied biologically related microstructures found in many pedogenic calcretes.

'Classical' vadose features such as meniscus and pendant cement geometries may also be used to assess whether the carbonate is vadose in origin. Any solution features in the vadose zone are likely to be vertically elongate, whereas those in the phreatic zone may be more horizonally elongate. Interestingly, the majority of the cavities in the massive dolomites in the Grès de Chaunoy are lens shaped and horizontal or subhorizontal. Illuvial concentrations of other soil components may be present in the vadose calcretes or dolocretes, such as illuviated clays. However, the absence of such features should not be used as a diagnostic criterion for groundwater forms.

Grains with multiple coatings are restricted to vadose profiles where the grains are able to move either in the soil (because of soil creep) or through fracture systems in the calcrete itself, to develop relatively even concentric coatings. Such coatings are commonly biological in origin and are indicative of the more biologically active soil zone (Wright & Tucker, 1991). In ferricretes, the morphology of nodules in pedogenic and groundwater types is quite different (see discussion in Wright *et al.*, 1992). Vadose nodules are typically more complex in form, structure and composition than groundwater nodules and this is likely to be the case in calcretes and dolocretes.

It would be reasonable to expect groundwater calcretes and dolocretes to have a more coarsely crystalline microstructure than pedogenic forms. This would result from more uniform conditions in the phreatic zone. This does not appear to be a reliable criterion and the factors controlling crystal forms are probably very complex. Note that the pedogenic dolocretes in the Grès de Chaunoy are mainly dolomicritic in composition, whereas the groundwater dolocretes exhibit a wider range of crystal sizes including spheroidal dolomite.

Carbon and oxygen stable isotopes can possibly be used to distinguish between the two calcrete/dolocrete types, but as shown in this study, the complexity of the bio- and geochemical/diagenetic systems may make any differences (or similarities) which occur difficult to interpret. Similarly, trace elements other than Sr may provide clues but criteria are not available.

One of the main problems in using these criteria, or any others, is that Cainozoic duricrusts are commonly polyphase, reflecting complex changes in landscape, climate, hydrology and subsequent diagenesis. Such complexity is actually the 'norm' and should be expected in ancient forms.

CONCLUSIONS

Thick dolomite horizons associated with fluvial channel sandstones in the Upper Triassic Grès de Chaunoy of the Paris Basin are interpreted as groundwater dolocretes. These carbonates developed along palaeodrainage systems under a semi-arid to arid climate. The dolocretes generally lack evidence of vadose processes, only exhibiting pendant or meniscus style cements in the upper parts of some profiles. The dolomite is highly replacive and displacive ranging from carbonate-cemented sandstones, through sandy dolomites, to almost pure dolostones.

Strontium and carbon isotope data of early diagenetic dolocrete cements indicate a continental origin of the porewater. Oxygen isotope data of authigenic silica associated with these dolomite cements also indicate ^{18}O-depleted, meteorically derived porewaters.

It would be surprising if similar groundwater cementation zones are not more widespread in the geological record. In Late Cainozoic drainage basins in many arid and semi-arid settings (e.g. Australia), such zones produce extensive gypcretes, calcretes, dolocretes and silcretes. Such groundwater systems represent a distinctive continental diagenetic facies.

ACKNOWLEDGMENTS

This paper is based on a PhD study of the senior author at the University of Berne under the supervision of Albert Matter. We are grateful to Elf-Aquitaine (France) for releasing core material and logistic data on the Keuper of the Paris Basin and granting permission to publish this paper. We thank Albert Matter, Olivier Brévart, Nigel Platt and Raymond Cussey for discussion and comments. This manuscript was carefully reviewed by Gregg Mack and Lynton Land and editorially handled by Stuart Burley. Their help is gratefully acknowledged. This study was supported by the Swiss National Science Foundation grant 20–26310.89. V.P.W. thanks British Petroleum for permission to use data collected while contracted to study Vert le Grand field. P.R.I.S. Contribution no. 241.

REFERENCES

ARAKEL, A.V. (1982) Genesis of calcrete in Quaternary soil profiles, Hutt and Leeman lagoons, Western Australia. *J. sedim. Petrol.*, **52**, 109–125.

ARAKEL, A.V. (1986) Evolution of calcrete in palaeodrainages of the Lake Napperby area, central Australia. *Palaeogeogr. Palaeoclim. Palaeoecol.*, **54**, 283–303.

ARAKEL, A.V. (1988) Carnotite mineralization in inland drainage areas of Australia. *Ore Geol. Rev.*, **3**, 289–311.

ARAKEL, A.V. (1991) Evolution of Quaternary duricrusts in Karinga Creek drainage system, central Australian groundwater discharge zone. *Aust. J. Earth Sci.*, **38**, 334–347.

ARAKEL, A.V., JACOBSEN, G., SALEHI, M. & HILL, C.M. (1989) Silicification of calcrete in palaeodrainage basins of the Australian arid zone. *Austr. J. Earth Sci.*, **36**, 73–89.

ARAKEL, A.V. & MCCONCHIE, D. (1982) Classification and genesis of calcrete and gypsite lithofacies in palaeodrainage systems of inland Australia and their relationship to carnotite mineralization. *J. sedim. Petrol.*, **52**, 1149–1170.

ARAKEL, A.V. & WAKELIN-KING, G.A. (1991) Chemical sedimentation and facies relationships in the Karinga Creek palaeo-drainage system, Amadeus Basin, Northern Territory. *Bull. Bur. Miner. Res.*, **236**, 581–589.

BREWER, R. (1976) *Fabric and Mineral Analysis of Soils*. Wiley, New York, 470 pp.

BULLOCK, P., FEDOROFF, N., JOGERIUS, A., STOOPS, G. & TURSINA, T. (1985) *Handbook of Soil Thin Section Description*. Waine Research Publ., UK, 152 pp.

CARLISLE, D. (1983) Concentration of uranium and vanadium in calcretes and gypcretes. In: *Residual Deposits* (Ed. by R. C. L. Wilson), *Spec. Publ. geol. Soc. Lond.*, **11**, 185–195.

CAVELIER, C. & LORENZ, J. (1987) Aspect et évolution géologiques du Bassin Parisien. *Mém. Bull. Inst. géol. Bassin Paris*, **6**, 1–271.

CERLING, T.E. (1984) The stable isotopic composition of modern soil carbonates and its relationship to climate. *Earth planet. Sci. Lett.*, **71**, 229–240.

CERLING, T.E. (1991) Carbon dioxide and the atmosphere: evidence from Cenozoic and Mesozoic paleosols. *Am. J. Sci.*, **291**, 337–400.

DEVER, L., FONTES, J.C. & RICHE, J. (1987) Isotopic approach to calcite dissolution and precipitation in soils under semi-arid conditions. *Chem. Geol. (Isotope Geosci. Sect.)*, **66**, 307–314.

EL-SAYED, M.I., FAIRCHILD, I.J. & SPIRO, B. (1991) Kuwaiti dolocrete: petrology, geochemistry and groundwater origin. *Sediment. Geol.*, **73**, 59–75.

ESPITALIÉ, J., MAXWELL, J.R., CHENET, Y. & MARQUIS, F. (1988) Aspects of hydrocarbon migration in the Mesozoic in the Paris Basin as deduced from an organic geochemical survey. *Org. Geochem.*, **13**, 467–481.

ESTEBAN, M. & KLAPPA, C.F. (1983) Subaerial exposure environment. In: *Carbonate Depositional Environments* (Ed. by P. A. Scholle, D. G. Bebout & C. H. Moore), *Mem. Am. Ass. petrol. Geol.*, **33**, 1–95.

FAURE, G. (1986) *Principles of Isotope Geology*. Wiley, New York, 589 pp.

FITZPATRICK, E.A. (1984) *Micromorphology of Soils*. Chapman & Hall, London, 433 pp.

FRIEDMAN, I. & O'NEIL, J.R. (1977) Compilation of stable isotope fractionation factors of geochemical interest. In: *Data of Geochemistry*, 6th edn (Ed. by M. Fleischer), *Prof. Pap. US geol. Surv.*, **440-KK**, 1–12.

GEISLER-CUSSEY, D. (1986) Approche sédimentologique et géochimique des mécanismes générateurs de formations évaporitiques actuelles et fossiles. Marais salants de Camargue et du Levant espagnol, Messinien méditerranéen et Trias lorrain. *Mém. Sci. Terre*, **48**, 1–268.

GOUDIE, A.S. (1983) Calcrete. In: *Chemical Sediments and Geomorphology: Precipitates and Residua in the Near-Surface Environment* (Ed. by A. S. Goudie & K. Pye), pp. 93–131. Academic Press, Orlando.

GUNATILAKA, A. (1989) Spheroidal dolomites—origin by hydrocarbon seepage? *Sedimentology*, **36**, 701–710.

JACOBSEN, G., ARAKEL, A.V. & CHEN YIJIAN (1988) The central Australian groundwater discharge zone: evolution of associated calcrete and gypcrete deposits. *Aust. J. Earth Sci.*, **35**, 549–565.

KHALAF, F.I. (1990) Occurrence of phreatic dolocrete within Tertiary clastic deposits of Kuwait, Arabian Gulf. *Sediment. Geol.*, **68**, 223–239.

KOEPNICK, R.B., DENISON, R.E., BURKE, W.H., HETHERINGTON, E.A. & DAHL, D.A. (1990) Construction of the Triassic and Jurassic portion of the Phanerozoic curve of seawater $^{87}Sr/^{86}Sr$. *Chem. Geol. (Isotope Geosci. Sect.)*, **80**, 327–349.

LAND, L.S. (1983) The application of stable isotopes to studies of the origin of dolomite and to problems of diagenesis of clastic sediments. In: *Stable Isotopes in Sedimentary Geology* (Ed. by M. A. Arthur, T. F. Anderson, I. R. Kaplan, J. Veizer & L. S. Land), *Soc. econ. Paleont. Miner. Short Course*, **10**, 1–22.

LIGANG, Z., JINGXIU, L., HUANBO, Z. & ZHENSHENG, C. (1989) Oxygen isotope fractionation in the quartz-water-salt system. *Econ. Geol.*, **84**, 1643–1650.

LUMSDEN, D.N. & CHIMAHUSKY, J.S. (1980) Relationship between dolomite nonstoichiometry and carbonate facies parameters. In: *Concepts and Models of Dolomitization* (Ed. by D. H. Zenger, J. B. Dunham & R. L. Ethington), *Spec. Publ. Soc. econ. Paleont. Miner.*, **28**, 123–137.

MACHETTE, M.N. (1985) Calcic soils of the southwestern United States. In: *Soils and Quaternary Geology of the Southwestern United States* (Ed. by D. L. Weide), *Spec. Pap. geol. Soc. Am.*, **203**, 1–21.

MACK, G.H., COLE, D.R., GIORDANO, T.H., SCHAAL, W.C. & BARCELOS, J.H. (1991) Paleoclimatic controls on stable oxygen and carbon isotopes in caliche of the Abo Formation (Permian), south-central New Mexico, U.S.A. *J. sedim. Petrol.*, **61**, 445–457.

MAIZELS, J.K. (1987) Plio-Pleistocene raised channel systems of the western Sharqiya (Wahiba), Oman. In: *Desert Sediments: Ancient and Modern* (Ed. by L. E. Frostick & I. Reid), *Spec. Publ. geol. Soc. Lond.*, **35**, 31–50.

MALIVA, R.G. & SIEVER, R. (1988) Pre-Cenozoic nodular cherts: evidence for opal-CT precursors and direct quartz replacement. *Am. J. Sci.*, **288**, 798–809.

MANN, A.W. & DEUTSCHER, R.L. (1978) Genesis principles for the precipitation of carnotite in calcrete drainages in Western Australia. *Econ. Geol.*, **73**, 1724–1737.

MANN, A.W. & HORWITZ, R.C. (1979) Groundwater calcrete deposits in Australia: some observations from Western Australia. *J. geol. Soc. Aust.*, **26**, 293–303.

MANZE, U. & BRUNNACKER, K. (1977) Über das Verhalten der Sauerstoff- und Kohlenstoff-Isotope in Kalkkrusten und Kalktuffen des mediterranen Raumes und der Sahara. *Z. Geomorph.* N.F., **21**, 343–353.

MÉGNIEN, C. (1980a) Synthèse géologique du Bassin de Paris, Vol. 1. Stratigraphie et Paléogéographie. *Mém. Bur. Rech. Geol. Min.*, **101**, 1–468.

MÉGNIEN, C. (1980b) Synthèse géologique du Bassin de Paris, Vol. 2. Atlas. *Mém. Bur. Rech. Geol. Min.*, **102**.

MÉGNIEN, C. (1980c) Tectogenèse du Bassin de Paris: étapes de l'évolution du bassin. *Bull. Soc. géol. France*, **12**, 669–680.

MILLIKEN, K.L. (1979) The silicified evaporite syndrome—two aspects of the silicification history of former evaporite nodules from southern Kentucky and northern Tennessee. *J. sedim. Petrol.*, **49**, 245–256.

MORA, C.I., DRIESE, S.G. & SEAGER, P.C. (1991) Carbon dioxide in the Paleozoic atmosphere: evidence from carbon-isotope compositions of pedogenic carbonate. *Geology*, **19**, 1017–1020.

NAYLOR, H., TURNER, P., VAUGHAN, D.J. & FALLICK, A.E. (1989) The Cherty Rock, Elgin: a petrographic and isotopic study of a Permo-Triassic calcrete. *Geol. J.*, **24**, 205–221.

NETTERBERG, F. (1980) Geology of southern African calcretes I. Terminology, description, macrofeatures and classification. *Trans. geol. Soc. South Africa*, **83**, 255–283.

PERRODON, A. & ZABEK, J. (1991) Paris Basin. In: *Interior Cratonic Basins* (Ed. by M. W. Leighton, D. R. Kolata, D. F. Oltz & J. J. Eidel), *Mem. Am. Ass. petrol. Geol.*, **51**, 633–679.

PELECHATY, S.M. & JAMES, N.P. (1991) Dolomitized Middle Proterozoic calcretes, Bathurst Inlet, Northwest Territories, Canada. *J. sedim. Petrol.*, **61**, 988–1001.

PURVIS, K. & WRIGHT, V.P. (1991) Calcretes related to phreatophytic vegetation from the Middle Triassic Otter Sandstone of South West England. *Sedimentology*, **38**, 539–551.

REEDER, R.J. (1983) Crystal chemistry of the rhombohedral carbonates. In: *Carbonates: Mineralogy and Chemistry* (Ed. by R. J. Reeder), *Rev. Miner.*, **11**, 1–47.

REEDER, R.J. & SHEPPARD, C.E. (1984) Variation of lattice parameters in some sedimentary dolomites. *Am. Miner.*, **69**, 520–527.

REEVES, C.C. (1983) Pliocene channel calcrete and suspenparallel drainage in West Texas and New Mexico. In: *Residual Deposits* (Ed. by R. C. L. Wilson), *Spec. Publ. geol. Soc. Lond.*, **11**, 179–183.

RICHTER, D.K. (1985) Die Dolomite der Evaporit- und der Dolcrete-Playasequenz im mittleren Keuper bei Coburg (NE-Bayern). *Neues Jb. geol. Paläont. Abh.*, **10**, 87–128.

RICOUR, J. (1960) La genèse des niveaux salifères. Cas du Trias francais. *Rev. géogr. phys. géol. dyn.*, **3**, 139–148.

SALOMONS, W., GOUDIE, A.S. & MOOK, W.G. (1978) Isotopic composition of calcrete deposits from Europe, Africa and India. *Earth Surf. Proc.*, **3**, 43–57.

SALOMONS, W. & MOOK, W.G. (1986) Isotope geochemistry of carbonates in the weathering zone. In: *Handbook of Environmental Isotope Geochemistry*, Vol. 2 (Ed. by P. Fritz & J. C. Fontes), pp. 239–269. Elsevier, Amsterdam.

SCHLESINGER, W.H. (1985) The formation of caliche in soils of the Mojave Desert, California. *Geochim. Cosmochim. Acta*, **49**, 57–66.

SEMENIUK, V. & MEAGHER, T.D. (1981) Calcrete in Quaternary coastal dunes in south western Australia: a capillary-rise phenomenon associated with plants. *J. sedim. Petrol.*, **51**, 47–68.

SIBLEY, D.F. & GREGG, J.M. (1987) Classification of dolomite rock textures. *J. sedim. Petrol.*, **57**, 967–975.

SMITH, A.G., HURLEY, A.M. & BRIDEN, J.C. (1981) *Phanerozoic Palaeocontinental World Maps.* Cambridge University Press, Cambridge, 102 pp.

SPÖTL, C. (1991) *Diagenesis and porewater evolution of the Keuper sediments in the subsurface of the Paris Basin (France).* PhD thesis, University of Berne, Switzerland, 255 pp.

SPY-ANDERSON, F.L. (1981) Dolocrètes et nodules dolomitiques. Résultats de la dolomitisation directe, en milieu continental, de sédiments terrigènes de la "formation bariolée supérieure" (Keuper) de la région des Vans (Ardèche, sud-est de la France). *Bull. Bur. Rech. géol. min.*, **1/3**, 195–205.

TALMA, A.S. & NETTERBERG, F. (1983) Stable isotope abundances in calcretes. In: *Residual Deposits* (Ed. by R. C. L. Wilson), *Spec. Publ. geol. Soc. Lond.*, **11**, 221–233.

TANDON, S.K. & NARAYAN, D. (1981) Calcrete conglomerate, a case-hardened conglomerate and cornstone—a comparative account of pedogenic and non-pedogenic carbonates from the continental Siwalik Group, Punjab, India. *Sedimentology*, **28**, 353–367.

THIRY, M. & MILNES, A.R. (1991) Pedogenic and groundwater silcretes at Stuart Creek Opal field, South Australia. *J. sedim. Petrol.*, **61**, 111–127.

VAHRENKAMP, V.C. & SWART, P.K. (1990) New distribution coefficient for the incorporation of strontium into dolomite and its implications for the formation of ancient dolomites. *Geology*, **18**, 387–391.

VEIZER, J. (1989) Strontium isotopes in seawater through time. *Ann. Rev. Earth planet. Sci.*, **17**, 141–167.

VOGEL, J.C. & VAN URK, H. (1975) Isotopic composition of groundwater in semi-arid regions of Southern Africa. *J. Hydrol.*, **25**, 23–36.

WATTS, N.L. (1980) Quaternary pedogenic calcretes from the Kalahari (southern Africa): mineralogy, genesis and diagenesis. *Sedimentology*, **27**, 661–686.

WRIGHT, V.P. (1990) A micromorphological classification of fossil and recent calcic and petrocalcic microstructures. In: *Soil Micromorphology: A Basic and Applied Science. Developments in Soil Science*, Vol. 19 (Ed. by L. A. Douglas), pp. 401–407. Elsevier, Amsterdam.

WRIGHT, V.P., SLOAN, R.J., VALERO GARCES, B. & GARVIE, L.A.J. (1992) Groundwater ferricretes from the Silurian of Ireland and Permian of the Spanish Pyrenees. *Sediment. Geol.*, **77**, 37–49.

WRIGHT, V.P. & TUCKER, M.E. (1991) Calcretes, Introduction. *Int. Ass. Sediment., Reprint Series*, **2**, 1–22.

WRIGHT, V.P. & VANSTONE, S.D. (1991) Assessing the carbon dioxide content of ancient atmospheres using palaeocalcretes: theoretical and empirical constraints. *J. geol. Soc. London*, **148**, 945–947.

(Manuscript received 28 April 1992; revision received 15 September 1992)

Mesogenesis (burial diagenesis)

Quartz-related mesogenesis

Quartz is the most common cement in sandstones. It forms typically as crystals that are continuous with detrital quartz grains (Waugh, 1970). The suggestion by this author, following seminal SEM analysis, that quartz cement growth is most rapid parallel to the c-axis has not been well developed and yet could have significance for the prediction of overall quartz cement growth rates.

Numerous sources of quartz cement have been advocated, including feldspar–clay reactions, stylolite-induced grain dissolution and then reprecipitation; grain-size-controlled redistribution of silica between facies in a given sandbody and even sources external to sandbodies. Although sandstones are obviously rich in quartz to start with, it has been suggested that silica moves from neighbouring mudstones into sandstones and crystallizes as quartz cement. Theoretical analysis by Mullis (1992) suggests that diffusion of silica from mudstone could only penetrate into the sandstone to within a few metres of the lithological margin. However, advective dispersion within the sandstone could plausibly redistribute mudstone-sourced silica throughout a sand body. Siliceous sponge spicules are proposed to be a source of quartz cement in shallow-marine sandstones (Vagle *et al.*, 1994). The sandstones studied by these authors have not been subjected to temperatures typically associated with quartz cementation (> 80–90°C), but still have abundant quartz cement suggesting that defining thermal–kinetic boundaries for quartz cement may be erroneous.

Dissolution of quartz can occur during burial and diagenesis, at least on a local scale. Quartz dissolution often coincides with carbonate cementation during mesogenesis and it is assumed that the two processes are related. Hurst (1981) studied quartz dissolution experimentally and concludes that, for surface reaction-controlled cases, there is a clear hierarchy of the relative solubility of quartz surfaces. Those with the highest surface energy and thus lowest natural stability should dissolve more easily than those with low surface energy. The least stable, and thus most soluble, include fracture surfaces whereas the most stable, and thus least soluble, include low Miller index, euhedral crystal faces, especially those parallel to the c-axis. Burley & Kantorowicz (1986) concur with Hurst in that quartz grains, in their combined empirical and experimental study of calcite cemented sandstones, show a clear preference for dissolution along fracture surfaces in comparison with most other types of quartz surface. However, this study also demonstrates the effect of carbonate mineralogy on quartz replacement. Quartz in dolomite and siderite cemented sandstones seems to display no preferential dissolution sites. The random nature of quartz dissolution in these cases is ascribed to a geological environment in which the overall progress of the reaction (quartz dissolution and influx of a component inducing carbonate cementation) was transport controlled rather than surface controlled. They conclude that different styles of quartz dissolution fabrics could be used potentially to reveal details of the process inducing carbonate cementation and quartz dissolution.

Reprinted from Sedimentology (1970) ***14****, 309–320*

FORMATION OF QUARTZ OVERGROWTHS IN THE PENRITH SANDSTONE (LOWER PERMIAN) OF NORTHWEST ENGLAND AS REVEALED BY SCANNING ELECTRON MICROSCOPY

B. WAUGH

Department of Geology, The University, Hull (Great Britain)

(Received December 23, 1969)
(Resubmitted March 5, 1970)

SUMMARY

The development of optically continuous quartz overgrowths is governed by the atomic structure and crystallographic orientation of the detrital quartz grains. Initial growth commences with the appearance of numerous oriented projections, with rhombohedral and prismatic form, on grain surfaces. Merging and overlap of the projections results in the formation of large crystal faces whose form is dependent upon the initial location of the projections with respect to the internal crystallograhic axes. Growth is particularly rapid along the direction of the *c*-axis. For unicrystalline quartz grains the ultimate growth phase is the production of polyhedral quartz crystals having the appearance of hexagonal dipyramids. In polycrystalline grains each quartz unit within a single grain develops a separate overgrowth, the form of which is similarly controlled by the internal structure of the individual quartz units. Hence, the completed overgrowth shows an irregular distribution of diversely oriented prism and rhombohedral faces.

OUTLINE PETROGRAPHY OF QUARTZ-CEMENTED PENRITH SANDSTONE

The Penrith Sandstone (Lower Permian) crops out in the valley of the River Eden in the counties of Cumberland and Westmorland, northwest England. It is a continental, red orthoquartzite, which was deposited as barchan sand dunes in an arid desert environment. The uniform red colouration of the sediment is due to the presence of a film of red iron oxide and clay mineral around the detrital grains.

Throughout the greater part of the outcrop the sandstone is tightly cemented by secondary quartz. This occurs as overgrowths of optically continuous quartz around the aeolian grains, the overgrowths invariably possessing a high degree of crystalline perfection. Indeed, the majority appear as almost perfect quartz crystals, between 1.0–1.5 mm in length, showing double pyramidal terminations to the prism faces. This feature of the rock was initially observed by HARKNESS (1862) and MURCHISON and HARKNESS (1864), who concluded that the crystals were in fact derived quartz crystals from the source rock. SORBY (1880), however, recognized that the crystal faces were clearly the product of secondary precipitation of quartz around detrital quartz grains.

Petrographically, two distinct structural classes of detrital quartz grain can be recognized in the sandstone: (*1*) unicrystalline grains, composed of a single unit of either unstrained or strained quartz; (*2*) polycrystalline grains, composed of several units of unstrained or strained quartz. This latter group includes derived vein quartz, metaquartzite, gneiss, schist, mylonite and strained grains with internal lineage boundaries (rather poorly defined boundaries delimiting zones of differing undulose extinction; grains with such boundaries may be described as semi-composite). In the polycrystalline grains each quartz unit possesses its own distinct crystallographic orientation, and the extinction position of any individual unit therefore differs markedly from that of an adjacent unit within the same grain.

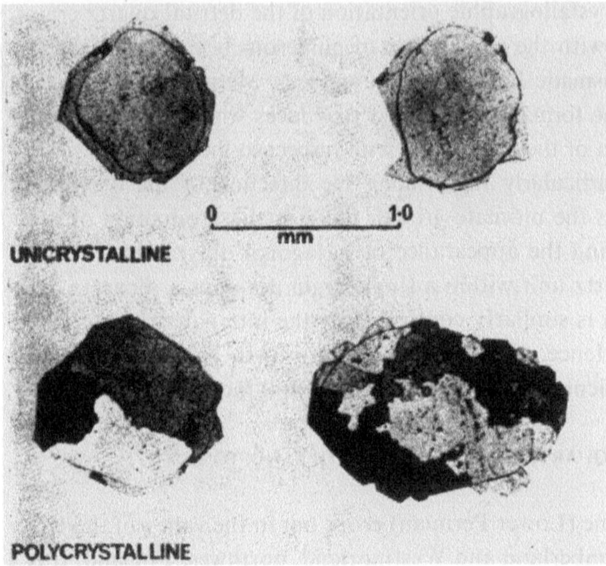

Fig.1. Thin-sections of unicrystalline and polycrystalline quartz grains with their corresponding overgrowths.

Optically continuous overgrowths are present on both the unicrystalline and polycrystalline grains (Fig.1). The maximum degree of crystalline perfection is attained by those on the unicrystalline grains, where the rhombohedral and prism faces are prominent. Polycrystalline grains display a similar overgrowth morphology, but in this case each individual quartz unit adjacent to the grain margin possesses a separate overgrowth. Further, on strained grains of both structural classes the overgrowths show the same intensity of straining as the detrital grains.

THE APPLICATION OF ELECTRON MICROSCOPY

In 1965, the author presented a preliminary account of an investigation to determine stages in the formation of quartz overgrowths in the Penrith Sandstone of northwest England. This investigation was primarily a transmission electron-microscope study of carbon-coated, gold-palladium shadowed replicas of overgrowth surfaces. Although successful in determining the general growth stages, the technique was restricted by a number of major factors: (*1*) Difficulty in producing a single, high-quality replica of a sufficiently representative surface area of the overgrowth. (*2*) As replicas, and not the grains themselves, are used it is not possible to determine such features as the orientation of the overgrowths with respect to the entire detrital grain surface, and the inter-relationships of the various growth stages on different sides of the same grain. (*3*) The mesh pattern of the copper grid supports, on which the replicas are placed, prevents photographing large areas of the overgrowths at comparatively low magnifications. Hence, there is the general difficulty to convincingly reproduce photographically the relationships of the growth stages on a single grain surface.

With the introduction of scanning electron microscopy, however, all these difficulties have readily been overcome. Consequently, it is now possible to present a more exhaustive account of the development of these naturally formed quartz overgrowths, and to differentiate between the structure of overgrowths produced on unicrystalline and polycrystalline quartz grains.

The scanning electron microscope has been described by OATLEY et al. (1965). The advantages of this instrument over the transmission electron microscope, for the type of study presented here, are numerous. For instance, replication is not involved, the sample being observed directly in the microscope, eliminating the major problems outlined above. Further, the magnification range can be varied between 20 × and 100,000 ×, the depth of focus is superior to that of an optical light microscope at comparable magnifications by a factor of at least 300, and the specimen can be viewed at any required orientation. One other feature is the ease at which stereomicrographic pairs can be prepared, by photographing the image twice and tilting the specimen by 5° between exposures. Such stereopairs enable detailed interpretation and measurement of surface features to be undertaken.

SPECIMEN PREPARATION FOR SCANNING ELECTRON MICROSCOPY

The preparation of specimens for viewing in the scanning electron microscope is a simple and quick process. Individual sand grains from the quartz cemented sandstone are separated from the host rock and mounted on standaıd aluminium stubs. The mounting medium used in this study was a dispersed silver in alcohol adhesive. Several grains can readily be mounted on a single stub. Alternatively, small fragments of rock, approximately 12 mm in diameter, can be mounted in an identical manner.

At this stage, non-conductive specimens are coated with a conductive material (e.g., carbon, aluminium, gold-palladium) in a vacuum coating unit, in order to achieve maximum specimen resolution. However, in this study coating was not necessary, probably because of the presence of the iron oxide coating to the quartz grains. Any "charging-up" of the specimens can usually be removed by spraying with an anti-static solvent.

With the exception of Fig.1 and 2, all the photographs presented are scanning electronmicrographs of overgrowths on single quartz grains. Stereopairs were prepared in all cases to facilitate interpretation.

THEORETICAL AND EXPERIMENTAL CONSIDERATIONS OF CRYSTAL GROWTH

It has long been recognized that epitaxial growth between two crystalline substances is only possible when these substances exhibit a comparable atomic structure. Thus, VAN DER MERWE (1949, p.201) has shown that, for an overgrowth to form on a crystalline substrate, there must first be formed an initial layer of regular atomic pattern (the monolayer), whose structure is similar to that of the substrate. Once such a layer becomes established its basic structure is preserved throughout the entire crystal lattice of the overgrowth. Further, BUCKLEY (1951, p.401) has noted that growth on the substrate commences simultaneously at several places on the same surface, and that it is the manner in which the new parts come into contact which determines the relative perfection of the final crystal. He maintains that perfection can only be achieved when there is a good fitting together of the new parts in all directions.

Experimental evidence, supporting the views of Van Der Merwe and Buckley, has been attained by ERNST and BLATT (1964), who successfully produced synthetic quartz overgrowths on strained and unstrained unicrystalline quartz grains. They demonstrated that growth commenced with the development of optically continuous rhombohedral projections of quartz on grain surfaces. Although no initial projections with a prismatic habit were observed, such forms have been recognized by HEALD (1950) and YEDLOSKY and DEAN (1961, fig.1) from natural occurrences. Continued growth, by merging and overlap of rhombohedra, resulted in the formation of larger faces with recognizable prismatic or rhombohedral form. The final growth stage of these larger faces produced polyhedral quartz crystals.

Considerable knowledge of the formation of quartz crystals has also been gained during the attempted production of perfect quartz crystals for industrial application. VAN PRAAGH (1947,1949) and THOMAS et al. (1949, fig.2) have reported on the successful growth of quartz crystals round cut spheres of quartz. In these experiments development was particularly rapid along the direction of the *c*-axis, leading to the formation of pronounced rhombohedral faces, whilst the prism faces remained comparatively small. The completed overgrowths consisted of a

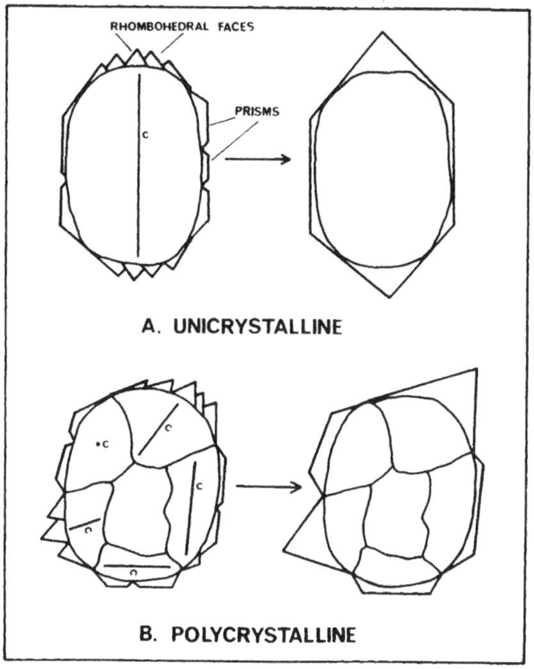

Fig.2. Theoretical development phases of quartz overgrowths on unicrystalline and polycrystalline grains, with respect to the orientation of the crystallographic *c*-axis.

cluster of rhombohedral faces at the distal ends of the grains, terminating barrel shaped prism faces.

Fig.2 demonstrates the theoretical development of overgrowths on both unicrystalline and polycrystalline grains with respect to the orientation of the crystallographic *c*-axis. In both structural types the prism faces are formed in the plane parallel to the *c*-axis and normal to the *a*-axes, coinciding with the equitorial regions of the grain. Conversely, the rhombohedral faces form in the plane normal to, and at the distal ends of, the *c*-axis.

However, in the polycrystalline grains, the complete overgrowth on any individual quartz unit could theoretically consist of a single crystal form. Thus, in those units where the *c*-axis is normal to the grain outline only rhombohedral faces may be present. Alternatively, where the *c*-axis is parallel to the grain outline prism faces will be dominant. Hence, the resultant overgrowth on a single polycrystalline grain will exhibit a complex of separate overgrowths of differing orientation and form.

THE DEVELOPMENT OF PENRITH SANDSTONE QUARTZ OVERGROWTHS

From these theoretical and experimental considerations it would appear reasonable to assume that naturally occurring overgrowths developed in a similar manner. However, in the majority of sandstones cemented by quartz overgrowths, only the end product of silification, i.e., polyhedral overgrowths, can usually be examined. Yet, in the Penrith Sandstone, the opportunity arises for a study of the complete development phases.

As this study was based on an examination of external overgrowth morphology, it was not possible to ascertain, with any degree of confidence, whether any of the overgrowths were developed on strained grains. Thin-sections of many grains were therefore prepared after a complete electron-microscope examination had been undertaken, in an attempt to relate any internal grain straining to a characteristic morphology of the overgrowth on such grains. In all cases, however, there was no apparent marked difference in the morphology of overgrowths formed on unstrained or strained grains. The same position exists also for the grains with internal lineage boundaries. In such grains ERNST and BLATT (1964) observed that, during overgrowth development, the initial rhombohedral projections are located on either side of the lineage boundary. Continued growth occurred by merging of the projections across the boundary, the position of the boundary being retained within the final overgrowth. Although this feature has not been recognized in electronmicrographs, it has been observed in many thin-sections of the Penrith Sandstone.

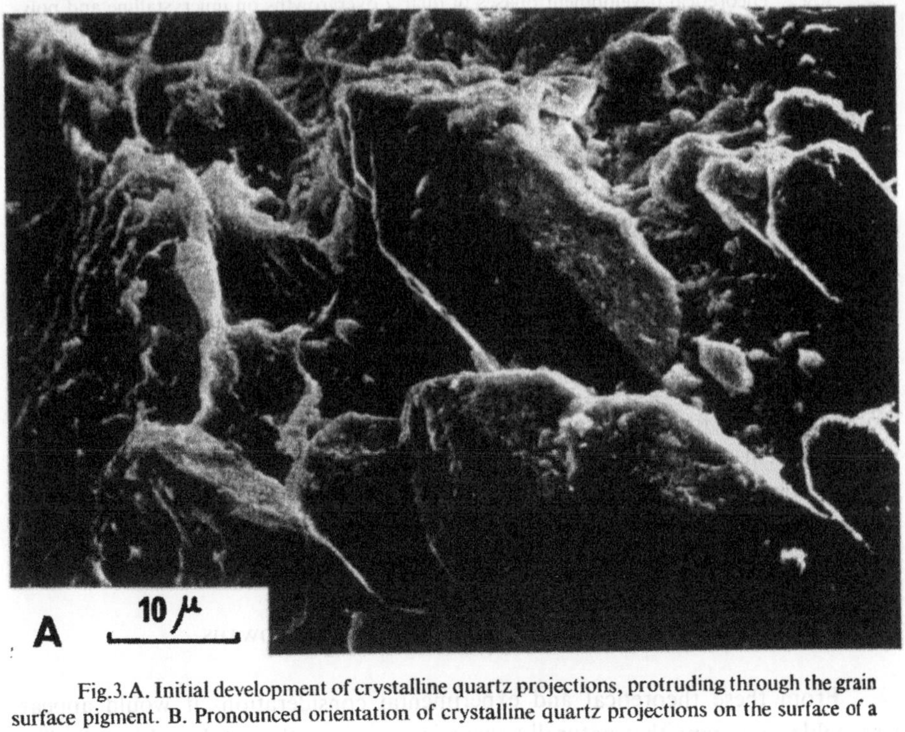

Fig.3.A. Initial development of crystalline quartz projections, protruding through the grain surface pigment. B. Pronounced orientation of crystalline quartz projections on the surface of a single unicrystalline grain.

Initial phase: unicrystalline and polycrystalline grains

As the Penrith Sandstone quartz grains are coated with a film of red iron oxide and clay mineral it is apparent that the silica-bearing solutions, which precipitated the quartz overgrowths, had adequate access through the film to bind the secondary quartz to the detrital grain. Hence, the initial phase of overgrowth development is recognized by the first appearance of crystalline quartz projections through the surface film. Fig.3A demonstrates this earliest stage, in which the projections are clearly seen protruding through the film, the projections themselves possessing an irregular coating of the surface pigment. The crystalline form of the projections is a combination of prism and rhombohedral faces, thus differing from those recognized by ERNST and BLATT (1964), where only rhombohedral projections were recorded. At this stage, although projections appear at several points on grain surfaces, they cover a comparatively small surface area.

Subsequent growth stages: unicrystalline grains

Following the initial phase, a period of rapid growth results in the entire surface becoming covered with projections (Fig.3B). Apart from the perfection in crystalline form, the most remarkable feature is the uniformity of their orientation, clearly reflecting epitaxial growth, coinciding with an identical atomic structure and

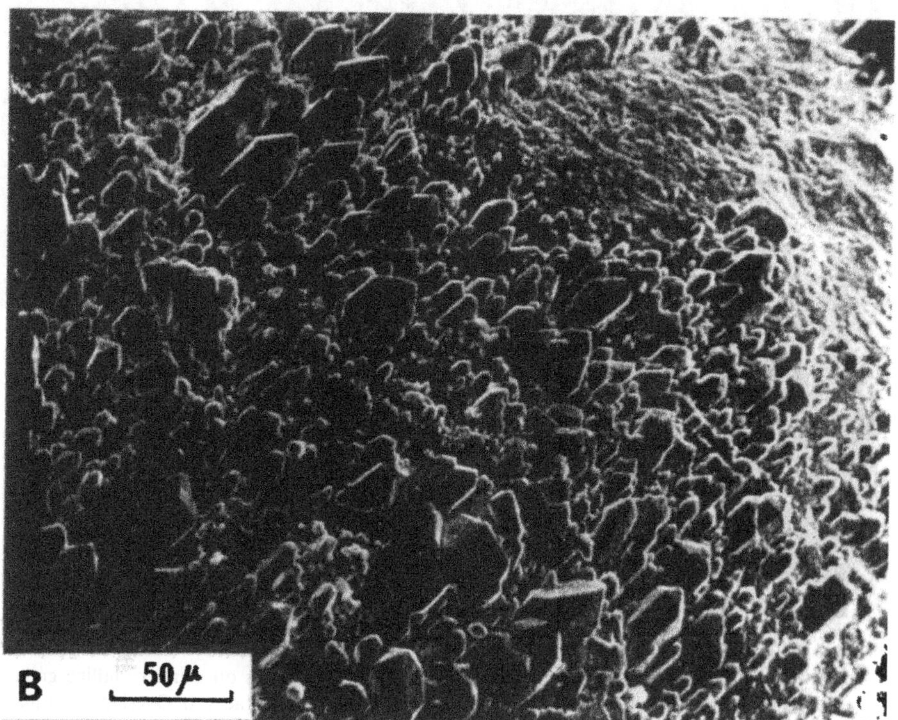

crystallographic orientation within the detrital grain. This phase, therefore, conforms to the pattern of crystal growth as advocated by VAN DER MERWE (1949), and it is conceivable that the crystalline projections are analogous to the monolayer of this author. Occasionally, isolated projections may show a divergence of up to 10° from the dominant orientation. Such a feature is probably the result of localized defects in the crystal lattice of the detrital grain. However, these misfits cannot be recognized in later growth stages.

With continued growth the second stage recognized by ERNST and BLATT (1964) becomes apparent. The crystalline projections become so numerous on grain surfaces that all subsequent growth is accomplished by merging and overlapping of adjacent projections, giving rise to the formation of larger crystal faces with a recognizable form. It is at this point where the location of the initial projections, with respect to the detrital grain crystal axes, becomes critical in determining whether prism or rhombohedral faces are formed.

Fig.4A demonstrates the concentration of numerous sets of rhombohedral faces at the distal end of the c-axis of a single quartz grain, where there is a marked grouping of oriented pyramidal projections. Overlap is at a fairly advanced stage in

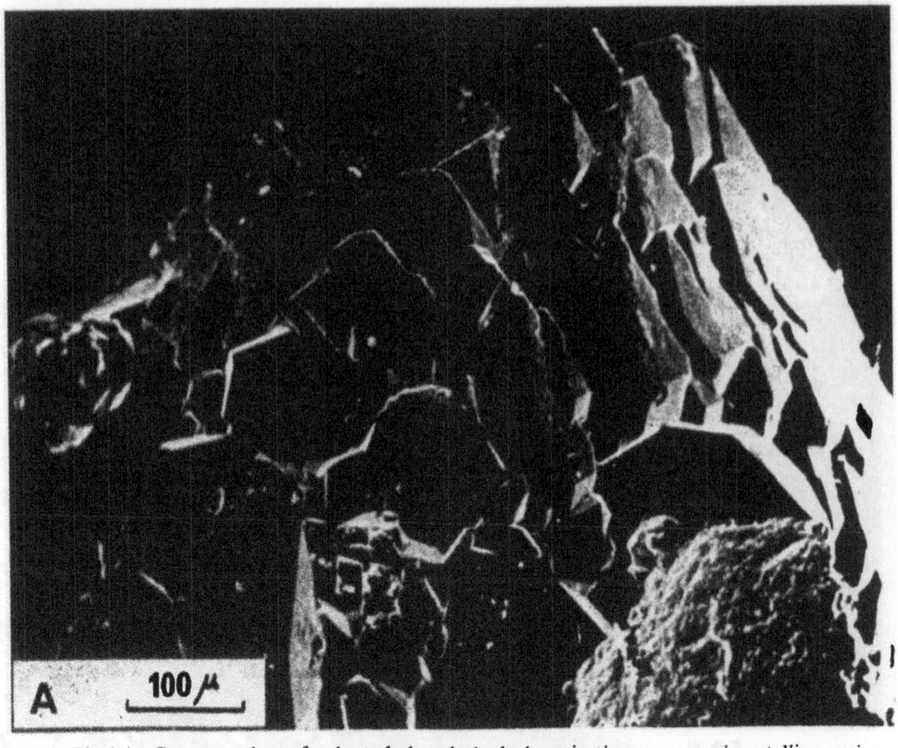

Fig.4.A. Concentration of oriented rhombohedral projections on a unicrystalline grain, with overlap of adjacent projections leading to the formation of larger faces. B. Overlap of prismatic projections on a unicrystalline grain, forming a larger prism face.

that adjacent sets have clearly merged to form much larger rhombohedral faces. The indication of the ultimate form of the pyramidal termination is readily apparent on this specimen.

An advanced stage in the formation of prism faces is represented in Fig.4B. Here, in an equitorial position on the quartz grain, considerable overlap of adjacent prismatic projections has resulted in the near completion of a single, large prism face, although the outline of the initial projections is still evident. Indeed, the preservation of these outlines assists in determining the orientation of the prism faces. It will be noticed that small, partially formed, rhombohedral faces occur at the distal ends of some of the prism faces, indicating that the *c*-axis in Fig.4B is trending towards the north west.

On individual quartz grains possessing both the growth stages shown in Fig.4 the rhombohedral faces are invariably larger and more pronounced than the prism faces, presumably due to preferential growth in the direction of the *c*-axis.

The final stage of overlapping produces almost perfectly formed, authigenic quartz crystals, having the general appearance of hexagonal dipyramids (Plate I). Close examination of the crystals reveals an inequality in the rate of growth of the crystal faces, in that one set of faces usually appears to be better developed than another. This particularly applies to the rhombohedral terminations (Plate IB,C)

PLATE I

Completed unicrystalline overgrowths, illustrating the double pyramidal terminations to the prism faces. Note the rounded outline of the detrital quartz grain in D.

where alternate faces are similar, but adjacent ones are different. In such cases the positive rhombohedral faces (10$\bar{1}$1) are usually larger than the corresponding negative faces (01$\bar{1}$1). Occasionally, as in the upper pyramidal termination of Plate IA, an equality of the growth rate may be observed. In Plate ID, although a very high degree of crystallinity is attained by the overgrowth, the rounded outline of part of the detrital grain, relatively free of overgrowth, is still visible.

Subsequent growth stages: polycrystalline grains

The development of quartz overgrowths on single units of the polycrystalline grains follows an identical pattern to that of the unicrystalline grains. The major difference, which is particularly well shown in the earlier development phases, is that the orientation of the crystalline projections on adjacent units within the same grain differs markedly. This feature reflects opposing orientations of the crystallographic axes of each quartz unit.

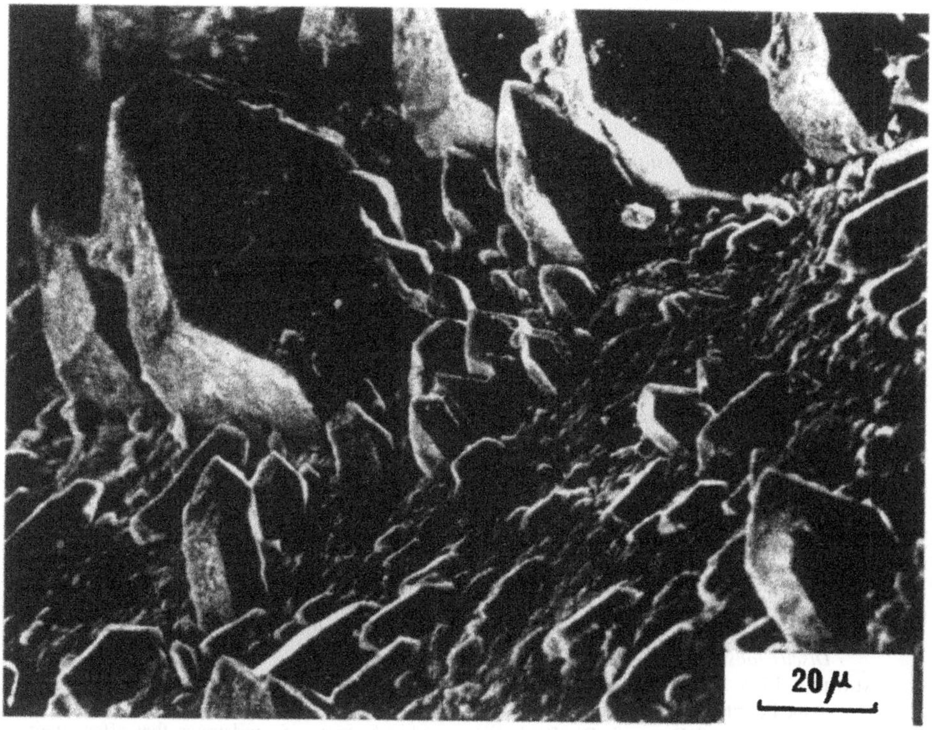

Fig.5. Polycrystalline grain with two distinct crystal forms and orientations of crystalline projections, corresponding to opposing crystallographic orientations of the two detrital quartz units.

For example, in Fig.5, two independent sets of projections are present, exhibiting different crystalline form and orientation. The dominant set, possessing the well developed rhombohedral faces, demonstrates growth in the direction of a *c*-axis oriented normal to the grain surface. Conversely, the second set displays the preferential development of prism faces with minor rhombohedral terminations, the *c*-axis of the detrital quartz unit being almost parallel with the grain surface.

The final form of the polycrystalline overgrowths, achieved by overlap of these early projections, is as illustrated in Fig.1 and Fig.2B.

ACKNOWLEDGEMENTS

The research was undertaken during the tenure of a Natural Environment Research Council Research Grant. Financial assistance from this source is gratefully acknowledged. The scanning electron microscopy was carried out on the Cambridge Instruments Stereoscan Mark IIa, in the Department of Geology, University of Leicester. The author is indebted to Professor P. C. Sylvester-Bradley for permission to use the instrument, and to Mr. G. McTurk for the excellent technical assistance in operating the electron microscope.

REFERENCES

Buckley, H. E., 1951. *Crystal Growth*. Wiley, New York, N.Y., 571 pp.

Ernst, W. G. and Blatt, H., 1964. Experimental study of quartz overgrowths and synthetic quartzites. *J. Geol.*, 72:461–470.

Harkness, R., 1862. On the sandstones and their associated deposits in the Vale of Eden, the Cumberland plain, and the southeast of Dumfrieshire. *Quart. J. Geol. Soc. London*, 18:205–218.

Heald, M. T., 1950. Authigenesis in West Virginia sandstones. *J. Geol.*, 58:624–633.

Murchison, R. I. and Harkness, R., 1864. On the Permian rocks of northwest England and their extension into Scotland. *Quart. J. Geol. Soc. London*, 20:144–165.

Oatley, C. W., Nixon, W. C. and Pease, R. F. W., 1965. Scanning electron microscopy. *Advan. Electron. Electron Phys.*, 21:181–247.

Sorby, H. C., 1880. On the structure and origin of non-calcareous stratified rocks. *Proc. Geol. Soc. London*, 36:46–92.

Thomas, L. A., Wooster, N. and Wooster, W. A., 1949. The hydrothermal synthesis of quartz. *Discussions Faraday Soc.*, 5:341–345.

Van der Merwe, J. H., 1949. Misfitting monolayers and oriented overgrowths. *Discussions Faraday Soc.*, 5:201–214.

Van Praagh, G., 1947. Synthetic quartz crystals. *Geol. Mag.*, 84:98–100.

Van Praagh, G., 1949. The hydrothermal crystallisation of vitreosil at constant temperature. *Discussions Faraday Soc.*, 5:338–341.

Waugh, B., 1965. A preliminary electron microscope study of the development of authigenic silica in the Penrith Sandstone. *Proc. Yorkshire Geol. Soc.*, 35:59–69.

Yedlosky, R. L. and Dean, J. R., 1961. Petrographic features of sandstones that affect their suitability for road material. *J. Sediment. Petrol.*, 31:372–389.

Reprinted from Sedimentology (1981) **28**, *451–459*

A scale of dissolution for quartz and its implications for diagenetic processes in sandstones

ANDREW R. HURST*

Sedimentology Research Laboratory, Department of Geology, University of Reading, Whiteknights, Reading RG6 2AB, U.K.

ABSTRACT

During diagenesis quartz grains undergo selective dissolution, controlled in location by the surface energy characteristics of the individual grains. Experimental etching in HF of isolated quartz grains reproduces comparable textures to those of natural occurrences. Some experimental results illustrate the specific effects of surface textures on their initial dissolution rates, so demonstrating the control surface energy variation has over dissolution. A hierarchy of grain surface characteristics, according to surface energies, provides a useful guide to the relative rates of dissolution during decomposition.

INTRODUCTION

In the study of dissolution rates of single crystals it is simple to establish that dissolving crystals do not dissolve at equal rates in all crystallographic directions. Low index faces (Miller indices) have slower dissolution rates than high index faces (Cabrera & Vermilyea, 1958). For quartz the pinacoid faces (high index faces representing termination in the *c*-axis direction) should dissolve more rapidly than the rhombohedral faces (low index faces parallel to the *c*-axis direction). Experimental dissolution of quartz spheres (Frondel, 1962) demonstrates this relationship between dissolution and face indices. Dissolution is most rapid in the *c*-axis direction, that is from the direction of growth of the pinacoid faces. From these simple crystallographic relationships, a hierarchy of textural characteristics can be constructed to illustrate their differing susceptibility to dissolution in terms of surface energy. This hierarchy for quartz grains (presented in Table 1), is numbered from 1 to 9 in a non-mathematical scale of increasing surface energy and decreasing crystallinity or crystal form. The numbering implies that if surface characteristics of no. 7 are present along with those of nos 1 and 2, surfaces of no. 7 will be preferentially dissolved when etched.

Such a scale allows predictions to be made about the location of areas of greatest dissolution of quartz grains. The implication of this is that dissolution and replacive textures formed during the diagenesis of quartz are not the result of a random process, but are controlled by the crystallographic properties of the quartz grains. A series of dissolution experiments were conducted to substantiate these theoretical predictions and in turn were compared with natural samples.

EXPERIMENTAL METHOD

Seven samples of disaggregated grains from lithified Jurassic sandstones were obtained by ultrasonic treatment in a water-bath. Non-quartz phases present in these disaggregated samples were removed following the methods of Sridhar, Jackson & Clayton (1975) and Jackson, Sayin & Clayton (1976). Each pure quartz sample was separated into size-fractions of >420, 250–420 and <250 μm by sieving. Nine of

* Present address: Geologisk Laboratorium, Statoil, Forus, N-4001 Stavanger, Norway.

0037-0746/81/0800-0451 $02.00
© 1981 International Association of Sedimentologists

Table 1. Classification of quartz grain surfaces with respect to surface morphology and surface energy

Scale	Face characteristics	References	
1	Low index faces	Cabrera & Vermilyea, 1958	
2	High index faces		
3	Face edges		
4	Junctions between edges (corners)		
5	Faceted areas	Cabrera & Vermilyea, 1958; Hurst, 1980	Increasing surface energy
6	Dissolution/replacement surfaces*		
7	Abraded surfaces		
8	Fractures†		
9	Disturbed surface layers	Lidström, 1968; Henderson *et al.*, 1970	

* Dissolution/replacement surfaces are areas of grains which have already undergone some dissolution or have been replaced by a cement which has subsequently been removed.

† Fractures can cleave along crystallographic planes so forming relatively stable surfaces. This is unusual in sedimentary quartz grains, conchoidal fractures predominating (see Iler, 1955, p. 258).

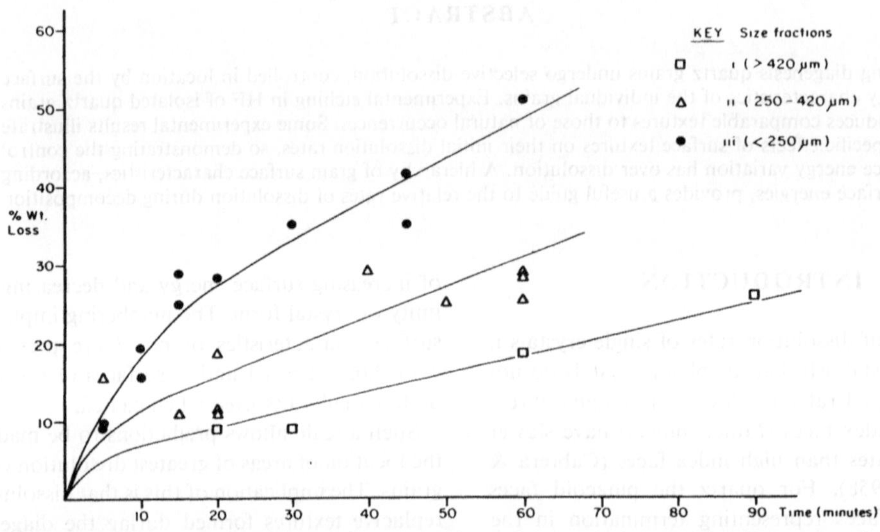

Fig. 1. Dissolution rate curves derived from the results of the preliminary etching. Data source is from Hurst (1980). The curves are hand-drawn as best fits the data.

the resulting size-fractions were sufficiently abundant to be sub-divided further. These were divided into thirds, weighed and etched in 40% hydrofluoric acid (HF) over time periods of 15, 20, 30, 45, 60 and 90 min. Reaction with HF was quenched with boric acid and samples were then washed repeatedly with distilled water before drying. Samples were re-weighed and percentage weight losses calculated. A second series of samples were prepared by the same method and etched over shorter time periods of 2, 5 and 15 min.

Surface textures of unetched and etched grains were studied using a Cam Scan electron microscope.

RESULTS

Figures 1 and 2 show the experimental results plotted graphically as rate curves. Full data tables are compiled in Hurst (1981). The three curves in Fig. 1 correspond closely to the three size-fractions used. All these curves are initially steep, levelling out after 15 min and showing relatively constant rates of dissolution thereafter. Dissolution was not extended to completion: therefore the expected sigmoidal curves such as presented by Henderson, Syers & Jackson (1970) were not obtained.

The short-period etching data is shown in Fig. 2 and Table 2. In general, the relationship between

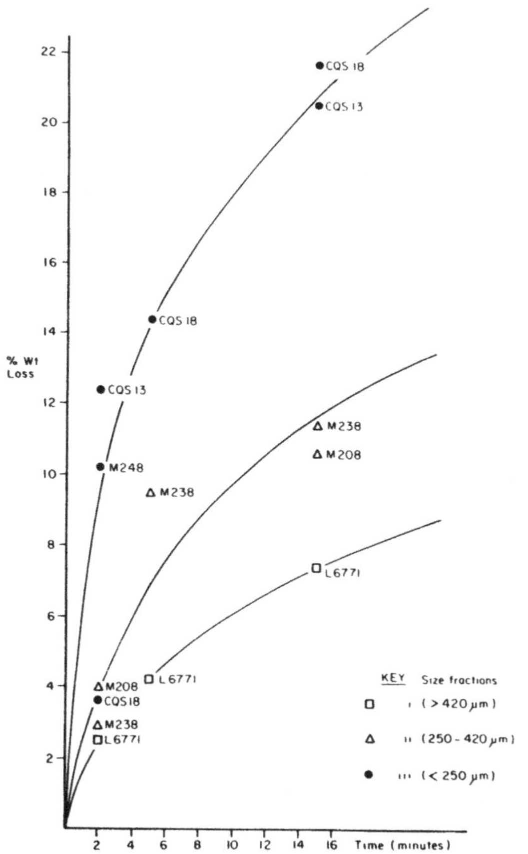

Fig. 2. Dissolution rate curves from short-period etching. See Table 2 for data.

Table 2. Percentage weight losses of samples used in short-period etching experiments

Sample	Etching time (min)		
	2	5	15
L 6771	2·5	4·2	7·4
M 208	4·0	5·75	10·6
M 238	2·9	9·5	11·4
CQS 18	3·6	14·4	21·8
CQS 13	12·4	—	20·6
M 248	10·2	—	—

grain-size and dissolution rate is retained during the short-period etching. However, after 2 min of etching these data show that the relationship between grain-size and dissolution rate is less clear. Sample CQS 18 shows an anomalously low initial dissolution rate (approx. 6·5%) compared with similar-sized fractions; samples L 6771 and M 238 show similar initial dissolution rates, although they are different size-fractions. After 5 min of etching all samples of the same size fractions show similar dissolution rates.

Surface textures prior to etching are described in Table 3 and illustrated in Fig. 3 (A–D).

Table 3. Description of grain surfaces with identification and estimation of their surface characteristics using the scale from Table 1 (IDX)

Sample	
CQS 18	All grains have a coating of small euhedral quartz grains (≤2 μm *a*-axes). The coating grains are not in optical continuity with each other or their host grains (Fig. 3A, B). IDX; ≤no. 4 = 70–85%, no. 6 = 15–30%
M 238	Syntaxial overgrowths are well developed with good crystal faces present (Fig. 3C). IDX; ≤no. 4 = 80–90%, no. 6 = 10–20%
M 208	As M 238 but with less overgrowths. IDX; ≤no. 4 = 60–70%, no. 6 = 30–40%
L 6771	Rounded grains with occasional small patches of syntaxial overgrowth. Surfaces have a fine crevassed network of grooves probably formed by a diagenetic dissolution process (Fig. 3D). IDX; nos 6 and 7 = 100%
M 248	As M 238 but finer grain-size and less authigenic faces. Some rounded grains. IDX; nos ≤ 4 = 50–70%, no. 6 = 30–50%
CQS 13	Some authigenic overgrowths present but generally grains have pitted surfaces. This pitting is interpreted as being a relict replacement feature after a replacive carbonate cement. IDX; no. ≤ 4 = 40–60%, nos 5 and 6 = 40–60%

DISCUSSION

Before the experimentally derived textures can be compared with those produced during diagenesis, it is important to establish that the rate-control of quartz dissolution is the same in both cases. Quartz disassociates in HF due to the high affinity of Si^{4+} to bond with F^- ions. In an excess of aqueous HF, the hexafluoro anions SiF_6^{2-} are formed (e.g. Cotton & Wilkinson, 1972, p. 319), following the reaction

$$SiO_2 + 6HF \rightarrow SiF_6^{2-} + 2H^+ + 2H_2O$$

at atmospheric pressure. This reaction is irreversible when HF is present in excess and therefore goes to completion. In these experiments, quartz dissolution is entirely dependent on the availability of fluoride ions and so problems with quartz saturation in the solvent are avoided. In general, quartz dissolution in natural systems will not be controlled by the avail-

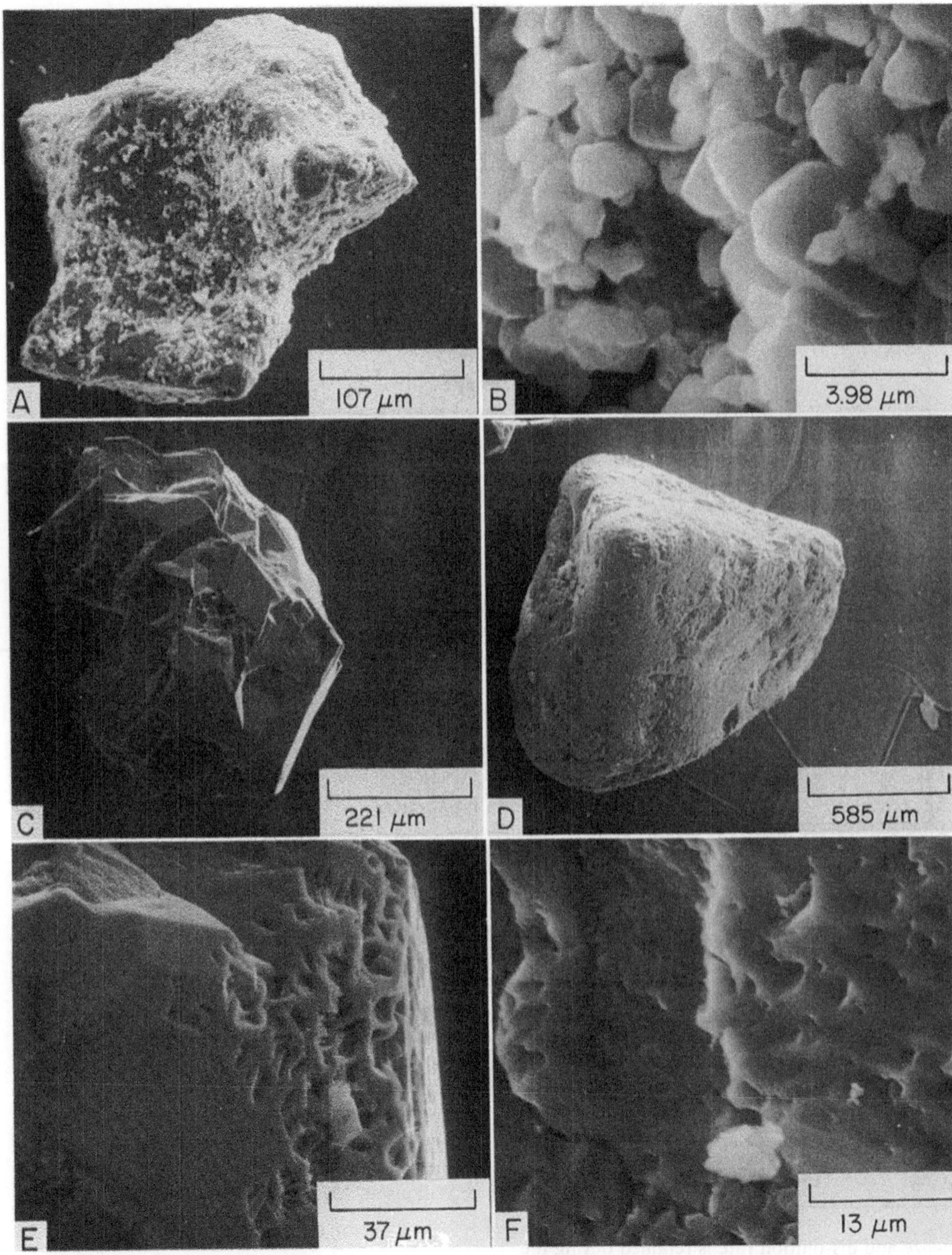

Fig. 3. Surface textures of grains shown by SEM. (A–B) CQS 18 single grain with fine crystalline quartz coating partly obscuring coarser authigenic overgrowths. (B) A close-up showing well-formed, but inequant, authigenic quartz lacking crystallographic continuity with each other and the host grain. (C) M 208 single grain well-developed syntaxial overgrowths forming large crystal faces. (D) L 6771 rounded grain with rare, small patches of authigenic overgrowth. (E) M 238 showing experimentally induced dissolution textures. In the discussion section concerning experimental anomalies the surface area calculation is based on 'typical' notch, of dimensions 5 × 2 × 5 μm, as illustrated here. (F) Naturally produced dissolution textures on quartz from the Brora Sandstone Member, Brora Arenaceous Formation, Oxfordian, Brora, N.E. Scotland.

ability of fluoride ions but by pH; quartz solubility rises rapidly at about pH 9 (Krauskopf, 1956).

Identification of the rate-control of dissolution, experimentally and during diagenesis, can be made by textural examination of the grains (Berner, 1978). Experimentally and naturally produced dissolution surfaces on quartz grains are shown in Fig. 3 (E, F) respectively. Clearly, both samples have textures produced by selective dissolution, the location of which is controlled by the crystallographic properties of the quartz. This type of dissolution is surface-reaction-controlled and is typical of slow dissolution rates of relatively insoluble minerals (e.g. Berner, 1978, table 1). Although one may have expected quartz to dissolve by a different mechanism in HF than during diagenesis, the textural evidence clearly shows that the rate-controlling mechanism is the same. Because of this similarity, the textures, the location of their development and their relative rates of development within either dissolution system are directly comparable. The actual rates of dissolution in both systems, which are almost certainly different, have no direct consequence on the textural comparisons made.

Having established the validity of textural comparison between the experimentally produced and natural samples, the results produced during the short-period etching (Fig. 2 and Table 2) can be considered in relation to dissolution in natural systems. If mineral grains can be considered as idealized spherical particles having a uniform surface energy at the solid–liquid interface, then the rate of dissolution can be related to the particle size by the Ostwald–Freundlich equation,

$$\ln \frac{S_r}{S_\infty} = \frac{2EV}{RT_r}$$

where S_∞ = solubility of an infinitely large particle, S_r = solubility of a small particle of radius r (Å), E = surface energy (Nm^{-1}), V = molar volume ($m^3\ kmol^{-1}$), R = gas constant (8318 $Nm\ kmol^{-1}\ K^{-1}$) and T = absolute temperature (K) (after Iler, 1955). However, the mineral grains illustrated in Figs 3 and 4 show that dissolution has been selective, and not uniform throughout the surface. This implies that the surface energy at the solid–liquid interface is not uniform and that the actual rate of dissolution is the sum of a number of rate components. For example, in the case of a perfect crystal form,

$$R_r = f(R_f, R_c, R_e)$$

where R_f = dissolution rate of faces, R_c = dissolution rate of corners and R_e = dissolution rate of edges. This relationship is not readily quantifiable for quartz. However, by reference to Table 1, it is easy to see that the rate of dissolution (R_r) of similar-sized grains will vary according to the proportions of faces, corners and edges present.

The hierarchy presented in Table 1 can be justified by combining experimental evidence with theoretical considerations. Problems of scale exist between what is a face and a facet on any one grain. However, this relationship is best understood by appreciating that as faces decrease in size, so their relative area of edges and corners increases in relation to the total face area, therefore increasing the relative abundance of high-energy areas. Corners and edges of crystals have higher surface energies than faces, as they are sites of kinks, growth steps and dislocations (e.g. Sunagawa, 1977). When etched, faceted areas can be demonstrated to undergo greater dissolution than adjacent faces (Hurst, 1980).

Abraded, and dissolution/replacement surfaces (nos 7 and 6, Table 1) both have increased surface energy compared with faces and faceted areas. This is difficult to prove mathematically, but the experimental work presented here and by Cabrera & Vermilyea (1958) shows that on dissolution, faceted areas develop into dissolution surfaces rather than vice versa. This implies that the development of dissolution surfaces represents an increase in surface energy. To quote Cabrera & Vermilyea, 'A dissolving crystal will ultimately not be bounded by low index faces but by curved or rough surfaces.' In the case discussed here, rough surfaces develop due to the selective nature of the dissolution reaction.

Fractures (no. 8) may cleave along crystallographic planes, so forming relatively stable surfaces. Sedimentary quartz grains often fracture conchoidally, cross-cutting crystallographic planes and so creating a relatively high surface energy (Iler, 1955).

Disturbed surface layers (or disrupted lattice layer quartz, Lidström, 1968) have a solubility between that of amorphous silica (120 ppm SiO_2) and crystalline quartz (6 ppm SiO_2) (Henderson *et al.*, 1970). No disturbed surface layers were encountered during the course of this work.

Experimental evidence

After 2 min of etching CQS 18 produced a low weight loss in comparison to other similar-sized fractions (Table 2). CQS 18 has authigenic coatings of fine crystalline quartz (Fig. 3B) which, by visual estimation, accounts for approximately 2% of the total grain volumes. This fine authigenic quartz does not completely cover all grain surfaces (Table 3). After 2 min of etching, CQS 18 has developed rounded, corroded textures (Fig. 4A) which appear to be the coarsest relicts of the original quartz coating. During dissolution, quartz was dissolved from all areas of the grains but selectively at higher rates from the areas of highest surface energy. Therefore, although undergoing dissolution, the coarsest authigenic faces have persisted longer than surrounding finer crystallites. Referring to Table 1, this behaviour exemplifies the relationship between nos 1 and 5 during dissolution. The relatively low initial

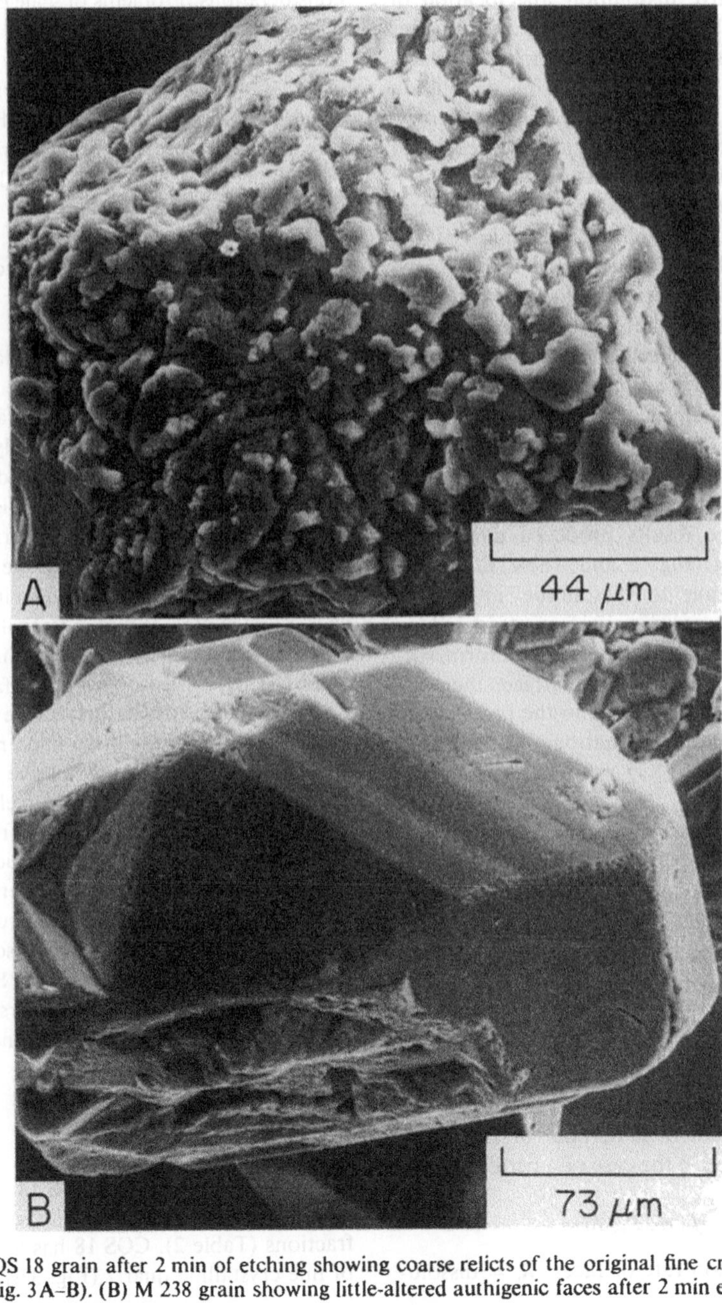

Fig. 4. (A) CQS 18 grain after 2 min of etching showing coarse relicts of the original fine crystalline quartz coating (see Fig. 3A–B). (B) M 238 grain showing little-altered authigenic faces after 2 min etching.

dissolution rate can be hypothesized as resulting from the presence of the coarser overgrowths impeding overall dissolution. After 5 min of etching, no traces of these overgrowths were found and the dissolution rate was comparable to that of other similar-sized fractions.

Samples M 238 and L 6771, although different size fractions, had similar weight losses after 2 min of etching. M 238 has well-developed, coarse, quartz overgrowths (Table 3) which after 2 min etching show only slight effects of the dissolution (Fig. 4 B). Conversely, L 6771 lacks authigenic overgrowths but has coarser, rounded grains (Fig. 3 D). From Table 1 it would be expected that the relatively low surface energy characteristics of M 238 ($>80\%$ of the grain area has characteristics $<$ no. 4) would tend to dissolve less readily than the relatively high energy faces of L 6771 (100% of the grain area has characteristics $\geqslant$ no. 6). After 5 min of etching, the surface characteristics of both samples are similar, and so the subsequent dissolution rate of the finer grain-size, M 238, is more rapid than for L 6771.

A simple calculation demonstrates the effectiveness of minor etched features on increasing total surface area. Let the total face area of a grain (excluding areas without coarse overgrowth) be composed of 10 faces to produce a surface area of $1{\cdot}6 \times 10^4$ μm². If, after 2 min of etching, 10 approximately equidimensional, rectangular notches (see Fig. 3 E), $5 \times 2 \times 5$ μm, develop on each face, the increase in area can be calculated:

$$10\,(2\,(5{\cdot}5)+2\,(2{\cdot}5)) = 700\ \mu\text{m}^2,$$

therefore on 10 faces the area increases by 7×10^3 μm². The significance of this increase of 44% total area is that, with only relatively minor etching on authigenic faces, the surface area increases very rapidly. With reference to the example of M 238, it is not unreasonable to expect that sample's dissolution rate to follow a higher rate once the authigenic faces have undergone even slight dissolution, due to the considerable increase in total surface area. In terms of textural development, it is important to note the way that dissolution selectively concentrates around corners and along edges of faces (Figs 3 E and 4 B), that is the areas of the grains with relatively high surface energy.

Application to diagenetic textures

These experimental examples illustrate the validity of using textural characteristics for predicting relative dissolution rates, or the selective dissolution

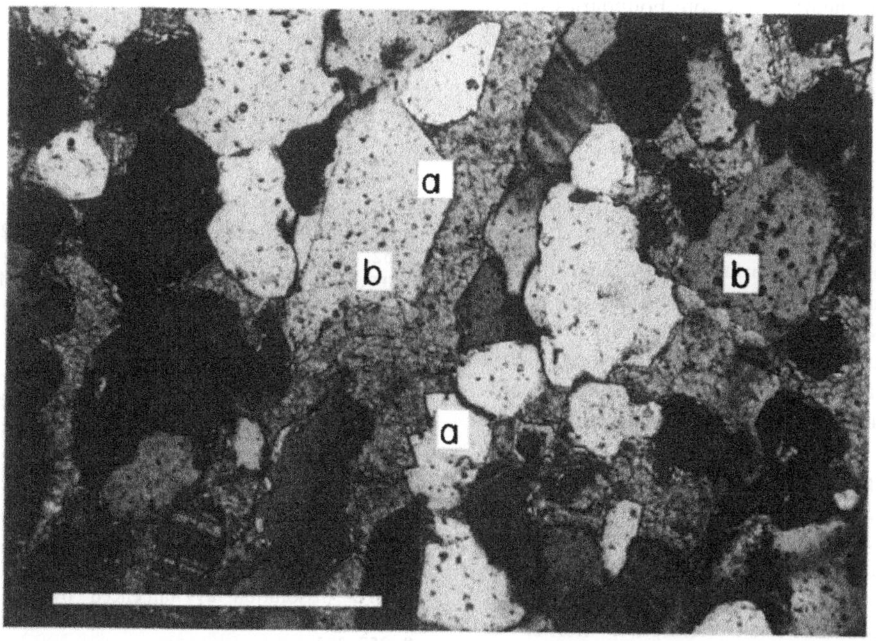

Fig. 5. Calcite-cemented sandstone under X-nicols. Areas marked 'a' represent stable euhedral quartz (low surface energy). Areas marked 'b' are areas where replacement of the quartz fabric has taken place. Scale bar = 1 mm.

of quartz surfaces. As dissolution is surface-reaction-controlled in both experimental and natural conditions, the results of these experiments can be applied to naturally occurring diagenetic textures in rocks. Calcite is observed as a replacive cement in some sandstones (Fig. 5). Replacive is here used descriptively, not genetically: the calcite cement occupies an area once occupied by quartz. Although it is impossible to state categorically that areas 'b' in Fig. 5 were not sites of large authigenic faces prior to calcite cementation, the significant feature is that faces 'a' have remained stable. Areas 'b' are irregular, unstable contacts between quartz and calcite, interpreted texturally as sites of instability between the two minerals (Spry, 1969). However, at sites 'a' little reaction has taken place. By analogy with the experimental results, this implies that the process of calcite cementation was selective in its location, as was the initial dissolution of quartz grains in HF.

If the quartz grains of a sandstone are similar to those samples studied here (i.e. contain a variety of surface textures), a corrosive pore-fluid or replacive cement will preferentially attack the rock fabric at the points of greatest surface energy. For example, consider two pores A and B within the same sandstone and with the same corrosive pore-fluid circulating through them: A has pore boundaries whose faces have relatively high surface energy (nos 5–7, Table 1); B has pore boundaries which have good crystal faces (nos. 1–2). The result of this corrosion would be to preferentially increase the volume of pore A, rather than pore B. When this relationship is extended throughout a sandstone fabric, it follows logically that the development of this corroded, secondary porosity will be located at primary pores which had the highest surface-energy faces forming their boundaries. Therefore, the degree of homogeneity of distribution of such a porosity or cement is a direct consequence of the preceding quartz textures. Clearly, this simple relationship is complicated if minerals, other than quartz, form a substantial part of the mineralogy. In the samples studied, feldspars rarely account for more than 5%, and micas never more than 2% of the bulk mineralogy. Feldspars and micas both have low-solubility products in natural waters, and petrographic evidence shows that their dissolution is surface-reaction-controlled (feldspars, Berner & Holdren, 1977; micas, Rimsaite, 1975). From these data for feldspars and micas, it would seem that the crystallographic parameters of cleavage and twin planes are of greater importance in controlling the location of dissolution than in the case of quartz. These characteristically weak planes in feldspars and micas may also have significant bearing on the location of secondary porosity.

To conclude, the experimental and natural examples presented herein justify the hierarchy presented in Table 1. This hierarchy may be further quantifiable by using more refined experimental techniques. The potential for predicting where dissolution or replacement will take place within a sandstone fabric is clearly of considerable impact, if the distribution of these features is of importance.

Further modelling of these relationships is required to define the useful limits of this work; however, these results provide a basis for re-examining the development of dissolution textures in sandstones.

ACKNOWLEDGMENTS

I would like to thank H. Irwin and A. Parker for constructive criticism of an early draft of this paper. D. Whitehead was a continual source of helpful criticism, assistance and discussion. R. Goldring is thanked for translating some of my ideas into English. This work was supported by NERC grant GT4/77/GS101.

REFERENCES

Berner, R.A. (1978) Rate control of mineral dissolution under earth surface conditions. *Am. J. Sci.* **278**, 1235–1252.

Berner, R.A. & Holdren, G.R. Jr. (1977) Mechanism of feldspar weathering: some observational evidence. *Geology*, **5**, 369–372.

Cabrera, N. & Vermilyea, D.A. (1958) The growth of crystals from solution. In: *Growth and Perfection of Crystals* (Ed. by R.H. Doremus, B.W. Roberts and D. Turnbull), pp. 393–410. Wiley, New York.

Cotton, F.A. & Wilkinson, G. (1972) *Advanced Inorganic Chemistry*, 3rd edn. Wiley (Interscience), New York.

Frondel, C. (1962) *Dana's The System of Mineralogy*, 7th edn, **111**, *Silica Minerals*. Wiley, London. 334 pp.

Henderson, J.H., Syers, J.K. & Jackson, M.L. (1970) Quartz dissolution as influenced by pH and the presence of a disturbed surface layer. *Israel J. Chem.* **8**, 357–373.

Hurst, A.R. (1980) *The diagenesis of Jurassic rocks of the Moray Firth, N.E. Scotland.* Unpublished Ph.D. Thesis, University of Reading.

ILER, R.K. (1955) *The Colloid Chemistry of Silica and Silicates.* Cornell University Press, Ithaca, New York.

JACKSON, M.L., SAYIN, M.L. & CLAYTON, R.N. (1976) Hexafluorosilicic acid reagent modification for quartz isolation. *Proc. Soil Sci. Soc. Am.* **40**, 958–960.

KRAUSKOPF, K.B. (1956) Dissolution and precipitation of silica at low temperatures. *Geochim. cosmochim. Acta,* **10**, 1–26.

LIDSTRÖM, L. (1968) Surface and bond forming of quartz and silicate minerals and their application in mineral processing techniques. *Acta polytech. scand.* **75** A, 149 pp.

RIMSAITE, J. (1975) Natural alteration of mica and reactions between released ions in mineral deposits. *Clays Clay Miner.* **23**, 247–255.

SPRY, A. (1969) *Metamorphic Textures.* Pergamon Press, Oxford. 350 pp.

SRIDHAR, K., JACKSON, M.L. & CLAYTON, R.N. (1975) Quartz oxygen isotope isotopic stability in relation to isolation from sediments and diversity of source. *Proc. Soil Sci. Soc.* **39**, 1209–1213.

SUNAGAWA, I. (1977) Natural crystallisation. *J. Crystal Growth* **42**, 214–223.

(*Manuscript received* 10 *March* 1980; *revision received* 30 *September* 1980)

Reprinted from Sedimentology (1986) ***33****, 587–604*

Thin section and S.E.M. textural criteria for the recognition of cement-dissolution porosity in sandstones

S. D. BURLEY* *and* J. D. KANTOROWICZ†

Department of Geology, The University, Hull HU6 7RX, U.K.

ABSTRACT

Cement-porosity relationships are described from the Lower Triassic Sherwood Sandstone Group and the Middle Jurassic Ravenscar Group in the United Kingdom. Calcite cemented sandstones display a variety of replacement textures, with preferential replacement of grains and of overgrowth faces with high free-surface energy. Dolomite and siderite cemented sandstones display similar textures but replacement is less specific and euhedral overgrowth surfaces are commonly embayed by carbonates. Examination of the more porous sandstones with the scanning electron microscope reveals a range of pitting and embayment textures in authigenic overgrowths and in detrital grains. These range from small 'v'-shaped notches and pits, through regular and irregular shaped embayments, into large depressions. These textures appear to be morphologically similar to the quartz surfaces seen in thin sections of carbonate cemented sandstones, and are interpreted to have been formed by the dissolution of pore-filling and grain replacive authigenic carbonates. This is confirmed by examination of experimentally exhumed overgrowth surfaces from carbonate cemented sandstones. These textures indicate that part of the intergranular porosity in these sediments is secondary in origin, and has been generated by the dissolution of carbonate cements. The identification of such textures may lead to a more confident interpretation of the nature of intergranular porosity in the subsurface.

INTRODUCTION

Schmidt, McDonald & Platt (1977), and Schmidt & McDonald (1979a, b), describe criteria for the recognition of secondary porosity in sandstones, and, more significantly, emphasize its widespread occurrence, particularly at depths where previous studies suggest porosity should be severely reduced or even eliminated by cementation and compaction. Such secondary porosity may be generated either through the dissolution of detrital framework components or authigenic cements. Secondary porosity which results from the dissolution of mineral cements, irrespective of whether they are replacive or purely pore-filling, is termed cement-dissolution porosity. The criteria presented by these authors for the recognition of cement-dissolution porosity are developed principally from thin section studies and are, in some instances, open to alternative interpretations. This paper describes additional textural criteria for the recognition of cement-dissolution porosity both in thin section and, more especially, with the scanning electron microscope (S.E.M.). The observations are thus intended to supplement and to refine the criteria developed by Schmidt *et al.* (1977), and to assist in the recognition of secondary porosity resulting from the dissolution of pore-filling cements which are peripherally replacive to the silicate framework. Modifications of earlier formed authigenic overgrowths produced *in situ* during diagenesis are stressed in this paper to avoid confusing cement-dissolution textures with features that may be produced on grain surfaces during transportation (see Krinsley & Doornkampf, 1973; Le Ribault, 1977).

Secondary porosity has been widely reported and extensively described in sandstones from diverse sedimentary basins in recent years. However, the

Present addresses: *Geologisches Institut, Universität Bern, Baltzerstrasse 1, CH-3012, Berne, Switzerland.

†Koninklijke/Shell Exploratie en Produktie Laboratorie, Volmerlaan 6, Rijswijk ZH, The Netherlands.

origin and extent of secondary porosity is strongly contested. Indeed, some authors disagree on the primary or secondary nature of intergranular pore space (see, e.g. Sommer, 1978; Schmidt & McDonald, 1979a; Bjørlykke, 1981, 1984; Giles & Marshall, 1983; Knox *et al.*, 1984). In this paper we do not intend to address the problem of the origin of secondary porosity in sandstones. Rather, we wish to highlight textural criteria for the recognition of secondary porosity which may be especially beneficial to the study of subsurface sandstones in which the easily recognizable features indicative of secondary porosity, such as grain dissolution, only rarely occur.

SEDIMENTOLOGICAL SETTING AND REGIONAL PATTERNS OF DIAGENESIS

The textures described herein are observed as a modification of both detrital grains and overgrowths from two Mesozoic sandstone units in the United Kingdom (Fig. 1). These are the Lower Triassic Sherwood Sandstone Group, a complex of red beds deposited largely by fluvial processes but with subordinate aeolian and lacustrine influences (Thompson, 1969, 1970; Audley-Charles, 1970; Pattison, Smith & Warrington, 1983; Thompson & Steel, 1983) and the Middle Jurassic Ravenscar Group, which accumulated in a series of fluvio-deltaic lobes intercalated with interdistributary bay, lagoonal and shallow-marine sediments (Hemingway, 1974; Livera & Leeder, 1981).

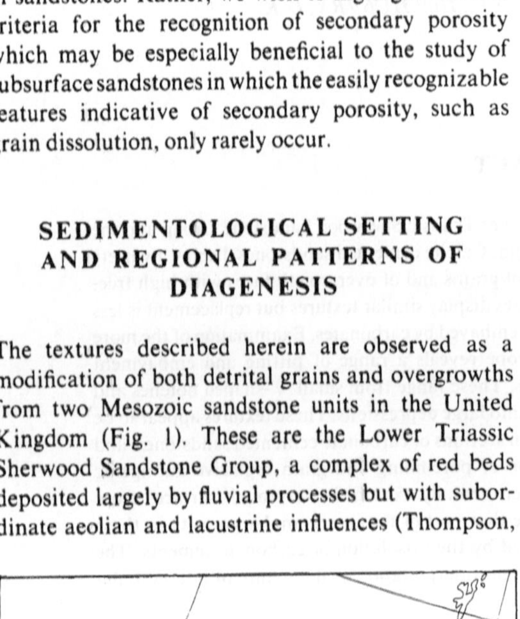

Fig. 1. Location of the study sections within the general framework of their Mesozoic sedimentary basins.

The Sherwood Sandstone exhibits a general south to north variation in detrital mineralogy from lithic arkoses to sublithic, subarkosic arenites, concomitant with a gradual change from proximal, coarse sandy and gravelly alluvium to distal sandy braided river deposits. Diagenetic mineralogies in the proximal and distal sandstones are significantly different. As representative examples, the Marchwood Borehole penetrated some 60 m of proximal Sherwood Sandstone alluvium on the eastern margin of the Wessex Basin, whilst wells drilled in the Irish Sea Basin encountered a 1 km thick distal sequence, the top 200 m of which were cored. Regional aspects of diagenetic mineralogies and textures of these sandstones are detailed in Burley (1984) and summarized in Figs 2 and 3. In essence, during an early phase of 'red bed type' diagenesis, comparable to that described for recent red beds (Walker, Waugh & Crone, 1978) these sandstones were extensively cemented with non-ferroan carbonates of pedogenic origin, and are analogous to modern caliche deposits. The early diagenetic mineralogy comprised quartz overgrowths, authigenic feldspar, mixed-layer illite-smectite clays and authigenic haematite together with the extensive carbonate cement. These authigenic minerals were associated with grain dissolution and clay replacement. As a result of burial, which reached in excess of 2 km in the Wessex Basin (Stoneley, 1982) and probably exceeded 3 km in the Irish Sea Basin (Colter, 1978), the originally smectitic clays that were present underwent conversion to illites which now contain no, or only a small (<5%), expanding component. Additional authigenic minerals precipitated under deep burial include further quartz overgrowths, ferroan carbonates, pyrite and minor barytes cement.

The petrography and diagenesis of the Middle Jurassic Ravenscar Group are described by Kantorowicz (1985), and summarized in Fig. 4. Diagenetic modifications involve the initial authigenesis of various clay minerals, quartz and carbonate cementation, carbonate dissolution and the precipitation of dickite. The Ravenscar Group currently crops out in the Cleveland Basin, to form the northern and western escarpment of the North Yorkshire Moors. During burial the Ravenscar Group underwent a double

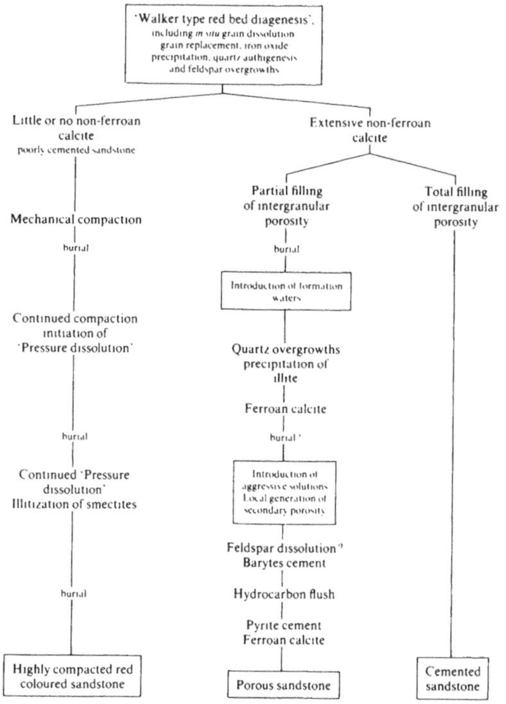

Fig. 2. Flow chart illustrating simplified schematic diagenetic pathways present in the Marchwood Borehole, Wessex Basin.

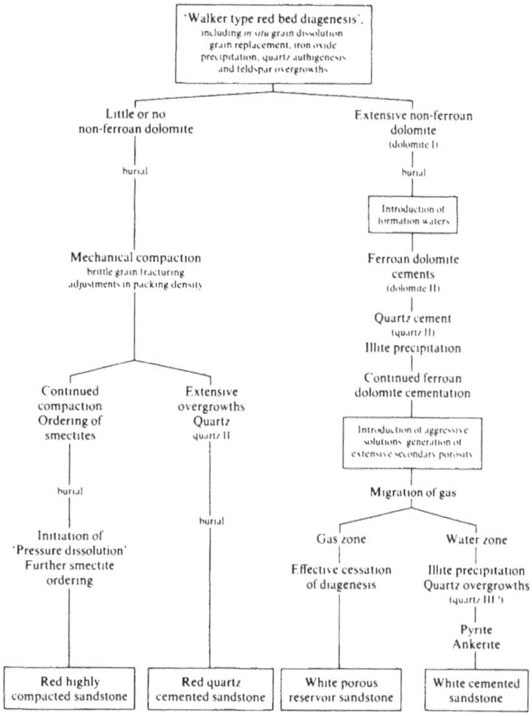

Fig. 3. Flow chart illustrating simplified schematic diagenetic pathways in the Sherwood Sandstone Group from the Irish Sea Morecambe Bay gas field.

inversion, being buried rapidly to 1·5 km by the mid-Cretaceous, uplifted to a depth of approximately 1 km in the late Cretaceous, reburied, and inverted again to outcrop today (Kent, 1980; Barnard & Cooper, 1983).

Methodology

Samples were taken from both outcrop and subcrop material to encompass all the variation in depositional environment, texture and grain size displayed by the studied sandstones. The Sherwood Sandstone Group was sampled in the B.G.S. Marchwood Borehole (SU 3991 1118) of the Wessex Basin, and in the Hydrocarbons GB Wells 110/3–3 and 110/2–1 in the Morecambe gas field, Irish Sea Basin, whilst the Ravenscar Group was sampled at outcrop on the North Yorkshire Moors.

Thin sections were prepared from samples routinely impregnated with blue-dyed epoxy-resin. Detailed petrographical work was undertaken to identify and to quantify both detrital and authigenic mineral components. Identification of carbonate cements was aided by thin section staining techniques (Dickson, 1965), and by quantitative chemical analysis with the electron microprobe. Following petrographical study, those samples exhibiting well developed porosity systems and relic carbonate cements were examined using a Cambridge Instruments Stereoscan S.E.M. equipped with a Link E.D.S. system to investigate the three-dimensional nature of cement-porosity relationships.

Selected sandstone samples with calcite and dolomite cement were carefully broken into fragments 5–10 mm in diameter and placed in a 10% HCl acid bath until effervescence ceased. The resulting isolated detrital grains were cleaned ultrasonically and examined with the S.E.M.

Terminology

In view of the wide range of surface textures observed on quartz grains and overgrowths in the studied sandstones, we have adopted an informal classification to describe replacement features. Our observations do not enable a unique genetic relationship to be

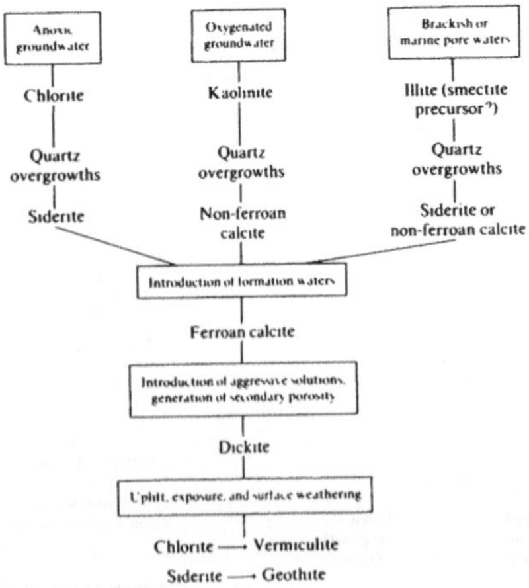

Fig. 4. Flow chart illustrating schematic diagenetic pathways in the Ravenscar Group sandstones (after Kantorowicz, 1985).

established between replacement features and particular carbonate cements. Accordingly, we have defined a hierarchy of replacement features based on size, and have employed morphological terminology to describe replacement features.

Pits or notches

These are the smallest corrosion features observed, typically being of between 1 and 2 μm in diameter, and having involved only limited removal of quartz (see, e.g. Figs 5H and 6H). In general, pits are highly irregularly shaped, whilst notches are distinctly 'v'-shaped and more uniform. Overgrowths are still recognizable as such, even when extensively pitted. Occasionally, larger rhomb-shaped notches occur which attain diameters of up to 10 μm.

Embayments

Larger or coalesced pits are termed embayments and are easily recognized in thin section (see Fig. 7E, F), as well as with the S.E.M. (see Figs 6E and 7C). Embayments are commonly of such large size as to penetrate through overgrowths into the detrital grain, and attain widths in excess of 20 μm.

Depressions

Where embayments have enlarged considerably, and have removed overgrowth surfaces to such an extent that they are no longer recognizable, we have used the term depression to describe the replacement texture. Here we infer that extensive peripheral replacement of both overgrowths and detrital grains has taken place, significantly reducing the silicate framework (see Fig. 7G).

Whilst sizes are quoted for the order of magnitude of the features observed in these sandstones, they do not imply an absolute quantification. The size of the surface textures developed depends both on the extent of dissolution and also on the size of the carbonate crystals precipitated. In other sandstones therefore, where larger carbonate crystals occur, morphologically similar textures may develop in which the absolute size of individual features differs significantly.

PETROGRAPHY

The proximal Sherwood Sandstone Group of the Marchwood Borehole

Calcite cements occur in virtually all of the samples studied in the Marchwood Borehole, and range in quantity from trace amounts to in excess of 30% according to thin section modal analysis. At least two separate generations of calcite cementation can be recognized (Fig. 2). An early, non-ferroan calcite is the dominant cement, and is quite distinct from later, poikilotopic ferroan calcite cements.

The non-ferroan calcite exhibits a variety of textures typical of recent caliche deposits, including nodular growth, displacive textures and radial rim cement fabrics. Additionally, where this non-ferroan calcite cement is abundant and occludes most of the intergranular pore space, the original detrital fabric is largely preserved with little or no evidence of compaction, indicating carbonate cementation took place prior to significant burial. The calcite is also highly corrosive with respect to both quartz and feldspar grains and, to a lesser extent, early quartz and feldspar overgrowths (Fig. 5A, B). There is a distinct tendency for calcite replacement to be more intense on those quartz surfaces with high free-surface energy (Hurst, 1981), being particularly well developed around grain peripheries, along fractures and between crystal boundaries in rock fragments (Fig. 9). Feldspar grains may also be extensively replaced (Fig. 5A), especially along cleavage planes, and are locally completely replaced. In such cases a 'floating' grain texture results and the former feldspar grain outlines can only be recognized by the presence of clay rims or relic overgrowths that have not been entirely replaced.

The ferroan calcite occurs as large, poikilotopic crystals which commonly nucleate on, and enclose, the non-ferroan calcite. According to modal analysis, ferroan calcite may amount up to 8% by area of the sandstones. This calcite is also replacive to both quartz and feldspar grains and, to a limited extent, their respective overgrowths.

Where early carbonate cements are not present or are very subordinate, a spectrum of textures can be recognized. At one extreme, the sandstones have undergone intense compaction and intergranular porosity has been reduced largely by grain-to-grain contact dissolution ('pressure dissolution'). At the other extreme the sandstones are characterized by open, well connected intergranular porosity, largely devoid of pore-lining or pore-filling authigenic clay minerals (Fig. 5C, D). Late ferroan calcite is sporadically developed in these porous sandstones and quartz overgrowths are present as up to 5% of modal analysis. Measured porosities reach 28% and permeabilities are greater than 1000 mD and locally reach 5000 mD (Knox *et al.*, 1984). There is an extreme heterogeneity of grain packing. Grain contacts vary from apparently 'floating' to straight or even locally sutured. Pore throats are often enlarged and oversized pores are common, the latter frequently being enclosed by insoluble clay rims (Fig. 5D). Grain outlines are highly irregular and both grains and overgrowths appear highly corroded.

The full extent of this corrosion is best illustrated with the S.E.M. In the porous zones, grain surfaces are covered with a complex pattern of angular ridges and depressions. In some cases, where more recognizably complete overgrowths occur, prismatic embayments can be observed penetrating the overgrowth and reaching the underlying grain (Fig. 5E). The lower surface of the larger depressions is covered by an intricate, interlocking pattern of irregular shaped pits and embayments (Fig. 5F, G). More rarely, the surface of the quartz overgrowth is observed to be covered with orientated 'v'-shaped notches (Fig. 5H).

The distal Sherwood Sandstone Group of the Irish Sea Basin Wells 110/3–3, 110/2–1

In contrast to the proximal Sherwood Sandstone, the dominant carbonate cements in the distal sequence are dolomite or ferroan dolomite. Studies of the adjacent, shallow-basin margin sequence in Cumbria (see Burley, 1984), indicate that non-ferroan dolomite cements, commonly displaying nodular growth textures, are abundant. In the deeply buried Sherwood Sandstone of the Irish Sea these early dolomites have recrystallized as iron-rich dolomites. Dolomite now occurs as large, poikilotopic crystals which are highly corrosive to the silicate framework and replace both quartz and feldspar grains and their respective early overgrowths (Fig. 6A, B). Cathodoluminescence petrography shows that the iron-bearing dolomites are zoned, and comprise an inner, brightly yellow-orange luminescing core enclosed in later red-luminescing dolomite or non-luminescent ankerite.

Prior to significant compaction or grain-to-grain contact dissolution, sandstones lacking dolomite cements were cemented with large, well developed syntaxial quartz overgrowths which locally total 19% of the whole rock. Further rhombic iron-rich dolomite and ankerite cements post-date the main phase of quartz authigenesis, partially occlude porosity and replace detrital grains and late quartz overgrowths.

As in the Marchwood Borehole, interbedded with these carbonate and quartz cemented horizons there are distinct zones of both tightly compacted and extremely porous, friable sandstones (see Fig. 3). The differentiation into compacted, cemented and porous zones is highly variable and irregular and may occur on a laminae scale within one thin section, or on a metre scale visible within the slabbed cores, and is apparently developed independently of depositional facies, grain size or texture. Where early dolomite cements were absent and subsequent quartz overgrowth cementation did not develop, mechanical compaction, through grain slippage and rotation, packing re-adjustment and brittle fracturing, together with chemical grain-to-grain contact dissolution, reduced intergranular porosity to negligible levels.

The highly porous zones, by contrast, have thin section porosities of up to 27% and measured porosities of up to 22%, with permeabilities of 1000 mD (Ebbern, 1981). In thin section, detrital quartz grains are extremely angular and exhibit irregular shapes. Grain packing is characteristically heterogeneous, oversized pores are common and pore throats are typically enlarged, resulting in floating grain textures (Fig. 6C, D). The S.E.M. confirms the presence of abundant corrosive textures on the surface of both detrital grains and overgrowths. Quartz overgrowths are present in all samples but are preserved to varying degrees. Where well developed, quartz overgrowths are modified by small, irregular embayments (Fig. 6E). More typical, however, is the presence of only small areas of overgrowths bounded partially by crystal faces and partially by irregular, corroded edges (Fig. 6F). The surface of the detrital grain is covered with innumerable, apparently random pits and embayments (Fig. 6G, H), which give the grain surface a highly corroded appearance. Overgrowth faces may be smooth but are commonly covered with small irregular notches.

The Middle Jurassic Ravenscar Group sandstones of the North Yorkshire Moors

A variety of carbonate cements are present in Ravenscar Group sandstones. Siderite occurs in the majority of samples from non-marine environments, comprising from a trace to as much as 30% of the rocks it cements (Fig. 7A). It is usually rhombic (Fig. 7B, D) but occasionally occurs as sphaerosiderite (Fig. 7A). Calcite, on the other hand, is well developed as a cement in marine sandstones but only occurs locally in non-marine sandstones. Ferroan calcite is more abundant than non-ferroan calcite but both occur as poikilotopic crystals. Both ferroan and non-ferroan calcite replace quartz selectively, but replace feldspar grains more extensively. In general, where carbonates occur, siderite appears to corrode and replace quartz grains and overgrowths to a greater extent than does calcite. The siderite is inferred to have been precipitated following authigenic quartz in these sandstones. Siderite is only rarely completely enclosed within quartz overgrowths and it is considered unlikely that the quartz overgrowths post-dated siderite precipitation and subsequently enclosed siderite rhombs. The highly replacive nature of the siderite is clearly demonstrated by the presence of siderite rhombs within detrital quartz grains.

Siderite replacement concentrates preferentially on the surfaces with relatively high free-surface energy (see Hurst, 1981). Also, siderite more frequently replaces the flat and relatively low free-energy faces of quartz overgrowths (Fig. 7A, B). Replacement of grains and overgrowths is manifested in both distinct rhomb-shaped 'notches' or larger, irregular embayments. Corrosion is typically intense, with siderite rhombs extensively replacing quartz overgrowths and penetrating through into the detrital grain. Once siderite has penetrated the overgrowth, replacement of the detrital grain is more extensive, opening out along the contact between the overgrowth and detrital grain, producing distinctive globular embayments (Fig. 7B). This preferential growth between grains and overgrowths indicates that the overgrowths are not necessarily in complete crystallographical continuity with the detrital grain. Siderite precipitation, therefore, accentuates planes of weakness, or indeed, follows enclosed dust rims or clay coatings which only allow local, rather than wholesale, attachment of the overgrowths to the grains. In sandstones with extensive siderite cement, peripheral replacement of the silicate framework may exceed 10% of the rock.

Fig. 5. Calcite cement-dissolution porosity textures.

(A) Extensive pore-filling and grain replacing non-ferroan calcite cement that effectively occludes all visible intergranular porosity. Peripheral replacement of quartz grains (R) has enlarged the intergranular areas. Detrital feldspar grains have undergone almost complete replacement (RF). Photomicrograph, PPL, Marchwood Borehole, 1689 m. Scale bar = 1.2 mm.

(B) Detail of detrital grain enclosed within non-ferroan calcite cement showing peripheral replacement of both grains (R) and relics of early quartz overgrowths (OG) enclosed within calcite. Photomicrograph, PPL, Marchwood Boreholes, 1689 m. Scale bar = 625 μm.

(C) Open grain framework with a well connected enlarged intergranular porosity system (P) containing relic non-ferroan calcite cement (c). Pore throats are unusually large and some detrital grains are apparently floating (FG). Abraded detrital grain margins appear highly corroded (A) and contain deep embayments whilst authigenic quartz overgrowth faces (OG) of low surface energy are stable. Partially dissolved rock fragments (RF) have collapsed and been compacted. Photomicrograph, PPL, Marchwood Borehole, 1687 m. Scale bar = 1 mm.

(D) Sandstone with large, oversized pores (OP) and common floating grain textures. Dark insoluble residue indicates the position of former calcite crystal boundaries. The fractured grain (F) was probably replaced along the high surface energy fracture by calcite prior to cement dissolution. Note the peripheral corrosion of detrital grain surfaces (C) and contrasting stable overgrowth faces (OG). Photomicrograph, PPL, Marchwood Borehole, 1687 m. Scale bar 1.2 mm.

(E) Typical surface texture of detrital quartz grain from the above sample as viewed with the S.E.M. Relic quartz overgrowth is clearly visible with corrosion edge (C) where former calcite crystal has replaced the overgrowth and detrital grain. Where corrosion has been more extensive only thin upstanding ridges (U) of the overgrowth remain. Scanning electron micrograph, Marchwood Borehole, 1687 m. Scale bar 20 μm.

(F) Corroded texture of the detrital grain surface. Embayments and depressions are highly irregular, exhibiting no preferred orientation or crystal shape. Scanning electron micrograph, Marchwood Borehole, 1687 m. Scale bar 10 μm.

(G) Detail of detrital grain surface with randomly orientated dissolution pits. Note the rounded appearance of both dissolution pits and grain surface. Scanning electron micrograph, Marchwood Borehole, 1687 m. Scale bar 5 μm.

(H) Detail of orientated 'v'-shaped notches covering the surface of authigenic quartz overgrowth. Scanning electron micrograph, Marchwood Borehole, 1687 m. Scale bar 4 μm.

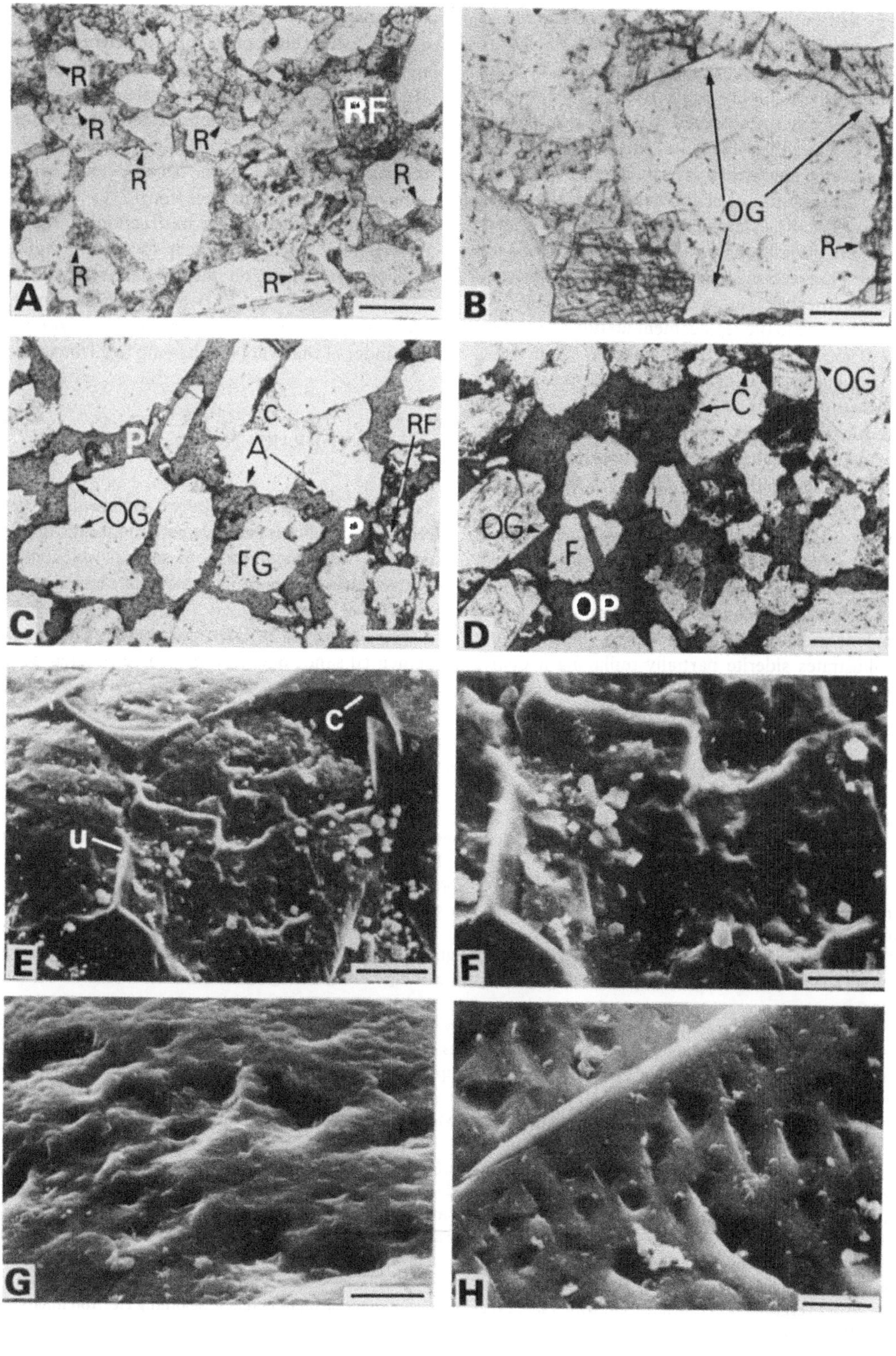
R
R
R
R
R
RF
R
R
R
A
OG
R
B
c
P
A
RF
OG
P
FG
C
C
OG
C
OG
F
OP
D
c
u
E
F
G
H

The distribution of porosity within Ravenscar Group sandstones is quite variable. Some samples are tightly cemented with carbonate cements and, where the cements are highly replacive, detrital silicate grains are commonly observed to be 'floating' or only in point contact. Elsewhere, and particularly in large, interconnected channel-facies sandstones, carbonate cements are subordinate or absent and porosity is well developed. In these samples the quartz overgrowth surfaces are often observed to be intensely embayed or pitted. Thin section observations show the presence of irregular, deep corrosion embayments (Fig. 7E, F), similar in shape to the replacement textures described above (Fig. 7A, B). These embayments are present, both on single crystals and at the juncture of overgrowths on different grains.

Examination with the S.E.M. reveals a wide range of textures in addition to those observed in thin section. These vary in size from small 'v'-shaped notches, to large embayments (Fig. 7H) which are probably analogous to those described above. Of these textures, only the smaller ones are regularly shaped. The 'v'-shaped notches average 1–2 μm in diameter whilst larger rhomb-shaped embayments are usually 10 μm long and around 5 μm wide (Fig. 7C, D). Figure 7(D) illustrates siderite partially replacing a quartz overgrowth. In this case the rhombic impression of the replacing phase can be recognized. Where larger embayments occur, only the edges have any morphologically distinct character. The inferred origin of these embayments (e.g. Fig. 7H) is by coalescence of smaller pits. Corrosion of quartz overgrowths in the Ravenscar Group is commonly so intense that only relics of the authigenic cements remain, and these display no recognizable crystal form. Pitting and corrosion embayments range in size from 2 to 5 μm wide notches, to almost complete destruction of grains and overgrowths. Pits also appear to occur irregularly, both on grains and on the crystal terminations of overgrowths. In Fig. 7(G) larger depressions have developed, preferentially removing the grain, the remainder of the overgrowth being free from attack.

EXPERIMENTAL EVIDENCE

To study the surface textures produced by grain-replacive carbonate cements both calcite and dolomite cements were removed from several well-cemented samples and the resulting loose grains examined with the S.E.M.

The surfaces of isolated quartz grains from calcite cemented sandstones are seen to be covered with an interlocking, anastomising network of upstanding ridges which form a prismatic box-work texture (Fig.

Fig. 6. Dolomite cement-dissolution porosity textures.
(A) Sandstone tightly cemented with dolomite. Note the presence of quartz overgrowth (og) enclosed within the dolomite cement and the enlarged intergranular areas. Photomicrograph, PPL, 110/3–3, 1067 m. Scale bar = 1.2 mm.
(B) Detail of quartz grains with overgrowths enclosed in dolomite cement. Both the overgrowths and detrital grains have suffered extensive peripheral replacement (R) by the dolomite cement. Photomicrograph, PPL, 110/3–3, 1067 m. Scale bar 625 μm.
(C) Thin section photomicrograph illustrating heterogeneous grain packing density with oversized pores (P), enlarged pore throats (T) and irregular distribution of quartz overgrowth cements (og). Photomicrograph, PPL, 110/3–3, 1092 m. Scale bar 900 μm.
(D) Interconnected, intergranular cement-dissolution porosity with enlarged pore throats (T). Detrital grains commonly exhibit floating textures and have extremely angular outlines. The distribution of quartz overgrowths (og) is extremely heterogeneous and although individual overgrowths are locally quite large, they are generally lacking. Photomicrograph, PPL, 110/3–3, 1075. Scale bar 300 μm.
(E) Irregularly shaped embayment penetrating quartz overgrowth. Scanning electron micrograph, 110/3–3, 1092 m. Scale bar 20 μm.
(F) Examination of the same sample as in (D) with the S.E.M. clearly demonstrates the extensive peripheral grain and overgrowth corrosion. The quartz overgrowth (og) is considered to have originally covered most of the detrital grain but is now deeply embayed. Scanning electron micrograph, 110/3–3, 1075 m. Scale bar 10 μm.
(G) Detail of corroded grain surface typical of grain in this sample. Occasional upstanding peaks (u) suggest that quartz overgrowths originally enclosed the grain surface, most of which is now covered with irregular corrosion pits. Scanning electron micrograph, 110/3–3, 1075 m. Scale bar 10 μm.
(H) Detail of overgrowth corrosion texture with a more regular development of small notches. Scanning electron micrograph, 110/3–3, 1075 m. Scale bar 4 μm.

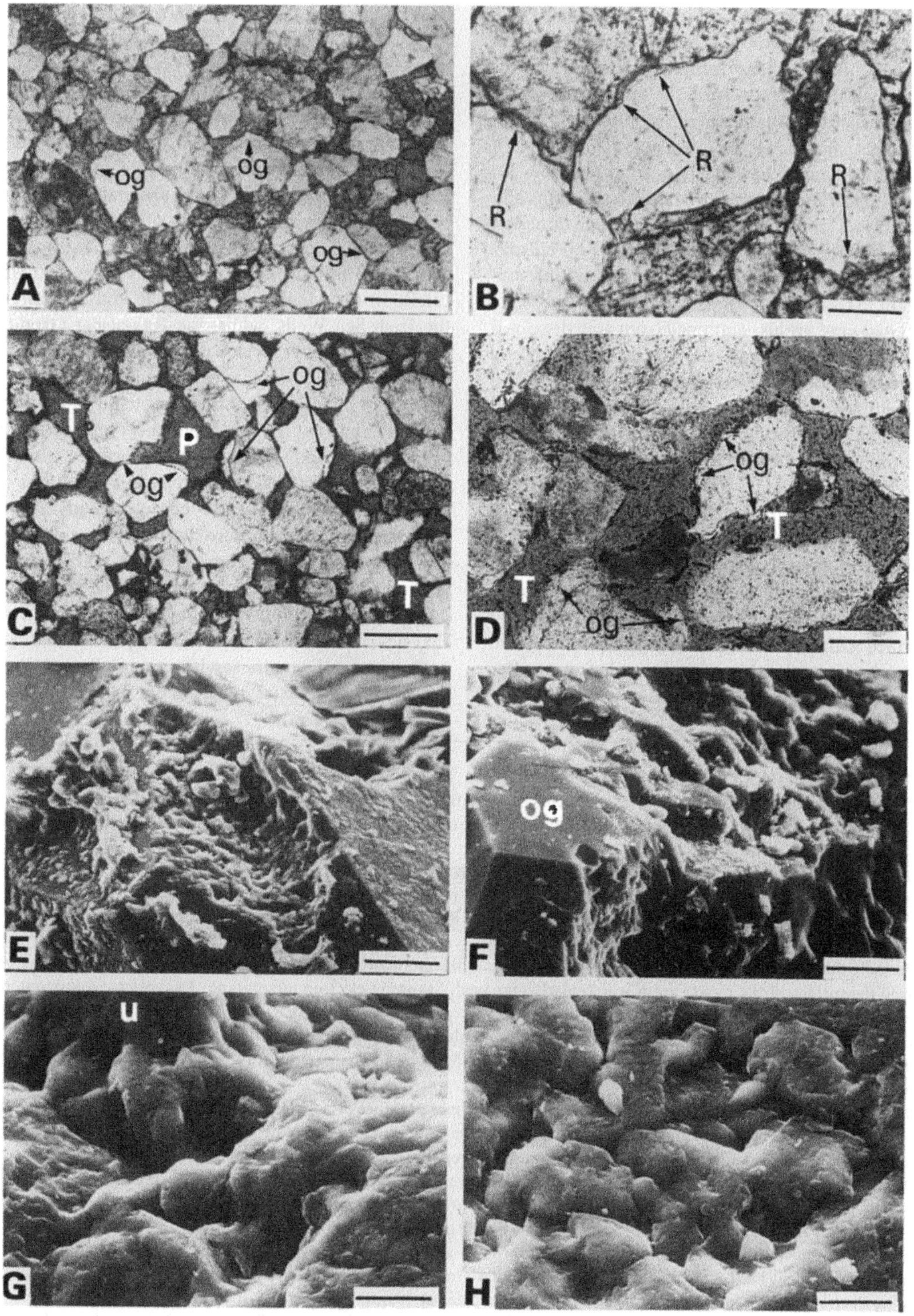
og
og
og
A
R
R
R
R
B
og
T
P
og
T
C
og
T
T
og
D
E
og
F
u
G
H

8A, B). Individual boxes vary in size between 10 and 4 μm, and fall within the size range of calcite crystals visible in thin section, forming radial grain-rim cements. Feldspar grains from the same acid treated samples also exhibit intense corrosion particularly in directions parallel to the cleavage (Fig. 8C–F). Corrosion features on feldspar grain surfaces in general appear more intense than in quartz grains and invariably exhibit a preferred dissolution direction relative to planes of high free-surface energy. Feldspar overgrowth faces, by contrast, appear more stable and resistant to corrosion.

Isolated sand grains, examined from acid dissolved dolomite cemented sandstones, are characterized by completely random and irregular surface features and grain outlines. Crystal faces are generally lacking and no regular pattern of etching or corrosion can be recognized (Fig. 8G, H). Grain surfaces are typically covered with abundant irregular depressions and ridges which often exhibit a stepped appearance.

As a control on the acid dissolution process quartz grains from porous Triassic sandstones exhibiting no corrosion textures in thin section or with the S.E.M., were bathed in dilute HCl for 5, 10 and 30 min, and examined with the S.E.M. following each immersion. There was no noticeable grain corrosion even after prolonged immersion.

INTERPRETATION OF THE DIAGENETIC TEXTURE AND POROSITY SYSTEM

The textures observed on detrital grain surfaces and seen to modify overgrowths in the Sherwood Sandstone and Ravenscar Groups result from the dissolution of replacive authigenic carbonates. By comparison with cemented intervals of the sandstones in the studied sections, the Sherwood Sandstone is considered to have been extensively cemented with calcite in the Marchwood Borehole and dolomite in the Irish Sea Wells, whilst the Ravenscar Group is considered to have been extensively cemented with siderite. These carbonate cements are considered to have been precipitated subsequent to an early quartz cement and to have peripherally replaced both overgrowths and detrital grain surfaces. In the Sherwood Sandstone, although carbonate-quartz overgrowth relationships are complex and textural evidence suggests that more than one phase of quartz and carbonate cementation may have taken place, it is clear that carbonates post-date early quartz authigenesis. Thin section textural evidence also indicates that siderite in the Ravenscar Group post-dates quartz authigenesis. Even in sandstones totally cemented with siderite, partial replacement of quartz over-

Fig. 7. Siderite cement-dissolution porosity textures.

(A) Pervasive pore-filling and grain replacing sphaerosiderite cement. Quartz overgrowths (og) are extensively replaced, whilst siderite occurs penetrating through the overgrowths and into the underlying grains. Photomicrograph, PPL, Ravenscar Group, outcrop. Scale bar 125 μm.

(B) Oxidized relics after partially dissolved rhombic siderite cement highlights the intense peripheral grain replacement (r). The siderite corrosion is irregular but non-specific and attacks both faces of low and high surface free energy. Embayments occur in several of the grains. Note the large, oversized pores (P). Photomicrograph, PPL, Ravenscar Group, outcrop. Scale bar 125 μm.

(C) Partially dissolved relics of siderite rhombs replacing authigenic quartz overgrowths (og). Where the siderite has been removed the surface of the quartz grain is irregularly embayed. Scanning electron micrograph, Ravenscar Group, outcrop. Scale bar 20 μm.

(D) Detail of *in situ* partially dissolved siderite rhombs (s) replacing quartz overgrowth (og). In this example the replacive siderite is controlling the shape of the resulting texture. Scanning electron micrograph, Ravenscar Group, outcrop. Scale bar 10 μm.

(E) Deep embayments in quartz overgrowths indicate of the former presence of pore filling and replacive siderite cement. The embayments clearly penetrate through the overgrowth into the underlying grain. Note the non-specific quartz overgrowth corrosion (C). Photomicrograph, PPL, Ravenscar Group, outcrop. Scale bar 70 μm.

(F) Deeply corroded embayment (C) passing through quartz overgrowth into the grain beneath. Distribution of embayments is extremely irregular, occurring on both detrital grain surfaces and overgrowth faces of low surface free energy. Photomicrograph, XP, Ravenscar Group, outcrop. Scale bar 80 μm.

(G) Extensive non-specific depression in quartz overgrowth and detrital grain indicative of siderite cement-dissolution porosity. Replacement has preferentially attacked the underlying grain leaving behind upstanding surfaces of overgrowths. Scanning electron micrograph, Ravenscar Group, outcrop. Scale bar 10 μm.

(H) Large corrosion embayments in quartz overgrowth. Scanning electron micrograph, Ravenscar Group, outcrop. Scale bar 20 μm.

og
og
A
r
r
P
B
og
C
S
og
S
D
C
E
C
F
G
H

growths is observed and it is conceptually difficult to infer siderite precipitation prior to quartz overgrowth cementation.

Corrosion features observed in highly porous sandstones from the Sherwood Sandstone and Ravenscar Groups are similar to surface textures present on isolated quartz grains revealed after artificial removal of carbonate cements. This supports the interpretation that the corrosion features result from the dissolution during diagenesis of replacive carbonates. Therefore, at least part of the present porosity system in these sandstones is secondary in origin. This porosity has been created by the removal of pore-filling and grain-replacive authigenic carbonates, and has the enlarged intergranular pore texture of Schmidt & McDonald (1979b). It is not, of course, possible to prove that all the present porosity in these sandstones is secondary in origin. The initial distribution of pore-filling carbonates may have been heterogeneous, with some relic primary intergranular porosity remaining in the cemented sandstones. Indeed, such relic primary intergranular porosity is probably important for introducing carbonate undersaturated solutions into the sandstones. The present porosity system in these sandstones is thus of hybrid origin. It comprises a combination of relic and rejuvenated primary intergranular porosity enhanced by the dissolution of replacive carbonate at the margins of adjacent grains and overgrowths. Moreover, there is clearly considerable variation in the morphology of the observed surface corrosion features. Those textures observed where calcite has been interpreted to have been removed are characterized by an anastomosing pattern of ridges and prismatic depressions. These are distinctly different from the more irregular, pervasive corrosion and deep embayments that are typical of surfaces where dolomite and siderite have been removed. Identification of the textures described herein allows the former presence of an authigenic intergranular cement that was peripherally replacive with respect to the silicate framework to be recognized but does not prove unequivocally that *all* the sandstone was previously cemented.

DISCUSSION

Carbonate replacement textures

Hurst (1981) proposed that quartz grains and their overgrowths are subject to selective dissolution during calcite replacement and erected a hierarchy of grain surface characteristics susceptible to dissolution based on surface-free energy. Surfaces of high-free energy such as disturbed surface layers, fractures or abraded detrital grain surfaces, are preferentially dissolved before surfaces of low-free energy such as authigenic overgrowth crystal faces. Such relationships are commonly observed in the calcite cemented Sherwood Sandstone and in those sandstones where calcite cements are considered to have been formerly present (Fig. 9). However, petrographical study suggests that both dolomite and siderite extensively corrode low

Fig. 8. Surface textures of quartz and feldspar grains after artificial removal of calcite and dolomite cements by acid dissolution.
(A) Detrital quartz grain after calcite cement removal covered with an interlocking, anastomising network of upstanding ridges which form a prismatic box-work texture. Scanning electron micrograph, Marchwood Borehole, 1687 m. Scale bar 25 μm.
(B) Detail of grain surface showing upstanding ridges. Note the straight edges to many of the ridges. Scanning electron micrograph, Marchwood Borehole, 1683 m. Scale bar 10 μm.
(C) Detrital feldspar grain from calcite cemented sandstone illustrating intense surface corrosion. Scanning electron micrograph, Marchwood Borehole, 1683 m. Scale bar 60 μm.
(D) Enlargement of surface in (C). Calcite crystals are interpreted to have precipitated on the lower part of the grain and grown towards the top dissolving the overgrowth (og) by a 'thin film' mechanism. Upstanding ridges of relic overgrowth are present only where calcite crystals did not coalesce. Scanning electron micrograph, Marchwood Borehole, 1683 m. Scale bar 25 μm.
(E) Detail of upstanding ridge (u), corrosion front (cf) and grain surface (gs). Both overgrowth and grain surface exhibit orientated corrosion textures. Upstanding ridge is 2–3 μm in height. Scanning electron micrograph, Marchwood Borehole, 1683 m. Scale bar 10 μm.
(F) Detail of orientated corrosion surface of feldspar overgrowth. Scanning electron micrograph, Marchwood Borehole, 1683. Scale bar 6 μm.
(G) Highly corroded quartz grain after dolomite cement dissolution showing irregular, angular corrosion texture. Note the stepped appearance of the dissolution features. Scanning electron micrograph, 110/3–3, 1107 m. Scale bar 25 μm.
(H) Detail of surface corrosion. Quartz grain surface is covered with irregular dissolution pits and relic ridges. Crystal outlines cannot be recognized. Scanning electron micrograph, 110/3–3, 1107 m. Scale bar 10 μm.

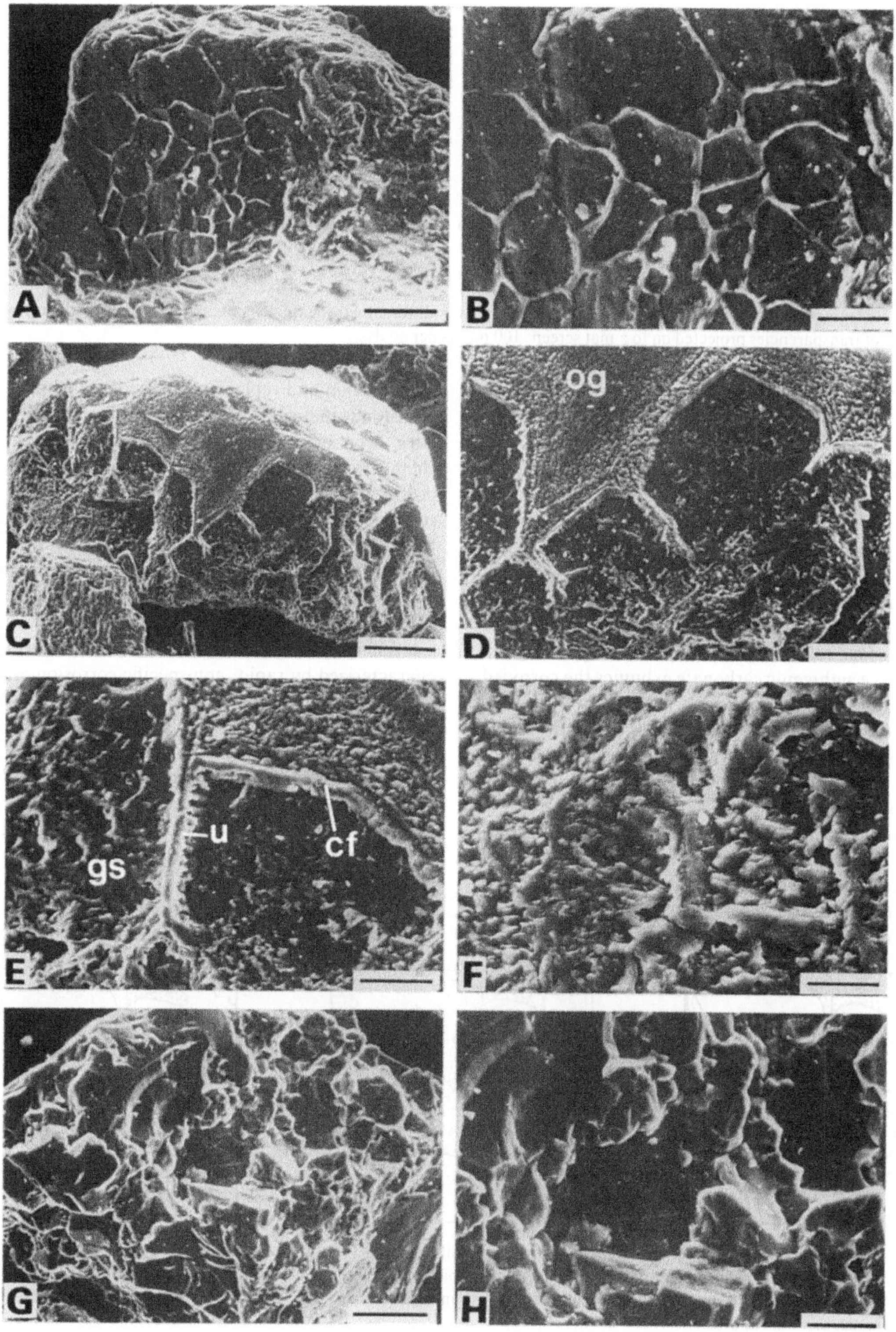
A
B
og
C
D
u
cf
gs
E
F
G
H

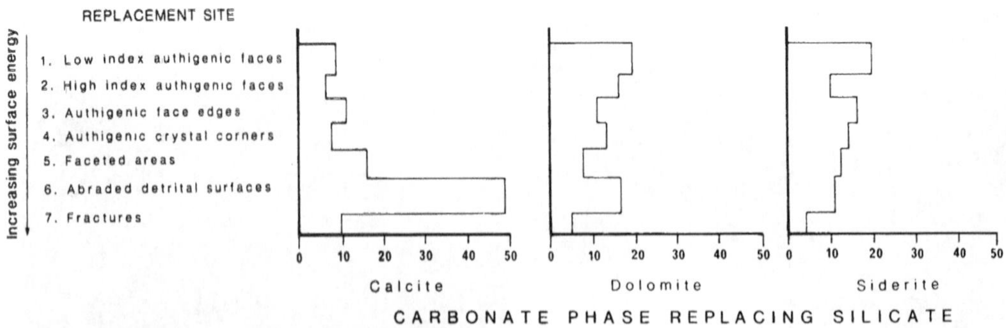

Fig. 9. Histograms showing distribution of replacement sites for different carbonate cements based on point count modal analysis of transparencies projected on to a mat screen: 100 points per analysis.

surface-free energy authigenic crystal faces and do not appear to be specific in the location of corrosion sites (Fig. 9).

The etching, embayment and partial replacement of quartz grains during carbonate precipitation is well known (Kashik, 1965; Blatt, Middleton & Murray, 1972; Friedman, Ali & Krinsley, 1976). Carbonates precipitated in sandstones, after quartz overgrowth authigenesis, are also commonly corrosive with respect to both the detrital grains and authigenic overgrowths. Subsequent removal of such pore-filling and grain replacive authigenic carbonate exhumes the corroded quartz grain and overgrowth topography (Fig. 10). Textures developed on the surfaces of quartz grains and overgrowths should, therefore, reflect the conditions in which dissolution was produced during carbonate precipitation.

The textures described from the Sherwood Sandstone and Ravenscar Groups provide evidence for the recognition of cement-dissolution porosity. These textures may also record the specific geochemical environment in which they formed during carbonate precipitation and replacement. Evidence from these sandstones indicates that the style of corrosion must have been determined by the specific pore water chemistry during reaction, all the minerals involved being carbonates. Following Berner (1978, 1980) two styles of dissolution reaction may take place, although Morse (1983) believes the processes may operate together. Dissolution reactions are either controlled by the rate of transport of ions to and away from the reacting surface, or by the reaction rate at the solid–solution interface. Transport controlled dissolution is characterized by rapid, non-specific corrosion and is typical of strongly concentrated solutions or highly soluble minerals. It gives rise to intense etching and corrosion of all the available sites. Surface reaction controlled dissolution, by contrast, is generally slow and more specific, being typical of slow dissolution of relatively insoluble minerals in solutions of low chemical reactivity. Surface reaction controlled dissolution thus tends to produce distinct crystallographically controlled features such as well defined notches.

Regular, straight edged shallow corrosion features

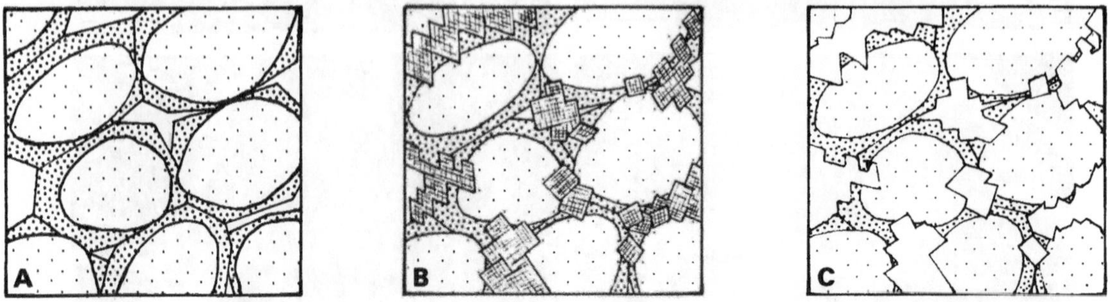

Fig. 10. Schematic development of corrosion textures on detrital grains and overgrowths. (A) Overgrowths enclosing detrital grains. (B) Extensive cementation of part of remaining pore space by pore-filling and grain replacive carbonate cement. (C) Dissolution of carbonate cement by aggressive pore-waters exhumes the corroded silicate framework topography and generates intergranular secondary porosity.

produced by calcite replacement of quartz grains and overgrowths are probably the result of surface reaction controlled dissolution. Dissolution appears to initiate through the development of small oriented 'v'-shaped notches and pits. These features may affect both overgrowths and detrital grain surfaces and appear to be crystallographically controlled by the dissolving mineral. Controlled dissolution results in the formation of larger etch pits which eventually coalesce until only upstanding ridges of the original grain surface and relic overgrowths remain. Such ridges are thought to represent the boundaries between individual calcite crystals which replaced the original quartz.

Corrosion during dolomite precipitation appears to be non-specific and readily strips quartz overgrowths almost completely from detrital grains. Large rhomb-shaped notches and isolated deep embayments, which penetrate through the overgrowth into the underlying detrital grain, were only observed with siderite corrosion. This irregular, more pervasive dissolution of the silicate framework associated with authigenic dolomite and siderite is more likely to be the result of transport controlled dissolution from chemically highly aggressive pore waters. The rhomb-shaped notches associated with siderite corrosion probably represent the initial stage of replacement. This is followed by more pervasive corrosion as the individual notches coalesce. In extreme cases of corrosion, orientated pits or notches are generally lacking and the shape of former carbonate crystal outlines can be recognized only rarely, corroded surfaces being characterized by an irregular network of dissolution pits and relic ridges.

The nature of authigenic mineral precipitation and continued crystal growth depends largely on the degree of pore water saturation (Berner, 1980). Strongly supersaturated solutions are required before seed crystal nucleation can occur. However, once nucleation is initiated, many nuclei are likely to form from highly supersaturated solutions. In contrast, crystal nucleation from sparingly saturated solutions is difficult, but once a crystal nucleus exists, growth on this nucleus is more likely than further nucleation.

In the sandstones described here, there is a relationship between the inferred style of quartz dissolution and both the type and texture of authigenic carbonates precipitated. Coarse, sparry or large poikilotopic calcite crystals are associated with surface reaction controlled dissolution. Numerous smaller crystals of both dolomite and siderite (and, to a lesser extent, calcite) are, in contrast, associated with transport controlled dissolution. There is also an apparent relationship between type of carbonate crystal morphology and the inferred pore water conditions in which the carbonates precipitated. Of course, crystal shape is a function of all the factors which influence nucleation and crystal growth. These include pressure, temperature and flow rate as well as pore water composition itself. Nonetheless, according to Folk (1974), large poikilotopic crystals (and presumably coarse, sparry cements) only form from sparingly supersaturated solutions. This is precisely the condition in which the crystallography of the dissolving mineral controls the resultant dissolution textures (Berner, 1980). More strongly supersaturated solutions may be invoked to explain the authigenesis of abundant, much smaller rhombs of dolomite and siderite. In such pore waters, reacting ions are concentrated at the pore walls and the crystallography of the precipitating mineral controls the resultant dissolution textures.

In the Ravenscar and Sherwood Sandstone Group sandstones, we recognize the widespread effects of both transport and surface reaction controlled dissolution. The observed relationships between the style of quartz dissolution and the conditions of carbonate precipitation suggest that the crystallography of the dissolving quartz grain or overgrowth only influences the style of dissolution when the pore water saturation is low. In strongly supersaturated solutions, the solution itself contains sufficient free energy to overcome bonds within the dissolving mineral. Consequently, the crystallography and free energy distribution of the dissolving mineral becomes subordinate as a control on the style of dissolution to the nature of the solution itself. The crystal shape of the precipitating carbonate is thus manifested as the surface texture of the dissolving mineral. Whilst this relationship holds good for stable silica minerals, such as quartz, with little free-energy anisotropy, more unstable silicate minerals, such as feldspar, with high free-energy anisotropy (cleavage directions, etc.), can be expected to exert more crystallographic influence on the dissolution texture regardless of pore-water composition.

Recognition of cement-dissolution porosity

Despite detailed and careful description of the textures characteristic of intergranular secondary porosity (Schmidt & McDonald, 1979b) there is still considerable debate over the recognition of secondary porosity created by the removal of intergranular cements. There is indeed, a lack of papers subsequently documenting examples of such intergranular secondary porosity (although see McDonald & Surdam, 1984). Probably one of the main reasons for the continued dispute is that cement-dissolution porosity typically closely mimics primary intergranular porosity. However, authigenic carbonates are generally corrosive with respect to the stable silicate framework of sandstones and will therefore, on dissolution, leave behind a record of their former presence. We emphasize that the textures described herein are observed to modify quartz overgrowths precipitated during diagenesis.

Our observations indicate that when the size of carbonate crystals precipitated around and partially replacing the silicate framework is significantly smaller than the diameter of the detrital grains, distinctive corrosion features are formed. As a result, the corrosive growth of carbonate crystals, presumably by a 'thin film' mechanism (Pettijohn, Potter & Siever, 1972), produces a replica of the carbonate crystals imprinted into the silicate mineral being replaced. If the corrosive carbonate mineral is precipitated as coarsely crystalline spar, or as large poikilotopic crystals, the resulting surface texture is likely to be less distinctive, and extreme care may be needed to recognize such corrosion textures, if indeed, they are present at all.

We therefore strongly advocate the routine use of the S.E.M., in conjunction with standard petrographical techniques, for recognizing and studying the textural relationships and distribution of secondary cement-dissolution porosity. Moreover, in supplementing the petrographical criteria detailed by Schmidt *et al.* (1977) and Schmidt & McDonald (1979b), we suggest that textures visible with the S.E.M. provide more convincing evidence for the presence of secondary cement-dissolution porosity (Fig. 11).

We have examined a wide variety of North Sea reservoir sandstones from Permian to Jurassic age and find widespread evidence for secondary cement-dissolution porosity. Moreover, examination of published S.E.M. photomicrographs of grain and overgrowth surfaces indicates that the textures we describe are present in sandstones of Palaeozoic to Mesozoic sandstones from diverse localities. The significance of these textures has not been realized or they have been interpreted differently (see, e.g. Antia & Whitaker, 1978; Werle & Schneider, 1978).

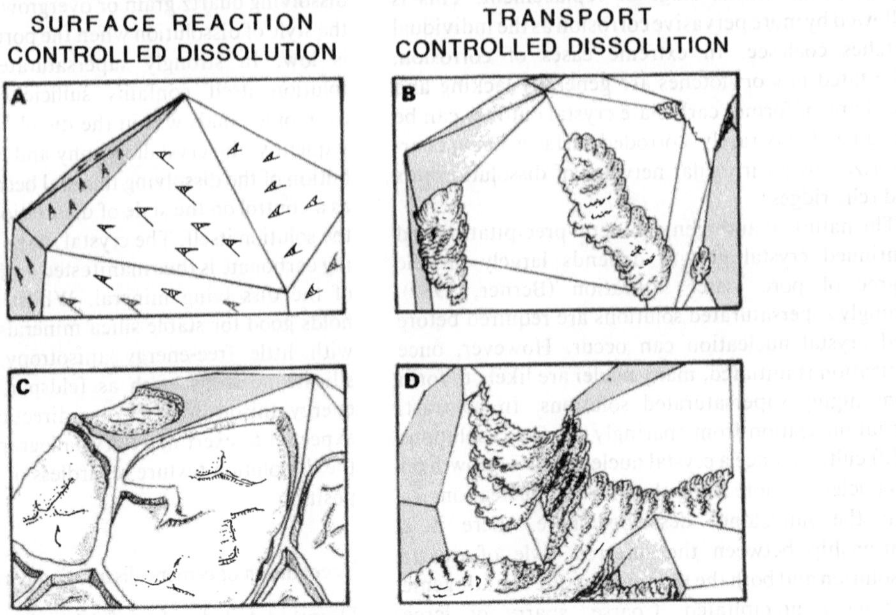

Fig. 11. Surface textural criteria for the recognition of cement dissolution porosity with the scanning electron microscope. (A) Regular, orientated notches in overgrowths. (B) Irregular pits and embayments developed initially at loci of high free surface energy. (C) Regular shaped embayments and large depressions almost completely stripping overgrowths from grain surface; regular pattern of upstanding ridges may remain. (D) Extensive, highly non-specific stripping of overgrowths producing large irregular depressions.

CONCLUSIONS

A variety of surface replacement textures are present on detrital grains and overgrowth faces in the Sherwood Sandstone and Ravenscar Groups that are interpreted to result from the exhumation of carbonate corrosion features. Notches, deep embayments and extensive corrosion depressions of authigenic overgrowths produced *in situ* during diagenesis must be observed to modify authigenic phases to document unequivocally the presence of secondary cement-dissolution porosity.

These surface textures may also record the specific geochemical environment in which they formed during carbonate precipitation and replacement. Surface reaction controlled dissolution of quartz, where the crystal size of the precipitated carbonate cement is significantly smaller than the detrital grain diameter, produces distinct textures on the surface of the detrital grain or overgrowth. Rapid transport controlled dissolution, from chemically highly reactive solutions or corrosion produced by carbonate crystals larger in size than the detrital grain diameter, produces irregular surface textures which may be more difficult to recognize.

ACKNOWLEDGMENTS

The research which forms the basis of this paper was carried out whilst the authors were in receipt of N.E.R.C. support grants and under the guidance of Dr B. Waugh at the University of Hull, England. However, our ideas have gradually crystallized since moving to new abodes and we are grateful for discussion with colleagues at the Geological Institute, University of Berne and Shell Research Laboratories, Rijswijk. Professor Albert Matter, Karl Ramseyer and Louis Macchi (who observed similar replacement textures in the Permian of western Cumbria) all provided useful stimulating discussion. Volkmar Schmidt and Charles Curtis suggested numerous improvements on the initial version of the manuscript whilst the referees, David Houseknecht and Steve Livera, clarified our argumentation with their critical reading. Thanks are also expressed to the numerous staff at B.G.S. Deep Geology and Hydrogeology Research Groups for their useful comments on early drafts of the manuscript.

British Gas and B.G.S. kindly made available material from the Irish Sea and Marchwood Boreholes respectively, and along with Shell International Petroleum Company, are thanked for permission to publish this paper. SDB gratefully acknowledges financial support from the Swiss National Foundation (project number 2.913–0.83). Andy Werthemann produced the photographic plates at the University of Berne, and Regina Gubler very kindly typed (and retyped) the manuscript.

REFERENCES

ANTIA, D.D.J. & WHITAKER, J.H. McD. (1978) A scanning electron microscope study of the genesis of the Upper Silurian Ludlow Bone Bed. In: *Scanning Electron Microscopy in the Study of Sediments* (Ed. by W. B. Whalley), pp. 119–136. Geo Abstracts, Norwich.

AUDLEY-CHARLES, M.G. (1970) Stratigraphical correlation of the Triassic rocks of the British Isles. *J. geol. Soc. London*, **126**, 19–47.

BARNARD, P. & COOPER, B.S.A. (1983) A review of geochemical data related to the North West European Gas Province. In: *Petroleum Geochemistry* (Ed. by J. Brooks). *Spec. Publ. geol. Soc. London*, **12**, 19–34. Blackwell Scientific Publications, Oxford.

BERNER, R.A. (1978) Rate control of mineral dissolution under earth surface conditions. *Am. J. Sci.* **278**, 1235–1252.

BERNER, R.A. (1980) *Early Diagenesis: a Theoretical Approach*. Princeton University Press, New Jersey.

BJØRLYKKE, K. (1981) Diagenetic reactions in sandstones. In: *Sediment Diagenesis* (Ed. by A. Parker and B. W. Sellwood). *NATO ASI, Series C, math. Phys. Sci.* **115**, 169–214.

BJØRLYKKE, K. (1984) Formation of secondary porosity: how important is it? In: *Clastic Diagenesis* (Ed. by D. A. McDonald and R. C. Surdam). *Mem. Am. Ass. Petrol. Geol.* **37**, 277–286.

BLATT, H., MIDDLETON, G. & MURRAY, R. (1972) *Origin of Sedimentary Rocks*. Prentice-Hall, New Jersey.

BURLEY, S.D. (1984) Patterns of diagenesis in the Sherwood Sandstone Group (Triassic), United Kingdom. *Clay Miner.* **19**, 403–440.

COLTER, V.S. (1978) Exploration for gas in the Irish Sea. In: *Key Notes for the MEGS* (Ed. by A. J. van Loon). *Geologie Mijnb.* **51**, 503–516. Amsterdam.

DICKSON, J.A.D. (1965) A modified staining technique for carbonates in thin section. *Nature*, **205**, 587.

EBBERN, J. (1981) The geology of the Morecambe Gas Field. In: *Petroleum Geology of the Continental Shelf NW Europe* (Ed. by L. V. Illing and G. D. Hobson), pp. 485–493. Institute of Petroleum, London.

FOLK, R.L. (1974) The natural history of crystalline calcium carbonate: effect of magnesium content and salinity. *J. sedim. Petrol.* **44**, 40–53.

FRIEDMAN, G.M., ALI, S.A. & KRINSLEY, D.A. (1976) Dissolution of quartz accompanying carbonate precipitation and cementation in reefs: example from the Red Sea. *J. sedim. Petrol.* **46**, 920–913.

GILES, M.R. & MARSHALL, J.D. (1983) Processes controlling secondary porosity development (abstract). In: *Patterns of mineral diagenesis on the N.W. European Shelf and their relations to facies and hydrocarbon accumulations.* 1983 Mineral Society Meeting, Cambridge.

HEMINGWAY, J.E. (1974) Jurassic. In: *The Geology and Mineral Resources of Yorkshire* (Ed. by D. H. Rayner and J. E. Hemingway), pp. 161–224. Yorkshire Geological Society.

HURST, A.R. (1981) A scale of dissolution for quartz and its implications for diagenetic processes in sandstones. *Sedimentology*, **28**, 451–460.

KANTOROWICZ, J.D. (1985) The petrology and diagenesis of Middle Jurassic clastic sediments Ravenscar Group, Yorkshire. *Sedimentology*, **32**, 833–853.

KASHIK, S.A. (1965) Replacement of quartz by calcite in sedimentary rocks. *Geochem. Inst.* **2**, 133–138.

KENT, P.E. (1980) Subsidence and uplift in East Yorkshire and Lincolnshire: a double inversion. *Proc. Yorks. geol. Soc.* **42**, 505–524.

KNOX, R.W.O'B., BURGESS, W., WILSON, K.S. & BATH, A. (1984) Petrological controls on porosity and permeability in Triassic Sandstones. *Clay Miner.* **19**, 441–456.

KRINSLEY, D.H. & DOORNKAMPF, J.L. (1973) *Atlas of Sand Surface Textures.* Cambridge University Press, 91 pp.

LE RIBAULT, L. (1977) *L'exoscopie des Quartz.* Masson, Paris, 150 pp.

LIVERA, S.E. & LEEDER, M.R. (1981) The Middle Jurassic Ravenscar Group ('Deltaic Series') of Yorkshire: recent sedimentological studies as demonstrated during a recent field meeting, 2–3 May 1980. *Proc. geol. Ass.* **92**, 241–250.

MCDONALD, D.A. & SURDAM, R.C. (Eds) (1984) Clastic diagenesis. *Mem. Am. Ass. Petrol. Geol.* **37**, 434 pp.

MORSE, J.W. (1983) The kinetics of calcium carbonate dissolution and precipitation. In: *Carbonates: Mineralogy and Chemistry* (Ed. by R. J. Reeder). *Rev. Mineral. Min. Soc. Am.* **11**, 227–264.

PATTISON, J., SMITH, D.B. & WARRINGTON, G. (1983) A review of late Permian and early Triassic biostratigraphy in the British Isles. In: *The Permian and Triassic Systems and their Mutual Boundaries* (Ed. by A. Logan and L. V. Hills). *Mem. Can. Soc. Petrol. Geol.* **2**, 220–260.

PETTIJOHN, F.J., POTTER, P.E. & SIEVER, R. (1972) *Sand and Sandstone.* Springer-Verlag, Berlin, 618 pp.

SCHMIDT, V. & MCDONALD, D.A. (1979a) The Role of Secondary Porosity in the course of sandstone diagenesis. In: *Aspects of Diagenesis* (Ed. by P. A. Scholle and P. R. Schluger). *Spec. Publ. Soc. econ. Paleont. Miner., Tulsa, No. 26*, 175–208.

SCHMIDT, V. & MCDONALD, D.A. (1979b) Texture and recognition of secondary porosity in sandstones. In: *Aspects of Diagenesis* (Ed. by P. A. Scholle and P. R. Schluger). *Spec. Publ. Soc. econ. Paleont. Miner., Tulsa, No. 26*, 209–225.

SCHMIDT, V., MCDONALD, D.A. & PLATT, R.L. (1977) Pore geometry and reservoir aspects of secondary porosity in sandstones. *Bull. Can. Petrol. Geol.* **25**, 271–290.

SOMMER, F. (1978) Diagenesis of Jurassic Sandstones in the Viking Graben. *J. geol. Soc. London*, **135**, 63–68.

STONELEY, R. (1982) The structural development of the Wessex Basin. *J. geol. Soc. London*, **139**, 545–554.

THOMPSON, D.B. (1969) Dome shaped aeolian dunes in the Frodsham Member of the so called 'Keuper' Sandstone Formation (Scythian-Ansian: Triassic) at Frodsham, Cheshire, England. *Sedim. Geol.* **3**, 263–289.

THOMPSON, D.B. (1970) Sedimentation of the Triassic (Scythian) Red Pebbly Sandstones in the Cheshire Basin and its margins. *Geol. J.* **7**, 183–215.

THOMPSON, D.B. & STEEL, R. (1983) Structures and Textures in the Triassic (Scythian?) braided stream conglomerates ('Bunter' Pebble Beds) in the Sherwood Sandstone Group, North Staffordshire. *Sedimentology*, **30**, 341–368.

WALKER, T.R., WAUGH, B. & CRONE, A.J. (1978) Diagenesis in the first cycle desert alluvium of Cenozoic age, southwestern U.S.A. and N.W. Mexico. *Bull. geol. Soc. Am.* **89**, 19–32.

WERLE, B. & SCHNEIDER, H.E. (1978) Scanning Electron Microscope observations of diagenesis in the Triassic sandstones of the Saar area, Germany. In: *Scanning Electron Microscopy in the Study of Sediment* (Ed. by W. B. Whalley), pp. 355–362. Geo Abstracts, Norwich.

(*Manuscript received 15 February 1985; revision received 19 July 1985*)

Reprinted from Sedimentology (1992) **39**, *99–107*

A numerical model for porosity modification at a sandstone–mudstone boundary by quartz pressure dissolution and diffusive mass transfer

ANDREW M. MULLIS

Department of Geology and Applied Geology, Lilybank Gardens, University of Glasgow, Glasgow G12 8QQ, UK

ABSTRACT

In many sandstones quartz cements significantly reduce porosity. The origin of these cements is often unclear. This paper investigates a possible mechanism for the generation of silica for quartz cement by pressure dissolution in interbedded mudstones.

Theoretical models of quartz pressure dissolution, including the effects of silica precipitation kinetics, show that the concentration of dissolved silica in the pore fluids of a compacting sediment increases with decreasing grain size and silica precipitation constant. Quartz precipitation is strongly inhibited by the presence of small amounts of clay within a sediment, suggesting that siltstones and quartz-rich mudstones which are undergoing pressure dissolution may act as a source of dissolved silica for export to nearby, coarser sediments.

A computational model for the diagenetic modification of a sandstone–mudstone interface due to pressure dissolution is described. Both sandstone and mudstone layers are assumed to be actively compacting by pressure dissolution, and mass transport by molecular diffusion is considered. As quartz precipitation in the mudstone layer is relatively slow compared to that in the sandstone, significant amounts of dissolved silica become available in the mudstone, and may be exported into the adjacent sandstone. In the absence of pore-fluid advection, this may result in the formation of extensive secondary quartz within the sandstone, close to the interface.

The volume of silica exported from the mudstone is limited by the length scale over which diffusion through the mudstone is effective. This is typically 3–5 m. The volume of silica available therefore suggests that extensive porosity modification within the adjacent sandstone can only occur close to the mudstone. Thus it is possible that thin sandstones could become cemented by slow diffusive transfer of silica, but that in thicker sandstones the silica may become dispersed by pore-fluid advection.

INTRODUCTION

Many sandstone formations contain quartz cements in quantities which are unlikely to have been derived locally from within the sandstone, either by pressure dissolution, or dissolution of detrital silicates such as K-feldspar. The most likely sources of this imported silica are adjacent or interbedded siltstones and mudstones (Land & Dutton, 1978; McBride, 1989). Direct evidence for this silica export from mudstones is provided by the excess quartz cementation found at the bed margin of some sequences, for example that reported by Füchtbauer (1983) for the Dogger Beta Formation of northern Germany. One source of mobile silica within mudstones is diagenetic reactions among clay minerals, such as the illitization of smectite (Boles & Franks, 1979). However, petrographical and geochemical analysis of Ordovician mudstones in Morocco by Evans (1989) suggest that pressure dissolution of quartz may also be an important source of dissolved silica within mudstones. By analysis of the bulk composition of mudstones within and surrounding concretions, Evans (1989) established that the mudstones may have exported 10–20% of their detrital quartz to adjacent sandstones. In this paper a computational model of an idealized sandstone–mudstone boundary will be used to demonstrate that the transport of silica from a siltstone or mudstone by quartz pressure dissolution is a natural consequence of the smaller grain size and the inhibition of silica precipitation within the mudstone.

Intergranular pressure dissolution, together with grain reorientation and mechanical deformation, is one of the principal mechanisms of clastic sediment compaction (Bjorlykke, 1983). In pressure dissolution, concentration of stress at grain-to-grain boundaries normal to lithostatic loading leads to increased solubility of material at the grain boundary. The most widely accepted model of pressure dissolution holds that this sets up a concentration gradient across the grain boundary, and that dissolved material is expelled by grain boundary diffusion (Weyl, 1959; Rutter, 1976). In order to facilitate reasonable diffusion rates at the low temperatures at which pressure dissolution is observed to operate, it is normally assumed that diffusion occurs through a thin water layer adsorbed onto the grain boundary (Rutter, 1976). Nevertheless, it is normally assumed within this model that grain boundary diffusion is the rate limiting process. A closely related model proposed by Raj & Chyung (1981) holds that the existence of a continuous thin aqueous film between two stressed surfaces is unlikely for materials with an aqueous wetting angle greater than zero. Instead they propose that the stress is supported by contacts between the two solids, with fluid at normal pressure occupying channels within the grain boundary, which provide a diffusion path away from the sites of dissolution. This is the so called 'island channel' model. The alternative to the grain boundary diffusion models (whether through an adsorbed thin film or through channels) is the undercutting model of Bathurst (1958), and its subsequent modifications. This model suggests that increased solubility at the grain contact leads to preferential dissolution at the rim of the grain-to-grain contact. This results in the grain contact undercutting the overlying grain, resulting in the eventual collapse of the grain contact. Modifications to the theory by Pharr & Ashby (1983) and Tada *et al.* (1987) suggest that the grain contact may deform plastically, rather than by brittle failure.

Numerous experiments have been carried in an attempt to elucidate the pressure dissolution mechanism. Recent work by Hickman & Evans (1991) on the compaction of halite lenses on flat silica surfaces in saturated brine found no evidence of island channel structures. Compaction experiments on quartz sand by Schutjens (1991) found that below 250°C compaction occurred by intergranular cracking, whereas above 300°C compaction occurred by pressure dissolution via a grain boundary diffusion mechanism. No evidence of undercutting was found. However, little petrographical evidence is available to suggest that natural sediments compact by intergranular microcracking, whereas there is a wealth of documented evidence of pressure dissolution features (e.g. concavo-convex grain contacts, sutured grain boundaries) in sediments which have never experienced temperatures in excess of 250°C. Consequently, it is suggested that the microcracking observed by Schutjens (1991) may possibly be the result of the rapid loading required to obtain measurable compaction in a reasonable time.

It has been assumed in this work that pressure dissolution proceeds via grain boundary diffusion through an adsorbed aqueous film. However, as the thermodynamic driving force is the same for all the pressure dissolution models discussed, the results presented are expected to be at least qualitatively similar if an alternative mechanism for solution transfer were assumed.

In a previous paper (Mullis, 1991) the effect of silica precipitation kinetics on the rate of intergranular pressure dissolution within a quartz sandstone was investigated. It was shown that grain size, a, and the precipitation constant for silica, k_-, can significantly affect both the rate of pressure dissolution and the concentration of dissolved silica in the interstitial pore fluids. In particular, it was shown that as the grain size and precipitation constant are decreased, the concentration of dissolved silica in the pore spaces, c', increases from its saturation value, c_0, determined by the pressure and temperature of the pore fluids, towards some higher limiting value determined by the increased solubility of quartz under the applied differential stress of lithostatic loading.

It was shown by Mullis (1991), using laboratory estimates of k_-, that for clean quartz sandstone, precipitation kinetics are unlikely to affect significantly the rate of quartz pressure dissolution. However, *in situ* estimates of k_- for the Jurassic Dogger Beta Formation of northern Germany made by Mullis (1991), from the data of Füchtbauer (1983), showed that in a geological setting, the value of k_- may be about three orders of magnitude smaller than in the laboratory, and in the range where precipitation kinetics dominate the rate of quartz pressure dissolution. It was suggested that this large difference between k_- for field and laboratory data was possibly due to the action of clays, which have been observed both in the field (Heald, 1956, 1965) and in the laboratory (Cecil & Heald, 1971) to inhibit strongly the precipitation of quartz. In a fine-grained sandstone or siltstone, containing a few per cent clay, we might thus expect k_- to be considerably smaller than in a clean sandstone, and hence to contain higher concentrations of dissolved silica than in a cleaner, medium- or coarse-grained sandstone. Where two such contrasting lithologies are in contact, quartz dissolved in the siltstone may be available for export to the sandstone, either by diffusion or by advection of pore fluids. It is the purpose of this paper to use theoretical models of an idealized sediment to investigate the extent to which silica dissolved in a silstone or mudstone by pressure dissolution can be exported by diffusion into an adjacent sandstone, and the ways in which this might alter the subsequent diagenetic history of the sandstone.

MATHEMATICAL FORMULATION

In Mullis (1991) the rate of pressure dissolution within a sediment of uniformly sized truncated spherical grains of radius a, in a simple cubic packing (see Fig. 1), was considered. It was shown that the rate at

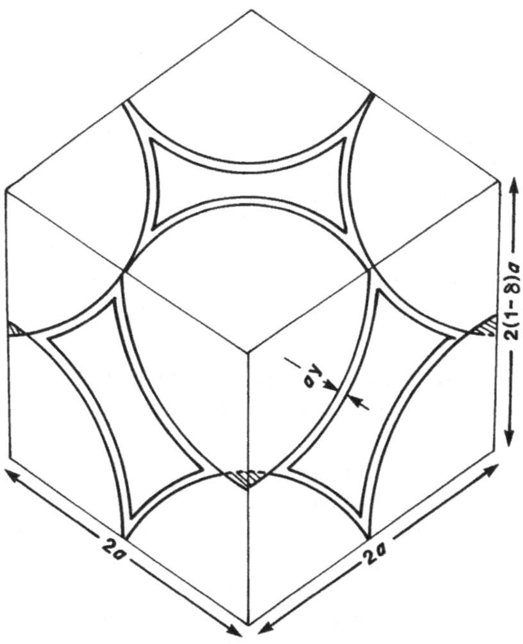

Fig. 1. A unit cell of cubically packed spheres, of radius a, compacting vertically under pressure dissolution by an amount $a\delta$. Material dissolved from the intergranular contacts is redeposited as a concentric rim of width ay on the free faces of the grains (not to scale).

which dissolved silica is supplied to unit volume of the solution, J_+, is

$$J_+ = \frac{\pi h D v c\left[\Sigma - P - \frac{RT}{v}\ln\left(\frac{c'}{c_0}\right)\right]}{a^3\phi RT(1-\delta)}, \qquad (1)$$

where h is the thickness of the adsorbed water layer between compacting grains, D is the diffusion coefficient for silica through an adsorbed water film, v is the molar volume of silica, Σ is the mean lithostatic load, P is the fluid pressure, R is the gas constant, T is the absolute temperature, ϕ is the porosity of the sediment and δ is the fractional compaction of the unit cell in the vertical (z) direction. The rate at which silica is precipitated out of supersaturated solution is given by Rimstidt & Barnes (1980), using first-order kinetics, as

$$J_- = -k'_-(c - c_0), \qquad (2)$$

where k'_- is a local precipitation rate constant given by

$$k'_- = k_-\frac{A/A_0}{M/M_0};$$

A is the wetted surface area of the quartz in the system and M is the mass of water in the system ($A_0 = 1$ m^2 and $M_0 = 1$ kg).

On the assumption that the pore fluids are static, the one-dimensional variation of the concentration of dissolved silica [$c(z,t)$] within a sediment in which a and k_- vary with z will be given by the diffusion equation, which including source and sink terms is

$$\frac{\partial c(z,t)}{\partial t} = \frac{D_0}{\phi}\frac{\partial}{\partial z}\left[\frac{\phi}{\theta^2}\frac{\partial c}{\partial z}\right] + J_+ - J_-, \qquad (3)$$

where D_0 is the diffusion constant for silica in free solution, θ^2 is the tortuosity of the diffusion path and t is time.

Due to the non-linear nature of Eq. (3), general analytical solutions are not available. However, the equation may be solved numerically, using finite difference methods.

Equation (3) has been used to model the diagenetic modification of an idealized two-layer sediment composed of truncated spherical grains in a cubic packing arrangement. Layer 1 ($0 < z \leqslant l_1$) represents a medium-grained sandstone as it receives dissolved silica from layer 2 ($-l_2 \leqslant z < 0$), an interbedded siltstone. The grain sizes have been taken as $a = 1{\cdot}77 \times 10^{-4}$ m for the sandstone, and $a = 2{\cdot}2 \times 10^{-5}$ m for the siltstone, these values being mean grain radii for medium-grained sand and coarse-grained silt as defined by the Udden–Wentworth size classification. The quartz precipitation constant has been taken as a variable in both layers.

The initial conditions for the solution of Eq. (3) are given by assuming no diffusion of material to occur for $t<0$. Under these conditions, the concentration of dissolved silica in each lithology is given by equating J_+ and J_- and solving iteratively for c'. Boundary conditions for the problem have been obtained by assuming $c'(z)$ to be symmetrical about $l_1/2$ for $0<z\leqslant l_1$, and $-l_2/2$ for $-l_2\leqslant z<0$, so that $\partial c/\partial z=0$, at these points. Strictly, this assumes a similar series of layers are present above and below the two under consideration. However, as the effects of the interface rarely extend more than 1–2 m into either layer, this does not affect the generality of the treatment.

The volume of quartz precipitated in unit time Δt is given by

$$V_{\text{dep}} = 8a^3\phi k_- v(c-c_0)\Delta t. \tag{4}$$

It is assumed that this forms a concentric rim of width ay around the existing grain. The surface area of the unit cell is given by

$$A = 4\pi a^2(1+y)^2 - 4\pi a^2(1+y) \times(\delta+y) - 8\pi a^2 y(1+y), \tag{5}$$

where the second term on the right is $8\times 1/4$ the surface area of a spherical cap of radius $a(1+y)$ and height $a(y+\delta)$, and the third term on the right is $16\times 1/4$ the area of a spherical cap of radius $a(1+y)$ and height ay. Combining Eqs (4) and (5), we obtain the change in y, Δy, in time Δt as

$$\Delta y = \frac{2avk_-(c'-c_0)(1-\delta)\phi\Delta t}{\pi(1+y)(1-\delta-2y)}. \tag{6}$$

The strain rate for the compacting sediment under the above assumptions is given by Mullis (1991) as

$$\frac{1}{z}\frac{\partial z}{\partial t} = \frac{4hDcv^2}{a^3RT(1-\delta)y^2}\left[\Sigma - P - \frac{RT}{v}\ln\left(\frac{c}{c_0}\right)\right]. \tag{7}$$

Thus the fractional compaction in time $t+\Delta t$ is given by

$$\delta'' = \delta' + \Delta t\frac{1}{z(1-\delta')}\frac{\mathrm{d}z}{\mathrm{d}t}, \tag{8}$$

where δ' is the fractional compaction at time t.

The porosity of the sediment is calculated from

$$\phi = \frac{V_{\text{cell}} - V_{\text{sol}}}{V_{\text{cell}}}, \tag{9}$$

where V_{cell} is the volume of the unit cell, given by

$$V_{\text{cell}} = 8a^3(1-\delta), \tag{10}$$

and V_{sol} is the volume of solid per unit cell, given by Mullis (1991) as

$$V_{\text{sol}} = \frac{2}{3}\pi a^3(2+6y-6y\delta-3y^2 - 4y^3 - 3\delta^2 + \delta^3 - 3\delta y^2). \tag{11}$$

The tortuosity of the diffusion path in a porous sediment, θ^2, is normally assumed to be related to the porosity via a power law relationship (e.g. Manheim & Waterman, 1974; Ullman & Aller, 1982). Here the variation in the tortuosity of the diffusion path, θ, as compaction proceeds has been modelled as

$$\theta^2 = \theta_0{}^2\left(\frac{\phi}{\phi_0}\right)^m, \tag{12}$$

where $\theta^2{}_0$ is the initial tortuosity and ϕ_0 is the initial porosity of 41·82%; this figure is appropriate for a cubic packing arrangement which has undergone 10% volumetric compaction in the manner described above. Ullman & Aller (1982) suggest that

$$\theta^2 \approx \frac{1}{\phi} \tag{13}$$

is a reasonable model for marine sandstones. Consequently, for the sandstone layer, values of $\theta_0=2{\cdot}5$ and $m=1$ have been adopted. These values are in agreement with the tortuosities measured by Manheim & Waterman (1974) for marine sandstones. For mudstones, Manheim & Waterman (1974) found that θ^2 may be approximately double that of a sandstone. Consequently, values of $\theta_0=5$ and $m=2{\cdot}5$ have been chosen, $m=2{\cdot}5$ satisfying the models of Lerman (1979) and Ullman & Aller (1982) for mudstone.

RESULTS OF THE NUMERICAL INVESTIGATION

The mathematical model has been used to investigate the diagenetic behaviour of a sandstone which is adjacent to a siltstone or quartz-rich mudstone containing a higher concentration of dissolved silica than the equilibrium concentration for the sandstone. As specific examples need to be considered when performing the numerical solution to Eq. (3), all the models discussed below assume conditions appropriate to ≈ 3 km burial. Pore fluid and lithostatic pressure have been estimated from ρgd, where ρ is density, g is acceleration due to gravity and d is depth of burial.

Taking $\rho=1000$ kg m^{-3} for water and $\rho=2650$ kg m^{-3} for the overlying rock column, we have $P=29{\cdot}4$ MPa and $\Sigma=78{\cdot}0$ MPa. Taking the geothermal gradient as 0·035 K m^{-1}, typical of rifted continental crust, and assuming a base temperature of 277 K, we have $T=378$ K. The diffusion constant for silica in sea water is given by Wollast & Garrels (1971) as 10^{-9} m^2 s^{-1} at 298 K. At the elevated temperature considered here, D_0 has been estimated by assuming the diffusion coefficient to be inversely proportional to the kinematic viscosity of water. The diffusion coefficient through an adsorbed water layer has been taken as $D=10^{-5}\,D_0$ (Rutter, 1976) and the thickness of the adsorbed layer has been taken as 3 nm (Tada *et al.*, 1987).

Initially, a model with a pure quartz mudstone is considered in order to investigate the effects of precipitation kinetics on the patterns of cementation that develop in the sandstone in the presence of an abundant supply of silica. However, pure quartz mudstones are rare, so we will then consider a mudstone composed of $\leqslant 50\%$ quartz. As the purpose of this paper is to demonstrate the effects of one process, not to produce a geochemical model of mudstone diagenesis, the non-quartz component of the mudstone has been assumed to be inert.

In all the cases modelled it has been assumed that pressure dissolution occurs between clean quartz grains. However, petrographically it has often been observed that thin clay layers between quartz grains promote pressure dissolution (Heald, 1956, 1965). This is thought to occur because the clays contain multiple adsorbed water layers which act as diffusion pathways away from the grain boundary. This is easily included in the model as an increased value of the product hD. As the effect of clay layers between grains is difficult to quantify, clean quartz contacts have been assumed. This will tend to make the model conservative, underestimating the importance of pressure dissolution as a source of aqueous silica.

Figure 2(a) shows the porosity of the sandstone, as a percentage of its initial porosity, in a 1-m-wide section adjacent to the sandstone–siltstone interface, for values of k_- in the range $2{\cdot}2\times10^{-9}$ s^{-1} to $6{\cdot}7\times10^{-11}$ s^{-1}, corresponding to values of $t_{1/2}$, defined in Mullis (1991) as

$$t_{1/2}=\frac{\ln 2}{k_-}, \tag{14}$$

of 10 and 330 years, respectively. Values of k_- smaller than this have not been considered as they require an

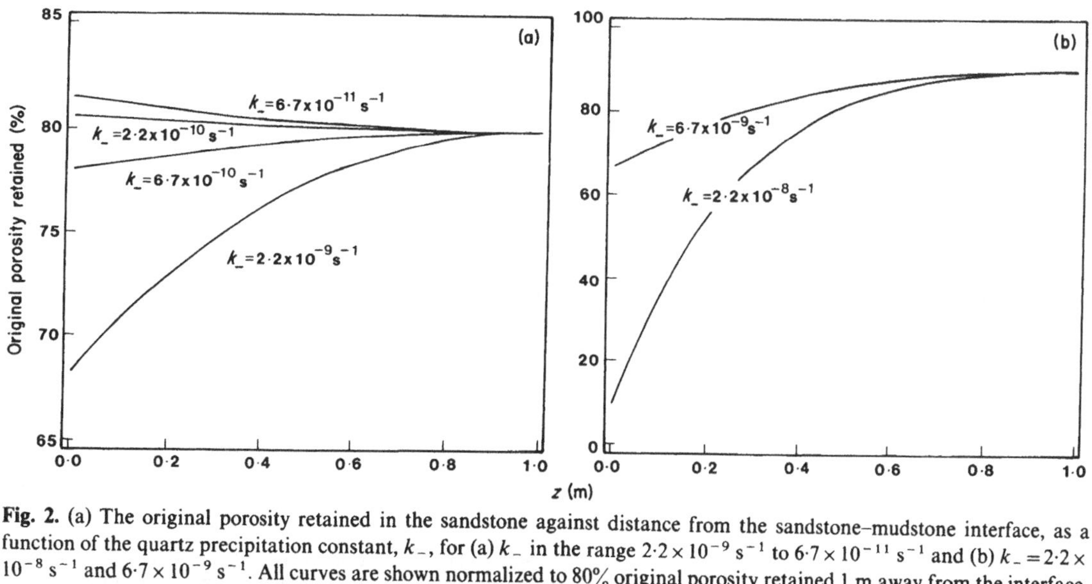

Fig. 2. (a) The original porosity retained in the sandstone against distance from the sandstone–mudstone interface, as a function of the quartz precipitation constant, k_-, for (a) k_- in the range $2{\cdot}2\times10^{-9}$ s^{-1} to $6{\cdot}7\times10^{-11}$ s^{-1} and (b) $k_-=2{\cdot}2\times10^{-8}$ s^{-1} and $6{\cdot}7\times10^{-9}$ s^{-1}. All curves are shown normalized to 80% original porosity retained 1 m away from the interface. The time taken for this degree of modification to take place is not constant with k_-, and the times corresponding to each curve are given in Table 1. Import of silica from the mudstone leads to increased growth of secondary quartz at the interface, but decreases the rate of compaction by pressure dissolution in this region. The balance between these processes, which is determined by the value of k_-, determines whether porosity is enhanced or occluded at the interface. In (b), the high value of k_- used in these models leads to the formation of extensive secondary quartz at the interface. This is particularly marked for $k_-=2{\cdot}2\times10^{-8}$ s^{-1}, where virtually all the original porosity has been destroyed adjacent to the mudstone.

unreasonably long time to produce significant compaction. Behaviour of the sandstone for values of k_- greater than $2 \cdot 2 \times 10^{-9}$ s^{-1} will be discussed below.

In the four models shown in Fig. 2(a) the value of k_- adopted for the siltstone is one-fifth of that assumed for the sandstone. Under these conditions the siltstone behaves as a reservoir of supersaturated aqueous silica, without significant precipitation of quartz occurring in the siltstone. For ease of comparison, all the models shown in Fig. 2(a) have been stopped when the porosity of the sandstone beyond the influence of the siltstone has lost 20% of its original porosity. The time taken for this 20% reduction in porosity is shown in Table 1.

It is apparent from Fig. 2(a) that although all four models under consideration show a net import of silica into the sandstone, the effect of this on porosity modification is dependent on the value of k_-. This difference may be understood with reference to Figs 3 & 4, which show the fractional compaction of the sandstone and the percentage of authigenic quartz in the mudstone, at the times given in Table 1. Increased concentrations of dissolved silica in the pore fluids adjacent to the mudstone promote precipitation of authigenic quartz, but also inhibit self-compaction through pressure dissolution. For high values of k_-, quartz precipitation is the dominant porosity-reducing agent in the sandstone near the sandstone–mudstone boundary, and the porosity of this zone drops below that of the bulk of the sandstone. As k_- decreases, self-compaction becomes more important as a porosity-reducing agent in the import zone, counteracting the effects of enhanced quartz precipitation, and ultimately leading to the porosity of the import zone being maintained above that of the bulk sandstone.

Figure 2(b) shows the predicted evolution of the sandstone–siltstone interface for values of k_- in the range $6 \cdot 7 \times 10^{-9}$ s^{-1} to $2 \cdot 2 \times 10^{-8}$ s^{-1}, higher than the values considered previously for k_-. Transport of silica away from the interface is strongly retarded by

Table 1. Times corresponding to the curves shown in Fig. 2 for 20% (*10%) of the original porosity to be removed, 1 m from the sandstone–mudstone interface, as a function of the quartz precipitation constant, k_-.

k_- (s^{-1})	t (Ma)
$2 \cdot 2 \times 10^{-8}$	2·92*
$6 \cdot 7 \times 10^{-9}$	1·51*
$2 \cdot 2 \times 10^{-9}$	3·82
$6 \cdot 7 \times 10^{-10}$	7·32
$2 \cdot 2 \times 10^{-10}$	18·29
$6 \cdot 7 \times 10^{-11}$	56·18

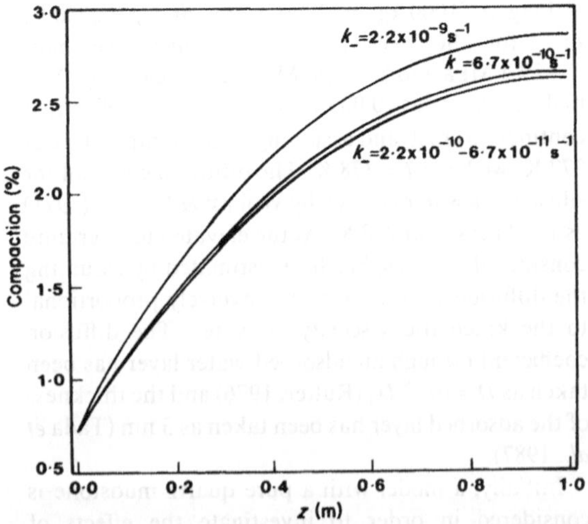

Fig. 3. The linear compaction in the sandstone against distance from the sandstone–mudstone interface, as a function of the quartz precipitation constant, k_-, for k_- in the range $2 \cdot 2 \times 10^{-9}$ s^{-1} to $6 \cdot 7 \times 10^{-11}$ s^{-1}. Curves are shown for times given in Table 1, which are not necessarily the same for each curve. For all values of k_-, compaction is strongly suppressed by the elevated silicia concentration at the sandstone–mudstone interface.

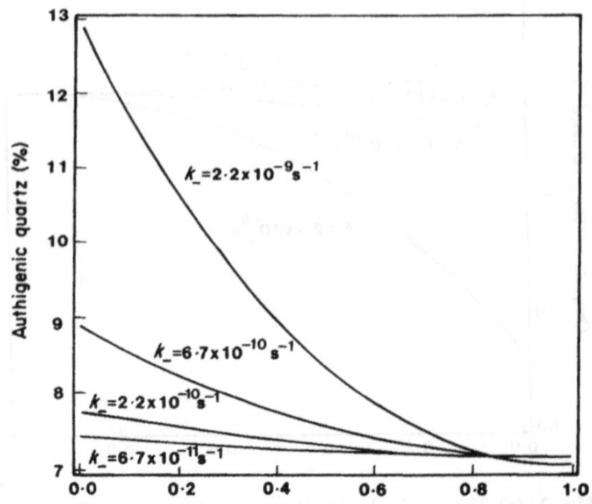

Fig. 4. The amount of secondary quartz in the sandstone against distance from the sandstone–mudstone interface, as a function of the quartz precipitation constant, k_-, for k_- in the range $2 \cdot 2 \times 10^{-9}$ s^{-1} to $6 \cdot 7 \times 10^{-11}$ s^{-1}. Curves are shown for times given in Table 1, which are not necessarily the same for each curve. All models show enhanced secondary quartz growth near the sandstone–mudstone interface, but this is particularly marked as the value of k_- increases.

rapid quartz precipitation, leading to extensive porosity modification in the interface region. This is particularly marked when $k_- = 2{\cdot}2 \times 10^{-8}\ s^{-1}$, where virtually all of the porosity has been lost at the interface.

In the models described above the source region for the dissolved silica has been taken as a pure quartz siltstone, in which k_- is much smaller than in the adjoining sandstone. Under such conditions, compaction of the siltstone is confined to a surface layer a few centimetres thick, which undergoes a large fractional compaction. More realistically, fine-grained quartz may represent $\leqslant 50\%$ of the siltstone and mudstone interval. In the models discussed below it has been assumed that quartz constitutes only a minor fraction, x, of the siltstone and mudstone lithologies. If the quartz is randomly distributed within the mudstone, approximately x^2 of the total rock may be composed of quartz grains in contact, with the potential for pressure dissolution. This fraction may be considerably higher if some grain sorting has occurred, so that the distribution of quartz is not random. The computational model described above is easily modified to account for this change by assuming that each actively dissolving unit cell is embedded in a matrix of $(1-x^2)/x^2$ inert cells. When an active unit cell reaches some predetermined, large amount of compaction, say 85%, it is transferred to inert status. This procedure is adopted purely to prevent the model calculating compaction rates for cells which are almost exhausted of silica, and can considerably increase the speed of execution of the model. It has further been assumed that precipitation of quartz within the mudstone is totally suppressed.

Figure 5 shows the porosity of the sandstone, in the vicinity of the sandstone–mudstone interface, as a percentage of the initial porosity. As before, all results are displayed at times corresponding to equal porosity loss ($\phi/\phi_0 = 80\%$) away from the sandstone–mudstone interface. The time taken for this porosity modification to occur in each case is given in Table 2. In this example $k_- = 2{\cdot}2 \times 10^{-9}\ s^{-1}$ has been adopted for all the models shown, which differ only in the quartz content of the mudstone. For a pure quartz source lithology, dissolution is confined to a thin boundary layer at the interface, and diffusion of silica through the siltstone is not important in restricting the movement of aqueous silica. However, for a siltstone and mudstone interval containing significant fractions of inert material, maintaining a silica concentration gradient across the mudstone is an important constraint on the volume of silica exported from the mudstone. The concentration of dissolved silica across the sandstone and mudstone layers is shown in Fig. 6, as a function of the quartz content of the mudstone which is available for pressure dissolution. Each curve has three distinct regions: a flat plateau, indicating no export of silica from regions distant from the sandstone; a region in which c drops linearly with z, as z approaches 0, which is the gradient dc/dz responsible for driving diffusion through the mudstone; and a convex downwards section, where silica is being precipitated out of solution. As a gradient dc/dz needs to be maintained to drive diffusion through the mudstone, the retreat of the active dissolution region into the mudstone leads to lower quartz supersaturations at the mudstone–sandstone interface, and hence lower silica fluxes into the sandstone.

Table 2. Times corresponding to the curves shown in Fig. 5 for 20% of the original porosity to be removed, 1 m from the sandstone–mudstone interface, as a function of the quartz content of the mud, x.

x (% SiO_2)	x^2 ($\%^2$ SiO_2)	t (Ma)
22·4	5·0	5·81
31·6	10·0	5·02
44·7	20·0	4·41
54·8	30·0	4·14

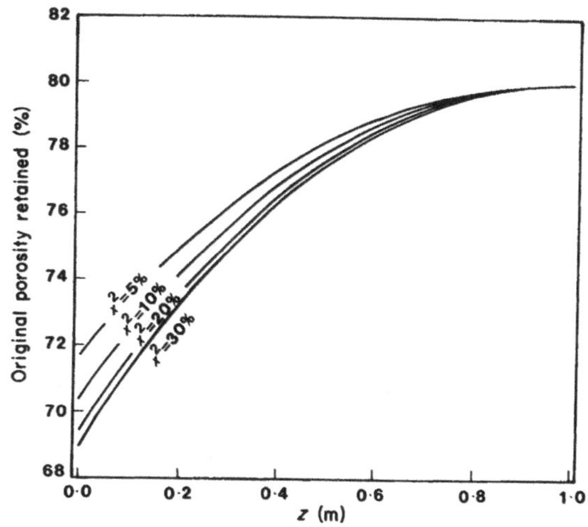

Fig. 5. The original porosity retained in the sandstone against distance from the sandstone–mudstone interface, as a function of the quartz content of the mudstone, x. All curves are shown normalized to 80% original porosity retained 1 m away from the interface. The time taken for this degree of modification to take place is not constant with x, and the times corresponding to each curve are given in Table 2.

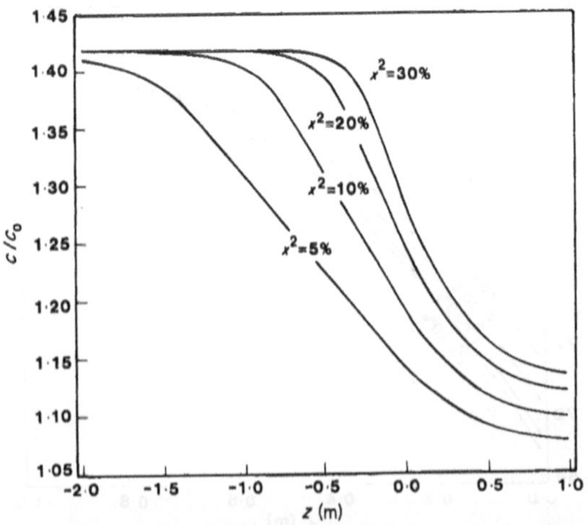

Fig. 6. The concentration of dissolved silica in the pore fluids of the sandstone and mudstone against distance from the sandstone–mudstone interface, as a function of the quartz content of the mudstone, x. Curves are shown for times given in Table 2, which are not necessarily the same for each curve.

DISCUSSION

The numerical models discussed above demonstrate that pressure dissolution and diffusion provide feasible mechanisms for the export of silica from a siltstone or mudstone into an adjacent sandstone. This process is summarized in Fig. 7. The importance and detectability of this silica export depends principally on the composition of the exporter lithology, and the speed of silica transport. For an approximately pure quartz source lithology, pressure dissolution is confined to a thin surface layer, which gradually recedes into the mudstone as the quartz is dissolved. This has the potential to release large volumes of silica into solution, although how far such a boundary might recede into the mudstone may be limited in practice. In addition, extensive quartz cements deposited close to the sandstone–siltstone interface will increase the tortuosity of the diffusion path and ultimately impede the flow of silica into the sandstone.

For a source lithology containing a significant percentage of other materials, such as clays, diffusion of silica through this material effectively limits silica export from quartz pressure dissolution to a zone 3–5 m thick. Under these conditions, the total volume of quartz exported will simply be a function of the volume of quartz available for dissolution in the first 3–5 m of mud. Also, because for silt-sized particles the principal control on both the rate of pressure dissolution and the concentration of dissolved silica in the pore fluids is the silica precipitation constant, the migration of dissolved silica from a clay-bearing mudstone to a clean quartz siltstone would be expected, even if there were no difference in grain size. This effect was reported in experiments by Rutter *et al.* (1986).

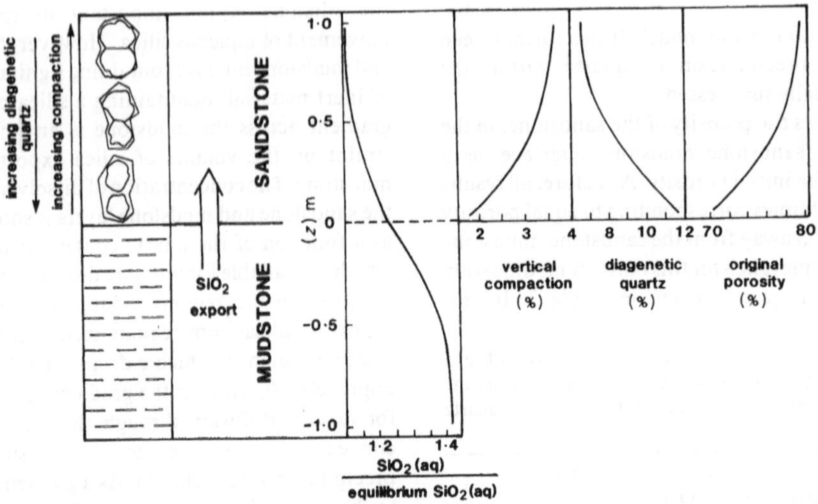

Fig. 7. Summary of the proposed porosity modification process. Aqueous silica in supersaturated solution is produced by intergranular pressure dissolution within a mudstone and is transported to an adjacent sandstone by molecular diffusion. Influx of aqueous silica across the boundary into the sandstone leads to enhanced precipitation of authigenic quartz and inhibits self-compaction by pressure dissolution. Curves shown correspond to a mudstone source containing 44·7% quartz by volume, 4·41 Ma after the onset of silica transport. k_- for the sandstone is $2{\cdot}2 \times 10^{-9}\ s^{-1}$.

In all the models described above, the supply of silica is thus limited. By considering a slow transport process, molecular diffusion, it has been demonstrated that this silica export may lead to significant porosity modification within a thin sandstone bed, or at the margins of a thicker sequence. However, if the pore fluids within the sandstone are not stationary, transport by advection-dispersion may be far more rapid than by molecular diffusion. If conditions exist to disperse rapidly the imported silica uniformly throughout a thick sandstone sequence, there may be no obvious features indicating silica import. Instead, a small uniform excess of authigenic quartz may be present throughout the sandstone sequence.

ACKNOWLEDGMENTS

This work was supported by NERC under grant No. GST 02/355. I am indebted to Stuart Haszeldine for helpful comments during the preparation of the manuscript, and E. H. Rutter and D. W. Houseknetch for their helpful reviews.

REFERENCES

BATHURST, R.G.C. (1958) Diagenetic fabrics in some British Dinantian limestones. *Lpool Manchr geol. J.*, **2**, 11–36.

BJORLYKKE, K. (1983) Diagenetic reactions in sandstones. In: *Proc. NATO Advanced Study Institute on Sediment Diagenesis* (Ed. by A. Parker & B. W. Sellwood), pp. 169–213. D. Riedel, Holland.

BOLES, J.R. & FRANKS, S.G. (1979) Clay diagenesis in Wilcox sandstones of SW Texas: implications of smectite diagenesis on sandstone cementation. *J. sedim. Petrol.*, **49**, 55–70.

CECIL, C.B. & HEALD, M.T. (1971) Experimental investigation of the effects of grain coatings on quartz overgrowth. *J. sedim. Petrol.*, **41**, 582–584.

EVANS, I.J. (1989) Geochemical fluxes during shale diagenesis, an example from the Ordovician of Morocco. In: *Water–Rock Interactions* (Ed. by J. Miles), pp. 219–222. Balkema, Rotterdam.

FÜCHTBAUER, H. (1983) Facies control on sandstone diagenesis. In: *Proc. NATO Advanced Study Institute on Sediment Diagenesis* (Ed. by A. Parker & B. W. Sellwood), pp. 269–288. D. Riedel, Holland.

HEALD, M.T. (1956) Cementation of Simpson and St. Peter sandstone in parts of Oklahoma, Arkansas and Missouri. *J. Geol.*, **64**, 16–30.

HEALD, M.T. (1965) Lithification of sandstones in West Virginia. *Bull. W. Va geol. Surv.*, **30**, 1–28.

HICKMAN, S.H. & EVANS, B. (1991) Experimental pressure solution in halite—the effect of grain/interphase boundary structure. *J. geol. Soc. London*, **148**, 549–560.

LAND, L.S. & DUTTON, S.P. (1978) Cementation of Pennsylvanian deltaic sandstone: isotopic data. *J. sedim. Petrol.*, **48**, 1167–1176.

LERMAN, A. (1979) *Geochemical Processes*. Wiley, New York, 481 pp.

MANHEIM, F.T. & WATERMAN, L.S. (1974) Diffusimetry (diffusion constant estimation) in sediment cores, by resistivity probe. *Initial Rep. Deep Sea Drill. Proj.*, **22**, 663–670.

MCBRIDE, E.F. (1989) Quartz cement in sandstones: a review. *Earth Sci. Rev.*, **26**, 159–166.

MULLIS, A.M. (1991) The role of silica precipitation kinetics in determining the rate of quartz pressure solution. *J. geophys. Res.*, **96**, 10 007.

PHARR, G.M. & ASHBY M.F. (1983) On creep enhanced by a liquid phase. *Acta metall.*, **31**, 129–138.

RAJ, R. & CHYUNG, C.K. (1981) Solution precipitation creep in glass ceramics. *Acta metall.*, **29**, 159–166.

RIMSTIDT, J.D. & BARNES, H.L. (1980) The kinetics of silica water reactions. *Geochim. Cosmochim. Acta*, **44**, 1683–1699.

RUTTER, E.H. (1976) The kinetics of rock deformation by pressure solution. *Phil. Trans. R. Soc.*, **A283**, 203–209.

RUTTER, E.H., MADDOCK, R.H., HALL, S.H. & WHITE, S.H. (1986) Comparative microstructures of natural and experimentally produced clay-bearing fault gouges. *Pure Appl. Geophys.*, **124**, 3–30.

SCHUTJENS, P.M.T.M. (1991) Experimental compaction of quartz sand at low effective stress and temperature conditions. *J. geol. Soc. London*, **148**, 527–539.

TADA, R., MALIVA, R. & SIEVER, R. (1987) A new mechanism for pressure solution in porous quartzose sandstone. *Geochim. Cosmochim. Acta*, **51**, 2295–2301.

ULLMAN, W.J. & ALLER, R.C. (1982) Diffusion coefficients in nearshore marine sediments. *Limonol. Oceanogr.*, **27**, 552–556.

WEYL, P.K. (1959) Pressure solution and the force of crystallization—A phenomenological theory. *J. geophys. Res.*, **64**, 2001–2025.

WOLLAST, R. & GARRELS, R.M. (1971) Diffusion coefficient of silica in sea water. *Nature Phys. Sci.*, **229**, 94.

(Manuscript received 13 March 1991; revision received 23 August 1991)

Reprinted from Sedimentology (1994) ***41****, 363–377*

Origin of quartz cements in some sandstones from the Jurassic of the Inner Moray Firth (UK)

GRETE BLOCK VAGLE,* ANDREW HURST† *and* HENNING DYPVIK‡

**BP Norway Limited U.A., Postboks 197, 4033 Forus, Norway*
†Production Geoscience Unit, Department of Geology & Petroleum Geology, University of Aberdeen, Aberdeen AB9 2UE, UK
‡Geology Department, University of Oslo, Blindern, Oslo, Norway

ABSTRACT

The extent of quartz cementation in shallow marine sandstones of the Brora Arenaceous Formation (Oxfordian) is closely related to the occurrence and abundance of *Rhaxella perforata* sponge spicules. Three cement morphologies are identified, chalcedonic quartz, microquartz and mesoquartz. Chalcedonic quartz forms matrix-supported cements which preserve moulds of *Rhaxella* spicules. Chalcedonic quartz crystals have inequant development of crystal faces, on average 0·1 μm in diameter, and are the first formed cement and reveal homogeneous dark grey tones on the SEM-CL/BEI. Microquartz forms 5–10 μm diameter crystals, which commonly grow on chalcedonic quartz substrates and show various grey tones under SEM-CL/BEI. Mesoquartz crystals grow in optical continuity with their host grains, have >20 μm *a*-axial diameter crystals, and exhibit distinctly zoned luminescence. Although no opaline silica is preserved, the quartz cement is interpreted to have formed from an opaline precursor. Detrital quartz has an average $\delta^{18}O$ composition of +12·2‰ and mesoquartz (syntaxial overgrowth) has an average $\delta^{18}O$ composition of +20·0‰. Estimates of the $\delta^{18}O$ compositions of microquartz and chalcedonic quartz are complicated by the problem of isolating the two textural types; mixtures of the two give consistently higher $\delta^{18}O$ compositions than mesoquartz, the higher estimate being +39·2‰. From oxygen isotope data the formation of quartz, microquartz and chalcedonic quartz is interpreted to have taken place between 35 and 71°C in marine derived pore waters. Organic and inorganic maturation data constrain the upper temperature limit to less than 60°C.

INTRODUCTION

Quartz is commonly the most abundant and frequently identified cement in sandstones. It is largely responsible for the consolidation of sandstones and ultimately the destruction of porosity in hydrocarbon reservoirs and aquifers. Quartz occurs at several stages of diagenesis and may have a variety of origins, e.g. biogenic, volcanogenic, from clay mineral transformations in adjacent shales, from sedimentary abrasion of quartz, grain to grain pressure solution, stylolitization, feldspar alteration, geothermal alteration, ascending solutions, weathering, circulating groundwater, etc. (McBride, 1989). Quartz cements frequently, but not always, occur as overgrowths on detrital quartz grains (Waugh, 1970). Quartz may also occur as a microcrystalline matrix, similar in appearance to chert or chalcedony, possibly being derived from opaline precursors (Williams *et al.*, 1985).

On the north-eastern coast of the Inner Moray Firth (Fig. 1), outcrops of marine, quartzose sandstones of Oxfordian age have quartz as the only common cement. This area has not been exposed to extensive burial diagenesis and clay mineralogical (Hurst, 1982) and vitrinite reflectance (Kolm Martens, 1988) data indicate a maximum burial temperature below 60°C. Thus, the outcrops afford an interesting opportunity to study the relationship between early diagenetic cements and depositional facies. The outcrops are of additional interest because they display facies similar to the main reservoir

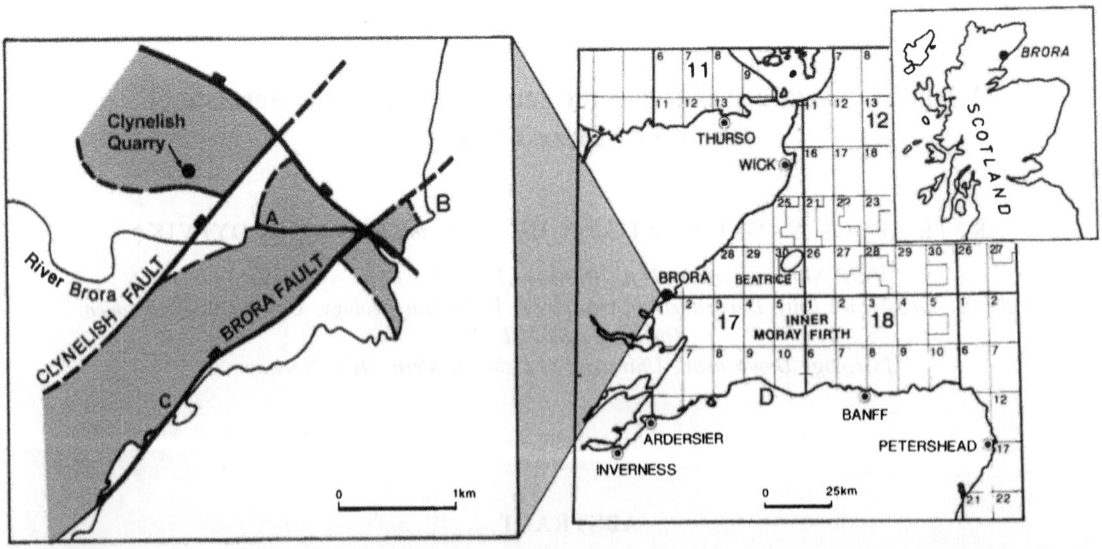

Fig. 1. Distribution map of the Brora Arenaceous Formation (stippled) outcrops and location of Clynelish Quarry, Fascally (A), Ardassie Point (B) and Strathsteven (C). Location of the Lossiemouth borehole (D) and the Beatrice Field are shown to the right.

sandstones in the Beatrice Oilfield (Fig. 1) and other prospective petroleum-areas within the Inner Moray Firth (Linsley *et al.*, 1980; Andrews & Brown, 1987).

MATERIALS AND METHODS

Petrography

Sedimentary description and sampling was made from several outcrops of the Brora Arenaceous Formation, and in particular at the Clynelish Quarry, Fascally and Strathsteven (Fig. 1). To provide comparison between the quartz cements at Brora with those elsewhere, seven additional samples were selected; five from exploration borehole 11/30-2 from the Beatrice Oilfield and two from the Lossiemouth borehole (Fig. 1; Hurst, 1985). Doubly polished thin sections impregnated with blue stained epoxy were made of 40 sandstone samples and examined with a polarizing microscope. All thin sections were point counted with at least 400 points per sample.

Scanning electron microscope (SEM) analysis was made using a Jeol JSM 840 with a Jeol Superprobe 733. Secondary electron image analysis (SEI) was carried out on freshly fractured, stub-mounted sample coated with gold. Some thin sections were carbon coated and examined on the SEM using back scattered electron imaging (BEI). Cathodoluminescence (CL) studies were made both using a Nuclide Corporation ELM 2 and the SEM Jeol Superprobe 733 with a photomultiplactor detector type R-268.

On the SEM, data from CL images and the BEI were combined to differentiate between detrital quartz, authigenic quartz, and porosity. SEM-CL images show luminescent detrital quartz and non-luminescent epoxy impregnated pores. BEI images show the same intensity of back scatterd electrons for authigenic quartz as for detrital quartz and low intensities for the porosity. Thus, by superimposing CL and BEI data, discrimination can be made between detrital quartz, authigenic quartz and porosity.

Stable isotope analysis

Quartz was isolated for stable isotope analysis using two independent methods.

Method 1. Five sandstone samples with high (approximately 37%) quartz cement contents were treated in an ultrasonic bath (Branso Sonifer S-75) for several hours to loosen authigenic silica from the detrital grains. Separation of the clastic grains (grain size 0·1–0·25 mm) from the authigenic silica was made by sieving the sandstones through a 63 μm mesh after ultrasonic treatment. Any clay minerals and carbonates present were removed by fusion with sodium pyrosulphate and HCl treatment (Syers *et al.*, 1968). Optical examination (SEI) of the authigenic quartz obtained showed textures similar to the quartz cement revealed by analysis of the bulk samples.

Method 2. A separate series of samples were disaggregated without attempting to isolate the authigenic and detrital quartz mechanically. These samples were separated into size fractions, and sequentially etched to decrease successively the proportion of authigenic quartz present following the procedure of Hurst (1981) and Lee & Savin (1985).

Extraction of oxygen from the quartz obtained from both preparation methods was done by a procedure similar to that of Clayton & Mayeda (1963), with a reproductivity of ± 0·5‰.

DATA

Sedimentology

A general coarsening upwards sequence is recognized within the Brora Arenaceous Formation which Sykes (1975) divided into three progressively coarser sandstones, the Fascally Sandstone, Clynelish Quarry Sandstone and Brora Sandstone members (Fig. 2). In the field, a general coarsening upwards sequence can be demonstrated, although the Fascally and Clynelish Quarry sandstones are often interbedded, and the contact between the Brora and Clynelish sandstones is difficult to observe (Hurst, 1993). Sykes (1975) assigned the generalized environmental interpretation of a coastal sand bar to the sequence. A more specific interpretation seems possible, and several characteristics typical of tidally dominated sand waves are recognized (Hurst, 1993). Sedimentological interpretation and correlation between outcrops is hindered by discontinuous exposure.

The Clynelish Quarry Sandstone Member is commonly extremely poorly consolidated, as is evidenced by the high porosities (average 35% from 47 core plugs of outcrops, Hurst, 1993). However, others, particularly from the Clynelish Quarry, are tightly cemented (Table 1), at outcrop resembling quartzites. At Clynelish Quarry, detailed sedimentary logging and petrographic inspection of thin sections reveal a relationship between spicule abundance and grain size. Spicules are most abundant in the finest grained intervals where an average of 7% spicules are present (Fig. 3).

Mineralogy

The highly quartzose nature (both detrital and diagenetic composition) of the sandstones (Table 1,

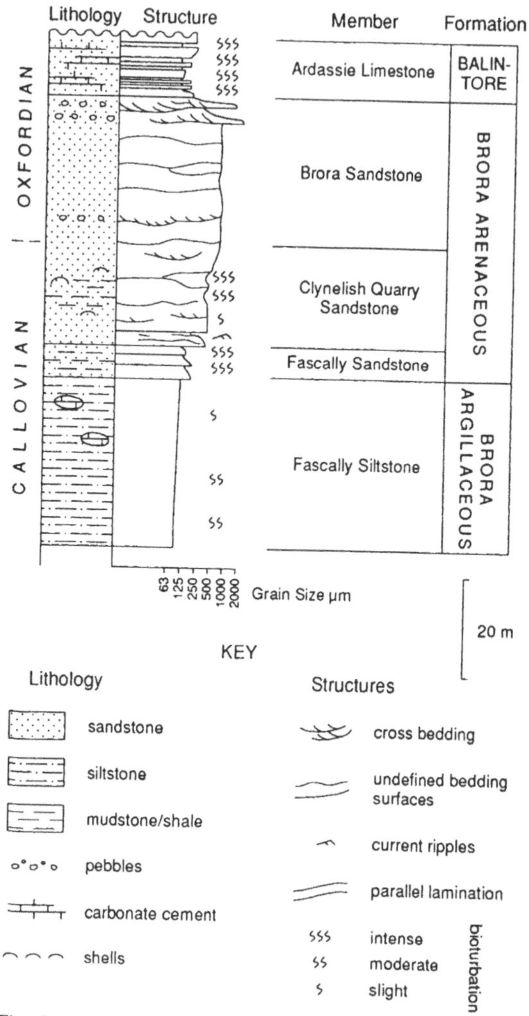

Fig. 2. General lithostratigraphy for the late Callovian to Oxfordian at Brora (modified from Hurst, 1993).

Fig. 4a) reflects both their mineralogically mature detrital compositions, and that quartz is the major cement present. Feldspar and rock fragments (gneiss, granites) are the only other mineral components regularly encountered during point counting, with traces of mica and kaolinite. No evidence is found for the presence of silica polymorphs other than quartz.

Quartz cementation at the Clynelish Quarry locality is invariably accompanied by the presence of fragments, or fossil moulds, most likely of *Rhaxella perforata* sponge spicules (Fig. 4b). The sandstones with extensive silica cementation have a high content

Table 1. Details of the studied sandstones. Values are percentages.

Sample	Locality	Grain size	Detrital quartz	Other detrital grains	Detrital matrix	Quartz cement	Other cements	Porosity	*Rhaxella*	Organic matter
Brora Sandstone Member										
4.11.13	Fascally	f	60·0	4·0	0·6	0·2	+	35·2	—	—
4.11.14	Fascally	f	67·5	4·5	1·2	+	0·2	26·6	—	—
4.11.16	Fascally	f	67·5	3·5	0·2	+	0·4	28·4	—	+
average			65·0	4·0	0·7	0·1	0·2	30·1	0·0	0·0
Clynelish Quarry Sandstone Member										
5.12.1	Strathsteven	f	59·8	6·3	1·4	1·0	1·0	30·5	—	—
5.12.2	Strathsteven	f	63·6	4·2	0·2	+	+	32·0	—	—
5.12.3	Strathsteven	f	62·0	4·2	1·6	+	0·2	32·0	—	—
5.12.4	Strathsteven	f	62·8	3·6	1·8	+	0·4	31·2	—	0·2
5.12.5	Strathsteven	f	63·5	5·7	1·0	0·4	+	29·4	—	—
average			62·3	4·8	1·2	0·3	0·3	31·0	0·0	0·0
6.13.1	Clynelish	f	71·3	1·5	—	21·0	+	6·4	—	+
6.13.2	Clynelish	f	44·0	4·9	0·8	40·4	+	(9·9)†	9·9	+
6.13.3	Clynelish	f	54·6	7·7	0·6	32·3	0·1	(1·6)	4·6	0·2
6.13.4	Clynelish	f	51·5	4·4	1·0	36·8	0·2	(6·1)	6·1	+
average			55·4	4·6	0·6	32·6	0·1	1·6	5·2	0·1
4.11.1	Fascally	f	66·8	4·8	0·8	+	+	27·6	—	+
4.11.2	Fascally	f	61·9	6·5	0·8	0·6	+	30·0	—	0·2
4.11.3	Fascally	f	61·4	6·0	0·8	0·4	0·4	31·0	—	—
4.11.4	Fascally	vf	47·2	7·6	10·8	0·2	1·2	32·0	—	1·0
4.11.5	Fascally	f	64·0	5·8	1·0	+	—	29·2	—	+
4.11.6	Fascally	f	63·4	7·2	1·0	0·2	0·2	28·0	—	+
4.11.7	Fascally	f	60·6	5·8	0·2	0·2	0·2	32·8	—	0·2
4.11.8	Fascally	vf	43·8	4·4	11·0	15·8	11·8	12·8	—	0·4
4.11.9	Fascally	f	43·9	4·5	26·4	0·3	11·7	12·8	—	0·4
4.11.10	Fascally	f	56·3	6·7	0·4	10·4	+	26·2	—	+
4.11.11	Fascally	f	57·8	5·0	2·2	8·8	0·2	26·0	—	+
4.11.12	Fascally	f	62·4	5·6	0·8	0·3	0·3	30·6	—	+
average			57·5	5·8	4·7	3·1	2·2	26·6	0·0	0·2

Table 1. *Continued.*

Sample	Locality	Grain size	Detrital quartz	Other detrital grains	Detrital matrix	Quartz cement	Other cements	Porosity	*Rhaxella*	Organic matter
Fascally Siltstone Member			Matrix*							
4.7.1	Cawcrask	vf	23·8	0·8	66·0	—		8·0	+	1·4
4.7.2	Cawcrask	vf	22·6	1·4	72·0	+		2·6	0·2	1·2
4.7.3	Cawcrask	vf	20·8	2·8	68·8	+		0·6	7·0	+
4.7.4	Cawcrask	vf	21·0	2·6	71·8	+		1·8	2·6	0·2
4.7.5	Cawcrask	vf	24·3	1·5	66·2	+		1·0	6·0	1·0
4.7.6	Cawcrask	vf	21·2	2·2	73·6	—		0·6	1·0	1·4
4.7.7	Cawcrask	vf	21·6	2·8	68·6	+		3·0	1·0	3·0
4.7.8	Cawcrask	vf	23·2	2·6	67·2	—		1·2	4·4	1·4
4.7.9	Cawcrask	vf	24·2	1·4	65·6	+		3·6	4·4	0·8
average			22·5	2·0	68·9	0·0		2·5	3·0	1·2
Fascally Sandstone Member										
4.7.10	Cawcrask	vf	28·8	2·6	56·4	+		0·4	8·8	3·0
4.7.11	Cawcrask	vf	25·4	4·4	64·4	—		4·4	—	1·4
4.7.12	Cawcrask	vf	29·9	2·6	49·1	+		0·6	16·4	1·4
4.7.13	Cawcrask	vf	43·2	4·4	42·2	+		4·6	3·6	2·0
4.7.14	Cawcrask	vf	34·4	3·4	49·2	—		0·2	12·0	0·8
4.7.15	Cawcrask	vf	37·0	4·2	53·2	+		2·4	0·8	2·4
4.7.16	Cawcrask	vf	41·6	5·6	41·6	+		9·2	—	2·0
4.7.17	Cawcrask	vf	37·8	3·6	49·6	+		3·8	+	5·2
4.7.18	Cawcrask	vf	16·6	2·0	58·0	+		+	21·6	1·8
4.7.19	Cawcrask	vf	28·0	4·0	55·2	+		6·4	5·0	1·4
4.7.20	Cawcrask	vf	29·6	3·7	38·0	+		4·3	22·8	1·6
4.7.21	Cawcrask	vf	35·0	2·9	51·4	+		7·3	2·6	0·8
4.7.22	Cawcrask	vf	20·8	2·4	63·2	+		6·0	6·8	0·8
4.7.23	Cawcrask	vf	39·4	5·4	31·4	—		22·8	—	1·0
4.7.24	Cawcrask	vf	32·0	1·4	54·4	+		9·8	—	1·4
average			32·0	3·5	50·5	0·0		5·5	6·7	1·8

Detrital fractions are divided into two categories, detrital quartz and detrital grains. Detrital matrix includes clay minerals, rare carbonates, pyrite, iron oxides and phosphates. Matrix* (Fascally Sandstone and Siltstone samples from Cawcrask) includes large proportions of detrital clay and authigenic carbonate cement. Note that *Rhaxella* spicules in samples 4.7.1–24 (Cawcrask) are replaced by calcite.

†Porosity in samples from Clynelish (in parentheses) is present as moulds of *Rhaxella* spicules.

+ = trace amount.

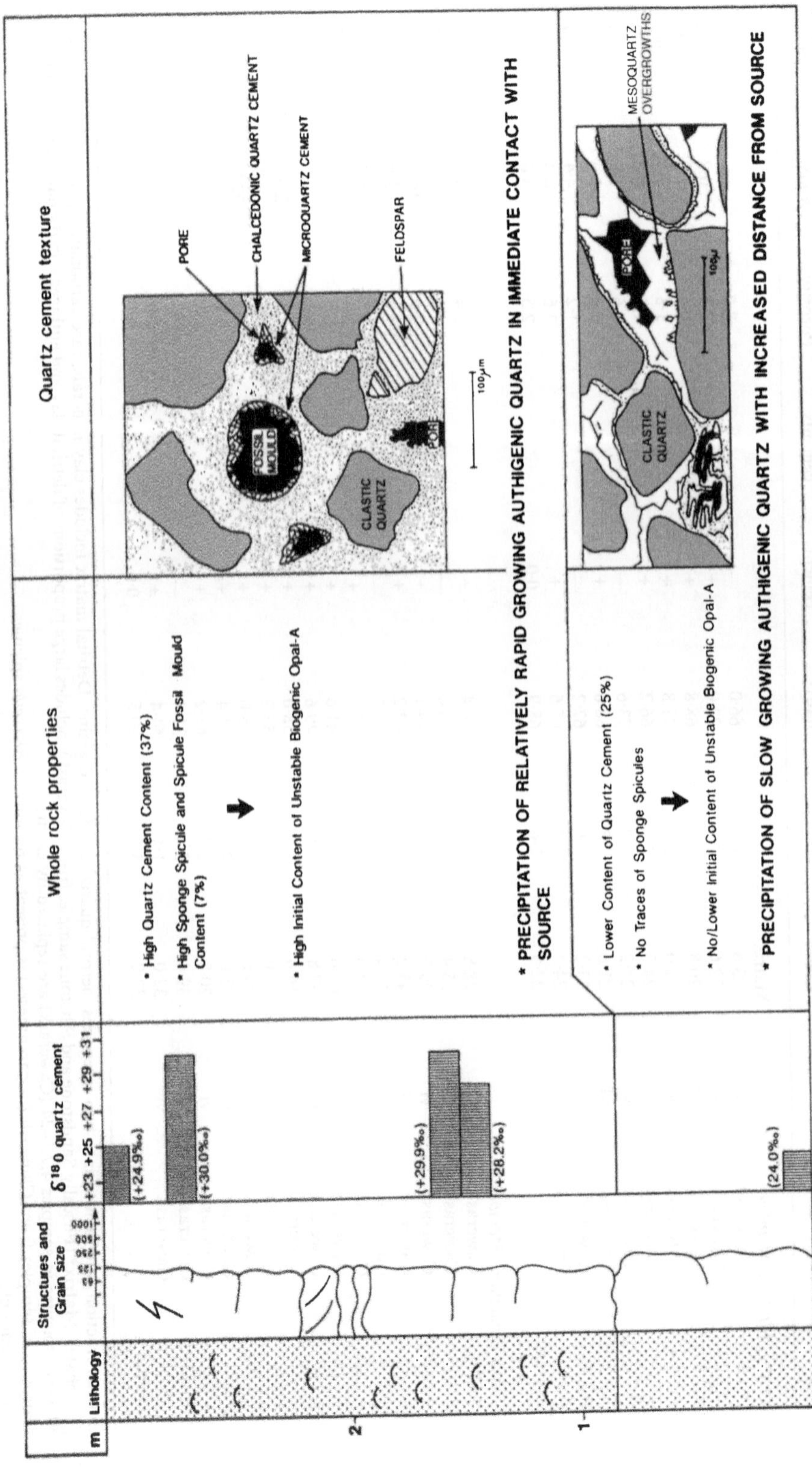

Fig. 3. Sedimentary log from the Clynelish Quarry Sandstone Member at Clynelish Quarry. Details of cement and spicule distribution and isotope data are summarized.

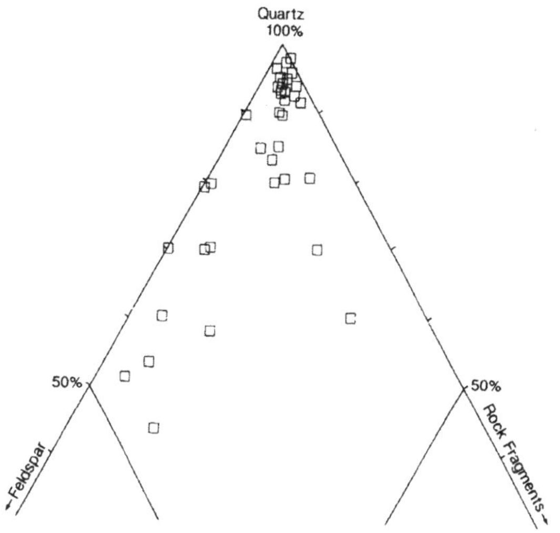

(a)

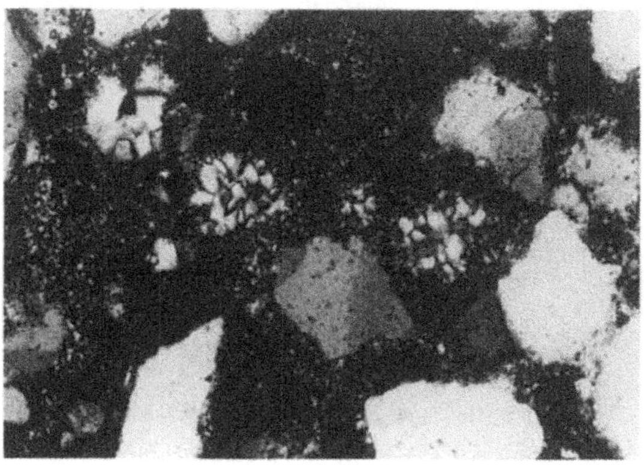

(b)

Fig. 4. (a) Composition of sandstones from the Brora Arenaceous Formation. (b) Thin section photomicrograph of a matrix supported, chalcedonic quartz cemented sandstone from the Clynelish Quarry. *Rhaxella perforata* sponge spicules are the approximately circular areas of radially orientated quartz crystals. Width of view= 640 μm.

of *R. perforata* (usually preserved as moulds; Table 1), while most sandstones lacking sponge spicules have very little quartz cement (normally less than 1% by volume). The *Rhaxella* spicules observed in the Fascally Siltstone and Sandstone members are replaced by calcite. It is assumed that SiO_2 from dissolution and replacement of the spicules has been exported since only minor amounts of quartz cement occur.

Petrography

Quartz cement occurs in three textural styles: chalcedonic quartz, microquartz and mesoquartz.

Chalcedonic quartz. Chalcedonic pore fillings occur where the individual quartz crystals have inequant development of crystal faces. Individual quartz crystals average 0·1 μm in diameter. The chalcedonic quartz smothers clastic grains, usually increasing in grain size away from the clastic grain boundaries (Fig. 5a). SEM-CL analysis of the cement reveals homogeneous, dark, grey tones.

Microquartz. Microquartz occurs as 5–10 μm diameter crystals, with little common orientation of neighbouring crystals (Fig. 5). Crystals are commonly twinned or occur in small clusters (Fig. 5d). Growth parallel to the *c*-axis is more rarely observed. Microquartz seems to have nucleated on the chalcedonic quartz cement and grows into open pore space, often filling sponge spicule fossil moulds, or fringing the walls of fossil moulds (Fig. 5c, d). In the SEM-CL this cement has a wider range of grey tones, noticeably lighter than the chalcedonic quartz.

Mesoquartz. Mesoquartz occurs as >20 μm *a*-axial diameter crystals, which often grow in optical continuity with host detrital grains. Mesoquartz is either directly attached to host detrital grains or is separated from them by thin layers of chalcedonic quartz and/or microquartz (Fig. 6). In CL, mesoquartz is characterized by zoning. In SEM-CL at least three individual zones are identified as approximately 2 μm thick, dark and light grey layers. In the cold cathode CL the zones luminesce with low intensities from black to red–brown. Using high magnification in the SEM-CL, three subzones can be identified in the most peripheral cement zones (Fig. 7).

Oxygen isotope analysis

Estimation of equilibration temperature is calculated using

$$10^3 \ln \alpha = 3{\cdot}38(10^6\, T^{-2}) - 3{\cdot}40,$$

an extrapolation of the Clayton *et al.* (1972) calibration, where α is the fractionation constant and *T* is the absolute temperature in degrees K. The curve of this equation coincides with the curve for fractionation between diatom frustules and water in the temperature range 0–30°C. Since no $\delta^{18}O$ values for

Fig. 5. (a) SEM photomicrograh showing chalcedonic quartz coating detrital grains which subsequently has been overgrown by microquartz. Scale bar = 10 μm. (b) SEM photomicrograph showing detrital grains within a matrix of microquartz and chalcedonic quartz. Moulds of sponge spicules are labelled 'a'. Scale bar = 100 μm. (c) View of chalcedonic quartz and microquartz matrix-supporting cement. Field of view is shown in (b). Scale bar = 10 μm. (d) Detail of the microquartz walls of a *Rhaxella perforata* sponge spicule mould. Field of view is shown in (c). Scale bar = 1 μm.

the pore water are available, the calculated precipitation temperature for the quartz cement can only be considered as an estimate. Two $\delta^{18}O$ values for the pore water are selected, − 1·2‰ SMOW, which is Shackleton & Kennett's (1975) value for preglacial ocean water, and − 2‰ SMOW, which is an average value for pore water in Jurassic sandstone reservoirs in the North Sea (Ekeberg & Aagaard, 1987). The results are interpreted to infer that quartz formed within a temperature range of 35–71°C (Table 2, Fig. 8).

As the different quartz cements have probably formed by successive recrystallization, each cement is likely to have precipitated at different temperatures, and indeed individual textural styles have probably formed at different temperatures. The petrographic evidence (Figs 4b and 5) for a biogenic origin for the silica cement is circumstantial. However, the moulds of *R. perforata*, which were originally opal-A, coincide with abundant quartz cement. opal-CT, the presumed intermediate between opal-A and quartz, is absent in the studied samples.

Estimation of the temperature of quartz precipitation (preparation method 1) is made from the $\delta^{18}O$ values from the oxygen isotope analysis of the quartz cement (Fig. 8). The sample dominated by

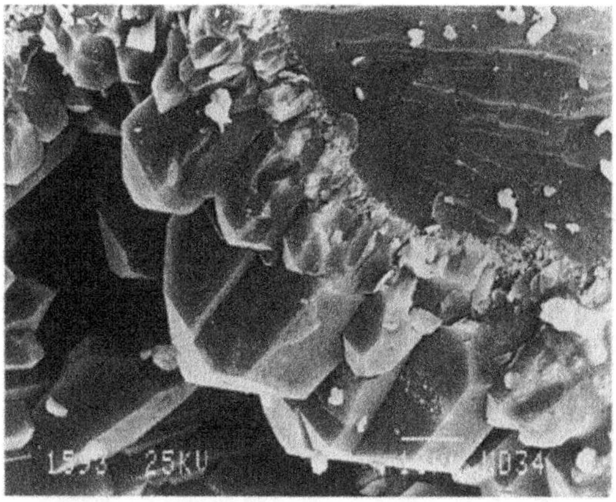

Fig. 6. Detrital quartz overgrown by progressively coarser mesoquartz overgrowths.

mesoquartz has the lowest value (+24‰). Mixtures of chalcedonic quartz and microquartz give somewhat higher values, an average of +28·3‰ (a range of +24·9 to +30·0‰). Using preparation method 2, $\delta^{18}O$ values obtained for mesoquartz overgrowths have an average composition of +20·0‰ ranging from +13·3 to +32·4‰ with one anomalously low value of +3·7‰ (Fig. 9). Two samples with microquartz and chalcedonic quartz had $\delta^{18}O$ compositions of +30·8 and +39·2‰. From duplicate samples containing mesoquartz compositional variations are recorded of up to 11·5‰ (average of 8·3‰ from four samples, Appendix Table A3). The $\delta^{18}O$ composition of detrital cores is lower than for diagenetic quartz and averages +12·2‰. Duplicate samples of detrital quartz have an average variability of 0·6‰ with a maximum recorded variability of 1·1‰ (Appendix Table A3).

DISCUSSION

Rhaxella perforata

Rhaxella sponge spicules (Hinde, 1893) are known from the Corallin and Portlandian beds in southern England in facies that vary from normal marine shallow shelf sediments (mudstone and muddy sandstones) to muddy biomicrites of hypersaline lagoonal origin (Wilson, 1968; Talbot, 1973). In the Moray Firth Basin, silty shales and fine grained sandstones containing *Rhaxella* spicules are abundant in the Upper Jurassic. The thickest of these 'spiculite sequences', sometimes referred to as the Alness Spiculite Member (mid-Oxfordian), has its greatest thickness in blocks 12/22 and 12/23 (Fig. 1), and is interpreted as large subtidal shoal deposits (Andrews & Brown, 1987).

In the onshore Jurassic at Brora, *Rhaxella* sponge spicules are found in shales, siltstones, fine grained sandstones and carbonates, but whole *R. perforata* sponges are not preserved. *Rhaxella* is not found in the coarser grained sandstones of the Brora Sandstone Member. The absence of well preserved sponges is indicative of rapid *post mortem* decomposition and sedimentary reworking of the spicules. Analogous spicule reworking is known from observations of the recent sponge family *Geodiidae* (Moore, 1955) which habitate hard substrates in low suspension, clean waters (Aarseth *et al.*, 1974). If *Rhaxella* had a similar habitat to the *Geodiidae*, it may be inferred that spicules occurring in the fine grained sandstones of the Clynelish Quarry Sandstone Member have been reworked from deeper parts of the shelf, and transported inshore by tidal currents.

A 3 m thick, overall fining upward section is identified from the petrographic study of samples (15 samples studied, four representative samples point counted) collected from the Clynelish Quarry, which is not easily discernable at outcrop (Fig. 3). The uppermost part of the sequence comprises very fine grained sandstone, with a typical content of approximately 7% siliceous sponge spicules or fossil moulds. The authigenic quartz content is very high, averaging around 37%. Clastic grains 'float' in a matrix of quartz cement dominated by chalcedonic quartz and microquartz (Fig. 4b), and a high depositional spicule content is inferred. Spicule content, and thus quartz cementation, is interpreted to be of depositional origin, spicules presumably having similar hydrodynamic behaviour to silt and very fine grained sand and so accumulating within the finest grained laminae. The coarser (but still classified as fine grained) sandstones (Table 1) have lower quartz cement contents, approximately 25%, and from their grain supported texture it is inferred that there was little detrital spicule content. Overgrowths of mesoquartz typify the coarsest grained (fine sand), spicule-free sandstones.

Textural variety

Crystal growth

Abundance of quartz cement generally coincides with depositional variations in *Rhaxella* content and it is

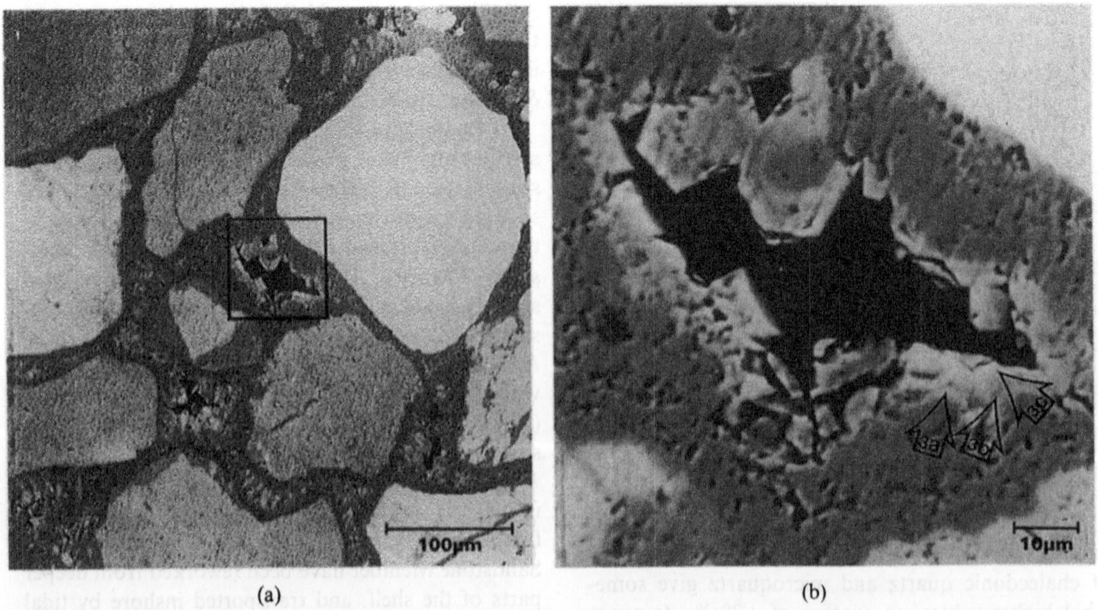

Fig. 7. (a) Combined BEI-CL micrograph of matrix supported quartz cement showing several generations of quartz growth, each with progressively coarser crystal size. (b) Detail of a single pore from the boxed area in (a) where three luminescent zones are identified within the mesoquartz; two light grey cements 3a and 3c, and a dark grey cement 3b. Similar zoning is apparent in mesoquartz when examined under cold cathode luminescence.

likely that a major part of the diagenetic mass transport of silica-rich solutions occurred only on a local scale. Calcite-replaced *Rhaxella* spicules within the Fascally Silstone and Sandstone members, both which have little quartz cement, provide some evidence. Although opal-CT is not detected, evidence of a biogenic origin (Fig. 5b, c), its early diagenetic 'floating' grain texture (Figs 4b and 5b), and distinctive isotopic character similar to chert support the inference that opal precursors were the source of the quartz cement. Preservation of spicule moulds (Fig. 5b, c) is interpreted to imply that local concentrations of *Rhaxella* spicules were probably sufficient for a grain supporting silica cement to precipitate prior to dissolution of all spicules (Fig. 3). Texturally similar *Rhaxella* moulds are described from the Corallian beds of Oxfordshire (Milodowski & Wilmot, 1984), differing only from the Clynelish Quarry Sandstone Member examples by the presence of opal-CT lepispheres rather than microquartz on the walls of the spicule moulds. By analogy, in the Clynelish Quarry Sandstone Member opal-CT may have pseudomorphed the spicules and subsequently recrystallized to quartz. Presumably, silica diagenesis in the Clynelish Quarry sandstones is more advanced than in the Corallian beds. Preservation of the *Rhaxella*

Table 2. Estimates of the temperature of quartz crystallization for $\delta^{18}O$ values of −1·2 and −2‰ relative to SMOW.

$\delta^{18}O$ (‰)	T (°C) ($\delta^{18}O_{water} = -1{\cdot}2$‰)	T (°C) ($\delta^{18}O_{water} = -2{\cdot}0$‰)
+24·9	65	60
+30·0	39	35
+29·9	40	36
+28·2	48	44
+24·0	71	66

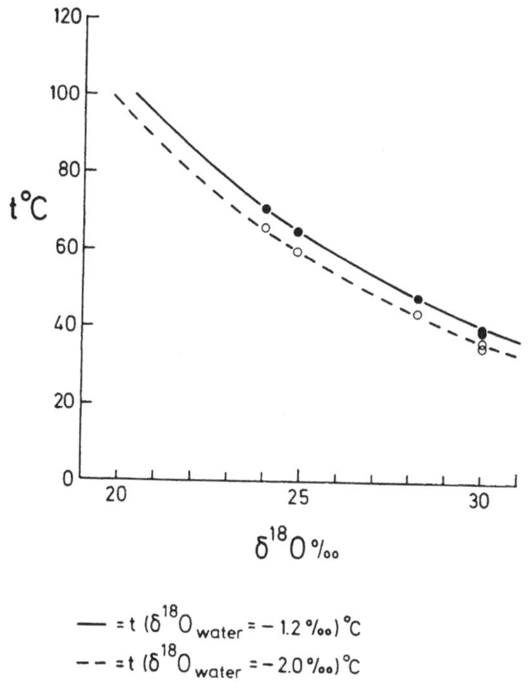

Fig. 8. Graphic representation of the $\delta^{18}O$ fractionation with change in temperature. The solid line is from Clayton *et al.* (1972) and the stippled line is from Labyerie (1974).

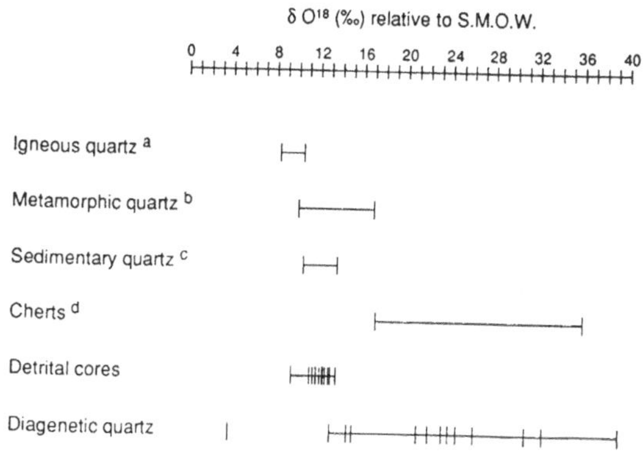

Fig. 9. Summary of the $\delta^{18}O$ compositions of quartz from detrital cores of grains and diagenetic fractions in this study. In this figure they are compared with: (a) igneous quartz (Taylor & Epstein, 1962); (b) regional metamorphic quartz (Garlick & Epstein, 1967); (c) quartz from sand and sandstones (Savin & Epstein, 1970); (d) Triassic to Miocene cherts (Knauth & Epstein, 1975). The detrital cores show values similar to sedimentary, igneous and metamorphic quartz, while the diagenetic quartz has ratios similar to cherts. See Appendix for further details.

morphology is indicative of early precompactional cementation. The high minus-cement porosity (up to 45%) in the chalcedony cemented sandstones is attributed to the post-depositional presence of a dense spicule matrix around the coarser detrital grains.

Chalcedonic quartz cement always occurs adjacent to the clastic grains (Fig. 5a), is often overgrown by microcrystalline quartz, and is interpreted to be the first of the three quartz cements to precipitate. The high specific surface area of the chalcedonic cement is indicative of a relatively rapid precipitation due to high degree of supersaturation with respect to quartz, which is interpreted to be related to opal-CT dissolution (Meyers, 1977). As increasing amounts of opal-CT dissolved and reprecipitated as chalcedonic quartz, the silica concentration of the pore water decreased and microquartz precipitated.

Microquartz crystals are often twinned, nucleate on substrates other than quartz, and have apparently chaotic *c*-axis orientation (Fig. 5d). The increased authigenic quartz grain size, from chalcedonic quartz to microquartz towards pore centres from the host detrital grains, is attributed to decreasing rates of crystal growth, which is in turn controlled by the gradual decrease in silica saturation as the pore water evolved through time. There is also some textural evidence for the progressive coarsening of authigenic quartz overgrowths (Fig. 6) by the Ostwald ripening process which is invoked to explain the progressive growth of larger overgrowths with lower specific surface area at the expense of smaller overgrowths with higher specific surface areas (cf. Williams & Crerar, 1985). Successively larger overgrowths appear more closely to mimic the orientations of the crystallographic axes of their host grains than their smaller precursor overgrowths (Fig. 6).

Luminescence

Detailed analysis of the origin of luminescence observed in quartz is beyond the scope of the present study. From analogy with earlier work, in which there is a consensus that variations in quartz luminescence are incompletely understood (Sprunt, 1981; Kearsley & Wright, 1988; Ramseyer *et al.*, 1988), it is inferred that the luminescent and weakly luminescent zones in the mesoquartz cement are probably indicative of changes in rates of cement precipitation through time, probably corresponding to changes of pore water chemistry or mode of growth. In this context, it is interesting to note that the chalcedonic quartz, interpreted texturally to have precipitated rapidly from highly silica-charged solutions, is uniformly non-luminescent, where as the microquartz, interpreted to have grown more slowly from pore waters with ever changing chemistry, has more varied luminescence. Zinkernagel (1978) suggested that defects in the quartz crystal lattice may cause variations in luminescence. Defects may be caused by substitution of Al^{3+} for Si^{4+} and natural growth defects such as dislocations.

Mesoquartz is the only diagenetic quartz with distinct zones parallel to the crystal faces (Fig. 7), which is consistent with slow crystal growth. The origin of the luminescent zones is not qualified, but slow crystal growth provides ample opportunity for a cement to be influenced by an evolving pore water. For example, metal cations which may act as luminescence activators/inhibitors (Fe^{3+}, Na^{+}, K^{+}, Ramseyer & Mullis, 1990) may vary in concentration in pore waters as heavy minerals dissolve during burial diagenesis.

Oxygen isotopes

Detrital quartz compositions have a very similar $\delta^{18}O$ compositional range to those reported for sedimentary quartz (Savin & Epstein, 1970) and metamorphic quartz (Garlick & Epstein, 1967). The similarity to the metamorphic quartz compositions (Fig. 9) probably reflects the provenance of the detrital quartz which is of polycyclical origin and ultimately derived from metamorphic basement. Variations in the isotopic composition of detrital grains (up to 1·1‰, sample L6771, Table A3) probably reflect the isotopic heterogeneity of the detrital quartz and will tend to become less if sample size is increased.

All varieties of diagenetic quartz have higher $\delta^{18}O$ compositions than the detrital cores. Microquartz and chalcedonic quartz have, in general, the highest values (up to +39·2‰) similar to the data of Knauth & Epstein (1975). The accuracy of estimates derived from method 2 is strongly influenced by the validity of the optical determination of proportions of quartz. The method is vindicated by producing reasonable results. However, the uncertainty associated with specific values is likely to be much larger than for the values of detrital quartz. This uncertainty is reflected in the results of analyses from duplicate samples of mesoquartz which, from the same sample but different subsamples, have compositional variations of up to 11·5‰, and average 8·3‰. Although some variation in the isotopic composition of diagenetic quartz is expected, we believe that the variations reported are largely a function of the inaccuracy of the visual estimates made by method 2.

Sudden changes in oxygen isotope signature are known to occur in intervals where opal-A to opal-CT, and opal-CT to quartz transitions occur (Murata & Randall, 1977). These step-wise changes are indicative of each silica polymorph retaining its isotopic signature until the burial temperature is sufficient to cause dissolution and neogenesis of the succeeding polymorph (Knauth & Epstein, 1975; Murata & Randall, 1977). Thus, the isotopic signature records the temperature and isotopic composition of the pore water at the time of reprecipitation. The oxygen isotopic signature at a given stratigraphic level does not necessarily reflect the maximum temperature to which that level has been exposed.

The $\delta^{18}O$ values of the quartz cements are interpreted to reflect the isotopic composition and the temperature of the pore water at the time of the opal-CT to quartz transformation, the final step in the formation of quartz. The reaction is reported to take place within a temperature range of 31–165°C (Murata & Randall, 1977; Pisciotto, 1981). This large temperature range for a specific diagenetic reaction is explained by the effects of particle size, shape and specific area on dissolution kinetics (Williams & Crerar, 1985). Varying proportions of other clastic minerals, especially those with high specific surface area, may, in addition to silica dissolution kinetics, contribute to causing the wide temperature range for this phase change (Williams *et al.*, 1985).

Constraining the temperature of formation for the quartz is difficult from the oxygen isotope data alone, both −2 and −1·2‰ formation water compositions give temperatures within the range 31–165°C. Burial temperatures of 65 and 71°C, estimated from a pore water with a ^{18}O of −1·2‰, are in excess of the burial temperatures estimated from organic and inorganic maturation parameters from the underlying Brora Argillaceous Formation; T_{max}=430°C with vitrinite reflectance of 0·35 from which an equivalent burial temperature of 30–50°C is derived (Kolm Martens, 1988). This confirms previous estimates of maximum burial temperatures of 50–60°C from clay mineral data (Hurst, 1982). From the temperature range indicated from vitrinite and clay mineral data, a −2‰ formation water composition appears likely.

Although recent sea water has a compositional range of −2·5 to +2·5‰ (Craig, 1961), −2‰, the oxygen isotope composition of North Sea formation waters (Ekeberg & Aagaard, 1989), is a low $\delta^{18}O$ composition for marine-derived pore water, and the influence of some meteoric water at the time of quartz precipitation cannot be discounted. The marine origin of the sandstones and their depositional isolation from sources of meteoric water may militate against extensive meteoric flushing, although the shallow burial history may have been conducive to meteoric water infiltration. Preglacial ocean water was probably 1·2‰ lower than recent marine ocean water (Shackleton & Kennett, 1975) so a value of −2‰ need not necessitate any meteoric influence. The transformation of opal-CT to quartz is interpreted to have occurred at the maximum burial temperature of the section. Textural and luminescence zoning of the authigenic quartz are indicative of several periods of quartz growth, and consequently the $\delta^{18}O$ signatures are probably averages of several different isotopic signatures that correspond to precipitation in various diagenetic environments.

CONCLUSIONS

Rhaxella sponge spicules formed a detrital matrix in the finest grained intervals of the Clynelish Quarry Sandstone Member which, during diagenesis, is assumed to have transformed from opaline silica to quartz. The transformation to quartz probably occurred at or near the maximum burial temperature for the section, oxygen isotope data indicating temperatures of recrystallization between 35 and 71°C, in marine-derived pore waters with a $\delta^{18}O$ composition of approximately −2‰ SMOW. Independent organic and inorganic maturation data may be interpreted to constrain the maximum burial temperature to <60°C.

Three morphological categories of quartz cement are identified, chalcedonic quartz, microquartz and mesoquartz. The dense chalcedonic quartz preserves the morphology of dissolved spicules and forms a floating grain texture for the detrital grains. Chalcedonic quartz, typically the first formed cement, is overgrown by microquartz, and is synonymous with the presence of *Rhaxella* spicules. Mixtures of chalcedonic quartz and microquartz have high $\delta^{18}O$ compositions, an average of +28·3‰ or greater. Mesoquartz is only common in the coarsest (fine sand) grained sandstones which contain few sponge spicules, and has average of $\delta^{18}O$ compositions of +20‰. Quartz cement distribution is largely a result of sedimentary processes which concentrated sponge spicules in the finest grained intervals of the sand body. Luminescent zonation is well developed in the mesoquartz, less so in the microquartz, and absent in the chalcedonic quartz. The origins of the luminescence are not resolved. Field, textural and isotopic data support that only local scale redistribution of silica occurred during diagenesis.

ACKNOWLEDGMENTS

Financial support for this work was provided by Det norske Vitenskaps Akademi and Statoil (G.B.V., H.D.), and in part by NERC grant GT4/77/GS101 (A.H.). Bård Flesjø is thanked for his contribution to sedimentological aspects of this work. Tony Fallick, Karl Ramseyer and Stuart Burley are thanked for their helpful criticism at various stages of completion of the manuscript. BP Norway Limited U.A. are acknowledged for their logistical support.

REFERENCES

AARSETH, I., BJERKELI, K., BJ RKLUND, K.R., B E, D., HOLM, J.P., LORENTZEN-STYR, T.J., MHYRE, L.A., UGLAND, E.S. & THIEDE, J. (1974) Late Quaternary sediments from Korsfjorden, western Norway. *SARISA*, **58**, 43–66.

ANDREWS, I.J. & BROWN, S. (1987) Stratigraphic evolution of the Jurassic, Moray Firth. In: *Petroleum Geology of North West Europe* (Ed. by J. Brooks & K. Glennie), pp. 785–795. Graham & Trotman.

CLAYTON, R.N. & MAYEDA, T.K. (1963) The use of bromine pentafluoride in the extraction of oxygen from oxides and silicates for isotopic analysis. *Geochim. Cosmochim. Acta*, **27**, 43–52.

CLAYTON, R.N., O'NEIL, J.R. & MAYEDA, T.K. (1972) Oxygen isotope exchange between quartz and water. *J. Geophys. Res.*, **77**, 3057–3067.

CRAIG, H. (1961) Isotopic variations in meteoric waters. *Science*, **133**, 1702–1703.

EKEBERG, P.K. & AAGAARD, P. (1989) Origin and evolution of formation water from oil fields in the Norwegian Shelf. *Applied Geochem.*, **4**, 131–142.

GARLICK, G.D. & EPSTEIN, S. (1967) Oxygen isotope ratios in co-existing minerals of regionally metamorphosed rocks. *Geochim. Cosmochim. Acta*, **31**, 181–214.

HINDE, G.J. (1893) A monograph of the British fossil sponges. *Palaeontogr. Soc. London*, 210–212.

HURST, A. (1981) A scale of dissolution for quartz and its implications for diagenetic processes in sandstones. *Sedimentology*, **28**, 451–459.

HURST, A. (1982) The clay mineralogy of Jurassic shales from Brora, NE Scotland. In: *International Clay*

Conference 1981 (Ed. by H. van Olphen & F. Veniale), pp. 677–684. Elsevier, Developments in Sedimentology 35.

HURST, A. (1985) Mineralogy and diagenesis of Lower Jurassic sediments of the Lossiemouth borehole, north-east Scotland. *Proc. York. geol. Soc.*, **45,** 189–197.

HURST, A. (1993) Bathonian to Oxfordian strata of the Brora area. In: *Excursion Guide to the Geology of E. Sutherland and Caithness* (Ed. by N.H. Trewin & A. Hurst), Scottish Academic Press.

KEARSLEY, A. & WRIGHT, P. (1988) Geological applications of scanning cathodoluminescence imagery. *Micro. Anal.* **7,** 49–51.

KNAUTH, P. & EPSTEIN, S. (1975) Hydrogen and oxygen isotope ratios in silica from the JOIDES Deep Sea Drilling Project. *Earth planet Sci. Lett.*, **25,** 1–10.

KOLM MARTENS, A. (1988) *Organic geochemical and sedimentological investigations of Jurassic shales from Brora, northeast Scotland.* Thesis, University of Oslo.

LABYERIE, L., JR (1974) New approach to surface seawater palaeotemperatures using $^{18}O/^{16}O$ in silica of diatom frustules. *Nature,* **248,** 40–42.

LEE, M. & SAVIN, S.M. (1985) Isolation of diagenetic overgrowths on quartz sand grains for oxygen isotopic analysis. *Geochim. Cosmochim. Acta,* **49,** 497–501.

LINSLEY, P.N., POTTER, H.C., MCNAB, G. & RACHER, D. (1980) The Beatrice field, Inner Moray Firth, U.K. North Sea. In: *Giant Oil and Gas Fields of the Decade, 1968–1978* (Ed. by M.T. Halbouty), *Mem. Am. Ass. petrol. Geol.*, **30,** 117–129.

MCBRIDE, E.F. (1989) Quartz cement in sandstones: a review. *Earth Sci. Rev.* **26,** 69–112.

MEYERS, W.J. (1977) Chertification in the Mississippian Lake Valley Formation, Sacremento Mountains, New Mexico. *Sedimentology,* **24,** 75–105.

MILODOWSKI, A.E. & WILMOT, R.D. (1984) Diagenesis, porosity and permeability in the Corallian Beds (Upper Oxfordian) from the Harwell research site, south Oxfordshire, U.K. *Clay Miner.*, **19,** 323–341.

MOORE, R.C. (1955) *Treatise on Invertebrate Palaeontology, E. Archaeocytha Porifera,* E41–E42. Geological Society of America.

MURATA, K.J. & RANDALL, R.G. (1977) Silica mineralogy and structure of the Monteray Shale, Temblor Range, California. *US geol. Surv. J. Res.*, **3,** 567–572.

PISCIOTTO, K.A. (1981) Diagenetic trends in the siliceous facies of the Monteray Shale in the Santa Maria Region, California. *Sedimentology,* **28,** 547–571.

RAMSEYER, K., BAUMANN, J., MATTER, A. & MULLIS, J. (1988) Cathodoluminescence colours of α-quartz. *Miner., Mag.*, **52,** 669–677.

RAMSEYER, K. & MULLIS, J. (1990) Factors influencing short-lived blue cathodoluminescence of α-quartz. *Am. Miner.*, **75,** 791–800.

SAVIN, S.M. & EPSTEIN, S. (1970) The oxygen isotope compositions of coarse grained sedimentary rocks and minerals. *Geochim. Cosmochim. Acta,* **34,** 323–329.

SHACKLETON, N.J. & KENNET, J.P. (1979) Palaeotemperature history of the Cenozoic and the initiation of Antarctic glaciation: oxygen and carbon isotopic analysis in DSDP sites 277, 179 and 281. *Initial Reports of the Deep Sea Drilling Project,* **24,** 313–324.

SPRUNT, E. (1981) Causes of quartz cathodoluminescence colours. *Scanning Electron Micrsc.* 525–535.

SYKES, R.M. (1975) The stratigraphy of the Callovian and Oxfordian stages (M-U Jurassic) in northern Scotland. *Scott. J. Geol.*, **11,** 51–78.

SYERS, J.K., CHAPMAN, S.L., JACKSON, M.L., REX, R.W. & CLAYTON, R.N. (1968) Quartz isolation from rocks, sediments and soils for determination of oxygen isotopic composition. *Geochim. Cosmochim. Acta,* **32,** 1022–1025.

TALBOT, M. (1973) Major sedimentary cycles in the Corallian Beds (Oxfordian) of southern England. *Palaeogeogr. Palaeoclimatol. Palaeoecol.*, **14,** 293–317.

TAYLOR, H.P., JR & EPSTEIN, S. (1962) Relationship between O^{18}/O^{16} ratios in coexisting minerals of igneous and metamorphic rocks. Part 1. Principles and experimental results. *Bull. geol. Soc. Am.*, **73,** 461–480.

WAUGH, B. (1970) Petrology, provenance and silica diagensis of the Penrith System (L. Permian) of NW England. *J. sedim. Petrol.*, **40,** 1226–1240.

WILLIAMS, L.A. & CRERAR, D.A. (1985) Silica diagenesis II. General mechanisms. *J. sedim. Petrol.*, **55,** 312–321.

WILLIAMS, L.A., PARKS, G.A. & CRERAR, D.A. (1985) Silica diagenesis I. Solubility controls. *J. sedim. Petrol.*, **55,** 301–311.

WILSON, R.C.L. (1968) Carbonate facies variation within the Osmington Oolite Series, southern England. *Palaeogeogr. palaeoclimatol. palaeoecol.*, **4,** 89–123.

ZINKERNAGEL, U. (1978) Cathodoluminescence of quartz and its application to sandstone petrology. *Contrib. Sedimentol.*, **8.**

(Manuscript received 3 April 1993; revision accepted 18 October 1993)

APPENDIX

Calculation of $\delta^{18}O$ compositions of detrital cores and diagenetic quartz shown in Fig. 9.

Compositions are estimated from bulk compositions of sequentially etched grains (Hurst, 1981). Relative abundances of detrital cores and their authigenic overgrowths were estimated by SEM examination of at least 12 grains per sample and are summarized in Table A1.

Values of $\delta^{18}O$ relative to SMOW were determined for each sample (i.e. a total of 29 samples) and are presented in Table A2. Conversion of data in Table A2 into compositions for detrital cores and authigenic quartz is done by solving simultaneous

Table A1. Relative abundance (%) of detrital quartz cores and diagenetic quartz overgrowths in samples from the Clynelish Quarry Sandstone Member (CQS), Lossiemouth borehole (L) and Beatrice Oilfield (M2) (see Hurst, 1980, 1985, for sample details).

		Subsample 1			Subsample 2			Subsample 3		
	Mesh	*a*	*b*	*c*	*a*	*b*	*c*	*a*	*b*	*c*
CQS18A	120	80	18	2	86	14	—	100	—	—
CQS 3H	120	80	17	3	94	6	—	100	—	—
CQS13A	120	92	5	3	97	3	—	100	—	—
L6771	40	100	trace	—	100	—	—	100	—	—
L6771	120	100	—	—	100	—	—	100	—	—
M238A	60	85	15	—	92	8	—	98	2	—
M208A	40	85	15	—	92	8	—	98	2	—
M243A	120	98	2	—	100	—	—	100	—	—
M248A	60	88	12	—	95	5	—	100	—	—
M238B	60	85	15	—	92	8	—	100	—	—

Mesh indicates the mesh size through which the sample grains passed. *a* detrital quartz; *b* mesoquartz; *c* microquartz and chalcedonic quartz.

Table A2. Isotopic analyses of subsamples 1, 2 and 3 from Table A1. All values are $\delta^{18}O$ relative to SMOW (‰).

	1	2	3
CQS18A	12·5	11·5	9·5
CQS 3A	14·3	12·0	11·2
CQS13A	13·6	20·0	22·3
L6771(40)	12·6	12·9	11·8
L6771(120)	11·5	11·8	—
M238A	13·7	12·9	12·4
M208A	14·2	12·3	13·7
M243A	13·1	13·6	13·1
M248A	13·7	13·4	12·4
M238B	12·8	11·7	12·4

equations based on the proportions of detrital quartz (*a*), mesoquartz (*b*) and microquartz/chalcedonic (*c*) (Table A1). An example calculation is given below.

Sample CQS18A

$$11{\cdot}5=0{\cdot}14b+0{\cdot}86a,$$
$$12{\cdot}5=0{\cdot}02c+0{\cdot}18b+0{\cdot}8a,$$

where $a=9{\cdot}5$. Thus, $b=23{\cdot}8$ and $c=30{\cdot}8$. Results are reported in Table A3. Values for detrital quartz (*a*) are derived from direct measurement on sample containing entirely detrital quartz or samples where only small quantites of authigenic quartz is present (samples M238A and M208A).

Values of $\delta^{18}O$ calculated by this method are very sensitive to minor variations in the estimation of proportions of detrital and diagenetic quartz. The validity of the method seems justified as the estimates of quartz proportions (Table A1) give reasonable values for $\delta^{18}O$ (Table A3). Only one sample (CQS13A) gives unreasonable (bulk $\delta^{18}O=20$‰) values for digenetic quartz. No source of analytical error was found.

Table A3. Estimates of the isotopic composition $\delta^{18}O$ of detrital (*a*), diagenetic (*b*) and microquartz/chalcedonic (*c*) quartz. Some samples have several values for detrital and diagenetic quartz. This is from duplicate analyses and samples where diagenetic quartz was present in more than one subsample.

	a	*b*	*c*
CQS18A	9·5	23·8	30·8
CQS 3A	11·2	24·5	39·2
CQS13A	13·6	?	?
L6771(40)	11.8 12.9 12.6	—	—
L6771(120)	11.5 11.8	—	—
M238A	12·4	21.0 22.0	—
M208A	12·1	14.6 26.1	—
M243A	13.1 13.6	13.1	—
M248A	12.4	23.2 32.4	—
M238B	12.4	3.7 15.1	—

Additional reference

Hurst, A. (1980) *The diagenesis of Jurassic rocks of the Moray Firth, NE Scotland.* PhD thesis, University of Reading.

Carbonate-cement-dominated mesogenesis

As well as forming during eogenesis, carbonate cements are common in sandstones during burial diagenesis. The origin of mesogenetic carbonate has been attributed to many sources, from simple recrystallization of biogenic debris and eogenetic cement through to the addition of both carbonate and divalent metals from external sources. The question of basin-scale fluid circulation is especially important for carbonate cements in sandstones, although an influx from fine-grained lithologies might require only localized diffusion or flow if mudstones and sandstones are intimately interbedded.

Zonation of mesogenetic ankerite and siderite (in terms of increasing iron and strontium content and increasingly radiogenic strontium) is assigned to sandstone formation waters that were controlled by minerals in neighbouring mudrocks (Mozley & Hoernle, 1990). These authors note that the mudstones thought to have controlled carbonate cementation in the sandstones had a similar diagenetic evolution of carbonate chemistry. Similarly, solutions responsible for the growth of ankerite in sandstones in a mixed sandstone–limestone sequence may have come from enclosing mudstones (Gawthorpe, 1987).

Reprinted from Sedimentology (1990) **37**, *817–836*

Geochemistry of carbonate cements in the Sag River and Shublik Formations (Triassic/Jurassic), North Slope, Alaska: implications for the geochemical evolution of formation waters

PETER S. MOZLEY* *and* KAJ HOERNLE

Department of Geological Sciences, University of California, Santa Barbara, CA 93106, USA

ABSTRACT

Carbonate cements (calcite, siderite, dolomite, and ankerite) formed throughout the diagenetic history of the Sag River and Shublik Formations. The trace element and isotopic geochemistry of these cements varies as a function of the timing of precipitation. Earliest calcites, formed prior to significant compaction of the sediment, are relatively enriched in Mg (up to 4·4 mol%), and have $^{87}Sr/^{86}Sr$ values (mean = 0·707898) compatible with the original marine pore waters. Later calcites are relatively Fe-rich (up to 5·0 mol%) and are characterized by increasing $^{87}Sr/^{86}Sr$ values (up to 0·712823) and Sr content with decreasing age. The Fe content of zoned siderite and dolomite/ankerite rhombs increases towards the outside of the rhombs (i.e. increasing Fe content with decreasing age).

These geochemical variations appear principally to result from changes in pore-water chemistry during diagenesis. The increase in $^{87}Sr/^{86}Sr$ and Sr content of the cements is most likely due to interaction between pore waters and ^{87}Sr-rich clay and possibly feldspar in Ellesmerian mudrocks (whole rock $^{87}Sr/^{86}Sr$ signatures for the mudrocks are >0·716). Pore-water Fe^{2+} concentration was probably controlled by diagenetic alterations involving Fe-bearing minerals (e.g. pyrite precipitation). A reconnaissance examination of carbonate cements in the overlying Kingak Shale indicates that similar alterations occurred in the Kingak.

The low $\delta^{18}O$ value of some calcite cements (−11·96‰ PDB) suggests that an influx of meteoric water may have occurred in the mid-Neocomian, though the low value could also result from an abnormally high geothermal gradient associated with mid-Neocomian rifting.

INTRODUCTION

Although the diagenesis and formation water chemistry of a number of units on Alaska's North Slope (Fig. 1) have been studied in detail (Melvin & Knight, 1984; Barnes, 1987; Eggert, 1987; Marinai, 1987; Kharaka & Carothers, 1988; Van de Kamp, 1988), the geochemical evolution of basin waters remains poorly understood. However, such information is vital to an accurate understanding of the hydrological and diagenetic history of the region. This paper addresses this aspect of North Slope basin history through a petrographical and geochemical analysis of carbonate cements in the Sag River and Shublik Formations (Late Triassic/Early Jurassic) in the National Petroleum Reserve, Alaska (NPRA).

The Sag River and Shublik Formations consist of organic-rich clastic and carbonate rocks that were deposited in a low-energy marine shelf environment (Jones & Speers, 1976). Because carbonate cements (calcite, dolomite, ankerite, and siderite) are abundant in both formations, and formed throughout much of the diagenetic history, analysis of the carbonate geochemistry has made it possible to infer the nature of changes in pore-water chemistry through time. These geochemical data indicate that dramatic variations occurred in the Fe^{2+}/Ca^{2+}, $Fe^{2+}/(Ca^{2+}+Mg^{2+})$, Sr^{2+}/Ca^{2+}, and $^{87}Sr/^{86}Sr$ ratios of the pore waters during diagenesis. In addition, a reconnaissance

*Present address: Geologisches Institut, Universität Bern, Baltzerstrasse 1, CH-3012 Bern, Switzerland.

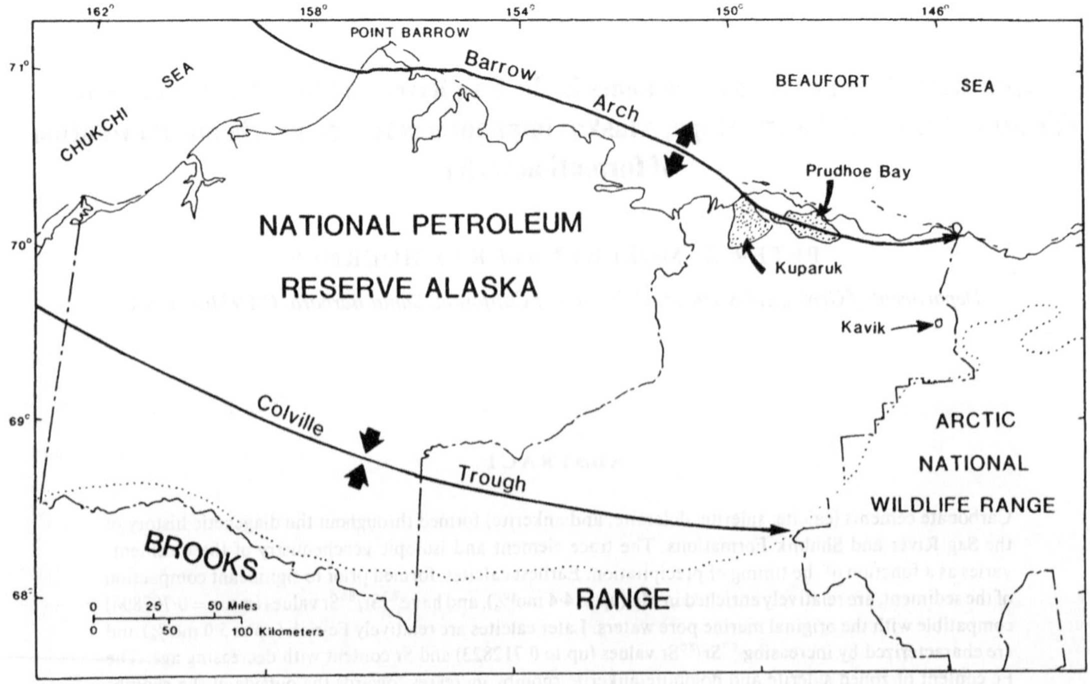

Fig. 1. Index map of the North Slope of Alaska showing major geographical and structural features. (Dotted line is northern limit of Brooks Range, stippled regions are major oil fields.)

examination of carbonate cements in sandstones within the overlying Kingak Shale (Jurassic) suggests that pore waters in the Kingak underwent similar alterations.

GEOLOGICAL SETTING

Stratigraphy

Regional

The post-Devonian sedimentary rocks of the North Slope are broadly divided into two major sedimentary packages (Fig. 2). The lowermost package, termed the Ellesmerian sequence, was derived from a continental landmass to the north of the present coastline. This sequence consists of Mississippian to Lower Cretaceous clastic and carbonate rocks deposited in non-marine to deep-marine environments. The Ellesmerian sequence averages about 1829 m in thickness, but locally may be up to three times this thick (Bird, 1985). The Sag River and Shublik Formations are part of this sequence of rocks. The Ellesmerian sequence is overlain by the Brookian sequence, a thick series (up to 6100 m; Bird, 1985) of predominantly clastic non-marine to deep-marine rocks of Early Cretaceous to Holocene age. These sediments were derived from rocks exposed in the Brooks Range.

Shublik Formation

The Shublik Formation is made up of phosphatic and glauconitic limestones, fine-grained sandstones, mudrocks, and rare ironstones. It ranges from 6·1 to 178 m in thickness (Bird, 1982). The Shublik conformably overlies the Ivishak Formation—a coarse-grained, fluvio-deltaic clastic unit that is the principal reservoir rock on the North Slope—except where the Ivishak has been removed by erosion near the Barrow Arch. The upper contact with the Sag River Formation is conformable (Bird, 1985). Palaeontological evidence indicates that the Shublik is Late Triassic (Carnian–Norian) in age (Jones & Speers, 1976; Dingus, 1984).

The Shublik was deposited in a zone of upwelling nutrient-rich waters in a marine shelf or shelf-slope environment (Dingus, 1984; Parrish, 1987). It is one of the most important hydrocarbon source rocks on the North Slope (Magoon & Bird, 1985).

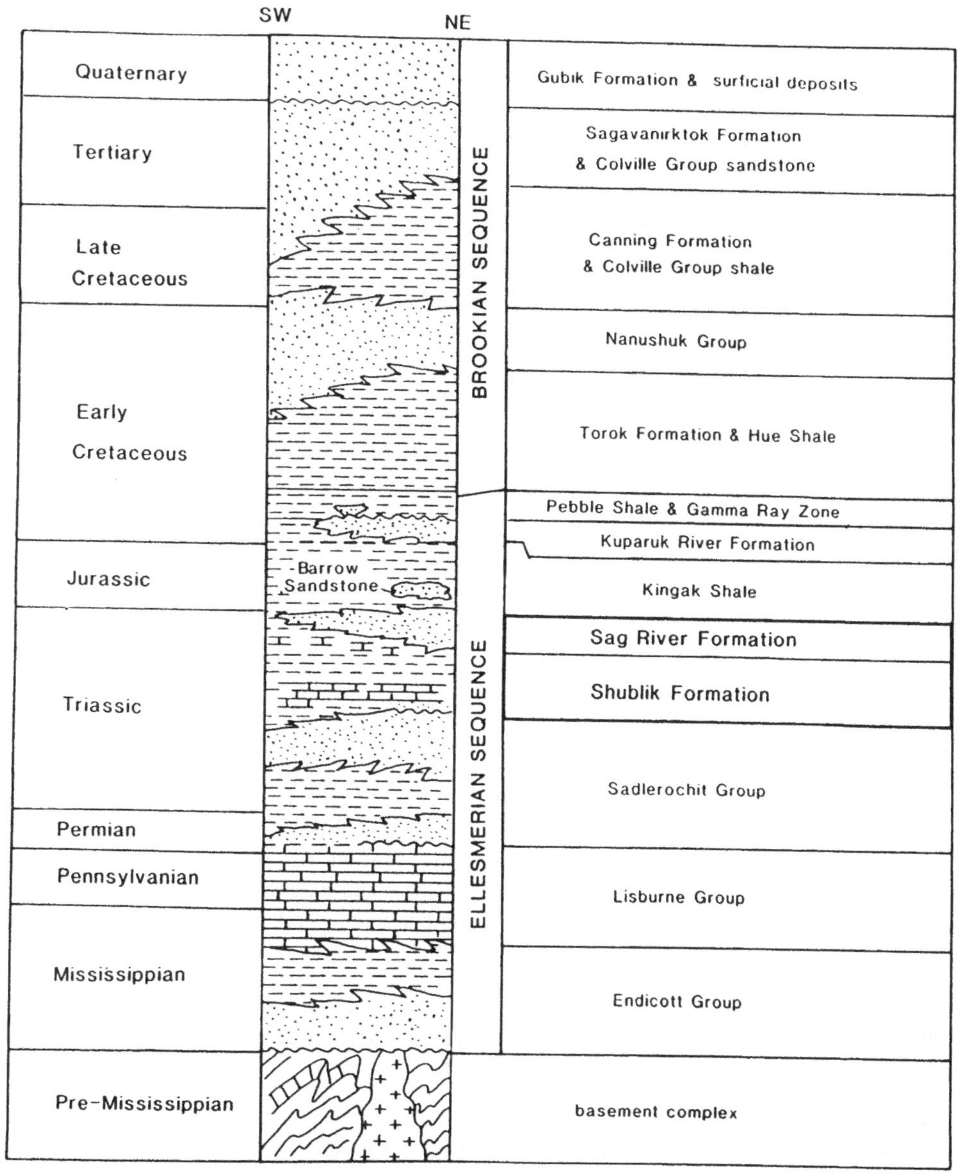

Fig. 2. Generalized stratigraphical column of the North Slope of Alaska. (Lithological code: brick pattern = limestone, dots = sandstone, dashes = mudrocks, complex = metamorphic.) After Bird (1985).

Sag River Formation

The Sag River Formation consists of fine-grained, glauconitic sandstones, mudrocks, and rare limestones and ironstones. In the National Petroleum Reserve Alaska (NPRA) it ranges from 15 to 82 m in thickness. The Sag River conformably overlies the Shublik Formation (Bird, 1985) and is disconformably overlain by the marine mudrocks of the Kingak Shale (Hubbard, Edrich & Rattey, 1987). It has been assigned a Late Triassic – earliest Jurassic (Norian–Hettangian) age (Jones & Speers, 1976).

The Sag River was deposited in a low-energy, offshore, marine-shelf environment (Jones & Speers,

1976; Barnes, 1987), and represents either partially reworked prodelta sands (Noonan, 1987) or shelf sandstones similar to the Cretaceous Ferron Sandstone in east-central Utah and modern sands offshore of Sapelo Island, Georgia (Barnes, 1987).

Nearly 4 billion barrels of oil and oil-equivalent gas are present in the Sag River Formation in the Prudhoe Bay field (Barnes, 1987), and in the Kavik field (Fig. 1) the Sag River also contains significant quantities of gas (Bird & Molenaar, 1987).

Basin evolution

The geological evolution of the North Slope has been discussed in detail elsewhere, and is only briefly covered here. Bird (1985) and Hubbard *et al.* (1987) are the principal sources for this discussion. The geological history of the North Slope can be summarized as follows.

(1) From early Mississippian to Early Jurassic the Ellesmerian sequence was deposited along the margin of a continental landmass to the north of the present-day coastline.

(2) From Early Jurassic to Early Cretaceous (mid-Neocomian), the region experienced regional extension, culminating in a rifting event that removed the northern landmass. Rift-margin uplift and subaerial erosion resulted in the formation of a 'break-up unconformity' (the mid-Neocomian unconformity) and the erosional truncation of Ellesmerian units. Meteoric water may have entered subaerially exposed permeable sandstone units such as the Ivishak Formation at this time (Kharaka & Carothers, 1988; Shanmugam & Higgins, 1988; Van de Kamp, 1988).

(3) By late Albian time, sediments shed from the rising Brooks Range to the south covered the Ellesmerian sequence. Continued sedimentation caused rapid subsidence and deep burial of Ellesmerian units. At this point hydrocarbon source rocks in the Colville Trough were buried to sufficient depths to generate oil; the earliest that hydrocarbon generation could have occurred in the region.

(4) From Albian to Recent, Brookian deformation and associated foredeep sedimentation migrated progressively north-eastward, with present-day deformation and sedimentation occurring north of the Barrow Arch in NE Alaska.

Burial history curves for representative NPRA wells are shown in Figure 3.

METHODS

Sampling

Samples of conventional core material from the Sag River and Shublik Formations (22 from the Sag River and 70 from the Shublik) in 10 wells (Fig. 4) were obtained from the US Geological Survey/NPRA core-repository in Menlo Park, California. In addition, six samples from the Kingak Shale that are lithological and stratigraphical equivalents of the Barrow Sandstone (K. Bird, pers. comm., 1985) were also obtained.

Analytical techniques

Petrographical analysis

The modal composition of the samples was determined by point counting (300 points for sandstones, 200 points for carbonates, 100 points for mudrocks). The mean grain size and degree of sorting were visually estimated using textural standards.

Microprobe analysis

The elemental composition of carbonate cements was determined on UNOCAL Science and Technology Division's JEOL Superprobe-733 (standard operating conditions: 15 kV acceleration potential, 20 nA sample current, 0·5–10 μm beam diameter) and on U.C. Santa Barbara's ARL EMX-SM (15 kV acceleration potential, 10 nA sample current, 5–10 μm beam diameter) using carbonate standards. The JEOL microprobe was used on samples for which the high-resolution back-scattered electron imaging system was needed to view diagenetic textures at high magnification.

O and C isotopes

Calcite. To avoid contamination by carbonate fossil fragments, pure calcite separates were obtained by hand-picking crushed samples under a binocular microscope. One sample (South Meade No. 1, 2692·5 m) was too fine grained for hand picking, making it necessary to obtain a bulk carbonate value. Because carbonate contaminants are present in low amounts relative to the calcite cement in this sample, the value is thought to represent the calcite cement accurately. Carbon dioxide gas for spectroscopic analysis was obtained by reacting the calcite with phosphoric acid at 25°C for 3 days. A phosphoric acid

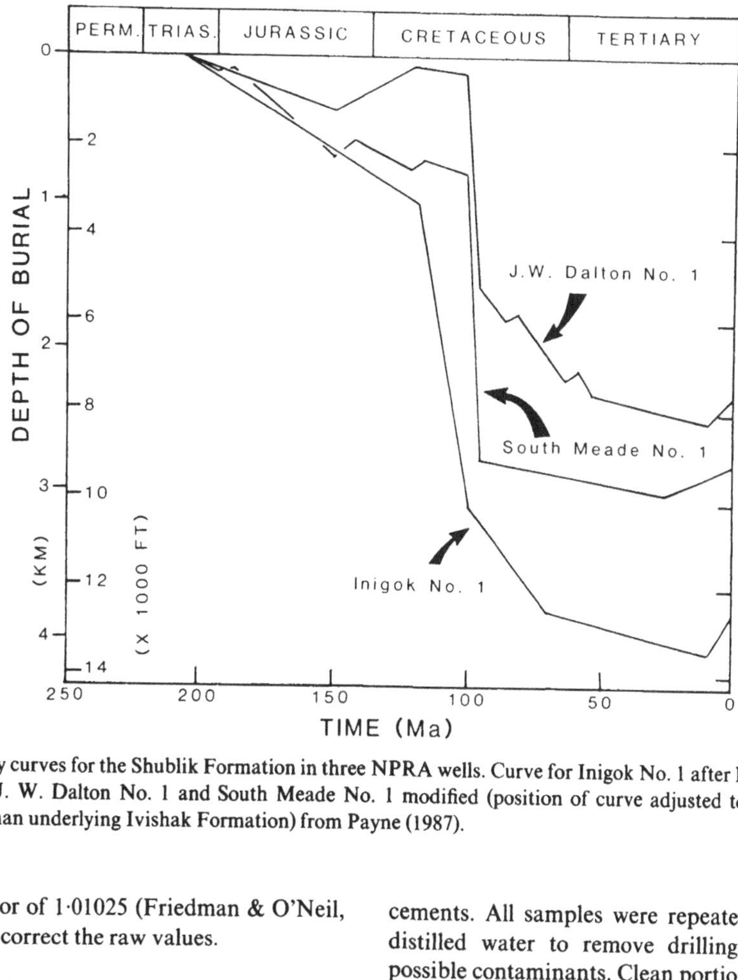

Fig. 3. Burial history curves for the Shublik Formation in three NPRA wells. Curve for Inigok No. 1 after Magoon & Claypool (1982); curves for J. W. Dalton No. 1 and South Meade No. 1 modified (position of curve adjusted to represent Shublik Formation rather than underlying Ivishak Formation) from Payne (1987).

fractionation factor of 1·01025 (Friedman & O'Neil, 1977) was used to correct the raw values.

Siderite. Siderite-rich samples containing little or no extraneous carbonate material were crushed and reacted directly with phosphoric acid. The samples were reacted in an oil bath at 50°C for 7 days to allow the reaction to go to completion. The siderite values were corrected using a 50°C phosphoric acid liberated CO_2–siderite fractionation factor of 1·01075 (Carothers, Adami & Rosenbauer, 1988).

Spectroscopic analysis and CO_2 extraction for carbonate cements were provided by the UNOCAL Science and Technology Division (Brea, California).

Sr isotopes

The $^{87}Sr/^{86}Sr$ ratio and Sr concentration were determined for selected samples. Because detrital carbonate material is common in these rocks, coarse-grained carbonate cements were preferentially sampled in order to facilitate mechanical separation of the cements. All samples were repeatedly washed with distilled water to remove drilling-mud and other possible contaminants. Clean portions of calcite were separated from crushed samples under a binocular microscope. One siderite sample and two of the pore-filling calcite samples were too fine-grained to be separated by hand, making it necessary to remove the carbonate through acid dissolution. The calcite was dissolved in 5 M acetic acid. Prior to dissolution, the clay fraction ($<2\,\mu m$ in size) was removed through repeated centrifugation of the samples in order to minimize possible exchange with clay mineral Sr (technique used by J. Schultz, 1985, U.C. Santa Barbara). The siderite sample (powdered) was first treated for 1 h with 5 M acetic acid to remove any possible calcite contamination. It was then dissolved in 2·5 M hydrochloric acid. The shale samples and a clay separate were dissolved in a mixture of hydrofluoric and perchloric acid.

Standard isotope dilution techniques were used to determine concentrations of Rb and Sr using ^{87}Rb- and ^{84}Sr-enriched spikes. Sr isotope ratios were

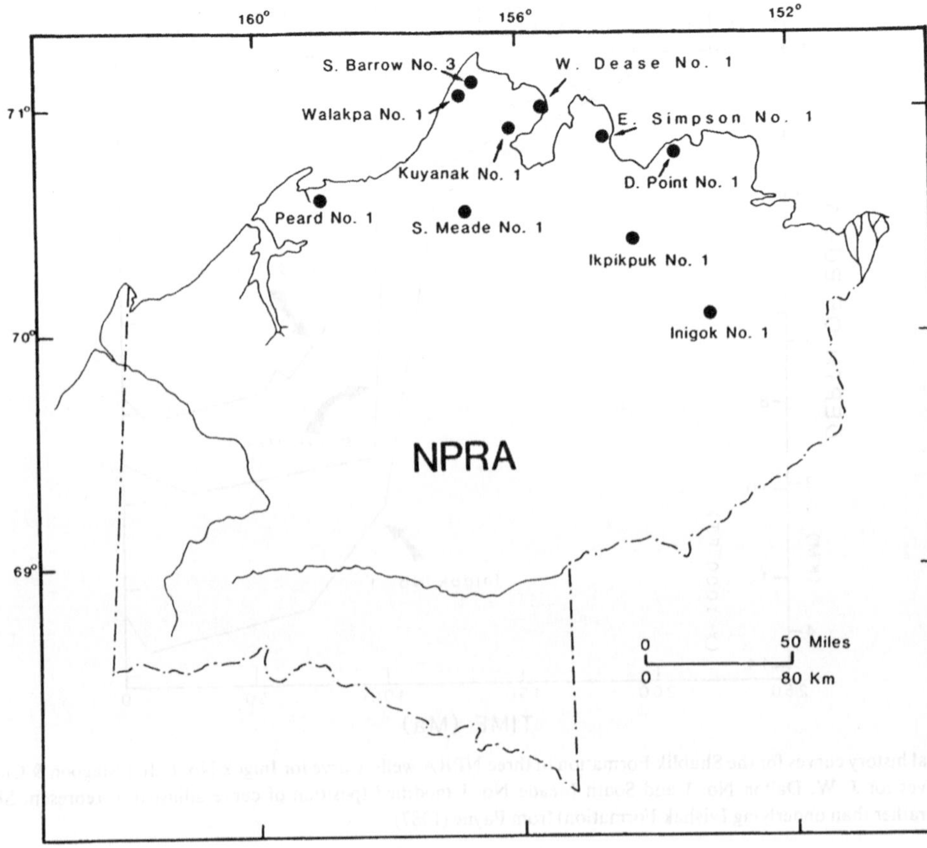

Fig. 4. Map of the National Petroleum Reserve, Alaska (NPRA), showing locations of wells sampled in this study.

determined on a Finnigan Mat 261 multiple collector mass spectrometer. Mass fractionation was corrected by normalizing $^{87}Sr/^{86}Sr$ ratios to 0·1194; all isotopic ratios were corrected to a value of 0·71025 for the NBS987 standard. A blank yielded a value of 4·39 pg. Sample weights ranged from 10 to 50 mg. Estimates of analytical precision based on repeated analysis are $\pm$0·00005 for $^{87}Sr/^{86}Sr$ ratios, and 0·5% for Sr and Rb concentrations. Errors in the concentration of Sr and Rb for the leached samples are estimated at 10% due to the difficulty in determining sample weights. All quoted errors are two standard deviations.

PETROLOGY

Sandstones

The sandstones are primarily fine-grained, moderately to well-sorted, sublitharenites (Fig. 5; classification of Folk, 1968). Framework grains consist mainly of monocrystalline quartz and chert, with lesser amounts of carbonate and argillaceous grains. Glauconite and fossil fragments are locally abundant, and mica and feldspar are present in trace amounts. The major non-framework components are calcite, siderite, and detrital clay (Fig. 6). Quartz overgrowths, pyrite, kaolinite, dolomite/ankerite, iron oxides, chlorite, and a fine-grained phosphatic matrix make up a significant portion of some samples (Fig. 6). Minor amounts of authigenic albite, carbonate-fluorapatite, barite, and residual oil are present locally.

Mudrocks

XRD analysis of the mudrocks indicates that they are composed of a mixture of quartz, illite/mica, and Fe-rich chlorite. Minor amounts of calcite, dolomite, siderite, feldspar, and pyrite are also present in many of the samples.

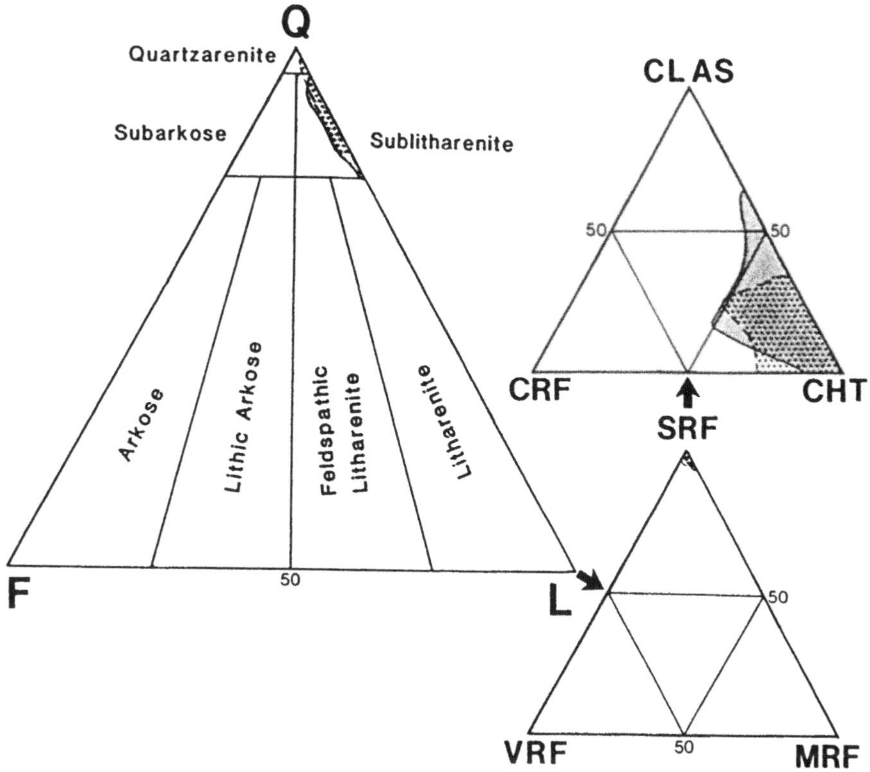

Fig. 5. Ternary diagrams illustrating the composition of framework grains in the Sag River and Shublik Formation sandstones. Grey pattern = Sag River ($n = 15$), stipple = Shublik ($n = 29$), Q = quartz, F = feldspar, L = lithic fragments, VRF = volcanic rock fragments, MRF = metamorphic rock fragments, SRF = sedimentary rock fragments, CRF = carbonate rock fragments, CHT = chert, CLAS = clastic rock fragments. Sandstone classification of Folk (1968).

Limestones

The limestones classify as arenaceous and fossiliferous grainstones, with lesser amounts of fossiliferous packstone, wackestone, and mudstone (classification of Dunham, 1963).

The major framework components are monocrystalline quartz and fossil fragments. Glauconite and phosphatic grains are locally abundant. Non-framework components consist mainly of calcite, with lesser amounts of phosphatic matrix, siderite, pyrite, and dolomite (Fig. 6). Small amounts of authigenic quartz, albite, barite, and haematite are present in a few samples.

CARBONATE PARAGENESIS

The Sag River and Shublik Formations have a complex paragenetic history, which ranges from very early authigenic minerals (such as glauconite and carbonate-fluorapatite) that must have formed near the sediment–water interface, to late-stage minerals (such as fracture-filling Fe-rich calcite) that formed subsequent to or during hydrocarbon generation in the basin (Fig. 7). Carbonate cements formed throughout this paragenetic history.

Siderite

Siderite principally occurs as small rhombs (generally 10–30 μm long axis) in intergranular areas, and as concretions (Fig. 8A). The concretions are common in both formations, and consist of an amalgamation of very fine-grained siderite rhombs.

Siderite appears to post-date glauconite, carbonate-fluorapatite, and pyrite, but pre-date most of the other authigenic minerals. Open packing and high minus-cement porosity (> 60%) in siderite-cemented regions,

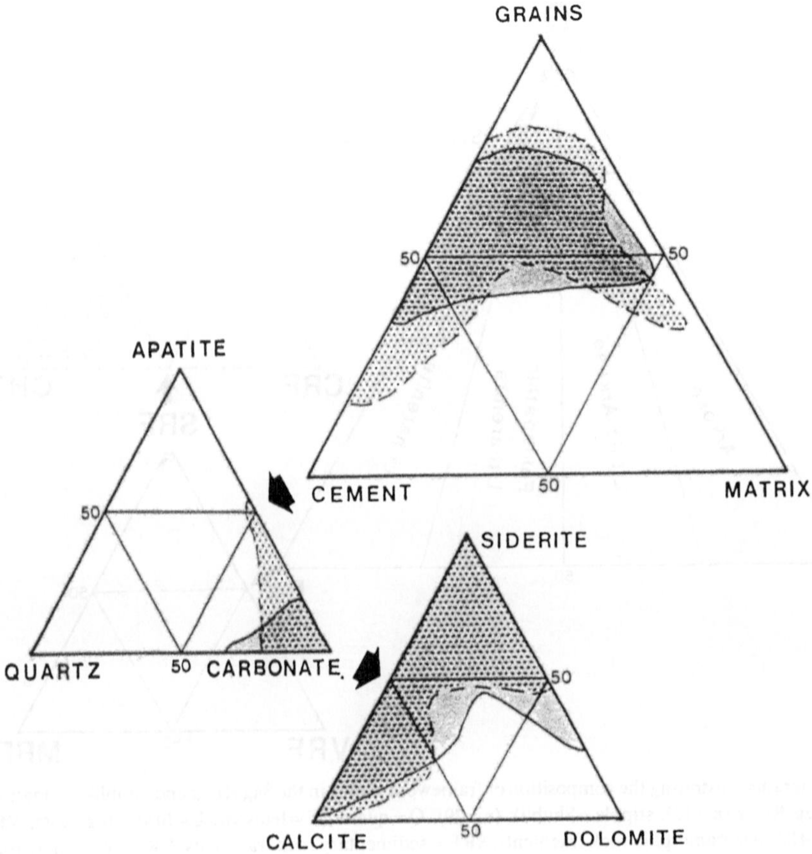

Fig. 6. Ternary diagrams illustrating the proportions of grains, cement, matrix, and major authigenic minerals in samples of the Sag River and Shublik Formations. Grey pattern = Sag River ($n = 18$), stipple = Shublik ($n = 43$).

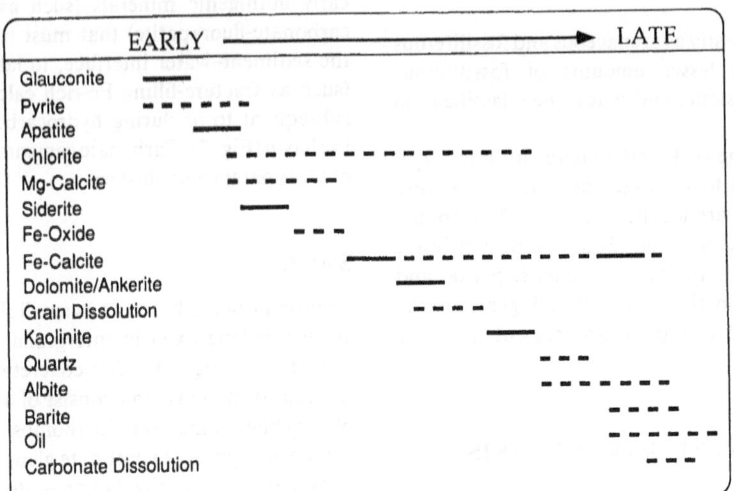

Fig. 7. Diagram illustrating generalized paragenetic sequence for the Sag River and Shublik Formations. Dashed lines show the range over which a given alteration may have occurred.

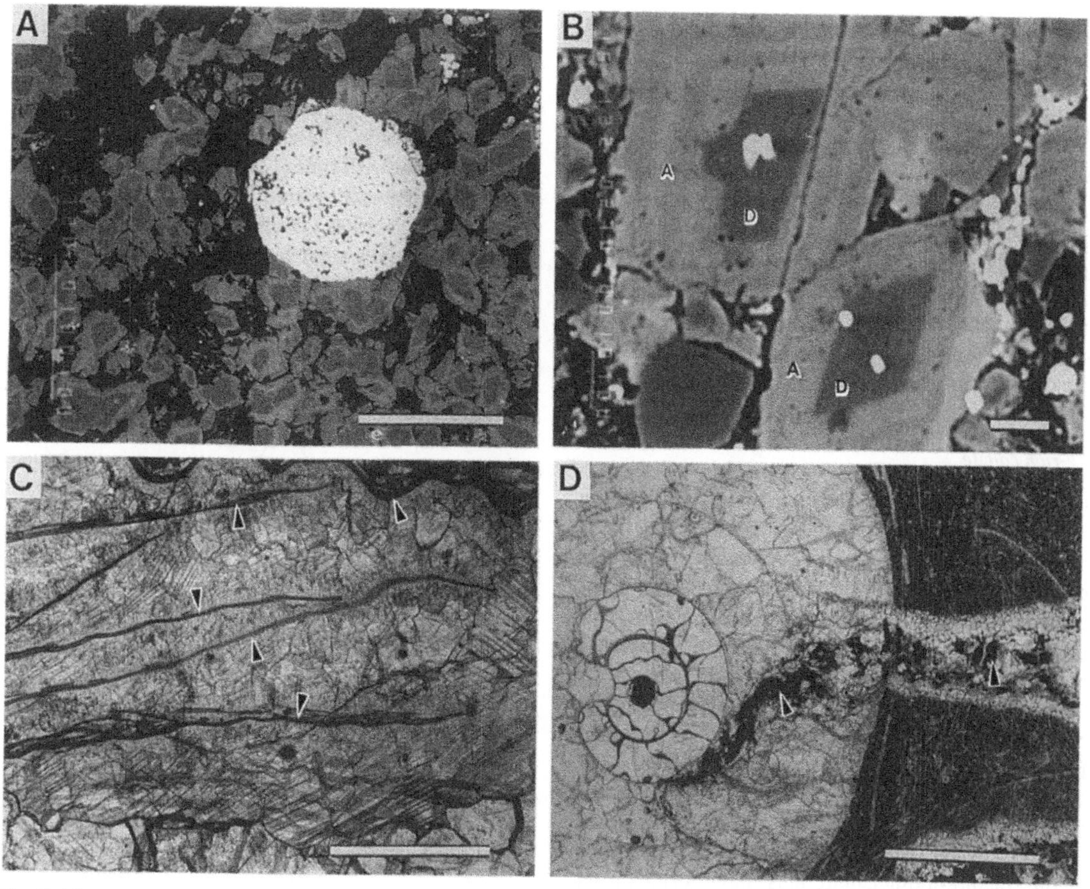

Fig. 8. Photomicrographs of carbonate cements. (A) Zoned intergranular siderite rhombs. The dark central portion of the rhombs contains less Fe and more Mg than the lighter outer portion. Framboidal pyrite (white) is also present. Back-scattered electrons, scale bar = 0·1 mm. (B) Zoned dolomite/ankerite rhombs. Dolomite (dark grey, D) makes up the centre of the rhombs, zoned ankerite (light grey, A) the outer portion. White areas are pyrite. Back-scattered electrons, scale bar = 0·01 mm. (C) Shell fragments (arrows) surrounded by pore-filling Mg-rich calcite. Plane light, scale bar = 1·0 mm. (D) Ammonite shell filled with pore-filling Mg-rich calcite. A fracture filled with Fe-rich calcite cross-cuts the Mg-rich calcite and shell. Inclusions of oil (black, arrows) are present in the fracture-filling calcite. Plane light, scale bar = 2·0 mm.

as well as deflection of mudrock laminae around siderite concretions, suggest that siderite formed prior to significant compaction.

Dolomite/ankerite

Dolomite and ankerite occur mainly as isolated rhombs. These rhombs are typically larger than those of siderite (20–75 μm long axis, vs. 10–30 μm for siderite), and are highly zoned, with both dolomite and ankerite layers present in the same rhomb (Fig. 8B). Often the rhombs appear to have originated as overgrowths on detrital dolomite grains.

Dolomite/ankerite rhombs formed subsequent to early Fe-rich calcite, but prior to quartz overgrowths and late Fe-rich calcite. Dolomite invariably preceded ankerite in rhombs containing both dolomite and ankerite.

Calcite

Calcite principally occurs as fracture-filling and poikilotopic pore-filling cement. Two chemically distinct types of calcite are present, one relatively enriched in Fe, the other in Mg (hereafter referred to as Fe-rich and Mg-rich calcite respectively). Mg-rich calcite only occurs in the Shublik Formation, whereas Fe-rich calcite is present in both units.

Mg-rich calcite

Mg-rich calcite is an early pore-filling cement. Rocks cemented by Mg-rich calcite have high minus-cement porosity (>40%) with no evidence for significant grain replacement, and appear to have experienced only minor compaction prior to cementation (Fig. 8C). The complete absence of quartz overgrowths from sandstones cemented by Mg-rich calcite indicates that the calcite pre-dates quartz overgrowth precipitation.

Fe-rich calcite

Fe-rich calcite occurs as pore- and fracture-filling cements that formed throughout much of the diagenetic history. Some of the pore-filling Fe-rich calcite occurs in samples that experienced only minor compaction prior to precipitation (i.e. minus-cement porosity >40% and undeformed ductile framework grains), whereas others fill pores in more compacted samples (i.e. minus-cement porosity 10–20%, highly deformed ductile framework grains). The fracture-filling calcite apparently formed very late in the diagenetic history. It cross-cuts both Fe- and Mg-rich calcite-cemented samples, and in places contains inclusions of oil (Fig. 8D)—an indication that some of the fracture-filling calcite must have formed after deep burial of the units in the Early Cretaceous when hydrocarbon generation occurred for the first time in the area.

The paragenetic relationship between Fe-rich calcite and quartz overgrowths confirms the presence of both early- and late-stage pore-filling calcite cements. The total absence of quartz overgrowths from Fe-rich calcite-cemented portions of some samples suggests that the calcite pre-dates overgrowth formation. In other samples, however, Fe-rich calcite clearly post-dates the overgrowths. Similarly, dolomite/ankerite rhombs both post- and pre-date Fe-rich calcite in some samples.

CARBONATE GEOCHEMISTRY

Elemental composition

Siderite

The siderite is extremely impure, with extensive substitution of Mg and, to a lesser extent, Ca for Fe in the siderite lattice (Fig. 9). Individual siderite rhombs are highly zoned (zonation can only be observed under u.v. light or through back-scattered electron imaging, Fig. 8A), a reflection of variations in the Fe/(Ca + Mg) ratio. The zonation almost always shows Fe-enrichment with time (i.e. Fe-poor rhomb centres, Fe-rich rhomb outer layers).

The impure nature of these siderites is characteristic of early diagenetic siderite from marine depositional environments, and probably results from the relatively low $Fe^{2+}/(Ca^{2+}+Mg^{2+})$ ratio of early marine pore waters (Mozley, 1989a).

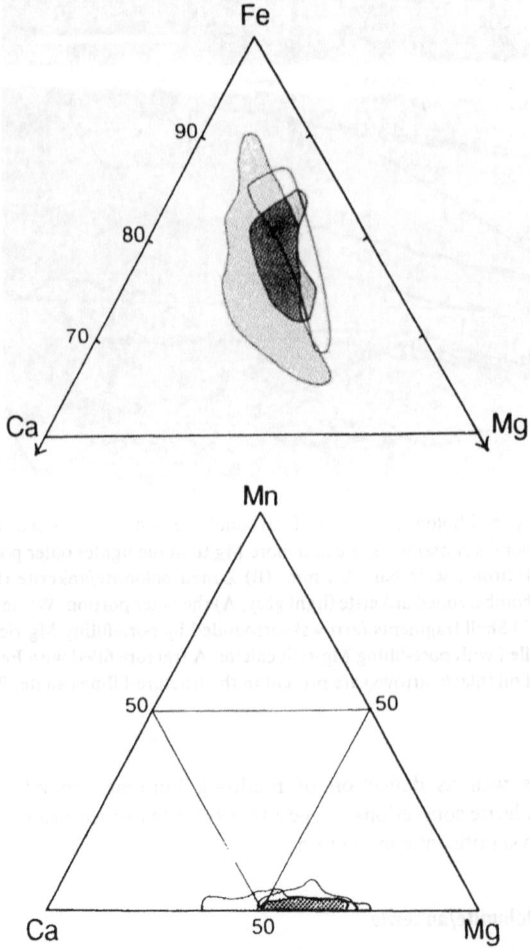

Fig. 9. Ternary diagrams illustrating the elemental composition of authigenic siderite in the Sag River and Shublik Formations, and Barrow Sandstone. (Top) Relative mole percentages of $FeCO_3$, $CaCO_3$, and $MgCO_3$; Ca and Mg corners = 60%, Fe corner = 100%. (Bottom) Relative mole percentages of $MnCO_3$, $CaCO_3$, and $MgCO_3$. Grey pattern = Sag River (n = 16), stipple pattern = Shublik (n = 34), no pattern = Barrow Sandstone (n = 14).

Dolomite/ankerite

Dolomite and ankerite occur in rhombs which range in composition from end-member dolomite to ankerite (Fig. 10). The ankerite is non-stoichiometric, containing excess Ca (57–58 mol% $CaCO_3$), and is strongly zoned (principally a variation in the Fe/Mg ratio of the ankerite).

Calcite

As previously mentioned, the calcite formed along distinct Mg- and Fe-rich compositional trends (Fig. 11).

Mg-rich calcite. The Mg-rich calcite contains 0·6–4·4 mol% $MgCO_3$ and 0–0·83 mol% $FeCO_3$. The relatively high Mg content probably results from the Mg-rich nature of early marine pore waters, whereas the low Fe content reflects low Fe^{2+} concentrations in the parent waters—the result of either precipitation under oxidizing conditions or contemporaneous precipitation of pyrite (see Coleman, 1985 for discussion of controls on composition of early carbonate minerals in organic-rich marine sediments). Alternatively, because early calcite cements in marine sediments often contain much higher Mg concentrations than these calcites (in 100–1000 m water depths, $MgCO_3$ content commonly 10–12 mol%, i.e. high-Mg calcite; Scoffin, 1987), they may have formed through the recrystallization of high-Mg calcite.

Fe-rich calcite. The Fe-rich calcite contains 0·5–5·0 mol% $FeCO_3$ and 0–1·64 mol% $MgCO_3$. The Fe-rich nature of the calcite reflects the presence of significant concentrations of Fe^{2+} and relatively low sulphide concentrations in the pore waters during precipitation.

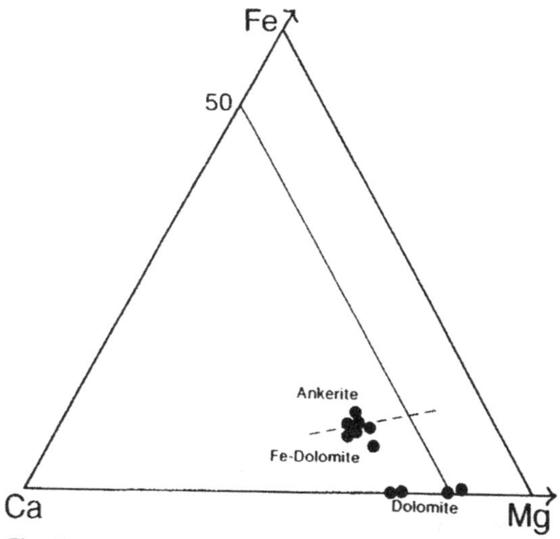

Fig. 10. Ternary diagram illustrating the elemental composition of dolomite and ankerite cements in the Sag River and Shublik Formations ($n=15$). Fe and Mg corners = 60%, Ca corner = 100%.

Isotope geochemistry

Oxygen and carbon

Oxygen isotopes. The siderite has a $\delta^{18}O$ value of

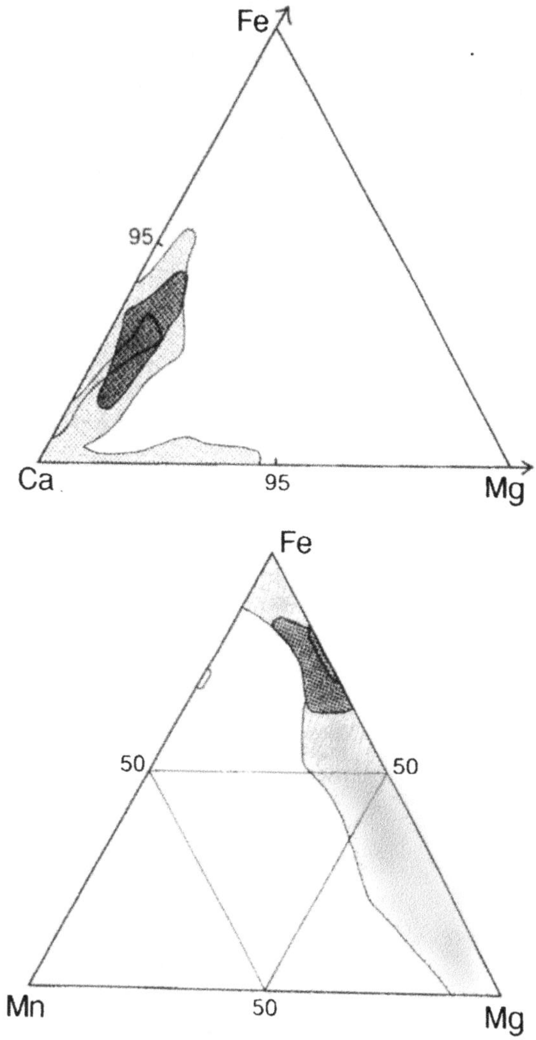

Fig. 11. Ternary plots illustrating the elemental composition of calcite cements from the Sag River and Shublik Formations, and Barrow Sandstone. (Top) Relative mole percentages of $CaCO_3$, $MgCO_3$, $FeCO_3$; Fe and Mg corners = 10%, Ca corner = 100%. (Bottom) Relative mole percentages of $MnCO_3$, $MgCO_3$, and $FeCO_3$. Grey pattern = Sag River ($n=23$), stipple pattern = Shublik ($n=69$), no pattern = Barrow Sandstone ($n=11$).

1·05‰ relative to PDB, whereas the calcites range from −3·08 to −12·57‰ (Table 1). The Mg-rich calcites have relatively high values of −3·08 and −4·62‰, whereas the Fe-rich calcites range from −5·77 to −11·96‰.

The $\delta^{18}O$ value for the siderite is consistent with precipitation from marine water at temperatures encountered at relatively shallow burial depths (i.e. around 26°C if water of $\delta^{18}O=0$‰ SMOW is assumed; calculated using the equation of Carothers, Adami & Rosenbauer, 1988).

The calcites have an 8‰ range in their $\delta^{18}O$ values. The highest values are compatible with precipitation from ocean water at temperatures encountered during shallow burial (i.e. 30–40°C if a water of $\delta^{18}O=0$‰ SMOW is assumed; Fig. 12). Higher temperatures of precipitation and/or ^{18}O-depleted parent waters (i.e. $\delta^{18}O<0$‰ SMOW) are necessary to explain the lower $\delta^{18}O$ values. In either case, the high temperatures or ^{18}O-depleted waters must have been present prior to significant burial (i.e. >1 km), because some low $\delta^{18}O$ cements occur in samples with high pre-cementation porosity (i.e. South Meade No. 1, 2692·5 m; −11·96‰ (PDB); around 35–40% minus-cement porosity with no evidence for significant grain replacement), and thus could not have formed subsequent to deep burial. Alternatively, these cements may have formed at depth in overpressured zones (i.e. compaction inhibited by high pore-fluid pressure). This seems unlikely, however, because overpressuring has never been reported for these units.

Carbon isotopes. The siderite has a very low $\delta^{18}C$ value of −11·50‰ relative to PDB. $\delta^{13}C$ values for the calcites range from +2·80 to −3·05‰ PDB (Table 1).

The extremely low value for the siderite indicates

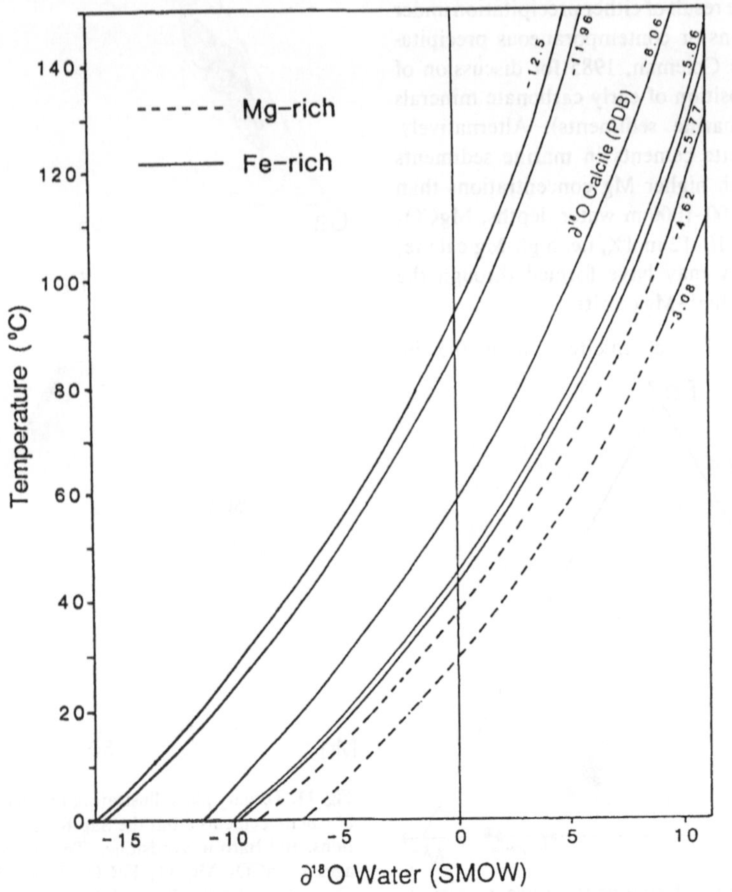

Fig. 12. Plot of $\delta^{18}O$ of pore water vs. precipitation temperature for calcite cements from the Sag River and Shublik Formations, and Barrow Sandstone. Diagram constructed using the equation $10^3 \ln \alpha = 2{\cdot}78\times10^6\,T^{-2}-2{\cdot}89$, where α is the isotope fractionation factor and T is temperature in kelvin (equation for H_2O–$CaCO_3$ fractionation, O'Neil *et al.*, 1969).

Table 1. Geochemical data for samples from the Barrow Sandstone, and Sag River and Shublik Formations in the National Petroleum Reserve, Alaska.

Well	Depth (m)	Unit	Material	$\delta^{13}C$ (‰PDB)	$\delta^{18}O$ (‰PDB)	$^{87}Sr/^{86}Sr$ (measured)	$^{87}Sr/^{86}Sr$ (replicate)	$^{87}Sr/^{86}Sr$ (initial)	Rb (p.p.m.)	Sr (p.p.m.)	Ca (mol%)	Mg (mol%)	Fe (mol%)	Mn (mol%)
IK	3134·0	SH	pf Mg-cal	−3·05	−3·08	0·707851 (34)	—	—	0·01	309·4	95·64	2·84	0·00	0·07
IK	3135·8A	SH	pf Mg-cal	−1·68	−4·62	0·707892 (39)	—	—	0·03	255·0	99·40	1·91	0·00	0·00
IK	3135·8B	SH	pf Mg-cal	—	—	0·707923 (15)	—	—	0·03	179·4	100·20	1·25	0·00	0·00
IK	3134·9	SH	pf Mg-cal	—	—	0·707707 (17)	0·707751 (27)	—	0·01	178·0	—	—	—	—
IK	3137·6	SH	pf Mg-cal	—	—	0·708119 (43)	—	—	0·02	608·0	95·95	2·09	0·15	0·07
PD	2523·4	BA	ff Fe-cal	−5·79	−12·57	0·711428 (34)	—	—	0·09	788·0	98·90	0·00	0·96	0·13
PD	2523·4	BA	conc sid	−13·40	0·88	0·708686 (13)	—	0·708592	0·75	68·7	6·57	11·15	82·02	0·25
PD	2582·0	SH	ff Fe-cal	—	—	0·711155 (48)	—	—	0·21	1428·0	95·58	0·00	2·42	0·16
PD	2583·8	SH	ff Fe-cal	−3·43	−8·06	0·711005 (07)	0·711050 (21)	—	0·76	1008·0	—	—	—	—
SB	777·2	KS	shale	—	—	0·722124 (10)	—	0·720672	61·07	344·5	—	—	—	—
SB	789·7	SR	conc sid	−11·50	1·05	—	—	—	—	—	—	—	—	—
SM	2692·5	SR	pf Fe-cal	−2·97	−11·96	0·710290 (10)	—	—	1·18	463·5	94·35	1·05	3·32	0·00
SM	2692·5	SR	clay sep	—	—	0·715238 (17)	0·715294 (44)	—	0·00	96·9	—	—	—	—
SM	2757·8	SH	ff Fe-cal	—	—	0·711177 (11)	—	—	0·27	1667·0	96·99	0·81	2·00	0·11
WD	1157·9	KS	shale	—	—	0·720594 (10)	—	0·716569	235·10	478·3	—	—	—	—
WD	1201·6	SH	shell	—	—	0·707847 (17)	0·707861 (37)	—	0·01	594·2	—	—	—	—
WD	1201·8	SH	ff Fe-cal	−1·46	−5·86	0·712753 (07)	—	—	3·40	2802·0	97·62	0·60	1·70	0·00
WD	1201·8	SH	pf Fe-cal	—	—	0·709473 (21)	—	—	0·17	473·6	95·89	1·11	2·95	0·00
WD	1202·0	SH	shell	—	—	0·707812 (10)	—	—	0·00	463·0	—	—	—	—
WD	1202·8	SH	ff Fe-cal	2·80	−5·77	0·712823 (11)	—	—	0·35	1603·0	97·98	0·37	1·57	0·07

Isotopic values represent a single analysis for a given sample, whereas elemental compositions are mean values of multiple spot analyses (see Mozley, 1988 for individual analyses). Numbers in parentheses are within-run statistical errors (2 SD) applicable to the last two digits; ages used in calculation of $^{87}Sr/^{86}Sr$ initials are: Shublik = 219 Ma, Sag River = 214 Ma, Kingak = 204 Ma, initials only listed where significantly different from measured value. Abbreviations: IK = Ikpikpuk No. 1, PD = Peard No. 1, SB = South Barrow No. 3, SM = South Meade No. 1, WD = West Dease No. 1, SH = Shublik; KS = Kingak Shak; BA = Barrow; SR = Sag River; pf = pore filling; ff = fracture filling; Fe-cal = Fe-rich calcite; Mg-cal = Mg-rich calcite; conc = concretionary; sid = siderite.

that much of it formed from waters enriched in carbon derived from the degradation of organic matter. Given the restricted chemical conditions under which siderite will precipitate (i.e. low Eh, very low sulphide concentrations), the low $\delta^{13}C$ values suggest precipitation either under suboxic conditions in the zone of Fe^{3+} reduction, or very early in the methanic zone (cf. Maynard, 1982; Gautier & Claypool, 1984; Walker, 1984; Curtis & Coleman, 1986).

Sr isotopes and Sr concentration

The $^{87}Sr/^{86}Sr$ ratio and Sr concentration of the calcite ranges considerably (Fig. 13, Table 1). Mg-rich samples have $^{87}Sr/^{86}Sr$ ratios ranging from 0·707707 to 0·708119 and Sr concentrations ranging from 178·0 to 608·0 p.p.m., whereas the Fe-rich samples vary more widely, ranging from 0·709473 to 0·712823 in $^{87}Sr/^{86}Sr$ and 463 to 2802·0 p.p.m. in Sr concentration.

There is a positive correlation between the $^{87}Sr/^{86}Sr$ ratio and Sr concentration of the calcites (Fig. 13). This relationship is consistent with a two-component mixing model (i.e. the mixing of Sr from two principal sources) involving a relatively low Sr concentration, low $^{87}Sr/^{86}Sr$ end-member and a relatively high Sr concentration, high $^{87}Sr/^{86}Sr$ end-member. Furthermore, the variation in $^{87}Sr/^{86}Sr$ and Sr concentration appears to have been time-dependent. Very early cements such as Mg-rich calcite have low concentrations and ratios, whereas demonstrably very late cements, such as fracture-filling calcite, have the highest concentrations and ratios. Calcites that precipitated between these end-members (i.e. Fe-rich pore-filling calcite) have intermediate values. Thus the amount of Sr and the $^{87}Sr/^{86}Sr$ ratios of the calcite increase systematically with decreasing age.

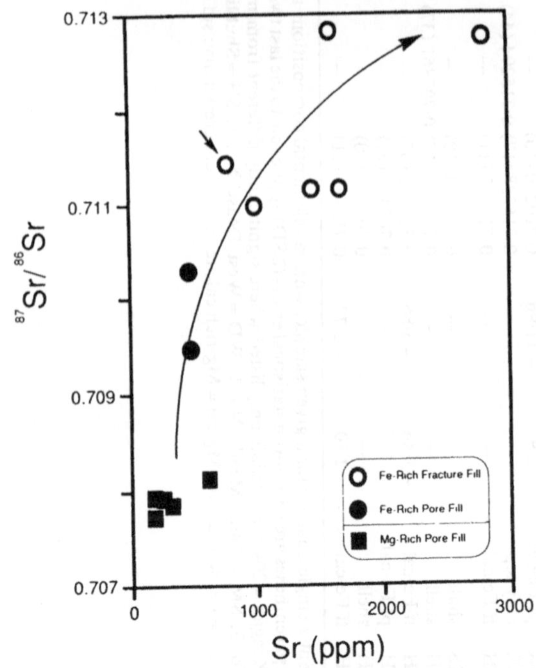

Fig. 13. Plot of $^{87}Sr/^{86}Sr$ vs. Sr concentration in calcite cements. Mg-rich calcites have relatively low Sr concentrations and $^{87}Sr/^{86}Sr$ ratios, whereas values for Fe-rich calcites are higher and range widely. The correlation coefficient for $^{87}Sr/^{86}Sr$ vs. 1/Sr is 0·83. Large arrow shows probable temporal trend. Small arrow = data for Barrow Sandstone sample.

The Mg-rich calcites have $^{87}Sr/^{86}Sr$ ratios within the range shown by Burke *et al.* (1982) for Triassic seawater. To confirm that Sag River and Shublik depositional water had a similar ratio to this 'typical' Triassic value, fragments of two pelecypod shells were analysed. The shell values (Table 1) agree with the Triassic values of Burke *et al.* (1982). Thus the Sr ratios of the Mg-rich calcite are consistent with precipitation from unaltered or only slightly altered Triassic seawater. The Fe-rich calcite cements, however, have ratios far too high to have been derived from these waters. More-radiogenic Sr (i.e. higher $^{87}Sr/^{86}Sr$) from another source (the upper end-member) must have been progressively added to the system to account for their values.

Given the very low Sr^{2+} content of most meteoric waters (Veizer, 1983; Faure, 1986) and the very high Sr content of many of the calcite samples, mixing with meteoric water cannot explain the high $^{87}Sr/^{86}Sr$ values.

The variation in $^{87}Sr/^{86}Sr$ and Sr concentration is, therefore, best explained as resulting from alteration of pore-water chemistry through interaction with rock components enriched in ^{87}Sr. There are a number of possible Sr sources in the section with which the pore water could have reacted.

(1) *Marine carbonate material.* Marine carbonate material is abundant throughout the Ellesmerian section, thus diagenetic alteration of this material could have provided Sr to formation waters. However, the $^{87}Sr/^{86}Sr$ ratios of marine carbonates are far too low to account for the high $^{87}Sr/^{86}Sr$ values observed in the late-stage carbonate cements (see data of Burke *et al.*, 1982).

(2) *Detrital mica.* Coarse detrital muscovite and biotite are present in sandstones and siltstones; however,

they are only present in minor amounts and there is no evidence that they have undergone alterations that contributed Sr to the pore waters.

(3) *Detrital feldspar*. Detrital feldspar is only present in Ellesmerian sandstones and siltstones in trace amounts, but some of the Ellesmerian shales contain 1–14% feldspar (Van de Kamp, 1988). In addition, Brookian sandstones and shales contain significant quantities of feldspar (2–8% for sandstones, Bartsch-Winkler, 1979; 7–15% for shales, Van de Kamp, 1988). However, because the increase in the $^{87}Sr/^{86}Sr$ ratio appears to have begun well before deposition of the Brookian sediments (i.e. some of the pore-filling calcites clearly formed prior to significant compaction, yet they have relatively high $^{87}Sr/^{86}Sr$ values), a unique Brookian source can probably be ruled out.

(4) *Clays*. Clays (principally Fe-rich chlorite and illite) are present as an intergranular matrix in sandstones and siltstones, and are the major constituent of the shales. A clay separate from a siltstone and whole-rock values for two shales from the Sag River and Shublik Formations that consist principally of clay, quartz, and possibly some feldspar were analysed for Sr; they all have $^{87}Sr/^{86}Sr$ ratios high enough to be the major source (range=0·715238–0·720672; Table 1).

Therefore, the most likely sources of the high $^{87}Sr/^{86}Sr$ are clay minerals and possibly feldspars in Ellesmerian sequence mudrocks. A variety of diagenetic alterations (e.g. dissolution or albitization of feldspar; dissolution, cation exchange reactions, or smectite to illite conversion in clays) could have released this Sr to interstitial waters.

Carbonates of the Barrow Sandstone

Marine sandstones that are lithological and stratigraphical equivalents of the Barrow Sandstone (K. Bird, pers. comm., 1985) occur in the Kingak Shale 12–24 m above the top of the Sag River Formation in several of the wells studied. The sandstones are compositionally and texturally comparable to those of the Sag River and Shublik Formations, and appear to have undergone a very similar diagenetic history (Mozley, 1988). A reconnaissance study of the geochemistry of carbonate cements in these rocks was conducted to see how they compare to those in the Sag River and Shublik Formations.

Siderite and calcite cements in the sandstones are very similar to those from the Sag River and Shublik Formations. The calcite appears to have developed along the same elemental and isotopic trends as the Fe-rich calcite of the Sag River and Shublik Formations (Figs 11 & 13; Table 1). Likewise, the elemental composition of the siderite is similar to that in the Sag River and Shublik Formations (Fig. 9), and has virtually the same $\delta^{18}O$ and $\delta^{13}C$ values (Table 1). The $^{87}Sr/^{86}Sr$ ratio obtained for concretionary siderite in the sandstones (see Mozley, 1989b for a discussion of the structure and mode of growth of these concretions) is higher than that of the Mg-rich calcite of the Shublik Formation, but lower than that of the Fe-rich calcite of all three units (Table 1). This is consistent with petrographical evidence indicating that siderite was formed prior to the Fe-rich calcite in all three units, but subsequent to the Mg-rich calcite of the Shublik Formation.

DISCUSSION

Analysis of the trace element and isotopic composition of the carbonate cements reveals considerable variation. A number of factors can influence the composition of carbonate cements, the most significant of which are pore-water composition, rate of precipitation, and recrystallization. The relative importance of these factors must be evaluated in order to interpret the geochemical data (Moore, 1989).

Recrystallization

Recrystallization of carbonate cements can significantly alter their composition. Two lines of evidence suggest that the Mg-rich calcite may be a product of recrystallization. (i) The very low compaction experienced by the samples prior to cementation indicates that precipitation must have occurred at very shallow burial depths (i.e. tens of metres), yet temperatures estimated from $\delta^{18}O$ values are too high (i.e. 30–40°C) for precipitation at such shallow depths if a water of $\delta^{18}O=0‰$ SMOW is assumed. (ii) Some of the Mg-rich calcites have an acicular crystal habit. Such crystal forms are commonly associated with high-Mg calcite and thus could represent an inherited texture from an initial high-Mg calcite that underwent conversion to low-Mg calcite.

Although the Mg-rich calcites may have undergone recrystallization, the preservation of complex compositional zonation in Fe-rich calcites, siderite, and dolomite/ankerite indicates that these minerals have not been recrystallized (such zonation was not observed in the Mg-rich calcites).

Variation in rate of precipitation

Because distribution coefficients for cations in calcite can be influenced by precipitation rate (Lorens, 1981; Given & Wilkinson, 1985a; Ten Have & Heijnen, 1985) some of the observed variations in carbonate composition may result from differing rates of precipitation.

Given & Wilkinson (1985a) indicated that acicular crystal habits, such as those seen in some of the Mg-rich calcites, may result from high rates of precipitation. Because high rates may favour incorporation of Mg (Given & Wilkinson, 1985a; a controversial topic, see Morse, 1985; Given & Wilkinson, 1985b), the Mg-rich nature of these calcites could in part result from high precipitation rates.

Although the composition of the Mg-rich calcite may have been influenced by a high rate of precipitation, the following points indicate that the variable composition of the Fe-rich calcite, dolomite/ankerite, and siderite does not result from variable rates of precipitation.

(1) The fracture-filling calcite cements have the highest percentages of Sr, but low percentages of Mg compared to earlier Fe-rich calcites. This is inconsistent with a kinetic control on calcite composition, because both Sr and Mg content should increase with increasing precipitation rate (Lorens, 1981; Given & Wilkinson, 1985a). Furthermore, the covariation of Sr concentration and $^{87}Sr/^{86}Sr$ values argues against a kinetic control of Sr concentration, because kinetic effects do not fractionate ^{87}Sr from ^{86}Sr.
(2) The Fe-content of early calcite, dolomite/ankerite, and siderite cements increased dramatically during precipitation. It is improbable that all three minerals would undergo changes in precipitation rate that affected their composition in a similar way.
(3) The crystal habit of these cements does not vary significantly. The Fe-rich calcites are characterized by equant forms, whereas dolomite/ankerite and siderite form small rhombs. Because changes in the rate of precipitation often result in very different crystal forms (Given & Wilkinson, 1985a), this suggests that radical shifts in precipitation rate did not occur.

Variation in pore-water composition

On the basis of the preceding dicussion, most of the observed variation in carbonate composition is inferred to have resulted from changes in pore-water composition during diagenesis.

Given that these are marine units, the original pore water was undoubtedly seawater. The $^{87}Sr/^{86}Sr$ ratios and textural relationships of the earliest of the carbonate cements, Mg-rich calcite, are compatible with precipitation from relatively unaltered Triassic seawater.

Following the precipitation of the Mg-rich calcite, the pore waters underwent major alterations which were recorded in later cements. Dramatic variations occurred in the Sr^{2+}/Ca^{2+}, $^{87}Sr/^{86}Sr$, Fe^{2+}/Ca^{2+}, and $Fe^{2+}/(Ca^{2+}+Mg^{2+})$ ratios.

As previously discussed, examination of Sr data for the calcite and siderite cements demonstrates that the Sr^{2+}/Ca^{2+} and $^{87}Sr/^{86}Sr$ ratios of the pore waters increased through time. These increases probably began fairly early, because siderite, an early cement that formed prior to significant compaction, has a $^{87}Sr/^{86}Sr$ ratio considerably higher than that of seawater.

This alteration appears to result from extensive interaction between pore water and ^{87}Sr-enriched clays and possibly feldspars in Ellesmerian mudrocks. Similar increases of pore-water Sr concentrations and $^{87}Sr/^{86}Sr$ ratios have been noted in studies of oil-field brines (Sunwall & Pushkar, 1979; Stueber, Pushkar & Hetherington, 1984; Chaudhuri, Broedel & Clauer, 1987; Russell, Cowart & Russell, 1988) and carbonate cements in other units (Stanley & Faure, 1979; Moore, 1985; Emery, Dickson & Smalley, 1987; Dutton & Land, 1988; Ruppel & Cander, 1988).

Systematic increases in the Fe content of calcite, dolomite/ankerite, and siderite cements indicate that the Fe^{2+}/Ca^{2+} and $Fe^{2+}/(Ca^{2+}+Mg^{2+})$ ratios dramatically increased during early diagenesis. The relatively low Fe content of fracture-filling Fe-rich calcites (the last stage of calcite cementation) indicates that following this initial increase, the Fe^{2+}/Ca^{2+} ratio of the waters gradually decreased. (These variations are clearly illustrated when $^{87}Sr/^{86}Sr$ is plotted against the Fe content of calcite cements, Fig. 14.)

The increase in the Fe^{2+}/Ca^{2+} and $Fe^{2+}/(Ca^{2+}+Mg^{2+})$ ratios may have occurred near the base of the sulphate reduction zone where Fe^{2+} previously kept at very low levels by pyrite precipitation is made available to interstitial waters (Gautier, 1985). This

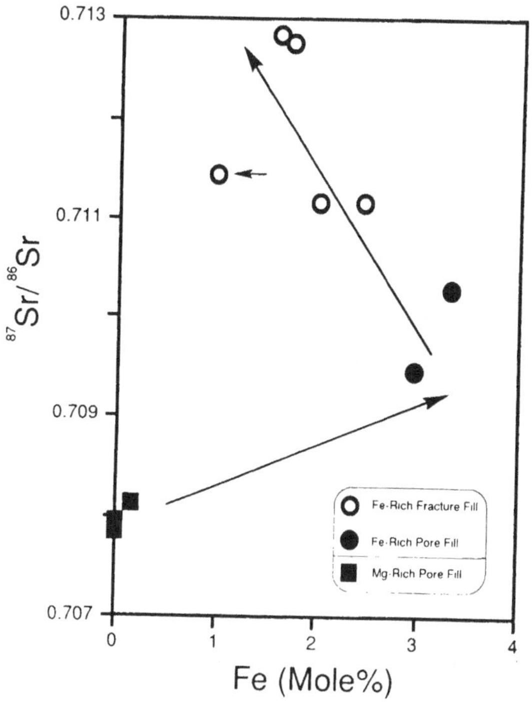

Fig. 14. Plot of $^{87}Sr/^{86}Sr$ vs. Fe content (mol%) of calcite. Mg-rich calcites are characterized by low $^{87}Sr/^{86}Sr$ ratios and low Fe percentages, whereas values for Fe-rich calcites are higher are more variable. Large arrows show probable temporal trends. Small arrow = Barrow Sandstone sample.

interpretation is supported by the presence of abundant early pyrite. The subsequent decrease in the Fe^{2+}/Ca^{2+} ratio may result from later diagenetic reactions such as precipitation of Fe-bearing minerals or exchange reactions with clays similar to those described by Drever (1971).

Several workers (Kharaka & Carothers, 1988; Shanmugam & Higgins, 1988; Van de Kamp, 1988) have suggested that permeable Ellesmerian units were flushed with meteoric water in the Early Cretaceous in association with the formation of the mid-Neocomian unconformity.

Although inconclusive, several observations indicate that the Sag River and Shublik Formations were affected by an incursion of meteoric water. Barnes (1987) showed that mineral leaching in the Sag River Formation in the Prudhoe Bay field resulted from the introduction of water at the mid-Neocomian unconformity. He concluded that this water was either meteoric water introduced during subaerial exposure in the mid-Neocomian, or water derived from the thick sequence of mudrocks that overlie the unconformity (D. Barnes, pers. comm., 1985).

In addition, some of the observed changes in carbonate composition could be the result of an incursion of meteoric water. The strongest evidence in favour of such an incursion is the small amount of compaction that rocks cemented by low-$\delta^{18}O$ calcite appear to have experienced prior to cementation (see oxygen isotopes section). The small amount of pre-cementation compaction shows that the calcite could not have formed subsequent to deep burial (i.e. >1 km), thus the low $\delta^{18}O$ values either reflect ^{18}O-depleted parent waters or anomalously high temperatures of precipitation (i.e. around 88°C) if a water of $\delta^{18}O = 0‰$ SMOW is assumed. (The rocks underwent deep burial for the first time during deposition of the Brookian sequence—subsequent to the mid-Neocomian uplift.) Unfortunately, however, the possibility that the rocks experienced unusually high temperatures prior to deep burial cannot be ruled out, because an extremely high geothermal gradient may have existed in the region in association with the Early Cretaceous rifting event (Magoon & Claypool, 1982).

CONCLUSIONS

(1) Carbonate cements (calcite, dolomite/ankerite, and siderite) formed throughout much of the diagenetic history. Early cements (e.g. Mg-rich calcite) probably formed near the sediment–water interface shortly after deposition, whereas some of the late-stage cements (e.g. Fe-rich fracture-filling calcite) formed during and/or subsequent to oil migration in the region (at least 100 Myr after deposition).

(2) The carbonate cements vary considerably in elemental and isotopic composition. This variation appears principally to reflect changes in the water chemistry through time.

(3) Early pore waters were of normal or slightly altered marine composition.

(4) The Fe^{2+}/Ca^{2+} and $Fe^{2+}/(Ca^{2+}+Mg^{2+})$ ratios of the pore waters increased dramatically during early diagenesis, and then decreased. The initial increase may have occurred near the base of the sulphate reduction zone, when significant quantities of Fe^{2+} became available to the pore waters for the first time.

(5) The Sr^{2+}/Ca^{2+} and $^{87}Sr/^{86}Sr$ ratios increased systematically during diagenesis, most likely as a result of rock–water interaction with ^{87}Sr-rich

clay and possibly feldspars in Ellesmerian sequence mudrocks.

(6) Low $\delta^{18}O$ values for calcite cements suggest that an incursion of meteoric water affected these rocks in the mid-Neocomian. However, because the region may have been affected by a high rift-related geothermal gradient at this time, the low values could also result from abnormally high temperatures of precipitation.

(7) Pore waters in sandstones in the lower portion of the Kingak Shale appear to have undergone a similar geochemical evolution to those of the Sag River and Shublik Formations.

ACKNOWLEDGMENTS

We are grateful to J. R. Boles and G. Tilton (U.C. Santa Barbara) for inspiration, discussions, and financial support throughout the study. J. Schultz (U.C. Santa Barbara) gave valuable advice on sample preparation and laboratory techniques for the Sr work. The geologists of the sedimentary geology group of the UNOCAL Science and Technology Division (particularly G. A. Crawford, A. Trevena, R. Cole, P. Lundegard, T. Elliot, and R. Clark) provided useful advice and access to their analytical facilities. K. Bird and I. Tailleur (U.S. Geological Survey, Menlo Park) answered innumerable questions concerning the geological setting and facilitated sampling of the NPRA core material. We thank S. Franks (ARCO Oil and Gas) and D. Barnes (Western Michigan University) for advice in the early stages of the study, and Albert Matter and Karl Ramseyer (Universität Bern) for useful discussions and encouragement during the preparation of the manuscript. The manuscript benefited from reviews by J. A. D. Dickson, J. C. Lacharpagne and K. Bird.

REFERENCES

BARTSCH-WINKLER, S. (1979) Textural and mineralogical study of some surface and subsurface sandstones from the Nanushuk Group, western North Slope, Alaska. In: *Preliminary Geologic, Petrologic, and Paleontologic Results of the Study of Nanushuk Group Rocks, North Slope, Alaska* (Ed. by T. S. Ahlbrandt). *U.S. geol. Surv. Circ.*, **794**, 61–76.

BARNES, D.A. (1987) Reservoir quality in the Sag River Formation, Prudhoe Bay Field Alaska: depositional environment and diagenesis. In: *Alaskan North Slope Geology*, Vol. I (Ed. by I. Tailleur & P. Weimer), pp. 85–94. Pacif. Sect. Soc. econ. Paleont. Miner., and Alaska Geol. Soc.

BIRD, K.J. (1982) Rock-unit reports of 228 wells drilled on the North Slope, Alaska. *U.S. geol. Surv., Open File Rept*, **82–278**, 106 pp.

BIRD, K.J. (1985) The framework geology of the North Slope of Alaska as related to oil-source rock correlations. In: *Alaska North Slope Oil/Rock Correlation Study* (Ed. by L. B. Magoon & G. E. Claypool). *Am. Ass. petrol. Geol. Studies in Geology*, **20**, 3–29.

BIRD, K.J. & MOLENAAR, C.M. (1987) Stratigraphy. In: *Petroleum Geology of the Northern Part of the Arctic National Wildlife Refuge, Northeastern Alaska* (Ed. by K. J. Bird & L. B. Magoon). *U.S. geol. Surv. Bull.*, **1778**, 37–60.

BURKE, W.H., DENISON, R.E., HETHERINGTON, E.A., KOEPNICK, R.B., NELSON, H.F. & OTTO, J.B. (1982) Variation of seawater $^{87}Sr/^{86}Sr$ throughout Phanerozoic time. *Geology*, **10**, 516–519.

CAROTHERS, W.W., ADAMI, L.H. & ROSENBAUER, R.J. (1988) Experimental oxygen isotope fractionation between siderite-water and phosphoric acid liberated CO_2–siderite. *Geochim. Cosmochim. Acta*, **52**, 2445–2450.

CHAUDHURI, S., BROEDEL, V. & CLAUER, N. (1987) Strontium isotopic evolution of oil-field waters from carbonate reservoir rocks in Bindley field, central Kansas, USA. *Geochim. Cosmochim. Acta*, **51**, 45–53.

COLEMAN, M.L. (1985) Geochemistry of diagenetic non-silicate minerals: kinetic considerations. In: *Geochemistry of Buried Sediments* (Ed. by G. Eglinton, C. D. Curtis, D. P. McKenzie & D. G. Murchison), pp. 39–54. The Royal Society, London.

CURTIS, C.D. & COLEMAN, M.L. (1986) Controls on the precipitation of early diagenetic calcite, dolomite, and siderite concretions in complex depositional sequences. In: *Roles of Organic Matter in Sediment Diagenesis* (Ed. by D. L. Gautier). *Spec. publs. Soc. econ. Paleont. Miner.*, **38**, 23–33.

DINGUS, A.S. (1984) *Paleoenvironmental Reconstruction of the Shublik Formation on the North Slope of Alaska*. MA thesis, University of California, Berkeley, 108 pp.

DREVER, J.I. (1971) Magnesium-iron replacement in clay minerals in anoxic marine sediments. *Science*, **172**, 1334–1336.

DUNHAM, R.J. (1963) Classification of carbonate rocks according to depositional texture. In: *Classification of Carbonate Rocks* (Ed. by W. E. Ham). *Mem. Am. Ass. petrol. Geol.*, **1**, 108–121.

DUTTON, S.P. & LAND, L.S. (1988) Cementation and burial history of a low-permeability quartzarenite, Lower Cretaceous Travis Peak Formation, East Texas. *Bull. geol. Soc. Am.*, **100**, 1271–1282.

EGGERT, J.T. (1987) Sandstone petrology, diagenesis, and reservoir quality, Lower Cretaceous Kuparuk River Formation, Kuparuk River Field, North Slope, Alaska [abstract]. In: *Alaskan North Slope Geology*, Vol. I (Ed. by I. Tailleur & P. Weimer), p. 108. Pacif. Sect. Soc. Econ. Paleont. Miner., and Alaska Geol. Soc.

EMERY, D., DICKSON, J.A.D. & SMALLEY, P.C. (1987) The strontium isotopic composition and origin of burial cements in the Lincolnshire Limestone (Bajocian) of central Lincolnshire, England. *Sedimentology*, **34**, 795–806.

FAURE, G. (1986) *Principles of Isotope Geology* (2nd edn). John Wiley and Sons, New York, 589 pp.

FOLK, R.L. (1968) *Petrology of Sedimentary Rocks*. Hemphill's Book Store, Austin, Texas, 182 pp.

FRIEDMAN, I. & O'NEIL, J.R. (1977) Compilation of stable isotope fractionation factors of geochemical interest (Data of Geochemistry, Sixth Edition, Chapter KK). *Prof. Pap. U.S. geol. Surv.*, **440-KK**, 12 pp.

GAUTIER, D.L. (1985) Interpretation of early diagenesis in ancient marine sediments. In: *Relationship of Organic Matter and Mineral Diagenesis* (Ed. by D. L. Gautier, Y. K. Kharaka & R. C. Surdam). *Soc. econ. Paleont. Miner. Short Course Notes*, **17**, 6–78.

GAUTIER, D.L. & CLAYPOOL, G.E. (1984) Interpretation of methanic diagnesis in ancient sediments by analogy with processes in modern diagenetic environments. In: *Clastic Diagenesis* (Ed. by D. A. McDonald & R. C. Surdam). *Mem. Am. Ass. petrol. Geol.*, **37**, 111–123.

GIVEN, R.K. & WILKINSON, B.H. (1985a) Kinetic control of morphology, composition, and mineralogy of abiotic sedimentary carbonates. *J. sedim. Petrol.*, **55**, 109–119.

GIVEN, R.K. & WILKINSON, B.H. (1985b) Kinetic control of morphology, composition, and mineralogy of abiotic sedimentary carbonates—reply. *J. sedim. Petrol.*, **55**, 921–926.

HUBBARD, R.J., EDRICH, S.P. & RATTEY, R.P. (1987) Geologic evolution and hydrocarbon habitat of the 'Arctic Alaska Microplate'. In: *Alaskan North Slope Geology*, Vol. II (Ed. by I. Tailleur & P. Weimer), pp. 797–830. Pacif. Sect. Soc. Econ. Paleont. Miner., and Alaska Geol. Soc.

JONES, H.P. & SPEERS, R.G. (1976) Permo-Triassic reservoirs on Prudhoe Bay Field, North Slope, Alaska. In: *North American Oil and Gas Fields* (Ed. by J. Braustein). *Mem. Am. Ass. petrol. Geol.*, **24**, 23–50.

KHARAKA, Y.K. & CAROTHERS, W.W. (1988) Geochemistry of oil-field water from the North Slope. In: *Geology and Exploration of the National Petroleum Reserve in Alaska, 1974–1982* (Ed. by G. Gryc). *Prof. Pap. U.S. geol. Surv.*, **1399**, 551–561.

LORENS, R.B. (1981) Sr, Cd, Mn and Co distribution coefficients in calcite as a function of calcite precipitation rate. *Geochim. Cosmochim. Acta*, **45**, 553–561.

MAGOON, L.B. & BIRD, K.J. (1985) Alaskan North Slope petroleum geochemistry for the Shublik Formation, Kingak Shale, Pebble Shale Unit, and Torok Formation. In: *Alaska North Slope Oil/Rock Correlation Study* (Ed. by L. B. Magoon & G. E. Claypool). *Am. Ass. petrol. Geol. Studies in Geology*, **20**, 31–48.

MAGOON, L.B. & CLAYPOOL, G.E. (1982) NPRA Inigok No. 1—use of Lopatin's method to reconstruct thermal maturity. *Pacif. Petrol. Geol. Newsletter*, **January**, 4–6.

MARINAI, R.K. (1987) Petrography and diagenesis of the Ledge Sandstone Member of the Triassic Ivishak Formation. In: *Petroleum Geology of the Northern Part of the Arctic National Wildlife Refuge, Northeastern Alaska* (Ed. by K. J. Bird & L. B. Magoon). *Bull. U.S. geol. Surv.*, **1778**, 101–116.

MAYNARD, J.B. (1982) Extension of Berner's 'new geochemical classification of sedimentary environments' to ancient sediments. *J. sedim. Petrol.*, **52**, 1325–1331.

MELVIN, J. & KNIGHT, A.S. (1984) Lithofacies, diagenesis and porosity of the Ivishak Formation, Prudhoe Bay Area, Alaska. In: *Clastic Diagenesis* (Ed. by D. A. McDonald & R. C. Surdam). *Mem. Am. Ass. petrol. Geol.*, **37**, 347–366.

MOORE, C.H. (1985) Upper Jurassic subsurface cements: a case history. In: *Carbonate Cements* (Ed. by N. Schneidermann & P. M. Harris). *Spec. publs Soc. econ. Paleont. Miner.*, **36**, 291–308.

MOORE, C.H. (1989) *Carbonate Diagenesis and Porosity*, Developments in Sedimentology 46. Elsevier, Amsterdam, 338 pp.

MORSE, J.W. (1985) Kinetic control of morphology, composition, and mineralogy of abiotic sedimentary carbonates—discussion. *J. sedim. Petrol.*, **55**, 919–921.

MOZLEY, P.S. (1988) *Petrography and Diagenesis of the Sag River and Shublik Formations in the National Petroleum Reserve, Alaska; and Topics in Siderite Geochemistry*. PhD thesis, University of California, Santa Barbara, 252 pp.

MOZLEY, P.S. (1989a) Relation between depositional environment and the elemental composition of early diagenetic siderite. *Geology*, **17**, 704–706.

MOZLEY, P.S. (1989b) Complex compositional zonation in concretionary siderite: implications for geochemical studies. *J. sedim. Petrol.*, **59**, 815–818.

NOONAN, W.G. (1987) Post-Ellesmerian depositional sequences of the North Slope subsurface. In: *Alaskan North Slope Geology*, Vol. I (Ed. by I. Tailleur & P. Weimer), pp. 459–478. Pacif. Sect. Soc. Econ. Paleont. Miner., and Alaska Geol. Soc.

O'NEIL, J.R., CLAYTON, R.N. & MAYEDA, T.K. (1969) Oxygen isotope fractionation in divalent metal carbonates. *J. Chem. Phys.*, **51**, 5547–5558.

PARRISH, J.T. (1987) Lithology, geochemistry, and depositional environment of the Triassic Shublik Formation, Northern Alaska. In: *Alaskan North Slope Geology*, Vol. I (Ed. by I. Tailleur & P. Weimer), pp. 391–396. Pacif. Sect. Soc. Econ. Paleont. Miner., and Alaska Geol. Soc.

PAYNE, J.H. (1987) Diagenetic variations in the Permo-Triassic Ivishak Sandstone in the Prudhoe Bay Field and central-northeastern National Petroleum Reserve in Alaska. In: *Alaskan North Slope Geology*, Vol. I (Ed. by I. Tailleur & P. Weimer), pp. 77–83. Pacif. Sect. Soc. Econ. Paleont. Miner., and Alaska Geol. Soc.

RUPPEL, S.C. & CANDER, H.S. (1988) Dolomitization of shallow-water platform carbonates by sea water and seawater-derived brines: San Andres Formation (Guadalupian), west Texas. In: *Sedimentology and Geochemistry of Dolostones* (Ed. by V. Shukla & P. A. Baker). *Spec. Publs Soc. econ. Paleont. Miner.*, **43**, 245–262.

RUSSELL, C.W., COWART, J.B. & RUSSELL, G.S. (1988) Strontium isotopes in brines and associated rocks from Cretaceous strata in the Mississippi Salt Dome Basin (southeastern Mississippi, U.S.A.). *Chem. Geol.*, **74**, 153–171.

SCOFFIN, T.P. (1987) *An Introduction to Carbonate Sediments and Rocks*. Blackie, Glasgow, 274 pp.

SHANMUGAM, G. & HIGGINS, J.B. (1988) Porosity enhancement from chert dissolution beneath Neocomian Unconformity: Ivishak Formation, North Slope, Alaska. *Bull. Am. Ass. petrol. Geol.*, **72**, 523–535.

STANLEY, K.O. & FAURE, G. (1979) Isotopic composition and sources of strontium in sandstone cements: the High Plains Sequence of Wyoming and Nebraska. *J. sedim. Petrol.*, **49**, 45–54.

STUEBER, A. M., PUSHKAR, P. & HETHERINGTON, E.A. (1984) A strontium isotopic study of Smackover brines and associated solids, southern Arkansas. *Geochim. Cosmochim. Acta*, **48**, 1637–1649.

SUNWALL, M.T. & PUSKHAR, P. (1979) The isotopic composition of strontium in brines from petroleum fields of southeastern Ohio. *Chem. Geol.*, **24**, 189–197.

TEN HAVE, T., & HEIJNEN, W. (1985) Cathodoluminescence activation and zonation in carbonate rocks: an experimental approach. *Geologie Mijnb.*, **64**, 297–310.

VAN DE KAMP, P. (1988) Stratigraphy and diagenetic alteration of Ellesmerian Sequence siliciclastic rocks, North Slope, Alaska. In: *Geology and Exploration of the National Petroleum Reserve in Alaska, 1974–1982* (Ed. by G. Gryc). *Prof. Pap. U.S. geol. Surv.*, **1399**, 833–854.

VEIZER, J. (1983) Chemical diagenesis of carbonates: theory and application of trace element technique: In: *Stable Isotopes in Sedimentary Geology* (Ed. by M. A. Arthur). *Soc. econ. Paleont. Miner. Short Course Notes*, **10**, 3-1 to 3-100.

WALKER, J.C.G. (1984) Suboxic diagenesis in banded iron formations. *Nature*, **309**, 340–342.

(Manuscript received 1 June 1989; revision received 17 March 1990)

Reprinted from Sedimentology (1987) **34**, *533–558*

Burial dolomitization and porosity development in a mixed carbonate-clastic sequence: an example from the Bowland Basin, northern England

ROBERT L. GAWTHORPE*

Department of Earth Sciences, The University of Leeds, Leeds LS2 9JT, England

ABSTRACT

Widespread dolomitization and leaching occur in the Asbian to Brigantian (Dinantian) sequence of the Bowland Basin. Within this mudrock-dominated succession, dolomite is developed in calcarenites and limestone breccia/conglomerates deposited in a carbonate slope environment (Pendleside Limestone) and also within graded quartz wackes deposited by density currents in a generally 'starved' basin environment (Pendleside Sandstone). The dolomitized intervals range in thickness from less than one metre to several tens of metres and have a stratabound nature.

All stages of calcite cement pre-date dolomitization and calcite veins are dolomitized. Dolomite crystals replace neomorphic spar and may also contain insoluble residues that were concentrated along stylolites. Thus dolomitization was a late stage process within the carbonate diagenetic sequence. A late-stage diagenetic origin is also indicated within the sandstones, with dolomite post-dating the development of quartz overgrowths.

Six main textural styles of dolomite are observed: (1) scattered; (2) mosaic; (3) subhedral to euhedral rhombic; (4) microcrystalline; (5) single crystal and (6) saddle. The style of dolomite developed is dependent on the host rock mineralogy, on whether it is space-filling or replacive and also on temperature. Chemically the dolomite varies from near stoichiometric compositions to ankeritic varieties containing up to 20 mole % $FeCO_3$. Generally the dolomites have isotopic compositions depleted in $\delta^{18}O$ compared to the host limestone, with similar or lighter $\delta^{13}C$ values.

Initial dolomite was of the scattered type, but with progressive replacement of the host a mosaic dolostone with a sucrosic texture was produced. There was a general increase in the Fe and Mn content and reduction in $\delta^{18}O$ ratio of the crystals during dolomitization. Leaching is restricted to partly dolomitized horizons, where calcite, feldspars, micas, ?clays and, to some extent, dolomite have been leached. This has produced biomouldic and vuggy secondary porosity within the carbonates, whereas in the sandstones honeycombed, corroded and floating grains associated with oversized pores occur. Porosity within both carbonates and sandstones is reduced by ferroan dolomite/ankerite cements.

Field, petrographic and chemical characteristics indicate that dolomitizing solutions were predominantly derived from the enclosing mudrocks (Bowland Shales) during intermediate/deep burial. Fluid migration out of the mudrocks would have been aided by dehydration reactions and overpressure, the fluids migrating along the most permeable horizons—the coarse grained carbonates and sandstones that are now dolomitized and contain secondary porosity.

INTRODUCTION

The carbonate minerals dolomite and ankerite are common phases in both sandstones and carbonates, and are of economic significance because of their common association with hydrocarbon reservoirs (Davies, 1979) and Pb–Zn mineralization (Macqueen, 1979). In sandstones and also in concretions within mudrocks, these carbonate phases are generally considered to have formed late in the diagenetic sequence, often in a 'deep burial' situation (e.g. Boles, 1978; Curtis, 1978, 1983a; Irwin, Curtis & Coleman, 1977; Kantorowicz, 1985). In recent years especially,

*Present address: BP Petroleum Development Ltd., Britannic House, Moor Lane, London EC2Y 9BU, England.

some occurrences of dolomite and ankerite within carbonates have also been ascribed to late diagenetic (burial) processes (e.g. Choquette, 1971; Dickinson & Coleman, 1980; Jodry, 1969; MacHargue & Price, 1982; Mattes & Mountjoy, 1980; Schofield & Adams, 1986; Wong & Oldershaw, 1981; Zenger, 1983). Both in these occurrences and in the case of dolomite/ankerite within sandstones, mudrocks have been cited as a possible source of ions for dolomite formation. However, doubts as to the amount of Mg^{2+} available from such a source have also been expressed, e.g. Land (1985). Of particular significance within sandstones has been the recognition of secondary porosity generated in these burial situations, again related to mudrock diagenesis (Curtis, 1978, 1983a; Franks & Forester, 1984; Hayes, 1979; Loucks, Dodge & Galloway, 1979; Milliken, Land & Loucks, 1981; Schmidt & McDonald, 1979). However, little attention has been paid to similar processes operating within carbonates.

The late Asbian to Brigantian succession of the Bowland Basin provides evidence of the relationships between, and origin of, deep burial dolomite/ankerite and secondary porosity in both sandstones and carbonates within a mudrock-dominated sequence. In particular, the sequence shows how deep burial diagenesis of a mixed terrigenous-carbonate sequence can produce extensive dolomitization associated with major secondary porosity development in platform margin carbonates.

METHODS

This paper records the results of an integrated field, petrographic and geochemical analysis of the dolomitization and associated leaching of Asbian to early Brigantian carbonates and sandstones. Outcrop and borehole core samples have together allowed basin-wide coverage of the carbonates; sandstone samples were principally obtained from the northern part of the study area, in particular from two boreholes (Fig. 1). Some 300 polished and standard thin sections were prepared from logged sections and were examined using a variety of techniques.

Chemical analyses were principally obtained by electron microprobe spot analysis, with additional data on whole rock chemistry provided by X-ray fluorescence (XRF). Microprobe analyses were carried out using a Link Systems 860–500 energy dispersive system fitted to a Jeol JXA-50A electron microprobe; all data were ZAF-corrected using Link ZAF4/FLS software. The operating conditions for analyses were an accelerating voltage of 20 kV, a beam current of 1 nA and a count time of 60 s. Element distribution maps were collected using Link System DIGIMAP software. In addition, backscattered and scanning electron microscopy (BSEM and SEM) were undertaken on the Jeol microprobe, with SEM analyses also performed on an ISI Super-I SEM. X-ray fluorescence analysis was carried out using a Phillips PW1400 spectrometer.

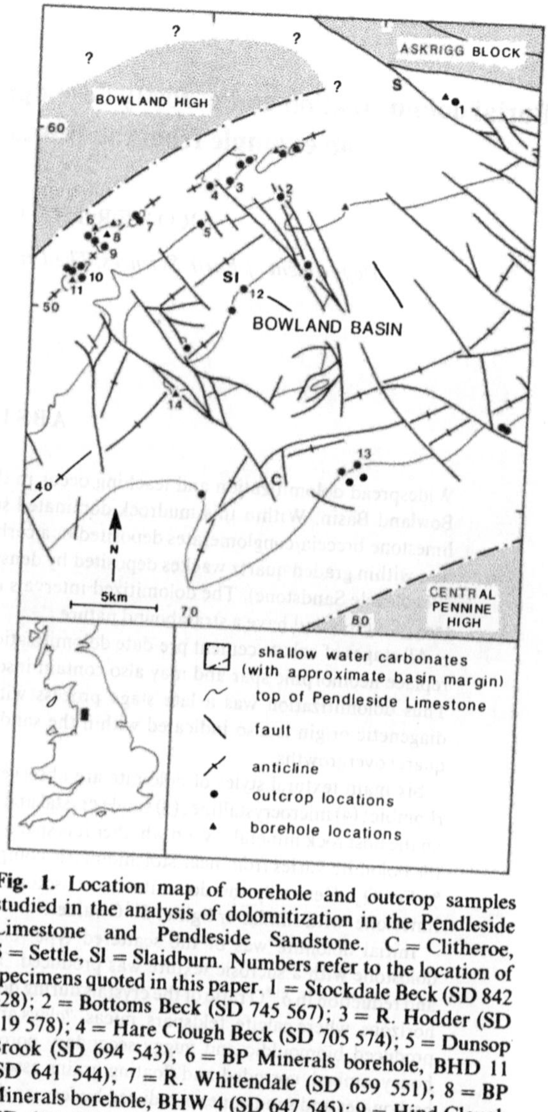

Fig. 1. Location map of borehole and outcrop samples studied in the analysis of dolomitization in the Pendleside Limestone and Pendleside Sandstone. C = Clitheroe, S = Settle, Sl = Slaidburn. Numbers refer to the location of specimens quoted in this paper. 1 = Stockdale Beck (SD 842 628); 2 = Bottoms Beck (SD 745 567); 3 = R. Hodder (SD 719 578); 4 = Hare Clough Beck (SD 705 574); 5 = Dunsop Brook (SD 694 543); 6 = BP Minerals borehole, BHD 11 (SD 641 544); 7 = R. Whitendale (SD 659 551); 8 = BP Minerals borehole, BHW 4 (SD 647 545); 9 = Hind Clough (SD 644 534); 10 = Penny Brook (SD 634 519); 11 = BP Minerals borehole, BHD 10 (SD 632 519); 12 = Skelshaw Brook (SD 724 506); 13 = Rad Brook (SD 788 428); 14 = BP Minerals borehole, MHD 14 (SD 674 456).

Cathodoluminescence (CL) was achieved using a Technosyn MkI cold cathode luminoscope. Operating conditions were an accelerating voltage of 2·2–2·5 kV and a gun current of 190–200 μA. Photographs were taken under these conditions on an Olympus OM2 attached to an Olympus POS microscope using automatic exposure and Ektachrome 160 tungsten film. Where possible cathodoluminescence observations and microprobe analyses were made on borehole core samples in order to reduce the effects of surface weathering. Carbon and oxygen stable isotope analyses were carried out by A. M. Boast at BP Research Centre, Sunbury.

GEOLOGICAL SETTING

The Bowland Basin was one of several shelf basins in northern England during the Dinantian (Lower Carboniferous), and contains a series of carbonates, terrigenous mudstones and rare sandstones. Facies analysis of the basin sequence indicates evolution from a carbonate ramp into a carbonate slope depositional environment between the Ivorian and early Brigantian, with two main episodes of tectonic activity which strongly influenced sedimentation (Gawthorpe, 1986). Post-early Brigantian sediments are predominantly terrigenous mudstones with local sandstones. Dinantian sedimentation was controlled by the extensional fault-controlled geometry of the basin, being closely related to a series of NE–SW-trending normal faults and NW–SE-trending transfer faults (Gawthorpe, 1987). The basin has a half graben morphology, controlled by a major basin-margin normal fault situated along the SE margin of the basin. Within the basin, antithetic and transfer faults produced a series of topographic highs and lows that affected facies and thickness variations from early Viséan to Namurian times. Carbonate sediment within the basin was derived from the NW margin of the basin, whereas terrigenous sediment had an extrabasinal source (mainly from the NE), and was transported axially along the basin. Post-Dinantian sediments are predominantly sandstones, reflecting the establishment of fluvio-deltaic environments over much of northern England. The original thickness of the Namurian and Westphalian strata is unknown because of uplift and erosion during ?Stephanian-Early Permian times, prior to the deposition of Permo-Triassic sandstones (Earp *et al.*, 1961). However, the thickness of Namurian and Westphalian sediments in adjacent areas indicates burial of the Dinantian sequence to a depth of approximately 3.5–4.5 km by Late Carboniferous times.

The dolomitized horizons described in this paper (the Pendleside Limestone and Pendleside Sandstone; Fig. 2) are of Asbian to Brigantian age and span the interval during which there was a transition from predominantly carbonate to terrigenous sedimentation. These horizons lie at the base of and within the Bowland Shales, a sequence up to 400 m thick (Fewtrell & Smith, 1980) composed mainly of terrigenous mudstones. The Pendleside Limestone was deposited in a carbonate slope environment. Thick deposits of limestone breccia/conglomerate facies of debris flow origin are developed along the northern margin of the basin, passing progressively into interbedded mudstones and graded calcarenites to the SE. Little has been published about the depositional history of the Pendleside Sandstone; however, the presence of sharp bases, rip-up clasts and the upward fining of individual beds suggest that they were event deposits, produced by deposition from turbidity currents (Wadge, Bateson & Evans, 1983). Both horizons show marked thickness and facies variations; for example, the Pendleside Sandstone is absent from some regions yet is up to 200 m thick in others. The greatest thickness of gravity flow-deposited carbonates and sands are located in the hangingwall of syn-sedimentary faults. Within the footwall of these same structures much thinner terrigenous mud-dominant sequences, commonly lacking gravity flow deposits, are present.

FACIES AND DIAGENETIC RELATIONSHIPS

Dolomitization of the Pendleside Limestone and Pendleside Sandstone occurs basin-wide, affecting intervals up to 50 m in thickness. Dolomitization is particularly intense towards the top of the Pendleside Limestone and is also associated with the upper and lower margins of the Pendleside Sandstone (Fig. 2). Transitions from dolomitized to undolomitized rock occur on a cm to m scale and often a spectrum of alteration fabrics can be observed in one thin section. In the field, dolostones are distinguished by their rusty orange-brown weathering, resulting from surface oxidation of Fe^{2+} within the dolomite lattice. In some instances original limestone textures have been pseudomorphed by dolomite whereas in others these textures have been completely destroyed. Another characteristic feature is the presence of mouldic porosity, with crinoid and shell moulds clearly visible.

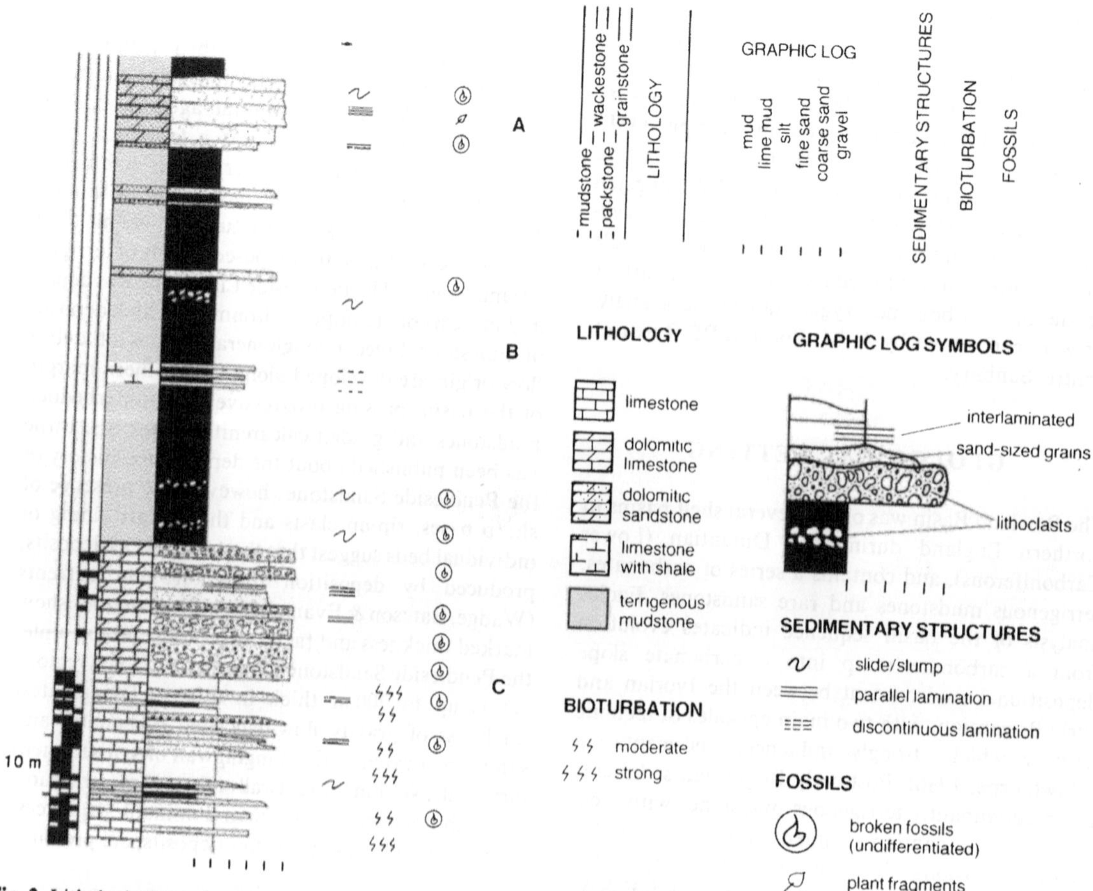

Fig. 2. Lithological log showing sedimentary and stratigraphic relationships between dolomitized intervals. A = Pendleside Sandstone; B = Bowland Shales; C = Pendleside Limestone. BP Minerals Borehole, BHD 11, location 6, Fig. 1.

Field recognition of partially dolomitized limestones and dolomitic sandstones is more difficult; differential weathering similar to that observed in the dolostones is the most distinctive feature. The morphology of dolomite bodies is difficult to determine owing to a combination of poor exposure, lateral variation within the sedimentary sequence and structural complexities. At outcrop, dolomite–limestone contacts are seen to transect bedding planes but are commonly restricted by mudstone beds, suggesting that the dolomitization has a stratabound character.

Dolomitization was a selective process; calcarenites and the lime sand matrix of limestone breccia/conglomerate beds were selectively replaced with lime mudstone/wackestone intervals commonly escaping alteration. This suggests that the permeability of the limestones was important in influencing the overall distribution of dolomite. However, within individual beds, lime mud matrix, foraminifera, peloids and microspar were preferentially replaced, with brachiopod fragments, coral and echinoderm debris and calcite spar being most resistant.

Dolomitization and associated leaching appear to have been late stage processes within the diagenetic sequence of both sandstones and carbonates. In particular, dolomite post-dates calcite veins and stylolites within the carbonates and quartz overgrowths in the sandstones. These features indicate that dolomite growth occurred in lithified or partially lithified sediment during burial.

Several additional authigenic phases can be seen to post-date the dolomite in secondary mouldic or vuggy pores. Kaolinite, often having a vermicular habit, is common within the sandstones and, to a lesser extent, the carbonates. Quartz is another authigenic phase that infills pore space remaining after dolomitization, particularly within the carbonates. In some instances quartz may extensively replace the dolomite. Bitumen, liquid hydrocarbons and base metal sulphides may also be present.

DOLOMITE TYPES

Six morphological types of dolomite have been recognized (Fig. 3); (1) scattered, (2) mosaic, (3) microcrystalline, (4) subhedral to euhedral, (5) large single crystal and (6) saddle.

Scattered dolomite

This type consists of 0·1–0·3 mm sized anhedral-to-subhedral crystals occurring either individually or in clusters up to 2 mm across that are distributed irregularly through the host limestone (Fig. 4a). Individual crystals contain submicroscopic inclusions as well as resolvable calcite and pyrite, and have sharp to undulose extinction under crossed polars. No strong zonation is observed under plane-polarized transmitted light, although some crystals may show diffuse cloudy cores and clear rims or vice versa. Cross-cutting relationships with allochems and spar in the host rock suggest this type of dolomite is replacive in origin. Dolomites within fine-grained hosts have a relatively uniform, subhedral form with well-developed compromise boundaries, whereas those in limestones consisting of various proportions of allochems, matrix and cement are more variable in morphology and size. In the case of the latter, the fine grain-size fraction (micrite, microspar, peloids, foraminifera, calcispheres) is preferentially replaced, commonly leaving crinoid, brachiopod and mollusc fragments, corals and cements unaltered. Where allochems are replaced, the dolomite often pseudomorphs the form of the allochem, or inclusions within the dolomite crystals form ghosts of the original fabric.

Cathodoluminescence (CL) reveals that this type of dolomite contains two main luminescent zones, both having a rather mottled luminescent character (Fig. 4b). The inner, dominant zone has a bright red luminescence whereas the outer zone is a very dull orange-brown. The contact between these two zones is generally sharp and planar, although highly irregular contacts (solution disconformities) have also been recognized. Locally a thin non-luminescent zone is present along the periphery of the dolomite crystals; this zone has an irregular contact with the underlying zone(s).

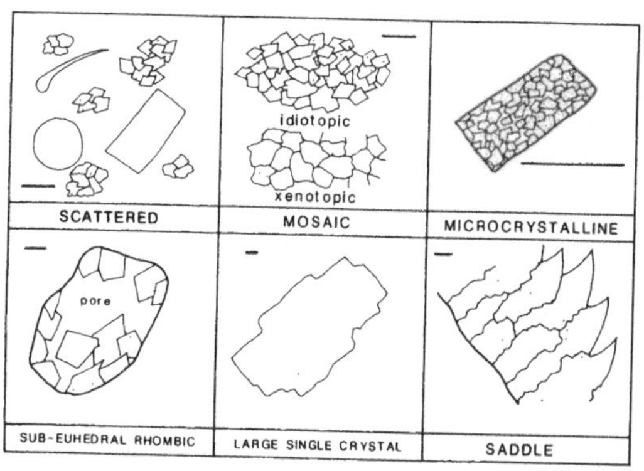

Fig. 3. Types of dolomite present in the Pendleside Limestone and Pendleside Sandstone. Scale bar approximately 100 μm in length.

Mosaic dolomite

This style of dolomite is volumetrically the most important within the carbonates, consisting of hand specimen- to outcrop-scale masses, or mosaics, of crystalline dolomite displaying a sucrosic texture. Individual crystals range in size from 150 μm to 700 μm and two end-member styles of mosaic dolomite can be recognized based on crystal form, the nature of intercrystalline boundaries and optical/CL characteristics. *Idiotopic mosaics* have a crystal size distribution between 100 μm and 250 μm, with local coarser crystals up to 700 μm, and have straight compromise boundaries. Individual crystals generally show unit extinction under crossed polars, have a subhedral to euhedral rhombic form (Fig. 4c) and contain variable amounts of inclusions, commonly ghosting the texture of the host limestone. Regular optical zonation is usually absent. The matrix present between the rhombs is usually dark brown in colour and of a translucent to opaque character; pyrite is a common constituent. *Xenotopic mosaics* are generally more coarsely crystalline, with crystal sizes from 200 μm to 600 μm (Fig. 4e). Individual crystals have an anhedral form, show sweeping extinction to varying degrees and are generally inclusion-rich, with pyrite a common inclusion. Intercrystalline boundaries are generally irregular and have a lobate character. This type of mosaic dolomite characteristically contains no intercrystalline matrix.

This style of dolomitization is replacive in origin, as indicated by the transection of grain and crystal boundaries within the host limestone by dolomite, the preservation of ghost features within the dolomite

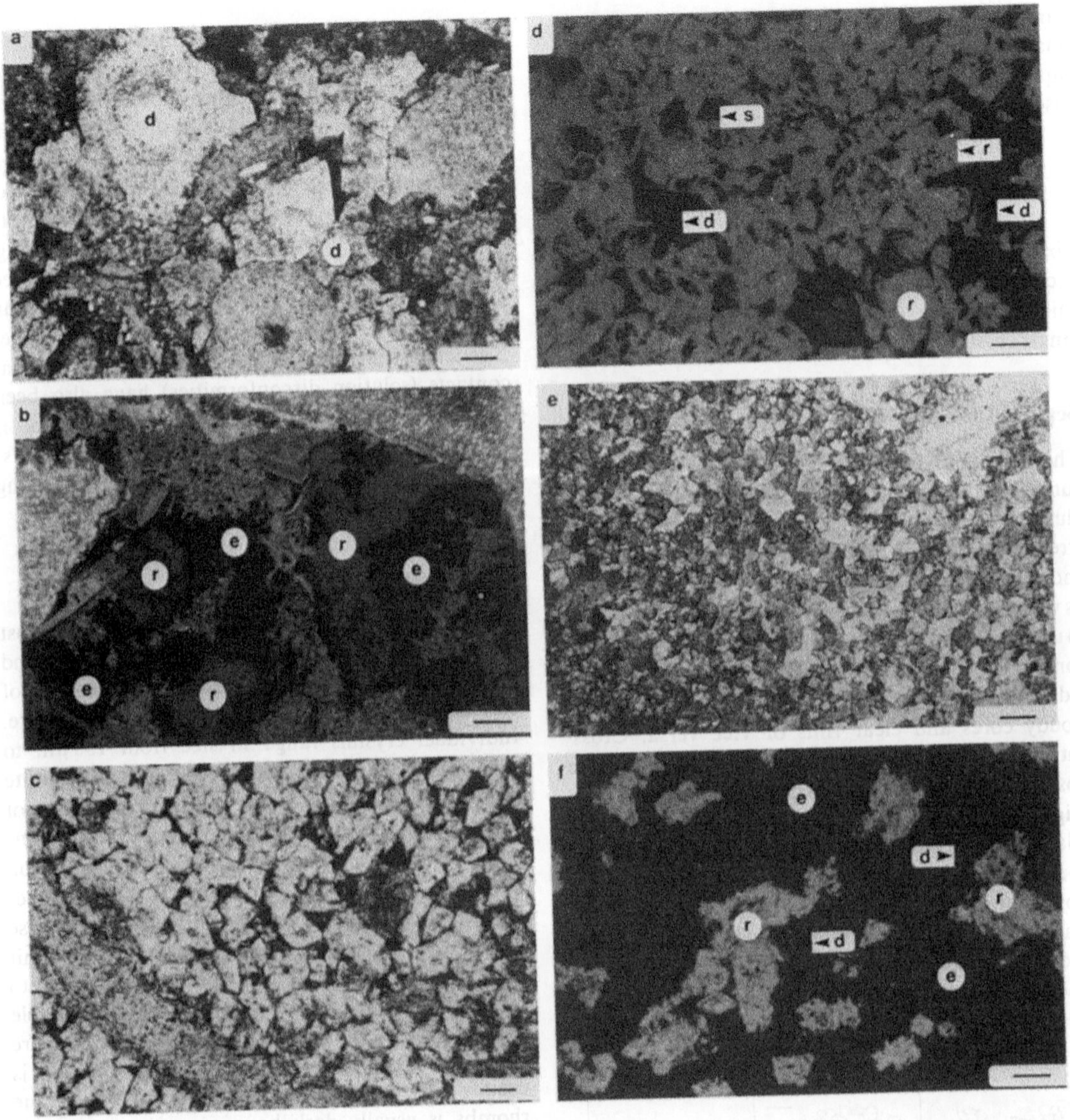

Fig. 4. (This page and facing page.) (a) Scattered dolomite within calcarenite. Dolomite (d) is preferentially replacing the finer grained carbonate constituents (micrite, microspar and microfossils). Scale bar = 100 μm, Plane polarised light (PPL). Pendleside Limestone, location 6, Fig. 1. (b) Scattered dolomite under cathodoluminescence (CL). Dolomite is composed of an inner red luminescent core (r) and an outer dull to non-luminescent. rim (e). Scale bar = 100 μm, CL. Pendleside Limestone, Location 3, Fig. 1. (c) Idiotopic mosaic dolomite showing sub- to euhedral crystals and well-developed compromise boundaries. Dolomitized shell fragment (lower left) is composed of ferroan dolomite. Scale bar = 100 μm, PPL. Pendleside Limestone, Location 3, Fig. 1. (d) Idiotopic mosaic dolomite under CL. The dominance of red luminescence (r; type 1 luminescence) is a characteristic feature, with dull to non-luminescent outer zones (d; type 2 luminescence). In this example a very thin red luminescent subzone (s) is present in the outer non-luminescent zone. Note the sub- to euhedral form. Scale bar = 100 μm, CL. Pendleside Limestone, location 3, Fig. 1. (e) Xenotopic mosaic dolomite characterized by anhedral crystals of dolomite with curved, irregular intercrystalline boundaries. Under crossed polarized light (XP) this type of dolomite often shows undulatory extinction. Scale bar = 300 μm, PPL. Pendleside Limestone location 9, Fig. 1 (f) Xenotopic mosaic dolomite under CL. Dull orange-brown luminescence (d; type 2 luminescence) and non-luminescent (e; type 3 luminescence) zones dominate with subordinate red (r; type 1 luminescence) cores. Scale bar = 100 μm, CL. Pendleside Limestone, Location 6, Fig. 1. (g) Microcrystalline dolomite with intercrystalline porosity. Note the sub- to euhedral form of individual crystals and the well

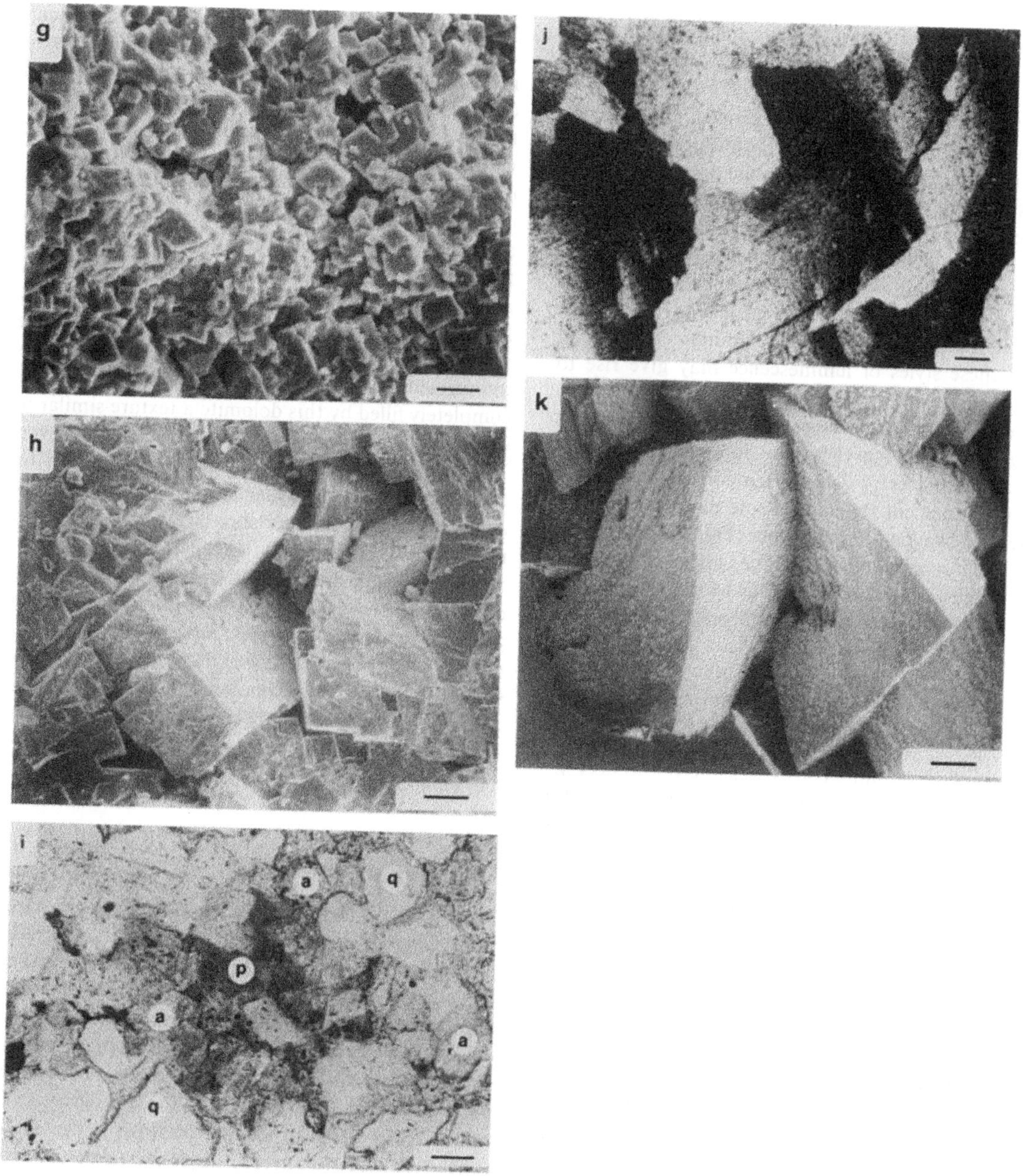

developed intercrystalline boundaries. Scale bar = 25 μm, Scanning electron photomicrograph (SEM). Pendleside Limestone, location 3, Fig. 1. (h) Sub- to euhedral rhombic dolomite partially occluding biomouldic porosity within dolostone. Scale bar = 100 μm, SEM. Pendleside Limestone, Location 6, Fig. 1 (i) Rhombic ankerite (a) partially filling secondary porosity (p) in sandstone. Note irregular 'corroded' morphology of quartz grains (q). Scale bar = 50 μm, PPL. Pendleside Sandstone, location c, Fig. 1. (j) Saddle dolomite showing undulatory extinction and irregular intercrystalline boundaries. Scale bar = 100 μm, XP. Pendleside Limestone, location 7, Fig. 1. (k) Saddle dolomite partially occluding secondary pore in calcarenite, showing well developed curved crystal faces. Small overgrowths on the dolomite are smithsonite. Scale bar = 200 μm, SEM. Pendleside Limestone, location 8, Fig. 1.

mosaics, the presence of calcite inclusions and the lack of fabric criteria commonly associated with space-filling cements (Bathurst, 1975).

Mosaic dolomites have a varied luminescence character, but a consistent CL stratigraphy can be recognized. Earliest formed mosaic dolomites have a mottled, bright red (type 1) luminescence (Fig. 4d), revealed in the cores of crystals and locally whole mosaic crystals. This is commonly succeeded by a diffuse, very dull orange-brown (type 2) zone, again with a mottled appearance (Fig. 4d, f). Slight variations in the intensity of luminescence within both these styles of luminescence may give rise to minor zonation. A third, non-luminescent (type 3) zone is present on the outer margins of mosaic crystals (Fig. 4f). Contacts between zones are generally irregular, and early zones often contain patches of later luminescent styles. Idiotopic mosaics are characterized by type 1 luminescence (Fig. 4d) with a thin, marginal zone of type 2 luminescence. In contrast, xenotopic mosaics are dominated by type 3 luminescence; earlier-formed type 1 and 2 zones are commonly highly irregular (Fig. 4f).

Microcrystalline dolomite

This type of dolomite consists of micron- to decimicron-sized anhedral to subhedral crystals (Fig. 4g) occurring in small clusters or incipient mosaics, and is generally associated with mosaic dolomite. The microcrystalline clusters have recognizable forms, including shell fragments and crinoid ossicles, with the internal structure of the parent fragment preserved as inclusions, or mimicked by the size distribution or the crystallographic orientation of the dolomite crystals. This type of dolomite is clearly replacive in origin. However, the shapes of crinoid and brachiopod fragments lacking preservation of internal structure can also be observed and, locally, biomouldic pores after such allochems may be lined or nearly totally filled by this style of dolomite. This evidence indicates a pore-filling origin. Clearly the microcrystalline dolomite may have more than one origin.

Little additional information is revealed under CL. In all sections analysed, microcrystalline dolomite crystals were non-luminescent.

Subhedral-to-euhedral rhombic dolomite

This type of dolomite consists of subhedral-to-euhedral rhombs 0·2–1·0 mm in size with sharp extinction. Individual crystals contain variable density of inclusions; in some cases variations in the amount of inclusions produce well-developed cloudy cores and clear rims or more complex zonation. Generally this type of dolomite occurs as clusters of subhedral crystals with well-developed compromise boundaries developed between individual rhombs (Fig. 4h). The clusters commonly have forms which suggest that they are pseudomorphs after brachiopod or echinoderm fragments. Sparry dolomite of this type is usually associated with mosaic dolomite and is seen to partly fill biomouldic pores in carbonates and to occlude secondary porosity in sandstones (Fig. 4i). Thus it is interpreted as a dolomite cement. When pores are completely filled by this dolomite a texture similar to euhedral mosaic dolomite is produced; however, subhedral to euhedral rhombs are characterized by coarser crystal size and pseudomorphic form. Cathodoluminescence characteristics (see below) also help to distinguish between the two types.

Generally euhedral pore-filling dolomite spar has a dull orange-brown (early) to non-luminescent (late) character under CL. Rare examples have cores of a bright-red luminescent style giving a CL stratigraphy similar to the mosaic dolomites. However, unlike the mosaic type, subhedral to euhedral sparry dolomites have much smaller red luminescent zones.

Large single crystal dolomite

This style of dolomite comprises mm-to-cm-sized single crystals of anhedral form. Crystals generally have sharp extinction and may or may not contain inclusions. Where present, inclusions are of either microcrystalline or subhedral rhombic dolomite or calcite. The morphology of these crystals suggests that single crystal dolomite replaces or fills biomouldic pores after crinoid fragments. Where inclusions of calcite are present, a replacive origin is indicated, a conclusion supported by the single-crystal nature and pseudomorphic form of the dolomite. However, this style of dolomite is also seen to partly fill biomouldic pores after crinoid fragments, as well as containing inclusions of pore-filling rhombic and microcrystalline dolomite.

Under CL large single sparry dolomite crystals show a mottled dull orange-brown and non-luminescent character. Where biomouldic pores are partly filled, a transition from mottled dull orange-brown (early) to non-luminescent (late) can be observed.

Saddle dolomite

Sparry crystals, 500 μm to several millimetres in size, with curved cleavage traces and crystal faces and sweeping extinction under crossed polars, represent another type of dolomite, and are usually associated with mosaic dolomite. This saddle dolomite (Radke & Mathis, 1980), also known as 'baroque dolomite' (e.g. Zenger, 1983) and 'white-sparry dolomite' (Beales, 1971), contains variable amounts of inclusions, which may give the crystals a zoned appearance and mainly consist of calcite, dolomite and/or fluid. Generally these saddle dolomites have a spear-shaped anhedral to euhedral form. They are present both as a replacive and a pore-filling phase. The replacive type is generally anhedral, forming hypidiotopic to xenotopic mosaics (Fig. 4k) and is generally inclusion-rich, with calcite and fluid inclusions predominant. Inclusions often preserve ghosts of the parent limestone texture. However, unlike most of the other types of replacive dolomite, the saddle type is not selective in its replacement. Biomouldic pores partially filled by saddle dolomite (Fig. 4k) indicate that it can occur as a cement phase also. These crystals are generally subhedral and rhombic to spear-shaped, with irregular compromise boundaries and commonly with well-developed growth bands (zones). Under CL, both replacive and pore-filling saddle dolomites are non-luminescent.

CHEMISTRY

There are no strict rules of nomenclature for dolomites in which the crystal lattice contains appreciable substitution by Fe and Mn. Carbonates with a dolomite structure containing variable amounts of Fe and Mn are variously referred to as ferroan dolomite and ankerite (ankerite being generally restricted to the more ferroan compositions). Here the term ankerite is restricted to Ca(Mg,Fe,Mn) carbonates with Mg/(Fe+Mn) <4:1 (Deer, Howie & Zussman, 1962). In the following account the term 'dolomite' is used to describe carbonates with a dolomite structure containing variable amounts of Fe and Mn unless further distinction is needed.

General characteristics

The dolomites vary widely in composition, from near stoichiometric $Ca_{1\cdot00}\ Mg_{1\cdot00}\ (CO_3)_2$ to phases containing appreciable Fe and/or Mn (Table 1). Iron substitution is the more common; the most ferroan composition containing 20 mole % $FeCO_3$ was observed in a dolomite of formula $Ca_{1\cdot02}\ Mg_{0\cdot51}\ Fe_{0\cdot40}\ Mn_{0\cdot07}\ (CO_3)_2$. The highest Mn content recorded was 15 mole % MnCO, occurring in a dolomite with a composition $Ca_{1\cdot04}\ Mg_{0\cdot64}\ Fe_{0\cdot02}\ Mn_{0\cdot30}\ (CO_3)_2$. This sample is, however, atypical, since it was taken from a dolomite vein associated with minor sphalerite. Typically Mn concentrations up to 3 mole % $MnCO_3$ are recorded. Most of the samples show a positive correlation between Fe and Mn substitution, although high Fe and low Mn concentrations and vice versa have been recorded. Magnesium-ion concentrations range from 25 mole % to 50 mole % $MgCO_3$, as a result of substitution of Fe and Mn for Mg within the dolomite lattice. Variation between 48 mole % and 54 mole % $CaCO_3$ occurs. A general trend from depletion to excess Ca is observed with increasing substitution of Fe and Mn into the dolomite lattice, although the whole range of Ca concentration has been recorded from ferroan dolomites with similar (Fe+Mn) contents. Other trace elements reported to substitute for the major elements within dolomite such as S, K, Al, Ba, Na, Sr, Pb and Zn (Morrow, 1982a, b), if present, were generally at concentrations below the detection limit of the microprobe. The generally low trace-element contents of whole rock samples (Table 2) are consistent with this.

Facies-related chemical variations

Dolomitization occurs within four facies associations of the Pendleside Limestone, representing basin margin and upper, mid and lower subenvironments of a carbonate slope environment, and also within the Pendleside Sandstone. Figure 5 relates the main variations in dolomite chemistry to the types of facies dolomitized. Only samples from totally dolomitized calcarenites and sandstones are presented in an attempt to constrain the data set, although interpretation of the results may be complicated by variations in chemistry during progressive dolomitization (see below). Four of the groups, the sandstones and basin margin, upper and lower slope carbonates, have quite tightly clustered compositions (Fig. 5a). The sandstones have excess Ca, with Fe and Mn ranging from approximately 12 mole % to 16 mole % $(Fe,Mn)CO_3$. Basin margin calcarenites have slightly excess Ca with $(Fe, Mn)CO_3$ approximately 10 mole %, whereas upper slope calcarenites are generally depleted in Ca and are ferroan dolomites with $(Fe,Mn)CO_3$ ranging from 5 to 8 mole %. The lower slope calcarenites have

Table 1. Representative electron microprobe analyses of dolomites from the Bowland Basin examined in this paper. Specimen number refers to location numbers on Fig. 1. + = Pendleside Sandstone; * = dolomite in vein; remaining analyses are from the Pendleside Limestone.

Mole %	4a	4b	5a*	5b*	5c*	6a+
$CaCO_3$	49.3	50.5	52.2	51.1	52.7	51.15
$MgCO_3$	44.5	45.9	45.5	45.6	45.4	33.9
$MnCO_3$	0.7	0.4	2.0	12.6	1.6	0.6
$FeCO_3$	5.5	3.2	0.3	0.7	0.3	14.0

Mole %	6b+	6c+	10a	10b	13a	13b
$CaCO_3$	52.6	53.1	50.7	51.1	49.3	49.2
$MgCO_3$	31.7	34.6	34.4	37.7	50.2	50.2
$MnCO_3$	0.5	0.5	1.8	2.6	0.0	0.1
$FeCO_3$	15.2	11.8	13.1	8.6	0.3	0.5

compositions nearest to stoichiometric dolomite, although they are depleted in Ca and have $(Fe,Mn)CO_3$ contents of <2 mole %. The mid-slope calcarenites have a wide spectrum of compositions covering almost the complete compositional range of the samples studied. Sandstone and lower slope facies have low Mn contents (<0·7 mole % Mn CO_3), the latter having a wide spread of Mn/Fe ratios despite the low Mn contents (0·3–2·0 mole %) (Fig. 5b). Both the upper slope and marginal facies have higher Mn/Fe ratios than the mid slope facies for a particular Mn content, suggesting enrichment in Mn within these facies. The marginal facies have some of the highest Mn contents, up to 2·3 mole % $MnCO_3$. Thus, with the exception of the lower slope facies, there appears to be a basinward increase in Fe and a counter gradient in Mn.

Chemical zonation

Because of the generally ferroan composition, staining of thin sections with potassium ferricyanide produces a turquoise stain on nearly all the dolomites and thus reveals little information regarding chemical zonation. However, some mosaic dolomite crystals display unstained cores with slightly turquoise rims and an increase in the intensity of the stain colour along crystal margins, indicating an increase in Fe substitution during crystal growth. This is supported by the CL characteristics of the dolomites; the transition from bright red, through dull orange-brown, to non-luminescent is consistent with increasing Fe content, reflecting the quenching effect of Fe on luminescence (Fairchild, 1983; Pierson, 1981). Element distribution maps (Ca, Mg, Fe and Mn) for an individual dolomite crystal (Fig. 6) also show this zoned character. Calcium shows a uniform distribution throughout the crystal (Fig. 6b) but Mg is depleted in the outer zone relative to the core of the crystal (Fig. 6c). This depletion is associated with enrichment in Fe and Mn (Fig. 6d, e), suggesting substitution of Fe and Mn for Mg. The boundary between the Fe- and Mn-poor core and the Fe- and Mn-rich rim is highly irregular and cuspate, in contrast to the zone boundaries within the rim, and is similar to the zone boundaries observed under CL. Within the Fe- and Mn-enriched rim two main zones can be identified; the inner zone is further enriched in Mn compared to the outer zone. Thus during progressive dolomitization there was a general increase in the degree of Fe and Mn substitution in the dolomite structure.

Quantitative variations in chemistry across several of the dolomite types are shown in Fig. 7. Dolomite in sandstone shows little variation in chemistry and has consistently ankeritic composition. Scattered dolomite within a partially dolomitized calcarenite similarly shows little evidence of zonation and is ferroan dolomite ($FeCO_3$ between 4·3 mole % and 6·5 mole % and between 0·8 mole % and 1·7 mole % $MnCO_3$). The other dolomite crystals show a well-zoned character with marked changes in chemistry (increase in Fe and Mn towards the crystal margins) and highly irregular, cuspate boundaries between zones (similar to those displayed in Fig. 6a). The cuspate boundaries may separate different types of dolomite (e.g. where euhedral and saddle dolomite overgrow corroded mosaic cores) or occur within dolomite crystals of mosaic type.

Thus during dolomitization there was an increase in Fe and Mn substitution, with abrupt changes in pore-fluid chemistry producing marked chemical zonation within individual dolomite crystals. This is

Table 2. Representative whole rock XRF analyses for the Pendleside Limestone. Oxides in wt. %; trace elements in ppm. LOI = Loss on ignition. ▲ = limestone; + = dolostones and dolomitic limestones. Specimen numbers refer to locations on Fig. 1.

	3a▲	3b▲	3c+	3d+	4a▲	4b+	9a▲	9b+	13a▲	13b+
SiO_2	3.65	2.62	4.09	4.08	0.73	4.78	2.47	4.03	4.77	3.91
TiO_2	0.01	0.02	0.02	0.01	0.01	0.01	0.00	0.02	0.01	0.01
Al_2O_3	0.17	0.40	0.09	0.06	0.05	0.11	0.06	0.06	0.09	0.05
Fe_2O_3	0.15	0.59	3.14	2.76	0.09	3.86	0.17	3.27	0.18	1.52
MnO	0.06	0.04	0.49	0.31	0.06	0.35	0.18	0.55	0.06	0.17
MgO	0.58	2.61	17.74	18.02	0.59	17.15	0.58	12.64	0.50	19.83
CaO	52.94	50.58	29.67	30.58	55.39	28.98	53.42	36.03	51.67	28.85
K_2O	0.00	0.07	0.00	0.00	0.00	0.00	0.00	0.01	0.00	0.01
P_2O_5	0.03	0.07	0.08	0.01	0.03	0.01	0.04	0.02	0.02	0.01
LOI	42.19	42.06	44.13	44.64	43.64	43.09	43.04	43.41	41.96	45.27
Total	99.80	99.07	99.45	100.48	100.55	98.34	99.97	100.05	99.26	99.64
Cr	0	0	0	0	0	0	0	0	0	0
Co	2	2	9	8	3	10	2	9	2	5
Ni	4	9	5	1	3	6	3	2	5	2
Cu	10	12	5	4	10	19	9	7	10	4
Zn	85	144	234	20	26	6645	9	13	17	26
Rb	0	5	1	1	0	1	1	1	1	2
Sr	401	524	83	59	386	51	420	67	757	42
Y	19	18	6	8	6	12	6	7	6	7
Zr	0	0	0	0	0	0	0	0	0	0
Nb	0	0	0	0	0	0	0	0	0	0
Pb	4	8	1	1	3	2	3	2	4	5
Ba	17	31	12	5	14	9	34	4	9	8
Nd	11	17	5	7	5	11	5	7	4	6
Ce	27	41	4	10	4	10	9	4	4	6
La	2	18	0	8	0	3	0	1	0	2
V	5	7	9	1	6	8	4	0	5	4
Sc	49	51	23	24	50	26	47	33	46	23

particularly evident where there is a change from zones with less than 5–7 mole % $FeCO_3$ to those in excess of this value. The contact between these zones is commonly highly irregular. No discrete chemical zonation is observed in dolomites containing <5–7 mole % $FeCO_3$, which are characterized by scattered and some mosaic types (Fig. 8), despite the recognition of zones under CL. Saddle, microcrystalline, single crystal and subhedral to euhedral rhombic types generally have in excess of 7 mole % $FeCO_3$ (Fig. 8), and commonly display discrete chemical zonation. Compared to mosaic and scattered dolomite these types of dolomite are generally enriched in Mn (Fig. 8). Mosaic dolomites show a wide range of compositions, with up to 10 mole % $FeCO_3$ and in excess of 2 mole % $MnCO_3$ (Figs 7, 8).

CARBON AND OXYGEN ISOTOPES

Samples of dolomite from both clastics and carbonates were analysed for stable carbon and oxygen isotopes, and in spite of the small number of samples ($n = 14$ for dolomite; $n = 2$ for calcite; Table 3), several observations can be made that help to constrain models for dolomitization. Two calcite samples from the host limestones have $\delta^{18}O$ and $\delta^{13}C$ values ($\delta^{18}O$ from $-7{\cdot}25‰$ to $-6{\cdot}32‰$ and $\delta^{13}C$ from $+1{\cdot}26‰$ to $+1{\cdot}40‰$), similar to those reported for Carboniferous marine carbonates (e.g. Dickson & Coleman, 1980). The dolomite samples from both the Pendleside Limestone and Pendleside Sandstone can be divided into two main groups based on their isotopic compositions (Fig. 9a). Group 1 dolomites have a range of $\delta^{13}C$ values from $+0{\cdot}87‰$ to $+2{\cdot}79‰$ and $\delta^{18}O$ from $-7{\cdot}21‰$ to $-3{\cdot}71‰$, while Group 2 dolomites have a range of $\delta^{13}C$ from $-2{\cdot}07‰$ to $-0{\cdot}76‰$ and $\delta^{18}O$ from $-9{\cdot}43‰$ to $-5{\cdot}55‰$. In addition, several dolomite samples from veins and vugs were analysed and have a range of $\delta^{13}C$ from $+1{\cdot}76‰$ to $+3{\cdot}18‰$ and $\delta^{18}O$ from $-11{\cdot}44‰$ to $-4{\cdot}68‰$

Calcite and dolomite precipitated under equilibrium conditions exhibit a temperature-dependent isotopic fractionation whereby dolomite is enriched in $\delta^{18}O$ by approximately 3‰ (e.g. Land, 1980). Thus Group 1 and 2 dolomites are generally depleted in $\delta^{18}O$ compared with calcite within the host limestone, suggesting an increase in temperature and/or depletion of the water in $\delta^{18}O$ between calcite cementation and dolomitization. As neither the $\delta^{18}O$ composition of the fluid or the temperature at which the dolomites were precipitated are known; the data cannot be interpreted uniquely.

Comparison of the $\delta^{18}O$ values of the dolomites

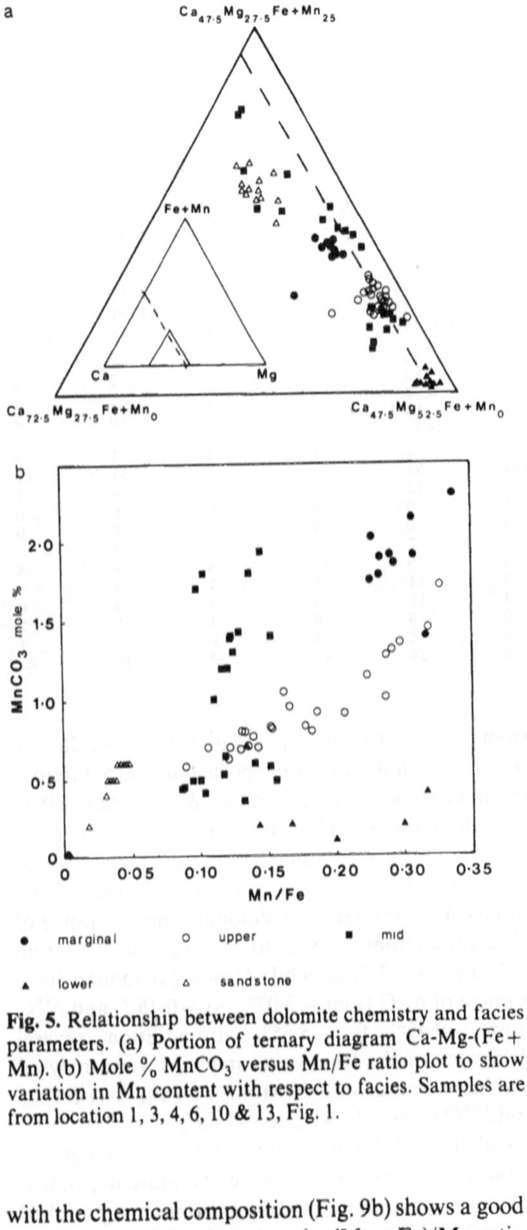

Fig. 5. Relationship between dolomite chemistry and facies parameters. (a) Portion of ternary diagram Ca-Mg-(Fe+Mn). (b) Mole % $MnCO_3$ versus Mn/Fe ratio plot to show variation in Mn content with respect to facies. Samples are from location 1, 3, 4, 6, 10 & 13, Fig. 1.

with the chemical composition (Fig. 9b) shows a good negative correlation between the (Mn + Fe)/Mg ratio and $\delta^{18}O$. In view of the chemical zonation (Figs 6, 7) present in the dolomites this suggests a progressive depletion in $\delta^{18}O$ during dolomitization. Such a conclusion is supported by comparing the values of $\delta^{18}O$ for mosaic dolomite with the $\delta^{18}O$ for grain replacements and pore fills that are paragenetically later (see below). The latter are depleted in $\delta^{18}O$ and have higher (Mn + Fe)/Mg ratios.

The range of carbon isotope values shown above can be interpreted in terms of the interaction of two main carbon sources: carbonate similar in composition to Carboniferous marine carbonates (i.e. $\delta^{13}C = +1‰$, e.g. Dickson & Coleman, 1980) and a source of isotopically light carbon. The majority of Group 1 dolomites and dolomite from veins have $\delta^{13}C$ values similar to those that would be expected if, during dolomitization, carbon was derived from the host limestone, with only very minor $\delta^{13}C$ enrichment (up to approximately 2‰, similar to the value predicted by Sheppard & Schwarcz, 1970). Group 2 dolomites, however, have lighter $\delta^{13}C$ values, indicating addition of $\delta^{13}C$-depleted carbon. If the carbon isotope data are viewed in terms of the facies relationships (Fig. 9c) of the dolomites, there is a trend of $\delta^{13}C$ depletion from proximal (limestone-dominant) facies, through distal limestone facies (interbedded limestones and mudstones), to the Pendleside Sandstone within the Bowland Shales (little marine carbonate, predominantly terrigenous clastics). The association of light $\delta^{13}C$ values in facies interbedded with organic-rich mudrocks suggests decarboxylation of organic matter as the main source of light carbon (e.g. Irwin, Curtis & Coleman, 1977; Curtis, 1978). Thus the differences in $\delta^{13}C$ between Group 1 and 2 dolomite values can be explained in terms of the interaction of an external, isotopically light carbon source derived from organic matter within the terrigenous mudrocks with a source of carbon derived from the dolomitized host. In the carbonate-dominated facies, the effects of the isotopically light CO_2 would be swamped by the large limestone-derived carbon source.

LEACHING AND POROSITY DEVELOPMENT

Development of volumetrically significant secondary porosity accompanied dolomitization within both sandstones and carbonates. In both cases this porosity was created as a result of leaching of grains and/or authigenic phases rather than by other processes such as grain shrinkage or fracturing. This secondary porosity development is considered to have been intimately associated with dolomitization, since dolomite pre- and post-dates dissolution and no secondary porosity generation occurred in undolomitized lithologies.

Within the dolomitized carbonates the most obvious effect of leaching is the production of fabric-selective mouldic porosity (Fig. 10a), most commonly after

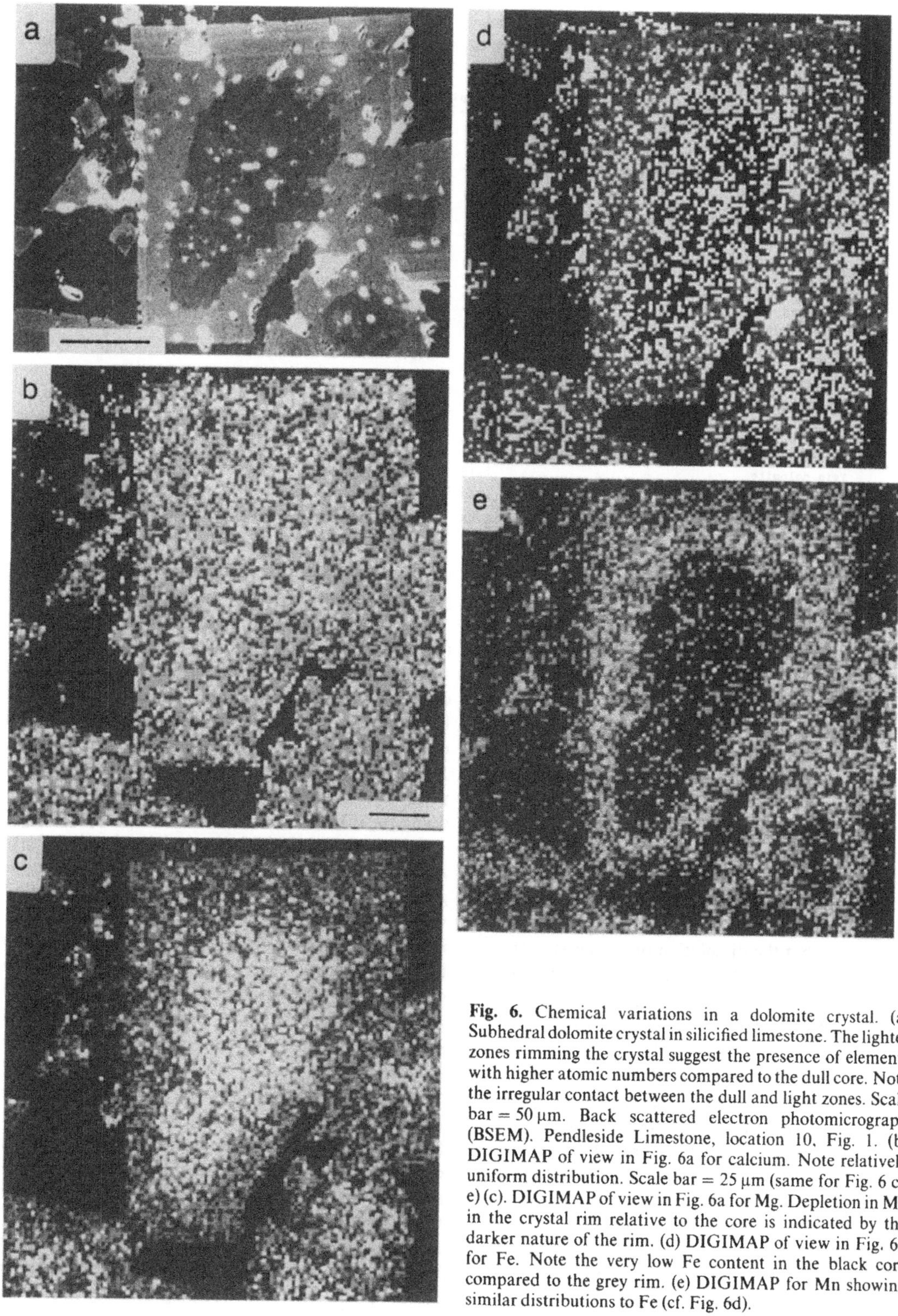

Fig. 6. Chemical variations in a dolomite crystal. (a) Subhedral dolomite crystal in silicified limestone. The lighter zones rimming the crystal suggest the presence of elements with higher atomic numbers compared to the dull core. Note the irregular contact between the dull and light zones. Scale bar = 50 μm. Back scattered electron photomicrograph (BSEM). Pendleside Limestone, location 10, Fig. 1. (b) DIGIMAP of view in Fig. 6a for calcium. Note relatively uniform distribution. Scale bar = 25 μm (same for Fig. 6 c–e) (c). DIGIMAP of view in Fig. 6a for Mg. Depletion in Mg in the crystal rim relative to the core is indicated by the darker nature of the rim. (d) DIGIMAP of view in Fig. 6a for Fe. Note the very low Fe content in the black core compared to the grey rim. (e) DIGIMAP for Mn showing similar distributions to Fe (cf. Fig. 6d).

Table 3. Isotopic and chemical analyses of dolomite and calcite from the Bowland Basin. * = Pendleside Limestone; + = Pendleside Sandstone; # = other dolomite. Isotopic analyses are quoted relative to the PDB standard. Specimen numbers refer to locations on Fig. 1.

Specimen No.	Description	Isotopic Ratio (per mil)		Mole Fraction				
		$\delta^{13}C$	$\delta^{18}O$	Ca	Mg	Mn	Fe	Sr
1*	Rhombic dolomite in silicified carbonate	2.45	-7.21	0.4680	0.4300	0.0136	0.0881	0.0007
2a*	Crinoid fragment	1.26	-6.32	0.9790	0.0200	0.0003	0.0003	0.0007
2b*	Rhombic dolomite in silicified carbonate	-0.76	-7.89	0.5210	0.4770	0.0003	0.0023	0.0002
3a*	Mosaic dolomite	0.87	-5.53	0.5180	0.4340	0.0078	0.0404	0.0002
3b*	Rhombic dolomite in secondary porosity	1.20	-5.49	0.5030	0.4240	0.0077	0.0653	0.0003
4a*	Saddle dolomite	1.56	-5.77	0.5090	0.4160	0.0099	0.0648	0.0003
4b*	Xenotopic mosaic dolomite	1.12	-4.68	0.5250	0.4540	0.0013	0.0206	0.0003
5a*	Mosaic dolomite	2.37	-3.71	0.5170	0.4420	0.0050	0.0367	0.0003
5b*	Vein dolomite with sphalerite cutting 5a	1.78	-4.68	0.5460	0.4120	0.0066	0.0354	0.0002
6a*	Idiotopic mosaic dolomite	2.79	-6.85	0.5520	0.4390	0.0015	0.0079	0.0003
6b+	Rhombic dolomite	-1.67	-9.43	0.5470	0.2540	0.0064	0.1925	0.0003
6c+	Rhombic dolomite	-2.07	-6.48	0.4550	0.3830	0.0014	0.1612	0.0009
6d+	Rhombic dolomite	-1.85	-5.55	0.4140	0.3980	0.0016	0.1876	0.0010
9#	Vein dolomite	3.18	-11.44	0.5040	0.3550	0.0107	0.1302	0.0002
12*	Sparry calcite cement	1.40	-7.25	1.0000	0.0010	0.0005	0.0005	0.0003
14#	Saddle dolomite	3.08	-8.31	0.5050	0.3320	0.0062	0.1561	0.0002

bioclasts (echinoderm and shell fragments). These mouldic pores are often enlarged, suggesting that leaching was not confined to the bioclasts but also affected the surrounding matrix (Fig. 10a). Although this type of porosity is clearly dependent on the size of the original grains, mouldic pores usually have diameters on a mm-to-cm scale. Another important type of secondary porosity developed within mosaic dolomites is non-fabric-selective vuggy porosity. Vuggy pores vary widely in both shape and size. Most commonly they have an irregular shape and are centimetre sized but vugs may, in some instances, be up to a metre in diameter. The presence of enlarged biomouldic pores may indicate that at least some of the vuggy porosity originated through dissolution of bioclasts and that the dissolution process was particularly effective in these areas, leaching the surrounding matrix extensively. Both mouldic and vuggy pores are generally partly to totally occluded by ferroan dolomite/ankerite of subhedral to euhedral rhombic, microcrystalline, saddle or single crystal form.

Intercrystalline porosity is generally of minor importance within the dolomitized carbonates, although it is present to a limited extent within euhedral mosaic and microcrystalline styles of dolomite (Fig. 4g). In the case of the microcrystalline dolomite it is not always clear whether this porosity is, in fact, reduced mouldic or truly intercrystalline. Within the dolostones where mouldic pores predominate, visual porosity typically ranges from zero to 15%; however, where vuggy porosity makes a significant contribution, porosities may exceed 30%.

The effects of leaching within dolomitized carbonates can also be observed where apparently no significant porosity is produced. Within mosaic dolomites highly irregular, often cuspate contacts between the zones of high and low Fe content observed under BSEM (Fig. 6a) and inferred from CL (Fig. 11a, b) suggest that dissolution/corrosion occurred prior to overgrowth of the more Fe-rich ferroan dolomite/ankerite.

Within the sandstones dissolution-related porosity is usually reduced or totally occluded by later authigenic phases. Where present it is commonly associated with the presence of dolomite. Dissolution of both interstitial cement/matrix and framework grains was the main process responsible for secondary porosity generation. Feldspars, mica, carbonate and ?clay minerals appear to have been the main phases selectively leached. Quartz may also have been affected since grains of quartz are often corroded along their boundaries, although this effect may have been produced by dissolution of a grain-corrosive (?carbonate) cement. In sandstones where porosity is only partly occluded by dolomite cement, the preserved pore system consists predominantly of irregular to elongate pores located between grains (Fig. 10b). Partial dissolution of framework grains produced intragranular porosity commonly associated with corroded and honeycombed plagioclase grains (Fig.

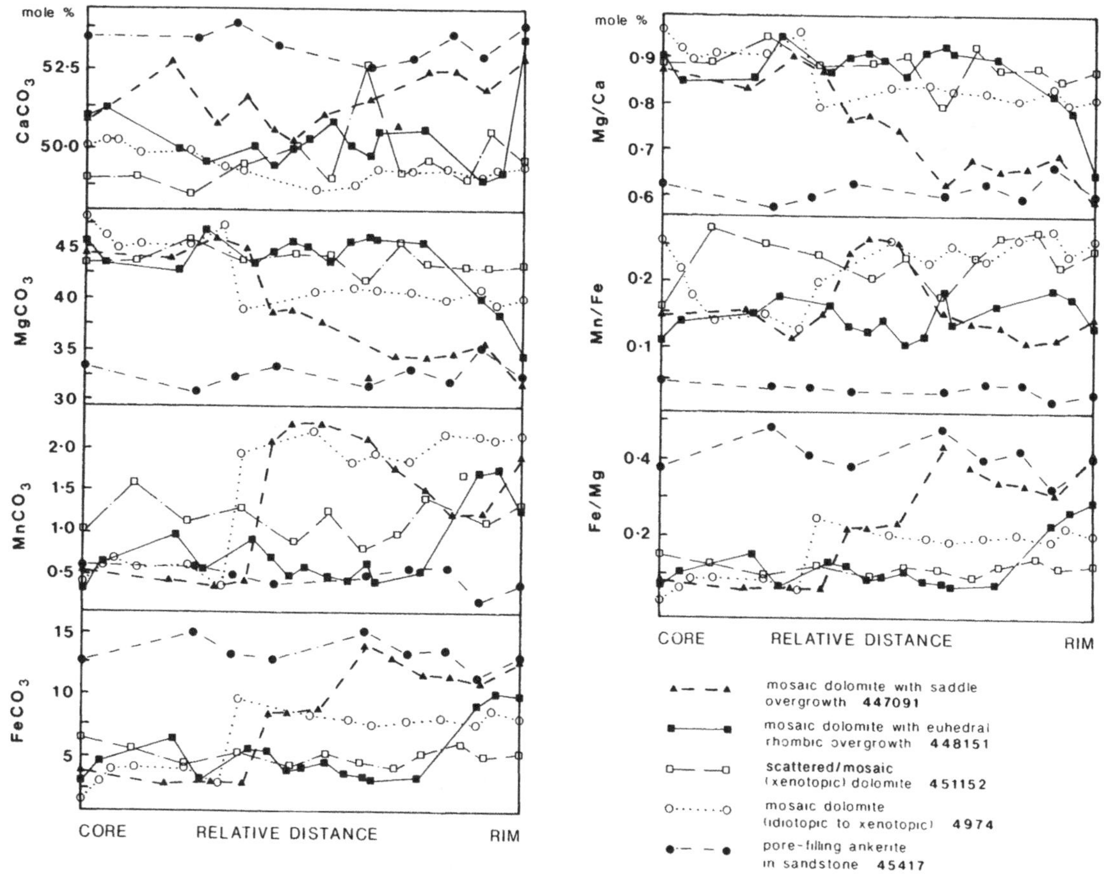

Fig. 7. Variation in chemistry across individual dolomite rhombs of various types or combinations of types. Numbers refer to individual specimens; 447091, location 10, Fig. 1; 448151, location 4, Fig. 1; 4974, location 3, Fig. 1; 451152 & 45417, location 6, Fig. 1.

10c). In sandstones with a high percentage of dolomite and little preserved porosity, the presence of leached grains, oversized pores and floating grains suggests that extensive dissolution has occurred prior to dolomite precipitation.

PARAGENESIS

In the Bowland Basin the transition from dolomitized to undolomitized carbonates and sandstones commonly occurs over distances of a few centimetres to metres and locally on the scale of a thin section. The spectrum of fabrics which can be observed allows constraints to be placed on dolomite paragenesis and the relationships of dolomitization, leaching and secondary porosity development. The dolomite stratigraphy provided by CL microscopy also helps to define the relationships between the various types of dolomite. Three main stages within the dolomitization–dissolution paragenetic sequence can be recognized; replacement, secondary porosity development and porosity occlusion (Fig. 12). The replacement stage within the carbonates is characterized by the development of mosaic dolomites. Initial dolomitization generally affected the fine-grained calcite constituents comprising the matrix between allochems (micrite and microspar), together with foraminifera, peloids, calcispheres and micritized grains. Scattered dolomite clusters, both anhedral and euhedral, characterize this initial stage of dolomitization. The dolomites formed during this early part of the replacement phase are generally low in Fe and Mn and commonly brightly luminescent. With increasing alteration, mosaic do-

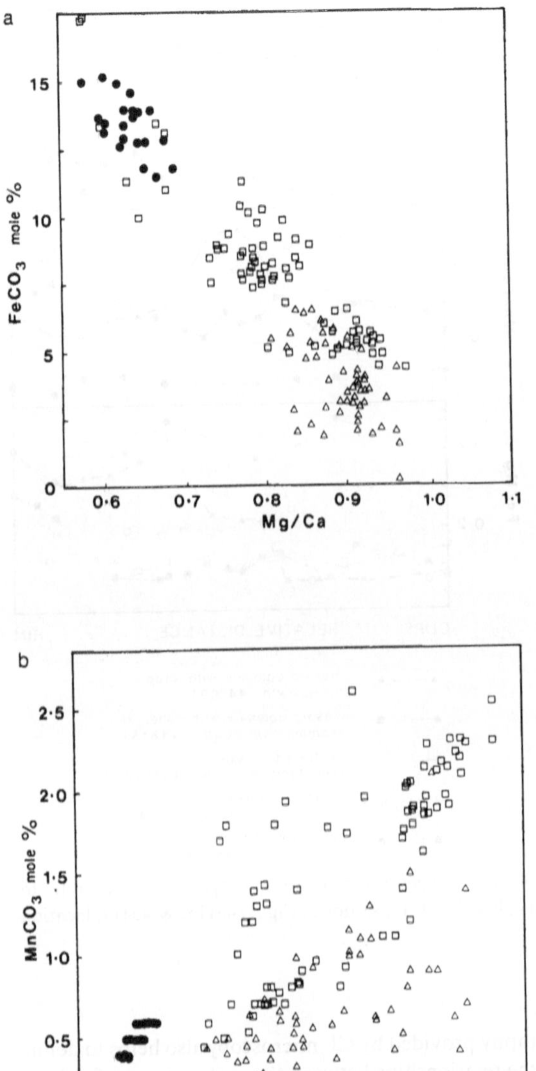

Fig. 8. Variation in chemistry with respect to dolomite paragenesis, subdivided into pre- and post-leaching. (a) Plot of mole % $FeCO_3$ versus Mg/Ca ratio. (b) Plot of mole % $MnCO_3$ versus Mn/Fe ratio. Note higher Fe and Mn contents of post-leaching dolomites. ● = dolomite in Pendleside Sandstone; □ = post-leaching, xenotopic mosaic (outer zones), euhedral rhombic & saddle dolomite (Pendleside Limestone). △ = pre-leaching, idiotopic and inner zones of xenotopic mosaics and scattered dolomite (Pendleside Limestone). Samples from locations 1, 3, 6 & 10, Fig. 1.

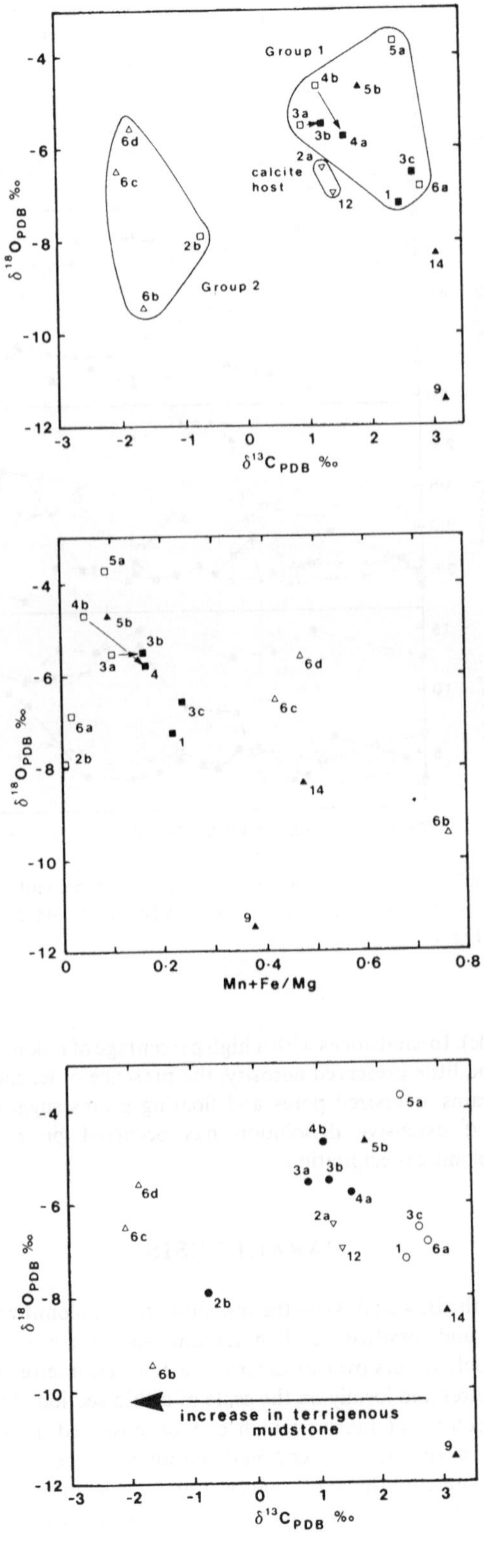

lomites formed, displaying a complex series of textures with one or both types of mosaic (xenotopic and idiotopic) developed. Under CL cores of brightly luminescent dolomite with well-developed crystal faces (characteristic of idiotopic dolomite) are commonly seen to be overgrown by anhedral, xenotopic-style dolomite. Thus idiotopic mosaics and diffuse dolomite clusters generally pre-date xenotopic mosaics.

During mosaic development Fe and Mn were incorporated to a greater extent , with the result that CL characteristics change to dull orange-brown or non-luminescent. More resistant components of the host (e.g. calcite spar and shell fragments) became dolomitized leaving only coarse sparry calcite. Other types of dolomite became apparent, with crinoid fragments commonly wholly or partially replaced by single crystal and euhedral sparry dolomite. Commonly other fragments as well as crinoid grains have also been replaced by microcrystalline dolomite. All these types of dolomite have dull to non-luminescent CL characteristics and contain >7 mole % $(Fe,Mn)CO_3$. The change in the type of dolomite preserves some of the original host texture. Ultimately this style of replacement led to the development of a dolostone. However, prior to complete dolomitization a marked change in the composition of pore fluids took place and was associated with major dissolution. This marks the second major stage in the paragenetic sequence, secondary porosity development. Analysis of specimens representing this part of the paragenetic sequence indicates that calcite was the main phase leached from the carbonates, although enlarged mouldic pores suggest pre-leaching dolomite may also have been dissolved. Generally porosity was devel-

Fig. 9. Isotopic and chemical variations of dolomites. (a) $\delta^{18}O$ versus $\delta^{13}C$ plot showing two main groups of dolomite in the Pendleside Limestone and Pendleside Sandstone. (b) $\delta^{18}O$ versus (Mn + Fe)/Mg ratio. Note the increase in (Mn + Fe)/Mg ratio and depletion in $\delta^{18}O$ with progressive dolomitization. Key for 9a and 9b: △ = dolomite in Pendleside Sandstone; ▲ = vein dolomite; □ = pre-leaching dolomite (Pendleside Limestone); ■ = post-leaching dolomite (Pendleside Limestone); ▽ = calcite host (grain and cement). (c) $\delta^{18}O$ versus $\delta^{13}C$ plot showing relationship between facies and isotopic composition. Note decrease in $\delta^{13}C$ with more distal facies (increase in interbedded mudstone). Key for 9c: ▲ = vein dolomite; ○ = dolomite in proximal (upperslope) calcarenites; ● = dolomite in distal (mid/ lower slope) calcarenites △ = dolomite in Pendleside Sandstone; ▽ = calcite host (grain and cement). Numbers refer to specimen numbers in Table 3 and location numbers in Fig. 1.

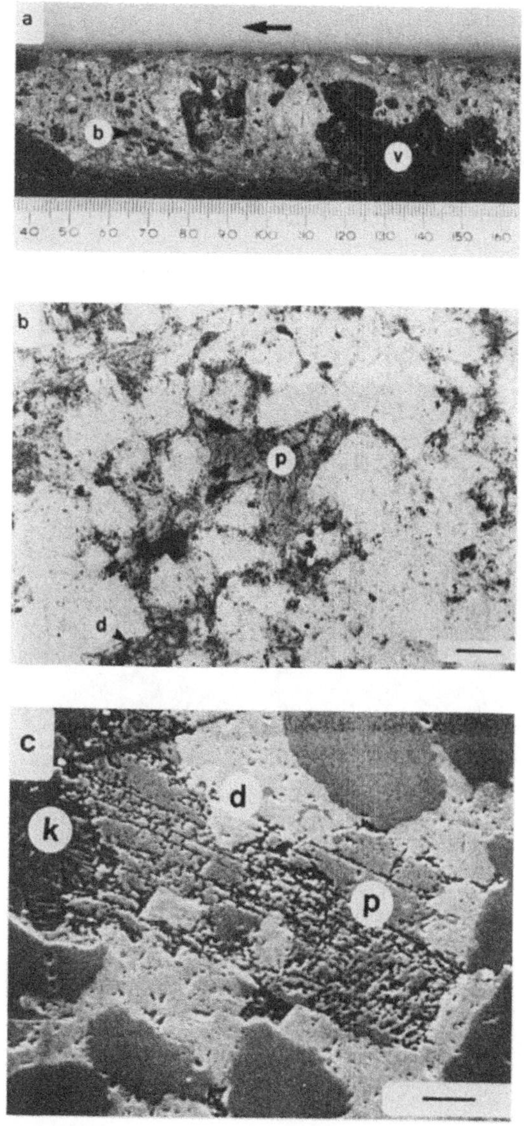

Fig. 10. (a) Core specimen showing secondary porosity developed within dolostone. Porosity consists of biomouldic porosity (b) after crinoid and brachiopod fragments and vuggy porosity (v), the latter possibly representing enlarged biomouldic porosity. Some pores are partially occluded by microcrystalline and euhedral rhombic dolomite. Scale is in mm. Pendleside Limestone, location 8, Fig. 1. (b) Secondary porosity (p) developed within sandstone, relict outlines of grains are marked by ?clay rims. Also note presence of dolomite/ankerite (d). Scale bar = 50 μm, PPL. Pendelside Sandstone, location 6, Fig. 1. (c) Honeycombed plagioclase grain (p) partially dissolved to yield intraparticle porosity. Kaolinite/dickite (k) can be seen filling secondary pore. Ferroan dolomite/ankerite (d) is also replacing the plagioclase. Scale bar = 100 μm, BSEM. Pendleside Sandstone, location 6, Fig. 1.

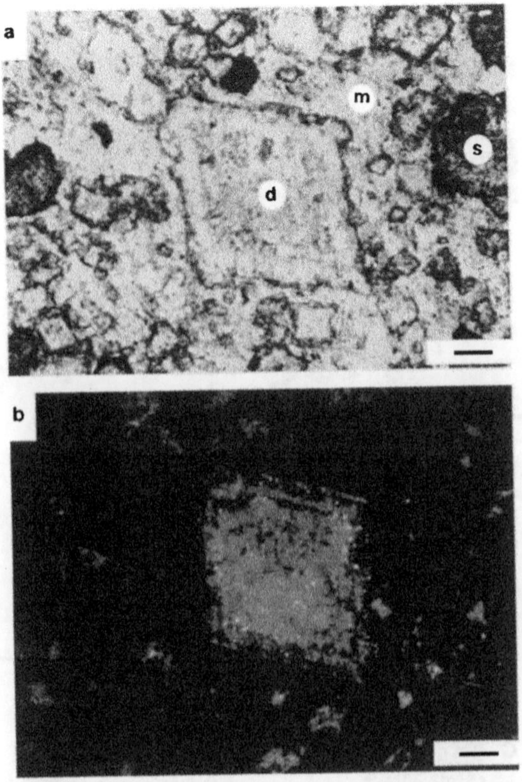

Fig. 11. (a) Euhedral rhombic dolomite crystal (d) in silicified limestone/dolostone (m) with disseminated sphalerite(s). Note cloudy core and clear rim habit. Scale bar = 100 μm, PPL. Pendleside Limestone, location 3, Fig. 1. (b) Same view as in Fig. 11a but under CL. The large rhomb in the centre of the field of view can be seen to be composed of a luminescent (red) core and non-luminescent rims. Note the highly irregular contact between these two types of luminescence and the presence of non-luminescent dolomite within the red core. These features suggest dissolution prior to precipitation of the extinct dolomite. Scale bar = 100 μm, CL. Pendleside Limestone, Location 3, Fig. 1.

oped as a result of dissolution of coarse bioclasts, crinoids and shell fragments, and coarse sparry cements. The presence of corroded cores within mosaic dolomites may reflect dissolution of matrix from between idiotopic mosaics and the generation of intercrystalline secondary porosity. In the sandstones, the most significant phases removed during leaching were carbonate grains and feldspars.

The third paragenetic stage is characterized by occlusion of this porosity by a variety of dolomite types, that may occur separately or together. Volumetrically the most important dolomite cements in both sandstones and carbonates are euhedral rhombic and saddle types. There appears to be no relationship between the type of mouldic pore or the type of pre-leaching dolomite and the nature of the dolomite cement phase, with the exception of large single crystal dolomite that preferentially infills biomouldic pores after crinoid ossicles. Initial growth of this type of dolomite may have been replacive in nature, in optical continuity with the calcite, and after dissolution of the calcite during leaching, porosity was occluded by syntaxial dolomite overgrowths. This is supported by the presence of calcite inclusions within this style of dolomite that reduces biomouldic pores. Indeed, other types of apparent dolomite cement (euhedral rhombic spar and microcrystalline), including the dolomites within the sandstones, may have similar complex replacive-cement histories.

ORIGIN OF THE BOWLAND DOLOMITES

Textural evolution and constraints on dolomitization processes

Several related factors were important in determining the texture of the dolomites. These are: 1) host rock mineralogy and texture; 2) chemistry of dolomitizing solutions; 3) availability of nucleation sites in the host; 4) temperature; and 5) time. Some of the textural features of the dolomites can be explained by the interaction of a local source of $CO^{2-}{}_3$ (Murray, 1964; Weyl, 1960) and Ca^{2+} with dolomitizing solutions which supplied Mg^{2+}, Fe^{2+} and Mn^{2+}. In this particular case the dolomitization reaction:

$$2CaCO_3 + Mg^{2+} \rightarrow CaMg(CO_3)_2 + Ca^{2+} \quad (1)$$

requires two moles of calcite to form one mole of dolomite. The excess carbonate may have been derived from dissolution of the unreplaced host. Such a mechanism could explain the increased porosity of the dolomites (Murray, 1964) and the cloudy core-clear rimmed character of some of the dolomite rhombs. However, although such a model is consistent with the similar carbon-isotope ratios of the dolomite and host calcite, several factors suggest that such a dolomitizing reaction was not generally applicable in the case of the Bowland Basin. Specifically, 1) the presence of calcite inclusions within many of the replacive mosaic dolomites, 2) dissolution of dolomite (evidenced by enlarged biomouldic pores and corroded CL zones), in addition to dissolution of calcite, to produce secondary porosity, 3) the presence of

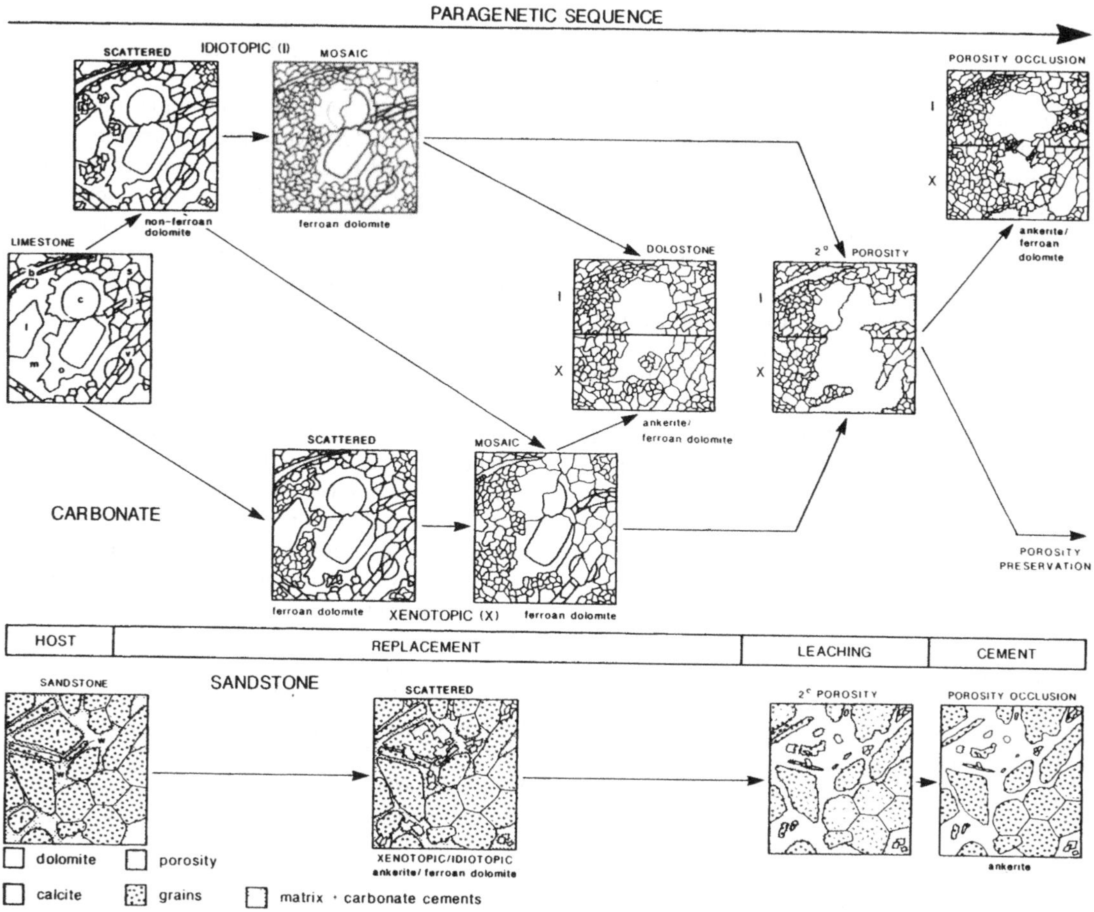

Fig. 12. Paragenetic relationships between the various types of dolomite, leaching and chemistry in the Pendleside Limestone and Pendleside Sandstone (see text for further explanation).

dolomite cements, and 4) the indication of addition of isotopically light carbon during dolomitization all suggest that dolomitization purely in terms of reaction (1) is not tenable. The apparent conservation of volume during dolomitization, as indicated by the preservation of host texture and lack of fracture or shrinkage porosity accompanying dolomitization, suggests that a volume-for-volume reaction may be applicable (Morrow, 1982a):

$$(2-x)\,CaCO_3 + Mg^{2+} + xCO^{2-} \rightarrow CaMg(CO_3)_2 + (1-x)CO^{2-}{}_3, \qquad (2)$$

(where $x = 0.25$ for the dolomitization of calcite).

The availability of nucleation sites appears to have been a particularly important factor in determining the location and texture of the dolomites. The high frequency of nucleation sites within fine-grained substrates (Sibley, 1982) is consistent with the selective, early replacement of the finer grained allochems, microspar and micrite within the host limestones. Sibley (*op. cit.*) suggested that pseudomorphic replacement of bioclasts in the Pliocene Seroe Domi dolomites was a result of dolomitization from solutions slightly undersaturated with respect to the substrate, with nucleation possibly enhanced by the development of kink sites on crystals as a result of the undersaturated conditions. Pseudomorphic and non-pseudomorphic replacement of shell fragments by microcrystalline dolomite within the Bowland sequence occurred prior to, or synchronous with the generation of biomouldic porosity and hence a similar mechanism may explain the origin of the pseudomorphic dolomite discussed in this study. Concurrent dissolution and replacement of the type described by Sibley (1982) may also explain the complex replacive and pore-filling textures of microcrystalline, euhedral rhombic and large single-crystal dolomite types. Thus when these are compared to the scattered and mosaic dolomites which contain calcite inclusions, it appears there was a progressive evolution of the dolomitizing fluids from a state of calcite saturation to undersaturation during progressive replacement.

In addition to constraining reaction mechanisms, some of the dolomite types help to place constraints on the temperature range within which dolomitization took place. Saddle dolomite is present as a replacive and pore-filling phase within both sandstones and carbonates. The warped crystal lattice of this type of dolomite has been attributed to a number of factors (Radke & Mathis, 1980), amongst which elevated temperature (>60°C) appears to be one of the most important (Gregg, 1983). Integration of crystal growth theory (Jackson, 1958a, b), experimental data and field examples (Gregg & Sibley, 1984), suggests that xenotopic dolomite textures are the result of growth at elevated temperatures in excess of 50°C. Thus the change from idiotopic to xenotopic mosaic dolomite within the Bowland sequence can be explained in terms of increasing temperature during progressive dolomitization.

In this way the textural characteristics of the dolomites from the Asbian to early Brigantian carbonates and clastics provide evidence concerning the nature of the dolomitization process. In particular, the dolomites appear to have formed at elevated temperatures (>50–60°C), indicating a burial origin. This is consistent with the replacement of calcite veins and the $\delta^{18}O$-depleted nature of the dolomite with respect to the host calcite, and also with the position of dolomite late in the paragenetic sequence of both sandstones and carbonates.

Chemical constraints

In order to produce dolomitization in the Bowland sequence, three basic conditions must have been fulfilled: 1) the source(s) of ions (Mg^{2+}, Fe^{2+}, Mn^{2+}, $CO^{2-}{}_3$) must have been sufficiently large to account for the observed extent of the dolomitized horizons; 2) fluids transporting these ions must have been capable of forming dolomite; 3) a mechanism for transporting the ions from their sources to the site of dolomitization must have existed. In addition, the following factors must be taken into account: 1) the generally highly ferroan nature of the dolomite; 2) the trend to higher Fe and Mn content with progressive dolomitization and $\delta^{18}O$ depletion; 3) the variations in isotopic ($\delta^{13}C$) and chemical composition associated with facies and geographic parameters; 4) the calcite-saturated to calcite-undersaturated evolution of the dolomitizing fluids; 5) a source of isotopically light carbon derived from adjacent mudrocks; and 6) similarities in the character of dolomitization in both sandstones and carbonates.

Model

The main features of the dolomitization model presented in this paper are summarized in Fig. 13. Potential sources of ions for dolomitization are the host rocks themselves and external sources. Whole rock XRF analyses of the host limestones show generally low concentrations of Mg, Mn and Fe and the cations involved in dolomitization are thus thought to have been derived largely from an external source, although carbon isotope values indicate a local source for at least some of the carbon. Whilst it could be argued that the elemental contents now recorded for the limestones indicate that they are already depleted in Fe, Mg and Mn, the presence of dolomite within both the carbonates and the sandstones (rocks of different depositional setting and chemical composition) is consistent with an external source of cations for dolomitization.

Further insight into potential sources for the dolomitizing solutions can be gained by considering the stratigraphic distribution of the dolomite. As already mentioned, dolomite is generally located towards the top of the Pendleside Limestone and at the margins of the Pendleside Sandstone, i.e. at the contacts with surrounding mudstone-dominated facies. This strongly suggests a mudrock source for the dolomitizing fluids.

Burial-related clay mineral transformations are a potential source of cations for dolomitization (MacHargue & Price, 1982). In particular the transformation of smectite through mixed-layer illite–smectite-to-illite with increasing depth of burial and temperature has been documented from a number of sedimentary basins (e.g. Burst, 1969; Foscolos & Kodama, 1974; Pearson, Watkins & Small, 1982; Perry & Hower, 1970), especially from the Tertiary of the Gulf Coast, U.S.A. Although the change from smectite to illite is complex, and several opinions exist as to the precise nature of the reaction(s) (Boles & Franks, 1979; Hower *et al.*, 1976; Nadeau *et al.*, 1984), it

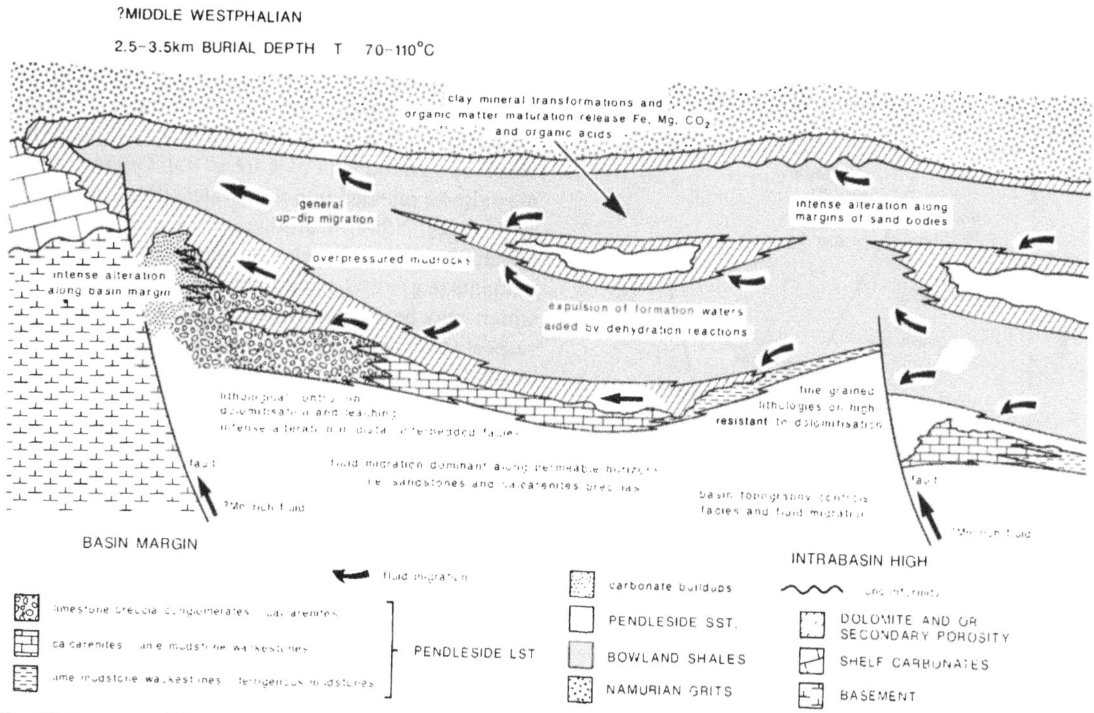

Fig. 13. Summary of the model for the origin of the late diagenetic dolomitization and porosity development in the Pendleside Limestone and Pendleside Sandstone as a result of expulsion of mudrock formation waters (see text for further explanation).

appears capable of releasing large amounts of cations, particularly Ca^{2+}, Mg^{2+}, Fe^{2+} and Si^{4+} into pore fluids. Boles & Franks (1979) suggested that Fe- and Mg-rich smectites do not react as readily as the more aluminous varieties and this may be one explanation of the increase in the ferroan nature of the Bowland dolomites with increasing replacement and cementation. Mass balance calculations (Boles & Franks, 1979; Boles, 1978) indicate that such transformations within Gulf Coast strata are more than capable of accounting for the volumes of ankerite and ferroan dolomite cement within the Tertiary sandstones.

Whole rock XRD analysis of samples from the Bowland Shales indicates the presence of illite. However, the amount of illite within the sequence has not been determined and the origin of the illite may be other than smectite transformation. These factors, together with the lack of data from mudrocks within the Bowland Basin (Worston and Bowland Shales) on the original percentage of smectite or its composition, make mass-balance calculations for the Bowland dolomites difficult to quantify. What is apparent from analysis of the generalized burial history of the Asbian/Brigantian sediments is that, during the Carboniferous, they underwent burial to sufficient depth for such clay mineral transformations to have occurred, probably during the Westphalian (Fig. 14). Additional support for burial dolomitization from mudrock-derived formation waters comes from the magnetic fabric of the dolostones and dolomitic limestones (Addison, Turner & Tarling, 1985). The dolostones have non-primary magnetic fabrics, interpreted to be the result of chemical remanent magnetization (CRM) generated during dolomitization, almost antiparallel to the primary remanence characteristic of the Pendleside Limestone. Addison *et al.* (1985) considered this normal CRM to have been acquired during the Westphalian, which would be consistent with the proposed timing of dolomitization based on the burial history of the sequence (Fig. 14).

No data are available concerning the possible release of Mn as a consequence of such reactions and the Mn contents of carbonates thought to be associated with shale-derived formation waters are not presented. Thus it is difficult to interpret the high Mn contents of dolomites replacing and/or cementing marginal carbonate facies. It is unlikely that the Mn could have been derived from the host limestone because of their

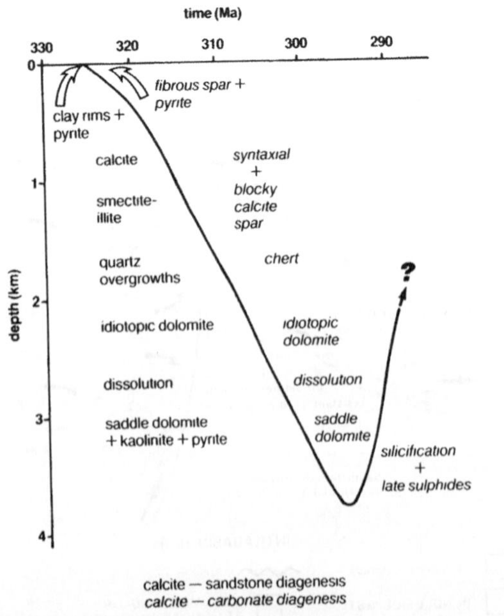

Fig. 14. Relationship between diagenesis of the Pendleside Limestone and Pendleside Sandstone and burial history. Depth range of diagenetic zones based on data derived mainly from studies on the Gulf Coast, U.S.A. (e.g. Boles & Franks, 1979, Milliken *et al.*, 1981) and burial curve calculated using the approach outlined by Perrier & Quiblier (1974) and Guidish *et al.* (1985). Thickness data form Earp *et al.*, (1961), Ramsbottom (1977) and Anderton *et al.* (1979); time scale from DeSouza (1982).

very low Mn contents. If Mn was also derived from the mudrocks the higher Mn contents further away from the source may be a result of the larger stability field of Mn compared to Fe in solution under reducing conditions. Alternatively the Mn may have been derived from a different source. High Mn contents are commonly associated with hydrothermal fluids within sedimentary sequences, and the presence of high Mn contents adjacent to fault zones and in vein dolomites may suggest a 'deep basinal' source. Manganese enrichment adjacent to fault zones within early- to mid-Dinantian sediments in Ireland (Tynagh) has been described by Russell (1975).

Organic matter, like clay minerals, undergoes a series of depth-related changes (Curtis, 1978, 1983a; Laplante, 1974; Tissot *et al.*, 1974), and at the elevated temperatures proposed for formation of the Bowland dolomites (>60°C) abiotic thermal decarboxylation reactions dominate the depth-related modification of organic matter (zone 4 of Curtis, 1978, 1983a). The result of such reactions is to release CO_2 into pore waters. Carbonates derived from such a source of CO_2 would have depleted $\delta^{13}C$ values. Hence mixing of carbon from the host limestone with isotopically light carbon derived from decarboxylation can explain the observed variations in $\delta^{13}C$ (Fig. 9c). Organic acids may also be released into pore waters (Surdam, Boese & Crossey, 1984; Carothers & Kharaka, 1978). Depending on the relative amounts of mineral oxidants (e.g. Fe_2O_3) and these organic species, pore waters may be supersaturated or undersaturated with respect to carbonates and potentially unstable aluminosilicate grains. Curtis (1983a) suggested that, because of a decrease in the amount and reactivity of species like Fe_2O_3 and the increased reactivity of organic matter with increasing burial, there will be a trend towards acidification of pore fluids with depth, potentially leading to major dissolution. The coincidence of the zones of smectite to illite transformation, release of CO_2 and organic acid production (Curtis, 1983a; Foscolos & Powell, 1979, 1980; Surdam *et al.* 1984) can account for the intimate association of dolomitization and secondary porosity generation within the Bowland carbonates and sandstones adjacent to thick mudrock sequences. Initially pore waters were saturated with respect to calcite and dolomite but, as the pore waters became increasingly acidic, carbonates and unstable framework grains (e.g. feldspars) within the sandstones and carbonates were dissolved. The combined effects of carbonate dissolution and the decrease in CO_2 and organic acid production towards the zone of oil generation (Curtis, 1983a; Surdam *et al.*, 1984) would have caused the pore waters to become less acidic and carbonate cements could therefore precipitate (Fig. 12). The occlusion of secondary porosity by kaolinite in both sandstones and carbonates was also a consequence of this reduction in acidity (Curtis, 1983b).

Although the necessary ions could have been derived from the adjacent mudrocks, and such a source can account for the observed geographic, textural and chemical characteristics of the Bowland dolomites, would the pore fluids produced within them have been capable of precipitating dolomite or dolomitizing limestone? Near-surface dolomitization is hindered by several kinetic factors (Folk & Land, 1975; Lippman, 1973; Morrow, 1982a) related to; 1) the problem of producing the highly ordered dolomite structure, 2) the highly hydrated nature of the Mg ion, and 3) the generally low concentration of CO_2 relative to Ca^{2+} and Mg^{2+}. The result is that the optimum conditions for near-surface dolomitization involve solutions with high Mg/Ca ratios (>1), high concentrations of CO_2 and low salinity (Morrow, 1982a). Subsurface fluids generally have low Mg/Ca ratios and high salinities (e.g. Hitchon, Billings & Klovan, 1971; Dickey, 1966, 1969) and by straight comparison with the near-surface environment such fluids would not be expected to dolomitize. In addition, Land (1985) argues against a burial origin for dolomitization on the grounds of an insufficient Mg source. However, the elevated temperatures of the burial environment may reduce the kinetic barriers that hinder dolomite formation at the surface; in particular, the increased reaction rates and reduction of the hydration barrier of the Mg ions with increasing temperature may allow dolomite formation at low Mg/Ca ratios and relatively high salinities (Usdowdki, 1968; Mattes & Mountjoy, 1980) and may explain the excess Ca in some of the dolomites. Other factors in the burial environment, especially the catalystic effect of organic acids and Fe (Carpenter, 1980; Gaines, 1980; Mattes & Mountjoy, 1980), may promote the formation of dolomite. Thus, although it is not clear whether or not mudrock-derived fluids would be capable of producing dolomite, dolomitization in the subsurface may be kinetically much easier than in near-surface environments, even from quite saline fluids.

Concomitant with the release of cations from the transformation of smectite to illite is the release of interlayer water (Burst, 1969). Although such dehydration processes may in detail involve several steps and may be dependent on the chemical composition of the clays (Foscolos & Powell, 1979, 1980), they provide a mechanism for the expulsion of mudrock pore waters from their source. The general lack of interbedded porous horizons within the mudrock source (Bowland Shales) suggests that they are likely to have been overpressured due to a combination of mineral dehydration reactions and aquathermal and compactional effects (e.g. Magara, 1978). The combination of these effects could have led to the migration of formation waters out of the mudrocks into more permeable horizons where flow would be concentrated. Such migration of pore waters can explain the location of the most intense dolomitization and secondary porosity generation in sandstones and coarse-grained carbonates adjacent to mudrocks. Preferential development in the coarser calcarenites and breccia/conglomerates rather than in the lime mudstones/wackestones is possibly the result of preservation of small amounts of primary porosity and/or higher permeabilities in the coarser grained lithologies. The development of dolomitized limestones, particularly with idiotopic dolomite textures, would have increased permeability and further concentrated fluid flow. This would explain the intimate association of intense carbonate dissolution and particularly dolomitized lithologies.

Areas of particularly intense dolomitization were thus controlled by the inter-relationship of facies mosaics and basin morphology. Facies patterns were influenced by the half-graben morphology of the basin and a series of antithetic and transfer faults related to the extensional tectonic regime operating during the Dinantian and early Namurian (Gawthorpe, 1987). The basement morphology probably also controlled later fluid migration. Iskenderian (1980) described a southwesterly increase in the degree of dolomitization in the south of the basin. This would coincide with a general up-dip migration of fluids towards the crest of an intrabasin basement structure controlled by movement along the Clitheroe Fault Zone.

On a basin-wide scale the half-graben nature of the basin, controlled by a normal fault along its southern margin, and the location of intense dolomitization along the NW margin are consistent with regional up-dip migration of mudrock-derived fluids from structural lows.

CONCLUSIONS

(A) Dolomitization and secondary porosity generation in the Pendleside Limestone and Pendleside Sandstone occurred at a similar time, late in the diagenetic sequence of these rocks, post-dating major cementation and calcite veining.

(B) Dolomite paragenesis is similar in both sandstones and limestones, involving a progressive increase in Fe and Mn substitution and depletion in $\delta^{18}O$ during dolomitization. Three stages in the paragenesis can be recognized; early replacement, followed by leaching and secondary porosity generation and, finally, cementation.

(C) The dolomite textures developed within the sequence provide information on the temperature of formation. In particular, the occurrence of saddle and xenotopic mosaic dolomite suggests elevated temperatures ($>60°C$). This is compatible with the depletion in $\delta^{18}O$ of the dolomites relative to the calcite host. This information, together with the paragenetic relations, indicates that the dolomite formed during moderate to deep burial diagenesis (mesogenetic zone).

(D) The location of highly dolomitized rocks and

variations in the chemistry of the dolomites suggest that fluids involved in dolomitization were derived from the adjacent mudrocks. A model involving release of Fe^{2+} and Mg^{2+} as a result of clay mineral transformations (such as smectite and mixed layer smectite-illite clays to illite) together with maturation of organic matter and the release of CO_2 and organic acids can account for the ions needed for dolomitization and leaching. Manganese may have been derived from a 'deep basinal' source, with migration of fluids to the site of dolomitization concentrated along fault zones.

(E) The coincidence during burial of the transformation of smectite to illite with maximum production of CO_2 and organic acids can explain the coincidence of dolomitization and secondary porosity generation. The increase in the Fe content of the dolomites during progressive dolomitization is thought to be related to the greater stability of smectites rich in Fe and Mg.

(F) Migration of formation waters out of the mudrocks would have been aided by dehydration reactions and the likely overpressured nature of the mudrocks. Fluid migration was concentrated along the more permeable carbonates and sandstones, i.e. the dolomitized intervals. The exact location of dolomitization was strongly influenced by the porosity and permeability of the host rocks and by fluid migration pathways. Fluid migration was probably controlled by the intrabasin structure of the Bowland Basin, since intense dolomitization and leaching are located along the margins of topographic highs within the basin where 'reactive' sediments were deposited.

ACKNOWLEDGMENTS

The author would like to thank Eric Condliffe, Mike Leeder, Rob Raiswell and Harry Clemmey for helpful discussions during the course of this work. Special thanks go to Jon Gluyas, Mark Boast and Max Coleman for discussing the isotopic data. Financial support from BP Minerals International Ltd is gratefully acknowledged.

REFERENCES

ADDISON, F.T., TURNER, P. & TARLING, D.H. (1985) Magnetic studies of the Pendleside Limestone: evidence for remagnetization and late diagenetic dolomitization during a post-Asbian normal event. *J. geol. Soc. London*, **142**, 983–994.

ANDERTON, R., BRIDGES, P.H., LEEDER, M.R. & SELWOOD, B.W. (1979) *A Dynamic Stratigraphy of the British Isles: a Study in Crustal Evolution*. George, Allen & Unwin, London.

BATHURST, R.G.C. (1975) *Carbonate Sediments and their Diagenesis. Developments in Sedimentology 12*, 2nd edn. Elsevier, Amsterdam.

BEALES, F.W. (1971) Cementation by white sparry dolomite. In: *Carbonate Cements* (Ed. by O. P. Bricker), pp. 339–346. Johns Hopkins University Press, Baltimore, Maryland.

BOLES, J.R. (1978) Active ankerite cementation in the subsurface Eocene of Southwest Texas. *Contr. Miner. Petrol.*, **68**, 13–22.

BOLES, J.R. & FRANKS, S.G. (1979) Clay diagenesis in Wilcox sandstones of S.W. Texas: implications of smectite diagenesis on sandstone cementation. *J. sedim. Petrol.*, **49**, 55–70.

BURST, J.F. (1969) Diagenesis of Gulf Coast clayey sediments and its possible relation to petroleum migration. *Bull. Am. Ass Petrol. Geol.*, **53**, 73–93.

CAROTHERS, W.W. & KHARAKA, Y.K. (1978) Aliphatic acid anions in oilfield waters—implications for origin of natural gas. *Bull. Am. Ass. Petrol. Geol.*, **62**, 2441–2453.

CARPENTER, A.B. (1980) The chemistry of dolomite formation 1: the stability of dolomite. In: *Concepts and Models of Dolomitization* (Ed by D. H. Zenger, J. B. Dunham & R. L. Ethington). *Spec. Publs Soc. econ. Paleont. Miner, Tulsa*, **28**, 111–121.

CHOQUETTE, P.W. (1971) Late ferroan dolomite cement, Mississippian carbonates, Illinois Basin, U.S.A. In: *Carbonate Cements* (Ed. by O. P. Bricker), pp. 339–346. John Hopkins University Press, Baltimore, Maryland.

CURTIS, C.D. (1978) Possible links between sandstone diagenesis and depth-related geochemical reactions in enclosing mudrocks. *J. geol. Soc. London*, **135**, 107–117.

CURTIS, C.D. (1983a) Geochemistry of porosity enhancement and reduction in clastic sediments. In: *Petroleum Geochemistry and Exploration of Europe* (Ed. by J. Brooks). *Spec. Publ. geol. Soc. Lond.*, **12**, 113–126.

CURTIS, C.D. (1983b) Link between aluminium mobility and destruction of secondary porosity. *Bull. Am. Ass. Petrol. Geol.*, **67**, 390–393.

DAVIES, G.R. (1979) Dolomite reservoir rocks: processes, controls, porosity development. In: *Geology of Carbonate Porosity. Am. Ass. Petrol. Geol.* [Continuing Education Course Note Series], **II**, C1–C17.

DEER, W.A., HOWIE, R.A. & ZUSSMAN, J. (1962) *Rock Forming Minerals, Vol. 5: Non-Silicates*. Longmans, London.

DESOUZA, H.A.F. (1982) Age data from Scotland and the Carboniferous time scale. In: *Numerical Dating in Stratigraphy* (Ed. by D. S. Odin), pp. 455–465. John Wiley & Sons, London.

DICKEY, P.A. (1966) Patterns of chemical composition in deep subsurface brines. *Bull. Am. Ass. Petrol. Geol.*, **50**, 2472–2478.

DICKEY, P.A. (1969) Increasing concentration of subsurface brines with depth. *Chem. Geol.*, **4**, 361–370.

DICKSON, J.A.D. & COLEMAN, M.L. (1980) Changes in carbon and oxygen isotope composition during limestone diagenesis. *Sedimentology*, **27**, 107–118.

EARP, J.R., MAGRAW, D., POOLE, E.G., LAND, D.H. & WHITEMAN, A.J. (1961) Geology of the country around

Clitheroe and Nelson. *Mem. Geol. Surv. U.K.* London, 346 pp.

FAIRCHILD, I.J. (1983) Chemical controls of cathodoluminescence of natural dolomites and calcites: new data and review. *Sedimentology*, **30**, 579–583.

FEWTRELL, M.D. & SMITH, D.G. (1980) Revision of the Dinantian stratigraphy of the Craven Basin, N. England. *Geol. Mag.*, **117**, 37–49.

FOLK, R.L. & LAND, L.S. (1975) Mg/Ca ratio and salinity, two controls over crystallization of dolomite. *Bull. Am. Ass. Petrol. Geol.*, **59**, 60–68.

FOSCOLOS, A.E. & KODAMA, H. (1974) Diagenesis of clay minerals from Lower Cretaceous shales of northeastern British Columbia. *Clays Clay Miner.*, **22**, 319–335.

FOSCOLOS, A.E. & POWELL, T.G. (1979) Catagenesis in shales and occurrence of authigenic clays in sandstones, North Sabine H-49 well, Canadian Arctic Islands. *Can. J. Earth Sci.*, **16**, 1309–1314.

FOSCOLOS, A.E. & POWELL, T.G. (1980) Mineralogical and geochemical transformations of clays during catagenesis and their relation to oil generation. In: *Facts and Principles of World Petroleum Occurrences. Mem. Can. Soc. Petrol. Geol., Calgary*, **6**, 153–172.

FRANKS, S.G. & FORESTER, R.W. 1984. Relationships among secondary porosity, pore-fluid chemistry and carbon dioxide, Texas Gulf Coast. In: *Clastic Diagenesis* (Ed D. A. McDonald & R. C. Surdam). *Mem. Am. Ass. Petrol. Geol.*, **37**, 63–80.

GAINES, A.M. (1980) Dolomitization kinetics: recent experimental studies. In: *Concepts and Models of Dolomitization* (Ed by D. H. Zenger, J. B. Dunham & R. L. Ethington). *Spec. Publs Soc. econ. Paleont. Miner., Tulsa*, **28**, 81–86.

GAWTHORPE, R.L. (1986) Sedimentation during carbonate ramp-to-slope evolution in a tectonically active area: Bowland Basin (Dinantian), N. England. *Sedimentology*, **33**, 185–206.

GAWTHORPE, R.L. (1987) Tectono-sedimentary evolution of the Bowland Basin, northern England, during the Dinantian. *J. geol. Soc. London*, **144**, 59–71.

GREGG, J.M. (1983) On the formation and occurrence of saddle dolomite—Discussion. *J. sedim. Petrol.*, **53**, 1025.

GREGG, J.M. & SIBLEY, D.F. (1984) Epigenetic dolomitization and the origin of xenotopic dolomite texture. *J. sedim. Petrol.*, **54**, 908–931.

GUIDISH, T.M., KENDALL, C.G.St.C., LERCHE, I., TOTH, D.J. & YARZAB, R.F. (1985) Basin evaluation using burial history calculations: an overview. *Bull. Am. Ass. Petrol. Geol.*, **69**, 92–105.

HAYES, J.B. (1979) Sandstone diagenesis—The hole truth. In: *Aspects of Diagenesis* (Ed by P. A. Scholle & P. R. Schluger). *Spec. Publs Soc. econ. Paleont. Miner., Tulsa*, **26**, 127–139.

HITCHON, B., BILLINGS, G.K. & KLOVAN, J.E. (1971) Geochemistry and origin of formation waters in the Western Canadian sedimentary basin III: Factors controlling chemical composition. *Geochim. cosmochim. Acta*, **35**, 567–598.

HOWER, J., ESLINGER, E.V., HOWER, M.E. & PERRY, E.A. (1976) Mechanism of burial and metamorphism of argillaceous sediment. I: Mineralogical and chemical evidence. *Bull. geol. Soc. Am.*, **87**, 725–737.

IRWIN, H., CURTIS, C.D. & COLEMAN, M.L. (1977) Isotopic evidence for source of diagenetic carbonates formed during burial of organic-rich sediments. *Nature*, **269**, 209–213.

ISKENDERIAN, F.A. (1980) *The sedimentology and diagenesis of the Pendleside Limestone Group in the Craven Basin of northern England.* Unpublished PhD thesis, University of Aston.

JACKSON, K.A. (1958a) Mechanisms of growth. In: *Liquid Metals and Solidification*, pp. 174–186. Am. Soc. Metals. Cleveland, Ohio.

JACKSON, K.A. (1958b) Interface structure. In: *Growth and Perfection of Crystals* (Ed by R. M. Doremus, B. W. Roberts & D. Turnbull), pp. 319–323. John Wiley & Sons, New York.

JODRY, R.L. (1969) Growth and dolomitisation of Silurian reefs, St. Clair Country, Michigan. *Bull. Am. Ass. Petrol. Geol.*, **52**, 957–981.

KANTOROWICZ, J.D. (1985) The origin of authigenic ankerite from the Ninian Field, U.K. North Sea. *Nature*, **315**, 214–216.

LAND, L.S. (1980) The isotopic and trace element geochemistry of dolomite: the state of the art. In: *Concepts and Models of Dolomitization* (Ed. by D. H. Zenger, J. B. Dunham & R. L. Ethington). *Spec. Publs Soc. econ. Paleont. Miner., Tulsa*, **28**, 87–110.

LAND, L.S. (1985) The origin of massive dolomite. *J. geol. Education*, **33**, 112–125.

LAPLANTE, R.E. (1974) Hydrocarbon generation in Gulf Coast Tertiary sediments. *Bull. Am. Ass. Petrol. Geol.*, **53**, 1281–1289.

LIPPMAN, F. (1973) *Sedimentary Carbonate Minerals.* Springer–Verlag, New York.

LOUCKS, R.G., DODGE, M.M. & GALLOWAY, W.E. (1979) Importance of secondary leached porosity in Lower Tertiary sandstone reservoirs along the Texas Gulf Coast. *Trans. Gulf-Cst Ass. Geol. Socs*, **29**, 16–171.

MACHARGUE, T.R. & PRICE, R.C. (1982) Dolomite from clay in argillaceous marine associated carbonates. *J. sedim. Petrol.*, **52**, 873–886.

MACQUEEN, R.W. (1979) Basemetal deposits in sedimentary rocks: some approaches. *Geosci. Can.*, **6**, 3–9.

MAGARA, K. (1978) *Compaction and Fluid Migration. Developments in Petroleum Science*, 9. Elsevier, Amsterdam.

MATTES, B.W. & MOUNTJOY, E.W. (1980) Burial dolomitization of the Upper Devonian Miette buildup, Jasper National Park, Alberta. In: *Concepts and Models of Dolomitization* (Ed by D. H. Zenger, J. B. Dunham & R. L. Ethington). *Spec. Publs. Soc. econ. Paleont. Miner., Tulsa*, **28**, 259–297.

MILLIKEN, K.L., LAND, L.S. & LOUCKS, R.G. (1981) History of burial diagenesis determined from isotopic geochemistry, Frio Formation, Brazoria Country, Texas. *Bull. Am. Ass. Petrol. Geol.*, **65**, 1397–1413.

MORROW, D.W. (1982a) Diagenesis 1. Dolomite Part 1: The chemistry of dolomitization and dolomite precipitation. *Geosci. Can.*, **9**, 5–13.

MORROW, D.W. (1982b) Diagenesis 2. Dolomite part 2: Dolomitization models and ancient dolostones. *Geosci. Can.*, **9**, 95–107.

MURRAY, R.C. (1964) Preservation of primary structures and fabrics in dolomite. In: *Approaches to Paleoecology* (Ed. by J. Imbrie & N. Newell), pp. 388–403. John Wiley, New York.

NADEAU, P.H., TAIT, J.M., MCHARDY, W.J. & WILSON, M.J. (1984) Interstratified XRD characteristics of physical mixtures of elemental clay particles. *Clay Miner.*, **19**, 67–76.

PEARSON, M.J., WATKINS, D. & SMALL, J.S. (1982) Clay diagenesis and organic maturation in Northern North Sea sediments, *Developments in Sedimentology*, **35**, pp. 665–675. Elsevier, Amsterdam.

PERRIER, R. & QUIBLIER, J. (1974) Thickness changes in sedimentary layers during compaction history; methods for quantitative evaluation. *Bull. Am. Ass. Petrol. Geol.*, **58**, 507–520.

PERRY, E.A. & HOWER, J. (1970) Burial diagenesis in Gulf Coast pelitic sediments. *Clays Clay Miner.*, **18**, 165–178.

PIERSON, B.J. (1981) The control of cathodoluminescence in dolomite by iron and manganese. *Sedimentology*, **28**, 601–610.

RADKE, B.M. & MATHIS, R.L. (1980) On the formation and occurrence of saddle dolomite. *J. sedim. Petrol.*, **50**, 1149–1168.

RAMSBOTTOM, W.H.C. (1977) Major cycles of transgression and regression (mesothems) in the Namurian. *Proc. Yorks. geol. Soc.*, **41**, 261–291.

RUSSELL, M.J. (1975) Lithogeochemical environment of the Tynagh base-metal deposit, Ireland, and its bearing on ore deposition. *Trans. Inst. Min. Metall.*, **84**, B128–B133.

SCHMIDT, V. & MCDONALD, D.A. (1979) The role of secondary porosity in the course of sandstone diagenesis. In: *Aspects of Diagenesis* (Ed. by P. A. Scholle & P. R. Schluger). *Spec. Publs Soc. econ. Paleont. Miner., Tulsa*, **26**, 175–208.

SCHOFIELD, K. & ADAMS, A.E. (1986) Burial dolomitisation of the Woo Dale Limestones Formation (Lower Carboniferous), Derbyshire, England. *Sedimentology*, **33**, 207–219.

SHEPPARD, S.M.F. & SCHWARCZ, H.P. (1970) Fractionation of carbon and oxygen isotopes and magnesium between coexisting metamorphic calcite and dolomite. *Contr. Miner. Petrol.*, **26**, 161–198.

SIBLEY, D.F. (1982) The origin of common dolomite fabrics: clues from the Pliocene. *J. sedim. Petrol.* **52**, 1087–1100.

SURDAM, R.C., BOESE, S.W. & CROSSEY, L.J. (1984) The chemistry of secondary porosity. In: *Clastic Diagenesis* (Ed. by D. A. McDonald & R. C. Surdam). *Mem. Am. Ass. Petrol. Geol.*, **37**, 127–150.

TISSOT, B., DURAND, B., ESPITALIE, J. & COMBAZ, A. (1974) Influence of nature and diagenesis of organic matter in formation of petroleum. *Bull. Am. Ass. Petrol. Geol.*, **58**, 499–506.

USDOWSKI, H.E. (1968) The formation of dolomite in sediments. In: *Recent Developments in Carbonate Sedimentology in Central Europe* (Ed. by G. Miller & G. M. Friedman), pp. 21–32. Springer–Verlag, Berlin.

WADGE, A.J., BATESON, J.H. & EVANS, A.D. (1983) Mineral reconnaissance surveys in the Craven Basin. Mineral Reconnaissance Programme. *Rep. Inst. geol. Sci. London*, **66**,

WEYL, P.K. (1960) Porosity through dolomitization: conservation-of-mass requirements. *J. sedim. Petrol.*, **30**, 85–90.

WONG, P.K. & OLDERSHAW, A. (1981) Burial cementation in the Devonian, Kaybob Reef Complex, Alberta, Canada. *J. sedim. Petrol.*, **51**, 507–520.

ZENGER, D.H. (1983) Burial dolomitization in the Los Burro Formation (Devonian), east-central California, and the significance of late diagenetic dolomitization. *Geology*, **11**, 519–522.

(Manuscript received on 28 April 1986; revision received on 4 September 1986)

Clay and aluminosilicate mineral-related mesogenesis

Clays, feldspars and other silicate minerals containing aluminium are involved in degrading reservoir quality in sandstones and so have been widely studied. The term greywacke has long been used by field geologists to describe a sandstone with abundant clay matrix. The origin of the matrix was debated because it was appreciated that sand and clay are not usually deposited at the same time. The experimental studies and conclusions of Whetten & Hawkins (1970) have stood the test of time and it is now held for many sandstones that the clay matrix is a diagenetic product of the breakdown of unstable detrital grains. Subsequent compaction of these clay-rich sand grains results in them being squeezed between more competent grains, and so would fill pores. Lee & De Vries Klein (1986) developed these ideas and concluded that the degree of clay and other aluminosilicate mineral diagenesis is controlled strongly by the original lithotype of the sandstone, the geothermal gradient and the time of sediment deposition relative to heat flow history. They report a complex series of reaction products in sandstones rich in volcanic rock fragments including zeolites and smectite. Iijima & Utada (1966) also examined zeolite diagenesis in clastic rocks rich in volcanic debris and consider that there are recognizable patterns of growth of different zeolite minerals as a function of depth. They assign specific families of zeolites to the eogenetic and mesogenetic regimes and even go as far as to assign laumontite to a range of burial depths between 3000 and 11 000 m. El Tabakh & Shreiber (1998) studied diagenesis in zeolite-bearing sandstones and concluded that some of the components were derived from outside the host sandstone, implying a degree of basin-scale movement involved in the diagenetic modifications.

Morad *et al.* (1989) studied K-feldspar cements and pseudomorphic replacements of unstable detrital feldspars in a series of sandstones deposited as alluvium at the base of the succession through to estuarine and marine sediments at the top in a graben margin. They concluded that the process operated by a localized feldspar dissolution and reprecipitation process. They also observed that the process was most prevalent in the stratigraphically younger marine sandstones, with the potassium, essential to the process, being sourced from mica and feldspar dissolution in the basal alluvial sands and carried by flowing water up into the marine facies.

Morton *et al.* (1989) studied the diagenesis of garnet, a common heavy mineral in sandstones. Garnets often display euhedral morphologies that could be taken to suggest garnet growth during diagenesis. However, Morton *et al.* prove that garnet undergoes crystallographically controlled dissolution that mimics diagenetic growth, but only under mesogenetic conditions.

Reprinted from Sedimentology (1970) ***15****, 347–361*

DIAGENETIC ORIGIN OF GRAYWACKE MATRIX MINERALS

JOHN T. WHETTEN AND JAMES W. HAWKINS JR.

Department of Geological Sciences, University of Washington, Seattle, Wash. (U.S.A.)
Geological Research Division, Scripps Institution of Oceanography, University of California, La Jolla, Calif. (U.S.A.)

(Received April 20, 1970)

(Resubmitted July 13, 1970)

SUMMARY

Phyllosilicates and zeolites grew in Columbia River sediments during hydrothermal experiments at relatively low temperatures and pressures. Although the new minerals may not be equilibrium assemblages, our results strongly support the idea that matrix minerals in graywackes may be the result of alteration of components thermodynamically unstable in the environment of diagenesis. Scanning electron micrographs show that the new minerals have formed as a mesh-like coating on original grains. The textural relationship of the new minerals to the original minerals resembles graywacke texture.

INTRODUCTION

"The dark fine grained matrix in which the sand grains are set is at once the essential characteristic of graywacke and the essence of the graywacke problem" (CUMMINS, 1962, p.52).

Graywackes—hard, dark-colored sandstones with abundant fine matrix—are relatively common sedimentary rocks in certain regions, but their origin, especially the origin of matrix, is a matter of continuing debate. The following points seem to us especially pertinent:

(*1*) Many graywackes were evidently deposited as deep-sea turbidite sands (PETTIJOHN 1957, pp.301–314), but some were deposited in a variety of other ways (CUMMINS, 1962, pp.55–57).

(*2*) The silt and clay content of many modern coarse turbidites was estimated by KUENEN (1966, p.269) to be about 1%. HUBERT (1964, p.781) stated that deep-sea sands and silts of the western North Atlantic consistently contain less than 5% clay, and none of those he examined had more than 12%. Turbidite sands analyzed from Leg 1 of the Deep Sea Drilling Project are apparently not distinguished by a large amount of clay matrix (BEALL and FISCHER, 1969, pp.565–570). Only rarely are sands found with *abundant* matrix of silt and clay suggestive of graywacke texture (HOLISTER and HEEZEN, 1964). SHEPARD (1964, p.18) concluded: "It seems likely that at present most (marine) sands are being transported in a form that prevents the incorporation of coarse sand in a muddy matrix."

(*3*) Not all ancient turbidites are graywackes. Some have little matrix, and in this respect are comparable to modern deep-sea sands (DOTT, 1963, p.124; OKADA, 1966).

(*4*) There is disagreement as to how modern deep-sea sands and silts are deposited—turbidity currents versus deep ocean-bottom currents—(HUBERT, 1964, pp.778–783), but no mechanism has been proposed that would be expected, as a rule, to deposit matrix-rich sediments. KUENEN (1966) demonstrated experimentally that coarse turbidites contain at most less than 10% silt and clay matrix. Finer turbidites may contain more.

(*5*) There is general agreement that graywacke matrix is "recrystallized", but recrystallized what? PETTIJOHN (1957, p.305) presented the view that ". . . all matrix minerals are, indeed, authigenic and are the result of reorganization of an *original interstitial mud*"[1]. C. Gilbert (in WILLIAMS et al., 1955, p.298) concurred: ". . . (matrix) probably formed from *original argillaceous detritus*[1] by compaction and recrystallization during long deep burial." On the other hand, CUMMINS (1962) suggested that the amount of matrix in graywacke sands was low at the time of deposition, but increased at the expense of sand grains by alteration of sand during diagenesis. These opposing ideas on the origin of graywacke matrix are, in our view, the nub of "the graywacke problem".

In this paper we present evidence that Cummins' view has merit, and suggest that it is the most reasonable and likely explanation for the orign of matrix in many graywackes.

The evidence comes from a series of hydrothermal experiments using typical samples of Columbia River bottom sediments as starting materials. With X-ray diffraction we have been able to observe changes in the sediments caused by elevated temperatures and pressures simulating moderate to deep burial. Phyllosilicates, the typical matrix minerals in graywackes, were readily formed in all samples, and electron micrographs show that the new minerals encrust or replace the original materials. Some data and conclusions presented elsewhere (HAWKINS and WHETTEN, 1969; WHETTEN et al., 1969) are included in the present review.

COLUMBIA RIVER SEDIMENTS

Bedload sediment in the Columbia River is derived from a lithologically heterogeneous drainage basin. However, the andesitic volcanoes of the Washington-Oregon Cascades are by far the largest contributors of bedload sediment. Smaller amount of sediment are derived from plutonic and metamorphic rocks (WHETTEN et al., 1969).

[1] Italics added.

The samples used in our experimental work are from the lower Columbia River, and consist of moderately to poorly sorted sands containing relatively little quartz and a large quantity of thermodynamically unstable clasts suggestive of a volcanic source terrane (Table I, II; Plate I). A distinctive characteristic is the large

TABLE I

MODAL ANALYSES OF THE MINERAL CONTENT OF COLUMBIA RIVER SEDIMENT SAMPLES *(in %)*

Constituent	*CC-43*	*CC-72*	*CC-148*
Quartz	15	8	5
Plagioclase	13	16	16
K-feldspar	7	4	1
Lithic fragments			
porphyritic glassy	61	63	68
porphyritic aphanitic	1	1	2
other	3	1	2
Mafic minerals	1	7	7
Heavy minerals ($\rho > 3.17$)			
Orthopyroxene	28	45	52
Clinopyroxene	31	21	27
Hornblende	33	8	14
Epidote	1	1	2
Garnet	6	22	4
Sphene	tr	1	—
Zircon	tr	1	1
Apatite	tr	—	—
Other	—	—	—

TABLE II

CHEMICAL ANALYSES OF COLUMBIA RIVER SEDIMENTS AND GRAYWACKES

Constituent	*Columbia River samples*			*Average Franciscan Formation*[1]	*Average graywacke*[2]
	CC-43	*CC-72*	*CC-148*		
SiO_2	67.53	67.3	63.22	67.5	66.7
TiO_2	0.71	0.9	0.85	0.5	0.6
Al_2O_3	14.10	13.4	16.13	13.5	13.5
Fe_2O_3	2.46	7.4	2.91	1.2	1.6
FeO	1.83	×[3]	2.52	3.0	3.5
MnO	0.07	0.1	0.10	0.1	0.1
MgO	1.67	3.5	2.67	2.2	2.1
CaO	3.36	3.5	5.03	2.4	2.5
Na_2O	3.08	3.5	3.82	3.6	2.9
K_2O	2.30	2.1	1.60	1.7	2.0
H_2O^+	1.03		0.70	2.5	2.4
H_2O^-	1.70		0.20	0.4	0.6
P_2O_5	0.18		0.29	0.1	0.2
CO_2				0.8	1.2

[1] BAILEY et al. (1964); [2] PETTIJOHN, 1963; [3] total Fe expressed as Fe_2O_3.

PLATE I

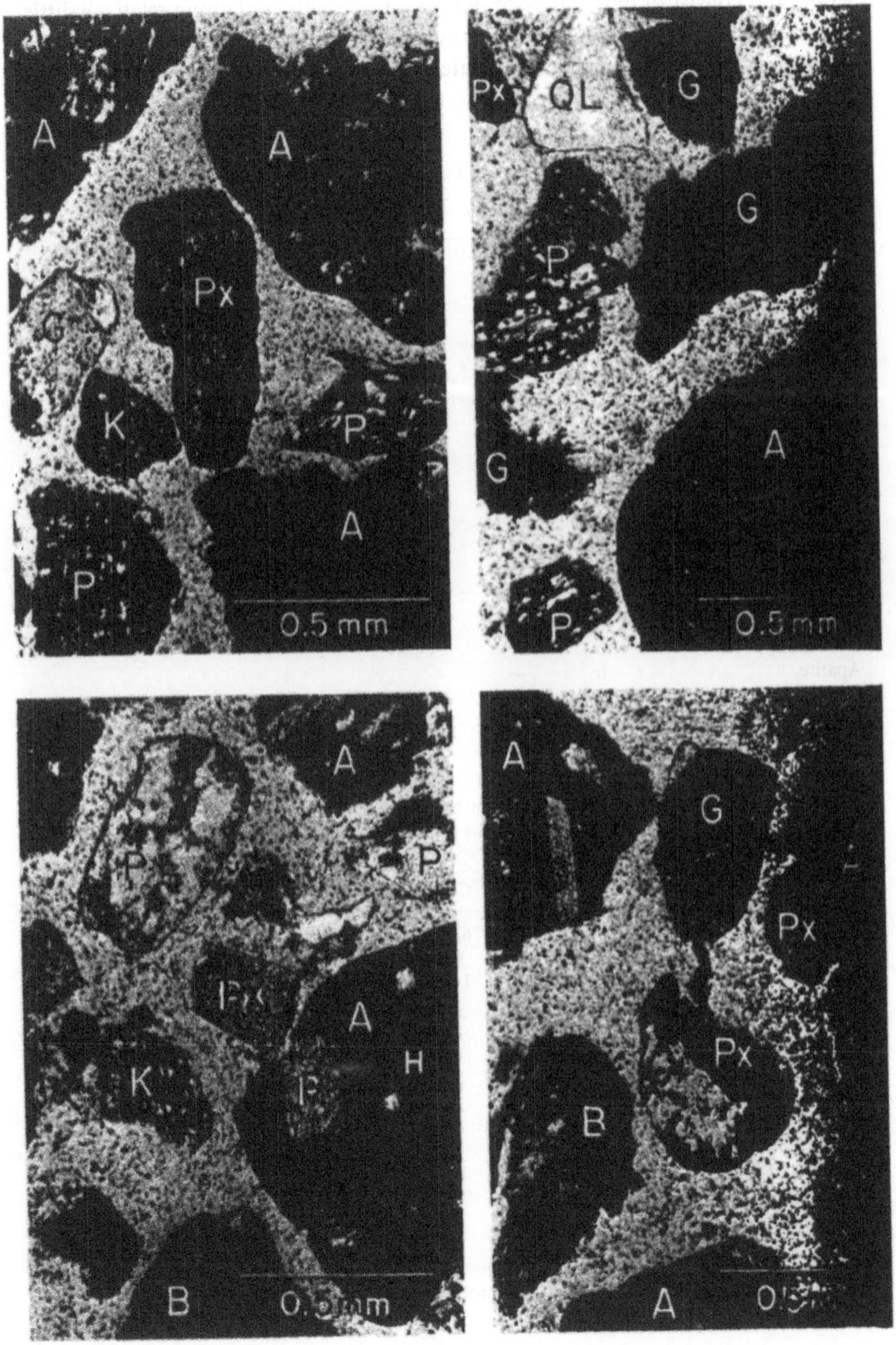

Photomicrographs of Columbia River sediment samples. Key to identified grains: *A* = andesite; *B* = basalt; *G* = devitrified glass; *H* = hornblende; *K* = K-feldspar; *P* = plagioclase; *Px* = pyroxene; *Q* = quartz; *QL* = quartzose lithic. Photographed in plane polarized light. (Samples: upper left and right: CC-148; lower left: CC-43; lower right: CC-72.)

quantity of devitrified volcanic glass, porphyritic volcanic fragments (Plate I, sample CC-148), and mafic minerals. Clay minerals (principally montmorillonite) can be detected by X-ray diffraction only after being concentrated from large volumes of sediment (KNEBEL et al., 1968).

Chemical analyses of Columbia River sediments are almost identical in composition to many graywackes but are low in silica and high in virtually all other major constituents compared to many sandstones (Table II). Particularly noteworthy is the high Na/K ratio of both graywackes and Columbia River sediments, a characteristic described as one of the most important attributes of graywackes (PETTIJOHN, 1963) and commonly believed, often erroneously, to be due to Na metasomatism. Sediments from the Columbia River vary only slightly in chemical composition along the river even though the mineral composition and the character of the lithic fragments are noticeably different (WHETTEN et al., 1969, p.1162).

We believe that Columbia River sediments are typical of a major class of clastic sediments deposited on the continental margin in a "eugeosynclinal" type environment. As such, they are useful materials for investigating diagenetic changes resulting from elevated temperature and pressure during laboratory experiments.

EXPERIMENTAL RESULTS

Three sediment samples were ground to silt-size (until all of the sample passed through a 200-mesh sieve) in order to increase reaction rates. Depending on the experiment, the samples were placed in either distilled water or artificial sea water in silver tubes and heated externally by resistance furnaces. Experiments were run for 14 to 60 days. P_{H_2O} was 1 and 1.5 kbar; run temperatures were 150 °C, 250 °C, and 300 °C. Minerals synthesized readily at 250 °C and 300 °C, but no new minerals were identified in the 150 °C-runs. The variability of P and T conditions and other experimental details are discussed elsewhere (HAWKINS and WHETTEN, 1969).

Minerals identified as products of the hydrothermal experiments are listed in Tables III and IV. Identifications were made by X-ray diffraction using Ni-filtered Cu$K\alpha$ radiation. Diffraction patterns of original and hydrothermally treated samples are given in Fig.1, 2, and electron micrographs in Plates II, III. The most distinctive change in appearance between treated and untreated samples is the apparent etching of grain surfaces and the formation of a "fibrous" coating on original grains.

The minerals identified as run products were not detected in the original material and are considered to be synthesized minerals. Some of these minerals, or their precursors (e.g., montmorillonite), were present in small amounts and may have served as nuclei for new mineral growth. Montmorillonite appeared as a new

TABLE III

RESULTS OF HYDROTHERMAL EXPERIMENTS AT 250 °C

Run number	Duration (days)	Pressure (kbar P_{H_2O})	Fluid phase	Synthesized products		
				CC-43	CC-72	CC-148
3	29	1	ASW	Mnt, Chl	Mnt	Mnt(?)
5	60	1	DW	Mnt	Mnt, Heu	
6	60	1	DW	Mnt(?)	Chl(?)	
8	30	1.5	DW	Mnt, K-M	Mnt	Chl

DW = distilled water; *ASW* = artificial seawater; *Chl* = chlorite; *Mnt* = montmorillonite; *K-M* = K-mica (illite); *Heu* = heulandite; (*?*) = quantities too small for certain identification.

TABLE IV

RESULTS OF HYDROTHERMAL EXPERIMENTS AT 300 °C, 1 kbar P_{H_2O} USING DISTILLED WATER

Run number	Duration (days)	Synthesized products		
		CC-43	CC-72	CC-148
1	30	Chl, Mnt, Cpt	Mnt	Mnt
7	60	Mnt	Mnt, Cpt, Rie	Mnt, Chl

Chl = chlorite; *Mnt* = montmorillonite; *Cpt* = clinoptilolite; *Rie* = riebeckite.

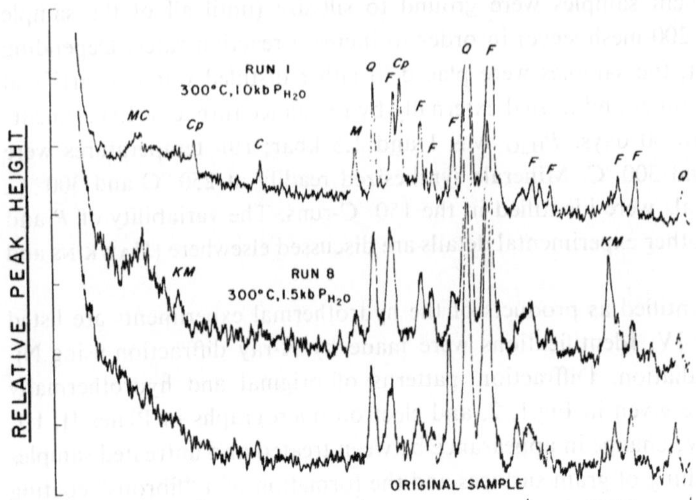

Fig.1. X-ray diffraction patterns of sample CC-43 using Ni filtered CuKα radiation, comparing original samples and results of hydrothermal experiments. Key: *C* = chlorite; *Cp* = clinoptilolite; *F* = plagioclase; *KM* = K-mica (illite); *M* = montmorillonite; *Q* = quartz. All samples run using approximately the same sample weight, surface area, and electronic settings.

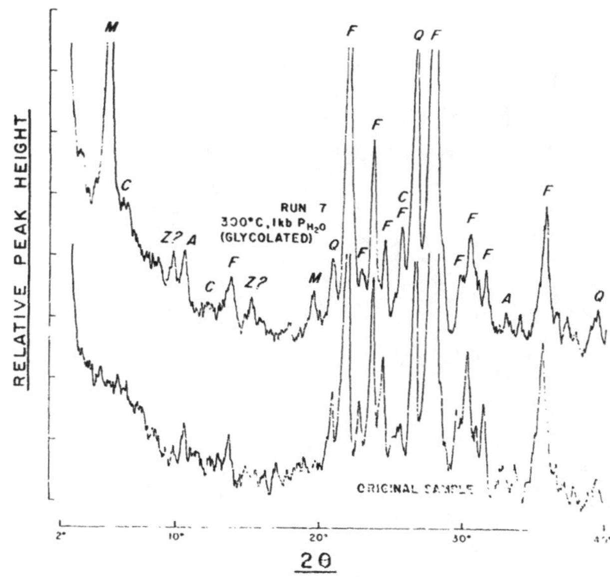

Fig.2. X-ray diffraction patterns of sample CC-148 using Ni filtered CuKα radiation, comparing original samples and results of hydrothermal experiments. Key: A = amphibole; C = chlorite; F = plagioclase; M = montmorillonite; Q = quartz; Z = zeolite. All samples run using approximately the same sample weight, surface area, and electronic settings.

mineral in nearly all the experiments above 200 °C. Other new minerals were chlorite, clinoptilolite, heulandite, an unidentified zeolite, and, in one sample, riebeckite (Table IV).

Kinetic factors, such as nucleation and diffusion rates, and the development of metastable phases exert a control on reaction rates. These factors probably explain the erratic production of zeolites and chlorite. The duration of the runs probably was too short to obtain equilibrium at the P and T used, and it seems likely that equilibrium conditions had to be overstepped considerably to produce the phases formed.

For example, clinoptilolite formed in sample CC-43 during 30 days at 1 kbar P_{H_2O} and 300 °C (run 1), but no clinoptilolite was detected after a 60-day run under the same conditions (run 7). Geologic evidence suggests that clinoptilolite may form at relatively low temperatures (e.g., 27°–55°C in the John Day Formation of Oregon; HAY, 1963, p.237), and that it usually cannot survive a depth of burial exceeding 3 km (HAY, 1966, pp.70,73). At Wairakei, New Zealand, heulandite (closely allied to clinoptilolite) reacts to form laumontite at about 200 °C under relatively low pressure (COOMBS, 1961, p.206). Therefore, 300 °C is probably above the stability field for clinoptilolite. In the 30-day run it may have formed metastably after overstepping the energy requirements for nucleation, but in the longer run it may have been replaced by another phase or phases.

PLATE II

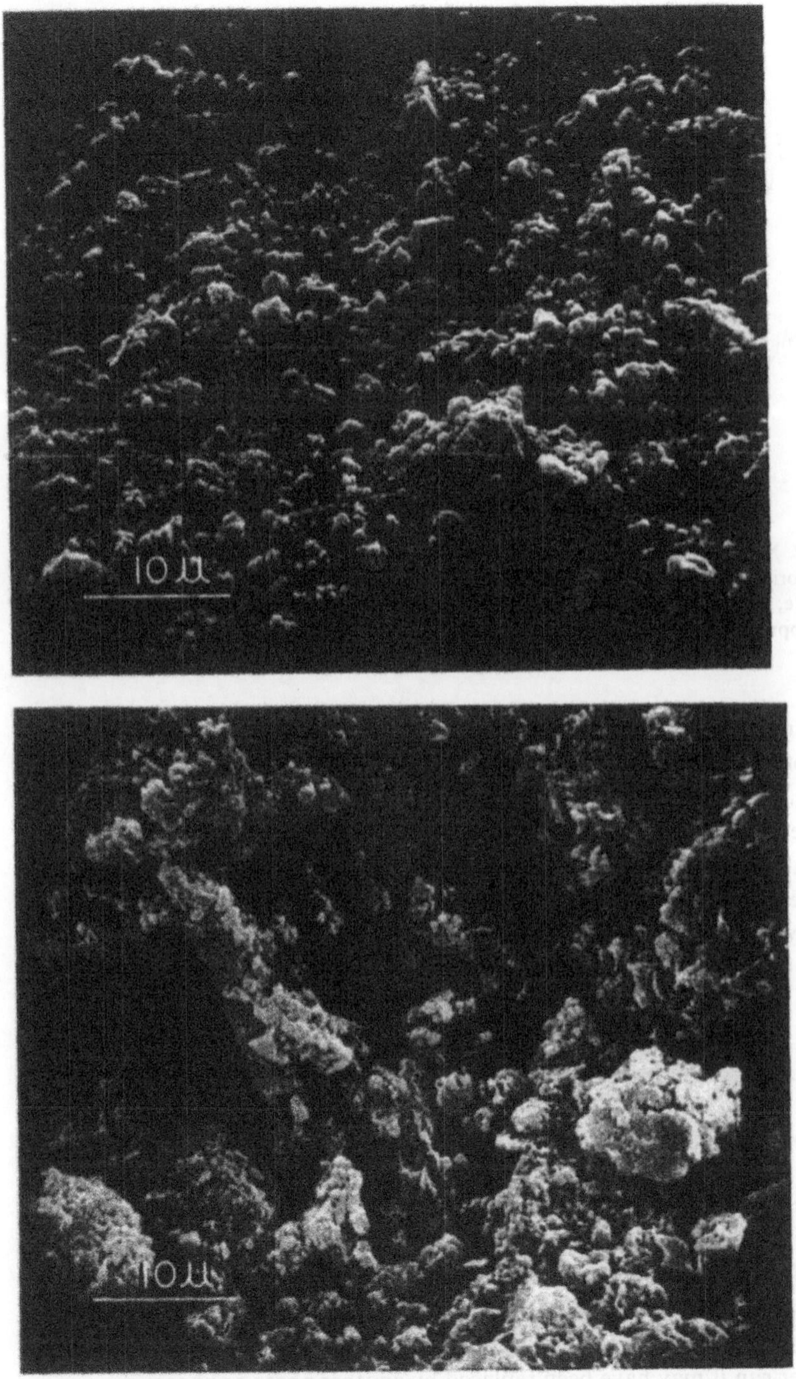

Electron micrograph (× 2000; taken with a scanning electron microscope) of original sample CC-43 (above) and synthesized products from run 8 (below). Note the fibrous mesh of new materials and the formation of a pitted surface on original materials in the treated sample.

PLATE III

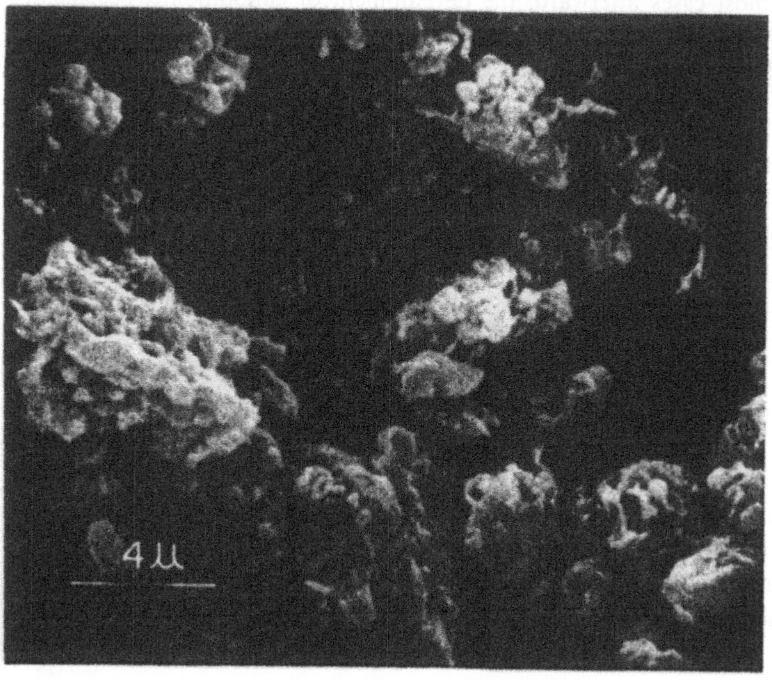

Electron micrograph (× 5000; taken with a scanning electron microscope) of original sample CC-43 (above) and synthesized products from run 8 (below). Note the fibrous mesh of new materials and the formation of a pitted surface on original materials in the treated sample.

The apparent lack of new minerals at 150 °C suggests that for the short time intervals used, temperature must be overstepped by perhaps as much as 50 °C in order to nucleate new phases.

The reactions that formed the new minerals were undoubtedly complex, but hydration and alteration of partially and completely devitrified lithic fragments in the original materials must have played an important role, especially in the formation of montmorillonite. Mafic minerals such as hornblende and pyroxene were present in the starting materials, but the mineral identification technique was not sensitive enough to determine whether or not they participated in mineral reactions. Certainly there is abundant petrographic evidence from sedimentary rocks that these minerals may be replaced by phyllosilicates (e.g., CROOK, 1960, p.546; HAY, 1963, p.222; WALKER, 1967) during intrastratal solution (PETTIJOHN, 1957, pp.674–679).

Some small changes in peak heights and composition (determined by X-ray diffraction) of feldspars were noted (HAWKINS and WHETTEN, 1969) but could have been caused by orientation factors and may not be significant. However, experimental work by HEMLEY (1962, p.196), HEMLEY and JONES (1964), and others indicates that at low temperatures feldspar alters to phyllosilicates; the reaction is controlled by temperature, pressure, and the alkali ion/H^+ ratio of the solutions, and is favored by high pH.

Although it was not possible to analyze the fluid phase either during the runs, or in most cases afterward, the development of clinoptilolite, heulandite, montmorillonite, and an amphibole with the X-ray diffraction characteristics of riebeckite (Table IV, run 7, sample CC-72) indicated a relatively high pH in runs in which distilled water was used. Clinoptilolite and montmorillonite coexist in the lower part of the John Day Formation (HAY, 1963, p.199) as well as in many other sedimentary sequences (HAY, 1966, p.73), and on the basis of HEMLEY'S (1962) experimental work, HAY (1963, p.237) concluded that solutions permeating the John Day pyroclastic deposits had high Si/Al and (Na + K)/H ratios. MILTON and EUGSTER (1959, p.141) indicated that a fluid phase with a high pH and low Po_2 produced the authigenic sodic amphiboles in the Green River Formation.

Other experiments are now in progress with the same starting materials, but using equipment for monitoring the chemistry of the fluid phase during the runs. Preliminary indications are that pH decreases to 3–4 when artificial sea water is used, and increases to 8–9 with distilled water. Apparently the pH of the fluid phases in the experiments described here followed the same pattern.

The fluid phase from samples of run 8 was analyzed semi-quantitatively by atomic absorption. All these samples had fluid phases with Na > 400 p.p.m.; relative concentrations were: CC-43 > CC-148 > CC-72. The relative K concentration was highest in sample CC-43, which had an illitic (K-mica) clay.

There were no known carbonates in the starting materials, yet zeolites were produced. This supports ZEN's (1961) hypothesis that zeolites form at relatively low aCO_2/aH_2O values.

Physical conditions of the experiments were approximately equivalent to burial to a depth of 3–4 km in areas with a geothermal gradient of 60°–80°C/km. The thermal gradients may be in excess of those in natural environments by as much as a factor of 2 because of the kinetic problems discussed above.

DISCUSSION

Our experiments demonstrate that minerals commonly found in the matrices of graywackes may have been sand grains that have undergone diagenetic transformation. Graywacke matrix need not be recrystallized interstitial mud nor argillaceous detritus. Electron micrographs show that the experimentally produced new minerals form as a coating on original minerals and produce a graywacke-like texture. Our experiments also indicate that these diagenetic processes may be essentially isochemical.

We readily acknowledge that our experiments are preliminary, and that others are needed. It is important to test different starting materials, and to determine the chemistry of the fluid phases during and after the experiments as we are now doing.

Petrographic evidence for the origin of matrix is not always decisive. For example, Plate IV shows photomicrographs of sedimentary rocks that have been called graywackes. Lithic fragments are easily recognized in sample O-349; hazy outlines of detrital lithic fragments can be seen in OM-5b; but the grain margins are much less apparent in OM-37a; and in AG-D-11-L there are few obvious detrital grains but abundant phyllosilicate minerals that form a "matrix". Chloritic matrix is most abundant in rocks with few recognizable clasts. Although these samples come from three unrelated units they illustrate an interesting progression in textural type. Using textural criteria alone, it is tempting to attribute such differences to differences in depositional conditions. Alternatively, thermodynamically unstable sand grains in these examples (consisting largely of volcanic fragments probably composed of both crystals and glass) become altered during diagenesis to produce matrix minerals. We believe our experimental results support the latter interpretation.

BRENCHLEY (1969) interpreted the matrix in Ordovician volcanic graywackes from North Wales as having a diagenetic origin. According to Brenchley (p.1299), matrix is absent where sand grains in the graywacke have been partly replaced by calcite and have a calcite cement; instead, the rock is a well-sorted sandstone. On the other hand, AUDLEY-CHARLES (1967, p.9) showed photomicrographs of graywacke from the Viqueque Formation of Timor that resemble those in Plate IV, and stated that "... post-depositional in situ diagenesis (to form matrix) is pal-

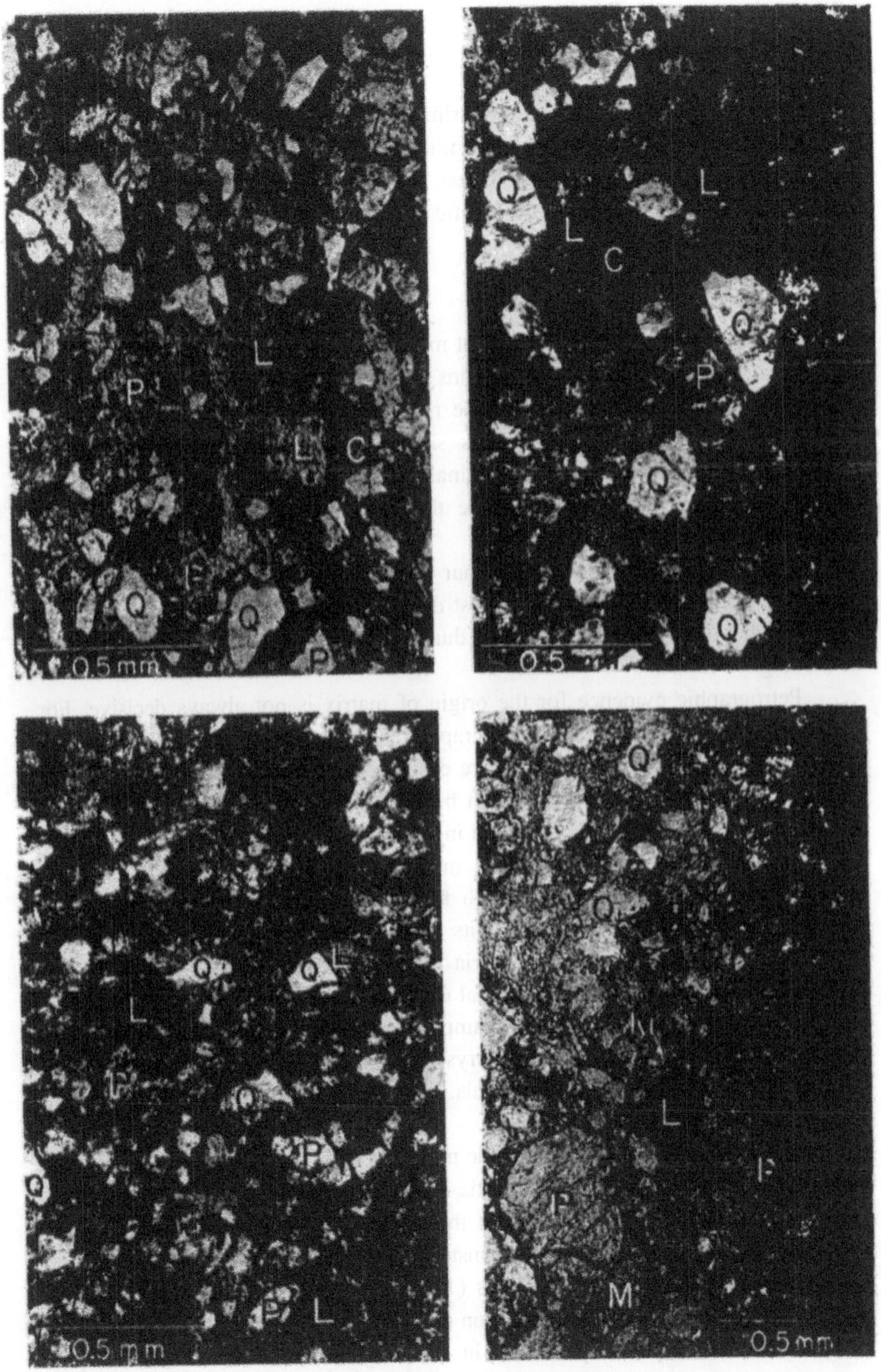

Photomicrographs of graywacke samples. Key to identified grains: C = chlorite; L = altered lithic fragments; M = "matrix" of chlorite and quartz; P = plagioclase; Q = quartz. Photographed in plane polarized light. Sample location: 0-439, Mt. Constitution, Orcas Island, Washington; OM-5b and OM-37a, Mt. Olympus, Clallam County, Washington; AG-D-11-L, Patton Escarpment, southern California borderland, dredged from 1500 m depth. (Samples: upper left O 349; upper right OM-5b; lower left OM-37a; lower right AG-D-11-L.)

pably impossible". However, Audley-Charles' evidence was vigorously disputed in a comment by RAHMANI (1968).

Another line of evidence for the diagenetic origin of matrix is that hornblende and biotite are altering intrastratally to produce Fe-rich clay matrix in Pleistocene and Pliocene arkosic sands of the Sonoran desert, and that the sedimentary sequence shows progressive alteration with age (WALKER, 1967). In suggesting this model as a general explanation for red beds, Walker (p.365–366) pointed to the paucity of mafic minerals in ancient red beds where they should occur as common accessory minerals.

A similar argument could be made regarding unstable clasts in graywackes. If it is accepted that Columbia River sediments are reasonably typical of the sediment types derived from a mountainous, lithologically heterogeneous source terrane, such sediments should appear in the geologic record. Where are ancient clastic sedimentary rocks with large quantities of mafic minerals and unstable lithic fragments preserved? Clearly, the unstable materials were not removed during weathering in the source area or during transportation (Table I). The absence in the geologic record of rocks resembling lithified Columbia River sediments is paralleled by the lack of modern sands that resemble, mineralogically and texturally, the presumed parent material of ancient graywackes.

There are several geologic factors bearing on these problems. Graywackes are typical of orogenic belts. Modern tectonic concepts (MORGAN, 1968) suggest that these belts are related to zones of subduction (NELSON and TEMPLE, 1969) or crustal consumption in which the sea floor and marine deposits are carried beneath continental areas. These zones are loci of volcanism and rapid sedimentation, the dominant sedimentary types being volcaniclastic sands of andesitic composition (DICKINSON, 1962; BAILEY et al., 1964). The rapid accumulation of thermodynamically unstable detritus in regions of active tectonism and volcanism favors diagenetic and low-grade metamorphic changes of these sediments. We believe that our experimental conditions may have duplicated the shallow-burial environment of some of these ancient areas of tectonism. There are numerous examples of graywackes such as those shown in Plate IV which have textures suggesting that hydration and recrystallization of unstable clasts is common. These interpretations and the results of our hydrothermal experiments lead us to conclude that the origin of graywacke matrix is largely the result of diagenetic processes and that CUMMINS' (1962) explanation for graywacke texture is essentially correct.

ACKNOWLEDGEMENTS

We thank R. G. Coleman, J. R. Curray, W. R. Dickinson, R. M. Garrels, J. R. Walker, A. L. Washburn, and E. L. Winterer for many helpful discussions. Financial support was provided by NSF Grants GA-634, GA-1406 and GA-15706.

AUDLEY-CHARLES, M. G., 1967. Greywackes with a primary matrix from the Viqueque Formation (Upper Miocene-Pliocene), Timor. *J. Sediment. Petrol.*, 37:5–11.

BAILEY, E. H., IRWIN, W. P. and JONES, D. L., 1964. Franciscan and related rocks, and their significance in the geology of western California. *Calif. Div. Mines Geol., Bull.*, 183:177 pp.

BEALL, JR., A. O. and FISCHER, A. G., 1969. Sedimentology. In: *Initial Reports of the Deep Sea Drilling Project, 1.* National Science Foundation, Washington, D.C., pp.521–593.

BRENCHLEY, P. J., 1969. Origin of matrix in Ordovician greywackes, Berwyn Hills, North Wales. *J. Sediment. Petrol.*, 39:1297–1301.

COOMBS, D. S., 1961. Some recent work on the lower grades of metamorphism. *Australian J. Sci.*, 24:203–215.

CROOK, K. A. W., 1960. Petrology of Parry Group, Upper Devonian-Lower Carboniferous, Tamworth-Nundle District, New South Wales. *J. Sediment. Petrol.*, 30:538–552.

CUMMINS, W. A., 1962. The greywacke problem. *Liverpool Manchester Geol. J.*, 3:51–71.

DICKINSON, W. R., 1962. Circum-Pacific andesite types. *J. Geophys. Res.* 73:2261–2269.

DOTT, JR., R. H., 1963. Dynamics of subaqueous gravity depositional processes. *Am. Assoc. Petrol. Geologists Bull.*, 47:104–128.

HAWKINS, J. W. and WHETTEN, J. T., 1969. Greywacke matrix minerals: hydrothermal reactions with Columbia River sediments. *Science*, 166:868–870.

HAY, R. L., 1963. Stratigraphy and zeolitic diagenesis of the John Day Formation of Oregon. *Geol. Sci.*, 42:199–262.

HAY, R. L., 1966. Zeolites and zeolitic reactions in sedimentary rocks. *Geol. Soc. Am., Spec. Papers*, 85:130 pp.

HEMLEY, J. J., 1962. Alteration studies in the systems Na_2O–Al_2O_3–SiO_2–H_2O and K_2O–Al_2O_3–SiO_2–H_2O, p. 196 in the Geol. Soc. Am. Abstracts for 1961. *Geol. Soc. Am., Spec. Papers*, 68:322 pp.

HEMLEY, J. J. and JONES, W. R., 1964. Chemical aspects of hydıothermal alteration with emphasis on hydrogen metasomatism. *Econ. Geol.*, 59:538–569.

HOLLISTER, C. D. and HEEZEN, B. C., 1964. Modern graywacke-type sands. *Science*, 146:1573–1574.

HUBERT, J. F., 1964. Textural evidence for deposition of many western North Atlantic deep-sea sands by ocean-bottom currents rather than turbidity currents. *J. Geol.*, 72:757–785.

KNEBEL, H. J., KELLEY, J. C. and WHETTEN, J. T., 1968. Clay minerals of the Columbia River: a qualitative, quantitative, and statistical evaluation. *J. Sediment. Petrol.*, 38:600–611.

KUENEN, PH. H., 1966. Matrix of turbidites: experimental approach. *Sedimentology*, 7:267–297.

MILTON, D. and EUGSTER, H., 1959. Mineral assemblages of the Green River Formation. In: P. H. ABELSON (Editor), *Researches in Geochemistry, 1.* Wiley, New York, N.Y., pp.118–150.

MORGAN, W. J., 1968. Rises, trenches, great faults, and crustal blocks. *J. Geophys. Res.*, 73:1959–1982.

NELSON, T. H. and TEMPLE, P. G., 1969. Global tectonics and migration of oceanic ridges and trenches. *EOS, Trans. Am. Geophys. Union*, 50:634–635.

OKADA, H., 1966. Non-greywacke "turbidite" sandstones in the Welsh geosyncline. *Sedimentology*, 7:211–232.

PETTIJOHN, F. J., 1957. *Sedimentary Rocks.* Harper, New York, N.Y., 2nd ed., 718 pp.

PETTIJOHN, F. J., 1963. Chemical composition of sandstones - excluding carbonate and volcanic sands. In: M. FLEISCHER (Editor), *Data of Geochemistry* (6th ed.)—*U.S. Geol. Surv., Profess. Papers*, 440-S:21 pp.

RAHMANI, R. A., 1968. Greywackes with a primary matrix from the Viqueque formation (Upper Miocene-Pliocene), Timor–comment on a Paper by M.G. Audley-Charles. *J. Sediment. Petrol.*, 38:271–273.

SHEPARD, F. P., 1964. Criteria in modern sediments useful in recognizing ancient sedimentary environments. In: L. M. J. U. VAN STRAATEN (Editor), *Deltaic and Shallow-Marine Deposits.* Elsevier, Amsterdam, pp.1–25.

Walker, T. R., 1967. Formation of red beds in modern and ancient deserts. *Geol. Soc. Am., Bull.*, 78:353–368.

Whetten, J. T., Kelley, J. C. and Hanson, L. G., 1969. Characteristics of Columbia River sediments and sediment transport. *J. Sediment. Petrol.*, 39:1149–1166.

Williams, H., Turner, J. and Gilbert, C., 1955. *Petrography: an Introduction to the Study of Rocks in Thin Sections.* Freeman, San Francisco, Calif., 406 pp.

Zen, E., 1961. The zeolite facies–an interpretation. *Am. J. Sci.*, 259:401–409.

Reprinted from Sedimentology (1972) ***19****, 141–143*

Discussion

DIAGENETIC ORIGIN OF GRAYWACKE MATRIX MINERALS: A DISCUSSION

J. P. B. LOVELL

Grant Institute of Geology, University of Edinburgh, Edinburgh (Great Britain)

(Received October 12, 1971)

"Petrographic evidence for the origin of matrix is not always decisive" write Whetten and Hawkins (1970, p.357). In the discussion of the textures shown in photomicrographs on their Plate IV they write: "Using textural criteria alone, it is tempting to attribute such differences to differences in depositional conditions. Alternatively, thermodynamically unstable sand grains in these examples . . . become altered during diagenesis to produce matrix minerals. We believe our experimental results support the latter interpretation."

I agree that petrographic evidence for the diagenetic origin of matrix may not be decisive, but it is in many cases at least highly suggestive. The ultimate proof of any theories of matrix origin must come from petrographic studies of ancient sedimentary rocks, especially regional studies such as those of Walker (1967) and Brenchley (1969) quoted by Whetten and Hawkins. The results of another such regional study are reviewed below, following an earlier short report (Lovell, 1969).

REGIONAL STUDY OF ORIGIN OF MATRIX

A study of the regional and local variation in the mineralogy of the sandstones of the Eocene Tyee Formation, Oregon Coast Range, indicates that the only statistically significant regional variations of abundant components are those of volcanic rock fragments, total rock fragments (which are largely volcanic) and matrix. The *total* of rock fragments plus matrix does not vary significantly across the region (Lovell, 1969, table 2). Rock fragments are more abundant in the south and southeast, while matrix is less abundant in these areas.

The following explanations of these facts are not mutually exclusive:

(*1*) Operator error. It is certainly difficult to distinguish rock fragments from matrix in many samples, but it is not difficult to distinguish either of them from other components of the rock. Double counts and other checks rule out the possibility that there was a systematic change in the identification of rock fragments and matrix which led to the regional variation under discussion.

(*2*) Difference in depositional conditions. From the start, rock fragments may have been more common in rocks deposited near the southern source, while the matrix, derived from the breakdown of rock fragments during transport, may have been more common distally. It is just conceivable that this is a special and coincidental case in which their total throughout the formation remains constant.

(*3*) Diagenesis. There are three main lines of evidence supporting this explanation:

(*a*) The partial coincidence between an area of low rock-fragment/matrix ratios, and an area of relatively abundant authigenic chlorite in the west centre of the outcrop area, where the Tyee Formation may have been more deeply buried than elsewhere, beneath more than 3000 m of later sediments and volcanics.

(*b*) The textural evidence seen in thin sections of disintegration of rock fragments, especially the abundant volcanic rock fragments.

(*c*) The third line of evidence comes from a study of the clay mineralogy of a low-matrix sample and a high-matrix sample, which were examined in two ways: (*i*) gentle disaggregation of the two rocks to study the existing matrices ($< 4\mu$ fractions), which were found to be dissimilar; (*ii*) grinding of the two rocks, followed by separation of the $< 4\mu$ fractions, which proved to have similar compositions.

These results support the indications from thin sections that the rock fragments in the low-matrix sample have altered mineralogically following burial, and would have formed matrix mineralogically similar to that in the high-matrix sample if disintegration had gone further.

(*4*) Regional variation in grain size. If the rocks sampled from the south and southeast were originally coarser than those sampled from other areas, the regional variation could be explained at least partly by the normal association between coarser-grained sandstones and abundant rock fragments, and finer-grained sandstones and abundant matrix (H. Füchtbauer, personal communication). The major difficulty in testing this hypothesis in the Tyee sandstones is diagenesis, which makes accurate analyses of the original grain size of a sample impossible. The data indicate that the present average grain size of the samples studied (c. 0.2 mm) does not change greatly over the region; no special effort in sampling in the field was needed to ensure this result, except in parts of the southern facies (Lovell, 1969).

CONCLUSIONS

I suggest that the evidence from the Tyee Formation supports the main conclusions of Whetten and Hawkins (1970), and those of the earlier proponents of a diagenetic origin of graywacke matrices such as Cummins (1962). This conclusion is hedged with some uncertainty about grain size variation. It is not incompatible with certain other suggestions concerning the origin of graywacke matrices (e.g., Füchtbauer, 1970, p.209).

I also suggest that the Tyee is an example of a formation, situated in an orogenic belt, in which "large quantities of mafic minerals and unstable lithic fragments" *are* preserved (in answer to the question put on p.359 of Whetten and Hawkins, 1970). This is presumably at least partly because of the notable lack of tectonic deformation. How this lack of deformation may in turn be related to "modern tectonic concepts" (Whetten and Hawkins, 1970, p.359) is another question, but it may well be that the ratio of rock fragments to matrix may provide a useful index of the relative depth of burial of parts of this and similar formations.

ACKNOWLEDGEMENT

I thank Professors H. Füchtbauer and F. L. Schwab for their criticisms of the manuscript.

REFERENCES

Brenchley, P. J., 1969. Origin of matrix in Ordovician greywackes, Berwyn Hills, North Wales. *J. Sediment. Petrol.*, 39:1297–1301.

Cummins, W. A., 1962. The greywacke problem. *Liverpool Manchester Geol. J.*, 3:51–72.

Füchtbauer, H., 1970. Lithification of clastic sediments. *Sedimentology*, 15:207–210.

Lovell, J. P. B., 1969. Tyee Formation: undeformed turbidites and their lateral equivalents: mineralogy and paleogeography. *Geol. Soc. Am. Bull.*, 80:9–22.

Walker, T. R., 1967. Formation of red beds in modern and ancient deserts. *Geol. Soc. Am. Bull.*, 78: 353–368.

Whetten, J. T. and Hawkins, J. W., Jr., 1970. Diagenetic origin of graywacke matrix minerals. *Sedimentology*, 15:347–361.

Reprinted from Sedimentology (1972) ***19***, *144–146*

DIAGENETIC ORIGIN OF GRAYWACKE MATRIX MINERALS: A REPLY

JOHN T. WHETTEN AND JAMES W. HAWKINS, JR.

Department of Geological Sciences, University of Washington, Seattle, Wash. (U.S.A.)
Geological Research Division, Scripps Institution of Oceanography, University of California, La Jolla, Calif. (U.S.A.)

(Received November 15, 1971)

We are pleased that Lovell's findings in the Tyee Formation (Lovell, 1969, 1972) support the hypothesis that graywacke matrix may result from the diagenetic alteration of thermodynamically unstable detrital clasts. We knew of Lovell's work and should have cited it in our paper (Whetten and Hawkins, 1970). We agree with him that regional studies of sedimentary rocks may be important in testing theories of matrix origin (Lovell, 1972). However, we believe that in the long run information from *ancient* sedimentary rocks is unlikely to shed much direct light on the matter, particularly if the hypothesis we advocate is correct in either the general sense or in special instances.

After all, regional petrographic studies of sedimentary rocks have been made for many years, yet the idea of a diagenetic origin for graywacke matrix was only recently stated (Cummins, 1962). If rocks are studied in which the matrix-producing diagenetic reactions have gone to completion, then there is no way to interpret the nature of the clastic grains that were originally present but subsequently destroyed. It is precisely for this reason that many geologists, relying strictly on petrographic evidence, have been lead into believing that much or all matrix is primary and detrital.

Alternatively, we suggest that emphasis should be put on petrologic studies of relatively young sediments in which diagenetic processes are still going on and the starting materials are still at least partially present. Pleistocene and Late Tertiary clastic sediments from a wide range of environments (including JOIDES cores) should be looked at. The only thorough study of diagenesis in such sediments that we are familiar with (Walker, 1967) is highly supportive of our views.

Finally, we would like to add a footnote to our 1970 paper and show some new experimental evidence for the diagenetic origin of matrix. Fig.1 is a scanning electron photomicrograph of a sand-size detrital hypersthene grain that was severely etched in a five-month run at 200°C and about 200 p.s.i. Montmorillonite was extensively produced during the run. Other hypersthene grains were also partially destroyed. A variety of grains probably participated in the reaction to form clay

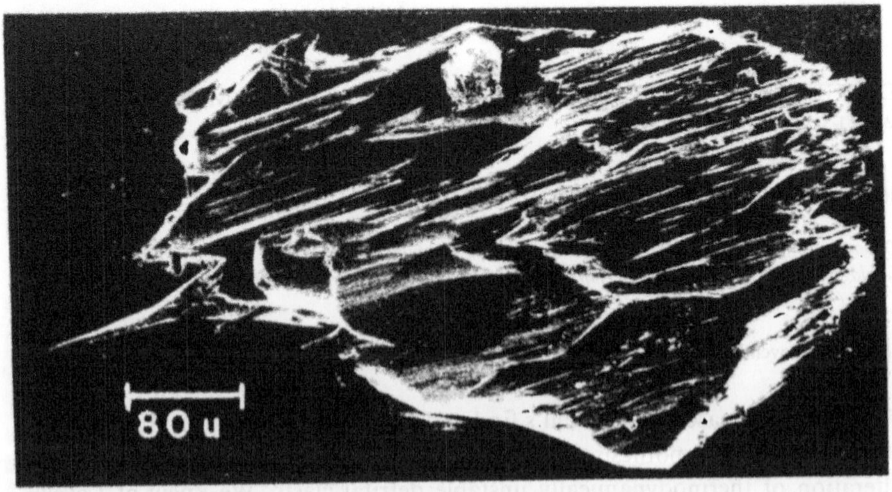

Fig. 1. Detrital hypersthene grain (coarse sand size) from the Columbia River. Grain was etched in five months in a brine solution composed of 62.3 g $NaCl$, 28.5 g $CaCl_2$, 16.3 g KCl, 3.6 g $MgCl_2$, and 7 ml HCl in 900 ml distilled water. P/T conditions were 200°C and about 200 p.s.i.

"matrix," but this is the first time we have been able to cite a specific mineral as a probable source. Füchtbauer (1970, pp.208–209) implies that fragments of devitrified volcanic glass formed our experimentally derived matrix, but now a wider range of minerals (including those of non-volcanic origin) must also be considered. Perhaps the most important point, with reference to our suggestion that relatively young sediments need study, is that had the run proceeded much longer (a year?) the grain shown in Fig. 1 would have been completely destroyed, leaving no direct evidence that it was ever present. The complete destruction of grains by solution or other processes is what makes a study of diagenesis in *ancient* sedimentary rocks difficult.

The Tyee is cited by Lovell (1972) as a formation where large quantities of mafic minerals and unstable lithic fragments *are* preserved; yet they cannot *all* be preserved, if some have been altered to matrix as Lovell believes. Hypersthene grains and a host of other minerals were probably destroyed. Unfortunately, few petrologic studies of modern detrital sands are available so that meaningful comparisons can be made with *ancient* sedimentary rocks like the Tyee.

We appreciate Lovell's supportive discussion. His work, the petrographic work of others (particularly with young sediments), and our own experimental evidence lead us to believe that texture and composition of clastic sedimentary rocks are determined as much by diagenesis as they are by the environment of deposition and source area conditions, a view somewhat different from the one recently expressed by Folk et al. (1970, pp.939–940).

REFERENCES

Cummins, W. A., 1962. The greywacke problem. *Liverpool Manchester Geol. J.*, 3:51–71.
Folk, R. L., Andrews, P. B. and Lewis, D. W., 1970. Detrital sedimentary rock classification and nomenclature for use in New Zealand. *N. Z. J. Geol. Geophys.*, 13(4):937–968.
Füchtbauer, H., 1970. Lithification of clastic sediments. *Sedimentology*, 15:207–210.
Loveli, J. P. B., 1969, Tyee Formation: undeformed turbidites and their lateral equivalents—mineralogy and paleogeography. *Geol. Soc. Am., Bull.*, 80:9–22.
Lovell, J. P. B., 1972. Diagenetic origin of greywacke matrix materials: a discussion. *Sedimentology*, 19:141–143.
Walker, T. R., 1967. Formation of red beds in modern and ancient deserts. *Geol. Soc. Am., Bull.*, 78:353–368.
Whetten, J. T. and Hawkins, J. W., 1970. Diagenetic origin of graywacke matrix minerals. *Sedimentology*, 15:347–361.

Reprinted from Sedimentology (1986) ***33***, *651–675*

Diagenesis of sandstones in the back-arc basins of the western Pacific Ocean

YONG IL LEE* *and* GEORGE DEVRIES KLEIN

Department of Geology, University of Illinois at Urbana-Champaign, 245 Natural History Bldg, 1301 W. Green Street, Urbana, IL 61801–2999, U.S.A.

ABSTRACT

Sandstones occur in back-arc basins of the western Pacific at DSDP sites 299 (Sea of Japan), 297 (northern Shikoku Basin), 445 and 446 (Daito-Ridge-and-Basin Province), 453 (Mariana Trough), 286 (New Hebrides Basin) and 285 (South Fiji Basin). These sandstones are dominantly volcaniclastic arenites derived from andesitic island arcs. The degree of sandstone diagenesis is dependent on original composition, burial rate, heat flow history of the basin, and timing of sandstone deposition with respect to rifting processes and associated high heat flow.

Sandstones containing a larger proportion of volcaniclastic components showed more diagenetic effects than sandstones containing a significant volume of other rock fragments and mineral components. Sandstones deposited during early stages of rifting (sites 445, 446) with a slow burial rate and high crustal heat flow showed the greatest degree of downhole diagenetic change. These diagenetic changes include early pore-space reduction and rim cementation by clay minerals followed later by calcite, and subsequent pore-fill cementation by clinoptilolite, heulandite, analcite and later calcite. Replacement of recognizable volcanic rock fragments by chert, calcite and zeolites was observed in the deepest part of the hole. Sandstones deposited after rifting under conditions of associated lower heat flow showed considerably less diagenetic changes, particularly if burial was rapid.

The high heat flow associated with earliest rifting, associated fluid circulation driven by thermal convection, and slow burial rate controlled the diagenetic history of these sandstones. Thus, timing of sandstone deposition with rifting stage and associated burial rates were key factors in controlling sandstone diagenesis in back-arc basins.

INTRODUCTION

Over the past decade, the study of sandstone diagenesis has undergone a rapid expansion. These studies demonstrated that factors influencing diagenetic changes in sandstones include original composition, burial depth, temperature, and pore-water chemistry (Carrigy & Mellon, 1964; Blatt, 1979; Hayes, 1979; Vavra, 1983). Sandstone diagenesis proceeds through several systematic steps starting with pore-space reduction by compaction, later rim cementation, followed by pore-fill cementation, and alteration and transformation of mineral phases with more-deeply buried sandstones (Wilson & Pittman, 1977). It is for this reason that porosity and permeability decrease downward in a borehole. Selective dissolution at depth causes development of secondary porosity within sandstones (Schmidt & McDonald, 1979).

In active continental margins, few studies of sandstone diagenesis, except one by Galloway (1974, 1979), were completed because most such studies focussed on framework components (Dickinson, 1982; Dickinson, Helmold & Stein, 1979; Dickinson & Valloni, 1980; Harrold & Moore, 1975; Valloni & Maynard, 1981). Sands and sandstones deposited in back-arc basins are derived mostly from andesitic volcanic island arcs (Karig, 1975; White et al., 1980; Klein & Lee, 1984; Klein, 1975, a, b, 1985). Little is known about the diagenesis of sandstones in this tectonic domain where extensional rifting and characteristic high heat flow should play an important role in diagenesis. Only one study of sandstone diagenesis

*Present address: Department of Geology, Seoul National University, San 56, Shinrim Dong, Kwanak-ku, Seoul 151, Korea.

was completed in a back-arc basin at DSDP site 445 on the southern edge of the Daito Ridge (Klein *et al.*, 1980a, b). There, a systematic change in diagenesis was observed downhole. The first change observed was rim cementation by clay minerals, followed downhole by pore-fill cementation by calcite and zeolites. In deeper horizons, volcaniclastic grains are replaced progressively by calcite, and at deeper levels by chert. Klein *et al.* (1980b) suggested that solution and replacement of volcaniclastic grains is more common at deeper stratigraphic intervals, whereas mineral precipitation and compaction are more characteristic at shallower elevations. No work has compared diagenetic changes in sandstones at other back-arc basin sites.

Statement of the problem

Despite this one study from site 445 (Klein *et al.*, 1980b), several problems remain. This paper focusses on the following problems in Pacific Ocean back-arc basins where sandstones are known to occur.

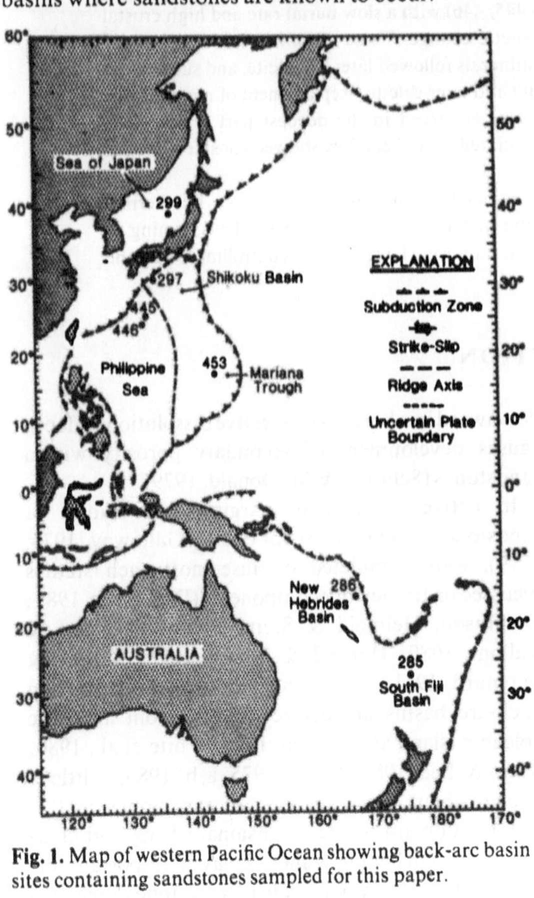

Fig. 1. Map of western Pacific Ocean showing back-arc basin sites containing sandstones sampled for this paper.

(1) What is the effect of sediment age on diagenesis, particularly where the sandstone framework consists of volcaniclastic grains that are reactive under low temperature and pressure in marine environments?

(2) How does original sandstone composition control diagenesis of sandstones of the same clan?

(3) How does the thermal history of the basement influence diagenesis of the overlying sediments within the basin particularly in areas of high heat flow?

(4) What is the effect of differing burial histories on sandstone diagenesis within a similar suite of rocks such as volcaniclastic sandstone?

(5) How does tectonic evolution of back-arc basins influence sandstone diagenesis?

Seventy-five samples from most sandstone intervals were obtained from DSDP sites 299, 297, 445, 446, 453, 286 and 285 (Fig. 1). Table 1 provides location data, basement ages and DSDP site report references. The reader is referred to those references and to Weissel (1981), Klein & Lee (1984, tables 3 and 4) and Klein (1985) for details concerning geology, rifting ages, ages of volcanism, sedimentology and correlation of sedimentation to tectonic and volcanic events.

SANDSTONE PETROLOGY

Methods

Laboratory analyses in this study included using polarizing microscopy, X-ray diffraction, scanning electron microscopy (SEM), and electron microprobe analysis of specific phases. Carbon and oxygen isotope analysis of cementing minerals was performed on samples obtained from site 445 (Lee, 1986).

Point-counting of 75 thin sections of sandstones from seven sites was completed to determine detrital and authigenic sandstone composition (Appendix, Tables 2 to 7). An average of 350 points were counted per section at sites 445 and 446, and an average of 230 points were counted at the other sites.

Petrologic summary

Sandstones at the seven sites listed in Table 1 range in particle size from fine silt to coarse sand. Most show poor sorting and grains are subrounded to angular in most samples, except at sites 299 and 297 where rounded grains are common.

Table 1. Location of DSDP sites discussed in this paper, basement ages (after Weissel, 1981) and depth interval of sandstones sampled in this study

Site	Latitude and longitude	Location	Depth interval of sandstones (below subbottom, m)	Basement age Myr BP	Site report reference
299	39° 29·69′N 137° 39·72′E	Toyama Fan, Sea of Japan	0–532	0–5	Karig *et al.* (1975)
297	30° 52·36′N 134° 09·89′E	Shikoku Basin	330–570	25–5	Karig *et al.* (1975)
445	25° 29·69′N 133° 12·49′E	Daito Basin	647–892	44–48	Klein *et al.* (1980)
446	24° 42·04′N 132° 46·49′E	Daito Basin	172–628	46–51	Klein *et al.* (1980)
453	17° 54·42′N 143° 40·95′E	Western Mariana Trough	303–443	4·7–5	Hussong *et al.* (1981)
286	26° 32·92′S 166° 22·18′E	New Hebrides Basin	206–649	37–45	Andrews *et al.* (1975)
285	26° 49·16′S 175° 48·24′E	South Fiji Basin	453–564	12–15	Andrews *et al.* (1975)

Detrital mineralogy is relatively simple, consisting mostly of plagioclase and recognizable volcanic rock and glass fragments with accessory quartz (Appendix, Tables 2 to 7). At site 297, quartz and sedimentary rock fragments are more common. The relative proportions of these constituents, especially quartz and volcanic glass and rock fragments, vary between sites and downhole. Sandstones were classified according to Folk (1968, fig. 2); most are litharenites with accessory feldspathic litharenites. Sandstone at sites 297 and 299 are the most mineralogically mature of our sample set (Fig. 2).

Almost all sandstones occurred on the tie-line between sedimentary and volcanic rock fragments in the rock fragment daughter triangle. These sandstones are classified as volcanic arenite, except at site 297 and three samples at site 446, which are sedarenites.

The principal framework component in these sandstones is volcanic rock fragments at nearly all of the sites we sampled (Appendix, Tables 2 to 7). Andesitic rock fragments are by far the most common volcanic rock fragment, although phyric basalt, microdolerite and aphyric basalt fragments were observed also, particularly at sites 445 and 446. Many volcanic rock fragments contain plagioclase lathes set in a groundmass of volcanic glass. The volumetric proportion of volcanic rock fragments changes downhole at three sites with a general increase downhole observed at site 299 (Table 2), and site 453 (Table 5), and a general decrease observed downhole at site 286 (Table 6).

Detrital volcanic glass is a major constituent in sandstones at most sites occurring both as framework grains and as matrix, especially at sites 453 (Table 5) and 286 (Table 6). Both fresh and altered glasses occur. Some shards are porous, rounded, etched and pitted by solution.

Plagioclase feldspar is a minor constituent in these sandstones. Most plagioclases are andesine, with bytownite observed at site 453 (Table 5). Alteration of plagioclases was observed to increase downhole, particularly at site 445 (Table 4).

At site 297 (Table 3), recognizable sedimentary rock fragments comprise the major detrital component. These are well-rounded and contain quartz and feldspar in an argillaceous matrix. The relatively larger volume of recognizable sedimentary and metamorphic rock fragments at site 297 was caused by Miocene uplift of Shikoku Island (Matsuda, Makamura & Sugimura 1967; Karig *et al.*, 1975; Sugi, Chinzei & Uyeda, 1983) which provided sediment to the northern Shikoku Basin. However, during the early Pliocene, the Nankai Trough formed and acted as a sediment trap, cutting off sediment yield to site 297.

Sandstone matrix at sites 299 and 297 (Appendix, Tables 2 and 3) consists mostly of terrigenous clay and volcanic glass. XRD analysis indicates that the clay fraction consists of mica, chlorite, kaolinite and smectite. Other framework components are listed in Tables 2 to 7 (see Appendix).

DIAGENETIC PHENOMENA IN BACK-ARC BASIN SANDSTONES

Although a large variety of diagenetic phenomena was observed at each site, most represent the end

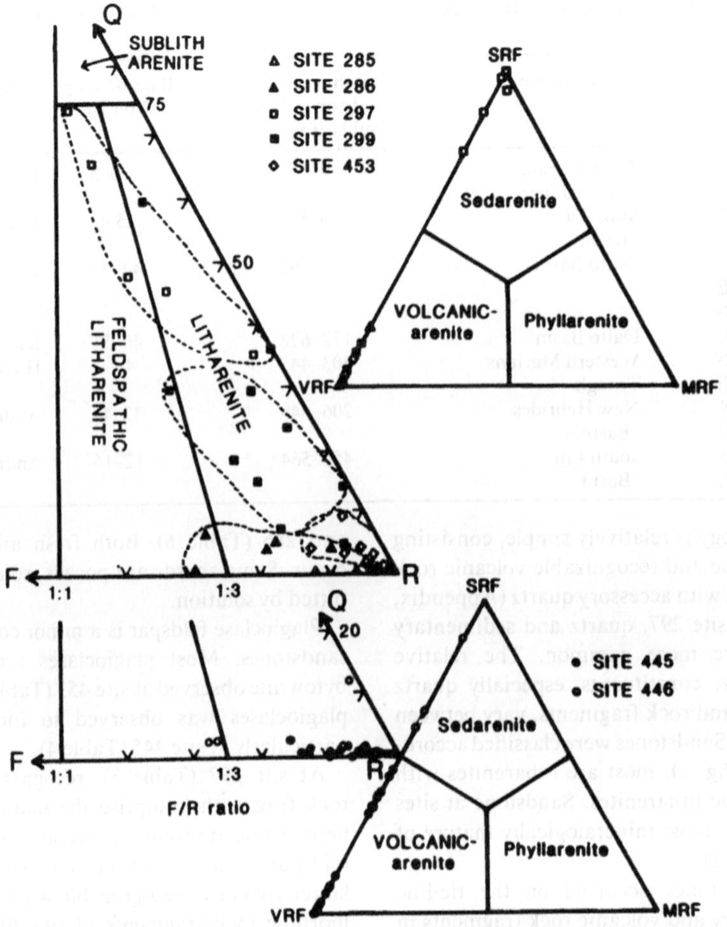

Fig. 2. Classification of sandstone from Pacific Ocean back-arc basins (classification after Folk, 1968).

product of cementation, complex alteration and replacement, and dissolution. Authigenic cements include anhydrite cement, micritic and sparry calcite rim and pore-fills (Fig. 3A, B), clay rims oriented normal to detrital grain surfaces (Fig. 3A, C), well-crystallized pore-filling zeolites including clinoptilolite, heulandite and analcite (Fig. 3C, D, E, F), and minor quantities of syntaxial feldspar and fossil overgrowths. Mechanically-compacted softer grains such as limestone and fossils also occur. Alteration and replacement phenomena include volcanic glass matrix altered to clay minerals and zeolites, cristobalite layers around volcanic rock fragments, and altered and chloritized volcanic rock fragments. These rock fragments also show replacement by chlorite, zeolites, chert and calcite, whereas plagioclase is replaced by calcite and sericite. Alteration of components such as limestone fragments and calcareous fossils is common. Cristobalite and mica, and quartz overgrowths were also observed.

Pore-lining and pore-filling cementation

Pore spaces in these sandstones contain a variety of cements ranging from clay coats on detrital grain surfaces to complex sequences of zeolites.

Clay minerals: Clay mineral rims are common on volcaniclastic grains. These coats are 30–50 μm thick. At site 297, no diagenetic smectite was observed. At site 445, smectite, chlorite and illite are oriented normal to detrital grain surfaces, and are absent at grain contacts suggesting an authigenic origin (Fig. 3A). X-ray diffraction (XRD) analysis indicated that smectite was 100% expandable, and chlorite is iron-rich. Minor quantities of discrete illite (observed in SEM) appears as a rim cement below a subbottom depth of 820 m.

Anydrite pore-fill cement: Anhydrite pore-fill cement, at site 299 (Fig. 4), shows an oval to elongate shape and develops independent of grain composition. Anhydrite pore-fill cement was observed also at site 297 showing radiating clusters of elongate, oval-shaped crystals.

Heulandite group minerals: Both clinoptilolite and heulandite were identified and classified according to Boles (1972).

Clinoptilolite: This mineral occurs at site 445 with heulandite, cristobalite, smectite and altered glass matrix (Fig. 3A). At site 446, it replaces both volcanic glass matrix and volcanic rock fragments.

In thin section or under the SEM, clinoptilolite occurs as tabular, lath-shaped crystals developing from alteration of volcanic glass. Most crystals display characteristic monoclinic symmetry and show well-developed cleavage. Some clinoptilolite crystals show evidence of dissolution. Electron microprobe analyses showed clinoptolite was characterized by Si/Al ratios greater than 4·00, and heat treatment at 550°C showed no change in this ratio (Lee, 1984).

Heulandite: The framework ion ratio of Si to Al (⟨4·0) in heulandite differs from this ratio in clinoptilolite, and heulandite is also thermally less stable. It occurs in sandstones at site 445 with an association identical to clinoptilolite (Fig. 3D). Electron microprobe analyses of some heulandites confirm the identification (Lee, 1984). Heat treatment at 550°C for one hour decreased peak intensity considerably.

Analcite: This mineral was observed at both sites 445 and 446. In SEM, analcite crystals show both cubo-octahedral and trapezohedral euhedral forms (Fig. 3E). Some analcite shows dissolution features (Fig. 3E, F). Some analcite forms pseudomorphs after clinoptilolite or heulandite and in volcanic rock fragments, it replaces plagioclase. Electron microprobe analyses of these analcites show Si/Al ratios from 2·08 to 2·82 (Lee, 1984).

Calcite: Both early and late calcite cements occur in sandstones at sites 445 and 446. The distinction between them depends on the relative timing of formation and occurrence. Early cements occur usually as pore rim and pore-fill cement and are both equant and sparry. Later calcite cement occurs only in deeper samples and fills any remaining pore space after the appearance of zeolite (Fig. 3A, B). Locally, this later calcite cement replaces zeolite cements, as well as volcanic rock fragments and plagioclases. Electron microprobe analyses indicate both types of calcite are iron-poor (Lee, 1984).

Cristobalite: This mineral forms spherulites or lepispheres up to 30 μm in diameter at site 445. It shows flat individual plates with ragged edges and occurs commonly with clinoptolite. Some cristobalite was precipitated on volcanic rock fragments and is overlain by pore-fill clinoptilolite or heulandite cement.

Replacement and dissolution phenomena

Several replacement, overgrowth and dissolution features occur commonly in these sandstones, including albtization of grains observed in two samples at site 299, erionite replacement of volcanic materials, dissolution of grains and cements, and replacement of grains by zeolite, calcite and chert.

Dissolution of grains: Dissolution of feldspar was observed rarely with plagioclase dissolution occurring along cleavage planes. Some volcanic glass fragments show pitted and etched surfaces under SEM indicating etching by pore solution. Volcanic glass matrix and groundmass of volcanic rock fragments are devitrified, and locally volcanic glass matrix is replaced by authigenic calcite. Partial dissolution of volcanic rock fragments was observed in several samples, including total grain dissolution (Fig. 3G) contributing to increased porosity. Augite shows dissolution with a saw-tooth shape.

Zeolite: Some zeolites occur as an alteration product of volcanic glass. Erionite was observed as bean-shaped bundles of needles with each bundle up to 8–10 μm long and about 5 μm thick. The sheaf of erionite needles encloses poikilitically another erionite sheaf, both with pore spaces lined with smectite. EDAX analysis of erionite shows the major elemental peaks to be K, Al, Si and Fe.

Authigenic feldspar: Authigenic albite and K-feldspar are present both as overgrowths on detrital feldspar grains and as rhombic crystals. The authigenic K-

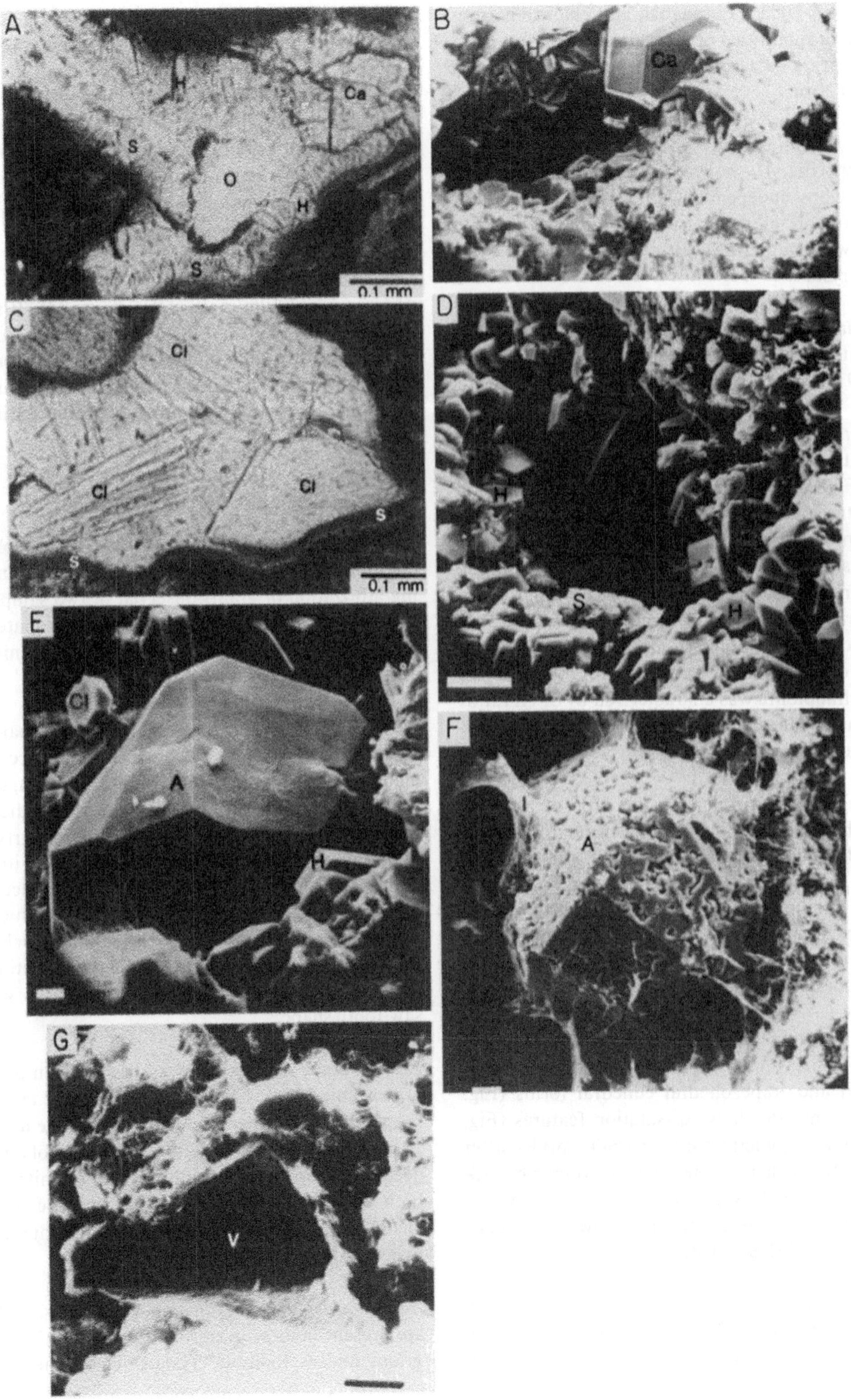
A
H
Ca
S
O
H
S
0.1 mm
B
Ca
C
Cl
Cl
Cl
S
S
0.1 mm
D
S
H
S
H
E
Cl
A
H
F
I
A
G
V

feldspar and albite overgrowths are clear and colourless, show uniform extinction and are in optical continuity with detrital feldspar. Some detrital feldspar underwent albitization, confirmed by electron microprobe analysis (Lee, 1984).

Chert and natrolite: Both occur as replacement minerals of volcanic rock fragments. Natrolite was observed only at site 445, and in thin section, natrolite occurs as a fibrous mineral. In natrolite, sodium is almost the only cation and the ratio of Si/AL ranges from 1·47 to 1·48 (Lee, 1984).

DEPTH DISTRIBUTION OF DIAGENETIC CHANGES

Depth variation of diagenetic changes differ from moderate to strong changes downhole at the seven sites we sampled. These data are shown in Table 2 to 6 (Appendix), and Figs 4 to 9.

Site 299: Both anhydrite pore-fill cement and cristobalite were observed from 80 to 140 m subbottom depth (Fig. 4).

Site 297: The only depth-dependent diagenetic changes observed at this site (Fig. 5) was the occurrence of anhydrite cement from 430 to 450 m subbottom depth.

Site 445: Major depth-dependent diagenetic changes were observed at site 445 (Fig. 6). There, smectite and chlorite rim cements were observed throughout, downhole, although chlorite is scattered in its distribution (Fig. 6). Early calcite rim cement and pore-fill cement also occurs throughout downhole (Fig. 6). The volume of early calcite cement increases from subbottom depths of 710 to 800 m, decreases from 800 to 830 m and is absent from 830 to 880 m (in which fossils are absent), but reappears at the bottom of this hole (Appendix, Table 4).

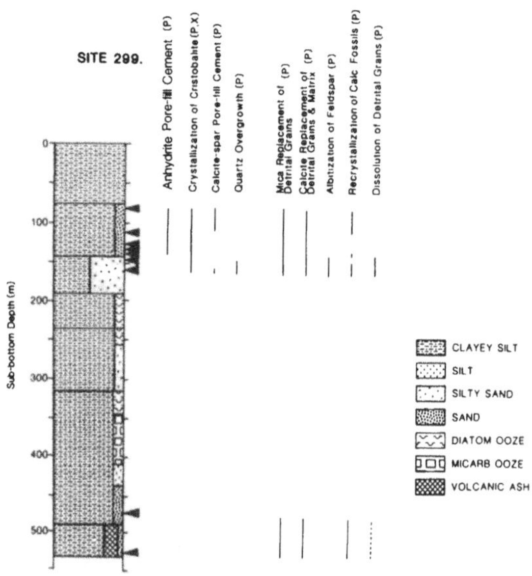

Fig. 4. Depth distribution of diagenetic changes in sandstones, site 299, Sea of Japan. Arrowheads show sample locations. (P-polarizing microscopy; X- X-ray diffraction analysis.)

Clinoptilolite pore-fill cement occurs above 670 m subbottom and in several deeper horizons (Fig. 6). Heulandite pore-fill cements was observed first at a subbottom depth of 660 m and occurs sporadically to a depth of 836 m (Fig. 6). Although some overlap occurs between these two minerals, heulandite pore-

---

Fig. 3. (OPPOSITE) **(A)** Photomicrograph showing sequence of cementation of smectite (S) rim cement, followed by heulandite (h) pore-fill cement, which in turn is followed by late sparry calcite (Ca) cement. 0 = open pore space. Site 445, sample 445–78–5(0–4), subbottom depth of 736·5 m. Plane light. **(B)** SEM photomicrograph showing heulandite (H) cement followed by growth of later sparry calcite (Ca) cement in remaining pore space. Site 445, sample 445–78–5(0–4), subbottom depth of 736·5 m. Bar scale is 10 μm. **(C)** Photomicrograph showing smectite (S) rim cement on volcaniclastic grain surface, and clinoptilolite (Cl) pore-fil cement showing different orientations. Site 445, sample 445–69–2(70–72), subbottom depth of 647·2 m. Plane light. **(D)** SEM photomicrograph showing heulandite (H) and smectite (S) pore-lining cement. Some heulandite crystals show ragged appearance suggesting partial dissolution after forming from volcanic glass precursor. Site 445, sample 445–85–1(18–20), subbottom depth of 797·20 m. Bar scale is 10 μm. **(E)** SEM photomicrograph showing analcite (A) crystal occupying centre of pore as a pore-fill cement. Dissolution features on clinoptilolite (Cl) and heulandite (h) suggests analcite developed at expense of both clinoptilolite and heulandite. Site 445, sample 445–88–2(24–26), subbottom depth of 827·24 m. Bar scale is 10 μm. **(F)** SEM photomicrograph showing analcite (A) pore-fill cement with partial dissolution, and illite (I) crystals. Site 445, sample 445–91–2(74–76), subbottom depth of 665·87 m. Bar scale is 10 μm. **(G)** SEM photomicrograph of secondary moldic porosity (V) caused by dissolution of volcanic rock fragment. Heulandite crystal shown inside moldic pore (lower centre right). Site 286, sample 286–32–2(78–81), subbottom depth of 588·8 m. Bar scale is 10 μm.

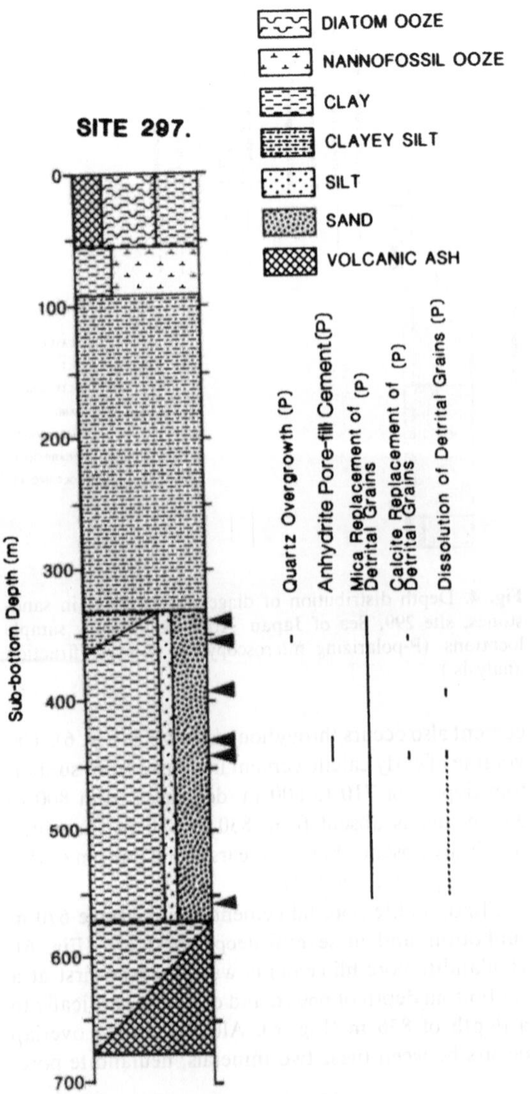

Fig. 5. Depth distribution of diagenetic changes in sandstones, site 297, northern Shikoku Basin. Arrowheads show samples locations. (Abbreviations as in Fig. 4.)

fill cement was precipitated later than clinoptilolite, because it occurs more often in pore centres, whereas clinoptilolite only occurs as pore rims.

At 730 m, analcite and later sparry calcite cements fills pore space completely. Analcite cement is observed as the major pore-filling cement below this depth. Later sparry calcite cement continues to the bottom of the hole.

Recrystallization of detrital grains such as calcareous fossils and limestone fragments occurs from 660 to 800 m (Appendix, Table 4). Grain alterations such

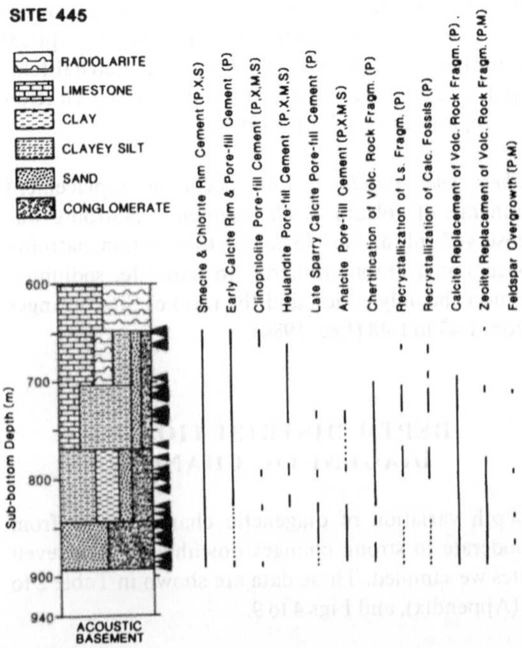

Fig. 6. Depth distribution of diagenetic changes, site 445, southern edge of Daito Ridge. Arrowheads show sample locations. (M = microprobe analysis; S = SEM determination; other abbreviations as in Fig. 4.)

as replacement of detrital grains by chert, zeolite and calcite occur below subbottom depths of 680 m.

Although a minor component, feldspar overgrowths are observed first at a depth of 710 m, and in several horizons below 790 m.

Site 446: A similar suite of downhole diagenetic changes were observed at site 446 (Fig. 7). However, the entire sandstone interval at site 446 occurs at a shallower depth below subbottom compared to site 445 and therefore diagenetic changes at site 446 show a more limited depth distribution. Smectite and chlorite occur in a depth range from 140 to 370 m. At 250 m subbottom, early calcite rim and pore-fill cements are observed first and persist downhole.

Clinoptilolite-heulandite occurs at 137 m subbottom and below 353 m. Analcite occurs from 250 to 270 m, and again below 372 m. Recrystallization and replacement of detrital grains occur mainly in sandstone below 300 m. Minor feldspar overgrowths are observed in above 260 m.

Site 453: Smectite rim cementation was observed throughout this site as the major diagenetic change

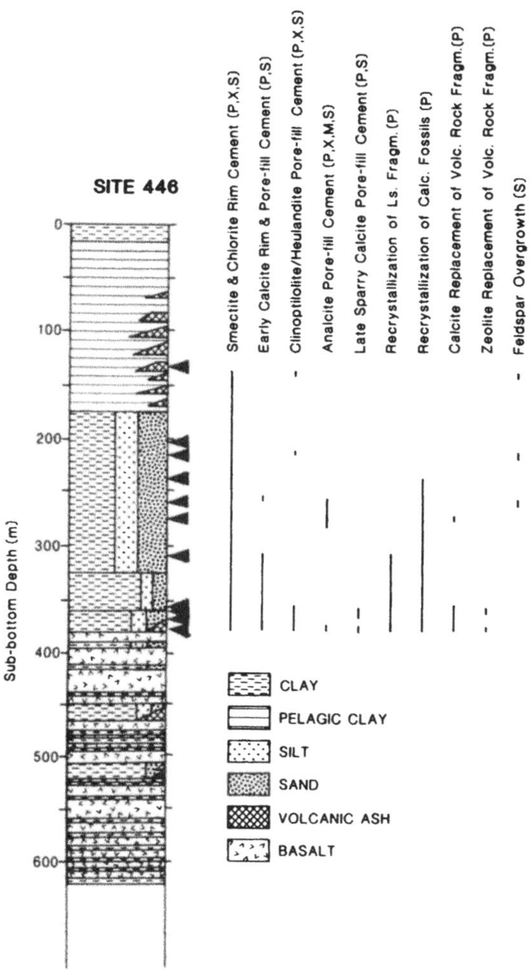

Fig. 7. Depth distribution of diagenetic changes in sandstones, site 446, Daito Basin. Arrowheads show sample locations. (Abbreviations as in Figs 4 and 6.)

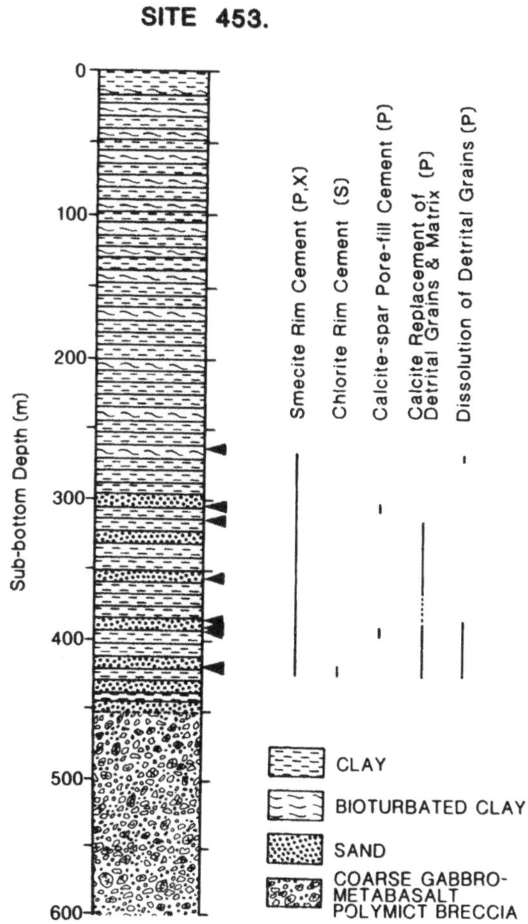

Fig. 8. Depth distribution of diagenetic changes in sandstones, site 453, Mariana Trough. Arrowheads show sample locations. (Abbreviations as in Figs 4 and 6.)

(Fig. 8). Honeycomb-structured chlorite rim cementation is observed at a depth of 420 m. Minor grain dissolution was observed mainly in subbottom depths below 380 m.

Site 286: Smectite rim cement and crystallization of cristobalite are observed throughout this site (Fig. 9). Sepioliote cement is detected in sandstone samples from 210 to 380 m subbottom. At 380 m subbottom, erionite pore-fill cement is observed. Replacement of detrital grains and matrix by calcite and dissolution of detrital grains are observed sparsely throughout. Recrystallization of calcareous components is observed in several samples in the upper half of the site.

MINERAL PARAGENESIS IN BACK-ARC BASIN DIAGENESIS

The depth distribution of diagenetic changes permit reconstruction of a mineral paragenetic sequence (Fig. 10). This paragenesis is based mostly on observations at sites 445, 446 and 286, where most diagenetic phenomena were observed. Most of the observed changes show a depth dependency downhole (Figs 4–9).

The first authigenic mineral phases observed downhole in thin section are smectite and chlorite rim cements, followed by precipitation of early calcite rim and pore-fill cement. This rim cementation was followed by precipitation of cristobalite and clinoptil-

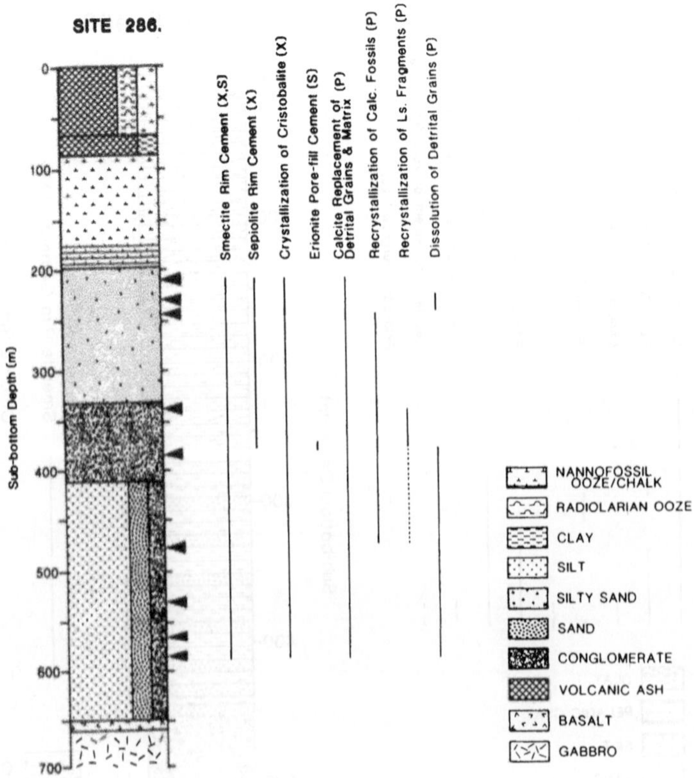

Fig. 9. Depth distribution of diagenetic changes in sandstones, site 286, New Hebrides Basin. Arrowheads show sample locations. (Abbreviations as in Figs 4 and 6.)

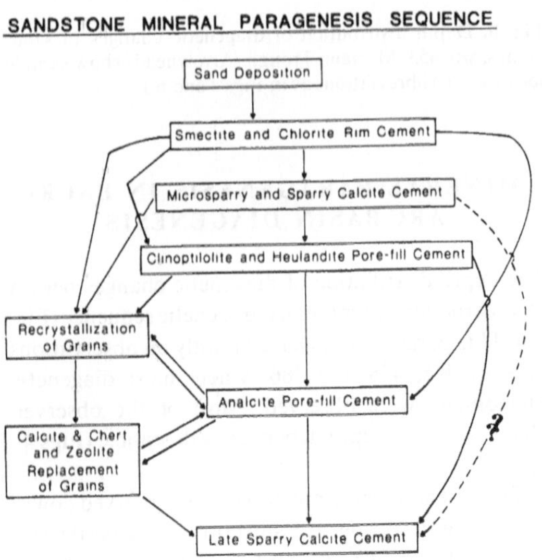

Fig. 10. Sandstone mineral paragenetic sequence in back-arc basins.

olite and heulandite cements filling interstices between smectite, chlorite and early calcite cement. Some zeolites appear to be a direct alteration product of volcanic glass. However, most of the clinoptilolite or heulandite show well-developed crystal outlines and were precipitated from solution on to smectite and chlorite rim cements, or on detrital grain surfaces, reducing porosity significantly. Locally, pores are filled completely by zeolites, in which case the contacts between crystals are sutured (Fig. 3C).

Downhole, clinoptilolite or heulandite are associated with analcite pore-fill cements. This analcite occurs in the pore center, was precipitated late, and shows well-developed euhedral crystals. In thin section, analcite shows some zoning at sites 445 and 446. Later calcite fills any pore space remaining after crystallization of analcite.

Mixtures of volcanic glass and terrigenous sediment comprise the matrices at sites 297 and 299. There diagenetic changes are rare, reaching no further than the earliest stages of the above paragenesis.

FACTORS CONTROLLING SANDSTONE DIAGENESIS IN PACIFIC BACK-ARC BASINS

Because major differences diagenetic changes in sandstones were observed between different sites and different basins, the relevant depositional and geodynamic history of each basin must be considered not only to interpret these diagenetic differences but also to determine the processes controlling diagenesis. A correlation diagram comparing timing of submarine fan and debris flow sandstone deposition, basin rifting and andesitic arc volcanism in each basin (Fig. 11) shows deposition of sandstones to differ with respect to timing of rifting and andesitic volcanic events. Sandstones deposited during active rifting are more likely to be influenced by coeval rates of high heat flow, but other variables such as original composition, sandstone age (Fig. 12) and burial history (Fig. 13) must also be considered.

Original sandstone composition

Original sandstone composition is known to play an important role in diagenesis (Carrigy & Mellon, 1964; Blatt, 1979; Hayes, 1979; Vavra, 1983) because different mineral components react differently to post-depositional alteration. Among rock fragments, basic

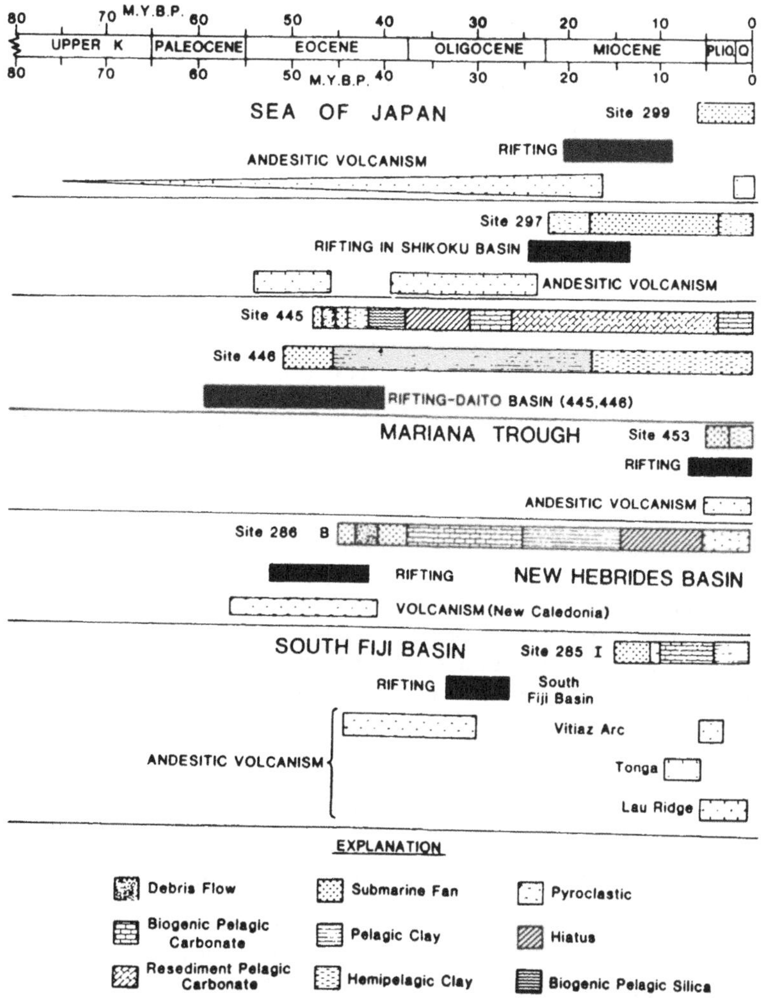

Fig. 11. Correlation diagram showing timing of sandstone deposition (as submarine fans and debris flows), other sediment deposition, rifting and andesitic arc volcanism in back-arc basins from which sandstones were sampled (after Weissel, 1981; Klein, 1985; Klein & Lee, 1984, tables 2, 3 and 4).

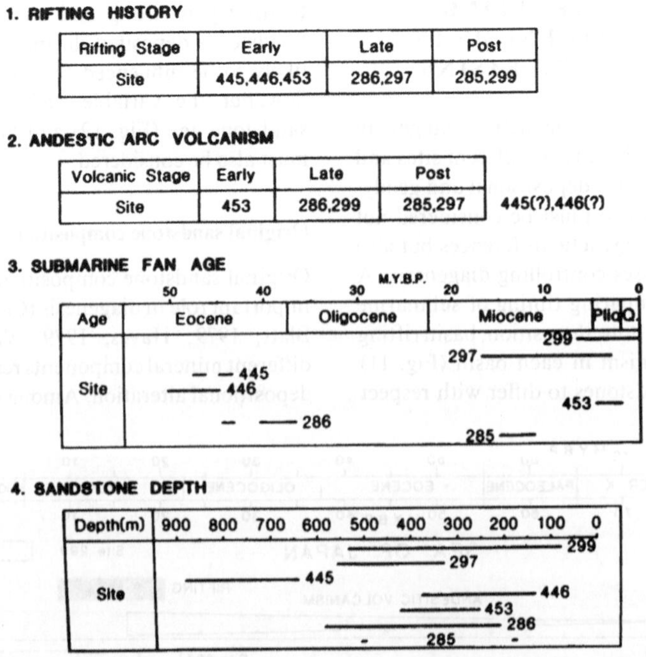

Fig. 12. Comparison of sandstone age, depth, rifting and arc volcanism, in back-arc basins, western Pacific (data from Klein, 1985).

volcanic rock fragments are altered most easily (Whetten & Hawkins, 1970) because of chemical instability, particularly as burial depth and temperature increase. These rock fragments are sources of Na,K,Al and Si. Both feldspar and volcanic rock fragments are replaced preferentially by authigenic minerals. The volume of authigenic components is negligible generally except at site 445 where it reaches 4%.

Sandstones deposited during active basin rifting (sites 453, 286, 445, 446) contain immature volcanic rock fragments and, therefore, more diagenetic alteration is to be expected in these sandstones. Sandstones at sites 297 and 299, which were deposited after rifting ceased, contain mineralogically more stable components (Fig. 2) and would be less altered diagenetically. The greater degree of diagenesis at sites 445 and 446, the moderate degree of diagenesis at site 286 and the minimal diagenetic changes at sites 297 and 299 confirm the interpretation that original composition is a significant variable in this case.

At site 297, sandstones contain significant amounts of quartz and argillaceous sedimentary rocks fragments which are less reactive (Fig. 2) and volcanic matrix is mixed with argillaceous material. Sandstones at site 297 are derived from Shikoku Island which is composed of Mesozoic and Cenozoic sedimentary rocks (Minato, Gorai & Hunahashi, 1965, Karig *et al.*, 1975, amongst others). At site 453, few diagenetic changes were observed because the main component is silicic volcanic glass which reacts more slowly with sea water than mafic glass; associated larger silica content retarded reactions between volcanic glass and pore fluids (Hawkins, 1981).

Hydrolysis of volcanic glass matrix also contributes to diagenesis by changing the composition of pore solutions. Sandstones at sites 445, 446, 286 and 453 contain a considerable volume of volcanic glass matrix (up to 61%), whereas the sandstone matrix at sites 297 and 299 consists of a mixture of such glass and argillaceous sediment. Alteration of matrix was observed to be greater, therefore, at sites 445, 446, 286 and 453.

Rifting history and heat flow

Rifting processes form back-arc basin by seafloor spreading processes similar to those of mid-ocean ridges (Weissel & Watts, 1975; Watts, Weissel & Larson, 1977; Weissel, 1981) and they are also characterized by high heat flow (Karig, 1971; Watanabe, Longseth & Anderson, 1977) which appears to

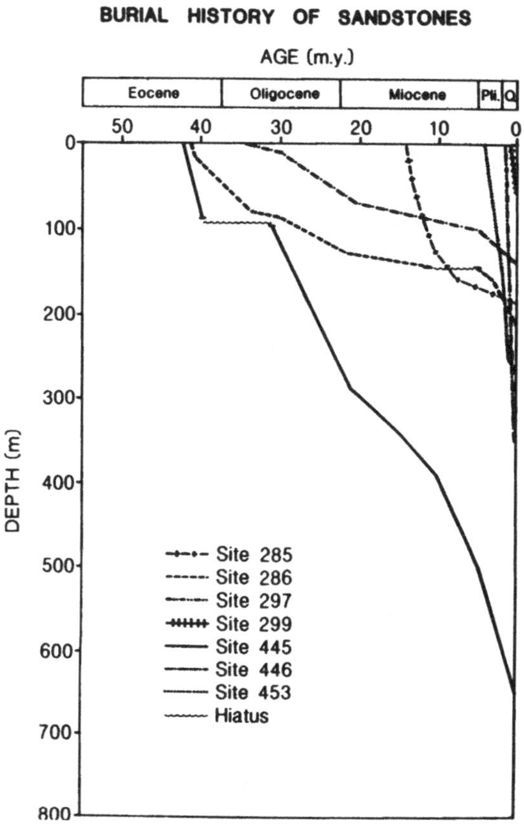

Fig. 13. Burial history curves of sandstones, back-arc basins, western Pacific Ocean.

depend on age (Karig, 1970, 1971). High heat flow is common in active basins, including those with a larger sediment thickness (Watanabe *et al.* 1977). Younger back-arc basins with low heat flow are also characterized by circulation of seawater into oceanic crust; this circulation causes lower heat flow (Watanabe *et al.*, 1977) especially where these basins contain a thin sediment cover.

Site 453 occurs in an actively-spreading basin, sites 297 and 299 in mature basins with high heat flow, and sites 445, 446, 286 and 285 are located within inactive basins with normal heat flow (Toksoz & Bird, 1977). When these basins rifted actively, each site was characterized by high heat flow. If sand was deposited in the basin during rifting, thermal effects from newly-formed basement (Sclater, Jaupart & Galson, 1980) are to be expected in the overlying sediment. Elevated temperatures during rifting increase reaction rates between sediment and pore water leading to thermal diagenesis.

Sandstone deposition occurred during early stages of basin rifting at sites 445, 446 and 453, and coevally with late stages of rifting at sites 286 and 297 (Klein, 1985; Figs 11 and 12). At sites 299 and 285, sandstone was deposited after rifting terminated. Therefore, thermal effects are expected to be greater in sandstones at sites 445, 446 and 453, and intermediate in sandstones deposited at sites 286 and 297. Minimal thermal diagenesis should be observed at sites 299 and 285. The observed sandstone diagenesis at sites 445, 446, 286 and 299 fit this prediction.

Sites 453 and 297 are exceptions to this prediction. Original sandstone composition limited diagenesis because of a larger quartz content at Site 297. Site 453 should show larger thermal influences because it occurs in the presently-active spreading Mariana Trough, but seawater penetrated through porous breccia beneath the sediment column and advected through it (Gieskes, 1981). This hydrothermal circulation in the basement might cause development of a cold-water barrier and a relatively low heat flow (2·4 HFU) at site 453, compared to its age and evolutionary stage (Abbott, Menke & Morin, 1983), thus accounting for the relatively little diagenetic change observed there.

Andesitic arc volcanism

Most sandstones in back-arc basins are derived from andesitic island arcs (Klein & Lee, 1984; Klein, 1985). However, these sandstones occur only when andesitic volcanism is associated with, or follows, basin rifting and if sediment yield from a large voicanic drainage area was caused by a large rate of tectonic uplift and associated increased denudation (Klein, 1984, 1985). The influence of andesitic volcanism on diagenesis is in controlling original sandstone composition, and timing of sandy sediment yield into the basins.

Sandstone age and burial depth

Both sandstone age and depth represent the burial factor in sandstone diagenesis. Sandstone age indicates the total time span during which diagenesis occurred since deposition, whereas burial depth is an independent indicator of the temperature and pressure to which these sediments are subjected. We observed that sandstones at sites 286, 445 and 446 are both older and experienced more diagenetic alteration when compared with sandstones at the other sites we examined. Burial depths differ between sites, however, suggesting that the observed diagenesis (Figs 4–9) may be caused also by different geothermal gradients.

The most common diagenetic products from volcanic sediments are smectite and zeolites and their occurrence is partly a function of age (Denoyer de Segonzac, 1970; Kastner & Stonecipher, 1978; Boles & Wise, 1978). Authigenic smectite in marine sediment forms, commonly, during early diagenesis. In deep-sea sediments, phillipsite is most abundant in Miocene or younger sediments, whereas clinoptilolite is most abundant in Eocene or older sediments (Boles & Wise, 1978; Kastner & Stonecipher, 1978) suggesting reaction rates of 10^6 yr or more. Moreover, the frequency of occurrence of analcite increases with age (Kastner & Stonecipher, 1978). It should be expected and we observed that sandstones containing clinoptilolite, heulandite and analcite at sites 445 and 446, and erionite at site 2896, are older than sandstones at the other sites in which zeolites were absent.

Construction of burial history curves (Fig. 13) showed that although sites 445 and 446 occur in the same basin, their burial histories differ because site 445 occurs on the edge of the Daito Basin, and site 446 occurs in its centre. Burial proceeded at slow to moderate rates at sites 286, 445 and 446 (Fig. 13), whereas burial was rapid at sites 297, 299, 453 and 285. Sandstones at sites 445, 446 and 286 were provided more time, therefore, to react with hydrothermally driven pore waters because compaction was minimal. At the other sites, rapid burial caused excess compaction, sealed off hydrothermal circulation and blanketed thermal influences from the basement. Burial rate therefore controlled directly the potential thermal diagenesis in these basins.

Palaeogeothermal gradient

Crystallization temperatures of zeolites observed at sites 445 and 446 appear to require a thermal regime in excess of normal geothermal gradients. Zeolites in deep-sea sediments cannot be attributed to deep burial metamorphism (Boles, 1981) within the shallow subbottom depths from which our samples came. Alkali clinoptilolite is stable at temperatures as high as 84°–91°C with excess opal-CT (low crystobalite) in Tertiary and Cretaceous silicic tuffs of marine deposits on land buried as deep as 3·5 km. Moreover, both alkali clinoptilolite and mordenite altered to analcite at 84°–91°C (Iijima & Utada, 1971; Iijima, 1975). Clinoptilolite-heulandite transforms to analcite experimentally at 100°C (Boles, 1971).

Both the first appearance of diagenetic analcite at a depth of 730 m, and the large volume of analcite observed at 850 m depth at site 445 requires expected temperatures of crystallization less than 38° and 45°C, assuming temperatures at the sediment–seawater interface were 2°C and a geothermal gradient of 5°C $(100\ m)^{-1}$ (based on isotopic analysis of later calcite cement; Lee, 1986). At site 446, diagenetic analcite occurs at a depth of 260 m, requiring a temperature of crystallization of 15°C for identical reasons. These crystallization temperatures at these shallow depths are considerably less than temperatures reported by Iijima (1975) and Boles (1971). Thus geothermal gradients at sites 445 and 446 must have been greater than estimated during early basin rifting in order to account for such high crystallization temperatures at such shallow depths.

Oxygen isotope measurements (Lee, 1986) of later calcite, clinoptilolite, heulandite, and analcite pore-fill cements also constrain temperatures of crystallization at site 445. These measurements were based on an extrapolation of alkali feldspar-water fractionation as per O'Neill & Taylor (1967) to oxygen isotopic analysis of phillipsite by Savin & Epstein (1970). The isotopic temperatures of crystallization for clinoptilolite and heulandite at site 445 (three samples) was 27°C, ranging from 23° to 62°C, whereas for two analcite samples it was 77° and 100°C. Isotopic temperatures of crystallization of later calcite pore-fill cements ranged from 32° to 44°C in seven samples.

At sites 445 and 446, the observed diagenetic products and their temperatures of crystallization suggest a significant influence of high heat flow from the basement. Although the Daito-Ridge-and Basin-Province is now an inactive back-arc basin (Toksoz & Bird, 1977), a high heat flow existed when the basin was spreading actively (Karig, 1975; Klein & Kobayashi, 1980).

Such high heat flow probably caused extensive hydrothermal circulation and advection of pore waters through these sandstones. Pore water advection through sediments in deep sea environments was recognized mostly from non-linear temperature gradients (Anderson, Hoburt & Langseth, 1979; Von Herzen, Crowe & Green, 1979), from pore water chemistry such as Ca^{++} and $^4He/^3He$ profiles (Abbott *et al.*, 1983; Sayles & Jenkins, 1982), and from observed average heat flow which was much less than predicted by theoretical conductive heat transfer (Anderson & Hobart, 1976; Parsons & Sclater, 1977). Hydrothermal circulation of pore water in sediments is present in areas with sediment less than 100–150 m thick in the Mariana Trough (Abbott *et al.*, 1983) and is known elsewhere in sediments up to 300 m thick or more (Von Herzen *et al.*, 1979; Sayles & Jenkins,

1982). The threshold thickness of the overlying sediment cover acts as a seal to inhibit hydrothermal circulation of sea water to the basement and ranges from 50 m (Karato & Becker, 1983) to 270 m (Becker *et al.*, 1983) depending on the permeability and rate of water movement through sediment (Abbott *et al.*, 1981). These sediment thickness limits were estimated from relatively less permeable sediments overlying relatively young ocean crust.

Sandstones at sites 445 and 446 are of comparable thickness and were relatively more permeable than deep-sea clays, silts or pelagic carbonates. Similar thermally-driven fluid circulation may account for sandstone diagenesis at sites 445 and 446. Elevated temperatures from advection from the basement may account also for anomalous zeolite temperatures of crystallization at the shallow depths discussed earlier. Temperatures of crystallization of later calcite may have been influenced by a lower temperature regime involving downwelling advection of ocean bottom waters.

Similar advection processes were suggested from isotopic and major cation analyses associated with a hiatus at DSDP site 323 (Anderson *et al.*, 1976; Lawrence *et al.*, 1979) although the sediment thickness there was only 60 m. Cooling of young oceanic crust by hydrothermal convection was completed essentially within 1–10 Myr depending on the spreading rate (Fehn & Cathles, 1979). A similar 10 Myr period of basement cooling seems reasonable for sites 445 and 446, particular in view of a known 8 Myr hiatus at site 445 about 400 m above acoustic basement. Therefore, thermal effects due to hydrothermal advection should occur throughout sandstones at both sites 445 and 446 and raise temperatures sufficiently to transform clinoptilolite and heulandite to analcite at the isotopic crystallization temperatures reported earlier, as well as temperatures of crystallization determined by Boles (1971), Iijima (1975) and Iijima & Utada (1971).

Depth-related temperature trends were compared in each back-arc basin (Fig. 14) and a geothermal gradient range was suggested for each site from its geodynamic history. Because heat flow data were available only for site 453 (Mariana Trough), an arbitrary geothermal gradient of 5°C $(100\ m)^{-1}$ (Fig. 4) was calculated using isotopic analysis of later calcite cement as a constraint. Sandstones at sites 285, 286, 297, 299 and 453 would be influenced by a smaller geothermal gradient suggesting less influence from excess temperature. Temperature trends for sites 445 and 446 were derived from the transformation from clinoptilolite/heulandite to analcite (Fig. 14). At site

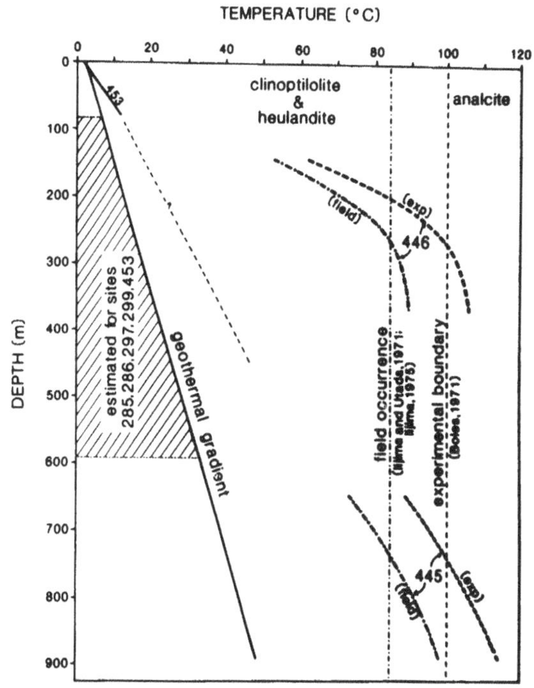

Fig. 14. Diagram showing temperature trend with depth for each DSDP site, back-arc basins, western Pacific. Curves for sites 445 and 446 are based on mineral transformation from clinoptilolite and heulandite to analcite according to both field and experimental occurrence and observations as discussed in text.

446, a steeper temperature trend with depth accounts for mineral reactions which occurred at depths shallower than site 445. The reported intrusion of 26 sills at site 446 (Klein *et al.*, 1980) may account for additional and shallower thermal effects. In summary, Fig. 14 suggests that sites 445 and 446 experience elevated temperatures well above average geothermal gradients during basin evolution, sandstone deposition and burial.

Temperature range

The upper limit of pore-water temperatures at sites 445 and 446 is estimated from the presence of authigenic minerals. Zeolite reactions from clinoptilolite-heulandite to analcite and the formation of albite suggest relatively high temperatures. When analcite is the only alkali zeolite present, it represents the highest temperature zeolite in the Na-K-Si system (Velde, 1977) and causes an association of analcite, albite, and potassium feldspar. Experimental studies show that anaclite-albite paragenesis is stable at 180°C at low pressure ($P_{H_2O} = P_{total}$) (Campbell & Fyfe, 1965; Liou, 1971; Thompson, 1971). An upper limit of 120°C at 1·1 kb lithospheric pressure was suggested for volcanic sediments in Japan (Iijima & Utada, 1971).

Coexisting clay minerals in our samples constrain this temperature range. The upper limit of smectite stability is suggested to range from 80° and 100°C (Steiner, 1968; Muffler & White, 1969; Perry & Hower, 1972). Therefore, volcaniclastic sandstones at sites 445 and 446 were altered within maximum temperatures ranging from 80° to 100°C.

CONCLUSIONS

Sandstone diagenetic reactions in back-arc basins of the western Pacific depend primarily on changes in heat flow during basinal rifting. Volcaniclastic sandstone deposition occurred during both early stages of rifting when heat flow was high (sites 445, 446 and 453), late stages of rifting when heat flow was less (site 286) and well after rifting ceased, when heat flow was normal (sites 285, 297 and 299). The greater degree of diagenesis at sites 445 and 446 is explained primarily from deposition during early rifting, and by heat flow modification, associated thermally-driven fluid circulation, and a slow rate of burial (Figs 13, 14 and 15). The moderate diagenetic influences observed at site 286 are explained in part by a moderate heat flow regime and slow burial rates (Figs 13 and 15). The minimal diagenetic influences observed at sites 453, 299, 297 and 285 appear to be influenced by a combination of low heat flow and rapid burial rates (Figs 13 and 15). The mineralogically more mature composition of sandstones at site 299 and 297 may account also for the minimal diagenesis observed there.

Most of these sandstones are volcanic arenites except at site 297 and some samples at site 446, which are sedarenite. These sandstones show large differences in diagenetic changes between sites and between basins. The greatest amount of diagenesis involving zeolites was observed at sites 445 and 446, intermediate diagenetic products involving erionite and smectite rim cementation were observed at site 286, and minimal diagenesis occurred at sites 453, 297 and 299 (Fig. 15). Downhole changes at sites 445 and 446 consist of pore-space reduction by mechanical compaction, rim cementation by clay minerals and early calcite, and pore-fill cementation by clinoptilolite, heulandite, analcite and later sparry calcite. In addition, grain alteration was observed with depth including recrystallization of detrital grains, such as calcareous fossil and limestone fragments, and detrital grain replacement by calcite, chert and zeolite. These diagenetic changes represent a depth-dependent mineral paragenetic sequence which was established from replacement and spatial observations at sites 445 and 446. Sandstones at sites 299, 297, 286 and 453 represent the earliest stages of this paragenetic sequence.

In terms of the major problems in sandstone diagenesis we identified in back-arc basins (p. 661), our study demonstrates that:

(1) Older and more deeply buried sandstones show a greater degree of diagenesis.
(2) Andesitic volcaniclastic sandstones are altered diagenetically far greater than associated mineralogically more mature sandstones.
(3) High rates of heat flow during the early rifting stage of basin evolution appear to account for most of the diagenetic changes we observed (see also (5) below).
(4) Sites characterized by slower rates of burial tend to show greater diagenetic alteration of sandstones.
(5) Contemporary deposition of sandstones during early stages of rifting in back-arc basins favours their diagenetic alteration because of associated high heat flow (see also (3) above).

ACKNOWLEDGMENTS

This paper is extracted partly from the senior author's Ph.D. dissertation (Lee, 1984). Research support was provided by grants from the University of Illinois Research Board and the National Science Foundation (Grant OCE-81-09447). The Deep Sea Drilling Project is also funded by the National Science Foundation through its grant C-482. The authors thank Drs R. L. Hay, A. T. Hsui, S. Marshak, P. A. Sandberg, C. L. Vavra and S. M. White for their comments on earlier versions of this manuscript.

REFERENCES

ABBOTT, D.H., MENKE, W., HOBART, M.A. & ANDERSON, R.N. (1981) Evidence for excess pore pressures in southwest Indian Ocean sediments. *J.Geophys. Res.* **86**, 1813–1827.

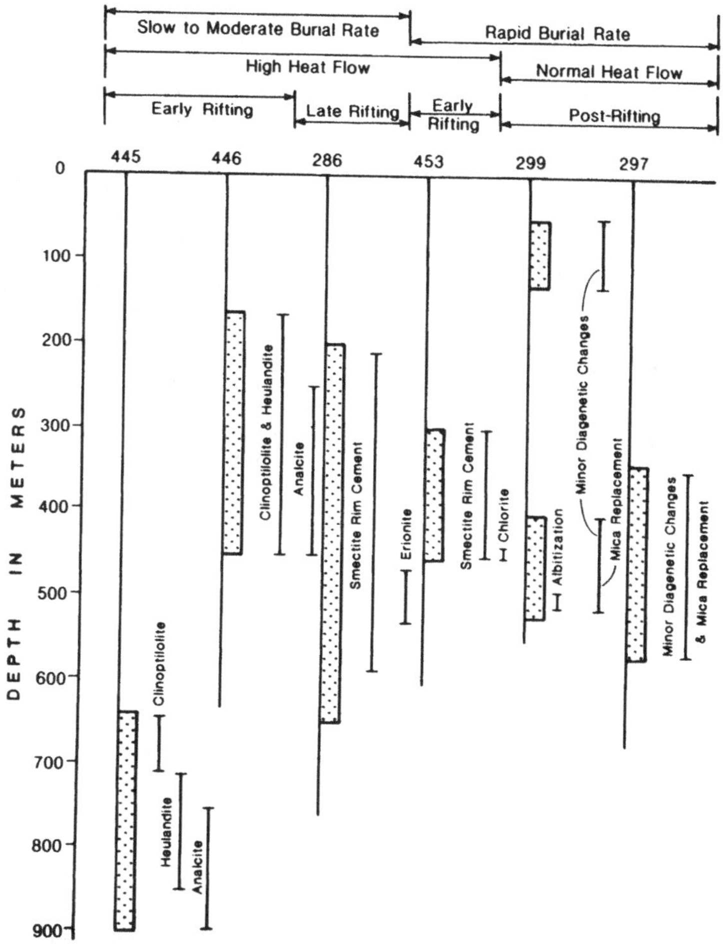

Fig. 15. Comparison of major diagenetic changes in each back-arc basin site containing sandstones with changing history of burial rate, heat flow and rifting history. Vertical line shows total depth of hole. Diagenesis at sites 445 and 446 was by zeolite mineralization in response to a combination of high heat flow during deposition and slow to moderate burial rates. Erionite diagenesis and smectite rim cementation at site 286 was associated with later stages of rifting and was caused by moderate heat flow and moderate burial rates. Rapid burial rates at sites 453, 299 and 297 appear to have prevented development of major diagenetic changes, although albitization of basal part of site 299 indicates initial high heat flow which was blanketed later by rapid burial.

ABBOTT, D.H., MENKE, W. & MORIN, R. (1983) Constraints upon water advection in sediments of the Mariana Trough. *J.Geophys. Res.* **88**, 1075–1093.

ANDERSON, R.N. & HOBART, M.A. (1976) The relation between heat flow, sediment thickness and age in the eastern Pacific. *J.Geophys. Res.* **81**, 2968–2989.

ANDERSON, R.N., HOBART, M.A. & LANGSETH, M.G. (1979) Convective heat transfer in oceanic crust and sediment in the Indian Ocean. *Science*, **204**, 828–832.

ANDERSON, T.F., DONNELLY, T.W., DREVER, J.I., GIESKES, J.M., KASTNER, M., LAWRENCE, J.R. & PERRY, E.A. (1976) Geochemistry and diagenesis of deep-sea sediments from Leg 35 of the Deep Sea Drilling Project. *Nature*, **261**, 473–476.

ANDREWS, J.E., PACKHAM, G.H. *et al.* (1975) *Initial Reports of the Deep Sea Drilling Project*, **30**, Washington, U.S. Government Printing Office, 753 pp.

BECKER, K., LANGSETH, M.G., VON HERZEN, R.P. & ANDERSON, R.N. (1983) Deep crustal geothermal measurements, Hole 504B, Costa Rica Rift. *J. geophys. Res.* **88**, 3447–3457.

BLATT, H. (1979) Diagenetic processes in sandstones. In: *Aspects of Diagenesis* (Ed. by Scholle, P.A. and Schluger, P.R.). *Spec. Publ. Soc. econ. Paleont. Miner., Tulsa*, **26**, 141–158.

BOLES, J.R. (1971) Synthesis of analcime from natural heulandite and clinoptilolite. *Am. Miner.* **56**, 1724–1734.

BOLES, J.R. (1972) Composition, optical properties, cell

dimensions, and thermal stability of some heulandite group zeolites. *Am. Miner.* **57**, 1463–1493.

BOLES, J.R. (1981) Zeolites in deep-sea sediments. In: *Mineralogy and Geology of Natural Zeolites* (Ed. by F. A. Mumpton). *Rev. Mineral.* **4**, 137–163. Mineralogical Society of America.

BOLES, J.R. & WISE, W.S. (1978) Nature and origin of deep-sea clinoptilolite. In: *Natural Zeolites: Occurrence, Properties, Use* (Ed. by L. B. Sand & F. A. Mumpton), pp 235–243. Pergamon Press, London.

CAMPBELL, A.S. & FYFE, W.S. (1965) Analcime-albite equilibria. *Am. J. Sci.* **263**, 807–816.

CARRIGY, M.A. & MELLON, G.B. (1964) Authigenic clay mineral cements in Cretaceous and Tertiary Sandstones of Albert. *J. sedim. Petrol.* **34**, 461–472.

DICKINSON, W.R. (1982) Composition of sandstones in circum-Pacific subduction complexes and fore-arc basins. *Bull. Am. Ass. Petrol. Geol.* **66**, 121–137.

DICKINSON, W.R., HELMOLD, K.P. & STEIN, J.A. (1979) Mesozoic lithic sandstones in central Oregon. *J. sedim. Petrol.* **49**, 501–516.

DICKINSON, W.R. & VALLONI, R. (1980) Plate tectonics and the provenance of sandstone in modern ocean basins. *Geology*, **8**, 82–86.

DENOYER DE SEGONZAC, G. (1970) The transformation of clay minerals during diagenesis and low-grade metamorphism. *Sedimentology*, **15**, 281–346.

FEHN, U. & CATHLES, L.M., (1979) An estimate of seawater circulation through oceanic crust (Abstract). *Eos*, **60**, 382.

FOLK, R.L. (1968) *Petrology of Sedimentary Rocks*. Hemphill, Austin, 170 pp.

GALLOWAY, W. (1974) Deposition and diagenetic alteration of sandstone in northeast Pacific arc-related basins: implications for graywacke genesis. *Bull. geol. Soc. Am.* **85**, 379–390.

GALLOWAY, W. (1979) Diagenetic control of reservoir quality in arc-derived sandstones: implications for petroleum exploration. In: *Aspects of Diagenesis* (Ed. by P. A. Scholle, P. R. Schluger). *Spec. Publ. Soc. econ. Paleont. Miner. Tulsa*, **26**, 251–262.

GIESKES, J.M. (1981) Deep sea drilling interstitial-water studies: implications for chemical alteration of the ocean crust Layer I and II: In: *The Deep Sea Drilling Project, A Decade of Progress* (Ed. by J. E. Warme, R. G. Douglas & E. L. Winterer). *Spec. Publ. Soc. econ. Paleont. Miner. Tulsa*, **32**, 149–167.

HARROLD, P.J. & MOORE, J.C. (1975) Composition of deep-sea sands from marginal basins of the northwestern Pacific. In: *Init. Rep. Deep Sea drill. Proj.* **31**, (Ed. by D. E. Karig, J. C. Ingle Jr, *et al.*), pp. 507–514. U.S. Government Printing Office, Washington.

HAWKINS, D.B. (1981) Kinetics of glass dissolution and zeolite formation under hydrothermal conditions. *Clays & Clay Miner.* **29**, 331–340.

HAYES, J.B. (1979) Sandstone diagenesis—the hole truth. In: *Aspects of Diagenesis* (Ed. by P. A. Scholle, & P. R. Schluger). *Spec. Publ. Soc. econ. Paleont. Miner.* **26**, 127–140.

HUSSONG, D.M., UYEDA, S. *et al.* (1981) *Init. Rep. Deep Sea drill. Proj.*, **60**, U.S. Government Printing Office, Washington, 929 pp.

IIJIMA, A. (1975) Effect of pore water to clinoptilolite-analcime-albite reaction series. *Univ. Tokyo, Faculty Sci. J.* Sec. 11, **19**, 133–147.

IIJIMA, A. & UTADA, M. (1971) Present-day zeolitic diagenesis of the Neogene geosynclinal deposits in the Niigata oil field, Japan. In: *Molecular Sieve Zeolites*—1 (Ed. by R. F. Gould) *Am. Chem. Soc. Adv. Chem. Ser.* **101**, 342–349.

KARATO, S. & BECKER, K. (1983) Porosity and hydraulic properties of sediments from the Galapagos center and their relation to hydrothermal circulation in the oceanic crust. *J. geophys. Res.* **88**, 1009–1017.

KARIG, D.E. (1970) Ridges and basins of the Tonga-Kermadec Island arc system. *J. geophys. Res.* **75**, 238–254.

KARIG, D.E. (1971) Origin and development of marginal basins in the western Pacific. *J. geophys. Res.* **76**, 2542–2561.

KARIG, D.E. (1975) Basin genesis in the Philippine Sea. In: *Init. Rep. Deep Sea drill. Proj.* **31**, (Ed. by D. E. Karig, J. C. Ingle Jr *et al*), pp. 875–880. U.S. Government Printing Office, Washington.

KARIG, D.E., INGLE, J.C. JR *et al.* (1975) *Init. Rep. Deep Sea drill. Proj.* **31**. U.S. Government Printing Office, Washington. 927 pp.

KASTNER, M. & STONECIPHER, S.A. (1978) Zeolites in pelagic sediments of the Atlantic, Pacific and Indian Oceans. In: *Natural Zeolites: Occurrence Properties, Use* (Ed. by L. B. Sand & F. A. Mumpton), pp. 199–220. Pergamon Press, New York.

KLEIN, G.DEV. (1975a) Sedimentary tectonics in southwest Pacific marginal basins based on Leg 30 Deep Sea Drilling Project cores from the South Fiji, Hebrides, and Coral Sea Basins. *Bull. geol. Soc. Am.* **86**, 1012–1018.

KLEIN, G.DEV. (1975b) Depositional facies of Leg 30 Deep Sea Drilling Project cores. In: *Init. Rep. Deep Sea drill. Proj.* **30** (Ed. by J. E. Andrews, G. H. Packham, *et al.*) pp. 423–442. U.S. Government Printing Office, Washington.

KLEIN, G.DEV. (1984) Relative rates of tectonic uplift as determined from episodic turbidite deposition in marine basins. *Geology*, **12**, 48–50.

KLEIN, G.DEV. (1985) The control of depositional depth, tectonic uplift, and volcanism on sedimentation processes in the back-arc basins of the western Pacific Ocean. *J. Geol.* **93**, 1–25.

KLEIN, G.DEV. & KOBAYASHI, K. (1980) Geological summary of the north Philippine Sea, based on Deep Sea Drilling Project Leg 58 results. In: *Init. Rep. Deep Sea drill. Proj.* **58** (Ed. by G. de V. Klein, K. Kobayashi *et al.*), pp. 951–961. U.S. Government Printing Office, Washington.

KLEIN, G.DEV, KOBAYASHI, K. *et al.* (1980a) *Init. Rep. Deep Sea drill. Proj.* **58**, U.S. Government Printing Office, Washington, 1017 pp.

KLEIN, G.DEV. & LEE, Y.I. (1984) A preliminary assessment of geodynamic controls on depositional systems and sandstone diagenesis in back-arc basins, western Pacific Ocean. *Tectonophys.* **102**, 119–152.

KLEIN, G.DEV., MCCONNVILLE, R.L., HARRIS, J.M. & STREFENSON, C.K. (1980b) Petrology and diagenesis of sandstones, Deep Sea Drilling Project Site 445, Daito Ridge. In: *Init. Rep. Deep Sea drill. Proj.* **58**. (Ed. by G. de V. Klein, K. Kobayashi, *et al.*), pp. 609–616. U.S. Government Printing Office, Washington.

LAWRENCE, J.R., DREVER, J.I., ANDERSON, T.F. & BRUCKNER, H.K. (1979) Importance of alteration of volcanic material in the sediments of DSDP Site 323: chemistry,

$^{18}O/^{16}O$ and $^{87}Sr/^{86}Sr$. *Geochim. cosmochim. Acta*, **43**, 573–588.

LEE, Y.I. (1984) *Petrology and diagenesis of medium-grained clastic sediments in the back-arc basins of the western Pacific Ocean.* Unpublished Ph.D. Dissertation, University of Illinois at Urbana-Champaign, 207 pp.

LEE, Y.I. (1986) Isotopic aspects of thermal and burial diagenesis of sandstones at DSDP Site 445, Daito Ridge, northwest Pacific Ocean. *Isotope Geosci.* (in press).

LIOU, J.C. (1971) Analcime equilibria. *Lithos*, **4**, 389–402.

MATSUDA, T., NAKAMURA, K. & SUGIMURA, A. (1967) Late Cenozoic orogeny in Japan. *Tectonophys.* **4**, 349–366.

MUFFLER, L.J.P. & WHITE, D.E. (1969) Active metamorphism in Upper Cenozoic sediments in the Salton Sea geothermal field and the Salton Trough, southeastern California. *Bull. geol. Soc. Am.* **80**, 157–182.

O'NEILL, J.R. & TAYLOR, H.P. (1967) The oxygen isotope and cation exchange chemistry of feldspars. *Am. Miner.* **52**, 1414–1437.

PARSONS, B. & SCLATER, J.G. (1977) An analysis of the variation of ocean floor bathymetry and heat flow with age. *J. geophys. Res.* **82**, 803–827.

PERRY, E.A. & HOWER, J. (1972) Late stage dehydration in deeply buried pelitic sediments. *Bull. Am. Ass. Petrol. Geol.* **56**, 2013–2021.

SAVIN, S.M. & EPSTEIN, S. (1970) The oxygen and hydrogen isotope geochemistry of ocean sediments and shales. *Geochim. cosmochim. Acta*, **34**, 43–63.

SAYLES, F.L. & JENKINS, W.J. (1982) Advection of pore fluids through sediments in the equatorial east Pacific. *Science*, **217**, 245–248.

SCHMIDT, V. & MCDONALD, D.A. (1979) The role of secondary porosity in the course of sandstone diagenesis. In: *Aspects of Diagenesis* (ed. by P. A. Scholle & P. R. Schluger) *Spec. Publ. Soc. econ. Paleont. Miner.* **26**, 175–208.

SCLATER, J.G., JAUPART, C. & GALSON, D. (1980) The heat flow through oceanic and continental crust and the heat loss of the earth. *Rev. Geophys. Space Phys.* **18**, 269–311.

STEINER, A. (1968) Clay minerals in hydrothermally altered rocks at Waikei, New Zealand. *Clays Clay Miner.* **16**, 193–213.

SUGI, N., CHINZEI, K. & UYEDA, S. (1983) Vertical crystal movements of northeast Japan since Middle Miocene. In: *Geodynamics of the Western Pacific–Indonesian Region* (Ed. by T. W. C. Hilde & S. Uyeda) *Am. Geophys Un. Geodyn. Ser.* **11**, 317–330.

THOMPSON, A.B. (1971) Analcite-albite equilibria at low temperature. *Am. J. Sci.* **271**, 79–92.

TOKSOZ, M.N. & BIRD, P. (1977) Formation and evolution of marginal basins and continental plateaus. In: *Island Arcs, Deep Sea Trenches and Back-arc Basins* (Ed. by M. Talwani & W. S. Pittman III). *Am. geophys. Union, Maurice Ewing Ser.* **1**, 379–393.

VALLONI, R. & MAYNARD, J.B. (1981) Detrital mode of Recent deep-sea sands and their relation to tectonic setting: a first approximation. *Sedimentology*, **28**, 75–83.

VAVRA, C.L. (1983) Mineral reactions and controls on zeolite facies alteration in sandstone in central Transantarctic Mountains, Antarctica (Abs). *Bull. Am. Ass. Petrol. Geol.* **67**, 563.

VELDE, B. (1977) *Clays and Clay Minerals in Natural and Synthetic Minerals.* Elsevier, Amsterdam, 218 pp.

VON HERZEN, R.P., CROWE, J. & GREEN, K. (1979) Fluid convection in eastern Pacific ocean crust (abstract). *Eos*, **60**, 382.

WATANABE, T., LANGSETH, M.G. & ANDERSON, R.N. (1977) Heat flow in back-arc basins of the western Pacific. In: *Island Arcs, Deep-Sea Trenches and Back-arc Basins* (Ed. by M. Talwani & W. S. Pittman, III). *Am. geophys., Union, Maurice Ewing Ser.* **1**, 137–161.

WATTS, A.B., WEISSEL, J.K. & LARSON, R.L. (1977) Sea-floor spreading in marginal basins of the western Pacific. *Tectonophys.* **37**, 167–181.

WEISSEL, J.K. (1981) Magnetic lineations in marginal basins of the western Pacific. *Phil. Trans. R. Soc. A*, **300**, 223–247.

WEISSEL, J.K. & WATTS, A.B. (1975) Tectonic complexities in the South Fiji Marginal Basin. *Earth planet. Sci. Lett.* **28**, 121–126.

WHETTEN, J.T. & HAWKINS, J.W. (1970) Diagenetic origin of graywacke matrix minerals. *Sedimentology*, **15**, 347–361.

WHITE, S.M., CHAMLEY, H., CURTIS, D.M., KLEIN, G.DEV. & MIZUNO, A. (1980) Sediment synthesis: Deep Sea Drilling Project Leg 58, Philippine Sea. In: *Init. Rep. Deep Sea drill. Proj.* **58**, (Ed. by G. de V. Klein, K. Kobayashi, *et al.*), pp. 964–1014. U.S. Government Printing Office, Washington.

WILSON, M.D. & PITTMAN, E.D. (1977) Authigenic clays in sandstones: recognition and influence on reservoir and paleoenvironmental analysis. *J. sedim. Petrol.* **47**, 3–31.

(*Manuscript received 2 April 1985; revision received 9 September 1985*)

Table 2. Modal analysis of sandstones,

Sample (interval in cm)	Sub-bottom depth (m)	Detrital							
		Quartz			Plagioclase	Orthoclase	Microcline	Volcanic rock fragment	Volcanic glass fragment
		Sharp ext.	Undulose ext.	Polyxls					
299-9-3 (124-130)	80·27	13·8	8·3		4·4	0·6		1·7	43·1
299-12-4 (48-54)	109·51	7·1	3·2		0·6			tr	60·3
299-14-5 (30-32)	129·81	21·5	12·7		3·5	0·9		tr	39·0
299-15-5 (63-69)	139·66	3·4	3·4	0·5	4·8	1·5		7·7	25·8
299-16-2 (42-44)	144·43	0·5	1·4	1·0	4·8	0·7		14·4	17·3
299-16-4 (10-24)	147·07	5·5	4·5	3·0	8·4	2·0		23·3	18·3
299-17-3 (48-53)	155·51	11·1	7·9	3·1	12·4	1·8	tr	22·1	11·5
299-35-1 (43-47)	475·45	2·9	0·4		3·8	0·4		26·3	37·1
299-38-5 (135-137)	529·86	7·7	7·3	1·4	3·2			50·5	tr

Table 3. Modal analysis of sandstones,

Sample (interval in cm)	Sub-bottom depth (m)	Detrital						
		Quartz			Plagioclase	Orthoclase	Microcline	Volcanic rock fragment
		Sharp ext.	Undulose ext.	Polyxls				
297-15-4(146-148)	339·47	3·2	4·6	1·4	2·3			0·5
297-16-2(49-53)	354·51	13·1	9·7	2·6	5·6	0·8		tr
297-17-2(74-79)	397·76	16·3	27·6	2·0	14·3	1·0	tr	1·0
297-18-2(10-15)	430·13	22·8	16·6	3·3	7·5	0·4	tr	tr
297-19-1(119-125)	447·72	23·7	9·0	11·3	6·0	0·4		1·1
297-22-2(139-141)	554·90	8·7	18·9		tr			12·6

site 299 (percentage based on 230 points)

Grains							Matrix	Cements		Other authigenic	
Sed. rock fragment	Horne-blende	Pyro-xene	Chert	Musco-vite	Glauco-nite	Fossil (foram %)		Clay	Anhy-drite	Chert	Calcite
2·8						8·8 (50)	13·7	0·5	1·2		1·0
		0·6			tr	7·7 (17)	19·7		0·8		tr
0·4		0·4					17·9	3·1	0·5		
tr			1·0		tr	6·7(0)	45·0	tr	0·9	0·5	
tr		0·5	0·5				59·6				
0·5	tr	1·5	4·5		0·5	0·5(0)	25·7			1·9	
0·8	tr	1·8	4·4	0·4	tr	1·3(0)	21·2				
	0·8	0·4	tr			1·3(0)	25·8				0·8
	0·9		tr	0·9		tr	28·2				

site 297 (percentage based on 230 points)

Grains							Matrix	Cements	
Sed. rock fragment	Met. rock fragment	Horn-blende	Pyroxene	Chert	Muscovite	Glauconite		Clay	Anhydrite
56·8					tr	tr	30·9		
22·0			0·4				43·3		
27·5	1·1			5·1		tr	4·1		
12·9				2·2			33·4		1·0
6·8	tr		tr	0·8	tr	tr	35·6	4·9	0·5
36·2		1·6	0·8		2·4		18·9		

Table 4. Modal analysis of sandstones,

Sample (interval in cm)	Sub-bottom depth (m)	Detrital Grains										
		quartz			Plagio-clase	Ortho-clase	Volcanic rock frag.	Lime-stone frag.	Horn-blende	Pyrozone	Chert	Fossil (foram %)
		Sharp ext.	Undulose ext.	Polyxls								
Site 445												
445-69-2(70-72)	647·20			tr	0·5		46·3	2·0	tr			30·6(95)
445-69-2(143-149)	647·93	0·3	0·6		2·8	0·6	67·5	1·1	tr		0·8	1·7(95)
445-71-1(90-91)	664·90		tr	tr	0·9		19·8	15·1	tr	tr	0·3	42·6(94)
445-71-2(37-40)	665·87			0·3	3·6	0·3	33·4	2·6	0·3			2·3(100)
445-74-1(93-97)	693·47	0·3			2·6		60·8	6·1	0·3			15·7(95)
445-75-3(23-25)	705·23	0·9		0·3	11·0	1·2	44·0	27·4	0·6	0·9	0·6	0·3(100)
445-75-6(97-100)	710·47	0·3			1·6	0·3	42·8	22·3	1·6	0·5	0·5	59·0(83)
445-76-1(23-28)	711·73	0·8			11·7		48·4	23·9	0·3	0·6		3·9(57)
445-76-1(122-125)	712·72				6·8		41·9	26·2	0·6	0·3	0·3	4·5(71)
445-77-4(99-104)	726·49			0·3	0·8		28·1	16·3				34·9(99)
445-78-3(58-63)	734·08			0·3	3·7		38·9	22·1		0·3		15·3(98)
445-78-5(0-4)	736·50			tr	tr		68·7	2·4				6·2(100)
445-79-2(143-148)	743·93				2·4		33·0	29·9				13·6(80)
445-82-4(1-5)	771·50			0·3	3·4		59·0	5·4	0·3			4·3(82)
445-83-4(9-13)	779·59			0·3	1·5		62·7	6·9		0·3		4·8(85)
445-84-2(124-126)	790·24		tr	tr	2·8		67·4	6·5				1·7(82)
445-85-1(18-20)	797·18	tr	tr	tr	2·1		71·9	6·8				3·0(81)
445-86-1(92-94)	797·92				3·0		88·1					tr
445-86-4(73-75)	811·73		0·3		2·2		84·8	1·4				0·5(83)
445-88-2(24-26)	827·24			0·3	1·7		78·0				1·2	tr
445-89-1(131-142)	836·31		tr		0·8		77·7					
445-90-2(11-14)	846·11			tr	2·3		78·5		0·6			tr
445-90-2(17-21)	846·17		0·3	tr	2·2		78·2				0·6	tr
445-90-2(22-26)	846·22		0·8		3·6		76·3	tr				tr
445-91-2(64-70)	856·14	0·3	tr		2·7		77·9	2·7				1·1(90)
445-91-4(84-88)	859·34	0·3			0·3		90·3	0·9				0·6(86)
445-92-3(40-48)	866·90		0·3	0·3	10·0		84·0	tr				
445-93-4(127-132)	878·77	tr	0·6	0·3	4·3		86·1	tr				
445-94-2(78-82)	884·78	0·3	0·9	0·6	9·9		80·8					
445-94-4(132-134)	888·32				0·3		13·6	7·3		0·3		69·1(100)
Site 446												
446-20-1(84-88)	137·34	0·9	0·3	0·3	13·4	0·6	39·1	0·9	0·6	3·1	0·3	0·3(62)
446-23-2(24-27)	201·76				4·6		60·8		4·1	6·9		
446-24-1(146-148)	211·97				0·5		64·5	0·5	1·1	4·3		
446-27-1(80-82)	239·81	0·5	0·5		4·9		59·4	1·3	3·6	3·1		0·5(100)
446-29-1(13-15)	258·13	2·6	1·3	0·4	12·2		49·1	0·4	2·6	2·2	tr	tr
446-30-4(97-102)	273·00	0·4	0·4		3·5		49·8		2·2	5·2		0·4(100)
446-34-4(31-35)	307·31				3·0		9·7	17·2	0·3	3·6		48·6(98)
446-38-cc(33-38)	353·00	0·8	0·4		1·2		28·9	21·1	tr	0·4		27·3(82)
446-39-2(84-88)	355·34						4·7	46·8		0·5		32·6(65)
446-40-1(81-86)	363·31				tr		17·9	27·6		tr		26·2(98)
446-41-1(71-75)	372·73	3·8	2·4		0·9		44·6			0·5		9·4(100)

sites 445 and 446 (percentage based on 350 points)

Matrix	Cements						Other Authigenic			Recrystallization		Remark
Volcanic glass	Calcite		Zeolite		Clay	Cristobalite	Zeolite	Calcite	Chert	Fossil	Limestone frag.	
	Early	Late	Clino-beaul.	Analcite								
14·1	1·6		5·6		0·8							
5·1			10·7		7·6		tr					
18·2	1·5		0·9		0·6	tr				2·4	tr	
53·4		tr	tr		4·3	tr						
12·5	2·7		3·8		0·6	tr		tr		4·6		
8·9			tr		tr				0·6		22·8	
9·6	3·9		tr		0·3			0·3		9·1	17·2	authi. Ab
8·6	1·4		1·4			tr			0·6	7·3	15·9	
	18·7		tr		tr		tr	tr	0·6	6·1	17·4	
	17·3				0·3			1·2	0·8	0·8	15·4	
	17·0				0·3			0·9	0·3	4·4	13·3	clinop.
tr		3·9	8·3	tr	8·6			1·1	0·9			↕
8·0	9·2		tr	tr	0·5			1·6	0·8	6·7	9·3	heolan.
15·0	8·5				1·1			1·7	0·3	7·7	33·3	
12·0	7·5		0·3		1·4		0·9	1·2		12·6	33·3	
	18·4		0·3	tr	0·6		0·6	1·4	0·3	7·1	17·8	authi. Ab, K-sp
	1·9	0·3	11·0	tr	1·1	tr	tr	1·1	0·9	12·3	18·9	authi. Ab, K-ap
4·6			tr	tr	1·4	tr		1·1	1·9			
6·1	1·1		1·1		0·3			0·8	1·4			
1·9	0·5	1·8	6·7	1·5	0·3	tr	1·1	2·9	2·0			↓small amount of illite
		7·0	10·0	tr	0·5		0·9	2·0				
		3·7		10·8	0·6		1·2	1·4	1·1			authi. Ab
		2·7		9·0	2·5		0·6	1·5	2·4			authi. Ab, K-sp
		8·0		6·6	0·3		1·1	2·6	0·8			authi. Ab
		4·6		8·7	1·1		0·5	0·3				
2·3			tr	2·3	0·3		2·4	0·6				
		tr		3·2	1·4		0·6	0·3	tr			authi. Ab
				3·6	1·6		0·9	1·4	1·2			
	0·3			1·7	1·3		tr	2·2	0·9			
	5·6		3·4					0·5				
30·4			9·1		0·3	tr						authi. K-sp
23·0					tr				0·4			
21·5			tr		7·5	tr						authi. K-sp
25·9									0·5	tr		
9·1	1·3			14·3	1·3	4·3						authi. K-sp
35·9				0·9	0·9			0·4		tr		
	17·5									8·2	24·0	
11·7	2·8	0·8	5·2		0·6			0·4		11·8	8·0	
	11·1	1·1	2·1				1·1	tr		17·7	7·9	
	8·4		17·3		1·1			1·5		18·5	9·1	
9·9		0·5	3·8	tr	16·4	tr	8·0			18·1		

Table 5. Modal analysis of sandstones,

Sample (interval in cm)	Sub-bottom depth (m)	Detrital: Quartz					
		Sharp ext.	Undulose ext.	Polyxls	Plagioclase	Orthoclase	Volcanic rock fragment
453–29–1(53–58)	266·05	1·8	1·8		1·4		26·2
453–33–2(63–69)	305·66		0·9		1·3	tr	3·9
453–34–2(30–35)	314·83	0·5			0·5		0·5
453–39–3(108–110)	364·59	1·4	0·9		1·9	tr	6·1
453–41–3(48–52)	383·00	2·8	0·5		4·2		6·1
453–42–1(116–122)	390·19	1·4	0·9		1·4		5·9
453–42–2(24–26)	390·75	1·7	0·4		5·1		22·8
453–45–3(92–97)	421·45	tr	2·4		3·9		30·9

Table 6. Modal analysis of sandstones,

Sample (interval in cm)	Sub-bottom depth (m)	Detrital: Quartz						
		Sharp ext.	Undulose ext.	Polyxls	Plagio-clase	Ortho-clase	Volcanic rock fragment	Volglass fragment
286–12–2(50–55)	208·53	0·9	2·8		13·4		21·7	15·2
286–13–2(101–109)	229·05		4·0		6·7		42·9	19·4
286–14–2(92–96)	246·94	0·4	1·6		6·1		22·2	8·9
286–19–1(105–110)	340·57		0·5		2·8		23·1	61·3
286–21–2(87–89)	379·88		tr		8·1		45·5	10·6
286–26–3(30–33)	475·82			0·5	22·6	0·5	32·1	21·3
286–29–3(124·127)	533·75	0·4	0·4		3·6		8·7	80·9
286–31–1(98–102)	568·50	tr			3·7		1·9	55·6
286–32–2(78–81)	588·80	1·3	2·1		13·8	0·4	10·4	53·8

site 453 (percentage based on 230 points)

Grains				Matrix	Cements		Other authigenic	
Volcanic glass fragment	Sed. rock fragment	Pyroxene	Fossil	Volcanic glass	Clay	Calcite	Chert	Calcite
6·3		0·5		61·9				
75·4	0·9	0·4	tr	17·2	tr	tr		tr
79·2	7·6	tr	tr	11·7			tr	tr
44·2	2·3	0·5		42·8				tr
61·0	0·5	1·4		23·5				
57·7	1·4	0·4		30·9		tr		tr
17·7		1·7		48·5	2·1			tr
26·6		1·5		34·6	tr			tr

site 286 (percentage based on 230 points)

grains							Matrix	Cement	Other authigenic	
Sed. rock fragment	Horn-blende	Pyroxene	Olivine	Hematite	Glauco-nite	Fossil (foram %)	Vol. glass	Cristobalite	Chert	Calcite
		2·7		tr		12·0(0)	42·4	1·4	0·9	
14·3	tr	1·2				2·8(tr)	7·9	0·8		tr
		tr		0·4	tr	0·8(tr)	56·1	3·7		tr
0·5		tr		0·9		0·9(50)	7·6	2·4		tr
0·4		2·0	tr	0·4		0·8(50)	31·4	0·8	tr	
0·5	tr	1·8	tr	0·5		tr	19·9	0·5		
		0·8				1·2(0)	3·9	tr		
							38·3	0·5		
		3·3					14·2	0·8		

Reprinted from Sedimentology (1989) ***36****, 635–650*

Diagenetic K-feldspar pseudomorphs in the Triassic Buntsandstein sandstones of the Iberian Range, Spain

SADOON MORAD

Department of Mineralogy and Petrology, Institute of Geology, Uppsala University, Box 555, S-751 22 Uppsala, Sweden

RAFAELA MARFIL *and* JOSÉ ANDRÉS DE LA PEÑA

Departamento de Petrologia y Geoqúimica, Universidad Complutense, 28040 Madrid, Spain

ABSTRACT

Early diagenetic K-feldspar in the Triassic Buntsandstein of the Iberian Range (Spain) occurs as pseudomorphs after detrital K-feldspar (Or<93) and plagioclase (Ab<96). These pseudomorphs are chemically pure (Or>99), untwinned, commonly heavily clouded by vacuoles and tiny inclusions, dark-luminescing and are composed of numerous fine euhedral crystals of K-feldspar. The latter property suggests that the pseudomorphs form via dissolution of detrital K-feldspar and plagioclase and precipitation of authigenic K-feldspar. X-ray diffraction analysis shows that the authigenic K-feldspar is intermediate microcline.

INTRODUCTION

Several papers have documented the albitization of detrital plagioclase (e.g. Land & Milliken, 1981; Boles, 1982; Gold, 1987) and K-feldspar (e.g. Middleton, 1972; Walker, 1984; Morad, 1986) in feldspathic sandstones from different parts of the world. These discoveries have further led some authors (e.g. McBride, 1985) to suggest that albitization of detrital feldspars should be expected in all deeply buried sandstones. However, sedimentary petrographers are seemingly still unaware that diagenetic K-feldspar pseudomorphs can form, by a process similar to albitization, via replacement of detrital K-feldspar and plagioclase by Or_{99-100}. Recognition of albite and K-feldspar pseudomorphs is of essential importance because they are commonly accompanied by significant modifications of mineralogy, porosity and permeability in sandstones.

The aim of this paper is to present a comprehensive description of textures (as revealed by light, scanning electron (SEM) and cathodoluminescence (CL) microscopy), chemistry (by means of an electron microprobe), and X-ray diffraction data on the authigenic K-feldspar.

GEOLOGICAL SETTING

The Buntsandstein of the Iberian Range (Fig. 1) represents the lower part of the Triassic Germanic sequence. Deposition of the Buntsandstein took place within a graben structure orientated NW–SE (Alvaro, Capote & Vegas, 1979). The basal alluvial deposit (the lower unit) is conglomeratic and grades upward into deltaic–estuarine (the middle unit) sandstones (Capote, Gomez, Rosell, Diaz, de la Torre, Sopena, Gabaldon, Ruiz & Yebenes, 1982; Arribas, 1984; Fig. 2). These sandstones in turn grade upward into tidal deposits composed of siltstones, shales and thin bedded micaceous sandstones (the upper unit; Fig. 2). The thickness of the Buntsandstein increases to the SE away from the Hesperian Massif (Fig. 1) and reaches a maximum thickness of about 1000 m. The most probable source rocks for the sandstones studied were the K-feldspar-rich gneisses of the Olla de Sapo

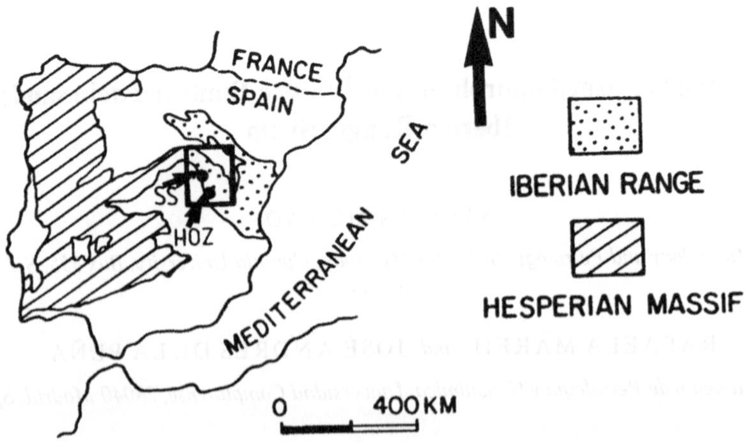

Fig. 1. Map of the Iberian Peninsula showing location of study area. SS and HOZ represents the locations of the borehole (at Sigüenza) and outcrop (at Barranco de la Hoz) samples, respectively.

Formation of the Hesparian Massif (Arribas, Marfil & de la Pena, 1985).

SAMPLES AND METHODS

Fifty samples collected from subsurface cores (SS samples) and from a representative outcrop section (Hoz samples; Fig. 1) were used in the present investigation. The samples cover the alluvial, deltaic–estuarine and tidal facies of the Buntsandstein (Fig. 2). Thin sections were prepared for all samples and examined by standard optical microscopy. Modal compositions were obtained by counting 500 points in each thin section. Electron microprobe analyses (EMP) were performed on carbon-coated thin sections using a Cameca Camebax BX50 equipped with three spectrometers and a backscatter electron detector (BSE). The correction program used was a Cameca PAP. Amounts of Or–Ab–An solid solutions were determined by the Cameca Geo-Multilabel program. Operating conditions during analyses were an accelerating voltage of 15 kV, a beam diameter of about 1 μm, a beam current of about 5–15 nA and a count time of 100 s. The standards used were natural orthoclase (K, Al, Si), albite (Na) and andradite (Ca). Element distribution maps were also obtained by the microprobe. A maximum counting time of about 7 min was used for oxide concentrations of less than 1·5 wt%. Carbon-coated thin sections were also examined by a Jeol scanning electron microscope (SEM) equipped with a BSE, a cathodoluminescence (CL) detector (Coy-Yll, 1969; Grant & White, 1978) and a Link Systems energy dispersive X-ray analyser (EDS). Thin sections were also examined by a commercial Nuclide Luminoscope and a luminescence microscope equipped with a hot-cathode (Zinkernagel, 1978; Matter & Ramseyer, 1985). Representative sample chips were coated with a thin layer of gold and examined by the SEM.

Hand-picked K-feldspar grains from four samples rich in diagenetic K-feldspar with small amounts of unaltered detrital K-feldspar were powdered, mixed with internal standard (silicon metal) and examined at room temperature by means of a Phillips X-ray diffractometer using CuKα radiation and a Ni-filter. The powders were scanned from $2\theta = 12°$ to $60°$ at a rate of $0{\cdot}3°\ 2\theta\ \text{min}^{-1}$. The unit cell parameters were calculated using a computer program (Ersson, N.-O., unpublished) and based on at least 14 sharp peaks. However, the problem of contamination by detrital feldspar is difficult to avoid and no estimate of the effect of such contamination has been attempted on the results presented here.

PETROGRAPHY OF THE BUNTSANDSTEIN SANDSTONES

Sandstones of the Buntsandstein are generally moderately to well-sorted arkoses to subarkoses. The grains are mostly subrounded and grain size ranges from 0·125 to 0·250 mm in the N and from 0·250 to 0·500 mm in the S parts of the Iberian Range. Thin-

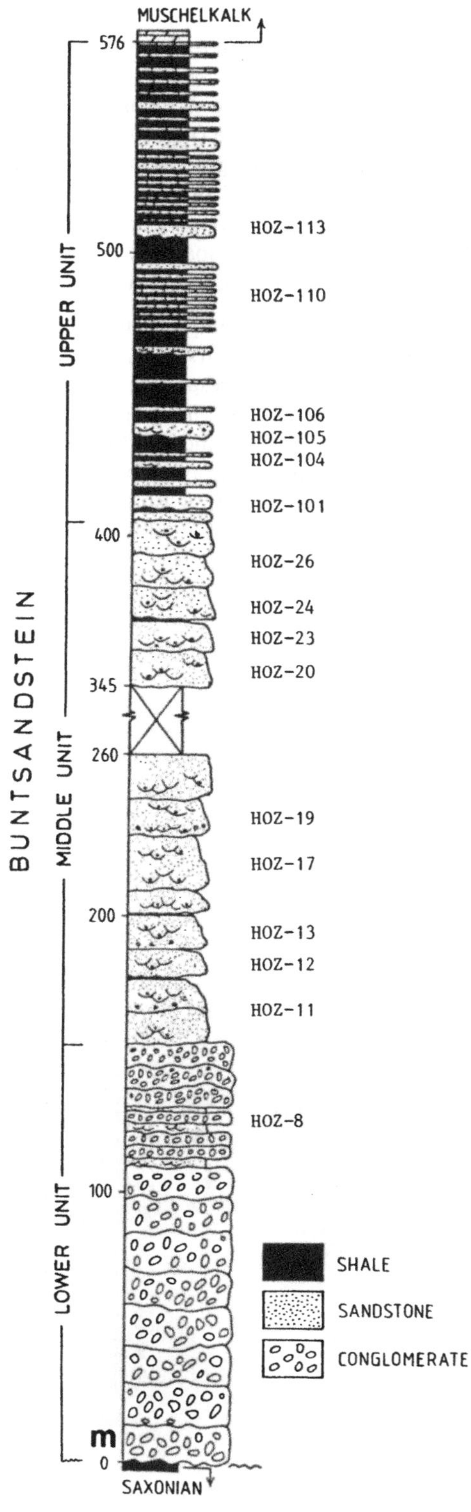

Fig. 2. Stratigraphic section of the Buntsandstein at Barranco de la Hoz.

section porosity ranges between about 5 and 15% by volume. Textural and mineralogical maturities increase from NW to SE within the basin (Arribas *et al.*, 1985). The framework grains are predominantly quartz (27–84% by volume) and K-feldspar (orthoclase, microcline and K-feldspar pseudomorphs; 11–30%) with smaller amounts of rock fragments (mostly slate, schist and gneisses), micas and chlorite (trace–15%). Minor amounts (< 4%) of plagioclase and heavy minerals (tourmaline, Ti-oxides, magnetite, haematite, pyrite, apatite, garnet and zircon) are also present. The framework minerals in the conglomeratic sandstones are composed of polycrystalline and monocrystalline quartz pebbles (up to about 70 mm in size) and sand grains and small amounts of heavy minerals and micas.

Major cementing minerals in sandstones of the Buntsandstein are carbonates (calcite, dolomite, ferroan dolomite, ankerite; 0–40%), quartz (0–4%), clay minerals (illite, mixed-layer illite–smectite, kaolinite and chlorite; trace–6%), iron oxides (0–3%), sulphates (anhydrite and barite; 0–3%) and K-feldspar (as overgrowths and, less commonly, intergranular crystals; 0–4%).

The sequence of major burial diagenetic events is outlined in Fig. 3. The detrital K-feldspars and micas in the lower parts of the stratigraphic section have been subjected to dissolution prior to and during the authigenesis of K-feldspar.

MINERALOGY, TEXTURE AND CHEMISTRY OF AUTHIGENIC K-FELDSPAR

Authigenic K-feldspar pseudomorphs occur in the transitional and marine facies sandstones of the Buntsandstein but are almost completely absent in sandstones which are cemented by early concretionary sulphate and calcite. These pseudomorphs were formed by replacement of twinned microcline, commonly untwinned orthoclase, and in a few cases, plagioclase.

Under the petrographic microscope, complete authigenic K-feldspar pseudomorphs appear untwinned, clouded by vacuoles and tiny inclusions (Fig. 4) and commonly display patchy or blocky to tabular extinction patterns. These optical characteristics are similar

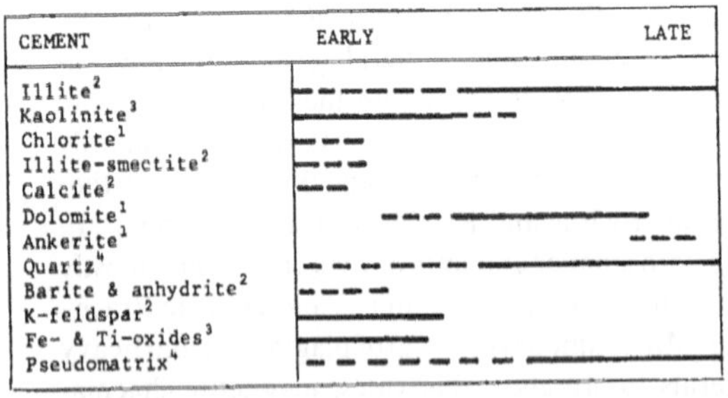

Fig. 3. Sequence of mineral diagenesis in the Buntsandstein. (1. in marine facies; 2, in transitional and marine facies; 3, in continental and transitional facies; 4, in all facies).

to those found in albitized feldspars (cf. Walker, 1984; Morad, 1986; Gold, 1987; Saigal *et al.*, 1988). K-feldspar pseudomorphs are best identified by dark cathodoluminescence (e.g. Kastner, 1971; Kastner & Siever, 1979; Saigal *et al.*, 1988). A dark greyish blue luminescence of the authigenic K-feldspar was detected by the hot-CL. Under the Nuclide Luminoscope the authigenic K-feldspar is completely dark. Detrital K-feldspars of the Buntsandstein display a bright blue luminescence colour under the hot-CL and the Nuclide Luminoscope.

Replacement by authigenic K-feldspar, which varies from partial to complete (Fig. 5), was guided by traces of cleavage planes (Fig. 6), by microfractures (Fig. 7) or by twin planes of the detrital feldspar (Figs 8a & b). Authigenic K-feldspar patches in partially pseudomorphosed feldspars are crystal clear (Fig. 8B). Figure 6 displays three interesting features:

1. The difference in chemical composition between the diagenetic and the K-feldspars ($Or_{99\cdot7}Ab_{0\cdot3}An_{0\cdot0}$ and $Or_{80\cdot9}Ab_{17\cdot8}An_{1\cdot3}$, respectively) can be detected by means of BSE imaging and X-ray elemental distribution of K and Na (Figs 6b & c). Authigenic K-feldspar is always almost pure end member (Or > 99 mol%; cf. Kastner & Siever, 1979) whereas the unaltered K-feldspars contain Na and Ca impurities (Table 1 and Fig. 9). Analyses of fresh detrital K-feldspar revealed compositions similar to relicts found in the pseudomorphs (Fig. 9). Thus the replacement by

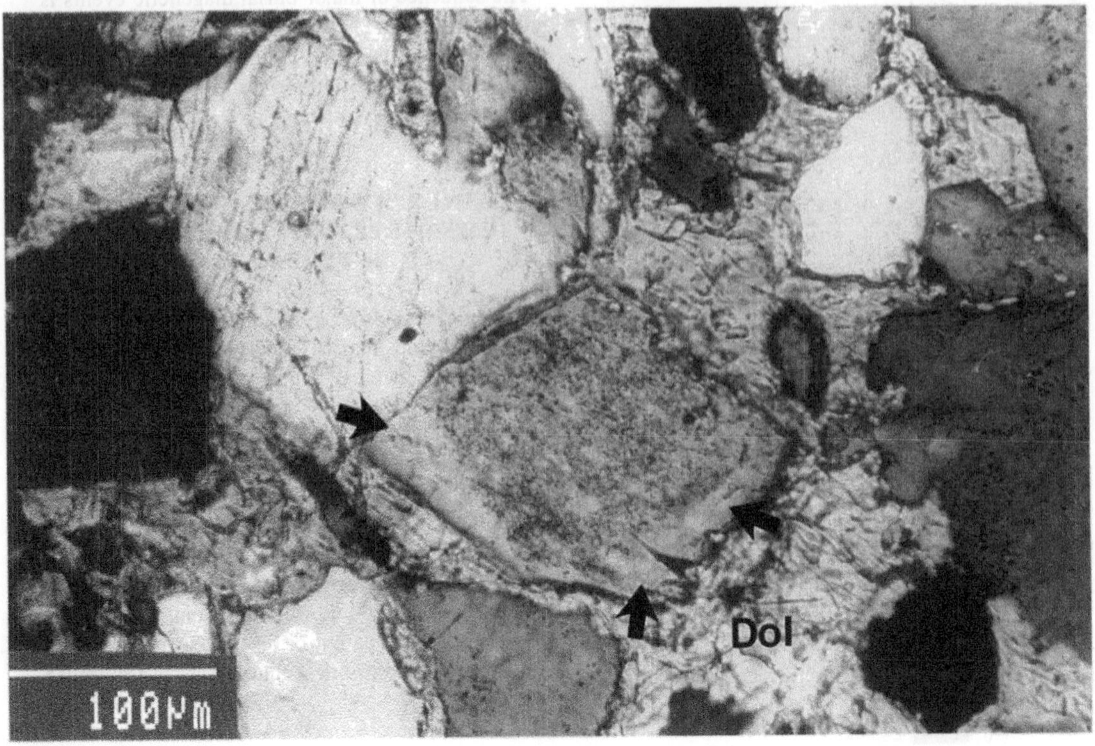

Fig. 4. A K-feldspar pseudomorph which is untwinned, clouded by tiny inclusions and vacuoles and has a clear K-feldspar overgrowth (arrows); cross-nicols.

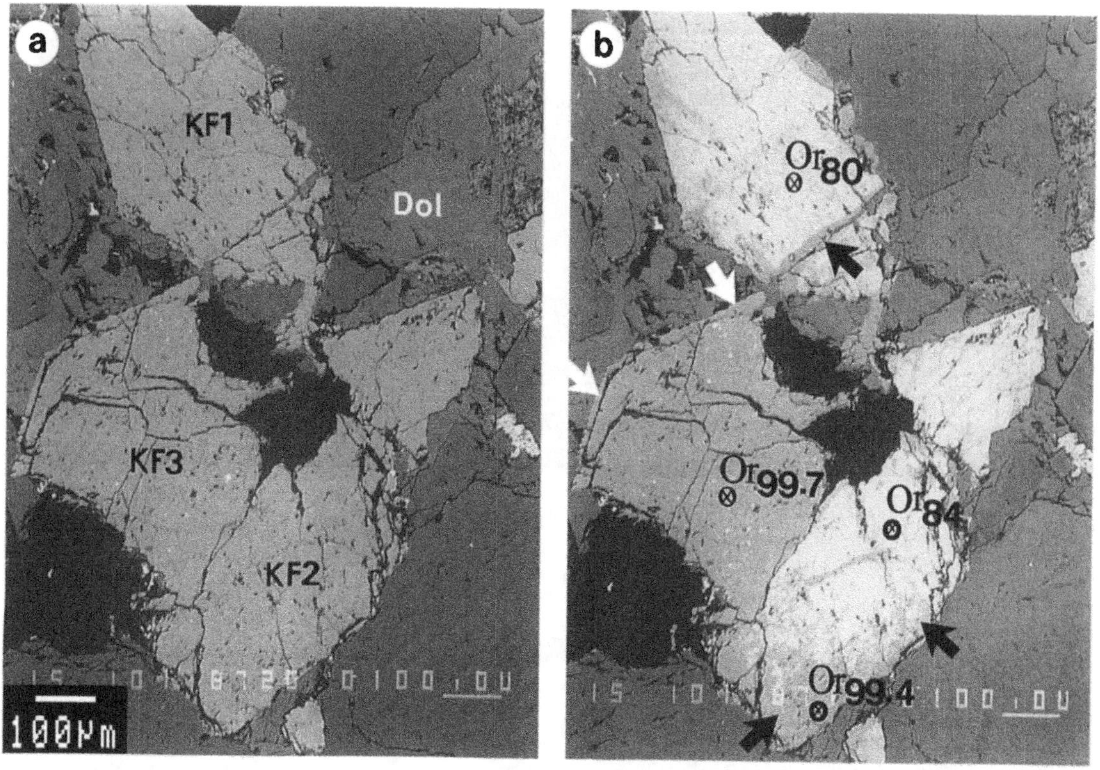

Fig. 5. (a) A BSE image of three K-feldspar grains; KF1, KF2 and KF3 embedded in dolomite cement (Dol); (b) A BSE/CL image of Fig. 5a showing that grain KF1 is just slightly replaced by authigenic K-feldspar compared to grain KF2, whereas grain KF3 is completely replaced by K-feldspar and has authigenic K-feldspar overgrowth (white arrows). The detrital K-feldspars have bright luminescence, whereas the diagenetic K-feldspars are dark luminescing. Black arrows show parts of grains KF1 and KF2 where replacement by K-feldspar occurred. Composition (mol% Or) is based on microprobe analyses.

authigenic K-feldspar was not controlled by detrital chemistry.

2. Less than 1 μm barite inclusions occur in planes parallel to the cleavage planes in the replaced K-feldspar (Fig. 6d). This feature is most common in the transitional facies sandstones (Fig. 2). Less than 1 μm Fe-oxide inclusions concentrated along microfractures in K-feldspar pseudomorphs were also observed.
3. Presence of numerous tiny pits which could be mistaken for inclusions when viewed under petrographic microscope (cf. Velbel, 1983). Relict detrital feldspars as well as K-feldspar overgrowths are almost completely devoid of inclusions and pits (Fig. 4).

A few of the detrital plagioclase grains are also replaced pseudomorphically by K-feldspar. Due to the small amounts (< 2%) of detrital plagioclase in the Buntsandstein this process is volumetrically not important. This replacement has resulted in textural, mineralogical and chemical features similar to those described above. The average composition of detrital plagioclase is $Ab_{94.1}An_{4.4}Or_{1.5}$.

Under the SEM, K-feldspar pseudomorphs contain abundant dissolution pits and discrete pseudorhombic and prismatic crystals (1–60 μm long) which are aligned parallel to one another, and have smooth surfaces (Fig. 10). Unaltered detrital K-feldspar and plagioclase show etched surfaces due to diagenetic dissolution. In many cases, the pseudomorphs are composed of many elongate, rod-like prismatic crystals (Fig. 11a) similar to those described by Burley (1984) in the Sherwood Sandstone Group (Triassic), of the United Kingdom. The surface texture of these latter pseudomorphs could explain their pitted nature in thin section (see Fig. 6). Dissolution voids, which vary greatly in amount, are commonly coated and

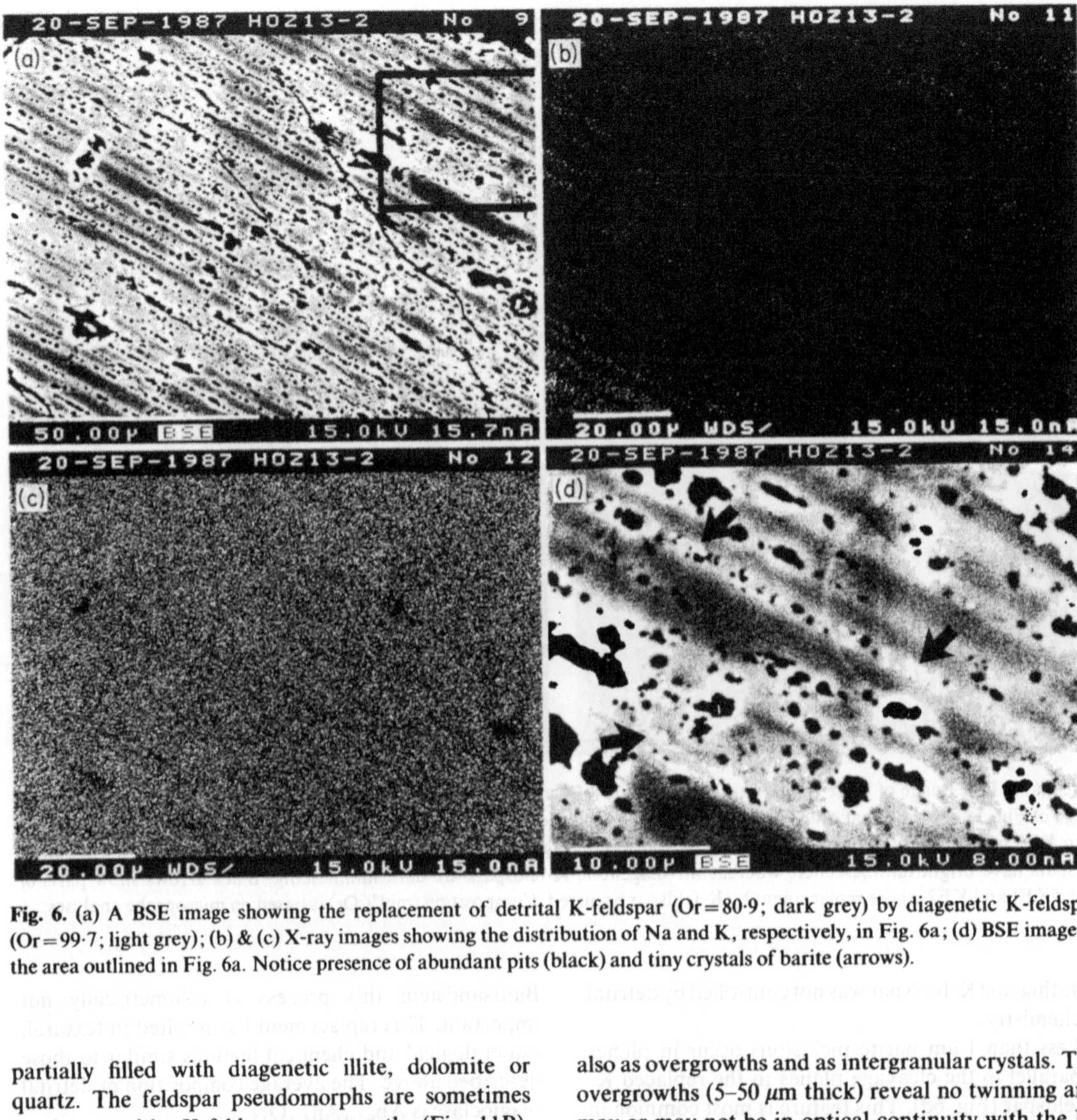

Fig. 6. (a) A BSE image showing the replacement of detrital K-feldspar (Or = 80·9; dark grey) by diagenetic K-feldspar (Or = 99·7; light grey); (b) & (c) X-ray images showing the distribution of Na and K, respectively, in Fig. 6a; (d) BSE image of the area outlined in Fig. 6a. Notice presence of abundant pits (black) and tiny crystals of barite (arrows).

partially filled with diagenetic illite, dolomite or quartz. The feldspar pseudomorphs are sometimes partly coated by K-feldspar overgrowths (Fig. 11B). In several instances, the K-feldspar pseudomorphs and the overgrowths appear as numerous flattened pseudorhombic crystals that overlap each other (Fig. 12). Similar crystals have been described by Waugh (1978) as combinations of the basal pinacoid (001) and prism (110) with a very thinly developed (010). Sometimes, authigenic K-feldspar has nucleated and grown as numerous parallel aligned crystals on severely etched detrital K-feldspar. Under the SEM, the authigenic (pure $KAlSi_3O_8$) and detrital (contain detectable amounts of Na) K-feldspars were differentiated based on numerous EDS analyses.

As pointed out above, authigenic K-feldspar occurs also as overgrowths and as intergranular crystals. The overgrowths (5–50 μm thick) reveal no twinning and may or may not be in optical continuity with the K-feldspar core. The SEM shows that the overgrowths rest on mixed-layer illite–smectite which coats the core or directly on the core. In many cases, K-feldspar overgrowths appear as pseudorhombic crystals (about 5–30 μm in size) which are randomly arranged or are almost parallel. Late diagenetic quartz envelops the K-feldspar overgrowth. The intergranular authigenic K-feldspar occurs as aggregates that consist of several euhedral crystals which vary considerably in size and shape.

X-ray diffraction data on the authigenic K-feldspar (Table 2) samples indicate that they are intermediate between microcline and orthoclase and have *b* and *c*

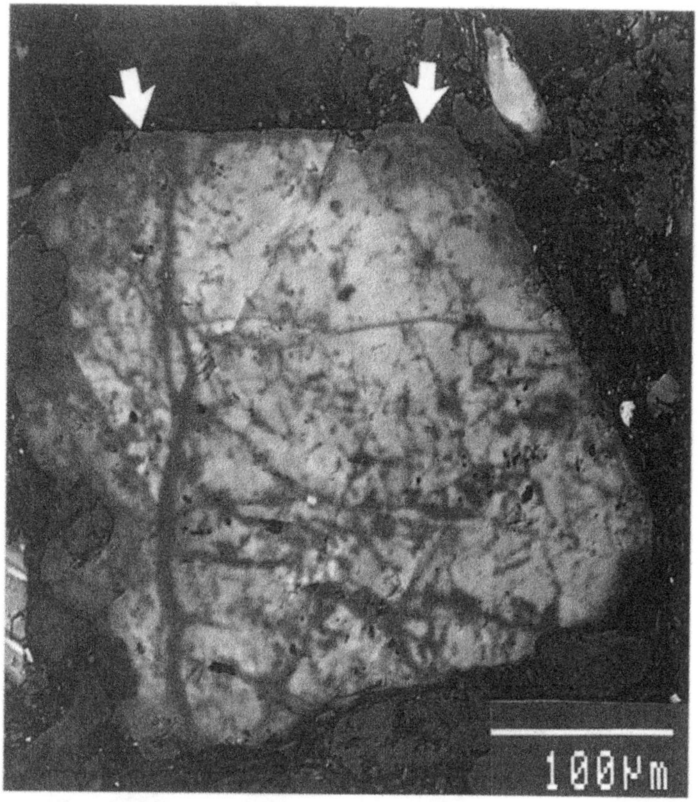

Fig. 7. A BSE/CL micrograph of a K-feldspar replaced along microfractures by dark luminescing diagenetic K-feldspar (dark). A thin layer of K-feldspar overgrowth also occurs (arrows).

cell edges that deviate from those expected for fully ordered microcline (Fig. 13). Two authigenic microclines analysed by Kastner (1971) are, on the basis of the *b* and *c* cell edges, fully ordered. The *b* and *c* dimensions depend on composition and on Al–Si order (Kroll & Ribbe, 1983). Incorporation of cations larger than Al results in an expansion of *b* and *c* cell edges (Martin & Falster, 1986; Stevenson & Martin, 1986). However, as the authigenic K-feldspars are pure the cell edge dimensions depend only on the Si/Al distribution (cf. Kastner & Siever, 1979). Kastner and Siever (1979) pointed out that authigenic K-feldspars are either highly ordered microcline, partially disordered adularia or orthoclase, or highly disordered sanidine. Ali & Turner (1982) found the pure K-feldspar overgrowths in Triassic sandstones from the United Kingdom to be characterized by −2V values between 20 and 30°, suggesting a potassian intermediate sanidine (cf. Burley, 1984).

DISCUSSION

Evidence of diagenetic origin of the K-feldspar pseudomorphs

Several lines of evidence suggest that the K-feldspar pseudomorphs of the Buntsandstein have been formed diagenetically. These are:

(1) Authigenesis of K-feldspar was restricted to the transitional and marine facies of the Buntsandstein (Fig. 2). This feature suggests that authigenesis of K-feldspar was favoured under a specific set of geochemical conditions (primarily a high $\alpha_{K^+}/\alpha_{H^+}$ ratio and a high $\alpha_{H_4SiO_4}$).

(2) The less common occurrence of diagenetic replacement by K-feldspars in early diagenetic concretionary cemented parts of the sandstones. Under such conditions the detrital feldspars were isolated from the geochemical system which favoured formation of K-feldspar.

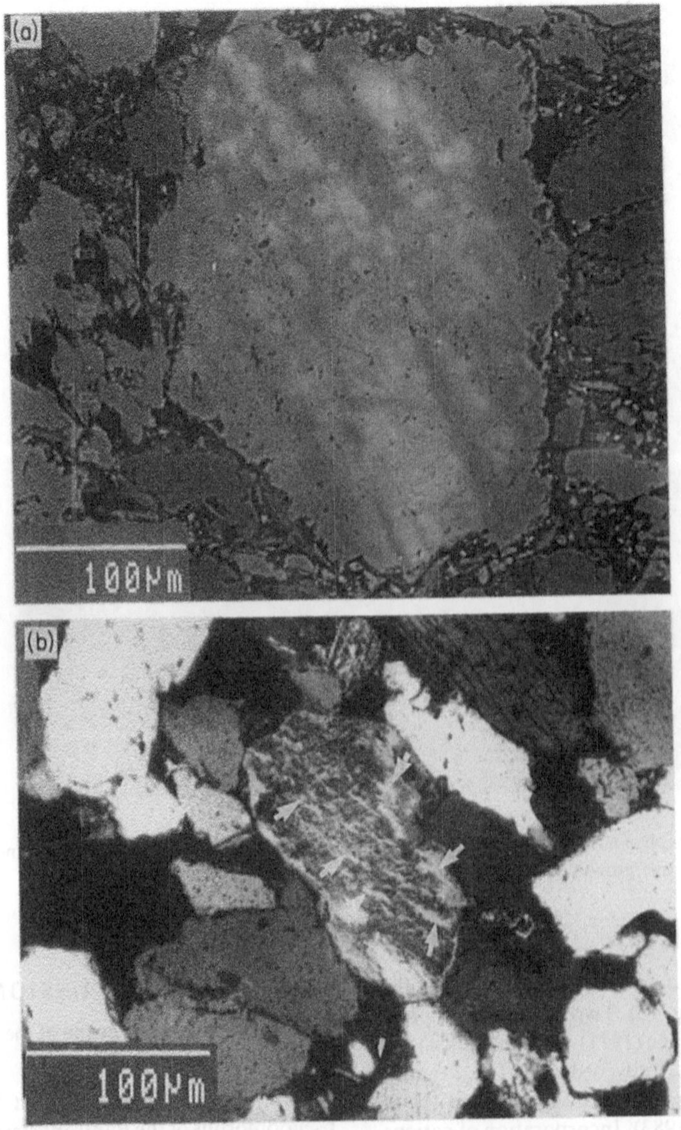

Fig. 8. (a) A BSE/CL micrograph showing the replacement of detrital microcline (bright) along the polysynthetic twinning planes by dark luminescing K-feldspar (dark). (b) Authigenic K-feldspar (light; arrows) which replaces detrital microcline along twinning and, probably, cleavage traces. Cross-nicols.

(3) The almost pure end-member composition also suggest a diagenetic origin for the K-feldspar pseudomorphs (Table 2; cf. Kastner, 1971; Stablein & Dapples, 1977; Kastner & Siever, 1979). Detrital K-feldspars of the Buntsandstein contain significant amounts of, particularly, albite solid solution.

(4) The dark cathodoluminescence of the K-feldspar pseudomorphs is also characteristic of a diagenetic origin (cf. Kastner, 1971; Kastner & Siever, 1979; Ali & Turner, 1982; Saigal *et al.*, 1988).

(5) Presence of microporosity (vacuoles) and inclusions in the K-feldspar pseudomorphs are characteristic of diagenetic feldspar pseudomorphs (Middleton, 1972; Ogunyomi, Martin & Hesse, 1981; Walker, 1984; Morad, 1986).

Table 1. Representative microprobe analyses of diagenetic and detrital feldspars in the Buntsandstein sandstones: DFC = detrital K-feldspar core, AFO = authigenic K-feldspar overgrowth, RDF = relict detrital feldspars, PF = diagenetic K-feldspar pseudomorphically replacing detrital feldspar.

Sample	Hoz 12		Hoz 105		Hoz 10		Hoz 101		Hoz 13		SS 17		Hoz 19		Hoz 17		Hoz 106	
	DFC 1	FOG 2	DFC 3	FOG 4	DFC 5	FOG 6	DF 7	PF 8	DF 9	PF 10	DF 11	PF 12	DF 13	PF 14	DF 15	PF 16	DF 17	PF 18
SiO_2	64·34	64·07	64·05	64·60	64·78	65·26	64·92	64·61	65·33	64·13	65·36	64·38	67·33	64·68	65·78	64·33	66·69	64·18
Al_2O_3	18·14	18·24	18·05	18·08	18·15	17·89	17·85	17·83	17·62	18·31	17·91	17·94	19·22	18·38	18·96	18·57	19·33	18·38
CaO	0·29	0·00	0·51	0·00	0·42	0·00	0·11	0·01	0·42	0·00	0·26	0·00	0·80	0·00	2·43	0·00	1·54	0·00
Na_2O	1·39	0·00	1·49	0·01	1·07	0·04	1·07	0·00	1·50	0·02	2·01	0·00	11·81	0·08	10·17	0·01	11·90	0·20
K_2O	14·14	16·86	14·49	16·60	14·73	16·98	15·21	16·64	14·23	16·78	13·93	16·74	1·14	16·53	2·68	16·41	0·54	16·12
Total	98·30	99·17	99·19	99·28	99·15	100·17	99·16	99·09	99·10	99·24	99·50	99·06	100·30	99·67	99·99	99·32	100·00	98·88
Number of ions on the basis of 8 (O)																		
Si	2·995	2·995	2·994	3·009	3·003	3·017	3·015	3·014	3·023	2·993	3·012	3·009	2·964	2·999	2·935	2·989	2·946	2·995
Al	0·994	1·005	0·995	0·992	0·992	0·975	0·977	0·980	0·961	1·007	0·973	0·988	0·997	1·005	0·997	1·017	1·006	1·011
Ca	0·014	0·000	0·025	0·000	0·021	0·000	0·006	0·000	0·020	0·000	0·013	0·000	0·038	0·000	0·116	0·000	0·073	0·000
Na	0·125	0·000	0·135	0·001	0·096	0·004	0·096	0·001	0·143	0·002	0·180	0·000	1·008	0·007	0·880	0·001	1·019	0·018
K	0·883	1·005	0·853	0·983	0·871	1·001	0·901	0·990	0·842	0·999	0·819	0·998	0·064	0·978	0·153	0·973	0·030	0·960
Σ	5·012	5·005	5·002	4·985	4·984	4·997	4·995	4·986	4·989	5·001	5·000	4·996	5·072	4·991	5·082	4·988	5·075	4·988
Molecular composition																		
Or	86·3	100·0	84·2	99·9	88·1	99·7	89·9	99·90	83·7	99·8	81·0	100·0	5·8	99·2	13·3	99·9	2·7	98·2
Ab	12·3	0·0	13·3	0·1	9·8	0·3	9·6	0·10	14·3	0·2	17·7	0·0	90·8	0·8	76·6	0·1	90·8	1·8
An	1·4	0·0	2·5	0·0	2·1	0·0	0·5	0·0	2·0	0·0	1·3	0·0	3·4	0·0	10·0	0·0	6·5	0·0

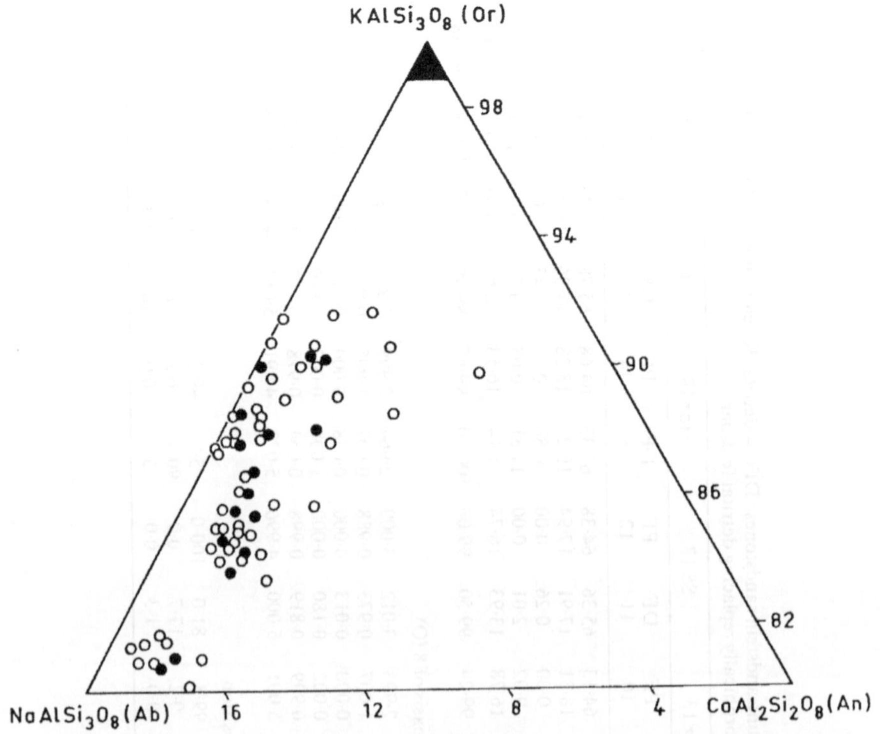

Fig. 9. Ternary plot of the end-member composition of detrital K-feldspars which are pervasively or slightly replaced by authigenic K-feldspar (open and solid circles respectively). Shaded area shows the composition field of authigenic K-feldspar (replacive and overgrowth types).

Fig. 10. A SEM micrograph of a K-feldspar pseudomorph composed of tiny euhedral crystals of diagenetic K-feldspar (e.g., arrows); RKF is relict detrital K-feldspar.

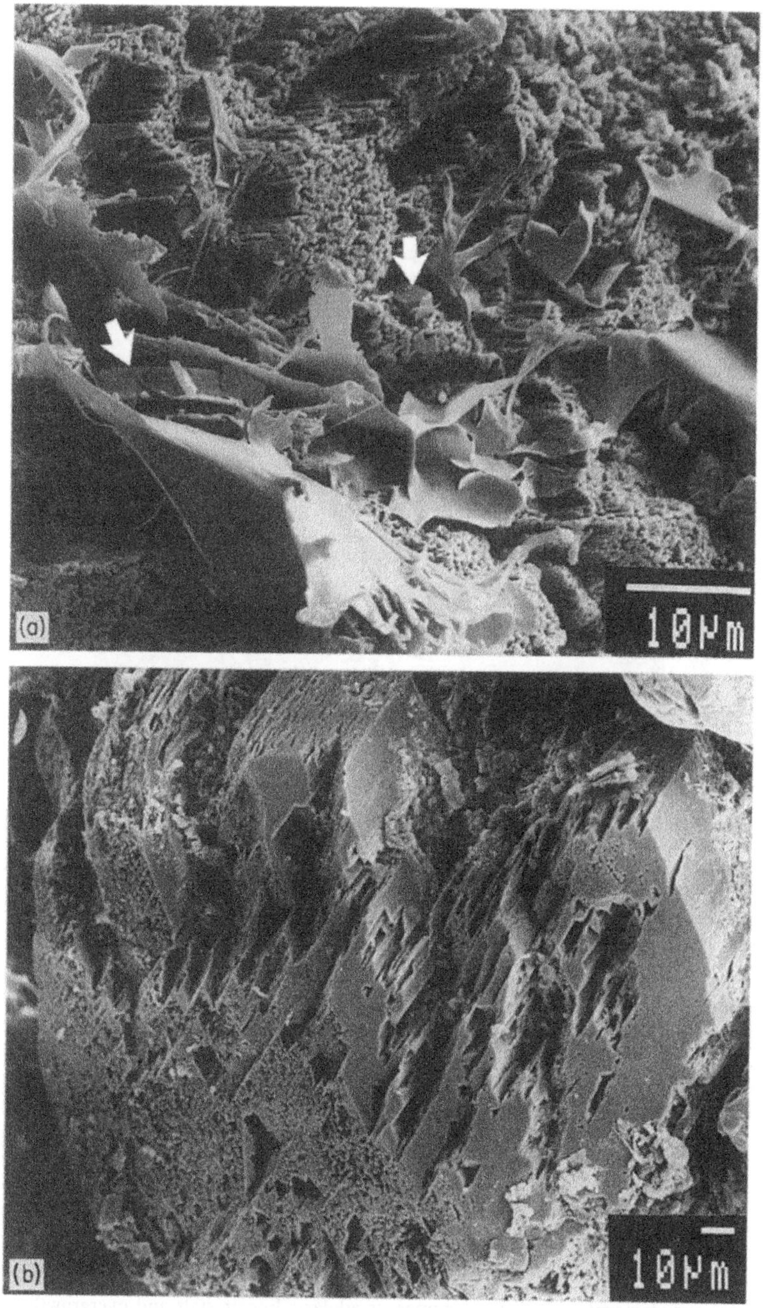

Fig. 11. (a) A K-feldspar pseudomorph composed of elongated prismatic crystals containing abundant intercrystalline micropores. Notice the presence of diagenetic illite flakes and discrete K-feldspar crystals (arrows). (b) A K-feldspar pseudomorph similar to that in Fig. 11a but which displays the presence of smooth-surfaced K-feldspar overgrowths. SEM micrographs.

Fig. 12. (a) A K-feldspar pseudomorph composed of numerous adularia-like crystals and overgrowth (arrows). (b) Higher magnification of the area outlined in Fig. 12a. SEM micrographs.

(6) The euhedral shapes and smooth surfaces of the K-feldspar crystals in the pseudomorphs and overgrowths are analogous with diagenetic albite pseudomorphs and overgrowths (see Walker, 1984; Morad, 1986), hence their diagenetic origin.

(7) Apparent absence of twinning is characteristic for diagenetic feldspar pseudomorphs (Middleton, 1972; Morad, 1986; Saigal *et al.*, 1988). This character is suggested here to be related to the fact that the textural framework of feldspar pseudo-

Table 2. Unit cell parameters of diagenetic K-feldspars.

Sample	a (Å)	b (Å)	c (Å)	α (°)	β (°)	γ (°)	v Å³
1 Hoz 10	8·5803	12·9723	7·2096	90·424	115·929	88·924	721·57
	0·0044	0·0038	0·0019	0·039	0·033	0·037	
2 Hoz 12	8·5628	12·9679	7·2076	90·282	115·989	89·165	719·32
	0·0045	0·0053	0·0022	0·060	0·043	0·085	
3 Hoz 101	8·5726	12·9704	7·2119	90·172	115·943	89·070	720·99
	0·0060	0·0046	0·0020	0·036	0·044	0·050	
4 SS 17	8·5657	12·9775	7·2110	90·228	116·009	88·952	719·71
	0·0047	0·0033	0·0013	0·029	0·024	0·026	
5 Kastner (1971)	8·5656	12·9632	7·2240	90·682	116·041	87·804	720·17
	0·0072	0·0024	0·0041	0·033	0·034	0·031	
6 Kastner (1971)	8·5852	12·9610	7·2212	90·537	115·956	87·984	722·01
	0·0026	0·0014	0·0017	0·016	0·014	0·015	

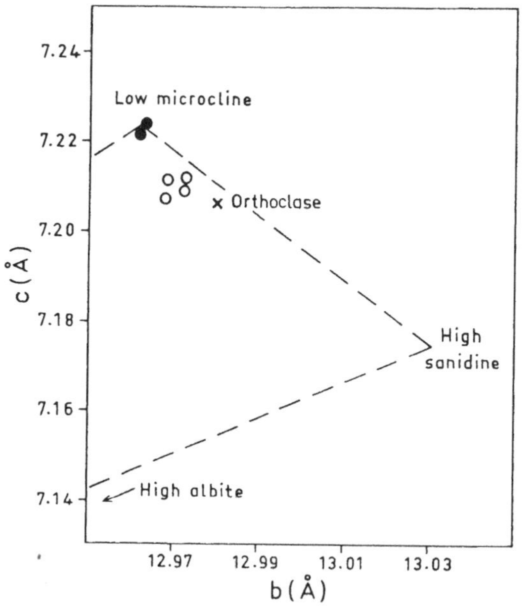

Fig. 13. Plots of the cell dimensions *b versus c* for diagenetic K-feldspars of the Buntsandstein (open circles) and diagenetic microclines of Kastner (1971). The coordinates of the end members are from Kroll & Ribbe (1983).

morphs is commonly comprised of numerous discrete crystals rather than a single crystal. Instead of twinning, some of the pseudomorphs are characterized by a patchy extinction pattern which is due to variations in the optical orientation of authigenic K-feldspar domains or the presence of relict detrital feldspar.

Process and conditions of K-feldspar authigenesis

The occurrence of diagenetic K-feldspar as numerous commonly parallel aligned crystals with variable amounts of intercrystalline porosity indicates that pseudomorphism occurs via a partial dissolution–precipitation process which promotes nucleation on numerous sites. De Vore (1959) and Chou & Wollast (1985), among many others, suggested that decomposition of feldspars is accomplished by several steps. The first step involves instantaneous exchange of alkali ions (in this case H^+ exchanged for K^+ and Na^+) followed by formation of Al–Si chain structures before diffusion of Si and Al as soluble ions occurs.

The replacive authigenic K-feldspar would thus have subsequently formed by nucleation on these Si–Al structural units. The intercrystalline porosity in the pseudomorphs has apparently acted as pathways for outwards diffusion of Na^+ and Ca^{2+} and inwards diffusion of K^+ to the detrital feldspar–pore solution interface. This has enhanced the penetrative replacement by K-feldspar which ultimately resulted in complete pseudomorphs. The guiding of the replacement process by grain fractures as well as twin and cleavage traces is firstly because these surfaces act as weakness planes along which pore fluids can penetrate the detrital feldspar, and secondly, is probably because these surfaces are characterized by greater free energy, thus greater dissolution rates.

Simple mass-balance calculations were made based on the average compositions of detrital and authigenic feldspars in order to obtain some idea about the amounts of K^+ needed to accomplish the pseudomorphism. The replacement of an average of about 8% detrital K-feldspar and 1% plagioclase requires the consumption of about 16 000 ppm K^+ (i.e. 40 pore volumes if sea-water is the only source of K^+). In these calculations it is assumed that sandstones at shallow burial had porosities of about 30%. The amounts of K^+ needed would thus increase if lower porosities are assumed. Ca^{2+} released upon replace-

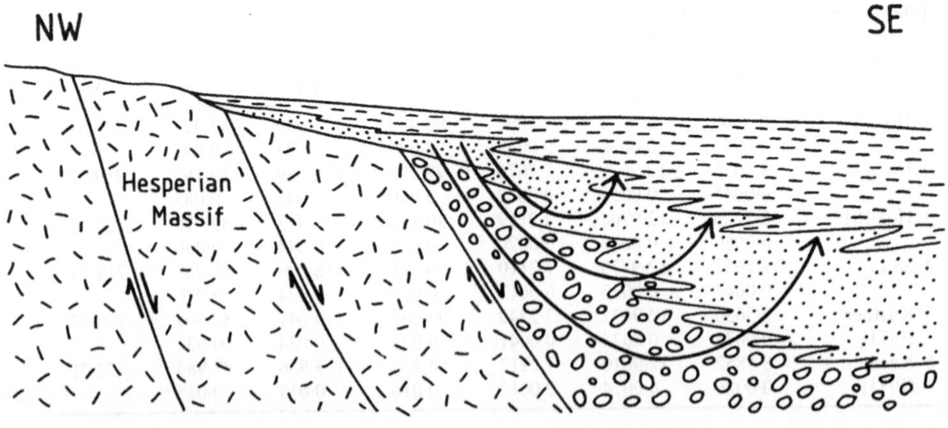

Fig. 14. Model for meteoric water recharge and flow into the Bundsandstein (not to scale).

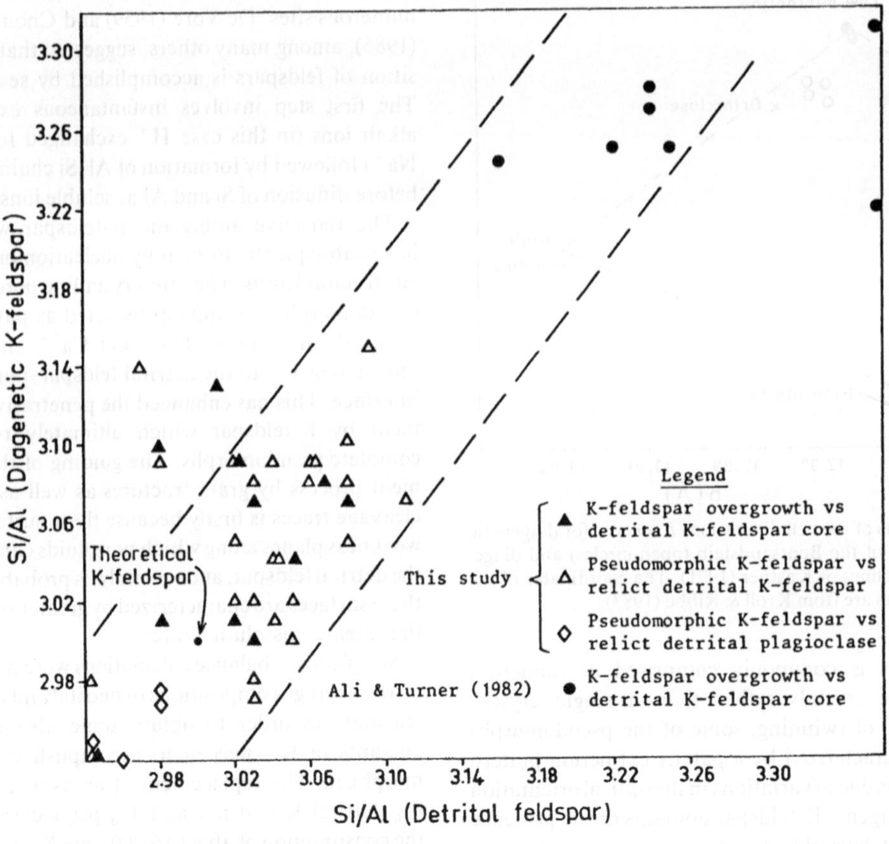

Fig. 15. Plot of the Si/Al ratios in diagenetic *versus* detrital feldspars. The correlation coefficient, based on data from the Bundsandstein, is +0·64.

ment of detrital K-feldspar and plagioclase by authigenic K-feldspar has been used in the authigenesis of dolomite. The fate of Na^+ is, so far, not well understood.

The diagenetic geochemical conditions which favoured formation of authigenic K-feldspar is presumably related to meteoric waters recharged in the permeable alluvial conglomeratic sediments at the base of fault escarpment (Fig. 14; cf. Kharaka & Perry, 1974; Dutton & Land, 1985). These waters evolved chemically ($\alpha_{K^+}/\alpha_{H^+}$ and αH_4SiO_4 increased) through reactions (almost complete leaching or kaolinitization) with detrital K-feldspar and micas (cf. Helgeson, Garrels & McKenzie, 1969; Drever, 1982). Authigenesis of K-feldspar has occurred at atmospheric *T–P* conditions (cf. Waugh, 1978; Stablein & Dapples, 1979; Ali & Turner, 1982).

The above hypothesis concerning the mechanism of pseudomorphism might explain the fairly good positive correlation between the Si/Al ratios in the detrital K-feldspar and the authigenic K-feldspar replacing it (Fig. 15) and the disordered nature of authigenic K-feldspar (Fig, 13). The latter is probably a property inherited from the parent detrital feldspar. For some kinetic reasons, the authigenic K-feldspar failed to reach full Al–Si ordering (cf. Ogunyomi *et al.*, 1981; Martin, 1984). One obvious feature revealed by Fig. 15 and Table 1 is that many of the Buntsandstein feldspars analysed and those presented by Ali & Turner (1982) deviate from the ideal stoichiometry (cf. Broxton, Bish & Warren, 1987) particularly in terms of the Si/Al ratio. The reason behind this discrepancy is not clear.

SUMMARY AND CONCLUSIONS

In this study we have described a previously unrecorded pseudomorphic replacement of detrital orthoclase, microcline and plagioclase by authigenic K feldspar. This process, which is principally analogous to the albitization process of detrital feldspars, is restricted to the deltaic–estuarine and tidal sandstone facies of the Buntsandstein. The pseudomorphism requires the consumption of K^+ which is suggested to be derived by meteoric waters that have evolved chemically through reaction with detrital micas and K-feldspars in the basal conglomeratic alluvial facies. Thus, K^+ was derived intraformationally.

K-feldspar pseudomorphs can be misinterpreted as detrital grains. However, these pseudomorphs are untwinned, dark-luminescing, pitted, and have almost an end-member composition (Or > 99 mol%). They are also vacoulized, commonly heavily clouded by tiny inclusions of barite and Fe-oxides and comprised of numerous parallel-aligned K-feldspar crystals. Detrital K-feldspars have variable amounts of Na and Ca (Or < 93 mol%) and detrital plagioclase is andesine to albitic (Ab < 96 mol%) in composition.

Formation of K-feldspar pseudomorphs, like albitization, is a dissolution–precipitation process. Replacement by authigenic K-feldspar is guided by grain fractures as well as cleavage and twin traces of the detrital feldspars. In contrast to albitization, replacement by K-feldspar has occurred at almost atmospheric *T—P* conditions.

ACKNOWLEDGMENTS

A substantial part of the SEM and cathodoluminescence work were performed at Saga Petroleum AS, Oslo under the supervision of Tor Mellem. We thank Hans Harryson who aided the microprobe analyses. Karl Ramseyer generously helped with the hot-cathode luminescence microscopy. We also thank A. A. Al Dahan, B. de la Cruz, S. D. Burley, Ö. Amcoff, P. B. Gold and K. Ramseyer for the critical reading of the manuscript. We thank A. Kaljusaar for drafting the figures and L. Wallsten for performing the XRD analyses. S. Morad thanks the Swedish Natural Science Research Council (NFR) for financial support. We regret the untimely death of A. Kaljusaar. R.M. and J.A.P. wish to thank Shell, Spain, J.E.N. and U.N.V.S.A. national companies for providing samples used in this study. O. Wallner took much care in preparing the thin sections.

REFERENCES

Ali, A.D. & Turner, P. (1982) Authigenic K-feldspar in the Bromsgrove Sandstone Formation (Triassic) of central England. *J. sedim. petrol.*, **52**, 187–197.

Alvaro, M., Capote, R. & Vegas, R. (1979) Un modelo de evolucion geotectonica para la Cadena Celtiberica. *Acta Geologica Hispanica*, **14**, 172–177.

Arribas, J. (1984) *Sedimentologia y Diagenesis del Bundsandstein y Muschelkalk de la Rama Argonesa de la Cordillera Iberica (Provincia de Soria y Zaragosa)*. Unpublished PhD Thesis, Universidad Complutense de Madrid, 354 pp.

Arribas, J., Marfil, R. & de la Peña, J.A. (1985) Provenance of Triassic feldspathic sandstones in the Iberian Range (Spain): Significance of quartz types. *J. sedim. Petrol.*, **55**, 864–868.

BLATT, H., MIDDLETON, G. & MURRAY, R. (1980) Origin of Sedimentary Rocks. Prentice-Hall, Englewood Cliffs, 782 pp.

BOLES, J.R. (1982) Active albitization of plagioclase, Gulf Coast Tertiary. *Am. J. Sci.*, **282**, 165–180.

BROXTON, D.E., BISH, D.L. & WARREN, R.G. (1987) Distribution and chemistry of diagenetic minerals at Yucca Mountain, Nye County, Nevada. *Clays Clay Miner.*, **35**, 89–110.

BURLEY, S.D. (1984) Patterns of diagenesis in the Sherwood Sandstone Group (Triassic), United Kingdom. *Clay Miner.*, **19**, 403–440.

CAPOTE, R., GOMEZ, J.J., ROSELL, J., DIAZ, M., DE LA TORRE, L.S., SOPENA, S., GABALDON, V., RUIZ, P. & YEBENES, A. (1982) *Evolucion Sedimentologica y Tectonica del Ciclo Alpino en el Terico Noroccidental de la Rama Castellana de la Cordillera Iberica.* Temas geologico-mineros IGME, 290 pp.

CHOU, L. & WOLLAST, R. (1985) Steady-state kinetics and dissolution mechanisms of albite. *Am. J. Sci.*, **285**, 963–993.

COY-YLL, R. (1969) Quelques aspects de la cathodoluminescence des mineraux. *Chem. Geol.*, **5**, 243–254.

DE LA PEÑA, J.A., ARRIBAS, J., DE LA CRUZ, B. & MARFIL, R. (1983) Diagenetic models of Permo–Triassic continental and transitional sandstones (red beds) in the Iberian Range, Spain. *4th Regional Meeting, Int. Ass. Sedimentol., Split, Yugoslavia, Abstr*, 137–139.

DE VORE, G.S. (1959) The surface chemistry of feldspars as an influence on their decomposition products. *Clays Clay Miner.*, **6**, 26–41.

DICKINSON, W.R., OJAKANGAS, R.W. & STEWART, R.J. (1969) Burial metamorphism of the late Mesozoic Great Valley sequence, Cache. *Bull. geol. Soc. Am.*, **80**, 519–526.

DREVER, J.I. (1982) *The Geochemistry of Natural Waters.* Prentice-Hall, Englewood Cliffs, 388 pp.

DUTTON, S.P. & LAND, L.S. (1985) Meteoric burial diagenesis of Pennsylvanian arkosic sandstones, southwestern Anadarko Basin, Texas. *Bull. Am. Ass. petrol. Geol.*, **69**, 22–38.

GOLD, P.B. (1987) Textures and geochemistry of authigenic albite from Miocene sandstones, Louisiana Gulf Coast. *J. sedim. Petrol.*, **57**, 353–362.

GRANT, P.R. & WHITE, S.H. (1978) Cathodoluminescence and microstructure of quartz. *Scanning Electron Microscopy* 1978, 789–794.

HELGESON, H.C., GARRELS, R.M. & MACKENZIE, T.A. (1969) Evaluation of irreversible reactions in geochemical processes involving minerals and aqueous solutions: II. Applications. *Geochim. Cosmochim. Acta*, **33**, 415–438.

KASTNER, M. (1971) Authigenic feldspars in carbonate rocks. *Am. Mineral*, **56**, 1403–1442.

KASTNER, M. & SIEVER, R. (1979) Low temperature feldspars in sedimentary rocks. *Am. J. Sci.*, **279**, 435–479.

KHARAKA, Y.K. & PERRY, F.A.F. (1974) The influence of geological membranes on the geochemistry of subsurface waters from Miocene sediments at Kettleman North Dome, California. *Wat. Resour. Res.*, **10**, 313–327.

KROLL, H. & RIBBE, P.H. (1983) Lattice parameters, composition and Al,Si order in alkali feldspars. In: *Feldspar Mineralogy* (Ed. by P. H. Ribbe). *Mineral. Soc. Am. Rev. Mineral.*, **2**, 57–99.

LAND, L.S. & MILLIKEN, K.L. (1981) Feldspar diagenesis in the Frio Formation, Brazoria County, Texas Gulf Coast. *Geology*, **9**, 314–318.

MARTIN, R.F. (1984) Patterns of albitization in the Montgenevre ophiolite, West Alps. *Bull. Miner.*, **107**, 345–356.

MARTIN, R.F. & FALSER, A.U. (1986) Proterozoic sanidine and microcline in pegmatite, Wausau complex, Wisconsin. *Can. Mineral.*, **24**, 709–716.

MATTER, A. & RAMSEYER, K. (1985) Cathodoluminescence microscopy as a tool for provenance studies of sandstones. In: *Provenance of Arenites* (Ed. by G. G. Zuffa), pp. 191–211. Reidel, Dordrecht.

MCBRIDE, E.F. (1985) Diagenetic processes that affect provenance determinations in sandstones. In: *Provenance of Arenites* (Ed. by G. G. Zaffa), pp. 95–113. Reidel, Dordrecht.

MIDDLETON, G.V. (1972) Albite of secondary origin in Charny sandstones, Quebec. *J. sedim. Petrol.*, **42**, 341–349.

MORAD, S. (1986) Albitization of K-feldspar grains in Proterozoic arkoses and greywackes from southern Sweden. *Neues. Jb. Miner. Mh.*, 145–156.

OGUNYOMI, O., MARTIN, R.F. & HESSE, R. (1981) Albite of secondary origin in Charny sandstones, Quebec: a re-evaluation. *J. sedim. Petrol.*, **51**, 597–606.

PETTIJOHN, F.J., POTTER, P.E. & SIEVER, R. (1973) *Sand and Sandstones.* Springer-Verlag, New York, 618 pp.

SAIGAL, G.C., MORAD, S., BJØRLYKKE, K., EGEBERG, P.K. & AAGAARD, P. (1988) Diagenetic albitization of detrital K-feldspars in Jurassic, Lower Cretaceous, and Tertiary clastic reservoir rocks from offshore Norway, I. textures and origin. *J. sedim. Petrol.*, **58**, 1003–1013.

STABLEIN, N.K. III & DAPPLES, E.C. (1977) Feldspars of the Tunnel City Group (Cambrian), western Wisconsin. *J. sedim. Petrol.*, **47**, 1512–1538.

STEVENSON, R.K. & MARTIN, R.F. (1986) Implications of the presence of amazonite in the Broken Hill and Geco metamorphosed sulfide deposits. *Can. Mineral.*, **24**, 729–745.

WALKER, T.R. (1984) Diagenetic albitization of potassium feldspar in arkosic sandstones. *J. sedim. Petrol.*, **54**, 3–16.

WAUGH, B. (1978) Authigenic K-feldspar in British Permo–Triassic sandstones. *J. geol. Soc. London*, **135**, 51–56.

ZINKERNAGEL, U. (1978) Cathodoluminescence of quartz and its application to sandstone petrology. *Contr. Sedimentol.*, **8**, 69 pp.

VELBEL, M.A. (1983) A dissolution–reprecipitation mechanism for the pseudomorphous replacement of plagioclase by clay minerals during weathering. *Science Geologiques Memoire*, **II**, 139–147.

(Manuscript received 5 May 1988; revision accepted 3 November 1988)

Reprinted from Sedimentology (1966) 7, *327–357*

ZEOLITES IN SEDIMENTARY ROCKS, WITH REFERENCE TO THE DEPOSITIONAL ENVIRONMENTS AND ZONAL DISTRIBUTION

AZUMA IIJIMA AND MINORU UTADA

Geological Institute, University of Tokyo, Hongo, Tokyo (Japan)
Institute of Earth Science, Faculty of Culture, University of Tokyo, Komaba, Tokyo (Japan)

(Received November 30, 1964)

SUMMARY

The authors have studied alterations of Cenozoic and Mesozoic pyroclastic rocks of Japan, which contain several kinds of zeolites in abundance. This paper summarizes zeolites in sedimentary rocks, with reference to the depositional environments and zonal distribution, by a survey of the literature in addition to the authors' data.

The zonal distribution of zeolites is recognized in buried sedimentary rocks as follows:

(*1*) Analcime, chabazite, natrolite and phillipsite	syngenetic or diagenetic origin
— about 0.2–1 km in depth —	diagenetic origin
(*2*) Fresh tuff	
Clinoptilolite, mordenite and erionite	
— about 1–3 km in depth —	
Analcime and heulandite	
— about 2–5 km in depth —	
Laumontite	
— about 3–11 km in depth —	

The zeolites in syngenetic or early diagenetic origin depend strongly upon a specific sedimentary environment. Phillipsite occurs largely in pelagic sediments of the younger geologic age. Analcime is found in saline-lake and terrestrial sediments in a warm, rather arid region, frequently associated with phillipsite, chabazite and natrolite. The zeolites are not influenced by the sedimentary environments but depend upon the depth of burial, i.e., increasing temperature and pressure. Most of clinoptilolite, mordenite and erionite, forming at a relatively shallow depth, occur only as an alteration product of acidic to intermediate volcanic glass and cement of the post-Jurassic pyroclastic rocks. Laumontite, forming at a greater depth, on the other hand, is widely distributed in the pre-Pliocene various sedimentary rocks.

Reprinted from Sedimentology (1998) **45**, *855–874*

Diagenesis of the Newark Rift Basin, Eastern North America

M. EL TABAKH* and B. C. SCHREIBER†
**School of Applied Geology, Curtin University, Perth, Western Australia 6001, Australia (E-mail: rtabakhm@cc.curtin.edu.au)*
†Department of Geology, Appalachian State University, 118 Rankin Science Building, Boone, NC, 28608, USA (E-mail: geologo@aol.com)

ABSTRACT

Late Triassic nonmarine strata in the Newark Rift Basin, eastern North America, accumulated in a subsiding half-graben prior to opening of the Atlantic Ocean. These strata consist, in ascending order, of the Stockton, Lockatong, and Passaic formations. Although different in specific lithology, these formations all exhibit diagenetic fabrics dominated by authigenic albite and analcime. These same minerals have a similar presence in Late Triassic (Newark Supergroup) strata of other rift-related basins to the north, suggesting that related authigenesis is not simply a result of local diagenetic factors.

The basal deposit, the Stockton Formation, is composed of fluvial sandstones and overbank mudstones, with nodular pedogenic calcite (calcrete). During burial, original micrite was first recrystallized into sparry calcite and then later subjected to partial replacement by authigenic albite, which is also present as overgrowths and void fillings in overbank mudstones. The Lockatong Formation contains organic-rich shales, carbonates, and evaporative mudstones deposited under cyclic conditions in laterally extensive lacustrine environments. Analcime comprises up to 40% of these strata by volume, occurring within the matrix, as a replacement of original carbonates and evaporites, and as fillings in macrovoids. The overlying Passaic Formation is made up of massive red mudstones, evaporites, and local calcareous lacustrine sequences. Evaporites are replaced by coarse-grained anhydrite together with some authigenic albite.

A central question concerns the source for the Na, Si, and Al required for albite and analcime authigenesis. It is suggested that, in addition to alteration of primary siliciclastic material, sodium in particular was supplied in two ways: (1) from high concentrations in original evaporative brines and groundwaters (Lockatong and Passaic formations); and (2) from dissolution of associated sodium-bearing evaporites (Lockatong and Passaic Formation) during diagenesis. It is proposed here that basin-sourced, Na-enriched brines circulated through the section over time. As albite is more stable at elevated temperatures relative to analcime, it developed in the lowermost strata of the basin (Stockton Formation). Analcime is more prevalent in the overlying Lockatong Formation.

Reprinted from Sedimentology (1989) ***36****, 927–942*

The origin of faceted garnets in sandstones: dissolution or overgrowth?

A. C. MORTON, G. BORG*, P. L. HANSLEY†, P. D. W. HAUGHTON**, D. H. KRINSLEY‡ *and* P. TRUSTY‡

*Stratigraphy and Sedimentology Research Group, British Geological Survey, Keyworth, Notts NG12 5GG, UK; *Bundesanstalt fur Geowissenschaften und Rohstoffe, P.O. Box 51 01 53, D-3000 Hannover 51, FRG; †Branch of Sedimentary Processes, US Geological Survey, Denver Federal Center, Denver, Colorado 80225-0046, USA; **Department of Geology, University of Glasgow, Glasgow G12 8QQ, UK; ‡Department of Geology, Arizona State University, Tempe, Arizona 85287-1404, USA*

ABSTRACT

Faceted garnets from a wide range of geological ages, environments and locations have been studied in polished grain mounts by a combination of backscattered electron microscopy and elemental mapping using energy-dispersive X-ray analysis. In all cases, the areas apparently showing positive relief on the faceted garnet surfaces are compositionally identical to the adjacent grain cores despite a wide variation in detrital garnet compositions. In one case, zoning within the grain core can be traced into the faceted areas on the grain surface. Thus, faceted areas must be considered to form part of the original detrital grains. Together with previously published studies on experimental garnet etching, thermodynamic conditions for garnet growth, textural relationships between faceted garnets and authigenic and detrital phases, and distribution of faceted garnets in the subsurface, this paper provides conclusive proof that faceted garnet surfaces form as a result of dissolution, not overgrowth.

INTRODUCTION

In recent years there has been a renewed interest in the origin of the facets that are commonly developed on the surfaces of detrital garnets in sandstones. Faceted garnets have only been reported from sandstones that have been subjected to temperatures and pressures typical of the diagenetic regime. Garnets in recent sediments such as soils may be pitted, but do not display crystallographically controlled facets (Velbel, 1984). The facets are positive features that are regularly distributed over grain surfaces, with faces that conform to crystallographic planes. These facets have previously been described as imbricate wedge markings (Rahmani, 1973) or as step-like (Zili & Zhengmou, 1987; Zili, 1988) or tile-like (Afanasev, 1985) features. Typical examples are shown in Figs 1–8. The earliest explanation of such facets was given by Sauer (1900), who interpreted them as overgrowths, but subsequently their development was ascribed to dissolution processes (e.g. Bramlette, 1929; Smithson, 1941; McMullen, 1959). The argument was reopened by Simpson (1976), who favoured the overgrowth hypothesis, and several authors (Howie, Simpson & Simpson, 1980, Mader, 1980; 1981; Maurer, 1982; Zili & Zhengmou, 1987; Zili, 1988) have subsequently followed this view. The same controversy has also been discussed by the Russian geological community: Preobrazhensky (1941), Serdyuchenko & Dobrotvorskaya (1949) and Pustowalloff (1955) suggested an authigenic overgrowth origin, a hypothesis that Sobolev, Vartanova & Shainyuk (1951) rejected on thermodynamic grounds. Subsequently, Zaporozhtzeva (1960) and Afanasev (1985) advanced the view that facets on garnet surfaces resulted from dissolution processes.

As Hansley (1987) has pointed out, no data supporting the overgrowth hypothesis have ever been published, with the main arguments in favour of this hypothesis being the crystallographic similarity between garnet facets and euhedral authigenic overgrowths on other mineral phases, such as quartz or feldspar. In contrast, a variety of lines of evidence have been cited in support of the etching hypothesis. These include the experimental simulation of garnet faceting, thermodynamic considerations, the distribution of faceted garnets in the subsurface, textural relationships between garnet and cementing minerals, and the pattern of compositional variation within individual faceted grains.

This paper is divided into two parts. In the first, the various lines of published evidence for the dissolution hypothesis are evaluated, and in the second, new data on intra-grain compositional variations are presented.

REVIEW OF PUBLISHED EVIDENCE

Experimental studies

Studies of the effects of corrosive solutions on garnet grains date as far back as the work of Bramlette (1929), who claimed to have produced facets on garnets using HF, but presented no photographic evidence. Similarly, McMullen (1959) reported that garnet dissolution in NaOH was associated with the formation of facets on grain surfaces, but again photographic proof was not given. Gravenor & Leavitt (1981) were the first to publish photomicrographs demonstrating that chemical corrosion does produce facets on garnet surfaces, although they admitted that the surface textures were not identical to those produced by natural processes. However, Hansley (1987) was able to simulate natural garnet facets from the Morrison Formation (San Juan basin, NW New Mexico) by subjecting garnets to oxalic acid solutions (0·1–1·0%) at elevated temperatures (75°C), thus corroborating other laboratory results that showed organic acids to be effective dissolution agents of aluminosilicate minerals (Surdam, Boese & Crossey, 1984; Huang & Keller, 1970). Recently, reactive short-chained organic acid anions have been detected in relatively high concentrations in many subsurface waters, particularly oilfield brines (Carothers & Kharaka, 1978). Thus, the experimental evidence indicates that naturally-occurring faceted garnets could have been produced by dissolution, although this clearly does not prove that all such occurrences were produced in this way.

Thermodynamic considerations

Thermodynamic constraints on the formation or authigenesis of garnets indicate that this group of silicates are typical high-temperature and/or high-pressure minerals associated mainly with metamorphic processes. It is, therefore, curious that relatively low-temperature diagenetic conditions are considered by many to favour garnet growth. The data summarized below on the stability and formation of garnet have been published relatively recently.

Pyrope is a high-pressure mineral, forming at above 800°C and over 14 kb pressure (Staudigel & Schreyer, 1977). Spessartine forms above 410°C at 0·05 kb (Matthes, 1961). Almandine can form at very low pressures, even at atmospheric pressure, but high temperatures (800°C to 1400°C) are required (Keesmann *et al.*, 1971). Almandine has been synthesized at 650°C and 4·31 kb (Holdaway & Lee, 1977). However, almandine–spessartine garnets can form at temperatures as low as 390°C at 0·2 kb, and crystals of mixed pyrope–almandine–spessartine composition have been synthesized experimentally at temperatures of 600°C to 800°C and pressures of 0·2 kb to 1·5 kb (Matthes, 1961). Experiments by Matthes (1961) showed that pyralspite garnets of the composition Pyr_{15}–Spe_5–Alm_{80} failed to crystallize at temperatures below 800°C and pressures below 1·5 kb. This result is of particular interest since faceted garnets (Fig. 6a) from unmetamorphosed Triassic sandstones of southern Germany are of the same composition (Borg, 1986). The maximum temperature affecting these sandstones was about 50°C.

Grossular forms above 470°C and 2 kb (Storre, 1970). Uvarovite may form at 855°C at surface pressure (Huckenholz & Knittel, 1975). Andradite forms above 400°C at 2 kb (Gustafson, 1974). Growth of andradite is possible at atmospheric pressure, but temperatures of more than 1137°C are required (Gustafson, 1974). Hydro-garnets (both hydro-grossular and hydro-andradite) can crystallize under surface pressure conditions at low temperatures, in the order of 50°C (Kuzel, 1968).

Interpretation of P, T constraints

Hydro-garnets are the only garnets that could be expected to form under diagenetic conditions, but to date such garnets have not been identified by the supporters of the overgrowth hypothesis, and no such compositions were detected in the study of faceted garnets discussed later in this paper. Elevated temperatures and pressures clearly promote the crystallization of garnet, and thus the likelihood of garnet authigenesis increases with burial depth. However, studies of the behaviour of garnets during deep burial show that garnet concentrations decrease as pore-fluid temperatures and burial depth increase within the diagenetic realm (Morton, 1984; Hansley, 1987).

However, there is a precedent for the formation of supposedly high-temperature phases as low-temperature diagenetic phases for overgrowths on both tourmaline and zircon have been reported from the sedimentary environment (Butterfield, 1936; Gautier, 1979; Morton & Humphreys, 1983). Thus, equilibrium thermodynamic evidence indicating that garnet overgrowth cannot occur during diagenesis cannot be regarded as proof without the supporting evidence presented here.

Distribution of faceted garnets in the subsurface

Although there are many papers that describe naturally occurring faceted garnets, there are comparatively few that discuss the relationship between the faceting and their distribution in the subsurface. Clearly, if faceting is the result of dissolution, then with continued and/or more intense corrosion, garnet should become reduced in proportion and ultimately disappear from the mineral assemblage. Such a relationship would not be expected if overgrowth processes were involved. However, to demonstrate that disappearance of garnet is the result of corrosion, it is necessary to prove that the loss is a secondary effect and not explicable in terms of changing provenance.

An early example is that given by Smithson (1941) who demonstrated that Middle Jurassic sandstones are devoid of garnet in a small area around the Cleveland anticline in North Yorkshire (UK). Entirely surrounding this area, which was the basin depocentre prior to inversion, is a zone in which highly faceted garnet is present but staurolite and kyanite are absent. With greater distance from the anticline, garnet lacks well-developed facets and coexists with staurolite and kyanite. Such a distribution cannot be explained by changing provenance. Thus the increasing degree of faceting and the ultimate removal of kyanite, staurolite and garnet must be explained in terms of corrosion, the degree of which intensified toward the original basin centre, due at least in part to increasing pore fluid temperature. Staurolite and kyanite, less stable minerals than garnet, were apparently dissolved at lower temperatures.

Another example with good geological control is the Eocene–Oligocene McKee Formation of Taranaki, New Zealand. Smale & Morton (1987) examined sandstones at burial depths ranging from 2135 m to 4141 m over a limited geographical area. In the relatively shallow sections, McKee sandstones contain common almandine–spessartine garnets lacking clear surface facets, but with increasing burial (and, consequently, at higher pore fluid temperatures) garnet surfaces become progressively more deeply faceted. At depths greater than 3600 m, garnet is entirely absent. There is no evidence for any change in sediment source, and thus the pattern, and the associated facet development, must be interpreted as the result of dissolution.

Other examples include the Upper Jurassic Morrison Formation of NW New Mexico and the Palaeocene Forties Formation of the central North Sea. Homogeneous sandstone without provenance variations in the Morrison Formation contains relatively garnet-poor zones with strongly faceted garnets sandwiched between garnet-rich zones containing garnets with no or only poorly-developed faceted surfaces (Hansley, 1987). Stable isotope data indicate that the zones with strongly faceted garnets have been flushed by warm (100°C to 130°C) deep-basin fluids; underlying and overlying strata containing unetched garnets show no evidence for the passage of warm fluids (Whitney & Northrop, 1987). In the Forties Formation, the most deeply buried sandstones at the centre of the basin (at about 3 km depth) are relatively depleted in garnet, and remaining garnet grains are strongly faceted (Morton, 1984). Both these examples show that depletion of garnet correlates with an increasing degree of faceting, indicating a dissolution mechanism. Furthermore, Morton (1987) has demonstrated that in the Forties Formation, preferential dissolution of the more soluble calcic garnets has taken place. However, in neither example does all the garnet disappear, although in the more deeply buried Jurassic of the central and northern North Sea areas, garnet does disappear entirely from the heavy mineral suites at depths between 3000 m and 3500 m (Morton, 1986). Similar relationships between burial depth and garnet disappearance are observed elsewhere. For example, McBride (1985) indicates that garnet disappears from US Gulf Coast Plio–Pleistocene sediments at burial depths in excess of 4 km. However, this account does not describe garnet surface textures or the possible role of provenance variations.

Other studies of garnet surface textures, such as those of Rahmani (1973) and Hemingway & Tamar-Agha (1975) also ascribe the progressive development of facets to increasing corrosion, but apart from the obvious analogy with the examples given above, the evidence given in these papers is not conclusive. Nevertheless, the examples discussed verify that well developed facets on garnet surfaces coincide with levels of relative garnet depletion, suggesting that dissolution is the causal mechanism. Furthermore, as discussed above, zones of extreme garnet faceting have been found adjacent to zones totally devoid of garnets, whose absence cannot be ascribed to provenance variations. Thus, the zone of extreme faceting must be regarded as a zone of advanced garnet dissolution in diagenetic conditions that fell just short of those causing total garnet loss.

Textural relationships

Because garnet is generally a very minor component of sandstones in volumetric terms, the relationship between faceted garnets and cements such as clay, carbonate or silica can only very rarely be observed. However, where it is possible to observe them, such relationships are vital in the resolution of the overgrowth/etching controversy. Only two such observations have been published, both very recently. The first was that of Hansley (1987), who showed that voids are present between authigenic grain-rimming chlorite and faceted garnet. This is clear proof that garnet dissolution took place after chlorite cementation, causing both a decrease in size of the grains and the generation of faceted surfaces. Salvino & Velbel (1989) observed similar phenomena in the Cambrian Munising Formation of Michigan (USA). Here, faceted garnets are associated with mouldic secondary porosity, indicating again that faceting is associated with a decrease, not an increase, in grain dimensions.

INVESTIGATION OF INTRA-GRAIN COMPOSITIONAL HETEROGENEITY

If garnet facets are the result of overgrowth and not dissolution, the positive areas on the grain surfaces should have a different chemical composition to the grain centres, as they would have formed under quite different physicochemical conditions. Two studies (Gravenor & Leavitt 1981; Borg 1986) reported electron microprobe analysis of various points within faceted garnet grains, and demonstrated that little or no compositional variation was present. However, Salvino & Velbel (1989) point out that it can be difficult to ascertain whether the microprobe has analysed the 'overgrowth' or the 'core'. Thus, microprobe analysis is not an ideal approach to the problem, and other methods that can provide information on the geographic variation in composition over entire garnet cross-sections need to be employed.

Combined BSE–EDX studies

Backscattered electron (BSE) imaging is an ideal method for evaluating compositional variations (Krinsley, Pye & Kearsley, 1983; Pye & Krinsley, 1984). The greyness of the BSE image is a function of the mean atomic weight of constituent elements in the analysed area, enabling changes in the ratios of light elements such as Mg and Ca to the heavy elements Mn and Fe to be rapidly assessed. Another approach is to use energy-dispersive X-ray analysis (EDX) to produce elemental dot-maps that show variations in distribution of the more abundant elements.

For this combined BSE–EDX study, faceted garnets were selected to represent the range of garnet compositions from a variety of geographical, stratigraphical and depositional settings. The sample set includes sandstones from the North Sea, the mid-Norway continental margin, southern Germany, New Mexico (USA) and New Zealand. Ages include Triassic, Jurassic, Palaeocene and late Eocene–Oligocene. Depositional environments range from fluvial and fluvio–deltaic to shallow and deep marine. Mg-, Ca-, Mn- and Fe-rich varieties are all represented. Sample details are given in Table 1.

All garnets were cut and polished, coated with carbon and examined in the backscatter mode under an ISI SS40 scanning electron microscope at Arizona State University. For BSE, an annular silicon solid state detector was used, placed perpendicular to the electron beam: working distance was 20 mm, and an acceleration voltage of 20 kV was used. Except for the garnets from southern Germany and New Mexico, which had already been studied in secondary electron mode (Borg, 1986; Hansley, 1987), garnets were also mounted and photographed in the secondary mode to demonstrate the presence of large scale facets. Here, the standard Everhart–Thornley secondary electron detector was used with an acceleration voltage of 20 kV; working distance was 40 mm.

Figures 1 to 7 show the results of these investigations. Although the faceted surfaces can be clearly discerned in thin section, neither the BSE nor the EDX techniques detected any changes in composition between the 'positive' faceted areas and the grain centres. This held true for garnets of all compositions,

Table 1. Sample details. For further information on geological setting, see Johnson & Stewart (1985) for samples 1–3, Gjelberg *et al.* (1987) for sample 4, Palmer (1985) for sample 5, Borg (1986) for sample 6, and Hansley (1986) for sample 7.

Sample & figure number	Location (borehole)	Depth (m)	Area	Formation & age	Depositional environment	Major garnet component
1	30/1–1	3123·3	Central North Sea	Forties (Palaeocene)	Submarine fan	Almandine Pyrope
2	21/18–2A	3013·3	Central North Sea	Fulmar (L. Jurassic)	Shallow marine	Almandine Pyrope
3	210/25–2	2619·4	Northern North Sea	Etive (M. Jurassic)	Shoreface	Almandine Grossular
4	6506/12–1	4009·0	Haltenbanken (mid-Norway)	Aldra (L. Jurassic)	Paralic	Almandine
5	McKee–4	2459·0	Taranaki (New Zealand)	McKee (Eocene–Oligocene)	Paralic	Spessartine Almandine
6	Funfbronn B3	106·2	Southern Germany	Keuper (Triassic)	Braided stream (arid)	Almandine
7	USGS S7	1019·3	NW New Mexico	Morrison (L. Jurassic)	Fluvial (arid)	Spessartine Almandine

whether they were Fe-Mg rich (Figs 1 & 2), dominantly Fe-rich (Figs 4 & 6), Fe-Ca rich (Fig. 3) or Fe-Mn rich (Figs 5 & 7). Consistent compositional similarity of garnet cores and facets cannot be reconciled with the overgrowth hypothesis. Even if, by some fortuitous chance, overgrowths in one example were to have the same composition as the core, it would be stretching the limits of belief to propose that in all cases garnet overgrowths had the same composition as their cores, despite the major variations in composition between the seven examples. Thus it must be concluded that, despite the wide geographical, environmental, stratigraphical and compositional variations present in the sample set, the facets must all have originated by dissolution processes.

Additional textural evidence for dissolution can be seen in Fig. 4a. This grain not only shows the typical faceted grain surface but also three parallel-oriented sub-grains, one long (x) and two short (y and z). EDX analysis indicated that these subgrains consist of TiO_2, probably rutile on crystallographic grounds. For the facets to have developed by overgrowth, it would be necessary to invoke crystal growth that engulfed three elongate rutile needles that already had a parallel orientation, a highly unlikely situation. The dissolution hypothesis, in contrast, would explain the relationship as the result of etching a garnet containing parallel inclusions of rutile.

Compositional zoning in detrital grains

The faceted grains analysed by combined BSE–EDX do not reveal significant internal compositional variation. As already discussed, this precludes an overgrowth origin for the facets, but also indicates that the degree of compositional zoning, if present, is below that which can be resolved by the technique. Quantitative electron microprobe analyses for cores and rims of the detrital garnets in Upper Triassic sandstones of southern Germany confirm that these are unzoned (Borg, 1986). The absence of zoning in these garnet populations is curious, considering that zoned garnets are widespread in many igneous and metamorphic source rocks. This may be in part a function of the small diameters of the garnet grains studied, generally *c.* 100 μm. In addition, the garnets analysed thus far may have been derived from high grade and/or slowly cooled source terrains in which diffusion may have eliminated original compositional variation. With this in mind, it is worthwhile considering garnets in metamorphic detritus stripped from a low grade polymetamorphic terrain.

Lower Old Red Sandstone (Early Devonian) alluvial sediments in the northwestern Midland Valley of Scotland contain abundant zoned almandine (Haughton, 1987; Haughton & Farrow, 1989). A typical detrital almandine is illustrated in Fig. 8a. The surface of the grain is characterized by well-developed facets. This grain was mounted in resin and ground to reveal a cross-section. Zoning maps of this cross-section (Fig. 8b) were made using a Geoscan 5 electron microprobe and a Link Systems energy-dispersive X-ray analyser.

The zoning maps show that there is a systematic variation in composition across the grain. There is a 2·5× increase in Mn and a concomitant fall in Fe, whereas Mg and Ca show only minor variation. The

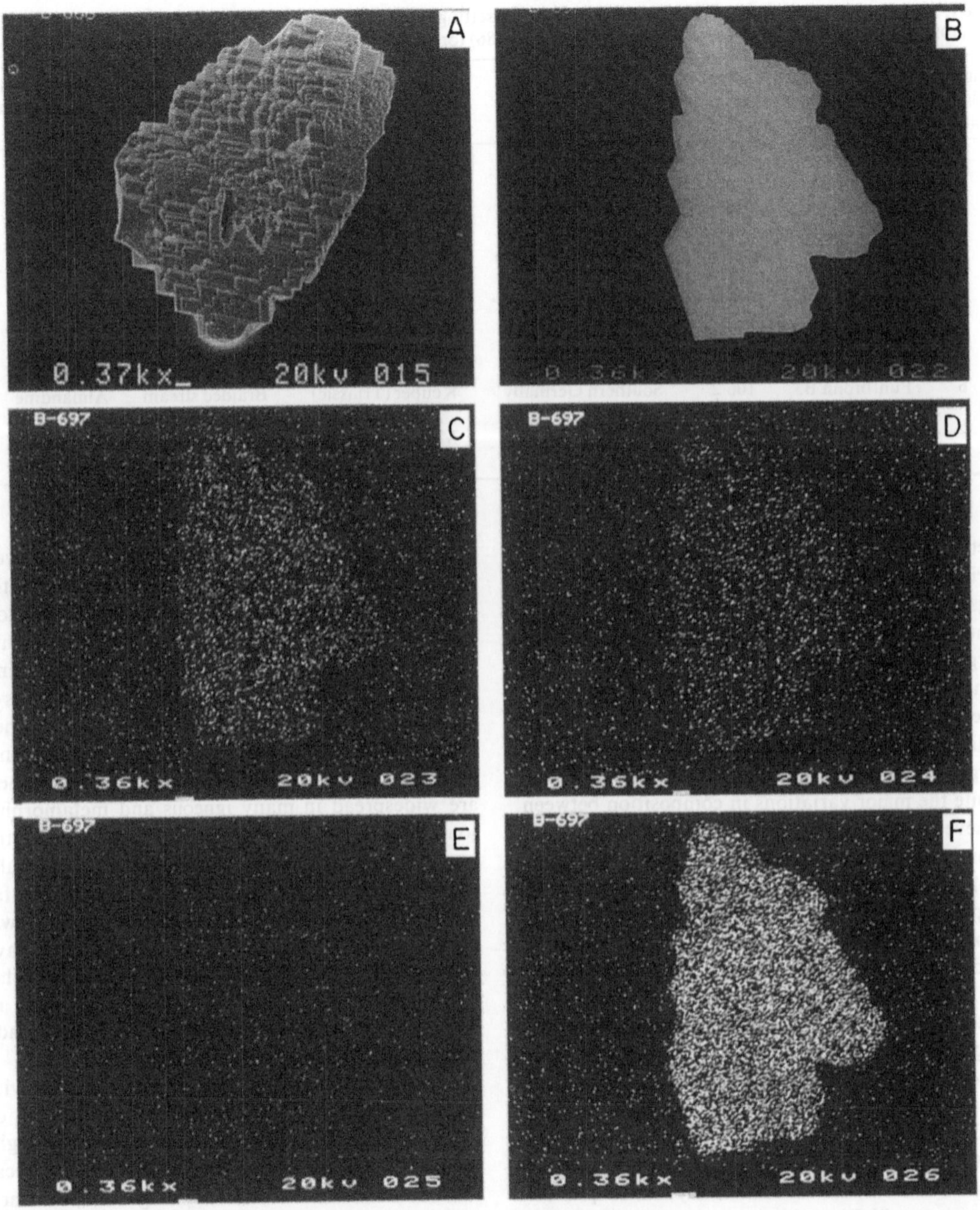

Fig. 1. Garnets from the Palaeocene Forties Formation in UK Central North Sea Well 30/1–1, 3123·3 m. Scale bar is 10 μm in all cases. (A) Typical surface textures viewed by secondary electron imaging. (B) Backscattered electron image of polished thin-section. Grain margin shows well-developed facets, yet no variations in grey level are observed. (C–F) Dot maps of, respectively, Mg, Ca, Mn and Fe contents of grain figured under BSE (Fig. 1b). Note lack of chemical variation.

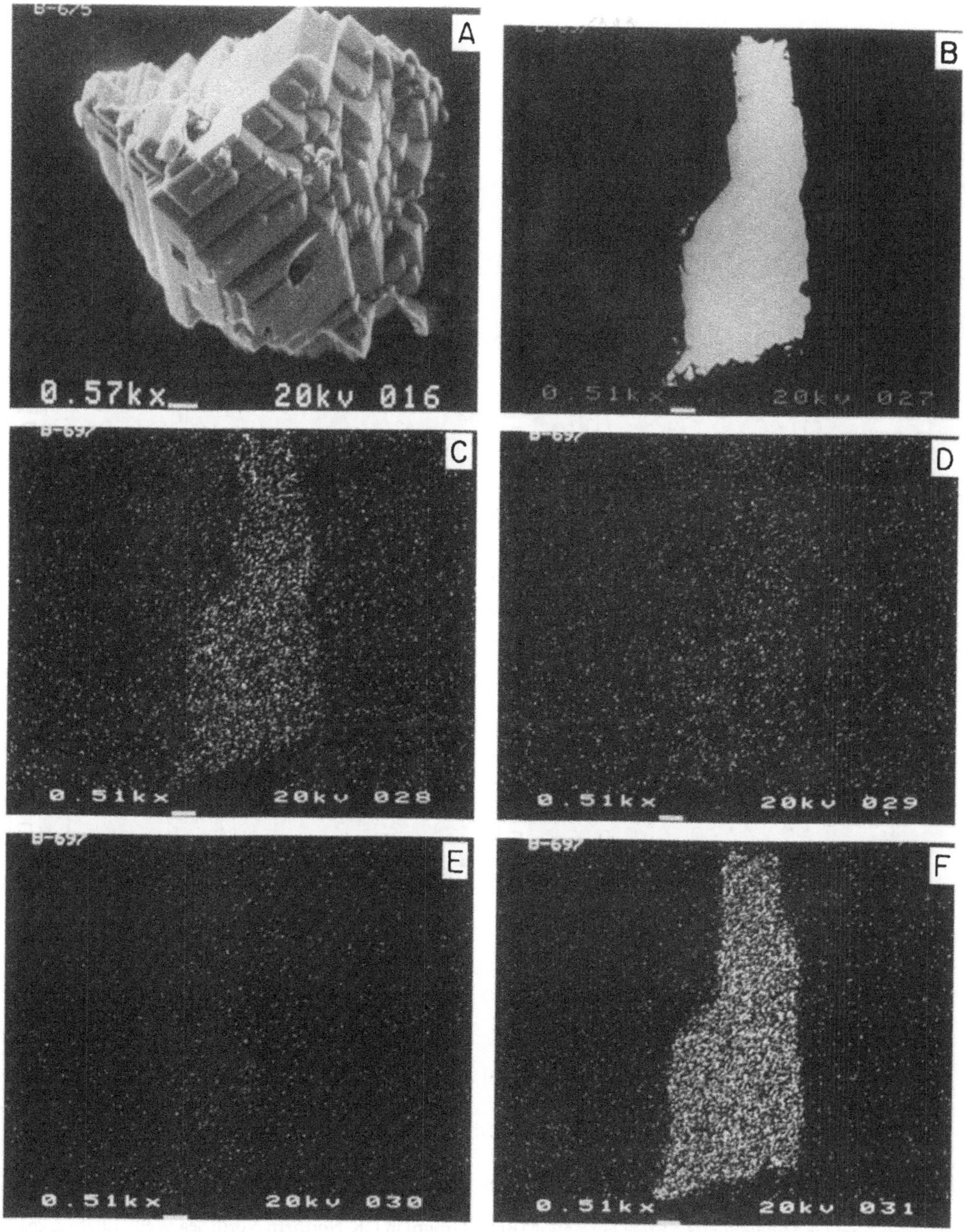

Fig. 2. Garnets from the Upper Jurassic Fulmar Formation in UK Central North Sea Well 21/18–2A, 3013·3 m. Scale bar is 10 μm in all cases. (A) Typical surface textures viewed by secondary electron imaging. (B) Backscattered electron image of polished thin-section. Grain margin shows well-developed facets, yet no variations in grey level are observed. (C–F) Dot maps of, respectively, Mg, Ca, Mn and Fe contents of grain figured under BSE (Fig. 2b). Note lack of chemical variation.

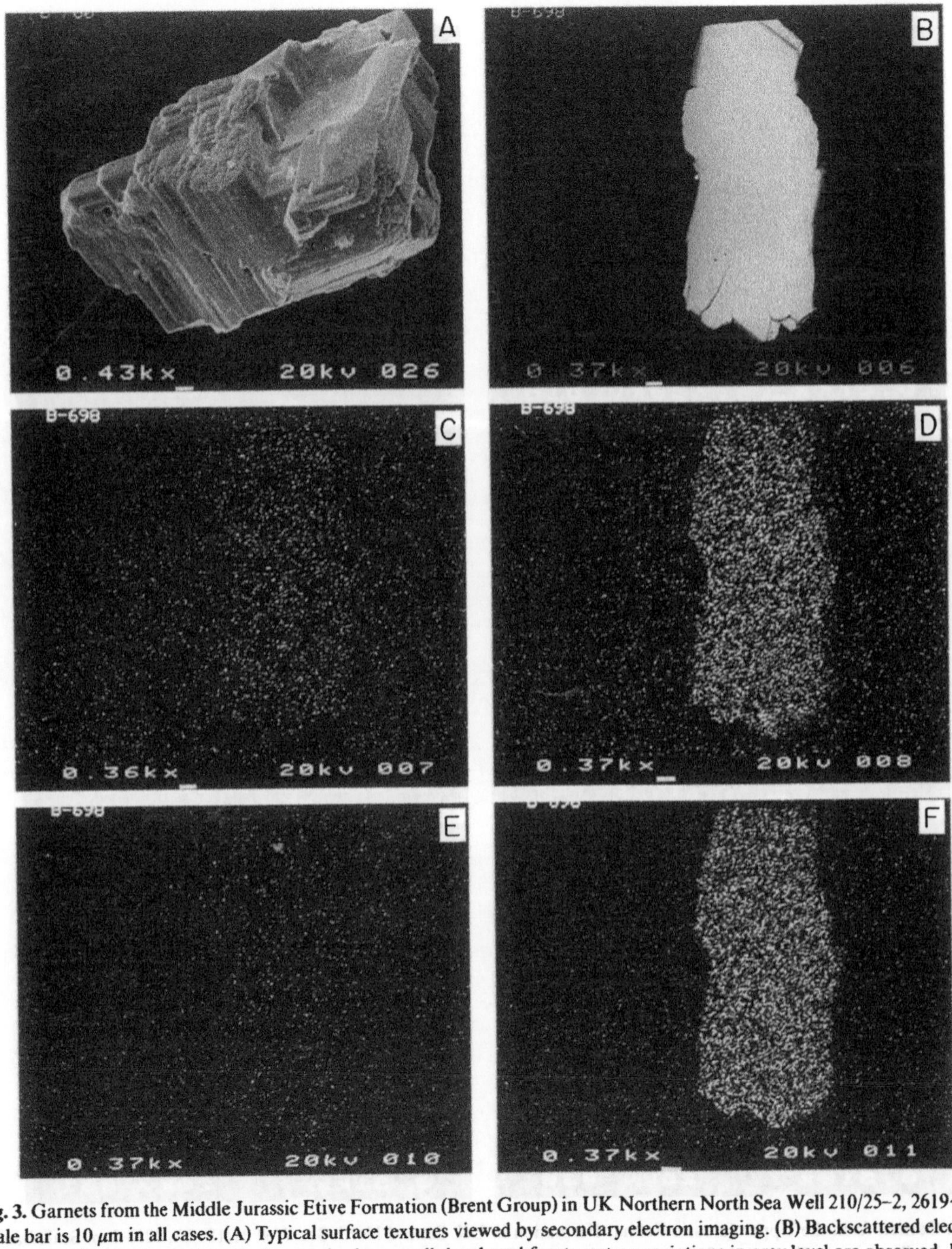

Fig. 3. Garnets from the Middle Jurassic Etive Formation (Brent Group) in UK Northern North Sea Well 210/25–2, 2619·4 m. Scale bar is 10 μm in all cases. (A) Typical surface textures viewed by secondary electron imaging. (B) Backscattered electron image of polished thin-section. Grain margin shows well-developed facets, yet no variations in grey level are observed. Edge effect visible where polished grain surface has positive relief with respect to embedding resin. (C–F) Dot maps of Mg, Ca, Mn and Fe contents of grain figured under BSE (Fig. 3b). Note lack of chemical variation.

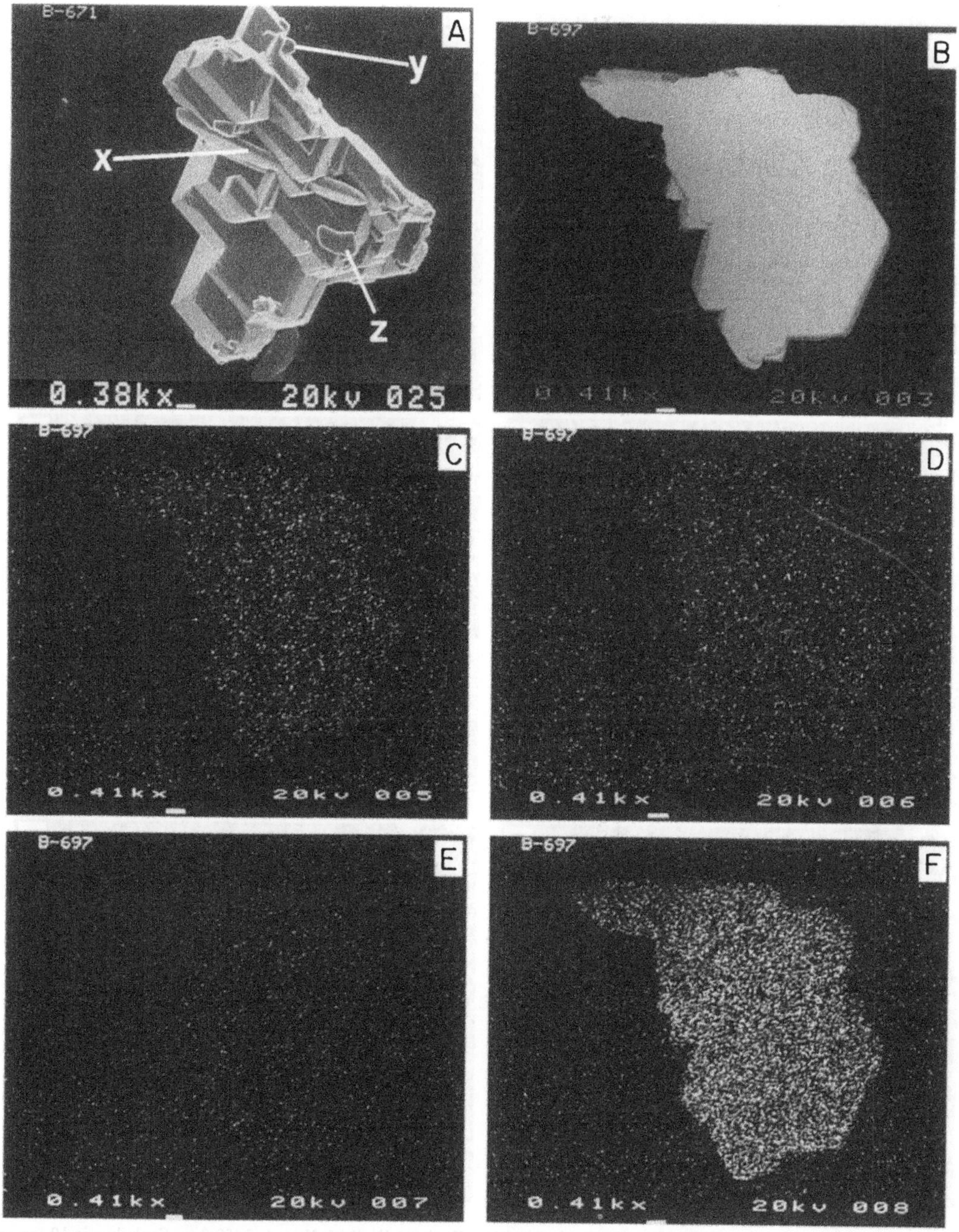

Fig. 4. Garnets from the Lower Jurassic Aldra Formation in Well 6506/12–1, 4009·0 m, from the Haltenbanken area, offshore mid-Norway. Scale bar is 10 μm in all cases. (A) Typical surface textures viewed by secondary electron imaging. Note parallel alignment of rutile inclusions x, y and z. (B) Backscattered electron image of polished thin-section. Grain margin shows well-developed facets, yet no variations in grey level are observed. Edge effect visible where polished grain surface has positive relief with respect to embedding resin. (C–F) Dot maps of, respectively, Mg, Ca, Mn and Fe contents of grain figured under BSE (Fig. 4b). Note lack of chemical variation.

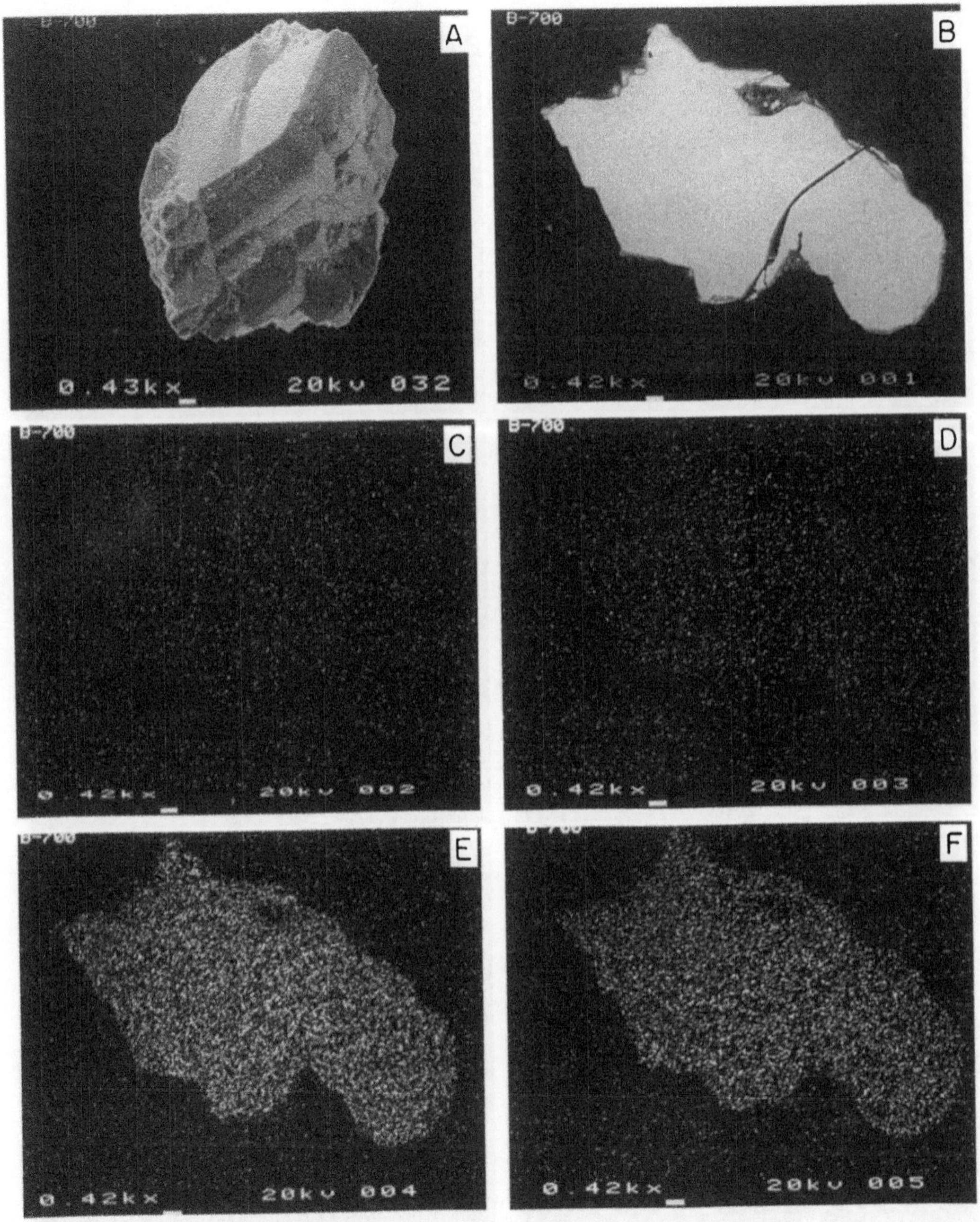

Fig. 5. Garnets from the Palaeogene McKee Formation in New Zealand Well McKee–4, 2459·0 m. Scale bar is 10 μm in all cases. (A) Typical surface textures viewed by secondary electron imaging. (B) Backscattered electron image of polished thin-section. Grain margin shows well-developed facets, yet no variations in grey level are observed. (C–F) Dot maps of, respectively, Mg, Ca, Mn and Fe contents of grain figured under BSE (Fig. 5b). Note lack of chemical variation.

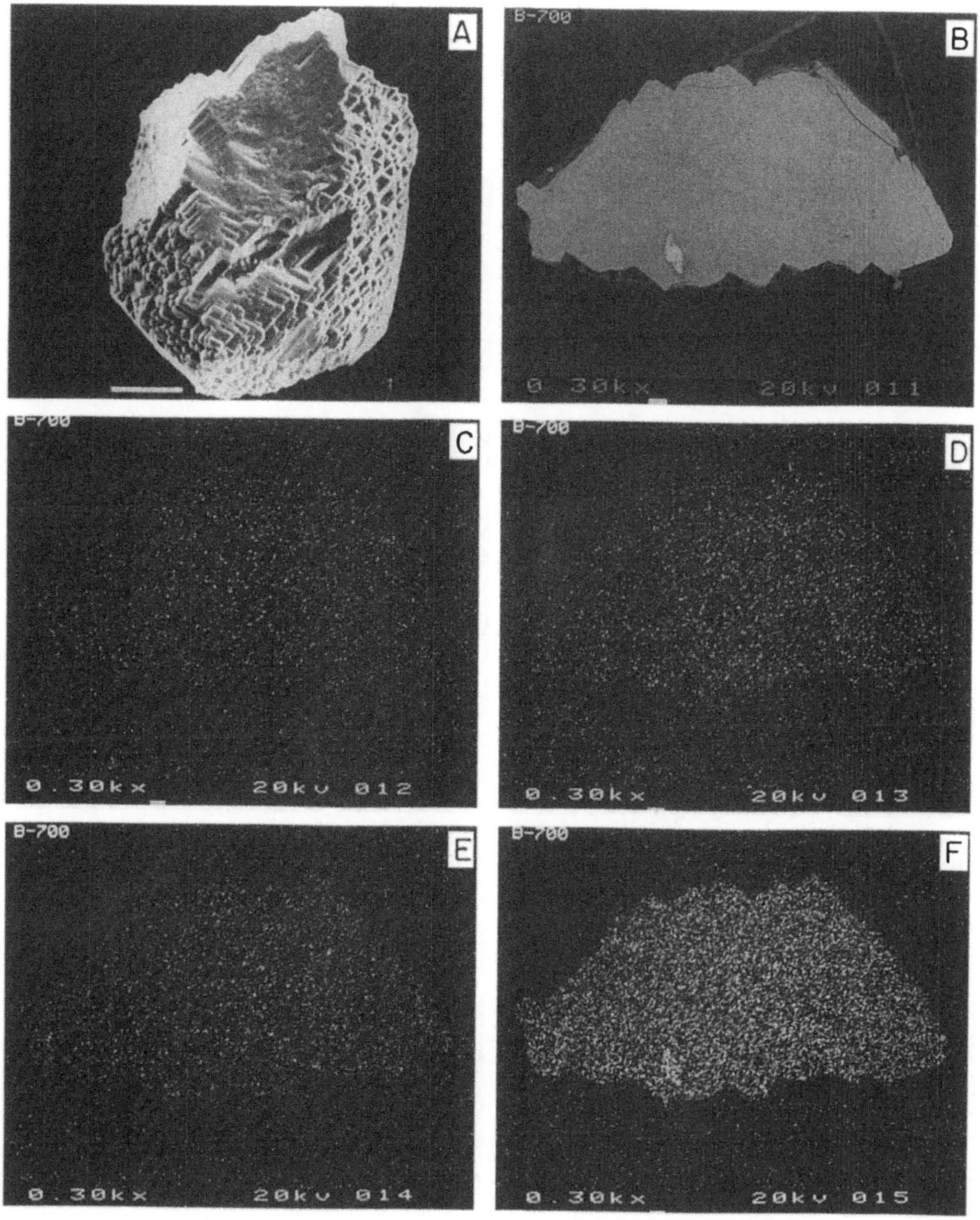

Fig. 6. Garnets from the Triassic Keuper Formation from southern Germany, Borehole Funfbronn B3, 106·2 m. Scale bar is 50 μm in (A), and 10 μm in the other cases. (A) Typical surface textures viewed by secondary electron imaging. (B) Backscattered electron image of polished thin-section. Grain margin shows well-developed facets, yet no variations in grey level are observed. Bright area is an Fe-rich inclusion (see Fe dot map, Fig. 6f). (C–F) Dot maps of, respectively, Mg, Ca, Mn and Fe contents of grain figured under BSE (Fig. 6b). Note lack of chemical variation.

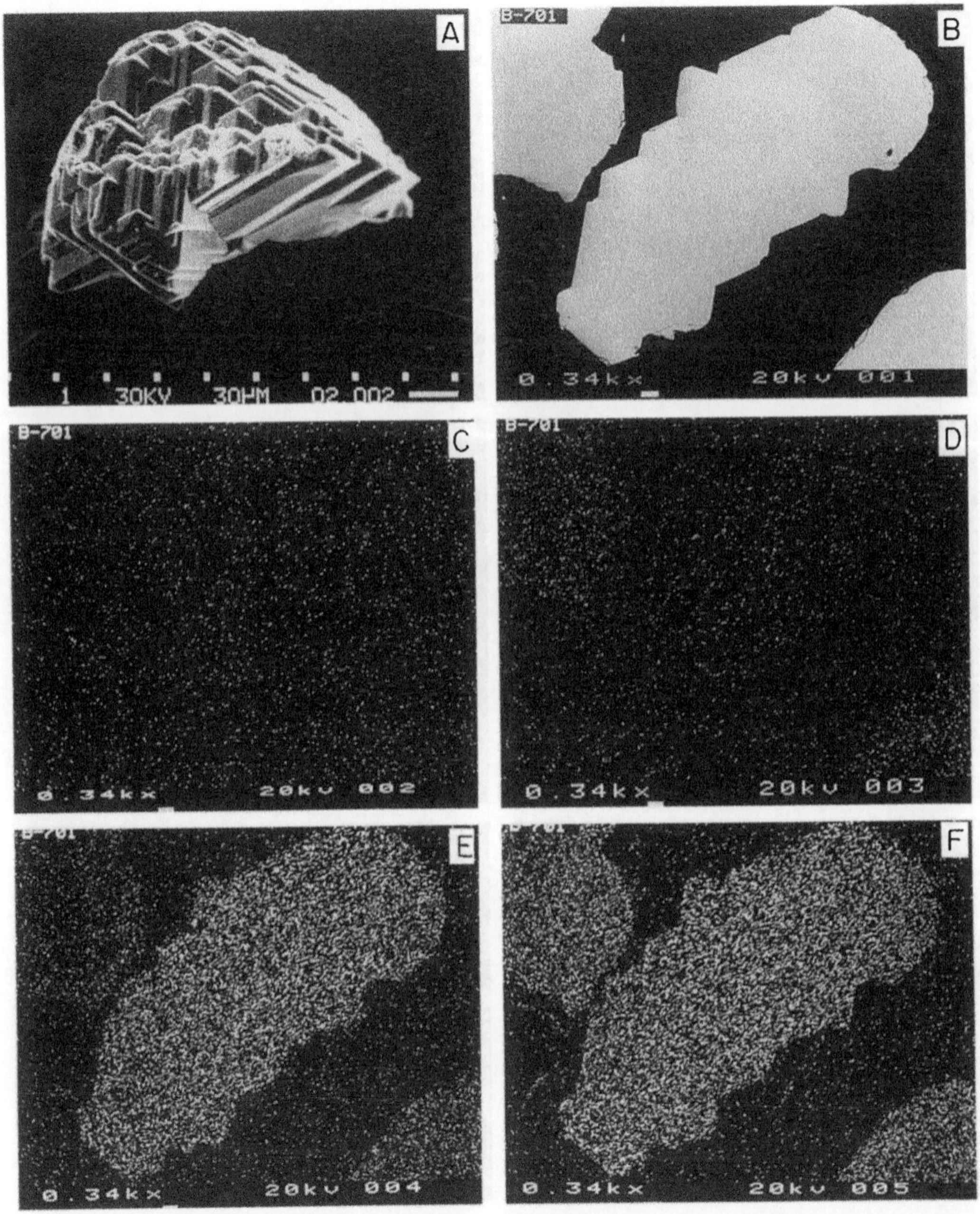

Fig. 7. Garnets from the Upper Jurassic Morrison Formation, NW New Mexico, USA, from USGS S7, 1019·3 m. Scale bar is 30 μm in (A), and 10 μm in the other cases. (A) Typical surface textures viewed by secondary electron imaging. (B) Backscattered electron image of polished thin-section. Grain margin shows well-developed facets, yet no variations in grey level are observed. (C–F) Dot maps of, respectively, Mg, Ca, Mn and Fe contents of grain figured under BSE (Fig. 3b). Note lack of chemical variation.

variation is not concentric about the grain core; rather, Mn concentrations increase steadily across the grain, from one rim, across the core, to the opposite rim. In normal or growth zoned metamorphic almandines, Mn is often concentrated at the core, resulting in typical bell-shaped profiles that have been attributed to progressive Mn depletion of the matrix in which the garnet grew. The detrital grain of Fig. 8 probably represents a fragment of a much larger concentrically-zoned crystal which was broken up during transport of the sediment from the source to the basin. This is consistent with the slight curvature of the Mn contours.

The most significant feature of the compositional zonation is that it can be traced right up to the grain rim where it is truncated by the faceted grain surface. Spots analysed on the positive areas of the grain boundary have compositions which conform to the zoning pattern established in the grain interior. This, together with the variation in composition around the grain circumference, cannot be explained if the facets formed as overgrowths. The retention of part of the original zoning profile is potentially of use in assessing the ultimate provenance of the grain. Zoning profiles, in addition to garnet compositions, can vary across a metamorphic source terrain. Establishing the pattern of zoning within grains may thus enhance the garnet geochemical technique, with the proviso that due consideration is given to the possible role of cross-section orientation in determining the zoning profile derived. Zoning is more likely to be encountered in relatively coarse detrital grains.

DISCUSSION

The geological literature documents the occurrence of faceted garnets in sandstones ranging in age from Cambrian to Tertiary that have been subjected to burial diagenesis. New data presented in this paper, together with the previously published data, now provide very strong evidence for garnet faceting by dissolution, rather than overgrowth. The arguments can be summarized as follows:

(1) Experiments with organic and inorganic solvents have successfully produced etched garnets, but only organic acids have produced facets with morphologies similar to naturally-occurring faceted garnets.

(2) Naturally-occurring faceted garnets are found in

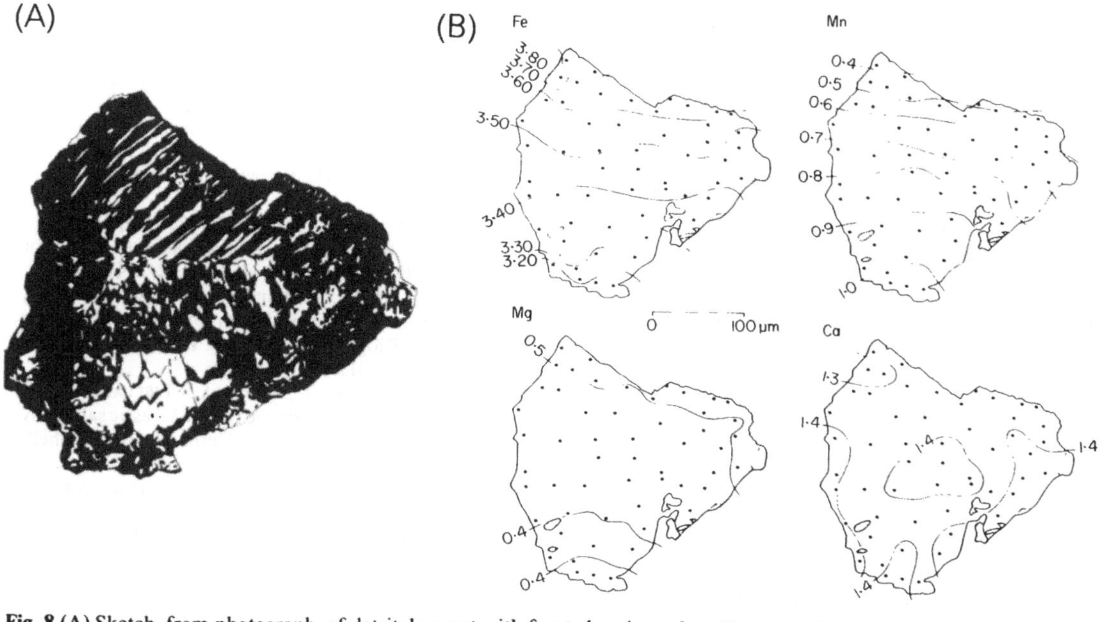

Fig. 8 (A) Sketch, from photograph, of detrital garnet with faceted grain surface. From the Lower Old Red Sandstone, near Comrie, Scotland (sample MV116·72). (B) Zoning maps contoured in cations per formula unit on the basis of 24 oxygens. Dots represent analysed points. Stippled areas are quartz inclusions.

zones of garnet depletion and adjacent to zones from which garnet has been entirely dissolved.

(3) Textural relationships, both between garnet and surrounding authigenic clay, carbonate and silica cements and between garnet grains and their inclusions, are not compatible with the overgrowth hypothesis.

(4) Thermodynamic data indicate that garnet overgrowth is not feasible in normal diagenetic conditions, unless the secondary material has the composition of a hydro-garnet. Such material was not recorded in this study; nor has it been reported elsewhere.

(5) In a wide variety of detrital garnets, compositional differences cannot be detected between grain centres and the positive areas at grain margins defined by the facets.

Based on the above observations, we believe that garnet overgrowths are unlikely to be generated in siliciclastic sequences under normal diagenetic conditions. We also believe that the overwhelming amount of evidence in favour of this view now places the onus of proof firmly with the proponents of the authigenic garnet hypothesis.

A final point concerns the implications for provenance studies that employ the garnet geochemical technique (e.g. Morton, 1985, 1987). This approach relies on rapid electron microprobe analysis of grain surfaces to delineate the compositional range of garnet suites. The value of this technique would clearly be doubtful if facets on garnets were actually overgrowths. This study verifies that many small detrital garnets (c. 100 μm diameter) show little difference in composition between grain margins and grain centres, an observation made earlier (Morton, 1985; Borg, 1986). When compositional zoning is present, variations are commonly not concentrically arranged around the grain core. Determination of a relatively large number of rim compositions from different grains should therefore bracket the range of compositional zoning.

CONCLUSIONS

Combined BSE–EDX studies and other evidence indicate that garnet facets are not overgrowth phenomena but are the result of dissolution processes. Dissolution is apparently facilitated by increased pore fluid temperature. Geological evidence suggests that increased pore fluid temperature is critical to the formation of faceted garnets, because none have been reported from soils or other recent sediments. Further research is required to determine the geochemistry of etching fluids; determine probable temperatures at which garnet dissolution occurs by reference to vitrinite reflectance, apatite fission track dating and stable isotope analysis of associated authigenic phases; and evaluate the role of pressure in garnet dissolution. In this way, etched garnet, together with other diagenetically-sensitive heavy minerals, may be used as indicators of diagenetic environment and thermal history.

ACKNOWLEDGMENTS

We are grateful to David Smale (New Zealand Geological Survey) for supplying garnets from the McKee Formation, Taranaki, New Zealand. We would also like to thank Maria Mange and Michael Velbel for thoughtful and constructive reviews. This paper is published with the approval of the Director, British Geological Survey (NERC).

REFERENCES

AFANASEV, V.P. (1985) The genesis of solution relief on garnets of the pyrope–almandine series. *Zapiski Vsgsoyuznogo Mineralogischeskogo Obshchestva*, **114 (1)**, 73–80.

BORG, G. (1986) Facetted garnets formed by etching—examples from sandstones of Late Triassic age, south Germany. *Sedimentology*, **33**, 141–146.

BRAMLETTE, M.N. (1929) Natural etching of detrital garnet. *Am. Miner.*, **14**, 336–337.

BUTTERFIELD, J.A. (1936) Outgrowths on zircon. *Geol. Mag.*, **73**, 511–516.

CAROTHERS, W.W. & KHARAKA, K. (1978) Aliphatic acid anions in oil-field waters—implications for origin of natural gas. *Bull. Am. Ass. Petrol. Geol.*, **62**, 2441–2483.

GAUTIER, D.L. (1979) Preliminary report of authigenic, euhedral tourmaline crystals in a productive gas reservoir of the Tiger Ridge Field, North-Central Montana. *J. sedim. Petrol.*, **49**, 911–916.

GJELBERG, J., DREYER, T., HØIE, A., TJELLAND, T. & LILLENG, T. (1987) Late Triassic to mid-Jurassic sandbody development on the Barents and mid-Norwegian shelf. In: *Petroleum Geology of north-west Europe*. Ed. by K.W. Glennie & J. Brooks, pp. 1105–1129. Graham & Trotman, London.

GRAVENOR, C.P. & LEAVITT, R.K. (1981) Experimental formation and significance of etch patterns on detrital garnets. *Can. J. Earth Sci.*, **18**, 765–775.

GUSTAFSON, W.I. (1974) The stability of andradite, hedenbergite and related minerals in the system Ca–Fe–Si–O–H. *J. Petrol.*, **15**, 455–496.

HANSLEY, P.L. (1986) Regional diagenetic trends and

uranium mineralization in the Morrison Formation across the Grants uranium region. In: *A Basin Analysis Case Study: The Morrison Formation, Grants Uranium Region.* (Ed. by C.E. Turner-Petersen, E.S. Santos & N.S. Fishman). *Am. Ass. Petrol. Geol. Studies in Geology*, **22**, 277–302.

HANSLEY, P.L. (1987) Petrologic and experimental evidence for the etching of garnets by organic acids in the Upper Jurassic Morrison Formation, northwestern New Mexico. *J. sedim. Petrol.*, **57**, 666–681.

HAUGHTON, P.D.W. (1987) Composition of Lower ORS detrital garnets–some preliminary observations. *Publ. Depart. Geology and Mineralogy, University of Aberdeen*, **6**, 56 (abstract).

HAUGHTON, P.D.W. & FARROW, C.M. (1989) Compositional variation in Lower ORS detrital garnets from the Midland Valley of Scotland and the Anglo–Welsh basin: implications for Early Devonian source areas, Acadian fault motions and Laurasian palaeogeography. *Geol. Mag.*, in press.

HEMINGWAY, J.E. & TAMAR-AGHA, M.Y. (1975) The effects of diagenesis on some heavy minerals from the sandstones of the Middle Limestone Group in Northumberland. *Proc. York. geol. Soc.*, **40**, 537–546.

HOLDAWAY, M.J. & LEE, S.M. (1977) Fe–Mg cordierite stability in high-grade pelitic rocks based on experimental, theoretical and natural observations. *Contr. Miner. Petrol.*, **63**, 175–198.

HOWIE, R.A., SIMPSON, I.M. & SIMPSON, G.S. (1980) The nature of the late Paleozoic glaciation in Gondwana as determined from an analysis of garnets and other heavy minerals: Discussion. *Can. J. Earth Sci.*, **17**, 957–958.

HUANG, W.H. & KELLER, W.D. (1970) Dissolution of rock-forming silicate minerals in organic acids: simulated first-stage weathering of fresh mineral surfaces. *Am. Miner.*, **55**, 2076–2094.

HUCKENHOLZ, H.G. & KNITTEL, D. (1975) Uvarovite: stability of uvarovite–grossularite solid solution at low pressure. *Contr. Miner. Petrol.*, **49**, 211–232.

JOHNSON, H.D. & STEWART, D.J. (1985) Role of clastic sedimentology in the exploration and production of oil and gas in the North Sea. In: *Sedimentology: Recent Developments and Applied Aspects.* (Ed. by P.J. Brenchley & B.P.J. Williams) *Spec. Publ. geol. Soc. London*, **18**, 249–310.

KEESMANN, I., MATTHES, S., SCHREYER, W. & SEIFERT, F. (1971) Stability of almandine in the system FeO–(Fe_2O_3)–Al_2O_3–SiO_2–(H_2O) at elevated pressures. *Contr. Mineral. Petrol.*, **31**, 132–144.

KRINSLEY, D., PYE, K. & KEARSLEY, A. (1983) Mudrocks examined by backscattered electron microscopy. *Geol. Mag.*, **120**, 109–114.

KUZEL, H.J. (1968) Uber die diadochie von Al^{3+}, Cr^{3+} und Fe^{3+} in $3CaO.Al_2O_3{\cdot}6H_2O$ oberhalb 50°C. *Neues J. Miner. Monatshefte*, **1968**, 87–96.

MCBRIDE, E.F. (1985) Diagenetic processes that affect provenance determinations in sandstones. In: *Provenance of Arenites.* (Ed. by G.G. Zuffa), pp. 95–113. Reidel, Dordrecht.

MCMULLEN, R.M. (1959) Etched detrital garnet from the Cardium Formation, Pembina area, central Alberta. *J. Alberta Soc. Petrol. Geol.*, **7**, 272–274.

MADER, D. (1980) Authigener granat im Bundsandstein der Westeifel. *Jahresberichte und Mitteilungen des Oberrheinischen Geologischen Vereines*, **62**, 217–227.

MADER, D. (1981) Diagenesis of the Bundsandstein (Lower Triassic) in western Eifel (Germany). *Neues J. Mineral. Abh.*, **142**, 1–26.

MATTHES, S. (1961) Ergebnisse zur granatsynthese und ihre beziehung zur naturlichen granatbildung innerhalb der pyralspit-gruppe. *Geochim. Cosmochim. Acta*, **23**, 233–294.

MAURER, H. (1982) Oberflächentexturen an schwermineralkörner aus der Unteren Süsswassermolasse (Chattien) der Westschweiz. *Eclog. geol. Helv.*, **75**, 23–31.

MORTON, A.C. (1984) Stability of heavy minerals in Tertiary sandstones from the North Sea Basin. *Clay Miner.*, **19**, 287–308.

MORTON, A.C. (1985) A new approach to provenance studies: electron microprobe analysis of detrital garnets from Middle Jurassic sandstones of the northern North Sea. *Sedimentology*, **32**, 553–566.

MORTON, A.C. (1986) Dissolution of apatite in North Sea Jurassic sandstones: implications for the generation of secondary porosity. *Clay Miner.*, **21**, 711–733.

MORTON, A.C. (1987) Influences of provenance and diagenesis on detrital garnet suites in the Paleocene Forties sandstone, central North Sea. *J. sedim. Petrol.*, **57**, 1027–1032.

MORTON, A.C. & HUMPHREYS, B. (1983) The petrology of the Middle Jurassic sandstones from the Murchison Field, North Sea. *J. Petrol. Geol.*, **5**, 245–260.

NICKEL, E.H. (1973) Experimental dissolution of light and heavy minerals in comparison with weathering and intrastratal solution. *Contr. Sedimentol.*, **1**, 1–68.

PALMER, J. (1985) Pre-Miocene lithostratigraphy of Taranaki Basin, New Zealand. *N. Z. J. Geol. Geophys.*, **28**, 197–216.

PREOBRAZHENSKY, I.A. (1941) On authigenous minerals and mineral formation. *Proc. Inst. Geol. Sci., Academy of Sciences, USSR, Petrography Series*, **40 (13)**, 41–53.

PUSTOWALLOFF, L.W. (1955) Uber sekundare veranderungen der sedimentgesteine. *Geol. Rdsch.*, **43**, 535–550.

PYE, K. & KRINSLEY, D. (1984) Petrographic examination of sedimentary rocks in the SEM using backscattered electron detectors. *J. sedim. Petrol.*, **54**, 877–888.

RAHMANI, R.A. (1973) Grain surface etching features of some heavy minerals. *J. sedim. Petrol.*, **43**, 882–888.

SALVINO, J.F. & VELBEL, M.A. (1989) Facetted garnets from sandstones of the Munising Formation (Cambrian), northern Michigan: petrographic evidence for origin via intrastratal solution. *Sedimentology*, in press.

SAUER, A. (1900) Granat authigenes hemingtheil im Bunter Keuper. *Bericht Oberrheinischen Geologischen Verhandlungen*, **33**, 42–46.

SERDYUCHENKO, D.P. & DOBROTVORSKAYA (1949) On certain new formations of minerals in sedimentary rocks. *Doklady Akademii Nauk SSSR Seriya Geologiya*, **69 (3)**, 423–425.

SIMPSON, G.S. (1976) Evidence of overgrowths on, and solution of, detrital garnets. *J. sedim. Petrol.*, **46**, 689–693.

SMALE, D. & MORTON, A.C. (1987) Heavy mineral suites of core samples from the McKee Formation (Eocene–Lower Oligocene), Taranaki: implications for provenance and diagenesis. *N. Z. J. Geol. Geophys.*, **30**, 299–306.

SMITHSON, F. (1941) The alteration of detrital minerals in the Mesozoic rocks of Yorkshire. *Geol. Mag.*, **78**, 97–112.

SOBOLEV, V.S., VARTANOVA, N.S. & SHAINYUK, A.I. (1951) The question of growth of grains of garnet and other metamorphic minerals in sedimentary rocks. *Zapiski Vsgsoyuznogo Mineralogischeskogo Obshchestva*, **80 (2)**, 122–128.

STAUDIGEL, H. & SCHREYER, W. (1977) The upper thermal stability of chlinochlore, $Mg_5Al\ (AlSi_3O_{10})\ (OH)_8$ at 10–35 kb pH_2O. *Contr. Miner. Petrol.*, **61**, 187–198.

STORRE, B. (1970) Stabilitatsbedingungen grossular-fuhrender paragenesen im system CaO–Al_2O_3–SiO_2CO_2–H_2O at elevated pressures. *Contr. Miner. Petrol.*, **31**, 132–144.

SURDAM, R.C., BOESE, S.W. & CROSSEY, L.J. (1984) The chemistry of secondary porosity. In: *Clastic Diagenesis*. (Ed. by D.A. McDonald & R.C. Surdam). *Mem. Am. Ass. Petrol. Geol.*, **37**, 127–149.

VELBEL, M.A. (1984) Natural weathering mechanisms of almandine garnet. *Geology*, **12**, 631–634.

WHITNEY, C.G. & NORTHROP, H.R. (1987) Diagenesis and fluid flow in the San Juan Basin, New Mexico—regional zonation in the mineralogy and stable isotope composition in sandstone. *Am. J. Sci.*, **287**, 353–382.

ZAPOROZHTZEVA, A.S. (1960) On the origin of stepped surfaces of disintegrated grains of garnet of the Cretaceous sediments of northern Yakutiya. *Doklady Akademii Nauk SSSR Seriya Geologiya*, **131 (2)**, 420–423.

ZILI, Z. (1988) Authigenic origin and formation temperature of step-like garnet from lower Tertiary sandstone, Shengli oil field, Shandong. *Acta Sedimentol. Sinica*, **6**, 13–20.

ZILI, Z. & ZHENGMOU, L. (1987) The burial diagenesis and reservoir assessment of Tertiary clastic rocks in Shengli Oil Field, Shandong. *Earth Sci. J. Wuhan Coll.*, **12**, 311–319.

(Manuscript received 22 July 1988; revision received 20 December 1988)

Effect of oil on sandstone mesogenesis

Sandstone diagenesis requires an aqueous medium because all reactions occur by dissolution and reprecipitation rather than as solid state processes. Consequently, oil emplacement into a sandstone is assumed to stop diagenesis. Reality is not quite so simple as this assertion because there is usually a thin film of water on mineral surfaces remaining after the emplacement of oil. How-ever, the observation that oil-bearing sandstones have undergone less mesogenetic transformation than the underlying aquifer (Yurkova, 1970) was used as early evidence that sandstone diagenesis is at least slowed down in the presence of oil.

Adding oil to sandstone can have more subtle effects on diagenesis. Post-emplacement biodegradation of oil typically leads to increased quantities of CO_2 in the remaining petroleum. Elevated CO_2 usually leads to increased acidity in the formation water. Biodegradation of oil is thus cited as the ultimate cause of localized enhancement of plagioclase dissolution (Ehrenberg & Jakobsen, 2001). It is not yet clear whether biodegradation of petroleum routinely induces mesogenetic alteration of sandstones, or whether this is an exceptional case.

Reprinted from Sedimentology (1970) **15**, *53–68*

COMPARISON OF POST-SEDIMENTARY ALTERATIONS OF OIL-, GAS- AND WATER-BEARING ROCKS

R. M. YURKOVA

Geological Institute, U.S.S.R. Academy of Sciences, Moscow (U.S.S.R.)

(Received March 6, 1970)

SUMMARY

One of the productive horizons (17th) of north Sakhalin has been studied. The areas within and beyond oil and gas pools, near the top and at the flanks of structures were compared with areas where the horizon under study was water-bearing. The rocks were least transformed by secondary processes within oil accumulations, where a sudden increase was observed in the content of feldspars and accessory minerals (epidote, ilmenite and sphene) which had not undergone intrastratal solution. Inside gas pools, however, no deceleration of epigenesis has been observed. This points to a more recent (post-Pliocene) accumulation of gas, whereas the first oil migration in these Miocene strata occurred in the Upper Pliocene, according to the hampered epigenesis.

INTRODUCTION

The inflow of oil which forces water out of the pore space of the reservoir rocks results in a drastic change of the conditions which certainly will affect the processes of secondary mineral genesis. These processes are known to take place in sedimentary rocks in the presence of pore water. It is to be expected that the expulsion of pore water by inflowing oil will slow down the secondary processes. The concept of "braking" of the authigenic mineral formation processes by oil present in the rocks was developed in a number of studies (ORLOVA, 1958; CHEPIKOV et al., 1959, 1960, 1967; FÜCHTBAUER, 1961; MILLOT, 1964; KLUBOVA, 1965; PEROZIO, 1965; PROZOROVICH, 1967, and others); it has been applied in petroleum geology for estimating the time of oil accumulation in reservoir rocks. The techniques used to assess oil accumulation time on the basis of epigenetic rock transformations are discussed by PROZOROVICH (1967).

This paper discusses the results of a comparative study of post-sedimentary alteration of reservoir rocks of the 17th productive horizon of north Sakhalin in respect of its distribution over a number of oil- and gas-bearing structures. The discussion is made in terms of the concept of oil retarding the secondary mineral formation processes.

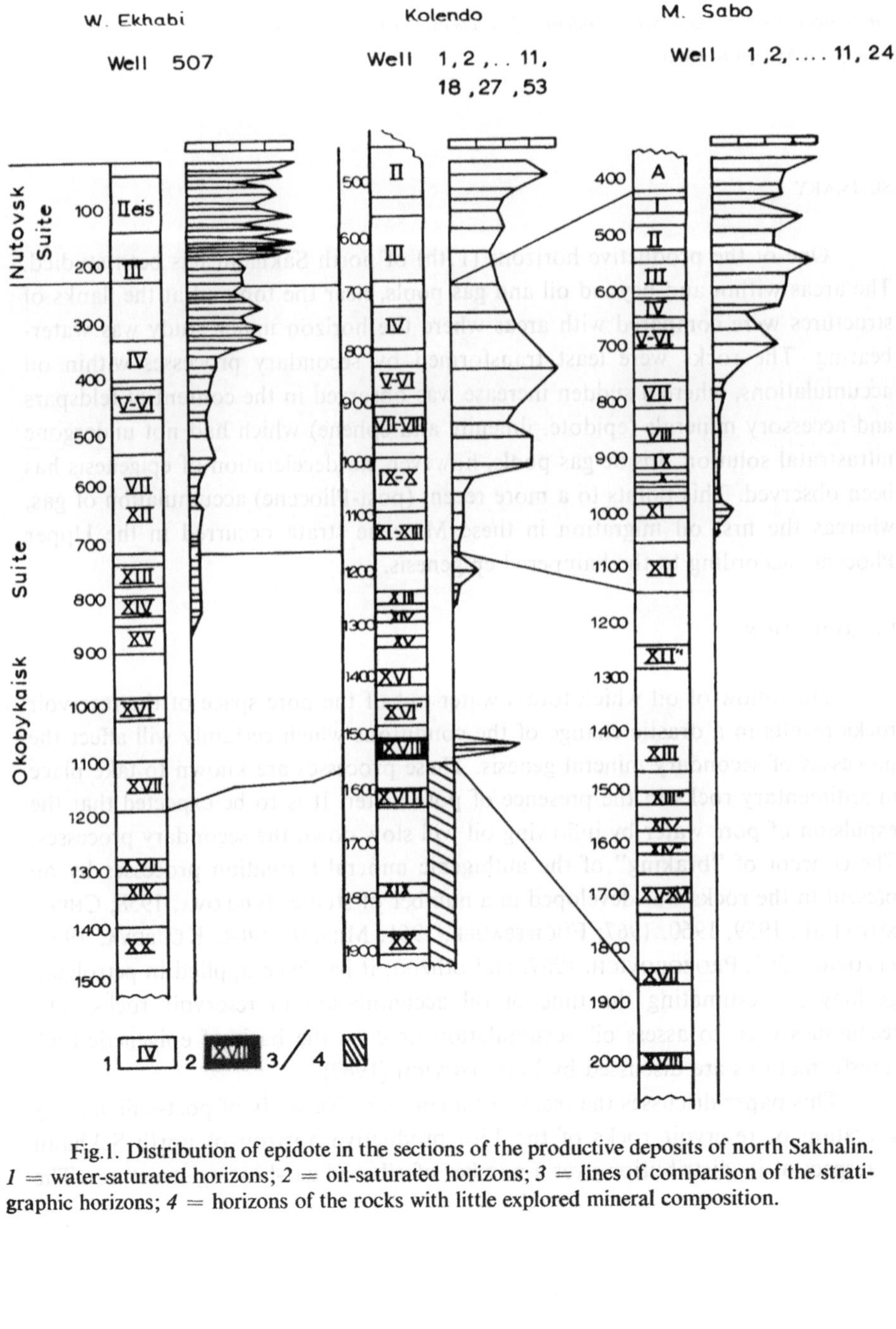

Fig.1. Distribution of epidote in the sections of the productive deposits of north Sakhalin. *1* = water-saturated horizons; *2* = oil-saturated horizons; *3* = lines of comparison of the stratigraphic horizons; *4* = horizons of the rocks with little explored mineral composition.

CHOICE AND DESCRIPTION OF THE OBJECT OF STUDY

The study of post-sedimentary change of the sandstones of the 17th horizon was preceded by an investigation of the composition of the rocks of the whole productive series and of the nature of their secondary transformation. It had been established that the sandstones and siltstones of the productive series are represented by quartz–feldspathic greywackes with polymineralllic (chlorite, kaolinite, hydromica, montmorillonite) clay matter and carbonate cement (YURKOVA, 1968). The mineral composition of these rocks, both the detrital and the secondary part, shows a regular variation across the section. A gradual impoverishment, down the section, of the composition of clastic components due to the disappearance and the replacement of unstable minerals (sphene, ilmenite, garnets, epidote, feldspars) and an expansion of the complex of authigenic minerals are clearly observable. This regularity is abruptly disturbed within the oil-pool boundaries (Fig.1). The disappearance of unstable components and the formation of authigenic rocks is hampered here as compared with the water-saturated areas of the oil-bearing series. This retardation is especially manifest in the deep productive horizons where some of the unstable components (sphene, epidote) disappear almost completely outside the oil accumulations. The 17th horizon was subjected to the study, because it formed in the period of the widespread Upper Miocene transgression and because it is well detectable in the Neogene sections, and contains the greatest amounts of gas and oil deposits as compared with the other horizons of the productive series (KLESHCHEV, 1966).

The horizon usually consists of some three or four sandstone beds separated by thin clay interlayers. The medium- or fine-grained sandstones frequently contain a considerable (up to 30%) admixture of silt. The horizon thickness varies within a wide range of 20–160 m, diminishing towards the crest. The anticlinal structures to which the rocks of this horizon were confined as well as data on its subsidence depth, temperature conditions, composition and mineralization of subsurface waters, and on the specific weight of the oil are given in Table I. According to some authors (ALEKSEICHIK et al., 1963; EVDOKIMOVA, 1963; SOLOMATINA, 1963; KLESHCHEV, 1966; etc.) the anticlinal structures of north Sakhalin being studied were formed simultaneously with the accumulation of the deposits. A majority of these were on the whole completely formed by the end of the Upper Miocene, except for the Kolendo fold, which somewhat lagged behind in its evolution (KLESHCHEV, 1966).

THE POST-SEDIMENTARY TRANSFORMATION OF THE SANDSTONES OF THE 17TH HORIZON

The post-sedimentary changes of the sandstones of the 17th horizon were considered separately for the various components: clastic rock-forming and acces-

TABLE I

DEPTH OF OCCURRENCE AND PROPERTIES OF UNDERGROUND WATERS AND OILS OF THE 17TH HORIZON (ACCORDING TO VARIOUS AUTHORS AND THE DATA FILED AT THE MOSCOW ALL-UNION RESEARCH INSTITUTE OF GEOLOGICAL EXPLORATION: AVERAGE VALUES)

Anticlinal structures	*Beds (local nomenclature)*	*Depth of occurrence of the bed roof (m)*	*Specific weight of oil (average minimal value in g/cm³)*	*Type of underground waters*	*Underground water mineralization (g/l)*	*Temperature (°C)*
Odoptu	XX	2,000–2,300	0.848	hydrocarbonaceous-sodic	24	—
Kolendo	XVII	1,360–1,560	0.878	hydrocarbonaceous-sodic	23	55
East Ekhabi (underthrust)	25	1,300–1,600	0.858	hydrocarbonaceous-sodic	25	50
East Ekhabi (thrust)						
northern part	XVII	700–1,400	0.848	hydrocarbonaceous-sodic	16	45
central part	XVII	400–800	0.864	hydrocarbonaceous-sodic	19	20
Ekhabi	XVII	750–900	0.865	hydrocarbonaceous-sodic	21	28
Kydylanyi	X	1,900–2,150	gas	hydrocarbonaceous-sodic	10	65
Tungor	XVII	1,840–1,960	gas	hydrocarbonaceous-sodic	24	60
Sabo	XVII	1,800–1,950	gas	hydrocarbonaceous-sodic	10	55
Maloye Sabo	XVIII	1,600–1,960	gas	hydrocarbonaceous-sodic	10	58
Maloye Sabo	XVI	1,400–1,680	water	hydrocarbonaceous-sodic	9	50

sory minerals, clay minerals and authigenic minerals. The reservoir rocks in the crest and in the lower part of the structure were compared with the non-productive sandstones of the 17th horizon.

Clastic rock-forming components

Quartz and rock fragments show no conspicuous post-sedimentation change. The feldspars in sandstones beyond the oil-pool outlines as well as in the area of gas-bearing structures and the structure of the unproductive 18th horizon are replaced either by calcite or kaolinite, rarely chlorite, but are left intact within oil-pool outlines. The difference in the feldspar content in rocks within and beyond oil pools is 7% versus 22%.

Biotite flakes are differently changed within and outside the oil pools. In the samples taken from oil pools all stages of the gradual transition of biotite are observed: from the intensively pleochroitic and highly bi-refringent mineral with a refractive index of $Nm = 1.638–1.650$ to a light greenish low bi-refringent nearly isotropic chlorite-like mineral with $Nm = 1.590–1.610$. About one-third of the biotite flakes have preserved intensive pleochroism and high bi-refringence. Beyond the oil-pool outline and within the 2nd marginal zone of the oil pool of the Kolendo deposit, montmorillonite and kaolinite are developed on the basis of biotite, in the sandstones. Little-changed biotite occurs here rather rarely.

Accessory minerals

Almost all samples that were studied show dentate and scalar forms of intrastratal solution of epidote, garnet, sphene, as well as transitions of sphene and ilmenite into leucoxene. Coarse sphene grains are for the most part broken into small fragments that are gradually enveloped by leucoxene films. These accessory minerals which are unstable during intrastratal solution are sharply increased in the sandstones within oil pools (Table II). The oil–water interface is the boundary of the abrupt increase in sphene, epidote, ilmenite and low-refraction garnets in the sandstones throughout the area of the majority of oil-bearing structures under consideration (Ekhabi, upper Ekhabi, Odopty).

In the Kolendo oil-bearing structure, the content of these minerals in sandstones shows two sharp changes from the border towards the crest: at the pool boundary and in the centre (see Table II; Fig.2). The highest contents of sphene, epidote, ilmenite and low-refraction garnets are observed in the sandstones of the first (central)zone. In the second marginal zone, the amount of these minerals is 2–4 times smaller, although it is still greater than the values observed beyond the oil-pool outlines (Table II; Fig.2). In the area of the gas-bearing structures and the structure with the unproductive 17th horizon the contents of the minerals being considered vary within rather narrow limits (0,1, 1–3, 3–12%) equal to the values indicated beyond the oil-pool outlines (Table II; Fig.3).

A pronounced increase in sphene, epidote, ilmenite, and low-refraction garnets in the sandstones within oil pools is established by the study of different heavy fractions: 0.25–0.1; 0.1–0.025; 0.1–0.01 mm. In the fraction 0.1–0.01 this increase is especially clear and judging by the results of graphic correlation it does not depend on the sandstone granulometry. The content of heavy fractions in sandstones within oil pools varies between 0.17 and 1.0% and beyond the pool outline between 0.16 and 0.86%. The arithmetic mean content in the former case is 0.68% (on the basis of 22 samples) and 0.56% in the latter case (on the basis of 20 samples only; consequently the differences are not significant). The fractions enriched with pyrite micronodules were taken into account in determining the mean figures. It appears probable that part of the unstable heavy minerals which were dissolved in the sandstones beyond the oil-pool outlines were substituted in

TABLE II

HEAVY-MINERALS FRACTION IN THE SANDSTONES OF THE 17TH HORIZON

Location of samples	*Number of samples*	*Allothigenic minerals in the order of their stability (after* PETTIJOHN, *1941)*				*Leucoxene*	*Fine-grained anatase assemblages*
		sphene	*epidote*	*ilmenite*	*garnets with N = 1.737–1.770*		
Within the outlines of a majority of the oil deposits considered and in the first zone of the Kolendo deposit	62	15–40	3–35	24–54	15–18	8–18	1–4
Second zone of the Kolendo deposit	29	3–9	2–8	7–16	10–15	15–35	1–7
Beyond the oil-pool outline	32	1–3	0–1	5–12	5–12	25–45	1–6
Within the outlines of the gas pools	17	1–2	0–1	3–12	4–12	25–38	1–7
Beyond gas-pool outlines	16	1–1	0–1	5–10	7–12	20–40	1–7
In the crest areas of the structure with the unproductive 17th horizon	7	1–3	0–1	5–10	4	25–38	2–4
On the flanks of the structure with the unproductive 17th horizon	7	1–4	0–1	5–12	6	23–38	2–4

Note: the contents of epidote, sphene and low-refraction garnets have been calculated in percentages of the sum of transparent minerals; those of ilmenite, leucoxene and anatase in percentages of the total amount of heavy minerals excluding pyrite. Fraction sizes: 0.25–0.01 mm.

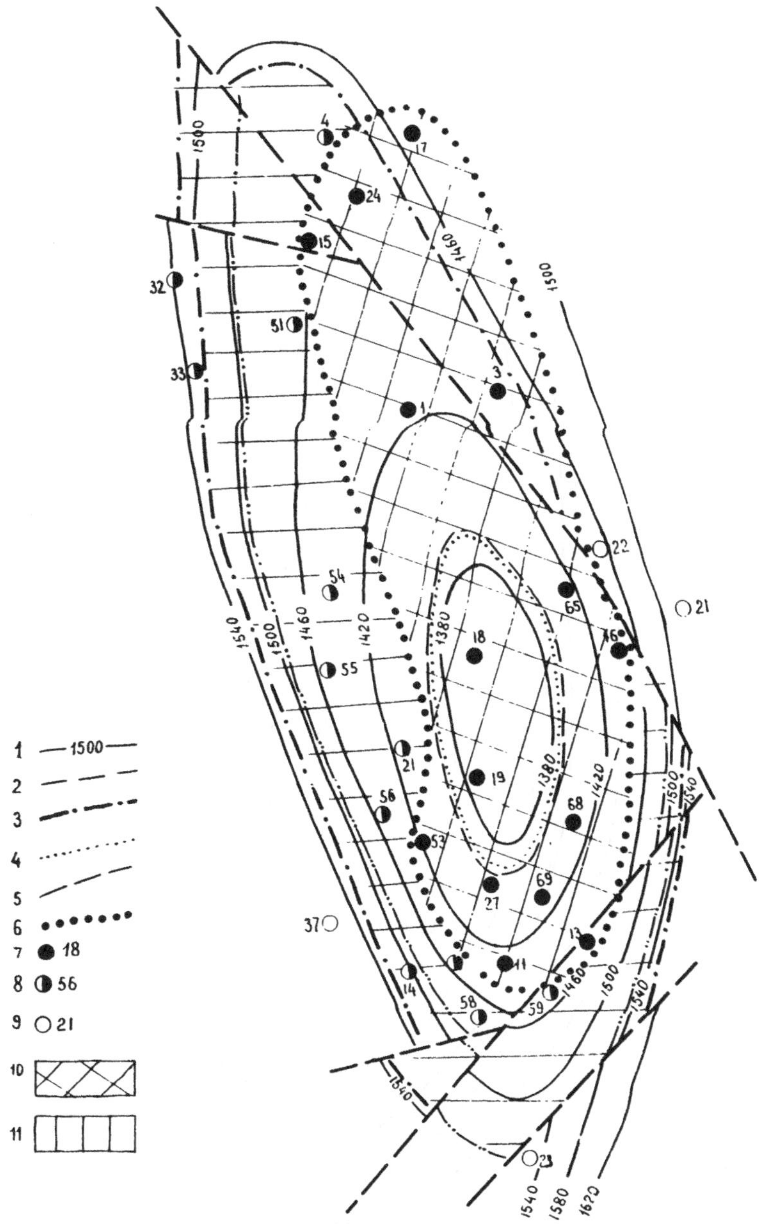

Fig.2. Structural map of the 17th horizon of the Kolendo deposit. *1* = contour lines of the upper surface of the 17th horizon; *2* = outlines of inner oil presence; *3* = outlines of outer oil presence; *4* = outline of presence of gas; *5* = tectonic disturbances; *6* = boundaries of the first zone (supposed outline of the deposit of the first phase of oil accumulation); *7*, *8*, *9* = sample locations (numbers on the right indicate the wells): *7* = high sphene content (15–40%); *8* = sphene content between 3 and 9 %; *9* = low sphene content (1–3 %); *10* = zone of sandstones abundant in sphene; *11* = zone of sandstones with 3–9 % sphene content.

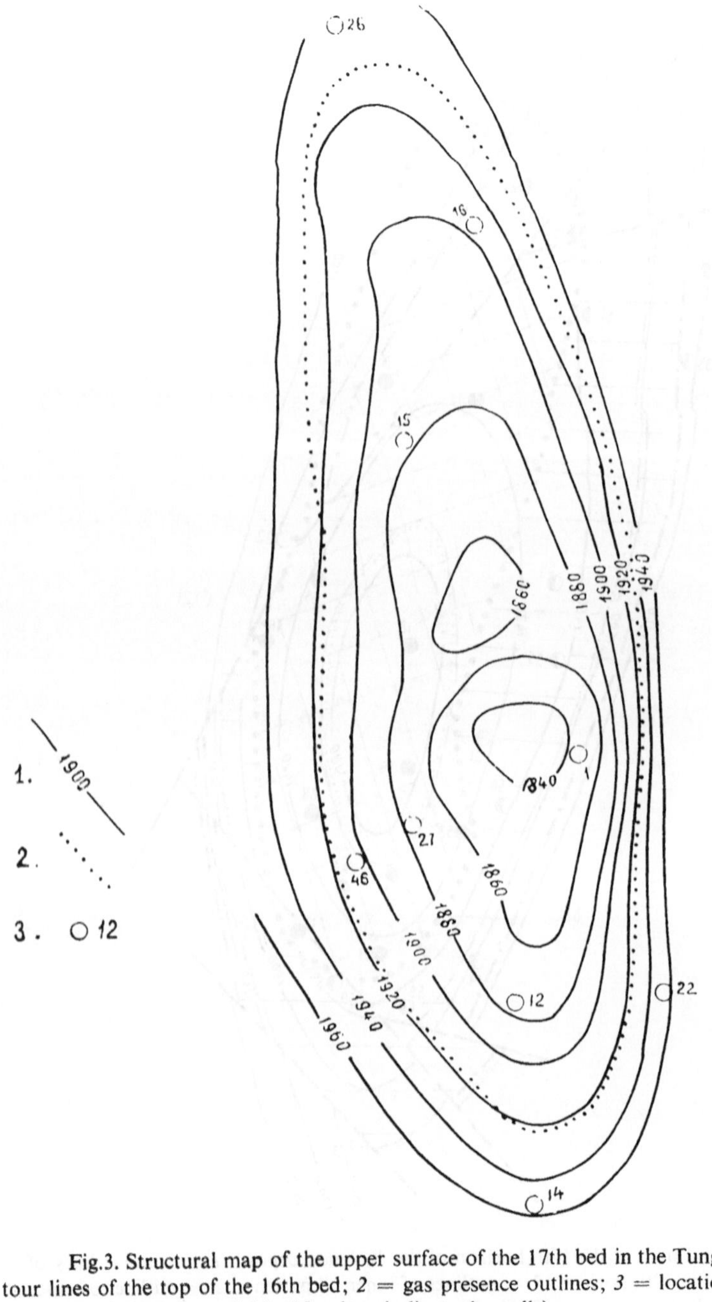

Fig.3. Structural map of the upper surface of the 17th bed in the Tungur pool. *1* = contour lines of the top of the 16th bed; *2* = gas presence outlines; *3* = location of samples with low content of sphene (1–3 %). (Numbers indicate the wells).

weight by leucoxene and fine-grained anatase. This is suggested by Table II which shows that the content of leucoxene and fine-grained anatase beyond the oil-pool outlines is 2–3 times as great as that within the pool (Table II).

TABLE III

X-RAY CHARACTERISTICS OF THE CLAY MATTER OF THE SANDSTONES OF THE 17TH HORIZON

Gas-bearing section (sample 1220)						*Oil-bearing section (sample 1108)*					
saturated		*annealed*		*column after HCl-treatment*		*saturated*		*annealed*		*saturated after LiCl-treatment*	
I	$\frac{d}{n}$ *(001)*	*I*	$\frac{d}{n}$ *(001)*	*I*	$\frac{d}{n}$ *(001)*	*I*	$\frac{d}{n}$ *(001)*	*I*	$\frac{d}{n}$ *(001)*	*I*	$\frac{d}{n}$ *(001)*
(1)	*(2)*	*(3)*	*(4)*	*(5)*	*(6)*	*(7)*	*(8)*	*(9)*	*(10)*	*(11)*	*(12)*
	14.6	5	14.2			10	17.6				
	10.0	10	10.1	1	9.8	1	14.2	3	13.3		
	7.1	–	–	6	7.1	1	10.0	9	9.9	6	9.8
	4.97	2	4.99	–	–	3	8.9	–	–	–	–
	54.74	–	–	–	–	2	7.1	–	–	2	7.1
	–	1	4.49	10	4.49	2	5.9	–	–	–	–
	4.27	1	4.26	5	4.14	1	4.96	3	4.99	–	–
0.5	4.01	1	4.04	–	–	1	4.74	–	–	2	4.7
–	–	0,5	3.80	1	3.84	3	4.46	1	4.49	3	4.4
9	3.56	0.5	3.50	4	3.54	1	4.26	2	4.24	3	4.2
4	3.34	9	3.34	8	3.36	–	–	1	3.78	–	–
1	3.23	2	3.21	4	3.20	4	3.59	1	3.47	1	3.49
1	3.04	0.5	3.05	2	2.98	5	3.34	10	3.34	8	3.34
			2.82								
1	2.82			1	2.75	0.5	3.25	3	3.20	4	3.17
1	2.56	1	2.58	9	2.57	2	2.97	–	–	3	3.12
			2.430								
0	52.485	0.5		3	2.480	1	2.55	1	2.59	1	2.56
									2.430		
2	2.388	–	–	4	2.337	1	2.451			1	2.43
1	2.294	–	–	2	2.294	1	2.229	1	2.270	1	2.35
1	2.140	1	2.132	1	2.199	1	2.144	1	2.120	1	2.11
–	–	–	–	2	2.124	2	1.980	1	1.990	2	1.971
3	2.001	2	1.994	1	1.990	1	1.818	1	1.818	2	1.796
–	–	1	1.886	1	1.889	1	1.663	1	1.672	2	1.684
1	1.796	1	1.799	2	1.818	1	1.549	1	1.545	2	11.642
3	1.663	2	1.663	4	1.663	1	1.501	0.5	1.508	2	1.540
1	1.542	0.5	1.533	2	1.535	2	1.378	1	1.378	4	1.493
1	1.494	1	1.497	6	1.488	–	–	–	–	–	–
–	–	–	–	1	1.453	–	–	–	–	–	–
1	1.372	1	1.380								
1	1.339										
1	1.304										

Clay minerals

Quasi-regular kaolinite occurs ubiquitously in the sandstone of the 17th horizon (Table III). It fills mainly the central parts of the pore space and possesses a flaky rosette-shaped or vermicular structure. In gas-bearing sandstones where kaolinite is predominant the coarsest flakes and assemblages are observed. Montmorillonite and hydromica occur within the oil-pool outlines in the pores of sandstones in the form of randomly oriented fine flakes. In the sandstones beyond the pool outline they frequently envelop clastic grains with fine brightly bi-refringent rims, whereas in the interior of the pores they show identical optical orientation with the flakes mentioned above. Montmorillonite is absent in the rocks of gas-bearing structures. Beyond the oil-pool outline, sandstone pores in the area of gas-bearing and unproductive structures contain authigenic light-green chlorite ($Nm = 1.616$) possessing a random aggregate, rosette-shaped or spherulitic structure.

Carbonates

The refractive index and the findings of the thermographic and chemical analyses suggest that the carbonate composition of the sand- and siltstones of the 17th horizon includes Fe-carbonate of the type of calciosideroplesite, calcite and ankerite (Table IV). Diagenetic pelitomorphic calciosideroplesite occurs ubiquitously in sandstones. It is developed on biotites, dispersed or accumulated in the pore interstices of rocks. Its content is especially high in the rocks within the oil-pool area (up to 15%). Calcite cement is developed beyond the oil-pool outline and in structures within the unproductive 17th horizon. Traces of fine-grained calcite occur also in the sandstone pores of the 2nd zone in the Loendo oil pool. Clear crystalline ankerite occurs only in sandstones of the gas-bearing structures.

Pyrite

Pyrite is present in sandstones of the 17th horizon either in the form of micronodules developed on vegetative relics and femic minerals or dispersed in the clay matter of the cement. Abundance of pyrite (up to 50–90%) of the heavy fraction is observed in the rocks above the peripheral area of the oil pool and along the outlines of the 2nd zone of the Kolendo pool.

Anhydrite

Spherulitic anhydrite assemblages are observed in single pores of the sand- and siltstones in the areas of both oil-bearing and gas-bearing structures, independent of rock position with respect to the pool outline. The pools with rocks containing anhydrite show an increased sulphur content, while the waters beyond the outline are marked by an increased content of sulphates (Table V).

TABLE IV

CHEMICAL COMPOSITION AND REFRACTION INDICES OF CARBONATES FROM THE SANDSTONES OF THE 17TH HORIZON

Chemical composition and refractive index	*First zone of the oil pool of Kolendo*		*Second zone of the oil pool of Kolendo*		*Beyond oil-pool outlines of Kolendo*		*Within gas pool of Tungur*	*Unproductive 16th bed of the Maloye–Sabo structure*	
								crest	*flanks*
Sample no.:	*1207/7*	*1207/2**	*1232/3*	*1232/3**	*1201*	*1160/2*	*1250*	*18.4*	*1125/2*
Insoluble residue	38.20	41.02	75.76	35.43	81.15	67.66	73.02	74.60	64.88
Al_2O_3	1.31	1.48	1.83	1.65	0.77	–	0.61	–	–
Fe_2O_3	0.72	1.47	2.01	0.44	0.29	0.57	1.92	0.36	0.41
FeO	20.88	20.03	6.08	24.66	2.16	1.92	2.52	0.89	1.45
CaO	3.18	2.97	1.48	3.66	4.95	13.03	5.03	10.66	16.28
MgO	3.73	4.20	1.06	3.97	0.53	1.49	2.55	–	–
MnO	0.43	0.37	0.16	0.45	0.04	0.11	0.08	0.14	0.07
CO_2	17.35	18.25	6.17	21.65	5.20	12.90	8.85	9.0	14.0
P_2O_5	–	–	0.11	–	0.01	0.01	0.02	0.01	0.01
Total	85.88	90.04	94.66	91.91	95.87	97.69	94.60	95.66	97.10
$CaCO_3$	13.2	11.8	16.5	12.2	76.1	80.6	48.5	92.0	92.3
$MnCO_3$	1.6	1.3	1.8	1.3	0.5	0.6	0.7	1.0	0.3
$MgCO_3$	7.8	14.7	15.2	12.1	4.5	10.4	28.8	–	–
$FeCO_3$	77.4	72.2	66.5	74.4	18.6	8.4	22.0	7.0	7.4
Refractive index	>1.78	>1.78	>1.78 1.659 ±0.002	>1.78	>1.78 1.659 ±0.002		1.690 ±0.002	1.659 ±0.002	

* Chemical compositions of the carbonates in the heavy fraction 0.25–0.01 mm; specific weight of the fraction: over 2.89

TABLE V

RELATIONSHIP OF SULPHUR CONTENT AND OCCURRENCE OF ANHYDRITE

Structures	*Maximum content of sulphur in oils of the 17th horizon*	*Content of sulphates in waters of the 17th horizon (mg/l)*	*Presence of anhydrite*
Kolendo	0.45	20–30	absent
Ekhabi	0.30	20–30	absent
Upper Ekhabi (thrust)	0.60	50–80	present
Upper Ekhabi (underthrust)	1.6	no data	present
Tungur	gas	50–80	present

DISCUSSION

The above data indicate that the sandstones of the 17th horizon are markedly different in the degree of post-sedimentary transformation (see Tables II, VI). The least affected by the secondary processes are sandstones within the oil pools. The post-sedimentary change of these rocks is manifested in: (*1*) a gradual transformation of biotites into a chlorite-like mineral and initial intrastratal solution of the unstable accessory minerals (epidote, sphene, low-refraction garnets, ilmenite); (*2*) recrystallization of fine-dispersed kaolinite into flaky kaolinite; and (*3*) formation of diagenetic sideroplesite, pyrite, and small quantities of leucoxene and fine-grained anatase.

Within the Kolendo oil field, such secondary alteration of sandstones is observed within the first (inner) zone only. The sandstones of the second zone contain, apart from the enumerated neogenetic minerals: montmorillonite and kaolinite developed on biotite; in pores, light-greenish chlorite aggregates, thin hydromica and montmorillonite films oriented around fragments; and rarely fine-grained calcite. The content of unstable heavy minerals such as sphene, epidote, ilmenite, low-refraction garnets is two or three times lower than in the inner zone.

The amount of leucoxene and fine-grained anatase, on the other hand, is that much greater (see Table II). The most affected by the secondary processes are the sandstones beyond the oil pools, in the area of gas-bearing structures and the structure with the unproductive 17th horizon. In addition to the neogenetic minerals enumerated above, here occur fine and medium grain-sized calcite and ankerite, and kaolinite developed on feldspars. The content of ilmenite and low-refraction garnet in the sandstones of these areas is much lower, while that of epidote and sphene falls down to 3–0%. The content of authigenic Ti-minerals increases (Table II). Judging by the composition and structure of the authigenic

TABLE VI

COMPOSITION AND THE SEQUENCE OF FORMATION OF AUTHIGENIC MINERALS IN THE SANDSTONES OF THE 17TH HORIZON

Stages of authigenic mineral formation	*Within the boundaries of a majority of the oil pools considered and in the 1st zone of the Kolendo oil pool*	*In the 2nd zone of the Kolendo oil pool*	*Beyond the oil pools and in the area of the structure with the unproductive 17th horizon*	*Within gas pools*	*Beyond gas pools*	*Allothigenic sphene content (%)*
Late diagenesis-initial epigenesis	pyrite, pelitomorphous calcio-sideroplesite, flaky kaolinite in pores, leucoxene, fine-grained anatase aggregate formations, chlorite formed on biotite					15–40
	first phase of oil accumulation		—			
Initial epigenesis		pyrite on the periphery of the 1st zone of the Kolendo oil pool		absent		
		leucoxene				2–3
		fine-grained anatase aggregates				
		montmorillonite and kaolinite developed on biotite				
		light-green chlorite in pores				
		rare growth of granular calcite			absent	
		second phase of oil accumulation		—		
			pyrite on the outlines of all oil pools		absent	1–3
			granular calcite		granular ankerite	
			kaolinite on feldspars			

minerals as well as by the level of preservation of clastic minerals that are unstable to intrastratal solution and replacement processes—recalling that petrographic-mineralogical criteria proposed by KOSSOVSKAYA (1962) and PEROZIO (1965, 1966) were used to assess the grade of epigenetic transformation—it may be asserted that the post-sedimentary change of the sandstones within a majority of oil pools and in the first zone of the Kolendo field are in the stage of late diagenesis–early initial epigenesis, while those outside the oil-pool outlines, and in the areas of oil-bearing structures and the unproductive 17th horizon structure belong to more advanced stages of initial epigenesis.

The investigations have thus established that the sandstones which are least affected by epigenesis are confined, the other geological and hydrogeological conditions being equal, to the areas within the outlines of the oil pools. This fact leads us to connect the delay in the secondary transformation of rocks with the presence of oil in these rocks, since according to the "braking" hypothesis oil is the most unfavourable medium for the processes of epigenetic solution and replacement of the unstable components and for authigenic mineral formation (see the Introduction).

On the other hand, the study has not shown any deceleration of the post-sedimentation processes in rocks occurring inside gas pools, which may be due to a recent accumulation of gas in the traps under investigation. The retardation of secondary transformation of oil-bearing rocks is indicated by various constituents of the detrital and cementing minerals, particularly by the clastic heavy minerals which proved to be more sensitive to intrastratal solution and replacement than biotite or the clay matrix. The most sensitive minerals seem to be ilmenite and especially sphene. Its content does not depend on the granulometric composition of the sandstones or on the position in the productive horizon section, nor—for most of the oil and gas pools considered, except that of Kolendo—on the location of the samples in the structure (see Fig.3). It appears possible to evaluate the retardation of the secondary change of sandstones observed in the oil pools by the sphene content in the heavy fractions of sandstones, which gives grounds for objective assessment of the time of oil accumulation in traps.

Probable geological interpretation of the findings

An attempt was made at delineating the phases and time of oil accumulation in the traps of the 17th horizon. The resulting oil-accumulation phases are given in Table VI. The evaluation was based on the assumption that sandstones showing an equal degree of post-sedimentary change over the area of a majority of oil pools considered were formed within the same (not very long) phase of oil accumulation. Moreover, post-sedimentary changes observed in the sandstones were assumed to have appeared prior to oil inflow. The presence of two sandstone zones differently affected by the secondary processes in the Kolendo pool was explained as a result of a two-phase oil accumulation in this trap. On the basis of identical post-sedimentary change of rocks it was assumed that oil accumulation in the first zone of the Kolendo structure (first phase) took place simultaneously with oil accumulation in most traps of the structures under consideration (Ekhabi, eastern Ekhabi, Odoptu).

The time spent by oil in the deposits was evaluated assuming a complete cessation of the secondary processes in sandstones after oil accumulation. A horizon or level was searched for in the continuous productive series section which possessed post-sedimentary changes equal or similar to those of the oil-saturated sandstones of the 17th horizon. If we assume that the sum of post-sedimentary changes of sandstones prior to oil inflow has been accumulating proportionally for a geological time interval, then the age of the sandstones of this horizon, and hence the time of their secondary transformation, may be set equal to the time that elapsed from the deposition of the oil-bearing rocks until the accumulation of oil in them. The post-sedimentary change of the sandstones within the oil pools of the 17th horizon corresponds to the secondary alteration of those rocks in the overlying unproductive horizons 1–4 (Ekhabi nomenclature). Time spent by oil in the traps of a majority of the structures under consideration and in the first zone of the Kolendo structure was assessed as the difference of the time of formation of the deposits of the 17th and 1st–4th horizons, taken on the same structure characterized by a relatively regular pattern (for instance, the Ekhabi structure). This difference between the second half of the Miocene and the Miocene–Pliocene boundary may be gauged as 1–3 million years. Therefore, the first phase of oil accumulation is referred to the Upper Pliocene, whereas the time of the second (more recent) oil accumulation phase may be evaluated as post-Pliocene. It should be noted that the oil accumulation time estimates given in the paper are based on a number of assumptions, such as the complete cessation of secondary mineralogenesis in rocks in the presence of oil, time-dependent post-sedimentary alteration of the rocks of the 17th and the 1 st–4th horizons, etc., so that they are tentative and to a great extent hypothetical. In spite of that, they are in good conformity with the oil accumulation time estimates given for the traps of the 17th horizon by Kovalchuk and Evdokimova (1967) who considered the history of the oil-pool formation from tectonic, lithological, hydrogeological and geochemical points of view. These authors associate the migration of hydrocarbons with the time of intensified tectonic movements in the Sakhalin folding phase at the end of the Pliocene and the beginning of the post-Pliocene.

REFERENCES

Alekseichik, S. N., 1963. Tektonika, istoriya geologicheskogo razvitiya i perspektivy neftegazonosnosti Sakhalina. (Tectonics, geological evolution history and gas and oil occurrence prospects of Sakhalin.) *Tr. Vses. Nauchn. Issled. Geologorazved. Inst.*, 217.

Chepikov, K. R., Ermolova, E. P. and Orlova, N. A., 1959. Epigennye mineraly kak pokazateli prikhoda nefti v peschanye promyshlennye kollektory. (Epigenetic minerals as indicators of oil migration to productive sand reservoirs.) *Dokl. Akad. Nauk S.S.S.R.*, 125 (5).

CHEPIKOV, K. R., ERMOLOVA, E. P. and ORLOVA, N. A., 1960. Epigennye mineraly v peschanykh porodakh produktivnykh gorizontov i ikh vliyanie na kollektornye svoistva. (Epigenetic minerals in sandy rocks of productive horizons and their impact on reservoir properties.) *Tr. Inst. Geol. Razrabotki Goryuch. Iskop., Akad. Nauk S.S.S.R.*, 1:240–246.

CHEPIKOV, K. R., ERMOLOVA, E. P. and ORLOVA, N. A., 1967. Nekotorye rezultaty izuchenia autigennykh mineralov s tselyu opredelenia otnositelnogo vremeni prikhoda nefti v porody-kollektory. (Some results of authigenic mineral studies towards relative timing of oil migration to reservoir rocks.) In: *Genezis Nefti i Gaza*. Nedra, Moscow, pp. 397–403.

EVDOKIMOVA, T. I., 1963. Tektonika Vostochno-Ekhabinskogo mestorozhdeniya i ee vliyanie na obrazovanie i razrushenie neftyanykh zalezhei. (Tectonics of the east Ekhabi deposit and its influence on formation and destruction of oil pools.) *Tr. Vses. Nauchn. Issled. Geologorazved. Inst.*, 224:109–117.

FÜCHTBAUER, H., 1961. Zur Quarzneubildung in Erdöllagerstätten. *Erdöl Kohle*, 14:169–173.

KLESHCHEV, V. A., 1966. Formirovanie i zakonomernosti razmeshcheniya mestorozhdenii nefti i gaza v verkhnemiotsenovykh otlozheniyakh severovostochnoi chasti ostrova Sakhalin (Formation and regularities of location of oil and gas pools in the Upper Miocene deposits of northeastern Sakhalin.) Avtoreferat dissertatsii na soiskanie uch. stepeni kand. g.-m. n., Moscow.

KLUBOVA, T. T., 1965. *Rol Glinistykh Mineralov v Preobrazovanii Organishcheskogo Veshchestva i Formirovanii Porovogo Prostranstva Kollektorov. (Role of Clay Minerals in Transformation of Organic Matter and Formation of Pore Interstices in Reservoirs.)* Nauka, Moscow.

KOSSOVSKAYA, A. G., 1962. Mineralogiya terrigennogo mezozoiskogo kompleksa Vilyuiskoi vpadiny i Zapadnogo Verkhoyanya (Mineralogy of the terrigenous Mesozoic complex of the Vilyuisk Trough and western Verchojanja.) *Tr. Geol. Inst., Akad. Nauk S.S.S.R.*, 63:66–68.

KOVALCHUK, V. S. and EVDOKIMOVA, T. P., 1967. Nekotorye zakonomernosti formirovaniya i razmeshcheniya gasoneftyanykh zalezhei Severo-Vostochnogo Sakhalina. (Some regularities of formation and location of gas and oil pools in northeastern Sakhalin.) In: *Genezis Nefti i Gaza*. Nedra, Moscow, pp. 661–666.

MILLOT, G., 1964. *Géologie des Argiles*. Masson, Paris, 499 pp.

ORLOVA, N. A., 1958. Structurnye osobennosti i petrograficheskii sostav neftenosnykh i vodonosnykh plastov v otlozheniakh devona i karbonata Tatarii, Bashkirii i Kuibyshevskogo Zavolzhya. (Structural features and petrographic composition of oil- and water-bearing beds in Devonian and Carboniferous deposits and the Kyubyshev Zavolzhye.) In: *Annotatsii Rabot po Geologii Nefti if Gaza za 1957*. Akad. Nauk S.S.S.R., Moscow, pp.18–19.

PEROZIO, G. N., 1965. Rentgenovskae izuchenie autigennykh mineralov indicatorov razlichnykh stadii preobrazovaniya granulyarnykh kollektorov nei primera izucheniya ryada mesto-rozhdenii Zapadno-Sibirskoi nizmennosti. In: *Voprosy Klassifikatsii i Metodika Izucheniya Kollektorov Nefti i Gaza.*

PEROZIO, G. N. and MANDRIKOVA, N. T., 1966. Izomorfizm v karbonatakh i ego znachenie dlya resheniya nekotorykh geneticheskikh voprosov. (Isomorphism in carbonates and its significance for solution of some genetic problems.) In: *Fizicheslie Metody Issledovaniya Mineralov Osadochnykh Porod.* Nauka, Moscow, pp.279–292.

PROZOROVICH, G. E., 1967. K metodike opredeleniya vremeni formirovaniya mestorozhdenii nefti po epigeneticheskim izmeneniyam produktivnykh plastov. (Method of determination of oil-pool formation time by epigenetic alteration of productive beds.) *Byull. Nauchn. Tekhn. Inform. Ser. Geol. Mestorozhdenii Polezn. Iskop.*, 2:4–10.

SOLOMATINA, L. N., 1963. Usloviya formirovaniya lokalnykh stuktur Erri i Sabo i priurochennykh k nim gazonosnykh zalezhei. (Conditions of formation of local structures of Erri and Sabo and gas and oil pools confined to them.) *Tr. Vses. Neft. Nauchn. Issled. Geologorazved. Inst.*, 224:81–90.

YURKOVA, R. M., 1968. Terrigennye mineraly assotsiatsii neogena Severnogo Sakhalina. (Terrigenous minerals of the Neogene assemblage of northern Sakhalin.) *Litologiya Polezn. Iskop.*, 5:30–42.

Reprinted from Sedimentology (2001) ***48****, 703–721*

Plagioclase dissolution related to biodegradation of oil in Brent Group sandstones (Middle Jurassic) of Gullfaks Field, northern North Sea

S. N. EHRENBERG* and K. G. JAKOBSEN†
**Statoil, N-4035 Stavanger, Norway (E-mail: sne@statoil.com)*
†Statoil, N-5020 Bergen, Norway (E-mail: kgja@statoil.com)

ABSTRACT

Brent Group sandstones from the north side of the Gullfaks Oilfield contain mostly 5–8% albitic plagioclase, whereas plagioclase is almost absent in the same strata in the southern part of the field. Absence of plagioclase throughout the entire vertical extent of the Brent Group in the southern wells seems to rule out provenance as the principal explanation for differing plagioclase content, which is therefore interpreted as the result of diagenesis. Hypotheses for the nature of the inferred leaching event include epigenetic meteoric diagenesis and mesogenetic release of acid components from clay minerals or kerogen, but these explanations are unable to account for the observed spatial distribution of the plagioclase-bearing and plagioclase-free sandstone intervals. However, overall correspondence between the area lacking plagioclase and oil compositions having both anomalously high CO_2 and organic geochemical indications of advanced biodegradation suggest a link between plagioclase dissolution and biodegradation of the present oil column. It is, therefore, proposed that acid components from biodegradation selectively reacted with albitic plagioclase to form kaolin, releasing sodium bicarbonate into the residual water. The plagioclase-free sandstones contain more kaolin than the plagioclase-bearing sandstones, as would be expected due to aluminium conservation. However, the wide and overlapping ranges of kaolin content in both groups suggest that most of the kaolin originated from processes other than biodegradation-driven plagioclase alteration, potentially including both epigenetic and mesogenetic dissolution, as well as deposition of detrital kaolin and kaolin precursors.

Keywords Biodegradation, diagenesis, feldspar, sandstone.

INTRODUCTION

The origin of abundant feldspar-dissolution effects and the ostensibly complementary kaolin cement typically present in shallow-marine and delta-plain sandstone reservoirs have been topics of long-standing interest and controversy (Schmidt *et al.*, 1977; Curtis, 1983; Bjørlykke, 1984, 1998; Surdam *et al.*, 1989). Eogenetic meteoric-water diagenesis has been cogently argued to be the main process responsible, based on mass-balance constraints, oxygen isotope data, and the observation of little to no feldspar dissolution or kaolin in sandstones from distal-shelf settings lacking the opportunity for contact with meteoric water during early burial (Bjørlykke, 1998). Nevertheless, some degree of additional feldspar dissolution could also take place during later burial due to release of acids by diagenesis of clay minerals and kerogen (Curtis, 1983; Bjørlykke, 1984). Such mesogenetic dissolution may be difficult to distinguish from eogenetic effects unless special spatial or petrographical relationships provide compelling evidence for timing, as in the examples reported by Ehrenberg (1991) and Ehrenberg *et al.* (1995).

The present study describes a set of mineralogical data from cores in the Gullfaks Field, where absence of plagioclase corresponds spatially to areas of biodegraded oil composition, suggesting mesogenetic timing of dissolution. However, not all observations are accounted for by this model. Although biodegradation appears to provide the most plausible explanation for the observed pattern of plagioclase occurrence overall, it does not readily explain the distribution of plagioclase in certain cores from the Rannoch Formation. There is also an unresolved problem involved with explaining how leaching can have removed plagioclase but not K-feldspar. Therefore, the goals of this report are to present the mineralogical data, to outline some alternative models explaining the observed patterns, and to discuss the constraints applicable to each alternative. These results may be useful for guiding other studies to recognize inorganic diagenetic effects of petroleum biodegradation, a relationship that should be widespread but seems not to have been noticed before now.

Nearshore-marine and deltaic sandstones of the Middle Jurassic Brent Group constitute the main reservoirs of Gullfaks Field (Figs 1–4; Eriksen *et al.*, 1987; Petterson *et al.*, 1990, 1992; Olaussen *et al.*, 1992; Tollefsen *et al.*, 1992). The Brent Group comprises five formations, the upper four of which have been sampled in the present study (Fig. 4). These strata underwent faulting, tilting and erosion during the Late Jurassic to Early Cretaceous Cimmerian tectonic episode (Karlsson, 1986). This tectonism created the fault blocks containing Gullfaks Field and also produced the unconformity truncating the upper surface of the Gullfaks reservoirs. The sandstone reservoirs are overlain and sealed by both Upper Jurassic shale (Heather Formation) and Lower to Upper Cretaceous shales and marls (overlying the Cimmerian unconformity). The burial history of the Brent Group of Gullfaks Field (burial curve in Olaussen *et al.*, 1992) involved initial shallow burial (<500 m) followed by Cimmerian uplift; gradually increasing burial from Late Cretaceous through Miocene time; and an additional ≈1 km of burial, beginning in Pliocene time, to the present maximum depth of 1·6–1·7 km below the sea floor at the top Brent reservoir. Pre-production reservoir fluid pressure in the Brent reservoirs is 310 bar; temperature is 72 °C; and formation water salinity is 4·4 wt% total dissolved solids.

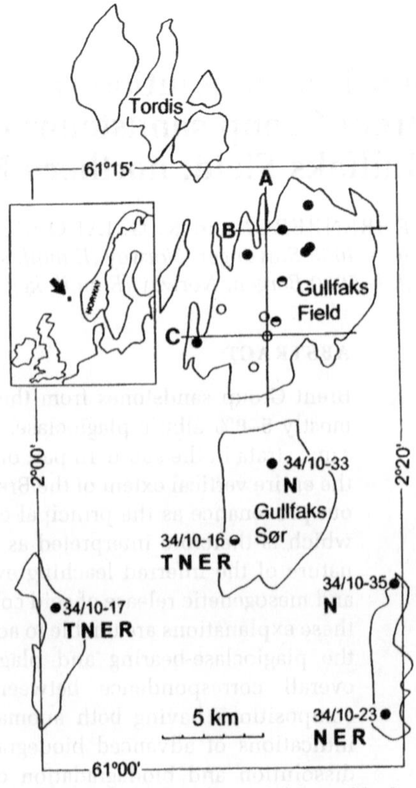

Fig. 1. Outlines of oil and gas fields in block 34/10, showing locations of wells where mineralogical data are available for cored intervals in Brent Group strata. Filled circles, wells in which plagioclase is mostly present in normal abundances; open circles, wells in which plagioclase is mostly absent; half-filled circles, wells where cored intervals are partly plagioclase-bearing and partly plagioclase-free. Letters under well numbers indicate the formations analysed in each well (T, Tarbert; N, Ness; E, Etive; R, Rannoch). Lines across Gullfaks Field show locations of structural cross-sections A, B, and C in Fig. 3.

ANALYTICAL METHODS

The term 'kaolinite' is used here to refer specifically to the polytype of that name, where identification has either been established analytically or is assumed based on shallow burial depth. According to the recommendation of Ehrenberg *et al.* (1993), the term 'kaolin' is used to refer to the kaolinite group minerals, where the polytype present is unspecified. A limited survey of the sample set in this study indicates that the kaolin present in Gullfaks Field consists entirely of kaolinite.

Two generations of bulk X-ray diffraction (XRD) data are available for the Gullfaks cores,

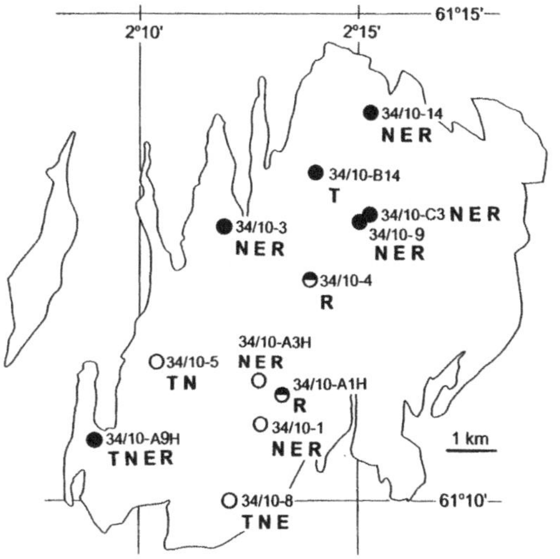

Fig. 2. Outline of Gullfaks Field (and Gullfaks West; left), showing locations of wells where mineralogical data are available for cored intervals in Brent Group strata. Filled circles, wells in which plagioclase is mostly present in normal abundances; open circles, wells in which plagioclase is mostly absent; half-filled circles, wells where the Rannoch Formation is plagioclase bearing in its upper part and plagioclase free in its lower part. Letters by well numbers indicate the formations analysed in each well (T, Tarbert; N, Ness; E, Etive; R, Rannoch).

both performed at Statoil: an older (pre-1986) set of 414 analyses from eight wells (34/10-14, -3, -5, -A3H, -1, -8, -4 and -A1H) and a newer (1994) set of 37 analyses from four wells (34/10-14, -B14, -8 and -A3H). Analytical procedures and correction factors used differ between these two data sets. The older data suffer from overestimation of feldspar content in some samples, apparently due to improper grinding technique (Ehrenberg, 1991). However, these analyses are nevertheless useful for distinguishing presence vs. absence of plagioclase and K-feldspar and for showing systematic differences in mineral abundances between sample sets run using the same analytical procedures. The 1994 analyses and additional analyses from wells near the Gullfaks Field (34/10-16, etc.; Fig. 1) were performed by P. H. Nadeau (Statoil). The 1994 analyses used random-powder mounts prepared by grinding 10 g of rock plus 15 mL of alcohol in a micronizer for 5 min, Cu radiation, and calculation of mineral abundances using intensity factors similar to Ramm (1991). Only analyses of whole core samples are included in this study.

The bulk chemical analyses also include two sets of data. Older (pre 1986) analyses were performed by X-ray fluorescence at Statoil by A. Rehkopff, whereas newer analyses were

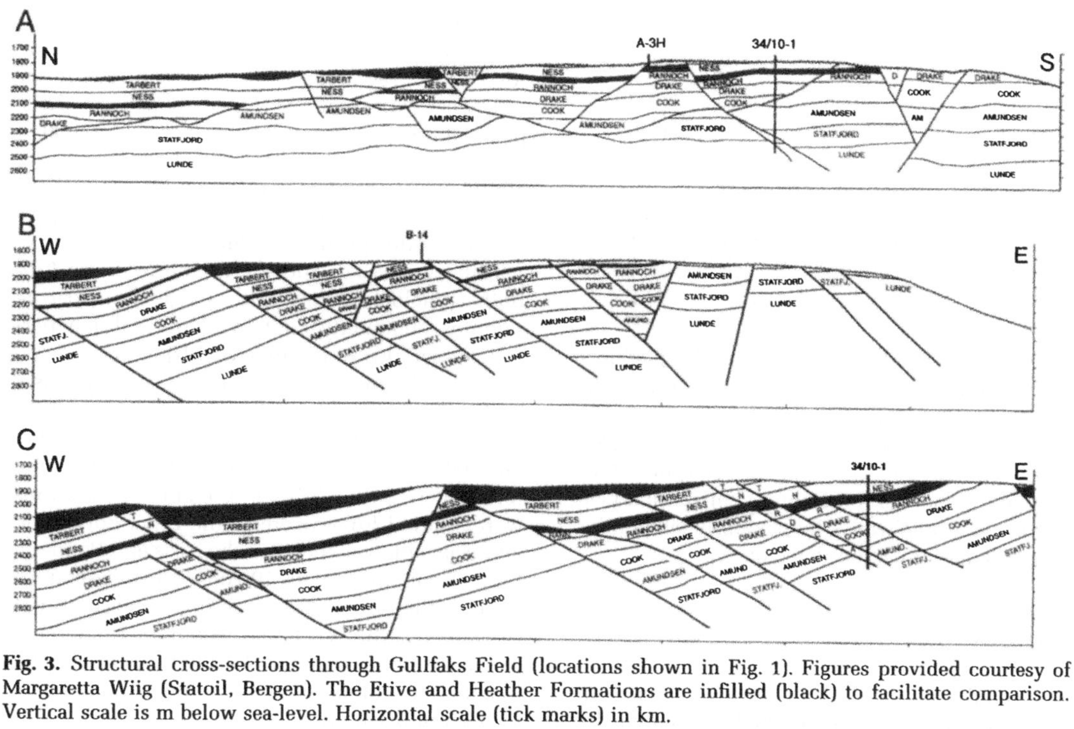

Fig. 3. Structural cross-sections through Gullfaks Field (locations shown in Fig. 1). Figures provided courtesy of Margaretta Wiig (Statoil, Bergen). The Etive and Heather Formations are infilled (black) to facilitate comparison. Vertical scale is m below sea-level. Horizontal scale (tick marks) in km.

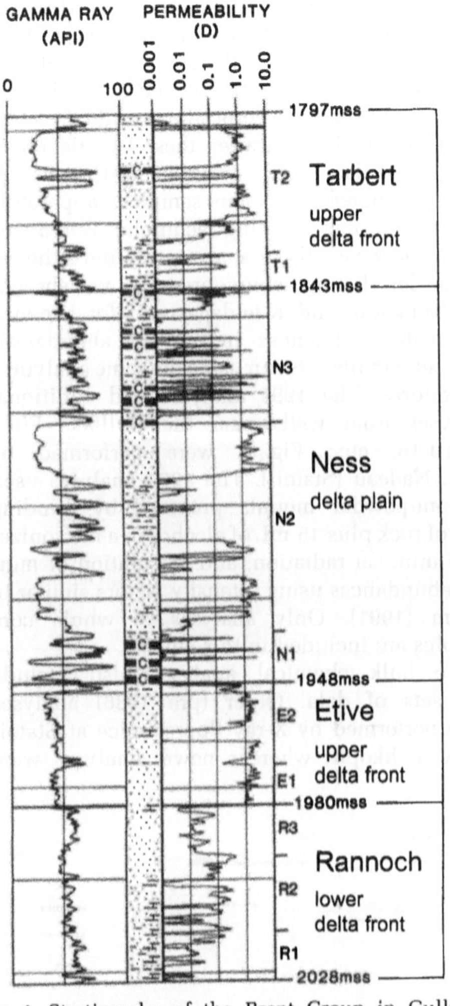

Fig. 4. Stratigraphy of the Brent Group in Gullfaks Field, as represented by well 34/10-8 ('C', coal beds). Depths in metres below mean sea-level. Modified from Eriksen *et al.* (1987).

performed by XRAL, Canada, using X-ray fluorescence and additional techniques described in Ehrenberg & Siring (1992).

Geochemical analyses of stock-tank oil (see Fig. 15) and oil and gas samples used for calculation of in-place reservoir fluid composition (see Fig. 16) were performed by gas chromatography using a flame-ionization detector.

RESULTS

The overall pattern of plagioclase occurrence in the Brent Group of Gullfaks Field is shown by the well symbols in Fig. 2 and in the histograms of Fig. 5. Vertical profiles of plagioclase content determined by XRD are shown for each well in Fig. 6. With the exception of well 34/10-A9H, plagioclase is mostly absent throughout the Brent Group in the southern part of Gullfaks Field, but is present in 'normal' amounts (generally 5–8 wt%, based on the newer XRD and bulk chemical data) in the northern part of the field.

Correlation of XRD plagioclase percentage with bulk-rock sodium content (Fig. 7) indicates that plagioclase in the Brent Group is mainly of Na-rich composition and that other Na-bearing minerals are present in only insignificant quantities. The low Na contents of all chemically analysed samples having zero XRD-measured plagioclase content confirms the reliability of the bulk XRD analyses for the purpose of recognizing plagioclase-free samples. However, the overall correlation between sodium content and bulk XRD percentage plagioclase is not as good in the earlier (pre-1986) data set as in the more recent (1994) XRD analyses (Fig. 7). Potassium, however, correlates poorly with K-feldspar (Fig. 8) because it is also present in mica and illitic clays. The albitic compositional range of the plagioclase is confirmed by electron microprobe analysis of plagioclase grains in representative samples from different formations (Fig. 9).

Petrographical observations also support the XRD data regarding plagioclase presence or absence. Thin sections of samples with XRD plagioclase reported contain plagioclase grains and commonly also have perthite grains with albitic lamellae intact (Fig. 10A). No plagioclase grains are apparent in thin sections of the samples with zero XRD plagioclase, and albitic lamellae of perthite grains are entirely dissolved (Fig. 10B). Both plagioclase-free and plagioclase-bearing samples commonly contain a few volume percentage moulds of dissolved feldspar grains.

Two wells in Fig. 2 are marked as plagioclase-bearing based on observations of stained thin sections. These two well sections were examined only late in the study as tests of the model being formulated. In well 34/10-9, eight thin sections were studied from a core in the Etive Formation and from side-wall cores in the Ness and Rannoch Formations. In well 34/10-A9H, 18 thin sections were studied from cores covering the entire Brent section. All thin sections examined from these two wells contain minor (4–6) to numerous (>20) plagioclase grains showing minor to moderate degrees of dissolution.

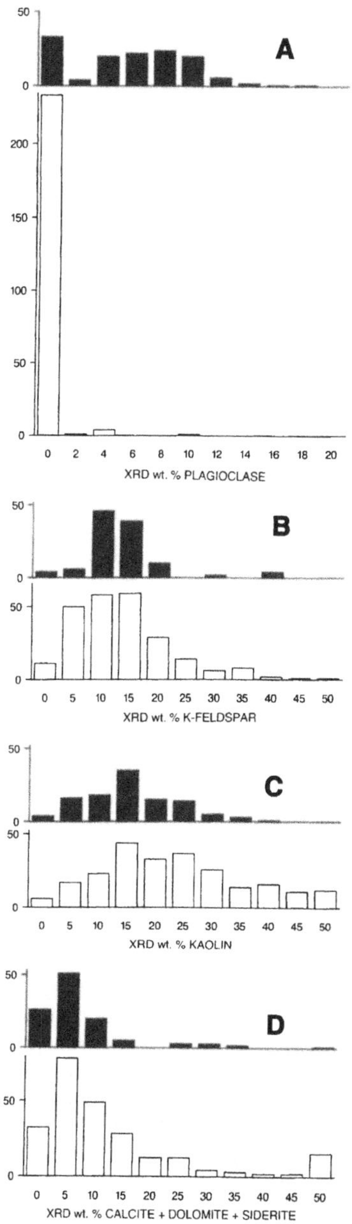

Fig. 5. Frequency (vertical axis) vs. mineral abundances determined by bulk XRD analysis in Brent Group sandstones from Gullfaks Field. Solid columns (133 samples) are samples from northern wells (filled circles in Fig. 2), where plagioclase is mostly present in normal abundances. Open columns (239 samples) are samples from southern wells (open circles in Fig. 2), where plagioclase is mostly absent. Wells 34/10-4 and 34/10-A1H (half-filled circles in Fig. 2) are not included. Both older and newer XRD data sets are included, as well as plagioclase abundances calculated from bulk-rock sodium content in well 34/10-C3.

Cathodoluminescence examination of representative Gullfaks sandstones shows that virtually all feldspar grains are brightly luminescent in various shades of blue, brown, or pink and are therefore detrital rather than authigenic. In general, detrital plagioclase and K-feldspar has bright cathodoluminescence, whereas authigenic feldspars are non-luminescent (Kastner, 1971; Marshall, 1988). The only non-luminescent feldspar present occurs as syntaxial overgrowths of non-luminescent K-feldspar on luminescent (detrital) K-feldspar grains.

Patterns of plagioclase occurrence can also be examined for each formation individually:

1 The Rannoch Formation is plagioclase bearing throughout in well 34/10-C3 and plagioclase free throughout in wells 34/10-1 and 34/10-A3H (Fig. 6). These wells thus conform to the overall pattern of plagioclase presence in the north and plagioclase absence in the south of Gullfaks Field. However, in wells 34/10-4 and 34/10-A1H, plagioclase is present in the upper part of the Rannoch Formation and absent from the lower part (Fig. 6). Further information is available from wells 34/10-3 and 34/10-14, where only the upper part of the Rannoch Formation is cored and is plagioclase bearing (Fig. 6).

2 The Etive Formation is plagioclase bearing in the northern well 34/10-14 and plagioclase free in the southern wells 34/10-1 and 34/10-A3H. However, in the otherwise plagioclase-free southern well 34/10-8, one sample from the very uppermost part of the Etive Formation and two samples from the immediately overlying base of the Ness Formation are plagioclase bearing. Furthermore, in well 34/10-3 the lower Etive is plagioclase bearing whereas the upper Etive is plagioclase free.

3 The Ness Formation is plagioclase bearing in the northern well 34/10-14 and is mostly plagioclase free in the southern wells 34/10-1, 34/10-8, 34/10-5 and 34/10-A3H, except for minor plagioclase in a few samples, some of which are calcite concretions. In well 34/10-3, the lower Ness is plagioclase bearing whereas the upper Ness is plagioclase free.

4 The Tarbert Formation is plagioclase bearing in the northern well 34/10-B14 and plagioclase free in the southern wells 34/10-5 and 34/10-8.

Further bulk XRD data are available from wells in separate structures in the southern part of block 34/10 (Fig. 1). Plagioclase is present in normal abundances throughout the cored intervals of the Brent Group in all of these wells except for 34/10-16, where plagioclase is present throughout the Rannoch and Etive Formations

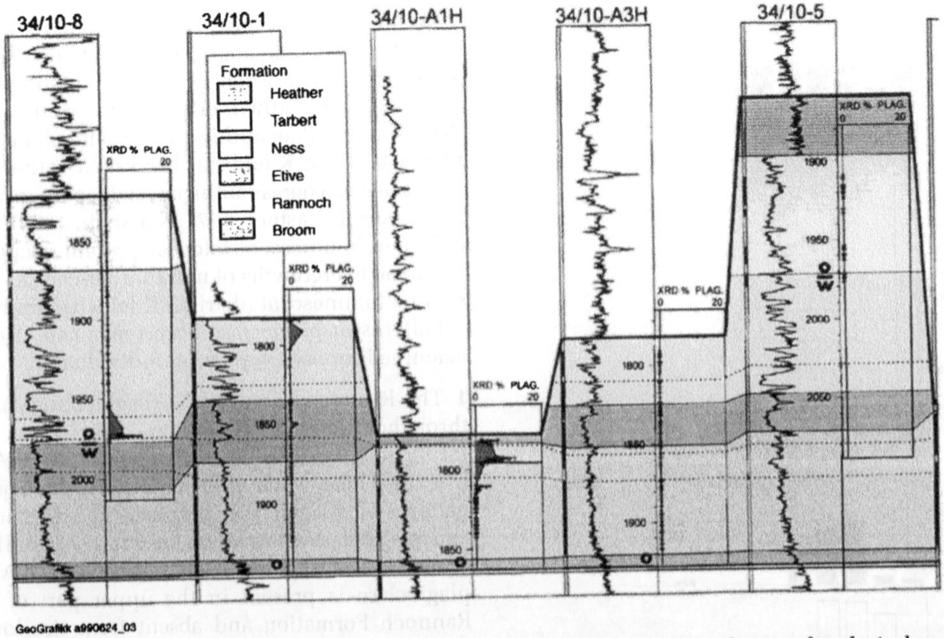

Fig. 6. Profiles of gamma ray activity (scale 0–150 API units) and plagioclase content (area under plagioclase curve is coloured red; '+' symbols at zero vertical reference line indicate analyses with zero plagioclase content) in the Brent Group in 10 wells from Gullfaks Field (Fig. 2) and one well (34/10–16) from Gullfaks Sør Field (Fig. 1). Cored intervals are indicated by solid bars at the left of the GR column. Base-Cretaceous unconformity is indicated by thick line at top of coloured section (Brent formations and Heather Formation). Depths are measured depth (m below kelly bushing), except for the deviated well 34/10-B14, where vertical depth is plotted. Plagioclase content was determined by XRD, except in well 34/10-C3, where weight percentage plagioclase was calculated from bulk sodium content (Fig. 7A). Dotted lines show positions of three interpreted time surfaces in Gullfaks wells.

and most of the Ness Formation, but is absent in the Tarbert Formation (Fig. 6).

The XRD analyses show that K-feldspar is present in almost all samples analysed and shows no correlation with plagioclase content (Figs 5 and 11). However, plagioclase-free samples may have somewhat lower K-feldspar on average, at least according to the newer set of bulk XRD analyses (Fig. 11). Similarly, there is no correlation between bulk-rock potassium and sodium contents (Fig. 12).

Kaolin content varies widely in both plagioclase-bearing and plagioclase-free sample groups, but there is an overall tendency in all four formations towards higher kaolin content in the plagioclase-free samples (Figs 5, 13 and 14).

INTERPRETATION

Significance of plagioclase absence

Variations in plagioclase content in Brent Group sandstones can reflect both primary variations in detrital sand composition and differences in diagenetic alteration. Depositional sand composition can be influenced by: (1) relative proportions of different provenance components (Dalland *et al.*, 1995); (2) hydrodynamic sorting processes (Odum *et al.*, 1976); and (3) varying degree of plagioclase alteration during sand transport to the site of final deposition (Johnsson & Meade, 1990). However, the present study is not concerned with interpreting the causes of varying plagioclase content within the plagioclase-bearing intervals; the problem of interest is the reason for the complete absence of plagioclase within virtually the entire Brent Group in the southern part of Gullfaks Field (Fig. 2). Two general explanations are possible: plagioclase absence may be either a primary depositional characteristic or an effect of diagenesis.

The first explanation to be considered is that plagioclase absence may be a primary depositional characteristic caused by varying mixtures of sands from plagioclase-rich and plagioclase-poor source terranes. For example, Dalland *et al.*

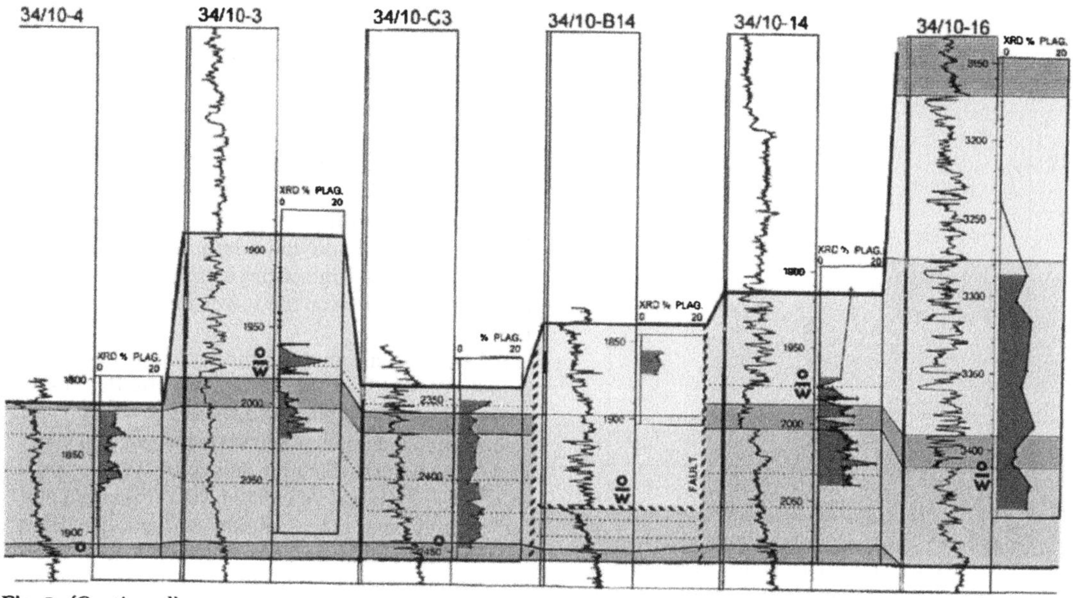

Fig. 6. (Continued).

(1995) showed that differences in plagioclase content in the Statfjord Formation of Gullfaks Field reflect differences in sand provenance. However, plagioclase variation in the Statfjord Formation is mainly vertical (stratified), whereas plagioclase variation in the Brent Group is mainly lateral. The lateral transition from plagioclase-bearing to plagioclase-free composition observed within the entire Brent Group, over a distance of only 5 km (Fig. 2), is unlikely to be the result of varying depositional sand composition because of lateral shifting of depositional centres during the course of Brent accumulation. Processes of sediment mixing and dispersal within deltaic

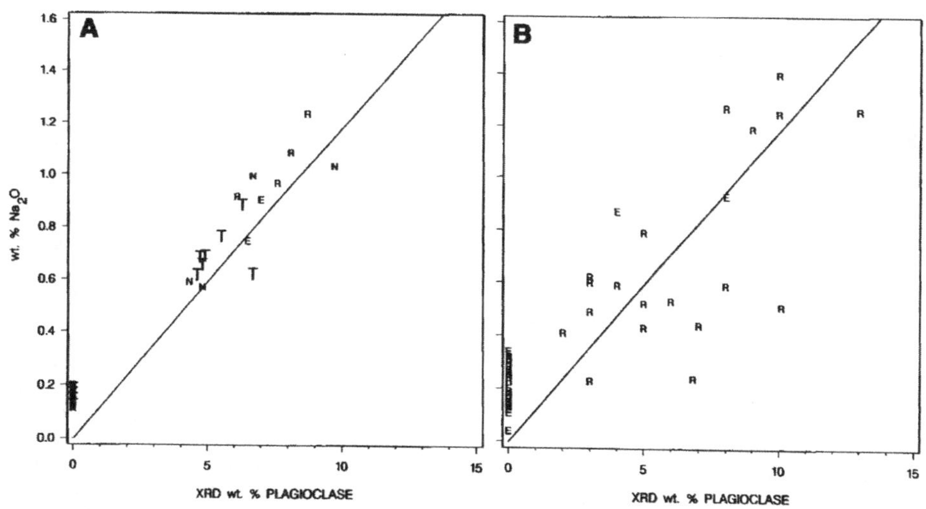

Fig. 7. Bulk-rock sodium (from X-ray fluorescence) vs. plagioclase content (from X-ray diffraction). (A) 1994 data set. (B) Pre-1986 data set. For the 1994 data the correlation has been improved using bulk chlorine analysis to correct for minor amounts of Na contained in salt precipitated from evaporation of trapped pore water. Diagonal line represents correlation for ideal albite. Plotting symbols indicate formations (T, Tarbert; N, Ness; E, Etive; R, Rannoch).

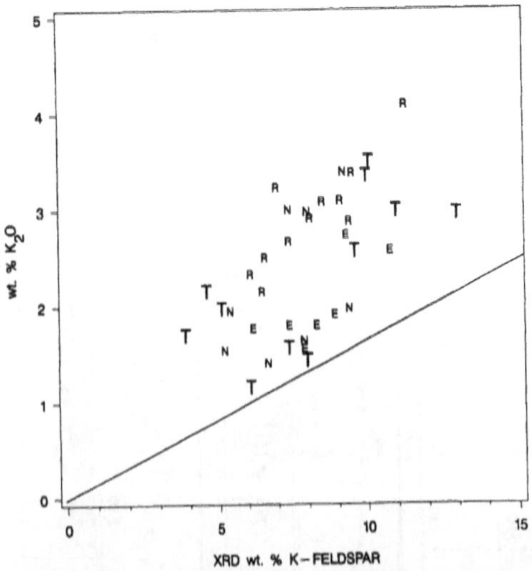

Fig. 8. Bulk-rock potassium vs. K-feldspar content determined by X-ray diffraction (1994 data set). The diagonal line shows the ideal relationship for stoichiometric K-feldspar as the only K-bearing mineral. The fact that the points plot above this line indicates the presence of other K-bearing minerals (mica, illite) and possibly some degree of analytical error (for example, under-estimation of percentage K-feldspar). Plotting symbols indicate formations (T, Tarbert; N, Ness; E, Etive; R, Rannoch).

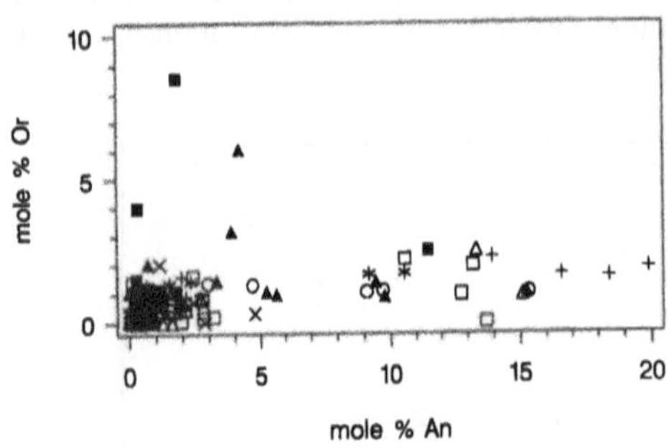

Fig. 9. Electron microprobe analyses of 125 different plagioclase grains in eight thin sections (two thin sections from each formation studied; plotting symbols indicate formation) from wells 34/10-14 and 34/10-B14. Analyses performed by T. Boassen at IKU Petroleum Research, Trondheim.

depositional systems the size of the palaeo-Brent delta (100–150 km wide; Graue *et al.*, 1987) operate over scales much larger than 5 km, such that provenance differences would not be expected to remain stationary in space throughout accumulation of successive depositional units. The occurrence of plagioclase throughout the Brent section in well 34/10-A9H (Fig. 2), documented by thin-section observations, also seems to rule out the possibility of a separate provenance for the southern part of the Gullfaks area. The second possibility thus appears unavoidable: that plagioclase absence is a diagenetic effect. Note that this does not exclude the possibility that some subordinate intervals of plagioclase-free sandstone could have had zero plagioclase content from the time of deposition, as discussed below specifically for the lower Rannoch Formation.

Additional support for the hypothesis of diagenetic plagioclase loss is provided by the observation that plagioclase is present (3–4 wt%) in two tightly calcite-cemented samples (1783·95 and 1790·95 m) from the otherwise plagioclase-free Ness Formation of well 34/10-1 (Fig. 6). It can be argued that the calcite cement predates plagioclase leaching and protected the plagioclase from contact with the pore waters responsible for dissolution. However, two other samples (1784·30 and 1784·60 m) between the above plagioclase-bearing depths are also tightly calcite cemented but lack plagioclase, possibly reflecting somewhat later timing of calcite cementation or less protective cement distribution. The calcite cement in these samples is blocky to poikilotopic with irregular to sweeping extinction, but geochemical data are not available. The data set also contains several other heavily calcite-cemented

Fig. 10. Photomicrographs showing feldspar and kaolin textures in typical Brent sandstones. (A) Perthite grain in a plagioclase-bearing sandstone (1988·3 m, 34/10-14; Ness Formation). K-feldspar lamellae are undissolved. Albite lamellae are mostly intact, but are dissolved near grain boundaries. Grains are coated by very finely crystalline, greenish clay. (B) Perthite grain in a plagioclase-free sandstone (1836·0 m, 34/10-A3H; Etive Formation). K-feldspar lamellae (stained yellow) are undissolved and bend around minor carbonate (?calcite) crystals (white). Albite lamellae are entirely dissolved. A grain-like mass of microporous kaolin is seen at the top centre. Grains are coated by dark brown (?oxidized), very finely crystalline siderite. (C) An undissolved K-feldspar grain in a plagioclase-free sandstone (1831·5 m, 34/10-A3H; Etive Formation); polarizing filters crossed. The grain is coated by a thin overgrowth of K-feldspar cement (optical orientation similar to the one twin set not at extinction). (D) A tightly packed kaolin aggregate (centre) is sandwiched between quartz grains. A 'halo' of microporous, vermicular kaolin (lower left and top right) surrounds the central aggregate where it is in contact with intergranular pore spaces (1836·0 m, 34/10-A3H; Etive Formation).

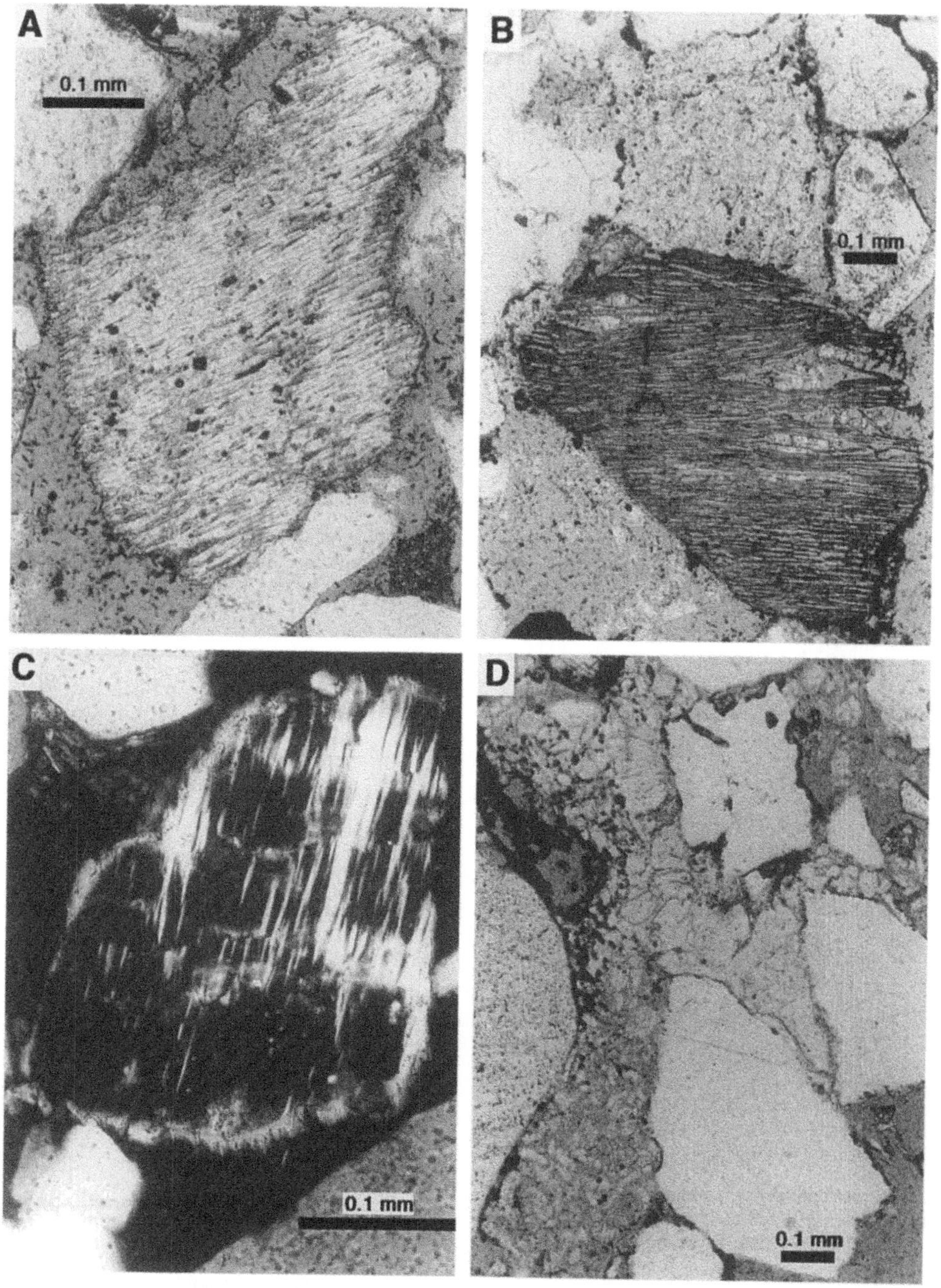

and plagioclase-free samples from plagioclase-free intervals (three Tarbert samples from 34/10 to five and three Rannoch samples from 34/10-A3H), which could again indicate later timing of calcite cementation or non-protective cement distribution.

Timing of plagioclase dissolution

Four alternative models may be considered regarding the possible nature and timing of post-depositional plagioclase loss:

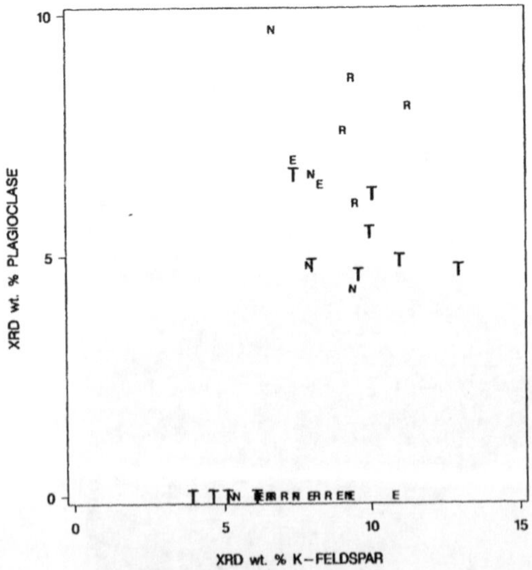

Fig. 11. Plagioclase vs. K-feldspar content (1994 data set). Plotting symbols indicate formations (T, Tarbert; N, Ness; E, Etive; R, Rannoch).

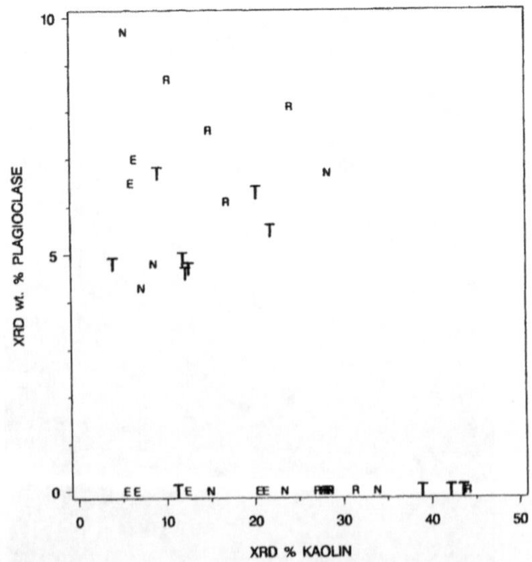

Fig. 13. Plagioclase vs. kaolin content (1994 data set). Plotting symbols indicate formations (T, Tarbert; N, Ness; E, Etive; R, Rannoch).

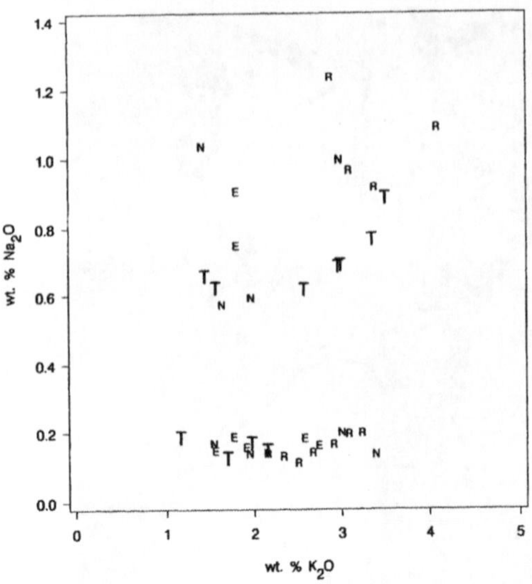

Fig. 12. Bulk-rock sodium vs. potassium content (1994 data set). Plotting symbols indicate formations (T, Tarbert; N, Ness; E, Etive; R, Rannoch).

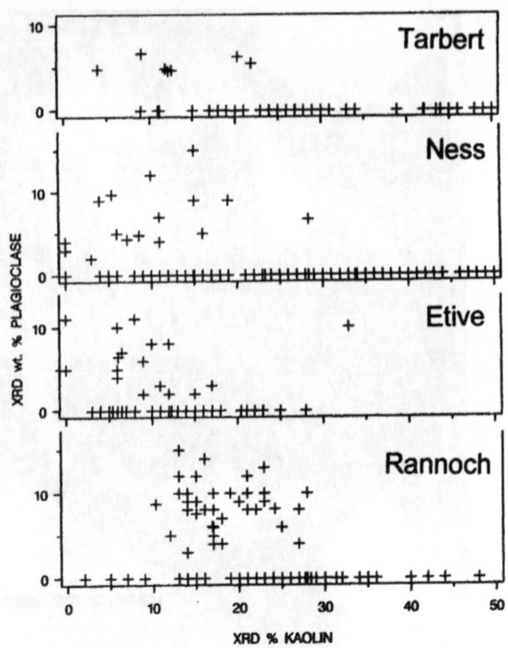

Fig. 14. Plagioclase vs. kaolin content, plotted separately for each formation (all data).

1 Leaching by meteoric water influx during or shortly after deposition of the Brent delta.
2 Shallow leaching by meteoric water influx during Cimmerian uplift and erosion.
3 Late leaching due to influx of acid components released by organic maturation, shale illitization, or addition of CO_2 released from deep-seated metamorphic processes.
4 Release of CO_2 and organic acids as by-products of biodegradation of the present oil column.

Model 1, leaching shortly after deposition, is particularly plausible for the Rannoch through Ness Formations because of the northward-prograding character of the lower part of the Brent succession (Fält *et al.*, 1989). Thus, strata further south would be closer to the emergent delta-plain setting where downward influx of aggressive soil-acid components from Ness peat swamps would have had greatest effect on underlying strata. However, an argument against this scenario is the observation of normal plagioclase contents in Brent cores from adjacent structures in the southern part of block 34/10 (Fig. 1), a notable exception being the Tarbert Formation in well 34/10-16 (Figs 1 and 6). With this exception, these cores indicate that the absence of plagioclase in the south side of Gullfaks Field is local rather than a widespread and intrinsic characteristic of the progradational lower part of the Brent delta. Another possible inconsistency with this model is the difficulty of explaining why coal-bearing lower-Ness strata in wells 34/10-3, 34/10-C3 and 34/10-14 are plagioclase bearing, if it is simply the water chemistry of the delta-plain setting that accounts for plagioclase removal.

A more serious problem is that the south-north division between leached and unleached sandstones persists also throughout the Tarbert Formation, which would not have been affected by acids from the underlying Ness delta plain environment. The Tarbert/Ness boundary marks a major transgression involving southward retreat of the Brent delta and reworking of underlying deposits (Graue *et al.*, 1987; Fält *et al.*, 1989). Thus the peat swamps overlying the Gullfaks area in Ness time would have been displaced far southward by Tarbert time.

Model 2, that leaching is related to the Cimmerian unconformity, could explain the observed pattern of leaching throughout the entire Brent stratigraphy, if it can be assumed that the southern part of Gullfaks Field experienced greater meteoric water influx than in the north. However, it would be expected according to this model that meteoric water influx, and therefore degree of feldspar dissolution, should increase with proximity to the Cimmerian unconformity, but this is not observed.The unconformity strikes roughly north–south (Fig. 3), such that the southern, plagioclase-free wells are not closer to the unconformity than the northern, plagioclase-bearing wells. The lack of any apparent correlation between plagioclase occurrence and vertical distance from the unconformity is also illustrated in Fig. 6, which does not support a causal relationship involving this surface. Absence of increase in kaolin content towards the unconformity has previously been documented within the Rannoch Formation of Gullfaks Field by Bjørkum *et al.* (1990), where it was pointed out that erosion may have removed rock faster that influx of meteoric water would have propagated a dissolution front into the sandstone.

Model 3, involving acid released during deeper burial (>1 km), would in general be expected to produce similar intensity of alteration throughout the Gullfaks structure unless the acid was supplied from a particular direction or area. One mechanism for focusing alteration towards one side of the Gullfaks structure would be if increased acidity were delivered along with the migration of oil from a particular direction. However, the probable filling direction of the Brent reservoirs in Gullfaks Field is up the structural dip, from the west (Larter & Horstad, 1992), and the main migration pathway is believed to be from the north–west (Thomas *et al.*, 1985; Karlsson, 1986; Horstad *et al.*, 1995), which would not tend to focus alteration in the southern part of the field. Furthermore, a number of formidable problems exist with generating and transporting sufficient reactive components to produce measurable degrees of feldspar alteration by the products of kerogen maturation (Barth & Bjørlykke, 1993). Another mechanism for focusing acid influx involves the release of acidity from shales by either clay mineral diagenesis or organic maturation (Curtis, 1983; Bjørlykke, 1984; Ehrenberg, 1991). However, this does not seem applicable to the Gullfaks case because there are no indications that the plagioclase-free intervals have greater proximity to shales (Fig. 6). Finally, the addition of CO_2 from deep-seated metamorphic or igneous processes could cause feldspar dissolution (Smith & Ehrenberg, 1989; Lundegard & Trevena, 1990), but there is no supporting evidence for this.

The final possibility, Model 4, is the preferred explanation for the observed pattern of plagioclase occurrence, because there is a correlation between this pattern and the degree of biodegradation of oils in the Gullfaks Brent reservoirs. Analyses of oil samples from different parts of the Gullfaks Brent system show that they belong to a single 'geochemical family', believed to have a common source and migration route (Horstad *et al.*, 1995). Oils from the nearby Tordis Field (Fig. 1) belong to this same family and represent an earlier position along the migration route. However, oils in the

Cook, Statfjord and Lunde Formations at Gullfaks Field and oils from the Brent and other reservoirs of Gullfaks Sør Field belong to distinctly different geochemical families.

Both stock-tank oils (Fig. 15) and calculated 'recombined' in-place oil compositions (Fig. 16A) show wide variations that are attributed to differing amounts of biodegradation in different locations within the Gullfaks Brent reservoir system (Larter & Horstad, 1992). The least biodegraded Gullfaks oils are similar to oil compositions from Tordis Field, where there are no signs of biodegradation. The recombined oil compositions have widely variable CO_2 content, with higher values showing a general correlation with the degree of biodegradation indicated by proportions of butane components (Fig. 16B). The higher CO_2 values lie well above the trend of CO_2 partial pressure vs. temperature for Norwegian-shelf sandstone reservoirs compiled by Smith & Ehrenberg (1989), whereas the lower-CO_2, less biodegraded compositions are close to this trend (Fig. 17), suggesting that biodegradation has disturbed the initial equilibrium condition of the mineral-CO_2 system.

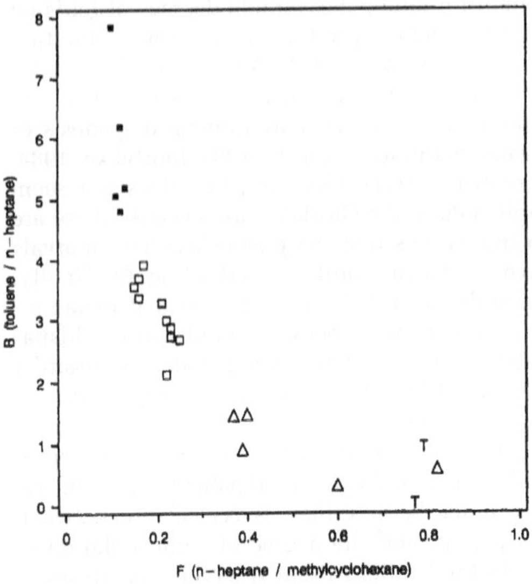

Fig. 15. Thompson indices (Thompson, 1987) measured on stock-tank oils from Brent Group reservoirs in Gullfaks Field and two locations in and near Tordis Field ('T' symbols). Symbols divide the 20 Gullfaks oils into three groups with differing degrees of biodegradation. Biodegradation increases towards the upper left, with decreasing abundance of normal-C_6 component (Halpern, 1995).

The above variations in oil composition show a geographical pattern in which the more biodegraded oils occur in the south-central part of Gullfaks Field and less altered oils occur in the flanking positions (Fig. 18). A similar pattern was reported by Larter & Horstad (1992). Comparison with Fig. 2 shows that the locations with more CO_2-rich, biodegraded oil compositions correspond to the area of plagioclase-free cored intervals. Thus, the apparent north–south contrast in plagioclase occurrence may reflect in part the limited geographical distribution of our mineralogical data. It is therefore proposed that the observed pattern of plagioclase absence (Fig. 2) is the result of biodegradation of oil, which occurred most intensely in the south-central area of the Gullfaks Brent reservoirs.

After this hypothesis was formulated, it was recognized that a possible test would be to determine whether plagioclase is present in the cores from well 34/10-A9H, which is located in the southern part of Gullfaks Field, has been extensively cored throughout the four main Brent formations, and contains oil showing low degree of biodegradation (Figs 15, 16 and 18). This test was carried out by staining a selection of the existing thin sections for K-feldspar, the feldspar grains remaining unstained thus being unambiguously recognizable as plagioclase. A set of thin sections from well 34/10-9 was also stained, as this well also contains oil showing only minor biodegradation. As described above, all thin sections examined from these wells contain minor to numerous plagioclase grains, confirming the prediction made from the biodegradation hypothesis.

Except for the two cores from the Rannoch Formation that are discussed below, the only data seemingly inconsistent with the biodegradation hypothesis are from well 34/10-3, where the oil is highly biodegraded (Fig. 18), but plagioclase is present throughout much of the cored interval (Fig. 2). However, closer inspection shows that these results may not be inconsistent after all. This is because the 34/10-3 oils are from drill-stem tests 2 and 3, with perforations at 1935–48 and 1895–1900 m, respectively, in the central to upper Ness Formation, where plagioclase is absent (Fig. 6). The plagioclase-bearing interval in this well is deeper than 1960 m (Fig. 6). Most of the plagioclase-bearing interval is in the water zone, and the upper oil-saturated part of the plagioclase-bearing interval is separated from the overlying, sampled portion of the oil column by a shaly, low-permeability barrier at 1953–60 m. It is

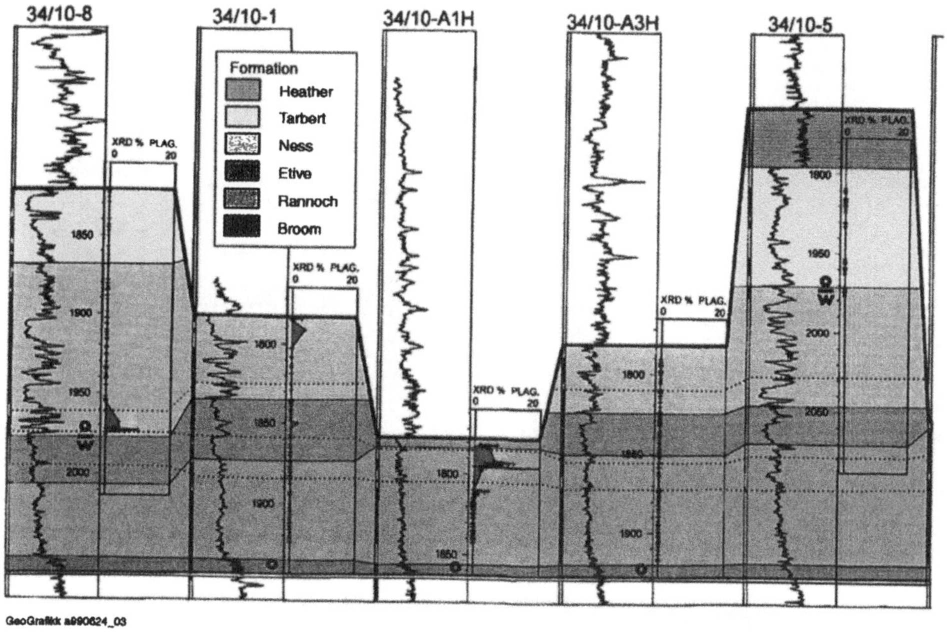

Fig. 6. Profiles of gamma ray activity (scale 0–150 API units) and plagioclase content (area under plagioclase curve is coloured red; '+' symbols at zero vertical reference line indicate analyses with zero plagioclase content) in the Brent Group in 10 wells from Gullfaks Field (Fig. 2) and one well (34/10–16) from Gullfaks Sør Field (Fig. 1). Cored intervals are indicated by solid bars at the left of the GR column. Base-Cretaceous unconformity is indicated by thick line at top of coloured section (Brent formations and Heather Formation). Depths are measured depth (m below kelly bushing), except for the deviated well 34/10-B14, where vertical depth is plotted. Plagioclase content was determined by XRD, except in well 34/10-C3, where weight percentage plagioclase was calculated from bulk sodium content (Fig. 7A). Dotted lines show positions of three interpreted time surfaces in Gullfaks wells.

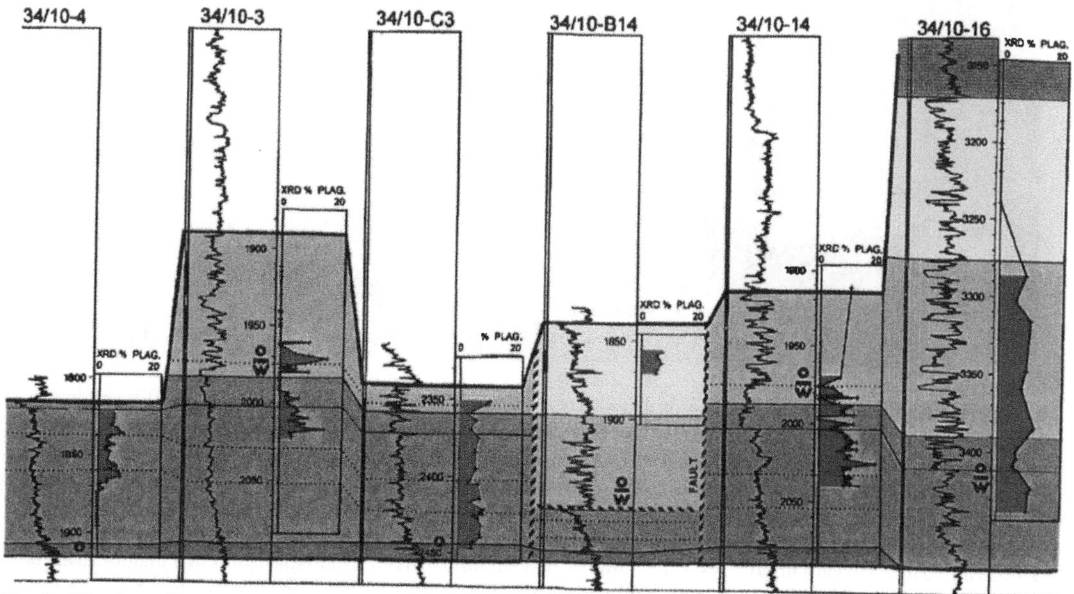

Fig. 6. (Continued).

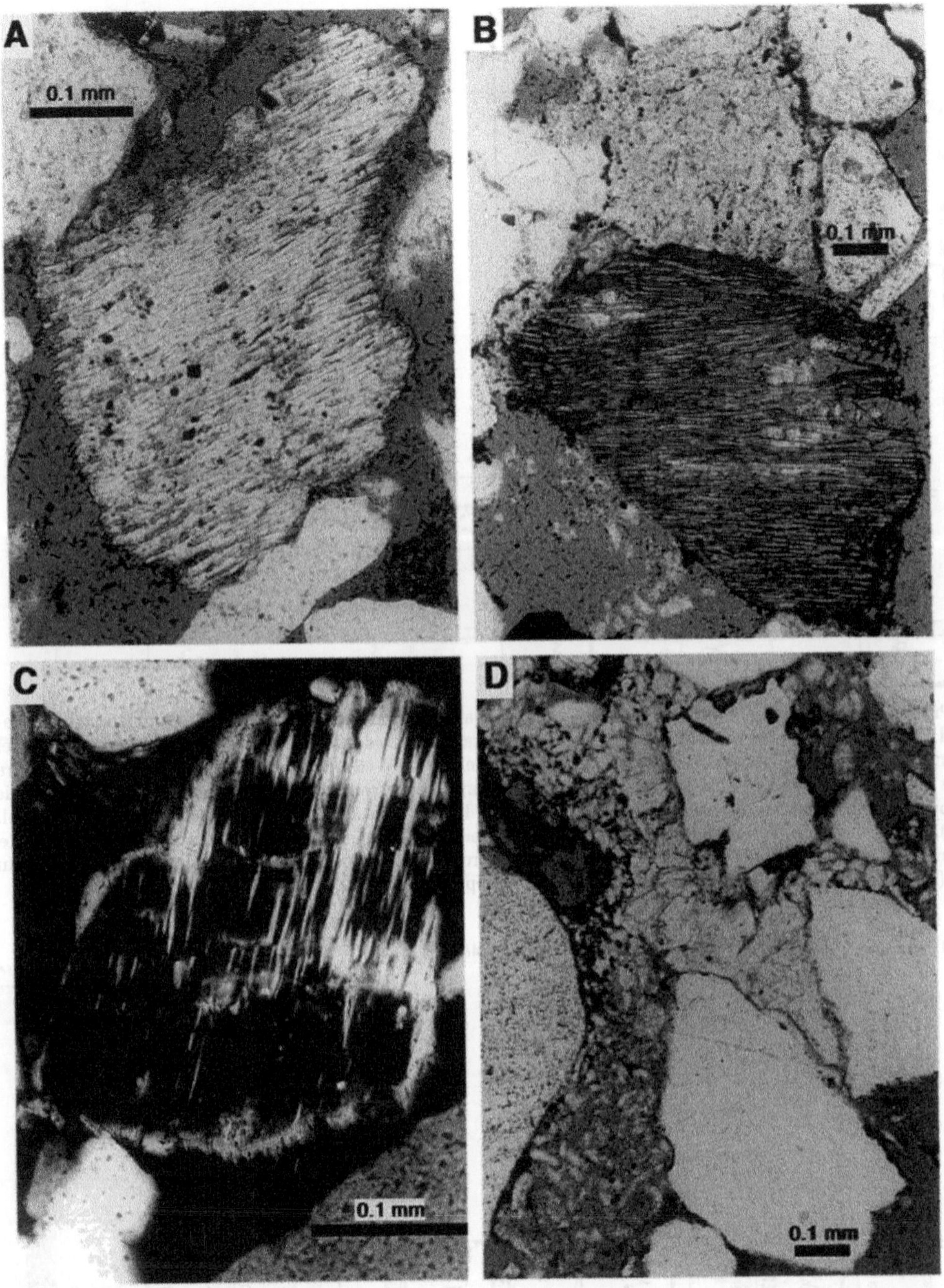

Fig. 10. Photomicrographs showing feldspar and kaolin textures in typical Brent sandstones. (A) Perthite grain in a plagioclase-bearing sandstone (1988·3 m, 34/10-14; Ness Formation). K-feldspar lamellae are undissolved. Albite lamellae are mostly intact, but are dissolved near grain boundaries. Grains are coated by very finely crystalline, greenish clay. (B) Perthite grain in a plagioclase-free sandstone (1836·0 m, 34/10-A3H; Etive Formation). K-feldspar lamellae (stained yellow) are undissolved and bend around minor carbonate (?calcite) crystals (white). Albite lamellae are entirely dissolved. A grain-like mass of microporous kaolin is seen at the top centre. Grains are coated by dark brown (?oxidized), very finely crystalline siderite. (C) An undissolved K-feldspar grain in a plagioclase-free sandstone (1831·5 m, 34/10-A3H; Etive Formation); polarizing filters crossed. The grain is coated by a thin overgrowth of K-feldspar cement (optical orientation similar to the one twin set not at extinction). (D) A tightly packed kaolin aggregate (centre) is sandwiched between quartz grains. A 'halo' of microporous, vermicular kaolin (lower left and top right) surrounds the central aggregate where it is in contact with intergranular pore spaces (1836·0 m, 34/10-A3H; Etive Formation).

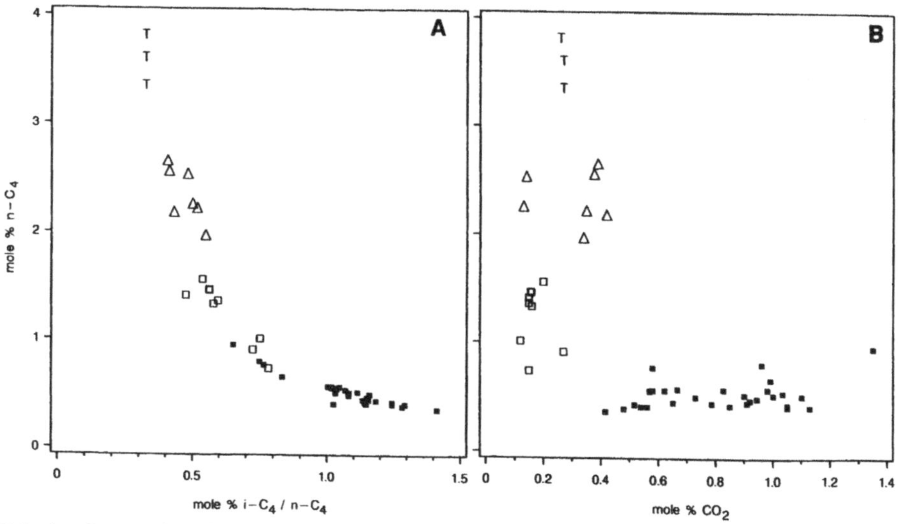

Fig. 16. Calculated 'recombined' in-place oil compositions from Brent Group reservoirs in Gullfaks Field and three locations in and near Tordis Field ('T' symbols). Symbols divide the Gullfaks oils into three groups with differing degrees of biodegradation, based on plot (B). Note that these groups do not exactly correspond with the three Gullfaks oil groups defined in Fig. 15 because the latter are defined for stock-tank oil compositions, whereas the present figure concerns calculated oil compositions at depth within the reservoir. Biodegradation increases with decreasing abundance of normal-C_4 component (Halpern, 1995), that is, towards the lower right in both plots.

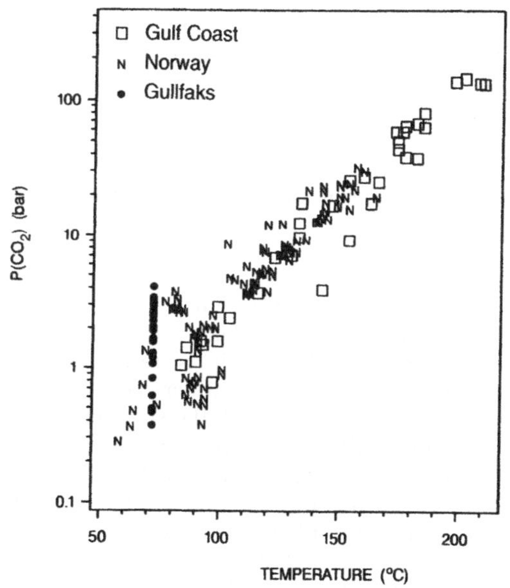

Fig. 17. $P(CO_2)$ (mol% $CO_2 \times$ fluid pressure) vs. temperature for hydrocarbon reservoirs from Fig. 2 of Smith & Ehrenberg (1989) and for Gullfaks Field.

possible therefore that the lower, plagioclase-bearing section in well 34/10-3 was not in communication with the corrosive fluid composition produced by biodegradation within the main part of the Ness Formation. A similar pattern is present in well 34/10-8 (Fig. 6), where plagioclase is absent throughout all of the Ness Formation except the lowermost 10 m, which is just above the oil/water contact and just below the same shaly barrier zone as in 34/10-3.

Biodegradation model

Biodegradation is known to be associated with increased acidity of petroleum, caused by generation of both CO_2 and organic acids as the products of bacterial metabolism (Mackenzie *et al.*, 1983; Behar & Albrecht, 1984; Jaffè & Gallardo, 1993). Increased acidity attending generation of CO_2 could cause plagioclase dissolution by the following reaction:

$$2CO_2 + 2NaAlSi_3O_8 + 3H_2O = Al_2Si_2O_5(OH)_4 + 4SiO_2 + 2Na^+ + 2HCO_3^- \quad (1)$$

whereby each mole of CO_2 supplied could account for one mole of plagioclase dissolved.

According to this reaction, a rather large amount of CO_2 is required to dissolve an average value of 7 wt% albite (Fig. 7A) from the sandstone. Using 30% porosity and grain density of 2·65, a cubic metre of Brent sandstone with 7 wt%

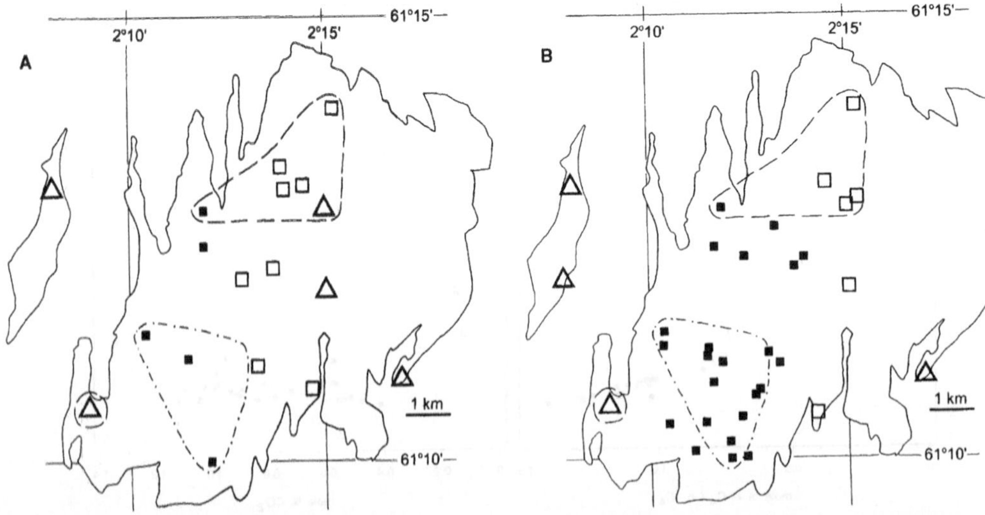

Fig. 18. Maps showing the geographical distribution of biodegradation groupings of oils from Brent Group reservoirs in Gullfaks Field. Dashed lines enclose areas with plagioclase-bearing cores (Fig. 2; excluding wells 34/10–4 and 34/10-A1H, where the Rannoch Formation comprises both plagioclase-bearing and plagioclase-free intervals). The dot-dash line encloses the area with plagioclase-free cores. (A) Thompson indices (Thompson, 1987) measured on stock-tank oils. Symbols correspond to the three groups defined in Fig. 15. The northernmost filled-square symbol represents well 34/10-3 (discussed in the text). (B) Calculated 'recombined' in-place oil compositions. Symbols correspond to the three groups defined in Fig. 16B. The northernmost filled-square symbol represents well 34/10-3 (discussed in the text).

albite contains roughly 500 mol of albite. Assuming 5% water saturation, 29° API oil and 0·3 wt% CO_2 in the oil (roughly 1 mole percentage), the same cubic metre of Brent sandstone from the oil zone would contain roughly 15 mol of CO_2. Therefore, the present CO_2 content of the oil in the southern part of Gullfaks Field represents only a small fraction of the total CO_2 required to produce the observed difference between the plagioclase-bearing and plagioclase-free cores using Eqn 1. Some of the apparently missing CO_2 would necessarily have been consumed by reaction with plagioclase, but it seems probable that the main reactants involved were the various organic acid species produced during biodegradation, rather than carbonic acid alone. Although there is no basis for independently estimating the amounts of CO_2 and organic acids produced at Gullfaks, the observed spatial correlation between plagioclase absence and oil biodegradation supports the contention that the products of biodegradation must have been sufficient to accomplish the required degree of plagioclase dissolution.

The problem of removing the sodium released by albite dissolution must also be considered. If the assumed 5% residual water volume were static, then dissolution of 7 wt% albite would result in an implausibly high concentration of $NaHCO_3$. Thus, 5000 g of water (5 volume percentage) plus 500 mol $NaHCO_3$ per cubic metre of sandstone reservoir gives a concentrated solution of roughly 46 wt% $NaHCO_3$. However, the process of biodegradation is widely regarded as requiring a continuous flow of water to supply nutrients and remove metabolic by-products (Hunt, 1979) and may therefore be dependent upon continuous flow of the residual water phase, whether or not biodegradation is mainly aerobic or anaerobic (Heider *et al.*, 1999). Therefore, using the model of Bjørkum *et al.* (1998), it is reasonable to postulate both significant water permeability through the Gullfaks caprock (Lower Cretaceous shale) and significant relative permeability for water in the oil zone, such that there was continuous upward flow of residual water through the oil column, of sufficient magnitude to support biodegradation and also remove the sodium released by plagioclase breakdown.

The Rannoch Formation

The Rannoch Formation in wells 34/10-4 and 34/10-A1H shows a pattern of plagioclase occurrence that appears inconsistent with the above general explanation of plagioclase absence due to biodegradation of oil. In these wells, plagioclase is present throughout the upper part of the Rannoch Formation and is absent throughout the lower part of the formation (Fig. 6). The Rannoch Formation throughout Gullfaks Field is characterized by a general trend of upward-increasing grain size and quartz content resulting in upward-increasing permeability and oil saturation (Fig. 4). Thus, dissolution of plagioclase in proportion to oil content might produce preferential leaching from the more oil-saturated upper part of the unit, but the observed pattern is difficult to account for.

A possible alternative is to postulate a unique plagioclase-deficient composition for the lower Rannoch Formation at the time of its deposition. This argument is complicated, however, because in well 34/10-C3, located further north in the field (Fig. 2), plagioclase is present throughout both the upper and lower Rannoch Formation (Fig. 6). It must be further postulated therefore that the lower Rannoch was deposited with plagioclase-deficient composition in the southern part of the field, but had normal plagioclase content when deposited further north. This is sedimentologically possible because of the northward progradational geometry of the Brent delta (Fig. 19). Thus, the earliest Rannoch deposits in Gullfaks Field (lower Rannoch in the south) could have been plagioclase deficient, and all later deposits (upper Rannoch in the south and time-equivalent lower Rannoch in the north, plus all overlying Brent strata) could have had normal depositional plagioclase content. Although this scenario offers a possible explanation consistent with the data, it is presently impossible to test. A possible test is to examine profiles of Sm–Nd isotopic analyses, an approach commonly used for detecting provenance differences (Mearns, 1989). However, the Nd data do not indicate any consistent differences in provenance between the upper and lower Rannoch intervals in either well 34/10-1 or 34/10-A1H (Fig. 19), which does not support the postulate of a provenance difference, but does not necessarily disprove it either (as different provenances can have either similar or different Nd isotopic composition).

Another perplexing observation is the presence of plagioclase in the uppermost Rannoch Formation in well 34/10-A1H, but the absence of plagioclase throughout the upper Rannoch in the nearby wells 34/10-1 and 34/10-A3H (Fig. 2). Because of the proximity of these wells, it is implausible to postulate a difference in sand provenance within the upper Rannoch Formation between well 34/10-A1H and the surrounding wells to explain this difference in plagioclase content. It seems equally doubtful that there should have been any significant difference in degree of biodegradation and consequent plagioclase leaching between these nearby locations, but this may be the only alternative available.

Despite the above problems with explaining plagioclase distributions in the Rannoch

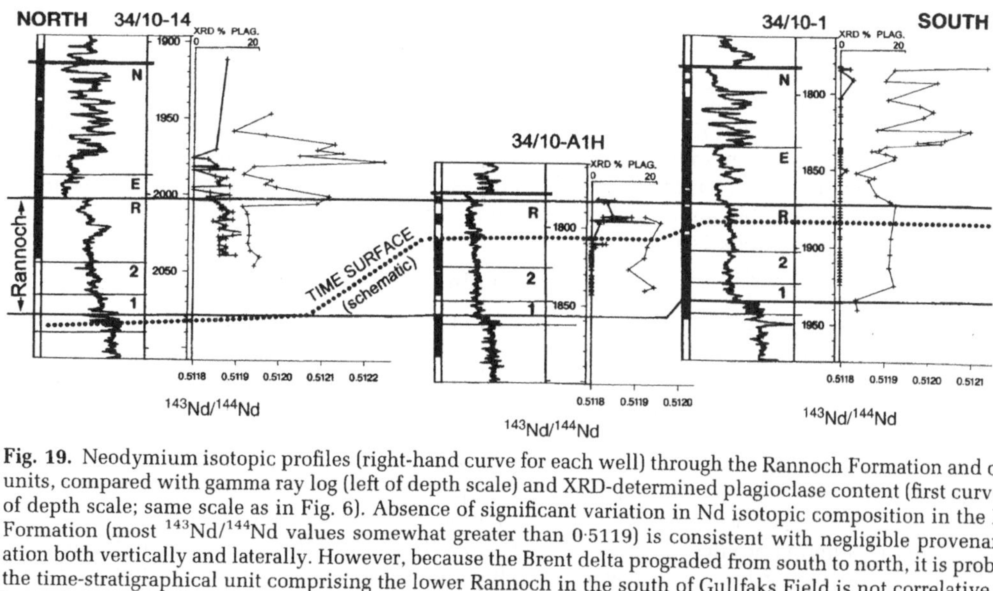

Fig. 19. Neodymium isotopic profiles (right-hand curve for each well) through the Rannoch Formation and overlying units, compared with gamma ray log (left of depth scale) and XRD-determined plagioclase content (first curve to right of depth scale; same scale as in Fig. 6). Absence of significant variation in Nd isotopic composition in the Rannoch Formation (most $^{143}Nd/^{144}Nd$ values somewhat greater than 0·5119) is consistent with negligible provenance variation both vertically and laterally. However, because the Brent delta prograded from south to north, it is probable that the time-stratigraphical unit comprising the lower Rannoch in the south of Gullfaks Field is not correlative with the lower Rannoch in the north side of the field (as illustrated by the dotted correlation line).

Formation in terms of the biodegradation model, it is not clear that this general model is necessarily invalidated by these inconsistencies.

K-feldspar

If plagioclase absence is caused by greater intensity of dissolution, then it might be expected that corresponding effects would also be observed for K-feldspar. However, this appears not to be the case. Plagioclase-free intervals have at most only moderately reduced K-feldspar content compared with plagioclase-bearing intervals (Figs 5 and 11), and only a few of the plagioclase-free samples also lack K-feldspar (Fig. 5). Furthermore, thin overgrowths of K-feldspar cement partly covering detrital K-feldspar grains are ubiquitous in both plagioclase-bearing and plagioclase-free sandstones (Fig. 10C). Similar overgrowths are typically the earliest form of K-feldspar to show extensive dissolution in Haltenbanken sandstones buried somewhat deeper than the Gullfaks cores (around 3·3–3·5 km below the sea floor; Ehrenberg & Nadeau, 1989). These observations indicate that K-feldspar was relatively resistant to dissolution under the conditions where biodegradation took place, which could reflect either thermodynamic or kinetic factors.

One possibility is greater thermodynamic stability of K-feldspar. However, based on an extensive suite of porewater analyses from North Sea sandstone reservoirs, Bjørlykke *et al.* (1995) reported that water compositions are undersaturated with respect to K-feldspar at depths greater than about 1·5 km. Slower kinetics of K-feldspar dissolution is the other possible explanation for preferential loss of albite. Both experimental and empirical geological data show that K-feldspar is more resistant to dissolution than sodic plagioclase at near-surface temperatures, but the degree to which these data are applicable under higher-temperature subsurface conditions is uncertain. Lasaga (1984) summarized experimental data on the rate of release of silica from different minerals under acidic surface conditions (25 °C, pH = 5) and reported the value for albite to be around seven times that of K-feldspar. Walderhaug (1998) tabulated other experimental results indicating that albite may dissolve about three times as fast as K-feldspar. A wealth of observational data on relative stability of Na-plagioclase and K-feldspar under surficial conditions have been contributed by studies of deep weathering profiles. Most such profiles show distinctly earlier loss of Na-plagioclase and bulk sodium content relative to K-feldspar and bulk potassium (Goldich, 1938; Wahlstrom, 1948; Grant, 1963; Bondam, 1967; Nesbitt & Young, 1984, 1989; Bain *et al.*, 1993; Morey & Setterholm, 1997). In general, sodic plagioclase is entirely lost before there is any reduction in K-feldspar content.

Whereas K-feldspar is more resistant to dissolution than sodic plagioclase in the near-surface, meteoric environment, this relationship is commonly reversed upon burial below a few kilometres depth (80–100 °C), where formation waters typically have moderate to high salinity. Here detrital Ca-bearing plagioclase recrystallizes to form authigenic albite and kaolin, and K-feldspar begins to be replaced by albite, the extent of reaction depending on the supply of Na^+ available from the formation water (Saigal *et al.*, 1988; Aagaard *et al.*, 1990; Morad *et al.*, 1990). Upon still deeper burial (>125–140 °C), K-feldspar reacts with kaolin to form illite, quartz and albite (Ehrenberg & Nadeau, 1989; Aagaard *et al.*, 1990). Where formation waters are Na-poor, however, albitization does not take place, and both K-feldspar and plagioclase dissolve to form illite and kaolin (Lundegard & Trevena, 1990).

Carbonates

Like K-feldspar, the various carbonate minerals siderite, dolomite and calcite show no difference in abundance between plagioclase-free and plagioclase-bearing intervals (Fig. 5) and thus appear to have been unaffected by the inferred dissolution process. However, this observation is not as troubling as the apparent lack of K-feldspar loss in the plagioclase-free intervals. As pointed out by Smith & Ehrenberg (1989), carbonate stability in the subsurface depends upon the fugacity of CO_2, which may be buffered by aluminosilicate phases. Addition of CO_2 can thus lead to carbonate precipitation rather than dissolution.

Kaolin

Bjørlykke *et al.* (1995) have argued that, even taking into consideration the possible effects of organic acid complexes, aluminium is effectively immobile on the scale of metres at the temperature and pH conditions expected at moderate depth (1–3 km) in siliciclastic sediments. Therefore, feldspar dissolution should result in local precipitation of Al-bearing products (initially kaolin) within the rock volume affected. If plagioclase alteration is the main source of kaolin in Gullfaks sandstones, then kaolin should be more abundant in strata where plagioclase has been entirely altered. Assuming conservation of aluminium, dissolution of each mole of albite should produce 0·5 mol of kaolin (Bjørlykke, 1984), equivalent to 0·6 wt% kaolin per weight percentage albite lost.

As noted above, there is a distinct tendency towards higher kaolin content in the plagioclase-free samples (Figs 5, 13 and 14). However, there is no linear correlation of increasing kaolin with decreasing plagioclase, and many plagioclase-free samples have very low kaolin content. The lack of such correlation can have various possible explanations; the most obvious of which is that depositional plagioclase content is likely to have been widely variable. Another factor is that the total kaolin content probably results from a combination of processes including, but not limited to, plagioclase alteration. For example, variable amounts of kaolin may be detrital (Bjørkum *et al.*, 1990); may have formed from poorly crystallized aluminosilicate weathering products (such as halloysite); or may have formed by eogenetic or mesogenetic dissolution of K-feldspar and mica. If much of the kaolin in a significant fraction of the Gullfaks sandstones is unrelated to plagioclase alteration, then any negative correlation between plagioclase and kaolin could be concealed by uncorrelated variations in the amount of nonplagioclase-related kaolin. Overall, the wide and overlapping ranges of kaolin content in both plagioclase-free and plagioclase-bearing sandstones suggest that much of the total kaolin content in both groups originated from processes other than post-depositional plagioclase alteration.

Petrographical observations indicate that much of the kaolin is not derived from feldspar dissolution, but formed from mica and lithic grains, either before sand deposition (detrital kaolin) or *in situ*. Similar observations were made for the Rannoch Formation by Bjørkum *et al.* (1990), but the same types of textures also dominate in most samples from the overlying Etive, Ness and Tarbert Formations. Some of the total kaolin has texture consistent with precipitation as the result of in-place feldspar dissolution, that is, as microporous clusters of euhedral crystals apparently unassociated with tightly packed, detrital-appearing kaolin/mica aggregates (Fig. 10B). However, in most samples, a large fraction of the kaolin occurs either (1) in tightly packed, grain-like aggregates commonly preserving areas of birefringent relict mica structure, or (2) in immediately surrounding 'halos' of highly microporous, euhedral kaolin booklets and vermicular crystals (Fig. 10D). The halo areas typically show a pattern of increasing intercrystal microporosity and uniformity of crystal size outwards from the detrital core areas, suggesting derivation of kaolin reactants from the tightly packed aggregates and an outward gradient of decreasing 'precipitation intensity'. It is unknown to what degree the tightly packed kaolin aggregates represent detrital kaolin as opposed to kaolinization of altered mica and lithic grains after sand deposition, but their textures indicate that they were not derived from feldspar grains.

Reservoir quality

Brent Group sandstones typically display wide ranges in porosity and permeability correlated with differences in depositional facies (petrologically expressed principally by variations in grain size and quartz content; Ehrenberg, 1997). For this reason, both the plagioclase-bearing and plagioclase-free sample groups from the present study show wide ranges of porosity and permeability. Within the newer dataset, where more comprehensive petrological data are available, no systematic differences in porosity or permeability are apparent between these groups. Therefore, it is unknown whether plagioclase alteration had any effect on reservoir quality. However, because such effects are not readily apparent, it is reasonable to assume that their influence is not significant.

CONCLUSIONS

1 Albitic plagioclase is absent throughout the Brent Group in most wells from the southern part of Gullfaks Field, but is present in normal amounts (5–8 wt%) in the same strata in north of the field. Correlation of bulk XRD plagioclase with bulk-rock sodium content and petrographical observations show that plagioclase absence is not an artefact of analytical error.

2 Plagioclase absence is attributed to dissolution due to increased acidity produced by oil biodegradation, which is especially advanced in the south-central part of Gullfaks Field.

3 Despite complete removal of plagioclase, both K-feldspar and carbonate minerals are essentially unaffected.

4 Kaolin is moderately more abundant in the plagioclase-free samples, but lack of correlation between increasing kaolin and decreasing plagioclase indicates that most kaolin originated from processes other than plagioclase dissolution. Textural observations also show that much of the kaolin present is not the product of feldspar dissolution.

Perhaps the main significance of this study is that it documents a case of compelling empirical evidence for an organic geochemical process driving a mesogenetic (burial) inorganic diagenetic reaction. Despite considerable discussion of possible links between organic and inorganic diagenesis (Surdam *et al.*, 1989), examples previously presented (e.g. Gautier, 1986) can mostly be classified as either concerning eogenetic (very shallow) processes or based on theoretical relationships subject to alternative interpretation not necessarily involving organic matter.

ACKNOWLEDGEMENTS

This paper is based in part upon an unpublished 1983 Statoil report 'Reservoir Sedimentology of the Brent Group, Gullfaks Field' by S. Olaussen, M. Eien and O. A. Malm, where the mineralogical trends discussed here were first recognized. We thank R. Patience for advice regarding biodegradation and J. Johannesen, J. Milter, K. K. Meisingset and A. Båtevik for information about Gullfaks oil compositions and well data. The manuscript was improved by suggestions from P. H. Nadeau, O. Walderhaug, P. A. Bjørkum, E. Graue, A. Brendsdal and *Sedimentology* referees R. S. Haszeldine and A. S. Trevena.

REFERENCES

Aagaard, P., Egeberg, P.K., Saigal, G.C., Morad, S. and **Bjørlykke, K.** (1990) Diagenetic albitization of detrital K-feldspars in Jurassic, Lower Cretaceous and Tertiary clastic reservoirs from offshore Norway, II. Formation water chemistry and kinetic considerations. *J. Sed. Petrol.*, **60**, 575–581.

Bain, D.C., Mellor, A., Robertson-Rintoul, M.S.E. and **Buckland, S.T.** (1993) Variations in weathering processes and rates with time in a chronosequence of soils from Glen Feshie, Scotland. *Geoderma*, **57**, 275–293.

Barth, T. and **Bjørlykke, K.** (1993) Organic acids from source rock maturation: generation potentials, transport mechanisms and relevance for mineral diagenesis. *Appl. Geochem.*, **8**, 325–337.

Behar, F.H. and **Albrecht, P.** (1984) Correlations between carboxylic acids and hydrocarbons in several crude oils. Alteration by biodegradation. *Org. Geochem.*, **6**, 597–604.

Bjørkum, P.A., Mjøs, R., Walderhaug, O. and **Hurst, A.** (1990) The role of the late Cimmerian unconformity for the distribution of kaolinite in the Gullfaks Field, northern North Sea. *Sedimentology*, **37**, 395–406.

Bjørkum, P.A., Walderhaug, O. and **Nadeau, P.H.** (1998) Physical constraints on hydrocarbon leakage and trapping revisited. *Petroleum Geosci.*, **4**, 237–239.

Bjørlykke, K. (1984) Formation of secondary porosity: how important is it? In: *Clastic Diagenesis* (Eds D.A. McDonald and R.C. Surdam), *AAPG Mem.*, **37**, 277–286.

Bjørlykke, K. (1998) Clay mineral diagenesis in sedimentary basins – a key to the prediction of rock properties. Examples from the North Sea Basin. *Clay Mineral.*, **33**, 15–34.

Bjørlykke, K., Aagaard, P., Egeberg, P.K. and **Simmons, S.P.** (1995) Geochemical constraints from formation water analyses from the North Sea and the Gulf Coast Basins on quartz, feldspar and illite precipitation in reservoir rocks. In: *The Geochemistry of Reservoirs* (Eds J.M. Cubitt and W.A. England), *Geol. Soc. Spec. Publ.*, **86**, 33–50.

Bondam, J. (1967) Undersøgelser vedrørende de geokemiske forhold i kaolinforekomsten ved Rønne på Bronholm. *Meddelelser Dansk Geol. Forening*, **17**, 297–356.

Curtis, C.D. (1983) Link between aluminum mobility and destruction of secondary porosity. *AAPG Bull.*, **67**, 380–393.

Dalland, A., Mearns, E.W. and **McBride, J.J.** (1995) The application of samarium-neodymium (Sm-Nd) provenance ages to correlation of biologically barren strata: a case study of the Statfjord Formation in the Gullfaks oilfield, Norwegian North Sea. In: *Non-Biostratigraphical Methods of Dating and Correlation* (Eds R.E. Dunay and E.A. Hailwood), *Geol. Soc. Spec. Publ.*, **89**, 201–222.

Ehrenberg, S.N. (1991) Kaolinized, potassium-leached zones at the contacts of the Garn Formation, Haltenbanken, mid-Norwegian continental shelf. *Mar. Petrol. Geol.*, **8**, 250–269.

Ehrenberg, S.N. (1997) Influence of depositional sand quality and diagenesis on porosity and permeability: examples from Brent Group reservoirs, northern North Sea. *J. Sed. Res.*, **67**, 197–211.

Ehrenberg, S.N. and **Nadeau, P.H.** (1989) Formation of diagenetic illite in sandstones of the Garn Formation, Haltenbanken, mid-Norwegian continental shelf. *Clay Mineral.*, **24**, 233–253.

Ehrenberg, S.N. and **Siring, E.** (1992) Use of bulk chemical analyses in stratigraphic correlation of sandstones: an example from the Statfjord Nord Field, Norwegian continental shelf. *J. Sed. Petrol.*, **62**, 318–330.

Ehrenberg, S.N., Aagaard, P., Wilson, M.J., Fraser, A.R. and **Duthie, D.M.L.** (1993) Depth-dependent transformation of kaolinite to dickite in sandstones of the Norwegian continental shelf. *Clay Mineral.*, **28**, 325–352.

Ehrenberg, S.N., Skjevrak, I. and **Gilje, A.E.** (1995) Asphaltene-rich residues in sandstone reservoirs of Haltenbanken province, mid-Norwegian continental shelf. *Mar. Petrol. Geol.*, **12**, 53–69.

Eriksen, T., Helle, M., Henden, J. and **Rognebakke, A.** (1987) Gullfaks. In: *Geology of the Norwegian Oil and Gas Fields* (Eds A.M. Spencer, E. Holter, C.J. Campbell, S.H. Hanslien, P.H.H. Nelson, E. Nysæther and E.G. Ormaasen), pp. 273–286. Norwegian Petroleum Society, Graham and Trotman, London.

Fält, L.M., Helland, R., Jacobsen, V.W. and **Renshaw, D.** (1989) Correlation of transgressive-regressive depositional sequences in the Middle Jurassic Bret/Vestland Group megacycle, Viking Graben, Norwegian North Sea. In: *Correlation in Hydrocarbon Exploration* (Ed. J.D. Collinson), pp. 191–200. Norwegian Petroleum Society, Graham and Trotman, London.

Gautier, D.L. (1986) Roles of organic matter in sediment diagenesis. *SEPM Spec. Publ.*, **38**, 203.

Goldich, S.S. (1938) A study in rock weathering. *J. Geol.*, **46**, 17–58.

Grant, W.H. (1963) Weathering of Stone Mountain Granite. *Clay Clay Mineral.*, **11**, 65–73.

Graue, E., Helland-Hansen, W., Johnsen, J., Lømo, L., Nøttvedt, A., Rønning, K., Ryseth, A. and **Steel, R.** (1987) Advance and retreat of the Brent Delta System, Norwegian North Sea. In: *Petroleum Geology of North West Europe* (Eds J. Brooks and K. Glennie), pp. 915–937. Norwegian Petroleum Society, Graham and Trotman, London.

Halpern, H.I. (1995) Development and applications of light-hydrocarbon-based star diagrams. *AAPG Bull.*, **79**, 801–815.

Heider, J., Spormann, A.M., Beller, H.R. and **Widdel, F.** (1999) Anaerobic bacterial metabolism of hydrocarbons. *FEMS Microbiol. Rev.*, **22**, 459–473.

Horstad, I., Larter, S.R. and **Mills, N.** (1995) Migration of hydrocarbons in the Tampen Spur area, Norwegian North Sea: a reservoir geochemical evaluation. In: *The Geochemistry of Reservoirs* (Eds J.M. Cubitt and W.A. England), *Geol. Soc. Spec. Publ.*, **86**, 159–183.

Hunt, J.M. (1979) *Petroleum Geochemistry and Geology*. W. H. Freeman, San Francisco, 617 pp.

Jaffè, R. and **Gallardo, M.T.** (1993) Application of carboxylic acid biomarkers as indicators of biodegradation and migration of crude oils from the Maracaibo Basin, western Venezuela. *Org. Geochem.*, **20**, 973–984.

Johnsson, M.J. and **Meade, R.H.** (1990) Chemical weathering of fluvial sediments during alluvial storage: the Macuapanim Island point bar, Solimes River, Brazil. *J. Sed. Petrol.*, **60**, 827–842.

Karlsson, W. (1986) The Snorre, Statfjord and Gullfaks oilfields and the habitat of hydrocarbons on the Tampen Spur, offshore Norway. In: *Habitat of Hydrocarbons on the Norwegian Continental Shelf* (Eds A.M. Spencer *et al.*), pp. 181–197. Norwegian Petroleum Society, Graham and Trotman, London.

Kastner, M.A. (1971) Authigenic feldspars in carbonate rocks. *Am. Mineral.*, **56**, 403–442.

Lasaga, A.C. (1984) Chemical kinetics of water–rock interactions. *J. Geophys. Res.*, **89**, 4009–4025.

Larter, S. and **Horstad, I.** (1992) Migration of petroleum into Brent Group reservoirs: some observations from the Gullfaks Field, Tampen Spur area North Sea. In: *Geology of the Brent Group* (Eds A.C. Morton, R.S. Haszeldine, M.R. Giles and S. Brown), *Geol. Soc. Spec. Publ.*, **61**, 441–452.

Lundegard, P.D. and **Trevena, A.S.** (1990) Sandstone diagenesis in the Pattani Basin (Gulf of Thailand): history of water–rock interaction and comparison with the Gulf of Mexico. *Appl. Geochem.*, **5**, 669–685.

Mackenzie, A.S., Wolff, G.A. and **Maxwell, J.R.** (1983) Fatty acids in some biodegraded petroleums. possible origins and significance. In: *Advances in Organic Geochemistry 1981* (Eds M. Bjorøy *et al.*), pp. 637–689. John Wiley & Sons, Chichester.

Marshall, D.J. (1988) *Cathodoluminescence of Geological Materials.* Unwin Hyman, Boston.

Mearns, E.W. (1989) Neodymium isotope stratigraphy of Gullfaks oilfield. In: *Correlation in Hydrocarbon Exploration* (Ed. J.D. Collinson), pp. 201–215. Norwegian Petroleum Society, Graham and Trotman, London.

Morad, S., Bergan, M., Knarud, R. and **Nystuen, J.P.** (1990) Albitization of detrital plagioclase in Triassic reservoir sandstones from the Snorre Field, Norwegian North Sea. *J. Sed. Petrol.*, **60**, 411–425.

Morey, G.B. and **Setterholm, D.R.** (1997) Rare earth elements in weathering profiles and sediments of Minnesota: implications for provenance studies. *J. Sed. Res.*, **67**, 105–115.

Nesbitt, H.W. and **Young, G.M.** (1984) Prediction of some weathering trends of plutonic and volcanic rocks based on thermodynamic and kinetic considerations. *Geochim. Cosmochim. Acta*, **48**, 1523–1534.

Nesbitt, H.W. and **Young, G.M.** (1989) Formation and diagenesis of weathering profiles. *J. Geol.*, **97**, 129–147.

Odum, I.E., Doe, T.W. and **Dott, R.H. Jr** (1976) Nature of feldspar-grain size relations in some quartz-rich sandstones. *J. Sed. Petrol.*, **46**, 862–870.

Olaussen, S., Beck, L., Fält, L.M., Graue, E., Malm, O.A. and **South, D.** (1992) Gullfaks Field – Norway East Shetland Basin, northern North Sea. In: *Treatise of Petroleum Geology, Atlas of Oil and Gas Fields, Structural Traps VI* (Eds N.H. Foster and E.A. Beaumont), pp. 55–83. American Association of Petroleum Geologists, Tulsa.

Petterson, O., Storli, A., Ljosland, E. and **Massie, I.** (1990) The Gullfaks Field: geology and reservoir development. In: *North Sea Oil and Gas Reservoirs II* (Eds A. Buller *et al.*), pp. 67–90. The Norwegian Institute of Technology, Graham and Trotman, London.

Petterson, O., Storli, A., Ljosland, E., Nygaard, O., Massie, I. and **Carlsen, H.** (1992) The Gullfaks Field. In: *Giant Oil and Gas Fields of the Decade 1978–88* (Ed. M.T. Halbouty), *AAGP Mem.*, **54**, 429–446.

Ramm, M. (1991) On quantitative mineral analysis of sandstones using XRD. *Dept. Geol. Oslo, Intern. Skr. Ser.*, **62**, 1–28.

Saigal, G.C., Morad, S., Bjørlykke, K., Egeberg, P.K. and **Aagaard, P.** (1988) Diagenetic albitization of detrital K-feldspars in Jurassic, Lower Cretaceous and Tertiary clastic reservoirs from offshore Norway, I. Textures and origin. *J. Sed. Petrol.*, **58**, 1003–1013.

Schmidt, V., McDonald, D.A. and **Platt, R.L.** (1977) Pore geometry and reservoir aspects of secondary porosity in sandstones. *Bull. Can. Petrol. Geol.*, **25**, 271–290.

Smith, J.T. and **Ehrenberg, S.N.** (1989) Correlation of carbon dioxide abundance with temperature in clastic hydrocarbon reservoirs: relationship to inorganic chemical equilibrium. *Mar. Petrol. Geol.*, **6**, 129–135.

Surdam, R.C., Crossey, L.J., Hagen, E.S. and **Heasler, H.P.** (1989) Organic–inorganic interactions and sandstone diagenesis. *AAPG Bull.*, **73**, 1–23.

Thomas, B.M., Møller-Pedersen, P., Whitaker, M.F. and **Shaw, N.D.** (1985) Organic facies and hydrocarbon distributions in the Norwegian North Sea. In: *Petroleum Geochemistry in Exploration of the Norwegian Shelf* (Ed. B.M. Thomas), pp. 3–26. Norwegian Petroleum Society, Graham and Trotman, London.

Thompson, K.F.M. (1987) Fractionated aromatic petroleums and the generation of gas-condensates. *Org. Geochem.*, **11**, 573–590.

Tollefsen, S., Graue, E. and **Svinddal, S.** (1992) The Gullfaks Field development: challenges and perspectives. *Soc. Petrol. Eng.*, **SPE 25054**, 377–393.

Wahlstrom, E.E. (1948) Pre-Fountain and Recent weathering on Flagstaff Mountain near Boulder, Colorado. *Geol. Soc. Am. Bull.*, **59**, 1173–1190.

Walderhaug, O. (1998) Chemical weathering at rock art sites in western Norway: which mechanisms are active and how can they be retarded? *J. Archaeol. Sci.*, **25**, 789–800.

Manuscript received 25 November 1999; revision accepted 15 December 2000.

Integrated time, temperature and water-composition analysis of sandstone mesogenesis

Sandstones in the mesogenetic regime have passed through the eogenetic regime. This apparent truism has proved worthwhile pursuing, in terms of overall understanding of mineralogy and fabric, as a way of developing holistic pictures of sandstone diagenesis. Many sandstones involve each of the common cements, such as quartz, carbonate and clay cements, as well as pyrite and sulphate cements. The growth of these cements may in part be the result of fluxes of formation water, so that identifying candidate water types that are potentially responsible for diagenetic modifications to sandstones has been an important aspect of the integrated study of sandstone diagenesis.

Baker (1991) considered that a combination of locally variable environment of deposition, and the consequent eogenetic modifications to the sand, influx of water from mudstones and the pervasive influx of meteoric water following maximum burial resulted in the formation of just about every type of carbonate cement, quartz, illite–smectite clay as well as pyrite. The combination of these cements controlled reservoir quality in the fluvio-deltaic Permian sandstones and only the careful unravelling of the complex series of diagenetic events allowed the prediction of reservoir quality trends.

Kantorowicz (1985) studied non-marine Mid-Jurassic sandstones from eo- through meso- to telogenesis, attributing early kaolinite and siderite to warm wet eogenesis. Subsequent mesogenesis involved quartz and ferroan calcite cementation owing to an influx of bicarbonate-rich waters. Telogenetic modifications resulted from oxidized meteoric water flushing and led to carbonate cement, feldspar dissolution and the conversion of diagenetic chlorite into expandable clays.

Morad *et al.* (1994) report eogenesis of fluvial Triassic sands dominated by kaolinite and K-feldspar cement and infiltrated clay. Initial mesogenetic alterations involved basinal brines and led to quartz, anhydrite and siderite cements. Continued burial resulted in dickite, chlorite, quartz and illite growth. The current formation waters were derived from older, deeper dolomitic mudstones. Dickite, common in these sandstones, is thermodynamically unstable in the presence of these formation waters, implying that the rate of dickite-consuming reactions may be very slow in sandstones.

Reprinted from Sedimentology (1991) ***38**, 819–838*

Diagenesis and reservoir quality of the Aldebaran Sandstone, Denison Trough, east-central Queensland, Australia

JULIAN C. BAKER*

Department of Geology and Mineralogy, University of Queensland, 4072, Queensland, Australia

ABSTRACT

The Lower Permian Aldebaran Sandstone is the principal hydrocarbon reservoir in the Denison Trough (Bowen Basin), east-central Queensland, Australia. It accumulated in a wide range of fluvio-deltaic and nearshore marine environments. Detailed petrological study of the unit by thin section, X-ray diffraction, scanning electron microscopy, electron microprobe and isotopic analysis reveals a complex diagenetic history which can be directly related to depositional environment, initial composition and burial–temperature history. Early diagenetic effects included the precipitation of pyrite, siderite and illite–smectite rims ($\delta^{18}O$ (SMOW) $= +8{\cdot}9$ to $+11{\cdot}3‰$). Deep burial effects included physico-chemical compaction and the formation of quartz overgrowths, siderite ($\delta^{13}C$ (PDB) $= -34{\cdot}0$ to $+11{\cdot}5‰$, $\delta^{18}O = -0{\cdot}7$ to $+22{\cdot}7‰$), illite/illite–smectite and ankerite ($\delta^{13}C = -9{\cdot}3$ to $-4{\cdot}9‰$, $\delta^{18}O = +7{\cdot}6$ to $+14{\cdot}4‰$). Involved fluids were in part 'connate meteoric' water derived from compaction of the underlying freshwater Reids Dome beds. Important post-maximum burial effects, controlled by deep meteoric influx from the surface, were ankerite and labile grain dissolution and formation of kaolinite ($\delta^{18}O = +7{\cdot}8$ to $+8{\cdot}9‰$, $\delta D = -115$ to $-99‰$), calcite ($\delta^{13}C = -9{\cdot}5$ to $+0{\cdot}9‰$, $\delta^{18}O = +9{\cdot}0$ to $+20{\cdot}0‰$) and dawsonite ($\delta^{13}C = -4{\cdot}0$ to $+2{\cdot}3‰$, $\delta^{18}O = +9{\cdot}8$ to $+19{\cdot}8‰$), the formation of dawsonite reflecting eventual stagnation of the aquifer. Entrapment of contained hydrocarbons was a relatively recent event which may be continuing today. Reservoir quality varies from marginal to good in the west to poor in the east, with predictable trends being directly linked to depositional environment and diagenesis.

INTRODUCTION

The Denison Trough is a deep, NNW-trending continental sub-basin situated along the western margin of the Bowen Basin in east-central Queensland (Fig. 1). Structurally, it consists of a series of N-trending half graben bounded by basement highs to the east (Comet Ridge) and west (Springsure Shelf and Nebine Ridge) (Brown *et al.*, 1983) (Fig. 2). The southern part of the trough is concealed beneath Jurassic sedimentary rocks of the Surat Basin. The trough contains up to 6500 m of Permo–Triassic siliciclastic sedimentary rocks folded into a series of large, N-trending, *en-echelon* anticlines and synclines. Closures along some anticlines now form traps for several economically significant hydrocarbon accumulations.

The Lower Permian Aldebaran Sandstone is an important hydrocarbon (gas) reservoir in the Denison Trough, and is the prime reservoir over all known hydrocarbon fields in the southwest of the trough (see Fig. 2). It accumulated in a wide variety of marginal marine and fluvio-deltaic environments created during the first regressive phase of a series of transgressive–regressive cycles of sediment accumulation in the trough. Seismically defined intraformational unconformities attest to its complex accumulation history involving syndepositional structuring and base-level fluctuations.

The purpose of this paper is to discuss the diagenesis of the Aldebaran Sandstone and existing links between depositional environment, diagenesis and reservoir

* Present address: Electron Microscope Centre, University of Queensland, 4072, Queensland, Australia.

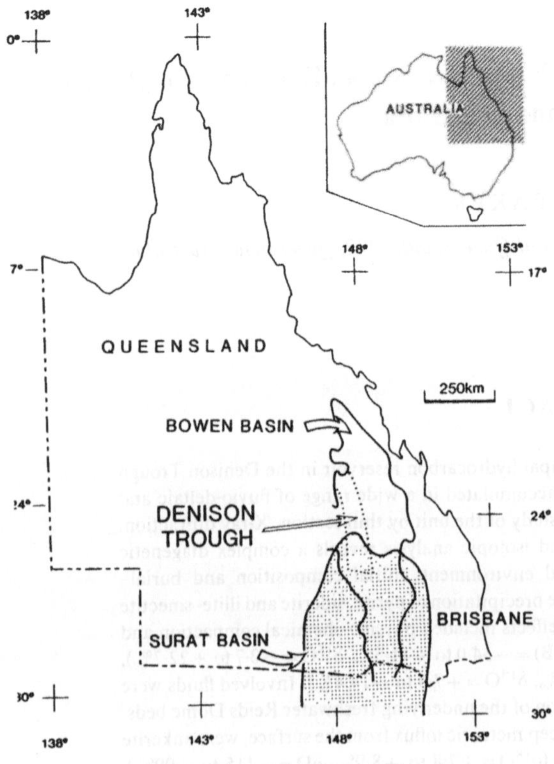

Fig. 1. Location of the Denison Trough.

quality which enable delineation of porosity and permeability trends within the unit and demarcation of favourable reservoir areas. Conclusions are also drawn on the timing of migration and entrapment of contained hydrocarbons. The discussion is largely based on results obtained from detailed petrological and geochemical studies of the detrital components and diagenetic minerals, interpreted in light of the known depositional framework and burial–temperature history of the unit.

STRATIGRAPHY

The Aldebaran Sandstone consists of a thick sequence of sandstones, conglomerates and minor mudrocks and coals which accumulated within, and eventually beyond, the Denison Trough during the Early Permian. It reaches a maximum thickness of over 700 m in the Consuelo Anticline area from where it thins towards the trough margins (Gray, 1976; Paten *et al.*, 1979). The Cattle Creek Formation, a thick sequence of mainly marine mudrocks, lies conformably beneath the Aldebaran Sandstone. In the far south, where the Cattle Creek Formation is absent, the Aldebaran Sandstone directly overlies thick mudrocks, sandstones and coals of the Reids Dome beds. Overlying are the Freitag and Ingelara formations, which consist of mainly marine sandstones and mudrocks respectively (Fig. 3). The Aldebaran Sandstone occurs virtually throughout the Denison Trough with exposure restricted to the Serocold, Springsure and Consuelo anticlines (see Fig. 2). Elsewhere it is covered by up to 1·4 km of Permo-Triassic strata of the upper Bowen Basin sequence, Jurassic sediments of the Surat Basin, and Cenozoic basalts and alluvium. Over the outcrop area, two subunits are recognized within the interval; an upper subunit comprising the larger part of the section and consisting of mainly medium to very coarse sandstones, pebbly sandstones and conglomerates, and a lower subunit consisting largely of fine to medium sandstones (e.g. Gray, 1976; Balfe, 1982). In the subsurface, three major stratigraphic breaks are recognized in the Aldebaran Sandstone on the basis of seismic and palynological evidence (Elliott, 1985).

DEPOSITIONAL FRAMEWORK

The coarsening-upward transition between the underlying Cattle Creek Formation and Aldebaran Sandstone represents the seaward progradation of delta and barrier-bar complexes diachronously over their prodelta and lower beachface counterparts. Two distinct deposystems were constituted by these regressive environments, the disposition of which was directly controlled by sites of sediment discharge into the trough.

Over the central–northern Denison Trough, facies relationships and palaeocurrent directions (Wilkinson, 1983; Parmenter, 1986; McLoughlin, 1988; J. J. Draper, unpublished data; C. R. Fielding, unpublished data) indicate that a deposystem consisted initially of a large, arcuate tide wave-dominated delta which prograded eastwards (transversely) into the trough from a single point of major sediment discharge located to the southwest of Springsure township (Fig. 4a). The coarse upper subunit in outcrop reflects the later establishment of a fan delta in response to continued tectonic activity (Fig. 4b). The fan delta terminated abruptly to the south and east where it entered the sea and became reworked. In the west, progradation of the delta/fan delta eventually led to the creation of purely fluvially controlled upper delta-

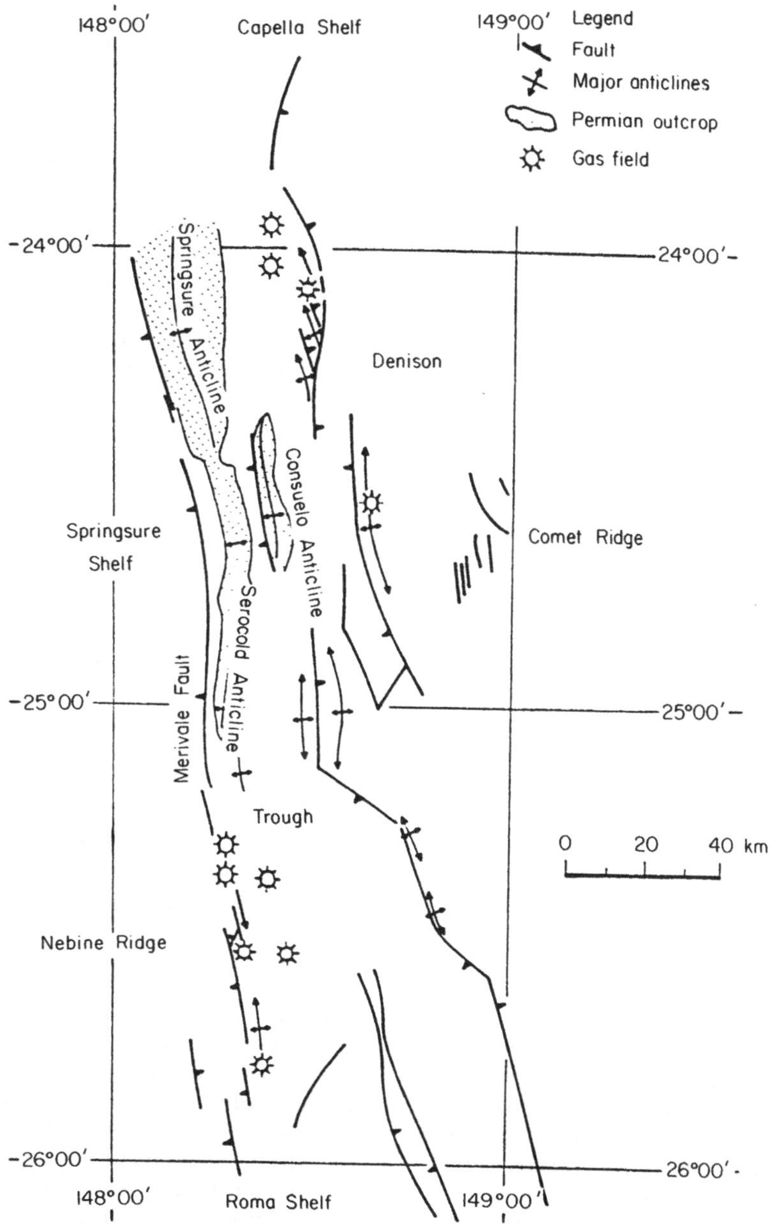

Fig. 2. Tectonic elements of the Denison Trough (from Brown *et al.*, 1983).

plain environments, whereas in the east, marginal marine–neritic conditions were sustained throughout accumulation of the unit.

To the south, remote from the delta/fan-delta complex, facies relationships (e.g. Draper & McClung, 1983) suggest that another deposystem consisted of mixed influenced, simple barriers/bars and associated tidal channels, deltas and flats which fringed the western margin of the trough. Eventually, they were replaced by unrestricted tide-dominated coastal environments. The pronounced increase in ice-rafted detritus in associated mudrocks towards the southwest (Draper, 1983) strongly suggests that sediments of the southern deposystem entered the trough from this

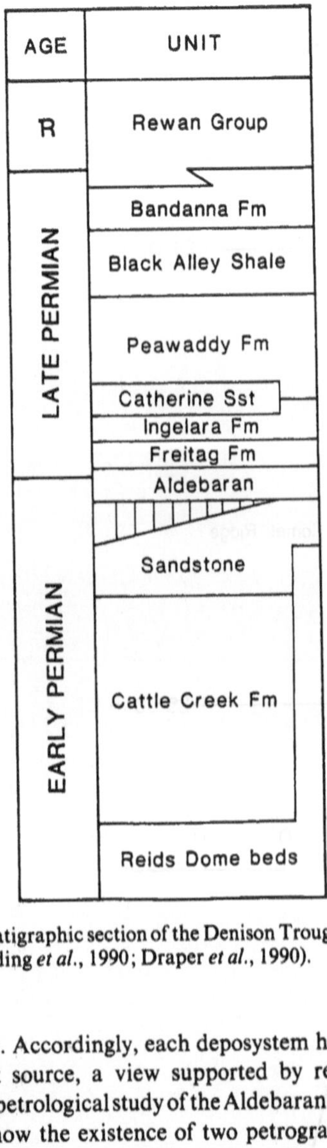

Fig. 3. Stratigraphic section of the Denison Trough (modified from Fielding *et al.*, 1990; Draper *et al.*, 1990).

direction. Accordingly, each deposystem had its own sediment source, a view supported by results of a detailed petrological study of the Aldebaran Sandstone which show the existence of two petrographic provinces within the unit which correspond areally to the two deposystems (Baker, 1989).

In the far south, small deltas prograded northwards (longitudinally) from the southern margin of the trough, and in the southeast, far removed from any major terrigenous sediment discharge, offshore environments were maintained in which synchronous coquinitic mudstones and sandstones accumulated below wave-base.

Accumulation of the Aldebaran Sandstone was interrupted by at least three erosional events, and finally ceased as a result of degradation of the source terrain and onset of a widespread marine transgression (Jackson *et al.*, 1980). This event is recorded by overlying sandstones and mudrocks of the Freitag and Ingelara formations.

BURIAL–TEMPERATURE HISTORY

Reconstruction of the burial–temperature history of the Aldebaran Sandstone was carried out using stratigraphic data and geothermometry techniques utilizing vitrinite reflectance and clay mineralogy.

The Denison Trough has experienced two major cycles of subsidence and uplift. Punctuated by several short periods of uplift, the first subsidence episode during the Early Permian to Late Triassic resulted in burial of the Aldebaran Sandstone to depths which areally varied between 1500 and 3000 m. Compression-induced positive inversion of the basin fill terminated this episode (Ziolkowski & Taylor, 1985), and by the earliest Jurassic had led to exposure of the unit over the present-day Permian outcrop area. Elsewhere it remained deeply buried. Renewed subsidence during the Early Jurassic heralded commencement of the second burial episode during which more than 1000 m of Surat Basin sediments accumulated over the Denison Trough until the Early Cretaceous. Late Cretaceous uplift, the second major uplift episode, has since led to removal of most of the Surat Basin section from over the Denison Trough. Permo-Triassic sediments, including the Aldebaran Sandstone, have been exhumed over the central–northern part of the trough as a consequence.

Considerably higher vitrinite reflectance profile gradients for similar reflectances in the Permo-Triassic section compared with those in the overlying Jurassic–Cretaceous section are interpreted to indicate that pre-Late Triassic palaeogeothermal gradients were not exceeded during the Jurassic and Cretaceous. Accordingly, since burial depths at the end of accumulation of the Surat Basin sequence were less than, or did not greatly exceed, those reached during the Late Triassic, maximum burial temperatures must have been attained in the Aldebaran Sandstone during the Late Triassic, i.e. at the end of the first subsidence episode. These maximum temperatures, as estimated from vitrinite reflectances using the modified model of Barker & Pawlewicz (1986), areally varied from about 90 to 150°C, and increased steadily eastwards across the trough. On the basis of data in Hoffman & Hower (1979) and Pollastro & Barker (1986), these temperature estimates are substantiated by the composition of mixed-layer illite–smectites in the associ-

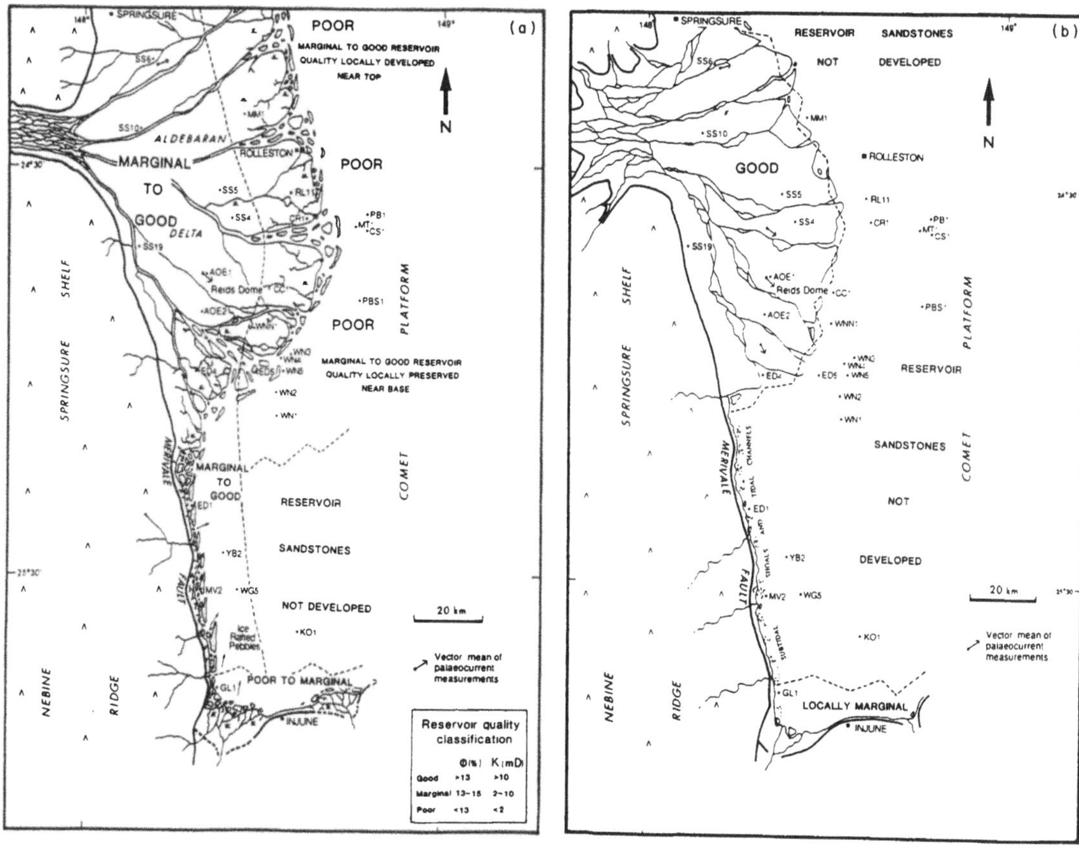

Fig. 4. Palaeogeography of the (a) lower and (b) upper Aldebaran Sandstone. Reservoir quality trends also shown.

ated mudrocks, although some now question the precision of illite–smectite composition as an absolute geothermometer (e.g. Hillier & Clayton, 1989; Freed & Peacor, 1989). Present-day formation temperatures range from about 20°C at the surface to a measured maximum of 75°C at a depth of 1400 m.

METHODOLOGY

Over 300 samples of sandstone, conglomerate and mudrock were collected from mainly the Aldebaran Sandstone in 43 boreholes and two areas of outcrop (Fig. 5). Thin sections were prepared for all sandstones and conglomerates using a blue impregnation resin to highlight porosity. Slides containing Ca, Fe and Mg carbonates were stained with alizarin red-S and potassium ferricyanide, and a number were stained with sodium cobaltinitrite to aid K-feldspar identification. Modal analyses were performed on most sandstones by counting 500 points per slide.

Orientated clay fraction X-ray diffraction (XRD) analyses were routinely carried out on each sample to enable precise identification of the clay minerals present. Superimposed traces for air-dried, glycolated (vapour, 12 h at 25°C) and heated (1 h at 375°C) orientated clay preparations additionally enabled an interpretation of the relative abundances of clay mineral species occurring in each sample using the method of Weir *et al.* (1975). Illite–smectite compositions were determined by comparing illite–smectite peak positions with those given in Hower (1981). Copper K_α radiation was used for all XRD analyses.

Scanning electron microscopy (SEM) was performed on 23 sandstone samples selected following the thin-section study. This work yielded important supplementary data on the nature of authigenic

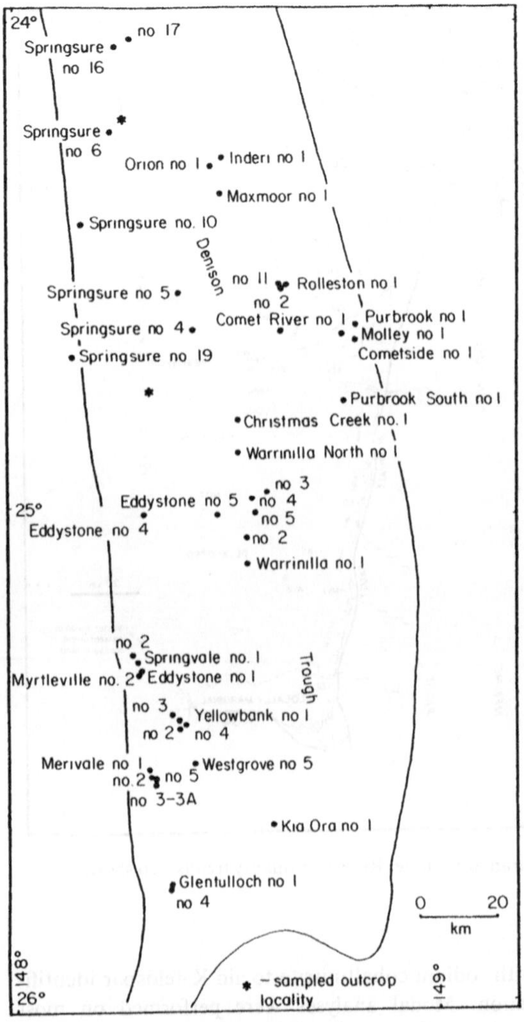

Fig. 5. Sampled borehole and outcrop localities.

minerals and porosity, and their spatio-temporal relationships.

Authigenic carbonate and clay compositions were confirmed using an electron microprobe, and carbon and oxygen isotope ratios were determined for all carbonate cements where they constituted greater than 3% of the rock and were dominated by a single mineral. In addition, oxygen and hydrogen isotopic compositions were determined for monomineralic authigenic clays occurring in 11 sandstones. The clays were separated by gently crushing the rock specimens and then, immediately after its placement into an ultrasonic bath, allowing the disaggregated material to settle in a column of distilled water containing 5 g l^{-1} Calgon (a clay anti-flocculant) for the appropriate times to obtain the $<2{\cdot}0\ \mu m$ (for kaolinite) and $<0{\cdot}5\ \mu m$ (for illite–smectite) size fractions; a procedure repeated three times. The standard isotopic analytical techniques of McCrea (1950) were used for the carbonates, and those of Clayton & Mayeda (1963) and Coleman *et al.* (1982) were used for the clays. All isotopic data are presented in per mil with respect to standard mean ocean water (SMOW) for hydrogen and oxygen (Craig, 1961), and the *Belemnitella americana* from the Peedee Formation (PDB) for carbon (Craig, 1957). The reproducibility of standard δD values was 3‰ or better, and that of standard $\delta^{13}C$ and $\delta^{18}O$ values was 0·2‰ or better. Phosphoric acid oxygen isotope fractionation factors for the carbonates used here are those reported by Sharma & Clayton (1965) and Rosenbaum & Sheppard (1986). In the absence of experimental data, the acid fractionation factor for dawsonite was assumed to be the same as for calcite. Mineralogical purity of the analysed clays was verified by XRD.

RESULTS

Mudrocks

The mudrock clay mineralogy is dominated by discrete illite, low-expandability (allevardite ordered) illite–smectite and kaolinite. Chlorite is locally present as a very minor constituent in the north. The proportion of illite and illite–smectite increases significantly at the expense of kaolinite east of the outcrop area. Constituent non-clay detrital minerals are quartz, K-feldspar, plagioclase and mica. Authigenic minerals detected are pyrite and siderite. Siderite, isotopically analysed in one sample (a marine mudstone), gave $\delta^{13}C$ and $\delta^{18}O$ values of $-34{\cdot}0$‰ and $+9{\cdot}2$‰ respectively.

Sandstones

Texture

Sampled sandstones range from fine to very coarse grained, but are nearly all medium grained. Lithologs and mudlogs show that much of the Aldebaran Sandstone is in fact composed of medium-grained sandstone, with coarser-grained sandstones more common in the upper subunit over the outcrop area where they are associated with pebbly sandstones and conglomerates.

Sandstones are poorly to very well sorted with the majority of reservoir sandstones well sorted except for those in the far south which are poorly sorted and commonly matrix supported rather than grain supported.

Framework grains

Sandstones from the Aldebaran Sandstone are quartz-rich. Approximately 75% of samples are sublitharenites according to the classification of Folk *et al.* (1970). The mean QFR ratio is 79:7:14. More feldspathic and lithic varieties are common over the southern and central-northern parts of the trough respectively. East of the outcrop area, sandstones are enriched in quartz by up to 16% compared with those in the west.

Quartz grains are mainly monocrystalline and non-undulatory. Feldspar types are orthoclase, microperthite, soda-rich plagioclase and microcline. K-feldspar dominates over plagioclase. Lithics are fragments of mainly acid to intermediate volcanics, fine-grained siliciclastics, low-grade metamorphics and granite. Accessories include volumetrically insignificant muscovite, biotite, zircon, tourmaline, rutile, garnet and apatite.

Diagenetic minerals

Quartz. Authigenic quartz occurs exclusively as syntaxial overgrowths, constituting up to 5% of some sandstones, but mainly constitutes less than 3%. Overgrowths are best developed in the highly quartzose sandstones located over the eastern side of the Denison Trough and on the Comet Ridge, and are lacking in those sandstones containing significant percentages of detrital matrix and in which framework grains are rimmed by authigenic clay.

Calcite. Although quite common, authigenic calcite is not a pervasive cement in the Aldebaran Sandstone. It typically occurs as a pore fill and grain replacement in well-defined, sporadically developed zones with vertical dimensions of up to several metres. Abundances range up to 30%. Where abundant, calcite typically forms large optically continuous, locally twinned patches which exhibit straight extinction and poikilotopically enclose the framework grains giving rise to lustre mottling in hand specimen. Probe analyses and stained thin sections revealed the presence of both non-ferroan and ferroan calcite varieties with FeO contents of up to 1·1%. MgO contents are invariably below 0·24% and MnO contents are mainly below 1·0%. Values of $\delta^{13}C$ and $\delta^{18}O$ range from −9·5 to +0·9‰ and +9·0 to +20·0‰ respectively.

Dawsonite. Dawsonite [$NaAlCO_3(OH)_2$], rare in sedimentary rocks worldwide, is widely distributed in the Aldebaran Sandstone where it partially infills pores and fractures, and replaces mainly labile framework grains. Unlike calcite it is extremely pervasive, occurring over large areas throughout all sandstones within sections ranging up to at least 170 m in thickness. Recorded abundances are up to 13·4%, with most values lying below 9%. Dawsonite is confined solely to sandstones of marine affinity, and has only been recorded elsewhere in the Denison Trough sequence in sandstones of the Freitag and Cattle Creek formations and uppermost Reids Dome beds. All these sandstones are closely associated with marine mudrocks. Dawsonite is absent in mudrocks and outcropping sandstones. Dawsonite crystals are characterized by their acicular habit although tabular forms were also noted. They typically form bundles, tufts, fans, rosettes and, less commonly, vermicules. In X-ray diffractograms, dawsonite is easily identified by its 110 peak at 0·566 nm. Although ranging from −4·0 to +2·3‰, $\delta^{13}C$ values generally show little variation, with the majority lying between −2·0 and +1·0‰. Values of $\delta^{18}O$ range from +9·8 to +19·8‰.

Ankerite. Very minor amounts (<1%) of ankerite are present in sandstones throughout the interval at depth. Abundances rarely exceed 4% and range up to a recorded maximum of 10·6%. Ankerite occurs as solitary or clustered rhombs with widths ranging up to 0·1 mm, and where more abundant, as small, irregular, coarsely crystalline or monocrystalline patches which infill pores and replace/corrode all other framework grains. The average composition of ankerites analysed with the microprobe is about $Ca_{0.53}Mg_{0.30}Fe_{0.16}Mn_{0.01}CO_3$. Values of $\delta^{13}C$ and $\delta^{18}O$ range from −9·3 to −4·9‰ and +7·6 to +14·4‰ respectively.

Siderite. Siderite in trace proportions is also very common in subsurface sandstones throughout the interval. Abundances greater than 2% and ranging up to a recorded maximum of 23% are very rare and spatially limited. For the most part, siderite is sucrosic or microgranular, and mainly occurs as a partial replacement of feldspars, rock fragments and mica. The average composition of the probed siderites is

$Fe_{0.66}Mg_{0.20}Ca_{0.11}Al_{0.02}Mn_{0.01}CO_3$. Both $\delta^{13}C$ and $\delta^{18}O$ values show considerable variability, ranging from −14·6 to +11·5‰ and −0·7 to +22·7‰ respectively. Except for one sample, all $\delta^{13}C$ values are negative.

Kaolinite. Authigenic kaolinite is very common in the Aldebaran Sandstone. Absolute abundances commonly exceed 8% and range up to 15% in the base of the unit in the south and throughout the interval over the outcrop area. In the east of the area, kaolinite abundance decreases dramatically with values seldom exceeding 2% and mainly lying below 1%. It typically forms randomly orientated, delicate booklet and accordian-like, loosely to densely packed aggregates of euhedral pseudo-hexagonal plates which line and infill scattered pores and replace labile grains. Plate widths are up to 50 μm. Deep subsurface kaolinite isotopically analysed has $\delta^{18}O$ values which lie in the narrow range from +7·8 to +8·9‰. Values of δD range from −115 to −99‰.

Illite–smectite/illite. Mixed-layer illite–smectite and discrete illite occur throughout most of the Aldebaran Sandstone at depth, with recorded combined abundances seldom exceeding 12%. In the south of the trough, abundances commonly exceed 8%, whereas over the central–northern parts of the trough, values are mainly below 5%. East of the outcrop area, illite–smectite and illite are the dominant and locally the only authigenic clays present in the sandstones. Here illite–smectite forms pervasive grain rims at the base of the Aldebaran Sandstone which consist of a meshwork of irregular crinkled flakes orientated perpendicular to grain surfaces. Many flakes have delicate, spiny projections or short lath-like digitate edges which commonly bridge gaps up to 20 μm wide near grain contacts in a web-like pattern. In addition to forming grain rims, authigenic illite–smectite, as well as illite, occurs as a product of labile grain alteration, typically forming in this situation a pore-filling cellular or honeycomb structure in which crystals are either sheet-like or more commonly exhibit no well-defined morphology. Lath or ribbon-like appendages are common. Nearly all illite–smectites analysed by XRD are allevardite ordered with between 65 and 85% illite interlayers (Fig. 6). Probe analysis of the illite–smectite rims confirmed the presence of K, Ca, Mg and Fe in addition to Si and Al. Isotopic analysis of the rims gave $\delta^{18}O$ values with little variability, ranging from +8·9 to +11·3‰.

Smectite. Minor amounts of smectite are present in a few samples originally situated in very close proximity to bentonite beds. Except for these occurrences, all smectite detected in sandstones is either infiltrated drilling mud or, in the case of the very shallow samples, related to near-surface weathering processes.

Other diagenetic clays locally present in trace amounts are glauconite and chlorite.

Sulphates/sulphides. Trace to minor amounts of barite and pyrite are restricted to sandstones which accumulated in marine environments.

Probe analyses of authigenic carbonates in the Aldebaran Sandstone are shown in Fig. 7, and a summary of the isotope results for carbonates and clays are presented in Figs 8 & 9 and Tables 1 & 2. Further details on these results can be found in Baker (1989).

PARAGENESIS

The relative timing of the major diagenetic minerals in the Aldebaran Sandstone was inferred from their textural relations as observed in thin section and under an SEM.

Illite–smectite rims pre-date physical compaction, underlie quartz overgrowths (Fig. 10a) and bound all carbonate cements. Accordingly, they are regarded as a very early precipitate. Siderite is enclosed within ankerite, and ankerite is enclosed within calcite (Fig. 10b), indicating that siderite was the earliest and calcite was the latest of these carbonates to precipitate. Dawsonite has grown on and around calcite, indicating that dawsonite is the latest carbonate, a fact substantiated by the occurrence in one sample of fracture-filling dawsonite cross-cutting an area extensively cemented with calcite (Fig. 10c). Dawsonite abuts euhedral quartz overgrowths (Fig. 10d) as do all the other carbonates with the exception of a very minor early siderite cement; relationships which suggest that the carbonates for the most part post-date quartz overgrowths. Kaolinite occupies spaces made available by ankerite dissolution (Fig. 10e) and is enclosed by calcite (Fig. 10f) indicating that it post-dates ankerite and pre-dates calcite. Kaolinite is also partly enclosed by quartz overgrowths (Fig. 10g) suggesting that either the two minerals are synchronous or that kaolinite pre-dates an episode of quartz overgrowth formation. Barite includes ankerite and replaces kaolinite indicating that it post-dates these

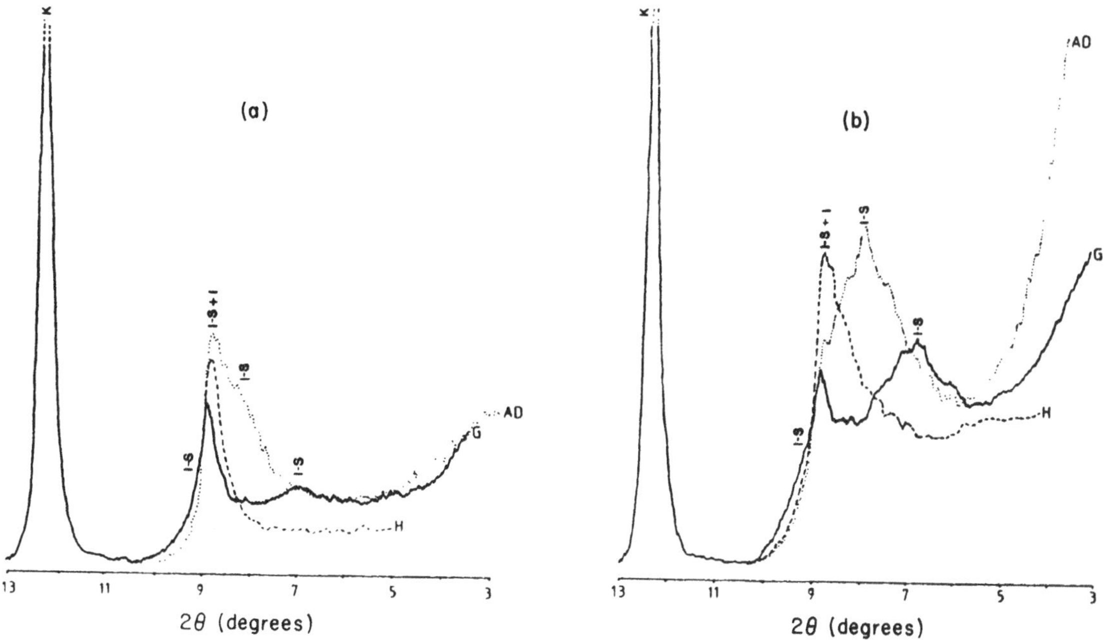

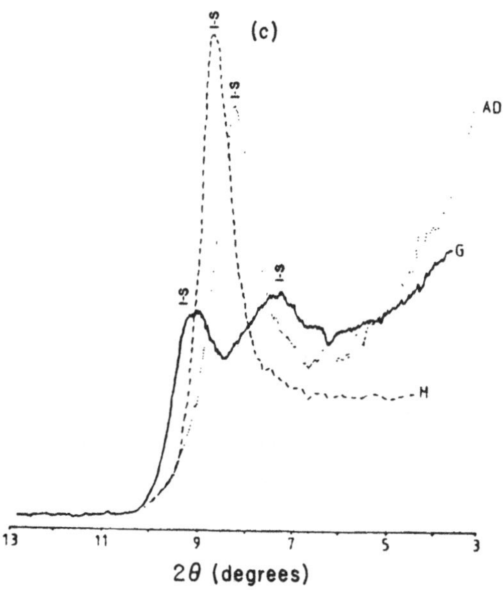

Fig. 6. Orientated clay fraction X-ray diffractograms for representative subsurface sandstones of the Aldebaran Sandstone from: (a) southern deposystem; Merivale no. 2, 1380·60 m; (b) outcrop area; Springsure no. 5, 557·45 m; (c) east of outcrop area; Warrinilla no. 4, 1423·64 m. I = discrete illite, I-S = allevardite ordered illite–smectite, K = kaolinite, AD = air dried, G = glycolated, H = heated.

minerals. Secondary grain dissolution pores are surrounded by welded quartz grains, detrital matrix and undeformed and intact illite–smectite rims (Fig. 10h) suggesting that they post-date compaction. Some are partially infilled by dawsonite, indicating formation prior to dawsonite precipitation.

DIAGENETIC HISTORY

Diagenesis is here divided into three regimes, eogenesis, mesogenesis and telogenesis, in accordance with the classification scheme of Schmidt & McDonald (1979). Eogenesis occurs at or near the sedimentation surface where fluid chemistries are mainly controlled by the depositional environment. Mesogenesis commences at greater depths where the sediment is effectively sealed from the predominant influence of surface agents. Telogenesis represents the regime at or near the surface following effective burial where surficial meteoric waters are brought into circulation displacing pre-existing pore fluids.

The major eogenetic mineral recognized in the Aldebaran Sandstone is grain rims of illite–smectite. Requiring high cation concentrations, which would

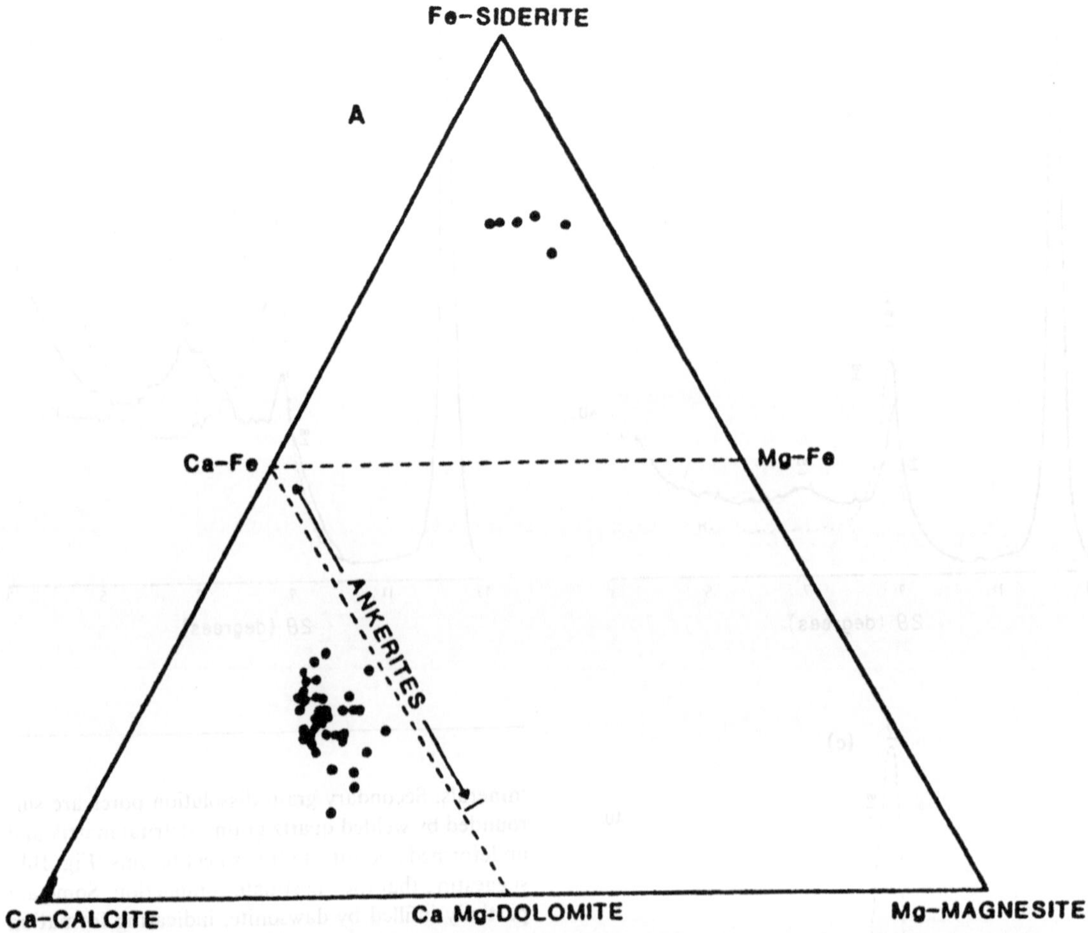

Fig. 7. Chemical composition of authigenic carbonates in the Aldebaran Sandstone in terms of FeO, CaO and MgO.

not have characterized groundwaters in the non-marine environments due to the humid palaeoclimate at the time of deposition, this mineral could only develop in sands which accumulated in normal marine environments. Other eogenetic minerals recognized in the marine sandstones are glauconite, pyrite and an early siderite. Despite the presence of intraformational unconformities and the potential for gravity-driven, lateral seaward migration of meteoric water from emergent intrabasinal areas to the west, marine sandstones show no evidence of early freshwater flushing. No eogenetic minerals could be recognized

Table 1. Authigenic carbonate isotopic compositions.*

	n	$\delta^{13}C$ (‰) (PDB)		$\delta^{18}O$ (‰) (SMOW)	
		Range	Mean ± SD	Range	Mean ± SD
Dawsonite	31	−4·0 to +2·3	−0·7 ± 1·7	+9·8 to +19·8	+13·9 ± 2·2
Calcite	24	−9·5 to +0·9	−5·7 ± 3·1	+9·0 to +20·0	+12·7 ± 2·4
Siderite	11	−34·0 to +11·5	−6·8 ± 11·0	−0·7 to +22·7	+11·0 ± 6·6
Ankerite	8	−9·3 to −4·9	−7·7 ± 1·5	+7·6 to +14·4	+10·8 ± 2·5

* Drs H. K. Herbert and H. R. Krouse are acknowledged for isotopic analysis of 40 of the samples.

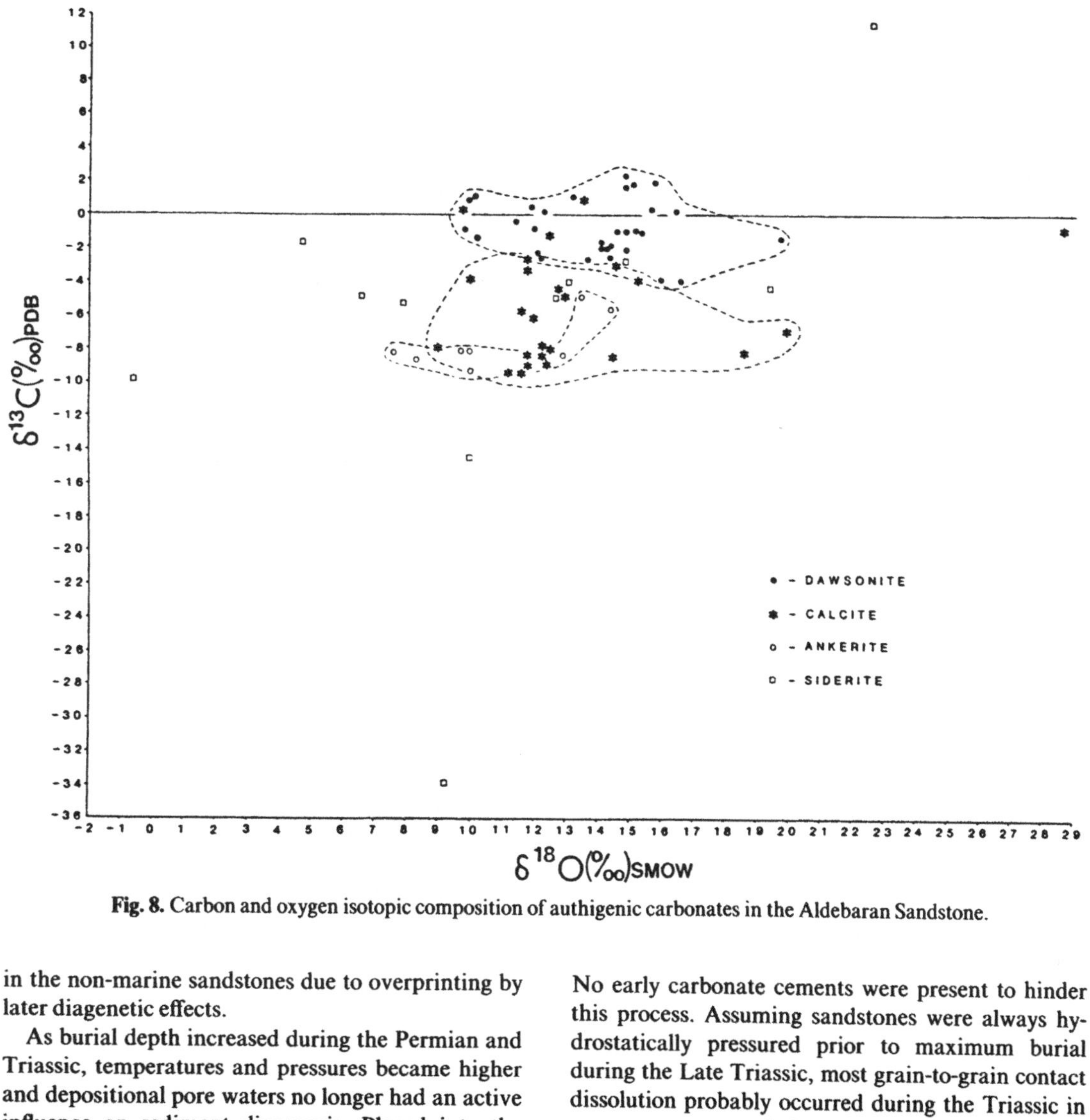

Fig. 8. Carbon and oxygen isotopic composition of authigenic carbonates in the Aldebaran Sandstone.

in the non-marine sandstones due to overprinting by later diagenetic effects.

As burial depth increased during the Permian and Triassic, temperatures and pressures became higher and depositional pore waters no longer had an active influence on sediment diagenesis. Placed into the mesogenetic regime, physico-chemical compaction became significant due to framework grain re-orientation and silica dissolution at contacts between quartzose grains (grain-to-grain contact dissolution). No early carbonate cements were present to hinder this process. Assuming sandstones were always hydrostatically pressured prior to maximum burial during the Late Triassic, most grain-to-grain contact dissolution probably occurred during the Triassic in response to accumulation of the thick Triassic units and concomitant substantial temperature increase.

Siderite (main generation), ankerite, illite–smectite/illite pore fill and quartz overgrowth formation, as well as compaction, probably took place in the

Table 2. Authigenic clay isotopic compositions.

	n	δD (‰) (SMOW)		$\delta^{18}O$ (‰) (SMOW)	
		Range	Mean ± SD	Range	Mean ± SD
Illite–Smectite	6	—	—	+8·9 to +11·3	+10·4 ± 1·0
Kaolinite	4	—	—	+7·8 to +8·9	+8·4 ± 0·6
Kaolinite	3	−115 to −99	−106 ± 8	—	—

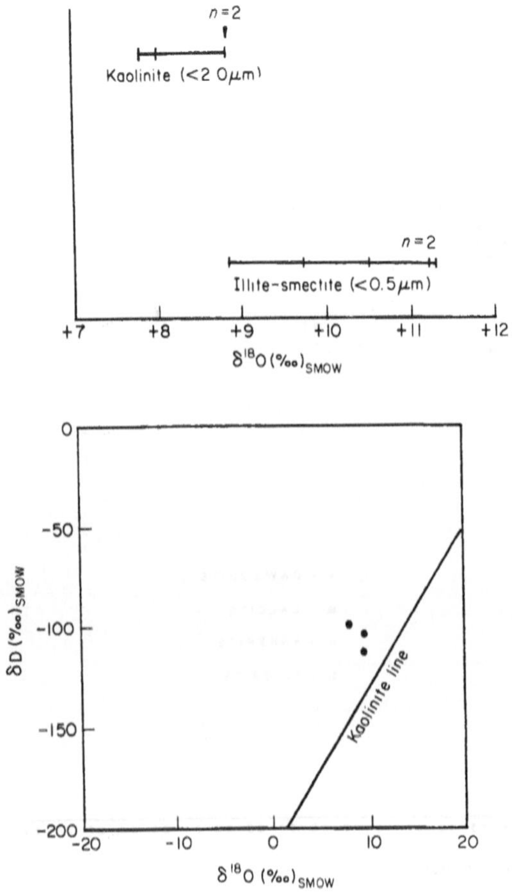

Fig. 9. Oxygen and hydrogen isotopic composition of authigenic clays in the Aldebaran Sandstone.

mesogenetic regime. Involved fluids would have been liberated from underlying and overlying marine mudrocks as a result of physical compaction (connate water) and, at temperatures above 70–80°C, the alteration to illite of mixed-layer clays (interlayer water in illite–smectites) (e.g. Burley, 1986). Meteoric water isotopic signatures of the siderites and ankerites (which show no evidence of having undergone later recrystallization) as well as illitized eogenetic illite–smectite grain rims (Fig. 11a–c) indicate that at least a proportion of water involved in the precipitation of carbonate and transformation of illite–smectite rims to a more illite-rich composition was expelled upward from the underlying Reids Dome beds, a thick (>2770 m) section of *freshwater* mudrocks, sandstones and coals. Connate water in this unit would have been originally highly depleted in ^{18}O, with $\delta^{18}O$ values probably around −17‰ (based on data in Bird & Chivas, 1988). Cognate sources of iron (volumetrically very minor detrital biotite), silica (quartzose grains) and calcium (carbonate fossils), and mixed-layer illite–smectites in associated mudrocks which released Si^{4+}, Ca^{2+}, Fe^{2+} and Mg^{2+} during their illitization (Boles & Franks, 1979) would have contributed the required ions for their development. The $\delta^{13}C$ values of siderite reflect carbon derived from both the fermentation and thermal decarboxylation of organic matter over a wide depth range (see Irwin *et al.*, 1977), whereas, assuming that ankerite is a deep burial cement, the $\delta^{13}C$ values of ankerite most likely reflect carbon derived from the thermal decarboxylation of organics augmented with carbon derived from the dissolution of marine carbonate fossils. Although the

Fig. 10. (a) Equant quartz overgrowth (QO) overlying illite–smectite (I-S) grain rim; Warrinilla no. 4, 1423·64 m. (b) Calcite (C) enclosing ankerite (A) indicating calcite is the later mineral; Westgrove no. 5, 889·10 m. (c) Microfault, with associated fracture gouge (FG) and partially infilled by dawsonite (D), cross-cutting sandstone extensively cemented by calcite (C). Relation indicates microfault formed after calcite and prior to dawsonite; Eddystone no. 5, 1073·14 m. (d) Pore-filling dawsonite (D), showing typical fibrous habit, abutting quartz overgrowths (QO); Eddystone no. 1, 902·68 m. (e) Ankerite–kaolinite relationship. Pore-filling/grain-replacement ankerite has undergone extensive dissolution (note remnant—A). Kaolinite (K) has precipitated in its place. Severely corroded and embayed margins of quartz grains (1), elongated areas between adjacent quartz grains (2), and presence of a small ragged patch of oxidized ankerite are strong evidence for ankerite decementation having occurred; Yellowbank no. 3, 1336·05 m. (f) Calcite (C) engulfing kaolinite (K) pore fill; Eddystone no. 1, 886·01 m. (g) Kaolinite (K) booklets engulfed by quartz overgrowth (QO). Relation interpreted to indicate that the two are coeval or overgrowth is the later phase; Eddystone no. 5, 1084·84 m. (h) Secondary pore defined by illite–smectite (I-S) rim which developed prior to dissolution of pore-precursor grain. Quartz overgrowths (QO) have since developed inside. Except for partial collapse produced artifically during sample preparation, pore has not been physically deformed indicating it and quartz overgrowths developed subsequent to compaction. Dawsonite (D) also present; Warrinilla no. 4, 1423·64 m. (a), (d), (f), (g) and (h) are scanning electron photomicrographs; scale bar in (b), (c) and (e) = 0·2 mm.

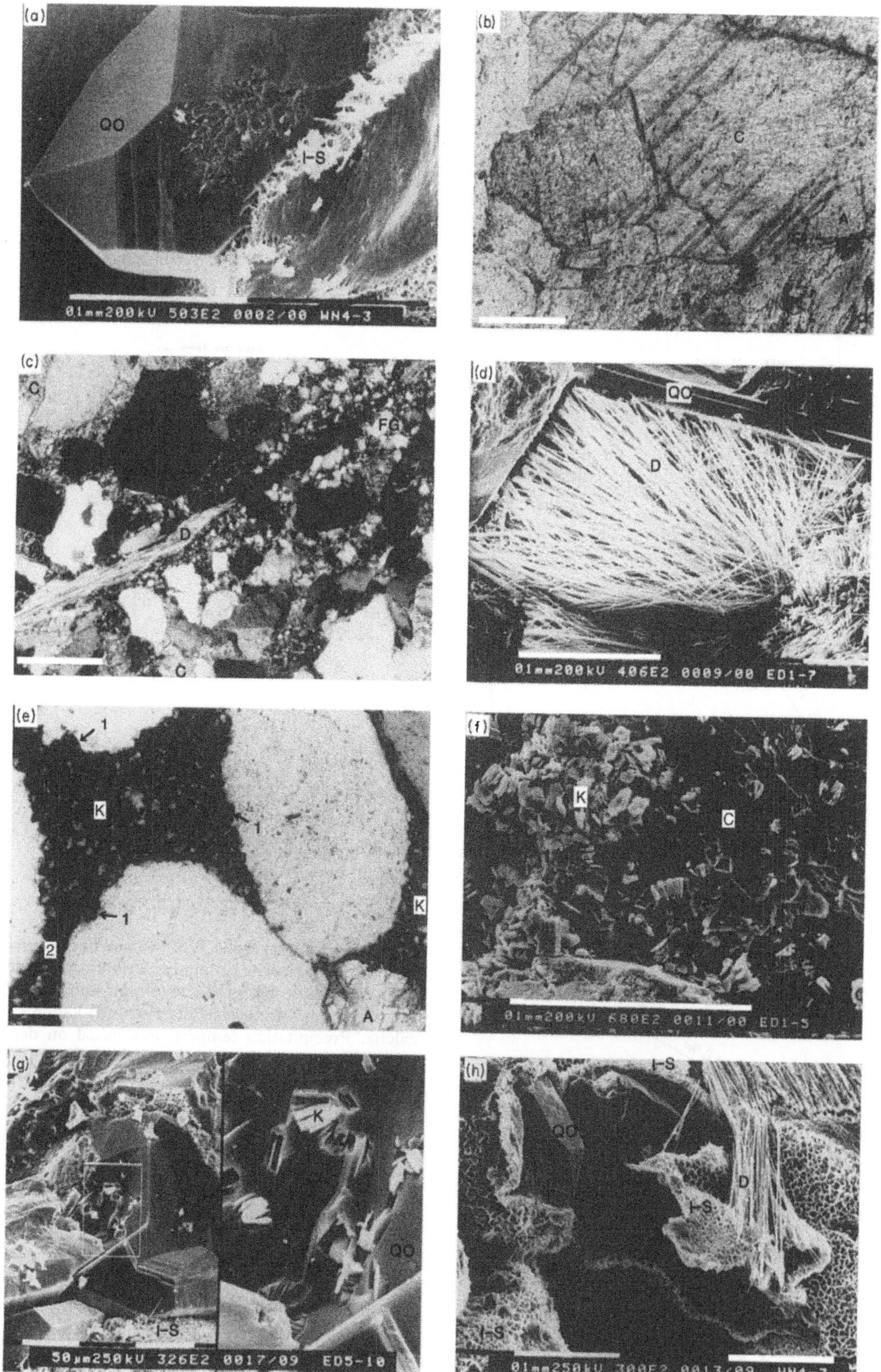
(a)
QO
I-S
01mm200kV 503E2 0002/00 WN4-3
(b)
A
C
A
(c)
C
FG
D
C
(d)
QO
D
01mm200kV 406E2 0009/00 ED1-7
(e)
1
K
1
1
K
2
A
(f)
K
C
01mm200kV 680E2 0011/00 ED1-5
(g)
K
QO
I-S
50µm250kV 326E2 0017/09 ED5-10
(h)
I-S
QO
D
I-S
I-S
01mm250kV 300E2 0013/09 WN4-3

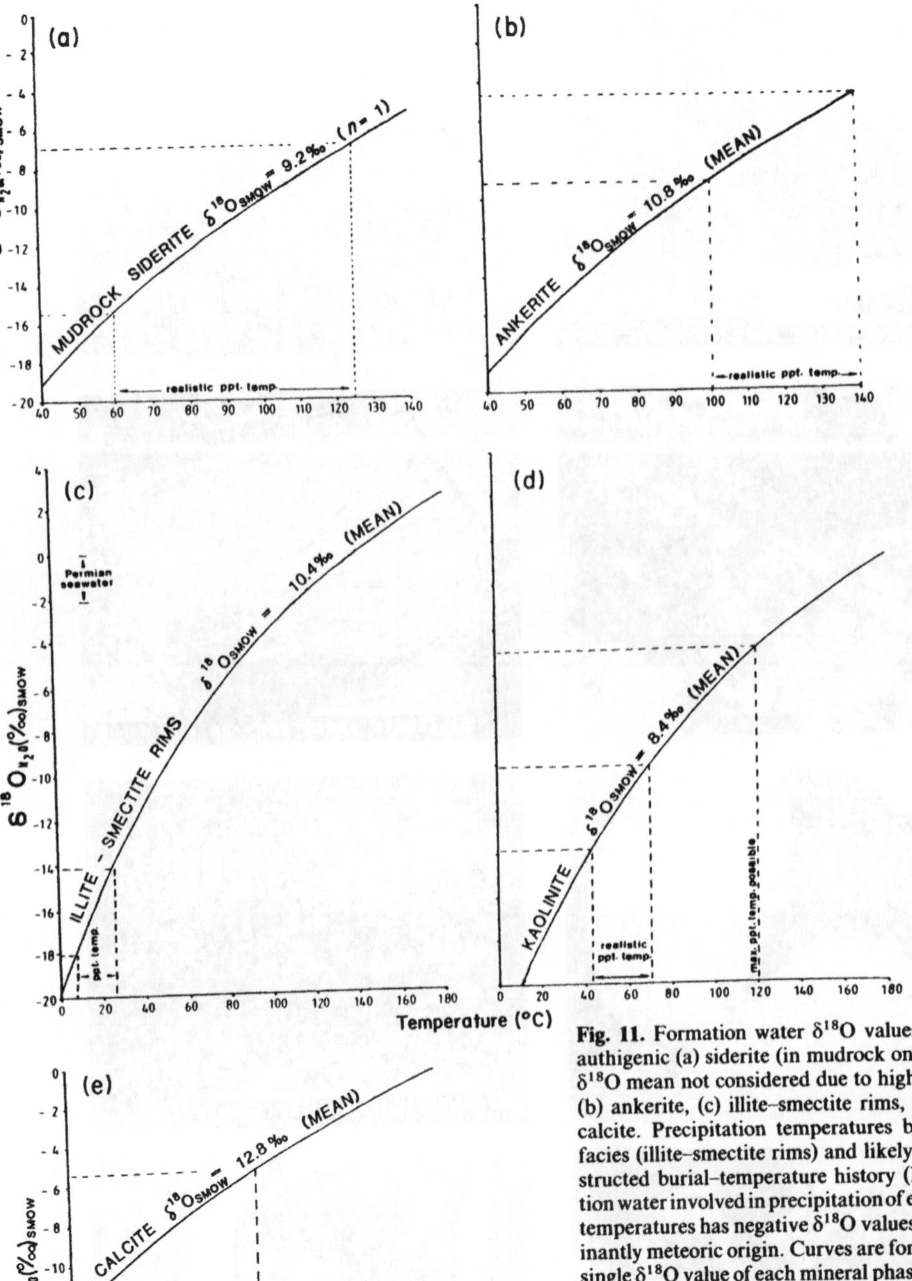

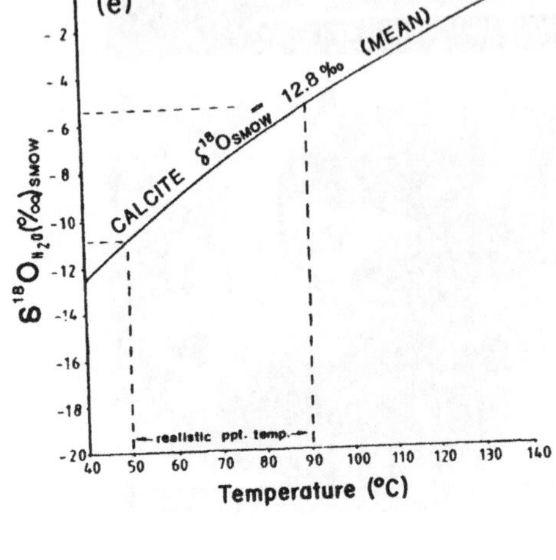

Fig. 11. Formation water $\delta^{18}O$ values vs. temperature for authigenic (a) siderite (in mudrock only; sandstone siderite $\delta^{18}O$ mean not considered due to high standard deviation), (b) ankerite, (c) illite–smectite rims, (d) kaolinite, and (e) calcite. Precipitation temperatures based on depositional facies (illite–smectite rims) and likely timing within reconstructed burial–temperature history (Baker, 1989). Formation water involved in precipitation of each phase at modelled temperatures has negative $\delta^{18}O$ values indicating a predominantly meteoric origin. Curves are for mean, or where $n = 1$, single $\delta^{18}O$ value of each mineral phase, and were calculated using the following mineral–water fractionation equations (with temperature, T, in degrees Kelvin): (a) $1000\ \ln_{\text{siderite–water}} = 3{\cdot}13 \times (10^6\ T^{-2}) - 3{\cdot}50$ (Carothers *et al.*, 1988); (b) $1000 \ln_{\text{dolomite–water}} = 3{\cdot}23 \times (10^6\ T^{-2}) - 3{\cdot}29$ (Golding & Wilson, 1983) (assumed for ankerite); (c) $1000\ \ln_{\text{illite/smectite–water}} = (2{\cdot}58 - 0{\cdot}19 \times I) \times 10^6\ T^{-2} - 4{\cdot}19$ (I = fraction of illite) (Savin & Lee, 1988); (d) $1000\ \ln_{\text{kaolinite–water}} = 2{\cdot}50 \times (10^6\ T^{-2}) - 2{\cdot}87$ (Land & Dutton, 1978); (e) $1000\ \ln_{\text{calcite–water}} = 2{\cdot}78 \times (10^6\ T^{-2}) - 2{\cdot}89$ (Friedman & O'Neil, 1977).

absence of mesogenetic secondary porosity created by expulsion of organic and carbonic acids from adjacent mudrocks could not be proved, it would seem unlikely that any such porosity was created in the Aldebaran Sandstone because of the neutralizing effect on any acids produced of abundant marine skeletal carbonate and feldspar present in these mudrocks (e.g. Bjørlykke, 1984).

The Aldebaran Sandstone underwent major uplift and deformation during the Late Triassic structuring episode. Thought to be linked to this event and attesting to freshwater flushing to depths of at least 700 m in the telogenetic regime is the widespread dissolution of ankerite and concomitant precipitation of kaolinite in areas vacated by the ankerite (Fig. 10e). This 'dissolution kaolinite' is highly depleted in D and ^{18}O, testifying to its origin from isotopically light waters rather than ^{18}O-rich water derived from clay reactions in mudrocks (Fig. 11d). Related also to this freshwater flushing episode was the creation of secondary pores through labile grain dissolution, formation of kaolinite by labile grain alteration, a second generation of quartz overgrowth development and, later, the precipitation of barite and calcite cements. Calcites are also relatively depleted in ^{18}O, reflecting precipitation from meteoric waters (Fig. 11e). These waters are thought to have originally entered the unit over the present-day outcrop area which, prior to its peneplanation by the earliest Jurassic, may have been a topographic high in which the unit was exposed. Calcite $\delta^{13}C$ values reflect the dominance of organic carbon and/or marine carbonate as carbon sources.

During re-burial by Jurassic and Lower Cretaceous sediments of the Surat Basin, burial depths and temperatures attained during this second burial phase were not sufficient to produce further release of connate and clay-mineral-bound water from the already well-compacted and illitized adjacent mudrocks. Consequently, the Aldebaran Sandstone remained in the telogenetic regime.

Over large parts of the Denison Trough south and east of the outcrop area, present-day formation water in the Aldebaran Sandstone, demonstrably of unmixed meteoric character on the basis of its isotopic composition (H. K. Herbert & H. R. Krouse, unpublished data), is enriched in sodium bicarbonate. Accordingly, in these areas there has been a dramatic change in water character following the freshwater flushing. Because these isotopically unmodified brines are considerably more depleted in D and ^{18}O than present-day surface waters in the Denison Trough, they may have been introduced during the Early Tertiary when the Australian continent was located at a much higher latitude and, as a consequence, rainfall over the Denison Trough was more depleted in D and ^{18}O because of the latitude effect (see Longstaffe, 1983). Therefore, on this basis, stagnation of the aquifer away from the outcrop area and the consequent enrichment of the formation fluids in sodium bicarbonate was a Tertiary event. Dawsonite precipitated as a result of this enrichment at temperatures similar to those existing in the unit today as shown by the close correspondence between borehole temperatures and dawsonite isotopic temperatures (Baker, 1989). Dawsonite $\delta^{13}C$ values average $-0{\cdot}5‰$, and are remarkably consistent over the whole basin, suggesting that the major carbon source was either magmatic carbon or marine skeletal carbonate.

The precipitation of dawsonite preceded the migration and entrapment of gaseous hydrocarbons as attested by the presence of the mineral throughout gas-bearing sandstones in quantities equal to that in the adjacent water-bearing sandstones. Accordingly, on this basis, which assumes diagenesis is inhibited or halted by the introduction of hydrocarbons (see below), because dawsonite precipitation was a late-stage event, hydrocarbon migration and entrapment must have also occurred very late in the diagenetic history, and may well still be continuing today. Hydrocarbon isotope data suggest that underlying coal seams of the Reids Dome beds were/are a major source for hydrocarbon gas occurring in the Aldebaran Sandstone (Baker, 1989).

It may be argued that gas emplacement pre-dated dawsonite precipitation, with the mineral being precipitated in the gas reservoir as a result of the upward diffusion of sodium and bicarbonate ions from below the gas–water contact. Diffusion is an ineffective mechanism of solute transfer on a scale of metres (Longstaffe, 1987), and therefore the high abundance of dawsonite throughout the gas-bearing sandstones (up to 13·4% of the bulk rock), equal to that in the adjacent water-bearing sandstones, seems to rule out this possibility. Moreover, it is difficult to conceive how irreducible water existing as pendular rings and thin adsorbed films (Chapman, 1982) could precipitate volumetrically significant amounts of dawsonite across large, mainly gas-filled pores as well as within framework grains. Another strong counter argument is based on the distribution of dawsonite. In most sections, the mineral is more abundant in the coarser, more-permeable sandstones, and in all sections is absent from the intercalated mudrocks. This indicates

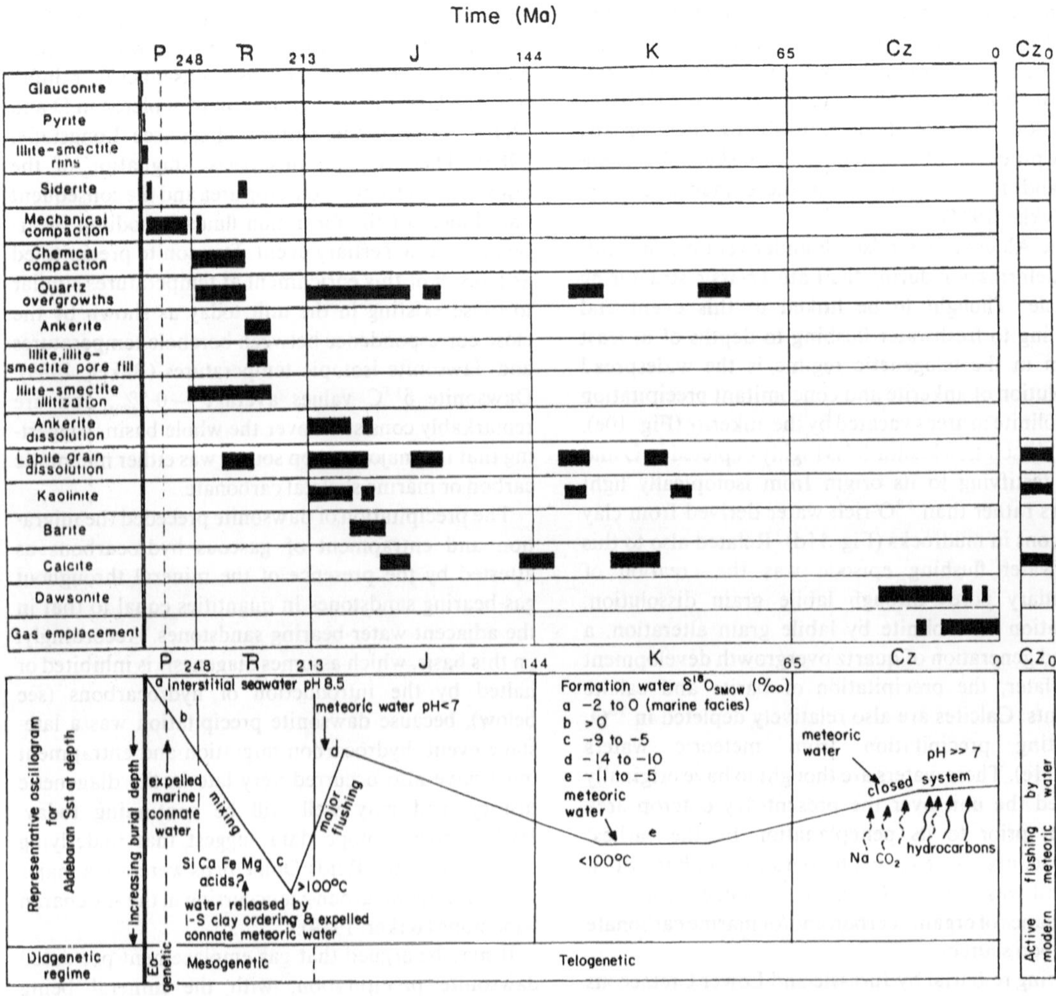

Fig. 12. Summary of diagenetic events in the Aldebaran Sandstone.

that precipitation was controlled entirely by flux of pore water. If ionic diffusion did operate, some dawsonite would be expected as a replacement in the mudrocks.

In the vicinity of the present-day outcrop area where active meteoric flushing still occurs, feldspar kaolinitization and dissolution are likely to be continuing.

A summary of the diagenetic history of the Aldebaran Sandstone is presented in Fig. 12.

RESERVOIR QUALITY

The principal hydrocarbon-producing interval of the Aldebaran Sandstone is a prominent basal sandstone section which developed within the southern deposystem. It can be subdivided into a lower and upper basal unit. The lower basal unit, up to 20 m thick, is widespread and accumulated in upper shoreface and beach environments. It has marginal to good reservoir quality with most porosity and permeability values (measured from core) in the range of 9–18% and 0·3–70 mD. The more restricted, slightly coarser, upper basal unit, up to about 12 m thick, accumulated in mainly back-barrier tidal delta, tidal channel and tidal flat environments. It has good reservoir quality (Fig. 13a) with most measured porosity and permeability values in the range of 12–22% and 10–400 mD. Good reservoir quality also characterizes the isolated medium- to coarse-grained tidal channel sandstones

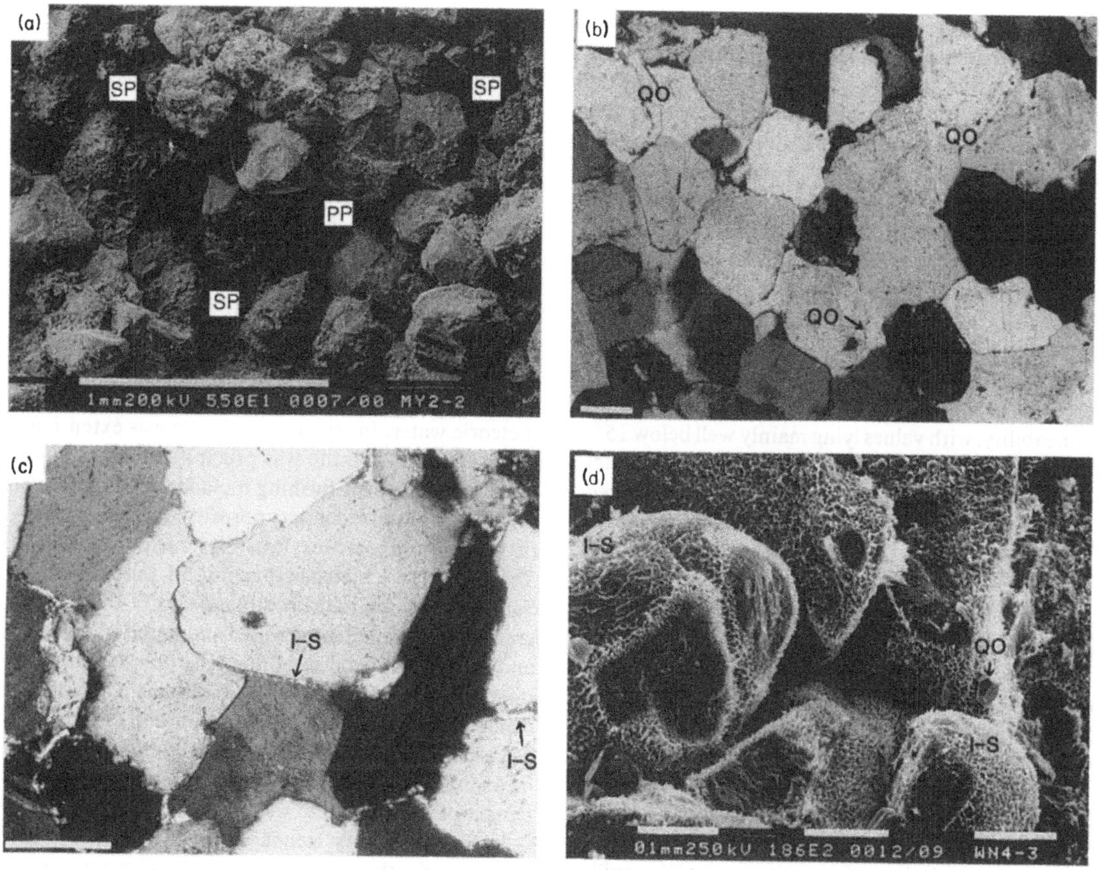

Fig. 13. (a) Sandstone with good reservoir quality from upper basal unit. Numerous primary (PP) and secondary (SP) pores are evident; Myrtleville no. 2, 687·95 m. (b) Texture representative of the Aldebaran Sandstone east of outcrop area. Primary porosity has been almost totally eliminated by quartz overgrowths (QO) and physico-chemical compaction; Rolleston no. 11, 973·20 m. (c) Promotion of grain-to-grain contact dissolution between quartz grains by illite–smectite (I-S) rims in a sandstone from east of the outcrop belt. Nearly all primary porosity has been obliterated; Christmas Creek no. 1, 1696·45m. (d) Thick illite–smectite (I-S) rims have preserved primary porosity by inhibiting grain-to-grain contact dissolution and quartz overgrowth development in sandstone from east of the outcrop area; Warrinilla no. 4, 1423·64 m. (a) and (d) are scanning electron micrographs, scale bars in (b) and (c) = 0·2 mm.

which locally occur higher in the section. Reflecting a major facies change, only poor to marginal reservoir quality appears to exist in the matrix-rich sandstones associated with the delta system which developed in the far south. Reservoir quality variations between these sandstone units primarily reflect grain-size differences, with highest porosities and permeabilities occurring in the coarsest units. Porosity reduction mainly by physico-chemical compaction and quartz overgrowth and carbonate precipitation, and porosity enhancement by labile (mainly feldspar) grain dissolution have generally enhanced variations caused by grain-size differences. Reservoir quality in drilled and prospective undrilled areas should differ little because stratigraphically equivalent sandstones accumulated in the same depositional environments, had similar pre-burial compositions, and were subject to the same diagenetic processes which controlled porosity reduction and enhancement.

Reservoir quality of sandstones that accumulated in the northern deposystem is highly variable. Over the outcrop area and areas to the west, it ranges from marginal to good, and deteriorates with depth as a result of the downward decrease in grain size from coarse sandstones and conglomerates of the upper fluvial subunit to fine and medium sandstones of the lower deltaic–nearshore marine subunit. Measured porosity and permeability values for the upper subunit mainly lie within the range 10–20% and 1–1000 mD, whereas those for the lower subunit lie within the range 12–22% and 0·5–50 mD. Diagenetic controls on reservoir quality are the same as those in the south. The section is non-prospective for hydrocarbons due to on-going meteoric flushing in this part of the basin.

East of the outcrop area, the effects of diagenesis, particularly grain-to-grain contact dissolution and silicification, have overshadowed any trends that may have arisen from textural variations. Nearly all sandstones, regardless of grain size and specific depositional environment, have very low porosity and permeability, with values lying mainly well below 15% and 10 mD. The extreme grain-to-grain contact dissolution and silicification which has obliterated nearly all primary porosity, and the lack of labile grain dissolution pores in the area reflect the eastward increase in detrital quartz caused by reworking in nearshore marine environments (Baker, 1989) and, locally, the presence of *thin* eogenetic illite–smectite grain rims which have promoted grain-to-grain contact dissolution (Fig. 13b, c). Relatively high maximum burial temperatures in this area may also be a contributing factor. In view of the depositional model developed for the Aldebaran Sandstone, reservoir quality in undrilled areas over this eastern region is expected to be, for the most part, poor, a noteworthy exception being the base of the unit over the Warrinilla area where *thick* illite–smectite rims have led to preservation of medium to good reservoir quality by inhibiting formation of quartz overgrowths and grain-to-grain contact dissolution (Fig. 13d). Recorded gas flows from this basal section undoubtedly reflect the presence of the grain-rimming illite–smectite.

Reservoir quality of the Aldebaran Sandstone is not related to present-day burial depth. For example, moderately good reservoirs exist at depths of around 1400 m in the south, whereas at about these depths east of the outcrop area, sandstones are tight. Accordingly, reservoir quality in this case cannot be predicted on the basis of depth, a fact hardly surprising considering the initial compositional heterogeneity and complex burial–temperature history of the unit.

A summary of regional reservoir quality trends in the Aldebaran Sandstone is included in Fig. 4.

CONCLUSIONS

The diagenetic history of the Aldebaran Sandstone is complex and can be directly linked to depositional environment, initial composition and burial–temperature history. Eogenetic effects included formation of glauconite, pyrite, siderite and illite–smectite rims. Mesogenetic effects were physico-chemical compaction and the formation of quartz overgrowths, siderite, illite/illite–smectite pore fill and ankerite. Involved fluids were a mixture of fresh and marine connate water and clay-mineral-bound water. Lacking major cognate sources of Fe and Mg, the ferroan carbonates may have incorporated cations released by the illitization of smectitic clays in the associated mudrocks. Late Triassic structuring subsequently placed the Aldebaran Sandstone in the telogenetic regime where it underwent flushing by deeply circulating meteoric water. In response, ankerite was extensively dissolved and kaolinite was precipitated in its place. Other effects of this flushing included the creation of grain-dissolution secondary porosity and precipitation of barite and calcite. Possibly during the Early Tertiary, active meteoric flushing of the Aldebaran Sandstone ceased to the south and east of the outcrop area. Trapped meteoric water stagnated chemically and became enriched in sodium bicarbonate. Dawsonite precipitated in response to this enrichment, an event which preceded the migration and entrapment of gaseous hydrocarbons as shown by the presence of dawsonite in gas-bearing sandstones in quantities equal to that in adjacent water-wet sandstones.

Reservoir quality trends in the Aldebaran Sandstone are clearly related to depositional environment and diagenesis. Depositional environment has had its most profound control on reservoir quality by way of dictating initial sediment composition and texture and pore fluid composition. This is shown by: (i) the dramatic detrital quartz increase east of the outcrop area due to reworking in nearshore marine environments (this has led to almost complete obliteration of primary porosity through grain-to-grain contact dissolution and silicification); (ii) differences in reservoir quality between the lower and upper basal units (southern deposystem) and lower and upper subunits (outcrop area) related to grain size which in turn has been controlled by depositional environment; and (iii) preservation of primary porosity and promotion of grain-to-grain contact dissolution by thick and thin eogenetic illite–smectite rims respectively which could only develop in marine environments. In the west, diagenesis has essentially enhanced reservoir quality variations caused by textural differences, whereas in the east, it has completely overprinted any facies-controlled reservoir quality variations that may have originally existed.

ACKNOWLEDGMENTS

This study forms part of a PhD research project supervised by Drs R. E. Chapman and C. R. Fielding. They are thanked together with Drs K. R. Martin and P. de Caritat and the two referees for their helpful reviews of earlier versions of this paper. I also thank the Department of Resource Industries, Queensland and AGL Petroleum for their support and co-operation. The valuable help and advice received from Dr S. D. Golding is greatly appreciated, and Drs H. K. Herbert and H. R. Krouse are gratefully acknowledged for providing unpublished relevant isotopic data. The research was supported by a University of Queensland Postgraduate Scholarship and a Commonwealth Postgraduate Award, and the preparation of this paper was greatly assisted by the Electron Microscope Centre, University of Queensland.

REFERENCES

Baker, J.C. (1989) *Petrology, diagenesis and reservoir quality of the Aldebaran Sandstone, Denison Trough, east-central Queensland*. PhD thesis, Queensland University.

Balfe, P.E. (1982) Permian stratigraphy of the Springsure–Arcturus Downs area. *Qld. Gov. Min. J.*, **83**, 133–155.

Barker, C.E. & Pawlewicz, M.J. (1986) The correlation of vitrinite reflectance with maximum temperature in humic kerogen. In: *Palaeogeothermics* (Ed. by G. Buntebarth & L. Stegena), pp. 79–93. Springer-Verlag, Berlin.

Bird, M.I. & Chivas, A.R. (1988) Stable isotope evidence for low-temperature kaolinitic weathering and post-formational hydrogen-isotope exchange in Permian kaolinites. *Chem. Geol.*, **72**, 249–265.

Bjørlykke, K. (1984) Formation of secondary porosity: how important is it? In: *Clastic Diagenesis* (Ed. by D. A. McDonald & R. C. Surdam), *Mem. Am. Ass. petrol. Geol.*, **37**, 277–286.

Boles, J.R. & Franks, S.G. (1979) Clay diagenesis in Wilcox sandstones of southwest Texas: implications of smectite diagenesis on sandstone cementation. *J. sedim. Petrol.*, **49**, 55–70.

Brown, R.S., Elliott, L.G. & Mollah, R.J. (1983) Recent exploration and petroleum discoveries in the Denison Trough, Queensland. *Aust. petrol. Expl. Ass. J.*, **23**, 120–135.

Burley, S.D. (1986) The development and destruction of porosity within Upper Jurassic reservoir sandstones of the Piper and Tartan Fields, Outer Moray Firth, North Sea. *Clay Miner.*, **21**, 649–694.

Carothers, W.W., Adami, L.H. & Rosenbauer, R.J. (1988) Experimental oxygen isotope fractionation between siderite-water and phosphoric acid liberated CO_2-siderite. *Geochim. Cosmochim. Acta*, **52**, 2445–2450.

Chapman, R.E. (1982) Effects of oil and gas accumulation on water movement. *Bull. Am. Ass. petrol. Geol.*, **66**, 368–378.

Clayton, R.M. & Mayeda, T.K. (1963) The use of bromine pentafluoride in the extraction of oxygen from oxides and silicates for isotopic analysis. *Geochim. Cosmochim. Acta*, **27**, 43–52.

Coleman, M.L., Shepherd, T.L., Durham, J.J., Rouse, J.E. & Moore, G.R. (1982) Reduction of water with zinc for hydrogen isotope analysis. *Analyt. Chem.*, **54**, 993–995.

Craig, H. (1957) Isotopic standards for carbon and oxygen and correlation factors for mass-spectrometric analysis of carbon dioxide. *Geochim. Cosmochim. Acta*, **12**, 133–149.

Craig, H. (1961) Standards for reporting concentrations of deuterium and oxygen-18 in natural waters. *Science*, **133**, 1833–1834.

Draper, J.J. (1983) The origin of pebbles in mudstones in the Denison Trough. In: *Proc. Permian Geology of Queensland, Symposium, Geological Society of Australia, Queensland Division*, pp. 305–316.

Draper, J.J. & McClung, G.R. (1983) Sedimentation and depositional environments of the Permian sequence in GSQ Eddystone 1, Denison Trough, Queensland. *Geol. Surv. Qld. Publ.*, **384**, 1–41.

Draper, J.J., Palmieri, V., Price, P.L., Briggs, D.J.C. & Parfrey, S.M. (1990) A biostratigraphic framework for the Bowen Basin. In: *Proc. Bowen Basin Symposium*. pp. 26–35.

Elliott, L.G. (1985) The stratigraphy of the Denison Trough. In: *Bowen Basin Coal Symposium. Geol. Soc. Aust. Abstracts*, **17**, 33–38.

Fielding, C.R., Falkner, A.J., Kassan, J. & Draper, J.J. (1990) Permian and Triassic depositional systems in the Bowen Basin. In: *Proc. Bowen Basin Symposium*, pp. 21–25.

Folk, R.L., Andrews, P.B. & Lewis, D.W. (1970) Detrital sedimentary rock classification and nomenclature for use in New Zealand. *N.Z. J. Geol. Geophys.*, **13**, 947–968.

Freed, R.L. & Peacor, D.R. (1989) Variability in temperature of the smectite/illite reaction in Gulf Coast sediments. *Clay Miner.*, **24**, 171–180.

Friedman, I. & O'Neil, J.R. (1977) Compilation of stable isotope fractionation factors of geochemical interest. In: *Data of Geochemistry*, 6th edn (Ed. by M. Fleischer), *Prof. Pap. US geol. Surv.*, **440-KK**, 12 pp.

Golding, S.D. & Wilson, A.F. (1983) Geochemical and stable isotope studies of the No. 4 Lode, Kalgoorlie, Western Australia. *Econ. Geol.*, **78**, 438–450.

Gray, A.R.G. (1976) Stratigraphic relationships of late Palaeozoic sediments between Springsure and Jericho. *Qld. Gov. Min. J.*, **77**, 146–163.

Hillier, S. & Clayton, T. (1989) Illite/smectite diagenesis in Devonian lacustrine mudrocks from northern Scotland and its relationship to organic maturity indicators. *Clay Miner.*, **24**, 181–196.

Hoffman, J. & Hower, J. (1979) Clay mineral assemblages as low grade metamorphic geothermometers: application to the thrust faulted disturbed belt of Montana, USA. *Spec. publs. Soc. econ. Paleont. Miner.*, **26**, 55–79.

Hower, J. (1981) X-ray diffraction identification of mixed-layer clay minerals. In: *Clays and the Resource Geologist* (Ed. by F. J. Longstaffe), *Miner. Ass. Canada, Short Course Handbook*, **7**, 39–59.

IRWIN, H., CURTIS, C. & COLEMAN, M. (1977) Isotopic evidence for source of diagenetic carbonates formed during burial of organic rich sediments. *Nature*, **269**, 209–213.

JACKSON, K.S., HAWKINS, P.J. & BENNETT, A.J.R. (1980) Regional facies and geochemical evaluation of the southern Denison Trough, Queensland. *Aust. petrol. Expl. Ass. J.*, **20**, 143–158.

LAND, L.S. & DUTTON, S.P. (1978) Cementation of a Pennsylvanian deltaic sandstone: isotopic data. *J. sedim. Petrol.*, **48**, 1167–1176.

LONGSTAFFE, F.J. (1983) Diagenesis 4. Stable isotope studies of diagenesis in clastic rocks. *Geosci. Can.*, **10**, 43–58.

LONGSTAFFE, F.J. (1987) Stable isotope studies in diagenetic processes. In: *Stable Isotope Geochemistry of Low Temperature Fluids* (Ed. by T. K. Kyser), *Miner. Ass. Canada, Short Course Handbook*, **13**, 187–257.

MCCREA, J.M. (1950) On the isotopic chemistry of carbonates and a palaeotemperature scale. *J. Chem. Phys.*, **18**, 849–857.

MCLOUGHLIN, S. (1988) Geology of the Inglis Dome, Denison Trough. *Dep. Geol. Mineral., Qld. Univ. Papers*, **12**, 229–264.

PARMENTER, D.S. (1986) *Permian geology of the Springsure-Rolleston area, Denison Trough, east-central Queensland.* BSc (Hons) thesis, Queensland University.

PATEN, R.J., BROWN, L.N. & GROVES, R.D. (1979) Stratigraphic concepts and petroleum potential of the Denison Trough, Queensland. *Aust. petrol. Expl. Ass. J.*, **19**, 43–52.

POLLASTRO, R.M. & BARKER, C.E. (1986) Application of clay-mineral, vitrinite reflectance and fluid inclusion studies to the thermal and burial history of the Pinedale Anticline, Green River Basin, Wyoming. In: *Roles of Organic Matter in Sediment Diagenesis* (Ed. by D. L. Gautier), *Spec. publs. Soc. econ. Paleont. Miner.*, **38**, 73–83.

ROSENBAUM, J. & SHEPPARD, S.M.F. (1986) An isotopic study of siderites, dolomites and ankerites at high temperatures. *Geochim. Cosmochim. Acta*, **50**, 1147–1150.

SAVIN, S.M. & LEE, M. (1988) Isotopic studies of phyllosilicates. In: *Hydrous Phyllosilicates (Exclusive of Micas)* (Ed. by S. W. Bailey), *Rev. Miner.*, **19**, 189–223.

SCHMIDT, V. & MCDONALD, D.A. (1979) The role of secondary porosity in the course of sandstone diagenesis. In: *Aspects of Diagenesis* (Ed. by P. A. Scholle & P. R. Schluger), *Spec. publs. Soc. econ. Paleont. Miner.*, **26**, 175–207.

SHARMA, T. & CLAYTON, R.N. (1965) Measurement of $^{18}O/^{16}O$ ratios of total oxygen carbonates. *Geochim. Cosmochim. Acta*, **29**, 1347–1353.

WEIR, A.H., ORMEROD, E.C. & MANSEY, I.M.I. (1975) Clay mineralogy of sediments of the western Nile Delta. *Clay Miner.*, **10**, 370–386.

WILKINSON, M.M. (1983) *Permian geology and environment of deposition of the Aldebaran Sandstone of the Serocold Anticline, east-central Queensland.* BSc (Hons) thesis, Queensland University.

ZIOLKOWSKI, V. & TAYLOR, R. (1985) Regional structure of the north Denison Trough. In: *Bowen Basin Coal Symposium. Geol. Soc. Aust. Abstracts*, **17**, 129–135.

(Manuscript received 25 October 1990; revision received 25 February 1991)

Reprinted from Sedimentology (1994) ***41****, 1253–1272*

Diagenesis and formation water chemistry of Triassic reservoir sandstones from southern Tunisia

S. MORAD,* H. N. BEN ISMAIL,† L. F. DE ROS,*[1] I. S. AL-AASM‡ *and* N-E. SERRHINI†

**Institute of Earth Sciences, Uppsala University, S-752 36 Uppsala, Sweden*
†Département de Géologie, Université de Tunis 2, Campus Universitaire 1060-Tunis, Tunisia
‡Department of Geology, University of Windsor, Windsor, Ontario N9B 3P4, Canada

ABSTRACT

The fluvial Triassic reservoir subarkoses and arkoses (2409·5–2519·45 m) of the El Borma oilfield, southern Tunisia, were subjected to cementation by haematite, anatase, infiltrated clays, kaolinite and K-feldspar at shallow burial depths from meteoric waters. Subsequently, basinal brines controlled the diagenetic evolution of the sandstones and resulted initially in the precipitation of quartz overgrowths, magnesian siderite, minor ferroan magnesite and anhydrite. The enrichment of siderite in ^{12}C isotope ($\delta^{13}C_{PDB} = -14{\cdot}5$ to -9‰) results from derivation of carbon from the thermal decarboxylation of organic matter. During further burial, the precipitation of dickite and pervasive transformation of kaolinite into dickite occurred, followed by the formation of microcrystalline K-feldspar and quartz, chlorite and illite, prior to the emplacement of oil. Present day formation waters are Na–Ca–Cl brines evolved by the evaporation of seawater and water/mineral interaction and are in equilibrium with the deep burial ($\leq 3{\cdot}1$ km) minerals. These waters are suggested to be derived from the underlying Silurian and Devonian dolomitic mudstones.

INTRODUCTION

Shallow, early diagenetic minerals in continental clastic sediments usually precipitate from meteoric pore waters, but mixing with marine waters occurs in nearshore/coastal sediments (Morad *et al.*, 1992). During burial diagenesis, mixing of these shallow diagenetic waters with basinal brines occurs, and hence profound changes in the formation water chemistry are common (Galloway, 1984; Hanor, 1987; Egeberg & Aagaard, 1989; De Ros *et al.*, 1994). These geochemical changes of the formation waters, combined with the influence of elevated temperatures and pressures, are expected to be reflected in the paragenetic sequence of the host sediments.

In the course of progressive burial diagenetic modifications of clay minerals in clastic sediments in several basins elsewhere, constraints have often been made only on the illitization and/or chloritization of smectite and kaolinite. Due to the difficulties encountered in distinction between kaolinite and dickite both microscopically and by X-ray diffraction analyses, the transformation of kaolinite into dickite has been poorly constrained and probably overlooked. Nevertheless, there have recently been several reports of this diagenetic transformation in sandstones (Ehrenberg *et al.*, 1993; McAulay *et al.*, 1993, 1994).

The aims of this paper are: (1) to unravel the relationship between the chemical/paragenetic characteristics of diagenetic minerals and the origin/geochemical evolution of formation waters in the Triassic clastic reservoirs of the El Borma Oilfield in southern Tunisia (Fig. 1); and (2) present textural evidence for the recognition of kaolinite to dickite transformation and to elucidate the conditions under which this transformation takes place in this geological setting. The recognition of burial diagenetic transformation of kaolinite into dickite is important as it

[1]Permanent address: Universidade Federal do Rio Grande do Sul, Instituto de Geociências, Av. Bento Gonçalves, 9500, Porto Alegre, RS, 91500, Brazil.

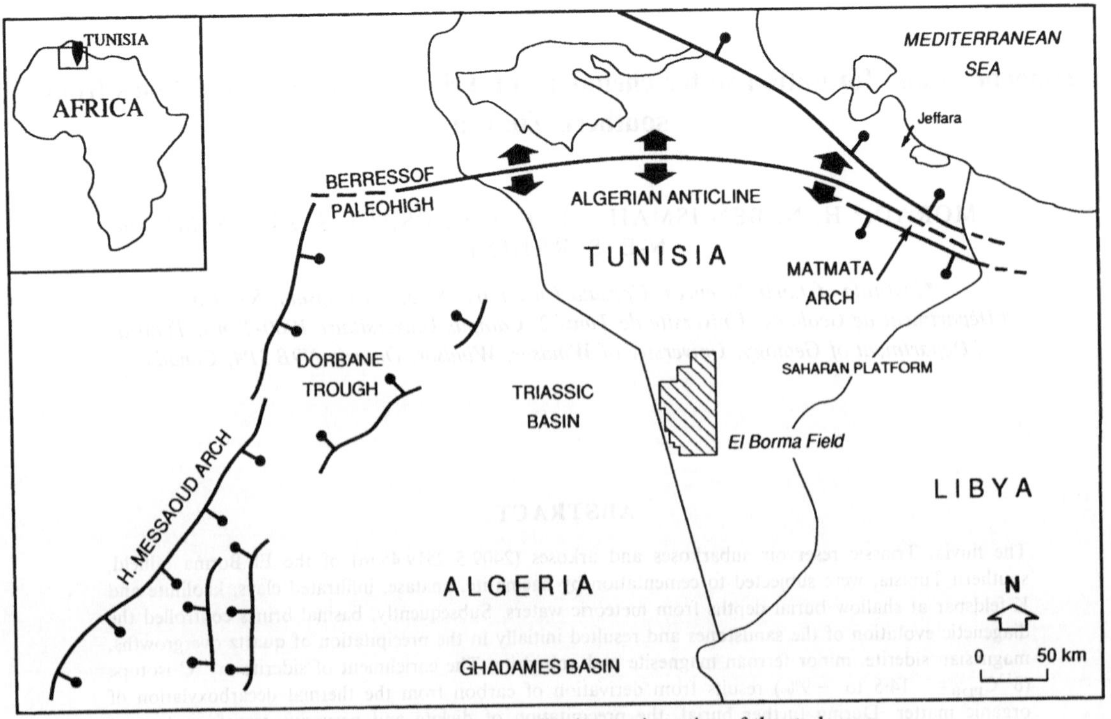

Fig. 1. Location map and a generalized tectonic setting of the study area.

is probably potentially a reliable palaeogeothermometer (Ehrenberg *et al.*, 1993).

GEOLOGICAL SETTING

El Borma Field is located on the tectonically stable Saharan platform. Epirogenic movements affected the platform and formed wide shallow basins and broads uplifts in Palaeozoic and Mesozoic time (Ben Ismail, 1991). The structure of the oilfield is an asymmetric anticline orientated SSE–NNW that formed in late Mesozoic to early Cenozoic time. North of El Borma, Triassic siliciclastic deposits are either thin or absent due to the formation of the Algerian anticline (Fig. 1) which acted as an east–west trending palaeohigh during early and middle Triassic time. West of El Borma, in eastern Algeria, the 3500 m deep Ghadames Basin (Fig. 1) is limited to the north by the Algerian anticline and to the west by the Hassi Messaoud palaeohigh. The latter is bounded by NNE trending faults which resulted from the late Carboniferous and Permian compressional tectonics with a shortening axis orientated at 120°. The faults reactivated as normal faults during the late Triassic tectonic extension and influenced the distribution of the Triassic reservoir sandstones in eastern Algeria. South of El Borma, the Hoggar shield (composed of metamorphic rocks) is located at a distance of 650 km in the central Sahara. Horsts and grabens, bounded by local N–S trending flexures and faults, extend from the Hoggar to the 'Triassic basin' close to the west of the El Borma area. Block movements were controlled by Precambrian basement faults that reactivated in Mesozoic time. Distribution of the Triassic sandstones is largely influenced by these structural elements.

In southern Tunisia, the Triassic sedimentary sequence (Fig. 2) comprises a lower clastic sequence (early to middle Triassic) that grades into a middle carbonate and, subsequently, an upper evaporite sequence (Ben Ismail, 1992). In the El Borma Field, the basal Triassic sandstones (early Carnian) are about 125 m thick and unconformably overlie Silurian and Devonian marine dolomitic mudstones (Fig. 3). The Triassic reservoir sandstones are subdivided into five reservoir intervals (A–E) separated from each other by mudstone beds. The basal E, C and D reservoirs are braided fluvial, whereas the B and A reservoirs are meandering fluvial in origin

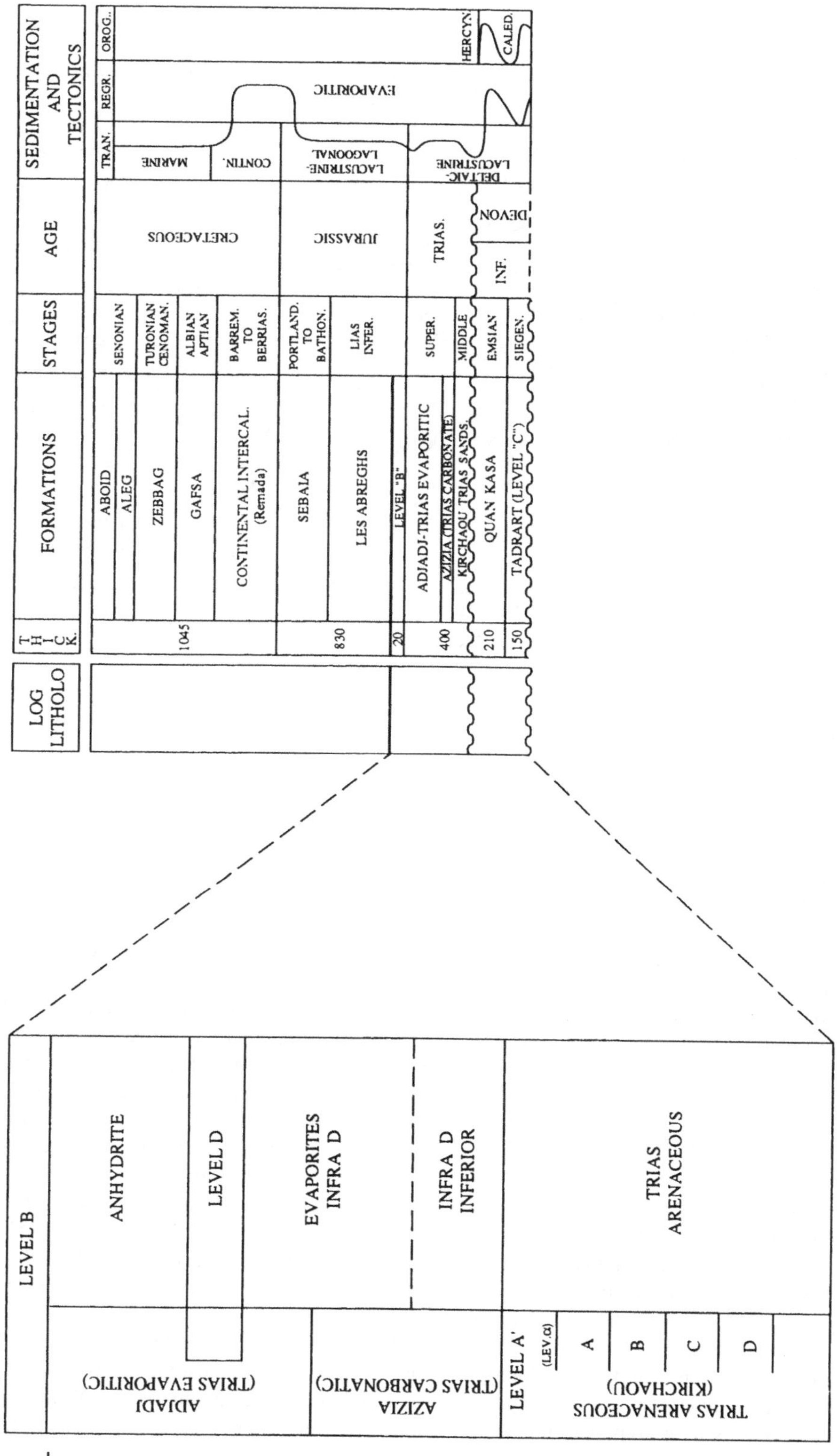

Fig. 2. A schematic stratigraphical column of southern Tunisia, detailing the Triassic sequence.

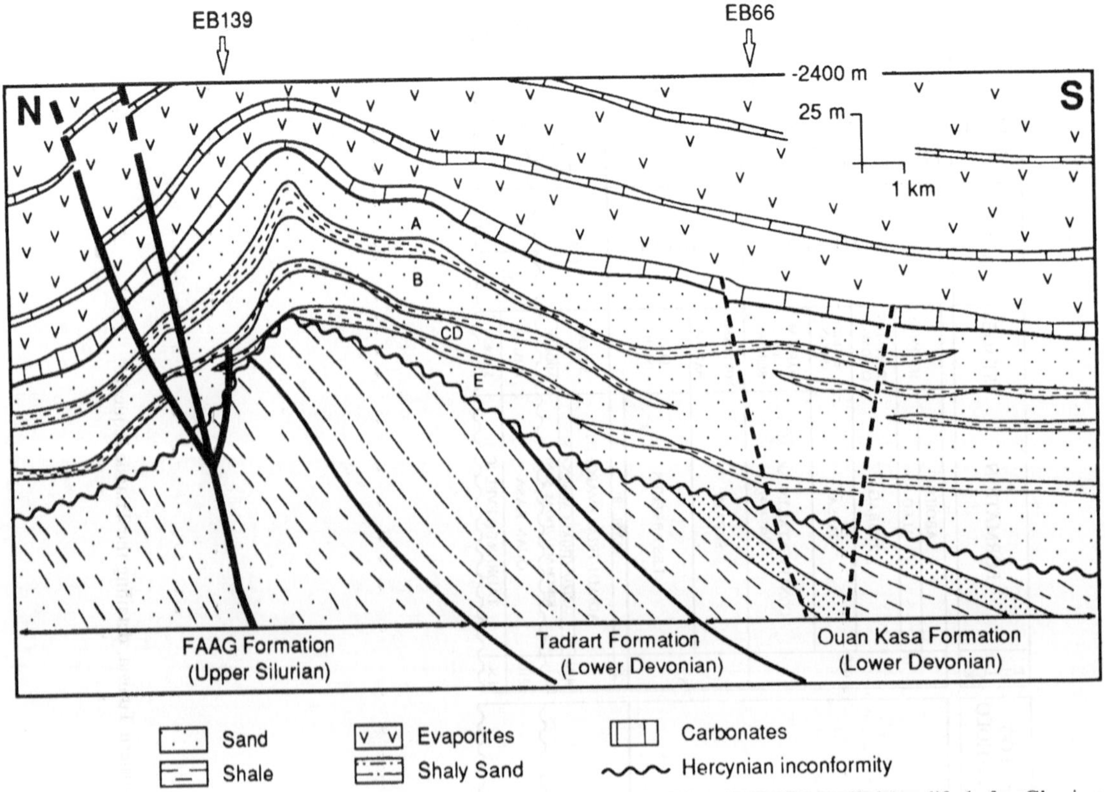

Fig. 3. A schematic structural cross section of the El Borma Field with location of the wells studied (modified after Ghenima, 1993).

(Serrhini, 1992). The poorly cored lowermost interval E fills the lows of the irregular Hercynian unconformity surface.

The burial history curve (Fig. 4) is typical for intracratonic basins where burial is gradual and continuous with late, relatively minor uplifts. The present day geothermal gradient of the area is 25–27°C km^{-1}.

ANALYTICAL TECHNIQUES

Sixty core samples (depths of 2409·5–2512·45 m; bottom hole temperatures ≈79–86°C) were selected from the five reservoir intervals in two boreholes (EB139 and EB66; Fig. 3). These two wells were chosen because they contain completely cored Triassic rocks. The samples were selected to cover all the variations in colour and apparent texture/lithology within the different reservoir facies. Epoxy-impregnated thin sections were prepared and examined with a standard petrographical microscope. The modal mineralogical compositions were determined by counting 300 points per thin section. The thin sections were coated with a thin layer of carbon for electron microprobe (EMP) analyses using a Cameca Camebax SX 50 instrument equipped with three crystal spectrometers and a back-scattered electron (BSE) detector. The operating conditions during analysis were an acceleration voltage of 15 kV, a measured beam current of 8 nA for siderite and 10 nA for the silicates, and a beam diameter of 1 μm. Clay minerals analysed were larger than ≈5 μm in size. The standards used were wollastonite (Ca and Si; 10 s), MgO (Mg; 10 s), $MnTiO_3$ (Mn; 20 s), haematite (Fe; 10 s), corundum (Al; 10 s), albite (Na; 5 s), strontianite (Sr; 10 s), barite (Ba; 10 s) and orthoclase (K; 5 s).

The less than 5 μm fraction was separated from five representative sandstone samples and kaolinite and dickite were distinguished by X-ray diffraction analyses using the non-basal reflections in randomly

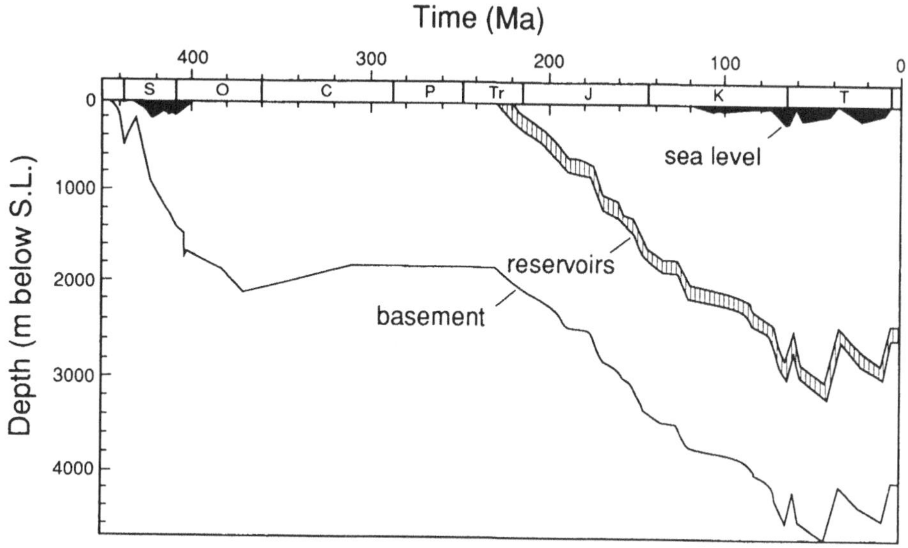

Fig. 4. Burial history diagram of the studied Traissic sandstones.

orientated samples (Ehrenberg *et al.*, 1993; McAulay *et al.*, 1993). Twenty-nine representative samples were coated with a thin layer of gold and examined with a JEOL JSM-T330 scanning electron microscope (SEM) at an accelerating voltage of 10 kV. Siderites in 20 samples were analysed for O and C isotope composition by treatment with 100% H_3PO_4 under vacuum at 50°C for 6 days (Al-Aasm *et al.*, 1990). The $\delta^{13}C$ and $\delta^{18}O$ in the evolved CO_2 gas were analysed using a SIRA-12 mass spectrometer. The results are reported in permil deviation from the PDB standard. Results were calculated using a fractionation factor of 1·009082 (Rosenbaum & Sheppard, 1986). The analytical reproducibility is better than ± 0·05‰ for $\delta^{13}C$ and $\delta^{18}O$.

Formation water chemical analyses were provided by the AGIP Oil Company. The $\delta^{18}O$ analyses of two formation water samples were performed by a Finnigan MAT 251 mass spectrometer using the CO_2 equilibration method.

TERMINOLOGY

Sandstone compositions are classified according to the scheme of McBride (1963). Clay mineral terminology follows the recommendations of AIPEA. The kaolin group mineral polytypes dickite and kaolinite are distinguished on the basis of powder mount XRD peaks as detailed in Bailey (1980) and SEM. Identification of illite and chlorite is made on the basis of a combination of textural, morphological, optical, EMP and XRD criteria. The term illite may include a swelling component.

Shallow burial meteoric diagenesis was used to indicate the interaction of depositional waters with host sediments. This category also includes contemporary recharge of fresh waters through fluvial sand-bodies. Fluids released due to compaction, along underseated faults and by decarboxylation reactions are described as deep burial diagenesis. Shallow and deep burial diagenetic regimes used here are equivalent to the terms eo- and mesodiagenesis (*sensu* Schmidt & McDonald, 1979), respectively.

DETRITAL MINERALOGY

The sandstones are grey, green to red in colour, fine to medium grained subarkoses and arkoses that contain carbonized plant debris. Modal analyses (Table 1) revealed that the most common framework mineral is monocrystalline quartz (≈30–45%, average 34%). Polycrystalline quartz forms ≈5–10% (average ≈8%) of the bulk sandstones. K-feldspars (≈7–12%; average ≈10%) dominate over plagioclase (trace to 3%; average 1·5%). The rock fragments (≈3–15%; average ≈9%) are mostly medium grade metamorphic in origin. Micas (trace to 6%) are dominated by muscovite. Pseudomatrix, mainly

Table 1. Representative and average ($n=60$) modal compositions of Triassic sandstones of EB139 wells of the El Borma oilfield, southern Tunisia.

Sample	EB13 2449·3	EB139 2453·9	EB139 2503·1	EB139 2508·1	EB139 2511·3	EB66 2417·1	EB66 2432·8	EB66 2434·4	EB66 2443·2	EB66 2445·7	EB66 2447·8	EB66 2448·6	Average
Detrital quartz	40·0	43·7	51·3	42·3	48·3	51·3	38·0	38·7	38·7	36·0	40·0	40·0	41·8
Quartz (monocrystalline)	34·3	34·0	44·7	37·0	41·0	39·3	33·0	32·7	30·3	29·0	35·0	30·3	34·1
Quartz (polycrystalline)	5·7	9·7	6·7	5·3	7·3	12·0	5·0	6·0	8·3	7·0	5·0	9·7	7·7
Detrital feldspar	10·3	7·7	8·7	7·0	6·3	6·7	11·7	10·0	10·7	10·7	10·7	11·3	9·8
Detrital K-felspar	10·0	7·0	8·7	7·0	5·0	6·7	11·7	10·0	10·7	10·7	10·3	11·3	9·5
Orthoclase	6·7	5·7	7·0	4·0	2·7	4·7	9·3	8·3	7·7	8·0	7·0	8·3	7·2
Microcline	3·0	1·3	1·7	3·0	2·0	1·0	2·3	1·7	3·0	2·7	2·7	2·3	2·0
Perthite	0·3	0·0	0·0	0·0	0·3	1·0	0·0	0·0	0·0	0·0	0·7	0·7	0·3
Detrital plagiocase	0·3	0·7	0·0	0·0	1·3	0·0	0·0	0·0	0·0	0·0	0·3	0·0	0·3
Plutonic r.f.	1·7	1·3	1·3	1·3	1·0	1·7	1·7	1·0	2·7	1·0	3·0	2·7	1·5
Metamorphic r.f.		4·7	3·3	0·7	1·0	3·7	12·0	7·3	10·0	8·0	9·7	9·0	6·7
Chert r.f.	0·3	1·3	1·0	1·0	1·0	0·0	0·0	1·0	0·0	0·7	0·0	0·3	0·5
Muscovite	2·3	0·3	1·0	0·3	2·7	0·0	2·0	2·0	0·7	1·0	1·7	0·7	1·6
Biotite	0·3	0·0	0·3	0·0	1·3	0·0	0·7	0·7	0·0	0·3	0·0	0·0	0·3
Heavy minerals	0·3	0·0	0·7	0·0	0·7	0·3	3·0	0·3	0·7	1·3	0·7	1·7	0·7
Mud intraclasts	0·0	1·0	0·0	0·7	1·0	1·3	0·0	0·0	3·3	0·3	0·0	0·0	0·5
Carbonate intraclasts	0·0	0·3	0·0	0·0	0·0	0·3	0·0	0·0	0·0	0·0	0·0	0·0	0·1
Clay peloids	0·0	0·3	0·3	0·0	0·0	0·3	0·0	0·0	0·0	0·0	0·0	0·0	0·1
Pseudomatrix	18·3	9·0	3·7	1·3	6·7	1·0	4·3	2·3	3·7	5·0	1·7	2·0	4·8
Total diagenetics	17·0	21·3	22·3	38·7	25·0	21·7	21·7	34·0	23·7	27·7	21·3	23·7	23·8
Clay coats	3·0	0·3	1·0	0·0	5·0	0·0	2·7	0·7	1·0	1·7	0·3	1·7	1·0
Haematite	0·3	0·0	0·3	0·0	0·7	0·3	1·3	0·0	0·3	0·7	0·0	0·3	0·2
Quartz overgrowth	3·7	8·7	2·0	6·0	4·7	6·7	6·7	2·3	3·0	6·0	6·7	5·3	5·0
Quartz discrete	0·3	1·0	0·3	1·0	0·0	2·0	0·0	0·0	0·7	0·3	0·3	0·0	0·3
K-feldspar overgrowth	1·3	1·3	0·3	4·0	0·3	3·3	4·3	3·3	1·3	4·0	4·3	2·7	2·3
K-feldspar discrete	0·3	1·0	0·3	0·7	0·0	0·7	0·7	0·0	0·3	0·3	0·7	1·0	0·2

Siderite/magnesite (poikilotopic)	1·0	3·3	2·3	1·7	5·0	3·7	0·0	0·0	0·0	0·0	0·0	0·0	1·5
Siderite/magnesite (blocky)	0·0	0·3	0·3	2·0	2·3	0·0	1·0	9·3	4·0	2·7	1·3	1·3	2·0
Siderite/magnesite (microcrystalline)	0·0	0·0	0·0	0·0	0·0	0·7	0·0	0·0	0·0	0·0	0·3	0·0	0·1
Siderite/magnesite (in grain)	4·3	2·0	3·7	1·7	0·7	2·0	1·3	12·7	4·3	4·3	3·0	0·7	3·6
Siderite/magnesite (in p.matrix)	1·7	1·0	1·3	1·7	0·7	1·0	0·7	4·0	2·7	2·0	0·0	0·0	1·5
Siderite/magnesite (in mica)	0·0	0·0	0·0	0·0	0·0	0·0	0·0	0·0	0·0	0·3	0·3	0·3	0·1
Kaolinite/Dickite (intergranular)	0·0	0·0	0·3	0·3	0·7	0·0	0·0	0·7	2·0	1·7	2·3	4·3	1·3
Kaolinite/Dickite (in grain)	0·0	0·0	1·7	0·0	1·3	0·0	0·0	0·7	1·7	2·3	1·0	4·0	1·7
Kaolinite/Dickite (in mica)	0·0	0·0	1·3	1·0	0·7	0·0	0·0	0·0	0·0	0·0	0·0	0·0	0·2
Kaolinite/Dickite (in p.matrix)	0·0	0·0	2·7	0·3	0·0	0·0	0·0	0·0	0·7	0·0	0·0	1·3	0·4
Illite (fibrous)	0·7	0·7	1·0	0·7	0·3	0·3	0·0	0·0	1·0	0·0	0·3	0·0	0·2
Anhydrite	0·0	1·7	2·7	12·3	1·7	0·0	0·0	0·0	0·0	0·0	0·0	0·0	1·4
Anhydrite (in grain)	0·0	0·0	0·0	4·0	0·0	0·0	0·0	0·0	0·0	0·0	0·0	0·0	0·2
Pyrite	0·0	0·0	0·3	0·7	0·7	0·3	0·0	0·0	0·0	0·3	0·0	0·0	0·1
Anatase	0·3	0·0	0·3	0·7	0·3	0·7	3·0	0·3	0·7	1·0	0·3	0·7	0·5
Macroporosity	2·3	9·0	6·0	6·7	5·0	11·7	5·0	2·7	6·0	8·0	11·3	8·7	7·8
Intergranular	2·0	8·0	4·3	6·0	4·7	9·7	4·3	2·3	5·0	5·0	10·0	5·7	6·4
Intragranular	0·3	1·0	1·7	0·7	0·3	2·0	0·7	0·3	1·0	3·0	1·3	3·0	1·4
Microporosity	abundant	common	common	rare	common	rare	common	common	common	common	common	abundant	common
Integranular volume	11·0	18·3	13·0	33·7	22·3	18·7	19·7	16·7	14·0	19·3	15·7	17·0	16·7
Grain volume	89·0	81·7	87·0	66·3	77·7	81·3	80·3	83·3	86·0	80·7	84·3	83·0	83·3
Grain replacement total	6·0	3·0	10·7	8·7	3·3	3·0	2·0	17·3	9·3	9·0	4·3	6·3	7·7
Packing	tight	loose	normal	loose	tight	normal	tight	loose	tight	tight	normal	normal	tight
Granulometry	fine	medium	fine	medium	fine	medium	fine	fine	fine	fine	medium	fine	fine
Sorting	moderate	moderate	moderate	good	good	good	good	moderate	moderate	good	moderate	good	good

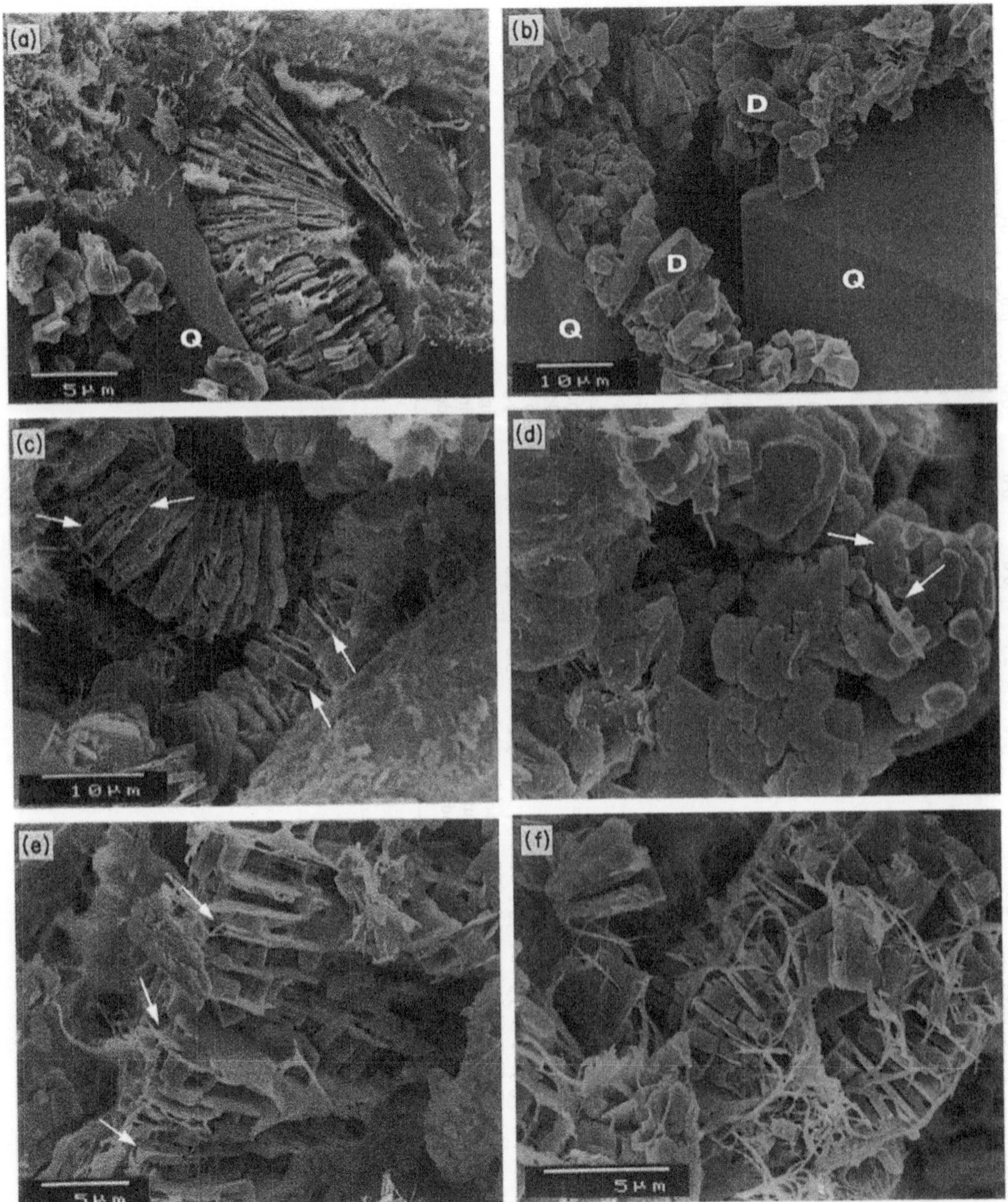

Fig. 5. SEM photomicrographs of: (a) parallel stacked, pseudohexagonal kaolinite crystals which are partially transformed into dickite and enveloped by diagenetic quartz overgrowths (Q); (b) authigenic dickite crystals (D) that post-date quartz overgrowths (Q); (c) kaolinite pervasively transformed into dickite with a few thin remnants of the original vermicular kaolinite (arrows); (d) a basal view of kaolinite transformed into dickite showing the etching of kaolinite remnants (arrows); (e) kaolinite crystals transformed into illite and dickite; note that illitization is restricted to the thin remnants of kaolinite (arrows); (f) thread-like illite that has grown on kaolinite crystals which are completely transformed into dickite.

formed by the deformation of mud intraclasts, forms between 1 and 18% (average ≈5%) of the bulk sandstones. The intergranular macroporosity (≈2–14%; average ≈6%) dominates over intragranular macroporosity (trace to 3%; average ≈1·5%).

DIAGENETIC MINERALOGY

Kaolinite and dickite

Kaolin (0–6%; average ≈2·5% of bulk sandstones; Table 1), which occurs as intergranular cement and as replacement of detrital feldspars, comprises numerous stacked pseudohexagonal crystals that often develop into vermicular aggregates. Detrital micas are kaolinized and display the typical expanded texture. Kaolinite is enveloped by quartz overgrowths (Fig. 5a). Dickite is texturally distinguished from kaolinite by its more euhedral blocky habit (cf. Tilley & Longstaffe, 1989; Ehrenberg *et al.*, 1993) and smooth surfaces. XRD analyses of randomly orientated powder samples revealed several strong reflections typical of dickite (4·13, 3·79, 2·50 and 2·327 Å) and, in one sample (EB-139, 2454·25 m), very weak kaolinite reflections (4·18, 2·495 and 2·339 Å). The XRD analyses of the orientated clay fractions revealed that the total kaolin varies between about 1 and 15% of the bulk clay fraction (Serrhini, 1992).

Dickite occurs in two modes: (1) as transformation of authigenic kaolinite, with the preservation or a slight disruption of the original vermicular stacking habit of the kaolinite and, less commonly, (2) as scattered crystals precipitated on the surface of authigenic quartz and in post-compactional intergranular pores (Fig. 5b). Replacement of kaolinite by dickite was recognized based on the vermicular stacking habit characteristic for kaolinite and the presence of etched remnants of kaolinite. Furthermore, kaolinite replaced by dickite occurs as sand sized patches with well defined outlines, probably related to earlier, complete replacement of detrital feldspars. Kaolinite transformed into dickite is commonly engulfed by quartz overgrowths. Authigenic dickite also occurs as randomly orientated blocky crystals that cover the quartz overgrowths. Transformation of kaolinized micas into dickite has occasionally caused a disruption of the original kaolinite layered texture. Pervasive to complete transformation of kaolinite into dickite occurs in samples from all facies and is accomplished through the growth of blocky, monoclinic dickite between the pseudohexagonal crystals of kaolinite. The kaolinite remnants are etched, due to partial dissolution (Fig. 5c,d), and/or selectively subjected to illitization (Fig. 5e). Dickite crystals are much thicker (≈ ≥1–8 μm) than kaolinite (<1 μm) and show no etching but are often covered by, and hence pre-date, authigenic illite (Fig. 5f).

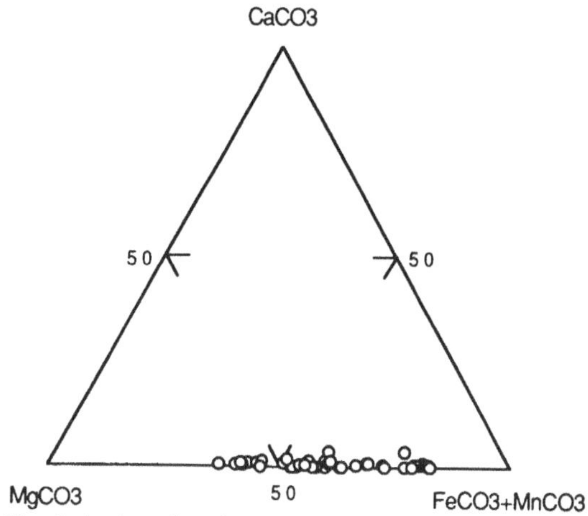

Fig. 6. A triangular plot showing the compositional ranges of siderite and magnesite.

Siderite and magnesite

The only carbonate cements detected in the sandstones (trace to 22%; average ≈7%; Table 1) are from the siderite–magnesite ($FeCO_3$–$MgCO_3$) series. They occur together in sandstones of all the reservoir intervals as finely crystalline (<50 μm) to coarsely crystalline (≈50–250 μm), occasionally poikilotopic pore filling and grain replacive cements.

The siderites are extremely Mg-rich ($MgCO_3$=18–47 mol%), being either sideroplesites or pistomesites (siderites having 5–30 and 30–50 mol% $MgCO_3$, respectively; Deer *et al.*, 1966) (Fig. 6). In some cases, the Mg content is up to 50–58 mol% and the phase can thus be termed breunnerite, which is a ferroan magnesite with 5–50 mol% $FeCO_3$ (Deer *et al.*, 1966). The substitution of siderite and magnesite by $MnCO_3$ is 1·8–4·5 mol%, whereas their substitution by $CaCO_3$ is between 0·2 and 2 mol%. Strontium is detected in trace amounts (<150 ppm) in a few of the siderites analysed.

The siderite crystals display complex zonation patterns in terms of Fe and Mg contents. The cores are richer in either iron or magnesium than the rims (Fig. 7a,d,f). There are no crystals showing systematic changes in Fe and/or Mg from core to rim; all have sharp alternating Fe-poor/Mg-rich (pistomesite to

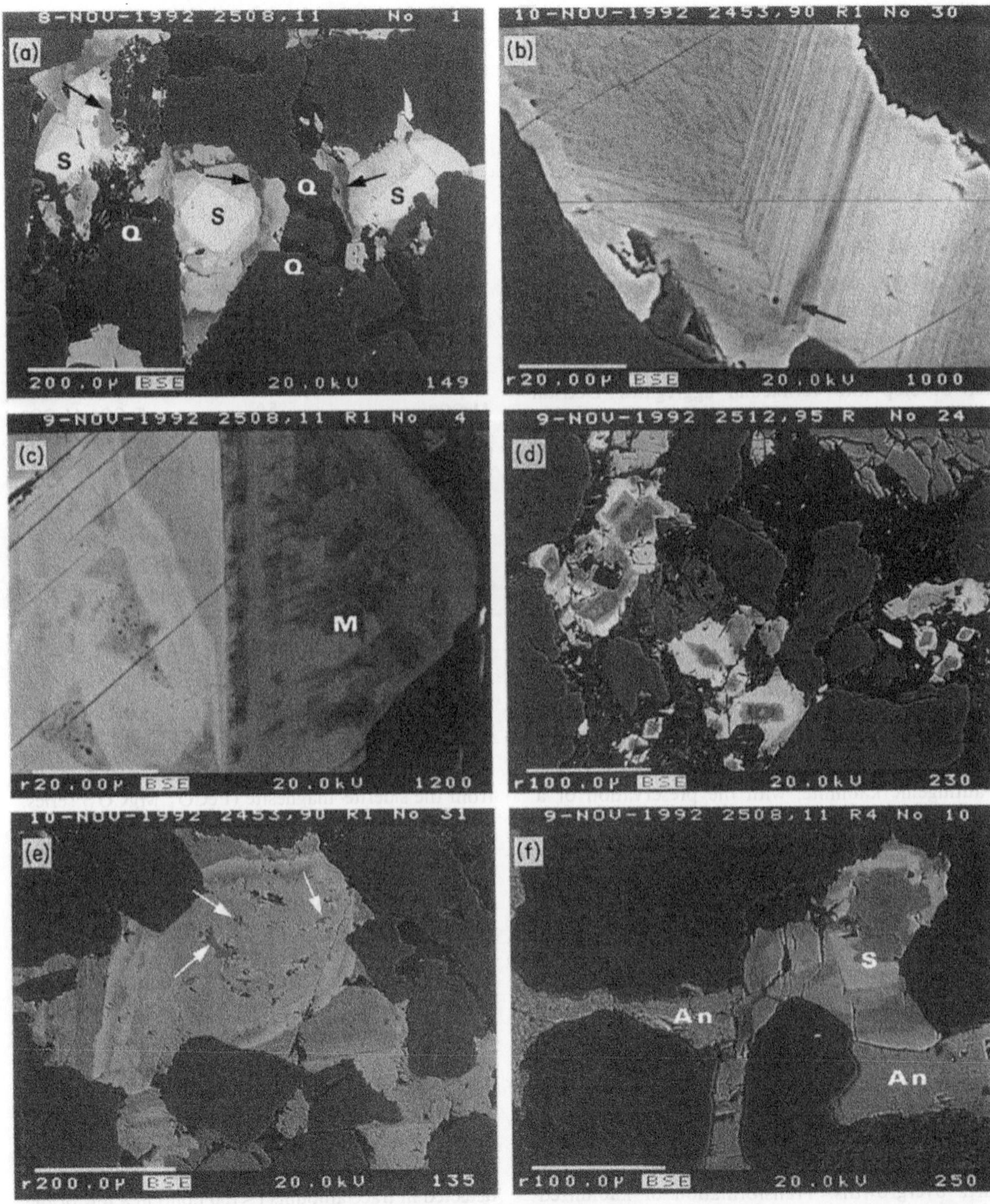

Fig. 7. (a) BSE image of pore filling magnesian siderite (S; $MgCO_3$=18–42 mol%) and ferroan magnesite (dark grey zones; arrows, $FeCO_3$=44 mol%); note the presence of abundant quartz overgrowths (Q). (b) BSE image showing the small scale zonation of siderite due to variations in the Mg for Fe substitution ($MgCO_3$=30–47 mol%); the arrow points to a ferroan magnesite zone ($FeCO_3$=40 mol%). (c) BSE image showing the dissolution of sideroplesite (light grey) and its replacement by pristomesite (dark grey). M, ferroan magnesite. (d) A BSE image showing the replacement of mud intraclasts by magnesian siderite crystals with $MgCO_3$ content that varies between 19 mol% (bright) and 40 mol% (grey). (e) A BSE image showing that siderite (light grey) filled the intergranular pores, corroded the quartz grains (dark grey), and almost completely replaced feldspar grains (see remnants; arrows); (f) A BSE image of zoned magnesian siderite (S) and anhydrite (An) filling intergranular pores. Dark grey grains are quartz.

breunnerite) and Fe-rich/Mg-poor (sideroplesite) zones. Some zones are extremely thin (1–2 μm; Fig. 7b). The breunnerite zones are both rare and very thin in all the siderite crystals. In all samples, the textural relationships suggest that the sideroplesite zones are corroded and replaced by the pistomesite (Fig. 7c).

In most samples, siderite has replaced the mud intraclasts (Fig. 7d) and grown displacively along the cleavage planes of severely dissolved and expanded micas. Thus, siderite formation seems to be closely related to the breakdown of clay minerals in the argillaceous fragments, and of micas. The replaced mud intraclasts (trace to 4%; average 1·5%; Table 1) are recognized by their oversized character as well as the presence of abundant clay minerals and silt sized quartz. These features are absent in the adjacent pore space.

Siderite partially fills the dissolution voids in detrital feldspars and slightly corrodes the grains (~1–13%; average ≈3·5%; Table 1). In some cases, the replacement of quartz and feldspar grains by siderite was pervasive, leading to the development of a texture resembling cemented oversized pores (Fig. 7e). Siderite abuts and slightly corrodes, and hence post-dates, quartz and K-feldspar overgrowths. Siderite is engulfed by, and hence pre-dates, anhydrite. However, in some sandstones, the textural relationship between siderite and anhydrite suggests precipitation at the same diagenetic stage (Fig. 7f).

The range of carbon and oxygen isotopic compositions of siderite in the Triassic sandstones is rather small: $\delta^{13}C_{PDB} = -14{\cdot}5$ to -9‰, $\delta^{18}O_{PDB} = -7{\cdot}8$ to $-2{\cdot}3$‰ (Fig. 8). To our knowledge, siderites displaying similar zonation patterns to those in the Triassic sandstones have not been published in the literature. However, two- and three-zoned magnesian siderite crystals have been reported in Cretaceous and Jurassic sandstone sequences by Mozley & Carothers (1992) and Macaulay *et al.* (1993), respectively. Purvis (1992) observed that in the Lower Permian Rotliegend sandstones, magnesian siderite crystals ($MgCO_3$ ≈5–50 mol%) have a uniform composition and show no or mottled zonation.

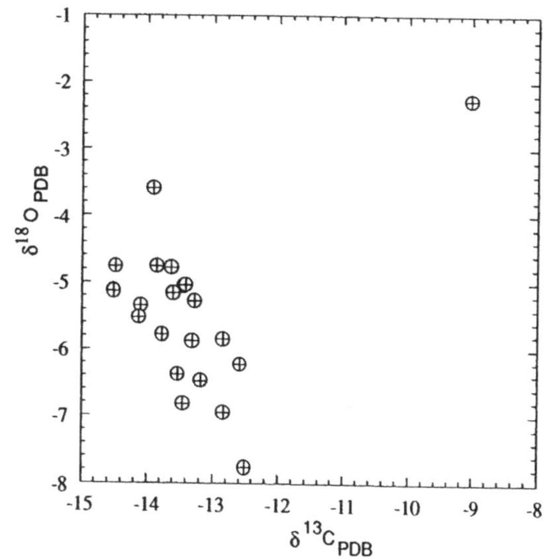

Fig. 8. The $\delta^{13}C$ versus $\delta^{18}O$ plot of siderite/magnesite.

Quartz

Quartz is the most common and often the most abundant cement (≈2–12%; average 5·3%; Table 1) in the sandstones. It occurs mainly as overgrowths (≈2–10%; average 5%) that envelop the quartz grains either partially or almost completely as continuous layers with euhedral outlines or as numerous small crystals. In some instances, the quartz overgrowths are coated with drusiform aggregates of prismatic quartz. The boundaries between the detrital quartz and the overgrowths are poorly defined, but are occasionally marked by haematite, fluid inclusions or illitic clay coatings (Fig. 9a). Microcrystalline quartz (≈3–10 μm across) occurs, together with fibrous illite, on quartz grains coated with infiltrated clay. Microcrystalline quartz (trace to 2%; average ≈0·3%), together with diagenetic chlorite, illite and K-feldspar, replaced the mud intraclasts.

K-feldspar

K-feldspar occurs (trace to 5%; average 2·5%; Table 1) in most sandstones and from all facies mainly as overgrowths (few micrometres to 100 μm in width) around detrital K-feldspars. Feldspar overgrowths are enclosed by, and thus pre-date, siderite, anhydrite and quartz cements. In some cases, the detrital feldspar cores have been dissolved partially to completely whereas the overgrowths as well as the clay coatings are preserved (Fig. 9b). The pores resulting from the dissolution of feldspars are filled with dickite, chlorite and illite. The feldspar overgrowths also reveal evidence of partial to pervasive dissolution (Fig. 9c). Microcrystalline (≈3–10 μm across), rhombohedral and prismatic K-feldspars have, together with illite and chlorite, replaced the mud intraclasts (Fig. 9d).

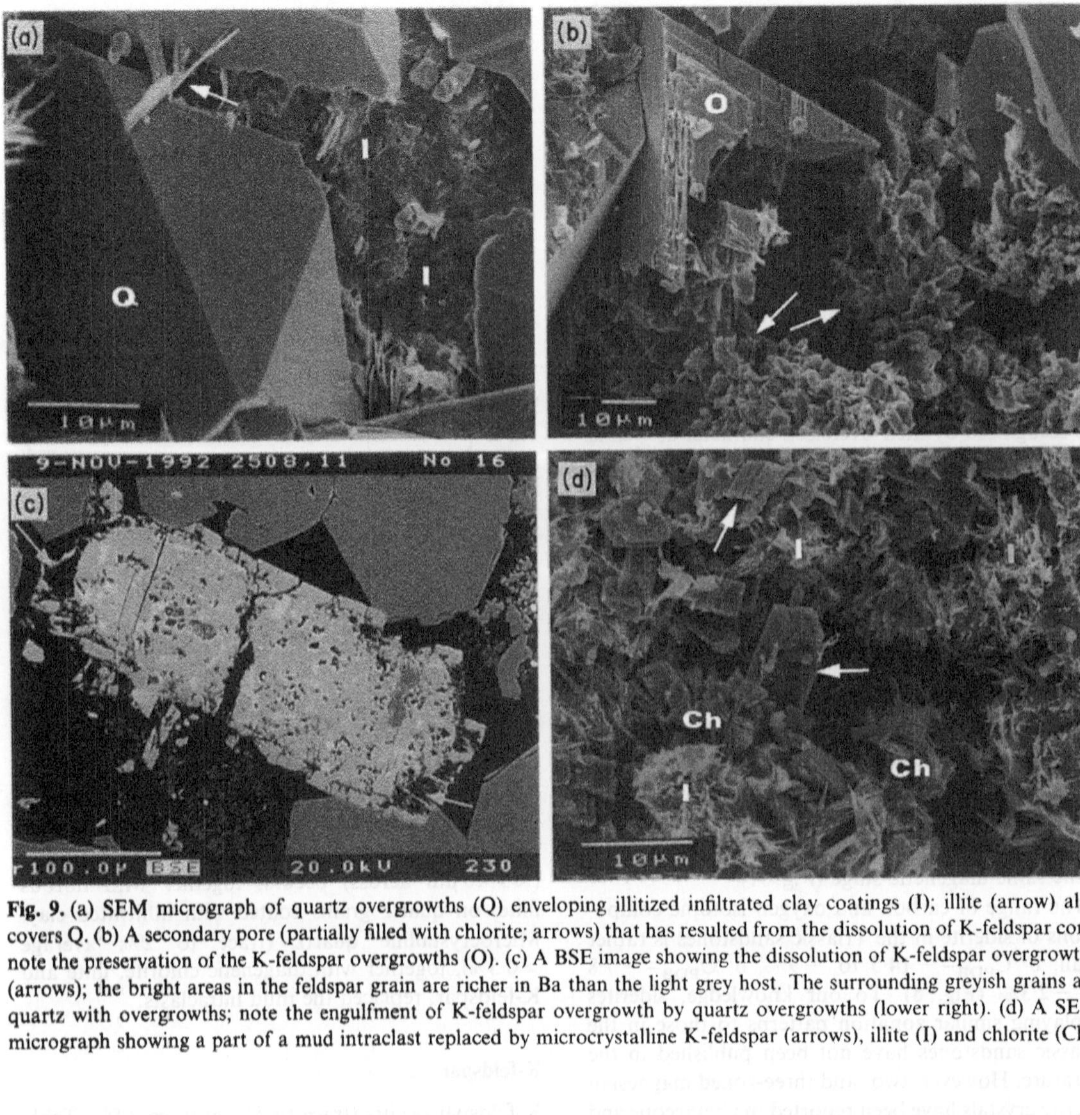

Fig. 9. (a) SEM micrograph of quartz overgrowths (Q) enveloping illitized infiltrated clay coatings (I); illite (arrow) also covers Q. (b) A secondary pore (partially filled with chlorite; arrows) that has resulted from the dissolution of K-feldspar core; note the preservation of the K-feldspar overgrowths (O). (c) A BSE image showing the dissolution of K-feldspar overgrowths (arrows); the bright areas in the feldspar grain are richer in Ba than the light grey host. The surrounding greyish grains are quartz with overgrowths; note the engulfment of K-feldspar overgrowth by quartz overgrowths (lower right). (d) A SEM micrograph showing a part of a mud intraclast replaced by microcrystalline K-feldspar (arrows), illite (I) and chlorite (Ch).

EMP analyses reveal that the authigenic feldspar is chemically pure, near stoichiometric $KAlSi_3O_8$ end-member, except that there is an excess of Al by 0·002–0·05 atoms. It should, however, be noted that this might be caused by errors in the analyses or is a consequence of assuming an ideal 8 oxygen atoms in the formula calculation. The detrital K-feldspars are often barian (BaO=0·0–2·2%) and contain variable amounts of albite solid-solution (Ab=2·5–13·6 mol%) but no anorthite (Ca).

Anhydrite

Poikilotopic anhydrite cement (0–12%; average ≈1·5%; Table 1) occurs in sandstones from all the reservoir intervals. It encloses the K-feldspar overgrowths with marginal replacement features. In any single sample, some of the anhydrite encloses considerable amounts of well developed, euhedral, quartz overgrowths whereas in other patches, quartz grains embedded in the anhydrite have developed only trace amounts of overgrowths. In the latter case, the anhydrite usually replaces the framework grains, leading to the development of cemented oversized spaces similar to what is seen with siderite (Fig. 7e). Quartz overgrowths are, however, much more abundant in areas which are uncemented by anhydrite and display a closer grain packing. This textural evidence indicates that: (a) poikilotopic anhydrite and quartz overgrowths have intermittently and alternately precipitated, and (b) precipitation of quartz overgrowths continued subsequent to the termination of anhydrite precipitation.

Infiltrated clays

Infiltrated clay mineral coatings (trace to 3%; average 1%; Table 1) comprise platelets tangentially deposited on the framework grains and show other textural features (e.g. detachment from grain surfaces) that match the criteria given by Moraes & De Ros (1990) for infiltrated clays. These coatings are texturally similar to the smectite with smooth surface texture described by Keller *et al.* (1986). However, due to its presence in trace amounts, smectite was not detected by XRD. Mechanically infiltrated clays are introduced during the near surface diagenesis of continental sediments subjected to episodic flooding under semiarid conditions (Walker *et al.*, 1978). As shown below, during burial diagenesis, the infiltrated clay coatings were subjected to partial to complete illitization and chloritization.

Illite

XRD analyses revealed that illite is the dominant clay mineral, varying in abundance between about 80 and 95% (Serrhini, 1992). Illite occurs as coatings of mat-like laths that lie parallel to the surfaces of detrital grains (Fig. 9a). These coatings were possibly derived from the partial to complete replacement of infiltrated smectite. This type of illite commonly has hair-like terminations (trace to 1%; average ≈0·2%; Table 1) that lie perpendicular to the grain surfaces (Fig. 9a). Illitized coatings are enveloped by quartz and feldspar overgrowths, and are thus very early in origin. Illite also occurs as hair- or thread-like crystals that have coated and/or intergrown with authigenic chlorite and dickite (Fig. 5f), on the surfaces of quartz overgrowths, and as replacement of authigenic kaolinite (Fig. 5e). Illite also occurs as sand sized patches with relatively well defined boundaries resulting from the replacement of compacted mud intraclasts. The amount of smectite mixed layers in the illite (Serrhini, 1992) varies between 5 and 20%. Variations in illite composition are minor. The average formula of illite calculated from EMP analyses of oxide weight% is: $K_{1.39}(Fe^{2+}{}_{0.54}Mg_{0.51}Al_{3.17})(Al_{1.17}Si_{6.83})O_{20}(OH)_4$.

Chlorite

Chlorite occurs as sand sized patches with relatively well defined boundaries resulting from the replacement of compacted mud intraclasts (Fig. 10b,c). Such a chlorite is often intergrown with variable amounts of authigenic illite and microcrystalline quartz and K-feldspar. Chlorite has also precipitated as interlocked pseudohexagonal crystals, occasionally with a rosette-like habit on top of the mat-like illite. This chlorite fills the intergranular and intragranular (after dissolved feldspars) pores and replaces the infiltrated clay coatings (Fig. 10d). XRD analysis of the clay fractions revealed trace to 5% chlorite (Serrhini, 1992). The chlorite has a small, broad 14 Å peak which was largely unaffected by glycolation, although a slight reduction in intensity was noted. Average EMP analyses of pure authigenic chlorite could not be obtained due to the intimate intergrowth with other clay minerals, K-feldspar and quartz.

Minor diagenetic minerals

Pyrite is very rare (0–0·7%; average 0·1%; Table 1), being observed only in a few samples. It occurs mainly as framboidal aggregates (4–10 μm in diameter) that lie on the surface of quartz overgrowths. Individual pyrite crystals in the framboids are 0·5–1 μm in diameter.

Microcrystalline haematite (0–0·7%; average 0·2%; Table 1) occurs beneath quartz and K-feldspar overgrowths, indicating the oxidizing conditions of the early continental meteoric diagenesis. Locally, finely crystalline haematite has recrystallized to euthedral pseudohexagonal specularite.

Ti oxides (0·3%; average 0·5%; Table 1) occur as pore lining anatase crystals that are enveloped by all other diagenetic minerals, and as an alteration product of detrital Fe–Ti oxides and biotite. In the latter case, anatase is associated with authigenic chlorite.

FORMATION WATER CHEMISTRY

The present day formation waters have a uniform composition (Table 2). They are extremely saline (TDS between 193 and 263 g l^{-1}) Na (58·5–67·5 g l^{-1})–Ca (19·4–25·3 g l^{-1})–Cl (143·9–161·7 g l^{-1}) brines and contain appreciable amounts of Mg (5567–7343 mg l^{-1}), K (1969–2430 mg l^{-1}), Sr (440–500 mg l^{-1}) and Fe (160–289 mg l^{-1}). Br concentration varies between 870 and 1217 mg l^{-1}. The

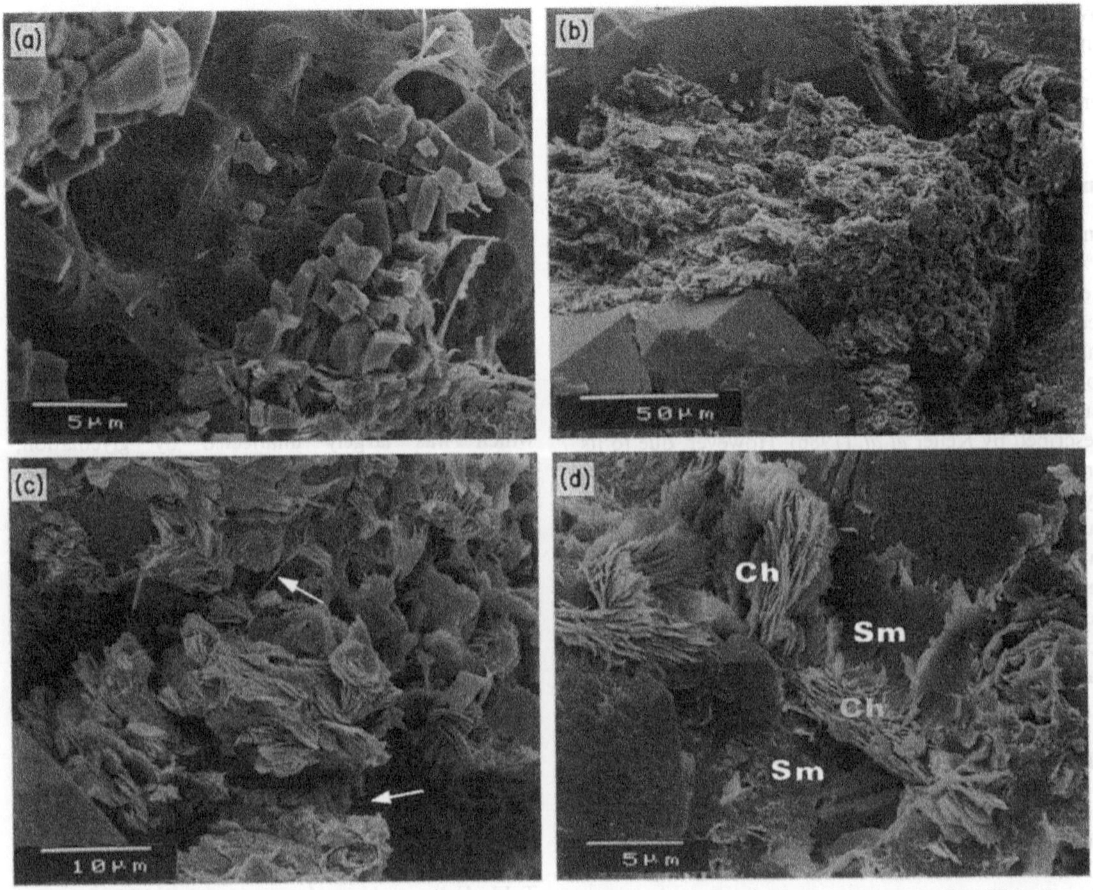

Fig. 10. SEM micrographs of: (a) mat-like illite with fibrous terminations that have resulted from the complete illitization of smectite grain coatings; (b) an altered mud intraclast; (c) Part of 10 (b) showing that the mud intraclasts have been replaced mainly by chlorite with smaller amounts of quartz (Q) and illite (arrows); (d) authigenic chlorite (Ch) that has replaced partially the infiltrated smectite coatings (Sm).

$\delta^{18}O_{SMOW}$ values of two formation waters are +0·44 and +0·65‰. The formation waters are characterized by relatively low pH (5·22–6·03) and hence low alkalinity (20–90 mg l^{-1}). The alkalinity is accounted for by mainly HCO_3^-. However, pH measured at low temperature and pressure is not representative for the *in situ* pH (Merino, 1975; Nesbitt, 1980; Reed & Spycher, 1984). This is due mainly to degassing of the water sample during collection and the temperature dependence of dissociation of weak acids and bases. The pH calculated based on the approach of Nesbitt (1980), which assumes that pH is controlled by the carbonate alkalinity and calcite solubility, varies between 4·38 and 5·57. The formation water samples contain small amounts of aliphatic acids, dominated by acetate (0·5–10 mg l^{-1}) and formate (0·5–14 mg l^{-1}), and traces of propionate and butirate.

DISCUSSION

Origin and evolution of formation waters

The Br^- versus Cl^- and the Br^- versus Na^+ values of the formation waters plot close to sea water evaporation lines, suggesting that these brines originally evolved by sea water evaporation (cf. Carpenter, 1978; Holser, 1979; Moore, 1983; Stoessell & Moore, 1983). As the studied sandstones are of fluvial origin, these brines have been derived from an external source displacing and/or mixing with the meteoric waters. The brines may have been

Table 2. Ionic concentrations (mg l^{-1}), temperature (°C), pressure (kg cm^{-2}) and measured pH of formation waters from the Triassic reservoirs of the El Borma Field.

Well	62	43	84	127	56	36
Na	61 838	62 694	59 417	59 071	59 121	58 657
K	2036	1969	2092	2029	2073	1996
SiO_2	27	20	27	26	23	22
Fe	200	193	186	182	160	179
Ca	21 033	21 048	19 400	19 580	20 885	19 361
Sr	500	478	452	448	466	467
Mg	6117	6934	5567	6004	6107	5934
Ba	9·5	11·0	10·0	6·0	6·0	8·3
Li	14	21	14	12	11	12
Br	1149	1161	1046	1127	1139	1132
Cl	155 308	155 095	147 334	144 859	144 217	143 932
SO_4	299	288	333	327	322	355
NH_4	129	134	118	145	145	146
HCO_3	n.d.	n.d.	n.d.	<10	<10	<25
H_3BO_3	205	152	195	233	230	233
Temperature	84	83	83	86	80	82
Pressure	193·7	213·6	235·0	211·8	243·1	243·1
pH	5·48	6·00	5·94	5·70	5·72	5·81

injected along faults from the underlying Silurian and Devonian marine dolomitic mudstones (see Fig. 3).

However, due to sediment–water interactions, several of the ions have either higher or lower concentrations relative to the sea water evaporation lines. For instance, the formation waters are depleted in SO_4^{2-} (212–346 mg l^{-1}), Mg^{2+} and K^+ relative to the seawater evaporation lines, as revealed from their plot versus Cl^- and Br^-. Depletion of these ions is due to the precipitation of anhydrite+minor pyrite (SO_4^{2-}), siderite+chlorite (Mg^{2+}) and illite+K-feldspar (K^+). Depletion in Mg^{2+} has probably also occurred due to dolomitization in the underlying Silurian and Devonian rocks. The Na^+ versus Cl^- values plot above the seawater evaporation line, indicating some dissolution of halite (cf. Kumar & Martinez, 1981).

Conversely, the formation waters are enriched in Ca^{2+} and Sr^{2+} relative to the normal seawater evaporation lines, probably due to the dolomitization in the Silurian and Devonian formations. No Sr was detected in the detrital K-feldspars by means of the EMP analyses. Detrital plagioclase is rare in the studied rocks and in the underlying formations and hence unimportant as a control on Ca^{2+} concentration. Albitization of plagioclase has been considered by several authors as a master control on the evolution of Ca-rich brine (Fisher & Boles, 1990; Land & Macpherson, 1992).

Paragenetic sequence

A paragenetic sequence has been constructed as is shown in Fig. 11. This sequence is based on the distribution pattern of diagenetic minerals and their textural relationships. However, due to the complex diagenetic pattern a complete knowledge of the timing and exact depths cannot be achieved for all the diagenetic effects observed. The paragenetic sequence has been controlled by shallow, meteoric diagenesis and, at greater depths, by basinal brines evolved from the evaporation of seawater and modified by water–rock interactions.

The meteoric diagenesis resulted in the infiltration of clay coatings, kaolinitization of feldspars and micas, and the precipitation of Fe and Ti oxides, kaolinite and K-feldspar overgrowths. The ions needed for the formation of feldspar overgrowths were mainly derived from the alteration of detrital micas and feldspars. Later, anhydrite and quartz overgrowths precipitated episodically and alternately. The precipitation of anhydrite, in particular, marked the onset of the influence of basinal brines on mineral diagenesis.

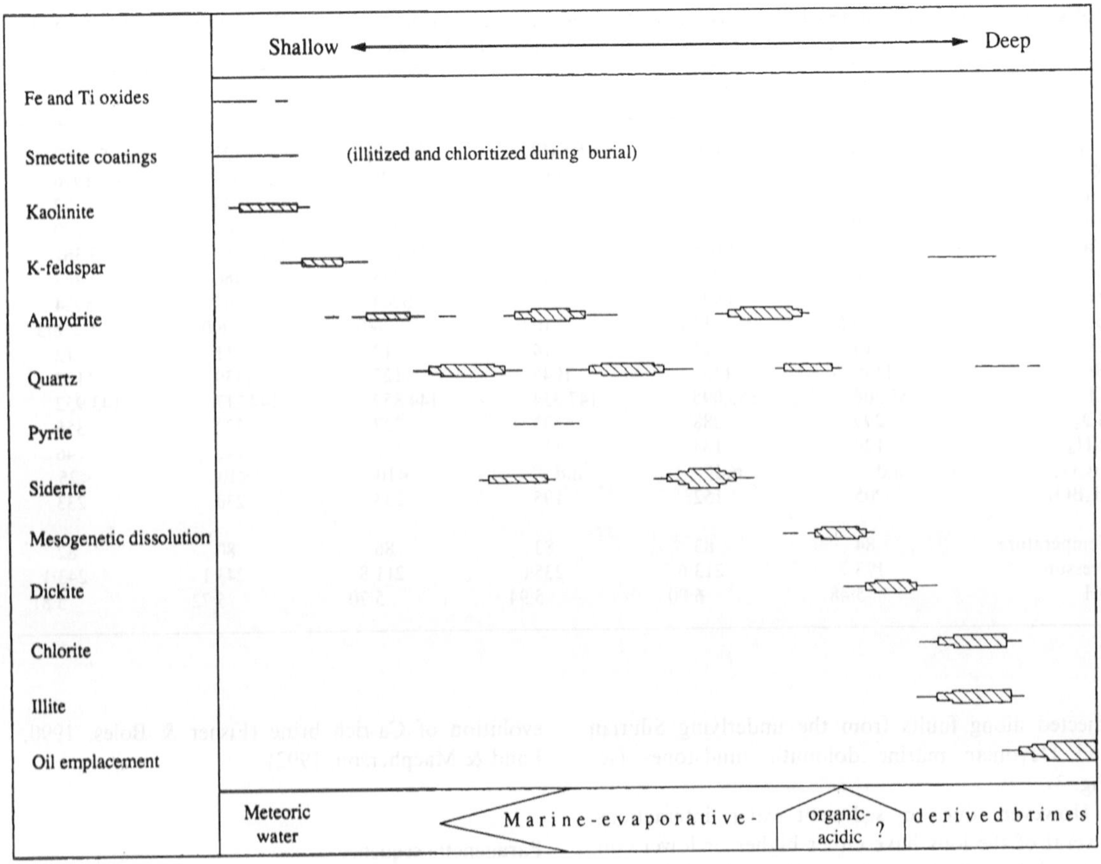

Fig. 11. Sequence of diagenesis in the Triassic sandstones and its possible relationship to changes in formation water chemistry.

Subsequently, the influence of basinal brines increased with increasing burial depths, and resulted in the precipitation of quartz overgrowths, siderite and anhydrite. As stylolites and intergranular pressure dissolution are rare in the Triassic sandstones, silica must have been derived from the brines and, internally, from the replacement of detrital quartz, rock fragments and feldspars by anhydrite and siderite.

The following diagenetic event was a slight dissolution of framework grains and cements. Subsequent to, or simultaneous with, this dissolution event, the precipitation of dickite and pervasive transformation of kaolinite into dickite took place. Upon further burial, the precipitation of microcrystalline K-feldspar, quartz, chlorite and illite occurred. The infiltrated clay mineral (smectite) coatings were illitized and chloritized whereas the mud intraclasts were replaced by illite, chlorite, microcrystalline quartz and K-feldspar. Therefore, it appears that many of the ions needed for the formation of deep burial (≤3·1 km; Fig. 4) minerals have been derived locally from within the sandstones (e.g. mud intraclasts, kaolinite, detrital feldspars and infiltrated clay coatings).

Clay mineral stability

Transformation of kaolinite into dickite occurs through a dissolution–precipitation process, as evidenced by the severe etching of the pseudohexagonal kaolinite remnants (Fig. 5c,d; cf. Ehrenberg *et al.*, 1993). Partial transformation of kaolinite into dickite did not disrupt the original vermicular stacking patterns of the kaolinite, yet the individual dickite crystals are blocky and considerably thicker.

The exact temperature of the kaolinite to dickite transformation is unknown. However, in North-Sea reservoir sandstones, transformation of kaolinite into dickite becomes prevalent at temperatures of about 80–130°C (Ehrenberg *et al.*, 1993; McAulay *et al.*, 1993, 1994). Tilley & Longstaffe (1989) concluded that diagenetic dickite in Lower Cretaceous sandstones of the Alberta Deep Basin formed at 150°C. The maximum burial of the Triassic sandstones was ≈3100 m (Fig. 4). Assuming the present-day geothermal gradient of 25–27°C km^{-1}, the maximum temperature which affected the sandstones is ≈100°C. This would mean that the transformation of kaolinite into dickite occurs at a temperature range similar to that estimated for the Brent Group reservoirs of the North Sea by McAulay *et al.* (1993, 1994).

If transformation of kaolinite into dickite occurred while the formation waters were still enriched in Mg and Fe (following the precipitation of siderite and magnesite), then the question that might arise is: why did kaolinite transform into dickite rather than chlorite? Although we have no proper answer to this question, transformation of kaolinite into dickite is probably related to an increase in a_{H^+} relative to $a_{Mg^{2+}}$ and $a_{Fe^{2+}}$. This chemical change in pore water chemistry might also have caused the burial dissolution of feldspars, siderite and anhydrite. Acidic fluids could have been derived from the maturation of organic matter in the source rocks (cf. Surdam *et al.*, 1984). If this is true, then there is probably a relationship between the timing of kaolinite to dickite transformation and burial porosity enhancement due to dissolution of framework grains and cement in reservoir sandstones. McAulay *et al.* (1993, 1994) observed that the transformation of kaolinite into dickite in the Brent Group occurs simultaneously with feldspar dissolution.

The present day formation waters fall close to the stability boundaries of illite–chlorite–K-feldspar in a_{K^+}/a_{H^+} versus $a_{Mg^{2+}}/a^2_{H^+}$ (Fig. 12), which is in agreement with our petrographical/mineralogical data on the late deep burial minerals. Although illite has crystallized on and between the dickite crystals (Fig. 5f), there is no conclusive textural evidence indicating a replacive relationship. However, flaky and fibrous illites have been preferably grown on the surfaces of etched kaolinite and/or occupy textural sites of kaolinite remnants in vermicular stacked kaolinite/dickite (Fig. 5e). These relationships, which are in accordance with the results of Ehrenberg *et al.* (1993), suggest that illite has preferentially replaced kaolinite whereas dickite is unaffected.

The greater stability of dickite is probably related to its better ordered crystal lattice compared with

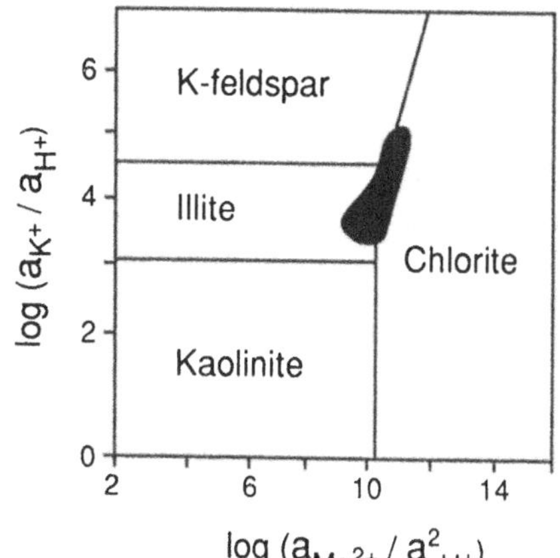

Fig. 12. Activity diagram (80°C, 300 bars; based on data from Helgeson *et al.*, 1978) showing the stability relationships between kaolinite, illite, chlorite and K-feldspar, and assuming quartz saturation. The formation waters plot along the stability fields of illite–chlorite–K-feldspar.

kaolinite. As discussed by Brindley (1980), in kaolinite, the vacant octahedral site occurs in the same relative position in each succeeding layer whereas in dickite these sites alternate between positions from one layer to the next. Kaolinite thus has one 1:1 layer per unit cell, while two layers are necessary to define the unit cell of dickite. Kaolinite has triclinic symmetry; the *a*/3 displacement of each succeeding 1:1 layer results in $b = 104{\cdot}8°$, and minor displacement along *b* results in a slight departure from 90° in *a*. Dickite is monoclinic; *b* is the same as for kaolinite, but there is no displacement along *b*. Kaolinite and dickite can display varying degrees of disorder resulting from random displacement of successive layers by 1/3*b*. In kaolinite the degree of such disorder can be extreme, but in dickite the degree of disorder is limited.

Conditions of carbonate authigenesis

The high Mg content of siderite is related to the high Mg concentration in the formation waters. The very low Ca content of the siderites (Fig. 6) is due to : (a) the preferable incorporation of Ca in anhydrite, and/or (b) the difference in ionic radius of Ca from Fe, which limits its substitution compared with Mg. Substitution of Ca for Fe in siderite is limited to 10–15 mol% $CaCO_3$ (Deer *et al.*, 1966). Mozley (1989) concluded that considerable substitution by $CaCO_3$ (up to 15 mol%) is characteristic for early diagenetic, marine siderites.

The stable isotopic data obtained are for bulk carbonate crystals and it is uncertain whether the strong fluctuations in the concentrations of Mg and Fe were accompanied by significant shifts in the isotopic composition of the formation waters too. The relatively late timing of precipitation (Fig. 11) coupled with the low $\delta^{13}C$ values of siderites indicate that dissolved carbon was derived mainly from the thermal decarboxylation of organic matter. Organic matter occurs in the interbedded mudstones and in the source rocks. The presence of only trace amounts or, more commonly, complete lack of pyrite in the rocks suggests that no significant amounts of dissolved carbon were supplied by sulphate reduction reactions, and that siderite precipitated under dominantly reducing non-sulphidic conditions.

Based on the fractionation equation of siderite–water given by Carothers *et al.* (1988), the $\delta^{18}O$ values indicate precipitation at temperatures between 42 and 74°C (mostly 55–60°C), assuming the $\delta^{18}O_{SMOW}$ values of present day formation waters (average +0·55‰). However, one might argue that at the time of siderite precipitation (shallower than dickite), the formation waters were probably influenced by meteoric fluids and thus had lower $\delta^{18}O$ values than present. This would mean that siderite has been precipitated at lower temperatures than those estimated above, which is unlikely when the relatively late timing of siderite formation is considered. Furthermore, siderites with the greatest $\delta^{18}O$ values are those displaying high intergranular volume and enclose minor amounts of quartz overgrowths, compared with siderites that have low $\delta^{18}O$ values. This confirms the role of temperature, rather than mixing with meteoric waters, in the $\delta^{18}O$ signatures of the siderite. Mozley & Carothers (1992) and Macaulay *et al.* (1993) observed that highly magnesian siderite in Cretaceous and Jurassic reservoir sandstones form at 77–86°C and 90°C, respectively.

The precipitation temperatures above were calculated assuming a fractionation relationship for pure siderite. However, due to the small size difference between Fe^{2+} and Mg^{2+}, the shift in fractionation related to the considerable $MgCO_3$ solid solution should be minor (Mozley & Carothers, 1992).

CONCLUSIONS

Mineral diagenesis during successive burial of the Triassic reservoir sandstones (bottom hole temperature ≈79–86°C; maximum burial temperature ≈100°C) of the El Borma oilfield has been accomplished as follows.

1. Shallow, early meteoric diagenesis that resulted in the infiltration of smectite clay coatings and the precipitation of Fe and Ti oxides, kaolinite and K-feldspar overgrowths.
2. During subsequent burial, the sediments were becoming influenced by basinal brines derived from the underlying Silurian and Devonian marine dolomitic mudstones. These brines were originally seawater which later evolved through evaporation and mineral interaction. At this stage the precipitation of quartz overgrowths, siderite and anhydrite occurred both episodically and alternately. However, the precipitation of quartz overgrowths, which continued after the termination of anhydrite precipitation, was followed by the precipitation of highly magnesian siderite (sideroplesite and pistomesite) and highly ferroan megnesite (breunnerite). The temperature of precipitation (based on $\delta^{18}O$ of siderite and of present day formation waters) varied between 42 and 74°C (dominantly 55–60°C).
3. During deep burial diagenesis (≤3·1 km; ≈100°C), the precipitation of authigenic dickite and pervasive to complete transformation of the shallow meteoric kaolinite into dickite occurred. Subsequently, the precipitation of microcrystalline quartz and K-feldspars, illite and chlorite, as well as illitization/chloritization of infiltrated clay coatings (smectite), kaolinite remnants and mud intraclasts occurred. Finally, the oil was emplaced.

Despite the fact that the present day formation waters fall close to the stability boundaries of illite–chlorite–K-feldspar, dickite has not been replaced by any of these minerals. Dickite is kinetically more stable than kaolinite, due to its more ordered crystal structure.

ACKNOWLEDGMENTS

We thank Hans Harryson for the help with the EMP analyses, P. K. Egeberg for calculating the ionic activities, Christina Wernström for drafting the figures, Christer Bäck and Bertel Giös for the photographic work, and H. Gaaya and M. Dridi for technical assistance. This research was funded by the Swedish Natural Science Research Council, NFR (to S.M.), the University of Tunis (to H.B.I.), the Canadian Natural Sciences and Engineering Research Council, NSERC (to I.S.A.), and Brazilian

National Research Council, CNPq (grant 200465/92.9-GL to L.F.D.R.). Constructive reviews by S. D. Burley, S. Ehrenberg and G. McAulay are gratefully acknowledged.

REFERENCES

AL-AASM, I.S., TAYLOR, B.E. & SOUTH, B. (1990) Stable isotope analysis of multiple carbonate samples using selective acid extraction. *Chem. Geol.*, **80**, 119–125.

BAILEY, S.W. (1980) Structures of Layer Silicates. In: *Clay Minerals and their X-ray Diffraction* (Ed. by G. W. Brindley and G. Brown), *Min. Soc. Monograph*, **5**, 1–124.

BEN ISMAIL, H.H. (1991) *Les Bassins Mesozóiques du Sud de la Tunisie: Stratigraphie Integreé, Caracteristiques Géophysiques et Evolution Geodynamique*. Thése es-Sciences, Université de Tunis.

BEN ISMAIL, H.H. (1992) Sequence stratigraphy and well log analysis in the Triassic sandstones of southern Tunisia (El Borma oil field). *29th Intern. Geol. Congr., Kyoto, Japan, Abs. Vol.*, p. 91.

BRINDLEY, G.W. (1980) Order-disorder in clay mineral structures. In: *Crystal Structure of Clay Minerals and Their Identification* (Ed. by G. W. Brindley and G. Brown), pp. 125–195. Mineralogical Society, London.

CAROTHERS, W.W., ADAMI, L.H. & ROSENBAUER, R.J. (1988) Experimental oxygen isotope fractionation between siderite–water and phosphoric acid liberated CO_2–siderite. *Geochim. Cosmochim. Acta*, **52**, 2445–2450.

CARPENTER, A.B. (1978) Origin and chemical evolution of brines in sedimentary basins. In: *Industrial Minerals Forum* (Ed. by K. S. Johnson and J. R. Russell), *Oklahoma Geological Survey Circular*, **79**, 60–77.

DEER, W.A., HOWIE, R.A. & ZUSSMAN, J. (1962) *Rock-Forming Minerals*, Vol. 5, *Non-Silicates*. Longmans, London.

DE ROS, L.F., ANJOS, S.M. & MORAD, S. (1994) Authigenesis of amphibole and its relationship to the diagenetic evolution of Lower Cretaceous sandstones of the Potiguar rift basin, northeastern Brazil. *Sediment. Geol.*, **88**, 253–266.

EGEBERG, P.K. & AAGAARD, P. (1989) Origin and evolution of formation waters from oilfields on the Norwegian Shelf. *Appl. Geochem.*, **4**, 131–142.

EHRENBERG, S.N., AAGAARD, P., WILSON, M.J., FRASER, A.R. & DUTHIE, D.M.L. (1993) Depth-dependent transformation of kaolinite to dickite in sandstones of the Norwegian Continental shelf. *Clay Miner.*, **28**, 325–352.

FISCHER, J. B. & BOLES, J. R. (1990) Water–rock interaction in Tertiary sandstones, San Joaquin California, USA. *Chem. Geol.*, **82**, 83–101.

GALLOWAY, W.E. (1984) Hydrogeologic regimes of sandstone diagenesis. In: *Clastic Diagenesis* (Ed. by D. A. McDonald and R. C. Surdam), *Mem. Am. Ass. petrol. Geol.*, **37**, 3–13.

GHENIMA, R. (1993) *Etude des Roches Méres Paléozoiques du Bassin de Ghedames. Modélisation de la Migration des Hydrocarbures et Application á l'Étude du Gisement d'El Borma*. Thése Doctorat, Institut Français du Pétrole, Université d'Orleans.

HANOR, J.S. (1987) Origin and migration of subsurface sedimentary brines. *Soc. Econ. Paleont. Mineral., Lecture Notes for Short Course, No. 21*.

HELGESON, H.C., DELANY, J.M., NESBITT, H.W. & BIRD, D.K. (1978) Summary and critique of the thermodynamic properties of rock-forming minerals. *Am. J. Sci.*, **278A**.

HOLSER, W.T. (1979) Trace elements and isotopes in evaporites. In: *Marine Minerals* (Ed. by R. G. Burns), *Reviews in Mineralogy*, Vol. 6, pp. 295–346. Mineral. Soc. Am., Washington.

KELLER, W.D., REYNOLDS, R.C. & INOUE, A. (1986) Morphology of clay minerals in the smectite-to-illite conversion series by scanning electron microscopy. *Clays Clay Miner.*, **34**, 187–197.

LAND, L.S. & MACPHERSON, G.L. (1992) Origin of saline formation waters, Cenozoic section, Gulf of Mexico sedimentary basin. *Bull. Am. Ass. petrol. Geol.*, **76**, 1344–1362.

KUMAR, M.B. & MARTINEZ, J.D. (1981) Character of brines from the Belle Isle and Weeks Island salt mines, Louisiana, U.S.A. *J. Hydrol.*, **54**, 107–140.

MACAULAY, C.I., HASZELDINE, R.S. & FALLICK, A.E. (1993) Distribution, chemistry, isotopic composition and origin of diagenetic carbonates: Magnus Sandstone, North Sea. *J. sedim. Petrol.*, **63**, 33–43.

MCAULAY, G.E., BURLEY, S.D., FALLICK, A.E. & KUSZNIR, N.J. (1994) Palaeohydrodynamic fluid flow regimes during diagenesis of the Brent Group in the Hutton-NW Hutton reservoirs: constraints from oxygen isotope studies of authigenic kaolin and reverse flexural modelling. *Clay Miner.*, in press.

MCAULAY, G.E., BURLEY, S.D. & JOHNES, L.H. (1993) Silicate mineral authigensis in the Hutton and NW Hutton fields: implications for subsurface porosity development. In: *Petroleum Geology of Northwest Europe* (Ed. by J. R. Parker), pp. 1377–1345. The Geological Society, London.

MCBRIDE, E.F. (1963) A classification of common sandstones. *J. sedim. Petrol.*, **33**, 664–668.

MERINO, E. (1975) Diagenesis in Tertiary sandstones from Kettleman North Dom, California-II. Interstitial solution: distribution of aqueous species at 100°C and chemical relation to diagenetic mineralogy. *Geochim. Cosmochim. Acta*, **39**, 1629–1645.

MORAD, S., MARFIL, R., AL-AASM, I.S. & GOMEZ-GRAS, D. (1992) The role of mixing-zone dolomitization in sandstone cementation: evidence from the Triassic Buntsandstein, the Iberian Range, Spain. *Sediment. Geol.*, **80**, 53–65.

MORAES, M.A.S. & DE ROS, L.F. (1990) Infiltrated clays in fluvial Jurassic sandstones of Reconcavo Basin, northeastern Brazil. *J. sedim. Petrol.*, **60**, 809–819.

MOORE, C.H. (1983) Chemical constraints and origins of four groups of Gulf Coast reservoir fluids. *Bull. Am. Ass. petrol. Geol.*, **67**, 869–906.

MOZLEY, P. (1989) Relationship between depositional environment and the elemental composition of early diagenetic siderite. *Geology*, **17**, 704–706.

MOZLEY, P. & CAROTHERS, W. W. (1992) Elemental and isotopic composition of siderite in the Kuparuk Formation, Alaska: effect of microbial activity and water/sediment interaction on early pore-water chemistry. *J. sedim. Petrol.*, **62**, 681–692.

Nesbitt, H.W. (1980) Characterization of mineral-formation water interactions in Carboniferous sandstones and shales of the illinois sedimentary basin. *Am. J. Sci.*, **280,** 607–630.

Purvis, K. (1992) Lower Permian Rotliegend sandstones, south North Sea: a case study of sanstone diagenesis in evaporite-associated sequences. *Sediment. Geol.*, **77,** 155–171.

Reed, M. & Spycher, N. (1984) Calculation of pH and mineral equilibrium in hydrothermal waters with application to the geothermometry and studies of boiling and dilution. *Geochim. Cosmochim. Acta,* **48,** 1479–1492.

Rosenbaum, J.M. & Shepard, S.M.F. (1986) An isotopic study of siderites, dolomites and ankerites at high temperatures. *Geochim. Cosmochim. Acta,* **50,** 1147–1150.

Schmidt, V. & McDonald, D.A. (1979) The role of secondary porosity generation in the course of sandstone diagenesis. In: *Aspects of Diagenesis* (Ed. by P. A. Scholle and P. R. Schluger), *Spec. Publ. Soc. econ. Paleont. Miner.*, **26,** 175–207.

Serrhini, N-E. (1992) Etude Sedimentologique et Diagenetique des Reservoirs Greseux du Trias dans les Sondages EB.70, EB.73 et EB.139 – Champ d'El Borma – (Tunisie Meridionale). *Memoire du Laboratoire de Geologie du Petrole et Bassins Sedimentaires no. 17*, Univ. Tunis II.

Stoessell, R.K. & Moore, C.H. (1983) Chemical constraints and origin of four groups of Gulf Coast reservoir fluids. *Bull. Am. Ass. petrol., Geol.*, **69,** 122–126.

Surdam, R.C., Boese, S.W. & Crossey, L.J. (1984) The chemistry of secondary porosity. In: *Clastic Diagenesis* (Ed. by D. A. Mcdonald and R. C. Surdam), *Mem. Am. Ass. petrol. Geol.*, **37,** 127–149.

Tilley, B.J. & Longstaffe, F.J. (1989) Diagenesis and isotopic evolution of porewaters in the Alberta Deep Basin: The Falher Member and Cadomin Formation. *Geochim. Cosmochim. Acta,* **53,** 2529–2546.

Walker, T.R., Waugh, B. & Crone, A.J. (1978) Diagenesis in first-cycle desert alluvium of Cenozoic age, southwestern United States and northwestern Mexico. *Bull. geol. Soc. Am.*, **89,** 19–32.

Reprinted from Sedimentology (1985) **32**, *833–853*

The petrology and diagenesis of Middle Jurassic clastic sediments, Ravenscar Group, Yorkshire

J. D. KANTOROWICZ*

Geology Department, University of Hull, Hull HU6 7RX, U.K.

ABSTRACT

The petrology of non-marine clastic sediments from the Middle Jurassic Ravenscar Group was investigated to identify and to assess the significance of possible controls on diagenesis. Diagenetic modifications took place in three broad regimes: within the depositional groundwater (eogenesis), during burial (mesogenesis), and during uplift and erosion (telogenesis).

Eogenesis involved the initial interaction of the original sedimentary assemblage with its depositional pore waters. Eogenetic processes were influenced strongly by bacterial degradation of organic matter present in the finer grained sediments. This bacterial activity caused a lowering of pore water pH, and subsequently reduced Eh. Feldspar dissolution and alteration of muscovite to kaolinite occurred. Quartz overgrowths and vermiform kaolinite precipitated in sands with oxygenated and mildly acidic pore waters, whereas chlorite and quartz overgrowths formed in sands with anoxic and neutral pore waters. This caused complete reduction of porosity in both finer-grained and texturally less-mature floodplain-facies sandstones. Channel-facies sandstones were cemented only with a thin veneer of quartz overgrowths, creating a rigid but still porous quartzose framework. Bacterial ferric-iron reduction throughout the water table subsequently established more ubiquitous conditions, raised pH, and caused rhombic siderite cementation within the remaining porosity.

Mesogenetic modifications, of the mineralogy formed at the surface, occurred when pore water composition changed during burial. Bicarbonate saturated solutions migrated through the surviving porosity, precipitating replacive, poikilotopic ferroan calcite. Carbonate dissolution occurred subsequently. Carbonate dissolution was possibly caused by the same solutions which then precipitated dickite and further authigenic quartz.

Telogenetic modifications took place during uplift and exposure. Groundwater reactions and subaerial exposure probably caused further carbonate and feldspar dissolution, as well as alteration of chlorite to interstratified chlorite/vermiculite.

It is concluded that the diagenesis of these non-marine clastic sediments reflects various factors, the complex interrelationship of which precludes the identification of any single factor as wholly controlling their diagenesis. Eogenetic modifications involved the interaction of the sediments and their pore waters. As a result of the metastability of the mineral assemblage established at this time, these modifications can still be recognized, despite the sediments' subsequent history.

INTRODUCTION

In recent publications a number of controls on the diagenesis of clastic sediments have been proposed, ranging in scale from local fluctuations in groundwater chemistry to plate tectonics. They include provenance (Stalder, 1975), depositional mineralogy (Walker, Waugh & Crone, 1978), depositional-water chemistry (Almon, Fullerton & Davies, 1976), changing pore-fluid composition (Burley, 1984), burial pressures and temperatures (Hutcheon, Oldershaw & Ghent, 1980), and exposure (Sommer, 1978). Of course, different factors may influence the diagenesis of a sedimentary rock at different times during its evolution (e.g. Glennie, Mudd & Nagtegaal, 1978). Within a sedi-

* Present address: Koninklijke/Shell Exploratie en Produktie Laboratorium, Volmerlaan 6, Rijswijk ZH, The Netherlands.

mentary sequence the diagenesis of mudrocks and of the enclosed sandstones may also be related (Boles & Franks, 1979).

In order to assess these possible controls on sandstone diagenesis, clastic sediments of known age, depositional environment and burial history are investigated. This paper reports the mineralogy and petrology of non-marine sandstones from the Middle Jurassic Ravenscar Group in Yorkshire, and elucidates the nature and origin of the diagenesis of these sediments.

The Ravenscar Group crops out on the North Yorkshire Moors (Fig. 1), and has been studied by a succession of geologists including Adam Smith, Sedgwick, Phillips and Murchison. The history of sedimentological interpretations of the Yorkshire 'deltaics', as they were formerly known, is summarized by Hemingway (1974), whilst more recent research is reviewed by Livera & Leeder (1981). The lithostratigraphy of Hemingway & Knox (1973) is given in Table 1. The sequence consists of shallow-marine clastics and carbonates, overlain by, and grading laterally into beach deposits, and both tidally-influenced and wholly non-marine alluvial sequences (Livera & Leeder, 1981). During burial these Middle Jurassic strata are believed to have suffered a 'double inversion'. Initially they were buried rapidly to almost 1500 m during the late Jurassic to early Cretaceous and uplifted to less than 1000 m during the mid-Cretaceous. The sequence was reburied to more than 2000 m during the Upper Cretaceous and Tertiary before renewed uplift and erosion (Kent, 1980; Barnard & Cooper, 1983).

Earlier published research into the diagenesis of these sandstones is reviewed by Hemingway (1974). More recently, Hemingway & Riddler (1982) related various aspects of the diagenesis of the Ravenscar Group and the Dogger Formation to local differences in the burial history of the basin. Consequently, although some aspects of the sequence's diagenetic history are known, there are to date no comprehensive, integrated studies to compare with the present work. The authigenic clay minerals of the Ravenscar Group have been described by Kantorowicz (1984). Siderite concretions and soil spheruliths in the Long Nab Member are described by Kantorowicz (1985).

TERMINOLOGY

Schemes for textural maturity and mineralogical classification of sandstones are modified from those proposed by Folk (1968) and McBride (1963) respectively. The terminology and classification used in the description and interpretation of porosity follows the schemes developed by Pittman (1979) and Schmidt & McDonald (1979b). Fine grained sediments are uniformly categorized as mudrocks (Spears, 1980). Clay mineral nomenclature follows AIPEA recom-

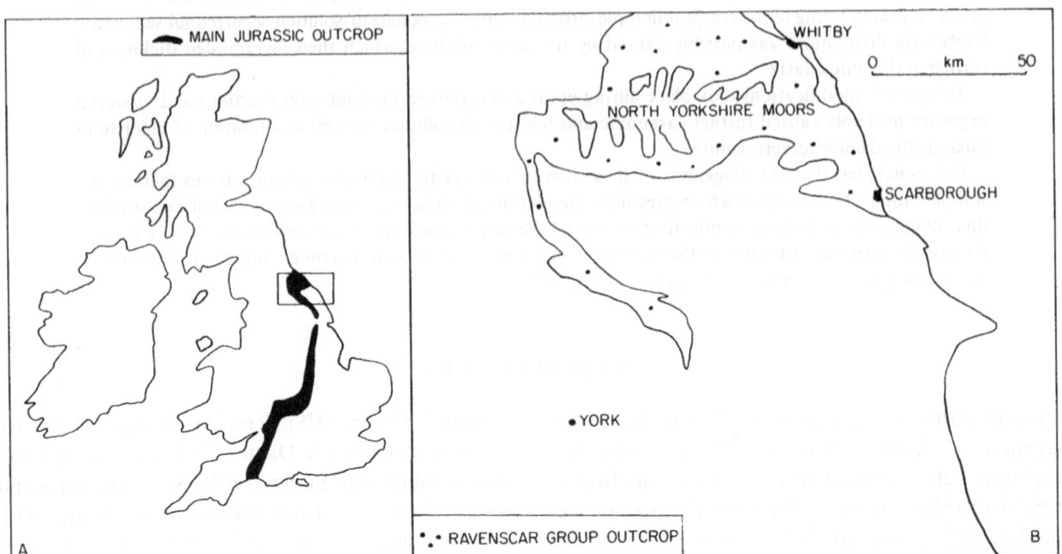

Fig. 1. (A) Jurassic outcrops of England with (B) inset. (B) Locality map illustrating the outcrop of the Ravenscar Group in Yorkshire.

Table 1. The lithostratigraphy of the Middle Jurassic Ravenscar Group on the Yorkshire Coast (modified after Hemingway & Knox, 1973)

Scalby Formation	Long Nab Member	
	Moor Grit Member	
Scarborough Formation (m)		
Cloughton Formation	Gristhorpe Member	
	Lebberston Member	Yons Nab Beds
		Millepore Beds (m)
	Sycarham Member	
Eller Beck Formation (m)		
Saltwick Formation		

(m)—marine.

mendations (Bailey, 1980), except that the term kandite is retained to emphasize the distinction between kaolinite and dickite, two different kandite minerals which occur in these rocks.

Diagenesis is the sum of those processes whereby an originally sedimentary assemblage attempts to reach and maintain an equilibrium with its environment. Diagenetic processes may be divided into three broad realms (after Schmidt & McDonald, 1979a). Eogenesis is the sum of those processes by which sediments move towards an equilibrium with their depositional pore waters. Mesogenesis is the regime in which diagenetic processes are unaffected by the influence of surface agents in the chemistry of the interstitial waters. Telogenesis results from the secondary introduction of surface waters, and includes weathering. These terms qualify the diagenetic events observed in these sediments and impart a genetic connotation to the inferred sequence. Although these terms have not been universally adopted they allow qualification beyond a purely relative description such as early or late.

METHODOLOGY

Blue-dye vacuum impregnated thin sections were made of all sandstone samples, and were stained for carbonates (Dickson, 1965). Basic petrography and modal analyses were supplemented with X-ray diffraction (XRD) to confirm and to quantify clay mineralogical identifications. Scanning electron microscopy (SEM) was used to elucidate the three-dimensional relationship between detrital and authigenic phases and porosity. Visual observations were supplemented with qualitative elemental analysis using a Link System 860 Energy Dispersive Analyser. This routine investigation was further enhanced by selective electron microprobe analysis, to give chemical analyses of minerals identified in this section. Selective cathodoluminescence (CL) was undertaken to clarify the sequences of diagenetic minerals observed (Ramseyer, 1983).

Stable isotope ratios were measured from carbonate cements and dickite, to investigate the changing conditions during the inferred sequence of authigenic precipitation. The purity of selected samples was checked by XRD. Samples were treated to remove free iron oxides (Mehra & Jackson, 1960) and organic compounds (Forester, Sandberg & Anderson, 1973). Dickite analysis followed the methodology of Clayton & Meyeda (1963). Duplicate samples were pre-heated to both 100°C and 250°C to remove interstitial and structural water, then reacted with 100% BrF_5 at 400°C for 12 hr. The oxygen evolved was reacted with a hot graphite rod to generate carbon dioxide. The carbon dioxide from both samples and a quartz standard reacted at the same time was analysed with a Micromass 903 Mass Spectrometer. Hydrogen isotopic compositions were not measured. Calcite was reacted with 100% phosphoric acid at 25°C for 2 hr, and siderite at 55°C for 96 hr. The carbon dioxide evolved was separated from water vapour in a refrigerated acetone trap (at −95°C), and analysed with a Micromass 903 Mass Spectrometer. The resulting raw data were corrected following the methods of Craig (1957) and Deines (1970) and are reported in conventional notation.

Calcite compositions have been calculated with a fractionation factor of 1·01025 (Sharma & Clayton,

1965). Coleman (1981, pers. comm.) and Gautier (1981, pers. comm.), both question the accuracy of the published siderite fractionation factor of 1·01169 at 25°C (Clayton, in Fritz *et al.*, 1971), and extrapolation from this value to 1·01121 at 59°C (Hangari, Ahmad & Perry, 1980). Because of this the data have been corrected with a fractionation factor of 1·01025, and are comparable with published calcite data.

One carbonate cemented sample (370) contained both siderite and calcite. Gases were extracted from the sample after 2 hr at 25°C and after a further 94 hr at 55°C. It is assumed that calcite reaction was completed after 2 hr and siderite reaction had not begun.

PETROGRAPHY

Three hundred samples were collected from three sedimentological facies at outcrop on the North Yorkshire Moors: channel- and floodplain-facies sandstones and floodplain-facies mudrocks. Floodplain-facies sandstones include crevasse- splay, sheetflood and levee deposits. The majority of the samples described here were collected from the Saltwick Formation, the Gristhorpe Member and the Scalby Formation (Table 1). Locality details are compiled in Kantorowicz (1982).

Channel-facies sandstones

Channel-facies sandstones crop out as individual or interconnected channels. These sandstones have elongated, ribbon-like form and are enclosed by floodplain-facies sediments. Texturally, the channel sandstones are generally medium- to fine-grained, supermature or mature quartz arenites and subarkoses (Figs 2 and 3A, B). Modal analyses are compiled in Table 2. Further data are recorded by Kantorowicz (1982). Authigenic phases comprise 10–30% of most samples, and up to 50% were extensive replacement occurs. Volumetrically, the most significant authigenic minerals are quartz overgrowths, siderite and dickite. Calcite, interstratified vermiculite/chlorite, chlorite and kaolinite, are present locally, but are not usually observed. Feldspar overgrowths, illite, brookite, anatase and pyrite also occur but comprise less than 1% of the rock on average. Porosity averages 15% but may be as high as 30%. Grain dissolution porosity occurs throughout and usually comprises 1% of each sample on average (Fig. 4B). Consequently, it only forms a small proportion of the total porosity in the more porous sandstones (e.g. Fig. 3A, B). Here, composite or enlarged intergranular porosity with open pore throats occurs characteristically. By contrast, although grain dissolution porosity also comprises 1% of the less porous sandstones, here less intergranular pore space remains and grain dissolution porosity becomes proportionally more significant. In samples with a higher percentage of authigenic minerals, especially pore-filling clays (Fig. 6A), a larger proportion of the porosity comprises microporosity.

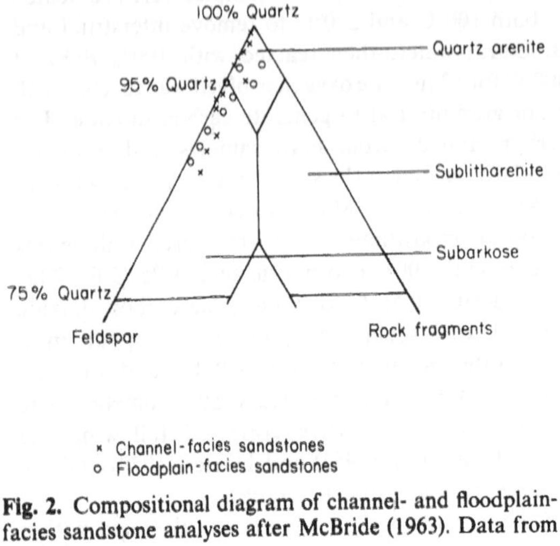

Fig. 2. Compositional diagram of channel- and floodplain-facies sandstone analyses after McBride (1963). Data from Table 2.

Syntaxial quartz overgrowths are ubiquitous as a thin cementing veneer on detrital quartz grains. They constrict and occasionally fully close pore throats. In general, they reduce rather than completely fill intergranular pore space (Fig. 3B). Quartz overgrowths may partially enclose authigenic grain-coating clays (Fig. 4D) and pore-filling dickite, but are absent where thicker clay coatings occur. The surfaces of overgrowths appear smooth at magnifications of up to at least × 500, but stepped overgrowths are also observed. CL observation sometimes reveals two episodes of overgrowth development, an inner bottle green luminescent zone, and an outer, non-luminescent zone. However, it was not possible to quantify separately the different episodes of authigenic quartz observed. Quartz overgrowths may be replaced by siderite (Fig. 6D) and calcite (Fig. 7A). Where quartz overgrowths line open pore spaces they are frequently observed to be corroded (Fig. 3C). This corrosion is recognized by small notches or pits 20–50 μm wide which disrupt crystal faces. These notches are occasionally rhomb-shaped. They are often coa-

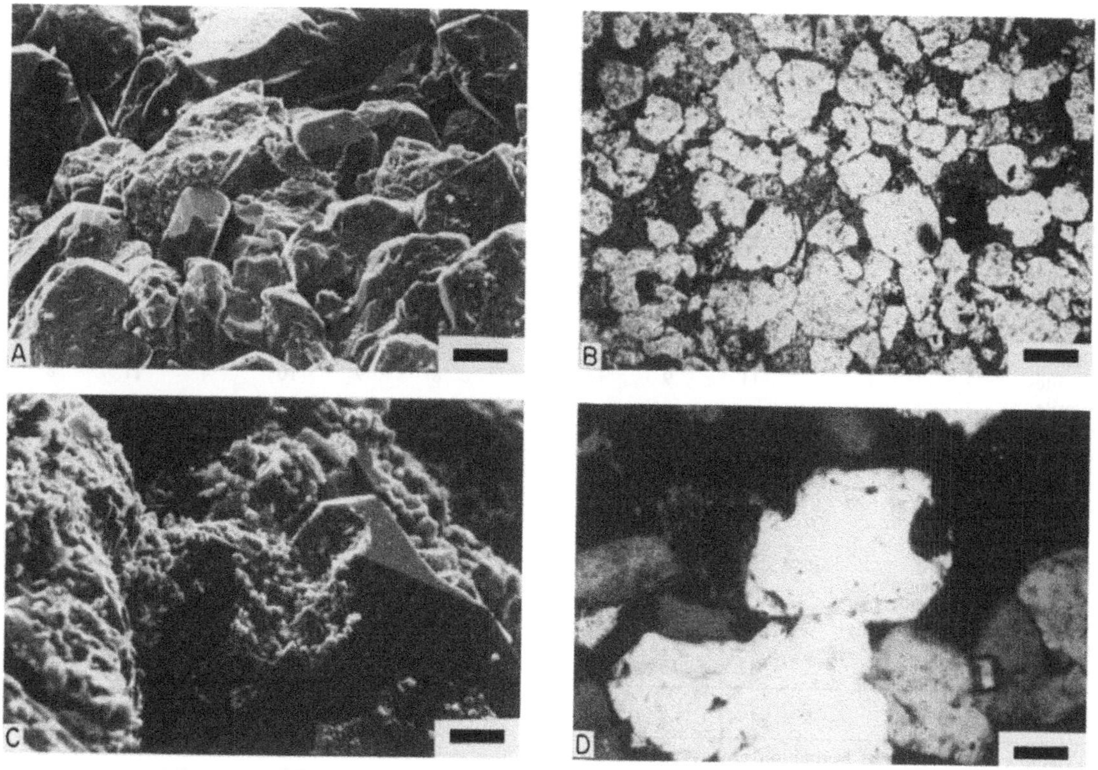

Fig. 3. (A) Scanning electron photomicrograph. Scale bar 27 μm. General view of a moderately well sorted fine to medium grained subarkose with extensive intergranular porosity. Pore throats are partially occluded by a thin veneer of quartz overgrowths which create a rigidly cemented framework. Sample 410, channel-facies sandstone, Long Nab Member, Boltby Reservoir. (B) Thin section photomicrograph. Scale bar 275 μm. Moderately well sorted fine to medium grained subarkosic channel facies sandstone with a characteristic veneer of quartz overgrowths, irregular and oversized pores, grain dissolution porosity and etched grain surfaces. The majority of grain to grain contacts are points, straight and long contacts only exist between overgrowths. Sample 606, channel-facies sandstone, Saltwick Formation, Eskdale. (C) Scanning electron photomicrograph. Scale bar 15 μm. Corroded and etched quartz overgrowth, with euhedral crystal faces and edges, abruptly intersected by a large, irregularly shaped pit. Replacement by the now removed mineral preferentially enlarged the exposed high free surface energy 'hole' in the crystal lattice rather than attacking the euhedral overgrowth surface. Sample 661, channel-facies sandstone, Saltwick Formation, Rosedale. (D) Thin section photomicrograph. Scale bar 75 μm. Medium grained quartz arenite with a moderately thick veneer of authigenic quartz. A larger and irregular shaped embayment has incised through the overgrowth and dust rim into the detrital grain. The corrosion embayment has been enlarged selectively whilst most euhedral faces remain unaffected. Sample 444, floodplain-facies sandstone, Saltwick Formation, Hayburn Wyke.

lesced into larger, irregular embayments, separated by relics of overgrowths, or are coalesced into depressions which obliterate completely euhedral crystal faces.

Sodium- and potassium-feldspar overgrowths occur on detrital sodium- and potassium-feldspar grains respectively. However, they usually comprise less than 1% of the rock. Their distribution and relationship with other authigenic phases is similar to that of quartz overgrowths. Microprobe analysis detected only stoichiometric overgrowths in these rocks (Table 3, Fig. 5) and, in spite of the common occurrence of dissolution (Fig. 4A, B), analysis revealed that the relic lamellae are also stoichiometric suggesting that feldspar dissolution is a congruent process. *In situ* replacement of detrital feldspar grains to kaolinite, dickite or illite was not observed.

Kandite minerals occur in three forms in these sediments (Kantorowicz, 1984). Kaolinite is found as an alteration product of muscovite, and as pore-filling vermiform aggregates. It is sometimes enclosed by siderite or calcite. Dickite is found as blocky aggre-

Table 2. Modal analyses of selected channel- and floodplain-facies sandstones from the Saltwick Formation

(a) Sample details: Saltwick Formation, Channel sandstones, Whitby

Sample number:	301	306	309	312	314	316	322	327	332	335
Quartz—simple	67	53	64	63	66	63	58	54	56	61
Quartz—polycrystalline	5	2	3	1	2	3	3	8	1	4
Rock fragments	—	1	1	1	—	—	—	—	—	—
Feldspars	2	4	1	6	2	3	4	3	3	4
Mica	—	—	—	1	—	—	—	—	—	—
Clay matrix	—	—	—	—	1	—	—	—	2	1
Opaque and heavy minerals	—	—	—	1	1	—	1	—	1	—
Fossils (plant debris)	—	—	—	—		—		—	—	
Illite		3	1		—	1	1			1
Kandites	11	9	14	8	8	9	10	4	11	8
Chlorite										
Quartz overgrowths	3	5	2	6	1	2	1	3	2	4
Feldspar overgrowths	—	—	—	—	—	—	—	—	—	—
Calcite									19	
Siderite (goethite)	—	6	—	2	—	4	14	12	4	3
Pyrite										
Intergranular porosity	5	13	13	11	18	13	7	14	—	11
Grain dissolution porosity	6	3	1	1	1	4	1	4	—	4
Total detritus	75	61	69	72	72	67	66	65	64	70
Total authigenics	14	23	17	16	9	16	26	19	36	16
Total porosity	11	16	14	12	19	17	8	18		15

(b) Sample details: Saltwick Formation, crevasse splay sandstones

Sample number:	317	320	331	350	358	359	552	563	573
Quartz—simple	62	71	70	59	44	59	71	70	53
Quartz—polycrystalline	2	2	1	—	2	—	1	—	2
Rock fragments	—	—	—	—	—	—	—	1	1
Feldspars	5	—	2	1	—	—	5	2	5
Mica	1	3	2	4	1	—	—	—	—
Clay matrix	—	6	10		—		—		—
Opaque and heavy minerals	—	1	1	—	—	—	1	—	—
Plant debris	—	—	—	—	—	—	—	—	—
Illite					1			4	3
Kandites	5	6	2	7	4		3	10	3
Chlorite									
Quartz overgrowths	9	3	4	7	1	14	10	6	3
Feldspar overgrowths	—	—	—	—	1	—	—	—	—
Calcite									
Siderite (goethite)	8	1	3	21	44	26	1	1	28
Pyrite	—							—	
Intergranular porosity	6	7	5	1	3	1	4	5	1
Grain dissolution porosity	2	—	—	—	—	—	—	1	—
Total detritus	71	83	86	64	47	59	79	73	61
Total authigenics	21	10	9	35	51	40	17	21	37
Total porosity	8	7	5	1	3	1	4	6	1

— denotes less than 1%.

(c) Sample details: Moor Grit Member, Bloody Beck					
Sample number:	540	544	584	587	590
Quartz—simple	76	66	70	76	76
Quartz—polycrystalline	—	2	1	1	—
Rock fragments	—	—	—	1	—
Feldspars	—	—	1	—	1
Mica					
Clay matrix					
Opaque and heavy minerals					
Illite					
Kandites	—	1	1	2	—
Chlorite	7	4	1	—	—
Quartz overgrowths	12	12	11	13	16
Feldspar overgrowths	—	—		—	
Calcite					
Siderite	1	2			
Pyrite			—		
Intergranular porosity	4	11	11	5	6
Grain dissolution porosity		2		2	
Total detritus	76	68	72	78	77
Total authigenics	20	19	13	15	16
Total porosity	4	13	11	7	6

gates irregularly distributed, plugging and filling composite enlarged-intergranular porosity (Fig. 6A, B). Dickite was not observed to be enclosed within carbonate cements.

Rhombic siderite is the commonest authigenic carbonate mineral present (Fig. 6C, D). Siderite rhombs often line pore space but total cementation is rare. The rhombs are typically 10–20 μm in length with few size variations observed. Siderite replaces both authigenic and detrital quartz and feldspar and encloses and replaces pyritized plant debris. Sphaerosiderite occurs as a pore-filling mineral which occasionally replaces quartz grains. Ferroan and non-ferroan calcite both occur as non-luminescent poikilotopic crystals which enclose, and replace, detrital and authigenic quartz (Fig. 7A), feldspar, kaolinite, altered muscovite and siderite. Carbonate cements were not observed enclosing dickite. Siderite is often oxidized to iron hydroxides.

Chlorite (Fig. 4D) and interstratified vermiculite/chlorite occur individually or together as thin- or irregularly developed grain-coatings, occasionally enclosed by quartz overgrowths. Both minerals occur associated with dickite but were not observed together with kaolinite.

A number of other authigenic minerals were observed in these sandstones. Authigenic illite occurs associated with detrital micas and, more rarely, growing on dickite. Brookite and anatase were observed in small pockets, often associated with partially dissolved detrital heavy minerals. Pyrite occurs in finer grained samples, where it is often associated with plant debris.

Floodplain-facies sandstones

These are laterally extensive, and have a sheet-like geometry, but are only rarely connected to larger sandbodies. The samples examined range from medium-grained to highly-micaceous siltstones, the majority being very fine or silty sandstones. Most are texturally mature, although finer samples are occasionally immature. The majority of the sandstones are subarkoses, with rarer quartz arenites (Table 2, Fig. 2). Porosity in floodplain-facies sandstones is generally lower than in the channels, consisting of irregularly distributed intergranular, and less significant grain-dissolution porosity. Porosity ranges from 1 to 14% with an average of 5%. Authigenic minerals comprise from 10 to 50% but on average constitute 20% of the rock (Table 2). Authigenic minerals include quartz overgrowths, dickite, kaolinite (Fig. 4C) and siderite. Potassium-feldspar overgrowths on orthoclase grains, illite, interstratified vermiculite/chlorite, pyrite, chlorite and brookite are all present occasionally.

Most sandstones are tightly cemented with one or more authigenic minerals. Variations of diagenetic modifications of these sediments may be related to

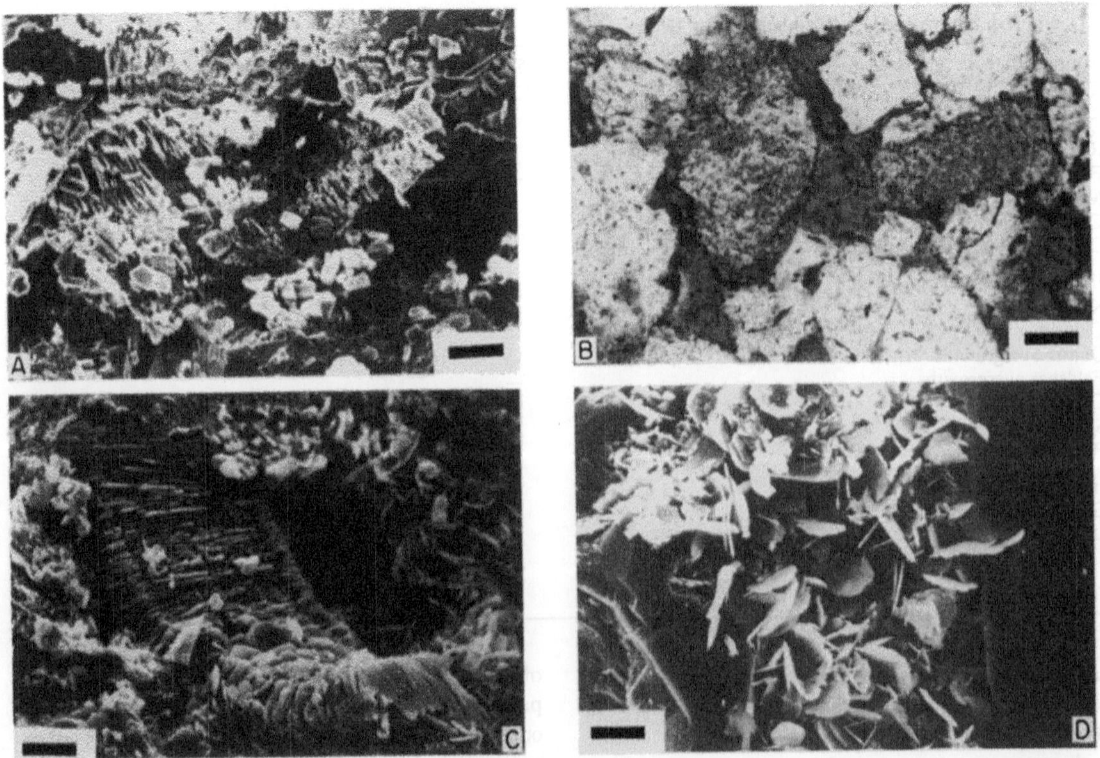

Fig. 4. (A) Scanning electron photomicrograph. Scale bar 27 μm. Partially dissolved feldspar grain. Dissolution has preferentially followed twin planes leaving a skeleton of original lamellae. In this example intergranular grain dissolution porosity is created but *in situ* alteration to kaolinite or illite is not observed. Note also that brittle fracturing of the relic grain has not occurred. Sample 567, channel-facies sandstone, Saltwick Formation, Eskdale. (B) Thin section photomicrograph. Scale bar 75 μm. Dissolution of potassium-feldspar grains. Dissolving grains exhibit a range from partial to complete grain dissolution with initial intergranular porosity developed parallel to twin lamellae, whilst only the rim of the central grain remains. Sample 606, channel-facies sandstone, Saltwick Formation, Eskdale. (C) Scanning electron photomicrograph. Scale bar 8 μm. Intergranular pore-filling kaolinite. The vermiform aggregate comprises numerous regularly arranged pseudo-hexagonal plates. Individual plates are thin but up to 40 μm wide. Sample 594, Floodplain-facies sandstone, Gristhorpe Member, Eskdale. (D) Scanning electron photomicrograph. Scale bar 8 μm. Thin grain-coating of authigenic chlorite. Plates, 5–10 μm in diameter. A quartz overgrowth occurs on the now obscured grain and partially encloses the chlorite. Sample 540, braided channel facies sandstone, Moor Grit Member, Bloody Beck.

grain size or matrix content. Fewer authigenic mineral occur in texturally immature samples, where intergranular porosity contains detrital clay (Fig. 7B). Mature, finer-grained samples are generally more tightly cemented than coarser ones (Fig. 7C, D). In finer-grained samples porosity is often completely filled with cement, whereas medium-grained samples, although equally rigidly cemented, were found to contain more intergranular porosity (e.g. Fig. 3D).

Qualitatively the paragenetic relationship between the authigenic minerals found in the floodplain-facies sandstones is similar to that in the channel-facies sandstones. However, quantitatively, the diagenetic modifications of the floodplain facies differ. Firstly, cementation by quartz overgrowths is more extensive (Fig. 7C, D), unless concentrations of matrix clay occur, in which case overgrowths are thin or absent (Fig. 7B). Secondly, kaolinite is more abundant than dickite, which only occurs rarely. Thirdly, rhombic siderite cementation is more common. Finally, calcite was not observed.

Floodplain-facies mudrocks

Floodplain-facies mudrocks are composed of micas, 'interstratified' mica/smectite with up to 10% smectite layers and kandites (Fig. 8). They also contain detrital

Table 3. Electron microprobe analyses of a sodium-feldspar grain and authigenic overgrowth. Sample 355, analysis numbers correspond to spots on Fig. 5

	1	2	3	4
SiO_2	67·45	66·84	68·49	63·34
TiO_2	n.d.	n.d.	n.d.	n.d.
Al_2O_3	19·39	18·81	19·45	19·27
FeO	n.d.	n.d.	0·27	n.d.
MgO	n.d.	n.d.	n.d.	n.d.
MnO	n.d.	n.d.	n.d.	n.d.
CaO	n.d.	n.d.	n.d.	n.d.
Na_2O	11·72	11·50	11·92	11·42
K_2O	0·13	0·14	0·19	n.d.
Total	98·74	97·28	100·31	94·03
Si	2·987	3·001	2·990	2·949
Al	1·012	0·995	1·000	1·057
Ca	0·000	0·000	0·000	0·000
Na	1·006	1·001	1·009	1·031
K	0·010	0·008	0·010	0·000
O	8·000	8·000	8·000	8·000

n.d. = not detected.

silt-sized quartz, organic matter, authigenic pyrite and siderite spheruliths.

STABLE ISOTOPE ANALYSIS

The results of these analyses are presented in Table 4 and the carbonate compositions in Fig. 9. Carbonate compositions are discussed relative to PDB and dickite relative to SMOW.

Siderite

The isotopic composition of the siderite samples averages $\delta^{13}C - 5·9‰$ PDB and $\delta^{18}O - 3·9‰$ PDB. This $\delta^{13}C$ composition is not consistent with a purely bacterial source of bicarbonate ($\delta^{13}C + 15$, or -20 to $-25‰$, Irwin, Curtis & Coleman, 1977) or with a seawater source ($\delta^{13}C$ 0‰), but is consistent with published data from freshwater cements and meteoric waters (Oana & Deevey, 1960; Murata, Friedman & Madsen, 1969; Tan & Hudson, 1974). Siderite is stable in reducing conditions (Berner, 1971). Hence, siderite formed in anoxic conditions, but from fresh water containing only subordinate amounts of bicarbonate produced during suboxic bacterial processes. It is well established that the $\delta^{18}O$ of meteoric water is more negative than that of seawater (Dansgaard, 1964). The $\delta^{18}O$ composition of these siderites could also reflect a fresh water source, being fractionated from Jurassic seawater ($\delta^{18}O - 1·2‰$ SMOW, Shackleton & Kennett, 1975).

Non-ferroan calcite

This cement also has a composition consistent with formation from fresh water.

Ferroan calcite

The $\delta^{13}C$ composition of the ferroan calcite samples ($-2·4$ to $-7·9‰$) is similar to the freshwater cements and so does not support a bacterial or wholly marine source of bicarbonate. However the $\delta^{18}O$ composition ($-11·9$ to $-13·8‰$, averaging $-12·6‰$), is significantly lighter than the non-ferroan calcite ($-5·7‰$), and the siderite (average $-3·9‰$).

The siderite and non-ferroan calcite formed from fresh water. The 6·9‰ break between the calcites and the 8·7‰ break between ferroan calcite and siderite do not indicate continued precipitation from fresh water. Pore-fluid composition changed at this time and ferroan calcite precipitated from formation water during burial.

Dickite

Dickite's isotopic composition (average $\delta^{18}O + 15·2‰$ SMOW) is relatively light. Assuming precipitation from Jurassic seawater ($\delta^{18}O$ $-1·2‰$) it would have formed at 87°C. It could equally have formed at lower temperatures but from isotopically light pore waters. It is suggested that dickite formed from hot or isotopically fractionated fluids in the subsurface, not at the earth's surface (cf. Dickson & Coleman, 1980).

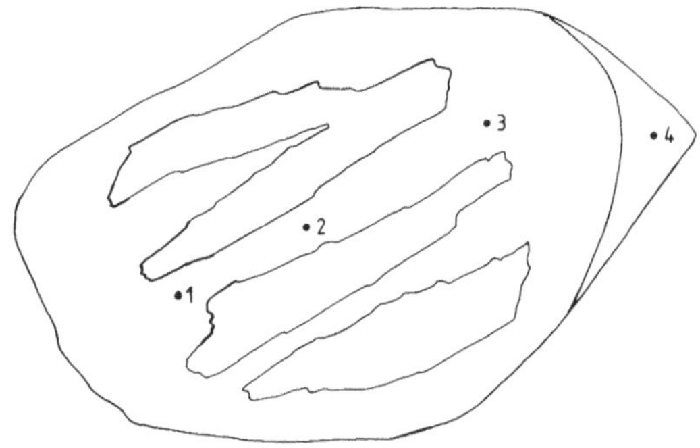

Fig. 5. Points of electron microprobe analysis of authigenic and detrital feldspar. Data from Table 3. Operating conditions were: beam voltage, 15 kV, specimen current 2·5 nA measured on cobalt, and beam width 1 μm. Minimum detection limits for the Geoscan are given by Dunham & Wilkinson (1978).

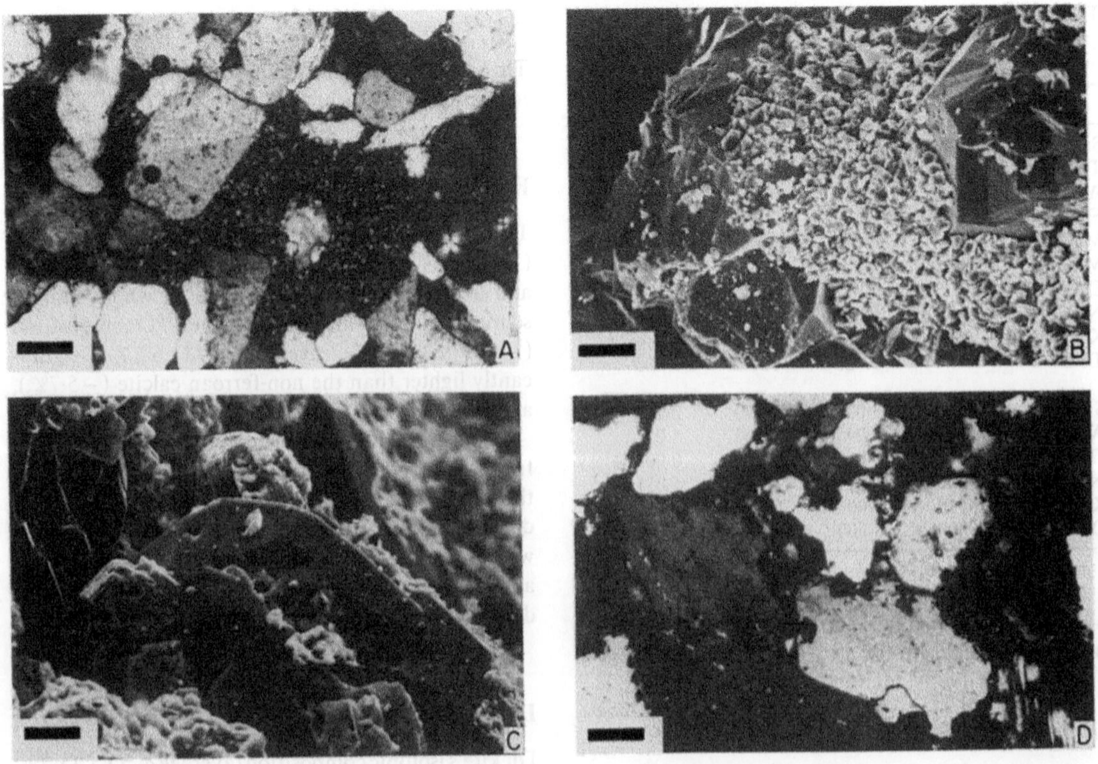

Fig. 6. (A) Thin section photomicrograph. Scale bar 75 μm. Pore-filling dickite. Numerous small blocky aggregates of dickite fill totally this enlarged intergranular pore spare. The pore walls are corroded and notched in places although these features are now infilled with dickite. Sample 472, braided channel facies sandstone, Moor Grit Member, Hundale Point. (B) Scanning electron photomicrograph. Scale bar 27 μm. Dickite occurs as an extensive intergranular pore filling mineral in this sample. Numerous short blocky aggregates occur, each comprising several thick pseudo-hexagonal plates, and up to 50 μm long but only 10–15 μm and one plate wide. Sample 471, braided channel facies sandstone, Moor Grit Member, Hundale Point. (C) Scanning electron photomicrograph. Scale bar 15 μm. Rhombic siderite replacing authigenic quartz. Sample 661, channel-facies sandstone, Saltwick Formation. Rosedale. (D) Thin section photomicrograph. Scale bar 60 μm. Siderite filling intergranular pore space and replacing detrital and authigenic quartz. Most extensive quartz replacement occurs where euhedral overgrowths are poorly developed or absent. Note the similar shape of the siderite embayments here and in Fig. 3(D). Sample 322, channel-facies sandstone. Saltwick Formation, Whitby.

Precipitation temperatures

If it is assumed that the carbonates precipitated from Jurassic seawater ($\delta^{18}O$ $-1{\cdot}2$ SMOW, Shackleton & Kennett, 1975), then the temperature at which the minerals formed can be estimated, using the following equations:

For calcites (Shackleton & Kennett, 1975)

$$T°C = 16{\cdot}9 - 4{\cdot}38\,(c - w) + 0{\cdot}10\,(c - w)^2.$$

For dickite (Dickson & Coleman's, 1980, modification of Savin & Epstein, 1970)

$$1000 \ln \alpha = 2{\cdot}58 \times 10^6 \times TK^{-2} - 3{\cdot}57.$$

The calculated temperatures are:

non-ferroan calcite	39°C,
ferroan calcite (average of 3)	87°C,
dickite (average of 5)	87°C.

Of course, during isotopic equilibrium precipitation, the oxygen isotope composition of any mineral is related to the composition of the pore waters from which it precipitates, as well as the temperature. As the isotopic composition is assumed, these temperatures are unlikely to be accurate but they do allow an assessment to be made of the changing conditions during mineral precipitation.

Fig. 7. (A) Thin section photomicrograph. Scale bar 75 μm. Ferroan calcite cemented subarkose with quartz overgrowths on the majority of grains, but only rarely *replaced* by the calcite. Calcite replaces preferentially grains or low order overgrowth surfaces. Sample 247, channel-facies sandstone, Moor Grit Member, White Nab Scarborough. (B) Thin section photomicrograph. Scale bar 75 μm. Poorly-sorted, fine-grained sandstone with intergranular pore space choked with matrix clay and silt sized quartz grains. Sample 320, floodplain-facies sandstone, Saltwick Formation, Whitby. (C) Scanning electron photomicrograph. Scale bar 27 μm. Extensively quartz overgrowth cemented sandstone. Overgrowths totally occlude porosity. Sample 452, floodplain-facies sandstone, Long Nab Member, Cromer Point. (D) Thin section photomicrograph. Scale bar 75 μm. Tightly quartz cemented sandstone. Cathodoluminescence reveals almost all contacts to be between overgrowths rather than detrital grains. Sample 452.

DIAGENESIS

The inferred sequence of diagenetic modifications of these sediments is summarized in Fig. 10. There are no differences between the paragenetic sequences for each sandstone facies, only in the quantities of authigenic minerals involved.

The first mineral reactions following deposition are interpreted to be the partial alteration of muscovite to kaolinite, the partial dissolution of feldspar, and the precipitation of quartz overgrowths, together with either kaolinite or chlorite. These reactions probably began in mildly acidic conditions, caused by aerobic bacterial degradation of organic matter in the finer grained sediments releasing bicarbonate into solution and lowering the pH (Curtis, 1978). The diagenetic modifications described above proceed at low pH (Fanning & Keramides, 1977; Huang, 1977; Velde, 1977). Alumina and silica released during these reactions caused pore waters to become saturated with respect to kaolinite and quartz (cf. Curtis & Spears, 1971; Tardy, 1971). Kaolinite precipitated ubiquitously whilst syntaxial quartz overgrowths only formed in the texturally mature sandstones. In other parts of the floodplain aerobic bacterial processes quickly removed oxygen and anaerobic bacterial activity began, reducing sulphates and then ferrous iron. Chlorite, rather than kaolinite, became stable under these conditions and began to form. Quartz overgrowths precipitating at this time enclosed chlorite (Fig. 4D).

Traces of various other authigenic minerals occur

Table 4. Stable isotope compositions of authigenic siderite, calcite and dickite. The siderite cements and non-ferroan calcite are also discussed in Kantorowicz (1985)

Description	Sample	$\delta^{13}C$ PDB	$\delta^{18}O$ PDB
Siderite cements	249	−8·0	−1·5
	263	−5·8	−7·7
	370	−3·9	−2·4
Average		−5·9	−3·9
Non-ferroan calcite	332	−7·6	−5·7
Ferroan calcite	250	−2·4	−13·8
	330	−7·9	−12·2
	370	−4·7	−11·9
Average		−5·0	−12·6

Dickite Sample	$\delta^{18}O$ SMOW	
	250°C	100°C
251	16·3	15·7
314	14·0	15·1
410	16·2	
480	15·1	14·4
502	14·2	
Average	15·2	15·1

predating carbonate cementation. Brookite and anatase precipitated locally, following the dissolution of detrital rutile (cf. Ixer, Turner & Waugh, 1979). The rare authigenic illite and potassium-feldspar overgrowths may have formed from potassium released during feldspar dissolution and retained in solution until pH rose, or from potassium in more brackish areas of the floodplain. Where pyrite occurs it records the bacterial reduction and removal from solution of sulphate ions, either released from organic matter during aerobic bacterial activity, or present in depositional pore waters in more brackish areas.

Siderite was the next major authigenic mineral to form, replacing and enclosing the minerals described above. Isotopic analysis suggests that it formed from fresh water containing a mixture of bicarbonate from the original meteoric water, and with subordinate amounts of bacterial bicarbonate, introduced during the processes described above. Non-ferroan calcite formed when oxygenated fresh water was introduced locally and siderite authigenesis was inhibited. Siderite's precipitation from fresh water constrains all the previously formed minerals to a similar origin.

Siderite precipitation began from strongly supersaturated solutions. Numerous crystal nuclei formed initially but, as the concentrations fell, slower growth led to the formation of well-formed rhombic euhedra. Precipitation began with crystals on the pore walls, often replacing the pore-lining phases (Fig. 6C, D). Continued growth in relatively stable pore water conditions is reflected by the uniform size of siderite rhombs. Siderite replacement of quartz and feldspar was not specific to grain surfaces, or grain margins with high surface-free energy, and once replacement had begun the embayments created continued to be preferentially enlarged (Fig. 6D). Siderite replacement of feldspar grains may have accentuated earlier dissolution along twin lamellae (cf. Hurst, 1981).

Ferroan calcite cementation succeeded siderite, but

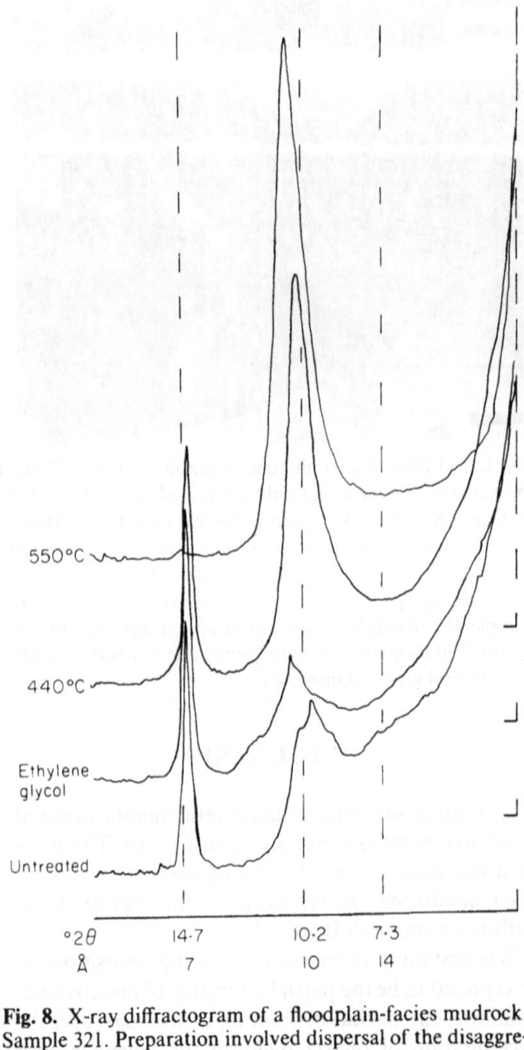

Fig. 8. X-ray diffractogram of a floodplain-facies mudrock. Sample 321. Preparation involved dispersal of the disaggregated mudrock in a sonic bath and sedimentation to separate the less than 2 μm e.s.d. fraction. This size fraction was redispersed in a sonic bath and a slurry pipetted on to warm unglazed ceramic tile. Runs were made from 2 to 60° with Co k_{α} radiation, at 40 kV, and 25 mA and with an Fe filter. Runs were made on untreated, glycolated, and 440 and 550°C heated samples.

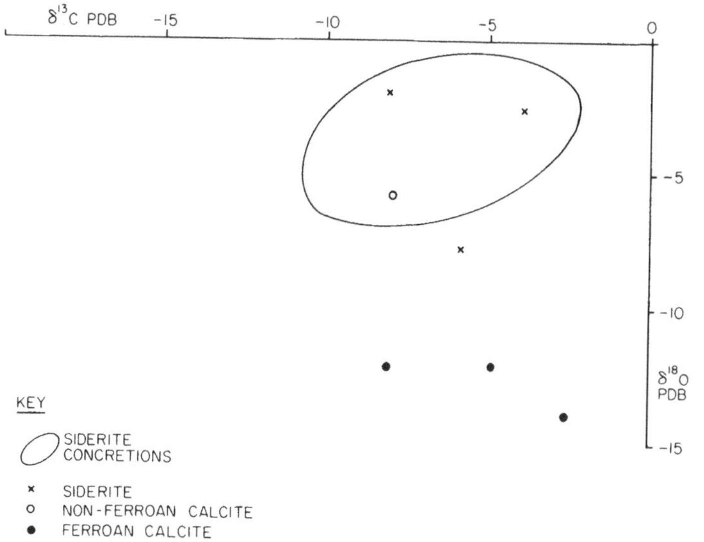

Fig. 9. Carbon and oxygen isotope composition of authigenic carbonate cements. Sample data from Table 4 and the field of siderite concretion analyses from an unpublished investigation by the author.

took place from pore waters introduced during burial, not from continuously evolving depositional waters. Ferroan calcite cementation, partial carbonate dissolution, and later dickite precipitation occurred during mesogenesis. Ferroan calcite replaced quartz overgrowths as well as partially filling the remaining porosity.

Where carbonate cements are absent, both quartz grains and quartz overgrowths are often observed to be corroded. It is inferred that the notches, embayments and irregular depressions on authigenic minerals were formed by the dissolution of a replacive mineral. Corrosion embayments occur where siderite replaces quartz overgrowths (Fig. 6C, D) and it is believed that dissolution of similar replacive siderite or calcite has created the textures now observed in porous sandstones (Fig. 3C, D). Similar textures were also observed on quartz overgrowths exhumed by HCl dissolution of siderite. Intergranular pore space, therefore, includes 'cement-dissolution porosity' (Schmidt & McDonald, 1979b). Large and irregularly shaped pores may have been created by the removal of carbonates, which had replaced detrital grains such as feldspars.

Dickite precipitated within this enlarged intergranular or composite pore space following the dissolution of carbonate cements. A model to explain coincident carbonate dissolution and kandite precipitation has been proposed by Curtis (1983). In this model alumina- and silica-rich low pH solutions cause carbonate dissolution and as pH rises kandite minerals precipitate. Alternatively, the dissolution of carbonates may have been unrelated to the subsequent precipitation of dickite. Dissolution may have been caused by the

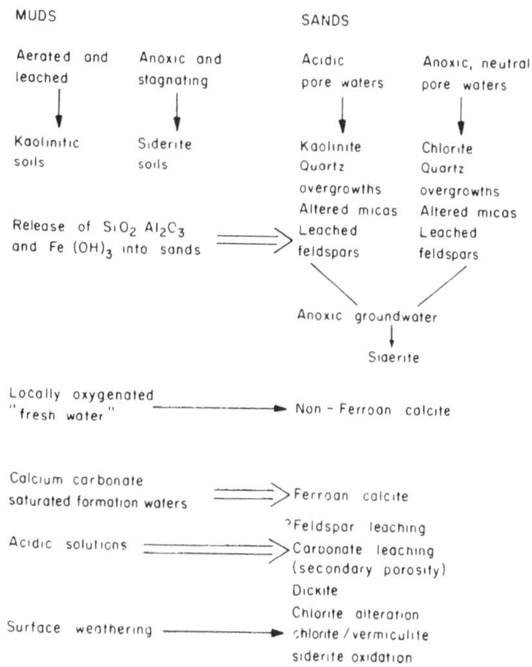

Fig. 10. Flow chart of sequence of diagenetic modifications of Ravenscar Group sediments.

introduction of carbonate undersaturated solutions (Cassan *et al.*, 1981). Further feldspar dissolution probably occurred also during the precipitation, in some sediments, but in feldspar-poor samples dickite is often quite abundant and the presence of feldspar grains was not a prerequisite for precipitation. It is likely that further secondary quartz precipitated from these solutions also. This explains the later quartz overgrowths observed with CL and may correspond to both the step-like growths seen on crystal faces with the SEM, and to the quartz occasionally observed to be enclosing dickite. Illitization of dickite and the precipitation of sodium-feldspar overgrowths probably occurred in alkaline pore waters during Tertiary burial. Illitization of dickite precludes the latter from a telogenetic origin during uplift (cf. Burley, 1984).

The final diagenetic modifications occurred during outcrop weathering. The alteration of chlorite to irregularly interstratified chlorite/vermiculite (Kantorowicz, 1984) oxidation of siderite and pyrite to iron oxides and hydroxides, and possibly continued dissolution of both carbonate cements and detrital and authigenic feldspars all may have occurred during Recent weathering.

In the mudrocks, the present assemblage contains only kandites and mica, with irregularly interstratified mica/smectites (Fig. 8). Processes during burial diagenesis have altered any discrete smectite present to an interstratified mica/smectite (Hower *et al.*, 1976).

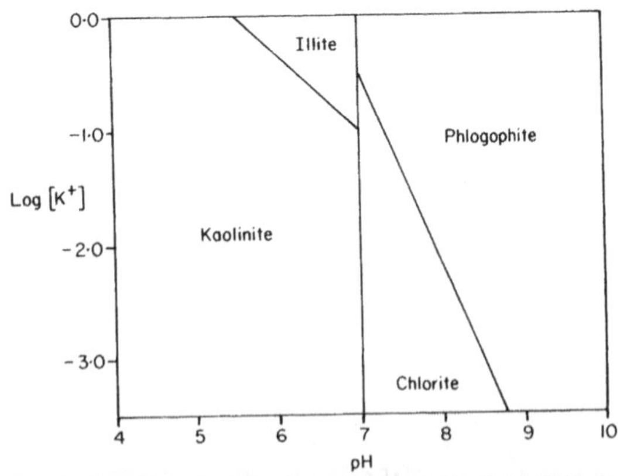

Fig. 11. Stability diagram for the system K_2O-MgO-FeO-Al_2O_3-SiO_2-H_2O at 25°C and 1 atm. with FeO, SiO_2 and H_2O in excess (after Garrels & Christ, 1965) illustrating the geochemical relationship between kaolinite and chlorite.

DISCUSSION

The diagenetic modifications of these Ravenscar Group sediments are recognized to have occurred within three realms: eogenesis, mesogenesis and telogenesis. In this discussion it is demonstrated that diagenesis is not controlled solely by any particular factor, such as detrital mineralogy or burial conditions. Within each of the three distinct realms diagenetic processes operated in an attempt to establish and maintain, an equilibrium between the sedimentary assemblage and its pore waters.

Eogenesis

The non-marine sediments of the Ravenscar Group were deposited in a range of poorly drained and well-drained floodplain environments. Diagenetic modifications of these sediments involved the interaction of the original sedimentary assemblage and its interstitial pore waters. The processes which resulted were an attempt to establish an equilibrium between the organic and inorganic constituents of the sedimentary assemblage and the depositional pore-water system.

Pore-water chemistry is often proposed as a fundamental diagenetic control. However, the amount of cement which can be deposited from a single pore volume is negligible. In order to precipitate quartz and siderite cements many pore volumes must have passed through the sediments prior to Upper Jurassic transgression. It was the interaction of this water with the sedimentary assemblage which most influenced diagenesis.

In these sediments the distribution of chlorite and kaolinite is mutually exclusive. Chlorite and kaolinite are stable in different pore-water conditions (Velde, 1977; Fig. 11), consequently depositional-water chemistry cannot have controlled mineral authigenesis (cf. Almon *et al.*, 1976).

In the wholly non-marine deposits the original pore waters may have been neutral or midly acidic (cf. Berner, 1971; Tardy, 1971; Gac & Tardy, 1980), whereas in the tidally influenced deposits pore waters may initially have been brackish. However, original depositional variations in pore-water chemistry were of little consequence, initial diagenetic modifications of the sediments being promoted by bacterial processes. Aerobic bacterial bicarbonate production would have lowered pH, and in dilute fresh water promoted dissolution of amorphous and relatively unstable detrital aluminosilicate minerals (cf. Curtis, 1978). In some parts of the floodplain a supply of fresh water maintained aerobic conditions. Here efficient drainage ensured that leaching established low ionic strength and oxidizing conditions suppressed ferrous

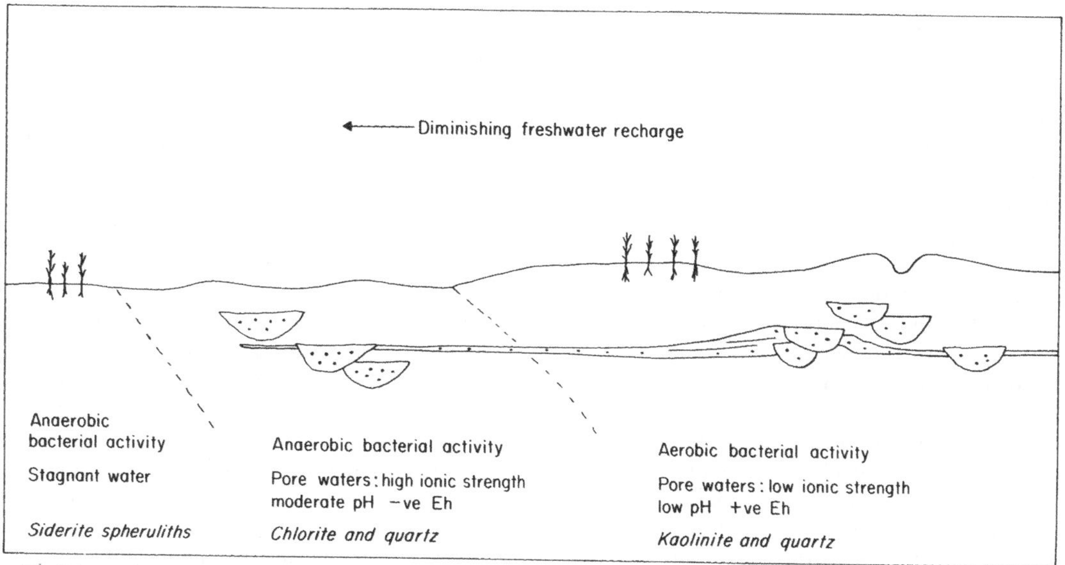

Fig. 12. Schematic eogenetic groundwater chemistry during initial modification of Ravenscar Group sediments. In the parts of the floodplain closest to fresh water recharge oxygenated conditions were maintained and kaolinite preceded siderite. Away from fresh water recharge pore waters rapidly became anoxic and chlorite formed. Meteoric and subordinate amounts of bacterial bicarbonate ions mixed during siderite cementation of the sands. In the muds bacterial processes dominated initially, and siderite soil spheruliths developed.

iron formation. With a supply of alumina and silica from mineral dissolution quartz overgrowths and kaolinite precipitated (Garrels & Christ, 1965; Curtis & Spears, 1971).

In those parts of the floodplain that lacked freshwater circulation, oxygen rapidly became depleted and anaerobic bacterial processes commenced. The scattered pyrite crystals record the reduction and removal of sulphate ions, before ferric iron reduction began. Subsequently, aluminosilicate diagenesis incorporated ferrous iron and chlorite precipitated (cf. Besly & Turner, 1983). Whilst no mineralogical sources of iron are present, iron oxides and hydroxides are common weathering products (e.g. Thomas, 1974) and may have been transported into the basin absorbed on detrital clay micelles (Greenland, 1975) or complexed with humic acid (Picard & Felbeck, 1976).

From the data available it is not possible to confirm whether the basin morphology influenced this groundwater division (Fig. 12). Pyrite occurs ubiquitously but in statistically insignificant quantities. It does not support any suggestion of local brackish pore water influences on clay mineral diagenesis. Rather, the mutually exclusive distribution of chlorite and kaolinite is inferred to be a consequence of the pore-water chemistry which was itself related to progressive interaction with organic matter in the sediment. The widespread siderite cementation of floodplain- and channel-facies sandstones was the final consequence of this interaction. Bacterial processes which were immediately effective in the muds, forming siderite spheruliths (Kantorowicz, 1985), progressively removed all oxygen and sulphate ions, reduced ferric iron to ferrous iron and raised pH. Siderite precipitation began when bicarbonate saturation occurred, and can, therefore, be related to the equilibrium of the original sedimentary assemblage and its pore waters.

Isotopic evidence of widespread siderite precipitation from fresh water confirms the geochemical discussion above. Sediment pore-water interaction culminated in uniform groundwater conditions, disturbed only locally where oxygenated, freshwater influx caused non-ferroan calcite precipitation. Oxygenated conditions had prevailed where kaolinite precipitated and anoxic conditions where chlorite formed. Reducing conditions were eventually initiated throughout the water table with the result that siderite succeeds both chlorite and kaolinite.

This progression towards uniform groundwater conditions supports the contention that sedimentary clastic assemblages are fundamentally unstable when deposited (Curtis, 1980). In these sediments eogenesis commenced with local modifications, as the least stable parts of the assemblage interacted. Bacterial processes were important locally in the muds influencing soil formation, whilst freshwater influence was more pronounced in the sands (Kantorowicz, 1985). Many of the initial processes involved were only effective within individual pores, for example, the replacement of rutile by brookite. Later modifications such as chlorite authigenesis involved less extreme disequilibrium, took longer and so only occurred under more established pore water conditions (cf. Walker *et al.*, 1978). The final eogenetic process, siderite precipitation, took place throughout.

Further variations in the abundance and distribution of authigeneic minerals result from the texture of the original sediments and to a certain extent their mineralogy. On a small scale, sediment texture controlled mineral authigenesis. Mature relatively well-sorted floodplain-facies sandstones contain quartz overgrowths, whilst those texturally immature sediments which contain large quantities of matrix clay do not (Fig. 7B). In spite of this, quartz sedimentation is often more *complete* in floodplain- than in channel-facies sandstones (Fig. 7C, D). This is because channel-facies sandstones are generally better sorted than the floodplain-facies deposits, and overgrowths occur as a thin, evenly distributed veneer (Fig. 3A). In the poorly sorted floodplain-facies sandstones, minus-cement porosities are lower, original pore geometry was more tortuous and so overgrowths more frequently both occlude and completely fill porosity. Quartz overgrowths are also more abundant in floodplain-facies sandstones, the result of silica being supplied both during leaching from above by meteoric waters and subsequently by the expulsion of quartz saturated pore waters from the surrounding muds during shallow-burial compaction (Meade, 1966).

Detrital mineralogy influenced eogenesis in two ways. Firstly, unstable minerals reacted with and affected pore water composition. Secondly, grains provided nuclei for the growth of authigenic minerals. The latter is often overlooked. Crystal growth on an existing nucleus is possible from saturated solutions, whilst supersaturation is required for seed-crystal formation and the degree of supersaturation will determine the form of initial crystal growth (Berner, 1980). It follows that lower concentrations are required for overgrowths to form on detrital quartz and feldspar grains than for other minerals, such as siderite or calcite to form discrete crystals. The limitation of a straightforward assessment of the detrital *mineralogy*, as a control on diagenesis, is in the nature of the most abundant authigenic minerals. These comprise silica, alumina, bicarbonate and iron. There are, however, no inferred mineralogical sources of carbon or iron in the original sediments (cf. Walker *et al.*, 1978). Detrital mineralogy *per se*, therefore, cannot be isolated as a control on the diagenesis of these sediments. Detrital mineralogy only influenced diagenesis when interacting with other factors.

Mesogenesis

Siderite was the last mineral to form during eogenesis. Subsequently pore water changed and ferroan calcite precipitated. This is inferred to mark the break in the paragenetic sequence between diagenetic processes which can be related to the chemistry of the depositional water system and processes which cannot. The previously established equilibrium became unbalanced, and further diagenetic processes took place in an attempt to restore equilibrium between the new pore waters and the eogenetic mineral assemblage. However, during ferroan calcite precipitation the metastability of most of the post-eogenetic assemblage also helped ensure that the effects of changing pore water composition were limited. During eogenesis, less stable minerals had been altered, or replaced, and stable authigenic minerals precipitated. These were metastable in the new pore water environment encountered during burial, where replacement of quartz by carbonate was the only widespread effect (Fig. 7A). A second change of pore water composition occurred when alumina- and silica-rich fluids were introduced into the larger interconnected sandbodies, precipitating dickite.

The original mineralogy of the coarse clastics was dominated by quartz and feldspar. Eogenetic quartz overgrowth cementation consequently created a rigid and stable quartzose framework which was unaffected by later physical and chemical changes (cf. Hutcheon *et al.*, 1980) This framework established an intergranular pore network within which both mineral precipitation and dissolution continued throughout diagenesis. In this case carbonates were precipitated and dissolved, and then dickite was precipitated.

The suggestion of a break between eogenetic siderite and non-ferroan calcite, and mesogenetic ferroan calcite, is supported by interpretation of the isotopic data presented above, as well as geochemical considerations. Poikilotopic ferroan calcites such as these are generally restricted to burial conditions and precipitation from solutions devoid of magnesium ions (Folk, 1974). Electron microprobe analyses of the eogenetic chlorite and siderite, reveal significant quantities of magnesium to have been present in the pore waters (Kantorwicz, 1984, 1985). Ferroan calcite precipitation, therefore, took place in pore waters with a different composition.

The inferred change from freshwater siderite and non-ferroan calcite, to formation water ferroan calcite, can only be confirmed isotopically because the succession of siderite by ferroan calcite would be expected during continuous precipitation (Curtis, 1978; Füchtbauer, 1980). However, whilst the difference in fractionation factors between calcite and water and siderite and water, is not known, it is unlikely to be as great as the 8·7‰ break found here (Coleman, 1981, pers. comm.).

Secondly, non-ferroan calcite is calculated to have formed at 39°C and ferroan calcite at 87°C. On this basis either a significant period of time elapsed and over 1 km of burial took place between non-ferroan and ferroan calcite precipitation, or the ferroan calcite episode formed from pore waters with a different isotopic composition. In either case a break exists between the two and the diagenetic realm changed from eogenesis to mesogenesis.

Maximum burial temperatures are interpreted to have been 100–105°C (Cooper, 1980, pers. comm.). Assuming a geothermal gradient of 3·2°/100 m 90°C burial temperatures could have been attained in the late Cretaceous or early Tertiary when the Ravenscar Group was buried to over 2 km below sea-level (Kent, 1980; Barnard & Cooper, 1983). During the Upper Jurassic and Lower Cretaceous the Ravenscar Group was buried to over 1500 m below sea-level. Elevated temperatures could only have been attained if an accelerated geothermal gradient of 6°C/100 m existed at this time.

The isotopic compositions of ferroan calcite and dickite suggest precipitation at similar temperatures (87°C). Three models may be considered in assessing the paragenetic relationship between the two minerals:

Model 1. If it is correct to assume that pore-water compositions were similar, then dickite precipitation may have been in equilibrium with carbonate dissolution (cf. Land & Dutton, 1978).

Model 2. Pore water compositions were not similar and changed again during burial.

Model 3. The temperatures could be coincidental, with the minerals forming at different times during burial.

In all these cases another break in the paragenetic sequence occurs.

If carbonate dissolution and clay replacement were in equilibrium (model 1) then the composition and the origin of the pore waters responsible may be assessed. Consequently, the likelihood of meteoric water involvement can be considered (cf. Sommer, 1978; Curtis, 1983). Using an estimated surface temperature of 25°C (Tan, Hudson & Keith, 1971; Irwin *et al.*, 1977; Marshall & Ashton, 1980) the composition of the water from which the non-ferroan calcite precipitated is recalculated as −3·9‰. This composition is lighter than estimated seawater compositions at the time but consistent with the established relationship between marine and fresh water (Dansgaard, 1964).

Using this composition the temperature of ferroan calcite and dickite precipitation can be recalculated as approximately 65°C. This is unrealistically high for precipitation from meteoric water, being a jump of 40°C from surface conditions. This would represent a depth of more than 1 km. In addition, water at 65°C in the subsurface, in equilibrium with the surrounding formation cannot be described as meteoric. In conclusion, irrespective of whether pore-water composition changed before dickite precipitated, dickite did not form from fresh water at the earth's surface.

An external source of the silica and alumina necessary for dickite precipitation can be inferred from modal analysis data from the Ravenscar Group which exhibit no positive correlation between dissolved feldspar and dickite (Table 2). The most abundant dickite was found in samples with abundant unaltered feldspar grains, and little evidence of grain-dissolution porosity. Concentrations of dickite also occur in samples with neither feldspar grains nor grain-dissolution porosity.

The corroded grains and etched overgrowths, which occur throughout the Ravenscar Group are features interpreted to be caused by cement dissolution (Schmidt & McDonald, 1979b). However, their presence does not prove that *all* the sediments were completely carbonate cemented. It is likely that part of the intergranular pore space within these rocks has never been cemented. Complete cementation would have occluded the conduits necessary for migrating pore waters which subsequently dissolved the carbonates. Precipitation and dissolution of a small amount of replacive carbonate cement has created the textures commonly identified to interpret porosity as wholly secondary. If carbonate dissolution and dickite precipitation were in equilibrium (model 1), then the reaction occurred in the subsurface, if not (models 2 and 3), this study has obtained no new data which could use to assess whether carbonate dissolution was caused by meteoric waters (Sommer, 1978), acidic solutions (Curtis, 1983), or simply bicarbonate undersaturated pore waters (Cassan *et al.*, 1981).

Reference should also be made to recent work by Morton (1984), who suggested, on the basis of heavy mineral analysis, that many sandstones which contain features indicative of secondary porosity cannot have been flushed with acids, but with solutions of approximately pH 8. If acidity is not invoked to explain carbonate dissolution one must consider thermal effects on solubility. Schmidt & McDonald (1979b) state that the zone of secondary porosity generation in the Gulf Coast is overpressured and that when solutions leak out into the overlying sediments pressure drops and calcite precipitates. It follows that the solutions in the overpressured zone, which have migrated up faults from deeper in the basin (Stanton, 1977), have cooled but pressure has not dropped. These solutions are likely to be undersaturated with respect to carbonates. It is possible that in similar circumstances introduction of non-acidic solutions could contribute to carbonate leaching.

Sandbody geometry and in particular the alluvial architecture of the original sediments, were also important controls on the distribution of authigenic modifications in the subsurface. Large and connected channel-facies sandstones, as well as nearby floodplain-facies deposits, are now relatively porous, but also contain significant quantities of dickite. Conversely, smaller, more isolated sandstone bodies, and especially thin floodplain-facies sandstones are often tighly cemented, or are less porous, and contain less dickite. It appears, therefore, that the larger sandbodies were conduits for pore-fluid migration in the subsurface, whilst the smaller sands remained unaffected by the activity of these solutions.

Telogenesis

At the present day the Ravenscar Group outcrops on the North Yorkshire Moors. Uplift into meteoric groundwaters and, finally, exposure to recent weathering, have affected the mineralogy of the sediments. The effects include clay alteration and the oxidation of ferrous iron-bearing minerals. Feldspar and carbonate dissolution may also have taken place. The effect of telogenesis on these sediment is the replacement of framework grains and now unstable clays, with more stable clays.

CONCLUSIONS

Diagenetic modification of non-marine clastic sediments from the Ravenscar Group has created a diverse collection of authigenic, unaltered, and altered detrital minerals. The processes involved reflect a number of influences which were important before, during, and after burial. These influences affected the accumulation of an initially unstable sedimentary assemblage and its subsequent modification to form the sedimentary rocks which outcrop today.

Eogenetic modification resulted from the attempted equilibration of depositional pore waters and detrital sediments. They were affected by the silica- and alumina-rich pore waters; the subarkosic mineralogy of the sediments; the texture of sandbodies throughout the floodplain; the quantities of iron and organic matter present in the sediments; and finally, the wet climate which frequently replenished the groundwater, and occasionally maintained a high water table.

On the floodplains, bacterial degradation of organic matter rapidly lowered groundwater pH. As a result, amorphous silica and alumina, as well as detrital feldspar grains began to dissolve and muscovite partially altered to kaolinite. Where meteoric water was continuously supplied to the sediments, Eh remained positive and vermiform aggregates of kaolinite precipitated. In parts of the groundwater, continued bacterial degradation of organic matter depleted oxygen and reduced ferric iron, a result of which was chlorite formation. Silica, both originally in solution and released during *in situ* reactions precipitated in the oxygenated and anoxic pore waters to form quartz overgrowths. Some of the least mature floodplain-facies sandstones were occluded completely, whereas most channel-facies sands were cemented into a rigid but porous quartzose framework. As bacterial ferric iron reduction commenced, pH rose and bicarbonate concentrations rose. Illite and potassium-feldspar overgrowths precipitated locally. Finally, the groundwater became anoxic throughout, and rhombic siderite precipitated in some of the remaining porosity.

During burial the remaining diagenetic modifications of the sandstones took place only in the larger and connected sandbodies. Ferroan calcite precipitated first but then carbonate cements were dissolved and dickite precipitated. A second episode of quartz overgrowths also formed in the subsurface.

Surface weathering has determined the most recent diagenetic modifications, with oxidation of siderite and pyrite, and possibly also dissolution of carbonates and aluminosilicates throughout the sediments. Weathering has caused the alteration of chlorite to a mixed-layer chlorite/vermiculite.

ACKNOWLEDGMENTS

This research was undertaken whilst in receipt of N.E.R.C. Grant GT4/79/GS/39 and the stable isotope analysis at the British Geological Survey (N.E.R.C.) Stable Isotopes Unit. I should like to thank the following: B. Waugh, who supervised this research; M. L. Coleman, and J. E. Rouse, who assisted in the isotope work; K. Ramseyer who kindly examined several samples with the luminoscope at Bern University; S. D. Burley, A. R. Hurst, and P. J. C. Nagtegaal who read early versions of the manuscript; and referees K. Bjørlykke and R. W. O'B. Knox.

The Director of the British Geological Survey and Shell International Petroleum Maatschappij are thanked for permission to publish this paper.

REFERENCES

ALMON, W.R., FULLERTON, L.B. & DAVIES, D.K. (1976) Pore space reduction in Cretaceous sandstones through chemical precipitation of clay minerals. *J. sedim. Petrol.* **46**, 89–96.

BAILEY, S.W. (1980) Summary of recommendations of AIPEA nomenclature committee. *Clay Miner.* **15**, 85–93.

BARNARD, P.C. & COOPER, B.S. (1983) A review of geochemical data related to the northwest European gas province. In: *Petroleum Geochemistry and Exploration of Europe* (Ed. by J. Brooks). Blackwell Scientific Publications, Oxford. *Spec. Publ. geol. Soc. London*, **12**, 19–33.

BERNER, R.A. (1971) *Principles of Chemical Sedimentology*. McGraw-Hill, New York.

BERNER, R.A. (1980) *Early Diagenesis; a Theoretical Approach*. Princeton University Press.

BESLY, B.M. & TURNER, P. (1983) Origin of red beds in a moist tropical climate (Etruria Formation, Upper Carboniferous, U.K.). In: *Residual Deposits: Surface Related Weathering Processes and Materials* (Ed. by R. C. L. Wilson). *Spec. Publ. geol. Soc. London*, **11**, 131–147. Blackwell Scientific Publications, Oxford.

BOLES, J.R. & FRANKS, S.G. (1979) Clay mineral diagenesis in Wilcox of south-west Texas: implications of smectite diagenesis in sandstone cementation. *J. sedim. Petrol.* **49**, 55–70.

BURLEY, S.D. (1984) Patterns of diagenesis in the Sherwood Sandstone Group (Triassic) U.K. *Clay Miner.* **19**, 403–440.

CASSAN, J.P., GARCIA PALACOIS, M.C., FRITZ, B. & TARDY, Y. (1981) Diagenesis of sandstone reservoirs as shown by petrographical and geochemical analysis of oil-bearing formations in the Gabon Basin. *Bull. Cent. Rech. Explor.-Prod. Elf-Aquitaine*, **5**, 113–135.

CLAYTON, R.N. & MAYEDA, T.K. (1963) The use of bromine pentafluoride in the extraction of oxygen from oxides and silicates for isotopic analysis. *Geochim. Cosmochim. Acta*, **27**, 43–52.

CRAIG, H. (1957) Isotopic standards for carbon and oxygen correction factors for mass spectrometric analysis for carbon dioxide. *Geochim. Cosmochim. Acta*, **12**, 133–149.

CURTIS, C.D. (1978) Possible links between sandstone diagenesis and depth related geochemical reactions occurring in enclosing mudstones. *J. geol. Soc. London*, **135**, 107–117.

CURTIS, C.D. (1980) Diagenetic alteration in black shales. *J. geol. Soc. London*, **137**, 189–194.

CURTIS, C.D. (1983) Geochemical studies on development and destruction of secondary porosity. In: *Petroleum Geochemistry and Exploration of Europe* (Ed. by J. Brooks). *Spec. Publ. geol. Soc. London*, **12**, 113–125. Blackwell Scientific Publications, London.

CURTIS, C.D. & SPEARS, D.A. (1971) Diagenetic development of kaolinite. *Clays Clay Miner.* **19**, 219–227.

DANSGAARD, W. (1964) Stable isotopes in precipitation. *Tellus*, **16**, 436–468.

DEINES, P. (1970) Mass spectrometer correction factors for the determination of small isotopic composition variations of carbon and oxygen. *Int. J. Mass Spectrom. Ion Phys.* **4**, 283–295.

DICKSON, J.A.D. (1965) A modified staining technique for carbonates in thin section. *Nature*, **205**, 587.

DICKSON, J.A.D. & COLEMAN, M.L. (1980) Changes in carbon and oxygen isotope composition during limestone diagenesis. *Sedimentology*, **27**, 107–118.

DUNHAM, A.C. & WILKINSON, F.C.F. (1978) Accuracy precision and detection limits of energy-dispersive electron microprobe analysis of silicates. *X-ray Spectrom.* **7**, 50–56.

FANNING, D.S. & KERAMIDES, V.Z. (1977) Micas. In: *Minerals in Soil Environments* (Ed. by J. B. Dixon and S. B. Weed), pp. 195–258. Soil Science Society of America, Madison, Wisconsin.

FOLK, R.L. (1968) *Petrology of Sedimentary Rocks*. Hemphills Bookstore, Austin, Texas.

FOLK, R.L. (1974) The natural history of crystalline calcium carbonate: effect of magnesium content and salinity. *J. sedim. Petrol.* **44**, 40–53.

FORESTER, R.M., SANDBERG, P.A. & ANDERSON, T.F. (1973) Isotopic variability of cheilostrome bryzoan skeleton. In: *Living and Fossil Bryozoa* (Ed. by G. P. Larwood), pp. 79–94. Academic Press, London.

FRITZ, P., BINDA, P.L., FOLINSBEE, F.E. & FROUSE, H.R. (1971) Isotopic composition of diagenetic siderites from Cretaceous sediments in western Canada. *J. sedim. Petrol.* **41**, 282–288.

FÜCHTBAUER, H. (1980) Experimental precipitation of ferroan calcite. *Int. Ass. Sedimentol. 1st Europ. Mtg Bochum*, pp. 170–171 (abstract).

GAC, J.Y. & TARDY, Y. (1980) Geochemistry of a tropical landscape on granitic rocks: the Lake Chad Basin. In: *Proceedings 3rd International Symposium on Water–Rock Interaction* (Ed. by A. Campbell), pp. 8–10. Alberta Research Council.

GARRELS, R.M. & CHRIST, G.L. (1965) *Solution, Minerals and Equilibria*. Harper & Row, New York.

GLENNIE, K.W., MUDD, G.C. & NAGTEGAAL, P.J.C. (1978) Depositional environment and diagenesis of Permian Rotliegendes Sandstones in Leman Bank and Sole Pit areas of the U.K. southern North Sea. *J. geol. Soc. London*, **135**, 25–34.

GREENLAND, D.J. (1975) Charge characteristics of some kaolinite-iron hydroxide complexes. *Clay Miner.* **10**, 407–416.

HANGARI, K.M., AHMAD, S.N. & PERRY (JR), E.C. (1980)

Carbon and oxygen isotope ratios in diagenetic siderite and magnetite from Upper Devonian ironstone, Wadi Shatti District, Libya. *Econ. Geol.* **75**, 538–545.

HEMINGWAY, J.E. (1974) Jurassic. In: *Geology and Mineral Resources of Yorkshire* (Ed. by D. H. Rayner and J. E. Hemingway), pp. 161–223. Yorkshire Geological Society, Leeds.

HEMINGWAY, J.E. & KNOX, R.W.O'B. (1973) Lithostratigraphic nomenclature of the Middle Jurassic strata of the Yorkshire Basin of north-east England. *Proc. Yorks. geol. Soc.* **39**, 527–535.

HEMINGWAY, J.E. & RIDDLER, G.P. (1982) Basin inversion in North Yorkshire. *Trans. Inst. Min. Metall. B* **91**, 175–186.

HOWER, J., ESLINGER, E.V., HOWER, M.E. & PERRY, E.A. (1976) Mechanism of burial and metamorphism of argillaceous sediment: I. Mineralogical and chemical evidence. *Bull. geol. Soc. Am.* **87**, 725–737.

HUANG, P.M. (1977) Feldspars, olivines, pryoxenes and amphiboles. In: *Minerals in Soil Environments* (Ed. by J. B. Dixon and S. B. Weed), pp. 553–602. Soil Science Society of America, Madison, Wisconsin.

HURST, A.R. (1981) A scale of dissolution for quartz and its implication for diagenetic processes in sandstones. *Sedimentology*, **28**, 451–460.

HUTCHEON, I., OLDERSHAW, A. & GHENT, E.D. (1980) Diagenesis of Cretaceous sandstones of the Kootenay Formation at Elk Valley (southeastern British Columbia) and Mt. Allen (southwestern Alberta). *Geochim. Cosmochim. Acta*, **44**, 1425–1435.

IRWIN, H., CURTIS, C.D. & COLEMAN, M.L. (1977) Isotopic evidence for source of diagenetic carbonates formed during burial of organic-rich sediments. *Nature*, **269**, 209–213.

IXER, R.A., TURNER, P. & WAUGH, B. (1979) Authigenic iron and titanium oxides in Triassic red beds (St. Bees Sandstone), Cumbria, Northern England. *Geol. J.* **14**, 179–192.

KANTOROWICZ, J.D. (1982) *Diagenetic modelling in Middle Jurassic clastic sediments from the Ravenscar Group, Yorkshire, and the Brent Group, Northern North Sea.* Unpublished Ph.D. Thesis, University of Hull.

KANTOROWICZ, J.D. (1984) The nature, origin and distribution of authigenic clay minerals from Middle Jurassic Ravenscar and Brent Group Sandstones. *Clay Miner.* **19**, 357–375.

KANTOROWICZ, J.D. (1985) Stable isotope analysis of authigenic carbonates from the Middle Jurassic Ravenscar Group, Yorkshire. In preparation.

KENT, P.E. (1980) Subsidence and uplift in East Yorkshire and Lincolnshire: a double inversion. *Proc. Yorks. geol. Soc.* **42**, 505–524.

LAND, L.S. & DUTTON, S.P. (1978) Cementation of a Pennsylvanian deltaic sandstone: isotopic data. *J. sedim. Petrol.* **48**, 1167–1176.

LIVERA, S.E. & LEEDER, M.R. (1981) The Middle Jurassic Ravenscar Group (Deltaic Series) of Yorkshire: Recent sedimentological studies as demonstrated during a recent Field Meeting, 2–3 May 1980. *Proc. geol. Ass.* **92**, 241–250.

MARSHALL, J.D. & ASHTON, M. (1980) Isotopic and trace element evidence of submarine lithification of hardgrounds in the Jurassic of eastern England. *Sedimentology*, **27**, 271–290.

MCBRIDE, E.F. (1963) A classification of common sandstones. *J. sedim. Petrol.* **33**, 664–669.

MEADE, R.H. (1966) Factors influencing the early stages of the compaction of clays and sands—review. *J. sedim. Petrol.* **36**, 1085–1101.

MEHRA, O.P. & JACKSON, M.L. (1960) Iron oxide removal from soils and clays by a dithionite-citrite system buffered with sodium bicarbonate. *Clays Clay Miner.* **7**, 317–327.

MORTON, A.C. (1984) Stability of detrital heavy minerals in Tertiary sandstones from the North Sea Basin. *Clay Miner.* **19**, 287–308.

MURATA, K.J., FRIEDMAN, I. & MADSEN, B.M. (1969) Isotopic composition of diagenetic carbonates in marine Miocene formations of California and Oregon. *Prof. Pap. U.S. geol. Surv. 614B*, 24 pp.

OANA, S. & DEEVEY, E.S. (1960) Carbon 13 in lake waters and its possible bearing on palaeolimnology. *Am. J. Sci.* **258A**, 253–272.

PICARD, G.L. & FELBECK (JR), G.T. (1976) The complexation of iron by marine humic acid. *Geochim. Cosmochim. Acta*, **40**, 1347–1350.

PITTMAN, E.D. (1979) Porosity, diagenesis and productive capability of sandstone reservoirs. In: *Aspects of Diagenesis* (Ed. by P. A. Scholle and P. R. Schluger). *Spec. Publ. Soc. econ. Paleont. Miner., Tulsa*, **26**, 159–173.

RAMSEYER, K. (1983) A new cathodoluminescence microscope and its application to sandstone diagenesis. *11th int. Cong. Sedimentology (I.A.S.)* Hamilton, p. 885 (abstract).

SAVIN, S. & EPSTEIN, S. (1970) The oxygen and hydrogen isotope geochemistry of clay minerals. *Geochim. Cosmochim. Acta*, **34**, 25–42.

SCHMIDT, V. & MCDONALD, D.A. (1979a) The role of secondary porosity in the course of sandstone diagenesis. In: *Aspects of Diagenesis* (Ed. by P. A. Scholle and P. R. Schluger). *Spec. Publ. Soc. econ. Paleont. Miner., Tulsa*, **26**, 175–207.

SCHMIDT, V. & MCDONALD, D.A. (1979b) Texture and recognition of secondary porosity in sandstones. In: *Aspects of Diagenesis* (Ed. by P. A. Scholle and P. R. Schluger). *Spec. Publ. Soc. econ. Paleont. Miner., Tulsa*, **26**, 209–225.

SHACKLETON, N.J. & KENNETT, J.P. (1975) Palaeotemperature history of the Cenozoic and the initiation of the Antarctic glaciation: oxygen and carbon isotope analysis in DSDP Sites 277, 279 and 281. In: *Init. Rep. Deep-Sea drill. Proj.* **29**, 743–755. Government Printing House, Washington, D.C.

SHARMA, T. & CLAYTON, R.N. (1965) Measurement of δ^{18}/δ^{16} ratios of total oxygen of carbonates. *Geochim. Cosmochim. Acta*, **29**, 1347–1354.

SPEARS, D.A. (1980) Towards a classification of black shales. *J. geol. Soc. London*, **137**, 125–129.

SOMMER, F. (1978) Diagenesis of Jurassic sandstones in the Viking Graben. *J. geol. Soc. London*, **135**, 63–67.

STALDER, P.J. (1975) Cementation of Pliocene–Quaternary fluviatile clastic deposits in and along the Oman Mountains. *Geologie Mijnb.* **54**, 148–156.

STANTON, G.D. (1977) Secondary porosity in sandstones of the Lower Wilcox (Eocene), Karnes County, Texas. *Trans. Gulf-Cst Ass. geol. Socs* **27**, 197–207.

TAN, F.C. & HUDSON, J.D. (1974) Isotopic studies of the

palaeoecology and diagenesis of the Great Estuarine Series (Jurassic) of Scotland. *Scott. J. Geol.* **10**, 91–128.

TAN, F.C., HUDSON, J.D. & KEITH, M.L. (1971) Jurassic (Callovian) paleotemperatures from Scotland. *Earth planet. Sci. Lett.* **9**, 421–426.

TARDY, Y. (1971) Characterisation of the principal weathering types of the geochemistry of waters from European and African crystalline massifs. *Chem. geol.* **7**, 253–271.

THOMAS, M.F. (1974) *Tropical Geomorphology. A Study of Weathering and Landform Development in Warm Climates.* MacMillan Press, London.

VELDE, B. (1977) *Clays and Clay Minerals in Natural and Synthetic Systems. Developments in Sedimentology* **21**, Elsevier, Amsterdam.

WALKER, T.R., WAUGH, B. & CRONE, A.J. (1978) Diagenesis in first cycle desert alluvium of Cenozoic age, southwestern United States and northwestern Mexico. *Bull. geol. Soc. Am.* **89**, 19–32.

Telogenesis (uplift-related diagenesis)

Telogenesis involves meteoric water infiltration into sandstones at erosional unconformities and has been thought capable of producing enhanced porosity. The distribution of kaolinite in relation to unconformities has been used to suggest that uplift related diagenesis has been responsible for kaolinite growth at the expense of other aluminosilicate minerals. Bjørkum *et al.* (1990), however, concluded that kaolinite present in rocks close to a basin-wide unconformity was detrital, rather than diagenetic kaolinite. Moreover, they claimed that the rate of telogenetic kaolinite growth could not keep pace with the rate of erosion. This conclusion urges caution on the use of unconformities as petroleum exploration targets, although the uncertainty in this debate survives unresolved.

Reprinted from Sedimentology (1990) **37**, *395–406*

The role of the late Cimmerian unconformity for the distribution of kaolinite in the Gullfaks Field, northern North Sea

PER ARNE BJØRKUM, RUNE MJØS, OLAV WALDERHAUG *and* ANDREW HURST*

*Rogaland Research Institute, PO Box 2503, Ullandhaug, 4004 Stavanger, Norway and *Statoil, PO Box 300, 4001 Stavanger, Norway*

ABSTRACT

A study of the Middle Jurassic Rannoch Formation in the Gullfaks Field shows that there is no relationship between the content and distribution of kaolinite and location relative to the late Cimmerian unconformity. From petrographic data most of the kaolinite is interpreted to be detrital, and only traces of authigenic kaolinite are observed. Mass-balance calculations are used to provide guidelines as to the likelihood of kaolinite being formed and preserved in sandstones under the unconformity. The result of the calculations shows that the propagation rate of the dissolution front in the sandstones was probably on average slower than the erosion rates during the formation of the late Cimmerian unconformity. Hence, transformation of significant amounts of feldspar and mica to kaolinite probably did not take place within the sandstones which at present underlie the unconformity in the Gullfaks Field. Periods of exposure during formation of unconformities may thus be less effective causes of kaolinitization in sandstones than has been assumed.

INTRODUCTION

Several hypotheses exist to explain the origin of kaolinite in Brent sandstone reservoirs in the North Sea. In the freshwater model (Hancock & Taylor, 1978; Sommer, 1978; Bjørlykke, Elverhøi & Malm, 1979) it has been proposed that the kaolinite is authigenic and formed by diagenetic degradation of detrital aluminosilicate minerals, mainly feldspar and mica, during meteoric-water flushing. Meteoric-water diagenesis may take place immediately after deposition in sandstones deposited in non-marine or paralic settings or, during subsequent uplift. While it is widely accepted, and theoretically justified, that kaolinite may form by degradation of mica and feldspars in meteoric water, the importance of the late Cimmerian unconformity in controlling the content and distribution of kaolinite within the Brent sandstones has not been convincingly demonstrated (Jeans, 1986).

The Brent Group sediments comprise deltaic deposits, divided into five formations: Broom, Rannoch, Etive, Ness and Tarbert (Vollset & Dore, 1984; Graue *et al.*, 1987). The Brent delta prograded northwards and deposited the wave- and storm-dominated lower to upper shoreface sandstones of the Rannoch Formation which were capped by the coarser grained, tidal inlet and upper shoreface/foreshore deposits of the Etive Formation. The Ness Formation is a delta plain sequence comprising sandstones, mudrocks and coals and is overlain by marginal and open-marine deposits of the Tarbert Formation which was deposited during the retreat of the Brent delta (Graue *et al.*, 1987).

The Middle Jurassic Brent Group was in most areas eroded and truncated during the Late Jurassic to Early Cretaceous. This erosional surface is commonly referred to as the 'late Cimmerian unconformity' (Ziegler, 1981; Rawson & Riley, 1982). In this paper the possible diagenetic effect of the late Cimmerian unconformity is studied by examining the kaolinite distribution in the Rannoch Formation sandstones of the Gullfaks Field, and by general mass-balance calculations. A generalized core log for the Gullfaks Field is presented in Fig. 1. The Gullfaks Field is located in the northern North Sea and is the most positive structural element in the Tampen Spur area

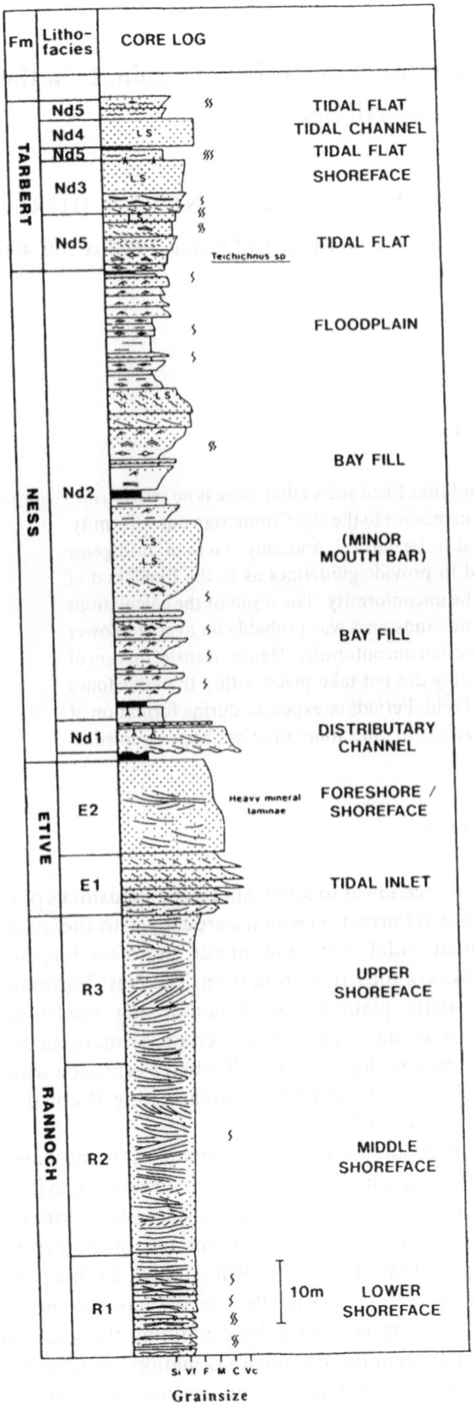

Fig. 1. Generalized core log for the Gullfaks Field.

(Erichsen *et al.*, 1987). The Etive, Ness and Tarbert formations are excluded from this study as diagenetic kaolinitization in these formations may have occurred in response to meteoric-water flushing both during and immediately after deposition, as well as during the late Cimmerian unconformity. Distinction between authigenic kaolinite formed by these separate events is difficult, if not impossible, using petrographic or geochemical methods. Since the Rannoch Formation was deposited in a lower-shoreface environment, it is less likely that this formation suffered syn- or early post-depositional kaolinitization by meteoric-water flushing.

Commonly, there is a positive correlation between the mica and kaolinite content of the Rannoch Formation. Bjørlykke & Brendsdal (1985) studied the Rannoch Formation in the Statfjord Field and concluded that the correlation between mica and kaolinite (XRD data) can mainly be explained by transformation of mica to kaolinite during meteoric-water flushing soon after deposition and at the late Cimmerian unconformity. Since the Rannoch Formation in the Gullfaks Field is truncated by the late Cimmerian unconformity in the well 34/10-4 (Fig. 2), any significant effect of the unconformity upon the kaolinite distribution may be tested.

SAMPLE MATERIAL AND METHODS

Seventy-five thin-sections from wells 34/10-1, 34/10-4 and 34/10-14 from the Rannoch Formation were point counted with 300 counts per thin-section. Seventy-seven samples were analysed by XRD (bulk analysis) and 17 samples were studied with an SEM equipped with an energy-dispersive X-ray analyser.

KAOLINITE DISTRIBUTION IN THE RANNOCH FORMATION

The sandstones of the Rannoch Formation in the Gullfaks Field consists of 30–40% quartz, 12–40% kaolinite, 5–20% mica, 10–15% K-feldspar, 0–10% plagioclase, 5–25% carbonate and traces of pyrite.

The kaolinite content in Rannoch Formation sandstones in the wells 34/10-1, 4 and 14 is plotted in Fig. 3. The kaolinite content is between 12 and 42% of the rock matrix. Although the unconformity is situated at the top of the Rannoch Formation in well 34/10-4, its kaolinite content is not higher than in well 34/10-1, and only slightly higher than in well 34/10-14 where the late Cimmerian unconformity is situated 90 and 70 m, respectively, above the top of the Rannoch Formation (Fig. 2). In both of these wells, shales of the Ness Formation lie between the unconformity and the top of the Rannoch Formation. Furthermore, the kaolinite content increases downwards in the Rannoch Formation in wells 34/10-1 and 34/10-4 (Fig. 3). The opposite trend would be expected if significant amounts of kaolinite were formed as a result of meteoric water moving downwards from the unconformity surface (discussed below). Even without petrographic evidence it seems unlikely that the kaolinite distribution in the Rannoch Formation is of a diagenetic origin related to the unconformity.

Several petrographic observations support the view that most of the kaolinite in the Rannoch Formation is detrital. The kaolinite occurs partly as grain-supporting clasts (Fig. 4) which have sharp boundaries, and have similar chemical compositions in their centres and edges (proven by back-scatter analysis on thin-sections). Furthermore, the size of the clay clasts is usually 1·2–2 times the size of the other framework grains (Fig. 4). The porosity within the clay clasts is estimated to be less than 10%, which is considerably lower than typical porosities in authigenic kaolinite (average porosity *c.* 50%; A. Hurst & P. Nadeau, unpublished observation), and well-developed crystal surfaces are not identified (Figs 5 & 6). Only traces of possible authigenic kaolinite (microporous pore-filling aggregates built up of stacks of pseudohexagonal platelets with well-developed crystal faces) are observed in each of the three wells.

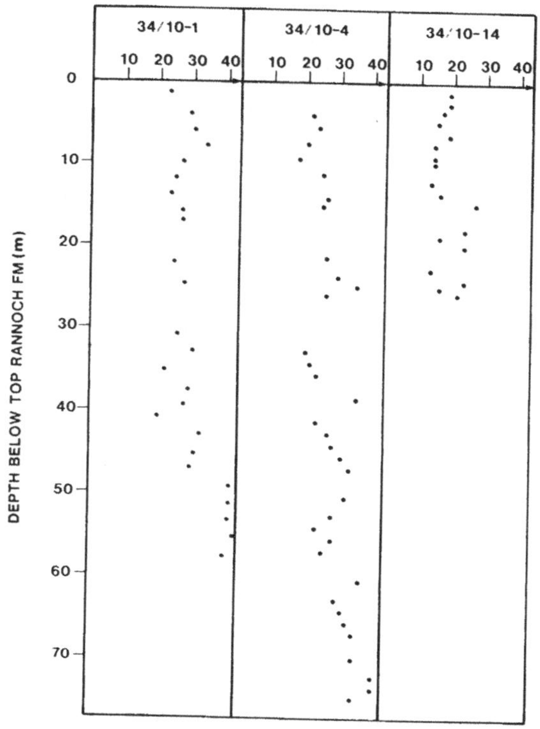

Fig. 3. The content of kaolinite (vol% determined by XRD) in the Rannoch Formation is shown for three wells in the Gullfaks Field. The kaolinite content increases downward in well 34/10-4, and the kaolinite content is about the same in wells 34/10-1 and 34/10-4 and only slightly lower in well 34/10-14. Only traces of authigenic kaolinite are observed in all three wells.

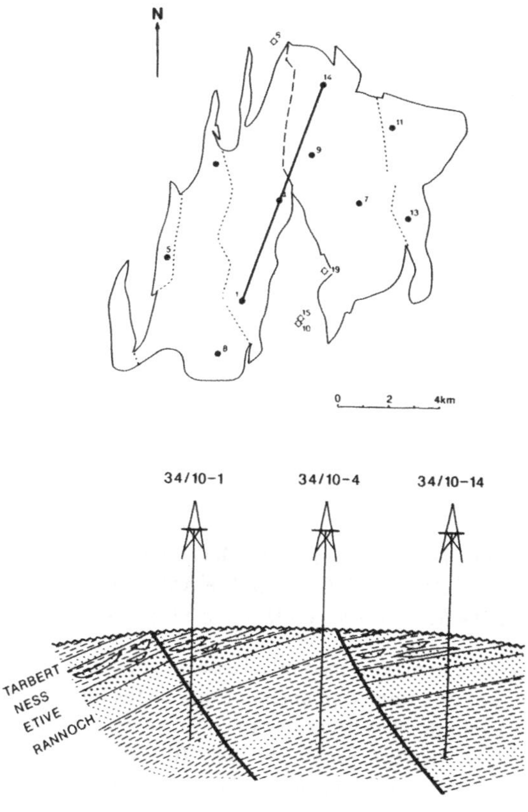

Fig. 2. Location of studied wells relative to the late Cimmerian unconformity. In wells 34/10-1 and 34/10-14, 90 and 70 m of sediments respectively are present between the top of the Rannoch Formation and the late Cimmerian unconformity. In well 34/10-4 the late Cimmerian unconformity is at the top of the Rannoch Formation.

Since only traces of possible authigenic kaolinite are observed, even in well 34/10-4 where the Rannoch Formation is located closest to the unconformity, we suggest that the unconformity has had a limited effect on the distribution of kaolinite in the Rannoch

Fig. 4. Kaolinitic clay clast (Ka) from the Rannoch Formation. The size of the clay clasts is typically 1·2–2 times (diameter) the size of the other framework grains. The element composition is the same in all parts of the clay clasts (back-scatter analysis on thin-sections). The clay clasts often show sharp grain boundaries, and well-developed pseudo-hexagonal kaolinite crystals are not observed, although less developed kaolinite crystals are seen in some cases. Insignificant authigenic modification of the clay clasts appears to have taken place after deposition. The white areas in the clay clast are carbonate cement. Width of photograph = 0·85 mm.

Formation in the Gullfaks Field. The increase in total kaolinite (XRD-data) downward in the Rannoch Formation in all three wells (Fig. 3) is interpreted simply as an increase in the detrital kaolinite content downsection, and that the kaolinite grains in the Rannoch Formation were already of kaolinitic composition when they were deposited. It is suggested that the positive correlation between the content of mica and the (detrital) kaolinite (Bjørlykke & Brendsdal, 1985) is due to their similar hydraulic behaviour during deposition.

The interpretation that the amount of kaolinite in the Rannoch Formation in the Gullfaks Field is controlled by sedimentary processes, and not the late Cimmerian unconformity, conflicts with the view advocated by Sommer (1978), Hancock & Taylor (1978) and Bjørlykke *et al.* (1979). Even if our distinction between detrital and authigenic kaolinite proves unreliable, the volume of kaolinite possible to generate by meteoric-water flushing can still be evaluated. In order to make this evaluation we quantified the processes that are important for the kaolinitization of sandstones during subaerial exposure. In our quantification we chose input parameters which favour the model advocated by Sommer (1978), Hancock & Taylor (1978) and Bjørlykke *et al.* (1979).

CONDITIONS AND PROCESSES OF IMPORTANCE FOR KAOLINITE FORMATION AND PRESERVATION

In order to make a theoretical assessment of the effect of the late Cimmerian unconformity on the formation of kaolinite, it is necessary to consider: (i) the water budget, (ii) the proton budget, (iii) the rates of dissolution of unstable minerals, and (iv) the effect of erosion.

(i) Water budget

The maximum water budget is determined by the annual rainfall. Most precipitation is re-introduced

Fig. 5. SEM micrograph of kaolinite observed in the Rannoch Formation. The kaolinite is texturally similar to the detrital kaolinite in Fig. 38 in Keller (1978). Width of photograph = 128 μm.

directly to the atmosphere by evaporation and organic respiration. Approximately 2% of the annual rainfall, worldwide on land, enters the oceans as groundwater (Lerman, 1979, p. 258). In areas where highly porous sediments are exposed, a higher fraction of the water is likely to enter the oceans as groundwater. In this paper we have assumed that 10% of the annual rainfall enters the oceans as groundwater.

The mean velocity U (m yr^{-1}) of the flow through a sandstone is

$$U = \frac{q'}{\varphi}, \quad (1)$$

where q' is the fraction of the annual rainfall (m yr^{-1}) that enters the sediment and φ is the porosity expressed as a decimal fraction.

The most uncertain assumption concerning the water budget is the fraction of the annual rainfall that enters sediments. Since data on the hydraulic head for the sandstones exposed during the formation of the late Cimmerian unconformity are lacking, we avoided calculations of the flow rates based on this approach. The flow rate used in our calculations should therefore be considered a 'guesstimate'.

(ii) Proton budget

Carbonic and organic acids are the main acids in weathering solutions. CO_2 in the atmosphere or in the soil air will dissolve in aqueous solutions following Henry's law:

$$m_{CO_2} = k_H P_{CO_2} = 10^{-1.5} \text{ mol kg}^{-1} \text{ bar}^{-1} P_{CO_2}, \quad (2)$$

where m_{CO_2} is the molality of CO_2 in the solution, k_H is the Henry's law constant and P_{CO_2} is the partial pressure of CO_2 in the gas phase. The CO_2 partial pressure in the atmosphere is $10^{-3.5}$ bar while the partial pressure in soil air may be higher, typically $10^{-2.0}$–$10^{-3.0}$ bar (Stumm & Morgan, 1981, p. 536). Since the concentration of CO_2 and other dissolved substances is low in meteoric water, the difference between concentrations given as molalities and concentrations given as molarities is negligible, and is disregarded in the following calculations.

The equilibrium concentration of CO_2 in water in contact with the atmosphere is, according to Eq. (2), $10^{-5.0}$ mol l^{-1}, while the equilibrium concentration of CO_2 in water in contact with soil air will vary between $10^{-3.5}$ and $10^{-4.5}$ mol l^{-1}.

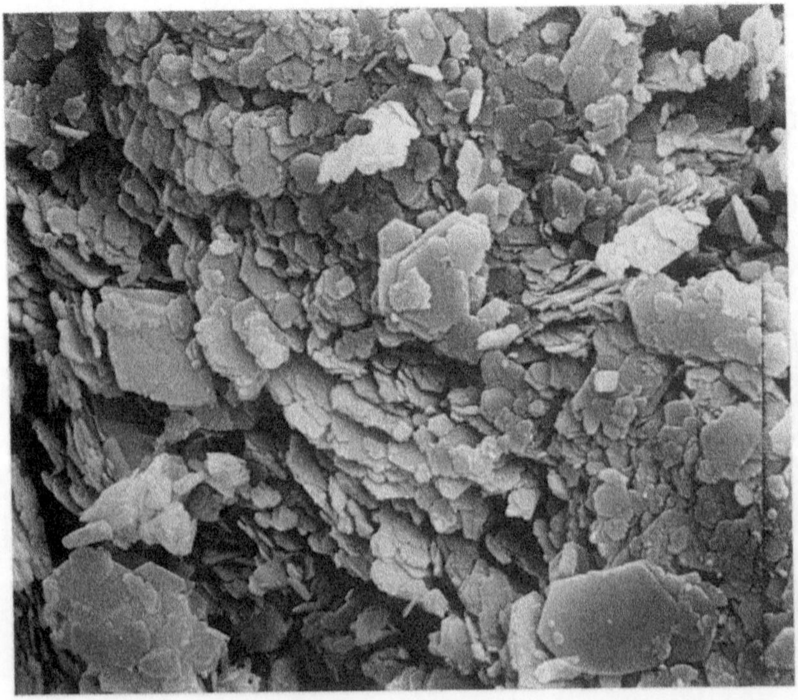

Fig. 6. SEM micrograph of kaolinite observed in the Rannoch Formation. The kaolinite is texturally similar to the detrital kaolinite in Fig. 39 in Keller (1978). Width of photograph = 36 μm.

For pH < 5, most of the dissolved CO_2 is in the form of hydrated molecular CO_2 (CO_2 (aq)). To act as an acid the hydrated CO_2 must first be converted to carbonic acid as follows:

$$H_2O + CO_2(aq) = H_2CO_3 = HCO_3^- + H^+. \quad (3)$$

In an open system, where the water is in equilibrium with a gas phase containing CO_2 at constant partial pressure, the total amount of dissolved carbon dioxide ($CO_2(aq) + H_2CO_3 + HCO_3^- + CO_3^{2-}$) increases with increasing pH (Stumm & Morgan, 1981, p. 179), since an increasing amount of the dissolved carbon dioxide will be in the form of carbonate and bicarbonate.

In a closed system, the total amount of dissolved carbon dioxide is constant for a given partial pressure of carbon dioxide (Stumm & Morgan, 1981, p. 173). Meteoric water penetrating soil profiles—with well-developed organic-rich horizons (Rose, Hawkes & Webb, 1979, Fig. 7.2)—dissolves a certain amount of carbon dioxide, and when the water has entered the C horizon where inorganic weathering dominates (Rose *et al.*, 1979, Fig. 7.2), the total amount of dissolved carbon dioxide is fixed. Hence, the total amount of dissolved carbon dioxide does not increase, even though the pH may increase due to interaction with minerals. As the pH increases, more of the total dissolved carbon dioxide is in the form of HCO_3^- and CO_3^{2-}. Consequently, carbon dioxide acts as a proton donor. The pH measured in soil waters is therefore not a good estimate of the total amount of protons available for mineral dissolution. In water containing $10^{-4.0}$ mol CO_2 l^{-1} the total amount of protons provided by the carbon dioxide as the pH increases from 5 to 7 is of the order of $10^{-4.0}$ mol l^{-1}, which is one order of magnitude higher than indicated by the pH alone.

The median concentration of organic components in shallow groundwaters is about 1 mg l^{-1} (Leehneer *et al.*, 1974; Thurman, 1984), with 20–40% of the dissolved carbon present as high molecular weight acids (Thurman, 1984). The median concentration of dissolved organic carbon in soil waters, however, is higher, ranging from 2 to 20 mg l^{-1} (Thurman, 1984). Organic acids may therefore act as pH buffers within the soil zone and the amount of protons provided by

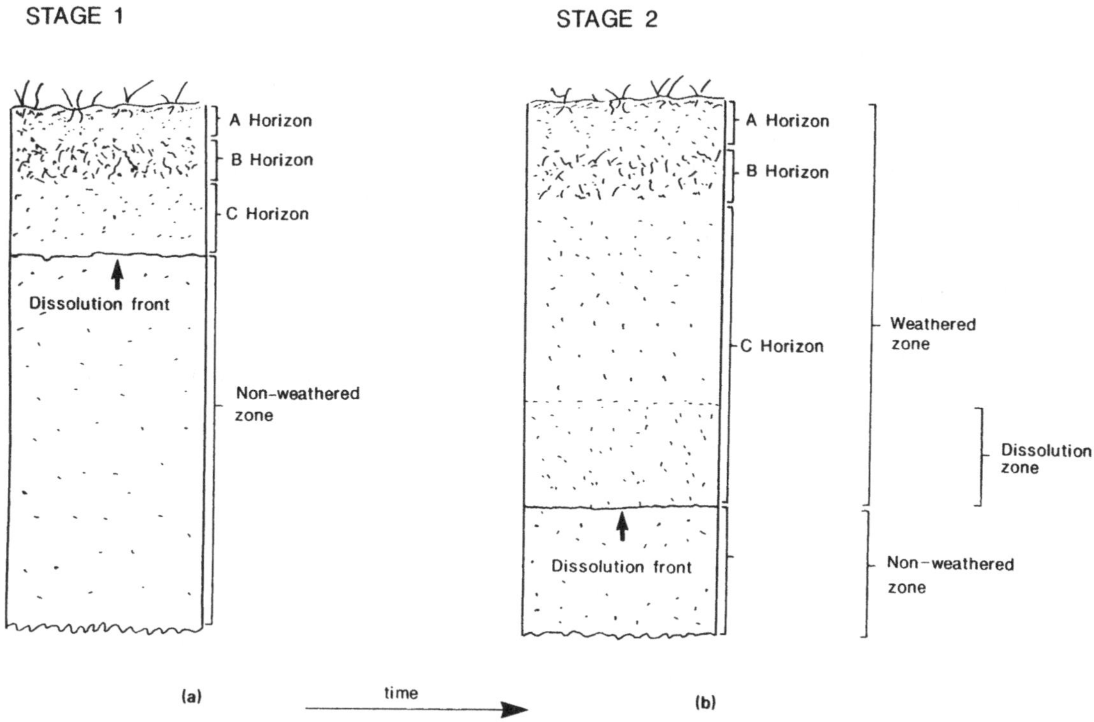

Fig. 7. A sketch of the weathering process associated with subaerial exposure. The *A horizon* is the zone of maximum biological activity, of eluviation, or both. The *B horizon* is the zone of illuviation. The A and B horizons are commonly referred to as the 'true soil' (Rose, Hawkes & Webb, 1979). The *C horizon* is the zone dominated by inorganic mineral decomposition or weathering (Rose *et al.*, 1979).

the organic acids may contribute somewhat to the dissolution of minerals. The amount of protons provided by the organic acids is less than the amount of protons provided by the carbon dioxide. In this study we have considered the contribution of organic acids to be an insignificant source for protons in comparison with CO_2.

(iii) Rate of mineral dissolution

Rates of mineral dissolution depend on the pH of the solution and the specific surface area of the minerals. The rates of dissolution for silicates are proportional to aH^+ at pH less than 7 (Lasaga, 1984). The dissolution-rate constants for K-feldspar and muscovite at 25°C and pH=5 are $5{\cdot}6\times10^{-13}$ mol m^{-2} s^{-1} (calculated from Busenberg & Clemency, 1976) and $8{\cdot}5\times10^{-14}$ mol m^{-2} s^{-1} (calculated from Lin & Clemency, 1981) respectively. The dissolution rate of plagioclase (albite) is $3{\cdot}9\times10^{-12}$ mol m^{-2} s^{-1} (Holdern & Berner, 1979). Apart from pH, the dissolution rate may be increased by complex-forming organic molecules (Lasaga, 1984). More work is needed, however, on the effect of organic acids on the dissolution rate of minerals (Lasaga, 1984).

(iv) Effect of erosion

Exposed sediments, or sedimentary rocks prone to meteoric-water penetration, are also prone to erosion. The rate of erosion depends on topography, climate, vegetation and the mechanical and chemical properties of the material being weathered. The diagenetic effect on the preserved sediments depends upon the relationship between the propagation rate of the dissolution front and the rate of erosion (Fig. 7). The *dissolution zone* is the zone where unstable minerals such as feldspar and mica dissolve because the pore-water is out of equilibrium with respect to these minerals. If these unstable minerals are completely altered in the upper part of the C horizon the descending pore-water will not start to be neutralized until it has penetrated down to the part of the sandstones where feldspar and mica are present (the dissolution zone in Fig. 7b). The dissolution front is the boundary between the dissolution zone and the non-weathered sandstone. The propagation rate of the dissolution front and the thickness of the dissolution zone together with the rate of erosion determine the weathered profile that may be observed in subaerially exposed sandstones. If the erosion rate exceeds the propagation rate of the dissolution front the weathered profile will not be preserved and any significant diagenetic effect (i.e. kaolinitization of feldspar and mica) on the subaerially exposed sandstone is unlikely to be observed.

MASS-BALANCE CALCULATIONS

If an annual precipitation of 3000 mm yr^{-1}, a porosity of 30% and that 10% of the annual precipitation enters the sediments as groundwater is assumed, the groundwater flow rate will, using Eq. (1), be 1·0 m yr^{-1}.

For a given flow rate, the diagenetic effect of the meteoric flow passing through the sandstones will depend upon the pH, the amount of CO_2, the mineralogy and the mineral dissolution rates. In the following calculations an average flow rate of 1 m yr^{-1} and a sandstone porosity of 30% is assumed. It is also assumed that the meteoric water has pH = 5 and contains $10^{-4\cdot0}$ mol l^{-1} of dissolved CO_2. The model sandstone is assumed to contain 15 vol% K-feldspar and 55 vol% quartz. Grain size is set to 0·5 mm and the grains are assumed to be spherical with smooth surfaces.

The K-feldspar is assumed to be transformed to kaolinite by the following equation (Bjørlykke, 1984):

$$\underset{\text{K-feldspar}}{2KAlSi_3O_8} + 2H_2CO_3 + H_2O = \underset{\text{Kaolinite}}{Al_2Si_2O_5(OH)_4} + 4SiO_2(aq) + 2HCO_3^- + 2K^+. \quad (4)$$

The amount of kaolinite formed is calculated by assuming conservation of aluminium. Dissolution of 1 mol K-feldspar will then give rise to 0·5 mol kaolinite.

Since each litre of pore-water contains $10^{-4\cdot0}$ mol dissolved CO_2 (discussed above), $10^{-4\cdot0}$ mol H^+ are available per litre of water (i.e. $10^{-4\cdot0}$ mol $- 10^{-7\cdot0}$ mol $\approx 10^{-4\cdot0}$ mol) for dissolution of K-feldspar assuming an increase in the pH to 7 during the dissolution process. After the available protons have been consumed the concentration of K^+ in the pore-water will then be approximately $10^{-4\cdot0}$ mol l^{-1} and the log $(aK^+/aH^+) \approx 3$. This is close to the equilibrium ratio for K-feldspar at 25°C in water slightly supersaturated with respect to silica (Fig. 1A of Bjørkum & Gjelsvik, 1988). By assuming the dissolution rate constant for K-feldspar presented above, a log (aK^+/aH^+) ratio of 3 is achieved within 1 yr. With an average flow rate of 1 m yr^{-1} the thickness of the dissolution zone (Fig. 7) will be approximately 1 m. In a sedimentary column which is 1 m thick, contains 15 vol% K-feldspar and covers 0·01 m^2, it takes approximately 45 000 pore volume exchanges to transform all the K-feldspar (13·4 mol) to kaolinite within the first metre below the surface. With an average flow rate of 1 m yr^{-1} it would therefore take approximately 45 000 years to transform all the K-feldspar within the first metre below the unconformity into kaolinite, i.e. 0·45 Myr to form about 7% kaolinite in the upper 10 m of the sandstone (the molar volumes of K-feldspar and kaolinite are approximately 110 and 100 ml respectively). Based on this we suggest that the propagation rate of the dissolution front is of the order of a few tens of metres per million years.

DISCUSSION

Mineral dissolution and kaolinite formation

The dissolution rate constant for muscovite is nearly one order of magnitude lower than for K-feldspar (Busenberg & Clemency, 1976; Lin & Clemency, 1981). Hence, it is expected that K-feldspar is preferentially dissolved with respect to muscovite. The specific surface area of muscovite may, however, be higher than for K-feldspar, and muscovite may therefore also undergo some dissolution. The dissolution rate of albite is seven times greater than the dissolution rate of K-feldspar. Hence, albite should dissolve faster than K-feldspar, and if K-feldspar is replaced by albite in our model sandstone, the thickness of the dissolution zone is reduced. The amount of kaolinite formed per unit time and the propagation rate of the dissolution front are not affected since these factors are controlled by the amount of protons provided by the water phase. In the Rannoch Formation the sandstones contain between 25 and 40% feldspar and mica. Hence, the amount of unstable minerals in the Rannoch Formation is approximately twice that in our model sandstone. This will reduce the propagation rate of the dissolution front by a factor of approximately two.

Also, due to the large amount of feldspar and mica and the small grain size (which increases the surface area), the thickness of the dissolution zone in the Rannoch Formation will also be less than in our model sandstone.

In sandstones containing a minor amount of unstable minerals the propagation rate of the dissolution front and the thickness of the dissolution zone will increase, and the zone of kaolinitization may reach several tens of metres below an unconformity (O. Walderhaug, & K. Bjørlykke, unpublished observation).

In the calculations a pH of 5 is assumed. Since pH increases as the K-feldspar (or mica) is dissolved, the rate of dissolution of the K-feldspar is retarded (Lasaga, 1984). This will tend to increase the thickness of the dissolution zone, but will not affect the total amount of kaolinite formed. Most of the protons (about 90%) that are provided by the carbon dioxide are, however, consumed before the pH has increased to values greater than 7, thus only minor amounts of K-feldspar will dissolve after the pore-water has reached a pH of 7. Also, the presence of organic acids will increase the dissolution rate of minerals (Lasaga, 1984) and so reduce the thickness of the dissolution zone.

Velbel (1986) concluded that dissolution rates for plagioclase in natural environments may be 1–3 orders of magnitude lower than laboratory rates. In our calculations we assumed that grains have smooth spherical surfaces and a grain size of 0·5 mm. It is well established that the surfaces of naturally weathered feldspar grains are not smooth, but rugose and often deeply etched. Generally, the surface area increases by one order of magnitude only after some 10% of the starting material has been dissolved; it may reach a maximum when 55% of the starting material has been consumed (Velbel, 1986). This increase in surface area is not taken into account in our calculations. Together with the increased dissolution rates caused by the possible presence of organic acids, the net effect on our calculation is likely to be minor. The fact that well-defined soil horizons do develop in sediments (Retallack, 1983) confirms that the thickness of the dissolution zone is commonly of the order of decimetres to metres, which is in agreement with the calculations presented in this paper.

The dissolution-rate constant for calcite is several orders of magnitude greater than for K-feldspar and mica (Berner & Morse, 1974; Busenberg & Clemency, 1976; Lin & Clemency, 1981). Hence, if calcite is present, it may dissolve preferentially to silicates, and only minor amounts of kaolinite may form due to meteoric-water influx in sediments containing calcite. Indeed, the pore-water may not reach equilibrium with the feldspars simply because there are not enough protons available. The dissolution rate of calcite may, however, be significantly affected by inhibiting agents present in natural environments, agents which are not taken into account in laboratory experiments (Berner, 1980, p. 189). Thus, presence of calcite in sediments may not always have significant effects on the dissolution rates of silicates.

Palaeoclimatic considerations

Although there has been a traditional assumption that the Jurassic climate of northern Britain/North Sea was humid, possibly tropical (Hallam, 1984), there is little concrete evidence to support this assumption, and in fact, a good deal of contradictory data (Hurst, 1985). Although the humidity increase from the Triassic into the Jurassic in northern Britain is well documented, a distinct seasonality has been identified, and a seasonal subtropical climate seems likely (Hudson, 1980; Hurst, 1985). Even more significant in terms of estimating the annual precipitation at the time of the Cimmerian unconformity, a marked decrease in humidity is believed to occur during deposition of the Upper Jurassic Kimmeridge Clay Formation and into the Lower Cretaceous (Hallam, 1984; Miller, 1990). Thus, the estimated annual precipitation used in our calculations (3000 mm yr^{-1}) is likely to be a considerable overestimate.

Erosion versus dissolution

In the North Sea several hundred metres of Jurassic sediments were eroded during formation of the late Cimmerian unconformity (Erichsen *et al.*, 1987), and in some places more than 1500 m of Jurassic and Triassic sediments have been removed (Karlsson, 1986). If it is assumed that erosion took place over a few million to a few tens of million years (the latter is a high estimate), the erosion rates were of the order of a few tens to several hundreds of metres per million years. Hence, the rate of erosion is at least equal to, or probably greater than, the rate of propagation of the dissolution front. In general, therefore, the sedimentary rocks underlying the late Cimmerian unconformity should not be expected to preserve any distinct diagenetic fingerprint attributable to the formation of the unconformity, except immediately below the unconformity.

In our calculations we assumed that the propagation rate of the dissolution front relative to the surface was equal to the propagation rate of the dissolution front in the sandstone. In a tilted fault block, the propagation rate of the dissolution front relative to the (horizontal) surface will be $V = V_d \sin \alpha$, where α is the dip angle and V_d is the propagation rate of the dissolution front in the exposed sandstone. Typical values for α in North Sea Jurassic sandstones are 5–10°; the propagation rate of the dissolution front relative to the surface will be 1/5–1/10 of the propagation rate of the dissolution front in the sandstone. Hence, the propagation rate of the dissolution front relative to the surface will be roughly 1/10 of our calculated values. Consequently, the propagation rate of the dissolution front relative to erosion rate can be regarded as an overestimate in the discussion above. This further reduces the probable preservation potential for the kaolinite formed by freshwater diagenesis during the late Cimmerian unconformity.

If the partial pressure of CO_2 in the soil profile is increased to $10^{-2\cdot0}$ bar, the amount of dissolved CO_2 is increased to $3\cdot3 \times 10^{-4\cdot0}$ mol l^{-1} and the dissolution capacity of the meteoric water triples. Even then the propagation rate of the dissolution front will be of the same order of magnitude or less than the average erosion rate. High partial pressure of carbon dioxide is only expected in well-developed soil profiles with organic-rich horizons. If erosion is relatively rapid, soil horizons tend to be thin and poorly developed (Kraus & Bown, 1986, p. 182). This suggests that the partial pressure of carbon dioxide in soil profiles developed during formation of the Cimmerian unconformity was probably low. If soil profiles containing organic-rich horizons did not develop, the amount of carbon dioxide dissolved in the pore-water will be $10^{-5\cdot0}$ mol l^{-1} (i.e. in equilibrium with the atmosphere) and the dissolution capacity of the pore-water penetrating the sandstones would have been one order of magnitude lower than in our calculations.

The comparison of erosion and dissolution rates supports the hypothesis that the late Cimmerian unconformity had limited effect on the overall distribution of kaolinite in the Jurassic sandstones of the North Sea. There may, however, be local areas where the erosion rate was lower and/or the water influx was higher, due to focusing of groundwater, such that the propagation rate of the dissolution front was higher than the rate of erosion. Within such areas, increased amounts of authigenic kaolinite may be observed down to several tens of metres below the unconformity. In sandstones with low contents of unstable minerals such as feldspar and mica, the propagation rate of the dissolution front may be significantly higher than in our calculations and the preserved weathered zone is therefore likely to be thicker.

Detrital versus authigenic kaolinite

The importance of petrographic differentiation between detrital and authigenic kaolinite is particularly important in the detrital kaolinite-rich Rannoch Formation. However, the fact that Rannoch sandstones were kaolinitic when deposited has no bearing on the relevance of our conclusions as detrital kaolinite in no way hinders formation of authigenic kaolinite. Ideally, the relationship between an unconformity and underlying kaolinitization in sandstones low in detrital kaolinite which were unaffected by immediate post-depositional meteoric-water flushing should be investigated. In some marine sandstones, which conform to these criteria, we observed only a very local (<10-m) zone of kaolinitization adjacent to unconformities. Similar observations on marine sandstones of the Magnus Field (D. Emery, K. Myers & R. Young, unpublished) confirm our observations and give support to our conclusions, as do data from the Snorre Field (P. A. Bjørkum, R. Knarud & M. Bergan, unpublished).

Closing remarks

Although the presented calculations can only be regarded as approximations to actual natural processes they do provide useful guidelines for evaluating the realism of particular diagenetic mechanisms. The pioneering work of Hancock & Taylor (1978) and Sommer (1978) on Brent Group diagenesis provided a burst of creativity by discussing several possible mechanisms for the formation of authigenic kaolinite. Unfortunately, there has been a tendency for workers to 'assume' that periods of unconformity produce large-scale meteoric-water infiltration causing kaolinitization without considering that the erosion associated with the unconformity will tend to remove the weathered sequence. Here, we evaluated data which are representative of the erosion rates, flow rates, dissolution rates, etc., during development of the Cimmerian unconformity, where possible, using data from the North Sea Jurassic. Our results give no credence to the unconformity as a significant cause of kaolinitization in the Rannoch sandstones, and show the importance of considering the rates of various processes associated with formation of unconformities. It seems appropriate to apply these types of simple quantifications whenever using diagenetic processes to explain mineral abundances and distributions.

CONCLUSIONS

In the Rannoch Formation the kaolinite content increases downward—i.e. away from the late Cimmerian unconformity—and from the petrographic data the kaolinite is interpreted to be detrital. Only traces of authigenic kaolinite are observed. Furthermore, the amounts of kaolinite in the Rannoch Formation in the three wells (Fig. 3) are similar. We suggest that formation of authigenic kaolinite by transformation of feldspar (or mica) due to meteoric-water penetration downwards from the late Cimmerian unconformity was insignificant in the Rannoch Formation sandstones which underlie the unconformity in the Gullfaks Field. This conclusion is probably applicable to other sandstones overlain by the Cimmerian unconformity.

Mass-balance calculations which include consideration of meteoric-water flow rates, amount of dissolved carbon dioxide, mineral dissolution rates and erosion rates, support the mineralogical interpretation. Results from the calculations show that the erosion rate was probably higher than the propagation rate of the dissolution front. It is therefore unlikely that much kaolinite formed during exposure is preserved in the sandstones underlying the late Cimmerian unconformity except immediately below the unconformity. In the light of these data, the significance of unconformities for kaolinitization in sandstones requires a re-evaluation.

ACKNOWLEDGMENTS

This work was funded by the Norwegian Department of Energy and is a part of the State Program for Improved Oil Recovery and Reservoir Technology. We thank Harry Shaw, Knut Bjørlykke, Dag Ormaasen, Michael Talbot and Henning Dypvik for constructive reviews of earlier drafts of this paper. Dominic Emery is thanked for his constructive refereeing comments and access to his as yet unpublished data. The XRD data and thin-sections were provided by Den Norske Stats Oljeselskap (Statoil) who are also acknowledged for supporting publication of this paper.

REFERENCES

BERNER, R. (1980) *Early Diagenesis—A Theoretical Approach.* Princeton University Press, NJ, 241 pp.

BERNER, R. & MORSE, J.W. (1974) Dissolution kinetics of calcium carbonate in seawater. IV. Theory of calcite dissolution. *Am. J. Sci.*, **274**, 108–134.

BJØRKUM, P.A. & GJELSVIK, N. (1988) An isochemical model for formation of authigenic kaolinite, K-feldspar and illite in sediments. *J. sedim. Petrol.*, **58**, 506–511.

BJØRLYKKE, K. (1984) Formation of secondary porosity: How important is it? In: *Clastic Diagenesis* (Ed. by D. A. McDonald and R. C. Surdam). *Mem. Am. Ass. Petrol. Geol.*, **37**, 277–288.

BJØRLYKKE, K. & BRENDSDAL, A. (1985) Diagenesis of the Brent Sandstones in the Statfjord Field, North Sea. In: *Roles of Organic Matter in Sediment Diagenesis* (Ed. by D. L. Gautier). *Spec. Publs Soc. econ. Paleont. Miner., Denver*, **38**, 157–167.

BJØRLYKKE, K., ELVERHØI, A. & MALM, A.D. (1979) Diagenesis in the Mesozoic sandstones from Spitsbergen and the North Sea: A comparison. *Geol. Runds.*, **68**, 1151–1171.

BUSENBERG, E. & CLEMENCY, C.V. (1976) The dissolution kinetics of feldspars at 25°C and at 1 atm CO_2 partial pressure. *Geochim. Cosmochim. Acta*, **40**, 41–50.

ERICHSEN, M., HELLE, M.H., HENDEN, J. & ROGNEBAKKEN, A. (1987) Gullfaks. In: *Geology of the Norwegian Oil and Gas Fields* (Ed. by A. M. Spencer *et al.*), pp. 273–286. Graham and Trotman, London, 493 pp.

GRAUE, E., HELLAND-HANSEN, W., JOHNSON, J., LOMO, L., NØTTVEDT, A., RØNNING, K., RYSETH, A. & STEEL, R. (1987) Advance and retreat of Brent Delta system, Norwegian North Sea. In: *Petroleum Geology of North West Europe* (Ed. by J. Brooks and K. W. Glennie), pp. 915–938. Graham and Trotman, London, 1219 pp.

HALLAM, A. (1984) Continental humid and arid zones during the Jurassic and Cretaceous. *Palaeogeogr. Palaeoclim. Palaeoecol.*, **47**, 195–223.

HANCOCK, N.J. & TAYLOR, A.M. (1978) Clay mineral diagenesis and oil migration in the Middle Jurassic Brent Sand Formation. *J. geol. Soc. London*, **135**, 69–72.

HOLDERN, G.R.J. & BERNER, R.A. (1979) Mechanism of feldspar weathering. I. Experimental studies. *Geochim. Cosmochim. Acta*, **43**, 1161–1171.

HUDSON, J.D. (1980) Aspects of brackish water facies and faunas from the Jurassic of NW Scotland. *Proc. geol. Ass.*, **91**, 99–105.

HURST, A. (1985) The implications of clay mineralogy to paleoclimate and provenance during the Jurassic in NE Scotland. *Scott. J. Geol.*, **21**, 143–160.

JEANS, C.V. (1986) Features of mineral diagenesis in hydrocarbon reservoirs. An introduction. *Clay Miner.*, **21**, 429–441.

KARLSSON, W. (1986) The Snorre, Statfjord and Gullfaks oil fields and the habitat of hydrocarbons on the Tampen Spur, offshore Norway. In: *Habitat of Hydrocarbons on the Norwegian Continental Shelf* (Ed. by A. M. Spencer *et al.*), pp. 181–198. Graham and Trotman, London, 354 pp.

KELLER, W.D. (1978) Classification of kaolin exemplified by their textures in scan electron micrographs. *Clays Clay Miner.*, **26**, 1–20.

KRAUS, M.J. & BOWN, T.M. (1986) Paleosols and time

resolution in alluvial stratigraphy. In: *Paleosols—Their Recognition and Interpretation* (Ed. by V. P. Wright), pp. 180–201. Princeton, NJ, 315 pp.

LASAGA, A.C. (1984) Chemical kinetics of water–rock interactions. *J. Geophys. Res.*, **89**, 4009–4025.

LEEHNEER, J.A., MALCOLM, R.L., MCKINLEY, P.W. & ECCLES, L.A. (1974) Occurrence of dissolved organic carbon in selected ground water samples in the United States. *J. Res. U.S. Geol. Surv.*, **2**, 361–369.

LERMAN, A. (1979) *Geochemical Processes—Water and Sediment Environments*. John Wiley, New York, 481 pp.

LIN, F. & CLEMENCY, C.V. (1981) The kinetics of dissolution of muscovites at 25°C and 1 atm CO_2 partial pressure. *Geochim. Cosmochim. Acta*, **45**, 571–576.

MILLER, R.G. (1990) A paleoceanographic approach to the Kimmeridge Clay Formation. In: *AAPG Proc. volume on deposition of organic facies* (Ed. by A. Y. Huc) (in press).

RAWSON, P.F. & RILEY, L.A. (1982) Late Jurassic–Early Cretaceous events and the 'Late Cimmerian Unconformity' in the North Sea. *Bull. Am. Ass. petrol. Geol.*, **66**, 2628–2648.

RETALLACK, G.J. (1983) A paleopedological approach to the interpretation of terrestrial sedimentary rocks: The mid-Tertiary fossil soils of Badlands National Park, South Dakota. *Geol. Soc. Am. Bull.*, **94**, 823–840.

ROSE, A.W., HAWKES, H.E. & WEBB, J.S. (1979) *Geochemistry in Mineral Exploration*. Academic Press, London, 657 pp.

SOMMER, F. (1978) Diagenesis of Jurassic sandstones in the Viking Graben. *J. geol. Soc. London*, **135**, 63–67.

STUMM, W. & MORGAN, J.J. (1981) *Aquatic Chemistry, An Introduction Emphasizing Chemical Equilibria in Natural Waters*. John Wiley, New York, 780 pp.

THURMAN, E.M. (1984) *Humic Substances. I. Geochemistry, Characterization and Isolation*. John Wiley, New York, 380 pp.

VELBEL, M.A. (1986) Influence of surface area, surface characteristics, and solution composition on feldspar weathering rates. In: *Geochemical Processes at Mineral Surfaces* (Ed. by J. A. Davis and K. F. Hayes), pp. 614–634. American Chemical Society Symposium Series 323, 683 pp.

VOLLSET, J. & DORE, A.G. (1984) A revised Triassic and Jurassic lithostratigraphic nomenclature for the Norwegian North Sea. *Nor. Petrol. Direct. Bull.*, **3**, 53 pp.

ZIEGLER, P.A. (1981) Evolution of sedimentary basins in North-West Europe. In: *Petroleum Geology of the Continental Shelf of North-West Europe* (Ed. by L. V. Illing and G. D. Hobson), pp. 3–39. Heyden & Son Ltd, London, 521 pp.

(Manuscript received 15 March 1989; revision received 16 October 1989)

Index

Page numbers in *italics* refer to figures; those in **bold** refer to tables.